1 I	2 II							3	4	5	6	7	8	9
1															

	1 H 1.008

2	3 Li 6.94	4 Be 9.01
3	11 Na 22.99	12 Mg 24.31

			3	4	5	6	7	8	9
4	19 K 39.10	20 Ca 40.08	21 Sc 44.96	22 Ti 47.90	23 V 50.94	24 Cr 52.00	25 Mn 54.94	26 Fe 55.85	27 Co 58.93
5	37 Rb 85.47	38 Sr 87.62	39 Y 88.91	40 Zr 91.22	41 Nb 92.91	42 Mo 95.94	43 Tc 98.91	44 Ru 101.07	45 Rh 102.91
6	55 Cs 132.91	56 Ba 137.34	71 Lu 174.97	72 Hs 178.49	73 Ta 180.95	74 W 183.85	75 Re 186.2	76 Os 190.2	77 Ir 192.2
7	87 Fr 223	88 Ra 226.03	103 Lr 257	104 Unq	105 Unp	106 Unh	107 Uns	108 Uno	109 Unn

57 La 138.91	58 Ce 140.12	59 Pr 140.91	60 Nd 144.24	61 Pm 146.92
89 Ac 227.03	90 Th 232.04	91 Pa 231.04	92 U 238.03	93 Np 237.05

			13 III	14 IV	15 V	16 VI	17 VII	18 VIII
								2 **He** 4.00
			5 **B** 10.81	6 **C** 12.01	7 **N** 14.01	8 **O** 16.00	9 **F** 19.00	10 **Ne** 20.18
			13 **Al** 26.98	14 **Si** 28.09	15 **P** 30.97	16 **S** 32.06	17 **Cl** 35.45	18 **Ar** 39.95
10	**11**	**12**						
28 **Ni** 58.71	29 **Cu** 63.54	30 **Zn** 65.37	31 **Ga** 69.72	32 **Ge** 72.59	33 **As** 74.92	34 **Se** 78.96	35 **Br** 79.91	36 **Kr** 83.80
46 **Pd** 106.4	47 **Ag** 107.87	48 **Cd** 112.40	49 **In** 114.82	50 **Sn** 118.69	51 **Sb** 121.75	52 **Te** 127.60	53 **I** 126.90	54 **Xe** 131.30
78 **Pt** 195.09	79 **Au** 196.97	80 **Hg** 200.59	81 **Tl** 204.37	82 **Pb** 207.19	83 **Bi** 208.98	84 **Po** 210	85 **At** 210	86 **Rn** 222

←—— Metals | Nonmetals

62 **Sm** 150.35	63 **Eu** 151.96	64 **Gd** 157.25	65 **Tb** 158.92	66 **Dy** 162.50	67 **Ho** 164.93	68 **Er** 167.26	69 **Tm** 168.93	70 **Yb** 173.04	Lanthanides
94 **Pu** 239.05	95 **Am** 241.06	96 **Cm** 247.07	97 **Bk** 249.08	98 **Cf** 251.08	99 **Es** 254.09	100 **Fm** 257.10	101 **Md** 258.10	102 **No** 255	Actinides

GENERAL CHEMISTRY

When mixed and heated, iron filings and sulfur form iron sulfide, with a vigorous evolution of heat.

GENERAL CHEMISTRY

P. W. ATKINS

Oxford University

SCIENTIFIC
AMERICAN
BOOKS Distributed by W. H. Freeman and Company

Cover image by Ken Karp.

Library of Congress Cataloging-in-Publication Data

Atkins, P. W. (Peter William), 1940–
  General chemistry.

  Includes index.
  1. Chemistry.   I. Title.
QD31.2.A75   1989   540   88-30580
ISBN 0-7167-1940-1

Printed in the United States of America

Scientific American Books is a subsidiary of Scientific American, Inc. Distributed by W. H. Freeman and Company, 41 Madison Avenue, New York, New York 10010

1 2 3 4 5 6 7 8 9 0 KP 7 6 5 4 3 2 1 0 8 9

PART I MATTER AND REACTIONS 1

1 Properties, Measurements, and Units 3

2 The Composition of Matter 41

3 Chemical Reactions 81

4 Reaction Stoichiometry 117

5 The Properties of Gases 151

6 Energy, Heat, and Thermochemistry 187

PART II ATOMS, MOLECULES, AND IONS 229

7 Atomic Structure and the Periodic Table 231

8 The Chemical Bond 277

9 The Shapes of Molecules 319

10 Liquids and Solids 351

11 The Properties of Solutions 393

CONTENTS IN BRIEF

PART III RATES AND EQUILIBRIUM 429

12 Chemical Kinetics 431

13 Chemical Equilibrium 477

14 Acids and Bases 511

15 Acids, Bases, and Salts in Water 545

16 Entropy, Free Energy, and Equilibrium 587

17 Electrochemistry 617

PART IV THE ELEMENTS 657

18 Hydrogen and the s-Block Elements 659

19 The p-Block Elements: I 691

20 The p-Block Elements: II 731

21 The d-Block Elements 761

22 Nuclear Chemistry 801

PART V ORGANIC CHEMISTRY **837**

23 The Hydrocarbons 839

24 Functional Groups and
   Biomolecules 871

APPENDIXES

1 Mathematical Information 911

2 Experimental Data 919

GLOSSARY **941**

**ANSWERS TO ODD-NUMBERED
NUMERICAL EXERCISES** **959**

ILLUSTRATION CREDITS **969**

INDEX **971**

PART I   MATTER AND REACTIONS   1

1   Properties, Measurements, and Units   3

The properties of substances   5

1.1   Physical and chemical properties   5

1.2   Pure substances and mixtures   9

Measurements and units   15

1.3   The International System of Units   16

1.4   Extensive and intensive properties   22

Using measurements   24

1.5   Unit analysis and conversion factors   24

*1.6   The reliability of measurements and
        calculations*   27

1.7   Significant figures in calculations   31

Summary   34

Exercises   35

TABLE OF CONTENTS

2   The Composition of Matter   41

Elements   42

2.1   The names and symbols of the elements   42

2.2   The periodic table   43

Atoms   46

2.3   The nuclear atom   47

2.4   The masses of atoms   51

2.5   Moles and molar mass   56

Compounds   59

2.6   Molecules and molecular compounds   59

2.7   Ions and ionic compounds   61

2.8   Chemical nomenclature   67

Summary   75

Exercises   76

3   Chemical Reactions   81

Chemical equations   82

3.1   Symbolizing reactions   82

3.2   Balancing equations   83

Precipitation reactions   86

3.3   Net ionic equations   87

*3.4   Using precipitation reactions in
        chemistry*   88

Acid-base reactions		90
3.5	Arrhenius acids and bases	90
3.6	The Brønsted definitions	92
3.7	Neutralization	94
Redox reactions		96
3.8	Electron transfer	96
3.9	Oxidizing and reducing agents	99
3.10	Half-reactions	105
Summary		110
Exercises		110

4	**Reaction Stoichiometry**	**117**
Interpreting stoichiometric coefficients		118
4.1	Mole calculations	118
4.2	Limiting reagents	123
4.3	Composition from measurements of mass	127
The stoichiometry of reactions in solution		132
4.4	Molar concentration	133
4.5	Volume of solution required for reaction	137
Titrations		139
4.6	Direct titrations	139
4.7	Indirect and back titrations	142
Summary		143
Exercises		144

5	**The Properties of Gases**	**151**
The gas laws		154
5.1	Pressure	154
5.2	The ideal gas	157
5.3	Using the ideal gas law	159
5.4	Real gases	164
The stoichiometry of reacting gases		166
5.5	The molar volume of a gas	166
5.6	Gaseous mixtures	170
The kinetic theory of gases		174
5.7	Molecular speeds	176
5.8	The liquefaction of gases	179
Summary		180
Exercises		181

6	**Energy, Heat, and Thermochemistry**	**187**
Energy and calorimetry		188
6.1	Energy	188
6.2	Enthalpy	195

The enthalpy of chemical change		205
6.3	Reaction enthalpies	206
6.4	Enthalpies of formation	214
The world's enthalpy resources		217
6.5	The enthalpy of fuels	219
6.6	The enthalpy of foods	220
Summary		222
Exercises		223

PART II ATOMS, MOLECULES, AND IONS   **229**

7	**Atomic Structure and the Periodic Table**	**231**
The structure of the hydrogen atom		232
7.1	Light	233
7.2	The spectrum of atomic hydrogen	236
7.3	Particles and waves	239
The structures of many-electron atoms		247
7.4	Orbital energies	247
7.5	The building-up principle	250
A survey of periodic properties		254
7.6	Blocks, periods, and groups	255
7.7	Periodicity of physical properties	256
7.8	Trends in chemical properties	264
Summary		268
Exercises		269

8	**The Chemical Bond**	**277**
Ionic bonds		278
8.1	The formation of ionic bonds	278
8.2	Variable valence	286
Covalent bonds		290
8.3	The electron-pair bond	291
8.4	Lewis structures of polyatomic molecules	295
8.5	Bond parameters	304
Charge distributions in compounds		307
8.6	Assessing the charge distribution	307
8.7	Polarization	310
Summary		313
Exercises		314

9	**The Shapes of Molecules**	**319**
Electron-pair repulsions		322

9.1	*The arrangement of electron pairs*	322
9.2	*Polar molecules*	328
	The orbital model of bonding	329
9.3	*Hybridization*	330
9.4	*Bonding orbitals*	332
	Molecular orbital theory	337
9.5	*Molecular orbitals*	339
9.6	*Bonding in Period 2 diatomic molecules*	341
9.7	*Orbitals in polyatomic molecules*	343
9.8	*A perspective on chemical bonding*	344
	Summary	345
	Exercises	346

10 Liquids and Solids **351**

	Forces between atoms, ions, and molecules	353
10.1	*Ion and dipole forces*	354
10.2	*Hydrogen bonding*	360
	The properties of liquids	361
10.3	*Viscosity and surface tension*	361
10.4	*Vapor pressure*	364
10.5	*Solidification*	368
	Solids	371
10.6	*X-ray diffraction*	372
10.7	*Metals and semiconductors*	374
10.8	*Ionic solids*	379
10.9	*Other types of solids*	381
	Summary	385
	Exercises	385

11 The Properties of Solutions **393**

	Measures of concentration	394
11.1	*Emphasizing the amount of solute*	394
11.2	*Emphasizing numbers of molecules*	396
	Solubility	400
11.3	*Saturation and solubility*	400
11.4	*The effect of pressure on gas solubility*	405
11.5	*The effect of temperature on solubility*	407
	Colligative properties	416
11.6	*The lowering of vapor pressure*	416
11.7	*Osmosis*	421
	Summary	424
	Exercises	424

PART III RATES AND EQUILIBRIUM **429**

12 Chemical Kinetics **431**

	The description of reaction rates	432
12.1	*Reaction rates*	432
12.2	*Rate laws and reaction order*	436
	Reaction mechanisms	449
12.3	*Elementary reactions*	449
12.4	*Chain reactions*	456
	Controlling rates of reactions	458
12.5	*Temperature dependence of reaction rates*	458
12.6	*Catalysis*	465
	Summary	469
	Exercises	470

13 Chemical Equilibrium **477**

	The description of chemical equilibrium	478
13.1	*Reactions at equilibrium*	478
13.2	*The equilibrium constant*	479
13.3	*Heterogeneous equilibria*	488
	Equilibrium calculations	491
13.4	*Specific initial concentrations*	492
13.5	*Arbitrary initial concentrations*	495
	The response of equilibria to the conditions	499
13.5	*The effect of added reagents*	499
13.7	*The effect of pressure*	500
13.8	*The effect of temperature*	502
	Summary	505
	Exercises	506

14 Acids and Bases **511**

	The definitions of acids and bases	512
14.1	*The Brønsted and Lewis definitions*	513
14.2	*Brønsted equilibria*	517
	Equilibria in solutions of acids and bases	521
14.3	*Ionization constants*	521
14.4	*Strong and weak acids and bases*	524
	Hydrogen ion concentration and pH	530
14.5	*The pH of solutions*	530
14.6	*Polyprotic acids*	536
	Summary	540
	Exercises	541

15	**Acids, Bases, and Salts in Water**	**545**
	Salts as acids and bases	546
15.1	*Ions as acids and bases*	547
15.2	*The pH of mixed solutions*	550
	Titrations and pH curves	552
15.3	*The variation of pH during a titration*	552
15.4	*Indicators and buffers*	560
	Solubility equilibria	566
15.5	*The solubility product*	567
15.6	*Precipitation reactions and qualitative analysis*	574
15.7	*Complex formation*	577
	Summary	580
	Exercises	581

16	**Entropy, Free Energy, and Equilibrium**	**587**
	The direction of spontaneous change	588
16.1	*Entropy and spontaneous change*	588
16.2	*The entropy change in the surroundings*	595
16.3	*The second law*	600
	Free energy	601
16.4	*Focusing on the system*	601
16.5	*Spontaneous reactions*	606
	Equilibria	608
16.6	*Free energy and composition*	609
16.7	*The equilibrium constant*	610
	Summary	611
	Exercises	612

17	**Electrochemistry**	**617**
	Electrochemical cells	619
17.1	*Cells and cell reactions*	620
17.2	*Practical cells*	625
	Thermodynamics and electrochemistry	628
17.3	*Cell potential and reaction free energy*	628
17.4	*The electrochemical series*	634
17.5	*The dependence of cell potential on concentration*	639
	Electrolysis	643
17.6	*The potential needed for electrolysis*	643
17.7	*The extent of electrolysis*	646
17.8	*Applications of electrolysis*	648
	Summary	651
	Exercises	652

PART IV	**THE ELEMENTS**	**657**

18	**Hydrogen and the *s*-Block Elements**	**659**
	Hydrogen	660
18.1	*The element hydrogen*	660
18.2	*Some important hydrogen compounds*	664
	Group I: the alkali metals	666
18.3	*The Group I elements*	667
18.4	*Some important Group I compounds*	671
	Group II: the alkaline earth metals	676
18.5	*The Group II elements*	676
18.6	*Some important Group II compounds*	680
	Summary	687
	Exercises	687

19	**The *p*-Block Elements: I**	**691**
	Group III: boron and aluminum	692
19.1	*The Group III elements*	692
19.2	*Group III oxides*	693
19.3	*Other important Group III compounds*	697
	Group IV: carbon and silicon	701
19.4	*The Group IV elements*	702
19.5	*Group IV oxides*	705
19.6	*Other important Group IV compounds*	711
	Group V: nitrogen and phosphorus	714
19.7	*The Group V elements*	715
19.8	*Hydrogen and halogen compounds*	716
19.9	*Group V oxides and oxoacids*	719
	Summary	724
	Exercises	725

20	**The *p*-Block Elements: II**	**731**
	Group VI: oxygen and sulfur	732
20.1	*The Group VI elements*	732
20.2	*Compounds with hydrogen*	735
20.3	*Some important compounds of sulfur*	739
	Group VII: the halogens	743
20.4	*The Group VII elements*	743
20.5	*Halides*	747
20.6	*Halogen oxides and oxoacids*	749
	Group VIII: the noble gases	751
20.7	*The Group VIII elements*	751
20.8	*Compounds of the noble gases*	753

Summary		754
Exercises		755

21 The d-Block Elements **761**

The d-block elements and their compounds 762
- 21.1 *Trends in properties* 763
- 21.2 *The elements scandium through nickel* 767
- 21.3 *The elements copper through mercury* 775

Complexes of the d-block elements 779
- 21.4 *The preparation and stability of complexes* 780
- 21.5 *The structures of complexes* 781
- 21.6 *Isomerism* 784

Crystal field theory 788
- 21.7 *The effects of ligands on* d *electrons* 788
- 21.8 *The electronic structures of many-electron complexes* 791

Summary 795
Exercises 796

22 Nuclear Chemistry **801**

Nuclear stability 802
- 22.1 *Nuclear structure and nuclear radiation* 802
- 22.2 *The identities of daughter nuclides* 805
- 22.3 *The pattern of nuclear stability* 809
- 22.4 *Nucleosynthesis* 812

Radioactivity 813
- 22.5 *Measuring radioactivity* 814
- 22.6 *The rate of nuclear disintegration* 815

Nuclear power 822
- 22.7 *Nuclear fission* 822
- 22.8 *Nuclear fusion* 828
- 22.9 *Chemical aspects of nuclear power* 829

Summary 832
Exercises 833

PART V ORGANIC CHEMISTRY **837**

23 The Hydrocarbons **839**

The alkanes 840
- 23.1 *Isomerism* 842
- 23.2 *Alkane nomenclature* 843
- 23.3 *The properties of alkanes* 849

The alkenes and the alkynes 851
- 23.4 *Alkene nomenclature* 852
- 23.5 *The carbon-carbon double bond* 853
- 23.6 *Alkene polymerization* 856
- 23.7 *Alkynes* 860

Aromatic hydrocarbons 862
- 23.8 *Arene nomenclature* 863
- 23.9 *Reactions of aromatic hydrocarbons* 865

Summary 866
Exercises 867

24 Functional Groups and Biomolecules **871**

The hydroxyl group 873
- 24.1 *Alcohols and ethers* 873
- 24.2 *Phenols* 877

The carbonyl group 879
- 24.3 *Aldehydes and ketones* 879
- 24.4 *Carbohydrates* 884

The carboxyl group 887
- 24.5 *Carboxylic acids* 887
- 24.6 *Esters* 889

Functional groups containing nitrogen 892
- 24.7 *Amines* 892
- 24.8 *Amino acids* 896
- 24.9 *DNA and RNA* 902

Summary 907
Exercises 907

Appendix 1 Mathematical Information **911**
- 1A: Scientific notation 912
- 1B: Logarithms 914
- 1C: Quadratic equations 915
- 1D: Graphs 915

Appendix 2 Experimental Data **919**
- 2A: Thermodynamic data at 25°C 920
  - 1 *Inorganic substances* 920
  - 2 *Organic compounds* 925
- 2B: Standard reduction potentials at 25°C 927
  - 1 *Potentials in electrochemical order* 927
  - 2 *Potentials in alphabetical order* 930
- 2C: Ground-state electron configurations 933
- 2D: The elements 934
- 2E: The top fifty chemicals produced in the United States 939

Glossary   941

**Answers to Odd-numbered
Numerical Exercises**   959

Illustration Credits   969

Index   971

1.1	Selecting convenient units	18
1.2	Converting between Celsius and Fahrenheit temperatures	20
1.3	Distinguishing extensive and intensive properties	23
1.4	Unit conversions	25
1.5	Converting derived units	26
1.6	Converting compound units	27
1.7	Counting significant figures	29
1.8	Adding and subtracting with the correct number of significant figures	32
1.9	Calculating mass percentage composition	33

2.1	Using the atomic number	50
2.2	Using mass numbers	54
2.3	Calculating the number of atoms in a sample	55
2.4	Using atomic weights to calculate the number of moles	58

WORKED EXAMPLES

2.5	Using molecular weight to calculate the number of moles	61
2.6	Working out the formula of an ionic compound	66
2.7	Writing the formula from the name	72
2.8	Naming molecular compounds	73

3.1	Balancing equations: I	85
3.2	Balancing equations: II	86
3.3	Using net ionic equations to balance chemical equations	88
3.4	Accounting for reactions in terms of proton transfer	95
3.5	Predicting the outcome of a neutralization	96
3.6	Using the electrochemical series	98
3.7	Calculating oxidation numbers	101
3.8	Identifying oxidizing and reducing agents	103
3.9	Balancing redox reactions by using half-reactions	107
3.10	Balancing redox equations using half-reactions	109

4.1 Calculating the number moles taking part in a reaction   119

4.2 Predicting the mass of a product   122

4.3 Limiting-reagent calculations   123

4.4 Identifying the limiting reagent: I   124

4.5 Identifying the limiting reagent: II   124

4.6 Calculating the percentage yield   126

4.7 Interpreting a combustion analysis   128

4.8 Calculating an empirical formula from a mass percentage composition   130

4.9 Determining a molecular formula from an empirical formula   132

4.10 Preparing a solution of specified concentration   135

4.11 Calculating the volume of solution to dilute   137

4.12 Calculating the volume of solution needed for a reaction   138

4.13 Measuring a concentration by titration   139

4.14 Determining concentration by redox titration   142

5.1 Interpreting a manometer reading   156

5.2 Using the ideal gas law   160

5.3 Calculating the density of a gas   161

5.4 Determining molecular weight from gas density   161

5.5 Calculating the volume at a new pressure   163

5.6 Calculating the volume at a new temperature   163

5.7 Calculating the volume after changes in both pressure and temperature   164

5.8 Using the van der Waals equation   165

5.9 Calculating the volume of a given mass of gas   167

5.10 Calculating the volume of gas produced by a given mass of reactant   168

5.11 Calculating the mass of reagent needed to react with a specified volume of gas   169

5.12 Calculating partial pressures   171

5.13 Measuring moles of gas produced by a reaction   173

5.14 Calculating a root-mean-square speed   175

5.15 Measuring molecular weight using Graham's law   178

6.1 Converting between calories and joules   190

6.2 Calculating the internal energy of an ideal gas   193

6.3 Calculating a change in internal energy   194

6.4 Calculating the work of expansion   197

6.5 Relating internal energy and enthalpy   198

6.6 Using the heat capacity   201

6.7 Calculating the heat produced by a reaction   202

6.8 Measuring the enthalpy of vaporization   203

6.9 Combining enthalpies of transition   205

6.10 Calculating the heat output of a fuel   207

6.11 Using enthalpy of combustion to calculate heat output   209

6.12 Combining reaction enthalpies   211

6.13 Using combustion reactions to reproduce a synthesis reaction   213

6.14 Using enthalpies of formation: I   216

6.15 Using enthalpies of formation: II   216

7.1 Calculating the wavelengths present in a spectrum   239

7.2 Using the uncertainty principle   241

7.3 Interpreting quantum numbers   244

7.4 Counting electrons in closed shells   249

7.5 Building up an electron configuration   250

7.6 Deducing electron configurations   253

7.7 Predicting the configurations of cations   254

7.8 Writing an electron configuration from a group number   256

7.9 Determining ionization energy   259

7.10 Accounting for the variation of ionization energies   261

7.11 Accounting for the variation in electron affinity   263

8.1 Using a Born-Haber cycle to account for compound formation   282

8.2 Predicting the compound formed by two elements   285

8.3 Interpreting the formulas of ionic compounds using electron dot diagrams   286

8.4 Judging whether bonding is ionic or covalent   292

8.5 Writing the Lewis structure of a diatomic ion   294

8.6 Writing Lewis structures for molecules 296

8.7 Writing resonance structures 298

8.8 Writing Lewis structures with expanded octets 300

8.9 Estimating a reaction enthalpy from bond enthalpies 305

8.10 Calculating the formal charge 308

8.11 Judging the plausibility of a structure 309

9.1 Predicting the shape of a simple molecule 324

9.2 Using VSEPR theory to predict shapes: I 325

9.3 Using VSEPR theory to predict shapes: II 326

9.4 Predicting the shape of a molecule with a multiple bond 327

9.5 Predicting the shape of a molecule using the shortcut 328

9.6 Predicting the polarity of a molecule 329

9.7 Selecting a hybridization scheme 331

9.8 Describing a structure in terms of a d-orbital hybridization 332

9.9 Analyzing the structure of a molecule in terms of σ and π bonds 335

9.10 Accounting for the absence of bonding 340

9.11 Deducing the electron configurations of Period 2 diatomic molecules 343

10.1 Estimating the strength of an ion-dipole interaction 355

10.2 Estimating the importance of dipole-dipole interactions in a gas 356

10.3 Estimating the strength of the London interaction in a gas 358

10.4 Accounting for a trend in boiling points 360

10.5 Assessing the rate of vaporization 365

10.6 Predicting the boiling point at a given pressure 367

10.7 Using Trouton's rule 367

10.8 Interpreting a phase diagram 370

10.9 Using the Bragg equation 373

10.10 Calculating the density of a metal 376

10.11 Counting the ions in a unit cell 380

11.1 Converting mass concentration to molar concentration 396

11.2 Calculating a mole fraction 397

11.3 Preparing a solution of a given molality 399

11.4 Converting between mole fraction and molality 399

11.5 Using Henry's law 406

11.6 Predicting the temperature dependence of solubility 409

11.7 Calculation of ion hydration enthalpy 412

11.8 Using the Born equation 413

11.9 Using Raoult's law 417

11.10 Using freezing-point depression to measure molar mass 420

11.11 Measuring molecular weight by osmometry 422

12.1 Expressing the reaction rate for different substances 433

12.2 Classifying reactions by order 439

12.3 Determining reaction order from experimental data 440

12.4 Using a rate law with a negative power 442

12.5 Calculating a concentration from a rate law 443

12.6 Measuring a rate constant 444

12.7 Using an integrated rate law 445

12.8 Using first-order half-lives 446

12.9 Using second-order half-lives 447

12.10 Deducing a rate law from a mechanism 455

12.11 Measuring an activation energy 460

12.12 Using the Arrhenius equation 461

12.13 Predicting the rate constant at one temperature from its value at another temperature 462

13.1 Calculating an equilibrium constant 481

13.2 Predicting the direction of reaction 484

13.3 Relating K_P and K_c 488

13.4 Calculating the concentration of a pure solid substance 489

13.5 Calculating one unknown concentration 492

13.6 Calculating the equilibrium composition for a decomposition reaction 493

13.7 Calculating a composition starting from stoichiometric proportions 495

13.8 Calculating an equilibrium composition starting from a mixture 496

13.9 Finding an equilibrium concentration by approximation 498

13.10 Predicting the effect of pressure 500

13.11 Predicting the effect of temperature 502

14.1 Identifying Lewis acids and bases 515

14.2 Identifying conjugate acids and bases 518

14.3 Predicting the acidities of solutions 519

14.4 Calculating the autoionization constant 520

14.5 Calculating pH 530

14.6 Calculating the pH of a solution of a base 533

14.7 Calculating the pH of a solution of a weak acid 534

14.8 Calculating the pH of a solution of a weak base 535

14.9 Calculating the pH of aqueous sulfuric acid 537

14.10 Calculating a true ionization constant 538

14.11 Calculating the carbonate ion concentration in carbonic acid 539

15.1 Calculating the pH of a salt solution: I 547

15.2 Calculating the pH of a salt solution: II 549

15.3 Calculating the pH before the equivalence point is reached in a strong acid–strong base titration 553

15.4 Calculating the pH after the equivalence point of a strong acid–base titration has been passed 554

15.5 Calculating the pH before the equivalence point is reached in a weak acid–strong base titration 556

15.6 Calculating the pH at the equivalence point of a weak acid–strong base titration 557

15.7 Calculating the pH at the equivalence point of weak base–strong acid titration 559

15.8 Calculating the pH of a buffer solution 565

15.9 Estimating buffer capacity 566

15.10 Writing solubility products 567

15.11 Predicting whether a salt will precipitate 575

15.12 Calculating solubility with complex formation taken into account 578

15.13 Calculating the concentration of a complex in solution 579

16.1 Using the Boltzmann formula for entropy 591

16.2 Calculating the entropy of a physical change 594

16.3 Calculating the standard reaction entropy 595

16.4 Predicting the boiling point of a substance 598

16.5 Calculating the minimum decomposition temperature 603

16.6 Calculating a standard free energy of formation 605

16.7 Judging the stability of a compound 606

16.8 Estimating the temperature at which a reaction can occur 608

16.9 Calculating an equilibrium constant 610

16.10 Calculating a vapor pressure 611

17.1 Specifying an electrode 622

17.2 Writing a cell diagram for a reaction 623

17.3 Deducing the reaction corresponding to a given cell 625

17.4 Writing a cell diagram using the electrochemical series 634

17.5 Predicting oxidizing power using the electrochemical series 635

17.6 Predicting the direction of a reaction with the electrochemical series 635

17.7 Predicting relative oxidizing strengths 636

17.8 Judging whether a metal can reduce an oxoacid 637

17.9 Calculating an equilibrium constant 640

17.10 Using the Nernst equation 641

17.11 Calculating the volume of gas produced in an electrolysis 646

17.12 Predicting the mass of an element produced by electrolysis 648

18.1 Judging the feasibility of reduction with hydrogen 661

18.2 Judging the reducing power of hydride ions 665

18.3 Accounting for an electrode potential 668

18.4 Accounting for a reaction rate 669

18.5 Predicting the properties of an element 679

18.6 Accounting for the bonding in beryllium hydride 682

18.7 Accounting for the effect of heat on hard water 685

19.1 Predicting group trends 693

19.2 Predicting the properties of alumina 696

19.3 Predicting the properties of boron halides 698

19.4 Predicting the properties of a group 701

19.5 Finding a method for preparing an oxide 705

19.6 Identifying the structure of a silicate 709

19.7 Accounting for a difference between carbon and silicon 713

19.8 Suggesting a preparation of phosphine 718

19.9 Predicting the result of a nonmetal halide reaction 718

19.10 Suggesting a preparation of a nitrogen oxide 722

20.1 Predicting the characteristics of oxygen 734

20.2 Accounting for a property of oxides 737

20.3 Predicting a reaction of sulfuric acid 742

20.4 Predicting the properties of a sulfur halide 742

20.5 Predicting the relative oxidizing abilities of the halogens 746

20.6 Predicting trends in oxoacid strength 749

21.1 Predicting the properties of titanium solutions 768

21.2 Suggesting a means of synthesizing a chromate 770

21.3 Accounting for the low reactivity of the coinage metals 775

21.4 Predicting the reactions of gold 778

21.5 Naming a coordination compound 783

21.6 Identifying optical isomerism 785

21.7 Calculating Δ_O from spectroscopy 790

21.8 Predicting the electron configuration of a complex 792

21.9 Predicting the magnetic properties of a complex 795

22.1 Identifying an element formed by α decay 806

22.2 Identifying an element formed by β decay 807

22.3 Identifying the products of other kinds of nuclear transmutations 808

22.4 Identifying the number of α and β emissions 810

22.5 Writing equations for nucleosynthesis reactions 813

22.6 Using the law of radioactive decay 818

22.7 Interpreting radiocarbon dating 821

22.8 Calculating the nuclear binding energy 826

22.9 Calculating the energy released during fission 827

23.1 Writing the formulas of isomeric molecules 842

23.2 Naming unbranched alkanes 844

23.3 Naming branched alkanes 847

23.4 Naming alkenes 852

23.5 Naming geometrical isomers 854

23.6 Naming derivatives of benzene 864

24.1 Classifying and naming alcohols 874

24.2 Predicting the product of a substitution reaction 877

24.3 Naming aldehydes and ketones 881

24.4 Naming carboxylic acids 888

24.5 Naming amines 893

24.6 Suggesting a synthesis of an amine 894

24.7 Judging whether a compound is optically active 897

TO THE INSTRUCTOR

I have aimed to write a text that conveys enthusiasm and gives pleasure while covering all the essentials of chemistry in a down-to-earth and useful way. I want—as I believe you want—your students to acquire a sense of the contribution that chemistry makes to our understanding of, control over, and respect for the world, an appreciation of chemical laws and patterns, and a degree of circumspection when confronting a social problem on which chemistry has a bearing. I have aimed to tell the reader a lot about chemistry, to express its logic, its excitement, its remarkable usefulness, and its dependence on judgment rather than pure deduction; and to show how to think scientifically. Above all, I have sought to transmit an *attitude* and to infuse the entire book with the sense of what I hope to achieve.

Through the years I have spent at work on this text, during which I have been in close contact with instructors around the world, and particularly in the United States, I have come to know and respect the enormous body of wisdom, experience, good practice, and professional judgment of those responsible for teaching General Chemistry, of other authors who have written texts for it, and of those who are merely concerned that it should be taught well. I have absorbed that experience, added my own, and in so doing have produced this text.

One of my express aims has been to avoid gimmickry, and to concentrate instead on presenting concepts in a logical, inviting, linear narrative rather than as an avenue of sideshows. The presentation is paced to give readers a sense of intellectual growth and evolving achievement, moving them from a state of almost total ignorance about chemistry to a stage at which they can deploy a variety of chemical concepts themselves.

The large-scale structure of the book is revealed by the names of the parts into which it is divided:

Part I   Matter and reactions
Part II   Atoms, molecules, and ions
Part III Rates and equilibrium
Part IV   The elements
Part V   Organic chemistry

This structure was adopted not to enforce a rigidity of organization (the part titles are easy enough to ignore) but rather to help the reader recognize that the subject *can* be broken into intellectually coherent blocks of knowledge. Another motive was

PREFACE

to give readers a sense of achievement (and in some cases, no doubt, a sense of relief) as the end of each part comes into sight. Schemes for providing structure, coherence, and relief have been incorporated on various scales throughout the text, down to the lowest level of subheadings.

The order of chapters has also been designed to reinforce the sense of structure and achievement. Specifically, the more observational topics—reactions, stoichiometry, and thermochemistry—are established early in the text, and then the more abstract topics relating to atoms and molecules are introduced. Nevertheless, I have not hesitated to use the concepts of atoms and molecules from the outset, since these are now familiar, in name at least, to just about everyone. Indeed, pictures of molecular models are scattered throughout the text, beginning in Chapter 1, usually alongside the first mention of an important molecular compound. The models make molecular structures familiar, prepare the way for organic chemistry, and foster chemical insight from the start. (It is an approach I used successfully in *Molecules*, a book I wrote for the Scientific American Library and intended for the nonchemist to enjoy.) Moreover, I have introduced the periodic table very early (in Chapter 2), since that is such a major component of the organization of chemical concepts and students should begin to use it as soon as possible. Chemical nomenclature and the various types of reactions are also discussed at the first opportunity (in Chapters 2 and 3). The sooner students acquire the basic language of chemistry, the sooner it will become second nature to them.

With the fundamentals established, the text moves into one of the heartlands of chemistry, the description of rates and equilibria, particularly as applied to the study of acids and bases. I have striven to show in this part of the book that it is possible to orchestrate the solution of an extraordinarily wide range of problems related to acids, bases, salts, and oxidizing and reducing reagents by making use of what is really a single idea. The attitude I want to convey is perhaps at its clearest here: that a simple idea richly deployed has immense power. The discussion of equilibria moves to its climax in Chapter 16, on entropy and free energy, where I figuratively pull back a curtain and introduce a new and powerful idea capable of explaining everything that has gone before. For this chapter, I have devised, after a lot of thought and advice, a new and more transparent nonmathematical approach to the introduction of entropy and free energy and their role in chemical thermodynamics.

In the remaining chapters, those dealing with the elements, including that remarkable element carbon, the text moves into another major component of General Chemistry, descriptive chemistry, the straightforward account of the physical and chemical properties of the elements and their compounds. These chapters represent a different cross section through chemistry, showing how the principles introduced in the first three parts enable us to understand the chemical personalities of the elements. They culminate at the point where organic chemistry merges with biochemistry, and the book ends with a tantalizing glimpse of chemicals coming to life.

I can point to a number of features that I have adopted to achieve my aims. Above all, I have concentrated on providing a clear statement of definitions and concepts. The written word is still the principal vehicle of communication in a textbook, and I have concentrated on making the exposition in this one clear, lively, and engaging. To supplement the text, I have developed an extensive, original, and systematic program of illustrations to help the reader imagine processes occurring at a molecular level and to develop insights that I hope will be retained long after the details of freshman chemistry have faded from memory. These diagrammatic representations of processes and structures are augmented by numerous photographs specially prepared for this text to familiarize the reader with the properties under discussion. The photographs promote an awareness of chemistry and chemicals that words alone cannot provide. Thus, through words and illustrations, this textbook spans the range of chemistry, from the once invisible atoms to the tangible everyday, forming an all-important bridge between understanding and experience.

Command of a concept comes from solving problems that relate to it. In common with all General Chemistry texts, this one emphasizes the importance of problem solving by offering numerous worked examples to illustrate the key ideas. My innovation lies in beginning every worked example with a "Strategy" section in which I suggest how the question might be tackled. Since it is important that students acquire a body of good sense about chemistry, many of the strategies start with a suggestion about how to guess the answer, or its order of magnitude if it is numerical. The rest of the strategy section then suggests ways of thinking about how the data should be deployed in order to arrive at the explicit answer. I am aware that there are probably as many ways of approaching a problem as there are instructors, and that in some instances you will feel more comfortable with approaches other than mine; but at least the student will find one perspective on each worked example and perhaps see how to approach similar cases.

What I have said so far may have given the impression that this book has an out-and-out "principles" orientation, with theory predominant. That is not so. I make no apology for the emphasis that I have put on the clear presentation of chemical theory, for this gives spine to the subject. However, it is not easy for a student to learn chemistry by putting either the principles or the descriptive chemistry in a wholly dominating position. These two deeply interwoven aspects of our subject grew up together, each one helping to edge the other forward, and I believe that we should emulate that symbiosis in our teaching. Students need organizational pegs to hang their knowledge on, and principles provide them. Therefore, while in the first three parts of the text I allow principles to steer the course, I use them as a framework within which to introduce a great deal of descriptive chemistry, as well as practical applications, social, medical, and environmental issues, and so on. I blend descriptive chemistry into the text with words and pictures wherever it seems appropriate. In many cases, I use an aspect of descriptive chemistry to introduce a new principle. All this I do with the conviction that it is imperative to show that chemistry is not cut-and-dried but is in fact a constant source of surprise. Once the principles have been firmly established, it is appropriate to allow descriptive chemistry to steer, and in the final two parts of the text, descriptive chemistry takes over, becoming, in the process, a way of showing how the principles work in practice. In these two parts, then, I weave a review of principles into the description of the elements and their reactions, often using the worked examples to show how to set about explaining a property or a trend.

There are a huge number of end-of-chapter exercises, most of them organized by topic. (Some suggestions for selecting exercises appropriate to particular needs are given in the *Instructor's Manual*.) In each chapter, the exercises range in difficulty from numerous drill questions to a selection of problems that are quite challenging. Although almost all are based strictly on the material dealt with in the chapter, a handful—two or three per chapter—invite the student to venture out from the bridgehead that the chapter has established. They are there for people who wish to stretch or be stretched. The most difficult are marked with an asterisk. In a somewhat different category are the exercises I have called "General" (so as not to light too bright a danger signal). The exercises in this section (unlike the exercises preceding them) draw on information from earlier chapters, requiring students to review and synthesize information from several places in the book. I have provided these in case you wish to use them to break down the artificial barriers between topics; they can easily be ignored.

A full package of supplements is available to accompany this text. I am fortunate to have been joined in my endeavor by a group of lively, experienced teachers who have worked hard and imaginatively to produce ancillaries that harmonize with my textbook yet also express their authors' vigorous personalities. They include

Study Guide, by David Becker, Oakland Community College
Instructor's Manual, by Joseph J. Topping, Towson State University
Solutions Manual, by Forrest C. Hentz, Jr. and G. Gilbert Long, North Carolina State University, and Joseph J. Topping, Towson State University
Laboratory Manual, by Alan Pribula, Towson State University
Computer Test Bank, by G. Gilbert Long, North Carolina State University
Video Demonstration for General Chemistry, by Ted Baldwin, Industrial Words and Images, with Dewey Carpenter, Louisiana State University, as chemical consultant.
Overhead Transparencies

Full details are available from the publisher.

TO THE STUDENT

I should like to start by introducing you to Moira Lerner, who has been my development editor from the beginning. She has made sure that what I write is suitable for the intended audience—you—and matches the kinds of courses you are just starting. In most books you will find editors thanked at the end of the preface, but Moira has played such a central role in the development of this text that she deserves a special mention at the outset. If you find that this text adds to your enjoyment of chemistry and presents topics in a way that you can understand, a lot of the credit should go to her.

Moira came to chemistry knowing perhaps even less about it than you do now. That is partly why it is relevant for me to mention her here. As we put the chapters together, she constantly questioned me about them: Why was I introducing that term now? Had I explained it? Why was it worth bothering with? Could I make it seem less boring? And so on. As time went on, I increasingly found her to be an ideal student, one who remained cheerful and hardworking, and who went on smiling as she bludgeoned her instructor—me—into clarity.

I hope you will enjoy chemistry both this year and when you need to draw on it again in the future. Take my word for it, *whatever* your chosen career, you *will* draw on chemistry. For some of you, it will be directly relevant, for you may become an academic or industrial scientist or a physician. If so, you will go on to build on this material in later courses and in your life's work. For others, it will be indirectly relevant, helping you to understand points that come up in different contexts. Thus, you will find that the knowledge of materials that chemistry provides will be useful should you become an engineer, a designer, a lawyer, an economist, or a painter, or take up a career in any of countless other professions. Even if you have no career, because you are out of work, too rich to bother, or in a penitentiary, chemistry will still touch almost every aspect of your life. It is responsible for the gasoline you use for travel, the construction materials in all the buildings around you, the fertilizers used to grow your food, the fabrics for your clothes, and the drugs to treat your injuries and diseases.

Chemistry is an intricate subject, because so many factors can influence the behavior of a substance. However, this book has been written with that intricacy in mind and with a concern that it should not overwhelm you. You will meet the ideas you need one by one and will gradually start seeing how they fit together. A part of the fascination of chemistry is knowing when one of many causes is likely to be dominant—when there is only one actor center stage—or when a whole chorus of effects must be taken into account. Similar problems of influences pulling in different directions are encountered in other aspects of life, and some of the attitudes you pick up here will be relevant in handling them, too. Chemistry shows how to unpick complex problems and teaches that many factors may be relevant to a rounded view of what is going on. In learning General Chemistry, you are actually training your mind to tackle a wide range of different tasks outside the subject itself.

A General Chemistry course looks at chemistry in a very broad way, introducing you to the principal ideas without getting too involved in detail. You can always study further if you want to find out more; but General Chemistry teaches you enough to understand in a general way the workings of the world as chemists see them.

You should pay particular attention to the worked examples in the text. There is a danger in treating a General Chemistry text like a novel and just gliding through rather than struggling with every page. I have tried to write the text so that you enjoy reading it, and want to turn the page, as in a good novel. However, it is up to you to make sure that you do in fact stop and pick a fight with each idea. Make yourself pause at the end of every subsection and decide whether you really understand the point being made. What is the message that I have tried to express? What are its logical consequences? The summaries at the end of the chapter should help you check whether the message that I think I have got across is the message that you have actually received.

Just as important, you should work carefully through the worked examples, which illustrate many of the key ideas and show the sorts of questions that you are likely to be asked in tests. This textbook is a bit different from others in that in each worked example I have explained how I myself would go about solving the problem. You may prefer to work it through in a different way—your instructor will be helpful here—but at least my remarks will help you get started (the hardest part of solving problems). In most cases (and this is where textbooks differ from real life), you will be able to work through a problem by thinking carefully about the material that has just been explained in the text. (Another very good source of information is the *Study Guide*, which in some cases will give you a different perspective on solving certain kinds of problems.) There are numerous exercises throughout the text (one in each worked example to give you an immediate self-test of whether you can do that sort of problem, and close to a hundred at the end of every chapter). You should do a lot of them: it is easy to convince yourself that you understand a concept, but doing the corresponding exercises is the only way to be sure you have truly grasped it.

Good luck with your course. Moira and I have done our best to provide you with a text that you will enjoy using and find easy to learn from. If you think I could have done better at a particular point, then write to me: keeping in touch with students and knowing their interests and difficulties is the best way of ensuring that a text is right.

ACKNOWLEDGMENTS

I have relied heavily on the collective wisdom of a large number of experienced instructors who have helped me develop my own understanding of what is needed in a text like this. Some have contributed countless pages of advice; others just a few comments that I have found myself hearing in my mind over and over again. Some of the most useful were the comments—on early drafts—that were most negative: how often I found myself agreeing with them, upon reflection. I hope that all these people will see that what was good has been retained and what was bad has been replaced by good. I should like to thank the following, purely in alphabetical order, for their help:

David L. Adams, Bradford College
John E. Adams, University of Missouri, Columbia
Martin Allen, College of St. Thomas (retired)
Norman C. Baenziger, University of Iowa
John E. Bauman, University of Missouri, Columbia
David Becker, Oakland Community College
James P. Birk, Arizona State University
Luther K. Brice, American University
J. Arthur Campbell, Harvey Mudd College
Dewey K. Carpenter, Louisiana State University
Geoffrey Davies, Northeastern University
Walter J. Deal, University of California, Riverside
John DeKorte, Northern Arizona University
Fred M. Dewey, Metropolitan State University
John H. Forsberg, St. Louis University
Marjorie H. Gardner, Lawrence Hall of Science

Gregory D. Gillespie, North Dakota State University
L. Peter Gold, Pennsylvania State University
Michael Golde, University of Pittsburgh
Thomas J. Greenbowe, Southeastern Massachusetts University
Robert N. Hammer, Michigan State University (consultant)
Joe S. Hayes, Mississippi Valley State University (retired)
Henry Heikkinen, University of Northern Colorado
Forrest C. Hentz, Jr., North Carolina State University
Jeffrey A. Hurlbut, Metropolitan State University
Earl S. Huyser, University of Kansas
Murray Johnston, University of Colorado, Boulder
Philip C. Keller, University of Arizona
Robert Loeschen, California State University, Long Beach
David G. Lovering, Royal Military College of Science, Shrivenham
James G. Malik, San Diego State University
Saundra McGuire, Cornell University
Amy E. Stevens Miller, University of Oklahoma
E. A. Ogryzlo, University of British Columbia
M. Larry Peck, Texas Agricultural and Mechanical University
Lee G. Pedersen, University of North Carolina, Chapel Hill
W. D. Perry, Auburn University
Everett L. Reed, University of Massachusetts
Don Roach, Miami-Dade Community College
E. A. Secco, St. Francis Xavier University
R. L. Stern, Oakland University
R. Carl Stoufer, University of Florida
Billy L. Stump, Virginia Commonwealth University
James E. Sturm, Lehigh University
James C. Thompson, University of Toronto
Donald D. Titus, Temple University
Joseph J. Topping, Towson State University
Patrick A. Wegner, California State University, Fullerton

One of the most formative episodes in the development of the text was a meeting in New Orleans with a group of advisors. Present in the group were David Becker, Dewey Carpenter, and Lee Pedersen, whom I have mentioned above, and also

Stephen J. Hawkes, Oregon State University
George F. Palladino, then at West Point Military Academy and now at the University of Pennsylvania
Duward F. Shriver, Northwestern University

Help and advice is too weak a term for what I have received from my publishers, Scientific American Books. The book was commissioned by Neil Patterson and John Staples and, after they had left the company, supported to the hilt by Linda Chaput and the team she has assembled, led, and inspired. Gary Carlson has put together the supporting package and has constantly helped me form my attitude—usually by forcing me to think again; Linda Davis has guided the development of this extraordinarily demanding text and has done so with sensitivity. Susan Moran was the project editor for much of the production, and whipped me along with charm, good humor, and (most important) efficiency as I sank for the third time in a sea of proofs. When she deservedly moved on to greater responsibilities, Georgia Lee Hadler, with whom I have worked enjoyably before, took over and continued what had become a happy tradition. Those talented designers Lynn Pieroni and Mike Suh have turned their graceful eyes to my manuscript, and have conjured from it a delightful page. Julia DeRosa and Janet Hornberger have made sure that even with an ocean between us the production bounded along with difficulties kept largely invisible from the author and that the execution of the text matched everyone's expectation. Bill Page ably coordinated the illustration program. I wish that the employees of York Graphic Services were less anonymous to me, but they will know (when they set this sentence) that I admire their accurate and skillful work. The same is true of the artists who developed the line illustrations from my sketches. Less invisible and anonymous were Ed Millman's contributions, for the nightmare-green ink of the copy editor brought uniformity to the text which otherwise it might have lacked. It is unusual to thank the sales representatives, for they might be thought to have a job to do only after my own is over. However, they did make a great contribution over the years of writing, for they brought me invaluable advice from the front and have helped me stay in touch with changing attitudes. It might seem that I have forgotten Moira Lerner, but remarks elsewhere show that I have not. She has been a slave driver, an inspiration, and a friend, and not such a bad development editor either.

My personal thanks go, too, to John Forsberg, Saint Louis University, who kept me in line with such good advice, and to O. C. Dermer, Oklahoma State University, who examined the chemistry with acuity and wisdom. Amy Stevens Miller and Debra Ewing, University of Oklahoma, were extraordinary: they checked and counterchecked all my numerical work and equations, all my calculations, and all my data, and I am deeply indebted to them. G. Gilbert Long and Forrest C. Hentz, Jr., North Carolina State University, Jack E. Powell, Iowa State University, and Jeffrey A. Hurlbut, Metropolitan State University, gave invaluable advice on the level, format, and content of the end-of-chapter exercises. The exceptionally fine photographs were taken for the text by Ken Karp under the guidance of Carlos Alva, Hunter College, and I enjoyed the sense of artistic and scientific creativity that they jointly brought to bear. Picture research was by Alice Dole, who discovered the ends of the earth as she trawled for the unusual among the mundane. Alison Marsland responded efficiently without audible complaint whenever I needed assistance, however demanding. Michael Clugston read the page proofs with his usual eagle eye, and Alexandra MacDermott read everything, often many times, and was a constant support and guide.

P. W. A.

GENERAL CHEMISTRY

MATTER AND REACTIONS

I

In this chapter we meet the basic language of chemistry, see a little of chemistry's method and aims, and learn the ground rules for reporting and using measurements. The illustration shows the Crab nebula—the remains of an exploding star first noticed in the year 1054. Such explosions scatter newly made elements through space and make them available on planets.

1 PROPERTIES, MEASUREMENTS, AND UNITS

The properties of substances

1.1 Physical and chemical properties

1.2 Pure substances and mixtures

Measurements and units

1.3 The International System of Units

1.4 Extensive and intensive properties

Using measurements

1.5 Unit analysis and conversion factors

1.6 The reliability of measurements and calculations

1.7 Significant figures in calculations

Chemistry begins in the stars. The stars are the source of the chemical elements, which are the building blocks of matter and the core of our subject. Within the stars, intense heat causes atoms of hydrogen, the smallest particles of the simplest element, to smash together, merge, and become atoms of other elements—carbon, oxygen, iron, and the rest. This merging releases even more heat, which generates starlight; so starlight, including sunlight, is a sign that the elements are still being formed.

Elements forged long ago inside ancient stars have found their way to earth. Many millions of years after a star is formed and it begins to cool, its outer layers may collapse, like a falling roof, into its exhausted core. This mighty quake produces such shock waves that the star shrugs off huge amounts of matter and sends it into space, in a huge explosion called a "supernova." Six such explosions have been detected in or near our galaxy in the past 1000 years, the most recent in 1987. The shock of the explosion raises the temperature in the star even higher than before: the Crab nebula (produced by the supernova of 1054) was visible in broad daylight for three weeks. At such high temperatures, even the heavy atoms in the star collide violently enough to merge and become still heavier atoms. The very heavy elements now found on earth, including uranium and gold, were made in this way.

Changes (we will come to know them as "chemical" changes) have taken place among the elements and converted them into the raw materials we find on earth; thus the elements formed in stars have become rocks, ocean, air, vegetation, and flesh. We humans have discovered how to change these raw materials into substances that are better suited to particular tasks.

In the early days of civilization, stone was replaced by metal as the material of which tools and weapons were made. At first, metals were blended together by accident. Then, as it was found that certain mixtures were easier to cast and mold, easier to use, or more durable, blending was done to achieve specific ends. The economic impact of the newly developed materials, together with humanity's spirit of inquiry, eventually gave rise to *science*, the systematically collected and organized body of knowledge based on experiment, observation, and careful reasoning. In particular, that spirit of inquiry gave rise to the branch of science known as chemistry. Greed admittedly drove that spirit originally, for chemistry sprang from the alchemists' vain struggle to convert lead into gold. Although the alchemists failed to copy the stars, in the course of their failure they discovered many conversions of matter that they could achieve. Thus they opened the door to modern chemistry.

Chemists have discovered through *experiments*—tests done under carefully controlled conditions—how to convert one kind of material into another. Their experiments, and the way they collect, organize, and use their observations and measurements, is where we begin our journey. In the course of it, in the remainder of the text, we meet the elements and see the combinations they form. We shall end it where chemistry merges with biology, with combinations of elements so elaborate that we regard them as being alive.

THE PROPERTIES OF SUBSTANCES

FIGURE 1.1 Sample of iron ore under a microscope.

Chemistry is the science of matter and the changes it can undergo. *Matter* is everything that takes up space. Matter includes the bricks and wood used for houses, the metal of airplanes, and the flesh and bone of human bodies. It includes water, air, earth, drugs, fertilizers, microchips, plastics, explosives, and food. Matter does not include light or abstract concepts like beauty (although chemistry, and chemicals, can be beautiful) because these do not take up space.

Chemists collect information about matter by making careful observations on a *sample*, a representative part of a whole. A sample of an ocean might be some seawater in a test tube; a sample of a rock might be a tiny fragment on a microscope slide (Fig. 1.1). As in everyday life, in chemistry we distinguish between different *substances*, or different kinds of matter. Water is one substance, iron another. We recognize different substances by their different *properties*, or distinguishing features. These properties include color, taste, smell, the ability to conduct electricity, and what happens when the substance is heated or mixed with other substances. The sweetening power of saccharin, for instance, was discovered by a dirty and careless chemist who licked his fingers in the laboratory and noticed a distinctive sweetness. (More chemists, though, have been made ill than have been made rich by potentially dangerous actions of this kind.)

1.1 PHYSICAL AND CHEMICAL PROPERTIES

In this section we start to sharpen concepts that are already familiar from everyday experience. That is a common feature of science: scientists accept much of everyday experience, express it precisely, and then explore the implications of what they have found out. Sharp and precise definitions provide an excellent basis for organizing observations and then making discoveries by noticing patterns or misfits in these patterns. The Russian chemist Dmitri Mendeleev made what is perhaps the greatest of all discoveries in chemistry, that the elements can be arranged as the periodic table (which we describe in the next chapter). Once he had recognized the pattern of the table, he went on to predict new elements by noting gaps in it.

We classify properties as either "physical" or "chemical" according to whether or not the formation of other substances is involved. That gold is yellow, conducts electricity, and melts at 1063°C are three of its physical properties because no new substance is formed. That natural gas—which is mainly methane—burns to produce carbon dioxide and water is one of methane's chemical properties, since it does involve the formation of a new substance.

Physical state and physical change. Often the first characteristic of a substance we notice is its *physical state*; that is, whether it is a solid, a liquid, or a gas at a particular temperature. We do this—barely con-

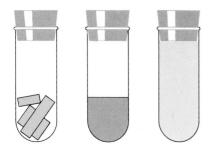

FIGURE 1.2 The three physical states of matter are easily distinguished by noting whether the substance retains its shape (solid), fills the bottom of the container (liquid), or fills the container completely however much is present (gas).

sciously, since it is so familiar a distinction—by noting which of the following physical properties the substance has (Fig. 1.2):

A **solid** is rigid and has a shape that is independent of the shape of its container.

Common substances that are solid at ordinary temperatures include brass, granite, quartz, and the elements copper, titanium, vanadium, and silicon. Many of the raw materials of the earth are solids, including its *minerals*, substances that are mined.

A **liquid** is fluid and takes the shape of the part of the container it occupies.

Common substances that are liquid at room temperature include gasoline and water. The only elements that are liquid at room temperature are mercury and bromine, but the element gallium melts to a liquid when held in the hand.

A **gas** is also fluid, but it fills the container it occupies and can easily be compressed into a smaller volume.

The term *vapor* is also widely used to mean a gas, but it usually signifies that the liquid state of the substance is also present, as in the water vapor that fills the space above the liquid in a kettle. Eleven elements are gases at room temperature. They include nitrogen and oxygen, the two principal components of the earth's atmosphere, and helium, neon, and argon, three of the so-called "noble gases."

The change of a substance from one of its physical states to another, as when liquid water freezes to ice, is called a *change of state*. Changes of state include *vaporization*, the formation of a gas or vapor, as when liquid water changes to water vapor during boiling. They also include *condensation*, the formation of a liquid from a gas, as when water vapor changes to a liquid, and *melting*, the formation of a liquid from a solid (see Table 1.1). One striking observation regarding some changes of state is that they occur at a single specific temperature. If we heat a solid, it remains a solid right up to its *melting point*, the temperature at which it melts; then some of it changes sharply to a liquid. Its temperature remains the same as we go on heating and more of it melts. Only when all the solid has melted does the temperature rise further. Similarly, when a liquid is cooled, it remains a liquid until its temperature has fallen to the *freezing point*, the temperature at which it freezes; then it begins to solidify. Once again, the temperature remains constant until all the liquid has frozen. The melting point is exactly the same as the freezing point, so ice melts to water at 0°C and water freezes to ice at 0°C.

Another sharp change occurs when a liquid is heated to its *boiling point*, the temperature at which it boils. Water does not boil until its temperature has reached 100°C, when it suddenly starts forming bubbles of vapor throughout its bulk. As in freezing and melting, the temperature of the boiling water remains constant until all the liquid has vaporized: a kettle of boiling water stays at 100°C until all the water is

TABLE 1.1 Changes of state

Initial state	Change	Final state
Solid	Melting or fusion	Liquid
Solid	Sublimation	Vapor
Liquid	Freezing	Solid
Liquid	Vaporization	Vapor
Vapor	Condensation	Liquid, solid

gone. Substances that boil at low temperatures are said to be *volatile*. Alcohol (ethanol) boils at 78°C and so is more volatile than water. Ether (diethyl ether) is more volatile still and boils at 34.5°C. The substances we ordinarily call gases are so volatile that their boiling points lie well below room temperature.

The sharpness of melting and boiling points, coupled with the fact that each substance melts and boils at a characteristic temperature, means that we can use these temperatures as guides to the identities of substances. One white powder looks very much like any other, but we can confirm that a white powder is aspirin, for example, by checking that it melts at 135°C. Several white substances may melt close to that temperature, but when it is also possible to check the temperature at which the sample boils (which is not possible with aspirin, because liquid aspirin decomposes before it boils), it can be identified with greater certainty. If each substance melted or boiled over a range of temperatures, this approach would not be feasible.

Chemical change. The conversion of one substance into another is called a *chemical change*. Chemical changes (which are more commonly called "chemical reactions") include the very complicated reactions that occur when a food is cooked and the substances that contribute to its flavor and aroma are formed. The extraction of metals like iron and copper from their *ores*, their natural mineral sources, is based on chemical changes, as is the production of synthetic fibers like nylon, acrylics such as Orlon, and polyesters such as Dacron from air, coal, and petroleum.

In some cases, chemical change can be brought about by passing an electric current through a substance, in the process called *electrolysis*. Electrolysis has been responsible for a number of important discoveries. For example, while the English chemist Humphry Davy was studying electrolysis systematically (in 1807), he discovered two new metals—sodium and potassium—within a few days of each other. Electrolysis is also the foundation of several great chemical industries, for it is used on a huge scale to convert aluminum oxide, which is mined as bauxite, to the metal aluminum (see Fig. 1.3). It is also used industrially to

FIGURE 1.3 Interior view of an aluminum plant showing the scale of the electrolysis operation. In the "pot room," electrolytic cells are connected in series.

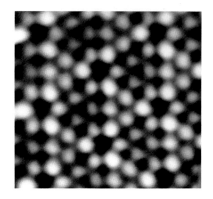

FIGURE 1.4 The fuzzy spheres in this photograph are individual silicon atoms. Advances in the design of microscopes—in this case an electron microscope— have only recently made pictures like this possible.

convert brine to sodium hydroxide and chlorine. Electrolysis is the only means we have for producing the gas fluorine, some of which is used for refining the uranium used in the nuclear power industry.

Elements and atoms. The single most important concept in chemistry is the <u>*atom*</u>, the smallest particle of an element that has the *chemical* properties characteristic of that element. An atom of gold has the chemical properties of gold, an atom of plutonium has the chemical properties of plutonium, and so on. The atoms of an element do not necessarily have the *physical* properties of the element: we cannot speak of a gold atom as "melting" or as being yellow. This is because many of the physical properties of substances are <u>*bulk properties*</u>, properties that depend on the collective behavior of large numbers of atoms. The melting point of a solid, for instance, depends on the collective behavior of the atoms as they change from an orderly, rigid arrangement in the solid to the disorderly, mobile collection typical of a liquid.

Atoms are far too small to be seen with the naked eye, but sophisticated microscopes can now make them out. A photograph of the surface of a sample of silicon is shown in Fig. 1.4; the fuzzy spheres are silicon atoms.

We now know of more than 100 different kinds of atoms. The precise number known in 1988 was 109, but new types of atoms are being made at about the rate of one every few years; by the time you read this we might know 110 or more. Having defined atoms, we can now define an <u>*element*</u> as a substance that consists of only one kind of atom. Hence there are about 100 different elements, each consisting of just one type of atom (Fig. 1.5a). The element hydrogen, for example, is a substance that consists only of hydrogen atoms, the element oxygen consists only of oxygen atoms, and so on. A list of the elements appears in Appendix 2D; there is no need to learn their names, or any of the other information given there, until we have discussed the elements at greater length in Chapter 2.

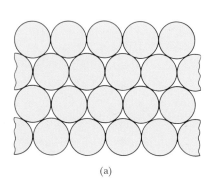
(a)

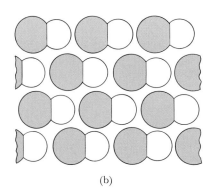
(b)

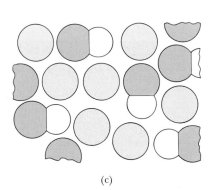
(c)

FIGURE 1.5 (a) An element consists of identical atoms (represented here by the yellow spheres). (b) A compound consists of different atoms bonded together in a strict ratio; here each molecule of a compound consists of one white atom bonded to one green atom. (c) A mixture is a simple intermingling of atoms in no set ratio; this is a mixture of the element and the compound.

Compounds and molecules. All matter is made up of atoms. A substance made up of atoms of at least two different elements in a definite ratio is called a <u>compound</u>. A compound is a *specific* combination of elements with a fixed composition. This is summarized by the following statement:

The **law of constant composition** states that a compound has the same composition whatever its source.

Water, for instance, is invariably made up of hydrogen and oxygen atoms, with two hydrogen atoms for each oxygen atom. It is therefore a compound of hydrogen and oxygen. Common table salt, sodium chloride, is a compound of the elements sodium and chlorine in which there is invariably one sodium atom for each chlorine atom. Methane is a compound of hydrogen and carbon, with four hydrogen atoms for each carbon atom. Carbon dioxide is a compound in which there are two (the meaning of the prefix *di-*) oxygen atoms for each carbon atom.

The different atoms in a compound are not just mixed together but are joined—the technical term is <u>bonded</u>—to one another in a specific way. In many compounds, the atoms are bonded together into definite, discrete groups called <u>molecules</u> (Fig. 1.5b). Just as an atom is the smallest particle with the chemical properties of an element, we can think of a molecule as being the smallest particle that has the characteristic chemical properties of a compound. We can speak, therefore, of molecules of water, methane, and carbon dioxide. These three molecules are shown in the models in the margin, where the black spheres represent carbon atoms, the white spheres hydrogen atoms, and the red spheres oxygen atoms. As we noted, a water molecule **(1)** consists of one oxygen atom bonded to two hydrogen atoms; a methane molecule **(2)** consists of one carbon atom bonded to four hydrogen atoms; and a carbon dioxide molecule **(3)** consists of a carbon atom bonded to two oxygen atoms. Sodium chloride, on the other hand, belongs to a different class of compounds (we discuss them in Chapter 2) and does not consist of individual molecules.

As is true of atoms and elements, the individual molecules of a compound do not necessarily have the same *physical* properties as the compound. We cannot think of a water molecule as freezing at 0°C or even as being a liquid. Water freezes when the water molecules in a bulk sample stick together to form an orderly, rigid structure. Above its freezing point, water consists of molecules that are able to move past each other so that the substance can flow.

1 Water

2 Methane

3 Carbon dioxide

1.2 PURE SUBSTANCES AND MIXTURES

From now on, when we use the term "substance" we shall always mean a single pure material. A substance may therefore be either an element or a compound. If it is an element (such as copper or bromine), all its atoms are the same. If it is a compound (such as water, sodium chloride, or benzene—a compound of carbon and hydrogen) it consists of atoms of specific elements bonded together in a definite ratio.

One very important task that chemists carry out is *chemical analysis*, the determination of the composition of a sample. Chemical analysis consists first in determining what elements are present and, if the substance is a compound, in what proportions. When the compound consists of molecules, a second goal of the analysis is to find the arrangement in which the atoms are bonded together. Chemical analysis is an important part of chemistry: once we know the composition of a substance and the arrangement of its atoms, we may be able to understand its properties, use it as a source of raw materials, or create more of it from other substances.

The first step in analysis is to decide whether a sample is a single substance (an element or compound) or a *mixture*, a mingling of different substances (Fig. 1.5c). We cannot do that simply by determining whether more than one element is present, because a compound consists of several elements but is a single substance. We might, however, be able to decide by making use of the differences between compounds and mixtures that are summarized in Table 1.2. The most important distinction at this stage is that a mixture normally exhibits a blend of the physical properties of its components, whereas the properties of a compound are usually strikingly different from the properties of its component elements. Water, a compound, is totally unlike the gases hydrogen and oxygen from which it is built. In contrast, a mixture of sugar and sand is both sweet (from the sugar) and gritty (from the sand). This suggests a critical test as to whether or not a sample is a mixture: see if its components can be separated by making use of their physical properties.

Techniques for separating mixtures. Some of the techniques that chemists use to separate mixtures are listed in Table 1.3. Three of the most important make use of three distinctive properties: the differing abilities of substances to dissolve, to vaporize, and to stick to surfaces.

In *filtration*, the sample is shaken with a liquid and then poured through a fine mesh. This technique depends on the different sizes of

TABLE 1.2 Differences between mixtures and compounds

Mixture	Compound
Can be separated using physical methods.	Cannot be separated using physical methods.
Composition is variable.	Composition is fixed.*
Properties are related to those of its components.	Properties are unlike those of its components.
Little heat is usually produced during formation.†	Considerable heat is usually produced during formation.‡

*There are exceptions: *nonstoichiometric compounds* have all the other characteristics of compounds, but have variable composition.
†There are exceptions: mixing sulfuric acid with water produces a lot of heat.
‡In some cases a chemical change is accompanied by considerable *absorption* of heat.

TABLE 1.3 Methods for separating mixtures

Method	Physical property used	Procedure
Centrifugation	Density	Rotation of liquid-solid mixture at high speed in a centrifuge; solid collects at bottom of sample tube
Filtration	Solubility	Pouring of solid mixture + liquid through a filter; solid trapped by filter
Recrystallization	Solubility	Slow crystallization of solid from solution
Distillation	Volatility	Boiling off the more volatile component of a liquid mixture
Chromatography	Ability to stick to surfaces	Passing of liquid or gaseous mixture over paper or through a column packed with material

particles. The technique can be used, for instance, to separate common salt from powdered glass: when a mixture of the two is shaken with water and poured through a filter paper, the glass particles are trapped and separated from the liquid. Filtration is often the first step in the treatment of domestic water supplies, since it removes grains of matter from the water.

The technique of *distillation* separates mixtures by vaporizing components of a mixture (Fig. 1.6). The technique depends on the different *volatilities* of the components, their readiness to vaporize. This technique can be used to remove water (which boils at 100°C) from salt,

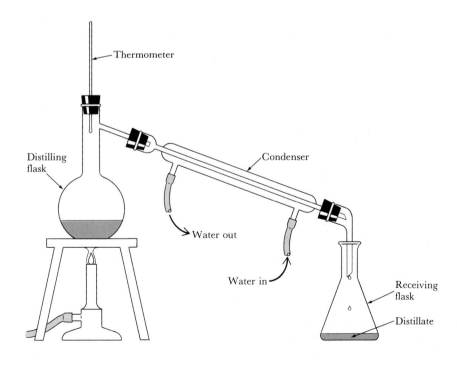

FIGURE 1.6 A simple distillation apparatus. The liquid mixture is heated, and the more volatile component boils off, is condensed, and is collected as the distillate.

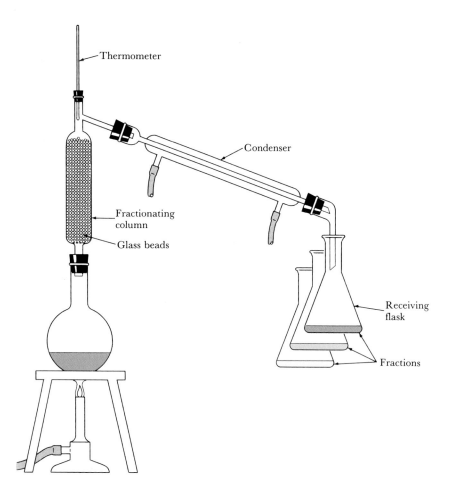

FIGURE 1.7 The apparatus used for fractional distillation includes a fractionating column packed with glass beads. The many condensations and vaporizations that occur on and from the surfaces of the beads ensure excellent separation of the components, which are collected as different "fractions."

Thermometer

Condenser

Fractionating column

Glass beads

Receiving flask

Fractions

FIGURE 1.8 The fractionating columns in an oil refinery work on a principle similar to that used in the laboratory, but on a much more substantial scale.

which boils at a much higher temperature and thus is left behind when the water is boiled off. It is most useful in the form of *fractional distillation* (Fig. 1.7). In this technique, which is used for separating liquid mixtures, the mixture is heated and the rising vapor passes up a column packed with glass beads. The vapor condenses on the beads and vaporizes again many times as it passes up the column, so that very effective separation is achieved. The first "fraction" to be collected is the most volatile of the mixed liquids; then, when that has been boiled off the sample, the next most volatile fraction comes through, and so on. Giant fractionating columns (Fig. 1.8) are used to separate the various fractions of petroleum, which range from the volatile liquids we use as gasoline (boiling at 50 to 200°C) and kerosene (175 to 325°C) to the heavier fractions that are used as diesel fuel (boiling above 275°C). The residue that remains behind after distillation is asphalt, which is used for surfacing highways.

Chromatography is a separation technique that relies on the different abilities of substances to *adsorb*, or stick to surfaces. There are various versions of the technique. In the simplest, the sample is washed along a strip of filter paper (Fig. 1.9). Substances that adsorb most weakly are washed along more quickly than others and, if they are colored, give rise to separate patches of color on the paper. (This is the origin of the name chromatography, which comes from the Greek words for "color

FIGURE 1.9 In paper chromatography the components of a mixture are separated by washing them along a paper—the "support"—with a liquid. This is the chromatogram of the components of a food coloring.

writing.") The location of a colorless substance is detected either by spraying the paper with a dye that attaches to the separated substances or by exposing it to ultraviolet light, which causes it to fluoresce (give out visible light).

In *gas-liquid chromatography* the sample is vaporized and carried in a stream of helium, the "carrier gas," through a long warmed tube. The tube is packed with aluminum oxide (or a similar unreactive substance) coated with the "stationary phase," a liquid of low volatility. The component that is adsorbed least strongly by the stationary phase emerges first from the far end (Fig. 1.10) and can be detected electronically. As time passes, the other components of the original mixture emerge, each detected and signaled by a peak on a chart. This chart, a record of the detector's output, is called a "chromatogram." It is a kind of fingerprint of the composition of the mixture. The different substances present can be collected as they emerge, or they can be identified from their positions in the chromatogram (Fig. 1.11). One use of gas chromatog-

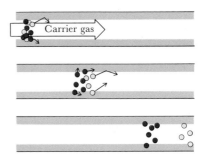

FIGURE 1.10 In gas-liquid chromatography a mixture is separated as its vapor is carried through a long coated tube. The components stick to the surface to different degrees and emerge at different times.

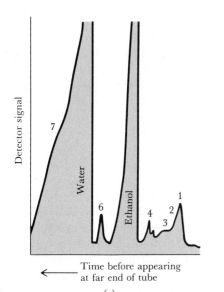

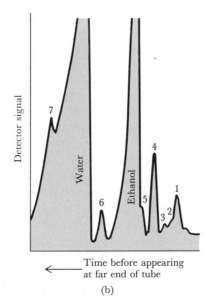

FIGURE 1.11 Gas chromatograms of (a) Scotch whisky and (b) bourbon whiskey, showing the components that contribute to their flavors. Mixing the components does not, it is said, recreate these flavors. Key: (1) Acetaldehyde, (2) formaldehyde, (3) ethyl formate, (4) ethyl acetate, (5) methanol, (6) propanol, (7) isoamyl alcohol.

FIGURE 1.12 This piece of rock is a heterogeneous mixture of many substances.

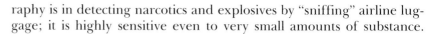

raphy is in detecting narcotics and explosives by "sniffing" airline luggage; it is highly sensitive even to very small amounts of substance.

Heterogeneous and homogeneous mixtures. We can see the different components of some mixtures with the unaided eye (Fig. 1.12). If we could magnify a small region of such a mixture so that individual atoms or molecules could be identified, we would see relatively large regions containing atoms or molecules of only one kind, and neighboring regions containing atoms or molecules of another kind. A mixture like this, a patchwork of different substances, is called "heterogeneous":

> A **heterogeneous mixture** is one in which the individual components, though mixed together, lie in distinct regions.

The regions in heterogeneous mixtures can be very small, perhaps only a few hundred atoms across. Among such mixtures are many of the rocks that form the landscape (Fig. 1.12). A mixture of sugar and sand, no matter how thoroughly mixed, is at the microscopic level a heterogeneous mixture.

In other mixtures we cannot make out distinct regions even with a microscope. Everywhere we look, on any scale of magnification, we would always find atoms or molecules of one component mingled among atoms or molecules of the other. Such mixtures are called "homogeneous":

> A **homogeneous mixture** is one in which the individual components are uniformly mixed, even on an atomic scale.

We cannot tell whether a sample is a homogeneous mixture or a pure substance until attempts have been made to separate it into different components. This is the case with air—a homogeneous mixture of nitrogen, oxygen, carbon dioxide, and several other colorless gases—for we cannot see that air is a mixture, even under a microscope. However, if we liquefy air we can separate its components by fractional distillation.

Homogeneous mixtures are also called *solutions*, although this name is often used only when there is much more of one component than another. Beer is a solution of alcohol in water, together with the substances that are responsible for its flavor. Seawater is a solution of salt (sodium chloride) and many other substances in water. The substance in excess—water, in these examples—is called the *solvent*, and the dissolved substances are the *solutes*.

We normally *dissolve* a solid (that is, form a solution in which it is the solute) by shaking it with the solvent or by stirring, as when we dissolve sugar in a cup of coffee. The opposite of dissolving is called either *crystallization*, when the solute slowly comes out of solution as crystals, or *precipitation* (Fig. 1.13), when it comes out of solution rapidly as a finely divided powder, a "precipitate." Crystallization often occurs when the solvent slowly evaporates, whereas precipitation is the result of a chemical reaction and is often almost instantaneous. The distinc-

FIGURE 1.13 Precipitation occurs when an insoluble compound is formed. Here yellow lead iodide, which is insoluble, precipitates when solutions of lead nitrate and potassium iodide are mixed.

tion between crystallization and precipitation actually is not very deep, because the powder is a mass of very small crystals.

Repeated dissolving and crystallization, the technique of _recrystallization_, is another means for separating substances, particularly impurities from a compound. The impurities stay in solution as the crystals form. This purification occurs on a grand scale when icebergs form from seawater: icebergs are composed largely of fresh water, as the salt remains behind in the ocean.

Beer and seawater are examples of _aqueous solutions_, solutions in which the solvent is water (_aqua_ is the Latin word for water). There are also many examples of _nonaqueous solutions_, those in which the solvent is not water. Gasoline is a nonaqueous solution of several solutes called "additives," which are added to the fuel to improve its properties. In "dry cleaning," grease and dirt on fabrics are dissolved in the solvent tetrachloroethylene, a compound of carbon and chlorine, which dissolves grease far better than water does.

Solutions need not be liquids. Solid homogeneous mixtures of metals are called _alloys_ and can be thought of as "solid solutions" of one or more metals in another metal. The alloy brass is a solid solution of zinc in copper, and pewter is a solid solution of the solutes antimony and copper in the solvent tin. The metal coating on the zinc core of a penny is an alloy of tin, zinc, and copper; the metal used for dimes and nickels is an alloy of nickel and copper.

MEASUREMENTS AND UNITS

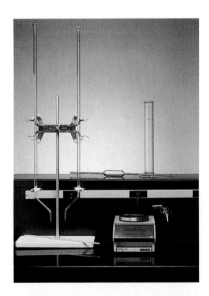

Many of the properties that we have discussed have been "qualitative," that is, purely descriptive. We have noted, for instance, that the metal used for dimes is an alloy of nickel and copper; but we have not mentioned the proportions in which they are mixed. In fact, many advances in science depend on "quantitative" observations, those that involve measurements and the reporting of numerical information (Fig. 1.14). Much of chemistry relies on the measurement of mass m, volume V, and temperature T:

The **mass** of a sample is the quantity of matter it contains.

The **volume** of a sample is the amount of space it occupies.

The **temperature** of a sample indicates how hot or cold it is.

Time (t) must also be measured when it is of interest to know how long it takes for the properties or composition of a sample to change.

More precisely, the mass of an object is a measure of its resistance to a change in its state of motion: a ball with a large mass takes more effort to throw than one with a small mass. It is also important to distinguish the mass of a sample from its "weight," which is the gravitational _pull_ on the sample. An astronaut has the same mass (is built from the same amount of matter) on earth, in space, and on the moon. The astro-

FIGURE 1.14 This photograph of a typical laboratory bench shows some of the measuring instruments often used by chemists. Clockwise from lower right: balance for weighing; thermometer for measuring temperature; two burets and a pipet for transferring volumes of liquids; graduated cylinder for measuring out volumes.

FIGURE 1.15 A very common measurement in a laboratory is that of the mass of a sample. This is done on a "balance" like the one shown here. The mass of the sample of powder (which is potassium chromate) is 10.140 g.

naut's weight is different in each of these places because the gravitational pulls are different. However, since weight is proportional to mass, so long as we make all our measurements in the constant gravity at the surface of the earth, we can measure mass by measuring weight. That is, we can measure the mass of a sample by comparing the pull exerted on it by the earth with the pull the earth exerts on a known mass. If the two pulls are the same, the masses are the same. This is the principle of the chemical balance and the origin of the term "weighing" for the measurement of mass (Fig. 1.15).

1.3 THE INTERNATIONAL SYSTEM OF UNITS

We measure a property of a sample by comparing it with a standard "unit" of that property. An example of a unit of length is "1 inch." The length of, say, a uranium rod may be reported as 5.0 inches only if everyone knows exactly what is meant by "1 inch" (Fig. 1.16). To make sure that everyone does, the different units are carefully defined at international meetings, and the definitions are then publicized around the world.

Scientists have found the *metric system* of units to be a convenient one. This system was introduced immediately after the French Revolution when, with revolutionary fervor, the French did away with their old units as well as their former leaders. The original metric system included what are now called the *base units*, from which all others can be constructed and which include units for mass, length, time, and temperature. Base units for electrical measurements have since been added, and the system has also been made more orderly. Together, the complete collection of rules, symbols, and definitions is now called the *International System of Units*. This is normally shortened to "SI," from the French *Système Internationale*. The great advantage of the system is that it lets scientists communicate easily with each other wherever they come from and whatever their field of interest. Chemists, for example, can communicate readily with biologists, engineers, physicists, and anyone else who needs their knowledge. The relationships between SI

SI base units

Property	Unit	Symbol
Mass	Kilogram	kg
Length	Meter	m
Time	Second	s
Temperature	Kelvin	K
Amount	Mole	mol
Electric current	Ampere	A
Luminous intensity	Candela	cd

FIGURE 1.16 The process of measurement consists of comparing an unknown with a standard and reporting the number of standard units involved in the property being measured. Here we are measuring the length of a uranium rod by comparison with units (inches) marked on a ruler. Its length is 5.0 inches.

TABLE 1.4 **Relations between units**

	Common unit	SI unit
Mass	1 ounce (oz)	28.35 grams (g)
	0.03527 oz	1 g
	1 pound (lb)	453.6 g
	2.205 lb	1 kilogram (kg)
	1 ton (2000 lb)	907.2 kg
Length	1 inch (in)	**2.54** centimeters (cm)
	0.3937 in	1 cm
	1 foot (ft)	**30.48** cm
	1 yard (yd)	0.914 meter (m)
	1.094 yd	1 m
	1 mile (mi)	1.6093 kilometers (km)
	0.6214 mi	1 km
Volume	1 liter (L)	**1000** cubic centimeters (cm^3)
	1 cubic foot (ft^3)	0.0283 cubic meter (m^3) (28.3 L)
	1 quart (qt)	946 cm^3 (0.946 L)
	1 gallon (gal)	3785 cm^3 (3.785 L)
Time	1 minute (min)	**60** seconds (s)
	1 hour (h)	**3,600** s
	1 day	**86,400** s

Entries in bold type are exact.

units and some older units (sometimes called "English units" and still widely used in the United States) are summarized in Table 1.4.

Mass. The first base unit we consider is the metric unit of mass, the *kilogram* (abbreviated kg). One kilogram (1 kg) is defined as the mass of a certain block of platinum kept at the International Bureau of Weights and Measures at Sèvres, just outside Paris.

Kilograms are fairly convenient units for everyday use. This book, for example, has a mass of about 1.5 kg. However, laboratory samples are typically much smaller, with masses of thousandths or ten-thousandths of a kilogram. Because it is sensible (and simpler) to use a unit that is similar in size to the thing being measured, smaller units of mass have been defined. One of these is the *gram* (g), which is exactly one-thousandth of a kilogram:

$$1 \text{ kg} = 1000 \text{ g}$$

A penny has a mass of 3 g, which is certainly more convenient to write (and remember) than 0.003 kg.

A still smaller unit of mass is the *milligram* (mg), which is defined as

$$1 \text{ mg} = \frac{1}{1000} \text{ g}$$

TABLE 1.5 SI prefixes

	Factor	Prefix*	Symbol
Multiples	1,000 (10^3)	kilo-	k
	1,000,000 (10^6, million)	mega-	M
	1,000,000,000 (10^9, billion)	giga-	G
Subdivisions	$\frac{1}{10}$ (10^{-1})	deci-	d
	$\frac{1}{100}$ (10^{-2})	centi-	c
	$\frac{1}{1,000}$ (10^{-3})	milli-	m
	$\frac{1}{1,000,000}$ (10^{-6}, millionth)	micro-	μ (Greek mu)
	$\frac{1}{1,000,000,000}$ (10^{-9}, billionth)	nano-	n
	$\frac{1}{1,000,000,000,000}$ (10^{-12}, trillionth)	pico-	p

*Other prefixes are also available, but are less commonly used.

The first samples of penicillin that were isolated in the early 1940s had masses of only a few milligrams.

The SI provides a number of prefixes like *kilo-* (k) and *milli-* (m) that may be applied to any unit. Each prefix multiplies its unit by some power of 10, from 10^{18} to 10^{-18}. The most important prefixes are listed in Table 1.5. The power-of-ten multiples in the table are expressed in "scientific notation." In that notation, 10^2 stands for $10 \times 10 = 100$, and 10^{-2} stands for $1/(10 \times 10) = \frac{1}{100}$. This notation is reviewed in Appendix 1A.

Length. The metric unit of length is the *meter* (m). Originally, the meter was defined as $\frac{1}{10,000,000}$ of the distance from the north pole to the equator. However, that distance was found to change slightly under the influence of the moon and the tides. Today the meter is defined as the distance light travels in just over $\frac{1}{300,000,000}$ second. (The precise definition is the distance it travels in exactly $\frac{1}{299,792,458}$ second.) This fraction was chosen so that the distance would be very close to the meter already in use.

The meter is a convenient measure of length for everyday use. Most people, for instance, are between 1.5 and 2 m tall. However, most laboratory samples have dimensions much smaller than 1 m, typically around 0.01 m. Therefore a more convenient unit for laboratory work is the *centimeter* (cm), with

$$1 \text{ cm} = \frac{1}{100} \text{ m}$$

A penny has a diameter of about 1 cm.

▼ **EXAMPLE 1.1** Selecting convenient units

When we later deal with atoms and molecules in more detail, we will find that their diameters are in the region of 10^{-10} m. What SI unit of length will be most convenient to use then?

STRATEGY The aim is to choose a unit comparable in size to the dimensions being reported. Therefore, we inspect Table 1.5 and look for a subdivision prefix that is close to 10^{-10}. If there are two candidates, it is often sensible to choose the one that avoids the use of decimal points.

SOLUTION Table 1.5 shows that nanometers (1 nm = 10^{-9} m) and picometers (1 pm = 10^{-12} m) might both be suitable. Since

$$1 \times 10^{-10} \text{ m} = 0.1 \times 10^{-9} \text{ m} = 0.1 \text{ nm}$$

and

$$1 \times 10^{-10} \text{ m} = 100 \times 10^{-12} \text{ m} = 100 \text{ pm}$$

it is probably more convenient to use picometers.

EXERCISE The diameter of a droplet of mist is about 10^{-7} m. What units would be most convenient to use when discussing it?

[*Answer*: nanometers]

▲

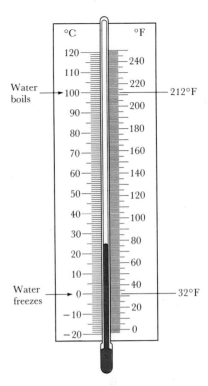

FIGURE 1.17 The Celsius (or centigrade) and Fahrenheit temperature scales. Note that the freezing point of water is 0°C (32°F), and its boiling point is 100°C (212°F).

Time. The SI unit of time is the *second* (s). (The name came about because the hour is first divided into *minute* parts called "minutes," which are then divided a *second* time into "seconds.") The second was originally defined as $\frac{1}{86,400}$ of the length of the day. However, the earth does not rotate at a constant rate; as a result, seconds were slightly longer on some days and shorter on others. The second is now defined as the duration of just under 10 billion (specifically, 9,192,631,770) oscillations of the light waves emitted by cesium atoms. This number of oscillations was chosen to make the time span very close to the second already in use.

Temperature. Temperatures are important in chemistry because they affect properties of substances, including physical state and the ability to undergo chemical change. Temperatures are often reported on the *Celsius scale*, named for Anders Celsius, the eighteenth-century Swedish astronomer who devised it. On the Celsius scale (which is also known as the "centigrade scale"), water freezes at 0°C ("zero degrees Celsius") and boils at 100°C. The Celsius scale and the everyday Fahrenheit scale (which is named for Daniel Fahrenheit, the German who proposed it in about 1714), on which water freezes at 32°F and boils at 212°F, are compared in Fig. 1.17.

We can convert a temperature on one scale to a temperature on the other with the formula

$$°F = 32 + \frac{9}{5} \times °C \tag{1}$$

This relation is one of many in chemistry that can be represented as a straight line on a graph (Fig. 1.18). The equation of a straight line is

$$y = a + bx$$

where a (the intercept) and b (the slope) are constants. Equation 1 has this form, with the intercept equal to 32 and the slope $\frac{9}{5}$. The graph can be quite useful for converting temperatures from one scale to the other, but for precise work Eq. 1 itself should be used.

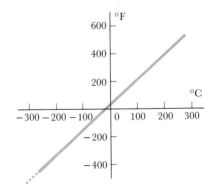

FIGURE 1.18 When temperatures on the Fahrenheit scale are plotted against those on the Celsius scale, a straight line is obtained. Notice that −40°C is the same as −40°F.

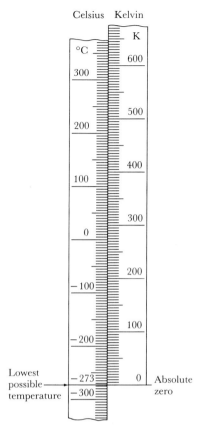

Celsius   Kelvin

FIGURE 1.19 The *Kelvin temperature scale* compared with the Celsius scale. The lowest possible temperature, *absolute zero*, is 0 on the Kelvin scale (0 K) and about −273 on the Celsius scale (−273°C).

▼ **EXAMPLE 1.2** **Converting between Celsius and Fahrenheit temperatures**

Express body temperature, about 99°F, on the Celsius scale.

STRATEGY Estimate the required temperature by referring to Fig. 1.17. Since Fahrenheit degrees are smaller than Celsius degrees, we expect temperatures on the Fahrenheit scale to be numerically greater than on the Celsius scale (so long as the temperature is greater than −40°C, where the scales coincide). For the precise value, we rearrange Eq. 1 to

$$°C = \frac{5}{9} \times (°F - 32)$$

and insert the Fahrenheit temperature.

SOLUTION A reading of 99 on the Fahrenheit scale is opposite 37 on the Celsius scale, so body temperature is about 37°C. From the equation:

$$°C = \frac{5}{9} \times (99 - 32)$$

$$= 37$$

That is, 99°F does correspond to 37°C.

EXERCISE Convert (a) 100°C to the Fahrenheit scale and (b) −10°F to the Celsius scale.

[*Answer:* (a) 212°F; (b) −23°C]

▲

In the International System, temperatures are reported on the *Kelvin scale*, named for Lord Kelvin, the Scottish physicist who invented it in 1848. The SI unit of temperature is called the *kelvin* (K); the word "degree" is not used. The kelvin is the same size as the Celsius degree, so there are 100 kelvins between the temperatures at which water freezes and boils. However, the zero on the Kelvin scale is 273.15 Celsius degrees below the freezing temperature of water. For most situations we can approximate this as 0 K = −273°C. This relationship is illustrated in Fig. 1.19, where the Kelvin scale begins opposite −273°C and has no negative values. The exact relation between the Kelvin and Celsius scales is

$$K = 273.15 + °C \tag{2}$$

This is the equation of a straight line with intercept 273.15 and unit slope (Fig. 1.20). On the Kelvin scale, water boils at 373 K (Fig. 1.21).

Kelvin chose −273°C for the starting point of his scale partly because (as we see in Chapter 5) the volume of a gas decreases as it is cooled and experiments suggested that the volume of any sample of gas reaches zero at −273°C. Since nothing can have a negative volume, −273°C is the *absolute zero* of temperature, the lowest possible temperature. Using a scale with 0 for that point therefore makes sense for scientific work, even though it leads to some unfamiliar numbers for common temperatures—such as 273 K for the freezing point of water and 373 K for its boiling point. The Kelvin scale, being more fundamental than the Celsius and Fahrenheit scales, has many advantages; whenever we use the symbol T in an equation in this text, we shall mean a temperature on the Kelvin scale.

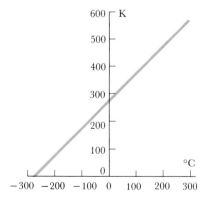

FIGURE 1.20 A plot of Kelvin temperatures against Celsius temperatures is another straight-line graph. The slope is unity because the kelvin has the same size as the Celsius degree.

Derived units. The volume V of a rectangular solid can be found by combining three measurements:

$$V = \text{length} \times \text{width} \times \text{depth}$$

Recall that we have defined the unit of length as the meter. This formula suggests that we can now define the unit of volume as the volume of a cube with each side equal to 1 m:

$$V = 1 \text{ m} \times 1 \text{ m} \times 1 \text{ m} = 1 \text{ m}^3$$

This unit of volume, the "cubic meter" (m^3), is an example of a *derived unit*, a combination of base units.

The cubic meter is a very large unit of volume for the samples we typically meet in the laboratory. Often a more convenient unit is the volume of a cube with sides 1 cm in length:

$$V = 1 \text{ cm} \times 1 \text{ cm} \times 1 \text{ cm} = 1 \text{ cm}^3$$

This unit of volume, the "cubic centimeter," is sometimes abbreviated cc (for cubic centimeter).

Although 1 cm^3 is a convenient unit for small volumes, it can be too small for describing the volumes of liquids and gases used in laboratories and in everyday life. The *liter* (L), defined as exactly 1000 cm^3, is therefore often used instead:

$$1 \text{ L} = 1000 \text{ cm}^3$$

The SI prefixes can be applied to the liter, and chemists find the *milliliter* (mL) particularly useful:

$$1 \text{ mL} = \frac{1}{1000} \text{ L}$$

FIGURE 1.21 The temperature of molten lava when it crashes into the sea at Surtsey, Iceland, is about 1000°C (about 1800°F or nearly 1300 K). The sea itself would be at about 5°C (41°F or 278 K) if it were not heated to the boiling point by the lava.

As 1 mL is exactly the same as 1 cm^3, it is a matter of choice which unit is used. In this book we use liters for the volumes of gases and large samples of liquids, milliliters for small liquid samples, and cubic centimeters for solids.

In deriving the unit of volume, we combined three measurements of the same kind (lengths). However, many properties are measured by combinations of different types of units. An example is *density d*, or mass per unit volume:

$$d = \frac{\text{mass}}{\text{volume}} = \frac{m}{V} \tag{3}$$

Substances of higher density (like lead) have a larger amount of matter in a given volume than substances of lower density (like aluminum).

We can derive the unit of density by thinking of a sample of unit mass (for convenience, 1 g) occupying unit volume (1 cm^3), so that

$$d = \frac{1 \text{ g}}{1 \text{ cm}^3} = 1 \text{ g/cm}^3$$

The derived unit of density is therefore the "gram per cubic centimeter" (g/cm^3). In practice we might find that the mass of a 2-cm^3 block of iron is 16 g, so that its density is

$$d = \frac{16 \text{ g}}{2 \text{ cm}^3} = 8 \text{ g/cm}^3$$

A similar derivation gives the "gram per milliliter" (g/mL) as the unit of density when volume is expressed in milliliters (as for liquids). The density of water at 25°C, for example, is 1.0 g/mL. The volumes of gases are normally reported in liters, so a convenient unit of density for them is the "gram per liter" (g/L). The density of air at sea level and 25°C is 1.2 g/L, about a thousand times less dense than that of water. Air is a much less "concentrated" form of matter than water.

1.4 EXTENSIVE AND INTENSIVE PROPERTIES

Chemists keep their collection of knowledge manageable by organizing it in several ways. Different branches of science often emerge from such organization (biology and the classification of living things into species is a notable example), for it enables patterns to be detected. The organization of knowledge is a first step in finding explanations and making predictions. One helpful way to classify properties is according to whether they are "intensive" or "extensive":

An **extensive property** depends on the size (the "extent") of the sample.

An **intensive property** is independent of the size of the sample.

The mass of a sample of sugar is an extensive property because the mass is greater when the sample is larger. The temperature of a sample of water taken from a thoroughly mixed heating tank is an intensive property because the same temperature is obtained whatever the size of the sample (Fig. 1.22). The density of a sample is intensive: although the mass and the volume increase as the size of the sample is increased, their *ratio* remains the same. Since the mass of a 2-cm^3 sample of lead (22.6 g) is twice the mass of a 1-cm^3 sample (11.3 g), their densities are equal (11.3 g/cm^3).

▼ **EXAMPLE 1.3** **Distinguishing extensive and intensive properties**

Which properties in the following statement are extensive, and which are intensive? "The density of water, which is a colorless liquid at room temperature, was determined by measuring the mass and volume of a sample."

STRATEGY Decide which properties depend on the size of the sample (extensive properties) and which do not (intensive properties). In some cases a property may be independent of size because an increase in one component property cancels an increase in another, as for density. In other cases, the property may, like temperature, be independent of the sample size by its nature.

SOLUTION Density, color, and temperature are all intensive. Mass and volume are extensive.

EXERCISE Classify the properties in the following statement: "Lead is a soft, dense metal with a low melting point."

▲ [*Answer*: Softness, density, and melting point are all intensive.]

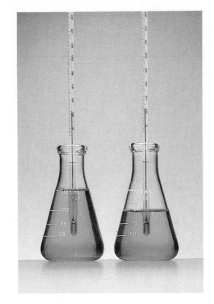

FIGURE 1.22 Mass is an extensive property, but temperature is intensive. These two samples of iron sulfate solution were both taken from the same well-mixed supply: they have different masses but the same temperature.

Intensive properties can be used to identify a substance, no matter what the size of the sample (Fig. 1.23). If a chemist finds that a metal object has a density of 11.3 g/cm^3, then it is certainly not aluminum (which has density 2.7 g/cm^3), but it *is* likely to be lead. On the other hand, an extensive property depends on the sample size and thus is not a guide to its identity: we cannot conclude that a lump of metal is aluminum simply by measuring its mass. It is strictly meaningless to

FIGURE 1.23 Some substances can be differentiated on sight, but some have similar appearances and can be identified only through testing. This student can probably recognize the blue crystals as copper sulfate. However, she needs to make careful tests to decide the identity of the white powder. It is in fact aspirin.

say, for example, that "lead is heavier than aluminum" because mass is extensive, and 1 m^3 of aluminum is much heavier than 1 cm^3 of lead. It *is* meaningful to say that "lead is denser than aluminum" because density is intensive, and any sample of lead is denser than any sample of aluminum. A feature of chemistry (and of science in general) is that it provides a language that helps to turn everyday expressions into precise, unambiguous statements.

USING MEASUREMENTS

Suppose the speed of a tennis ball is reported as 20 meters per second (m/s), and that of a car as 55 miles per hour (mi/h). Which is moving faster? To answer this kind of question we have to compare like with like, which means we must express the two speeds in the same units. This requires conversion from one set of units to the other. We also need to make conversions between units when a property is reported in one set of units but the information is needed in another. For example, we might know the mass and volume of a sample of gasoline in pounds and gallons, but need to find its density in grams per milliliter.

1.5 UNIT ANALYSIS AND CONVERSION FACTORS

While traveling in Canada, we see that our destination is 200 kilometers (200 km) away. How can we convert that distance to miles? We could begin by writing

$$\text{Distance in miles} = \text{distance in kilometers} \times \text{number of miles per kilometer}$$

The number of miles per kilometer is an example of a *conversion factor*, a ratio that converts values from one unit to another. Since Table 1.4 gives

$$1 \text{ km} = 0.6214 \text{ mi}$$

we know that there are 0.6214 mi in each kilometer. Hence,

$$\text{Distance in miles} = 200 \text{ km} \times \frac{0.6214 \text{ mi}}{1 \text{ km}}$$
$$= 124 \text{ mi}$$

In calculations of this kind, we use the following rules for rounding:

Round x00000 . . . through x49999 . . . to x.

Round x50000 . . . through x99999 . . . to $x + 1$.

Thus, 1.23 can be rounded to 1.2, and 1.2348 can be rounded to 1.235. (We explain later how to choose the number of figures to keep.) These rules also mean that 1.449 should be rounded in one step to 1.4, not first to 1.45 and then to 1.5. In complicated calculations we leave the rounding for the final step.

Unit analysis. The calculation we have just performed is an example of *unit analysis*, a technique for making conversions between units by using conversion factors. To construct a conversion factor we first identify the information required, as in "the number of miles per kilometer." Then we use a relation like those in Table 1.4 to express the information numerically. The relation 1 km = 0.6214 mi may be rearranged into

$$1 = \frac{0.6214 \text{ mi}}{1 \text{ km}}$$

which shows that there is 0.6214 mile per kilometer. The fact that dividing 0.6214 mi by 1 km gives 1 simply means that 0.6214 mi and 1 km are equivalent lengths. The term on the right is the conversion factor we used in the example.

If we were converting from miles to kilometers, we would need the number of kilometers per mile:

Distance in kilometers = distance in miles
× number of kilometers per mile

In this case we would rearrange 1 km = 0.6214 mi into

$$\frac{1 \text{ km}}{0.6214 \text{ mi}} = 1$$

which shows again that 1 km is the same as 0.6214 mi. The term on the left is the conversion factor we would need for this conversion.

A second key feature of unit analysis is that we multiply and divide the units in the same way as the numbers. When we converted from kilometers to miles, we canceled the km in the numerator and the denominator, leaving the unit miles (mi). This is a good check on the conversion; if the units do not come out right, the wrong factor has been used.

▼ EXAMPLE 1.4   Unit conversions

An automobile fuel tank is filled with 15 gallons (gal) of gasoline at a service station. How many liters is that?

STRATEGY   Since a liter is smaller than a gallon, we expect that there will be more than 15 liters (L). For the precise answer, we need to calculate

Volume in liters = volume in gallons × number of liters per gallon

We find the conversion factor, the number of liters per gallon, from Table 1.4.

SOLUTION   From Table 1.4, 1 gal = 3.785 L, so the conversion factor is

$$1 = \frac{3.785 \text{ L}}{1 \text{ gal}}$$

Therefore,

$$\text{Volume in liters} = 15 \text{ gal} \times \frac{3.785 \text{ L}}{1 \text{ gal}}$$

$$= 57 \text{ L}$$

EXERCISE   Calculate the mass in ounces (oz) of a 250-g package of breakfast cereal.

[*Answer*: 8.8 oz]

Converting derived units. The same principles apply to the conversion of derived units. Suppose we want to compare the speeds mentioned earlier, the 20 m/s of a tennis ball and the 55 mi/h of a car, and we decide to convert 20 m/s to miles per hour. The relation we need is

Speed in miles per hour = speed in meters per second
× number of miles per meter
× number of seconds per hour

Note that more complex conversions like these require two or more conversion factors. Although they can be set up in a line so that all the multiplications and cancellations are done in one step, there is nothing wrong in proceeding step by step. Indeed, that sometimes gives greater insight into the significance of each conversion. However, in this case, we will demonstrate the "one-line" approach. We find the two conversion factors from the relations

$$1 \text{ mi} = 1.609 \text{ km} = 1609 \text{ m} \qquad 1 \text{ h} = 3600 \text{ s}$$

which we rearrange to

$$\frac{1 \text{ mi}}{1609 \text{ m}} = 1 \qquad 1 = \frac{3600 \text{ s}}{1 \text{ h}}$$

Then

$$\text{Speed in miles per hour} = \frac{20 \text{ m}}{1 \text{ s}} \times \frac{1 \text{ mi}}{1609 \text{ m}} \times \frac{3600 \text{ s}}{1 \text{ h}}$$

$$= 45 \text{ mi/h}$$

The ball is traveling at 45 mi/h, so it is moving more slowly than the car.

▼ EXAMPLE 1.5 Converting derived units

How far in meters can an automobile travel on 1 mL of gasoline if its consumption is 18 mi/gal?

STRATEGY Expect a short distance, perhaps only a few meters, because 1 mL is a very small fraction of the fuel an automobile normally carries. We can find the answer once we know the consumption in meters per milliliter. Therefore, we begin by converting 18 mi/gal to meters per milliliter. This is done by setting up the relation

Consumption in meters per milliliter = consumption in miles per gallon
× number of meters per mile
× number of gallons per milliliter

and finding the conversion factors from Table 1.4.

SOLUTION From Table 1.4,

$$1 \text{ mi} = 1609 \text{ m} \qquad 1 \text{ gal} = 3.785 \times 10^3 \text{ mL}$$

which we rearrange to

$$1 = \frac{1609 \text{ m}}{1 \text{ mi}} \qquad \frac{1 \text{ gal}}{3.785 \times 10^3 \text{ mL}} = 1$$

Then

$$\text{Consumption in m/mL} = \frac{18 \text{ mi}}{1 \text{ gal}} \times \frac{1609 \text{ m}}{1 \text{ mi}} \times \frac{1 \text{ gal}}{3.785 \times 10^3 \text{ mL}}$$

$$= 7.7 \text{ m/mL}$$

(The m in mL does not cancel the m in the numerator: one stands for "milli" and the other for "meter.") Since the car's consumption is 7.7 m/mL, it can travel nearly 8 m on 1 mL of fuel. This result may give you a clearer picture of the rate at which fuel disappears from the tank as you travel along a highway.

EXERCISE Express the density of ice, 0.92 g/cm^3, in pounds per cubic foot (lb/ft^3), given that 1 ft^3 = 28.3 L.

[*Answer:* 57 lb/ft^3]

▲

▼ **EXAMPLE 1.6** **Converting compound units**

Tire pressures are often reported in pounds per square inch (lb/in^2). Express 25 lb/in^2 in kilograms per square meter.

STRATEGY Treat the multiple unit (square inches) as a product, and use conversions from Table 1.4 to set up the string of conversion factors that will give us the required units. Since we need one conversion factor for each power of the unit (in^2, for instance, requires two), we convert a power of a unit by the conversion factor raised to that power.

SOLUTION From Table 1.4, we need

$$1 \text{ lb} = 453.6 \text{ g}$$

$$= 0.4536 \text{ kg}$$

$$1 \text{ in} = 2.54 \text{ cm}$$

$$= 2.54 \times 10^{-2} \text{ m}$$

Then, after arranging these as conversion factors, we have

$$\text{Pressure} = 25 \frac{\text{lb}}{\text{in}^2} \times \frac{0.4536 \text{ kg}}{1 \text{ lb}} \times \left(\frac{1 \text{ in}}{2.54 \times 10^{-2} \text{ m}} \right)^2$$

$$= \frac{25 \times 0.4536}{(2.54 \times 10^{-2})^2} \frac{\text{lb} \times \text{kg} \times \text{in}^2}{\text{in}^2 \times \text{lb} \times \text{m}^2}$$

$$= 18 \times 10^3 \text{ kg/m}^2$$

EXERCISE Express 6.5 g/mm^2 in oz/ft^2.

[*Answer:* 21 × 10^3 oz/ft^2]

▲

1.6 THE RELIABILITY OF MEASUREMENTS AND CALCULATIONS

An important aspect of the public and international character of science is the meticulous care with which scientists report and use measurements. A simple example of how information should *not* be used is

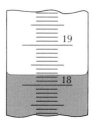

FIGURE 1.24 This volume of liquid could be reported as 18.26 mL, but the last digit is uncertain.

the following calculation of the density of sodium chloride (table salt), given that the mass of a sample is 2.5 g and its volume is 1.14 cm^3:

$$d = \frac{2.5 \text{ g}}{1.14 \text{ cm}^3} = 2.19298 \text{ g/cm}^3$$

But why is this incorrect? As you will see, the answer hinges on the significance of "2.5 g" and "1.14 cm^3."

Significant figures. Suppose we are measuring the volume of a liquid in a buret, and the level of the liquid is as shown in Fig. 1.24. We could report it as 18.26 mL, but the last digit, the 6, is little more than a guess. Different people looking at the same setup might report the volume as 18.25 mL or 18.27 mL or even 18.24 mL. Everyone would agree that it was 18.2*something* mL, but they would probably disagree about the *something*. Even if we averaged the readings to, say, 18.26 mL, the last digit would be uncertain.

Scientists would report the range of uncertainty in this measurement by writing 18.26 ± 0.02 mL. This notation indicates that the true value is estimated to lie between (18.26 − 0.02) mL = 18.24 mL and (18.26 + 0.02) mL = 18.28 mL. However, because this method is cumbersome, the convention is often adopted that the last digit in any reported measurement is uncertain by ±1. We would interpret a reported 18.26 mL, for example, as meaning that the true volume lies between 18.25 mL and 18.27 mL. As you can see, this convention may underestimate the range within which the true value lies, so it must be used cautiously.

The number of digits in a measurement, up to and including the first uncertain digit (the 6 in 18.26 mL), are called the *significant figures* in the measurement. There are four significant figures in a report of 18.26 mL. In a volume reported as 18 mL (signifying that it lies between 17 mL and 19 mL), there are two significant figures.

The zeros in a measurement such as 22.0 mL, 80.1 g, or 0.0025 kg sometimes cause a problem with regard to their significance. The trailing zero in 22.0 mL is significant because it indicates that the volume lies between 21.9 mL and 22.1 mL. Hence 22.0 mL has three significant figures. The zero in 80.1 g counts as an ordinary digit, so 80.1 g also has three significant figures. However, the leading zeros in 0.0025 kg are not significant. We can see this by changing the units from kilograms to grams and noting that 0.0025 kg is exactly the same as 2.5 g, a measurement with two significant figures.

To find the number of significant figures in a measurement with leading zeros, rewrite it in scientific notation with one nonzero digit to the left of the decimal point. Then all the digits in the number multiplying the power of 10 are significant figures. For example, by writing 0.0025 kg as 2.5×10^{-3} kg, we see that it has two significant figures—just as when we wrote it as 2.5 g. This also works for other measurements: the volume 22.0 mL would be rewritten as 2.20×10^1 mL, and so it has three significant figures.

Special care must be taken with a measurement reported as a round number of tens (say, 30 mL or 200 g). Are the zeros significant, implying that the mass lies between 199 g and 201 g? Or are they space fillers, implying that the mass lies between 190 g and 210 g, or even between 100 g and 300 g? One way of avoiding ambiguity is to report the measurement in scientific notation. Then 200 g would be written as 2.00×10^2 g or 2.0×10^2 g or 2×10^2 g, depending on the uncertainty of the measurement. Another way of showing that the zeros are significant, but which is only rarely used, is to write the decimal point with no zeros after it. Then 200. g means the same as 2.00×10^2 g: a value with three significant figures. Throughout this text, you should assume (unless told otherwise) that all zeros at the ends of numbers are significant.

▼ **EXAMPLE 1.7** **Counting significant figures**

Determine the number of significant figures in (a) 50.00 g; (b) 0.00501 m; (c) 0.00100 mm.

STRATEGY Write each measurement in scientific notation, keeping all trailing zeros but not leading zeros. Count the number of digits in each.

SOLUTION In the following, the significant figures are in bold type.
(a) **50.00** g = **5.000** $\times 10^1$ g; *four* significant figures
(b) 0.00**501** m = **5.01** $\times 10^{-3}$ m; *three* significant figures
(c) 0.00**100** mm = **1.00** $\times 10^{-3}$ mm; *three* significant figures

EXERCISE Determine the number of significant figures in (a) 2.1010 kg; (b) 100.000°C; (c) 0.0000001 K.

[*Answer:* (a) 5; (b) 6; (c) 1]

▲

We can now see why the density of 2.19298 g/cm³ that we calculated above is misleading. The mass of the sample, which we are told is 2.5 g, actually lies somewhere in the range 2.4 g to 2.6 g. The volume, 1.14 cm³, lies somewhere in the range 1.13 cm³ to 1.15 cm³. The density therefore lies somewhere between that calculated from the smallest mass and largest volume,

$$d = \frac{2.4 \text{ g}}{1.15 \text{ cm}^3} = 2.1 \text{ g/cm}^3$$

and that calculated from the largest mass and smallest volume,

$$d = \frac{2.6 \text{ g}}{1.13 \text{ cm}^3} = 2.3 \text{ g/cm}^3$$

Quoting the density as 2.2 g/cm³ (signifying that it lies in the range 2.1 g/cm³ to 2.3 g/cm³) is justified. Reporting it as 2.19298 g/cm³ is not.

Integers and definitions. The results of counting are *exact*. There is no uncertainty in the report "12 eggs": it means exactly 12, not 12 ± 1. The *integer* (whole number) 12 could in fact be taken to mean 12.000 . . . with the zeros continuing forever, and we could think of it as having an infinite number of significant figures. The values printed in bold type in Table 1.4 are also exact, even though some have digits to the right of a decimal point. They are *definitions*, and the relation 1 in = 2.54 cm defines an inch exactly. In this case 2.54 is shorthand for 2.5400 . . . with the zeros continuing forever. In contrast, 1 lb = 453.6 g is only an approximation to the exact (defined) value, 1 lb = 453.59237 g. The 273.15 in Eq. 2, the relation between Kelvin and Celsius temperatures, is also exact.

Errors in measurements. All measurements are uncertain, but some are more uncertain than others. Suppose we measure the mass of a sample of magnesium several times and get the following results:

2.5124 g   2.5122 g   2.5122 g   2.5125 g   2.5123 g

These measurements are all very close to the average value (2.5123 g) and there is little *random error*—the variation from measurement to measurement, which sometimes gives a high value and sometimes a low one. When the random error is small we say that measurements are *precise*. The five measurements above are indeed precise, and we could feel reasonably safe in reporting the mass as 2.5123 g, with five significant figures.

However, if there were a draft in the room, causing the balance to behave erratically while we were making the measurements, we might have recorded the following values:

2.5218 g   2.6214 g   2.5123 g   2.4134 g   2.4926 g

The random error here is large, and although the average value is the same as before, the measurements are imprecise. In this case we might feel safe in reporting the mass as 2.5 g, with only two significant figures. In other words, the lower the precision (the greater the random error), the smaller the number of significant figures that are justified.

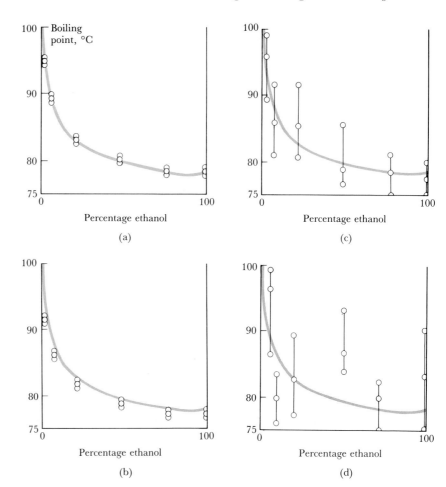

FIGURE 1.25 These diagrams illustrate the difference between accuracy and precision. The graphs are of the boiling point of a mixture of ethanol and water; the curve shows the true values and the circles are measurements. (a) Precise and accurate measurements; (b) precise but inaccurate measurements; (c) accurate but imprecise measurements; (d) inaccurate and imprecise measurements.

Now suppose that there is a speck of dust of mass 0.0100 g on the balance pan. Our measurements might be

$$2.5224 \text{ g} \quad 2.5222 \text{ g} \quad 2.5222 \text{ g} \quad 2.5225 \text{ g} \quad 2.5223 \text{ g}$$

with an average value 2.5223 g. The measurements are precise, but there is a *systematic error*—an error that appears in every measurement and does not average out. Measurements without a systematic error are said to be *accurate*; those with a systematic error are *inaccurate*. That is, accurate measurements are close to the actual value, and inaccurate measurements are not. The accuracy of a measurement depends on both the quality of the apparatus and the skill of the person using it.

We can summarize these remarks as follows:

Precise measurements have small random error and are reproducible in repeated trials.

Accurate measurements have small systematic error and give a result close to the actual value being measured.

These ideas are illustrated in Fig. 1.25.

1.7   SIGNIFICANT FIGURES IN CALCULATIONS

We shall often need to calculate the value of a property from other available information, called *data*. A simple example is the calculation of the density, where we used the mass and volume of the sample as data. In such cases we need to decide how many significant figures in the calculated value are justified by the number of significant figures in the data. In this section we discuss simple rules that *suggest* the correct number. These rules are designed to keep us from writing 2.19298 g/cm^3 for the density of a sample of sodium chloride with data as imprecise as a mass measured as 2.5 g and a volume measured as 1.14 cm^3.

Addition and subtraction.   In addition and subtraction, *the number of decimal places in the result should be the same as the smallest number of decimal places in the data.* Suppose we need the total volume of three blocks of copper with volumes 1.12, 1.4, and 2.000 cm^3. Then according to this rule, only one digit should follow the decimal point in the sum:

$$
\begin{array}{l}
1.12 \quad \text{cm}^3 \\
1.4 \\
\underline{2.000} \\
4.520 \text{ cm}^3 \text{: round to } 4.5 \text{ cm}^3
\end{array}
$$

This rule is based on the fact that the range of uncertainty of the least precise measurement (here 1.4 cm^3) is so great that it determines the precision of the sum. The rule is only a rough guide, and it may in some cases give too optimistic a range of uncertainty. The only reliable procedure, the one used by professional scientists, is to calculate the least and greatest values from the data, and to report the full range of uncertainty.

▼ **EXAMPLE 1.8** **Adding and subtracting with the correct number of significant figures**

Report (a) the total volume of a sample of water prepared by adding 25.6 mL to 50 mL; (b) the temperature in kelvins corresponding to the boiling point of sulfur, 444.674°C.

STRATEGY Add or subtract as required, identify the data value with the smallest number of digits following the decimal point, and round the result to that number.

SOLUTION (a) The smallest number of decimal places in the data is 0 (in 50 mL), so the total volume is

$$25.6 \text{ mL} + 50 \text{ mL} = 75.6 \text{ mL: round to } 76 \text{ mL}$$

(b) The conversion formula of Eq. 2 is

$$K = °C + 273.15$$

where 273.15 is exact (in other words, it has an infinite number of digits after the decimal point). Therefore, we round to the number of digits in the given Celsius temperature. Since

$$444.674 + 273.15000 \ldots = 717.824$$

the boiling point of sulfur is 717.824 K.

EXERCISE Report (a) the total mass of a sample prepared from 1.001 g of sugar, 2.05 g of salt, and 5.0 g of water; (b) the Celsius temperature corresponding to the melting point of iron, 1813 K.

▲   [*Answer*: (a) 8.1 g; (b) 1540°C]

Multiplication and division. In multiplication and division, *the number of significant figures in the result should be the same as the smallest number of significant figures in the data.* This rule leads to the proper precision in the sodium chloride density example: for a mass of 2.5 g and a volume of 1.14 cm³, the least number of significant figures is two and

$$d = \frac{2.5 \text{ g}}{1.14 \text{ cm}^3} = 2.19298 \text{ g/cm}^3\text{: round to } 2.2 \text{ g/cm}^3$$

The basis for this rule is the same as for the previous rule. The widest range of uncertainty in the data dominates and leads to a precision that is *approximately* that given by the rule. Again, the rule can be optimistic, and a more reliable procedure is to calculate and report the full range of uncertainty.

Multiplication by an integer is a disguised form of addition. The total length of two 5.11-cm rods is 2 × 5.11 cm, which actually means

$$5.11 \text{ m} + 5.11 \text{ m} = 10.22 \text{ m}$$

The number of decimal places in the result is therefore obtained from the rule for addition. The same holds for division by an integer, so

$$\frac{10.22 \text{ m}}{2} = 5.11 \text{ m}$$

An example: mass percentage composition. As an example of the usefulness of measurement concepts, we can apply them to the composition of mixtures. Ideally, composition should be reported in terms of *intensive* properties, so that the results apply to samples of any size. In contrast, most cooking recipes are *extensive*, because they are designed to lead to a cake or dish of predetermined size.

A convenient way of reporting the composition of a mixture intensively is to state the "mass percentage" (mass %) of each substance present:

The **mass percentage** of a substance A is the mass of A present in a sample, expressed as a percentage of the sample's total mass:

$$\text{Mass } \% \; A = \frac{\text{mass of } A \text{ in sample} \times 100\%}{\text{total mass of sample}}$$

Mass percentage is intensive, because the ratio does not change as the size of the sample increases. A table of alloys might state that a certain type of brass is 35% zinc by mass. We would then know that a 100-g sample contains 35 g of zinc and 65 g of copper. A 10.0-kg brass candlestick made of the same alloy would contain 3.5 kg of zinc and 6.5 kg of copper.

▼ **EXAMPLE 1.9** **Using mass percentage composition**

One meteorite is known to have a mass percentage composition of 26% nickel and 74% iron (Fig. 1.26). Calculate the mass of a sample of the meteorite that would contain 55 g of nickel.

STRATEGY We can use the mass percentage composition to set up a conversion factor between the mass of nickel required and the mass of sample that should be used. Since we know that 100 g of meteorite contains 26 g of nickel, we base the conversion factor on

$$100 \text{ g meteorite} = 26 \text{ g nickel}$$

SOLUTION The conversion we require is

$$\text{Mass of meteorite} = \text{mass of nickel}$$
$$\times \text{ grams of meteorite per gram of nickel}$$
$$= 55 \text{ g nickel} \times \frac{100 \text{ g meteorite}}{26 \text{ g nickel}}$$
$$= 2.1 \times 10^2 \text{ g meteorite}$$

EXERCISE The composition of commercial cold remedy is 87% aspirin, 7% caffeine, 6% vitamin C. What mass of the mixture should you weigh out in order to obtain 5 g of caffeine?

[*Answer*: 7×10 g]

▲

▼ **EXAMPLE 1.10** **Using mass percentage composition and density**

A sample of bronze consists of 82% copper and 18% tin by mass and has a density of 8.7 g/cm^3. What volume of sample contains 75 g of tin?

STRATEGY First, we can calculate the mass of bronze that contains the required mass of tin, proceeding as in Example 1.9. We can then convert that mass of bronze to the corresponding volume of sample by using the density as a conversion factor.

FIGURE 1.26 Nickel-steel was known even in prehistoric times. The head of this Chinese dagger is a nickel-steel meteorite. The dagger dates from the Chou dynasty, which began in 1027 B.C.

SOLUTION The first conversion uses

$$100 \text{ g bronze} = 18 \text{ g tin}$$

and gives

$$\text{Mass of bronze} = 75 \text{ g tin} \times \frac{100 \text{ g bronze}}{18 \text{ g tin}} = 417 \text{ g bronze}$$

The density tells us that

$$8.7 \text{ g bronze} = 1 \text{ cm}^3 \text{ bronze}$$

and we find (rounding to two significant figures)

$$\text{Volume of bronze} = 417 \text{ g bronze} \times \frac{1 \text{ cm}^3 \text{ bronze}}{8.7 \text{ g bronze}} = 48 \text{ cm}^3$$

EXERCISE A sample of iron ore consists of 57% by mass of iron and has a density of 5.0 g/cm^3. What volume of the ore contains 1.0 kg of iron?

[*Answer*: 3.5×10^2 cm^3]

SUMMARY

1.1 Chemistry makes progress through careful observation of the properties of samples of **matter,** the material of the universe. It deals with **substances,** which are different pure kinds of matter, and which may be either elements or compounds. **Atoms** are the smallest particles with the chemical properties of an element. A substance is an **element** if all its atoms are identical. A substance is a **compound** if it consists of different atoms that are present in fixed proportions and bonded together. In many cases, compounds consist of **molecules,** which are definite groupings of atoms and are the smallest particles that show the chemical properties of the compound. Properties are classified as either **physical,** if they do not involve a chemical change, or **chemical,** if they do. The former include the **physical state** of the substance (whether it is a **gas, liquid,** or **solid**) and the temperatures at which **changes of state** (**melting, freezing, vaporization,** and **condensation**) occur. The chemical properties of a substance include behavior in **electrolysis,** in which an electric current is passed through the substance.

1.2 The composition of a sample is obtained by **analysis,** the first step of which is to check if a sample is **pure**—a single substance—or a **mixture,** consisting of one or more intermingled substances. Mixtures can be separated by techniques that depend on the different physical properties of the components, including **filtration, distillation, chromatography,** and **recrystallization.** Mixtures may be **homogeneous** or **heterogeneous;** the former type includes **solutions** and **alloys.**

1.3 Many of the observations made in chemistry are **measurements,** or numerical observations. Four measurements frequently used are those of **mass, volume, time,** and **temperature.** The results of measurements are reported using the **International System of Units** (SI units). The **base units** of the system include the **kilogram** for mass, the **meter** for length, the **second** for time, and the **kelvin** for temperature. **Derived units** for measuring other properties are obtained from these base units. Multiples and subdivisions of base and derived units are constructed with prefixes that denote different powers of 10 (see Table 1.5).

1.4 A property that depends on the size of the sample is called an **extensive** property; one that is independent of size is called an **intensive** property. Intensive properties are often more useful for identifying a substance.

1.5 Measurements are converted from one set of units to another using **unit analysis,** in which a property expressed in one set of units is multiplied by a **conversion factor;** the units are multiplied and divided like numbers.

1.6 Great care must be used not to give the impression that a reported value is more reliable than is the case. The convention is adopted that the last digit of a numerical value is taken to be uncertain to the extent of ± 1. The number of digits up to and including the first uncertain digit is the number of **significant figures** in a measurement. If in a series of measurements there is only a small **random error,** a variation from measurement to measurement, then the measurements are said to be **precise;** the greater the number of significant figures, the greater the precision. A measurement is **accurate** if it is close to the true value, which implies that there is no **systematic error** (an error that does not average to zero).

1.7 When **data,** or supplied information, are used in a calculation, the result must not give a false impression of its reliability. Rules have been devised for estimating the number of significant figures justified by the data, but they must be used cautiously. Measurements of the composition of mixtures are reported intensively as **mass percentages,** the mass expressed as a percentage of the whole.

EXERCISES

The exercises (apart from those in the General section) are paired, so that each even-numbered exercise is closely related to the odd-numbered exercise that precedes it. Answers to the odd-numbered exercises are listed at the back of the book. Treat all zeros at the ends of numbers as significant unless otherwise stated. An asterisk (*) is a warning that the exercise is a little more difficult than the rest.

Physical and Chemical Properties

In Exercises 1.1 to 1.4, classify the properties and changes as either physical (P) or chemical (C).

1.1 (a) Color; (b) melting point; (c) flammability; (d) volatility.
1.2 (a) Density; (b) freezing point; (c) edibility; (d) tendency to rust.

1.3 (a) Melting; (b) evaporating; (c) burning; (d) rusting.
1.4 (a) Condensing; (b) boiling; (c) freezing; (d) digesting.

In Exercises 1.5 and 1.6, underline all the *physical* properties and changes in the given statements.

1.5 "The temperature of the land is an important factor for the ripening of crops, because it affects the evaporation of water and the humidity of the surrounding air."
1.6 "Coffee beans are ground into a powder, and some of the substances present are dissolved in hot water; other substances that are present are destroyed at the high temperatures sometimes used."

In Exercises 1.7 and 1.8, identify the pure substances (P) and mixtures (M).

1.7 (a) Ink; (b) coffee; (c) gold; (d) diamond.
1.8 (a) A soft drink; (b) gasoline; (c) chalk; (d) seawater.

In Exercises 1.9 and 1.10, state the physical property on which each separation method depends.

1.9 (a) Distillation; (b) chromatography; (c) evaporation to dryness; (d) sorting by hand.
1.10 (a) Filtration; (b) recrystallization; (c) gold panning; (d) gasoline refining.

In Exercises 1.11 and 1.12, suggest a method for separating each mixture.

1.11 (a) Acetic acid and water; (b) a solution of sugar in water; (c) the substances responsible for the color of a rose; (d) air.
1.12 (a) A mixture of chalk and table salt; (b) a mixture of chalk, table salt, and sugar; (c) the components of natural gas; (d) the substances responsible for the odor of spearmint.

The International System of Units

Exercises 1.13 and 1.14 give practice in the use of metric prefixes. In each case, complete the equation.

1.13 (a) 250 mL = _____ cm^3
(b) 250 mL = _____ L
(c) 28.3 g = _____ kg
(d) 25.4 mm = _____ cm
1.14 (a) 100 cm^3 = _____ L
(b) 50.00 mL = _____ L
(c) 0.454 kg = _____ g
(d) 0.0123 g = _____ mg

In Exercises 1.15 and 1.16, rewrite the statements using the units indicated in brackets.

1.15 A sample of tin of area 1 cm^2 [mm^2] was placed on a small block of lead of volume 1 cm^3 [m^3]. The two metals were placed on the bottom of a 1-m^3 [cm^3] tank and 10 mL [m^3] of acid was poured in.
1.16 A minute sample of the metal iridium of volume 1 mm^3 [cm^3] was recovered from land of area 1 km^2 [m^2]. The chemist doing the analysis wanted to determine the amount to be found in 1 m^3 [mm^3] of soil, and so analyzed a 0.010-m^3 [mL] sample.

In Exercises 1.17 and 1.18, complete the conversions using information from Table 1.4.

1.17 (a) 1.0 cm^2 = _____ in^2
(b) 1.0 yd^2 = _____ m^2
(c) 1.0 acre = 4840 yd^2 = _____ m^2
(d) 1.0 hectare = 100 m × 100 m = _____ acre
1.18 (a) 1.0 in^2 = _____ mm^2
(b) 1.0 L = _____ ft^3
(c) 1.0 barrel = 42 gal = _____ L
(d) 1.0 acre = _____ km^2

In Exercises 1.19 and 1.20, write the numbers using scientific notation.

1.19 (a) 10,000; (b) 186,000; (c) 0.000000001; (d) 55 billion; (e) one ten-thousandth.
1.20 (a) 100,000; (b) 0.00454; (c) 1989; (d) 100 billion; (e) three ten-billionths.

In Exercises 1.21 and 1.22, complete each equation using scientific notation.

1.21 (a) 1 pm = _____ m
(b) 1 nm = _____ cm
(c) 10 g = _____ mg
(d) 100 cm = _____ km
1.22 (a) 100 pm = _____ nm
(b) 1 angstrom = 10^{-8} cm = _____ pm
(c) 0.454 kg = 4.54 × _____ mg
(d) 500 nm = 5.00 × _____ mm

In Exercises 1.23 and 1.24, rewrite each statement using scientific notation and converting to the unit indicated in brackets.

1.23 (a) It is often necessary to know the number of atoms in an area of 1 nm^2 [m^2].

(b) The volume of a solid sample is typically 1 cm^3 [mm^3].

(c) The volume of a certain gas sample is 1 L [dm^3].

(d) The maximum loading of a certain shelf is 6 kg/m^2 [g/cm^2].

1.24 (a) The area occupied by an atom on a surface is about 5×10^4 pm^2 [nm^2].

(b) The smallest particle found in a sample of river water was 100 nm^3 [m^3].

(c) The size of a typical lab flask is 250 mL [dm^3].

(d) A certain measurement gave the result 5 kg$^2 \cdot$ m/s^2 [g$^2 \cdot$ cm/s^2].

Temperature Conversions

In Exercises 1.25 and 1.26, convert the temperatures. State the significance of each temperature identified by an asterisk.

1.25 (a) *98.6°F = _____ °C = _____ K

(b) *−40°C = _____ °F = _____ K

(c) *0 K = _____ °F

(d) *212°F = _____ K

1.26 (a) −269°C = _____ K

(b) 1000°C = _____ °F

(c) 80°F = _____ °C = _____ K

(d) 32°C = _____ °F = _____ K

In Exercises 1.27 and 1.28, rewrite the statements using the temperature scales indicated in brackets.

1.27 (a) The melting point of gold is 1063°C [°F, K].

(b) Sulfur boils at 445°C [°F, K].

(c) Neon boils at −411°F [K].

(d) The temperature of space is 2.7 K [°C].

1.28 (a) Propane boils at −42°C [°F, K].

(b) Properties are often reported at 298 K [°C, °F].

(c) The element iridium boils at 4550°C [°F].

(d) Helium boils at 4 K [°C, °F].

Derived Units

In Exercises 1.29 and 1.30, state the unit in which each property would be expressed if the base units were taken as (1) the kilogram, meter, second, and kelvin; (2) the gram, centimeter, second, and kelvin. (Some of the combinations may seem odd, but we see later that they have an important role in chemistry.)

1.29 (a) Speed (= distance per unit time); (b) acceleration (= speed divided by time); (c) kinetic energy

[= $\frac{1}{2}$ mass times (speed)^2]; (d) force (= mass times acceleration).

1.30 (a) Density (= mass per unit volume); (b) momentum (= mass times speed); (c) pressure (= force per unit area); (d) energy (= force times distance).

Extensive and Intensive Properties

In Exercises 1.31 and 1.32, state whether each property is extensive (E) or intensive (I).

1.31 (a) The color of copper sulfate; (b) the mass of steel; (c) the cost per gallon of gasoline; (d) the density of lead.

1.32 (a) The density of water; (b) the temperature of boiling sodium; (c) the volume occupied per unit mass of butter; (d) the cost of gasoline.

In Exercises 1.33 and 1.34, which statements are meaningful as they stand?

1.33 (a) Gasoline costs more than water.

(b) Iron is heavier than water.

(c) Molten iron is hotter than melting ice.

(d) Hydrogen is lighter than air.

1.34 (a) Oil is less dense than water.

(b) Both copper sulfate and the sky are blue.

(c) Water has a higher boiling point than liquid air.

(d) Steam occupies a greater volume than water.

Unit Conversion

In Exercises 1.35 to 1.40, complete the equations using the information in Tables 1.4 and 1.5. Make sure that your answers have the correct number of significant figures.

1.35 (a) 15 L/s = _____ gal/s

(b) 5 lb/ft^3 = _____ kg/ft^3

(c) 65 kg/cm^2 = _____ lb/cm^2

(d) 2×10^2 mm^2/s = _____ in^2/s

1.36 (a) 50 in^3/min = _____ cm^3/min

(b) 100 kg/ft^2 = _____ ton/ft^2

(c) 10 acres/day = _____ m^2/day

(1 acre = 4840 yd^2)

(d) 5 kg $\cdot$ m^2/s^2 = _____ g $\cdot$ cm^2/s^2

1.37 (a) 15 L/s = _____ L/h

(b) 5 lb/ft^3 = _____ lb/m^3 = _____ lb/cm^3

(c) 65 kg/cm^2 = _____ kg/in^2

(d) 11.3 g/cm^3 = _____ g/m^3 = _____ g/ft^3

1.38 (a) 50 in^3/min = _____ in^3/s

(b) 100 kg/ft^3 = _____ kg/m^3 = _____ kg/cm^3

(c) 10 acres/day = _____ acre/s

(d) 8.9×10^3 kg/m^3 = _____ kg/cm^3 = _____ kg/in^3

1.39 (a) 65 kg/cm^2 = _____ lb/in^2

(b) 11.3 g/cm^3 = _____ lb/ft^3 = _____ ton/ft^3

(c) 15 L/s = _____ gal/h

(d) 3.0×10^8 m/s = _____ mi/h

1.40 (a) 10 acres/day = _____ m²/s
= _____ mm²/ms

(b) 2698 kg/m³ = _____ g/cm³

(c) 1.00 ft³/min = _____ gal/h = _____ cm³/s

(d) 2.0 kg/cm² = _____ g/nm²

1.41 Express the diameter of the earth, 1.3×10^4 km, in miles.

1.42 Express the surface area of the earth, 5.1×10^{14} m², in square miles.

1.43 Convert the density of copper, 8.9 g/cm³, into pounds per cubic foot.

1.44 Convert the density of gold, 19.3 g/cm³, into (a) pounds per cubic foot; (b) kilograms per cubic meter.

1.45 One "light-year" is the distance light travels in 1 year and is equal to 9.46×10^{15} m. (a) How far is that in miles? (b) How far (in meters) is (i) 1 light-second; (ii) 1 light-nanosecond? (c) The nearest star is about 4 light-years away; how long would it take a space ship traveling at 20,000 mi/h to reach it?

1.46 In astronomy, distances are often quoted in "parsecs" (pc), for which 1 pc = 3.08×10^{16} m. (a) Express 1 pc in (i) light-years; (ii) miles. (b) How tall is a 5-ft 7-in person in parsecs?

In Exercises 1.47 and 1.48, express the statement in the appropriate SI units.

1.47 "A typical young 5-ft 10-in male weighs 154 lb and needs to consume about 2.0 oz of protein a day."

1.48 "A typical young 5-ft 4-in, 120-lb female needs to consume about 1.6 oz of protein a day."

1.49 The normal recommended consumption of vitamin C for a young adult is 60 mg/day. In the course of a year, to how much does that correspond in (a) grams; (b) ounces?

1.50 The recommended *maximum* intake of copper by an adult is 3 mg/day. In the course of a lifetime of 75 years, to how much does that correspond in (a) grams; (b) ounces?

1.51 Gold ore typically contains 10 g of gold per 1000 kg of ore, and the current value of gold is $500 per ounce. What mass of ore in tons must be refined to obtain $1 million worth of gold?

1.52 The element manganese occurs in seawater to the extent of about 9.5 tons/mi³. What mass of manganese (in micrograms) is present in a 250-mL sample of seawater?

Precision and Accuracy

In Exercises 1.53 and 1.54, state the number of significant figures in each measurement.

1.53 (a) 2.00 g; (b) 202 m; (c) 200. m; (d) 0.0200 s.

1.54 (a) 3.00100 g; (b) 0.001 K; (c) 2.998×10^8 m/s; (d) 1001 m.

1.55 Shown below are two vessels graduated in milliliters. Report the volumes, using the correct number of significant figures.

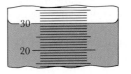

1.56 Repeat Exercise 1.55 for the vessels shown below.

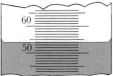

Reliable Calculations

In Exercises 1.57 and 1.58, round each number to the stated precision (SF means "significant figures").

1.57 (a) 2.49 K to two SF; (b) 2.49 K to one SF; (c) 2.998×10^8 m/s to two SF; (d) 9.649×10^4 A · s to one SF.

1.58 (a) 4.48 kg · m/s² to two SF; (b) 4.48 kg · m/s² to one SF; (c) 9.649×10^4 A · s to two SF; (d) 9.27×10^{-24} g to one SF.

In Exercises 1.59 to 1.72, complete each calculation to the correct number of significant figures.

1.59 The masses of copper, zinc, and manganese in a sample of an alloy were measured as 1.11, 1.1, and 1 g, respectively. What is the total mass of the three metals?

1.60 A forensic scientist collected three samples from the scene of a crime. Their masses were 0.110, 0.111, and 0.0001 g. What is the total mass?

1.61 The length of a piece of wood is 100.0 cm, and its thickness and width are 2.54 and 1.04 cm, respectively. What is the sum of the three dimensions?

1.62 A pharmacist made up a capsule containing 0.110 g of one drug, 0.111 g of another, and 0.0005 g of a third. What is the total mass in the capsule?

1.63 2.001 μg of a rare and expensive substance was removed from a sample of mass 1.78×10^{-4} g. How much is left?

1.64 An observation was made on a spacecraft 7.77 million mi from the earth. During the next 2 h it moved 3.33×10^4 mi closer to the earth. What then was its new distance from this planet?

1.65 If 800. tons of concrete is added to 1000. tons, and then 1 ton is removed, how much is left?

1.66 If 8.0×10^3 spectators left a crowd of 8.0×10^4, how many are left?

1.67 The speed of light is 2.998×10^8 m/s. (a) How far can it travel in 62.0 s? (b) How far can it travel in 62 s?

1.68 What is the total mass of two samples if one has a mass of 1.001 g and the other, of density 0.658 g/mL, has a volume of 2.51 cm^3?

1.69 Many equations in physics depend on the value of 4π. What is its value when π is approximated by (a) 3.142; (b) 3.141593?

1.70 A particular mathematical expression is the form $\pi + \pi/10 + \pi/100 + \pi/1000$. What is its value when π is approximated by (a) 3.14; (b) 3.1416?

1.71 The concentration of magnesium in seawater is 1.3 mg/L. What mass of magnesium is present in 10.0 L of seawater?

1.72 The concentration of gold in seawater is 0.011 μg/L. What mass of gold (in kilograms) is present in the Pacific Ocean, of volume 6.96×10^{11} km^3?

Mass Percentage Composition

1.73 Brass is an alloy of copper and zinc. Calculate the mass percentage composition of a brass formed by alloying 525 g of copper with 255 g of zinc.

1.74 Bronze is an alloy of copper and tin (and sometimes other elements). Calculate the mass percentage composition of a bronze formed by alloying 75 g of tin with 345 g of copper.

*__1.75__ If 55 g of tin were added to the brass of Exercise 1.73, what would be the mass percentage composition of the resulting mixture?

*__1.76__ If 55 g of zinc were added to the bronze of Exercise 1.74, what would be the mass percentage composition of the resulting mixture?

General

1.77 A powdered pharmaceutical mixture consisted of 2.5% by mass of the pain-relieving drug codeine. What mass of the mixture should be prescribed if the patient is to receive a dose of 0.50 g of codeine?

1.78 A contraceptive pill contains 0.050% of the active agent ethynodiol diacetate. What mass of pills contains 5.0 grams of the compound?

1.79 "Bell metal" is an alloy consisting of 78.0% by mass copper and 22.0% by mass tin. A smelter plans to produce 850 tons of the alloy annually and to obtain the tin from an ore composed 33.0% by mass of the mineral cassiterite. Cassiterite consists of 78.0% by mass tin. How many tons of ore must be added to the planned production?

1.80 "Zinc solders" may be used for joining aluminum. One manufacturer produces a zinc solder that has mass composition 75.0% tin, 1.5% aluminum, and the balance zinc. Zinc is obtained from the mineral sphalterite, which contains 67% zinc, and a typical ore contains only 5.0% of sphalerite. What mass of ore is needed to produce 100 kg of the zinc solder?

1.81 "German silver" is an alloy composed 52.0% by mass of copper, 26.0% of zinc, and 22.0% of nickel. Its density is 8.45 g/cm^3. What volume of the alloy contains 82.0 g of copper?

1.82 Sea water contains 0.0065% by mass bromine and its density is 1.028 g/mL. How many liters of seawater would be needed for the production of 100 kg of bromine by a plant that extracted it with complete efficiency?

1.83 The volume of 1.240 g of sugar was measured and found to be 0.781 cm^3. (a) What is the density of sugar? (b) How could the volume of a sample of sugar be measured?

1.84 The density of alcohol is 0.79 g/mL. What is the mass of 1 qt of pure alcohol?

1.85 The density of dry air is 1.2 g/L. What is the mass of air (in kilograms) in a room measuring 5 m by 5 m by 3 m?

1.86 The surface area of the earth, including the oceans, is 5.10×10^{14} m^2, and the population is 5×10^9.

(a) What is the average area available to each person? (b) About 70% of the surface is water; what is the average land area available per person?

1.87 The population of Manhattan is 5×10^6 on a working day, and its land area is 57 km^2. What is average area available to each inhabitant?

1.88 Pure gold is called "24-karat gold," where each "karat" is $\frac{1}{24}$ part. Thus 18-karat gold is a mixture of 18 parts gold by mass, and 6 parts of another substance (usually silver). Calculate the mass percentage composition of (a) 18-karat gold; (b) 14-karat gold.

1.89 "Solders" are usually alloys of tin and lead. What masses are present in a 100-g piece of solder of mass composition 48% tin?

1.90 The mass of a nickel coin is 5.0 g, and the metal is an alloy of 25% nickel and the remainder copper. What masses of the two elements are present in the coin?

1.91 A food scientist was investigating the effect of alcoholic beverages and prepared a mixture of 20.0 g of alcohol with 50.0 g of water. Calculate (a) the mass percentage composition of the mixture; (b) the mass of alcohol in 1.0 kg of the mixture; (c) the mass of water per kilogram of alcohol.

1.92 Vinegar is largely dilute acetic acid. A scientist examining its effect on the taste of food prepared a mixture resembling vinegar by mixing 7.5 g of acetic acid with 25.0 g of water. Calculate (a) the mass percentage composition of the mixture; (b) the mass of acid in 1.0 kg of the mixture; (c) the mass of water per kilogram of the acid.

1.93 Seawater is a very complex mixture, but its main component other than water is sodium chloride (table salt). A scientist investigating the corrosive action of seawater on steel prepared a solution resembling it by dissolving 5.24 g of table salt in 50.0 g of water. What are (a) the mass percentage composition of the solution; (b) the mass of solute per kilogram of solution; (c) the mass of solute per kilogram of solvent?

1.94 Many foods are sweetened with sugar, but fears have been expressed that high concentrations of sugar can cause cancer. This relation has been investigated by exposing living cells to solutions containing high concentrations of sugar. As part of the investigation, a scientist prepared a solution containing 15 g of sugar in 250 g of water. What are (a) the mass percentage composition of the solution; (b) the mass of solute per kilogram of solution; (c) the mass of solute per kilogram of solvent?

1.95 What mass of table salt must be dissolved in 100 g of water to prepare a solution that (a) is 10.0% salt by mass; (b) contains 15.0 g of salt per kilogram of solution; (c) contains 15.0 g of salt per kilogram of solvent?

1.96 What mass of sugar must be dissolved in 250 g of water to prepare a solution that (a) is 5.00% sugar by mass; (b) contains 10.0 g of sugar per kilogram of solution; (c) contains 10.0 g of sugar per kilogram of solvent?

This chapter introduces the chemical elements and their atoms. As well as learning a little about their internal structure, we see how atoms form two important classes of compounds and how these compounds are named. We also meet the "mole," a unit used throughout chemistry. The uses to which a knowledge of the composition of matter is put include the design and manufacture of microchips, such as this one, which shows the element silicon deposited on sapphire.

THE COMPOSITION OF MATTER

2

Elements
2.1 The names and symbols
    of the elements
2.2 The periodic table

Atoms
2.3 The nuclear atom
2.4 The masses of atoms
2.5 Moles and molar mass

Compounds
2.6 Molecules and
    molecular compounds
2.7 Ions and ionic
    compounds
2.8 Chemical nomenclature

Matter is built from the 100 or so different kinds of atoms that correspond to the 100 or so known elements. Atoms themselves have a definite structure, and atoms of different elements are built from different combinations of the same *subatomic particles*, particles smaller than atoms. Scientists are beginning to understand that all the rich diversity of the world around us can be traced back to a handful of different components. Whatever the material—sugar, chalk, seawater, a rock, a rose, or part of a brain—we can follow its composition back to atoms and beyond, back to the subatomic particles. We trace some of the connecting links in this chapter.

ELEMENTS

One hundred nine elements have been discovered (Fig. 2.1), and each one has been given a name and a symbol. Although 109 may seem a large number of elements to study, they can be organized into a pattern that makes it very easy to learn their properties. Moreover, for the purposes of chemistry, all the elements can be regarded as being constructed from only three subatomic particles.

2.1 THE NAMES AND SYMBOLS OF THE ELEMENTS

The names of some elements are ancient, because those elements have been known for a long time. This is the case for copper, a name derived from "Cyprus," where it was once mined, and gold, from the Old English word for "yellow." Elements identified more recently have been named, in a more or less random fashion, by their discoverers. In some cases the name takes note of a characteristic property. Chlorine is a yellow-green gas, and its name is derived from the Greek word for "yellow-green." For others, the chemist seems to have turned poet. Vanadium, which forms attractively colored compounds, is named

FIGURE 2.1 Samples of some common elements. Clockwise from the orange liquid bromine are mercury, iodine, cadmium, phosphorus, and copper.

FIGURE 2.2 The element vanadium is named after the Scandinavian goddess of beauty because of the attractive colors of some of its compounds.

after Vanadis, the Scandinavian goddess of beauty (Fig. 2.2). Some elements are named in honor of people or places. These include americium, berkelium, californium, einsteinium, and curium. The most recently discovered elements have not yet been given their final names. They are known by temporary names such as unnilennium (for element 109, un standing for 1, nil for 0, and enn for 9) until scientists agree on permanent names.

We represent elements by _chemical symbols_ made up of letters. Many of these symbols are the first, or first two, letters of the element's name:

Hydrogen	H	Carbon	C	Nitrogen	N	Oxygen	O
Helium	He	Aluminum	Al	Nickel	Ni	Silicon	Si

Note that the first letter of a symbol is always uppercase, and the second letter is always lowercase (as in Ni, not NI). Some elements have symbols formed from the first letter of the name and a later letter:

Magnesium	Mg	Chlorine	Cl	Zinc	Zn	Plutonium	Pu

The rest—fortunately only the eleven listed in Table 2.1—have symbols taken from their Latin or German names. Appendix 2D lists the names and symbols of all the elements.

2.2 THE PERIODIC TABLE

The known elements are also listed inside the front cover of this book, in a special arrangement called the "periodic table." Since this table is very important (as will become clear as we go on), it is a good idea to become familiar with it as early as possible. We will see that an element's location in the table is an excellent guide to its properties and the kinds of compounds it forms. It is therefore wise to note the position of every element you meet for the first time, and to note which elements are its neighbors, since it is likely to resemble them.

TABLE 2.1 Elements with symbols taken from Latin and German names

Element	Symbol	Origin of symbol
Antimony	Sb	_Stibium_
Copper	Cu	_Cuprum_
Gold	Au	_Aurum_
Iron	Fe	_Ferrum_
Lead	Pb	_Plumbum_
Mercury	Hg	_Hydrargyrum_
Potassium	K	_Kalium_
Silver	Ag	_Argentum_
Sodium	Na	_Natrium_
Tin	Sn	_Stannum_
Tungsten	W	_Wolfram_

FIGURE 2.3 The structure of the periodic table, showing the names of some of the regions and groups. Currently, chemists are considering a recommendation to change the numbering of the groups to 1 through 18, as shown toward the top.

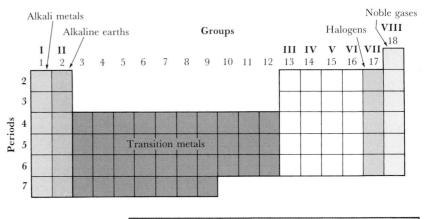

FIGURE 2.4 The alkali metals react with water with increasing violence from lithium to cesium. (a) Lithium reacts quietly. (b) Sodium reacts vigorously enough for the resulting heat to melt the metal. (c) Potassium reacts vigorously enough to ignite hydrogen produced from the water.

Groups and periods. The vertical columns of the periodic table are called *groups*; the horizontal rows are called *periods* (Fig. 2.3). An element and its *congeners*—the other members of its group—are generally very similar and show a gradation in properties. The properties of sodium (Na), for example, are a good clue to the properties of its congeners in Group I, namely lithium (Li), potassium (K), rubidium (Rb), and cesium (Cs). These five elements, which are collectively called the *alkali metals*, are all soft, silvery metals. All melt at low temperatures: lithium at 180°C, potassium at 64°C, and cesium at 30°C. They all produce hydrogen gas when in contact with water, lithium gently but cesium with explosive violence (Fig. 2.4). All the alkali metals are stored under oil to keep them out of contact with air and moisture.

Next to the alkali metals of Group I are the *alkaline earths* of Group II, which include calcium (Ca), strontium (Sr), and barium (Ba). They resemble the alkali metals in several ways but react with water less vigorously. Calcium, strontium, and barium react strongly enough to

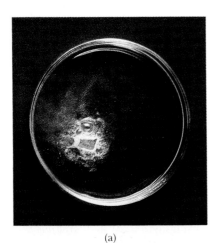

(a)

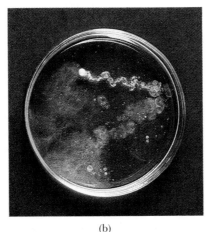

(b)

(c)

release hydrogen from the water, but magnesium (Mg) does so only if it is heated; beryllium (Be) does not react with water even if it is red hot.

At the other end of the table, in Group VIII, are the elements known as the *noble gases*. They are so called because they form very few compounds; they are chemically aloof. Indeed, until the 1960s they were called the "inert gases" because it was thought that they formed no compounds at all. Since then, a few dozen compounds of krypton (Kr), xenon (Xe), and radon (Rn) with oxygen and fluorine have been prepared. All these Group VIII elements are colorless, odorless gases. They are "monatomic" gases, in the sense that they exist as independent single atoms.

Next to the noble gases are the *halogens* of Group VII, most importantly fluorine (F), chlorine (Cl), bromine (Br), and iodine (I). We will see that many of the properties of the halogens show a smooth variation from fluorine to iodine. Fluorine is a pale yellow, almost colorless gas, chlorine a yellow-green gas, bromine a red-brown liquid, and iodine a purple-black solid (Fig. 2.5).

The elements in the part of the table that lies between Groups II and III are called the *transition metals*. They include the important structural elements titanium (Ti) and iron (Fe) and the *coinage metals* copper (Cu), silver (Ag), and gold (Au). The transition metals take their name from their role as a transition between the chemically active metals of Groups I and II and the much less active metals of Groups III and IV.

Metals and nonmetals. A *metal* is a substance that conducts electricity, has a metallic luster, and is malleable and ductile. A *malleable* substance (from the Latin word for "hammer") is one that can be hammered into thin sheets (Fig. 2.6). A *ductile* substance (from the Latin word for "drawing out") is one that can be drawn out into wires. Sodium, for example, conducts electricity, has a luster when freshly cut, is mallea-

(a)

(b)

FIGURE 2.6 One characteristic of a metal is that it can be deformed by hammering. Gold can be hammered into such thin sheets that they are almost transparent. This is gold foil photographed (a) against a light and (b) in reflected light.

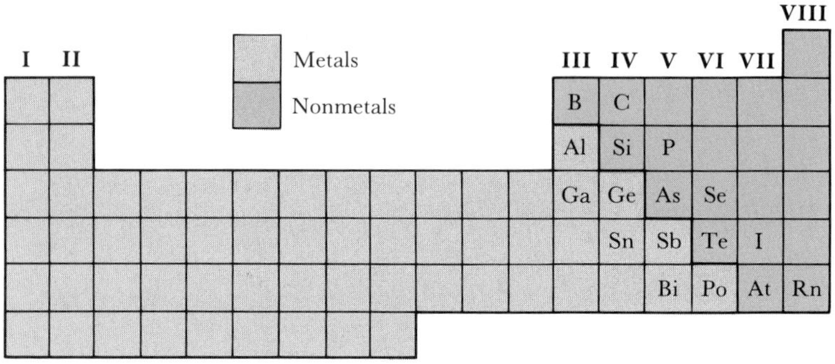

FIGURE 2.7 The distribution of metals and nonmetals in the periodic table.

Metals	Nonmetals

(periodic table diagram showing: B, C, Al, Si, P, Ga, Ge, As, Se, Sn, Sb, Te, I, Bi, Po, At, Rn as nonmetals in the upper right)

ble, and is reasonably ductile. A *nonmetal* is a substance that does not conduct electricity very well and is neither malleable nor ductile. Sulfur is a brittle yellow solid that does not conduct electricity, cannot be hammered into thin sheets, and cannot be drawn out into wires. All gases are nonmetals.

A striking feature of the periodic table becomes clear when we mark the positions of the metals and nonmetals (Fig. 2.7). We find that the metallic elements occur to the lower left of the table, whereas the nonmetallic elements occur to the upper right. One of the first applications of the periodic table now becomes clear: with a glance at the table we can tell whether an element is a metal or a nonmetal. We may never have heard of thallium (Tl) before, but its position in the periodic table shows us that it is a metal. We can even guess that, because it follows the transition metals, it is not a very reactive metal.

ATOMS

The first experimental evidence for the existence of atoms was assembled in 1805 by John Dalton, a scruffy but learned English schoolteacher (Fig. 2.8). His *atomic hypothesis* was that

1. All the atoms of a given element are identical.

2. The atoms of different elements have different masses.

3. A compound is a specific combination of atoms of more than one element.

4. In a chemical reaction, atoms are neither created nor destroyed but exchange partners to produce new substances.

A "hypothesis" is a suggestion that is put forward to account for a series of observations. Dalton based his hypothesis on measurements of the masses of elements that formed combinations. We do not need to go through the same arguments today because experimental techniques have progressed since Dalton's day, and the evidence for atoms is now much more direct. For instance, as we can see in Fig. 1.4, it is now possible to photograph individual atoms. Other support comes from a

FIGURE 2.8 John Dalton (1776–1844), the English schoolteacher who used experimental measurements to argue that matter consisted of atoms. This very early photograph—a "daguerrotype"—was made in 1842.

newly invented technique that, in effect, feels the surface of a solid and shows the position of its atoms as bumps (Fig. 2.9).

But what are atoms? Are atoms truly "uncuttable," as Dalton thought and as their name suggests (the word *atom* comes from the Greek words meaning "not cuttable"), or can they be broken apart into subatomic particles? If the latter is the case and we can discover how an atom is built, we might even be able to take a lead atom apart and reassemble it as gold! Even if that turns out to be impossible, we should at least be able to account for the existence of different elements and understand why they have different properties.

2.3 THE NUCLEAR ATOM

Experiments carried out at the end of the nineteenth century and early in the twentieth showed that atoms *are* built from smaller particles. They showed that atoms do have an internal structure made up of "electrons," "protons," and "neutrons."

Electrons. Just before the end of the nineteenth century (in 1897), the British physicist J. J. Thomson (Fig. 2.10) was studying the effect of high voltages on gases. He applied a potential difference (a "voltage") between two electrodes (or metal contacts) in a glass tube containing a small amount of gas. When he did so, he noticed that light was emitted from a location close to the "cathode," or negatively charged electrode (Fig. 2.11). His observations suggested that a stream of particles moved from the cathode toward the other electrode and caused the gas to emit light as they passed through it. Thomson called the stream of particles *cathode rays*. He found that they caused a spot of light where they hit a specially treated screen, and that he could cause the spot to move by placing electrically charged plates or a magnet near the path of the

FIGURE 2.9 Atoms show up as bumps on the surface of a solid when a technique called "surface tunneling microscopy" is used. This is an image of the surface of silicon. The cliff is one atom high.

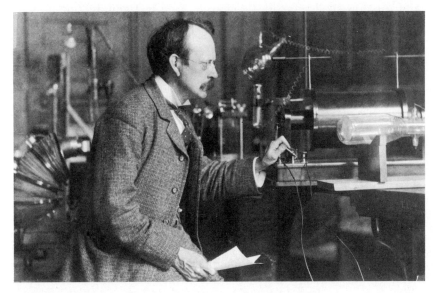

FIGURE 2.10 Joseph John Thomson (1856–1940), the discoverer of the electron, with the apparatus he used.

FIGURE 2.11 A close-up view of the glow near a cathode; the apparatus is much like that used by Thomson in the experiments that led to the discovery of the electron.

TABLE 2.2 **Properties of subatomic particles**

Particle	Symbol	Charge*	Mass, g
Electron	e⁻	−1	9.109×10^{-28}
Proton	p	+1	1.673×10^{-24}
Neutron	n	0	1.675×10^{-24}

*Charges are given as multiples of 1.60×10^{-19} C.

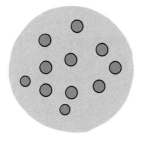

FIGURE 2.12 Thomson proposed that an atom consists of electrons (blue circles) in a jellylike positively charged substance.

FIGURE 2.13 Ernest Rutherford (1871–1937), who was responsible for many discoveries concerning the structure of the atom and its nucleus.

rays. He also found that the properties of the rays were the same no matter what metal he used for the electrodes.

The technological outcome of Thomson's discovery was the invention of the "cathode-ray tube" used in television. Each time you see a television picture you are watching a (sometimes) more entertaining version of Thomson's experiment. The scientific outcome of his work was the discovery that cathode rays are streams of negatively charged particles coming from inside the atoms that make up the electrodes. These particles are now called *electrons* and denoted e⁻. The fact that the same particles are obtained from electrodes made from any metal suggests that electrons are part of the makeup of all atoms.

Thomson found that he could deduce some of the properties of electrons from the way the spot moved when he altered the electric charge on the "deflecting plates" or the strength of the magnet. He was able to measure the value of e/m, the ratio of the size of the electron's charge e to its mass m. Later workers, most notably the American Robert Millikan, devised experiments to measure the mass and charge separately. By about 1910 they knew that an electron's mass is only 9.11×10^{-28} g. This makes it by far the lightest of the subatomic particles of interest in chemistry (see Table 2.2). The electric charge of an electron is 1.60×10^{-19} C, where C is the abbreviation for the SI unit of charge, the coulomb. We shall not use this numerical value until much later, and for now shall refer to the charge of an electron as "one unit" of negative charge.

Nuclei. Ordinary matter is neither attracted toward nor repelled by charged electrodes. This tells us that atoms do not have an electric charge; they are electrically "neutral." Thomson's experiment, however, showed that atoms contain electrons, which are negatively charged particles. Atoms must therefore also contain enough positive charge to cancel the negative charges of the electrons. Thomson suggested that the electrons were scattered through a jellylike positively charged substance (Fig. 2.12). However, this view was overthrown by the results of an experiment suggested by the New Zealander Ernest Rutherford (Fig. 2.13).

Rutherford was interested in the observation that some elements, including radium, emit streams of particles called *alpha particles* (α particles). By studying how α particles behave in the presence of electrically charged plates and magnets, he identified these particles as helium atoms that have lost all their electrons. According to Thomson's view, α particles should be blobs of positively charged jelly. Two of Rutherford's students, Hans Geiger and Ernest Marsden, carried out an experiment which Rutherford thought up to introduce Marsden to a particular piece of apparatus. They shot α particles through a very thin piece of gold foil, only a few atoms thick (like the one shown in Fig. 2.6), and observed the positions of flashes of light where the particles hit a screen (Fig. 2.14). Since they thought the atoms in the foil were all spheres of jelly, they expected the α particles to travel straight through. That is, they expected the α particles to be deflected from their paths only very slightly.

What they actually observed astonished them. Although most of the α particles were deflected only slightly, about 1 in 20,000 was deflected through more than 90°. Sometimes an α particle bounced straight back in the direction from which it had come. "It was almost as incredible," said Rutherford, "as if you had fired a 15-inch shell at a piece of tissue paper and it had come back and hit you."

The explanation had to be that atoms and α particles are not soft blobs of jelly but are, or at least contain, hard particles that can bounce off each other. Rutherford suggested a structure, called the _nuclear atom_, to account for this conclusion. He proposed that all the positive charge and most of the mass of an atom is concentrated in a minute lump, the _atomic nucleus_, and that the electrons move around it (Fig. 2.15a). According to Rutherford's model, an atom is almost all empty space, like the solar system, with the nucleus taking the place of the sun and the electrons the planets. Rutherford's suggestion accounts well for Geiger and Marsden's observations (Fig. 2.15b). The α particles are the minute nuclei of helium atoms. They pass through the gold atoms, which are mostly empty space, unless they score a direct hit on one of the minute gold nuclei. This happens very rarely, so most α particles are undeflected. However, about 1 in 20,000 scores a direct hit on a gold nucleus. The positive charge of the particle is then strongly repelled by the positive charge of the nucleus, and the α particle is deflected through a large angle, like a colliding billiard ball.

Measurement of the deflection angles gives information about the charges and diameters of atomic nuclei. Rutherford concluded that the nucleus of a gold atom carries nearly 100 units of positive charge (the value is now known to be 79) and that it is a minute, dense sphere about 10^{-14} m in diameter, or around $\frac{1}{10,000}$ the diameter of the atom itself. If we were to think of an atom as magnified to the size of a football stadium, the nucleus would be the size of a fly at the stadium's center.

Protons and atomic number. Thomson showed that an atom contains electrons. Rutherford showed that these electrons surround a minute, positively charged central nucleus. Nuclear physicists have continued the process of separating atoms into simpler components. They have found that atomic nuclei are built from two kinds of particles called _nucleons_. One kind of nucleon is called the _proton_ (p), and the other the _neutron_ (n). A proton is 1836 times heavier than an electron; it has a positive charge that exactly cancels an electron's negative charge. A neutron has almost the same mass as a proton but, as its name suggests, is electrically neutral.

The number of protons in an atomic nucleus is called the _atomic number Z_ of the element. Henry Moseley, a young British scientist, was

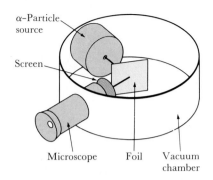

FIGURE 2.14 The experimental arrangement used by Geiger and Marsden. The α particles came from a sample of the radioactive gas radon. Their deflections were measured by observing flashes of light ("scintillations") where they struck a zinc sulfide screen.

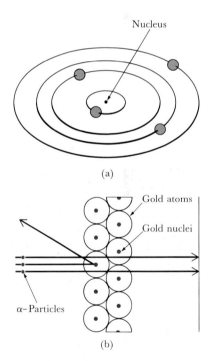

FIGURE 2.15 (a) Rutherford proposed that an atom consists of a minute, heavy central nucleus with electrons surrounding it like planets around the sun. The nucleus is far smaller than can be shown here: atoms are mostly empty space. (b) Rutherford's model explains the deflection of some α particles.

the first to measure atomic numbers accurately (shortly before he was killed in action in World War I). Moseley knew that when elements are bombarded with rapidly moving electrons they emit "x-rays." These rays are like light, but can pass through many substances. He found that the type of x-rays emitted by an element depends on its atomic number, and by observing the x-rays, he was able to determine the value of Z. We now know the atomic numbers of all the elements. They are given in the periodic table inside the front cover. In most versions of the periodic table, the data are arranged as shown in the margin. (The meaning of "atomic weight" will be explained later in the chapter.)

Atomic number →	1
Symbol ⟶	H
Atomic weight ⟶	1.008

Since an atom is electrically neutral, the number of protons in its nucleus must be the same as the number of electrons outside its nucleus. Therefore, counting the number of protons in the nucleus is an indirect way of counting the number of electrons. Since for hydrogen $Z = 1$, we know at once that a hydrogen atom has one electron. A gold atom, with $Z = 79$, has 79 electrons around its nucleus. For uranium $Z = 92$, so we know that every uranium atom has 92 electrons. The atomic number of an element is sometimes added as a subscript to the left of its chemical symbol, as in $_1$H and $_{92}$U.

▼ **EXAMPLE 2.1** **Using the atomic number**

The number of copper atoms in a 3.0-g lump of copper (approximately the mass of a penny) is 2.8×10^{22}. How many electrons are present in the sample? How much do they contribute to its mass?

STRATEGY We can calculate how many electrons are present from the total number of atoms and the number of electrons in each atom. The number of atoms present is given. The number of electrons in each atom is equal to the atomic number of copper, which is given in the periodic table. The total mass of the electrons is the product of the total number of electrons and the mass of a single electron, which is given in Table 2.2.

SOLUTION Since for copper $Z = 29$, each atom has 29 electrons. The total number of electrons is therefore

$$29 \text{ electrons per atom} \times (2.8 \times 10^{22} \text{ atoms}) = 8.1 \times 10^{23} \text{ electrons}$$

The mass of one electron is 9.11×10^{-28} g, so the total mass of the electrons present is

$$(8.1 \times 10^{23} \text{ electrons}) \times (9.11 \times 10^{-28} \text{ g}) = 7.4 \times 10^{-4} \text{ g}$$

That is, the electrons account for only 0.74 mg; the copper nuclei are responsible for almost all the mass.

EXERCISE Calculate (a) the number of electrons and (b) the total mass of electrons in 1.0 kg of iron, in which there are 1.08×10^{25} atoms.

[*Answer*: (a) 2.8×10^{26} electrons; (b) 0.26 g]

▲

We can now see that, in the periodic table, the elements are arranged in order of increasing atomic number. Since the table also shows how the properties of the elements repeat periodically, it follows that these properties vary periodically as the number of protons and electrons increases. This *periodicity* of the elements, or cyclic repetition of properties as Z increases, is one of the most remarkable features of matter. It is explained in Chapter 7.

2.4 THE MASSES OF ATOMS

Table 2.2 and Example 2.1 show that since the electron mass is almost negligible, the total mass of an atom is approximately the sum of the masses of the protons and neutrons in its nucleus. We know the mass of a single proton or neutron (it is listed in Table 2.2). We also know how many protons are present in each atom (from Z). Therefore, we should be able to deduce the number of neutrons present in an atom by measuring the mass of the atom and deciding how many neutrons are needed, in addition to the protons, to make up that mass.

When Dalton first proposed his atomic hypothesis, the masses of individual atoms could not be measured directly. He did try to deduce their *relative* masses in a series of clever experiments based on the masses of elements that combined together. He deduced (in 1803) that an atom of copper is about 60 times heavier than an atom of hydrogen, which is correct. Hence, he knew that a 1-g sample of copper contains only $\frac{1}{60}$ the number of atoms as a 1-g sample of hydrogen. However, because the electrolysis of water always produces a mass of oxygen eight times greater than that of hydrogen, he also thought that an atom of oxygen is about eight times heavier than hydrogen, which is wrong. (It is 16 times heavier; he did not know that there are not one but two H atoms for each O atom in water.)

Dalton's values were corrected later in the nineteenth century, and chemists have long been able to determine the number of atoms in a sample from its mass. It was not until earlier in this century, however, that they could weigh individual atoms directly, accurately, and precisely.

Mass spectrometry. The masses of atoms are now measured with a _mass spectrometer_. Atoms or molecules of the element—either a gaseous element (neon, for instance) or the vapor of a liquid or solid element (such as mercury or zinc)—are fed into the spectrometer's "ionization chamber." There they are exposed to a beam of rapidly moving electrons. When one of these electrons collides with an atom, it can knock an electron out of it, leaving the atom with a positive charge and thus creating a positive "ion":

An **ion** is an electrically charged atom or group of atoms.

(The name "ion" comes from the Greek word for "go," because of the way charged particles *go* either toward or away from a charged electrode.) The positive ions are accelerated out of the chamber by a high voltage applied between two plates. The speed attained by the ions depends on their mass, with lighter ions reaching higher speeds than heavy ones.

As an ion passes between the two curved charged plates, its path is bent by an amount that depends on its speed (and hence on its mass). The voltage between the plates is slowly changed, and a signal is produced when it is just strong enough to bend the path so that the ions arrive at the detector (Fig. 2.16). The mass of the ion is then calculated from the voltage needed to produce a signal; the mass of the atom is the sum of the masses of the ion and its missing electron. A hydrogen

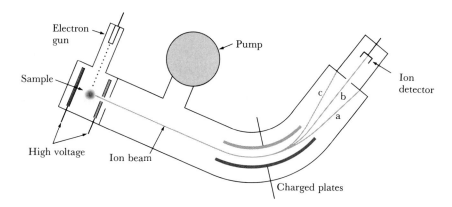

FIGURE 2.16 As the voltage between the charged plates of the mass spectrometer is changed, the path of the ions moves from *a* to *c*. When the path is at *b*, the ion detector sends a signal to a recorder.

atom, for example, is found to have mass 1.67×10^{-24} g, and a carbon atom 1.99×10^{-23} g, about 12 times greater.

Isotopes and mass number. The measurement of precise atomic masses led to a major discovery. One of Dalton's assumptions was that all the atoms of a given element are identical. However, when most elements are studied with a mass spectrometer, atoms with slightly different masses are detected even though a chemically pure sample is being used. In a sample of pure neon, for example, most of the atoms have mass 3.32×10^{-23} g, which is about 20 times that of a hydrogen atom. However, some have mass 3.65×10^{-23} g, about 22 times heavier, and a few have mass 3.49×10^{-23} g, about 21 times heavier (Fig. 2.17). All three types of atom have the chemical properties of neon and are called "isotopes" of neon:

> **Isotopes** are atoms of an element that have the same atomic number but different atomic masses.

The name comes from the Greek words for "equal place"; it signifies that although the atoms have different masses, they belong to an element that occupies one place in the periodic table.

It is easy to explain the existence of isotopes by assuming that the atomic nucleus of a given element has a fixed number of protons but a variable number of neutrons. Neutrons add to the mass of an atom, but they do not affect the number of electrons it needs for electrical neutrality. Hence the different numbers of neutrons in the isotopes of an element change their masses but not their chemical properties. The *mass number A* of an atom is the total number of nucleons (protons plus neutrons) in its nucleus. Isotopes of a given element all have the same atomic number but different mass numbers.

The proton, the neutron, and an atom of the most common isotope of hydrogen all have about the same mass. Therefore, an isotope of mass number *A*, with *A* protons and neutrons in its nucleus, is about *A* times heavier than a hydrogen atom. In the case of neon, since the mass spectrometer gives the masses of the isotopes as about 20, 21, and 22 times that of a hydrogen atom, their mass numbers must be 20, 21, and 22, respectively. Since each nucleus has 10 protons (because $Z = 10$ for neon), the isotopes must have 10, 11, and 12 neutrons respectively (Fig. 2.18).

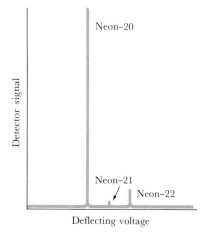

FIGURE 2.17 The signal obtained from a mass spectrometer shows that a sample of neon consists of atoms with different masses. Neon-20, an atom of neon about 20 times heavier than hydrogen, is the most abundant.

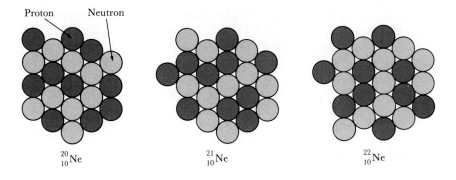

Proton   Neutron

$^{20}_{10}$Ne $^{21}_{10}$Ne $^{22}_{10}$Ne

FIGURE 2.18 The nuclei of isotopes have the same numbers of protons (red circles) but different numbers of neutrons (gray circles).

An isotope is named by writing its mass number after the name of the element, as in neon-20, neon-21, and neon-22. Its symbol is obtained by writing the mass number as a superscript to the left of the chemical symbol, as in ^{20}Ne, ^{21}Ne, and ^{22}Ne.

Hydrogen has three isotopes (see Table 2.3). The most common (^{1}H) has no neutrons, so its nucleus consists of a lone proton. The other two isotopes are less common but nevertheless are so important that they

TABLE 2.3   Some common isotopes

Name	Atomic number Z	Number of neutrons	Mass number A	Mass	Abundance, %	Symbol
Hydrogen	1	0	1	1.674×10^{-24} g 1.008 amu	99.985	^{1}H
Deuterium	1	1	2	3.344×10^{-24} g 2.014 amu	0.015	^{2}H or D
Tritium	1	2	3	5.008×10^{-24} g 3.016 amu	*	^{3}H or T
Carbon-12	6	6	12	1.9926×10^{-23} g 12 amu exactly	98.90	^{12}C
Carbon-13	6	7	13	2.159×10^{-23} g 13.00 amu	1.10	^{13}C
Oxygen-16	8	8	16	2.655×10^{-23} g 15.99 amu	99.76	^{16}O
Chlorine-35	17	18	35	5.807×10^{-23} g 34.97 amu	75.77	^{35}Cl
Chlorine-37	17	20	37	6.138×10^{-23} g 36.97 amu	24.23	^{37}Cl
Cobalt-60	27	33	60	—	*	^{60}Co
Strontium-90	38	52	90	—	*	^{90}Sr
Gold-197	79	118	197	3.271×10^{-22} g 197.0 amu	100	^{197}Au
Uranium-235	92	143	235	3.902×10^{-22} g 235.0 amu	0.72	^{235}U
Uranium-238	92	146	238	3.953×10^{-22} g 238.05 amu	99.27	^{238}U

*Unstable, radioactive element.

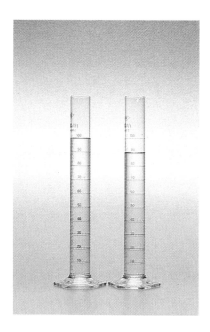

FIGURE 2.19 These two samples, both of which have a mass of 100 g, illustrate that heavy water (D_2O) is denser than ordinary water.

are given special names and symbols. One (^2H) is called deuterium (D), and the other (^3H) is tritium (T). A deuterium atom, with a nucleus that consists of one proton tightly bound to one neutron, is about twice as heavy as an ordinary hydrogen atom (more precisely, 1.998 times as heavy) and, in combination with oxygen, gives "heavy water." Heavy water, which is used in some nuclear reactors, has a density of 1.11 g/mL at 20°C, 11% greater than that of ordinary water (Fig. 2.19).

▼ **EXAMPLE 2.2** Using mass numbers

How many neutrons are present in atoms of (a) uranium-235; (b) ^{60}Co?

STRATEGY The number identifying the isotope gives the mass number A, the total number of protons and neutrons. The atomic number Z (and hence the number of protons) is unique to the particular element. The difference $A - Z$ is the number of neutrons.

SOLUTION (a) As the mass number of uranium-235 is 235, the total number of nucleons is $A = 235$. The periodic table shows that uranium has $Z = 92$, so its nucleus contains 92 protons. The number of neutrons is then $235 - 92 = 143$.
(b) ^{60}Co (or cobalt-60) has $Z = 27$ and $A = 60$. Therefore, the number of neutrons in its nucleus is $60 - 27 = 33$.

EXERCISE How many neutrons are present in an atom of (a) carbon-12; (b) ^{35}Cl?

[*Answer*: (a) 6; (b) 18]

▲

Isotopic abundance. A sample of an element (such as the chlorine produced by the electrolysis of sodium chloride) is typically a mixture of isotopes. The <u>abundance</u> of an isotope is the percentage of that isotope (in terms of <u>numbers of atoms</u>) present in a sample of the element:

$$\text{Abundance of X} = \frac{\text{number of atoms of isotope X}}{\text{total number of atoms in sample of X}} \times 100\%$$

We can find the number of atoms of each isotope present with a mass spectrometer, since the heights of the peaks on a mass spectrum are proportional to the numbers of atoms with each mass.

The <u>natural abundance</u> of an isotope is its abundance in a naturally occurring sample. The natural abundance of neon-20 is 91%, which means that 91 atoms out of 100 in a naturally occurring sample are neon-20. The natural abundance of uranium-235, the uranium isotope widely used as a nuclear fuel, is only 0.7%, so only 7 atoms out of every 1000 in a block of uranium are uranium-235. Some natural abundances are listed in Table 2.3. As illustrated in Fig. 2.20, some elements have very few isotopes, but others are rich mixtures of several.

Atomic mass units and atomic weights. The atomic masses in Table 2.3 are all very small, lying in the range 10^{-24} to 10^{-22} g. Since it is convenient to have a unit that results in simpler numerical values, atomic masses are often reported as multiples of the "atomic mass unit" (amu):

An **atomic mass unit** is $\frac{1}{12}$ the mass of one atom of carbon-12.

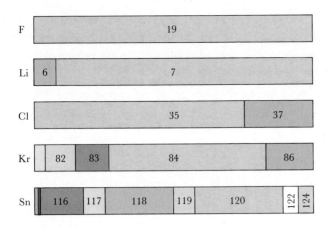

FIGURE 2.20 The abundances of the isotopes of some common elements, shown schematically: the lengths of the bars are proportional to the percentages of the isotopes in a naturally occurring mixture.

Since the mass of a single carbon-12 atom is 1.9926×10^{-23} g, and 1 amu is one-twelfth that mass, it follows that

$$1 \text{ amu} = 1.6605 \times 10^{-24} \text{ g}$$

The mass of any atom can be expressed in atomic mass units by using this relation as a conversion factor (as we illustrate in Example 2.3); some values are included in Table 2.3. Atomic masses lie in the range 1 amu (for hydrogen) to about 250 amu for the heaviest elements (uranium and the synthetic elements following it in the periodic table).

A natural sample of an element is a mixture of isotopes with different atomic masses. The *average* mass of the atoms in a naturally occurring sample is called the *atomic weight* (AW) of the element (Fig. 2.21). Although this is a poor name (because weight is quite different from mass, as explained in Chapter 1), it has had a long history and is widely used. Thus, in a sample of chlorine gas, both chlorine-35 and chlorine-37 atoms are present in their natural abundances, 75.8% ^{35}Cl and 24.2% ^{37}Cl. Since the masses of the two isotopes are 34.97 and 36.97 amu, respectively, the average mass of the atoms in the sample, which is the atomic weight of chlorine, is

$$AW = \frac{75.8}{100} \times 34.97 \text{ amu} + \frac{24.2}{100} \times 36.97 \text{ amu}$$

$$= 35.5 \text{ amu}$$

Atomic weights are shown in the periodic table and listed in Appendix 2D. We shall use them frequently in the remainder of this text.

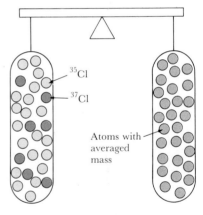

FIGURE 2.21 The atomic weight of chlorine is the average mass of the atoms in a naturally occurring sample, with the abundances of the isotopes taken into account.

▼ **EXAMPLE 2.3** **Calculating the number of atoms in a sample**

When 19.24 g of water is electrolyzed, 2.15 g of hydrogen gas is produced. Calculate the number of hydrogen atoms in 19.24 g of water.

STRATEGY Since no atoms are created or destroyed during a chemical change, the number of hydrogen atoms in 19.24 g of water is equal to the number in 2.15 g of hydrogen gas. This number is found by setting up the conversion

$$\text{Number of H atoms} = \text{mass of H in grams} \times \text{number of H atoms per gram}$$

and deducing the number of H atoms per gram from the atomic weight of hydrogen (obtained from the periodic table).

SOLUTION Since the atomic weight of hydrogen is 1.008 amu,

$$\text{Average mass of H atom} = 1.008 \text{ amu} \times \frac{1.6605 \times 10^{-24} \text{ g}}{1 \text{ amu}}$$

$$= 1.674 \times 10^{-24} \text{ g H}$$

That is,

$$1 \text{ H atom} = 1.674 \times 10^{-24} \text{ g H}$$

and the conversion factor from mass to number of atoms is

$$\frac{1 \text{ H atom}}{1.674 \times 10^{-24} \text{ g H}} = 1$$

Therefore,

$$\text{Number of H atoms} = 2.15 \text{ g H} \times \frac{1 \text{ H atom}}{1.674 \times 10^{-24} \text{ g H}}$$

$$= 1.28 \times 10^{24} \text{ H atoms}$$

EXERCISE Calculate the number of oxygen atoms in the sample (17.10 g of oxygen gas was produced), given that the average mass of one atom is 2.66×10^{-23} g. Do you have any comment regarding these results?

▲   [*Answer*: 6.43×10^{23} O atoms; exactly half the number of H atoms]

2.5 MOLES AND MOLAR MASS

The calculation in Example 2.3 shows that even a small sample of hydrogen contains a vast number of atoms. Similarly, when we pick up a penny we are holding about 10^{22} atoms of various kinds, which is more than the number of stars in the visible universe. A unit has been invented to express numbers as big as this and to make calculating the numbers of atoms in samples much more convenient.

The mole. The "mole" (mol) is the unit chemists use for expressing large numbers of atoms, ions, and molecules. The name comes—somewhat jokingly—from the Latin word for "massive heap."

One mole is the number of atoms in exactly 12 g of carbon-12.

Although 1 mol is defined in terms of carbon atoms, the unit applies to any object, just as 1 dozen means 12 of anything. The mass of one ^{12}C atom is exactly 12 amu, or $12 \times (1.6605 \times 10^{-24} \text{ g})$, so the number of atoms in 12 g of carbon-12 is

$$\text{Number of } ^{12}\text{C atoms} = \frac{\text{mass of sample}}{\text{mass of 1 } ^{12}\text{C atom}}$$

$$= \frac{12 \text{ g C atoms}}{12 \times 1.6605 \times 10^{-24} \text{ g}}$$

$$= 6.022 \times 10^{23} \text{ C atoms}$$

That is, 1 mol of atoms (of any element) is 6.022×10^{23} atoms of the element. The same is true of 1 mol of any object—atoms, ions, or molecules: 1 mol always means 6.022×10^{23} objects.

The number 6.022×10^{23} is called *Avogadro's number* N_A, in honor of the nineteenth century scientist Amedeo Avogadro (Fig. 2.22), who helped to establish the fact that atoms exist. Its value should be committed to memory. Several awe-inspiring illustrations can be devised to help visualize the enormous size of this number. For instance, an Avogadro's number of soft-drink cans would cover the surface of the earth to a height of over 200 mi.

To obtain 1 mol of any element, we must weigh out a mass in grams equal to its atomic weight in atomic mass units. For example, if we want 1.000 mol of magnesium (atomic weight of 24.31 amu), we weigh out 24.31 g of magnesium. The justification for this simple procedure is as follows. Let the atomic weight of the element be x amu. Then the number of atoms in a sample of mass x g is

$$\text{Number of atoms} = \frac{\cancel{x} \, \cancel{g}}{\cancel{x} \times 1.6605 \times 10^{-24} \, \cancel{g}}$$
$$= 6.022 \times 10^{23}$$

which is Avogadro's number. From now on, therefore, we know that to measure out 1.000 mol Cu atoms, we must measure out 63.54 g of copper; to measure out 1.000 mol P atoms, we must weigh out 30.97 g of phosphorus; and so on. The "massive heaps" corresponding to 1 mol of atoms of some elements are shown in Fig. 2.23; they have all been obtained by weighing out an amount of the element in grams equal to its atomic weight in atomic mass units.

Molar mass. The mass per mole of atoms of an element is called the *molar mass* of the atoms (the word *molar* is often used to mean "per mole"). Since the mass of 1 mol of Mg atoms is 24.31 g, their molar mass is 24.31 g/mol. Similarly, the molar mass of copper atoms is 63.54 g/mol. It is very easy to write down the molar mass of any ele-

FIGURE 2.22 Lorenzo Romano Amedeo Carlo Avogadro, count of Quaregna and Cerreto (1776–1856).

FIGURE 2.23 Each sample consists of 1 mol of atoms of the element. Clockwise from the upper right are 207 g of lead, 32 g of sulfur, 201 g of mercury, 64 g of copper, and 12 g of carbon.

ment's atoms by finding its atomic weight in the periodic table. If the atomic weight of the element is x amu, then the molar mass of the atoms is x g/mol.

The molar mass acts as a basis for the conversion factor between the mass of a sample of an element and the number of moles of atoms present. For example, for magnesium we can write

$$1 \text{ mol Mg atoms} = 24.31 \text{ g Mg}$$

and use this relation to construct a conversion factor. If we want to know how many moles of Mg atoms there are in 10 mg of magnesium, we form the conversion factor

$$\frac{1 \text{ mol Mg atoms}}{24.31 \text{ g Mg}} = 1$$

This states that 1 mol of Mg atoms is equivalent to 24.31 g of magnesium. Then we carry out the conversion:

$$\text{Moles of Mg atoms} = 10 \times 10^{-3} \text{ g Mg} \times \frac{1 \text{ mol Mg atoms}}{24.31 \text{ g Mg}}$$

$$= 4.1 \times 10^{-4} \text{ mol Mg atoms}$$

When small amounts of material are involved, the millimole (mmol; 1 mmol = 10^{-3} mol) can be a very useful unit. In our magnesium example, the result could be reported as 0.41 mmol Mg atoms.

▼ **EXAMPLE 2.4** **Using atomic weights to calculate the number of moles**

In an electrolysis of water, 2.53 g of hydrogen gas was formed. To how many moles of hydrogen atoms (produced as H_2 molecules) does that correspond?

STRATEGY This is a straightforward conversion-factor calculation:

$$\text{Moles of H atoms} = \text{mass of H} \times \text{moles of H atoms per gram of H}$$

The conversion factor is obtained from the atomic weight, using the information that

$$1 \text{ mol H atoms} = 1.008 \text{ g H}$$

SOLUTION The conversion factor is

$$\frac{1 \text{ mol H atoms}}{1.008 \text{ g hydrogen}} = 1$$

Therefore,

$$\text{Moles of H atoms} = 2.53 \text{ g H} \times \frac{1 \text{ mol H atoms}}{1.008 \text{ g H}}$$

$$= 2.51 \text{ mol H atoms}$$

That is, the electrolysis produces 1.51×10^{24} H atoms as H_2 molecules.

EXERCISE The same experiment produced 20.04 g of oxygen gas. How many moles of oxygen atoms were produced (as O_2 molecules)? Do you have any comment?

▲   [*Answer*: 1.25 mol O atoms; twice as many H atoms as O atoms]

COMPOUNDS

Now we turn our attention from elements to compounds. We have noted that a compound consists of a specific combination of atoms, such as the two hydrogen atoms per oxygen atom of water or the four hydrogen atoms per carbon atom of methane. In this part of the chapter we extend that concept; we begin by distinguishing between "molecular" and "ionic" compounds:

A **molecular compound** is a compound that consists of molecules.

An **ionic compound** is a compound that consists of ions.

Molecular compounds include water, methane, and ammonia, as well as sulfur dioxide, cane sugar (sucrose), benzene, and aspirin. Ionic compounds consist of positively and negatively charged ions held together by the attraction between their opposite electric charges. They include sodium chloride, which consists of "sodium ions" (positively charged sodium atoms) and "chloride ions" (negatively charged chlorine atoms). Each small crystal of the compound is a huge collection of sodium and chloride ions. When you take a pinch of salt, you are picking up crystals consisting of great numbers of ions held together by this ionic attraction.

1 Hydrogen cyanide

2.6 MOLECULES AND MOLECULAR COMPOUNDS

Recall that we introduced the "molecule" as the smallest particle of a compound that has the chemical properties of the compound. In terms of atoms,

2 Ammonia

A **molecule** is a definite and distinct group of bonded atoms.

Some models of molecules are shown in Section 1.1; three more are shown in the margin here: hydrogen cyanide **(1)**, ammonia **(2)**, and ethanol **(3)**. As before, the black spheres represent carbon, white hydrogen, and red oxygen. In addition, blue spheres represent nitrogen.

Molecular formulas. The number of atoms of each element in a molecule is given as a subscript in the "molecular formula" of the compound:

3 Ethanol

A **molecular formula** is a combination of chemical symbols and subscripts showing the number of atoms of each element present in a molecule.

The molecular formulas of hydrogen cyanide, ammonia, and ethanol are HCN, NH_3, and C_2H_6O, respectively. Glucose **(4)** consists of much more complicated molecules than these. Its molecular formula $C_6H_{12}O_6$ shows that it consists of six C atoms, twelve H atoms, and six O atoms.

4 Glucose

5 Carbon monoxide

6 Carbon dioxide

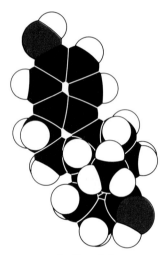

7 Estradiol

8 Phosphorus

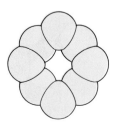

9 Sulfur

A *diatomic molecule* is a molecule that consists of only two atoms. The lethal gas carbon monoxide **(5)**, which is produced when gasoline burns in a limited supply of air, consists of diatomic molecules of formula CO: each molecule has one carbon atom and one oxygen atom. A *polyatomic molecule* is one that consists of more than two atoms (*poly* comes from the Greek word for "many"). Carbon dioxide **(6)**, the gas produced when gasoline burns in plenty of air, consists of polyatomic molecules (in this case "triatomic" molecules) of formula CO_2. The female sex hormone estradiol **(7)** is much larger: its molecular formula is $C_{18}H_{24}O_2$, which shows that it consists of 18 C atoms, 24 H atoms, and two O atoms.

Some of the elements exist as molecules. Hydrogen gas, for example, consists of diatomic H_2 molecules. The nitrogen and oxygen of the air exist as diatomic N_2 and O_2 molecules. Elemental phosphorus **(8)** exists as a white waxlike solid known as "white phosphorus," which consists of P_4 molecules. Sulfur **(9)** is made up of S_8 molecules in which eight S atoms form a crownlike ring.

Molecular weight. The *molecular weight* (MW) of a compound is the average mass of one of its molecules. It is calculated by adding together the atomic weights of the elements present. Thus the molecular weight of water is

$$MW(H_2O) = 2 \times AW(H) + AW(O)$$

$$= 2 \times 1.008 \text{ amu} + 16.00 \text{ amu}$$

$$= 18.02 \text{ amu}$$

By the same argument that we used for atoms, we see that if we need 1.000 mol of H_2O molecules for an experiment, then we must weigh out 18.02 g of water. A vending machine that delivers 100 g (about 100 mL) of a dilute aqueous solution like coffee or cola delivers about $5\frac{1}{2}$ mol of H_2O molecules. Figure 2.24 shows 1 mol of molecules of some common compounds.

FIGURE 2.24 Each sample contains 1 mol of molecules of a molecular compound. From left to right are 18 g of water, 46 g of ethanol, 180 g of glucose, and 342 g of sucrose.

The molar mass of molecules (like that of atoms) can be written from the molecular weight. If the molecular weight of the compound is x amu, then the molar mass (or mass per mole) of the molecules is x g/mol. The molar mass of H_2O, for instance, is 18.02 g/mol. As in the case of atoms, we use the molar mass to set up a conversion factor between the mass of compound in a sample and the number of moles of molecules in the sample. We can write, for example, that

$$1 \text{ mol } H_2O = 18.02 \text{ g } H_2O$$

and use it to write a conversion factor between grams of water and moles of H_2O molecules.

▼ **EXAMPLE 2.5** **Using molecular weight to calculate the number of moles**

The molecular formula of ordinary alcohol, which chemists call "ethanol," is C_2H_6O. How many moles of ethanol molecules are present in a 100-g sample?

STRATEGY The number of moles is found by setting up the conversion

Moles of C_2H_6O
$$= \text{mass of ethanol} \times \text{moles of } C_2H_6O \text{ per gram of ethanol}$$

We can set up the conversion factor once we know the molecular weight of ethanol. This is the sum of the atomic weights of the elements present in the molecule. Their values are given in the periodic table.

SOLUTION The molecular weight of ethanol is

$$MW(C_2H_6O) = 2 \times AW(C) + 6 \times AW(H) + AW(O)$$

$$= 2 \times 12.01 \text{ amu} + 6 \times 1.008 \text{ amu} + 16.00 \text{ amu}$$

$$= 46.07 \text{ amu}$$

Its molar mass is therefore 46.07 g/mol, and

$$1 \text{ mol } C_2H_6O = 46.07 \text{ g } C_2H_6O$$

The conversion factor from mass to moles is then

$$\frac{1 \text{ mol } C_2H_6O}{46.07 \text{ g } C_2H_6O} = 1$$

so

$$\text{Moles of } C_2H_6O \text{ molecules} = 100 \text{ g } C_2H_6O \times \frac{1 \text{ mol } C_2H_6O}{46.07 \text{ g } C_2H_6O}$$

$$= 2.17 \text{ mol } C_2H_6O$$

EXERCISE Calculate the number of moles of sucrose ($C_{12}H_{22}O_{11}$) molecules in a 3.0-g cube of sugar.

[*Answer:* 8.8×10^{-3} mol $C_{12}H_{22}O_{11}$]

▲

2.7 IONS AND IONIC COMPOUNDS

We have seen that although atoms are electrically neutral, they can be transformed into electrically charged particles called ions. One method of *ionizing* atoms, or converting them into ions by removing electrons, is

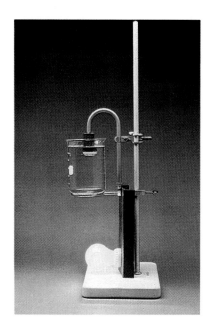

FIGURE 2.25 Pure water is a poor conductor of electricity, as is shown by the very dim glowing of the bulb in the circuit. However, when salt is dissolved in the water, the sodium and chloride ions carry the electric current, and the solution is a moderately good conductor.

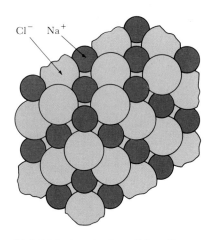

Cl⁻   Na⁺

FIGURE 2.26 In solid sodium chloride, an ionic compound, the ions are held together by the attraction of their opposite charges. The red spheres represent sodium ions, the green spheres chloride ions. The ions become free to move when salt dissolves in water.

to expose them to a beam of rapidly moving electrons, as in a mass spectrometer. The electrons in the beam collide with electrons in the atoms, knocking electrons out of the atoms and causing those atoms to become positively charged. Another method is to pass an electric current through a gas, as in a fluorescent lamp or a neon sign.

The fact that ionic compounds consist of ions can be demonstrated with the apparatus of Fig. 2.25. When two charged electrodes are immersed in a beaker of pure water, very little current flows through the circuit because pure water, a molecular compound, is a poor conductor of electricity. However, if sodium chloride is dissolved in the water, a current does flow; it is carried by sodium ions moving toward one electrode and by chloride ions moving toward the other. These ions are *not* formed when the compound dissolves, but rather they already exist in the solid (Fig. 2.26) . They become free to move when they are no longer packed tightly together in a solid but instead are surrounded by the liquid. A solution of an ionic compound, such as sodium chloride, is an example of an "electrolyte solution":

An **electrolyte solution** is a solution that conducts electricity.

A **nonelectrolyte solution** does not conduct electricity.

A solution of cane sugar (sucrose) in water is a nonelectrolyte solution, because sucrose molecules are electrically neutral and do not conduct electricity.

In the presence of charged electrodes, oppositely charged ions move in opposite directions. The ions are called either "cations" or "anions," depending on their charge:

A **cation** is a positively charged ion.

An **anion** is a negatively charged ion.

(The prefixes cat- and an- come from the Greek words for "down" and "up," which reflect two opposite directions.) In sodium chloride, the sodium ions are the cations and the chloride ions are the anions.

Cations. It is easy to explain the existence of ions in terms of the nuclear atom. Since an electron has one unit of negative charge, removing one electron from a neutral atom leaves behind a cation with one unit of positive charge (Fig. 2.27). Each electron that an atom loses increases its overall positive charge by one unit. A sodium cation is a sodium atom that has lost one electron; it is denoted Na^+. When a calcium atom loses two electrons, it becomes the calcium ion Ca^{2+} (read "calcium two-plus"), an ion of "charge number" +2. When an aluminum atom loses three electrons, it becomes the aluminum ion Al^{3+}, an ion of charge number +3.

Some of the cations formed by the elements are listed in Table 2.4. A very useful rule, which illustrates another application of the periodic table, is that in Groups I, II, and III of the table, the group number is the maximum (and sometimes the only) charge number a cation can

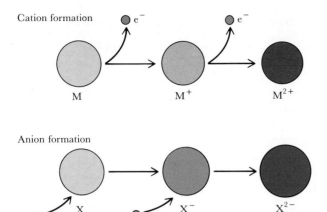

Cation formation

M → M⁺ → M²⁺

Anion formation

X → X⁻ → X²⁻

FIGURE 2.27 A cation of an element M is formed by the loss of one or more electrons, and an anion of an element X by the gain of electrons.

have. Thus, indium is in Group III and forms In^{3+} ions (and also In^+); barium is in Group II and forms only Ba^{2+} ions.

Table 2.4 shows that some atoms can lose different numbers of electrons. An iron atom, for instance, can lose two electrons to become Fe^{2+} or three electrons to become Fe^{3+} (Fig. 2.28). Copper is another element that behaves in this way: it can lose either one electron to form Cu^+ or two to become Cu^{2+}. This variability of charge is most pronounced among the transition metals, but the heavier elements in Groups III and IV also show it (as we saw for indium).

Anions. If an atom gains an extra electron, it acquires a single negative charge and becomes an anion. Each electron that an atom gains increases its overall negative charge by one unit. A chloride ion is a chlorine atom that has gained an additional electron; it is denoted Cl^-. When an oxygen atom gains two electrons, it becomes O^{2-}, an ion of charge number -2. When a nitrogen atom gains three electrons, it becomes N^{3-}, an ion of charge number -3.

TABLE 2.4   The cations formed by some elements*

I	II	Transition metals			III	IV
Li^+	Be^{2+}					
Na^+	Mg^{2+}				Al^{3+}	
K^+	Ca^{2+}	Fe^{2+}, Fe^{3+}	Cu^+, Cu^{2+}	Zn^{2+}	Ga^{3+}	
Rb^+	Sr^{2+}		Ag^+	Cd^{2+}	In^+, In^{3+}	Sn^{2+}, Sn^{4+}
Cs^+	Ba^{2+}		Au^+, Au^{3+}	Hg_2^{2+}, Hg^{2+}	Tl^+, Tl^{3+}	Pb^{2+}, Pb^{4+}

*The most important polyatomic cation is NH_4^+, the ammonium ion.

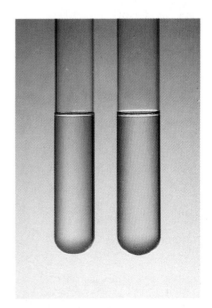

FIGURE 2.28 Iron is an example of an element that can form ions with different charges. Solutions containing Fe^{2+} are usually pale green, and solutions containing Fe^{3+} are usually yellow-brown.

TABLE 2.5 Some anions formed by the elements*

IV	V	VI	VII
C^{4-}, C_2^{2-} Carbide	N^{3-} Nitride	O^{2-} Oxide	F^- Fluoride
CN^- Cyanide	N_3^- Azide	O_2^- Superoxide	
NCO^- Cyanate	NH_2^- Amide	O_2^{2-} Peroxide	
NCS^- Thiocyanate		O_3^- Ozonide	
		OH^- Hydroxide	
	P^{3-} Phosphide	S^{2-} Sulfide	Cl^- Chloride
		HS^- Hydrogen sulfide	
			Br^- Bromide
			I^- Iodide
			I_3^- Triiodide

*Note that the arrangement of this table corresponds to the location of the elements in the periodic table.

Some of the anions formed by the elements are listed in Table 2.5. It is helpful to note that for an element at the right of the periodic table, the charge number of the anion it forms is equal to the group number minus 8. Thus oxygen, in Group VI, forms the oxide ion O^{2-} with charge number $6 - 8 = -2$. Phosphorus, in Group V, forms the phosphide ion P^{3-} with charge number $5 - 8 = -3$.

Polyatomic ions. By a _polyatomic ion_ we mean a bonded group of atoms with an overall positive or negative charge. Since polyatomic ions are like charged molecules, they are often called "molecular ions." It is important, however, to understand that the corresponding uncharged molecules might not exist. A simple example of a polyatomic cation is the ammonium ion NH_4^+ **(10)**, which occurs in ammonium chloride. As its formula shows, the ammonium ion consists of one N atom and four H atoms, but it has one fewer electron than required for electrical neutrality. The neutral molecule NH_4 is unknown.

The most common polyatomic anions are the _oxoanions_, polyatomic anions that contain oxygen. They include the carbonate (CO_3^{2-}), nitrate (NO_3^-), nitrite (NO_2^-), and sulfate (SO_4^{2-}) anions. The carbonate ion **(11)**, for example, consists of one C atom, three O atoms, and an additional two electrons. The corresponding neutral molecule CO_3 is unknown. The nitrite ion consists of one N atom, two O atoms, and one additional electron. The corresponding neutral molecule (nitrogen

10   Ammonium ion

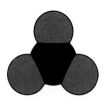

11   Carbonate ion

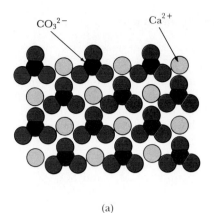

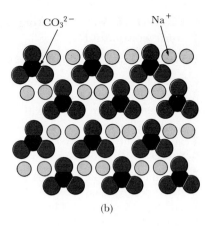

(a)   (b)

FIGURE 2.29 The formula unit of a compound indicates the relative number of atoms of each element present. (a) In $CaCO_3$ the Ca, C, and O atoms are present in the proportions $1:1:3$ (the latter two as the triangular carbonate ion). (b) In Na_2CO_3 the proportions of Na, C, and O are $2:1:3$.

dioxide, NO_2) is known and is a dark brown gas that contributes to the color of smog.

Formula units and formula weight. We symbolize an ionic compound by giving a chemical formula that shows the *relative* number of atoms of each element in the compound. In sodium chloride, for instance, there is one Na^+ ion for each Cl^- ion, so its formula is NaCl. In sodium carbonate there are two Na^+ ions per carbonate ion CO_3^{2-}, so its formula is Na_2CO_3 (Fig. 2.29). When a subscript has to be added to a polyatomic ion, the ion is written within parentheses, as in $(NH_4)_2SO_4$; here the parenthesis shows that there are two NH_4^+ ions for each SO_4^{2-} ion in the compound ammonium sulfate. We omit the charges on the ions when writing a formula, but it must be remembered that the compound consists of ions.

The relative numbers of cations and anions in an ionic compound are easy to determine, once we know the charges of the ions present. Since the compound as a whole is electrically neutral, the charges of the cations must be canceled by the charges of the anions (and vice versa). In sodium chloride, the charges are canceled if there is one Cl^- present for each Na^+ ion, which gives NaCl as the formula (Fig. 2.30). In magnesium chloride the charge of one Mg^{2+} ion is canceled by two Cl^- ions, so its formula is $MgCl_2$. In ammonium sulfate the double charge of SO_4^{2-} is canceled by two NH_4^+ ions, so its formula must be $(NH_4)_2SO_4$.

Formula unit of charge
$-1 +1 = 0$

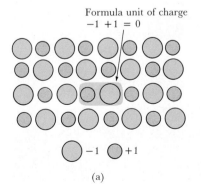

Formula unit of charge
$3 \times (-2) + 2 \times (-3) = 0$

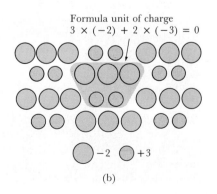

$\bigcirc -1$ $\bigcirc +1$

(a)

$\bigcirc -2$ $\bigcirc +3$

(b)

FIGURE 2.30 An ionic compound must be electrically neutral. These diagrams illustrate how this occurs in the case of (a) NaCl, where there is one cation for each anion, and for (b) Al_2O_3, where there are two Al^{3+} cations for three O^{2-} anions.

▼ EXAMPLE 2.6 Working out the formula of an ionic compound

Give the formula of aluminum sulfate.

STRATEGY The charges on the aluminum and sulfate ions are given in Tables 2.4 and 2.7 (below). One way to decide how many anions are needed to cancel the charge of the cation is to divide the number of positive charges on the cation by the number of negative charges on the anion. If the result is not a whole number, find the number of anions needed to cancel the charges of two (or, if necessary, three) cations.

SOLUTION The two ions in the compound are Al^{3+} and SO_4^{2-}. The three positive charges of Al^{3+} require three negative charges, or one and one-half SO_4^{2-} ions. The six positive charges of two Al^{3+} ions therefore require three SO_4^{2-} ions, which we denote $(SO_4)_3$ in the formula. The formula is therefore $Al_2(SO_4)_3$.

EXERCISE Write the formula for magnesium phosphate.

[*Answer*: $Mg_3(PO_4)_2$]

▲

A *formula unit* is a group of ions that matches the formula of the compound. Thus, the formula unit of NaCl consists of one Na^+ ion and one Cl^- ion; the formula unit of $(NH_4)_2SO_4$ consists of two NH_4^+ ions and one SO_4^{2-} ion. The *formula weight* (FW) of an ionic compound is the mass of one formula unit. Like molecular weight, it is obtained by adding together the atomic weights of the elements present in the unit. For magnesium chloride,

$$FW(MgCl_2) = AW(Mg) + 2 \times AW(Cl)$$

$$= 24.31 \text{ amu} + 2 \times 35.45 \text{ amu}$$

$$= 95.21 \text{ amu}$$

Once we know this formula weight, we also know that to obtain 1.00 mol of $MgCl_2$ (actually 1.00 mol of Mg^{2+} ions and 2.00 mol of Cl^- ions), we must weigh out 95.21 g of magnesium chloride. The "massive heaps" corresponding to 1 mol of formula units of some common ionic compounds are shown in Fig. 2.31.

FIGURE 2.31 Each sample contains 1 mol of an ionic compound. From left to right are 58 g of sodium chloride, 100 g of calcium carbonate, 278 g of iron sulfate heptahydrate, and 78 g of sodium peroxide.

The molar masses of formula units (like those of atoms) can be written from their formula weights. If the formula weight of the compound is x amu, then its molar mass (or mass per mole) is x g/mol. The molar mass of $MgCl_2$, for instance, is 95.21 g/mol. As in the case of atoms and molecules, the molar masses of formula units are used to set up conversion factors between the mass of compound in a sample and the number of moles of units in the sample. We can write, for example, that

$$1 \text{ mol } MgCl_2 = 95.21 \text{ g } MgCl_2$$

and use this to form a conversion factor. Suppose, for example, that we want to know how many grams of magnesium chloride to measure out to obtain 0.223 mol of $MgCl_2$. We first form the conversion factor

$$1 = \frac{95.21 \text{ g } MgCl_2}{1 \text{ mol } MgCl_2}$$

and then write

$$\text{Mass of } MgCl_2 = 0.223 \text{ mol } MgCl_2 \times \frac{95.21 \text{ g } MgCl_2}{1 \text{ mol } MgCl_2}$$

$$= 21.2 \text{ g } MgCl_2$$

That is, we weigh out 21.2 g of magnesium chloride.

2.8 CHEMICAL NOMENCLATURE

Many compounds have at least two names. A *common name* is one that has become familiar from everyday use but gives little or no clue to the compound's composition. Such names may have come into use before the composition was discovered; they include "water," "salt," "sugar," and "quartz." A *systematic name* is a name that reveals which elements are present (and, in some cases, how the atoms are arranged) and is constructed according to particular rules. The systematic name for table salt, "sodium chloride," shows at once that it is a compound of sodium and chlorine. A systematic name specifies a compound exactly. This is useful when the compound is unfamiliar or new, and it is essential when the compound has no common name. The systematic naming of compounds follows the rules of chemical *nomenclature*; like the defining of SI units, it is an international activity.

A distinction is traditionally made between "organic" and "inorganic" compounds. *Organic compounds* are compounds containing the element carbon, and usually hydrogen. They include methane, propane, glucose, and millions of other substances. These compounds are called "organic" because it was once incorrectly believed that they could be formed only by living organisms. All other compounds are classed as *inorganic compounds*; these include water, calcium sulfate, ammonia, silica, hydrochloric acid, and many more. In addition, some very simple carbon compounds, particularly carbon dioxide and the carbonates, which include chalk (calcium carbonate), are treated as inorganic compounds. We shall not deal with organic nomenclature at this stage, but concentrate here on naming inorganic compounds.

Naming cations. The names of cations are formed simply by adding "ion" to the name of the element, as in "sodium ion." Except in special circumstances, certain elements always form cations of one characteristic charge type. Thus, potassium is always present as K^+ in any compound, zinc is always present as Zn^{2+}, and aluminum is always present as Al^{3+}. The common elements that behave in this way include

Element	Characteristic ionic charge
Group I (the alkali metals)	+1
Group II (the alkaline-earth metals)	+2
Zinc and cadmium	+2
Aluminum	+3

When an element can form more than one kind of cation, such as copper forming both Cu^+ and Cu^{2+}, we need to distinguish between them. The most straightforward way of doing so is to use the *stock number*, a roman numeral equal to the number of electrons lost by the atom. Then Cu^+ is called the copper(I) ion and Cu^{2+} is called the copper(II) ion. Similarly, Fe^{2+} is the iron(II) ion and Fe^{3+} is the iron(III) ion.

The transition metals form so many different kinds of ions that it is almost always essential to show their charges and to include a roman numeral in the names of transition-metal compounds. However, the general aim in chemical nomenclature is to be unambiguous but brief, so when a transition element does in fact form only one kind of cation, there is no need to include a stock number in the names of that element's compounds. The most common example is silver: "silver chloride" always means "silver(I) chloride," even though some silver(II) and silver(III) compounds are known. Most scandium compounds contain Sc^{3+}, and there is no need to specify them as scandium(III).

In a somewhat older system of nomenclature, ions with different charges are named by first identifying the *stem* of the name of the element (its name with any *-ium* ending removed) and then adding a new *suffix* (or ending). The stem from chromium, for example, is *chrom-*. The ion with the lower charge is then denoted by the suffix *-ous*, giving chromous for Cr^{2+}, and the one with the higher charge is denoted by the suffix *-ic*, giving chromic for Cr^{3+}. To make matters worse, for some elements the Latin form of the name (Table 2.1) is used to form the stem. For instance, a compound used to provide fluoride ions in toothpaste, tin(II) fluoride, is often called stannous fluoride. Some common examples of these names and their translation into modern nomenclature are given in Table 2.6. We shall not use this system, but we must be aware of it since it is still quite commonly used.

Naming anions. The names of monatomic anions are formed by adding the suffix *-ide* to the stem of the element:

Element	Stem	Ion	
Fluorine	Fluor-	Fluoride ion	F^-
Oxygen	Ox-	Oxide ion	O^{2-}
Nitrogen	Nitr-	Nitride ion	N^{3-}

TABLE 2.6 **Names of cations with different charge numbers**

| Element | Ion | Names | |
		Old style	New style
Cobalt	Co^{2+}	Cobaltous	Cobalt(II)
	Co^{3+}	Cobaltic	Cobalt(III)
Copper	Cu^{+}	Cuprous	Copper(I)
	Cu^{2+}	Cupric	Copper(II)
Iron	Fe^{2+}	Ferrous	Iron(II)
	Fe^{3+}	Ferric	Iron(III)
Lead	Pb^{2+}	Plumbous	Lead(II)
	Pb^{4+}	Plumbic	Lead(IV)
Manganese	Mn^{2+}	Manganous	Manganese(II)
	Mn^{3+}	Manganic	Manganese(III)
Mercury	Hg_2^{2+}	Mercurous	Mercury(I)
	Hg^{2+}	Mercuric	Mercury(II)
Tin	Sn^{2+}	Stannous	Tin(II)
	Sn^{4+}	Stannic	Tin(IV)

There is no need to worry about alternatives because the monatomic anions of an element can have only one charge number. The ions formed from the halogens are collectively called *halide* ions, and include fluoride (F^-), chloride (Cl^-), and bromide (Br^-) ions.

The nomenclature of oxoanions (anions containing at least one oxygen atom) has given chemists a lot of trouble because there are so many of them. The names and symbols of the more common oxoanions are summarized in Table 2.7 and should be learned now. The role of the "acids" listed in the table will be explained shortly. The general rule is that the names of oxoanions are formed by adding the suffix *-ate* to the stem of the name of the element that is not oxygen:

Element	Stem	Oxoanion
Carbon	Carbon-	Carbonate ion CO_3^{2-}
Sulfur	Sulf-	Sulfate ion SO_4^{2-}
Chlorine	Chlor-	Chlorate ion ClO_3^{-}

The problem with oxoanions is that a given element can often form a variety of oxoanions with different numbers of oxygen atoms; nitrogen, for example, forms NO_2^- and NO_3^-. In such cases, the ion with the larger number of oxygen atoms is given the suffix *-ate*, and the one with the smaller number of oxygen atoms is given the suffix *-ite* (think of a mite as being small):

Element	Stem	Oxoanions	
Nitrogen	Nitr-	Nitrate ion	NO_3^{-}
		Nitrite ion	NO_2^{-}
Sulfur	Sulf-	Sulfate ion	SO_4^{2-}
		Sulfite ion	SO_3^{2-}

TABLE 2.7   Common oxoanions and their parent acids*

Transition metals		IV	V	VI	VII
		H_2CO_3 Carbonic acid	HNO_3 Nitric acid	H_2O Water	
		HCO_3^- Hydrogen carbonate (bicarbonate)	NO_3^- Nitrate	OH^- Hydroxide	
		CO_3^{2-} Carbonate		O^{2-} Oxide	
			HNO_2 Nitrous acid		
			NO_2^- Nitrite		
$Cr_2O_7^{2-}$ Dichromate	MnO_4^- Permanganate	H_3PO_4 Phosphoric acid	H_2SO_4 Sulfuric acid	$HClO_4$ Perchloric acid	
CrO_4^{2-} Chromate	MnO_4^{2-} Manganate	$H_2PO_4^-$ Dihydrogen phosphate	HSO_4^- Hydrogen sulfate (bisulfate)	ClO_4^- Perchlorate	
		HPO_4^{2-} Hydrogen phosphate	SO_4^{2-} Sulfate		
		PO_4^{3-} Phosphate			
		H_3PO_3 Phosphorous acid†	H_2SO_3 Sulfurous acid	$HClO_3$ Chloric acid	
		HPO_3^{2-} Phosphite	HSO_3^- Hydrogen sulfite (bisulfite)	ClO_3^- Chlorate	
			SO_3^{2-} Sulfite		
			$S_2O_3^{2-}$ Thiosulfate	$HClO_2$ Chlorous acid	
				ClO_2^- Chlorite	
				$HOCl$ Hypochlorous acid	
				OCl^- Hypochlorite	
				BrO_3^- Bromate	
				IO_3^- Iodate	

*Note that the arrangement of this table corresponds to the location of the elements in the periodic table.
†Phosphorous acid, H_3PO_3, can lose up to two hydrogen atoms.

Some oxoanions have relatives with an even smaller proportion of oxygen. Their names are formed by adding the prefix *hypo-* (from the Greek word for "under") to the stem with the -ite suffix, as in the hypochlorite ion ClO^-. Other oxoanions have a higher proportion of oxygen than the -ate oxoanions and are named with the prefix *per-* added to the -ate form of the name (*per* is the Latin word for "all over," suggesting that the element's ability to combine with oxygen is finally satisfied). An example is the perchlorate ion ClO_4^-. Chlorine forms oxoanions that span the range:

Perchlorate ion	ClO_4^-	highest oxygen content
Chlorate ion	ClO_3^-	
Chlorite ion	ClO_2^-	
Hypochlorite ion	ClO^-	lowest oxygen content
Chloride ion	Cl^-	No oxygen

Some important oxoanions contain hydrogen. Of these, the most common is the hydroxide ion OH^-; others include versions of the oxoanions mentioned above (see Table 2.7). These are named by adding "hydrogen" to the name of the oxoanion, as in

Hydrogen carbonate (bicarbonate) ion	HCO_3^-
Hydrogen sulfate (bisulfate) ion	HSO_4^-
Hydrogen sulfite (bisulfite) ion	HSO_3^-

(The older names—bicarbonate, bisulfate, and bisulfite—are still quite widely used.) When two hydrogen atoms are present, the prefix *di-* (from the Greek word for "two") is added, as for the dihydrogen phosphate ion $H_2PO_4^-$. Prefixes for showing numbers of atoms are listed in Table 2.8.

Naming ionic compounds. The name of an ionic compound is built from the names of the ions present in the compound, with the word "ion" omitted and in the order cation anion. Typical names include sodium chloride (a compound containing Na^+ and Cl^- ions), ammonium nitrate (containing NH_4^+ and NO_3^- ions), and calcium hydrogen carbonate (Ca^{2+} and HCO_3^- ions). The copper chloride that contains Cu^+ ions is called copper(I) chloride and the chloride that contains Cu^{2+} ions is called copper(II) chloride.

Prefixes are not normally used in naming ionic compounds. A compound such as $CuCl_2$ is called simply copper(II) chloride, not copper dichloride. There is no ambiguity regarding the number of Cl^- ions present. The same is true of $CaCl_2$, which is called simply calcium chloride, and of Al_2O_3, aluminum oxide. However, there are some exceptions among the common names; the most important is MnO_2, manganese(IV) oxide, which is commonly called manganese dioxide.

Prefixes are also used in naming *hydrates*, inorganic compounds with H_2O molecules present in the solid. For example, blue copper(II) sulfate crystals do not consist only of Cu^{2+} and SO_4^{2-} ions; each pair of ions is accompanied by five H_2O molecules. The overall formula unit is therefore written $CuSO_4 \cdot 5H_2O$, in which the 5 is understood to multi-

TABLE 2.8 Some prefixes used for naming compounds

Prefix	Meaning
Mono-	1
Di-	2
Tri-	3
Tetra-	4
Penta-	5
Hexa-	6
Hepta-	7
Octa-	8
Nona-	9
Deca-	10

FIGURE 2.32 Blue copper sulfate crystals ($CuSO_4 \cdot 5H_2O$) lose water above 150°C and form the white anhydrous powder ($CuSO_4$). The color is restored when water is added. This is a test that can be used to identify water.

ply the entire H_2O unit. This hydrate is copper(II) sulfate pentahydrate. The water in the crystal is called "water of hydration." It can be driven off by heating the crystals (Fig. 2.32), which then crumble to the white powder of "anhydrous" copper(II) sulfate $CuSO_4$. *Anhydrous* means "without water." When water is poured on the powder, the blue color is restored as the pentahydrate reforms.

Sodium carbonate is another example of a compound that can exist in various hydrated forms. Two of the most important are the decahydrate $Na_2CO_3 \cdot 10H_2O$, which is the compound used as "washing soda," and the anhydrous form Na_2CO_3. The latter, which is called "soda ash," is used in large amounts in glassmaking.

▼ **EXAMPLE 2.7   Writing the formula from the name**

Write the formula of the compound nickel(II) chloride hexahydrate.

STRATEGY   Identify the chemical symbols of the ions specified in the name, using Table 2.7 for the oxoanions if necessary (it should not be necessary in this simple case). Ensure that the formula unit is electrically neutral by selecting the appropriate number of each ion. For the hydration, refer to Table 2.8 for the meaning of the prefix.

SOLUTION   Nickel(II) indicates the presence of Ni^{2+}. Electrical neutrality is obtained with two Cl^- ions, giving $NiCl_2$. The prefix hexa- denotes 6; so the compound is $NiCl_2 \cdot 6H_2O$.

EXERCISE   Write the formula for magnesium perchlorate hexahydrate.
[*Answer*: $Mg(ClO_4)_2 \cdot 6H_2O$]
▲

Naming molecular compounds.   The names of many molecular compounds are formed as though the compound were ionic. This is an advantage in that we need not know whether a compound is ionic or molecular before naming it. It is a disadvantage in that we cannot tell from the name whether a compound is ionic or molecular. That knowledge comes with experience, but a good guide is that compounds containing only nonmetals (i.e., elements on the right of the periodic table or hydrogen) are likely to be molecular. A compound is likely to be ionic if one of the elements is a metal, especially an element from Group I or II.

A *binary compound* is one that consists of only two elements, as in HCl, H_2O, and CO_2. Except in a few cases, the molecular formula of a binary compound of hydrogen and a nonmetal has the H atom named first and written first:

Hydrogen chloride   HCl
Hydrogen sulfide H_2S

Some binary compounds of hydrogen are so common that they have nonsystematic names that have to be learned. These are given in Table 2.9.

Most other common binary molecular compounds contain at least one element from Group VI or VII. These elements are named sec-

TABLE 2.9 Some common names for molecular compounds

Formula*	Systematic name†	Common name
H_2O		Water
H_2O_2		Hydrogen peroxide
NH_3		Ammonia
N_2H_4		Hydrazine
NH_2OH		Hydroxylamine
PH_3		Phosphine
NO	Nitrogen monoxide	Nitric oxide
N_2O	Dinitrogen oxide	Nitrous oxide

*For historical reasons, the molecular formulas of binary hydrogen compounds of Group V elements are written with the Group V element first.
†The compounds for which no systematic names are given do have such names, but they are almost never used.

ond, and the number of atoms of each type is indicated with a prefix from Table 2.8:

Phosphorus trichloride PCl_3 Dinitrogen oxide N_2O
Sulfur hexafluoride SF_6 Dinitrogen pentoxide N_2O_5

The prefix mono- is never used with the first element and only very rarely with the second: the only common example of its use is in "carbon monoxide." Binary compounds of oxygen are named as "oxides" unless the other element is fluorine:

Chlorine dioxide ClO_2
Oxygen difluoride OF_2

Some oxides have common names that are still widely used; these are listed in Table 2.9 and should also be memorized.

▼ **EXAMPLE 2.8   Naming molecular compounds**

Name the compound with molecular formula N_2O_3.

STRATEGY   The compound is a nitrogen oxide. In Table 2.8, find the prefix for 2 and attach it to -nitrogen, and the prefix for 3 and attach it to -oxide.

SOLUTION   The prefix for 2 is di- and that for 3 is tri-; the compound is therefore dinitrogen trioxide.

EXERCISE   Name the molecular compounds SF_4 and S_2Cl_2.
▲   [*Answer*: Sulfur tetrafluoride; disulfur dichloride]

Acids.   Both HCl and H_2S are examples of an important class of compounds called <u>acids</u>. For now we shall take an acid to be a compound that contains hydrogen and can release hydrogen ions H^+ in water. (Acids are very important in chemistry, and starting in Chapter 3 we discuss them much more fully.) Hydrogen chloride is an acid because it dissolves in water to give a solution of hydrogen cations and chloride anions known as "hydrochloric acid." Methane (CH_4) contains hydrogen, but as it does not release hydrogen ions in water it is not an acid. A clue that a substance is an inorganic acid is that its formula begins with H, as in HCl and H_2SO_4.

Formally, the "hydrogen halides"—HF, HCl, HBr, and HI—are named like any other binary compounds of nonmetals: hydrogen fluoride, hydrogen chloride, and so on. However, aqueous solutions of these compounds have special names, formed by adding the prefix *hydro-* and the suffix *-ic acid* to the name of the element. Thus, an aqueous solution of HCl is generally called hydrochloric acid. These acids are the *parent acids* of the halides in the sense that the halide ions are left when the hydrogen ion has been released.

Element	Stem	Ion	Parent acid
Bromine	Brom-	Bromide ion Br^-	Hydrobromic acid HBr
Iodine	Iod-	Iodide ion I^-	Hydroiodic acid HI

Among the most important acids are the *oxoacids*, the acids containing oxygen. These molecular compounds are the parent acids of the oxoanions; they consist of molecules in which all the negative charges of an oxoanion are canceled by bonded hydrogens, as in H_2SO_4 and H_3PO_4. An oxoacid is named by replacing the *-ate* suffix of the anion with *-ic acid*; thus, the parent acid of the carbonates is H_2CO_3, carbonic acid. The more important oxoanions and oxoacids are listed in Table 2.7.

When both -ate (larger number of O atoms) and -ite (smaller number of O atoms) oxoanions exist, the parent acids are named as the *-ic acid* and the *-ous acid,* respectively:

> -ic oxoacids are the parent acids of -ate oxoanions.
> -ous oxoacids are the parent acids of -ite oxoanions.

Thus, we have sulfuric acid (H_2SO_4) as the parent of the sulfate anion (SO_4^{2-}), and sulfurous acid (H_2SO_3) as parent of the sulfite anion (SO_3^{2-}). This rule carries over into the parent acids of the per- and hypo- oxoanions: $HClO_4$, the parent acid of the perchlor*ate* ion, is called perchlor*ic* acid; HClO, the parent acid of the hypochlor*ite* ion, is called hypochlor*ous* acid. The complete range of chlorine acids is therefore

Perchlorate ion	ClO_4^-	$HClO_4$	perchloric acid (highest oxygen content)
Chlorate ion	ClO_3^-	$HClO_3$	chloric acid
Chlorite ion	ClO_2^-	$HClO_2$	chlorous acid
Hypochlorite ion	ClO^-	HClO	hypochlorous acid (lowest oxygen content)
Chloride ion	Cl^-	HCl	hydrochloric acid (no oxygen)

In this chapter we have taken matter apart, as far as chemists need to go; we have encountered elements and compounds, atoms, ions, and molecules, electrons and nuclei. We have now met the building blocks of matter. It is time to see how to use these blocks to build new materials.

SUMMARY

2.1 The classification of matter into ever more fundamental components is summarized in the figure below. Every element has a name and a **chemical symbol.**

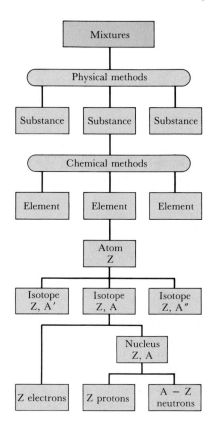

2.2 The elements are often listed in an arrangement called the **periodic table.** The columns of the table are called **groups,** and the rows are **periods.** Some of the groups have special names, such as the **alkali metals** of Group I, the **alkaline earths** of Group II, the **halogens**

of Group VII, and the **noble gases** of Group VIII. The **transition metals** are at the center of the table, between Groups II and III. Elements with similar characteristics are close to each other in the periodic table. All the **metals,** for instance, lie to the left of the **nonmetals,** which are located at the upper right.

2.3 According to Dalton's **atomic hypothesis,** all the atoms of a given element are the same, and compounds are specific combinations of atoms. Atoms are made up of **subatomic particles,** particles smaller than atoms, of which the most important in chemistry are the **electron,** the **proton,** and the **neutron.** An atom consists of a central **nucleus** surrounded by electrons. The **atomic number,** the number of protons in the nucleus, identifies the element and determines the number of electrons in the atom.

2.4 The **mass number,** which is the total number of **nucleons** in the nucleus, determines the mass of the atom. Individual atomic masses are measured with a **mass spectrometer** and may be expressed in **atomic mass units.** Most elements have **isotopes,** atoms that differ in mass because they have nuclei with different numbers of neutrons (but the same number of protons). Most natural samples of elements are actually mixtures of isotopes; the average atomic mass of such samples is called the **atomic weight** of the element.

2.5 The very large numbers of atoms in typical samples are expressed more conveniently in terms of the **mole.** The **molar mass** of the atoms of an element, or the mass per mole of atoms, is the basis for a conversion factor between mass and number of moles in that mass.

2.6 Molecular compounds consist of molecules; the **molecular weight** is the sum of the atomic weights of the elements in the compound. A molecular compound is symbolized by a **molecular formula,** showing the actual numbers of atoms in each molecule.

2.7 Atoms may lose or gain electrons to become **ions** and may form **ionic compounds.** Positively charged ions are called **cations,** and negatively charged ions are **anions.** Important polyatomic anions include the **oxoanions,** anions formed by the loss of hydrogen from their **parent acids,** the **oxoacids.** An ionic compound is symbolized by a formula showing the relative numbers of cations and anions in the compound.

2.8 The rules for naming compounds systematically are the basis of **chemical nomenclature.** The names of cations are listed in Table 2.6, and those of simple anions in Table 2.5. The names of oxoanions are listed in Table 2.7. For transition elements, the **Stock number,** the charge, must always be given.

EXERCISES

The Names of Elements and the Periodic Table

In Exercises 2.1 to 2.4, give the name of each element and state which group it belongs to in the periodic table.

2.1 (a) Li; (b) Ga; (c) Xe; (d) Tl.

2.2 (a) Fr; (b) Ge; (c) Sb; (d) Te.

2.3 (a) Ru; (b) Rh; (c) Re; (d) Rn.

2.4 (a) Pd; (b) Pt; (c) Pu; (d) Pa.

In Exercises 2.5 to 2.8, give the symbol and, where relevant, the group number of each element.

2.5 (a) Arsenic; (b) antimony; (c) astatine; (d) gold.

2.6 (a) Bromine; (b) barium; (c) bismuth; (d) boron.

2.7 (a) Tungsten; (b) tantalum; (c) terbium; (d) thulium.

2.8 (a) Curium; (b) californium; (c) cadmium; (d) chromium.

2.9 Give the names and symbols of four metallic elements in Groups III through VIII of the periodic table.

2.10 Give the names and symbols of four nonmetallic elements in Groups III through VIII of the periodic table.

The Nuclear Atom

In Exercises 2.11 to 2.14, state the numbers of protons, neutrons, and electrons in each atom or ion.

2.11 (a) Carbon-12; (b) oxygen-16; (c) uranium-235; (d) uranium-238.

2.12 (a) Deuterium; (b) fluorine-19; (c) lead-208; (d) lead-204.

2.13 (a) $^1H^+$; (b) $^4He^{2+}$; (c) $^{37}Cl^-$; (d) $^{32}S^{2-}$.

2.14 (a) $^{23}Na^+$; (b) $^{27}Al^{3+}$; (c) $^{16}O^{2-}$; (d) $^{31}P^{3-}$.

2.15 Calculate the total mass of (a) neutrons, (b) protons, and (c) electrons in 1.00 g of iron-57. (d) What fraction of your own mass is due to the neutrons in your body, if we suppose you to consist principally of water?

2.16 Calculate the total mass of (a) neutrons, (b) protons, and (c) electrons in 1.00 g of uranium-235. Sup-

pose that an automobile can be considered to be pure iron-57; (d) what fraction of its mass is due to its neutrons?

Atomic Masses

2.17 The masses of some atoms are given in Table 2.3. Use that information to calculate the number of atoms in (a) 1.0 mg of gold; (b) 1.0 L of oxygen which has a mass of 1.4 g.

2.18 Calculate the number of atoms in (a) 12 g of carbon; (b) 1.0 L of hydrogen, which has a mass of 90 mg.

2.19 Calculate (a) the mass of a single water molecule and (b) the number of molecules in 100 g of water.

2.20 Methane is the principal component of natural gas. Calculate (a) the mass of a single methane molecule and (b) the number of molecules in 1.0 L of methane, which has a mass of 0.72 g.

2.21 Calculate the total mass of copper in 1.0 g of copper(II) sulfate, $CuSO_4 \cdot 5H_2O$.

2.22 Calculate the total mass of nickel in 1.0 g of nickel(II) sulfate, $NiSO_4 \cdot 6H_2O$.

2.23 What mass of iron contains as many iron atoms as there are (a) carbon atoms in 12 g of carbon; (b) chromium atoms in 25 g of chromium?

2.24 What mass of copper contains as many copper atoms as there are (a) carbon atoms in 12 g of carbon; (b) nickel atoms in 10 g of nickel?

2.25 Estimate the total mass of electrons in your body by assuming that you are mainly water.

2.26 The mass of the copper in the Statue of Liberty is 2.5×10^5 kg. What is the total mass of the electrons in the statue?

2.27 In mass spectrometry it is common to make the approximation that the mass of a cation is the same as the mass of the parent atom. To how many significant figures must the mass of a hydrogen atom be reported before the difference between its mass and that of a hydrogen cation is significant? The mass of a hydrogen atom is 1.67×10^{-27} kg. What is the difference between the mass of a hydrogen cation and that of a hydrogen atom, expressed as a percentage of the atomic mass?

2.28 The mass of an iron atom is 9.27×10^{-26} kg. What are the differences between this mass and those of the two most common iron ions, expressed as a percentage of the atomic mass?

Atomic Mass Units

2.29 The mass of one atom of helium-3 is 3.02 amu. What is that in grams? How many atoms of helium are there in a sample of mass 1.0 mg?

2.30 The mass of one atom of deuterium is 2.01 amu. What is that in grams? How many deuterium (D_2) molecules are there in 1.0 mg of the gas?

2.31 The mass of one atom of chlorine-35 is 5.81×10^{-26} kg. What is that in atomic mass units? What would be the mass of a chlorine-35 atom on a scale on which carbon-12 is exactly 12?

2.32 The mass of one atom of fluorine-19 is 3.15×10^{-26} kg. What is that in atomic mass units? What would be the mass of a fluorine-19 atom on a scale on which carbon-12 is exactly 12?

Moles

In Exercises 2.33 and 2.34, express the given numbers as moles.

2.33 (a) 6.02×10^{23} hydrogen atoms; (b) 1.2×10^{24} electrons; (c) 300 million people, the approximate population of the United States; (d) 10^{22} stars, the number of stars in the observable universe.

2.34 (a) 6.02×10^{23} hydrogen molecules; (b) 3×10^{20} protons; (c) 1 trillion (10^{12}) grains of sand, about a thousand metric tons; (d) 10^{11} brain cells, about the number inside your head.

In Exercises 2.35 and 2.36, express the given numbers of moles as numbers of objects.

2.35 (a) 1.00 mol O_2 molecules; (b) 0.50 mol Na^+ ions; (c) 1.0 mmol C atoms; (d) 2.0 mol e^-.

2.36 (a) 1.00 mol O atoms; (b) 0.25 mol SO_4^{2-} ions; (c) 1.5 mmol Al atoms; (d) 2.0 mol $C_6H_{12}O_6$ (glucose) molecules.

Atomic Weights

2.37 Calculate the atomic weight of natural carbon, which consists of 98.89% carbon-12 (of mass 12 amu exactly) and 1.11% carbon-13 (of mass 13.003 amu).

2.38 Calculate the atomic weight of natural boron, which consists of 19.78% boron-10 (of mass 10.013 amu) and 80.22% boron-11 (of mass 11.093 amu).

2.39 Calculate the atomic weight of lithium given that it consists of lithium-6 (of mass 6.02 amu) in 7.42% abundance and lithium-7 (7.02 amu) in 92.48% abundance.

2.40 Lithium-6 is being partially drained from naturally occurring lithium samples because it is needed by the nuclear industry. What would be the atomic weight of a sample in which the lithium-6 abundance has been reduced to 5.22%? What are the percentage abundances of lithium-6 and lithium-7 in a sample with atomic weight 6.80 g per mole of Li atoms?

Isotopic Abundance

2.41 The principal isotopes of the yellow solid element sulfur, their abundances, and their masses are given below. How many atoms are there in a 1.000-g sample of sulfur?

Isotope	Abundance, %	Mass, amu
Sulfur-32	95.0	31.97
Sulfur-33	0.8	32.97
Sulfur-34	4.2	33.97

2.42 The principal isotopes of the gas krypton, their abundances, and their masses are given below. How many atoms of krypton are there in 0.00300 mg of the gas (which is enough to occupy about 1 mm³ under normal conditions)?

Isotope	Abundance, %	Mass, amu
Krypton-78	0.3	77.92
Krypton-80	2.3	79.92
Krypton-82	11.6	81.91
Krypton-83	11.5	82.91
Krypton-84	56.9	83.91
Krypton-86	17.4	85.91

In Exercises 2.43 and 2.44, calculate the mass of the sample you should prepare to obtain each of the given numbers of atoms:

2.43 (a) 1.00 mol of C atoms as graphite; (b) 0.50 mol of Cl atoms as gaseous chlorine (Cl_2); (c) 1.5 mmol of Pt atoms as the metal; (d) 10.0 mol of S atoms as S_8 molecules.

2.44 (a) 1.00 mol of C atoms as diamond; (b) 2.5 mmol of P atoms as P_4 molecules; (c) 25 mol of Fe atoms as iron metal; (d) 1.0 mmol of Pu atoms as the metal.

2.45 If the cost of gold bullion is \$480 per troy ounce (31.10 g), how much does a gold atom cost?

2.46 The cost of a particular 1.00-karat diamond, which consists of the element carbon, is \$3500. How much does each carbon atom cost? (1 karat = 200 mg)

Molecular and Formula Weights

In Exercises 2.47 and 2.48, calculate the molecular weights of the given molecular compounds.

2.47 (a) Sulfur hexafluoride; (b) ammonia; (c) acetylene, C_2H_2; (d) estradiol, $C_{18}H_{24}O_2$.

2.48 (a) Phosphorus pentachloride; (b) hydrogen sulfide; (c) oxalic acid, $H_2C_2O_4$; (d) testosterone, $C_{19}H_{28}O_2$.

In Exercises 2.49 and 2.50, calculate the formula weights of the given compounds.

2.49 (a) Silver chloride, AgCl; (b) potassium cyanide, KCN; (c) ammonium sulfate, $(NH_4)_2SO_4$; (d) copper(II) bromide, $CuBr_2 \cdot 4H_2O$.

2.50 (a) Magnesium fluoride, MgF_2; (b) manganese nitrate, $Mn(NO_3)_2 \cdot 4H_2O$; (c) ammonium dihydrogen phosphate, $(NH_4)H_2PO_4$; (d) lanthanum sulfate, $La_2(SO_4)_3 \cdot 9H_2O$.

In Exercises 2.51 and 2.52, calculate the number of moles of molecules in each of the given samples.

2.51 (a) 1.00 g of benzene (C_6H_6); (b) 1.00 g of sucrose ($C_{12}H_{22}O_{11}$).

2.52 (a) 100 g of water; (b) 1.0 kg of ethanol (C_2H_6O).

2.53 Calculate the number of moles of (a) AgCl units, (b) Ag^+ ions, and (c) Cl^- ions in 1.0 g of silver chloride.

2.54 Calculate the number of moles of (a) KCN units, (b) K^+ ions, and (c) CN^- ions in 1.0 g of potassium cyanide.

2.55 A chemist measured out 1.0 g of copper(II) bromide, $CuBr_2 \cdot 4H_2O$. How many Br^- ions are present in the sample?

2.56 A chemist prepared a sample by mixing 1.0 g of ammonium phosphate, $(NH_4)_3PO_4 \cdot 3H_2O$, with 1.0 g of potassium phosphate, K_3PO_4. How many NH_4^+ ions are in the sample?

In Exercises 2.57 and 2.58, determine what mass of each substance you would need to measure out to obtain the given number of moles.

2.57 (a) 1.0 mol of H_2O molecules; (b) 2.0 mol of Fe atoms; (c) 1.5 mol of UO_3 formula units; (d) 0.10 mol of $C_2H_4O_2$ (acetic acid) molecules.

2.58 (a) 1.0 mol of H_2S molecules; (b) 2.0 mol of Al atoms; (c) 1.0 mol of $AuCl_3 \cdot 2H_2O$ units; (d) 1.0 mol of C_6H_6 (benzene) molecules.

In Exercises 2.59 and 2.60, express the given masses as moles of the specified objects.

2.59 (a) 1.0 kg of water as H_2O molecules; (b) 100 g of ethanol as C_2H_6O molecules; (c) 1.0 g of silver nitrate as $AgNO_3$ formula units; (d) 3.0 g of sucrose as $C_{12}H_{22}O_{11}$ molecules.

2.60 (a) 200 mg of hydrogen as H atoms; (b) 200 mg of hydrogen as H_2 molecules; (c) 1.0 g of calcium carbonate as $CaCO_3$ formula units; (d) 3.0 g of glucose as $C_6H_{12}O_6$ molecules.

Chemical Nomenclature

Name the ions in Exercises 2.61 to 2.64, giving both the older and the newer names where they exist.

2.61 (a) Fe^{2+}; (b) H^+; (c) Au^{3+}; (d) Pb^{2+}.
2.62 (a) Sn^{2+}; (b) Pb^{4+}; (c) Hg_2^{2+}; (d) Mn^{2+}.

2.63 (a) F^-; (b) H^-; (c) C_2^{2-}; (d) CN^-.
2.64 (a) Se^{2-}; (b) P^{3-}; (c) HSe^-; (d) OI^-.

In Exercises 2.65 and 2.66, name the oxoanions.

2.65 (a) HCO_3^-; (b) NO_2^-; (c) BrO_3^-; (d) ClO_2^-.
2.66 (a) HSO_3^-; (b) AsO_4^{3-}; (c) $H_2AsO_4^-$; (d) OF^-.

In Exercises 2.67 to 2.70, name the compounds.

2.67 (a) CsI; (b) $CuCl_2$; (c) Ag_2SO_4; (d) $(NH_4)_2HPO_4$.
2.68 (a) RbF; (b) $AuCl_3$; (c) $Al(OH)_3$; (d) $(NH_4)HSO_3$.
2.69 (a) SF_4; (b) SiO_2; (c) NF_3; (d) XeF_4.
2.70 (a) $SiCl_4$; (b) N_2O_4; (c) ClO_2; (d) XeO_3.

In Exercises 2.71 and 2.72, propose chemical formulas for the given compounds.

2.71 (a) Magnesium oxide (magnesia); (b) calcium phosphate (the major inorganic component of bones); (c) aluminum sulfate; (d) calcium nitride.

2.72 (a) Silver chloride; (b) mercury(I) chloride (calomel); (c) calcium phosphide; (d) iron(III) selenide.

In Exercises 2.73 and 2.74, propose molecular formulas for the given compounds.

2.73 (a) Selenium trioxide; (b) carbon tetrabromide; (c) carbon disulfide; (d) iodine heptafluoride.

2.74 (a) Sulfur dichloride; (b) dinitrogen monoxide; (c) chlorine pentafluoride; (d) dinitrogen pentoxide.

In Exercises 2.75 and 2.76, name the acids.

2.75 (a) HF; (b) H_2SeO_4; (c) H_2SeO_3; (d) HOBr.
2.76 (a) $HClO_2$; (b) $HClO_3$; (c) H_3PO_3; (d) H_2XeO_4, where XeO_4^{2-} is the xenate ion.

General

*2.77 Calculate the average density of a single carbon atom by assuming that it is a uniform sphere of radius 77 pm. (The volume of a sphere is $\frac{4}{3}\pi r^3$, where r is its radius.) Express the answer in grams per cubic centimeter. The density of diamond, a crystalline form of carbon, is 3.5 g/cm^3. What does that suggest about the way the atoms are packed together in diamond?

*2.78 Assume that the entire mass of an atom is concentrated in its nucleus, a sphere of radius 1.5×10^{-5} pm. What is the density of a carbon nucleus? What would be the radius of the earth if its matter were compressed to the same density? (Its actual radius is 6.4×10^3 km, and its average density 5.5 g/cm^3.)

The change of one substance into others by chemical reaction is the essence of chemistry. This chapter explains how chemists describe reactions symbolically, and how they organize their subject by classifying reactions into different types according to certain characteristic features. The illustration shows the corrosion that formed on the Titanic during its immersion. The reactions responsible for corrosion are among those we describe in this chapter.

3

CHEMICAL REACTIONS

Chemical equations

3.1   Symbolizing reactions
3.2   Balancing equations

Precipitation reactions

3.3   Net ionic equations
3.4   Using precipitation
      reactions in chemistry

Acid-base reactions

3.5   Arrhenius acids and
      bases
3.6   The Brønsted
      definitions
3.7   Neutralization

Redox reactions

3.8   Electron transfer
3.9   Oxidizing and
      reducing agents
3.10 Half-reactions

Chemists have developed a powerful language for discussing reactions that uses formulas and equations. Although this may seem dry, the language allows us to perform a wide range of important calculations that provide information about chemical changes. This formalism also allows us to group reactions into classes, three of which we discuss in this chapter. As so often happens in chemistry, the classification scheme then helps us to recognize patterns that make the subject much easier to understand and use. Moreover, once we have identified the different types of reaction likely to occur when substances are mixed or heated, we can begin to make predictions about the outcomes in specific cases.

CHEMICAL EQUATIONS

The starting materials in a chemical reaction are called the *reactants*. The substances formed as a result of the reactions are called the *products*. A chemical reaction can therefore be summarized as

$$\text{Reactants} \longrightarrow \text{products}$$

Chemists speak of "reagents" as well as of reactants: a *reagent* is a substance that reacts with a wide variety of other substances, and for that reason is regularly stocked in chemistry laboratories (Fig. 3.1). Hydrochloric acid is one example. A reactant, however, is a substance taking part in a specified reaction.

3.1 SYMBOLIZING REACTIONS

A simple example of a reaction is the formation of water from hydrogen and oxygen, which we may symbolize as

$$\text{Hydrogen} + \text{oxygen} \longrightarrow \text{water}$$

This reaction is an example of a *synthesis*, a reaction in which a substance is formed from simpler starting materials (in this case, elements). Another type of reaction is *decomposition*, in which a substance is broken down into simpler substances. An example is the reaction that occurs when chalk is heated to 800°C (Fig. 3.2):

$$\text{Calcium carbonate} \longrightarrow \text{calcium oxide} + \text{carbon dioxide}$$

This *thermal decomposition*, a decomposition brought about by heat, is typical of most carbonates.

Chemists report a reaction by stating its *chemical equation*, which is an expression showing the chemical formulas of the reactants and products. As we shall see, for the reactions described above, they write

$$2H_2 + O_2 \longrightarrow 2H_2O$$

$$CaCO_3 \xrightarrow{\Delta} CaO + CO_2$$

Note that we distinguish a "chemical reaction," the actual chemical change, from a "chemical equation," its representation in terms of chemical symbols. The symbol Δ (the Greek letter delta) signifies that

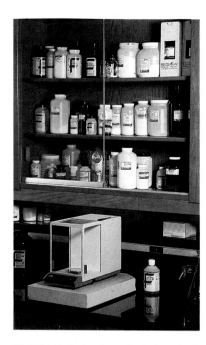

FIGURE 3.1 A typical chemical laboratory is equipped with a wide range of reagents that can be used as reactants.

the reaction occurs at an elevated temperature (in this case, at about 800°C). The numbers multiplying entire chemical formulas, so that $2H_2O$ stands for $2 \times (H_2O)$, are called the *stoichiometric coefficients* of the substances. This awkward name comes from the Greek words for "element" and "measure." These coefficients are included to ensure that the same number of atoms of each element appears in the reactants and in the products: a chemical reaction cannot create or destroy atoms. In the water synthesis, for example, there are four hydrogen atoms on the left side of the arrow (among the reactants) and four on the right side of the arrow (among the products); there are two oxygen atoms on each side of the arrow as well. A chemical equation therefore takes into account Dalton's view that a chemical change is a *rearrangement* of atoms between different partners (Fig. 3.3). It summarizes the identities of the substances in the reaction (the qualitative information) as well as the fact that the number of each type of atom is preserved (the quantitative information).

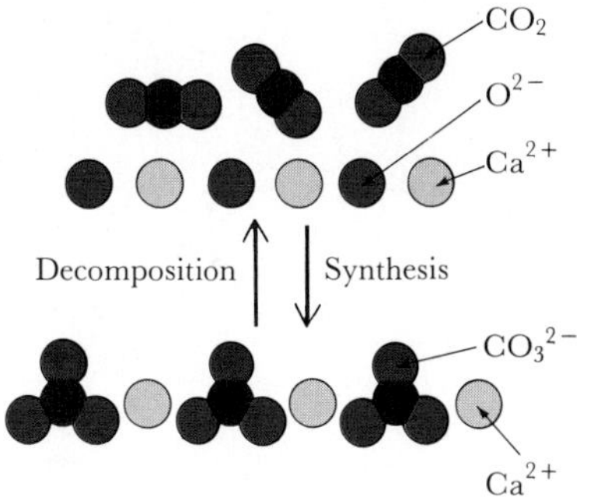

A chemical equation shows even more information if it also includes the physical state (solid, liquid, or gas) of each substance taking part. This is done by adding one of the following labels:

$$(s) \text{ for solid} \qquad (l) \text{ for liquid} \qquad (g) \text{ for gas}$$

The label (*aq*) is used to show that a substance is in aqueous solution (dissolved in water). The full-dress versions of the equations given above are then

$$2H_2(g) + O_2(g) \longrightarrow 2H_2O(l)$$

for the synthesis if liquid water (as distinct from water vapor or ice) is the product, and

$$CaCO_3(s) \xrightarrow{\Delta} CaO(s) + CO_2(g)$$

for the decomposition.

3.2 BALANCING EQUATIONS

A chemical equation is *balanced* when the same number of atoms of each element appears on both sides of the arrow. (Those discussed so far have been balanced.) Simple reactions can often be balanced almost at a glance; others need more work.

FIGURE 3.2 In a decomposition, a substance changes into less complex substances, which may be elements. The reverse of decomposition is synthesis, the formation of a substance from simpler starting materials.

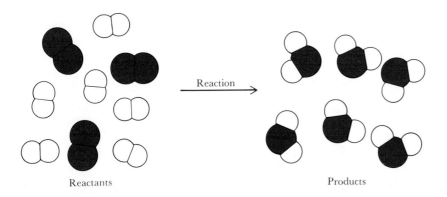

Reactants

Reaction

Products

FIGURE 3.3 A chemical reaction is a rearrangement of the atoms present in the reactants into the groupings that are characteristic of the products.

To formulate—and balance—a chemical equation, we begin by writing the *skeletal equation*, an equation that summarizes the qualitative information about the reaction, as in

$$H_2 + O_2 \longrightarrow H_2O \qquad \triangle$$

Hydrogen + oxygen forms water

(We shall use the international *hazard* road sign $\triangle$ to warn that an equation is not balanced.) Then we choose coefficients that balance all the elements:

$$2 \times (H_2) + O_2 \longrightarrow 2 \times (H_2O)$$

An equation must never be balanced by changing the *subscripts* in the chemical formulas. That would suggest that different substances were taking part in the reaction. Changing H_2O to H_2O_2 in the original equation and writing

$$H_2 + O_2 \longrightarrow H_2O_2$$

certainly results in a balanced equation. However, this balanced equation is a summary of a *different* reaction—the formation of hydrogen peroxide from its elements. Writing

$$2H + O \longrightarrow H_2O$$

also gives us a balanced equation, but one that expresses the reaction between hydrogen and oxygen *atoms*, not the molecules that are the actual starting materials.

In many cases, when we multiply one formula by a coefficient in order to balance a particular element, the balance of other elements is upset. This can be very frustrating when an equation contains a number of different elements. It is therefore wise to reduce the amount of work by working systematically. One straightforward procedure is to balance the element that occurs the least number of times first, and then to work up to the element that occurs the most times.

We can illustrate the method by balancing the equation for a *combustion*, a reaction that occurs when an element or compound burns in oxygen (Fig. 3.4). When organic compounds containing only carbon, hydrogen, and oxygen burn in a plentiful supply of air, the only products are carbon dioxide and water. It follows that if the organic compound is the methane (CH_4) of natural gas, the skeletal equation is

$$CH_4 + O_2 \longrightarrow CO_2 + H_2O \qquad \triangle$$

Since C and H occur in two formulas and O occurs in three, we begin with C and H. The C atoms are already balanced. We balance the H atoms by multiplying H_2O by 2 to give four H atoms on each side:

$$CH_4 + O_2 \longrightarrow CO_2 + 2H_2O \qquad \triangle$$

Now only O remains to be balanced. Since there are four O atoms on the right but only two on the left, the O_2 should be multiplied by 2. The result is

$$CH_4 + 2O_2 \longrightarrow CO_2 + 2H_2O$$

FIGURE 3.4 A methane flame gives out heat as a result of the combustion of methane to carbon dioxide and water. The flame is blue because the heat generated by the reaction excites the C_2 and CH units formed during the combustion, which then give off colored light.

The equation is now balanced, as we can verify by counting the number of atoms of each element on each side of the arrow. At this stage we should specify the states. If water vapor is the product, we write

$$CH_4(g) + 2O_2(g) \longrightarrow CO_2(g) + 2H_2O(g)$$

In some cases, balancing leads to fractional coefficients, as in the equation for the combustion of butane (C_4H_{10}):

$$C_4H_{10}(g) + \tfrac{13}{2}O_2(g) \longrightarrow 4CO_2(g) + 5H_2O(g)$$

Since it is much more convenient to deal with whole numbers, it is usually sensible (but not necessary) to clear the fractions by multiplying the *entire* equation by a numerical factor. Here we can clear the fraction $\tfrac{13}{2}$ by multiplying through by 2, obtaining

$$2C_4H_{10}(g) + 13O_2(g) \longrightarrow 8CO_2(g) + 10H_2O(g)$$

▼ **EXAMPLE 3.1 Balancing equations: I**

The purple solid potassium permanganate ($KMnO_4$) is produced from the mineral pyrolusite (manganese dioxide, MnO_2) in a two-stage process (Fig. 3.5). The first stage is the reaction of the pyrolusite with aqueous potassium hydroxide and oxygen to give dark green potassium manganate (K_2MnO_4) and water. Write the balanced equation for this stage of the process.

STRATEGY We start by writing the skeletal equation. Then it is usually a good idea to spend a moment trying to balance it by inspection. If the correct coefficients are not immediately obvious, work through the elements as suggested above.

SOLUTION The skeletal equation is

$$MnO_2 + KOH + O_2 \longrightarrow K_2MnO_4 + H_2O \qquad \triangle$$

Assuming we cannot balance it by inspection, we note that Mn, K, and H each occur in two formulas, and O occurs in five. The Mn atoms are balanced (one on each side). We balance the two K atoms on the right by multiplying KOH by 2:

$$MnO_2 + 2KOH + O_2 \longrightarrow K_2MnO_4 + H_2O \qquad \triangle$$

Hydrogen is balanced, so we move on to oxygen. On the right are five O atoms; there would be five on the left if O_2 were multiplied by $\tfrac{1}{2}$:

$$MnO_2 + 2KOH + \tfrac{1}{2}O_2 \longrightarrow K_2MnO_4 + H_2O$$

This is balanced, as we can check by counting the atoms of each element on each side. Finally, we multiply through by 2 to clear the fraction and add the state labels:

$$2MnO_2(s) + 4KOH(aq) + O_2(g) \longrightarrow 2K_2MnO_4(aq) + 2H_2O(l)$$

EXERCISE In the next stage of the process, chlorine gas is bubbled through the potassium manganate solution, and potassium permanganate and potassium chloride are produced. Write the balanced equation for this stage of the process.

[*Answer*: $2K_2MnO_4(aq) + Cl_2(g) \longrightarrow 2KMnO_4(aq) + 2KCl(aq)$]

(a)

(b)

(c)

FIGURE 3.5 (a) The starting materials for the production of potassium permanganate are the white potassium hydroxide and manganese dioxide. (b) In this laboratory version of the process, the first step converts the dioxide to potassium manganate (green) by reaction with oxygen and potassium hydroxide solution. (c) In the final step, the green solution of manganate ions is converted to the purple solution of permanganate ions using chlorine.

▼ **EXAMPLE 3.2** Balancing equations: II

A solution of nitric acid (HNO_3) and the gas nitric oxide (NO) are produced when the gas nitrogen dioxide (NO_2) is bubbled through water. Write the chemical equation for the process.

STRATEGY We write the skeletal equation and then balance it, using the order that causes least disturbance to the elements already balanced.

SOLUTION The skeletal equation is

$$NO_2 + H_2O \longrightarrow HNO_3 + NO \qquad \triangle$$

Since H occurs in two formulas, N in three, and O in four, we balance them in that order. We can balance the two H atoms on the left by multiplying HNO_3 by 2:

$$NO_2 + H_2O \longrightarrow 2HNO_3 + NO \qquad \triangle$$

The three N atoms on the right are balanced by multiplying NO_2 by 3:

$$3NO_2 + H_2O \longrightarrow 2HNO_3 + NO$$

This also balances the O atoms (and does not disturb the H balance). The complete equation is therefore

$$3NO_2(g) + H_2O(l) \longrightarrow 2HNO_3(aq) + NO(g)$$

EXERCISE The colorless gas boron trifluoride reacts with water to produce white crystals of boric acid (H_3BO_3) and a solution of tetrafluoroboric acid (HBF_4). Write the chemical equation for the reaction.

▲   [*Answer*: $4BF_3(g) + 3H_2O(l) \longrightarrow H_3BO_3(s) + 3HBF_4(aq)$]

PRECIPITATION REACTIONS

A soluble ionic compound dissolves in water to give an electrolyte solution (Section 2.7). An example is a solution of sodium chloride in water, which contains Na^+ cations and Cl^- anions moving freely and generally independently through the solvent. A solution of silver nitrate is also an electrolyte, but the ions present are Ag^+ cations and NO_3^- anions. Now suppose we mix the two electrolyte solutions. We immediately get a white precipitate, which analysis shows is the almost insoluble compound silver chloride. The remaining solution contains Na^+ cations and NO_3^- anions because sodium nitrate ($NaNO_3$) is soluble in water.

The reaction we have just described is an example of a *precipitation reaction* (Fig. 3.6), a reaction in which a solid product is formed when

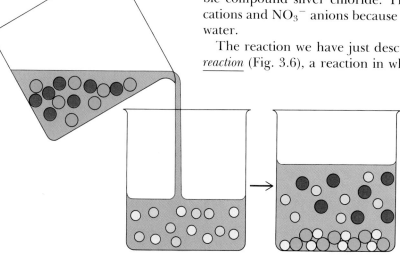

FIGURE 3.6 In a precipitation reaction, two soluble substances react to produce an insoluble precipitate of a new substance. A second, soluble product is usually left in solution.

two electrolyte solutions are mixed. In our case, the reaction is

$$AgNO_3(aq) + NaCl(aq) \longrightarrow AgCl(s) + NaNO_3(aq)$$

| Electrolyte solution | + | electrolyte solution | $\longrightarrow$ | precipitate + | electrolyte solution |

A colorful example is the formation of the insoluble yellow solid lead chromate ($PbCrO_4$, the artist's pigment "chrome yellow") when solutions of lead nitrate and potassium chromate are mixed (Fig. 3.7):

$$Pb(NO_3)_2(aq) + K_2CrO_4(aq) \longrightarrow PbCrO_4(s) + 2KNO_3(aq)$$

3.3 NET IONIC EQUATIONS

The essential character of a precipitation reaction becomes clear when we write its chemical equation in a form that shows the ions explicitly. For the silver chloride precipitation, in terms of the ions present in the solution we have

$$Ag^+(aq) + \cancel{NO_3^-(aq)} + \cancel{Na^+(aq)} + Cl^-(aq) \longrightarrow$$
$$AgCl(s) + \cancel{Na^+(aq)} + \cancel{NO_3^-(aq)}$$

Since the $Na^+(aq)$ and $NO_3^-(aq)$ ions appear as both reactants and products, they play no direct role in the reaction. That is, they are *spectator ions*, ions that are present but remain unchanged. The equation of the *net ionic reaction*, the net change, is obtained by canceling the spectator ions, which leaves

$$Ag^+(aq) + Cl^-(aq) \longrightarrow AgCl(s)$$

This equation focuses on the essential feature of the reaction (Fig. 3.8). It shows that, in the precipitation, the Ag^+ ions provided by one solution combine with the Cl^- ions provided by the other and precipitate as an insoluble solid. The precipitation of the insoluble solid is sometimes referred to as the "driving force" for the reaction.

FIGURE 3.7 Lead chromate (also known as "chrome yellow") is used in the paint with which traffic lanes are marked. It is manufactured in an industrial version of the laboratory preparation, by precipitation from lead nitrate and potassium chromate solutions. (The lead nitrate solution is the colorless solution in the flask.)

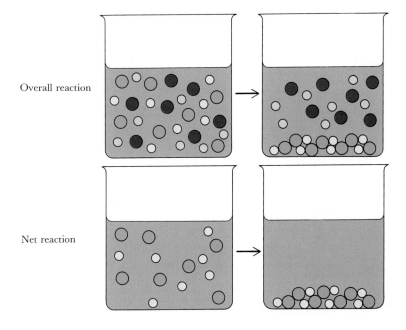

Overall reaction

Net reaction

FIGURE 3.8 By removing the spectator ions from a chemical equation, we can focus on the essential process—the net ionic reaction.

▼ EXAMPLE 3.3 Using net ionic equations to balance chemical equations

Write the balanced equation for the precipitation of barium iodate when barium nitrate and ammonium iodate solutions are mixed.

STRATEGY We begin by balancing the skeletal net ionic equation, for once that is done the spectator ions can be added in equal numbers on each side of the arrow to give the overall equation. It is unnecessary to balance the equation element by element because the polyatomic anions stay intact and can be balanced as though they were elements. Refer to Table 2.7 for the formulas of anions.

SOLUTION The skeletal equation showing that barium iodate precipitates when barium ions and iodate ions are mixed is

$$Ba^{2+}(aq) + IO_3^-(aq) \longrightarrow Ba(IO_3)_2(s)$$ ⚠

This is easily balanced by inspection, since the numbers of Ba^{2+} and IO_3^- ions must be the same on both sides:

$$Ba^{2+}(aq) + 2IO_3^-(aq) \longrightarrow Ba(IO_3)_2(s)$$

Now we add the spectator ions. One starting material is barium nitrate, for which two NO_3^- ions must be included on the left (and therefore on the right). The other is ammonium iodate, for which $2NH_4^+$ must be added to the left, and therefore to the right:

$$Ba^{2+}(aq) + 2NO_3^-(aq) + 2NH_4^+(aq) + 2IO_3^-(aq) \longrightarrow$$
$$Ba(IO_3)_2(s) + 2NH_4^+(aq) + 2NO_3^-(aq)$$

This is the same as

$$Ba(NO_3)_2(aq) + 2NH_4IO_3(aq) \longrightarrow Ba(IO_3)_2(s) + 2NH_4NO_3(aq)$$

EXERCISE Write the balanced equation for the reaction in which silver nitrate reacts with sodium chromate to give a precipitate of silver chromate.

▲ [*Answer*: $2AgNO_3(aq) + Na_2CrO_4(aq) \longrightarrow Ag_2CrO_4(s) + 2NaNO_3(aq)$]

3.4 USING PRECIPITATION REACTIONS IN CHEMISTRY

If a compound MX is insoluble in water, the net ionic equation for its precipitation is

$$M^+(aq) + X^-(aq) \longrightarrow MX(s)$$

An example is

$$Ag^+(aq) + I^-(aq) \longrightarrow AgI(s)$$

Therefore, if we can find two soluble compounds MA and BX such that A^- and B^+ act as spectator ions, then the general form of the precipitation reaction will be

$$MA(aq) + BX(aq) \longrightarrow MX(s) + BA(aq)$$

as in

$$AgNO_3(aq) + KI(aq) \longrightarrow AgI(s) + KNO_3(aq)$$

Precipitation reactions are one kind of _double replacement_ reaction (or "metathesis," from the Greek words for "exchange"), a reaction in which atoms or ions exchange partners. By choosing the starting compounds carefully, we can use a precipitation reaction to produce a desired insoluble compound, and then separate it from the reaction mixture by filtration (Fig. 3.9).

To be able to predict what will happen when we mix solutions, and to choose compounds that will give a desired product, we need to know whether compounds are soluble or insoluble. The empirical (observation-based) _solubility rules_ given in Table 3.1 summarize the solubility patterns of a range of common compounds in water. We see, for instance, that PbI_2 is insoluble. Thus, it is formed whenever solutions containing Pb^{2+} and I^- ions are mixed:

$$Pb^{2+}(aq) + 2I^-(aq) \longrightarrow PbI_2(s)$$

Since $Pb(NO_3)_2$ is soluble (like all nitrates) and KI is soluble (like all common Group I compounds), we can predict that mixing solutions of lead(II) nitrate and potassium iodide will result in the precipitation of lead(II) iodide:

$$Pb(NO_3)_2(aq) + 2KI(aq) \longrightarrow PbI_2(s) + 2KNO_3(aq)$$

The K^+ and NO_3^- ions are spectators and remain in solution.

Alternatively, we could form lead(II) iodide by mixing lead(II) perchlorate and sodium iodide, both of which are soluble:

$$Pb(ClO_4)_2(aq) + 2NaI(aq) \longrightarrow PbI_2(s) + 2NaClO_4(aq)$$

Now the Na^+ and ClO_4^- ions are the spectators, and the net ionic reaction is the same as before.

Because the hydroxides of Group I are soluble but those of other elements are not (with the exception of the moderately soluble alkaline

(a)

(b)

FIGURE 3.9 Two precipitation reactions for preparing compounds. (a) In one, sodium sulfide is added to a lead nitrate solution, giving an immediate precipitate of black lead sulfide. (b) In the other, cobalt(II) chloride and sodium carbonate solutions are mixed, giving a precipitate of cobalt(II) carbonate.

TABLE 3.1 Solubility rules

Rule	Exceptions*
Soluble compounds	_Compounds containing:_
Compounds of Group I elements	
Ammonium compounds	
Chlorides, bromides, iodides	Ag^+, Hg_2^{2+}, Pb^{2+}
Nitrates, chlorates, perchlorates	
Acetates	(Ag^+), (Hg_2^{2+})
Sulfates	Sr^{2+}, Ba^{2+}, Pb^{2+} (Ca^{2+}), (Ag^+)
Sparingly soluble and insoluble compounds	
Carbonates, phosphates, oxalates, chromates	Group I elements, NH_4^+
Sulfides	Group I elements, NH_4^+
Hydroxides, oxides	Group I elements (Ca^{2+}), (Sr^{2+}), (Ba^{2+})

*Parenthesis denotes that a compound is moderately soluble.

earths), we can predict from the solubility rules that a solid hydroxide is obtained as a precipitate in reactions like

$$FeCl_3(aq) + 3KOH(aq) \longrightarrow Fe(OH)_3(s) + 3KCl(aq)$$

In this case the "driving force" for the reaction is the precipitation of the insoluble hydroxide.

ACID-BASE REACTIONS

In Section 2.8 we defined an acid as a compound that contains hydrogen and releases hydrogen ions in water. This definition was proposed by the Swedish chemist Svante Arrhenius toward the end of the nineteenth century. We shall begin this discussion of acids and bases with the outdated but still quite useful Arrhenius definitions, and then turn to the more modern definitions.

3.5 ARRHENIUS ACIDS AND BASES

An *Arrhenius acid* is a compound that contains hydrogen and releases hydrogen ions (H^+) in water. Hydrogen chloride and nitric acid are thus Arrhenius acids. An *Arrhenius base* is a compound that produces hydroxide ions (OH^-) in water. Sodium hydroxide is such a base, since it consists of Na^+ and OH^- ions and releases them when it dissolves in water. Aqueous solutions of bases are called *alkalis*.

Arrhenius acids and bases are recognized by their effect on the color of certain dyes known as "indicators." One of the most famous of these is litmus, a vegetable dye obtained from lichen. Aqueous solutions of acids turn litmus red; aqueous solutions of bases turn it blue (Fig. 3.10).

Acids and bases are produced in huge amounts by industry (Table 3.2). Sulfuric acid, for example, is manufactured in greater ton-

FIGURE 3.10 The acidities of various household products can be demonstrated by adding an indicator (red cabbage in this case) and noting the resulting color. Red indicates acidic, blue basic. From left to right, the household products are lemon juice, soda water, 7-Up, vinegar, ammonia, Drano, milk of magnesia, and detergent in water. Note that ammonia and Drano are such strong bases they destroy the red cabbage dye, so no color is visible.

TABLE 3.2 The top twelve chemicals manufactured in the United States, classified (where appropriate) as acid or base

Substance	Rank	Production, $\times 10^9$ kg	Class
Sulfuric acid	1	35	Acid
Nitrogen	2	20	
Ethylene	3	16	
Ammonia	4	15	Base
Oxygen	5	15	
Lime	6	14	Base
Sodium hydroxide	7	10	Base
Chlorine	8	10	
Phosphoric acid	9	10	Acid
Propylene	10	7	
Sodium carbonate	11	8	Base*
Urea	12	7	Base

*For reasons explained in Chapter 14, a solution of sodium carbonate is basic.

nage than any other chemical; phosphoric acid ranks 9th, nitric acid is 13th, and hydrochloric acid is 28th. Sulfuric acid was until recently so important to industry that its annual production was used as a measure of a nation's degree of industrialization and commercial prosperity. However, most sulfuric acid is now used to make fertilizers from phosphates, so its production is becoming more a measure of *agricultural* activity (Fig. 3.11). Bases too are produced on an enormous scale. Ammonia ranks fourth in annual tonnage among industrial chemicals; however, it is first according to the number of molecules manufactured, because the NH_3 molecule is so light. (Its molar mass is 17 g/mol, compared with sulfuric acid's 98 g/mol.) Sodium hydroxide ranks seventh in annual tonnage.

We can generally recognize which compounds are acids because their formulas are written with the acid hydrogen atoms at the front of the molecular formula. Thus, we can immediately recognize that HCl, H_2CO_3, and HSO_3^- are acids but that CH_4 and NH_3 are not. The molecular formula of acetic acid is best written $HC_2H_3O_2$ because it has only one acidic hydrogen atom.

3.6 THE BRØNSTED DEFINITIONS

In 1923, the Danish chemist Lars Brønsted proposed the improved definitions of acids and bases that we shall use from now on:

A **Brønsted acid** is a proton donor.

A **Brønsted base** is a proton acceptor.

The same definitions were proposed independently by the English

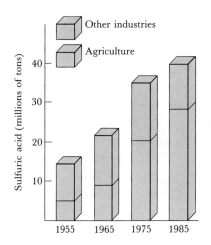

FIGURE 3.11 The uses of sulfuric acid have shifted over the past 30 years from mainly industrial to mainly agricultural.

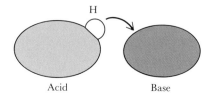

FIGURE 3.12 A Brønsted acid is a proton donor; a Brønsted base is a proton acceptor.

1 Hydronium ion

chemist Thomas Lowry, and the theory based on them is widely called the *Brønsted-Lowry theory* of acids and bases (Fig. 3.12). In the context of Brønsted acids and bases, a "proton" means a hydrogen ion (H^+). Thus the definitions focus attention on the transfer of H^+ from a donor molecule or ion (the acid) to an acceptor molecule or ion (the base).

Brønsted acids and bases. The Arrhenius acid HCl is an acid according to the Brønsted definition: when it dissolves in water, it donates a proton to a neighboring water molecule. That is,

$$HCl(g) + H_2O(l) \longrightarrow H_3O^+(aq) + Cl^-(aq)$$

Since proton donation results in the formation of ions (HCl, for example, becoming Cl^-), we call it *ionization*, and say that HCl is ionized when it dissolves in water. An H_2O molecule that has accepted a proton from a proton donor becomes a *hydronium ion* H_3O^+ (**1**). It is best to think of the hydronium ion as a polyatomic cation (the oxygen analog of NH_4^+), not just a hydrated proton.

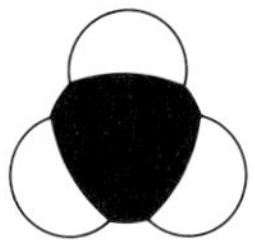

When sodium hydroxide dissolves in water, the hydroxide ions it provides can accept protons from any hydronium ions present:

$$H_3O^+(aq) + OH^-(aq) \longrightarrow 2H_2O(l)$$

Since OH^- accepts a proton, it is a Brønsted base. Hence anything that contains OH^- ions (such as sodium hydroxide) behaves as a base. Compounds that do not contain OH^- ions may also be bases. An oxide ion, for instance, is a Brønsted base. When an ionic oxide (such as CaO) dissolves in water, the negative charge of the O^{2-} anions attracts protons from H_2O molecules, and the oxide ion is converted to a hydroxide ion (Fig. 3.13):

$$O^{2-}(aq) + H_2O(l) \longrightarrow 2OH^-(aq)$$

The strong ability of an O^{2-} ion to attract protons accounts for our never seeing aqueous solutions of ionic oxides: all soluble oxides immediately acquire a proton and form hydroxides. Calcium oxide ("lime" or "quicklime") reacts so vigorously with water that the wooden boats once used to transport it sometimes caught fire. It is more safely transported as "slaked lime," $Ca(OH)_2$, which is lime with its thirst for water slaked.

Ammonia is also a proton acceptor and hence a Brønsted base:

$$NH_3(aq) + H_2O(l) \longrightarrow NH_4^+(aq) + OH^-(aq)$$

This behavior resembles that of the oxide ion (Fig. 3.14). However, whereas all the O^{2-} ions are converted to OH^- ions in water, only a small proportion (typically 1 in 100) of the NH_3 molecules are converted to NH_4^+ ions. Since only a small proportion of the ammonia molecules exist as ions in solution, ammonia is an example of a "weak electrolyte":

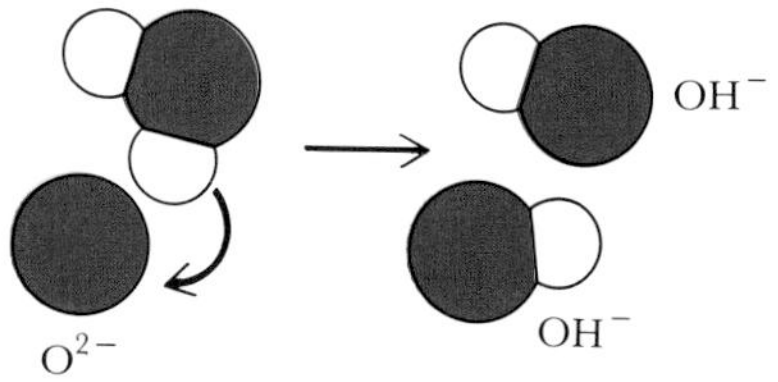

A **weak electrolyte** is a substance that dissolves to give a solution in which only a small proportion of the solute molecules are ionized.

The solution itself is called a "weak electrolyte solution." Most solutions of acids (such as acetic acid, oxalic acid, and sulfurous acid) are weak

FIGURE 3.13 An oxide ion in water is a strong base that attracts a proton from a neighboring water molecule to produce a hydroxide ion.

electrolyte solutions, with only a small proportion of the acid molecules converted to their anions and H_3O^+ ions. A few salts also form weak electrolyte solutions: two examples are mercury(II) chloride and lead(II) acetate, neither of which is fully ionized in water. For writing net ionic equations, it is normally appropriate to write all strong electrolytes as their individual ions. In contrast to weak electrolytes are compounds such as HCl that are "strong electrolytes":

A **strong electrolyte** is a substance that dissolves to give a solution in which almost all the solute molecules are ionized.

The solution itself is called a "strong electrolyte solution." Only a few common acids are strong electrolytes: they are $HClO_4$, HNO_3, HCl, HBr, HI, and H_2SO_4 (its first proton only). They should be memorized now. For writing net ionic equations it is normally appropriate to write all weak electrolytes in their nonionized form.

Acids, bases, and the periodic table. The alkali metal and alkaline earth metal oxides and hydroxides (CaO and KOH, for example) are bases. This is easy to understand on the grounds that O^{2-} ions and OH^- ions are both strong proton acceptors. In contrast to the basic character of these oxides, carbon dioxide combines with water to give H_2CO_3, an acid, and sulfur trioxide combines to give H_2SO_4, another acid. These H_2CO_3 and H_2SO_4 molecules can release protons partly because the oxygen is not present as ions and so attracts protons less strongly than in the ionic oxides.

We classify oxides as *basic oxides* if they are bases, and as *acidic oxides* if they react with water to give acids. Acidic oxides are not themselves acids (they have no hydrogen to donate), but they become acids when they react with water. Basic oxides are ionic solids (such as CaO and Na_2O) and hence are formed by metals. The acidic oxides are all molecular compounds (such as CO_2 and SO_3) and hence are formed by the nonmetals. The correlation between an element's being a metal and forming a basic oxide on the one hand, and its being a nonmetal and forming an acidic oxide on the other, is confirmed when we note the locations in the periodic table of the elements that form the two classes of oxide (Fig. 3.15). We find that elements that form basic oxides lie to the left of a diagonal line, and those that form acidic oxides lie to its right. Approximately the same diagonal separates the metals from the nonmetals (see Fig. 2.7).

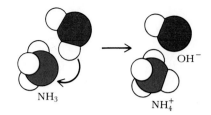

FIGURE 3.14 An NH_3 molecule behaves like an oxide in water, accepting a proton from a neighboring water molecule. However, only a tiny fraction of the available NH_3 molecules are converted to NH_4^+ ions.

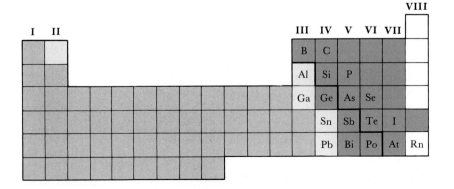

FIGURE 3.15 The locations in the periodic table of elements that have acidic and basic oxides. Acidic oxides are in pink, basic oxides and hydroxides in blue. The amphoteric oxides are marked in yellow.

If we know that an element is a metal, we can now predict that it forms basic oxides and hydroxides. Similarly, if we know that an element is a nonmetal, we can predict that its oxides are acidic. We would thus predict (rightly) that the oxide and hydroxide of barium are basic, whereas the oxides of selenium are acidic.

The classification of oxides is less clear for elements close to the dividing diagonal. Elements in that region of the periodic table (which, we have seen, exhibit a blend of metallic and nonmetallic character) have oxides that show both acidic and basic properties. These oxides are called "amphoteric" (from the Greek word meaning "both"). An *amphoteric* substance is one that reacts with both acids and bases. Note that the definition does not say that an amphoteric substance *is* an acid or a base but simply that it *reacts* with them in some way. Aluminum oxide is an example of an amphoteric oxide, since it reacts with acids and with bases. Solid aluminum oxide reacts very slowly, so the reactions that show its amphoteric character are best observed with freshly precipitated aluminum hydroxide (prepared by adding ammonia to a solution containing aluminum ions). The hydroxide reacts with acid:

$$Al(OH)_3(s) + 3HCl(aq) \longrightarrow AlCl_3(aq) + 3H_2O(l)$$

It also reacts with base:

$$Al(OH)_3(s) + NaOH(aq) \longrightarrow Na[Al(OH)_4](aq)$$

The product here is sodium aluminate; $[Al(OH)_4]^-$ is the aluminate ion. The hydroxide of aluminum's Group II neighbor beryllium behaves in the same way and so is also amphoteric. It reacts with acids to give solutions of Be^{2+} ions and with bases to give solutions of the beryllate ion, $[Be(OH)_4]^{2-}$.

3.7 NEUTRALIZATION

A characteristic reaction of hydrochloric acid and sodium hydroxide is the "neutralization" of one by the other to form a "neutral" compound, sodium chloride, which is neither an acid nor a base:

$$HCl(aq) + NaOH(aq) \longrightarrow NaCl(aq) + H_2O(l)$$

Compounds that can be formed in this way take their name from "salt" itself and are called *salts*. The cation of the salt is provided by the base, and the anion is provided by the acid. Examples include potassium nitrate (from KOH and HNO_3) and calcium sulfate (from $Ca(OH)_2$ and H_2SO_4).

Proton transfer. The essential feature of a neutralization reaction is *proton transfer*, and in particular the transfer of a proton from H_3O^+ to OH^-. This can be seen most clearly by writing the neutralization reaction given above as a net ionic equation. In water, HCl exists as H_3O^+ ions and Cl^- ions, and NaOH exists as Na^+ ions and OH^- ions; in terms of ions the overall equation is

$$H_3O^+(aq) + Cl^-(aq) + Na^+(aq) + OH^-(aq) \longrightarrow$$
$$Na^+(aq) + Cl^-(aq) + 2H_2O(l)$$

The Na^+ and Cl^- ions are unchanged by the reaction and hence are spectators, so the net ionic equation is

$$H_3O^+(aq) + OH^-(aq) \longrightarrow 2H_2O(l)$$

The "driving force" of a neutralization reaction is the transfer of a proton from a hydronium ion to a hydroxide ion, with the formation of water molecules (Fig. 3.16).

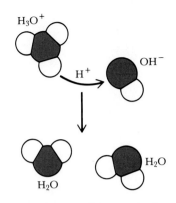

FIGURE 3.16 The net ionic reaction between an acid and a base is the formation of water by the transfer of protons from hydronium ions to hydroxide ions.

▼ **EXAMPLE 3.4** **Accounting for reactions in terms of proton transfer**

When an acid reacts with a carbonate or bicarbonate, carbon dioxide is formed. Write the net ionic equations that show how H_2CO_3 is formed by proton transfer and then decomposes into CO_2 and H_2O.

STRATEGY Write the net ionic equation for the reaction, and identify the Brønsted base that acts as the proton acceptor.

SOLUTION In the overall reaction

$$NaHCO_3(aq) + HCl(aq) \longrightarrow NaCl(aq) + CO_2(g) + H_2O(l)$$

the net ionic equation is

$$HCO_3^-(aq) + H^+(aq) \longrightarrow CO_2(g) + H_2O(l)$$

The proton is present as a hydronium ion. We can express this by adding one more H_2O molecule to both sides of the equation:

$$HCO_3^-(aq) + H_3O^+(aq) \longrightarrow CO_2(g) + 2H_2O(l)$$

The HCO_3^- ion is the Brønsted base. In the reaction, H_3O^+ ions donate protons to HCO_3^- ions, forming H_2CO_3 molecules:

$$HCO_3^-(aq) + H_3O^+(aq) \longrightarrow H_2CO_3(aq) + H_2O(l)$$

The H_2CO_3 decomposes into carbon dioxide and water. Similarly, the carbonate accepts two protons and reacts as follows:

$$CO_3^{2-}(aq) + 2H_3O^+(aq) \longrightarrow CO_2(g) + 3H_2O(l)$$

EXERCISE Write the net ionic equation for the action of acids on solid sulfides.

[*Answer*: $S^{2-}(s) + 2H_3O^+(aq) \longrightarrow H_2S(g) + 2H_2O(l)$]

▲

Chemical applications of neutralization reactions. A substance such as sodium hydroxide is the "base" of a range of salts; different salts are obtained from it by neutralization with different acids. (This is the origin of the name "base.")

$$HCl(aq) + NaOH(aq) \longrightarrow NaCl(aq) + H_2O(l)$$

$$H_2SO_4(aq) + 2NaOH(aq) \longrightarrow Na_2SO_4(aq) + 2H_2O(l)$$

In each case, the driving force of the reaction is the same, while the salt is the combination of spectator ions left behind after protons have been transferred from H_3O^+ ions to OH^- ions.

In a neutralization reaction, the anion of the salt is provided by the acid (so hydroiodic acid will produce iodides). The cation of the salt is provided by the base (so cesium hydroxide will produce cesium salts). It follows that when we want to use an acid-base neutralization to prepare a compound, we must choose an acid and a base that provide the spectator ions corresponding to the salt we require.

▼ **EXAMPLE 3.5** **Predicting the outcome of a neutralization**

How could we prepare rubidium sulfate, Rb_2SO_4, by neutralization?

STRATEGY Identify the base and the acid needed to prepare the salt. The cation (Rb^+) is the spectator ion supplied by the base, and the anion (SO_4^{2-}) is the spectator ion supplied by the acid.

SOLUTION A base containing Rb^+ is rubidium hydroxide. An acid that provides SO_4^{2-} ions is sulfuric acid. The neutralization reaction is then

$$H_2SO_4(aq) + 2RbOH(aq) \longrightarrow Rb_2SO_4(aq) + 2H_2O(l)$$

Chapter 4 describes how to determine the correct amounts of acid and base to use.

EXERCISE Suggest a preparation of ammonium sulfate.

[*Answer*: $2NH_3(aq) + H_2SO_4(aq) \longrightarrow (NH_4)_2SO_4(aq)$]

▲

REDOX REACTIONS

We have seen that reactions between acids and bases involve the transfer of protons from one molecule or ion to another. Another important class of reactions involves the transfer of *electrons*. We shall see that reactions of this kind include the combustion of methane,

$$CH_4(g) + 2O_2(g) \longrightarrow CO_2(g) + 2H_2O(g)$$

the reaction between magnesium and chlorine,

$$Mg(s) + Cl_2(g) \longrightarrow MgCl_2(s)$$

the reaction between a metal and an acid to produce hydrogen,

$$Zn(s) + 2HCl(aq) \longrightarrow ZnCl_2(aq) + H_2(g)$$

and the reaction used to obtain iron from its ore (Fig. 3.17),

$$Fe_2O_3(s) + 3CO(g) \xrightarrow{\Delta} 2Fe(l) + 3CO_2(g)$$

3.8 ELECTRON TRANSFER

The reaction between magnesium and oxygen (Fig. 3.18) is

$$2Mg(s) + O_2(g) \longrightarrow 2MgO(s)$$

It amounts, in essence, to the loss of electrons from Mg atoms (to form Mg^{2+} ions) and the gain of electrons by O atoms in O_2 (to form O^{2-} ions). The reaction does not actually take place by a simple electron transfer, because complicated events occur as the metal is converted to the oxide. However, the overall effect is that of an electron transfer. Similarly, in the reaction between magnesium and chlorine, the overall outcome is that of a transfer of electrons, in this case from Mg atoms to Cl atoms:

$$Mg(s) + Cl_2(g) \longrightarrow \underbrace{Mg^{2+}(s) + 2Cl^-(s)}_{MgCl_2(s)}$$

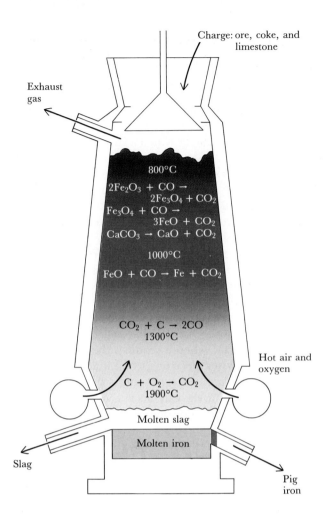

Charge: ore, coke, and limestone

Exhaust gas

800°C

$2Fe_2O_3 + CO \rightarrow$
$2Fe_3O_4 + CO_2$
$Fe_3O_4 + CO \rightarrow$
$3FeO + CO_2$
$CaCO_3 \rightarrow CaO + CO_2$

1000°C

$FeO + CO \rightarrow Fe + CO_2$

$CO_2 + C \rightarrow 2CO$
1300°C

Hot air and oxygen

$C + O_2 \rightarrow CO_2$
1900°C

Molten slag

Molten iron

Slag

Pig iron

FIGURE 3.17 The reduction of iron ore takes place inside a blast furnace containing a charge of the ore with coke and limestone. Different reactions occur in different regions when the blast of air and oxygen is admitted. The ore, an oxide, is reduced to the metal by reaction with the carbon monoxide produced in the furnace.

Oxidation and reduction. The reaction between a substance and oxygen was originally known as "oxidation." The meaning of that term has been extended to cover any reaction that involves loss of electrons:

Oxidation is electron loss.

When magnesium reacts with oxygen, the magnesium is oxidized. Magnesium is also oxidized when it reacts with chlorine—even though that reaction does not involve oxygen.

The name "reduction" originally meant reduction of the oxygen content of compounds, and in particular the extraction of metals from their ores, as in

$$Fe_2O_3(s) + 3CO(g) \xrightarrow{\Delta} 2Fe(l) + 3CO_2(g)$$

An essential feature of this reaction is the conversion of Fe^{3+} ions to Fe atoms by electron gain. The modern meaning of the term "reduction" builds on this feature:

Reduction is electron gain.

FIGURE 3.18 An example of oxidation: magnesium burning brightly in oxygen. Magnesium also burns in water and carbon dioxide, so magnesium fires are very difficult to extinguish.

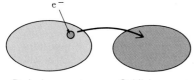

Reducing agent   Oxidizing agent

FIGURE 3.19 Oxidation is electron loss, and reduction is electron gain. The electron donor is the reducing agent, and the acceptor is the oxidizing agent.

TABLE 3.3 A fragment of the electrochemical series*

Oxidizing agent	$\xrightarrow{\text{Reduced}}$ $\xleftarrow{\text{Oxidized}}$	Reducing agent
Strongly oxidizing		
F_2		F^-
MnO_4^- (acidic)		Mn^{2+}
Cl_2		Cl^-
$Cr_2O_7^{2-}$ (acidic)		Cr^{3+}
NO_3^- (acidic)		NO
Ag^+		Ag
Fe^{3+}		Fe^{2+}
Cu^{2+}		Cu
SO_4^{2-}		H_2SO_3
H_3O^+		H_2
Fe^{2+}		Fe
Zn^{2+}		Zn
Mg^{2+}		Mg
Na^+		Na
		Strongly reducing

*A more extensive (and quantitative) list is given in Appendix 2B.

The Fe^{3+} ions in the oxide are reduced to Fe atoms. The O atoms in the O_2 that reacts with magnesium are reduced to O^{2-} ions. The Cl atoms in the Cl_2 that reacts with magnesium are reduced to Cl^- ions.

Oxidation combined with reduction. Oxidation (electron loss) must be accompanied by reduction (electron gain) because one substance cannot lose electrons in a chemical reaction unless another substance gains them. Therefore, any reaction in which a substance is oxidized involves the reduction of another substance. Since we cannot speak of oxidation without reduction, reactions equivalent to the transfer of electrons are called <u>*redox reactions*</u>, for *red*uction-*ox*idation reactions.

The substance that causes oxidation is called the *oxidizing agent*; oxygen and chlorine are two of the oxidizing agents we have met so far. The substance that causes reduction is called the *reducing agent*; thus carbon monoxide is the reducing agent in the reduction of iron oxide. In terms of electron transfer, the oxidizing agent is the substance that removes electrons from the substance being oxidized (Fig. 3.19). The reducing agent is the substance that supplies electrons to the substance being reduced. Since an oxidizing agent gains electrons when it acts, it is reduced in the reaction. Oxygen that reacts with magnesium, for instance, is reduced to O^{2-} when it oxidizes the magnesium. Similarly, since a reducing agent loses electrons when it acts, it is oxidized in the reaction. The reducing agent magnesium is oxidized to Mg^{2+}.

In principle, any substance can accept or donate electrons, but whether or not it does so depends on the other substance taking part in the reaction. For example, nitrogen can be reduced to ammonia by hydrogen, or oxidized to nitric oxide by oxygen. The relative strengths of substances as oxidizing and reducing agents can be discussed quantitatively, and we shall do so in Chapter 17. For the present, we need only realize that substances may be arranged in order of their oxidizing strength. This ordering is called the *electrochemical series*, a part of which is shown in Table 3.3. Since an oxidizing agent can oxidize a reducing agent that lies below it in the series, we can use the series to predict the outcomes of redox reactions.

▼ EXAMPLE 3.6 Using the electrochemical series

What will happen when a piece of iron is placed in a silver nitrate solution?

STRATEGY We have to consider whether Fe can reduce the Ag^+ ions in the solution to Ag atoms; if it can, metallic silver will be deposited. This is equivalent to deciding whether Ag^+ ions can *oxidize* Fe to Fe^{2+} ions, for if Ag^+ acts as an oxidizing agent, it will be reduced to Ag. The Ag^+ ion can oxidize the Fe to Fe^{2+} if it lies higher than Fe in the electrochemical series, for then the driving force of the reaction will favor the oxidation of iron.

SOLUTION Since the oxidizing agent Ag^+ lies above the reducing agent Fe, the iron will be oxidized, Ag^+ ions will be reduced, and silver will be deposited in the reaction

$$Fe(s) + 2Ag^+(aq) \longrightarrow Fe^{2+}(aq) + 2Ag(s)$$

EXERCISE What will happen to the iron when potassium permanganate which has been made acid with hydrochloric acid is added to an iron(II) sulfate solution?

[*Answer*: Fe(II) is oxidized to Fe(III) and $Mn^{2+}(aq)$ is formed.]

The roles of the oxidizing and reducing agents in redox reactions resemble those of the acids and bases in acid-base reactions. However, there is an important difference. In an acid-base reaction, there is an actual transfer of a proton from the acid to the base. In a redox reaction, the outcome of the reaction amounts to the transfer of an electron, but that transfer might involve a much more complex process than the simple migration of an electron from one substance to another (Fig. 3.20). It may, for example, be brought about by the transfer of an ion or a group of atoms.

3.9 OXIDIZING AND REDUCING AGENTS

We can tell when an acid-base neutralization has occurred by examining the equation of the reaction to see if a hydrogen ion has been transferred. How, though, do we know whether electrons have been transferred and that a redox reaction has taken place? In what sense has electron transfer taken place in the combustion of methane,

$$CH_4(g) + 2O_2(g) \longrightarrow CO_2(g) + 2H_2O(l)$$

or in the reaction

$$UF_4(s) + F_2(g) \longrightarrow UF_6(s)$$

which is used in refining uranium? The problem is even more severe for reactions with very complicated equations, such as the one between potassium permanganate and oxalic acid ($H_2C_2O_4$; **2**), in the presence of acid:

$$2KMnO_4(aq) + 6HCl(aq) + 5H_2C_2O_4(aq) \longrightarrow$$
$$2MnCl_2(aq) + 10CO_2(g) + 8H_2O(l) + 2KCl(aq)$$

What, if anything, has been oxidized, and what has been reduced?

Acids as oxidizing agents. In simple cases we can check whether the charge number of an element has increased as a result of electron loss (oxidation) or decreased by electron gain (reduction). If zinc is converted to Zn^{2+} ions, we know that it has been oxidized. This conversion takes place when an acid acts on zinc, in the reaction

$$Zn(s) + 2HCl(aq) \longrightarrow ZnCl_2(aq) + H_2(g)$$

for example. We can easily see that this reaction is a redox reaction, especially if we write the net ionic equation:

$$Zn(s) + 2H^+(aq) \longrightarrow Zn^{2+}(aq) + H_2(g)$$

The Zn becomes Zn^{2+}, so it is oxidized by the hydrogen ion. The hydrogen ion $H^+(aq)$—which is actually present as the hydronium ion—gains an electron, forms an H atom, and bubbles out of the solution as H_2 molecules. The hydrogen ion is therefore reduced (by the zinc) to molecular hydrogen. In reactions of this kind, the hydrogen ion supplied by an acid is the oxidizing agent; the hydrogen ion is a sufficiently powerful oxidizing agent to oxidize any metal below hydrogen in the electrochemical series.

Metals that lie above hydrogen in the electrochemical series are not oxidized by hydrogen ions. Copper, for example, lies above hydrogen

FIGURE 3.20 Many reactions are redox reactions, following the same general pattern as a reaction with oxygen or with hydrogen. The reaction that occurs when bromine is poured on red phosphorus (which is shown here) is also a redox reaction.

2 Oxalic acid

and is unaffected by hydrochloric acid. However, the anion of an oxoacid may be a more powerful oxidizing agent than a hydrogen ion and may be able to oxidize what the hydrogen ion cannot. This is the case with nitric acid: the nitrate ion lies above copper in the series and hence can oxidize that metal. When copper is heated in dilute nitric acid, a blue solution of copper nitrate and the colorless gas nitric oxide are produced:

$$3Cu(s) + 8HNO_3(aq,\ dilute) \longrightarrow 3Cu(NO_3)_2(aq) + 2NO(g) + 4H_2O(l)$$

(Later in this chapter we shall balance some more complicated equations like this one.) The net ionic equation is

$$3Cu(s) + 8H^+(aq) + 2NO_3^-(aq) \longrightarrow 3Cu^{2+}(aq) + 2NO(g) + 4H_2O(l)$$

The charge number of the copper is increased to $+2$, so it has been oxidized. The NO_3^- ion is reduced to NO. If concentrated nitric acid is used instead, the products include nitrogen dioxide (Fig. 3.21):

$$Cu(s) + 4HNO_3(aq,\ conc) \longrightarrow Cu(NO_3)_2(aq) + 2NO_2(g) + 2H_2O(l)$$

The net ionic equation,

$$Cu(s) + 4H^+(aq) + 2NO_3^-(aq) \longrightarrow Cu^{2+}(aq) + 2NO_2(g) + 2H_2O(l)$$

shows that in this case the NO_3^- ion is reduced only to NO_2. Sulfuric acid is also an oxidizing agent, and when hot it is reduced by copper to sulfur dioxide:

$$Cu(s) + 2H_2SO_4(aq,\ conc) \xrightarrow{\Delta} CuSO_4(aq) + SO_2(g) + 2H_2O(l)$$

Oxidation numbers. When the Stock number (Section 2.8) of an element is given, we can check it to see whether an atom or ion has lost or gained electrons. For example, the conversion of an iron(II) compound to an iron(III) compound is an oxidation of Fe^{2+} to Fe^{3+}. An actual example is the oxidation of iron(II) nitrate by potassium permanganate in acidic solution:

$$5Fe^{2+}(aq) + MnO_4^-(aq) + 8H^+(aq) \longrightarrow$$
$$5Fe^{3+}(aq) + Mn^{2+}(aq) + 4H_2O(l)$$

An increase in the charge number of a monatomic ion—Fe^{2+} to Fe^{3+}, for example—is an oxidation. A decrease—as in Br to Br^-—is a reduction. Chemists have found a way of assigning to any atom in any kind of compound an *effective* charge number, called its *oxidation number* (ON) and defined so that an increase in its value corresponds to oxidation and a decrease corresponds to reduction. Oxidation numbers are a kind of theoretical version of the litmus test for acids and bases, but they are used to detect oxidizing and reducing agents instead. (There are actual indicators for oxidation and reducing agents: these are described in Section 4.6.)

An oxidation number is assigned by use of the rules listed in Table 3.4. The rules must be applied in the order given, and the process must stop as soon as an oxidation number has been obtained (because a later rule might contradict an earlier one). The rules are based on some of the ideas chemists have developed about the way atoms in molecules share electrons—a point that is explained in Chapter 8. Some insight into the significance of oxidation numbers comes from

(a)

(b)

FIGURE 3.21 (a) Copper reacts with dilute nitric acid to give blue copper(II) nitrate and nitric oxide. (b) With concentrated nitric acid, nitrogen dioxide is also produced; the blue nitrate solution is turned green by this brown gas.

TABLE 3.4 Oxidation numbers

Work through the following rules *in the order given*. Stop as soon as an oxidation number has been assigned.

1. The sum of the oxidation numbers (ONs) of all the atoms in the molecule or ion is equal to its total charge

2. For atoms in their elemental form, ON = 0

3. For elements of Group I,   ON = +1
   Group II,   ON = +2
   Group III (except boron),   ON = +3 for M^{3+}
                                                ON = +1 for M^+
   Group IV (except carbon, silicon),   ON = +4 for M^{4+}
                                                ON = +2 for M^{2+}

4. For hydrogen, ON = +1 in combination with nonmetals
                 ON = −1 in combination with metals

5. For fluorine, ON = −1 in all compounds

6. For oxygen, ON = −1 in peroxides (O_2^{2-})
               ON = $-\frac{1}{2}$ in superoxides (O_2^-)
               ON = $-\frac{1}{3}$ in ozonides (O_3^-)
               ON = −2 unless combined with F

noting that the number assigned to an element in a compound is equal to the charge that the atom would have if every O atom in the compound were present as an O^{2-} ion (Fig. 3.22). For example, if the oxygen atom in H_2O is treated as an O^{2-} ion, then for overall charge neutrality each H atom must be treated as an H^+ ion; thus the oxidation number of H in H_2O is +1. In the chlorate ion ClO_3^- we must treat (for present purposes only) each O atom as being O^{2-}. That adds up to six negative charges in all. Since the overall charge number of the ion is −1, the effective charge of the Cl atom must be +5, so its oxidation number is +5. The rules in the table simply generalize this approach to compounds that do not contain oxygen.

For a monatomic cation, the oxidation number is equal to the charge number (and hence also equal to the Stock number of the element). For example, the oxidation number of Ni^{2+}, the nickel(II) ion, is +2, and the oxidation number of nickel metal is 0. In the conversion from Ni to Ni^{2+}, the oxidation number of nickel increases to +2, which corresponds to oxidation.

▼ EXAMPLE 3.7 Calculating oxidation numbers

Calculate the oxidation numbers of the atoms in (a) SO_2; (b) SO_4^{2-}.

STRATEGY Since the sulfur atom has more oxygen atoms attached to it in the sulfate ion than in sulfur dioxide, we expect that it will have a higher oxidation number in the ion. One approach is to treat each O atom as O^{2-}, and to find the charge that the S atom needs to produce the overall charge of the atom or ion. More formally, we can work *in order* through the rules in Table 3.4. We shall use the latter procedure, for practice with the rules.

SOLUTION (a) SO_2: By rule 1, the sum of the oxidation numbers of the atoms in the compound is zero. Rules 2 to 5 are not relevant. According to

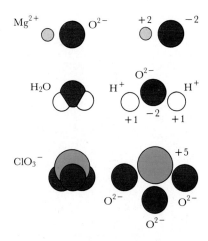

FIGURE 3.22 The oxidation number of an element is the charge it would have if each O atom present in the compound were an O^{2-} ion. In MgO, the oxidation number of Mg is +2, the same as its actual charge. The oxidation number of H in H_2O is +1, and that of Cl in ClO_3^- is +5.

rule 6, each oxygen has ON $= -2$. Hence, by rule 1, we know that ON(S) $+ 2 \times (-2) = 0$, where ON(S) is the oxidation number of sulfur. Then ON(S) $- 4 = 0$, so we must have ON(S) $= +4$. (b) SO_4^{2-}: By rule 1, the sum of the oxidation numbers of the atoms in the ion is -2. Rules 2 to 5 are not relevant. By rule 6, each oxygen has ON $= -2$. Hence, by rule 1, we have ON(S) $+ 4 \times (-2) = -2$. Thus, ON(S) $= +6$. As anticipated, the sulfur is more highly oxidized in the sulfate ion than in sulfur dioxide.

EXERCISE Calculate the oxidation numbers of the elements in (a) H_2S; (b) PO_4^{3-}; (c) NO_3^-.
[*Answer*: (a) ON(H) $= +1$, ON(S) $= -2$; (b) ON(P) $= +5$, ON(O) $= -2$; (c) ON(N) $= +5$, ON(O) $= -2$]

▲

3 Ethanol

4 Acetic acid

FIGURE 3.23 Another example of a redox reaction: in this one, gaseous hydrogen is burning in an atmosphere of bromine to give hydrogen bromide.

Oxidation number and oxygen content. The oxidation numbers of elements other than oxygen increase as the number of oxygen atoms present in a compound increases. The oxidation number of nitrogen in NO is $+2$; in NO_2 it is $+4$; and in NO_3^- it is $+5$. In the oxidation of copper by nitric acid, the oxidation number of the nitrogen decreases from $+5$ in the nitrate ion to $+2$ (if NO is produced) or $+4$ (if NO_2 is produced). In both cases, therefore, the nitrate ion is reduced.

The relation between oxidation number and oxygen content is consistent with the historical use of the term oxidation to mean an increase in the oxygen content of a compound (as when Mg is oxidized to MgO). That definition remains very useful in organic chemistry, where many oxidation reactions actually involve the replacement of hydrogen atoms by oxygen atoms. For example, the souring of wine is caused by the oxidation of ethanol (C_2H_6O; **3**) to acetic acid ($C_2H_4O_2$; **4**):

$$C_2H_6O(aq) + O_2(g) \longrightarrow C_2H_4O_2(aq) + H_2O(l)$$

In this reaction the oxygen content of ethanol is increased when it is converted to acetic acid by the oxidizing agent—the oxygen of the air.

Identifying oxidizing and reducing agents. We can find out whether electron transfer has occurred in a reaction by examining the oxidation numbers of all the elements taking part. If there is any change of oxidation numbers, the reaction is redox; if not, it is not redox. If the reaction is redox, the oxidizing agent is the substance that contains the element that is reduced. The reducing agent is the substance that contains the element that is oxidized (Fig. 3.23). Some examples of redox reactions are given in Table 3.5. The table also shows the oxidizing and reducing agents in each case.

In the uranium fluoride reaction, for example, the oxidation numbers are as follows:

$$\overset{+4\;\;-1}{UF_4} + \overset{0}{F_2} \longrightarrow \overset{+6\;-1}{UF_6}$$

They change, so the reaction is redox. The oxidation number of uranium increases from $+4$ to $+6$, implying that it has been oxidized and hence that UF_4 is the reducing agent. The oxidation number of fluorine decreases from 0 in F_2 to -1 in the compound, showing that F is reduced and hence that F_2 is the oxidizing agent.

TABLE 3.5 Common redox reactions

Reaction	Oxidizing agent	Reducing agent
$2H_2(g) + O_2(g) \longrightarrow 2H_2O(l)$	$O_2(g)$	$H_2(g)$
$N_2(g) + 3H_2(g) \longrightarrow 2NH_3(g)$	$N_2(g)$	$H_2(g)$
$UF_4(s) + F_2(g) \longrightarrow UF_6(s)$	$F_2(g)$	$UF_4(s)$
$Fe^{2+}(aq) + Ce^{4+}(aq) \longrightarrow Fe^{3+}(aq) + Ce^{3+}(aq)$	$Ce^{4+}(aq)$	$Fe^{2+}(aq)$
$Zn(s) + Cu^{2+}(aq) \longrightarrow Zn^{2+}(aq) + Cu(s)$	$Cu^{2+}(aq)$	$Zn(s)$
$6Fe^{2+}(aq) + Cr_2O_7{}^{2-}(aq) + 14H^+(aq) \longrightarrow$ $\qquad 6Fe^{3+}(aq) + 2Cr^{3+}(aq) + 7H_2O(l)$	$Cr_2O_7{}^{2-}(aq)$	$Fe^{2+}(aq)$
$I_2(aq) + 2S_2O_3{}^{2-}(aq) \longrightarrow 2I^-(aq) + S_4O_6{}^{2-}(aq)$	$I_2(aq)$	$S_2O_3{}^{2-}(aq)$

▼ EXAMPLE 3.8 Identifying oxidizing and reducing agents

Chromate ions react with chloride ions to produce chlorine gas and chromium(III) ions. Identify the oxidizing and reducing agents.

STRATEGY Since the chromate ion ($CrO_4{}^{2-}$; Table 2.7) has so many oxygen atoms, we anticipate that it is an oxidizing agent; this is confirmed by its high position in the electrochemical series. We should check this by using the rules of Table 3.4 to calculate changes in the oxidation numbers of chromium and chlorine, and thus identify the oxidized (ON increased) and reduced (ON decreased) elements.

SOLUTION In the chromate ion, the rules require that $ON(Cr) + 4ON(O) = -2$, with $ON(O) = -2$; hence $ON(Cr) = +6$. In $Cr^{3+}(aq)$, we have $ON(Cr) = +3$. Therefore, the chromium is reduced in the reaction. Since it is a part of the chromate ion, that ion is the oxidizing agent. The oxidation number of chlorine changes from -1 in Cl^- to 0 in Cl_2, showing that the chloride ion is oxidized and hence that the chloride ion is the reducing agent.

EXERCISE In basic solution, iodate ions react with $[Cr(OH)_4]^-(aq)$ ions to form iodide ions and chromate ions. Identify the reducing and oxidizing agents.

▲   [*Answer:* IO_3^- is the oxidizing agent, $[Cr(OH)_4]^-$ the reducing agent]

In the net ionic equation for an acid-base neutralization, the oxidation numbers are as follows:

$$\overset{+1 \ -2}{H_3O^+}(aq) + \overset{-2 \ +1}{OH^-}(aq) \longrightarrow \overset{+1 \ -2}{2H_2O}(l)$$

They do not change, so the reaction is not a redox reaction. Since the same net ionic equation applies to all acid-base neutralizations, the same is true of all of them. This confirms that acid-base neutralizations and redox reactions are two distinct classes of reactions.

Disproportionation and self-oxidation. A special type of redox reaction is one in which a single element is both reduced and oxidized, as in the reaction used for preparing nitric acid. The only element in the equation to undergo a redox change is nitrogen, which changes from +4 to +5 and +2:

$$\overset{+4}{3NO_2}(g) + H_2O(l) \longrightarrow \overset{+5}{2HNO_3}(aq) + \overset{+2}{NO}(g)$$

Some nitrogen atoms are therefore oxidized, and some are reduced. Such a "self-redox" reaction is called a "disproportionation":

A **disproportionation** is a redox reaction in which a single substance is simultaneously oxidized and reduced.

Another example is provided by the chlorites, which have a tendency to disproportionate in aqueous solution, especially when heated:

$$\overset{+3}{3ClO_2^-}(aq) \longrightarrow \overset{+5}{2ClO_3^-}(aq) + \overset{-1}{Cl^-}(aq)$$

In some compounds, an anion oxidizes or reduces its own cation. Both *self-oxidation* and *self-reduction* occur when ammonium nitrate is heated to prepare dinitrogen oxide (nitrous oxide):

$$\overset{-3}{N}H_4\overset{+5}{N}O_3(s) \longrightarrow \overset{+1}{N_2O}(g) + 2H_2O(l)$$

In this reaction the nitrate anion is the oxidizing agent, and the ammonium cation is the reducing agent. The nitrogen in NH_4^+ is oxidized, and the nitrogen in NO_3^- is reduced. This is the reverse of disproportionation: the nitrogen atoms initially have different oxidation numbers, but in the product their oxidation numbers are the same.

Oxidizing ability and oxidation number. In general, an element can have a range of oxidation numbers, up to a maximum equal to its group number, depending on the other elements present in the compound. The maximum oxidation number of sulfur (Group VI) is +6. Nitrogen (Group V) has one of the widest ranges of oxidation numbers of any element, taking on all whole-number values from −3 (in NH_3) to +5 (in HNO_3). (We discuss the reasons for the range and the maximum value in Chapter 8.) The oxidizing and reducing abilities of a compound depend on the oxidation numbers of the elements it contains. An oxoanion, for example, cannot be oxidized if the central element already has its maximum oxidation number.

Compounds in which an element has an oxidation number low in its range are often good reducing agents. This follows from the element's ability to be oxidized to a higher oxidation number. An example is provided by sulfur, which has oxidation numbers ranging from −2 to +6. In hydrogen sulfide the oxidation number of sulfur is −2, so we expect that compound to be a reducing agent and to be readily oxidized. This is its function in the "Claus process" for recovering sulfur from natural gas and petroleum:

$$\overset{-2}{2H_2S}(g) + \overset{+4}{SO_2}(g) \longrightarrow \overset{0}{3S}(s) + 2H_2O(l)$$

Compounds in which an element has an oxidation number in the middle of its range tend to undergo disproportionation. This is possible because both higher and lower oxidation numbers can be reached. Copper has oxidation numbers 0, +1, and +2, and copper(I) ions disproportionate in solution:

$$2\overset{+1}{Cu^+}(aq) \longrightarrow \overset{+2}{Cu^{2+}}(aq) + \overset{0}{Cu}(s)$$

Only insoluble copper(I) salts can be prepared in an aqueous environment, for soluble copper(I) salts disproportionate immediately.

Compounds in which an element has an oxidation number high in its range tend to be good oxidizing agents (Table 3.3), since the element can be reduced to a lower oxidation number. An example is the purple permanganate ion (MnO_4^-), in which the oxidation number of Mn is +7. The yellow chromate ion (5), in which chromium has oxidation number +6, is a powerful oxidizing agent and a useful reagent in the laboratory. The CrO_4^{2-} ion is stable in basic solutions, but when acid is added to a solution of a chromate it is converted (without change of oxidation number) to the orange dichromate ion (6) by loss of water (Fig. 3.24):

$$2CrO_4^{2-}(aq) + 2H^+(aq) \longrightarrow Cr_2O_7^{2-}(aq) + H_2O(l)$$

The oxidizing ability of oxoanions is often greater in acidic solution than in basic solution.

5 Chromate ion

6 Dichromate ion

3.10 HALF-REACTIONS

We can show the reduction and oxidation steps of a redox reaction separately by expressing the overall equation as the sum of two imaginary _half-reactions_. The equation for the half-reaction in which zinc is oxidized is

$$Zn(s) \longrightarrow Zn^{2+}(aq) + 2e^-$$

(The state of the electrons is not given because they are "in transit" and do not have a definite state.) The equation for the half-reaction in which hydrogen ions are reduced is

$$2H^+(aq) + 2e^- \longrightarrow H_2(g)$$

The overall equation is their sum:

$$Zn(s) + 2H^+(aq) + 2e^- \longrightarrow Zn^{2+}(aq) + 2e^- + H_2(g)$$

Note that the charges balance in the equations for the two half-reactions (they sum to the same number—here zero—on both sides of the arrow).

Redox reactions in photography. Half-reactions are a _conceptual_ way of breaking redox reactions into simpler steps. However, in some redox reactions, an electron is actually released and then captured by another substance elsewhere.

One example is in black-and-white photography. The emulsions used to coat photographic film contain silver bromide with small amounts of silver iodide. Photographic printing papers are coated with silver chloride. Each of these emulsions contains microscopic crystals,

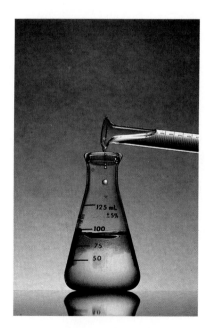

FIGURE 3.24 The chromate ion CrO_4^{2-} is yellow. When acid is added to a chromate solution, the ions form dichromate ions $Cr_2O_7^{2-}$, which are orange.

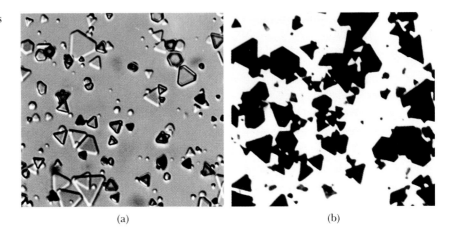

FIGURE 3.25 Microphotographs of the silver halide crystals in a photographic emulsion (a) before and (b) after development, at a magnification of 1800×.

(a)   (b)

or "grains," of the silver halide (Fig. 3.25). These grains are typically about 500 nm in diameter and are held in gelatin that has been coated onto the flat film or paper.

The reaction that records the image is an example of a *photochemical reaction*, a reaction caused by light. When the emulsion is exposed to light, an electron is driven out of a halide ion wherever the light falls:

$$Br^- \xrightarrow{\text{light}} Br + e^-$$

This half-reaction, the loss of an electron from the anion, is an oxidation. The liberated electron wanders through the grain and attaches to a nearby silver cation:

$$Ag^+ + e^- \longrightarrow Ag$$

This step, the gain of an electron by the cation, is a reduction. The equation for the overall photochemical reaction is the sum of the equations for the half-reactions. It is a redox reaction in which a silver cation is reduced by a bromide anion:

$$Ag^+ + Br^- \xrightarrow{\text{light}} Ag + Br$$

The redox reaction occurs only where the light falls, and it results in the formation of small clusters of silver atoms within grains. There must be about four or more silver atoms in a cluster for a grain to be developable, and grains that have this minimum number form the "latent image."

To develop the film, the silver ions remaining in the grains that contain the latent image must be reduced, but not those in the unexposed grains. This is achieved with a mild reducing agent, typically the organic compound hydroquinone ($C_6H_6O_2$; **7**). Developing *amplifies* the latent image, because the grains containing four or so silver atoms are converted to microscopic particles of metallic silver containing around 10^{10} atoms. To complete developing, the excess silver halide is dissolved in aqueous sodium thiosulfate $Na_2S_2O_3 \cdot 5H_2O$, or photographer's "hypo" (a shortening of its old name). Thiosulfate anions attach to the remaining silver ions, forming the soluble $[Ag(S_2O_3)_2]^{3-}$ anion, but not to the metallic silver that forms the image. The soluble silver is then washed away, leaving the metallic silver behind and the film no longer sensitive to light.

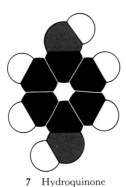

7 Hydroquinone

Balancing redox equations using half-reactions. Redox reactions are often difficult to balance. However, the process is made simpler by balancing the equation for each half-reaction separately, and then adding the two balanced half-equations together. It is necessary, in the combining, to ensure that all the electrons released in the oxidation are used in the reduction.

▼ **EXAMPLE 3.9 Balancing redox reactions by using half-reactions**

Use half-reactions to balance the chemical equation for the oxidation of aluminum to alumina (Al_2O_3).

STRATEGY Write the skeletal (unbalanced) equations for the oxidation and reduction half-reactions separately, balance them individually, and add enough electrons to each to balance the charges. Form the overall equation by adding together the two half-reactions, after first making sure that all the electrons released in the oxidation are taken up in the reduction.

SOLUTION We balance the charges in the oxidation half-reaction $Al(s) \rightarrow Al^{3+}(s)$ by adding three electrons to the right-hand side, obtaining

$$Al(s) \longrightarrow Al^{3+}(s) + 3e^-$$

We balance the charges in the reduction half-reaction $O_2(g) \rightarrow 2O^{2-}(s)$ by adding four electrons to the left-hand side, getting

$$O_2(g) + 4e^- \longrightarrow 2O^{2-}(s)$$

The numbers of electrons will match if we multiply the first equation through by 4, and the second by 3. Then we can form the overall reaction by adding the two half-equations:

$$
\begin{array}{ll}
4Al(s) & \longrightarrow 4Al^{3+}(s) \qquad\qquad + 12e^- \\
3O_2(g) + 12e^- \longrightarrow & \qquad\qquad 6O^{2-}(s) \\
\hline
4Al(s) + 3O_2(g) + \cancel{12e^-} \longrightarrow \underbrace{4Al^{3+}(s) + 6O^{2-}(s)}_{2Al_2O_3(s)} + \cancel{12e^-}
\end{array}
$$

The overall redox equation is therefore

$$4Al(s) + 3O_2(g) \longrightarrow 2Al_2O_3(s)$$

EXERCISE Balance the reaction in which P_4 reacts with O_2 to form P_4O_{10}.

[*Answer*: $P_4(s) + 5O_2(g) \longrightarrow P_4O_{10}(s)$]

▲

When a redox reaction involves oxoanions, oxidation numbers can act as a useful guide to the two half-reactions. To illustrate the steps involved in balancing such a reaction, we shall consider the reaction between potassium permanganate ($KMnO_4$) and oxalic acid ($H_2C_2O_4$) in a solution made acidic with hydrochloric acid (Fig. 3.26). As in more elementary cases, we begin by writing the net ionic skeletal (unbalanced) equation. In this case, the reaction produces Mn^{2+} ions and carbon dioxide, most of which stays in solution.

$$MnO_4^-(aq) + H_2C_2O_4(aq) \longrightarrow Mn^{2+}(aq) + CO_2(aq) \qquad \triangle$$

We do not include the hydrochloric acid at this stage, but keep the H^+ ions it provides in reserve and use them to balance the equation as they become necessary. Now we get down to the business of balancing the equation. This involves six steps.

FIGURE 3.26 The addition of potassium permanganate solution to oxalic acid solution causes a redox reaction that is used in analysis. While some oxalic acid is present, the permanganate is reduced to very pale pink Mn^{2+} ions. When all the oxalic acid has been oxidized, the addition of permanganate turns the solution purple (as shown here) because the MnO_4^- ions survive.

1. *Identify the oxidizing and reducing agents.* This we do by writing the oxidation numbers of the elements:

$$\overset{+7}{MnO_4^-}(aq) + \overset{+3}{H_2C_2O_4}(aq) \longrightarrow \overset{+2}{Mn^{2+}}(aq) + \overset{+4}{CO_2}(aq) \quad \triangle$$

Oxygen and hydrogen can normally be ignored because their oxidation numbers usually remain unchanged (but keep an eye open for peroxides, listed in Table 3.4, and other less common states of oxygen, as well as for hydrogen gas). The oxidation numbers show that MnO_4^- is the oxidizing agent (the oxidation number of Mn is reduced from $+7$ to $+2$ in the reaction) and that oxalic acid is the reducing agent (the oxidation number of C is increased from $+3$ to $+4$, so it undergoes oxidation).

2. *Write the skeletal equation for the oxidation half-reaction.* Since the oxalic acid is oxidized, the equation is

$$H_2C_2O_4(aq) \longrightarrow CO_2(aq) \quad \triangle$$

3. *Balance the oxidation half-equation, ignoring any charges.* For oxoanions in *acidic* solution, we balance H and O by including $H^+(aq)$ or $H_2O(l)$ or both in the equation. In *basic* solution, we balance with $OH^-(aq)$ and $H_2O(l)$. In the acidic solution of this example, we use the H^+ ions provided by the acid and write

$$H_2C_2O_4(aq) \longrightarrow 2CO_2(aq) + 2H^+(aq) \quad \triangle$$

4. *Balance the electric charges by adding electrons.* In this example, we need two electrons on the right to cancel the two positive charges, so the balanced half-equation is

$$H_2C_2O_4(aq) \longrightarrow 2CO_2(aq) + 2H^+(aq) + 2e^-$$

5. *Repeat Steps 2 to 4 for the reduction half-reaction.* The skeletal reduction half-equation is

$$MnO_4^-(aq) \longrightarrow Mn^{2+}(aq) \quad \triangle$$

The presence of O on the left suggests that we need H_2O on the right to balance it; however, as that also introduces H atoms, we also need H^+ ions on the left. Ignoring charges, we balance the atoms as

$$MnO_4^-(aq) + 8H^+(aq) \longrightarrow Mn^{2+}(aq) + 4H_2O(l) \quad \triangle$$

Since the charge of Mn on the left is $-1 + 8 = +7$, but on the right $+2$, we need five electrons on the left:

$$MnO_4^-(aq) + 8H^+(aq) + 5e^- \longrightarrow Mn^{2+}(aq) + 4H_2O(l)$$

6. *Combine the two half-equations by matching the number of electrons and adding.* The number of electrons released in the oxidation will

match the number taken up in the reduction if we multiply the first half-reaction by 5 and the second by 2:

$$5H_2C_2O_4(aq) \longrightarrow 10CO_2(aq) + 10H^+(aq) + 10e^-$$
$$2MnO_4^-(aq) + 16H^+(aq) + 10e^- \longrightarrow 2Mn^{2+}(aq) + 8H_2O(l)$$
$$2MnO_4^-(aq) + 5H_2C_2O_4(aq) + 6H^+(aq) \longrightarrow 2Mn^{2+}(aq) + 8H_2O(l) + 10CO_2(aq)$$

This is the fully balanced net ionic equation.

▼ **EXAMPLE 3.10** **Balancing redox equations using half-reactions**

Sodium hypochlorite in a solution that also contains sodium hydroxide reacts with chromium(III) hydroxide to produce sodium chromate and sodium chloride. Write the balanced equation.

STRATEGY Since chromate ions (CrO_4^{2-}) contain more oxygen than $Cr(OH)_3$, we anticipate that a redox reaction is taking place in which ClO^- ions act as oxidizing agents. Begin by writing the net ionic skeletal equation. Then work through the step-by-step procedure, using H_2O and OH^- for the balancing.

SOLUTION The skeletal net ionic equation and oxidation numbers are

$$\overset{+3}{Cr}(OH)_3(s) + \overset{+1}{Cl}O^-(aq) \longrightarrow \overset{+6}{Cr}O_4^{2-}(aq) + \overset{-1}{Cl}^-(aq) \qquad ⚠$$

As anticipated, the hypochlorite ion is the oxidizing agent. Its reduction half-reaction is

$$ClO^-(aq) \longrightarrow Cl^-(aq) \qquad ⚠$$

Using $H_2O(l)$, $OH^-(aq)$, and e^-, we balance this equation to

$$ClO^-(aq) + H_2O(l) + 2e^- \longrightarrow Cl^-(aq) + 2OH^-(aq)$$

The oxidation half-reaction is

$$Cr(OH)_3(s) \longrightarrow CrO_4^{2-}(aq) \qquad ⚠$$

which we balance to

$$Cr(OH)_3(s) + 5OH^-(aq) \longrightarrow CrO_4^{2-}(aq) + 4H_2O(l) + 3e^-$$

The number of electrons supplied by the oxidation reaction is matched to the number used up in the reduction by multiplying the equations by 2 and 3, respectively. The overall equation is the sum of the matched half-reaction equations; after the canceling of duplications, that is

$$2Cr(OH)_3(s) + 3ClO^-(aq) + 4OH^-(aq) \longrightarrow$$
$$2CrO_4^{2-}(aq) + 3Cl^-(aq) + 5H_2O(l)$$

The complete equation is obtained by adding $7Na^+$ ions to each side:

$$2Cr(OH)_3(s) + 3NaClO(aq) + 4NaOH(aq) \longrightarrow$$
$$2Na_2CrO_4(aq) + 3NaCl(aq) + 5H_2O(l)$$

EXERCISE When hypochlorite ions (ClO^-) are added to plumbite ions ($[Pb(OH)_3]^-$) in basic solution, lead(IV) oxide is formed. Write the balanced equation.

[*Answer*: $[Pb(OH)_3]^-(aq) + ClO^-(aq) \longrightarrow$
$PbO_2(s) + Cl^-(aq) + OH^-(aq) + H_2O(l)$]

▲

This chapter has concentrated on the qualitative aspects of reactions. The three types of reactions introduced here have key features in common, as is explained in later chapters. Precipitation reactions are dealt with in Chapter 15. Acids and bases are so important that we shall spend two chapters, Chapters 14 and 15, on their properties. Moreover, we have not discussed how to determine what substances can oxidize or reduce a given substance. This point is taken up in Chapter 17, where we also see how redox reactions are used to generate electricity.

Now we are ready to discuss some quantitative features and to start making numerical predictions about what happens when reagents are mixed.

SUMMARY

3.1 and 3.2 A chemical reaction may be described by writing its **chemical equation.** This equation summarizes qualitative information about the identities of the **reactants,** or starting materials, and the **products,** or materials formed. When it is **balanced** by including the **stoichiometric coefficients,** it also expresses the preservation of the atoms of each reactant element.

3.3 In a **precipitation reaction** a solid precipitate is formed when two solutions are mixed. The essential feature of a precipitation reaction (and other reactions involving ions) is made clear in the **net ionic equation,** written by canceling the **spectator ions**—the ions left unchanged by the reaction.

3.4 A precipitation reaction is a special case of a metathesis or **double-replacement** reaction, one in which atoms or ions exchange partners. The **solubility rules,** a summary of the solubilities of common compounds, are useful in deciding what is likely to be the product of a precipitation reaction.

3.5 and 3.6 The original **Arrhenius definitions** of an **acid** as a source of hydrogen ions and a **base** as a source of hydroxide ions have been superseded by the **Brønsted definitions,** which concentrate on **proton transfer:** the **proton donor** is the acid, and the **proton acceptor** is the base. Metals have **basic oxides,** and nonmetals have **acidic oxides.** Some oxides, those of elements on the frontier between metals and nonmetals, are **amphoteric,** reacting with both acids and bases.

3.7 The product of a neutralization of an Arrhenius acid and an Arrhenius base is a **salt** in which the cation is supplied by the base and the anion by the acid. In Brønsted terms, **neutralization** in water involves the transfer of a proton from the **hydronium ion** to the hydroxide ion with the formation of water.

3.8 Redox reactions involve the transfer of electrons from the **reducing agent** to the **oxidizing agent. Oxidation** is equivalent to electron loss, and **reduction** to electron gain. The oxidizing strengths of substances are listed in order in the **electrochemical series.**

3.9 Which substance has been oxidized and which reduced in a redox reaction can be determined by calculating the **oxidation numbers,** or effective charges, of the elements taking part. If no oxidation number changes, the reaction is not redox. A special case of a redox reaction is **disproportion,** in which the same element is simultaneously oxidized and reduced.

3.10 An overall redox reaction is conceptually equal to a **reduction half-reaction** and an **oxidation half-reaction;** the equation for the overall reaction is the sum of the equations for the half-reactions.

EXERCISES

Balancing Equations

In Exercises 3.1 and 3.2, write and balance the equation for each reaction. Do not include state symbols in equations in these two exercises.

3.1 (a) The reaction of sodium with water, which leads to the evolution of hydrogen and the formation of sodium hydroxide

(b) The reaction between sodium oxide and water, which produces sodium hydroxide

(c) The burning of lithium in nitrogen, in which lithium nitride is formed

(d) The reaction of calcium with water, in which hydrogen is evolved and calcium hydroxide is formed

3.2 (a) The heating of sodium peroxide (Na_2O_2) with sodium metal, in which sodium oxide is formed

(b) The photosynthesis reaction in which carbon dioxide and water are combined to form glucose ($C_6H_{12}O_6$) and oxygen

(c) The heating of metallic iron with antimony sulfide (Sb_2S_3), in which elemental antimony and iron(II) sulfide are formed

(d) The reaction of aluminum with boron oxide, in which elemental boron and aluminum oxide are formed

3.3 In one stage of ironmaking, the iron ore (Fe_2O_3) reacts with carbon monoxide to form Fe_3O_4 and carbon dioxide. Write the equation for this reaction.

3.4 The *overall* outcome of all the reactions involved in ironmaking is the conversion of iron ore and carbon monoxide into iron and carbon dioxide. Write the equation for the overall reaction.

In Exercises 3.5 to 3.12, balance the equations for the given reactions. Do not include state symbols in equations in these exercises.

3.5 The reaction of boron trifluoride with sodium borohydride ($NaBH_4$) leads to the formation of sodium tetrafluoroborate ($NaBF_4$) and the gas diborane (B_2H_6):

$$BF_3 + NaBH_4 \longrightarrow NaBF_4 + B_2H_6$$

3.6 One route to the production of sulfur, and to the removal of sulfur dioxide fumes from waste gases, is to cause hydrogen sulfide to react with sulfur dioxide:

$$H_2S + SO_2 \longrightarrow S + H_2O$$

3.7 Diborane burns in oxygen, with the formation of boron oxide (B_2O_3) and water:

$$B_2H_6 + O_2 \longrightarrow B_2O_3 + H_2O$$

3.8 Iodic acid can be prepared by the action of sulfuric acid on an iodate:

$$Ba(IO_3)_2 + H_2SO_4 \longrightarrow HIO_3 + BaSO_4$$

3.9 One route to the formation of chromium(III) chloride is to pass chlorine gas over a mixture of chromium(VI) oxide and carbon:

$$CrO_3 + C + Cl_2 \longrightarrow CrCl_3 + CO$$

3.10 The compound of formula $Sb_4O_5Cl_2$, which has been investigated because of its interesting electrical properties, can be prepared by warming a mixture of antimony(III) oxide and antimony(III) chloride:

$$Sb_2O_3 + SbCl_3 \longrightarrow Sb_4O_5Cl_2$$

3.11 Hydrofluoric acid is used to etch glass. It produces a frosted surface because it attacks glass, which can be represented by a silica formula unit (SiO_2), and forms silicon tetrafluoride:

$$HF + SiO_2 \longrightarrow SiF_4 + H_2O$$

3.12 As the pentahydrate $Na_2S_2O_3 \cdot 5H_2O$, sodium thiosulfate forms the large white crystals used as photogra-

pher's "hypo"; it can be prepared by bubbling oxygen through a solution of sodium polysulfide (Na_2S_5):

$$Na_2S_5 + O_2 \longrightarrow Na_2S_2O_3 + S_2$$

In Exercises 3.13 to 3.20, balance the given equations. The reactions are more complicated, but their equations can be balanced in the usual way, and some of the correct coefficients are already provided. Do not include state symbols in these exercises.

3.13 Potassium thiosulfate is used to remove any excess chlorine from fibers and fabrics that have been bleached with that gas:

$$Cl_2 + K_2S_2O_3 + H_2O \longrightarrow KHSO_4 + HCl$$

3.14 The red solid S_4N_2, which melts to a red liquid at 25°C, can be prepared by the reaction between the mercury compound $Hg_5(NS)_8$ and disulfur dichloride (S_2Cl_2) in carbon disulfide solution at 20°C:

$$Hg_5(NS)_8 + 4S_2Cl_2 \longrightarrow Hg_2Cl_2 + HgCl_2 + 4S_4N_2$$

3.15 The first stage in the manufacture of nitric acid from ammonia produced by the Haber synthesis is its oxidation to nitric oxide. A mixture of ammonia and air is passed over heated platinum:

$$NH_3 + O_2 \longrightarrow NO + H_2O$$

3.16 Sodium thiosulfate is oxidized to sodium trithionate ($Na_2S_3O_6$), and sodium sulfate is formed as well, when hydrogen peroxide is added to a solution of the thiosulfate:

$$2Na_2S_2O_3 + H_2O_2 \longrightarrow Na_2S_3O_6 + Na_2SO_4 + H_2O$$

3.17 When a mixture of nitric and hydrochloric acids is poured on copper, a complicated reaction occurs. Brown fumes of nitrogen dioxide are formed, and the solution turns blue-green as copper forms copper(II) chloride:

$$Cu + HNO_3 + HCl \longrightarrow CuCl_2 + NO + H_2O$$

3.18 Diphosphates, which are based on the ion $P_2O_7^{4-}$, can be produced by metathesis:

$$Na_4P_2O_7 + AgNO_3 \longrightarrow Ag_4P_2O_7 + NaNO_3$$

3.19 Until the 1960s, there were no known compounds of xenon. Now many are known, and chemists are still investigating their properties. It is found that alkaline solutions of potassium hydrogen xenate ($KHXeO_4$) decompose slowly, forming the colorless compound potassium perxenate (K_4XeO_6), a very strong oxidizing agent, and evolving bubbles of oxygen:

$$8KHXeO_4 + KOH \longrightarrow$$
$$5Xe + 3K_4XeO_6 + O_2 + H_2O$$

3.20 The gas nitric oxide can be prepared by the action of dilute sulfuric acid on a nitrite:

$$NaNO_2 + H_2SO_4 \longrightarrow NO + HNO_3 + H_2O + Na_2SO_4$$

In Exercises 3.21 and 3.22, write the equation for the combustion of each compound.

3.21 (a) Ethane (C_2H_6); (b) glucose ($C_6H_{12}O_6$); (c) toluene (C_7H_8); (d) tristearin (an animal fat, $C_{57}H_{110}O_6$).
3.22 (a) Benzene (C_6H_6); (b) sucrose ($C_{12}H_{22}O_{11}$); (c) naphthalene ($C_{10}H_8$); (d) triolein (a vegetable oil, $C_{57}H_{104}O_6$).

When nitrogen is present in a compound along with C and H, combustion leads to the formation of nitrogen gas as well as carbon dioxide and water. In Exercises 3.23 and 3.24, write the equation for the combustion of each compound.

3.23 (a) Methylamine (CH_5N); (b) the amino acid glycine ($C_2H_5O_2N$); (c) the analgesic (pain-killer) Tylenol® ($C_8H_9O_2N$); (d) caffeine ($C_8H_{10}O_2N_4$), the stimulant in coffee.
3.24 (a) Ethylamine (C_2H_7N); (b) the analgesic phenacetin ($C_{10}H_{13}O_2N$); (c) the psychoactive drug sold as Methedrine® ("speed," $C_{10}H_{15}N$); (d) nicotine ($C_{10}H_{14}N_2$), the main stimulant in tobacco.

Precipitation Reactions

In Exercises 3.25 and 3.26, state which of the given substances are likely to be soluble in water.

3.25 (a) $AgNO_3$; (b) $Ca(CH_3CO_2)_2$; (c) Na_2SO_4; (d) $AgCl$.
3.26 (a) Na_2S; (b) $Zn(CH_3CO_2)_2$; (c) $Fe(OH)_3$; (d) $CuBr_2$.

In Exercises 3.27 and 3.28, determine which combinations would result in precipitation reactions. For each of those that would, write the net ionic equation and identify the product that would be obtained as a precipitate.

3.27 (a) $FeCl_3(aq)$ mixed with $NaOH(aq)$; (b) $AgNO_3(aq)$ mixed with $KI(aq)$; (c) $Pb(NO_3)_2(aq)$ mixed with $NaCH_3CO_2(aq)$.
3.28 (a) $NH_4NO_3(aq)$ mixed with $Ca(CH_3CO_2)_2(aq)$; (b) $NaClO_4(aq)$ mixed with $Pb(NO_3)_2(aq)$; (c) $Na_2SO_4(aq)$ mixed with $BaCl_2(aq)$.

Acids and Bases

In Exercises 3.29 and 3.30, identify each substance as either an acid (A) or base (B).

3.29 (a) $Mg(OH)_2$; (b) $HOCl$; (c) HNO_2; (d) $B(OH)_3$.
3.30 (a) HN_3; (b) H_2CO_3; (c) NH_3; (d) CH_3NH_2.

In Exercises 3.31 and 3.32, state whether the oxides of each element are likely to be acidic, basic, or amphoteric.

3.31 (a) Mg; (b) P; (c) Be; (d) At.
3.32 (a) Fr; (b) Ra; (c) In; (d) Si.

In Exercises 3.33 and 3.34, state which acid and base you would use to prepare each salt. Write the equation for the reaction in each case.

3.33 (a) Potassium bromide; (b) copper(II) nitrate; (c) calcium acetate; (d) sodium sulfate.
3.34 (a) Cesium nitrate; (b) iron(II) sulfate; (c) magnesium oxalate; (d) sodium dihydrogen phosphate.

In Exercises 3.35 and 3.36, state the reaction you would expect, if any, when a dilute acid acts on each substance.

3.35 (a) Zinc; (b) magnesium; (c) copper; (d) aluminum.
3.36 (a) Potassium bromide; (b) iron; (c) gold; (d) calcium.

Electron-Transfer Half-reactions

In Exercises 3.37 and 3.38, write the equations for the reduction and oxidation half-reactions for each given redox reaction.

3.37 (a) The displacement of silver from solution by copper:

$$Cu + 2Ag^+ \longrightarrow Cu^{2+} + 2Ag$$

(b) The oxidation of iron(II) ions by lead(IV):

$$2Fe^{2+} + Pb^{4+} \longrightarrow 2Fe^{3+} + Pb^{2+}$$

(c) The decomposition of hydrogen chloride into its elements:

$$2HCl \longrightarrow H_2 + Cl_2$$

3.38 (a) The oxidation of tin(II) by iron(III):

$$Sn^{2+} + 2Fe^{3+} \longrightarrow Sn^{4+} + 2Fe^{2+}$$

(b) The displacement of copper from solution by aluminum:

$$2Al + 3Cu^{2+} \longrightarrow 2Al^{3+} + 3Cu$$

(c) The reaction of heated iron with chlorine:

$$2Fe + 3Cl_2 \longrightarrow 2Fe^{3+} + 6Cl^-$$

In Exercises 3.39 and 3.40, balance each of the given half-reaction equations by adding the appropriate number of electrons, and state whether the half-reaction is an oxidation or a reduction.

3.39 (a) $AgCN \longrightarrow Ag + CN^-$
(b) $ClO_3^- + H_2O \longrightarrow ClO_2^- + 2OH^-$
(c) $2F^- + H_2O \longrightarrow F_2O + 2H^+$
3.40 (a) $3I^- \longrightarrow I_3^-$
(b) $2NO_2^- + 2H_2O \longrightarrow N_2O_2^{2-} + 4OH^-$
(c) $SeO_4^{2-} + H_2O \longrightarrow SeO_3^{2-} + 2OH^-$

3.41 One stage in the extraction of cobalt from ore involves the treatment of a basic solution of cobalt(II) ions with sodium hypochlorite, which precipitates solid cobalt(III) hydroxide. Write the balanced redox and half-reaction equations.

3.42 One stage in the extraction of gold from rocks involves dissolving the metal from the rock with basic sodium cyanide that has been thoroughly aerated. This results in the formation of $[Au(CN)_2]^-$ and OH^- ions. The next stage is to precipitate the gold with zinc dust. This forms $Zn(CN)_4^{2-}$. Write the balanced overall and half-reaction equations for both stages.

Oxidation Numbers

In Exercises 3.43 to 3.48, give the oxidation numbers of the elements other than hydrogen and oxygen (except where requested) in each species.

3.43 (a) Cl^-; (b) Cl_2; (c) OCl^-; (d) $HOCl$.
3.44 (a) N_2; (b) NO; (c) NO_2^-; (d) HNO_2.
3.45 (a) IO_3^-; (b) IO_2^-; (c) N_2H_4; (d) N_3^-.
3.46 (a) ClO_4^-; (b) SO_4^{2-}; (c) HSO_3^-; (d) $S_2O_3^{2-}$.

3.47 (a) The manganate ion; (b) the dichromate ion; (c) the superoxide ion O_2^- (give the ONs of the oxygens in this case).
3.48 (a) The permanganate ion; (b) the chromate ion; (c) the ozonide ion O_3^- (give the ONs of the oxygens).

In Exercises 3.49 and 3.50, identify each change of oxidation number in the given reactions.

3.49 (a) $Cl_2(g) + 2Br^-(aq) \longrightarrow 2Cl^-(aq) + Br_2(aq)$, a reaction used to prepare bromine from brine
(b) $Pb(s) + NaNO_3(s) \longrightarrow PbO(s) + NaNO_2(s)$, a method of preparing nitrites
(c) $Br_2(l) + 3F_2(g) \longrightarrow 2BrF_3(g)$, a method of preparing the "interhalogen compound" bromine trifluoride
(d) $3NO_2(g) + H_2O(l) \longrightarrow 2HNO_3(aq) + NO(g)$, one stage in the Ostwald process for preparing nitric acid
3.50 (a) $KCl(l) + Na(g) \longrightarrow NaCl(s) + K(g)$, a reaction used to prepare potassium metal
(b) $Cl_2(g) + 2NaOH(aq) \longrightarrow$
$$NaCl(aq) + NaOCl(aq) + H_2O(l),$$
the reaction used to prepare hypochlorites
(c) $NO(g) + O_3(g) \longrightarrow NO_2(g) + O_2(g)$, a reaction that destroys ozone in the upper atmosphere
(d) $2Al(s) + 2OH^-(aq) + 6H_2O(l) \longrightarrow$
$$2[Al(OH)_4]^-(aq) + 3H_2(g),$$
a reaction showing aluminum's amphoteric character

Oxidizing and Reducing Agents

In Exercises 3.51 to 3.54, identify the oxidizing and reducing agents in each reaction.

3.51 (a) $Mg(s) + 2HCl(aq) \longrightarrow MgCl_2(aq) + H_2(g)$, a simple way of preparing hydrogen in the laboratory

(b) $B_2O_3(s) + 3Mg(s) \longrightarrow 2B(s) + 3MgO(s)$, a reaction used to prepare the element boron
(c) $2H_2S(g) + SO_2(g) \longrightarrow 3S(s) + 2H_2O(l)$, a reaction used to produce sulfur from the hydrogen sulfide obtained from oil and natural gas wells
3.52 (a) $6Li(s) + N_2(g) \longrightarrow 2Li_3N(s)$, a reaction showing how lithium resembles magnesium
(b) $2Al(s) + Cr_2O_3(s) \longrightarrow Al_2O_3(s) + 2Cr(s)$, an example of a "thermite reaction" used to obtain some metals from their ores
(c) $2XeF_2(aq) + 2H_2O(l) \longrightarrow$
$$2Xe(g) + 4HF(aq) + O_2(g),$$
a typical reaction of xenon difluoride

3.53 $3As_2S_3(s) + 10NO_3^-(aq) + 10H^+(aq) + 4H_2O(l)$
$$\longrightarrow 6H_3AsO_4(aq) + 9S(s) + 10NO(g)$$
3.54 $3P(s) + 2H^+(aq) + 5NO_3^-(aq) + 2H_2O(l) \longrightarrow$
$$3H_2PO_4^-(aq) + 5NO(g)$$

In Exercises 3.55 and 3.56, decide whether you should choose an oxidizing agent or a reducing agent to bring about each change.

3.55 (a) Obtain bromate ions from bromine in basic solution.
(b) Obtain sulfate ions from dithionate ions $(S_2O_6^{2-})$.
(c) Obtain nitric oxide from potassium nitrate solution.
3.56 (a) Obtain chlorine dioxide from sodium chlorate solution.
(b) Obtain sodium sulfide from sodium sulfate.
(c) Obtain manganese dioxide from a manganese(II) salt solution.

3.57 (a) Can Ag^+ ions oxidize Cu to Cu^{2+}?
(b) Can Cu^{2+} ions oxidize zinc to Zn^{2+}?
(c) Can permanganate ions be used to oxidize Cr^{3+} ions to dichromate ions in acidic solution?
3.58 (a) Can silver release hydrogen from hydrochloric acid?
(b) Can acidified dichromate ions be used to oxidize Mn^{2+} ions to permanganate ions?
(c) What happens when a piece of zinc is placed into a solution of Fe^{3+}?

Balancing Redox Equations

In Exercises 3.59 and 3.60, balance the half-reaction equations, all of which occur in acidic solution, and state whether each is an oxidation or reduction.

3.59 (a) $VO^{2+}(aq) \longrightarrow V^{3+}(aq)$
(b) $PbSO_4(s) \longrightarrow PbO_2(s) + HSO_4^-(aq)$
(c) $H_3PO_3(aq) \longrightarrow H_3PO_4(aq)$
3.60 (a) $MnO_2(s) \longrightarrow Mn^{2+}(aq)$
(b) $H_2C_2O_4(aq) \longrightarrow 2CO_2(g)$
(c) $B(s) \longrightarrow H_3BO_3(aq)$

In Exercises 3.61 and 3.62, balance the half-reaction equations, all of which occur in basic solution, and state whether each is an oxidation or reduction.

3.61 (a) $ClO^-(aq) \longrightarrow Cl^-(aq)$
(b) $Se(s) \longrightarrow SeO_3^{2-}(aq)$
(c) $IO_3^-(aq) \longrightarrow IO^-(aq)$

3.62 (a) $HPbO_2^-(aq) \longrightarrow Pb(s)$
(b) $2SO_3^{2-}(aq) \longrightarrow S_2O_4^{2-}(aq)$
(c) $BH_4^-(aq) \longrightarrow H_2BO_3^-(aq)$

In Exercises 3.63 to 3.70, balance each equation by the half-reaction method. (A) indicates an acidic solution, (B) a basic solution.

3.63 (a) $Cu(s) + NO_3^-(aq) \longrightarrow$
$$Cu^{2+}(aq) + NO(g) \quad (A)$$
(b) $ZnS(s) + O_2(g) \longrightarrow ZnO(s) + SO_2(g)$
(c) $Cl_2(g) \longrightarrow HOCl(aq) + Cl^-(aq) \quad (A)$

3.64 (a) $H_2S(g) + Cl_2(g) \longrightarrow S(s) + Cl^-(aq) \quad (A)$
(b) $S_2O_4^{2-}(aq) \longrightarrow$
$$S_2O_3^{2-}(aq) + HSO_3^-(aq) \quad (A)$$
(c) $H_2O_2(aq) \longrightarrow O_2(g) \quad (A)$

3.65 (a) $Cl^-(aq) + O_2(g) \longrightarrow OCl^-(aq) \quad (B)$
(b) $Tl_2O_3(s) + NH_2OH(aq) \longrightarrow$
$$TlOH(s) + N_2(g) \quad (B)$$
(c) $MnO_4^-(aq) + Br^-(aq) \longrightarrow$
$$MnO_2(s) + BrO_3^-(aq) \quad (B)$$

3.66 (a) $Br_2(l) \longrightarrow BrO_3^-(aq) + Br^-(aq) \quad (B)$
(b) $Ag(s) + CN^-(aq) + O_2(g) \longrightarrow$
$$[Ag(CN)_2]^-(aq) \quad (B)$$
(c) $P_4(s) \longrightarrow H_2PO_2^-(aq) + PH_3(g) \quad (B)$

3.67 (a) $Cr(OH)_3(s) + OCl^-(aq) \longrightarrow$
$$CrO_4^{2-}(aq) + Cl^-(aq) \quad (B)$$
(b) $MnO_4^-(aq) + H_2SO_3(aq) \longrightarrow$
$$Mn^{2+}(aq) + HSO_4^-(aq) \quad (A)$$
(c) $As_2S_3(s) + NO_3^-(aq) \longrightarrow$
$$H_3AsO_4(aq) + S(s) + NO(g) \quad (A)$$

3.68 (a) $Pb(s) + PbO_2(s) + SO_4^{2-}(aq) \longrightarrow$
$$PbSO_4(s) \quad (A)$$
(b) $MnO_4^-(aq) + CH_3OH(aq) \longrightarrow$
$$Mn^{2+}(aq) + HCOOH(aq) \quad (A)$$
(c) $I_2(s) + NO_3^-(aq) \longrightarrow IO_3^-(aq) + NO(g) \quad (A)$

3.69 (a) $H_2O_2(aq) + Cl_2O_7(g) \longrightarrow$
$$ClO_2^-(aq) + O_2(g) \quad (B)$$
(b) $MnO_4^{2-}(aq) \longrightarrow$
$$MnO_2(s) + MnO_4^-(aq) \quad (B)$$
(c) $CrI_3(s) + H_2O_2(aq) \longrightarrow$
$$CrO_4^{2-}(aq) + IO_4^-(aq) \quad (B)$$

*3.70 (a) $[Pb(OH)_4]^{2-}(aq) + ClO^-(aq) \longrightarrow$
$$PbO_2(s) + Cl^-(aq) \quad (B)$$
(b) $MnO_2(s) \longrightarrow MnO_4^{2-}(aq) + H_2(g) \quad (B)$
(c) $FeO_4^{2-}(aq) \longrightarrow Fe^{3+}(aq) + O_2(g) \quad (B)$

3.71 The hydrogen sulfite ion is a moderately strong reducing agent and, depending on the conditions, is oxi-dized to either hydrogen sulfate ion or dithionate ion, $S_2O_6^{2-}(aq)$. Write the equations for the half-reactions corresponding to the two oxidations. The reaction of hydrogen sulfite ions with iodine (to form iodide and hydrogen sulfate ions) is used to determine their concentration. Write the overall equation for this reaction.

3.72 The hydrogen sulfite ion can act as an oxidizing agent in the presence of strong reducing agents, sodium amalgam (sodium dissolved in mercury; only the sodium is involved) forming dithionite ion ($S_2O_4^{2-}$) and sodium ions. Write the corresponding equations for the half-reactions in basic solution.

*3.73 The thiosulfate ion ($S_2O_3^{2-}$) is a moderately strong reducing agent. It is used to determine the concentration of iodine using a reaction in which the latter is converted to iodide ions, and the thiosulfate to tetrathionate ions ($S_4O_6^{2-}$). Write the equations for the half-reactions and the overall reaction.

3.74 Stronger oxidizing agents than iodine oxidize thiosulfate ion to sulfate in acid solution. This reaction is used in the commercial bleaching of fabrics, to remove any excess chlorine from the fibers. Write the equations for the half-reactions and the overall reaction.

General

In Exercises 3.75 and 3.76, classify each reaction as precipitation (P), acid-base neutralization (N), or redox (R). Identify the oxidizing and reducing agents in each redox reaction, write the net ionic equation for each precipitation reaction, and identify the acid and the base in each acid-base neutralization.

3.75 (a) The production of sulfur trioxide from sulfur dioxide in the manufacture of sulfuric acid:
$$2SO_2 + O_2 \longrightarrow 2SO_3$$

(b) The reaction that occurs when ammonia burns in air:
$$4NH_3 + 3O_2 \longrightarrow 2N_2 + 6H_2O$$

(c) The reaction used to produce elemental phosphorus from its oxide:
$$P_4O_{10} + 10C \longrightarrow P_4 + 10CO$$

(d) The formation of insoluble silver chromate from soluble sodium chromate:
$$2AgNO_3(aq) + Na_2CrO_4(aq) \longrightarrow$$
$$Ag_2CrO_4(s) + 2NaNO_3(aq)$$

*3.76 (a) The reaction used in measuring the concentration of carbon monoxide in a gas stream:
$$5CO + I_2O_5 \longrightarrow I_2 + 5CO_2$$

The amount of iodine formed is a measure of the amount of carbon monoxide that was passed through the solution of I_2O_5.

(b) The reaction in which the amount of iodine formed by the reaction in (a) is measured:

$$I_2 + 2S_2O_3{}^{2-} \longrightarrow 2I^- + S_4O_6{}^{2-}$$

(c) The test for bromide ions in solution, in which a dense, creamy precipitate of silver bromide results when a drop of the solution is added to silver nitrate solution:

$$AgNO_3(aq) + Br^-(aq) \longrightarrow AgBr(s) + NO_3{}^-(aq)$$

(d) The heating of uranium tetrafluoride with magnesium, one stage in the purification of uranium metal:

$$UF_4 + 2Mg \longrightarrow U + 2MgF_2$$

In this chapter we see how to interpret chemical equations quantitatively. The techniques we learn here enable us to predict the amount of product formed in a reaction, calculate the required amounts of reactants, and use measurements of mass and volume to discover the compositions of compounds. Calculations like these are used to decide how much hydrogen and oxygen were needed to launch the space shuttle Atlantis, *seen here at the start of its October 1985 flight.*

REACTION STOICHIOMETRY

Interpreting stoichiometric coefficients

4.1 Mole calculations
4.2 Limiting reagents
4.3 Composition from measurements of mass

The stoichiometry of reactions in solution

4.4 Molar concentration
4.5 Volume of solution required for reaction

Titrations

4.6 Direct titrations
4.7 Indirect and back titrations

A balanced equation is a quantitative statement about a reaction; hence, it can be used to make numerical predictions about the outcome of the reaction. In particular, we can use a chemical equation to predict how much product will result when a given mass of starting material reacts completely. This is exactly the kind of information chemists need when they are designing chemical plants, analyzing the suitability of fuels, or working out how much reactant to use in an experiment. The same kinds of calculations can also be used to analyze the compositions of samples and to determine the chemical formulas of compounds.

INTERPRETING STOICHIOMETRIC COEFFICIENTS

The chemical equation for the reaction of hydrogen and oxygen to produce water is

$$2H_2(g) + O_2(g) \longrightarrow 2H_2O(l)$$

This shows that for every two H_2 molecules and one O_2 molecule that react, two H_2O molecules are formed:

$$2 \text{ H}_2 \text{ molecules} + 1 \text{ O}_2 \text{ molecule} \longrightarrow 2 \text{ H}_2\text{O molecules}$$

We can multiply this equation through by Avogadro's number. It then follows that for every 2 moles of H_2 molecules (12.04×10^{23} molecules) that react, 1 mole of O_2 molecules also react, and 2 moles of H_2O molecules are formed:

$$2 \text{ mol H}_2 \text{ molecules} + 1 \text{ mol O}_2 \text{ molecules} \longrightarrow$$
$$2 \text{ mol H}_2\text{O molecules}$$

That is, the stoichiometric coefficients—the numbers multiplying the chemical formulas in a chemical equation—tell us how many moles of each substance react or are produced in the reaction (Fig. 4.1).

4.1 MOLE CALCULATIONS

The interpretation of stoichiometric coefficients as numbers of moles opens up the path to a wide range of calculations (Fig. 4.2). We discuss the general strategy here and show two of the calculations. In later chapters, we shall see how the same strategy is used in other cases.

The numbers of moles taking part in a reaction. Suppose we wanted to know how much water is formed when 0.20 mol of H_2 molecules react with oxygen. If we knew the number of moles of H_2O produced per mole of H_2 molecules, we could use that information as a conversion factor to calculate the answer:

$$\text{Moles of H}_2\text{O} = \text{moles of H}_2 \times \text{moles of H}_2\text{O per mole of H}_2$$

The necessary information is found in the equation for the reaction:

$$2H_2(g) + O_2(g) \longrightarrow 2H_2O(l)$$

Since this shows that 2 mol of H_2O molecules are formed when 2 mol of H_2 molecules react, we can write

$$2 \text{ mol H}_2 = 2 \text{ mol H}_2\text{O}$$

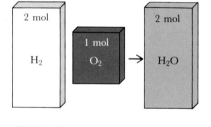

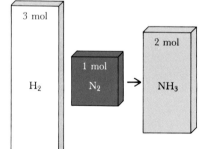

FIGURE 4.1 The stoichiometric coefficients in a chemical equation show the number of moles of each reactant that take part in a reaction, and the number of moles of each product that is formed when the reactants undergo complete reaction. These two illustrations interpret the reactions $2H_2 + O_2 \rightarrow 2H_2O$ and $3H_2 + N_2 \rightarrow 2NH_3$.

FIGURE 4.2 The large, round structures are ammonia storage tanks in a modern plant for producing ammonia by the Haber synthesis. The quantities of hydrogen and nitrogen that must be supplied to produce a given amount of ammonia can be calculated using the techniques described in this chapter.

The conversion factor from moles of H_2 molecules to moles of H_2O molecules is therefore

$$1 = \frac{2 \text{ mol } H_2O}{2 \text{ mol } H_2}$$

When we say that the term on the right is equal to 1, we mean that 2 mol of H_2O is equivalent to 2 mol of H_2 in this reaction. We use this factor just as though we were converting units, so in the present case

$$\text{Moles of } H_2O = 0.20 \text{ mol } H_2 \times \frac{2 \text{ mol } H_2O}{2 \text{ mol } H_2}$$

$$= 0.20 \text{ mol } H_2O$$

We now know that when 0.20 mol of H_2 (or 0.40 g of hydrogen gas) reacts with oxygen to form water, 0.20 mol of H_2O (or 3.6 g of water) is formed.

▼ EXAMPLE 4.1 Calculating the number of moles taking part in a reaction

How many moles of N_2 are needed to produce 5.0 mol of NH_3 molecules in the Haber synthesis?

STRATEGY In all stoichiometry calculations, begin by writing down the balanced equation; then identify the conversion that is needed, and obtain the required factor from the stoichiometric coefficients in the equation.

SOLUTION The equation for the synthesis is

$$N_2(g) + 3H_2(g) \longrightarrow 2NH_3(g)$$

To find the conversion factor required, we note that

$$\text{Moles of } N_2 = \text{moles of } NH_3 \times \text{moles of } N_2 \text{ per mole of } NH_3$$

Thus we need the number of moles of N_2 that react per mole of NH_3 produced. From the chemical equation we write

$$1 \text{ mol } N_2 = 2 \text{ mol } NH_3$$

and rearrange it to

$$\frac{1 \text{ mol N}_2}{2 \text{ mol NH}_3} = 1$$

We then use this factor to find

$$\text{Moles of N}_2 = 5.0 \text{ mol NH}_3 \times \frac{1 \text{ mol N}_2}{2 \text{ mol NH}_3}$$

$$= 2.5 \text{ mol N}_2$$

EXERCISE How many moles of NH_3 molecules can be produced from 2.0 mol of H_2 molecules if all the H_2 reacts?

[*Answer*: 1.3 mol NH_3]

▲

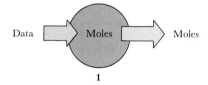

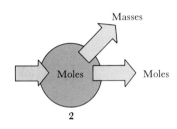

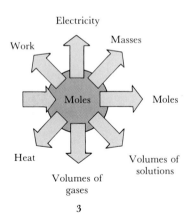

The masses taking part in a reaction. We have just seen that the number of moles of reactants can be related to the number of moles of products in a reaction. We have already seen (in Section 2.5) how to use molecular weights and molar masses to convert between moles and mass. By combining the two calculations, we obtain a way of calculating the *masses* of substances produced in reactions from the masses of the starting substances. That is, we can convert the masses of the reactants to moles, use the chemical relations between moles to convert from moles of reactants to moles of products, and then convert from moles of products to masses of products.

The central importance of the mole in chemistry is illustrated by the diagrams in the margin. Diagram **1** summarizes the first kind of calculation we did, where we used the relation between moles of reactants and moles of products. Diagram **2** summarizes the calculation we have just discussed and shall use soon. It shows that the data—which might be moles reacting, masses reacting, or other information about the reactants—can also be used to calculate the masses of the substances produced or consumed. As we proceed, the number of arrows in the diagram will grow; by Chapter 17 we shall have reached diagram **3,** in which the mole serves as pivot for a wide range of calculations.

At this stage we concentrate on calculations dealing with masses. As an example of the type of problem we are considering here, suppose that we want to know the mass of iron that can be obtained from 10.0 kg of hematite ore (Fe_2O_3) by the reduction

$$Fe_2O_3(s) + 3CO(g) \longrightarrow 2Fe(s) + 3CO_2(g)$$

(See Fig. 4.3.) We can find the mass of iron produced by first converting the mass of ore to a number of moles of Fe_2O_3, then using the chemical equation to convert moles of Fe_2O_3 to moles of Fe atoms, and finally converting moles of Fe atoms to mass of iron.

First, we convert the mass of ore to moles of Fe_2O_3:

Moles of Fe_2O_3
      = mass of Fe_2O_3 × moles of Fe_2O_3 per gram of Fe_2O_3

To complete this step we use the molar mass of Fe_2O_3,

$$1 \text{ mol Fe}_2O_3 = 159.70 \text{ g Fe}_2O_3$$

FIGURE 4.3 Iron ore is gouged out of the hillside by buckets like this. We could determine the mass of iron that can be produced from each load with a stoichiometric calculation.

which gives

$$\text{Moles of Fe}_2\text{O}_3 = 10.0 \times 10^3 \text{ g Fe}_2\text{O}_3 \times \frac{1 \text{ mol Fe}_2\text{O}_3}{159.70 \text{ g Fe}_2\text{O}_3}$$

$$= 62.6 \text{ mol Fe}_2\text{O}_3$$

Now we do a mole-to-mole conversion of the kind illustrated in Example 4.1. We begin by noting that

Moles of Fe = moles of Fe_2O_3 × moles of Fe per mole of Fe_2O_3

and use the chemical equation to find that

$$1 \text{ mol Fe}_2\text{O}_3 = 2 \text{ mol Fe}$$

Therefore,

$$\text{Moles of Fe} = 62.6 \text{ mol Fe}_2\text{O}_3 \times \frac{2 \text{ mol Fe}}{1 \text{ mol Fe}_2\text{O}_3}$$

$$= 125 \text{ mol Fe}$$

Finally, we convert from moles of Fe atoms to mass of iron using the molar mass of iron atoms in the form

$$1 \text{ mol Fe} = 55.85 \text{ g Fe}$$

This gives

$$\text{Mass of Fe} = 125 \text{ mol Fe} \times \frac{55.85 \text{ g Fe}}{1 \text{ mol Fe}}$$

$$= 7.0 \times 10^3 \text{ g}$$

Hence 7.0 kg of iron can be obtained from the 10.0 kg of ore.

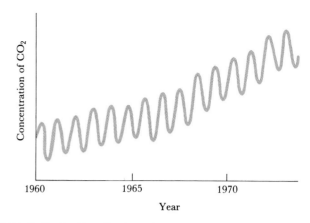

▼ **EXAMPLE 4.2** **Predicting the mass of a product**

An environmental problem currently receiving wide attention is the increase of carbon dioxide in the atmosphere (Fig. 4.4). Calculate the mass of carbon dioxide produced when 100 g of propane (C_3H_8) is burned.

STRATEGY Since an O atom is eight times as heavy as the two H atoms it replaces, the mass of CO_2 produced will be quite a lot larger than the 100 g of fuel. Begin the calculation by writing the equation for the combustion; this will provide the conversion from moles of C_3H_8 to moles of CO_2. Work through the string of conversions needed to proceed from mass to moles of reactant, then from moles of reactant to moles of product, and finally from moles of product to mass of product. Construct the conversion factors from the molecular weights and the chemical equation.

SOLUTION The combustion equation is

$$C_3H_8(g) + 5O_2(g) \longrightarrow 3CO_2(g) + 4H_2O(g)$$

The first conversion required is from mass of fuel to moles of C_3H_8 (which has molar mass 44.09 g/mol):

$$\text{Moles of } C_3H_8 = 100 \text{ g } C_3H_8 \times \frac{1 \text{ mol } C_3H_8}{44.09 \text{ g } C_3H_8}$$

$$= 2.27 \text{ mol } C_3H_8$$

Since the chemical equation implies that

$$1 \text{ mol } C_3H_8 = 3 \text{ mol } CO_2$$

the second conversion is

$$\text{Moles of } CO_2 = 2.27 \text{ mol } C_3H_8 \times \frac{3 \text{ mol } CO_2}{1 \text{ mol } C_3H_8}$$

$$= 6.81 \text{ mol } CO_2$$

Finally, since 1 mol CO_2 = 44.01 g CO_2, we have

$$\text{Mass of } CO_2 = 6.81 \text{ mol } CO_2 \times \frac{44.01 \text{ g } CO_2}{1 \text{ mol } CO_2}$$

$$= 3.00 \times 10^2 \text{ g } CO_2$$

That is, the combustion produces 300 g of carbon dioxide.

EXERCISE Calculate the mass of ammonia produced when 50 tons of hydrogen are used in its synthesis.

[*Answer*: 280 tons]

4.2 LIMITING REAGENTS

We could now calculate the mass of oxygen that a given mass of fuel needs if all of it is to burn to carbon dioxide and water. However, suppose we supply a smaller amount of oxygen than is needed. Then less of the fuel can burn, and less of the product will form. Since in this situation the oxygen limits the amount of product formed, it is called the "limiting reagent" (Fig. 4.5):

The **limiting reagent** in a particular experiment is the reactant that governs the maximum amount of product that can be formed.

An analogy is a hamburger stand that has a large supply of rolls but not much meat. The maximum number of hamburgers it can produce is then limited by the amount of meat available, so the meat is the limiting reagent. When there is no more meat, the excess bread remains unused (Fig. 4.6).

FIGURE 4.5 A smokey flame is the sign that insufficient oxygen is being provided to burn all the fuel. Therefore, oxygen is the limiting reagent. The white incandescence is due to white-hot carbon particles produced by the incomplete combustion.

▼ **EXAMPLE 4.3** **Limiting-reagent calculations**

How much octane (C_8H_{18}) will remain if 100 g are supplied and the maximum amount is burned in 150 g of oxygen (the limiting reagent)?

STRATEGY The mass remaining is the difference between the mass supplied and the mass that burns. We can find the mass that burns by calculating the number of moles of O_2 molecules supplied, and then the number of moles of C_8H_{18} molecules that amount of O_2 can oxidize. The string of conversions

Mass of $O_2 \longrightarrow$ moles of $O_2 \longrightarrow$ moles of $C_8H_{18} \longrightarrow$ mass of C_8H_{18}

proceeds in the same way as in Example 4.2.

SOLUTION The combustion equation is

$$2C_8H_{18}(l) + 25O_2(g) \longrightarrow 16CO_2(g) + 18H_2O(g)$$

Since 1 mol O_2 = 32.00 g O_2, 150 g of oxygen corresponds to 150/32.0 = 4.69 mol O_2. From the chemical equation, the relation between moles of O_2 and moles of C_8H_{18} is

$$2 \text{ mol } C_8H_{18} = 25 \text{ mol } O_2$$

so that

$$\text{Moles of } C_8H_{18} = 4.69 \text{ mol } O_2 \times \frac{2 \text{ mol } C_8H_{18}}{25 \text{ mol } O_2}$$

$$= 0.375 \text{ mol } C_8H_{18}$$

Since 1 mol C_8H_{18} = 114.22 g C_8H_{18}, this amount corresponds to 43 g C_8H_{18}. The mass of octane remaining unburned is therefore 100 g − 43 g = 57 g.

EXERCISE When potassium nitrate is heated with lead, lead(II) oxide and potassium nitrite are formed. In a certain experiment, 10.0 g of lead (the limiting reagent) was heated with potassium nitrate. Calculate the maximum mass of potassium nitrite that could have been formed.

[*Answer*: 4.11 g]

▲

Limiting reagent

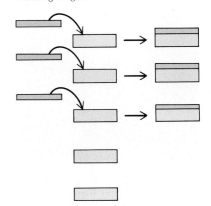

FIGURE 4.6 The role of a limiting reagent: only as much product can be produced as the limiting reagent allows.

Deciding which is the limiting reagent. Unless we supply exactly the right amounts of reactants, one of them is always the limiting reagent. We decide which one that is by calculating the amount of product that can be formed from each of the amounts of reactants supplied. Then the substance that leads to the smallest amount of product is the limiting reagent. If we found that we had enough bread to produce 100 hamburgers but enough meat to produce only 75, then the meat would be the limiting reagent.

▼ **EXAMPLE 4.4** **Identifying the limiting reagent: I**

Suppose that 2 mol of H_2O and 5 mol of NO_2 are supplied in the reaction used to produce nitric acid from nitrogen dioxide,

$$3NO_2(g) + H_2O(l) \longrightarrow 2HNO_3(l) + NO(g)$$

Which is the limiting reagent?

STRATEGY As explained above, we calculate the moles of HNO_3 that can be obtained by complete reaction of each reactant. The limiting reagent is the reactant that produces the least HNO_3.

SOLUTION From the chemical equation, we see that

$$3 \text{ mol } NO_2 = 2 \text{ mol } HNO_3$$

and

$$1 \text{ mol } H_2O = 2 \text{ mol } HNO_3$$

The 2 mol of H_2O can produce

$$\text{Moles of } HNO_3 = 2 \text{ mol } H_2O \times \frac{2 \text{ mol } HNO_3}{1 \text{ mol } H_2O}$$

$$= 4 \text{ mol } HNO_3$$

The 5 mol of NO_2 can produce

$$\text{Moles of } HNO_3 = 5 \text{ mol } NO_2 \times \frac{2 \text{ mol } HNO_3}{3 \text{ mol } NO_2}$$

$$= 3\tfrac{1}{3} \text{ mol } HNO_3$$

The available NO_2 produces the smaller amount of HNO_3, so it is the limiting reagent.

EXERCISE A 4-mmol (0.1-g) slice of sodium metal is dropped into a 5-mmol (0.1-mL) drop of water. Which is the limiting reagent in the reaction $2Na(s) + 2H_2O(l) \rightarrow 2NaOH(aq) + H_2(g)$?

[*Answer:* Na]

▲

FIGURE 4.7 Calcium carbide is a dirty-white solid formed from limestone and carbon. When, as here, water is poured on it, acetylene, which burns with a brilliant white flame, is produced.

▼ **EXAMPLE 4.5** **Identifying the limiting reagent: II**

Calcium carbide (CaC_2) reacts with water to form calcium hydroxide and the flammable gas acetylene (C_2H_2). This reaction was once used in vehicle lamps, because the acetylene burns with a bright white flame (Fig. 4.7). Today it is a potentially important route from inorganic raw materials to organic synthetics. Which is the limiting reagent when 100 g of water react with 100 g of calcium carbide?

STRATEGY We are given masses, not moles. However, we can use the same strategy as in Example 4.4 if we first convert from mass to moles. We need to calculate the amount of C_2H_2 that can be formed by each reactant if it reacts completely. Then the limiting reagent is the one that produces the smaller amount of acetylene.

SOLUTION The chemical equation is

$$CaC_2(s) + 2H_2O(l) \longrightarrow Ca(OH)_2(s) + C_2H_2(g)$$

which tells us that

$$1 \text{ mol } CaC_2 = 1 \text{ mol } C_2H_2$$

and

$$2 \text{ mol } H_2O = 1 \text{ mol } C_2H_2$$

Since 1 mol CaC_2 = 64.10 g CaC_2, the number of moles of CaC_2 supplied in 100 g of the carbide is

$$\text{Moles of } CaC_2 = 100 \text{ g } CaC_2 \times \frac{1 \text{ mol } CaC_2}{64.10 \text{ g } CaC_2}$$

$$= 1.56 \text{ mol } CaC_2$$

This can produce

$$\text{Moles of } C_2H_2 = 1.56 \text{ mol } CaC_2 \times \frac{1 \text{ mol } C_2H_2}{1 \text{ mol } CaC_2}$$

$$= 1.56 \text{ mol } C_2H_2$$

The number of moles of H_2O supplied is

$$\text{Moles of } H_2O = 100 \text{ g } H_2O \times \frac{1 \text{ mol } H_2O}{18.02 \text{ g } H_2O}$$

$$= 5.55 \text{ mol } H_2O$$

This can produce

$$\text{Moles of } C_2H_2 = 5.55 \text{ mol } H_2O \times \frac{1 \text{ mol } C_2H_2}{2 \text{ mol } H_2O}$$

$$= 2.78 \text{ mol } C_2H_2$$

Since the available carbide produces less acetylene, it is the limiting reagent.

EXERCISE Which is the limiting reagent when 1 ton of hydrogen reacts with 8 tons of nitrogen in the synthesis of ammonia?

[*Answer*: Hydrogen]

▲

Reaction yield. The chemical equation for the combustion of octane (C_8H_{18}), a representative compound in the mixture sold as gasoline, is

$$2C_8H_{18}(l) + 25O_2(g) \longrightarrow 16CO_2(g) + 18H_2O(g)$$

Using this equation, we can calculate that when 100 g of octane burns in a plentiful supply of oxygen (so that oxygen is not the limiting reagent), 308 g of carbon dioxide should be produced. This 308 g is the "theoretical yield" of CO_2:

The **theoretical yield** of a product is the maximum mass that can be obtained from a given mass of a specified reactant.

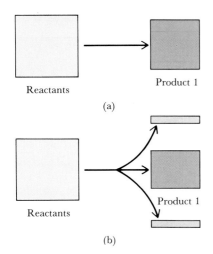

(a)

(b)

FIGURE 4.8 (a) A reaction's yield of products might be 100% if only one reaction takes place. However, if a reactant can take part in several reactions simultaneously (b), the yield will be less than 100% because other products will be formed.

When octane burns in a restricted supply of oxygen, carbon monoxide is formed as well as carbon dioxide:

$$2C_8H_{18}(l) + 17O_2(g) \longrightarrow 16CO(g) + 18H_2O(g)$$

It is not unusual for reactants to take part in several different reactions at the same time; then a single equation is an incomplete description of the changes taking place. When other reactions occur along with the reaction of interest, or when for some reason that reaction does not run its full course, the actual mass of a particular product formed (CO_2, for instance) is less than the theoretical yield (Fig. 4.8). We then speak of the "percentage yield" of the product:

The **percentage yield** of a product is the percentage of its theoretical yield achieved in practice:

$$\text{Percentage yield} = \frac{\text{mass produced}}{\text{theoretical yield}} \times 100\%$$

Suppose we find that in an actual combustion of 100 g of octane, only 92 g of carbon dioxide is produced. Then

$$\text{Percentage yield of } CO_2 = \frac{92 \text{ g } CO_2}{308 \text{ g } CO_2} \times 100 = 30\%$$

If the percentage yield is very close to 100%, we say that the reaction is "complete" or that it is "quantitative."

▼ **EXAMPLE 4.6 Calculating the percentage yield**

When 100 g of water was poured on 100 g of calcium carbide, 28 g of acetylene was produced. What was the percentage yield of acetylene in the reaction?

STRATEGY We need to know both the actual yield and the theoretical yield of C_2H_2. The actual yield is given as data. The theoretical yield is obtained from the stoichiometry of the reaction. This was calculated in Example 4.5, where we found that calcium carbide is the limiting reagent in this situation and that 100 g of CaC_2 produces a theoretical yield of 1.56 mol of C_2H_2. We must calculate the actual yield as a percentage of this figure (Fig. 4.9).

SOLUTION Since 1 mol C_2H_2 = 26.04 g C_2H_2, the theoretical yield of acetylene from 100 g of the carbide is

$$\text{Mass of } C_2H_2 = 1.56 \text{ mol } C_2H_2 \times \frac{26.04 \text{ g } C_2H_2}{1 \text{ mol } C_2H_2}$$

$$= 40.6 \text{ g } C_2H_2$$

Since the actual yield is 28 g C_2H_2,

$$\text{Percentage yield of } C_2H_2 = \frac{28 \text{ g } C_2H_2}{40.6 \text{ g } C_2H_2} \times 100\% = 69\%$$

EXERCISE When 10.0 g of lead was heated with 24.0 g of potassium nitrate, 3.8 g of potassium nitrite was formed. Calculate the percentage yield of potassium nitrite.

[*Answer*: 92%]

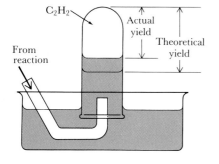

FIGURE 4.9 If calcium carbide reacted to give the stoichiometric amount of acetylene, we would expect the larger volume. However, in practice, a smaller volume is obtained and the yield is less than 100%.

▲

If the percentage yield of product in each step of a long series of reactions is low, very little of the final product will be formed from even a large amount of starting material. The fact that a low percentage yield of ammonia is obtained in the Haber process is of the greatest economic importance, because ammonia is the starting point for many industrial processes and is vital to agriculture. One of the achievements of chemistry—one that we shall examine in later chapters—has been to find ways of improving percentage yields so that costs can be minimized and useful substances made more generally available.

4.3 COMPOSITION FROM MEASUREMENTS OF MASS

We have already stressed (in Section 2.6) that a compound is a *specific* combination of elements with a fixed composition. This is summarized in the following statement:

Law of constant composition: A compound has the same composition whatever its source.

It follows that the composition of a compound is a good guide to its chemical identity. Several techniques have been devised for determining mass percentage composition; the one described here makes use of the calculations introduced in this chapter. It is one example of *gravimetric analysis*, the analysis of composition by measurement of mass.

Combustion analysis. When a sample of an organic compound burns in an unrestricted supply of oxygen, any carbon it contains is converted into carbon dioxide. Since oxygen is in excess, the carbon in the compound limits the amount of carbon dioxide formed. Moreover, since each atom of carbon ends up in one molecule of carbon dioxide, the combustion has the form

$$\ldots C \ldots + O_2(g) \longrightarrow CO_2(g)$$

for every C atom in the molecule. The relation between the carbon in the compound and in the product CO_2 is then

$$1 \text{ mol C} = 1 \text{ mol CO}_2$$

Hence, by measuring the number of moles of CO_2 produced, we can find the number of moles of C atoms that were in the sample. The technique of discovering the composition of a sample by examining the products of its combustion is called *combustion analysis*.

Similarly, when a sample of a compound burns in a plentiful supply of oxygen, all the hydrogen atoms it contains are converted into water:

$$4(\ldots H \ldots) + O_2(g) \longrightarrow 2H_2O(l)$$

(See Fig. 4.10.) Therefore, if we measure the number of moles of H_2O molecules produced when the compound burns, we can find the number of moles of H atoms that were in the sample, since

$$4 \text{ mol H} = 2 \text{ mol H}_2O$$

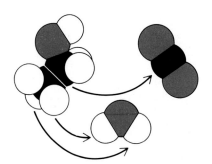

FIGURE 4.10 During combustion in a plentiful supply of oxygen, each C atom produces a CO_2 molecule, and each two H atoms produce one H_2O molecule.

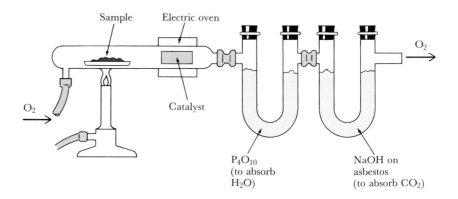

FIGURE 4.11 The apparatus used for a combustion analysis. The masses of CO_2 and H_2O produced are obtained from the masses of the collecting tubes before and after the experiment. (The catalyst ensures that any CO produced is oxidized to CO_2.)

So long as the heat is not too intense, any nitrogen in the sample is released as nitrogen gas. Since

$$2(\ldots N \ldots) \longrightarrow N_2(g)$$

we can find the moles of N atoms that were present in the sample from

$$2 \text{ mol N} = 1 \text{ mol } N_2$$

In practice, the combustion of nitrogen compounds leads to a mixture of nitrogen oxides as well as nitrogen gas. These are reduced back to nitrogen by passing the gas over hot copper.

Many organic compounds contain only C, H, and O, so their only products of combustion are CO_2 and H_2O. The moles of H_2O and CO_2 are measured by burning the sample in a stream of oxygen in the apparatus shown in Fig. 4.11. The gas stream passes through a tube containing phosphorus(V) oxide (P_4O_{10}, widely called "phosphorus pentoxide" because its formula was once written P_2O_5); the P_4O_{10} acts as a "drying agent," a substance that absorbs water. We can find how much water is produced by noting the difference between the masses of the drying agent before and after the experiment; this mass of water can be converted to moles using the molar mass of H_2O. After it passes through the drying agent, the stream is passed over asbestos coated with sodium hydroxide, which absorbs the carbon dioxide. The increase in the mass of the coated asbestos gives the mass of carbon dioxide produced, and hence the number of moles of CO_2 produced. Since only C, H, and O were present in the original compound, we can calculate the mass of oxygen originally present (and hence the number of moles of O atoms) by subtracting the masses of H and C atoms that were present from the original mass of the sample.

▼ EXAMPLE 4.7 Interpreting a combustion analysis

A combustion analysis was carried out on a 1.621-g sample of ethanol, which is known to contain only C, H, and O. The masses of water and carbon dioxide produced were 1.902 g and 3.095 g, respectively. What is the mass percentage composition of ethanol?

STRATEGY We need to find the masses of C, H, and O in the sample. We can find the number of moles of C and H atoms from the number of moles of CO_2 and H_2O produced in the combustion. These values can be converted to masses using the molar masses. The mass of oxygen in the sample can be obtained by subtraction.

SOLUTION The numbers of moles of CO_2 and H_2O (of molar masses 44.01 g/mol and 18.02 g/mol, respectively) produced were

$$\text{Moles of } CO_2 = 3.095 \text{ g } CO_2 \times \frac{1 \text{ mol } CO_2}{44.01 \text{ g } CO_2}$$

$$= 0.07032 \text{ mol } CO_2$$

$$\text{Moles of } H_2O = 1.902 \text{ g } H_2O \times \frac{1 \text{ mol } H_2O}{18.02 \text{ g } H_2O}$$

$$= 0.1056 \text{ mol } H_2O$$

The numbers of moles of C and H atoms in the original sample were therefore

$$\text{Moles of C} = 0.07032 \text{ mol } CO_2 \times \frac{1 \text{ mol C}}{1 \text{ mol } CO_2}$$

$$= 0.07032 \text{ mol C}$$

$$\text{Moles of H} = 0.1056 \text{ mol } H_2O \times \frac{4 \text{ mol H}}{2 \text{ mol } H_2O}$$

$$= 0.2112 \text{ mol H}$$

Hence the total masses of C and H atoms (of molar masses 12.01 g/mol and 1.008 g/mol) in the sample were

$$\text{Mass of C} = 0.07032 \text{ mol C} \times \frac{12.01 \text{ g C}}{1 \text{ mol C}}$$

$$= 0.8445 \text{ g C}$$

$$\text{Mass of H} = 0.2112 \text{ mol H} \times \frac{1.008 \text{ g H}}{1 \text{ mol H}}$$

$$= 0.2129 \text{ g H}$$

These two elements account for 1.057 g of the sample. The mass of O atoms present was therefore 1.621 g − 1.057 g = 0.564 g. The mass percentage composition is

$$\text{Mass \% C} = \frac{0.8445 \text{ g C}}{1.621 \text{ g}} \times 100\% = 52.10\% \text{ C}$$

$$\text{Mass \% H} = \frac{0.2129 \text{ g H}}{1.621 \text{ g}} \times 100\% = 13.13\% \text{ H}$$

$$\text{Mass \% O} = \frac{0.564 \text{ g O}}{1.621 \text{ g}} \times 100\% = 34.8\% \text{ O}$$

EXERCISE When 0.528 g of sucrose, a compound of carbon, hydrogen, and oxygen, was burned, 0.306 g of water and 0.815 g of carbon dioxide were formed. Calculate the mass percentage composition of sucrose.

[*Answer*: 42.1% C, 6.48% H, 51.4% O]

▲

Empirical formulas. It is often more helpful to report the composition of a compound in terms of the numbers of atoms of each element present, rather than in terms of mass percentages. Suppose we know

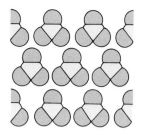

(a) XY_3 molecules

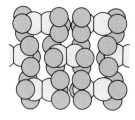

(b) X_2Y_6 molecules

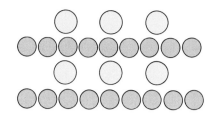

(c) XY_3 ionic compound

FIGURE 4.12 The molecular compounds (a) and (b) and the ionic compound (c) all have the empirical formula XY_3, since in each case there are three Y atoms for each X atom.

that a sample of a particular substance consists of 23% phosphorus and 77% chlorine by mass. Since a 100-g sample of the compound contains 23 g of phosphorus (of molar mass 30.97 g/mol) and 77 g of chlorine (of molar mass 35.45 g/mol), the number of moles of the atoms of each element present are

$$\text{Moles of P} = 23 \text{ g P} \times \frac{1 \text{ mol P}}{30.97 \text{ g P}}$$

$$= 0.74 \text{ mol P}$$

$$\text{Moles of Cl} = 77 \text{ g} \times \frac{1 \text{ mol Cl}}{35.45 \text{ g Cl}}$$

$$= 2.2 \text{ mol Cl}$$

Therefore, the atoms of the two elements are present in the ratio

$$\frac{\text{Number of Cl atoms}}{\text{Number of P atoms}} = \frac{2.2 \text{ mol}}{0.74 \text{ mol}} = 3.0$$

and for each P atom in the compound there are three Cl atoms.

The relative numbers of atoms in a sample are reported as the "empirical formula" of the compound:

An **empirical formula** is the simplest chemical formula that shows the relative numbers of atoms of each element in a compound.

In our example compound there are three Cl atoms for each P atom, so its empirical formula is PCl_3. We cannot yet conclude that PCl_3 is the *molecular* formula, for the same ratio of atoms would be obtained if the compound consisted of P_2Cl_6 molecules, P_3Cl_9 molecules, and so on, or if it were ionic (Fig. 4.12).

▼ **EXAMPLE 4.8** Calculating an empirical formula from a mass percentage composition

The mass percentage composition of vitamin C is 40.9% C, 4.57% H, and 54.5% O. What is its empirical formula?

STRATEGY We can calculate the number of moles of the atoms of each element in a 100-g sample using the mass percentage composition and the molar masses of the elements. The empirical formula is written by expressing the ratios of these numbers in a chemical formula using the smallest possible whole numbers.

SOLUTION The numbers of moles of atoms present in a 100-g sample are

$$\text{Moles of C} = 40.9 \text{ g C} \times \frac{1 \text{ mol C}}{12.01 \text{ g C}}$$

$$= 3.41 \text{ mol C}$$

$$\text{Moles of H} = 4.57 \text{ g H} \times \frac{1 \text{ mol H}}{1.008 \text{ g H}}$$

$$= 4.53 \text{ mol H}$$

$$\text{Moles of O} = 54.5 \text{ g O} \times \frac{1 \text{ mol O}}{16.00 \text{ g O}}$$

$$= 3.41 \text{ mol O}$$

Dividing through by the smallest number of moles (3.41) gives

$$\text{C: } 1.00 \qquad \text{H: } 1.33 \qquad \text{O: } 1.00$$

and multiplication by 3 (to clear the fraction 1.33, or $\frac{4}{3}$) gives

$$\text{C: } 3.00 \qquad \text{H: } 4.00 \qquad \text{O: } 3.00$$

showing that there are 3 C atoms and 4 H atoms for every 3 O atoms in the compound. Hence, the empirical formula of vitamin C is $C_3H_4O_3$.

EXERCISE The mass percentage composition of thionyl difluoride is 37.25% S, 18.59% O, and 44.16% F. Calculate its empirical formula.

[*Answer*: OSF_2]

▲

An empirical formula unit is the group of atoms, in kind and number, that appears in the empirical formula. For example, the empirical formula unit of the substance with empirical formula PCl_3 consists of one P atom and three Cl atoms.

To decide on the *molecular* formula of a compound, we measure its molecular weight and see how many empirical formula units are needed to account for it. The most precise method for measuring the molecular weights of simple molecules (those containing fewer than about 30 atoms) is mass spectrometry, the technique introduced in Section 2.4 for measuring the masses of atoms. Polyatomic ions (such as $C_2H_6O^+$ if the sample is ethanol) are formed by bombarding the sample vapor with electrons in the ionization chamber; the ions so formed are accelerated by a potential difference. Their masses are then calculated from the strength of the deflecting electric field needed for the detector to give a signal (Fig. 4.13). Other methods that use simpler apparatus are also available. They are sometimes more suitable than mass spectrometry, especially if the compound is not volatile. Some are described later in this chapter, and others in Chapters 5 and 11.

The molecular weight of ethanol has been found to be 46.07 amu. The empirical formula weight, the mass of one empirical formula unit, is

$$\text{FW}(C_2H_6O) = 2 \times 12.01 \text{ amu} + 6 \times 1.008 \text{ amu} + 16.00 \text{ amu}$$

$$= 46.07 \text{ amu}$$

Since the formula weight is the same as the molecular weight, we conclude that the molecular formula of ethanol is also C_2H_6O. On the

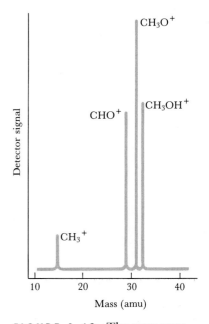

FIGURE 4.13 The mass spectrum of methanol (which is simpler than that of ethanol) and the identities of the polyatomic ions responsible for the signal. The CH_3OH molecule breaks up into the various ions when it is ionized.

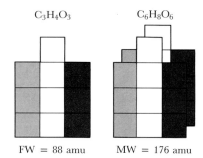

C3H4O3   C6H8O6

FW = 88 amu   MW = 176 amu

FIGURE 4.14 We can deduce the molecular formula of a compound (in this case, vitamin C) by deciding how many empirical formula units are needed to account for the molecular weight of the compound. Here two $C_3H_4O_3$ units are needed.

other hand, although the empirical formula of vitamin C is $C_3H_4O_3$, corresponding to a formula weight of 88.06 amu, its molecular weight is found to be 176.12 amu. This shows that each molecule consists of two $C_3H_4O_3$ formula units (Fig. 4.14), and hence that the molecular formula of vitamin C is $C_6H_8O_6$.

In general, then,

$$\text{Number of formula units per molecule} = \frac{\text{molecular weight}}{\text{empirical formula weight}}$$

The formula weight of P_2O_5 is 141.90 amu, but its molecular weight is known to be 283.80 amu; hence each molecule consists of

$$\frac{283.80 \text{ amu}}{141.90 \text{ amu}} = 2 \text{ formula units}$$

▼ **EXAMPLE 4.9 Determining a molecular formula from an empirical formula**

The molecular weight of ethyl butanoate, a compound that contributes to the flavor of pineapple, is 116 amu. Its empirical formula, determined from its mass percentage composition, is C_3H_6O. What is its molecular formula?

STRATEGY We have to decide how many empirical formula units are needed to account for the measured molecular weight. Therefore, we must calculate the empirical formula weight and compare it with the molecular weight.

SOLUTION The empirical formula weight is

$$FW(C_3H_6O) = 3 \times AW(C) + 6 \times AW(H) + AW(O)$$
$$= 3 \times 12.01 \text{ amu} + 6 \times 1.008 \text{ amu} + 16.00 \text{ amu}$$
$$= 58.08 \text{ amu}$$

The measured molecular weight is 116 amu, so each molecule must consist of two formula units. Therefore, the molecular formula is $2 \times C_3H_6O$, or $C_6H_{12}O_2$.

EXERCISE The molecular weight of glucose is 180 amu, and its empirical formula is CH_2O. Deduce its molecular formula.

[*Answer*: $C_6H_{12}O_6$]

▲

THE STOICHIOMETRY OF REACTIONS IN SOLUTION

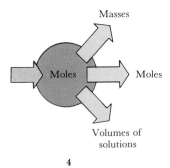

Masses

Moles   Moles

Volumes of solutions

4

Chemists very often study reactions in solution. To do this they need to know how to calculate the *volumes* of solutions that react together. For example, they may want to know the volume of an acid that would be required to neutralize a given volume of a base. In addition, since a solution is a homogeneous mixture (not a compound) of variable composition, its *concentration*, the amount of solute per liter of solution, must also be taken into account. Like calculations involving masses that react, calculations involving volumes of solutions that react are conveniently done via moles. That is, we are now going to deal with the new arrow in diagram **4**.

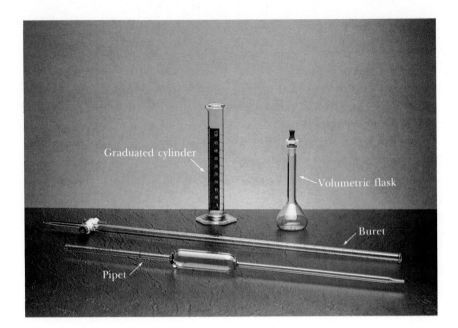

FIGURE 4.15 The glassware used for measuring the volumes of solutions in a chemical laboratory.

The apparatus used for measuring volumes of liquids was mentioned in Section 1.3 and is shown in Fig. 4.15. A graduated cylinder is used to measure volumes of liquids that do not need to be known very precisely. A buret is used to deliver an arbitrary but precise volume of solution into a flask. A pipet is used to deliver a definite, fixed volume. A volumetric flask is used to prepare a solution of known volume.

4.4 MOLAR CONCENTRATION

The link between the volume of a solution and the number of moles of solute it contains is its "molar concentration":

> The **molar concentration** of a solution is the number of moles of solute it contains per liter of solution.

Molar concentration is widely called "molarity" and denoted [X], where X is the chemical formula of the solute. It follows from the definition that

$$[X] = \frac{\text{moles of X}}{\text{liters of solution}}$$

Other measures of concentration are introduced in Chapter 11, but this is all we need for now.

Molar concentration is defined in terms of the volume of *solution*, not the volume of solvent used to prepare the solution. This makes sense, because we need to know what volume of the solution—solute plus solvent—to measure out so as to supply a given amount of solute. The usual way to prepare an aqueous solution of known concentration is to weigh out a known mass of solute into a volumetric flask, dissolve it in a little water, and then fill the flask up to the mark with water (Fig. 4.16).

(a)   (b)   (c)

FIGURE 4.16 The steps involved in making up a solution of known concentration (here, of potassium permanganate). (a) A known mass of solute is washed into a graduated flask. (b) Some water is added to dissolve it. (c) Finally, water is added up to the mark.

The unit of molar concentration is mol/L ("moles per liter"). However, as this unit is used so frequently, chemists denote it with the symbol M, which is read "molar." A solution prepared by dissolving 0.010 mol of sucrose molecules ($C_{12}H_{22}O_{11}$) in enough water to give 100 mL of solution would have a molar concentration of

$$[C_{12}H_{22}O_{11}] = \frac{0.010 \text{ mol } C_{12}H_{22}O_{11}}{0.100 \text{ L solution}}$$

$$= 0.10 \ M \ C_{12}H_{22}O_{11}(aq)$$

We shall sometimes report concentrations using the unit mM ("millimolar"), which is equal to $10^{-3} \ M$. A 0.0050 M HCl(aq) solution, for example, can be reported as 5.0 mM HCl(aq).

The relation between volume of solution and amount of solute. The concentration of a solution is actually a conversion factor that relates the volume to the moles of solute it contains. For example, if we want to know the number of moles of glucose molecules in 15 mL (15×10^{-3} L) of the 0.010 M $C_{12}H_{22}O_{11}(aq)$ solution, we write

Moles of $C_{12}H_{22}O_{11}$ = volume of solution
$\times$ moles of $C_{12}H_{22}O_{11}$ per liter of solution

Since "moles of $C_{12}H_{22}O_{11}$ per liter of solution" is the molar concentration of glucose, we have

Moles of $C_{12}H_{22}O_{11}$ = 15×10^{-3} L solution $\times \dfrac{0.010 \text{ mol } C_{12}H_{22}O_{11}}{1 \text{ L solution}}$

$$= 1.5 \times 10^{-4} \text{ mol } C_{12}H_{22}O_{11}$$

In general,

$$\text{Moles of X} = \text{volume of solution} \times [\text{X}] \qquad (1)$$

▼ **EXAMPLE 4.10** **Preparing a solution of specified concentration**

Calculate the mass of potassium permanganate needed to prepare 250 mL of a 0.0380 M $KMnO_4(aq)$ solution.

STRATEGY To calculate the required mass of $KMnO_4$, we need to know the number of moles of $KMnO_4$ formula units in 250 mL of the solution, and then convert this number of moles to mass of $KMnO_4$ using the molar mass (158.04 g/mol).

SOLUTION The number of moles of $KMnO_4$ in 250 mL of the solution is

Moles of $KMnO_4$ = volume of solution × moles of $KMnO_4$ per liter

$$= 250 \times 10^{-3} \text{ L solution} \times \frac{0.0380 \text{ mol } KMnO_4}{1 \text{ L solution}}$$

$$= 9.50 \times 10^{-3} \text{ mol } KMnO_4$$

The mass of $KMnO_4$ present is

Mass of $KMnO_4$ = moles of $KMnO_4$ × mass per mole of $KMnO_4$

$$= 9.50 \times 10^{-3} \text{ mol } KMnO_4 \times \frac{158.04 \text{ g } KMnO_4}{1 \text{ mol } KMnO_4}$$

$$= 1.50 \text{ g } KMnO_4$$

That is, 1.50 g of potassium permanganate should be dissolved in enough water to prepare 250 mL of solution.

EXERCISE Calculate the mass of glucose needed to prepare 150 mL of 0.44 M $C_6H_{12}O_6(aq)$ solution.

[*Answer*: 12 g]

▲

Transferring a specified amount of solute. Once we know the concentration of a solution, we can determine what volume of that solution will supply a specific number of moles of solute. Suppose, for example, that we want 0.760 mmol of $KMnO_4$ for a certain oxidation reaction, and we have available the 0.0380 M $KMnO_4(aq)$ solution prepared in Example 4.10. We set up the conversion

Volume of solution = moles of $KMnO_4$
       × liters of solution per mole of $KMnO_4$

The conversion factor is given by the molar concentration expressed in the form

$$0.0380 \text{ mol } KMnO_4 = 1 \text{ L solution}$$

and rearranged to

$$1 = \frac{1 \text{ L solution}}{0.0380 \text{ mol } KMnO_4}$$

Then the required volume of solution is

$$\text{Volume of solution} = 0.760 \times 10^{-3} \text{ mol } KMnO_4$$
$$\times \frac{1 \text{ L solution}}{0.0380 \text{ mol } KMnO_4}$$
$$= 20.0 \times 10^{-3} \text{ L solution}$$

FIGURE 4.17 A micropipet is used to measure out very small volumes of solutions.

That is, we should run 20.0 mL of the permanganate solution (from a buret) into the flask to give us 0.760 mmol of $KMnO_4$.

This calculation also brings out one of the advantages of using solutions in chemistry. It shows that we can transfer a small amount of a substance (in this case, less than 1 mmol of $KMnO_4$) quite readily and precisely by first dissolving it in a manageable amount of solvent, and then measuring out the volume that corresponds to the desired amount of solute. It is also very much easier to transfer some substances when they are dissolved than when they are not. For instance, it is much easier to transfer HCl and NH_3 as their solutions than as the gases themselves.

Dilution. Another advantage of using solutions is that they may be used to transfer not just small but *very* small amounts of a substance. This is important when only very little of the substance is available, perhaps because of its rarity (as for moon dust) or its cost (as for a new pharmaceutical compound). It can also be important when a sample being examined contains a minute amount of the substance of interest, as in the analysis of a contaminated food for traces of poison.

Suppose we were investigating a rare compound and needed to add 0.010 mmol (1.0×10^{-5} mol) of $KMnO_4$ to a sample to oxidize it. One method would be to use 0.26 mL of the 0.0380 M $KMnO_4(aq)$ solution. A well-equipped modern research laboratory has an instrument called a "micropipet" (Fig. 4.17) that can measure out very tiny volumes of liquid accurately, perhaps as little as 10^{-3} mL (1 microliter), about the volume of a pinhead. However, for many applications there is a much simpler approach that does not require such expensive equipment. This is simply to *dilute*, or reduce the concentration of, the solution by adding solvent. Then a larger volume of solution contains the same number of moles of solute and the transfer can be carried out accurately using an ordinary pipet or buret. If our original 0.0380 M solution were diluted a hundredfold, then 0.010 mmol of $KMnO_4$ formula units would be contained in 26 mL of solution, which could be measured out easily and accurately in the normal way.

To use dilution, we need to know how to reduce the concentration of a solution from an initial value $[X]_i$ to some final target value $[X]_f$. Suppose we transfer a volume V_i of the original solution to a volumetric flask. The number of moles of solute we have transferred is

$$\text{Moles of X} = V_i \times [X]_i$$

Now let us add enough solvent to increase the volume to V_f. The new molar concentration is this same number of moles divided by the new volume:

$$[X]_f = \frac{V_i \times [X]_i}{V_f}$$

This can be rearranged into the more easily remembered form

$$V_i \times [X]_i = V_f \times [X]_f \qquad (2)$$

That is,

Initial volume × initial concentration

= final volume × final concentration

▼ EXAMPLE 4.11 Calculating the volume of solution to dilute

What volume of 0.0380 M KMnO$_4$ solution should be used to prepare 250 mL of a 2.50 mM KMnO$_4$(aq) solution?

STRATEGY Decide which volume and concentration data are available, identify the one unknown, and rearrange Eq. 2 to give that unknown in terms of the available data.

SOLUTION The data are

> Initial volume V_i = ?
> Final volume V_f = 250 mL
> Initial concentration [KMnO$_4$]$_i$ = 0.0380 M KMnO$_4$
> Final concentration [KMnO$_4$]$_f$ = 2.50 × 10^{-3} M KMnO$_4$

We rearrange Eq. 2 to

$$V_i = \frac{V_f \times [\text{KMnO}_4]_f}{[\text{KMnO}_4]_i}$$

and substitute the available data to get

$$V_i = \frac{250 \text{ mL} \times 2.50 \times 10^{-3} \ M \ \text{KMnO}_4(aq)}{0.0380 \ M \ \text{KMnO}_4(aq)}$$

$$= 16.4 \text{ mL}$$

That is, 16.4 mL of the original solution should be measured into a 250-mL volumetric flask using a buret, and water added up to the mark (Fig. 4.18).

EXERCISE Calculate the volume of 0.0155 M HCl(aq) that should be used to prepare 100 mL of a 0.523 mM HCl(aq) solution.

[*Answer*: 3.37 mL]

▲

FIGURE 4.18 A convenient procedure for transferring very small amounts of substance is to use dilution. In this illustration 16.4 mL of permanganate solution, prepared as shown in Fig. 4.16, has been transferred into a 250-mL flask from a buret (left). Water is then added up to the mark (right). A 25-mL sample of this solution contains only 0.0625 mmol KMnO$_4$.

The use of techniques like dilution gives chemists very precise control over the quantities of substances, even in very small amounts. Pipeting 25.0 mL of the 2.50 mM KMnO$_4$(aq) solution prepared in the example corresponds to transferring as little as 0.0625 mmol of KMnO$_4$, or only 9.88 mg of the permanganate. Furthermore, solutions can be stored in concentrated form to save space (and hence money), and then diluted to whatever concentration is appropriate for their intended use.

4.5 VOLUME OF SOLUTION REQUIRED FOR REACTION

We are now at the stage where we can calculate the volume of one solution that is needed to react with a given volume of another solution. This kind of calculation is important in acid-base neutralizations, where we may be interested in knowing the volume of acid to add to a solution of a base so as to produce a salt. It is also needed in redox chemistry, where we may be interested in knowing how much oxidizing agent we should add to a solution to make a particular amount of product. There are other applications as well, including measuring the concentration of an unknown solution, discovering the identity of the solute, and measuring molecular weights.

The general strategy for this type of calculation is to compute the number of moles of one reactant from the volume and concentration of its solution. That number of moles is then converted to moles of the other reactant (the stoichiometric coefficients are used to construct the conversion factor). The final step is to convert the number of moles of the second reactant to the required volume of its solution, using as a conversion factor the molar concentration of that solution. The chain of conversions is therefore

$$\text{Volume of solution A} \xrightarrow{\text{concentration of A}} \text{moles of substance A}$$
$$\xrightarrow{\text{stoichiometry}} \text{moles of substance B}$$
$$\xrightarrow{\text{concentration of B}} \text{volume of solution B}$$

▼ **EXAMPLE 4.12 Calculating the volume of solution needed for a reaction**

Calculate the volume of 0.250 M HCl(aq) needed to neutralize 25.00 mL of 0.100 M NaOH(aq).

STRATEGY The pivot of the calculation is the number of moles of HCl that react with a given number of moles of NaOH, so we set up the calculation using moles. The number of moles of NaOH available is found from the molar concentration of the NaOH solution and its volume. The number of moles of HCl to which this corresponds is given by the chemical relation between HCl and NaOH. The volume of acid containing that number of moles of HCl is then found from the molar concentration of the acid.

SOLUTION The number of moles of NaOH in the solution is

Moles of NaOH = volume of solution × moles of NaOH per liter of solution

In a 0.100 M NaOH solution, we have

$$0.100 \text{ mol NaOH} = 1 \text{ L solution}$$

$$\text{Moles of NaOH} = 25.00 \text{ mL solution} \times \frac{0.100 \text{ mol NaOH}}{1 \text{ L solution}}$$

$$= 2.50 \text{ mmol NaOH}$$

Now the chemical equation

$$\text{HCl}(aq) + \text{NaOH}(aq) \longrightarrow \text{NaCl}(aq) + \text{H}_2\text{O}(l)$$

shows that 1 mol HCl = 1 mol NaOH, so we require

$$\text{Moles of HCl} = 2.50 \text{ mmol NaOH} \times \frac{1 \text{ mol HCl}}{1 \text{ mol NaOH}}$$

$$= 2.50 \text{ mmol HCl}$$

The volume of solution containing this number of moles of HCl is obtained from the HCl concentration,

$$0.250 \text{ mol HCl} = 1 \text{ L solution}$$

which gives

$$\text{Volume of solution} = 2.50 \text{ mmol HCl} \times \frac{1 \text{ L solution}}{0.250 \text{ mol HCl}}$$

$$= 10.0 \text{ mL solution}$$

EXERCISE Calculate the volume of 0.220 M NaOH(aq) that should be added to 25.0 mL of 0.743 M H$_2$SO$_4$(aq) to neutralize it completely.

[*Answer*: 169 mL]

▲

TITRATIONS

The techniques described in this chapter are widely used to determine the compositions of mixtures, the concentrations of solutions, and the chemical formulas of compounds. This aspect of chemistry is called *volumetric analysis*, or chemical analysis by measurement of volume. The process of analyzing composition by measuring the volume of one solution needed to react with another solution is called *titration* (the name comes from the French word for "assay," or test of quality). When one solution is an acid and the other a base, we speak of an "acid-base titration." When one solution contains a reducing agent and the other an oxidizing agent, we speak of a "redox titration."

4.6 DIRECT TITRATIONS

In a titration, we typically are trying to find the concentration of one of the two solutions. The solution of unknown concentration is called the *analyte*, and we assume we have a known volume of that solution in a flask. The solution of known concentration in the buret is the *titrant* (Fig. 4.19). In a direct titration we slowly add titrant to the analyte and measure the volume of titrant needed to react with all the analyte. The point at which exactly the right volume of titrant has been added so that the reaction is just complete is called the *equivalence point* of the titration. To obtain accurate results, it is important to use a reaction that is quantitative (that is, one that has 100% yield).

Acid-base titrations. In an acid-base titration the equivalence point is reached when exactly enough acid has been added to neutralize the base. The equivalence point can be detected with an indicator such as litmus or phenolphthalein, which changes color as the solution changes from basic to acidic (Fig. 4.20). The *end point* of the titration is the point at which enough titrant has been added so that the indicator color is halfway between its acid and base colors. If the indicator has been well chosen, the end point is almost the same as the equivalence point, so detecting the end point is the same as detecting the equivalence point. (The correct choice of acid-base indicator is an important aspect of titration, one we deal with in detail in Chapter 15.)

At the equivalence point of the titration, we know the volumes of the base and acid that have reacted and the concentration of one of them (the titrant). Our aim is to find the concentration of the other. To do so, we follow the usual route for stoichiometric calculations, using the relation between moles of reactants as the pivot of the calculation.

▼ **EXAMPLE 4.13** Measuring a concentration by titration

A 25.0-mL sample of vinegar (a dilute solution of acetic acid, CH_3COOH) was titrated with 0.500 *M* NaOH(*aq*). The equivalence point was reached when 38.1 mL of the base had been added. Find the concentration of acetic acid in the vinegar.

STRATEGY We know the volume and concentration of the titrant, and hence we can calculate the number of moles of NaOH used in the titration. The neutralization reaction lets us convert this number of moles to the number of moles of CH_3COOH in the 25.0 mL of analyte. Knowing the number

FIGURE 4.19 The arrangement typically used for a titration.

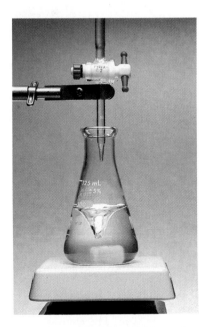

FIGURE 4.20 An acid-base titration. In this titration the indicator is phenolphthalein.

of moles contained in that volume, we can calculate the initial molar concentration of the acetic acid.

SOLUTION The number of moles of NaOH used in the titration is

$$\text{Moles of NaOH} = 38.1 \times 10^{-3} \text{ L} \times \frac{0.500 \text{ mol NaOH}}{1 \text{ L}}$$

$$= 1.905 \times 10^{-2} \text{ mol NaOH}$$

The neutralization reaction is

$$CH_3COOH(aq) + NaOH(aq) \longrightarrow NaCH_3CO_2(aq) + H_2O(l)$$

$$1 \text{ mol } CH_3COOH = 1 \text{ mol NaOH}$$

Then the number of moles of acetic acid used is

$$\text{Moles of } CH_3COOH = 1.905 \times 10^{-2} \text{ mol NaOH} \times \frac{1 \text{ mol } CH_3COOH}{1 \text{ mol NaOH}}$$

$$= 1.905 \times 10^{-2} \text{ mol } CH_3COOH$$

(We do the rounding at the end of the calculation.) Since this number of moles was originally in 25.0 mL of solution, the concentration of the acid was

$$[CH_3COOH] = \frac{1.905 \times 10^{-2} \text{ mol } CH_3COOH}{25.0 \times 10^{-3} \text{ L}}$$

$$= 0.762 \text{ mol/L } CH_3COOH$$

That is, the vinegar is $0.762 \ M \ CH_3COOH(aq)$.

EXERCISE A sample of hydrochloric acid was prepared that contained 0.72 g of hydrogen chloride in 250 mL of solution. It was used to titrate 25.0 mL of a solution of calcium hydroxide, and the equivalence point was reached when 15.1 mL of acid had been added. What was the concentration of the calcium hydroxide solution?

[*Answer*: 24 mM Ca(OH)$_2$(aq)]

Redox titrations. In a redox titration, the equivalence point occurs when enough oxidizing agent has been added to oxidize all the reducing agent. It can be detected by watching for the color change of a *redox indicator*, a substance that changes color when it is converted from its oxidized to its reduced state. An example is ferroin, a complex compound of iron, which changes from an almost colorless pale blue in its oxidized form to red in its reduced form. The equivalence point can sometimes be detected without using an indicator, if the reduced and oxidized forms of one of the reagents have different colors (Fig. 4.21). This is the case when iron(II) is oxidized to iron(III) in solution by permanganate ions. The permanganate ion is purple, but in acid solution it is reduced to the pale pink Mn^{2+} ion in the reaction

$$MnO_4^-(aq) + 8H^+(aq) + 5e^- \longrightarrow Mn^{2+}(aq) + 4H_2O(l)$$

and the iron is oxidized in the reaction

$$Fe^{2+}(aq) \longrightarrow Fe^{3+}(aq) + e^-$$

The overall reaction is

$$5Fe^{2+}(aq) + MnO_4^-(aq) + 8H^+(aq) \longrightarrow$$
$$5Fe^{3+}(aq) + Mn^{2+}(aq) + 4H_2O(l)$$

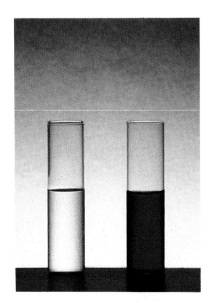

FIGURE 4.21 A redox titration of oxalic acid with potassium permanganate. Before the equivalence point (left), the permanganate is converted to the very pale pink Mn^{2+} ion. The equivalence point is recognized by noting when the added permanganate survives as permanganate (right).

TABLE 4.1 **Some permanganate titrations**

Substance analyzed	Oxidation half-reaction	Comment
$Fe^{2+}(aq)$	$Fe^{2+}(aq) \longrightarrow Fe^{3+}(aq) + e^-$	
$Br^-(aq)$	$2Br^-(aq) \longrightarrow Br_2(g) + 2e^-$	Titrate in boiling $2\ M\ H_2SO_4(aq)$.
$H_2O_2(aq)$	$H_2O_2(aq) \longrightarrow O_2(g) + 2H^+(aq) + 2e^-$	Titrate in $1\ M\ H_2SO_4(aq)$.
As(III)	$H_3AsO_3(aq) + H_2O(l) \longrightarrow$ $H_3AsO_4(aq) + 2H^+(aq) + 2e^-$	Titrate in $1\ M\ HCl(aq)$.
Oxalic acid	$C_2H_2O_4(aq) \longrightarrow$ $2CO_2(g) + 2H^+(aq) + 2e^-$	*

*Mg^{2+}, Ca^{2+}, Sr^{2+}, Ba^{2+}, Zn^{2+}, Co^{2+}, Pb^{2+}, Ag^+, and other elements with insoluble oxalates may be analyzed by precipitating the metal oxalate, dissolving it in acid, and titrating the oxalate so produced. The reduction half-reaction of MnO_4^- in acid solution is

$$MnO_4^-(aq) + 8H^+(aq) + 5e^- \longrightarrow Mn^{2+}(aq) + 4H_2O(l)$$

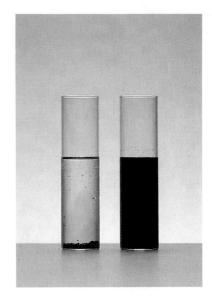

FIGURE 4.22 Iodine is only slightly soluble in water (left), but much more soluble in aqueous potassium iodide (right), owing to the formation of the I_3^- ion.

Since the iron ions are nearly colorless, the solution becomes purple immediately after the equivalence point is reached, when the added permanganate ions start to survive. Some uses of permanganate in redox titrations are summarized in Table 4.1.

Another widely used redox technique is *iodimetry*, a redox procedure in which the oxidizing agent is iodine. Some applications of the technique are summarized in Table 4.2; the terms used in the table are explained in more detail below. Iodine itself is only slightly soluble in water, but it dissolves readily in a potassium iodide solution because triiodide ions (I_3^-) are formed:

$$I_2(s) + I^-(aq) \longrightarrow I_3^-(aq)$$

The triiodide ion can be thought of as a store of I_2 molecules, ready to be released when the solution acts as an oxidizing agent (Fig. 4.22). In iodimetry, the indicator is usually starch, which turns blue in the presence of I_2 (Fig. 4.23).

TABLE 4.2 **Some applications of iodimetry**

Substance analyzed	Oxidation half-reaction	Comment*
As(III)	$H_3AsO_3(aq) + H_2O(l) \longrightarrow$ $H_3AsO_4(aq) + 2H^+(aq) + 2e^-$	D
SO_2	$SO_2(aq) + 2H_2O(l) \longrightarrow$ $SO_4^{2-}(aq) + 4H^+(aq) + 2e^-$	B
H_2S	$H_2S(aq) \longrightarrow S(s) + 2H^+(aq) + 2e^-$	B†
Glucose	$\dots CHO(aq) + 3OH^-(aq) \longrightarrow$ $\dots CO_2^-(aq) + 2H_2O(l) + 2e^-$	B

*Iodine is titrated using the reaction $I_3^-(aq) + 2S_2O_3^{2-}(aq) \longrightarrow 3I^-(aq) + S_4O_6^{2-}(aq)$. D: Use direct titration with I_3^-. B: Add excess I_3^- and back-titrate with thiosulfate.
†Use the H_2S determination indirectly for the analysis of Zn^{2+}, Cd^{2+}, Hg^{2+}, Pb^{2+}, and other insoluble sulfides: precipitate the sulfide; filter, wash, and treat the precipitate with acid; and then back-titrate as for $H_2S(aq)$.

FIGURE 4.23 Starch is commonly used as an indicator for the presence of iodine. The iodine molecules lie inside spirals of starch molecules and give it a dark blue color.

The same type of calculation applies to redox titrations as to acid-base titrations. The only difference is that the conversion factor between the two reactants is obtained from the redox equation.

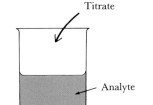
Titrate

Analyte

(a)

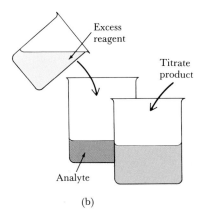
Excess reagent

Titrate product

Analyte

(b)

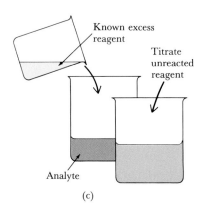
Known excess reagent

Titrate unreacted reagent

Analyte

(c)

FIGURE 4.24 (a) A *direct titration:* the analyte is titrated directly with the titrant. (b) An *indirect titration:* the analyte is converted with 100% yield to a second analyte, which is then titrated. (c) A *back-titration:* the analyte is the limiting reagent and the excess titrant remaining after the analyte-titrant reaction is what is titrated.

▼ **EXAMPLE 4.14** **Determining concentration by redox titration**

A 50.00-mL sample of iron(II) sulfate was titrated with 0.1000 M cerium(IV) sulfate solution, in which the cerium becomes reduced to cerium(III). The equivalence point, detected with ferroin, was reached when 32.00 mL of the cerium solution had been added. What was the concentration of the iron(II) solution?

STRATEGY We can calculate the number of moles of Ce(IV) used in the titration from the volume and concentration of the cerium(IV) sulfate solution. That number of moles can be converted to the number of moles of Fe(II) in the analyte by use of the redox equation. Finally, we can calculate the initial concentration of the analyte by dividing the number of moles of Fe(II) in it by its volume.

SOLUTION The number of moles of Ce(IV) used is

$$\text{Moles of Ce(IV)} = 32.00 \times 10^{-3} \text{ L} \times \frac{0.1000 \text{ mol Ce(IV)}}{1 \text{ L}}$$

$$= 3.200 \times 10^{-3} \text{ mol Ce(IV)}$$

The redox reaction is

$$Fe^{2+}(aq) + Ce^{4+}(aq) \longrightarrow Fe^{3+}(aq) + Ce^{3+}(aq)$$

so that

$$1 \text{ mol Fe}^{2+} = 1 \text{ mol Ce}^{4+}$$

It follows that the analyte contained 3.200×10^{-3} mol of Fe(II). Since its volume was 50.00 mL, its molar concentration was

$$[\text{Fe(II)}] = \frac{3.200 \times 10^{-3} \text{ mol Fe(II)}}{50.00 \times 10^{-3} \text{ L}}$$

$$= 6.400 \times 10^{-2} \text{ mol/L Fe(II)}$$

That is, the iron(II) sulfate solution is 64.00 mM FeSO$_4$(aq).

EXERCISE A 0.1500 M solution of oxalic acid was prepared, and a boiling 25.00-mL acidified sample of the solution was titrated with 19.85 mL of potassium permanganate solution. What was the concentration of the permanganate solution? The redox equation is given in Table 4.1.

[*Answer:* 0.07557 M KMnO$_4$(aq)]

▲

4.7 INDIRECT AND BACK TITRATIONS

A second general strategy for doing acid-base and redox titrations is *indirect titration.* In this technique, the first step is to add an excess of a reagent to the analyte (which therefore becomes the limiting reagent). Then the amount of product of the analyte-reagent reaction is measured by direct titration and interpreted in terms of the amount of analyte that must have been present in the original sample (Fig. 4.24).

Suppose we wanted to use iodimetry to measure the concentration of chlorine in drinking water. We could take, say, a 25-mL sample, and add to it an excess of potassium iodide solution; some of the iodide ions would be oxidized by the chlorine to iodine (Fig. 4.25):

$$Cl_2(aq) + 2I^-(aq) \longrightarrow 2Cl^-(aq) + I_2(aq)$$

(The iodine would be present as triiodide ions.) Then, since in this reaction 1 mol Cl_2 = 1 mol I_2, the concentration of I_2 in the resulting solution is an indirect measure of the concentration of Cl_2 in the sample. We would find the concentration of I_2 by direct titration.

The titration of iodine is almost always done using a solution of sodium thiosulfate (the thiosulfate ion is $S_2O_3^{2-}$; **5**), which reduces the iodine back to iodide and, in the process, is itself oxidized to the tetrathionate ion ($S_4O_6^{2-}$; **6**):

$$I_3^-(aq) + 2S_2O_3^{2-}(aq) \longrightarrow 3I^-(aq) + S_4O_6^{2-}(aq)$$

$$1 \text{ mol } I_3^- = 2 \text{ mol } S_2O_3^{2-}$$

The equivalence point is detected as the loss of color of a starch indicator that is added in the final stages of the titration.

In an indirect titration—which may be redox or acid-base—the amount of reagent added (the KI solution in the chlorine determination) is unimportant so long as it is in excess. In a *back-titration* an excess of reagent is also added, but the amount added must be known since the procedure involves measuring the amount of the added reagent that remains after the rest has reacted with the analyte (Fig. 4.24). For example, the concentration of glucose in a solution can be determined by oxidizing it with excess iodine (in solution as triiodide ions):

$$C_6H_{12}O_6(aq) + I_3^-(aq) + 3OH^-(aq) \longrightarrow$$
$$C_6H_{11}O_7^-(aq) + 2H_2O(l) + 3I^-(aq)$$

Then the iodine that remains (as I_3^-) is titrated in the normal way with sodium thiosulfate. Since we know the amount of iodine added, and we have found the amount remaining, we can work out the concentration of glucose that must have been present initially.

5 Thiosulfate ion

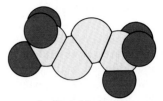

6 Tetrathionate ion

FIGURE 4.25 When chlorine gas is bubbled through a solution containing iodide ions it oxidizes them to iodine, which colors the solution brown.

SUMMARY

4.1 The quantitative uses of chemical equations are based on the interpretation of stoichiometric coefficients as moles of reacting substances. In stoichiometric calculations, the general strategy is to convert from the given data to moles of reactants or products, then to convert from moles of one substance to moles of another, and finally to convert from moles to the property of interest.

4.2 The amount of the **limiting reagent** available for reaction determines the maximum amount of product

that can be formed, and hence determines the **theoretical yield** of the reaction, the yield expected on the basis of the chemical equation. However, other reactions taking place may reduce the actual yield below the theoretical yield; the ratio of actual yield to theoretical is reported as the **percentage yield.**

4.3 The concept of a limiting reagent is the basis of **combustion analysis,** an example of **gravimetric analysis,** for the determination of mass percentage composition. The composition of a substance is often reported as an **empirical formula,** the simplest formula showing the proportions of atoms present. The empirical formula is used to obtain the molecular formula by comparing its formula weight with the molecular weight measured in another experiment.

4.4 and 4.5 The link between a volume of solution and the number of moles it contains is the **molar concentration,** the number of moles of solute per liter of solution. Solutions are very useful in chemistry, partly because, through **dilution** or the reduction of concentration, we can transfer very small fractions of a mole of solute. The volume of one solution needed to react with another is obtained by converting each volume to moles of solute and using the reaction stoichiometry to decide the number of moles, and hence the volume, needed.

4.6 The composition of mixtures, solutions, and compounds may be determined by **volumetric analysis,** of which the most important technique is the **titration** of an **analyte** with a **titrant.** The normal procedure is to measure one solution into another until the **equivalence point,** the point at which enough titrant has been added to react with all the analyte, is reached; this point can be shown by a color change of either the reagents or an added **indicator.** In **acid-base titrations** an acid reacts with a base; in **redox titrations** an oxidizing agent reacts with a reducing agent. A particularly useful oxidizing agent is iodine, which (in solution as the triiodide ion) is used in **iodimetry.**

4.7 Either type of titration may be performed as a **direct, indirect,** or **back-titration** by titrating the analyte, the product of the analyte and a reagent, or an excess of added reagent, respectively.

EXERCISES

Mole Calculations

4.1 How many moles of H_2 molecules are needed to convert 5 mol of O_2 molecules to water?

4.2 How many moles of H_2 molecules are needed to convert 5 mol of O_2 to hydrogen peroxide?

4.3 Calculate the number of moles of molecules produced when 2 mol of TNT molecules explode in the reaction

$$4C_7H_5O_6N_3(s) + 21O_2(g) \longrightarrow$$
$$28CO_2(g) + 10H_2O(g) + 6N_2(g)$$

4.4 Calculate the number of moles of molecules produced when 2 mol of nitroglycerin molecules explode in the reaction

$$4C_3H_5O_9N_3(l) \longrightarrow$$
$$6N_2(g) + 12CO_2(g) + 10H_2O(g) + O_2(g)$$

In Exercises 4.5 and 4.6, write the relation, in the form a mol A = b mol B, for the numbers of moles of the underlined substances taking part in each reaction.

4.5 (a) The neutralization of hydrochloric acid by a base:

$$\underline{HCl}(aq) + \underline{NaOH}(aq) \longrightarrow NaCl(aq) + H_2O(l)$$

(b) The neutralization of sulfuric acid by a base:

$$\underline{H_2SO_4}(aq) + 2\underline{NH_3}(aq) \longrightarrow (NH_4)_2SO_4(aq)$$

(c) The reaction of nitrogen dioxide and water, used in the manufacture of nitric acid:

$$3\underline{NO_2}(g) + H_2O(l) \longrightarrow 2\underline{HNO_3}(aq) + NO(g)$$

(d) The oxidation of iodide ions to iodine, used in the measurement of iodide concentration:

$$2KMnO_4(aq) + 16HCl(aq) + 10\underline{KI}(aq) \longrightarrow$$
$$2MnCl_2(aq) + 8H_2O(l) + 5\underline{I_2}(aq) + 12KCl(aq)$$

4.6 (a) The precipitation of silver chloride, used in the measurement of chloride-ion concentrations:

$$\underline{AgNO_3}(aq) + \underline{KCl}(aq) \longrightarrow AgCl(s) + KNO_3(aq)$$

(b) The neutralization of phosphoric acid with a base:

$$\underline{H_3PO_4}(aq) + 3\underline{NaOH}(aq) \longrightarrow Na_3PO_4(aq) + 3H_2O(l)$$

(c) The oxidation of hydrogen sulfide by sulfur dioxide:

$$2\underline{H_2S}(g) + SO_2(g) \longrightarrow 3\underline{S}(s) + 2H_2O(l)$$

(d) The oxidation of hydrogen sulfite ions to hydrogen sulfate ions by dichromate ions in acidic solution:

$$\underline{Na_2Cr_2O_7}(aq) + 3\underline{NaHSO_3}(aq) + 8HCl(aq) \longrightarrow$$
$$2CrCl_3(aq) + 3NaHSO_4(aq) + 2NaCl(aq) + 4H_2O(l)$$

4.7 Calculate the number of moles of CO_2 molecules produced when 1.0 mol of C_8H_{18} (octane) molecules burns in air.

4.8 Calculate the number of moles of H_2O molecules produced when 1.0 mol of NH_3 molecules burns in air to nitrogen and water.

4.9 How many moles of O_2 should be supplied to burn 1 mol of C_3H_8 (propane) molecules in a camping stove?

4.10 How many moles of O_2 molecules should be supplied to burn 1 mol of CH_4 molecules in a domestic furnace?

4.11 When the white crystalline solid ammonium nitrate is heated, it gives the colorless gas dinitrogen oxide and water. How many moles of N_2O molecules are produced when 0.10 mol of NH_4NO_3 is decomposed?

4.12 Ozone reacts with aqueous potassium iodide solution to form oxygen, iodine, and potassium hydroxide solution. How many moles of KI formula units are needed to react with 0.20 mol of O_3 molecules?

4.13 Sodium thiosulfate ($Na_2S_2O_3$), photographer's "hypo," reacts with unexposed silver bromide in the film emulsion to form sodium bromide and a compound of formula $Na_5[Ag(S_2O_3)_3]$. How many moles of $Na_2S_2O_3$ formula units are needed to make 0.10 mol of AgBr soluble?

4.14 Sodium thiosulfate ($Na_2S_2O_3$) is used in the laboratory to measure the concentration of iodine in solutions. In the reaction between the two substances, sodium iodide and sodium tetrathionate ($Na_2S_4O_6$) are produced. If 0.15 mol of $Na_2S_2O_3$ formula units were needed to react with all the iodine, how many moles of I_2 molecules were present initially?

Mass Calculations

4.15 Carbon dioxide is formed when carbonates are heated. This process is used industrially to produce lime (CaO) from limestone ($CaCO_3$). Calculate the mass of carbon dioxide produced when 10.0 g of calcium carbonate are decomposed.

4.16 A typical computation in the iron industry involves finding the mass of iron that can be obtained from a given mass of iron oxide (Fe_2O_3) when it is reduced to the element. What is the maximum mass of iron that can be obtained from 100 kg of the oxide? Is magnetite (Fe_3O_4) a richer source of iron?

4.17 Calculate the mass of alumina (Al_2O_3) produced when 100 g of aluminum burns in oxygen.

4.18 Calculate the mass of magnesium oxide (MgO) produced when 100 g of magnesium burns in oxygen.

4.19 "Slaked lime," $Ca(OH)_2$, is formed from "quicklime" (CaO) by adding water. What mass of water is needed to convert 10 kg of quicklime to slaked lime? What mass of slaked lime is produced?

4.20 What mass of carbon is needed to produce 100 kg of iron from iron(III) oxide in a reaction that produces (a) carbon monoxide; (b) carbon dioxide? What mass of iron ore is needed in each case?

4.21 In ironmaking, quicklime (CaO) combines with silica (SiO_2) that is present as an impurity in the iron ore and produces a molten slag of formula $CaSiO_3$. What mass of slag is produced from 1.0 metric ton of silica (1 metric ton is 1000 kg)? What mass of quicklime is needed? The quicklime is produced by the action of heat on limestone ($CaCO_3$) in the furnace. What mass of limestone is needed?

4.22 Other impurities present in iron ore include alumina (see Exercise 4.17). The quicklime combines with alumina too, and the slag then contains $Ca(AlO_2)_2$. What mass of alumina (Al_2O_3) can be removed by 1.0 metric ton of limestone?

4.23 Camels store the fat tristearin ($C_{57}H_{110}O_6$) in the hump. As well as being a source of energy, the fat is a source of water, because when it is used the reaction

$$2C_{57}H_{110}O_6(s) + 163O_2(g) \longrightarrow 114CO_2(g) + 110H_2O(l)$$

takes place. What mass of water is available from 1.0 kg of fat?

4.24 The souring of wine occurs when ethanol (C_2H_6O) is oxidized by oxygen to acetic acid ($C_2H_4O_2$) and water. By how much would the mass of (a) a closed bottle and (b) an open bottle of wine containing 75 g of ethanol increase if the ethanol were completely converted into acetic acid?

4.25 When a hydrocarbon burns, water is produced as well as carbon dioxide. (For this reason, clouds of condensed water droplets are often seen coming from automobile exhausts, especially on a cold day.) The density of gasoline is 0.8 g/mL. Assuming it to be all octane, C_8H_{18} (it is not, but octane is representative of the complex mixture sold as gasoline), calculate the mass of water produced when 1 gal of gasoline is burned.

4.26 The density of oak is 0.7 g/cm^3. Assuming oak is a substance with the formula $C_6H_{12}O_6$, calculate the mass of water produced when an oak log 10 cm by 10 cm by 15 cm is burned. Would the result be different if wood were represented by the formula $C_{6n}H_{12n}O_{6n}$, with n large?

4.27 The compound diborane (B_2H_6) was at one time considered for use as a rocket fuel. How many metric tons of liquid oxygen would a rocket have to carry to burn 10 metric tons of diborane completely? (The products of the combustion are B_2O_3 and H_2O.)

4.28 The solid fuel in the booster stage of the space shuttle is a mixture of ammonium perchlorate and aluminum powder, which react as follows:

$$6NH_4ClO_4(s) + 10Al(s) \longrightarrow$$
$$5Al_2O_3(s) + 3N_2(g) + 6HCl(g) + 9H_2O(g)$$

What mass of aluminum should be mixed with 5.0 × 10^3 kg of ammonium perchlorate, if the reaction proceeds as stated?

Limiting Reagents

4.29 What is the maximum mass of methane (CH_4) that can be burned if only 1.0 g of oxygen is available?

4.30 What is the maximum mass of propane (C_3H_8) that can be burned if only 1.0 g of oxygen is available?

4.31 What is the maximum mass of glucose ($C_6H_{12}O_6$) that can be burned in 10 g of oxygen?

4.32 What is the maximum mass of sucrose ($C_{12}H_{22}O_{11}$) that can be burned in 10 g of oxygen?

4.33 A solution containing 5.0 g of silver nitrate was mixed with another containing 5.0 g of potassium chloride. Which was the limiting reagent for the precipitation of silver chloride?

4.34 A solution containing 1.0 g of barium chloride was mixed with another containing 0.50 g of sodium sulfate. Which was the limiting reagent for the precipitation of barium sulfate?

4.35 Phosphorus trichloride reacts with water to form phosphorous acid and hydrogen chloride:

$$PCl_3(l) + 3H_2O(l) \longrightarrow H_3PO_3(aq) + 3HCl(g)$$

Which is the limiting reagent when 25 g of the trichloride is mixed with 10 g of water? What mass of phosphorous acid can be produced?

4.36 When xenon difluoride (XeF_2) dissolves in water, it slowly reacts with the water to produce xenon gas, hydrogen fluoride, and oxygen:

$$2XeF_2(s) + 2H_2O(l) \longrightarrow 2Xe(g) + 4HF(g) + O_2(g)$$

Which is the limiting reagent when 1.0 g of the difluoride is dissolved in 50 g of water? What mass of hydrogen fluoride can be produced?

Reaction Yield

4.37 The theoretical yield of sodium perxenate (Na_4XeO_6) in a certain reaction was 1.25 g, but only 1.21 g was obtained. What was the percentage yield of the reaction?

4.38 The theoretical yield of ammonia in an industrial synthesis was 550 tons, but only 480 tons was obtained. What was the percentage yield of the reaction?

4.39 A compound of formula AH (where A stands for the remainder of the molecule) and molar mass 231 g/mol reacts with another of formula BOH and molar mass 125 g/mol to give a compound AB. In a preparation of AB, 2.45 g of AH reacted and resulted in the formation of 2.91 g of AB. What was the percentage yield of the reaction?

4.40 A compound of formula ANH_2 and molar mass 168 g/mol reacts with another of formula BOH and molar mass 150 g/mol to give a compound ANHB. It was found that 1.15 g of ANH_2 resulted in the formation of 1.55 g of ANHB. What was the percentage yield of the reaction?

Chemical Composition from Measurements of Mass

In Exercises 4.41 and 4.42, calculate the empirical formula of each compound from its mass percentage composition.

4.41 (a) Benzoic acid: 69% C, 5% H, 26% O
(b) Aniline: 77.4% C, 7.6% H, 15.0% N

4.42 (a) Aspirin: 60.0% C, 4.5% H, 35.5% O
(b) Urea: 20.0% C, 6.7% H, 26.6% O, 46.7% N

In Exercises 4.43 and 4.44, calculate the molecular formula of each compound from its empirical formula and molecular weight.

4.43 (a) Decene: CH_2, 140 amu; (b) benzene: CH, 78 amu; (c) glucose: CH_2O, 180 amu; (d) nicotine: C_5H_7N, 162 amu.

4.44 (a) Hydrazine: NH_2, 32 amu; (b) trichlorobenzene: C_2HCl, 181 amu; (c) crocetin, a compound responsible for the color of some flowers: C_5H_6O, 328 amu.

4.45 A sample of the colorless liquid benzene, which is a compound of carbon and hydrogen, was analyzed, and a 0.456-g sample was found to contain 0.421 g of carbon. Calculate the empirical formula of benzene.

4.46 The colorless liquid toluene, which is used as an industrial solvent and is present in gasoline, is a compound of carbon and hydrogen. On analysis, 0.357 g of toluene was found to contain 0.326 g of carbon. What is the empirical formula of toluene?

4.47 Oxalic acid occurs in rhubarb. Analysis of a sample of the acid showed that it contained 27% C, 2.2% H, and 71% O. What is its empirical formula?

4.48 Fructose is a type of sugar that occurs in fruit and is the principal compound responsible for the sweetness of honey. Analysis of a sample of fructose showed that it contained 40% C, 7% H, and 53% O. Calculate its empirical formula.

4.49 In an experiment, 4.14 g of the element phosphorus combined with chlorine to produce 27.8 g of a white solid compound. What is the empirical formula of the compound?

4.50 A chemist found that 4.69 g of sulfur combined with fluorine to produce 15.81 g of a gas. What is the empirical formula of the gas?

4.51 A forensic chemist analyzed a sample of plastic found at the scene of a crime and found it to have mass percentage composition 86% C and 14% H. What is the empirical formula of the plastic?

4.52 Teflon® is the Du Pont trade name for a class of synthetic fluorocarbon plastics. (Fluorocarbons are compounds of fluorine and carbon.) A chemist from a rival laboratory analyzed a sample and found it to have mass percentage composition 24% C and 76% F. What is the empirical formula of the plastic?

4.53 The empirical formula of the white crystalline solid naphthalene, which was once widely used as "mothballs," was found to be C_5H_4, and a separate determination of its molecular weight gave the value 128 amu. What is the molecular formula of naphthalene?

4.54 The empirical formula of the organic compound responsible for the red color of ripe tomatoes was found to be C_5H_7. Its molecular weight was measured as

537 amu. What is the molecular formula of the compound?

4.55 The empirical formula of the female sex hormone estrone was found to be $C_9H_{11}O$, and its molecular weight is 270 amu. What is the molecular formula of estrone?

4.56 A chemist interested in the chemistry of natural products found that cocaine, which is obtained from the leaves of the coca shrub and was present in the original formulation of Coca-Cola®, has the empirical formula $C_{17}H_{21}O_4N$. Its molecular weight is 303 amu. What is the molecular formula of cocaine?

4.57 In a combustion analysis of a 0.152-g sample of the artificial sweetener aspartame, it was found that 0.318 g of carbon dioxide, 0.084 g of water, and 0.0145 g of nitrogen were produced. What is the empirical formula of aspartame? Mass spectrometry gave its molecular weight as 294 amu. What is its molecular formula?

4.58 The bitter compound quinine is a component of tonic water and is used as protection against malaria. When a sample of mass 0.487 g was burned, 1.321 g of carbon dioxide, 0.325 g of water, and 0.421 g of nitrogen were produced. Its molecular weight was found to be 324 amu. Determine the empirical and molecular formulas of quinine.

4.59 The stimulant in coffee and tea is caffeine, a substance of molecular weight 194 amu. When 0.376 g of caffeine was burned, 0.682 g of carbon dioxide, 0.174 g of water, and 0.110 g of nitrogen were formed. Determine the empirical and molecular formulas of caffeine, and write the equation for its combustion.

4.60 Nicotine, the stimulant in tobacco, causes a very complex set of physiological effects on the body. It is known to have molecular weight 162 amu. When a 0.395-g sample was burned, 1.072 g of carbon dioxide, 0.307 g of water, and 0.068 g of nitrogen were produced. What are the empirical and molecular formulas of nicotine? Write the equation for its combustion.

Concentration and Dilution

4.61 A 1.230-g sample of sodium chloride was dissolved in enough water to make 100.0 mL of solution. What is the molar concentration of the solution?

4.62 A 1.661-g sample of glucose ($C_6H_{12}O_6$) was dissolved in enough water to make 50.00 mL of solution. What is the molar concentration of the solution?

4.63 A chemist studying the properties of photographic emulsions decided to prepare a 0.100 M $AgNO_3(aq)$ solution by adding silver nitrate (one of the few soluble silver compounds) to a 50.00-mL volumetric flask and then adding water up to the mark. What mass of silver nitrate should be used?

4.64 Silver chloride, which is insoluble in water, can be prepared from a silver nitrate solution by adding a solu-

tion of a soluble chloride. A chemist decided to use potassium chloride, and wanted to prepare a 0.050 M KCl(aq) solution by adding potassium chloride to a 250.0-mL volumetric flask and then adding water up to the mark. What mass of potassium chloride should be used?

4.65 A chemist measured 23.60 mL of 0.0500 M KCl(aq) from a buret into a flask. How many moles of Cl^- ions were transferred?

4.66 A chemist measured 17.8 mL of 0.0152 M KMnO$_4$(aq) from a buret into a flask. How many moles of permanganate ions were transferred?

4.67 A student prepared a solution of sodium hydroxide by adding 1.577 g of the solid to a 250.0-mL volumetric flask and adding water up to the mark. Some of the solution was transferred into a buret. How much should you tell the student to run into a flask in order to transfer (a) 1.0 mmol of NaOH formula units; (b) 1.0 mmol of OH^- ions? What inaccuracy is there in the procedure (think about the chemical properties of NaOH)?

4.68 A student investigating the properties of carbonate ions made up a solution of sodium carbonate by adding 5.150 g of the anhydrous solid to a 250.0-mL volumetric flask and adding water up to the mark. What volume of the solution contains (a) 4.20 mmol of Na_2CO_3 formula units; (b) 4.20 mmol of CO_3^{2-} ions? (c) Suppose the student had mistakenly used 5.150 g of the $Na_2CO_3 \cdot 10H_2O$ instead of the anhydrous solid, but transferred the same volumes of solution. What amounts (in moles) of Na_2CO_3 formula units and CO_3^{2-} ions would be contained in the same volumes of solution?

4.69 What volume of a 0.126 M NaOH(aq) solution should be diluted to 500.0 mL to produce a 1.45×10^{-2} M NaOH(aq) solution?

4.70 What volume of a 15.9 M HNO$_3$(aq) solution should be diluted to 250.0 mL to produce a 1.00 M HNO$_3$(aq) solution?

4.71 A solution was prepared by weighing 5.820 g of potassium chloride into a volumetric flask, dissolving, and adding water up to 250.0 mL. Then 25.00 mL of this solution was transferred to a 250.0-mL flask and water was added up to the mark. What is the concentration of the second solution?

4.72 A solution was prepared by weighing 6.120 g of sodium acetate (NaC$_2$H$_3$O$_2$) into a 500.0-mL flask, dissolving, and adding water up to the mark. Then 50.00 mL of this solution was diluted to 250.0 mL by the addition of water. What is the concentration of the final solution?

Acid-Base Titrations

4.73 A solution was prepared by dissolving 10.00 g of sodium hydroxide in enough water to make 250.0 mL of solution. It was found that 25.00 mL of this solution required 19.40 mL of hydrochloric acid to reach the equivalence point. Calculate the molar concentration of the acid.

4.74 A solution was prepared by dissolving 4.05 g of sodium hydroxide in enough water to make 150.0 mL of solution. It was found that 25.00 mL of this solution required 14.84 mL of hydrochloric acid to reach the equivalence point. Calculate the molar concentration of the acid.

4.75 The sodium hydroxide solution of Exercise 4.73 was used in a titration with sulfuric acid (H_2SO_4). It was found that 25.00 mL of the solution was required to neutralize 17.25 mL of the acid completely. Calculate the molar concentration of the acid.

4.76 The sodium hydroxide solution of Exercise 4.74 was used in a titration with phosphoric acid (H_3PO_4). It was found that 25.00 mL of the solution was required to neutralize 8.80 mL of the acid completely. Calculate the molar concentration of the acid.

4.77 A solution of hydrochloric acid was prepared by measuring 10.00 mL of the concentrated acid into a 1.000-L volumetric flask and adding water up to the mark. Another solution was prepared by adding 0.530 g of anhydrous sodium carbonate to a 100.0-mL volumetric flask and adding water up to the mark. Then, 25.00 mL of the latter solution was pipetted into a flask and titrated with the diluted acid. The equivalence point was reached after 26.50 mL of acid had been added. What is the molar concentration of the original (concentrated) acid?

4.78 A solution of phosphoric acid was prepared by measuring 10.00 mL of the concentrated acid into a 1.000-L volumetric flask and adding water up to the mark. Another solution was prepared by adding 0.871 g of anhydrous sodium carbonate to a 100.00-mL volumetric flask and adding water up to the mark. Then 25.00 mL of the latter solution was pipetted into a flask and titrated with the dilute acid to the $H_2PO_4^-$ equivalence point. The volume of acid required was 28.98 mL. What was the molar concentration of the original (concentrated) acid?

Redox Titrations

4.79 A chemist prepared 250.0 mL of a solution containing 5.23 g of sodium thiosulfate in the form of crystals of composition $Na_2S_2O_3 \cdot 5H_2O$. When this solution was titrated against 25.00 mL of an iodine solution, equivalence was reached after the addition of 12.80 mL. What is the molar concentration of the iodine solution? What is its concentration in grams per liter?

4.80 A chemist dissolved 0.852 g of potassium permanganate in enough water to make 250.0 mL of solution. This solution was used to titrate 25.00 mL of an acidified hydrogen peroxide solution, the reaction being

$$5H_2O_2(aq) + 2KMnO_4(aq) + 3H_2SO_4(aq) \longrightarrow$$
$$2MnSO_4(aq) + K_2SO_4(aq) + 8H_2O(l) + 5O_2(g)$$

Equivalence was reached after 20.20 mL of the permanganate solution had been added. What is the molar concentration of the peroxide solution? What is its concentration in grams per liter?

4.81 Hydrazinium sulfate ($N_2H_5 \cdot HSO_4$) is oxidized by iodine to nitrogen. Calculate the concentration of a solution of hydrazine sulfate for which 18.78 mL of 0.0543 M $I_3^-(aq)$ was required to titrate 25.00 mL of the solution. The reaction is

$$N_2H_5^+ + 2I_2 \longrightarrow N_2 + 5H^+ + 4I^-$$

***4.82** The concentration of nitrite ions in a solution can be determined by back-titration using Ce(IV) as oxidizing agent. In one experiment, a 0.254-g sample containing sodium nitrite and inert materials was treated with 25.00 mL of a 0.122 M $Ce^{4+}(aq)$ solution. The excess cerium(IV) was then back-titrated with 13.20 mL of 0.0154 M $Fe^{2+}(aq)$ solution. Calculate the mass percentage of $NaNO_2$ in the sample. The reaction is

$$2Ce^{4+} + NO_2^- + H_2O \longrightarrow 2Ce^{3+} + NO_3^- + 2H^+$$

***4.83** The concentration of hydrocyanic acid (HCN) can be measured by titration with iodine, using the reaction

$$HCN(aq) + I_3^-(aq) \longrightarrow ICN(aq) + 2I^-(aq) + H^+(aq)$$

In one experiment, 15.00 mL of hydrocyanic acid was titrated with 8.75 mL of $I_3^-(aq)$ solution. The same triiodide solution had been titrated against an acidified solution containing 1.453 g of As_4O_6, which in solution forms H_3AsO_3; 11.42 mL of the triiodide solution were required. The reaction is

$$H_3AsO_3(aq) + I_3^-(aq) + H_2O(l) \longrightarrow$$
$$H_3AsO_4(aq) + 3I^-(aq) + 2H^+(aq)$$

What was the concentration of the hydrocyanic acid?

***4.84** Formalin, a liquid used to preserve biological specimens, is an aqueous solution of formaldehyde (CH_2O). Its concentration may be determined by oxidation to formate ions (HCO_2^-) with excess I_3^- ions, followed by back-titration with sodium thiosulfate. In a certain experiment, 25.00 mL of 0.0436 M $I_3^-(aq)$ solution was added to a 10.00-mL formalin solution. The excess iodine was then titrated with 12.56 mL of 0.0655 M $Na_2S_2O_3(aq)$ solution. What was the concentration of the formalin solution?

General

***4.85** The concentration of phosphite ions in a solution can be measured by oxidation to phosphate using iodine, followed by back-titration with sodium thiosulfate. A 1.00-mL sample of aqueous phosphorous acid was diluted to 100.00 mL, and 25.00 mL of the diluted solution was oxidized with 25.00 mL of 0.0548 M $I_3^-(aq)$ solution. The excess iodine was back-titrated with 8.34 mL of 0.0743 M $Na_2S_2O_3(aq)$ solution. What was the concentration of the original phosphorous acid?

*__4.86__ The concentration of a potassium permanganate solution can be measured by titration with As_4O_6 in acid solution, which it oxidizes to arsenic acid. It was found that when 0.1423 g of the oxide was used to prepare 50.00 mL of aqueous solution, 32.50 mL of a permanganate solution was needed to reach the equivalence point. What are the oxidation and reduction half-reactions and the overall reaction? What is the concentration of the permanganate solution? (See Exercise 4.83.)

*__4.87__ A vitamin C tablet was analyzed to determine whether it did in fact contain, as the manufacturer claimed, 1 g of the vitamin. A tablet was dissolved in enough water to make a 100.00-mL solution, and a 10.00-mL sample was titrated with iodine (as potassium triiodide). It was found that 10.1 mL of 0.0521 M $I_3^-(aq)$ was needed to reach equivalence. Given that 1 mol vitamin C = 1 mol I_3^-, is the manufacturer's claim correct? The molecular weight of vitamin C is 176 amu.

__4.88__ The mass percentage of sulfur in a fuel may be determined by burning the fuel in oxygen and passing the resulting SO_2 and SO_3 into dilute hydrogen peroxide, where they form sulfuric acid, the concentration of which may be measured volumetrically. In one experiment, 8.54 g of a fuel was burned, and the resulting sulfuric acid was titrated with 17.54 mL of 0.100 M $NaOH(aq)$. What is the mass percentage of sulfur in the fuel?

This chapter introduces the properties of the simplest state of matter—the gaseous state. It shows how the response of gases to changes in pressure and temperature can be predicted, and it extends the stoichiometric calculations of Chapter 4 to the volumes of gases taking part in reactions. In addition, we see how the properties of gases can be understood in terms of a simple but powerful molecular model. The illustration shows one of the most important gases of all: the atmosphere of the earth.

THE PROPERTIES OF GASES

The gas laws

5.1 Pressure
5.2 The ideal gas
5.3 Using the ideal gas law
5.4 Real gases

**The stoichiometry of
reacting gases**

5.5 The molar volume of a gas
5.6 Gaseous mixtures

The kinetic theory of gases

5.7 Molecular speeds
5.8 The liquefaction of gases

The study of gases is central to the study of chemistry. Their importance to the economy can be judged from the fact that they include half the top ten industrial chemicals (see Appendix 2E). Among them are nitrogen, which is extracted from the atmosphere and reduced to ammonia, and oxygen, which is used for steelmaking. Oxygen's role in steelmaking is to oxidize the impurities left in the crude iron obtained from blast furnaces; about 1 ton of oxygen is needed to produce 1 ton of steel. Third in annual tonnage is the gas ethylene (C_2H_4), much of which is used to make the plastic polyethylene. Air, that most important of gaseous mixtures (Table 5.1), affects everything we do. Eleven of the elements are gases under normal conditions, as are hundreds of compounds (Fig. 5.1 and Table 5.2).

The main goal of this chapter is to explain how reaction stoichiometry is used to predict the volumes of gases that take part in reactions. However, since the volume of a gas changes sharply when the temperature and pressure change, we cannot make reliable predictions until we know how to take these factors into account. We deal with this aspect of gases first.

TABLE 5.1 The composition of dry air at sea level

| Constituent | Molecular weight,* amu | Percentage composition | |
		Volume	Mass
Nitrogen, N_2	28.02	78.09	75.52
Oxygen, O_2	32.00	20.95	23.14
Argon, Ar	39.95	0.93	1.29
Carbon dioxide, CO_2	44.01	0.03	0.05

*The average molecular weight of air molecules, allowing for the different abundances, is 28.96 amu.

FIGURE 5.1 Most gases are colorless, but here we see three with characteristic colors: pale yellow-green chlorine, purple iodine vapor, and brown nitrogen dioxide.

TABLE 5.2 The preparation and properties of some gases

Gas	Preparation and properties	Gas	Preparation and properties
Monatomic gases			
Helium, He	Distillation of liquefied natural gas. Colorless; odorless; insoluble; forms no compounds; condenses at $-269°C$ (4 K). Used in balloons, gas chromatography, and for low-temperature research.	Neon, Ne	Distillation of liquid air. Colorless; odorless; insoluble; forms no compounds; condenses at $-246°C$ (27 K). Used in neon lights.
Diatomic gases			
Oxygen, O_2	Distillation of liquid air $$2KClO_3(s) \xrightarrow{\Delta,\ MnO_2} 2KCl(s) + 3O_2(g)$$ Colorless; highly reactive; life-supporting; not very soluble in water; condenses at $-183°C$ to pale blue liquid.	Chlorine, Cl_2	Electrolysis of molten sodium chloride. Pale yellow-green; reactive; reacts with water forming HOCl; condenses at $-34°C$ to yellow-green liquid. Used as a water-purifying agent and in many industrial processes, such as the production of PVC (polyvinyl chloride).
Nitrogen, N_2	Distillation of liquid air. Colorless; not very reactive; not very soluble in water; condenses at $-196°C$. Used for making fertilizers and other nitrogen-containing compounds (including plastics) and for providing inert atmospheres.	Carbon monoxide, CO	Combustion of carbon or carbon compounds in a restricted supply of oxygen. Water-gas reaction: $$C(s) + H_2O(g) \longrightarrow H_2(g) + CO(g)$$ Action of sulfuric acid on formic acid: $$H_2CO_2(l) \xrightarrow{\Delta,\ conc\ H_2SO_4} CO(g) + H_2O(l)$$ Colorless; flammable; very poisonous; almost insoluble in water. Used in steelmaking.
Fluorine, F_2	Electrolysis of molten KF/HF mixture. Almost colorless; highly reactive; reacts with water, forming HF; condenses at $-188°C$. Used for isotope separation (UF_6) and for producing fluorocarbons, including Teflon®.		
Triatomic gases			
Hydrogen sulfide, H_2S	Present in natural gas $$FeS(s) + 2HCl(aq) \longrightarrow FeCl_2(aq) + H_2S(g)$$ Colorless; rotten-egg odor; intensely poisonous; very soluble in water to give an acid solution; condenses at $-60°C$. Used in chemical analysis to produce insoluble sulfides. Used as source of sulfur and to manufacture sulfuric acid.	Sulfur dioxide, SO_2	$$S_8(s) + 8O_2(g) \longrightarrow 8SO_2(g)$$ $$SO_3^{2-}(aq) + 2HCl(aq) \longrightarrow 2Cl^-(aq) + H_2O(l) + SO_2(g)$$ Colorless; choking, pungent odor; poisonous; soluble in water to give acidic solution (sulfurous acid, H_2SO_3); condenses at $-10°C$. Used in the production of sulfuric acid.
Carbon dioxide, CO_2	Product of fermentation and combustion $$CO_3^{2-}(s\ or\ aq) + 2HCl(aq) \longrightarrow 2Cl^-(aq) + CO_2(g) + H_2O(l)$$ Colorless; odorless; soluble in water to give an acid solution (carbonic acid); condenses directly to a solid at $-78°C$. Used naturally in photosynthesis; solid carbon dioxide is a freezing agent.	*Tetraatomic gas*	
		Ammonia, NH_3	Haber synthesis (Section 13.8) $$N_2(g) + 3H_2(g) \xrightarrow{\Delta,\ pressure} 2NH_3(g)$$ Colorless; pungent; very soluble in water to give a basic solution; condenses at $-33°C$.

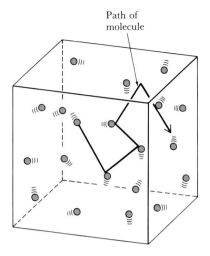

Path of
molecule

FIGURE 5.2 A gas may be pictured as a collection of widely spaced molecules in continuous, chaotic motion.

THE GAS LAWS

We have defined a gas to be the state of matter in which a substance fills any container it occupies. In molecular terms, a gas is a collection of widely separated molecules in chaotic, random motion (Fig. 5.2). This description is reflected in the name "gas," which is derived from the same root as the word *chaos*. The wide separation of their molecules accounts for the fact that gases are highly *compressible*, or readily confined to a smaller volume by a piston in a confining cylinder. A gas that is free to expand and contract also responds strongly to changes in temperature, occupying a larger volume as it is heated, and a smaller one as it is cooled. These responses are expressed quantitatively by the laws described in this section.

5.1 PRESSURE

Everyone who has ever pumped up a bicycle tire or squeezed an inflated balloon has experienced the opposing force arising from the "pressure" of the confined air. Scientists define the pressure P as the force exerted per unit area:

$$P = \frac{\text{force}}{\text{area}}$$

A gas exerts pressure on an object as a result of the collisions that the molecules make with the object's surface (Fig. 5.3). A vigorous, chaotic storm of molecules results in a strong force and hence a high pressure; a gentler storm of molecules results in a low pressure.

Barometers and the units of pressure. The pressure exerted by the atmosphere is measured with a *barometer*. This instrument was invented in the fifteenth century by Evangelista Torricelli, a student of Galileo. Torricelli (whose name incidentally means "little tower") formed a little tower of mercury by sealing a long glass tube at one end, filling it with the liquid, and inverting it into a beaker (Fig. 5.4). The column of mercury fell until the pressure exerted by the liquid matched the pressure exerted by the atmosphere outside. In this type of barometer, the final height of the column is proportional to the atmospheric pressure* and hence can be used to measure it.

   The height of the liquid column in a mercury barometer on a typical day at sea level is about 760 mm (corresponding to the 30 in referred to in weather forecasts). Hence a pressure of 760 mm of mercury (written 760 mmHg) corresponds to normal atmospheric pressure at sea level. The millimeter of mercury is almost exactly equal to the pressure unit "Torr," so the two can be used interchangeably: atmospheric pressure is about 760 Torr at sea level on an average day. A low-pressure gas, perhaps one being used to study reactions high in the atmosphere where the pressure is very low, might have a pressure of only a few

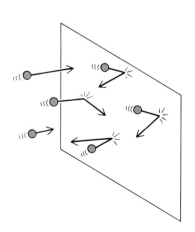

FIGURE 5.3 As the molecules of a gas collide with the walls of their container, they exert a force on it. The average force per unit area is the pressure of the gas.

*The precise relation is $P = dgh$, where P is the atmospheric pressure, d the density of the liquid, g the acceleration of gravity (9.81 m/s²), and h the height of the column.

Torr (a few millimeters of mercury). The atmospheric pressure at the cruising altitude of commercial aircraft (10 km) is about 200 Torr.

Since chemists often deal with substances at atmospheric pressure, they also use the unit called the "atmosphere" (atm):

$$1 \text{ atm} = 760 \text{ Torr exactly}$$

This relation is used as a conversion factor between Torr and atmospheres. For instance, a pressure of 500 Torr corresponds to

Pressure in atmospheres = pressure in Torr × atmospheres per Torr

$$= 500 \text{ Torr} \times \frac{1 \text{ atm}}{760 \text{ Torr}}$$

$$= 0.658 \text{ atm}$$

A very-low-pressure region of the atmosphere (an "area of low pressure" on a weather chart, Fig. 5.5) typically has a pressure of about 0.98 atm, and the eye of a hurricane might reach as low as 0.90 atm. A typical summer anticyclone, a region of high pressure, is at about 1.03 atm.

Pressures are also reported in other units, including the SI unit pascal (Pa). Table 5.3 lists some conversion factors between units.

Manometers. One instrument that is used to measure the pressure of a gas confined within a container is called a _manometer_. The name is derived from the Greek word for "thin," since manometers are commonly used for measuring pressures slightly below atmospheric—"thin" atmospheres. The gas sample presses on one surface of the liquid in the manometer tube, and the atmosphere presses on the other (Fig. 5.6). A balance is reached when the difference (left) or sum (right) of the pressures of the sample and the column of liquid matches the outside pressure. A liquid less dense than mercury is normally used, since otherwise small differences of pressure would result in height differences too

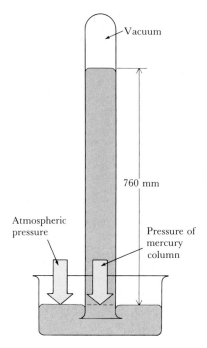

FIGURE 5.4 In a mercury barometer the pressure of a column of mercury is balanced against the pressure of the atmosphere; the height of the column is thus proportional to the atmospheric pressure. The space above the mercury is a vacuum, so it adds no pressure.

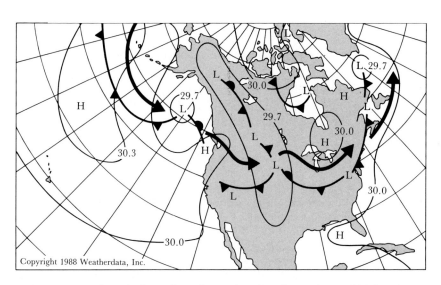

FIGURE 5.5 A typical weather chart, showing the regions of high pressure and low pressure. On this chart, pressure is given in inches of mercury.

TABLE 5.3 Pressure units and conversion factors

SI unit: the pascal (Pa)

$$1 \text{ Pa} = 1 \text{ kg/(m} \cdot \text{s}^2) = 1 \text{ N/m}^2$$

Conventional units*

1 bar = 100 kPa

1 atm = 101.325 kPa

1 atm = 760 Torr

1 atm = 760 mmHg
   = 14.70 lb/in^2

1 Torr = 133.3 Pa

1 mmHg = 133.3 Pa

*Exact relations are shown in bold type. In elementary work, 1 Torr is considered the same as 1 mmHg, and the two can be used interchangeably.

FIGURE 5.6 In a manometer the pressure of the sample is balanced against the sum (or difference) of the pressures of the atmosphere and the liquid. This shows a mercury manometer, but larger height differences are obtained with manometers that use water.

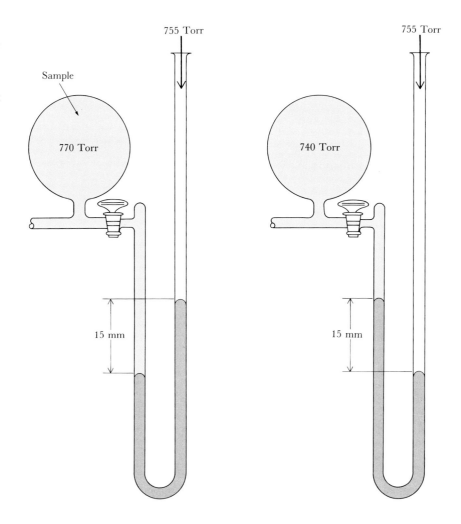

small to detect. Since water is 13.6 times less dense than mercury, a pressure difference of 10.0 Torr (10.0 mmHg) shows up as a 136-mm difference in the heights of the columns of a water manometer.

▼ EXAMPLE 5.1 Interpreting a manometer reading

A glass bulb containing argon was connected to a water manometer. The pressure of the gas caused the liquid to rise on the atmosphere side until it was 10.6 cm higher than on the sample side. The external pressure that day was 755.0 Torr. What was the pressure of the argon?

STRATEGY Because the argon pushed the column out against atmospheric pressure, the pressure of the argon was greater than atmospheric (greater, that day, than 755.0 Torr). The solution therefore hinges on being able to convert 10.6 cm of water to an equivalent number of Torr. That pressure is then added to the external pressure to give the pressure of the argon. As stated above, mercury is 13.6 times as dense as water. Therefore, a pressure that can hold up a 1.00-mm column of mercury can hold up a 13.6-mm (1.36 cm) column of water. This implies the relation

$$1.36 \text{ cm water} = 1.00 \text{ Torr}$$

We use this relation to convert the given height of water in the manometer to Torr.

SOLUTION The pressure due to the column of water is

$$\text{Pressure} = \text{height of water} \times \text{Torr per centimeter of water}$$

$$= 10.6 \text{ cm water} \times \frac{1.00 \text{ Torr}}{1.36 \text{ cm water}}$$

$$= 7.79 \text{ Torr}$$

The pressure of the argon is therefore

$$P = 7.79 \text{ Torr} + 755.0 \text{ Torr}$$

$$= 762.8 \text{ Torr}$$

EXERCISE In the same laboratory at the same time, a sample of neon gave a reading of 15.8 cm, with the water column on the atmosphere side lower than that on the sample side. What was the pressure of the neon?

[*Answer:* 743.4 Torr]

▲

5.2 THE IDEAL GAS

The gaseous state was the first state of matter to be examined scientifically. The first measurements were made by the Anglo-Irish scientist Robert Boyle in the seventeenth century. Technological motives for studying gases led to more discoveries when, in the eighteenth century, people began to fly in balloons. Their interest in ballooning stimulated the French scientists Jacques Charles and Joseph-Louis Gay-Lussac to make measurements of how the temperature of a gas affected its pressure, volume, and density. In fact, Charles built the first hydrogen balloon, which became known as the *charlière*. Gay-Lussac established a world altitude record of 23,018 feet in 1804.

Boyle's law. The experiments of Boyle, Charles, Gay-Lussac, and Avogadro in the nineteenth century showed that the pressure P, volume V, and temperature T of a fixed mass of gas all depend on each other, and a change in any one affects the others. The volume of a sample of gas decreases as the pressure is increased, in a manner summarized by "Boyle's law":

Boyle's law: At constant temperature and for a fixed mass of gas, the volume is inversely proportional to the pressure:

$$V \propto \frac{1}{P} \quad \text{or} \quad PV = \text{constant}$$

Thus, for example, doubling the pressure reduces the volume by half (Fig. 5.7). Gases respond readily to pressure because there is so much space between the molecules that they can easily be confined to a smaller volume. Boyle's law applies whatever the gas—argon, carbon dioxide, water vapor, or any other.

Since Boyle's law is the equation of a straight line passing through the origin (Section 1.3), we can write it as

$$V = 0 + \text{constant} \times \frac{1}{P}$$

$$y = a + \quad b \quad \times x$$

Hence it can be tested by plotting experimental values of V against $1/P$ and seeing how close the result is to a straight line. As Fig. 5.8 shows,

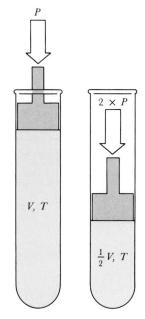

FIGURE 5.7 When pressure is applied to a sample of gas at constant temperature, its volume is decreased. Doubling the pressure reduces the volume of the gas by half.

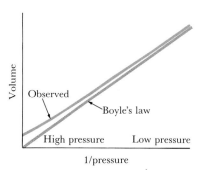

FIGURE 5.8 Boyle's law, $V \propto 1/P$, suggests that a plot of the volume of a real gas against the inverse of its pressure should be a straight line. Experiments show that gases obey this rule well at low pressures.

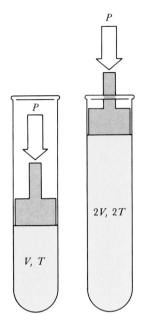

FIGURE 5.9 When a gas is heated at constant pressure, it expands. Doubling the temperature on the Kelvin scale doubles its volume.

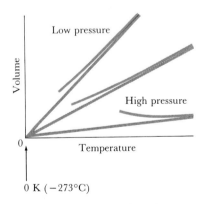

FIGURE 5.10 Charles's law, $V \propto T$, implies that a plot of the volume of a gas against temperature should be a straight line. Note that all the lines extrapolate to $V = 0$ at $T = 0$ (corresponding to −273°C); however, this is well below the temperatures at which Charles's law is valid.

the law is reasonably well obeyed at and below about 1 atm pressure. For reasons we shall examine later, it does poorly at high pressures and low temperatures.

Charles's law. The volume of a gas sample increases when it is heated, in a manner summarized by "Charles's law":

> **Charles's law:** The volume of a fixed mass of gas at constant pressure is proportional to its absolute temperature:
>
> $$V \propto T \qquad \text{or} \qquad V = \text{constant} \times T$$

Doubling the temperature from 293 K (20°C) to 586 K (313°C) doubles the volume of any gas that is free to expand against a constant pressure (Fig. 5.9). Like Boyle's law, Charles's law is obeyed at and below pressures of about 1 atm, but not at high pressures or low temperatures.

Charles's law is also the equation of a straight line:

$$V = 0 + \text{constant} \times T$$
$$y = a + \quad b \quad \times x$$

It can therefore be tested by plotting the volume of a sample of gas against the temperature (Fig. 5.10); the result should be a straight line which, when extended by extrapolation, passes through $V = 0$ at $T = 0$ (at −273°C). This is how the absolute zero of temperature was first identified: extrapolated plots of Charles's law passed through zero volume at −273°C no matter which gas was used in the experiment. Since the same value was observed for all gases, and since a volume cannot be negative, the graphs suggested that −273°C is the lowest temperature that can be reached by any substance. Kelvin then set the zero of his scale at this lowest possible temperature.

Avogadro's law. Avogadro's contribution to the gas laws was to suggest that the volume occupied by a gas was a measure of the number of molecules present, independent of their identity:

> **Avogadro's law:** The volume of a sample of gas at a given temperature and pressure is proportional to the number n of moles of molecules in the sample:
>
> $$V \propto n \qquad \text{or} \qquad V = \text{constant} \times n$$

Hence, doubling the number of moles of gas doubles the volume of the gas if its pressure and temperature are held constant (Fig. 5.11).

We can use the data in Table 5.4 to judge the validity of Avogadro's view that the volume occupied by a given number of moles of gaseous molecules is independent of their chemical identity. The volumes occupied by 1 mol of various kinds of molecules are only approximately the same at normal pressures (near 1 atm), but they all approach the same value as the pressure is decreased.

The ideal gas law. Avogadro's, Boyle's, and Charles's laws are now known to be three special cases of a single law, the "ideal gas law":

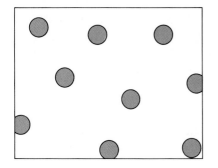 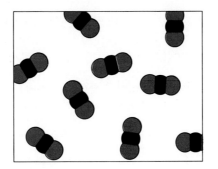

Ideal gas law: $$PV = nRT \qquad (1)$$

This law holds approximately for all gases; the constant R, called the *gas constant*, has the same value for every gas (it is "universal"). Once R has been measured for one gas, for instance by measuring n, P, V, and T for a given sample and calculating

$$R = \frac{PV}{nT}$$

the same value may be used for any gas. (In practice, other methods are used to measure R more precisely, including, as we shall see, relating it to the speed of sound in a gas.) It is found that at 0°C (273.15 K) and 1.00 atm pressure, 1.00 mol of a gas occupies 22.414 L. Hence

$$R = \frac{1.00 \text{ atm} \times 22.414 \text{ L}}{1.00 \text{ mol} \times 273.15 \text{ K}}$$

$$= 0.08206 \text{ L} \cdot \text{atm/(K} \cdot \text{mol)}$$

(This result is read as "0.08206 liter-atmospheres per kelvin-mole".) Table 5.5 gives values of R for pressure in other units.

The hypothetical gas that obeys the ideal gas law exactly at all temperatures and pressures is called an *ideal gas*. Although no real gas is ideal (for reasons explained below), the law is a good description of most gases at and below about 1 atm of pressure. Moreover, all gases are found to obey the law more and more closely as the pressure is reduced to very low values. It is normally acceptable for basic applications to assume that gases are ideal.

5.3 USING THE IDEAL GAS LAW

The ideal gas law is an example of an "equation of state":

An **equation of state** is a mathematical relation linking the pressure, volume, and temperature of a substance and the amount present.

Equations of state are very useful because they make it possible to calculate any one property once we know the other three. In the case of the ideal gas equation of state (Eq. 1), for example, if we know the number of moles in a sample of a gas, its temperature, and its volume,

TABLE 5.4 Molar volumes of gases at 0°C and 1 atm pressure

Gas	Volume, L/mol
Ammonia	22.40
Argon	22.09
Carbon dioxide	22.26
Nitrogen	22.40
Oxygen	22.40
Hydrogen	22.43
Helium	22.41
Ideal gas	22.414

TABLE 5.5 The gas constant in various units

R	When P is in
0.08206 L · atm/(K · mol)	Atmospheres
62.37 L · Torr/(K · mol)	Torr
8.314 J/(K · mol)	Pascals

then we can calculate its pressure from

$$P = \frac{nRT}{V}$$

whatever its identity.

▼ **EXAMPLE 5.2** **Using the ideal gas law**

Calculate the pressure inside a television picture tube, given that its volume is 5.0 L, its temperature is 23°C, and it contains 0.010 mg of nitrogen.

STRATEGY Assemble the data in the form required by the equation. The temperature must be in kelvins, and the mass of nitrogen must be converted to moles of N_2 molecules.

SOLUTION The temperature corresponds to $T = 296$ K. The number of moles of N_2 molecules is

$$n = 0.010 \times 10^{-3} \text{ g } N_2 \times \frac{1 \text{ mol } N_2}{28.02 \text{ g } N_2}$$

$$= 3.6 \times 10^{-7} \text{ mol } N_2$$

Hence, the pressure is

$$P = \frac{3.6 \times 10^{-7} \text{ mol}}{5.0 \text{ L}} \times \frac{0.08206 \text{ L} \cdot \text{atm}}{\text{K} \cdot \text{mol}} \times 296 \text{ K}$$

$$= 1.7 \times 10^{-6} \text{ atm}$$

This very low gas pressure (corresponding to a very small number of molecules striking the walls of the tube) is needed so that the electrons in the beam are not deflected by collisions with gas molecules, for that would give a blurred, dim picture.

EXERCISE Calculate the pressure exerted by 1.0 g of carbon dioxide in a 1.0-L flask at 300°C.

[*Answer*: 1.1 atm]

▲

The density of a gas. The ideal gas law is used to calculate how other physical properties vary with temperature and pressure. In this respect the law provides a simple example of how to rearrange and modify an equation for use in calculating a required property. We shall use it to find the density d, or mass per unit volume, of a gas in terms of its temperature and pressure. The calculation leads to a method for measuring the molecular weights of volatile substances.

As in the stoichiometry calculations of Chapter 4, the mole is the link between what we are given and what we want to know. Our aim is to convert mass per unit volume to moles per unit volume, and then to use the ideal gas law to write the number of moles in terms of the pressure and temperature. First we express the mass of the sample in terms of the number of moles of molecules, using the molar mass M of the gas molecules:

Mass of sample = number of moles in sample × mass per mole

$$= n \times M$$

Then we use this result to express the density in terms of moles:

$$d = \frac{\text{mass of sample}}{\text{volume of sample}} = \frac{n \times M}{V}$$

Finally we relate the number of moles to the pressure, using the ideal gas law in the form

$$\frac{n}{V} = \frac{P}{RT}$$

On substituting this relation into the equation for d, we get

$$d = P \times \frac{M}{RT} \qquad (2)$$

This equation shows that the density of a gas increases as the pressure is increased. More precisely, it shows that the density is proportional to P, so doubling the pressure doubles the density. The equation also shows how the density depends on temperature: so long as the pressure is constant (which means allowing the gas to expand as we heat it), the density decreases as the temperature is raised (Fig. 5.12).

FIGURE 5.12 The decrease in density with increasing temperature is used to provide the buoyancy of a hot-air balloon.

▼ EXAMPLE 5.3 Calculating the density of a gas

Estimate the density of air at 20°C and 1.0 atm pressure by supposing that it is largely nitrogen.

STRATEGY Assemble the data needed for Eq. 2; remember to convert the temperature to kelvins.

SOLUTION The molar mass of N_2 is 28 g/mol, and 20°C corresponds to 293 K. Since at this temperature $RT = 24.0$ L · atm/mol,

$$d = 1.0 \;\cancel{\text{atm}} \times \frac{28 \text{ g}}{1 \;\cancel{\text{mol}}} \times \frac{1 \;\cancel{\text{mol}}}{24.0 \text{ L} \cdot \cancel{\text{atm}}}$$

$$= 1.2 \text{ g/L}$$

That is, 1 L of air at sea level has a mass of 1.2 g, nearly a thousand times less than that of 1 L of water (of mass 1 kg).

EXERCISE Calculate the density of carbon dioxide at 1.0 atm and 25°C.
[*Answer*: 1.8 g/L]
▲

For a given pressure and temperature, Eq. 2 shows that gases with greater molecular weight have greater density. That is reasonable; since the same number of molecules are present in the same volume, heavier molecules mean there is a greater total mass in that volume (Fig. 5.13). The next example shows how the measurement of density is used to determine the molecular weight of a gas.

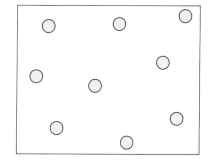

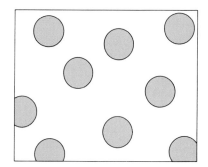

FIGURE 5.13 At the same temperature and pressure, a molecule occupies the same volume in any gas; hence the greater the molecular weight, the greater the density of the gas.

▼ EXAMPLE 5.4 Determining molecular weight from gas density

The organic compound geraniol, a component of oil of roses, is used in perfumery. At 260°C it was found that the density of the geraniol vapor is 0.480 g/L when the pressure is 103 Torr. Calculate the molecular weight of geraniol.

STRATEGY The molecular weight in atomic mass units is the same as the molar mass in grams per mole, so we can find the molecular weight by measuring the molar mass. The data are

$$d = 0.480 \text{ g/L} \qquad T = 533 \text{ K } (260° \text{ C})$$
$$P = 103 \text{ Torr} \qquad M = ?$$

and the unknown M can be found by rearranging Eq. 2 into

$$M = \frac{RT}{P} \times d$$

and substituting the data. We must, however, use the value of R (from Table 5.5) for pressures given in Torr, $R = 62.37 \text{ L} \cdot \text{Torr}/(\text{K} \cdot \text{mol})$.

SOLUTION Substitution of the data gives

$$M = \frac{62.37 \text{ L} \cdot \text{Torr}}{\text{K} \cdot \text{mol}} \times \frac{533 \text{ K}}{103 \text{ Torr}} \times \frac{0.480 \text{ g}}{1 \text{ L}}$$

$$= 155 \text{ g/mol}$$

Hence the molecular weight of geraniol is 155 amu.

EXERCISE The oil produced from eucalyptus leaves contains the organic compound eucalyptol. At 190°C, a sample of eucalyptol vapor had a density of 0.400 g/L at a pressure of 60.0 Torr. Calculate the molecular weight of eucalyptol.

▲ [*Answer*: 193 amu]

The dependence of volume on the conditions. The ideal gas law is widely used to calculate the changes that a sample undergoes when any of the conditions P, V, n, or T changes. This could be important, for instance, in the design of an ammonia synthesis plant, because the pressure of nitrogen changes when it is heated to the temperature required by the process. In many cases of this kind it turns out that we are dealing with a given sample of gas, and therefore one that contains a fixed number of moles. When that is so, we can use a simplified version of the ideal gas law.

Suppose that a sample initially has pressure P_i, volume V_i, and temperature T_i; then according to Eq. 1,

$$\frac{P_i V_i}{T_i} = nR$$

Now suppose that the pressure of the same sample is changed to P_f, with volume V_f, and temperature T_f; then we can also write

$$\frac{P_f V_f}{T_f} = nR$$

Since the number of moles of gas is unchanged, the right-hand sides of these two equations are the same. Hence we can equate the left-hand sides, obtaining

$$\frac{P_i V_i}{T_i} = \frac{P_f V_f}{T_f} \tag{3}$$

We can rearrange this equation to express any one of the six quantities in terms of the other five.

▼ EXAMPLE 5.5 Calculating the volume at a new pressure

A sample of gas occupies 1.00 L under 760 Torr pressure. Calculate the volume it would occupy at the same temperature when the pressure is 690 Torr (as in the eye of a hurricane).

STRATEGY Since the second pressure is lower, we anticipate that the gas will occupy a greater volume. We collect the data, making sure that there is only one unknown (if there is more than one unknown, we must obtain the missing information from tables of data). We rearrange Eq. 3 to give this unknown in terms of the other quantities, cancel the quantities that are unchanged, and then substitute the known values.

SOLUTION The data are

$$P_i = 760 \text{ Torr} \qquad V_i = 1.00 \text{ L} \qquad T_i = T_f$$
$$P_f = 690 \text{ Torr} \qquad V_f = ?$$

Since the unknown is V_f, we rearrange Eq. 3 to

$$V_f = V_i \times \frac{P_i}{P_f} \times \frac{T_f}{T_i}$$

The temperatures cancel, leaving

$$V_f = 1.00 \text{ L} \times \frac{760 \text{ Torr}}{690 \text{ Torr}} = 1.10 \text{ L}$$

As anticipated, the volume has increased.

EXERCISE What pressure is needed to confine a sample of argon into 300 mL if it occupies 500 mL when the pressure is 750 Torr?

[*Answer*: 1250 Torr]

▼ EXAMPLE 5.6 Calculating the volume at a new temperature

A hot-air balloon rises because the hot air inside the balloon is less dense than the cooler air outside. Calculate the volume that a sample of air would occupy at 40°C if it occupies 1.00 L at 20°C and the same pressure.

STRATEGY We anticipate that, because volume increases with temperature, the volume is greater at 40°C than at 20°C. We collect the data, making sure that there is only one unknown, rearrange Eq. 3 to give this unknown in terms of the other quantities, cancel the quantities that are unchanged, and then substitute the known values.

SOLUTION The data are

$$P_i = P_f \qquad V_i = 1.00 \text{ L} \qquad T_i = 293 \text{ K } (20°C)$$
$$V_f = ? \qquad T_f = 313 \text{ K } (40°C)$$

As the unknown is V_f, we rearrange Eq. 3 to

$$V_f = V_i \times \frac{P_i}{P_f} \times \frac{T_f}{T_i}$$

The pressures cancel, leaving

$$V_f = 1.00 \text{ L} \times \frac{313 \text{ K}}{293 \text{ K}} = 1.07 \text{ L}$$

That is, the sample occupies 1.07 L, an increase of 7%. This implies that air is 7% less dense at the higher temperature.

EXERCISE To what temperature should a sample of gas be cooled from 25°C to reduce its volume to half its initial value, at constant pressure?

[*Answer*: −124°C]

▼ EXAMPLE 5.7 Calculating the volume after changes in both pressure and temperature

At sea level, at a temperature of 20.2°C and a pressure of 755 Torr, the volume of a certain sample of air was 100 L. Calculate the volume of the sample at an altitude of 10 km (approximately the cruising altitude of a commercial jet), where the temperature is −50.5°C and the pressure is 225 Torr.

STRATEGY The decrease in temperature reduces the volume occupied by the gas, but the decrease in pressure increases the volume. The outcome depends on which effect dominates. The 70-K decrease in temperature is a change of 24%, and the 530-Torr decrease in pressure is a change of 70%. We therefore suspect that the pressure decrease is the more important effect, and that there will be a net increase of volume. As before, we begin by collecting the data, making sure that there is only one unknown. We rearrange Eq. 3 to give this unknown in terms of the other data, cancel the properties that are unchanged, and substitute the known values.

SOLUTION The data are

$$P_i = 755 \text{ Torr} \qquad V_i = 100 \text{ L} \qquad T_i = 293.4 \text{ K } (20.2°C)$$
$$P_f = 225 \text{ Torr} \qquad V_f = ? \qquad T_f = 222.7 \text{ K } (−50.5°C)$$

Since the unknown is V_f, Eq. 3 gives

$$V = 100 \text{ L} \times \frac{755 \text{ Torr}}{225 \text{ Torr}} \times \frac{222.7 \text{ K}}{293.4 \text{ K}}$$

$$= 255 \text{ L}$$

The volume more than doubles. As we expected, the decrease in pressure has a greater effect than the decrease in temperature.

EXERCISE Calculate the final pressure of a sample of gas that occupies 500 mL at 2.3 atm and 100°C and is compressed into 200 mL and cooled to 25°C.

[*Answer*: 4.6 atm]

▲

5.4 REAL GASES

As we have seen, real gases do not behave exactly as predicted by the ideal gas law. The deviations arise from *intermolecular forces*, the attractions and repulsions between gas molecules (Fig. 5.14). These interactions become important at high pressures, when the molecules are closer together. The evidence for attraction is the existence of liquids and solids, which consist of molecules held together by that attraction. The evidence for repulsion between molecules that are in very close proximity is the difficulty of compressing liquids and solids, which suggests that their molecules resist being pressed together.

The relative importance of the attractions and repulsions is shown by the curve in Fig. 5.15. The distance below the horizontal axis represents the strength of attraction, and the distance above it represents the strength of repulsion. There is no significant attraction at separations greater than about 10 molecular diameters, because the molecules are then too far apart to influence each other. The attraction increases as the molecules approach and influence each other more strongly. However, when the molecules come into contact, the attraction is overcome by a stronger repulsion. This repulsion increases sharply and produces

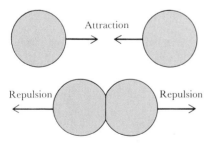

FIGURE 5.14 Molecules attract each other at distances up to several molecular diameters. However, as soon as they come into contact, they repel each other strongly.

a very strong resistance to the molecules moving any closer together than just touching.

Many suggestions have been proposed for changing the ideal gas law to accommodate the effect of intermolecular forces. One of the earliest and most useful of these improved equations of state was proposed by Johannes van der Waals, a nineteenth-century Dutch scientist:

Van der Waals equation:

$$\left(P + \frac{an^2}{V^2}\right) \times (V - nb) = nRT \qquad (4)$$

The constant a represents the effect of attractions, and b the effect of repulsions. The constant b is a measure of the volume of the gas molecules themselves; it is zero in an ideal gas, in which the pointlike molecules have zero volume. Some experimental values for a and b are listed in Table 5.6. The two constants are obtained experimentally by adjusting their values until the van der Waals equation fits the observed dependence of the pressure on the volume, temperature, and amount of gas present.

▼ **EXAMPLE 5.8** Using the van der Waals equation

In a plant for the production of polyethylene, 10 kg of ethylene is compressed to 50 L at 20°C. What pressure does it exert?

STRATEGY So much gas in such a volume is likely to exert too great a pressure for the ideal gas law to be a good approximation. The problem then becomes one of using the van der Waals equation. Our goal is to calculate the pressure, so our first job is to rearrange Eq. 4 into an expression for P:

$$P = \frac{nRT}{V - nb} - \frac{an^2}{V^2}$$

We are given T (20°C corresponds to 293 K) and V (50 L), and we can calculate n from the mass of ethylene (10 kg) and its molecular weight, 28.05 amu. In addition, we need a and b, which are given in Table 5.6. It is a good idea to do such lengthy calculations term by term, and then bring the results together at the end.

SOLUTION The number of moles of ethylene present is

$$n = 10 \times 10^3 \text{ g C}_2\text{H}_4 \times \frac{1 \text{ mol C}_2\text{H}_4}{28.05 \text{ g C}_2\text{H}_4} = 357 \text{ mol C}_2\text{H}_4$$

so

$$nRT = 357 \text{ mol C}_2\text{H}_4 \times \frac{0.08206 \text{ L} \cdot \text{atm}}{1 \text{ K} \cdot \text{mol C}_2\text{H}_4} \times 293 \text{ K}$$

$$= 8580 \text{ L} \cdot \text{atm}$$

From Table 5.6, $a = 4.47$ L$^2 \cdot$ atm/mol^2 and $b = 0.057$ L/mol. Then

$$V - nb = 50 \text{ L} - 357 \text{ mol} \times 0.057 \text{ L/mol}$$

$$= 29.7 \text{ L}$$

$$a \times \left(\frac{n}{V}\right)^2 = \frac{4.47 \text{ L}^2 \cdot \text{atm}}{1 \text{ mol}^2} \times \left(\frac{357 \text{ mol}}{50 \text{ L}}\right)^2$$

$$= 228 \text{ atm}$$

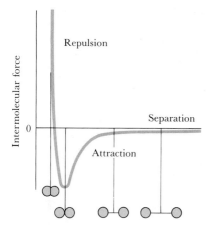

FIGURE 5.15 This curve shows how the intermolecular forces vary with separation. As two molecules approach (read from right to left), they attract each other increasingly strongly, as shown by the dropping of the curve well below zero. Then, when they are close enough to touch, they repel strongly, as shown by the rapid rise of the curve above zero.

TABLE 5.6 Van der Waals constants for some gases

	a, L$^2 \cdot$ atm/mol^2	b, L/mol
Air	1.4	0.039
Ammonia	4.17	0.037
Argon	1.35	0.032
Carbon dioxide	3.59	0.043
Ethylene	4.47	0.057
Helium	0.034	0.024
Hydrogen	0.244	0.027
Nitrogen	1.39	0.039
Oxygen	1.36	0.032

Combining these results as required by the van der Waals equation gives

$$P = \frac{8580 \text{ L} \cdot \text{atm}}{29.7 \text{ L}} - 228 \text{ atm}$$

$$= 61 \text{ atm}$$

The ideal gas law would predict 172 atm for this same pressure.

EXERCISE Calculate the pressure exerted by 100 g of carbon dioxide confined to a 500-mL flask at 25°C.

[*Answer*: 64 atm]

THE STOICHIOMETRY OF REACTING GASES

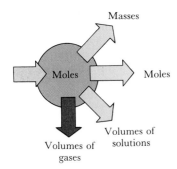

We saw in Section 4.1 how to calculate the number of moles of product when we know the number of moles of reactant or mass of reactant. If the product were a gas, and we knew the volume taken up by each mole of product molecules, we could go on to predict the volume of gas produced. In fact, we can add another arrow to the stoichiometry diagram (**1**)—and calculate reacting volumes for any reaction involving gases—if we can make the conversion

Volume of gas = number of moles × volume per mole

5.5 THE MOLAR VOLUME OF A GAS

The *molar volume* V_m of a substance (*any* substance, not only a gas) is the volume it occupies per mole of molecules:

$$V_m = \frac{\text{volume occupied}}{\text{moles of molecules}} = \frac{V}{n}$$

According to Avogadro's law, a gas consisting of 1 mol of X molecules occupies the same volume as one consisting of 1 mol of Y molecules, whatever the identities of substances X and Y. Therefore, the molar volume of O_2 should be the same as that of CO_2 or any other gas at the same temperature and pressure. The data in Table 5.4 and Fig. 5.16 confirm that this is approximately true under normal conditions.

Standard temperature and pressure. We can calculate the molar volume of an ideal gas as

$$\frac{V}{n} = \frac{RT}{P}$$

by substituting the appropriate values of P and T (Fig. 5.17). This equation shows that the molar volume decreases (the sample is compressed) as the pressure is increased at constant temperature. It also shows that the molar volume increases as the temperature is increased at constant pressure; that is, the increase in temperature causes the gas to expand.

To avoid having to compute the molar volume each time a calculation is done or reported, chemists generally report properties at an internationally agreed-upon *standard temperature and pressure* (STP), defined as 0°C (273.15 K) and 1 atm. At STP, the molar volume of most

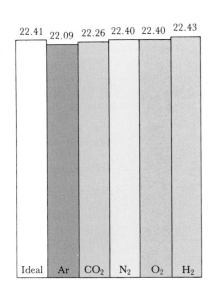

FIGURE 5.16 The molar volumes of several gases at the same temperature and pressure (at 1 atm pressure and 0°C, in this case) are all very similar.

Temperature (K)

	50	100	200	300	700	800	1000
0.1	41.0	82.1	164	246	576	657	821
1.0	4.1	8.2	16.4	24.6	57.5	65.7	82.1
3.0	1.4	2.8	5.6	8.2	19.2	21.9	27.4
10.0	0.41	0.82	4.64	2.46	5.75	6.57	8.21
30.0	0.14	0.28	0.56	0.82	1.92	2.19	2.74
100	0.04	0.08	0.16	0.25	0.58	0.66	0.82
300	0.01	0.03	0.06	0.08	0.19	0.22	0.27

Pressure (atm)

FIGURE 5.17 The molar volumes (in L/mol) of an ideal gas at different pressures and temperatures.

gases is approximately 22.4 L (about the volume of a cube of side 1 ft; Fig. 5.18); it is precisely this value if the gas is ideal. We can confirm this with the ideal gas equation:

$$V_m = \frac{RT}{P} = \frac{0.08206 \text{ L} \cdot \text{atm}}{1 \text{ K} \cdot \text{mol}} \times \frac{273.15 \text{ K}}{1.00 \text{ atm}}$$

$$= 22.4 \text{ L/mol}$$

Since 1.00 mol of gas molecules of any kind—argon, nitrogen, and so on—occupies 22.4 L at STP, we can write

$$1 \text{ mol Ar} = 22.4 \text{ L Ar}$$

$$1 \text{ mol N}_2 = 22.4 \text{ L N}_2$$

and so on, for any gas at STP. These relations can then be used to form the conversion factor between the number of moles of gas molecules and the volume the gas occupies at STP. If we want the volume at any other temperature and pressure, we use the ideal gas law.

The volume occupied by a given mass of gas. Sometimes we need the volume that a given *mass* of gas occupies at STP. The conversion to moles is the key to this calculation: we convert the mass of the sample to a number of moles, and then use the molar volume to convert this number of moles to liters.

▼ **EXAMPLE 5.9** **Calculating the volume of a given mass of gas**

Calculate the volume occupied by 10 g of CO_2 at STP.

STRATEGY Once we know the number of moles of CO_2 in the sample, we can use the molar volume to convert to liters. Therefore, we convert mass to moles first, and then moles to liters.

FIGURE 5.18 The molar volume of an ideal gas at 0°C and 1 atm.

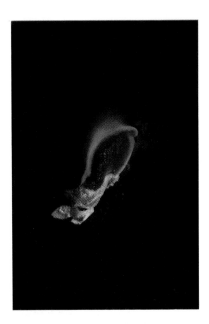

FIGURE 5.19 Sulfur burns with a blue flame and produces the dense gas sulfur dioxide.

SOLUTION We convert mass to moles using the molar mass of CO_2 (44.01 g/mol):

$$\text{Moles of } CO_2 = 10 \text{ g } CO_2 \times \frac{1 \text{ mol } CO_2}{44.01 \text{ g } CO_2}$$

$$= 0.227 \text{ mol } CO_2$$

Now we convert from moles to liters at STP:

$$\text{Volume of } CO_2 = 0.227 \text{ mol } CO_2 \times \frac{22.4 \text{ L } CO_2}{1 \text{ mol } CO_2}$$

$$= 5.1 \text{ L } CO_2$$

EXERCISE Calculate the volume occupied by 1.0 kg of hydrogen at STP.
[*Answer*: 1.1×10^4 L H_2]

The volumes of gases taking part in reactions. The molar volume is the last link in the chain we need to calculate the volumes of reacting gases. An example of the kind of calculation we can now do is to find the volume of sulfur dioxide produced by burning a specified mass of sulfur in the reaction

$$S_8(s) + 8O_2(g) \longrightarrow 8SO_2(g)$$

(see Fig. 5.19). As in the stoichiometry calculations of Chapter 4, we convert the given mass of sulfur to moles of S_8, then use the chemical equation to convert moles of S_8 to moles of SO_2, and finally (the new step in this chapter), convert moles of SO_2 to liters of the gas. This is illustrated in the following example.

▼ **EXAMPLE 5.10 Calculating the volume of gas produced by a given mass of reactant**

Calculate the volume of sulfur dioxide produced at STP by the combustion of 10 g of sulfur.

STRATEGY We follow the strategy set out above, converting from mass to moles of S_8, from S_8 to SO_2, and finally from moles of SO_2 to volume.

SOLUTION The conversion from mass of sulfur to moles of S_8 (of molar mass 256.48 g/mol) is

$$\text{Moles of } S_8 = 10 \text{ g } S_8 \times \frac{1 \text{ mol } S_8}{256.48 \text{ g } S_8}$$

$$= 0.0391 \text{ mol } S_8$$

Since

$$1 \text{ mol } S_8 = 8 \text{ mol } SO_2$$

when all the sulfur has been burned, we shall have

$$\text{Moles of } SO_2 = 0.0391 \text{ mol } S_8 \times \frac{8 \text{ mol } SO_2}{1 \text{ mol } S_8}$$

$$= 0.313 \text{ mol } SO_2$$

At STP, the molar volume of SO_2 is 22.4 L/mol; hence this number of moles of SO_2 molecules occupies

$$\text{Volume of SO}_2 = 0.313 \text{ mol SO}_2 \times \frac{22.4 \text{ L SO}_2}{1 \text{ mol SO}_2}$$

$$= 7.0 \text{ L SO}_2$$

EXERCISE Calculate the volume of acetylene (C_2H_2) produced at STP when 10 g of calcium carbide reacts completely with water.

[*Answer*: 3.5 L C_2H_2]

▼ **EXAMPLE 5.11** **Calculating the mass of reagent needed to react with a specified volume of gas**

A team working on a submarine design was investigating the use of potassium superoxide (KO_2) for purifying the air. The superoxide acts by combining with carbon dioxide and releasing oxygen:

$$4KO_2(s) + 2CO_2(g) \longrightarrow 2K_2CO_3(s) + 3O_2(g)$$

(see Fig. 5.20). Calculate the mass of KO_2 needed to react with 50 L of carbon dioxide at STP.

STRATEGY The overall logic of the calculation is the same as above, but it is carried out in reverse: we convert from volume of CO_2 to moles of CO_2, then from moles of CO_2 to moles of KO_2, and finally from moles of KO_2 to mass of KO_2 using its molar mass (71.1 g/mol).

SOLUTION The three conversions are based on the following equivalencies:

1 mol CO_2 = 22.4 L CO_2, obtained from the molar volume at STP
4 mol KO_2 = 2 mol CO_2, obtained from the chemical equation
1 mol KO_2 = 71.1 g KO_2, obtained from the molar mass of KO_2

The first conversion, from liters to moles of CO_2, is

$$\text{Moles of CO}_2 = 50 \text{ L CO}_2 \times \frac{1 \text{ mol CO}_2}{22.4 \text{ L CO}_2}$$

$$= 2.23 \text{ mol CO}_2$$

(We shall round at the end.) The second, from CO_2 to KO_2, is

$$\text{Moles of KO}_2 = 2.23 \text{ mol CO}_2 \times \frac{4 \text{ mol KO}_2}{2 \text{ mol CO}_2}$$

$$= 4.46 \text{ mol KO}_2$$

The third conversion, from moles to mass of KO_2, is

$$\text{Mass of KO}_2 = 4.46 \text{ mol KO}_2 \times \frac{71.1 \text{ g KO}_2}{1 \text{ mol KO}_2}$$

$$= 317 \text{ g KO}_2$$

which we round to 320 g of KO_2.

EXERCISE Calculate the volume of carbon dioxide at STP needed to make 1.0 g of glucose ($C_6H_{12}O_6$) by photosynthesis.

[*Answer*: 0.75 L CO_2]

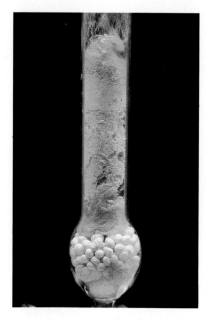

FIGURE 5.20 When carbon dioxide is passed over yellow potassium superoxide, it reacts to form colorless potassium carbonate. This reaction is used to remove CO_2 from air in confined spaces.

The molar volumes of gases, around 22.4 L/mol at STP, are much larger than those of liquids and solids, which are more typically 20×10^{-3} L/mol. The molar volume of liquid water, for instance, is 18 mL/mol. In other words, a mole of molecules in a gas occupies about a

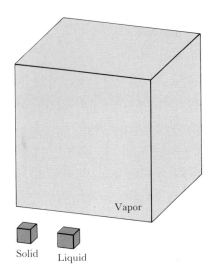

FIGURE 5.21 The molar volumes of solid, liquid, and gaseous water at STP.

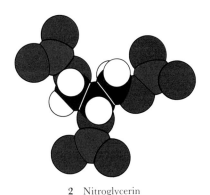

2   Nitroglycerin

thousand times the volume of a mole of molecules in a liquid or solid (Fig. 5.21). It follows that when liquids or solids react to form a gas in a reaction, there is about a thousandfold increase in volume. The increase may be even greater if several gas molecules are produced from each reactant molecule (Fig. 5.22).

One consequence of such an increase in volume is the explosive action of nitroglycerin ($C_3H_5N_3O_9$; **2**). When subjected to a shock wave from a detonator, it decomposes into many small molecules:

$$4C_3H_5N_3O_9(l) \longrightarrow 6N_2(g) + O_2(g) + 12CO_2(g) + 10H_2O(g)$$

Hence 4 mol of $C_3H_5N_3O_9$ molecules, corresponding to a little over 500 mL of the liquid, produce 29 mol of gas molecules of various kinds, giving a total of about 600 L of gas at STP. The pressure wave from the sudden 1200-fold expansion provides the destructive shock of the explosion. The detonator, which is typically lead azide, $Pb(N_3)_2$, works on a similar principle; it releases a large volume of nitrogen gas when it is struck:

$$Pb(N_3)_2(s) \longrightarrow Pb(s) + 3N_2(g)$$

5.6   GASEOUS MIXTURES

What we know about the properties of pure gases can be adapted very easily to describe the behavior of mixtures of gases. This should not be surprising. We have seen that the physical properties of gases are largely independent of their identities. Hence it is unimportant whether or not all the molecules in a sample are the same, and we can expect a mixture of gases to behave much like a single pure gas.

Dalton's law. Suppose that when a certain amount of oxygen is introduced into a container, it results in a pressure of 100 Torr. Suppose

FIGURE 5.22 An explosion caused by the ignition of coal dust. The shock wave is created by a tremendous expansion of volume as large numbers of gas molecules are formed.

SOLUTION We convert mass to moles using the molar mass of CO_2 (44.01 g/mol):

$$\text{Moles of } CO_2 = 10 \text{ g } CO_2 \times \frac{1 \text{ mol } CO_2}{44.01 \text{ g } CO_2}$$

$$= 0.227 \text{ mol } CO_2$$

Now we convert from moles to liters at STP:

$$\text{Volume of } CO_2 = 0.227 \text{ mol } CO_2 \times \frac{22.4 \text{ L } CO_2}{1 \text{ mol } CO_2}$$

$$= 5.1 \text{ L } CO_2$$

EXERCISE Calculate the volume occupied by 1.0 kg of hydrogen at STP.
[*Answer*: 1.1×10^4 L H_2]

The volumes of gases taking part in reactions. The molar volume is the last link in the chain we need to calculate the volumes of reacting gases. An example of the kind of calculation we can now do is to find the volume of sulfur dioxide produced by burning a specified mass of sulfur in the reaction

$$S_8(s) + 8O_2(g) \longrightarrow 8SO_2(g)$$

(see Fig. 5.19). As in the stoichiometry calculations of Chapter 4, we convert the given mass of sulfur to moles of S_8, then use the chemical equation to convert moles of S_8 to moles of SO_2, and finally (the new step in this chapter), convert moles of SO_2 to liters of the gas. This is illustrated in the following example.

▼ **EXAMPLE 5.10 Calculating the volume of gas produced by a given mass of reactant**

Calculate the volume of sulfur dioxide produced at STP by the combustion of 10 g of sulfur.

STRATEGY We follow the strategy set out above, converting from mass to moles of S_8, from S_8 to SO_2, and finally from moles of SO_2 to volume.

SOLUTION The conversion from mass of sulfur to moles of S_8 (of molar mass 256.48 g/mol) is

$$\text{Moles of } S_8 = 10 \text{ g } S_8 \times \frac{1 \text{ mol } S_8}{256.48 \text{ g } S_8}$$

$$= 0.0391 \text{ mol } S_8$$

Since

$$1 \text{ mol } S_8 = 8 \text{ mol } SO_2$$

when all the sulfur has been burned, we shall have

$$\text{Moles of } SO_2 = 0.0391 \text{ mol } S_8 \times \frac{8 \text{ mol } SO_2}{1 \text{ mol } S_8}$$

$$= 0.313 \text{ mol } SO_2$$

At STP, the molar volume of SO_2 is 22.4 L/mol; hence this number of moles of SO_2 molecules occupies

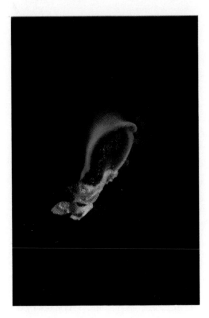

FIGURE 5.19 Sulfur burns with a blue flame and produces the dense gas sulfur dioxide.

FIGURE 5.17 The molar volumes (in L/mol) of an ideal gas at different pressures and temperatures.

Temperature (K)

Pressure (atm)	50	100	200	300	700	800	1000
0.1	41.0	82.1	164	246	576	657	821
1.0	4.1	8.2	16.4	24.6	57.5	65.7	82.1
3.0	1.4	2.8	5.6	8.2	19.2	21.9	27.4
10.0	0.41	0.82	4.64	2.46	5.75	6.57	8.21
30.0	0.14	0.28	0.56	0.82	1.92	2.19	2.74
100	0.04	0.08	0.16	0.25	0.58	0.66	0.82
300	0.01	0.03	0.06	0.08	0.19	0.22	0.27

gases is approximately 22.4 L (about the volume of a cube of side 1 ft; Fig. 5.18); it is precisely this value if the gas is ideal. We can confirm this with the ideal gas equation:

$$V_m = \frac{RT}{P} = \frac{0.08206 \text{ L} \cdot \text{atm}}{1 \text{ K} \cdot \text{mol}} \times \frac{273.15 \text{ K}}{1.00 \text{ atm}}$$

$$= 22.4 \text{ L/mol}$$

Since 1.00 mol of gas molecules of any kind—argon, nitrogen, and so on—occupies 22.4 L at STP, we can write

$$1 \text{ mol Ar} = 22.4 \text{ L Ar}$$

$$1 \text{ mol N}_2 = 22.4 \text{ L N}_2$$

and so on, for any gas at STP. These relations can then be used to form the conversion factor between the number of moles of gas molecules and the volume the gas occupies at STP. If we want the volume at any other temperature and pressure, we use the ideal gas law.

The volume occupied by a given mass of gas. Sometimes we need the volume that a given *mass* of gas occupies at STP. The conversion to moles is the key to this calculation: we convert the mass of the sample to a number of moles, and then use the molar volume to convert this number of moles to liters.

▼ **EXAMPLE 5.9** **Calculating the volume of a given mass of gas**

Calculate the volume occupied by 10 g of CO_2 at STP.

STRATEGY Once we know the number of moles of CO_2 in the sample, we can use the molar volume to convert to liters. Therefore, we convert mass to moles first, and then moles to liters.

FIGURE 5.18 The molar volume of an ideal gas at 0°C and 1 atm.

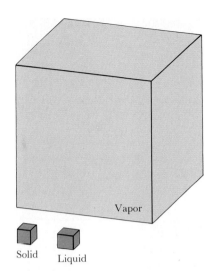

FIGURE 5.21 The molar volumes of solid, liquid, and gaseous water at STP.

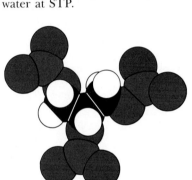

2 Nitroglycerin

FIGURE 5.22 An explosion caused by the ignition of coal dust. The shock wave is created by a tremendous expansion of volume as large numbers of gas molecules are formed.

thousand times the volume of a mole of molecules in a liquid or solid (Fig. 5.21). It follows that when liquids or solids react to form a gas in a reaction, there is about a thousandfold increase in volume. The increase may be even greater if several gas molecules are produced from each reactant molecule (Fig. 5.22).

One consequence of such an increase in volume is the explosive action of nitroglycerin ($C_3H_5N_3O_9$; **2**). When subjected to a shock wave from a detonator, it decomposes into many small molecules:

$$4C_3H_5N_3O_9(l) \longrightarrow 6N_2(g) + O_2(g) + 12CO_2(g) + 10H_2O(g)$$

Hence 4 mol of $C_3H_5N_3O_9$ molecules, corresponding to a little over 500 mL of the liquid, produce 29 mol of gas molecules of various kinds, giving a total of about 600 L of gas at STP. The pressure wave from the sudden 1200-fold expansion provides the destructive shock of the explosion. The detonator, which is typically lead azide, $Pb(N_3)_2$, works on a similar principle; it releases a large volume of nitrogen gas when it is struck:

$$Pb(N_3)_2(s) \longrightarrow Pb(s) + 3N_2(g)$$

5.6 GASEOUS MIXTURES

What we know about the properties of pure gases can be adapted very easily to describe the behavior of mixtures of gases. This should not be surprising. We have seen that the physical properties of gases are largely independent of their identities. Hence it is unimportant whether or not all the molecules in a sample are the same, and we can expect a mixture of gases to behave much like a single pure gas.

Dalton's law. Suppose that when a certain amount of oxygen is introduced into a container, it results in a pressure of 100 Torr. Suppose

also that when an amount of nitrogen at the same temperature is introduced instead, it results in a pressure of 200 Torr. John Dalton—whose contribution to atomic theory is described in Chapter 2—wondered what the total pressure would be if the two gases were present simultaneously. He made some fairly crude measurements and concluded that the total pressure of both gases in the same container would be 300 Torr, the sum of the individual pressures.

Dalton summarized his observations in terms of the "partial pressures" of gases:

> The **partial pressure** of a gas in a mixture is the pressure it would exert if it alone occupied the container.

In our example, the partial pressures of the oxygen and nitrogen in the mixture are 100 Torr and 200 Torr, respectively, since those are the pressures the gases exert when each one is alone in the container. Dalton then described the behavior of gaseous mixtures by his "law of partial pressures":

> **Dalton's law of partial pressures:** The total pressure of a mixture of gases is the sum of the partial pressures of its components.

Dalton's law is illustrated in Fig. 5.23.

The partial pressure P_A of a gas A depends on the number n_A of moles of A present, the temperature, and the total volume the mixture occupies. It is calculated as though no other gases were present, and hence can be found from the ideal gas law in the form

$$P_A = \frac{n_A RT}{V} \qquad (5)$$

▼ EXAMPLE 5.12 Calculating partial pressures

1.00 g of air consists of approximately 0.76 g of nitrogen and 0.24 g of oxygen. Calculate the partial pressures and the total pressure when this sample occupies a 1.00-L vessel at 20°C.

STRATEGY Collect the given data, and decide whether there are enough to calculate the unknown quantities using Eq. 5. If there is more than one unknown, it may be necessary to convert another item of data, or to obtain additional information from tables.

SOLUTION The data needed for the use of Eq. 5 are

$$P_{N_2} = ? \qquad n_{N_2} = ? \qquad V = 1.00 \text{ L} \qquad T = 293 \text{ K } (20°\text{C})$$
$$P_{O_2} = ? \qquad n_{O_2} = ? \qquad V = 1.00 \text{ L} \qquad T = 293 \text{ K } (20°\text{C})$$

The numbers of moles n_{N_2} and n_{O_2} can be obtained from the masses and molar masses (28.02 g/mol for N_2 and 32.00 g/mol for O_2):

$$\text{Moles of } N_2 = 0.76 \text{ g } N_2 \times \frac{1 \text{ mol } N_2}{28.02 \text{ g } N_2}$$

$$= 0.0271 \text{ mol } N_2$$

$$\text{Moles of } O_2 = 0.24 \text{ g } O_2 \times \frac{1 \text{ mol } O_2}{32.00 \text{ g } O_2}$$

$$= 0.00750 \text{ mol } O_2$$

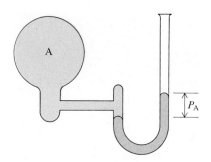

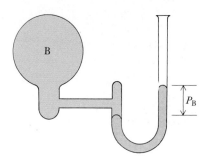

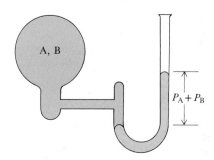

FIGURE 5.23 According to Dalton's law, the total pressure of a mixture of gases is the sum of the *partial pressures* of the components, the pressures they exert when each occupies the container alone (at the same temperature).

Then, from Eq. 5, we have

$$P_{N_2} = 0.0271 \text{ mol } N_2 \times \frac{0.08206 \text{ L} \cdot \text{atm}}{1 \text{ K} \cdot \text{mol}} \times \frac{293 \text{ K}}{1.00 \text{ L}}$$

$$= 0.65 \text{ atm } N_2$$

$$P_{O_2} = 0.00750 \text{ mol } O_2 \times \frac{0.08206 \text{ L} \cdot \text{atm}}{1 \text{ K} \cdot \text{mol}} \times \frac{293 \text{ K}}{1.00 \text{ L}}$$

$$= 0.18 \text{ atm } O_2$$

The total pressure is the sum of these partial pressures:

$$P = 0.65 \text{ atm} + 0.18 \text{ atm} = 0.83 \text{ atm}$$

EXERCISE The composition of dry air is given more precisely in Table 5.1. Calculate the partial pressures of all the components listed there and the total pressure when 1.00 g of dry air is confined to a 500-mL container (such as a bicycle tire) at 25°C.

[*Answer*: $P_{N_2} = 1.32$ atm; $P_{O_2} = 0.354$ atm; $P_{Ar} = 0.0158$ atm; $P_{CO_2} = 5.56 \times 10^{-4}$ atm; $P = 1.69$ atm]

Measuring moles of gas collected over water. One example of the kind of stoichiometry problem we need to be able to solve arises when we collect a gas over water (Fig. 5.24). This is a common way of collecting a not very soluble gas, for example, oxygen, that has been prepared in the laboratory. The gas inside the inverted collecting bottle is a mixture of water vapor and oxygen; knowing the total volume of gas collected, we need to determine the number of moles of O_2 produced by the reaction.

We might, for example, be interested in the number of moles of O_2 that can be obtained by heating a sample of potassium chlorate:

$$2KClO_3(s) \xrightarrow{\Delta} 2KCl(s) + 3O_2(g)$$

(In practice this reaction is carried out in the presence of a little manganese dioxide, which acts as a "catalyst," a substance that helps the reaction to run quickly and smoothly.) If we could measure the volume of gas collected and the partial pressure of the O_2, then we could use

$$n_{O_2} = \frac{P_{O_2} V}{RT}$$

to find the number of moles of O_2 produced. The problem is to determine P_{O_2}, since we can measure only the *total* pressure P, the sum of the partial pressures of the oxygen and the water vapor. We measure P by adjusting the height of the bottle so that the levels of the water inside and outside the bottle are equal (Fig. 5.25). Then the total pressure inside the bottle is equal to the atmospheric pressure, which can be measured with a barometer. Since the total pressure (P) of the gas inside the bottle is the sum of the partial pressures of the oxygen (P_{O_2}) and the water vapor (P_{H_2O}),

$$P_{O_2} = P - P_{H_2O}$$

The partial pressure of water vapor above liquid water is called its *vapor pressure* (a property that we explore in much more detail in Section 10.4). For the present all we need know is that the vapor pressure of water depends on the temperature as listed in Table 5.7. At room tem-

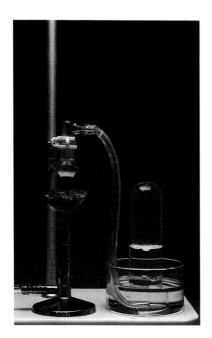

FIGURE 5.24 Gases produced by reactions are often collected over water in a "gas bottle." In this experiment, potassium chlorate is being heated with a small amount of black manganese dioxide, and the oxygen produced is being collected.

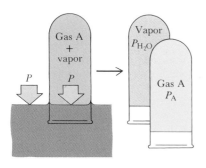

FIGURE 5.25 The partial pressure of gas A can be measured if the heights of water inside and outside the bottle are equalized, so that the total pressure inside is equal to the atmospheric pressure. Then the vapor pressure of the water may be subtracted from the atmospheric pressure to give the pressure due to the gas alone.

perature the vapor pressure of water is only about 20 Torr—less than 3% of atmospheric pressure—so the partial pressure of the oxygen in the collecting bottle is not very different from the total pressure.

▼ **EXAMPLE 5.13** **Measuring moles of gas produced by a reaction**

Some potassium chlorate was heated with a little manganese dioxide as catalyst on a day when the atmospheric pressure was 755.2 Torr, and a volume of 370 mL of gas was collected in a gas bottle over water at 20°C. How many moles of O_2 were produced?

STRATEGY We can calculate the number of moles using the ideal gas law once we know the partial pressure, the volume, and the temperature. The volume and temperature are given. According to Dalton's law, the partial pressure of the oxygen is the difference between the total pressure and the vapor pressure of water, which can be obtained from Table 5.7.

SOLUTION The partial pressure of the oxygen in the bottle is

$$P_{O_2} = 755.2 \text{ Torr} - 17.5 \text{ Torr} = 737.7 \text{ Torr}$$

Since at 20°C (293 K),

$$RT = \frac{62.37 \text{ L} \cdot \text{Torr}}{1 \text{ K} \cdot \text{mol}} \times 293 \text{ K}$$

$$= 1.83 \times 10^4 \text{ L} \cdot \text{Torr/mol}$$

we have, for the oxygen,

$$n_{O_2} = P_{O_2} V \times \frac{1}{RT}$$

$$= 737.7 \text{ Torr} \times 0.370 \text{ L} \times \frac{1 \text{ mol}}{1.83 \times 10^4 \text{ L} \cdot \text{Torr}}$$

$$= 1.49 \times 10^{-2} \text{ mol}$$

That is, 14.9 mmol of O_2 molecules are produced.

EXERCISE When a small piece of zinc was dissolved in dilute hydrochloric acid, 446 mL of hydrogen was collected over water at 25°C and 760 Torr. Calculate the number of moles of H_2 produced.

[*Answer*: 17.7 mmol H_2]

▲

Partial pressure and mole fraction. A very useful measure of the compositions of mixtures is the "mole fraction" (Fig. 5.26):

The **mole fraction** of a substance in a mixture is the number of moles of the substance in the mixture, expressed as a fraction of the total number of moles in the mixture.

The mole fraction applies equally to solid, liquid, and gaseous mixtures, but in this chapter we apply it only to gases. For a "binary mix-

TABLE 5.7 **The vapor pressure of water**

Temperature, °C	Vapor pressure, Torr
0	4.48
2	5.29
4	6.10
6	7.01
8	8.05
10	9.21
12	10.52
14	11.99
16	13.63
18	15.48
20	17.54
25	23.76
30	31.82
40	55.32
50	92.51
60	149.38
70	233.7
80	355.1
90	525.8
100	760.0

FIGURE 5.26 The mole fraction x_A of a substance A in a mixture is the number of moles of A expressed as a fraction of the total number of moles in the mixture.

$x_A = 0.33$

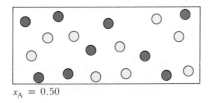
$x_A = 0.50$

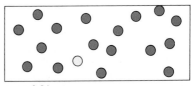
$x_A = 0.94$

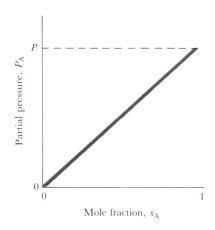

FIGURE 5.27 The partial pressure P_A of a gas in a mixture is proportional to its mole fraction x_A.

ture," a mixture of two substances, the mole fractions x_A and x_B of the two components A and B are

$$x_A = \frac{n_A}{n_A + n_B} \qquad x_B = \frac{n_B}{n_A + n_B} \qquad (6)$$

The mole fraction of A is zero when no A molecules are present, and 1 when the sample is pure A. If half the molecules are A, then the mole fraction of A is 0.5.

The partial pressure of a gas A can be calculated from its mole fraction as

$$P_A = x_A \times P \qquad (7)$$

where P is the total pressure of the mixture. To show that this is so, we first apply Eq. 5 to each substance in the mixture:

$$P_A = n_A \times \frac{RT}{V} \qquad P_B = n_B \times \frac{RT}{V}$$

The total pressure is the sum of both the partial pressures, so

$$P = P_A + P_B = (n_A + n_B) \times \frac{RT}{V}$$

Dividing the first equation by the last then gives us

$$\frac{P_A}{P} = \frac{n_A}{n_A + n_B}$$

The term on the right is x_A, so

$$\frac{P_A}{P} = x_A \qquad \text{or} \qquad P_A = x_A \times P$$

Equation 7 shows that the partial pressure of gas A is zero if its mole fraction x_A is zero (no molecules of A are present). If only A molecules are present ($x_A = 1$), the partial pressure of A is the same as the total pressure. If half the molecules are A ($x_A = 0.5$), its partial pressure is half the total pressure (Fig. 5.27).

THE KINETIC THEORY OF GASES

A gas is a collection of molecules that are in continuous, chaotic motion and, except during collisions, are widely separated from each other (Fig. 5.28). A molecule may be traveling through space one moment, but a fraction of a second later it might collide with another molecule. After the collision it travels off in a different direction, only to collide almost at once with yet another molecule. This description is the basis of the *kinetic theory* of gases (from the Greek word for "move"), which is based on three simple assumptions:

1. A gas consists of a collection of molecules in continuous random motion.

2. The molecules are infinitely small pointlike particles that move in straight lines until they collide.

3. Gas molecules do not influence each other except during collisions.

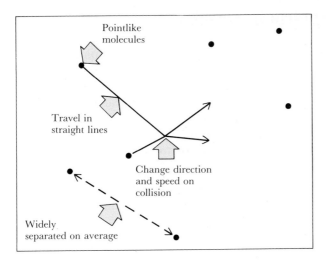

FIGURE 5.28 The assumptions
of the kinetic theory of gases.
These assumptions account for
the properties of an ideal gas.

Despite this simplicity, the theory leads to an interpretation of the bulk
properties of the gas in terms of the properties of its molecules. In
particular, it leads to the conclusion that the pressure and volume of a
gas are related by the equation

$$PV = \tfrac{1}{3}nMv^2 \tag{8}$$

where n is the number of moles of gas molecules present, M is their
molar mass, and v is their *root-mean-square speed* (or "rms speed"), defined as the square root of the average of the squared speeds of all the
molecules in the sample:

$$v = \sqrt{\frac{1}{k}(v_1{}^2 + v_2{}^2 + \cdots + v_k{}^2)}$$

The greater the average speed of the molecules in the gas, the greater
the rms speed.

▼ EXAMPLE 5.14 Calculating a root-mean-square speed

Traffic police measured the speeds of five cars as 55.1, 60.5, 62.5, 68.2, and
70.1 mi/h. What is their root-mean-square speed and their average speed?

STRATEGY To find the rms speed, we form the sum of the squares of the
speeds, divide by the number of cars, and then take the square root. The
average speed is the sum of the speeds divided by the number of cars.

SOLUTION The sum of the squares of the speeds is

$$(55.1)^2 + (60.5)^2 + (62.5)^2 + (68.2)^2 + (70.1)^2 = 2.017 \times 10^4 \ (\text{mi/h})^2$$

Division by 5 gives 4.034×10^3 (mi/h)^2, and taking the square root yields
63.5 mi/h as the rms speed. The sum of the speeds is

$$55.1 + 60.5 + 62.5 + 68.2 + 70.1 = 316.4 \ \text{mi/h}$$

Division by 5 gives 63.3 mi/h for the average speed.

EXERCISE Four cars have speeds 34.1, 36.2, 38.3, and 38.3 mi/h. What is
their root-mean-square speed and their average speed?

[*Answer*: 36.8 mi/h; 36.7 mi/h]

5.7 MOLECULAR SPEEDS

Gas molecules travel at about the speed of sound. The reasoning behind this remark is that sound is a wave of pressure carried through a substance by its molecules. Since the motion of the sound wave depends on the speed at which the molecules can move, it is likely that the speed of sound in a gas is close to the average speed of its molecules. The speed of sound in air is about 300 m/s (about 700 mi/h), so we can expect average molecular speeds to be near that value.

Speed and temperature. The kinetic theory of gases gives Eq. 8, from which we can find PV. However, we also know that $PV = nRT$. Hence we can equate these two expressions, obtaining

$$\tfrac{1}{3}nMv^2 = nRT$$

Dividing both sides by n and rearranging them gives

$$v^2 = \frac{3RT}{M}$$

Finally, taking the square root of both sides gives a relation between the root-mean-square speed of the molecules and the temperature:

$$v = \sqrt{\frac{3RT}{M}} \tag{9}$$

In passing we note that the rms speed depends on R. Hence, as mentioned in Section 5.2, measuring the speed of sound in a gas is one way of measuring the gas constant.

Equation 9 shows that the rms speed and average speed of the molecules of a gas increase as the temperature increases. In particular, it shows that the speed of molecules in a gas is proportional to the square root of the temperature (in kelvins). Doubling the temperature of *any* gas (from 200 K to 400 K, for example) increases the rms speed of its molecules by a factor of $\sqrt{2} = 1.4$. When the temperature of *any* gas is raised from 20°C (293 K) to 1000°C (1273 K), the rms speed of its molecules increases by a factor of

$$\sqrt{\frac{1273 \text{ K}}{293 \text{ K}}} = 2.08$$

Detailed calculations show that the rms speed of oxygen molecules at 20°C is about 500 m/s (nearly 1100 mi/h), so in a furnace where the temperature is close to 1000°C, the rms speed must be around 1000 m/s, or 2200 mi/h.

Speed and molecular weight. Equation 9 shows that, for a given temperature, the higher the molecular weight of the gas, the lower the average speed of its molecules. Thus carbon dioxide molecules travel, on average, more slowly than oxygen molecules. The equation also shows that the rms speed of the molecules in a gas is inversely proportional to the *square root* of the molar mass. Since the molar masses of CO_2 and O_2 are 44.01 and 32.00 g/mol, respectively, it follows that the rms speeds of these molecules are in the ratio

$$\sqrt{\frac{32.00}{44.01}} = 0.85$$

At any given temperature, then, the rms speed of CO_2 molecules is only 85% of that of O_2 molecules. A sample of damp air at 25°C can therefore be pictured as a storm of molecules, with the CO_2 molecules, the heaviest molecules present, lumbering along on average at about 920 mi/h while H_2O molecules, the lightest present, zip around at 1440 mi/h or so. Some rms molecular speeds are shown in Fig. 5.29.

Diffusion. The spreading of one substance through another substance is called *diffusion*. A solid diffuses into a liquid as it dissolves, and a solid can diffuse slowly into another solid where they are in contact. Here we shall concentrate on the diffusion of one gas into another. This occurs whenever two or more gases are introduced into the same region of space or are separated by a porous barrier. The gases diffuse because both types of molecules move chaotically and quickly mingle together.

The diffusion of one gas into another contributes to the transport of pheromones (chemical signals between animals) and perfumes through air. It helps to keep the composition of the atmosphere approximately constant, since gases in abnormally high concentrations diffuse away. Even in the absence of wind, diffusion helps to disperse gases leaking from chemical plants.

The rate of diffusion of a gas depends on the speed of its molecules: the faster they move, the more quickly they mingle. We have seen that, so long as the temperature is constant, the average speed of the molecules in a gas is inversely proportional to the square root of their molecular weight. Therefore, the rates of diffusion, as measured by the number of moles of gas molecules diffusing per second, should show the same dependence (Fig. 5.30). This is the content of a law proposed by Thomas Graham, a nineteenth century Scottish chemist:

Graham's law: The rate of diffusion of a gas is inversely proportional to the square root of its molecular weight.

Graham's law has been used to measure the molecular weights of volatile compounds, but more accurate methods (particularly mass spectrometry, discussed in Section 2.4) are now available. Since the rate of diffusion is inversely proportional to the time a given amount of gas takes to diffuse, Graham's law implies that the time t required for diffusion is *directly* proportional to the square root of the molecular weight:

$$t \propto \sqrt{MW}$$

The ratio of the times two different gases require to diffuse when they are at the same temperature and pressure is therefore

$$\frac{t_A}{t_B} = \sqrt{\frac{\text{molecular weight of gas A}}{\text{molecular weight of gas B}}} \qquad (10)$$

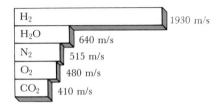

H_2	1930 m/s
H_2O	640 m/s
N_2	515 m/s
O_2	480 m/s
CO_2	410 m/s

FIGURE 5.29 The average speeds of H_2 and of the molecules in air at 20°C. Hydrogen is included to emphasize how much faster light molecules travel than heavy ones.

FIGURE 5.30 The plug on the left is soaked in hydrochloric acid, and that on the right in aqueous ammonia. Reaction occurs where the two diffusing gases meet. The reaction occurs closer to the HCl end, which has the more slowly diffusing molecules.

▼ **EXAMPLE 5.15 Measuring molecular weight using Graham's law**

Volatile liquids sold as "Freons" are used in refrigeration units; they are compounds of carbon, chlorine, and fluorine. It was found that a certain amount of a Freon gas required 186 s to diffuse through a porous plug when the pressure was held constant. The same amount of carbon dioxide required 112 s. Calculate the molecular weight of the Freon.

STRATEGY Since the Freon takes longer to diffuse, we can guess that its molecular weight is greater than that of CO_2 (which is 44.01 amu). Since the data give the time for carbon dioxide to diffuse, and its molecular weight is known, the only unknown is the molecular weight of the Freon. This is obtained by rearranging Eq. 10 to

$$\text{Molecular weight of Freon} = \text{MW}(CO_2) \times \left(\frac{t_{Freon}}{t_{CO_2}}\right)^2$$

and substituting the data.

SOLUTION The rearranged equation gives

$$\text{MW(Freon)} = 44.01 \text{ amu} \times \left(\frac{186 \text{ s}}{112 \text{ s}}\right)^2$$

$$= 121 \text{ amu}$$

This value is consistent with the molecular formula CCl_2F_2 for this Freon (which is sold as "Freon-12").

EXERCISE Carvone is the flavor component of spearmint. The time required for a certain number of moles of carvone molecules in vapor form to diffuse through a porous plug was 186 s. Under the same conditions, the same number of moles of argon atoms diffused in 96 s. Calculate the molecular weight of carvone.

[*Answer*: 150 amu]

▲

Graham's law explains why some isotope separation plants are so large and difficult to hide. Nuclear power generation depends on our being able to separate uranium-235 from the much more abundant uranium-238. One separation technique uses a series of reactions to convert uranium ore to the volatile solid uranium hexafluoride. Then UF_6 vapor is allowed to diffuse through a series of porous barriers. The molecules containing uranium-235, being lighter than those containing uranium-238, diffuse more quickly and thus may be separated from the rest. However, since the ratio of the molecular weights of $^{238}UF_6$ to $^{235}UF_6$ is only 1.008, the ratio of their diffusion rates is only $\sqrt{1.008} = 1.004$, so very little separation occurs at each diffusion stage. To improve the separation efficiency, the vapor is passed through many stages, with the result that the plants must be very large. The original plant in Oak Ridge, Tennessee had 4000 diffusion stages (Fig. 5.31), and covered an area of 43 acres.

The Maxwell distribution of speeds. The *average* speeds (more precisely, the rms speeds) of gas molecules are given by Eq. 9. However, like cars in traffic, individual molecules have speeds that vary over a very wide range. Moreover, as in a traffic collision, two molecules colliding almost head on might be brought almost to a standstill. Then, in the next instant, one of these molecules might be struck to one side by

FIGURE 5.31 The individual diffusion stages in the original uranium-235 diffusion plant at Oak Ridge, Tennessee. There are thousands in the entire plant.

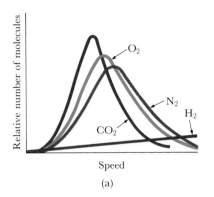

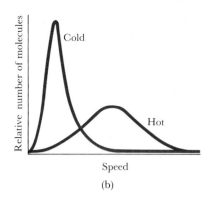

FIGURE 5.32 (a) The range of molecular speeds for several gases, as given by the Maxwell distribution. All the curves correspond to the same temperature. The greater the molecular weight, the narrower the spread of speeds. (b) This diagram also shows the Maxwell distribution, but now the curves correspond to a single substance at different temperatures. The higher the temperature, the broader the spread of speeds.

another and move off at the speed of sound. An individual molecule undergoes several billion changes of speed each second.

The percentage of the molecules of a gas sample that are moving at each speed at any instant is called the *distribution* of molecular speeds for that gas. The general formula was first calculated by the Scottish scientist James Maxwell, whose conclusions are summarized in Fig. 5.32. The graphs in Fig. 5.32a show that the heaviest molecules (CO_2) travel with speeds close to their average value. A light molecule (H_2) not only has a higher average speed, but the speeds of many individual molecules are very different from the average speed. This implies that many light molecules are likely to have such high speeds that they escape from the gravitational pull of small planets and go off into space. One consequence is that hydrogen molecules and helium atoms, which are both very light, are very rare in the earth's atmosphere.

The graph in Fig. 5.32b shows that the spread of the speeds widens as the temperature increases. At low temperatures most molecules of a gas have speeds close to the average for the gas. At high temperatures a high proportion have speeds widely different from their average. This has important consequences for the rates of chemical reactions, as we shall see in Section 12.6.

5.8 THE LIQUEFACTION OF GASES

At low temperatures, gas molecules move so slowly that intermolecular attraction may cause one molecule to be captured by another and stick to it instead of moving freely. When the temperature is reduced to below the boiling point of the substance, the gas condenses to a liquid: the molecules move too slowly to escape from each other, and the entire sample condenses to a jostling crowd of molecules held together by the attractions between them.

The simplest method of liquefying a gas is to immerse a sample in a bath kept at a lower temperature than the boiling point of the gas. Chlorine (with boiling point $-35°C$) condenses to a liquid when a sealed tube of the gas is immersed in a bath cooled with solid carbon dioxide (Fig. 5.33). Solid carbon dioxide is readily available commercially as "dry ice." It is so called because it evaporates directly to a gas without first forming a liquid. Temperatures as low as about $-77°C$ can be reached by adding chips of solid carbon dioxide to a low-freezing-point liquid such as acetone.

FIGURE 5.33 Chlorine can be liquefied at atmospheric pressure by cooling it to below $-35°C$.

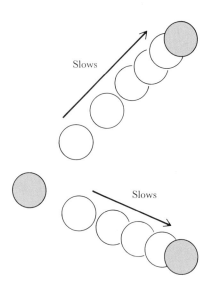

Gases can also be liquefied by making use of the relation between temperature and molecular speed. Since lower average speed corresponds to lower temperature, slowing the molecules is equivalent to cooling the gas. Molecules can be slowed by making use of the attractions between them, exactly as the gravitational pull of the earth slows a ball that has been thrown into the air. An expanding gas is like a vast number of balls separating from a vast number of earths, and although the attractions are not gravitational they have the same effect: increasing the average separation of molecules in a gas lowers their average speed (Fig. 5.34). In other words, as a gas expands, it cools. This is called the *Joule-Thomson effect* in honor of James Joule and William Thomson (later Lord Kelvin of the temperature scale), who first studied it.

The Joule-Thomson effect is used in some commercial refrigerators to liquefy gases. The gas to be liquefied is compressed and then allowed to expand through a small hole, the "throttle." The gas cools as it expands, and the cooled gas is circulated past the incoming compressed gas. This cools the incoming gas before it expands and cools still further. As the gas is continually recompressed and recirculated, its temperature progressively falls until finally it condenses to a liquid. If the gas is a mixture like air, the liquid it forms can be fractionally distilled to separate its components. This is the technique used for harvesting nitrogen, oxygen, neon, argon, krypton, and xenon from the atmosphere.

FIGURE 5.34 Cooling by the Joule-Thomson effect is a result of the molecules traveling more slowly as they climb away from each other's attraction.

SUMMARY

5.1 Gases exert a **pressure,** or force per unit area, that is measured with either a **barometer** (for the atmosphere) or a **manometer** (for the pressure inside a container). Gas pressure arises from the impact of gas molecules on surfaces.

5.2 and 5.3 The relation between the pressure, volume, number of moles, and temperature of a substance is called its **equation of state.** At low pressures, all gases obey the same equation of state—the **ideal gas law** ($PV = nRT$). Special cases of this equation are **Boyle's law** for the dependence of volume on pressure, **Charles's law** for the dependence of volume on temperature, and **Avogadro's law** for the relation between volume and number of moles of gas present. A hypothetical gas that obeys the ideal gas law exactly is called an **ideal gas.** All gases behave ideally at very low pressures, and approximately ideally at normal pressures and temperatures. The ideal gas law provides a way of calculating the molecular weight of a gas from its density, and of predicting the changes that occur when the pressure and temperature are varied.

5.4 Real gases behave differently from an ideal gas, especially at high pressure. The behavior of a real gas is described by the **van der Waals equation,** which takes into account the effects of **intermolecular forces**—the attractions and repulsions among gas molecules.

5.5 At **standard temperature and pressure** (0°C and 1 atm) all gases have very nearly the same **molar volume,** the volume per mole of molecules. This molar volume can be used as a conversion factor to relate the number of moles of gas molecules in a sample to the volume it occupies. It is also the conversion factor needed to extend the stoichiometric calculations of Chapter 4 to the calculation of the volumes of gases taking part in reactions. When volumes at other temperatures and pressures are needed, they are obtained using the ideal gas law.

5.6 The pressure of a mixture of gases is described by **Dalton's law of partial pressures,** which treats each gas in a mixture as though it alone were present in the container. The partial pressure of a gas in a mixture is proportional to its **mole fraction,** the number of moles of the gas present, expressed as a fraction of the total number in the mixture; it can be calculated from the ideal gas law.

5.7 The **kinetic theory of gases** supposes that the molecules of a gas are in chaotic random motion, have negligible size, and move independently in straight lines except during collisions. The theory leads to the conclusion that the **root-mean-square speed** of the molecules in a gas is proportional to the square root of their temperature and inversely proportional to the square root of

their molecular weight. The mobility of gas molecules accounts for the rapidity with which they **diffuse,** or spread, through substances. The kinetic theory accounts for **Graham's law,** which states that the rate of diffusion of a gas is inversely proportional to the square root of its molecular weight. The proportion of molecules having a particular speed at any temperature is given by the **Maxwell distribution.**

5.8 Gases are liquefied either by cooling in a cold bath or by making use of the **Joule-Thomson effect,** the cooling caused by expansion.

EXERCISES

Pressure

In Exercises 5.1 to 5.4, complete the given conversions using Table 5.4.

5.1 (a) 1 bar = _____ atm
  (b) 1 Torr = _____ mmHg
  (c) 1 N/m^2 = _____ Torr
  (d) 1 mmHg = _____ Pa
5.2 (a) 1 Pa = _____ Torr
  (b) 1 kPa = _____ N/m^2
  (c) 1 atm = _____ mbar
  (d) 1 Torr = _____ Pa

5.3 (a) 760 Torr = _____ atm
  (b) 1 bar = _____ kPa
  (c) 101 kPa = _____ Torr
  (d) 14 lb/in^2 = _____ kPa
5.4 (a) 1 kbar = _____ Torr
  (b) 1000 atm = _____ kbar
  (c) 1000 atm = _____ lb/in^2
  (d) 14 lb/in^2 = _____ N/m^2

For Exercises 5.5 to 5.10, note that the pressure exerted by a column of liquid or gas of height h and density d is given by the product gdh, where $g = 9.81$ m/s^2 is the acceleration of gravity. The density of water is 1 g/cm^3; the density of mercury is 13.6 g/cm^3. In Exercises 5.5 and 5.6, express the given pressures in the units specified.

5.5 (a) The pressure (in atmospheres and Torr) exerted by a 1.0-m-high column of water.
  (b) The total pressure (in pounds per square inch and atmospheres) 2 m below the surface of a swimming pool.
5.6 (a) The pressure (in atmospheres and pounds per square inch) exerted by a 1.0-m-high column of mercury.
  (b) The pressure difference (in atmospheres and Torr) between the top and bottom of a 15-cm vertical drinking straw full of a waterlike liquid. Suppose you

could suck the liquid up to a height of 100 cm using your lungs; how high could you suck a column of mercury? (Do not check the answer experimentally, because inhaling mercury is very dangerous.)

5.7 What is the difference in atmospheric pressure between the roof and ground at the World Trade Center in New York (1377 ft high) when the average air density is 1.2 g/L?
5.8 What is the difference in pressure between the surface and the bottom of a 6-ft-deep swimming pool?

5.9 Atmospheric pressure is typically 1.0 atm at sea level. The air density is 1.2 g/L. What is the mass of a column of air, of cross-sectional area 1 m^2, extending upward from sea level?
5.10 The atmospheric pressure at the top of Mt. Everest is 0.27 atm, and its summit is 29,000 ft above sea level. Estimate the mass of a column of air of cross-sectional area 1 m^2, extending (a) above and (b) below its summit.

In Exercises 5.11 and 5.12, calculate the pressure in Torr of a sample of gas when the external pressure is 752.8 Torr and a water manometer reads as given (the density of water at the temperature of the experiment is 0.9978 g/cm^3).

5.11 (a) Sample side is 15.0 cm lower than atmosphere side.
    (b) Sample side is 10.2 cm higher than atmosphere side.
5.12 (a) Sample side is 12.5 cm higher than atmosphere side.
    (b) Sample side is 15.2 cm lower than atmosphere side.
5.13 A manometer was constructed using benzene (density 0.879 g/mL) instead of water (density 0.997 g/mL). What height would the column in the benzene manometer be if the height in a water manometer was 10.0 cm?

5.14 A manometer was constructed using heavy water

(density 1.104 g/mL) instead of ordinary water (density 0.997 g/mL). What height would the column in the heavy-water manometer be if the height in an ordinary-water manometer was 15.7 cm?

The Ideal Gas Law

In Exercises 5.15 and 5.16, calculate the final volume if the pressure is changed as indicated while the temperature is held constant.

5.15 (a) The pressure on 1.00 L of hydrogen is increased from 1.00 atm to 2.00 atm.

(b) The pressure on 500 mL of air is decreased from 753 Torr to 693 Torr.

5.16 (a) The pressure on 500 mL of air is decreased from 1.00 atm to 0.90 atm.

(b) The pressure on 100 mL of carbon dioxide is increased from 165 Torr to 760 Torr.

In Exercises 5.17 and 5.18, calculate the pressure needed to produce the given final volume while the temperature is held constant.

5.17 (a) Expand 1.00 L of air at 1.00 atm to 2.00 L.

(b) Compress 450 mL of hydrogen at 760 Torr to 350 mL.

(c) Compress 600 mL of helium at 0.900 atm to 350 mL.

5.18 (a) Compress 1.00 L of air at 1.00 atm to 500 mL.

(b) Expand 350 mL of hydrogen at 760 Torr to 500 mL.

(c) Expand 1.00 mL of helium at 0.10 atm to 1.00 L.

In Exercises 5.19 and 5.20, calculate the initial volume of gas needed to produce each final volume, assuming no change in temperature.

5.19 (a) Produce 1.00 L of air at 10.0 kPa from air at 1.00 atm.

(b) Produce 10.0 mL of argon at 1000 atm from argon at 753.5 Torr.

5.20 (a) Produce 10.0 mL of helium at 1.0 kbar from helium at 99.8 kPa.

(b) Produce 10 L of air at 1.0×10^{-6} Torr from air at 1.02 atm.

For Exercises 5.21 and 5.22, suppose that the pressure is held constant in a certain industrial process. Calculate the final volume when the temperature of each sample of gas is changed as indicated.

5.21 (a) 1.00 L of air at 20.0°C is heated to 25.0°C.

(b) 250 mL of argon at −50.0°C is heated to 1000°C.

5.22 (a) 100.00 mL of hydrogen at 20.0°C is cooled to 0.0°C.

(b) 250 mL of air at 0.0°C is heated to 100°C.

In Exercises 5.23 and 5.24, calculate the Celsius temperature required to produce each final volume, assuming that the pressure and mass are held constant.

5.23 (a) Produce 100.0 mL of air from 200.0 mL at 25.0°C.

(b) Produce 50.00 mL of helium from 51.00 mL at 20.0°C.

5.24 (a) Produce 1.00 L of air from 100.0 mL at 25.0°C.

(b) Produce 50.0 mL of helium from 49.0 mL at 20.0°C.

In Exercises 5.25 and 5.26, determine whether or not the given data confirm Charles's law.

5.25

	Temperature, °C							
	−10	0	10	20	30	40	50	60
Volume of sample 1, mL	120	125	129	134	138	143	147	152
Volume of sample 2, mL	100	104	108	111	115	119	123	127

5.26

	Temperature, °C								
	−30	−20	−10	0	20	40	60	80	100
Volume of sample 1, mL	92	96	100	104	111	119	127	134	142
Volume of sample 2, mL	66	69	71	74	79	85	90	96	101

In Exercises 5.27 and 5.28, calculate the final pressure when the given samples are heated at constant volume.

5.27 (a) A sample of air at 760.0 Torr and 0.0°C is heated to 25.0°C.

(b) A sample of helium at 25.0°C and 1.00 atm is cooled to 4.0 K.

5.28 (a) A sample of air at 760 Torr and 20.0°C is cooled to 0.00°C.

(b) A sample of air at 25.0°C and 690 Torr is heated to 1000°C.

In Exercises 5.29 and 5.30, calculate the pressure (in Torr and atmospheres) of the given samples.

5.29 (a) A 250.0-mL flask containing 0.100 mol of Ar atoms at 30.0°C.

(b) A 100-mL flask containing 0.010 g of carbon dioxide at 100.0°C.

5.30 (a) A 1.0-L flask containing 0.200 mol of Ne at 25°C.

(b) A 250-mL flask containing 0.150 g of ammonia at 60.0°C.

In Exercises 5.31 and 5.32, calculate the mass of the given gas needed to produce the given conditions in a 250.0-mL flask.

5.31 (a) Air at 1.00 atm and 20.0°C.
(b) Carbon dioxide at 10.0 Torr and 100.0°C.
5.32 (a) Argon at 1.00 Torr and 0.00°C.
(b) Hydrogen at 1.0 Pa pressure at 25°C.

In Exercises 5.33 and 5.34, calculate the final pressure of the gas.

5.33 1.00 L of air at 760 Torr and 20.0°C that is cooled to −20.0°C and compressed to 750 mL.
5.34 500.0 mL of helium at 1.00 atm and 0.00°C that is heated to 100.0°C and compressed to 300.0 mL.

In Exercises 5.35 and 5.36, calculate the final volume of a 100.0-mL sample of air that is subjected to the given changes.

5.35 Originally at 0.0°C and 1.00 atm, it is heated to 25°C, the pressure is increased to 2.00 atm, and then it is heated again to 100.0°C.
5.36 Originally at 20.0°C and 759 Torr, it is heated to 100.0°C, the pressure is decreased to 200 Torr, then it is heated to 1000°C, and finally the pressure is decreased to 100 Torr.

In Exercises 5.37 and 5.38, calculate the molecular weight and deduce the molecular formula of the named compound, using the information given.

5.37 0.125 g of benzene vapor (empirical formula CH) at 60.0°C gives rise to a pressure of 133 Torr in a 250-mL flask.
5.38 0.155 g of naphthalene vapor (empirical formula C_5H_4) at 100°C gives rise to a pressure of 113 Torr in a 250-mL flask.

Real Gases

In Exercises 5.39 and 5.40, use the van der Waals equation to calculate the pressure of each gas. Also calculate the percentage difference between the calculated pressure and that predicted by the ideal gas law.

5.39 (a) 1.00 g of carbon dioxide in a 500-mL container at 10°C.
(b) 1.00 mol of "air molecules" contained in 22.4 L at 0.0°C.
5.40 (a) 1.0 g of ammonia in a 250.0-mL container at 25°C.
(b) 1.00 mol of helium atoms contained in 2.24 L at 0.0°C.

Molar Volumes

In Exercises 5.41 to 5.44, calculate the volume occupied at STP by each sample of gas.

5.41 (a) 1.00 mol of H_2 molecules; (b) 2.00 mol of O_2 molecules; (c) 0.152 mol of "air molecules"; (d) 0.11 mol of Ar atoms.
5.42 (a) 1.00 mol of He atoms; (b) 3.00 mol of NH_3 molecules; (c) 2.11 mol of C_2H_4 (ethylene) molecules; (d) 0.116 mol of H atoms.

5.43 (a) 1.00 g of oxygen; (b) 1.00 g of hydrogen; (c) 5.00 g of carbon dioxide; (d) 5.00 g of ammonia.
5.44 (a) 1.00 g of ethylene; (b) 1.00 g of sulfur dioxide; (c) 1.00 g of air; (d) 1.00 kg of nitric oxide.

In Exercises 5.45 and 5.46, calculate the number of moles of molecules present in each volume of gas at STP.

5.45 (a) 100 mL of nitrogen; (b) 100 mL of ammonia; (c) 1.0 m³ of air; (d) 1.0 cm³ of helium.
5.46 (a) 100 mL of carbon dioxide; (b) 100 mL of sulfur dioxide; (c) 1.0 L of ozone; (d) 250 mL of hydrogen sulfide.

Gaseous Mixtures

In Exercises 5.47 and 5.48, calculate the partial pressures and the total pressure (in Torr) of each mixture. Every mixture is held in a 1.00-L container at 0.0°C.

5.47 (a) 0.010 mol of N_2 and 0.030 mol of H_2.
(b) 0.001 mol of N_2, 0.003 mol of H_2, and 0.002 mol of NH_3.
(c) 0.112 g of argon and 0.112 g of neon.
(d) 1.51 mg of oxygen, 1.01 mg of carbon monoxide, and 1.05 mg of carbon dioxide.
5.48 (a) 0.002 mol of NH_3 and 0.003 mol of SO_2.
(b) 0.002 mol of H_2, 0.002 mol of Cl_2, and 0.004 mol of HCl.
(c) 0.105 g of methane and 0.105 g of carbon dioxide.
(d) 1.00 mg of argon, 2.00 mg of neon, and 3.00 mg of xenon.

5.49 A sample of damp air in a 1.00-L container exerts a pressure of 762.0 Torr at 20.0°C, but when it is cooled to −10.0°C the pressure falls to 607.1 Torr as the water condenses. What mass of water is present?
5.50 1.00 g of water was added to a 2.00-L flask containing dry carbon dioxide at 20.0°C and 505 Torr pressure; the flask was then heated to 200.0°C. What was the final pressure?

5.51 The mass percentage composition of the atmosphere of Mars is 95% carbon dioxide, 3% nitrogen, and 2% other gases, principally argon. What are the partial pressures of these components at the surface, where the total pressure is 5 Torr?
5.52 The mass percentage composition of the atmosphere of Venus is 97% carbon dioxide, 3% nitrogen, and very little of anything else. What are the partial pressures of these components at the surface, where the total pressure is 90 atm?

5.53 A barometer could be made with water, but it would need to be nearly 11 m high. Moreover, since water is more volatile than mercury, the "vacuum" at the top of the closed tube would be imperfect and would depress the water column slightly. What height would the water reach on a day when the atmospheric pressure is 1.000 atm and the temperature is 25°C?

5.54 What correction should be made to allow for the vapor pressure of mercury at 25°C (0.22 Pa) when we use a mercury barometer?

The Volumes of Reacting Gases

5.55 Calculate the volumes of hydrogen and oxygen produced at STP when 5.2 g of water is electrolyzed.

5.56 Calculate the mass of water that can be synthesized from 500 L of hydrogen at STP.

5.57 What volume of carbon dioxide at STP is produced when excess dilute hydrochloric acid is added to 5.0 g of calcium carbonate? The equation is

$$2HCl(aq) + CaCO_3(s) \longrightarrow CO_2(g) + CaCl_2(s) + H_2O(l)$$

5.58 What volume of dinitrogen oxide (N_2O) is produced at STP when 3.67 g of ammonium nitrate is heated? The equation is

$$NH_4NO_3(s) \longrightarrow N_2O(g) + 2H_2O(g)$$

5.59 One industrial process for producing sulfur makes use of the reaction between sulfur dioxide and hydrogen sulfide. What volume of sulfur dioxide at STP is needed to prepare 100 kg of sulfur? What volume is needed if it is supplied at 10 atm and 150°C?

5.60 In the Haber synthesis of ammonia, what volume of hydrogen at STP must be supplied to produce 1000 kg of ammonia? What volume is needed at 200 atm and 400°C?

5.61 Carbon dioxide (from the air) and ammonia (from the Haber process) react at 185°C and 200 atm pressure to produce urea, $CO(NH_2)_2$, and water. Urea is increasingly being used as a fertilizer. Under the given conditions, what volume of ammonia will produce 100 kg of urea?

5.62 Xenon (Xe) and fluorine (F_2) combine when heated to 350°C, forming the colorless solid xenon tetrafluoride (XeF_4). (The reaction also produces XeF_2 and XeF_6, but ignore those here.) What volumes of xenon and fluorine are required at STP for the formation of 100 g of the tetrafluoride, if we assume 100% yield?

In Exercises 5.63 to 5.66, calculate the number of moles of gas produced in the reaction after correcting for the vapor pressure of water.

5.63 In an electrolysis experiment, 220 mL of hydrogen was collected at 20°C when the external pressure was 720.2 Torr.

5.64 Potassium chlorate was heated in the presence of manganese dioxide, and 120 mL of oxygen was collected over water at 22°C when the external pressure was 762.1 Torr.

5.65 Dilute hydrochloric acid was poured on some zinc, and 154 mL of hydrogen was collected over water at 18°C when the external pressure was 743.2 Torr.

5.66 The removal of oxygen and carbon dioxide from air left a mixture of nitrogen and argon. Collected over water, the mixture occupied a volume of 1.25 L when the external pressure was 761.1 Torr.

Molecular Speeds

In Exercises 5.67 and 5.68, calculate the rms speed of each molecule, given that the rms speed of H_2 molecules at 25°C is 1920 m/s.

5.67 (a) H_2 molecules at 0°C; (b) H_2O molecules at 25°C; (c) CO_2 molecules at 0°C; (d) Br_2 molecules at 25°C.

5.68 (a) H_2 molecules at 1000°C; (b) C_6H_6 molecules at 25°C; (c) H_2O molecules at 0°C; (d) D_2O molecules at 0°C.

In Exercises 5.69 and 5.70, suggest the order of increasing speed of sound in the given gases.

5.69 Hydrogen, carbon dioxide, and argon.

5.70 Water vapor, chlorine, and air.

In Exercises 5.71 and 5.72, calculate the relative speeds of the given molecules.

5.71 (a) H_2 at 100.0°C relative to H_2O at 0.0°C.
(b) D_2O at 100.0°C relative to D_2 at 0.0°C.
(c) D_2O relative to H_2O.

5.72 (a) I_2 vapor relative to F_2.
(b) HI relative to DI.
(c) He on the sun's surface (at 6500°C) relative to He on the earth's surface (at 15°C).

Diffusion

In Exercises 5.73 and 5.74, calculate the time needed for each gas or vapor to diffuse through a porous plug.

5.73 (a) CO_2, (b) H_2O, (c) C_6H_6, and (d) I_2, given that under the same conditions the same amount of Ar takes 147 s to diffuse.

5.74 (a) Ar, (b) H_2, (c) O_2, and (d) UF_6, given that under the same conditions the same amount of carbon dioxide takes 115 min to diffuse.

5.75 One of the components of pineapple flavor is ethyl butyrate, which a combustion experiment showed to have the empirical formula C_3H_6O. Its vapor took 162 s to diffuse through a certain porous plug; the same number of carbon dioxide molecules diffused through the same plug in 100 s under the same conditions. What is the molecular formula of the compound?

5.76 A compound of empirical formula C_2H_3 took 349 s to diffuse through a porous plug, and the same number of argon atoms took 210 s to diffuse through the same plug under the same conditions. What is the molecular formula of the compound?

In Exercises 5.77 and 5.78, calculate the molecular weight and molecular formula of the phosphorus(III) compound.

5.77 When 1.04 g of phosphorus was burned in a restricted supply of oxygen, the mass of the phosphorus(III) oxide produced was 1.85 g. A certain amount of the vapor of that phosphorus(III) oxide took 302 s to diffuse, whereas the same number of carbon dioxide molecules took only 135 s under the same conditions.

5.78 When 2.36 g of phosphorus was burned in chlorine, 10.5 g of phosphorus(III) chloride was produced. The chloride's vapor took 1.8 times as long to diffuse as the same number of carbon dioxide molecules under the same conditions.

General

5.79 At 180.0°C, 27.20 mg of the antimalarial drug quinine in a 250.0-mL container gave rise to a pressure of 9.48 Torr. In a combustion experiment, 1.72×10^{-2} g of quinine produced 4.70×10^{-2} g of carbon dioxide, 1.15×10^{-2} g of water, and 1.51 mg of nitrogen. What is the molecular formula of quinine?

5.80 At 280.0°C and in a 500.0-mL container, 0.115 g of eugenol, the compound responsible for the odor of cloves, caused a pressure of 48.3 Torr. In a combustion experiment, 1.88×10^{-2} g of eugenol burned to 5.00×10^{-2} g of carbon dioxide and 1.24×10^{-2} g of water. What is the molecular formula of eugenol?

5.81 Yellow sulfur consists of S_8 molecules. When a sample of sulfur was vaporized, it gave rise to a pressure that was about three times larger than that expected, and as heating continued the pressure rose more quickly than the ideal gas law predicts. Suggest an explanation.

5.82 The pressure of a sample of hydrogen fluoride was found to be lower than expected, and to rise more quickly than the ideal gas law predicts. Suggest an explanation.

This chapter explores one of the roles of energy in chemistry. We concentrate here on the heat given out or absorbed during chemical and physical changes, and how it is described, measured, and predicted. We also extend the stoichiometry calculations to deal with the heat outputs and requirements of reactions. The illustration shows the burning off of waste gases from a blast furnace. Calculations like the ones described in this chapter let us assess the energy requirements of industrial and biological processes.

6

ENERGY, HEAT, AND THERMOCHEMISTRY

Energy and calorimetry

6.1 Energy
6.2 Enthalpy

The enthalpy of chemical change

6.3 Reaction enthalpies
6.4 Enthalpies of formation

The world's enthalpy resources

6.5 The enthalpy of fuels
6.6 The enthalpy of foods

When steam condenses to water, it releases enough heat to scald. In contrast, to vaporize water requires heat: we must, for example, supply heat to a kettle of water in order to boil the water away. Many chemical reactions produce heat, which often is simply allowed to go to waste but may be used to provide warmth or power an engine. Some chemical reactions take in heat as they occur. One of the most striking examples is the reaction that occurs when the solids barium hydroxide, $Ba(OH)_2 \cdot 8H_2O$, and ammonium thiocyanate, NH_4SCN, are mixed together in a flask:

$$Ba(OH)_2 \cdot 8H_2O(s) + 2NH_4SCN(s) \longrightarrow$$
$$Ba(SCN)_2(aq) + 2NH_3(g) + 10H_2O(l)$$

So much heat is required that the reaction freezes any water that happens to be on the outside of the flask (Fig. 6.1).

In summary, heat is a product of many reactions, and it is used—like a reactant—in others. This suggests that we might be able to add one more arrow to the mole diagram (**1**) and use relations between moles to make predictions about heat. This aspect of stoichiometry is called *thermochemistry*. Thermochemical calculations allow us to begin with given masses of reactants as data, express the masses as moles, and then predict the amount of heat the reaction will release or absorb. By analogy with our discussions in the last two chapters, we can anticipate that calculations of this kind depend on knowing how much heat is produced when 1 mol of a reactant is consumed, and then setting up the conversion

Heat produced = moles of reactant
× heat produced per mole of reactant

Once we know the conversion factor, we can calculate the heat produced when any number of moles (and hence any mass or volume) of reactant is consumed.

ENERGY AND CALORIMETRY

Thermochemistry is a branch of *thermodynamics*, the study of the transformation of energy from one form to another. We therefore begin by discussing some of the properties of energy and the means by which energy changes are measured.

6.1 ENERGY

By "energy" we mean the following:

Energy is the capacity to do work or supply heat.

A wound-up spring possesses energy, because as it unwinds it can do work such as raising a weight against the pull of gravity. A fuel possesses energy because as it burns it supplies heat to its surroundings. A fuel can also be used to do the work of moving a vehicle along a highway.

FIGURE 6.1 The reaction between ammonium thiocyanate and barium hydroxide draws in a lot of heat; here it causes water to freeze on the outside of the flask.

Atoms and molecules act as stores of energy. In reactions in which the products have more energy than the reactants, energy must be supplied for the products to form. That is the case in photosynthesis, in which energy supplied by the sun is used to build carbohydrates (typically glucose, $C_6H_{12}O_6$) from carbon dioxide and water:

$$6CO_2(g) + 6H_2O(l) + \text{energy} \longrightarrow C_6H_{12}O_6(s) + 6O_2(g)$$

In other reactions the products have a lower energy than the reactants, and the excess energy is released when the reaction takes place. This is the case when carbohydrates burn, either in a controlled way, as in respiration—when food is oxidized in living cells—or in an uncontrolled way, as in a forest fire (Fig. 6.2):

$$C_6H_{12}O_6(s) + 6O_2(g) \longrightarrow 6CO_2(g) + 6H_2O(l) + \text{energy}$$

Kinetic energy. One way in which energy is stored is as *kinetic energy* E_K, the energy of motion. The kinetic energy of a single molecule of mass m is related to its speed v by

$$E_K = \tfrac{1}{2}mv^2$$

The higher its speed, the greater its kinetic energy. An ideal gas consists of many moving molecules, and its total energy is the sum of all the individual kinetic energies. Raising the temperature of the gas, which increases the average speed of the molecules, increases its energy. This stored energy is released when the gas cools and the molecules move more slowly again.

We can use the expression for kinetic energy to derive the SI unit of energy. The kinetic energy of a mass of 2 kg traveling at 1 m/s is

$$E_K = \frac{1}{2} \times 2 \text{ kg} \times (1 \text{ m/s})^2$$

$$= 1 \text{ kg} \cdot \text{m}^2/\text{s}^2$$

FIGURE 6.2 Uncontrolled combustion takes place during a forest fire: the cellulose of the vegetation is oxidized back to the carbon dioxide and water from which it was originally formed by photosynthesis.

The energy unit $1 \text{ kg} \cdot \text{m}^2/\text{s}^2$ (kilogram meter-squared per second-squared) is used so often that it is given a special name, the *joule* (J):

$$1 \text{ J} = 1 \text{ kg} \cdot \text{m}^2/\text{s}^2$$

The name honors the nineteenth century English scientist James Joule, a student of Dalton's, who made a major contribution to our understanding of energy. Each beat of a human heart uses about 1 J of energy as it drives the blood through the body. To lift this book from the floor to the table, we need to use about 10 J of energy. To throw it across the room at 5 m/s requires nearly 15 J.

A nonSI energy unit still widely used in chemistry is the *calorie* (cal), from the Latin word for "heat":

$$1 \text{ cal} = 4.184 \text{ J exactly}$$

One calorie is the amount of energy required to increase the temperature of 1 g of water by 1°C.

▼ EXAMPLE 6.1 Converting between calories and joules

How many joules of energy are required to heat 100 g of water from 20°C to boiling (100°C)?

STRATEGY We know that 1 cal increases the temperature of 1 g of water by 1°C, so we can readily find the number of calories needed to heat 100 g to 100°C. Then we can use the conversion factor to change the result to joules. Since 1 J is *smaller* than 1 cal, we expect more joules than calories.

SOLUTION Heating 100 g of water through 80°C from 20°C to boiling requires 80×100 cal, or 8.0 kcal. In joules, this becomes

$$\text{Energy required} = 8.0 \times 10^3 \text{ cal} \times \frac{4.184 \text{ J}}{1 \text{ cal}}$$

$$= 3.3 \times 10^4 \text{ J}$$

That is, about 33 kJ is needed.

EXERCISE How much energy (in joules) is released when 25 g of water cools from 60°C to 20°C?

[*Answer*: 4.2 kJ]

▲

The kinetic energy of a gas molecule at room temperature is only about 6×10^{-21} J. Since there are 6×10^{23} molecules in a 1-mol sample, the total kinetic energy of 1 mol of molecules is about 4 kJ (4×10^3 J). If we could harness all that energy, we could use it to raise about 400 copies of this book from the floor to a table. We shall soon see that energies of chemical reactions are often of the order of kilojoules and megajoules (1 MJ = 10^6 J).

Potential energy. The second means by which energy is stored is as *potential energy* (E_P), the contribution to the energy from the interactions between particles. Two nitrogen atoms have a much higher po-

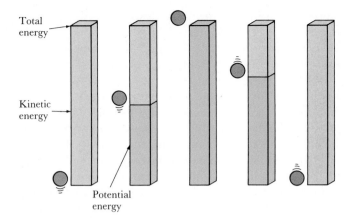

Total energy

Kinetic energy

Potential energy

FIGURE 6.3 During the flight of a ball, its kinetic energy is converted to potential energy. At the highest point of its flight, its speed is zero and all its energy is potential energy. Just before it reaches the ground again, it has the same kinetic energy as it was given initially. The total energy remains the same throughout the flight.

tential energy when they are far apart than when they are close together as parts of an N_2 molecule. Likewise, a baseball has a higher potential energy when it is high in the air than when it is on the surface of the earth.

Just as we must supply a lot of energy to throw a ball high in the air against the pull of gravity, we must supply a lot of energy to separate an N_2 molecule into atoms. The actual amount is 1.6×10^{-18} J; if we multiply that by Avogadro's number, we find that it takes 960 kJ to break 1 mol of N_2 molecules (28 g of nitrogen gas) into atoms. This is five times more than is needed to separate 1 mol of F_2 molecules into atoms, which partly explains why nitrogen is such an inert gas and fluorine is so reactive: it is much more difficult to get nitrogen atoms from N_2 than to get fluorine atoms from F_2. Some idea of the size of 960 kJ is that the same amount of energy supplied to a 100-g ball could raise the ball to a height of about 600 miles or accelerate it to 10,000 mi/h.

Total energy and internal energy. A ball thrown upward has a high speed and high kinetic energy initially, but it slows as it rises. That is, its kinetic energy decreases as its potential energy increases (Fig. 6.3). Although the amounts of the two kinds of energy are changing—with one decreasing and the other increasing—at every stage the *total energy E* is constant. This constancy of total energy is summarized as follows:

Law of conservation of energy: Energy can be neither created nor destroyed.

This law implies that the total energy of the universe remains the same as the ball rises, stops, and falls again. However, it implies much more, for it also implies that the total energy of the universe is the same before, during, and after *any* change, including a chemical reaction. The universe has a definite amount of energy which is the same for all time (or so we believe). Throughout history, people have tried to build "perpetual-motion machines," engines that run without using fuel (and thus without using energy). The law of conservation of energy implies that perpetual motion is impossible, because such machines would create energy from nothing. The fact that no one has ever succeeded in

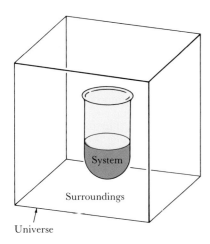

FIGURE 6.4 The "system" is the sample or reaction mixture that we are interested in. Outside the system are the "surroundings." The system and surroundings together form the "universe."

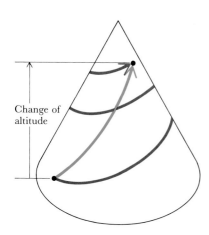

FIGURE 6.5 The altitude of someone on a mountain is a kind of state property: the same change of altitude results, whatever the route between the two points.

building a perpetual-motion machine (despite many attempts and false claims) is evidence for the validity of the law.

In thermochemistry, the total energy of all the particles in a sample is called its *internal energy E*. The sample is normally called the *system* (Fig. 6.4), and we shall use that term from now on. Everything outside the system is called the *surroundings*. If we think of the system as the reaction mixture itself, then the surroundings are the flask and the atmosphere or the water bath in which the flask is immersed. The system and its surroundings together make up the *universe*. In principle, the "universe" is the actual universe—everything there is—but in practice we normally take it to include only the more immediate surroundings.

Since energy cannot be created or destroyed, the internal energy of a system is constant so long as it is isolated from its surroundings—isolated in the sense that energy cannot enter or escape from the system. This is summarized by the "first law" of thermodynamics, a restatement of the law of conservation of energy in thermodynamic terms:

> **First law of thermodynamics:** The internal energy of an isolated system is constant.

We can change the internal energy of a system by supplying energy from the surroundings (by heating the system, for instance) or by allowing some energy to escape to the surroundings (by allowing the system to cool). However, the law of conservation of energy applies to the universe as a whole; hence, when the internal energy of the system increases by a certain amount, the total energy of the surroundings must decrease by exactly the same amount, and vice versa.

State properties. The internal energy of a sample of a substance depends on its physical state (whether it is gas, liquid, or solid), the temperature, and the pressure it experiences. It also depends on the size of the sample. In other words, internal energy is an *extensive* property, so 100 g of iron has twice the internal energy of 50 g of iron at the same temperature and pressure.

The internal energy of a sample of a substance is the same at a given temperature and pressure whatever may have happened to the sample in the past and whatever its origin. A 100-g sample of water at 25°C and 1 atm pressure has exactly the same internal energy whether it has been freshly synthesized from hydrogen and oxygen or obtained by distillation from a solution. The internal energy of the sample is also the same whether the sample has been kept at constant temperature or heated and then cooled back to 25°C. This is all summarized by saying that internal energy is a "state property":

> A **state property** is a property of a substance that is independent of how the substance was prepared.

An everyday example of a "state property" is altitude, or height above sea level (Fig. 6.5): once we arrive at the top of a mountain, our altitude is the same no matter how we got there. Other state properties are pressure, volume, and temperature.

Calculating the internal energy of an ideal gas

Calculate the internal energy of 1 mol of an ideal gas at 25°C.

STRATEGY In an ideal gas there are no intermolecular interactions, so the potential energy of the molecules is zero. Since the kinetic energy of each molecule of mass m is $\frac{1}{2}mv^2$, the internal energy (the total energy) of 1 mol of the gas is this quantity multiplied by Avogadro's number. The value of v is given by Eq. 9 in Section 5.7, and $N_A \times m$ is the molar mass M of the gas.

SOLUTION The internal energy is

$$E = N_A \times \tfrac{1}{2}mv^2$$

$$= M \times \frac{1}{2} \times \frac{3RT}{M}$$

$$= \tfrac{3}{2}RT$$

Substituting $R = 8.314$ J/(K · mol) and $T = 298$ K yields

$$E = \frac{3}{2} \times 8.314 \text{ J/(K · mol)} \times 298 \text{ K}$$

$$= 3.72 \text{ kJ/mol}$$

EXERCISE How much energy is required to heat the ideal gas from 25°C to 100°C?

[*Answer*: 0.94 kJ/mol]

When one substance changes into another, the internal energy of the system changes as well. Because internal energy is a state property, the change in its value that results from the physical or chemical change is independent of how the change came about. The starting substance has a certain internal energy, and so does the final substance; the change in internal energy is equal to the difference. This is like the difference in altitude between two points on a mountain being independent of the path taken to go from one to the other (Fig. 6.6). For example, we might measure the change of internal energy when 100 g of water is vaporized at a certain temperature and pressure, and find that it is 50 kJ. If we then measured out another 100 g of water, electrolyzed it, and then discharged a spark through the resulting hydrogen and oxygen to produce water vapor at the same temperature and pressure, we would find that the overall change of internal energy for the second sample is also 50 kJ.

Changes in a state property X from an initial value X_i to a final value X_f are denoted by the Greek uppercase delta (Δ), with

$$\Delta X = X_f - X_i$$

A change in internal energy is therefore symbolized ΔE. If the internal energy decreases during a change, so that E_f is less than E_i, then ΔE is negative; if E increases, then ΔE is positive. Since E_i and E_f are the same for every path from the initial state to the final state, ΔE is the same for every path between the two states.

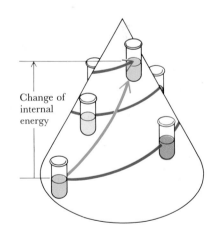

Change of internal energy

FIGURE 6.6 The overall change in internal energy is independent of the path taken from initial to final state. The two different paths represented here (one a direct change, and one a series of changes) lead to the same change in internal energy.

▼ EXAMPLE 6.3 Calculating a change in internal energy

What is the change in internal energy per mole of an ideal gas when it is heated from 25°C to 50°C in a sealed container?

STRATEGY We know from Example 6.2 how to calculate the internal energy of a gas at any temperature T; the required energy change is the difference between the internal energies at the two given temperatures.

SOLUTION Using the formula $E = \frac{3}{2}RT$, we find

$$\Delta E = \tfrac{3}{2}RT_f - \tfrac{3}{2}RT_i$$

$$= \frac{3}{2} \times 8.314 \text{ J/(K} \cdot \text{mol)} \times (323 \text{ K} - 298 \text{ K})$$

$$= +0.31 \text{ kJ/mol}$$

The plus sign shows that the internal energy has increased.

EXERCISE Calculate the change in energy per mole of an ideal gas that is cooled from 50°C to 25°C.

[*Answer*: −0.31 kJ/mol]

Heat. We can change the internal energy of a system in several ways. In this chapter we shall be concerned mainly with the transfer of energy as "heat":

Heat is energy that is transferred as the result of a temperature difference between a system and its surroundings.

If the system has a higher temperature than its surroundings, energy flows out of the system as heat, and the internal energy of the system decreases. The escaping energy increases the *thermal motion*, or random, chaotic motion, of the atoms in the surroundings (Fig. 6.7). If the surroundings are a gas, like the air, the increased thermal motion takes the form of faster disorderly motion of the colliding molecules. If the system has a lower temperature than its surroundings, energy flows in as heat and its internal energy rises. This decreases the thermal motion of the atoms in the surroundings.

The amount of heat produced or absorbed by a reaction depends on how the reaction takes place. We can illustrate this by considering the reaction between hydrogen and oxygen. Normally, this reaction produces a great deal of heat. However, we could carry out the reaction in a "fuel cell," a device that generates electricity from a chemical reaction and has been used on some space missions. Both reactions release the same amount of energy (the change of internal energy, a state property, is independent of how the reaction takes place). However, in the fuel cell, the released energy is primarily in the form of electrical power, and much less energy is obtained as heat (Fig. 6.8).

Suppose we carry out a reaction in such a way that the only energy transfer between the system and the surroundings is in the form of heat. Then by measuring the flow of heat into or out of the system, we can find the change of internal energy that has occurred. If we denote

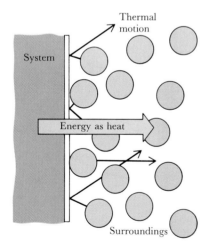

FIGURE 6.7 When energy leaves a system as the result of a temperature difference between the system and its surroundings (that is, as heat), it stimulates random thermal motion of the atoms in the surroundings.

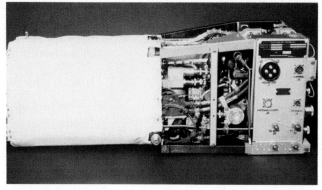

FIGURE 6.8 Energy can be released from a reaction in a number of ways; in particular, different quantities of heat may be produced, even though the change of internal energy is the same. The combustion of hydrogen on the left produces more heat than the same reaction in the NASA fuel cell on the right.

the amount of energy that enters a system as heat by q, then as long as no other changes occur,

$$\Delta E = q$$

If we found that 10 kJ had entered the system as heat during some reaction (Fig. 6.9), we could conclude that its internal energy had increased by 10 kJ, so $\Delta E = +10$ kJ. If we found that during another reaction 10 kJ of energy had left the system as heat, q would be negative and we would write $\Delta E = -10$ kJ (the minus sign indicating a decrease in internal energy).

We can detect the flow of heat out of a system by observing the temperature of its surroundings. (If the reaction flask is immersed in a water bath, then this is the temperature of the water in the bath.) If the temperature of the surroundings rises, we can conclude that heat has been released. This is the case in the combustion of methane,

$$CH_4(g) + 2O_2(g) \longrightarrow CO_2(g) + 2H_2O(l) + \text{energy}$$

The reaction releases energy because the products CO_2 and H_2O have lower combined internal energy than the reactants CH_4 and O_2. If the temperature of the surroundings goes down (as in the barium hydroxide–ammonium thiocyanate reaction of Fig. 6.1), we know that heat has been absorbed by the system. In the methane combustion, energy leaves the system, so its internal energy has decreased. In the barium hydroxide reaction, energy enters the system, so its internal energy has increased.

6.2 ENTHALPY

The relation $\Delta E = q$ applies when heat is the only form in which energy leaves or enters a system. Many chemical reactions, however, are car-

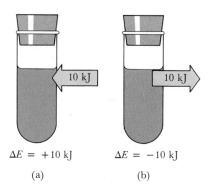

$\Delta E = +10$ kJ $\Delta E = -10$ kJ

(a)   (b)

FIGURE 6.9 (a) When 10 kJ of energy enters a system as heat and no other changes take place, the internal energy increases by 10 kJ. (b) If 10 kJ leaves the system, the internal energy decreases by the same amount.

2 Octane

ried out in flasks that are open to the atmosphere and hence subjected to a constant pressure and free to change their volume; then energy can escape in forms other than as heat. We can illustrate this with the combustion of octane (C_8H_{18}; **2**), a component of gasoline:

$$2C_8H_{18}(l) + 25O_2(g) \longrightarrow 16CO_2(g) + 18H_2O(g)$$

Since 34 mol of gas molecules are formed for every 25 mol of O_2 molecules used, at constant pressure the products occupy a much greater volume than the reactants. When octane is burned in a container that is open to the atmosphere, these products push back the atmosphere as they are formed (Fig. 6.10). Since this expansion occurs against the steady pressure of the atmosphere, it uses energy. Hence the total change in internal energy is greater than the change we detect as heat. That is, when the system is free to expand (or contract), the heat output alone is not an accurate measure of the change in the system's internal energy.

Expansion work. We can calculate the energy used to make room for the products of a reaction by noting that, in physics, *work w* is defined as the product of the distance moved and the force opposing the motion:

$$w = \text{distance} \times \text{opposing force}$$

This expression implies that we have to do a lot of work to travel a long distance against a strong opposing force (like bicycling against the wind). When a system expands, the opposing force arises from the pressure of the surroundings (usually the atmosphere) on the system. We can imagine the reaction as taking place inside a container fitted with a piston of area A, which is pushed out through a distance d as the product gases form (Fig. 6.11). Since the pressure on the piston is the force per unit area (see Section 5.1), the force is the pressure times the area, and we can write

$$w = \text{distance} \times (\text{area} \times \text{pressure})$$
$$= d \times A \times P$$

The product $d \times A$ is the change in volume of the system, so

$$w = P \times \Delta V$$

If the volume does not change ($\Delta V = 0$), no work is done. With the pressure in atmospheres and the volume in liters, the work is calculated in liter-atmospheres (L · atm). It is converted to joules using the equation

$$1 \text{ L} \cdot \text{atm} = 101 \text{ J}$$

This follows from the definition of 1 atm in Table 5.4, so that

$$1 \text{ L} \cdot \text{atm} = 10^{-3} \text{ m}^3 \times [101 \times 10^3 \text{ kg}/(\text{m} \cdot \text{s}^2)]$$
$$= 101 \text{ kg} \cdot \text{m}^2/\text{s}^2$$

and the definition of the joule as $1 \text{ kg} \cdot \text{m}^2/\text{s}^2$.

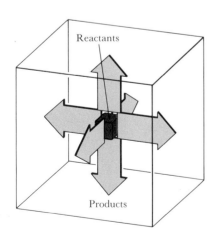

FIGURE 6.10 As gaseous products are formed, they must push back the atmosphere. This uses energy, so less is available as heat.

▼ **EXAMPLE 6.4** **Calculating the work of expansion**

Calculate the work that must be done at STP to make room for the products of the octane combustion

$$2C_8H_{18}(l) + 25O_2(g) \longrightarrow 16CO_2(g) + 18H_2O(g)$$

when 2 mol of C_8H_{18} is burned and all the gases are ideal.

STRATEGY We know that 25 mol of gas molecules are replaced by 34 mol of gas molecules in this reaction, a net increase of 9 mol. Since we know the molar volume of an ideal gas at STP (22.4 L/mol), we can calculate the change in volume and hence the work needed for the expansion.

SOLUTION The change in volume is

$$\Delta V = 9 \text{ mol} \times 22.4 \text{ L/mol} = 202 \text{ L}$$

Since the external pressure is 1.0 atm, the work required is

$$w = P \times \Delta V = 1.0 \text{ atm} \times 202 \text{ L} = 2.0 \times 10^2 \text{ L} \cdot \text{atm}$$

This is equivalent to

$$w = 2.0 \times 10^2 \text{ L} \cdot \text{atm} \times \frac{101 \text{ J}}{1 \text{ L} \cdot \text{atm}}$$

$$= 2.0 \times 10^4 \text{ J}$$

That is, the expansion requires 20 kJ of energy.

EXERCISE Calculate the work needed to make room for the products in the combustion of 1 mol of glucose to carbon dioxide and water vapor at STP.

[*Answer*: 14 kJ]

▲

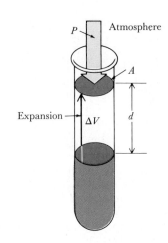

FIGURE 6.11 To calculate the work needed to make room for gases produced in a reaction, we imagine the reaction as taking place inside a cylinder fitted with a piston. The piston is pushed out a distance d against an opposing force $P \times A$ arising from the atmospheric pressure.

Heat output at constant pressure. When a system undergoes some kind of change but has a fixed volume, the heat output is equal to the change in internal energy ($q = \Delta E$). The heat output of a system free to expand or contract under the steady pressure of the atmosphere is not equal to ΔE, since energy must be used to make room for any products. However, we can use the heat output at constant pressure to define a property of the system called its "enthalpy change" ("enthalpy" comes from the Greek words for "heat inside"):

The **enthalpy change** ΔH of a system is equal to its heat output at constant pressure:

$$\Delta H = q \quad \text{at constant pressure}$$

Note that this defines only the *change* of enthalpy of a system. The enthalpy itself, designated H, is defined as

$$H = E + PV$$

Thus, if we know the internal energy of a system, we can calculate its enthalpy by adding the product of its pressure and volume ($P \times V$) to E. However, in chemistry we never need to know the enthalpy of a system but only the changes in the enthalpy when reactions and other

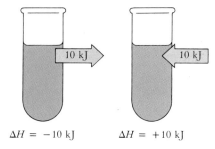

$\Delta H = -10 \text{ kJ}$ $\Delta H = +10 \text{ kJ}$

FIGURE 6.12 This system is open to the atmosphere, and hence free to expand or contract to keep the pressure constant. When 10 kJ of energy leaves the system as heat, the enthalpy decreases by 10 kJ. If 10 kJ enters this system, the enthalpy rises by the same amount.

transformations take place. The more useful relation in chemistry, therefore, is the one identifying the enthalpy change with the heat output at constant pressure. If we measure a heat output of 10 kJ when a certain mass of sulfur burns to form sulfur dioxide at constant pressure, we know that the enthalpy of the system has decreased by 10 kJ ($\Delta H = -10$ kJ) during the reaction (Fig. 6.12). Conversely, if we know that the enthalpy change is $\Delta H = -10$ kJ when a certain amount of nitrogen and hydrogen react to give ammonia, then we also know that the reaction will release 10 kJ of energy as heat into the surroundings if it takes place at constant pressure.

The values of ΔH and ΔE differ by the amount of work that needs to be done to make room for the products of the reaction. Since that work is equal to $P \, \Delta V$, we have

$$\Delta H = \Delta E + P \, \Delta V \qquad (1)$$

For example, if $\Delta E = -100$ kJ in a certain combustion reaction, but 10 kJ of work needs to be done to make room for the products, the change of enthalpy is

$$\Delta H = -100 \text{ kJ} + 10 \text{ kJ} = -90 \text{ kJ}$$

Hence, only 90 kJ of energy can be released as heat if the reaction is carried out at constant pressure, whereas 100 kJ can be released at constant volume.

▼ **EXAMPLE 6.5** **Relating internal energy and enthalpy**

The heat output of the reaction $N_2(g) + 3H_2(g) \rightarrow 2NH_3(g)$ when 1 mol of N_2 reacts at constant volume and at 25°C is 41 kJ. What heat output would be obtained if the same reaction took place at a constant pressure of 1.0 atm?

STRATEGY The heat output at constant pressure is equal to the enthalpy change; therefore we must calculate ΔH. To do so, we need the values of ΔE and $P \, \Delta V$. The change in internal energy is equal to the heat output at constant volume, which is given. The value of ΔV can be found from the change in the number of moles of gas molecules and the molar volume of an ideal gas at 25°C, which is 24.5 L/mol.

SOLUTION Since 41 kJ of heat is lost at constant volume, $\Delta E = -41$ kJ. In the reaction, 4 mol of gas molecules combines to form 2 mol, so there is a net *reduction* of 2 mol and hence a *decrease* in volume:

$$\Delta V = -2 \text{ mol} \times 24.5 \text{ L/mol} = -49.0 \text{ L}$$

It follows that

$$P \, \Delta V = 1.0 \text{ atm} \times (-49.0 \text{ L}) = -49 \text{ L} \cdot \text{atm}$$

This is equivalent to

$$P \, \Delta V = -49 \text{ L} \cdot \text{atm} \times \frac{101 \text{ J}}{1 \text{ atm}} = -4.9 \text{ kJ}$$

Therefore, the change in enthalpy is

$$\Delta H = \Delta E + P \, \Delta V$$
$$= -41 \text{ kJ} + (-4.9 \text{ kJ}) = -46 \text{ kJ}$$

That is, when the reaction takes place at constant pressure, it releases 46 kJ of heat. This is greater than the amount released at constant volume because at constant pressure the system contracts.

EXERCISE   At constant volume, the reaction of 1 mol of C to form carbon monoxide in the reaction $2C(s) + O_2(g) \rightarrow 2CO(g)$ releases 113.0 kJ of heat. What will it release at a constant pressure of 1.0 atm and 25°C?

▲ [*Answer:* 110.5 kJ]

Like overall changes of internal energy, overall changes of enthalpy are the same whatever path is taken between the initial and final states of the system. In other words, enthalpy is a state property. Enthalpy changes are often quite similar in magnitude to changes in internal energy (as we have seen, they typically differ by a few kilojoules). However, enthalpy changes are more useful in chemistry, because chemical reactions are commonly carried out in systems open to the atmosphere and hence are free to expand or contract.

Exothermic and endothermic reactions.   Some reactions and physical changes give out heat when they take place, and others take heat in:

An **exothermic process** is a process that gives out heat.

An **endothermic process** is a process that takes in heat.

In terms of the change of enthalpy, exothermic reactions are those for which ΔH is negative (because heat leaves the system and lowers the enthalpy; Fig. 6.13). Endothermic reactions are those for which ΔH is positive (because heat enters the system and increases the enthalpy).

All combustions are exothermic. The *thermite reaction*, the reduction of a metal oxide by aluminum, as in

$$2Al(s) + Fe_2O_3(s) \longrightarrow Al_2O_3(s) + 2Fe(s)$$

is spectacularly exothermic; once started, it produces enough heat to melt the iron (Fig. 6.14). The vaporization of water, like all vaporizations, is endothermic because heat is taken in as the liquid changes to vapor. Dissolving may be either slightly exothermic or slightly endothermic, depending on the solute. When ammonium nitrate dissolves in water, the process is quite strongly endothermic. This property is the basis of the "instant cooling packs" in some first-aid kits. These draw in heat when a tube of the salt is broken inside a bag of water and the salt dissolves. The reaction between barium hydroxide and ammonium thiocyanate (Fig. 6.1) is also endothermic.

Calorimetry.   We can measure the internal-energy and enthalpy changes that accompany a reaction by measuring the heat released or absorbed. If the reaction takes place at constant volume (in a sealed container), the heat output is equal to the change in the internal energy of the system. If it occurs at constant pressure (open to the atmosphere), the heat output is equal to the change in enthalpy:

$$\text{At constant volume,} \quad q = \Delta E$$
$$\text{At constant pressure,} \quad q = \Delta H \tag{2}$$

FIGURE 6.13   The decomposition of ammonium dichromate to chromium(III) oxide is so exothermic that, once ignited, it produces a miniature volcano.

FIGURE 6.14   The thermite reaction is another highly exothermic reaction—one that can melt the metals it produces.

FIGURE 6.15 A bomb calorimeter. The combustion is started with an electrically ignited fuse, and the temperature of the entire assembly is monitored.

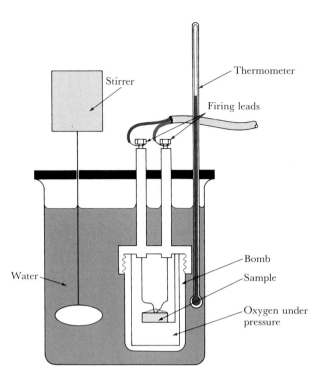

Thermometer

Firing leads

Stirrer

Bomb

Water

Sample

Oxygen under pressure

The values of ΔE and ΔH, and hence the values of q in these two cases, are related by Eq. 1.

Heat is measured with a *calorimeter*. A simple calorimeter consists of a Styrofoam cup with a loose-fitting lid through which a thermometer is inserted. (The lid is loose-fitting so that the contents are under constant atmospheric pressure.) When reactants are placed in the cup, the heat produced by the ensuing reaction increases the temperature of the reaction mixture. By measuring this increase, we can determine the heat output. Since the reaction occurs at constant pressure, the heat output gives ΔH.

A more complicated calorimeter, the *bomb calorimeter* (Fig. 6.15), is often used to study combustions. The sample is weighed into a small bucket and sealed into the central container, the "bomb." Next, the bomb is filled with compressed oxygen and immersed in the water bath. The sample is then ignited electrically, and the change in the temperature of the water is measured. Since the volume of the sealed bomb is constant, the heat of reaction is equal to ΔE.

Heat capacity. We now consider how to use the change of temperature of the calorimeter to arrive at the heat output of the reaction. For any object or substance, a temperature rise ΔT and the heat q responsible for that rise are proportional to each other. Hence we can write

$$q = C \times \Delta T \qquad (3)$$

The constant C is called the *heat capacity*; in this case, it depends on the size of the calorimeter and the materials from which it is made. The greater the heat capacity of any substance, the greater the amount of heat needed to produce a given temperature rise in that substance. A substance for which 1 J of heat brings about a temperature rise of 1 K has a heat capacity of 1 J/K (1 joule per kelvin). The calorie is so defined that 1 g of water has a heat capacity of 1 cal/°C.

The heat capacities of some common substances are given in Table 6.1. Since heat capacity is an extensive property, the table gives the *specific heat capacity* c (also often called just the "specific heat"), which is the heat capacity per gram. Since the heat capacity of a sample is the product of its specific heat capacity and the mass of the sample, Eq. 3 can be written

$$q = c \times \text{mass} \times \Delta T \qquad (4)$$

▼ EXAMPLE 6.6 Using the heat capacity

How much heat is required to heat 500 g of water from 20.0°C to boiling (100.0°C)?

STRATEGY We need the heat capacity of the water, which is obtained by multiplying the specific heat capacity by the mass of the sample. Then the heat required is the product of this heat capacity and the temperature rise.

SOLUTION The heat capacity of 500 g of water, with specific heat 4.18 J/(K · g), is

$$C = 500 \text{ g} \times 4.18 \text{ J/(K} \cdot \text{g)} = 2.09 \text{ kJ/K}$$

The heat required to bring about a temperature rise of 80.0 K is therefore

$$q = 2.09 \text{ kJ/K} \times 80.0 \text{ K} = 167 \text{ kJ}$$

EXERCISE Calculate the heat required to increase the temperature of a 1.0-kg granite block from 20°C to 100°C.

[*Answer*: 64 kJ]

▲

TABLE 6.1 Specific heat capacities of some materials

Substance	Specific heat capacity, J/(K · g)
Air	1.01
Benzene	1.05
Brass	0.37
Copper	0.38
Ethanol	2.42
Glass (Pyrex)	0.78
Granite	0.80
Marble	0.84
Polyethylene	2.3
Stainless steel	0.51
Water	4.18

The heat released by a reaction taking place inside a calorimeter increases the temperature of the *entire* calorimeter, which includes the reaction mixture, its container, and the surrounding water bath (Fig. 6.16). We therefore need to know the heat capacity of the entire calorimeter if we are to use the observed temperature change to calculate the heat produced by the reaction. This heat capacity could be calculated from the masses and specific heats of all the parts of the calorimeter. However, it is normally more accurate to observe the temperature rise caused by a reaction of known heat output, and use that to determine C for the calorimeter. This is an example of *calibration*, the interpretation of an observation by comparison with known information. For example, suppose a test sample increased the temperature of the calorimeter by 5.2 K. Also, under identical conditions, a sample producing 80.0 kJ of heat was found to raise the temperature of the same calorimeter by 8.4 K. This latter experiment allows us to write

$$80.0 \text{ kJ} = 8.4 \text{ K}$$

which we can use as a conversion factor to interpret the 5.2 K temperature rise of the sample:

$$\text{Heat output} = 5.2 \text{ K} \times \frac{80.0 \text{ kJ}}{8.4 \text{ K}} = 50 \text{ kJ}$$

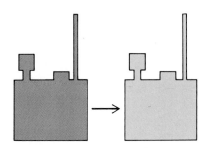

FIGURE 6.16 When a reaction takes place in a calorimeter, the temperature of the entire apparatus increases (shown here by a change of color). Therefore, to interpret the temperature increase, we need to know the heat capacity of the entire apparatus.

We could also calibrate the calorimeter by passing a known electric current through a heater inside the calorimeter for a known time and measuring the temperature rise. The energy supplied by the heater is

Heat in joules = current in amperes × voltage × time in seconds

Knowing that, we can calculate the heat capacity of the calorimeter from the temperature rise we observe. Alternatively, the heat supplied by an electric heater is its power in watts (W) multiplied by the time in seconds during which it is turned on:

Heat in joules = power in watts × time in seconds

In the latter calculation, we use $1 \text{ J} = 1 \text{ W} \cdot \text{s}$.

▼ **EXAMPLE 6.7 Calculating the heat produced by a reaction**

When 50.0 mL of 1.0 M NaOH(aq) and 50.0 mL of 1.0 M HCl(aq) were mixed in a simple calorimeter at constant pressure, the temperature rose from 20.1°C to 26.5°C. The heat capacity of the calorimeter was measured while it contained 100 mL of sodium chloride solution (which is the final product of this reaction) and was found to be 440.2 J/K. Calculate the heat produced by the neutralization and the change of enthalpy of the reaction.

STRATEGY Since the reaction occurs at constant pressure, the heat output is equal to the enthalpy change. The temperature increase is given, so we can use Eq. 3 to convert the temperature increase to a heat output.

SOLUTION The temperature rise of 6.4°C implies that $\Delta T = +6.4$ K. From Eq. 3, the heat produced by the reaction is

$$q = C \times \Delta T$$

$$= 440.2 \text{ J/K} \times 6.4 \text{ K} = 2.8 \text{ kJ}$$

Since the heat is lost from the reaction mixture, $\Delta H = -2.8$ kJ.

EXERCISE When 1.56 g of ethanol was burned in a calorimeter of total heat capacity 7.21 kJ/K, the temperature increased from 24.8 to 30.6°C. Calculate the enthalpy change of the combustion reaction.

[*Answer*: −41.8 kJ]

▲

The enthalpy of physical change. The enthalpy of any substance increases when it is heated. We can normally detect this increase as a rise in temperature. For example, if 10 kJ of heat is supplied to 100 g of water, the enthalpy of the water increases by 10 kJ. Since the heat capacity of 100 g of water is 0.42 kJ/K, this increase in enthalpy appears as a temperature rise of about 24 K. Note that while temperature is an intensive property, enthalpy is *extensive*. To produce the same 24-K temperature rise in 200 g of water, we would have to increase its enthalpy by twice as much, supplying 20 kJ of energy as heat.

If we continue to add heat to the water sample, its temperature continues to rise. At 100°C the water starts to boil if the external pressure is 1 atm. Now, however, as we continue to heat the sample, and hence continue to increase its enthalpy, the temperature of the water stops rising. The energy we are supplying is now being used to overcome the attractions between water molecules, and to vaporize the water. As we noted in Section 1.1, this behavior is characteristic of all changes of state: whenever a liquid freezes or boils, its temperature remains unchanged until the transition is complete. Since heat has been supplied to boil away the water, the enthalpy of the vapor at 100°C must be higher than that of liquid at the same temperature. That is why steam can cause severe scalding; when it comes into contact with the skin, a large *reduction* of enthalpy occurs as it condenses back to water, and the heat released (about 2 kJ per gram of water) damages the tissues.

The difference in enthalpy per mole of molecules between the vapor and liquid states of a substance is called the *enthalpy of vaporization* ΔH_{vap} of that substance. For water at 100°C,

$$\Delta H_{vap} = H_{gas} - H_{liquid} = +40.7 \text{ kJ/mol}$$

Values for some other substances are given in Table 6.2. The absorption of heat from skin when perspiration vaporizes is one of the body's strategies for keeping us cool. Vaporization also accounts for the action of cosmetic "cold creams," which are mixtures of oil and water. They feel cool because the water evaporates after they are applied to the skin.

▼ **EXAMPLE 6.8** **Measuring the enthalpy of vaporization**

An electric heater was used to heat 250 g of water to its boiling point in a simple calorimeter open to the atmosphere. Heating was then continued until 35 g of water had vaporized. From the voltage and current ratings of the heater and the time involved, it was calculated that the vaporization required 77 kJ of heat. Calculate the enthalpy of vaporization of water at 100°C.

STRATEGY We can anticipate an *increase* in enthalpy (positive ΔH), since heat is absorbed by the sample as it vaporizes. Because vaporization occurs at constant pressure, the heat absorbed by 35 g of water is equal to the increase in its enthalpy. We can convert this to an enthalpy change per mole of H_2O molecules by using a conversion factor formed from the molar mass of H_2O (18.02 g/mol).

SOLUTION From the data, we know that the change in enthalpy of the water is +77 kJ when 35 g is vaporized. The number of moles of H_2O in 35 g of water is

$$\text{Moles of } H_2O = 35 \text{ g } H_2O \times \frac{1 \text{ mol } H_2O}{18.02 \text{ g } H_2O}$$

$$= 1.94 \text{ mol } H_2O$$

The enthalpy change per mole of H_2O is therefore

$$\Delta H_{vap} = \frac{+77 \text{ kJ}}{1.94 \text{ mol}} = +40 \text{ kJ/mol}$$

EXERCISE The same heater was used to heat a quantity of benzene to 80°C, its boiling point. Then heating was continued, during which time 71 g of boiling benzene were evaporated (in a hood, because benzene is carcinogenic and toxic) and 28 kJ of heat was supplied. Calculate the enthalpy of vaporization of benzene, C_6H_6, at its boiling point.

[*Answer:* +31 kJ/mol]

▲

The enthalpy change that accompanies melting, the *enthalpy of melting*, is

$$\Delta H_{melt} = H_{liquid} - H_{solid}$$

TABLE 6.2 Standard enthalpies of physical change*

Substance	Formula	Freezing point, K	ΔH°_{melt}, kJ/mol	Boiling point, K	ΔH°_{vap}, kJ/mol
Acetone	CH_3COCH_3	177.8	5.72	329.4	29.1
Ammonia	NH_3	195.3	5.65	239.7	23.4
Argon	Ar	83.8	1.2	87.3	6.5
Benzene	C_6H_6	278.7	9.87	353.3	30.8
Ethanol	C_2H_5OH	158.7	4.60	351.5	43.5
Helium	He	3.5	0.02	4.22	0.08
Methane	CH_4	90.7	0.94	111.7	8.2
Methanol	CH_3OH	175.5	3.16	337.2	35.3
Water	H_2O	273.2	6.01	373.2	40.7 (44.0 at 25°C)

*Values correspond to the transition temperature. The superscript ° signifies that the change takes place at 1 atm and that the substance is pure. Note that, generally, the higher the freezing point, the higher the enthalpy of melting.

It is also called the *enthalpy of fusion*, fusion being another name for melting. The enthalpy of melting of water is +6 kJ/mol, showing that 6 kJ of energy must be supplied as heat to melt 18 g of ice at 0°C. Melting is endothermic, because heat is absorbed when a substance melts and its molecules break away from their neighbors. Some values are given in Table 6.2.

The *enthalpy of freezing* is the negative of the enthalpy of melting: the same quantity of heat is released when a liquid freezes as is absorbed when it melts. For water at 0°C, the enthalpy of freezing is −6 kJ/mol. Hence, to freeze 18 g of water at 0°C in a refrigerator, 6 kJ of energy must be removed as heat. This is so in general: the enthalpy change for the reverse of *any* process (any chemical reaction or physical change) is the negative of the enthalpy change for the original process:

$$\Delta H(\text{forward change}) = -\Delta H(\text{reverse change})$$

Sublimation is the direct vaporization of a solid to a gas without the formation of a liquid. The *enthalpy of sublimation* is the enthalpy change

FIGURE 6.17 The temperature of a solid rises as heat is supplied. At the melting point, the temperature remains constant while the supplied heat is used to melt the sample. When enough heat has been supplied to melt all the solid, the temperature begins to rise again. A similar pause occurs at the boiling point.

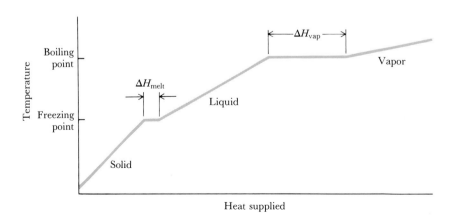

per mole of molecules that accompanies sublimation:

$$\Delta H_{\text{sub}} = H_{\text{gas}} - H_{\text{solid}}$$

Sublimation is always endothermic (ΔH is positive), because heat is absorbed when a substance changes from the solid state, in which the molecules are packed close together, to the gaseous state, in which they move about freely.

▼ **EXAMPLE 6.9** **Combining enthalpies of transition**

At 25°C the enthalpy of melting of sodium metal is +2.6 kJ/mol, and the enthalpy of vaporization of liquid sodium is +98 kJ/mol. Calculate the enthalpy of sublimation of the solid at this temperature.

STRATEGY Because enthalpy is a state property, the enthalpy change for the transition from solid to vapor is independent of the path taken. Therefore, we look for a path for which we know the enthalpy change at each stage. One such path is *solid → liquid* followed by *liquid → vapor*. The enthalpy change in the first step is the enthalpy of melting; that in the second step is the enthalpy of vaporization. The overall enthalpy change is the sum of the two **(3)**.

SOLUTION The enthalpy of sublimation is the sum

$$\Delta H_{\text{sub}} = \Delta H_{\text{melt}} + \Delta H_{\text{vap}}$$

$$= +2.6 \text{ kJ/mol} + 98 \text{ kJ/mol}$$

$$= +101 \text{ kJ/mol}$$

EXERCISE The enthalpy of sublimation of iodine is +57.3 kJ/mol, and its enthalpy of vaporization is +41.8 kJ/mol. Calculate its enthalpy of freezing.
[*Answer:* −15.5 kJ/mol]

▲

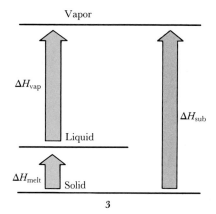

A graph showing the variation of the temperature of a sample as it is heated is called its *heating curve*. As we see in Fig. 6.17, the temperature of a solid rises steadily up to its melting point as it is heated. The temperature remains constant at the melting point as the heat supplies the energy necessary to form the liquid (the enthalpy of melting). Then it rises again up to the boiling point. The temperature now remains constant as the heat supplies the energy necessary to form the vapor (the enthalpy of vaporization). After the sample has evaporated completely, the temperature rises again as heating continues.

THE ENTHALPY OF CHEMICAL CHANGE

When 1 mol of CH_4 molecules (16 g of methane) burns in air under certain conditions, 890 kJ of heat is produced. That is, when the reaction

$$CH_4(g) + 2O_2(g) \longrightarrow CO_2(g) + 2H_2O(l)$$

takes place, the total enthalpy of the system (the reaction mixture) decreases by 890 kJ for each mole of CH_4 molecules that react. This decrease of 890 kJ is the *reaction enthalpy* ΔH, the change of enthalpy for the reaction *exactly as the chemical equation is written*, which implies that 1 mol of CH_4 has been burned:

$$CH_4(g) + 2O_2(g) \longrightarrow CO_2(g) + 2H_2O(l) \qquad \Delta H = -890 \text{ kJ}$$

If the same reaction is written with the coefficients all multiplied by 2, implying that 2 mol of CH_4 molecules have been burned, the reaction enthalpy is twice as great:

$$2CH_4(g) + 4O_2(g) \longrightarrow 2CO_2(g) + 4H_2O(l) \qquad \Delta H = -1780 \text{ kJ}$$

6.3 REACTION ENTHALPIES

Once we know the reaction enthalpy, we can use the methods of Chapter 4 to calculate the enthalpy change when any amount of reactant is involved. For example, suppose we want to know the heat output resulting from the combustion of 150 g of methane (Fig. 6.18). We begin by converting the given mass to moles of CH_4, as follows:

$$\text{Moles of } CH_4 = \text{mass of } CH_4 \times \text{moles of } CH_4 \text{ per gram}$$

$$= 150 \text{ g } CH_4 \times \frac{1 \text{ mol } CH_4}{16.04 \text{ g } CH_4}$$

$$= 9.35 \text{ mol } CH_4$$

Then we use the reaction enthalpy to calculate the heat output of a reaction in which 9.35 mol of CH_4 is burned:

$$\text{Heat output} = \text{moles of } CH_4 \times \text{heat output per mole of } CH_4$$

The reaction enthalpy tells us that

$$1 \text{ mol } CH_4 = 890 \text{ kJ}$$

which produces the conversion factor

$$1 = \frac{890 \text{ kJ}}{1 \text{ mol } CH_4}$$

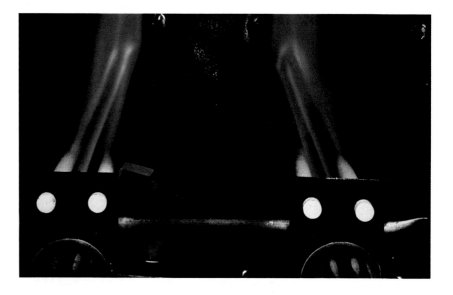

FIGURE 6.18 The interior of a furnace burning natural gas: the enthalpy change in the combustion reaction releases heat that is used to raise the temperature of the surrounding water.

Therefore,

$$\text{Heat output} = 9.35 \text{ mol } CH_4 \times \frac{890 \text{ kJ}}{1 \text{ mol } CH_4}$$

$$= 8.32 \times 10^3 \text{ kJ}$$

That is, the combustion provides 8.32 MJ of heat.

▼ EXAMPLE 6.10 Calculating the heat output of a fuel

The combustion of propane is described by the equation

$$C_3H_8(g) + 5O_2(g) \longrightarrow 3CO_2(g) + 4H_2O(l) \qquad \Delta H = -2220 \text{ kJ}$$

Calculate the mass of propane that must be burned to obtain 350 kJ of heat, which is just enough to heat 1 L of water from room temperature (20°C) to boiling.

STRATEGY As in all stoichiometric calculations, we convert to moles and use the mole relations given by the chemical equation. Since we are given the heat output, the first step is to convert that to moles of propane:

$$\text{Moles of } C_3H_8 = \text{heat output} \times \text{moles of } C_3H_8 \text{ per joule}$$

Since we are asked for the mass of propane needed, this number of moles of C_3H_8 molecules must be converted to a mass of propane:

$$\text{Mass of } C_3H_8 = \text{moles of } C_3H_8 \times \text{mass of } C_3H_8 \text{ per mole}$$

The first conversion factor comes from the reaction enthalpy. The second comes from the molar mass of propane (44.09 g/mol).

SOLUTION The two conversion factors are obtained from

$$1 \text{ mol } C_3H_8 = 2220 \text{ kJ}$$

and

$$1 \text{ mol } C_3H_8 = 44.09 \text{ g } C_3H_8$$

Therefore, the first conversion (from joules to moles) is

$$\text{Moles of } C_3H_8 = 350 \text{ kJ} \times \frac{1 \text{ mol } C_3H_8}{2220 \text{ kJ}}$$

$$= 0.158 \text{ mol } C_3H_8$$

The second conversion (from moles to grams) is

$$\text{Mass of } C_3H_8 = 0.158 \text{ mol } C_3H_8 \times \frac{44.09 \text{ g } C_3H_8}{1 \text{ mol } C_3H_8}$$

$$= 6.95 \text{ g } C_3H_8$$

EXERCISE The reaction for the burning of butane is

$$2C_4H_{10}(g) + 13O_2(g) \longrightarrow 8CO_2(g) + 10H_2O(l) \qquad \Delta H = -5754 \text{ kJ}$$

What mass of butane supplies 350 kJ of heat?

[*Answer*: 7.07 g]

▲

Standard reaction enthalpies. The heat given out or taken in by a reaction depends on the physical states of the reactants and products. Less heat is produced by the combustion of methane if the product water is formed as a vapor rather than as a liquid:

$$CH_4(g) + 2O_2(g) \longrightarrow CO_2(g) + 2H_2O(g) \qquad \Delta H = -802 \text{ kJ}$$

$$CH_4(g) + 2O_2(g) \longrightarrow CO_2(g) + 2H_2O(l) \qquad \Delta H = -890 \text{ kJ}$$

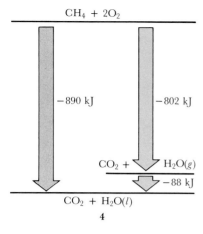

$CH_4 + 2O_2$

-890 kJ

-802 kJ

$CO_2 + $ $H_2O(g)$

-88 kJ

$CO_2 + H_2O(l)$

4

The enthalpy of water vapor is 44 kJ/mol higher than that of liquid water (see Table 6.2); hence 88 kJ less heat (for 2 mol of H_2O) is released in the reaction if the vapor is formed rather than the liquid **(4).**

Because enthalpy changes depend on temperature and pressure, chemists find it convenient to report reaction enthalpies based on an internationally agreed-upon set of standard conditions. They report values for reactions in which the reactants and products are in their "standard states":

The **standard state** of a substance is its pure form at 1 atm pressure.

The standard state of liquid water at some specified temperature is pure water at that temperature and 1 atm pressure. The standard state of ice at some temperature is pure ice at that temperature and 1 atm pressure. A reaction enthalpy based on such standard states is called a "standard reaction enthalpy" $\Delta H°$ (the superscript ° always denotes a "standard" value) and defined as follows:

A **standard reaction enthalpy** is a reaction enthalpy that is measured when reactants in their standard states change to products in their standard states.

Standard reaction enthalpies can be reported for any temperature, but data are usually given at 25°C (more precisely, at 298.15 K). All the values used in this text are for this temperature unless noted otherwise.

The standard reaction enthalpy for the combustion of methane is the change in enthalpy when pure methane and pure oxygen gases, each at 1 atm pressure, react to form pure carbon dioxide and pure water, both at that same pressure (Fig. 6.19). If the product is liquid water, we have

$$CH_4(g) + 2O_2(g) \longrightarrow CO_2(g) + 2H_2O(l) \qquad \Delta H° = -890 \text{ kJ}$$

If the product is water vapor, then

$$CH_4(g) + 2O_2(g) \longrightarrow CO_2(g) + 2H_2O(g) \qquad \Delta H° = -802 \text{ kJ}$$

FIGURE 6.19 The standard reaction enthalpy is the difference between the enthalpy of the *pure* reactants at 1 atm pressure and that of the pure products at the same pressure and the specified temperature (typically but not necessarily 25°C). The values are expressed on a scale on which O_2 is zero; this point is explained later.

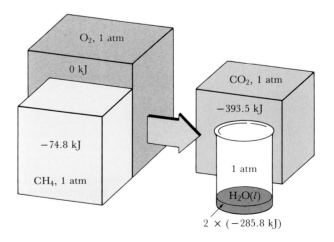

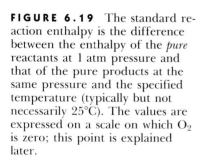

TABLE 6.3 Standard enthalpies of combustion at 25°C

Substance*	Formula	ΔH_c°, kJ/mol
Acetylene	$C_2H_2(g)$	-1300
Benzene	$C_6H_6(l)$	-3268
Carbon	$C(s)$, graphite	-394
Carbon monoxide	$CO(g)$	-283
Ethanol	$C_2H_6O(l)$	-1368
Glucose	$C_6H_{12}O_6(s)$	-2808
Hydrogen	$H_2(g)$	-286
Methane	$CH_4(g)$	-890
Methanol	$CH_4O(l)$	-726
Octane	$C_8H_{18}(l)$	-5471
Propane	$C_3H_8(g)$	-2220
Sucrose	$C_{12}H_{22}O_{11}(s)$	-5645
Toluene	$C_7H_8(l)$	-3910
Urea	$CH_4ON_2(s)$	-632

*C is converted to CO_2, H to H_2O, and N to N_2. More values are given in Appendix 2A.

Standard enthalpy of combustion. The reaction enthalpies of combustions are important for judging the suitability of fuels and for measuring the enthalpies of other kinds of reactions. Their values are usually reported as "standard enthalpies of combustion" ΔH_c°:

The **standard enthalpy of combustion** of a substance is the change in enthalpy per mole of the substance when it burns completely under standard conditions.

The complete combustion of hydrocarbons produces carbon dioxide and water. An example is the combustion of propane,

$$C_3H_8(g) + 5O_2(g) \longrightarrow 3CO_2(g) + 4H_2O(l) \qquad \Delta H^\circ = -2220 \text{ kJ}$$

which shows that the standard enthalpy of combustion of propane is

$$\Delta H_c^\circ = \frac{-2220 \text{ kJ}}{1 \text{ mol } C_3H_8} = -2220 \text{ kJ/mol } C_3H_8$$

Many standard enthalpies of combustion have been measured, and some are listed in Table 6.3. All the values are negative, since all combustions release heat and are exothermic.

▼ **EXAMPLE 6.11** Using enthalpy of combustion to calculate heat output

Gasoline, though a mixture, is thermochemically similar to pure octane. Calculate the heat released when 1.0 L of gasoline (of density 0.80 g/mL) burns completely under standard conditions at 25°C.

STRATEGY To calculate the heat output, we need the number of moles of C_8H_{18} burned, since then we can use the enthalpy of combustion (Table 6.3). The number of moles is obtained from the mass of the octane, which in turn we find from the volume and density, both of which are given.

SOLUTION The three conversion factors we need are obtained from

$$1.0 \text{ mL octane} = 0.80 \text{ g } C_8H_{18}$$

$$1 \text{ mol } C_8H_{18} = 114.2 \text{ g } C_8H_{18}$$

$$1 \text{ mol } C_8H_{18} = 5471 \text{ kJ}$$

The mass of octane burned is

$$\text{Mass of } C_8H_{18} = 1.0 \text{ L } C_8H_{18} \times \frac{0.80 \text{ g}}{1.0 \times 10^{-3} \text{ L}}$$

$$= 8.0 \times 10^2 \text{ g } C_8H_{18}$$

This corresponds to

$$\text{Moles of } C_8H_{18} = 8.0 \times 10^2 \text{ g } C_8H_{18} \times \frac{1 \text{ mol } C_8H_{18}}{114.2 \text{ g } C_8H_{18}}$$

$$= 7.0 \text{ mol } C_8H_{18}$$

The heat output is therefore

$$\text{Heat output} = 7.0 \text{ mol } C_8H_{18} \times \frac{5471 \text{ kJ}}{1 \text{ mol } C_8H_{18}}$$

$$= 3.8 \times 10^4 \text{ kJ}$$

The combustion is exothermic, and it releases 38 MJ of heat into the surroundings. This is enough to heat more than 120 L of water from room temperature to boiling.

EXERCISE The density of ethanol is 0.79 g/mL. Calculate the heat produced when 1.0 L burns under standard conditions at 25°C.

[*Answer*: 23 MJ]

▲

2C + O₂ + O₂

(b) −221 kJ

2CO + O₂

−787 kJ

(c) −566 kJ

2CO₂

5

Hess's law. Because enthalpy is a state property, its value changes by the same amount whatever the path from given reactants to specified products. We used this property in Example 6.9 to calculate the enthalpy of sublimation of a solid. Now we apply the same argument to chemical changes. For example, consider the oxidation of carbon to carbon dioxide:

$$2C(s) + 2O_2(g) \longrightarrow 2CO_2(g) \qquad \text{(a)}$$

The enthalpy change that results from this reaction is the same (provided the conditions of temperature and pressure are the same) as the total enthalpy change involved in first forming carbon monoxide,

$$2C(s) + O_2(g) \longrightarrow 2CO(g) \qquad \text{(b)}$$

and then oxidizing it to carbon dioxide (**5**):

$$2CO(g) + O_2(g) \longrightarrow 2CO_2(g) \qquad \text{(c)}$$

The two-step formation of carbon dioxide via reactions (b) and (c) is an example of a *reaction sequence*, a series of reactions in which products of one reaction take part as reactants in the next. The equation for the *overall reaction*, the net outcome of the sequence, is the sum of the equations of the individual steps:

$$2C(s) \qquad\qquad + \quad O_2(g) \longrightarrow 2CO(g) \qquad\qquad\qquad (b)$$
$$2CO(g) + \quad O_2(g) \longrightarrow \qquad\qquad\qquad 2CO_2(g) \qquad (c)$$
$$\overline{2C(s) + 2CO(g) + 2O_2(g) \longrightarrow 2CO(g) + 2CO_2(g)}$$
$$2C(s) \qquad + 2O_2(g) \longrightarrow 2CO_2(g) \qquad\qquad (a)$$

Since the overall equation (a) is the sum of the individual equations (b) and (c), the reaction enthalpy of (a) is the sum of the individual reaction enthalpies of (b) and (c):

$$2C(s) + O_2(g) \longrightarrow 2CO(g) \qquad \Delta H° = -221.0 \text{ kJ} \qquad (b)$$

$$2CO(g) + O_2(g) \longrightarrow 2CO_2(g) \qquad \Delta H° = -566.0 \text{ kJ} \qquad (c)$$

$$\Delta H°(\text{overall}) = -221.0 \text{ kJ} + (-566.0 \text{ kJ}) = -787.0 \text{ kJ}$$

The same conclusion—that reaction enthalpies can be combined like the individual reaction equations—applies to any reaction that can be considered to take place in several steps. This is summarized as follows (Fig. 6.20):

Hess's law: A reaction enthalpy is the sum of the enthalpies of any sequence of reactions (at the same temperature and pressure) into which the overall reaction may be divided.

Germain Hess was a Swiss chemist who proposed the law in 1840. We now know that it is a consequence of the fact that enthalpy changes are independent of the path by which specified reactants are transformed into specified products. In other words, Hess's law is a straightforward consequence of the first law of thermodynamics and the conservation of energy.

The intermediate reactions in a sequence need not be reactions that can actually be carried out in the laboratory. That is, the intermediate reactions may be "hypothetical." For example, the reaction

$$2C(s) + 2H_2(g) \longrightarrow C_2H_4(g)$$

cannot be carried out in practice. It can, nevertheless, be used in a reaction sequence to determine the enthalpy of an overall reaction. The only requirement is that the individual chemical equations in the sequence must balance and must add up to the equation for the reaction of interest. Subject to those conditions, a reaction enthalpy can be calculated from any convenient sequence.

▼ **EXAMPLE 6.12 Combining reaction enthalpies**

Calculate the standard reaction enthalpy for the incomplete combustion of octane to carbon monoxide and water.

STRATEGY We can anticipate that the enthalpy of the partial combustion of octane to carbon monoxide is smaller (less negative) than the enthalpy of the complete combustion to carbon dioxide: less heat is given out when the oxidation terminates at a less fully oxidized product. We know (from Table 6.3) the enthalpy of combustion to the dioxide,

$$2C_8H_{18}(l) + 25O_2(g) \longrightarrow 16CO_2(g) + 18H_2O(l) \qquad (a)$$

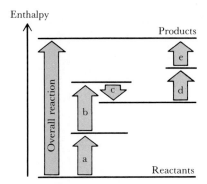

FIGURE 6.20 If the overall reaction can be broken down into a series of steps, then the corresponding overall reaction enthalpy is the sum of the reaction enthalpies of those steps. None of the steps need be a reaction that can actually be carried out in the laboratory.

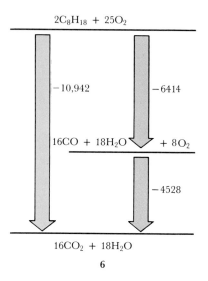

$2C_8H_{18} + 25O_2$

$-10,942$ -6414

$16CO + 18H_2O$ $+ 8O_2$

-4528

$16CO_2 + 18H_2O$

6

We want the enthalpy for the combustion to the monoxide,

$$2C_8H_{18}(l) + 17O_2(g) \longrightarrow 16CO(g) + 18H_2O(l) \qquad (b)$$

This suggests that we should subtract reaction (b) from reaction (a) to see what reaction enthalpy should be combined with the known value to give us the unknown.

SOLUTION   The subtraction of reaction (b) from reaction (a) gives

$$2C_8H_{18}(l) + 25O_2(g) \longrightarrow 16CO_2(g) \qquad\qquad\quad + 18H_2O(l) \qquad (a)$$
$$2C_8H_{18}(l) + 17O_2(g) \longrightarrow \qquad\qquad 16CO(g) + 18H_2O(l) \qquad (b)$$

$$8O_2(g) \longrightarrow 16CO_2(g) - 16CO(g)$$

Eliminating $-16CO(g)$ on the right of the arrow by adding $16CO(g)$ to each side gives

$$16CO(g) + 8O_2(g) \longrightarrow 16CO_2(g) \qquad (c)$$

This is the equation for the combustion of carbon monoxide, and its enthalpy is given in Table 6.3. The enthalpies of the three reactions are related in the same way as the chemical equations themselves **(6)**:

$$\Delta H°(a) - \Delta H°(b) = \Delta H°(c)$$

From Table 6.3:

$$\Delta H°(a) = 2 \text{ mol } C_8H_{18} \times (-5471 \text{ kJ/mol } C_8H_{18})$$

$$= -10,942 \text{ kJ}$$

$$\Delta H°(c) = 16 \text{ mol CO} \times (-283 \text{ kJ/mol CO})$$

$$= -4528 \text{ kJ}$$

Therefore,

$$\Delta H°(b) = \Delta H°(a) - \Delta H°(c)$$

$$= -10,942 \text{ kJ} - (-4528 \text{ kJ}) = -6414 \text{ kJ}$$

Since reaction (b) corresponds to the combustion of 2 mol of C_8H_{18} molecules,

$$\Delta H_c° = \frac{-6414 \text{ kJ}}{2 \text{ mol } C_8H_{18}} = -3207 \text{ kJ/mol } C_8H_{18}$$

As anticipated, this reaction is less exothermic than the complete combustion.

EXERCISE   Calculate the standard reaction enthalpy for the incomplete combustion of propane to carbon monoxide and water.

[*Answer:* $-1371 \text{ kJ/mol } C_3H_8$]

▲

We can use the same procedure to calculate the enthalpies of a wide range of reactions. In all such calculations, the skill to develop is the ability to find a reaction sequence that leads to the required overall reaction. Combustion reactions are a very useful source of data, because they result in CO_2 and H_2O (in the combustion of hydrocarbons), as well as N_2 (if nitrogen is present) and SO_2 (from any sulfur); these products act as a pool of atoms that can be used for rebuilding other compounds (Fig. 6.21). This is illustrated by the following example.

▼ EXAMPLE 6.13 Using combustion reactions to reproduce a synthesis reaction

Use standard enthalpies of combustion to calculate the standard reaction enthalpy for the synthesis of propane from its elements.

STRATEGY We must decide how to reproduce the overall equation

$$3C(s) + 4H_2(g) \longrightarrow C_3H_8(g) \qquad (a)$$

as sums and differences of combustion equations. One approach is to write the combustion equations we know we are likely to need (those of C and H_2), add them together, and subtract reaction (a). The resulting equation can often be rearranged into another combustion equation for which an enthalpy is listed in Table 6.3.

SOLUTION The combustion of 3 mol of C atoms is described by the equation

$$3C(s) + 3O_2(g) \longrightarrow 3CO_2(g) \qquad (b)$$

The combustion of 4 mol of H_2 has the equation

$$4H_2(g) + 2O_2(g) \longrightarrow 4H_2O(l) \qquad (c)$$

The sum of these two equations is

$$3C(s) + 4H_2(g) + 5O_2(g) \longrightarrow 3CO_2(g) + 4H_2O(l) \qquad (d)$$

Subtraction of reaction (a) from reaction (d) leaves

$$
\begin{aligned}
3C(s) + 4H_2(g) + 5O_2(g) &\longrightarrow 3CO_2(g) + 4H_2O(l) \quad (d)\\
3C(s) + 4H_2(g) &\longrightarrow \qquad\qquad\qquad\quad C_3H_8(g) \quad (a)\\
\hline
5O_2(g) &\longrightarrow 3CO_2(g) + 4H_2O(l) - C_3H_8(g)
\end{aligned}
$$

By adding C_3H_8 to both sides, we can rearrange this result to

$$C_3H_8(g) + 5O_2(g) \longrightarrow 3CO_2(g) + 4H_2O(l) \qquad (e)$$

which is the equation for the combustion of propane. The reaction enthalpies are added and subtracted in the same way as the equations:

$$\Delta H°(b) + \Delta H°(c) - \Delta H°(a) = \Delta H°(e)$$

The unknown, $\Delta H°(a)$, is therefore given by

$$\Delta H°(a) = \Delta H°(b) + \Delta H°(c) - \Delta H°(e)$$

From Table 6.3,

$$\Delta H°(b) = 3 \text{ mol C} \times (-394 \text{ kJ/mol C}) = -1182 \text{ kJ}$$

$$\Delta H°(c) = 4 \text{ mol H}_2 \times (-286 \text{ kJ/mol H}_2) = -1144 \text{ kJ}$$

$$\Delta H°(e) = 1 \text{ mol C}_3H_8 \times (-2220 \text{ kJ/mol C}_3H_8) = -2220 \text{ kJ}$$

Hence,

$$\Delta H°(a) = (-1182 \text{ kJ}) + (-1144 \text{ kJ}) - (-2220 \text{ kJ})$$

$$= -106 \text{ kJ}$$

This shows that the synthesis reaction is exothermic, releasing nearly 106 kJ of heat for each mole of propane molecules formed.

EXERCISE Calculate the standard reaction enthalpy for the synthesis of 1 mol of $C_6H_6(l)$ from its elements.

[*Answer*: +50 kJ]

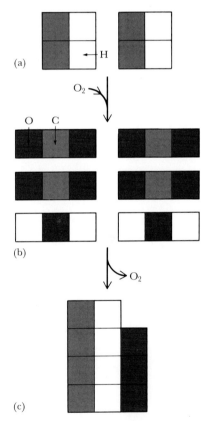

FIGURE 6.21 Combustions provide a pool of hydrogen, carbon, oxygen, and nitrogen atoms from which new compounds can be built. The reactants (a) combine with oxygen to give (b) CO_2 and H_2O molecules, which can be used (c) to form other molecules using the same C and H atoms.

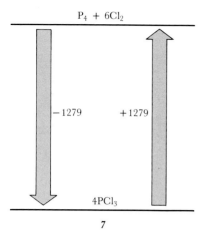

$P_4 + 6Cl_2$

-1279 $+1279$

$4PCl_3$

7

In some applications of Hess's law, it is necessary to use the reverse of a reaction for which $\Delta H°$ is listed. Since the enthalpy change for any reverse process is the negative of the change for the forward process, the reverse-reaction enthalpy is obtained simply by changing the sign of the forward-reaction enthalpy (**7**). As an example,

$$P_4(s) + 6Cl_2(g) \longrightarrow 4PCl_3(l) \qquad \Delta H° = -1279 \text{ kJ}$$
$$4PCl_3(l) \longrightarrow P_4(s) + 6Cl_2(g) \qquad \Delta H° = +1279 \text{ kJ}$$

A reaction that is exothermic in one direction is endothermic in the reverse direction.

6.4 ENTHALPIES OF FORMATION

We compute only ΔH (and not H) because only *changes* of enthalpy are useful in chemistry. In everyday life, we speak of altitudes above sea level without knowing (or always caring about) the actual altitudes above the center of the earth. In chemistry, without knowing the actual enthalpy of a substance, we can report the enthalpies of compounds as a kind of "thermochemical altitude" above a "thermochemical sea level."

Standard enthalpies of formation. If we let the elements define our thermochemical sea level, then any compound is at a thermochemical altitude given by the enthalpy change that occurs when it is formed from its elements (Fig. 6.22). We therefore can define a "standard enthalpy of formation" $\Delta H_f°$ as follows:

> The **standard enthalpy of formation** of a compound is the standard reaction enthalpy per mole of formula units for the compound's synthesis from its elements in their most stable form at 1 atm pressure.

Values for a number of compounds are listed in Table 6.4, which is extracted from a longer table in Appendix 2A. All the tabulated data in this text are for substances at 25°C. The most stable form of hydrogen at 25°C and 1 atm pressure is the gas; the most stable form of bromine is the liquid; that of iron is the solid. Since graphite is the most stable form of carbon at 25°C and 1 atm pressure, enthalpies of formation for organic compounds are based on their synthesis from graphite.

Enthalpies of formation are reported in kilojoules per mole of the compound. As an example, in the synthesis of liquid water,

$$2H_2(g) + O_2(g) \longrightarrow 2H_2O(l)$$

$\Delta H° = -571.6$ kJ and 2 mol of H_2O molecules are formed; hence, the standard enthalpy of formation of liquid water is

$$\Delta H_f° = \frac{-571.6 \text{ kJ}}{2 \text{ mol } H_2O} = -285.8 \text{ kJ/mol } H_2O$$

A common means of determining an enthalpy of formation when the compound cannot be synthesized directly is to use combustion enthalpies. This is illustrated for propane in Example 6.13.

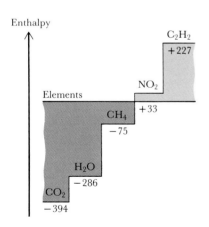

Enthalpy

C_2H_2 $+227$

NO_2
Elements $+33$

CH_4
-75

H_2O
-286

CO_2
-394

FIGURE 6.22 The elements define a thermochemical "sea level" for reporting the enthalpies of compounds. Some compounds have enthalpies of formation that are positive; others have negative enthalpies of formation.

TABLE 6.4   Standard enthalpies of formation at 25°C

Substance*	Formula	ΔH_f°, kJ/mol	Substance*	Formula	ΔH_f°, kJ/mol
Inorganic compounds			*Inorganic compounds*		
Ammonia	$NH_3(g)$	−46.11	Sulfur trioxide	$SO_3(g)$	−395.72
Ammonium nitrate	$NH_4NO_3(s)$	−365.56	Sulfuric acid	$H_2SO_4(l)$	−813.99
Carbon dioxide	$CO_2(g)$	−393.51	Water	$H_2O(l)$	−285.83
Carbon disulfide	$CS_2(l)$	+89.70		$H_2O(g)$	−241.82
Carbon monoxide	$CO(g)$	−110.53			
Dinitrogen oxide	$N_2O(g)$	+82.05	*Organic compounds*		
Dinitrogen tetroxide	$N_2O_4(g)$	+9.16	Acetylene	$C_2H_2(g)$	+226.73
Hydrogen chloride	$HCl(g)$	−92.31	Benzene	$C_6H_6(l)$	+49.0
Hydrogen fluoride	$HF(g)$	−271.1	Ethane	$C_2H_6(g)$	−84.68
Hydrogen sulfide	$H_2S(g)$	−20.63	Ethanol	$C_2H_6O(l)$	−277.69
Nitric acid	$HNO_3(l)$	−174.10	Ethylene	$C_2H_4(g)$	+52.26
Nitric oxide	$NO(g)$	+90.25	Glucose	$C_6H_{12}O_6(s)$	−1268
Nitrogen dioxide	$NO_2(g)$	+33.18	Methane	$CH_4(g)$	−74.81
Sodium chloride	$NaCl(s)$	−411.15	Methanol	$CH_4O(l)$	−238.66
Sulfur dioxide	$SO_2(g)$	−296.83	Sucrose	$C_{12}H_{22}O_{11}(s)$	−2222

*A longer list is given in Appendix 2A.

Combining standard enthalpies of formation.   Standard enthalpies of formation can be used to calculate the standard enthalpy of *any* reaction. Suppose, for example, that we are interested in the reaction in which dinitrogen tetroxide forms nitrogen dioxide:

$$N_2O_4(g) \longrightarrow 2NO_2(g) \qquad (a)$$

This equation can be expressed as the *difference* between the equations for the formation of the product and the formation of the reactant,

$$N_2(g) + 2O_2(g) \longrightarrow 2NO_2(g) \qquad (b)$$
$$N_2(g) + 2O_2(g) \longrightarrow \qquad\qquad N_2O_4(g) \qquad (c)$$
$$\overline{\qquad\qquad \longrightarrow 2NO_2(g) - N_2O_4(g)}$$

which becomes reaction (a) when we add N_2O_4 to both sides. It follows **(8)** that the standard reaction enthalpy associated with reaction (a) is the difference between the standard reaction enthalpies associated with reactions (b) and (c). We can obtain the latter from the standard enthalpies of formation of the two compounds (Table 6.4):

$$\Delta H^\circ(b) = 2 \text{ mol } NO_2 \times (+33.18 \text{ kJ/mol } NO_2) = +66.36 \text{ kJ}$$

$$\Delta H^\circ(c) = 1 \text{ mol } N_2O_4 \times (+9.16 \text{ kJ/mol } N_2O_4) = +9.16 \text{ kJ}$$

The reaction enthalpy for reaction (a) is therefore the difference

$$\Delta H^\circ(a) = 66.36 \text{ kJ} - 9.16 \text{ kJ} = +57.20 \text{ kJ}$$

This same approach can be used for any reaction. For a reaction

$$\text{Reactants} \longrightarrow \text{products}$$

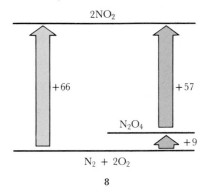

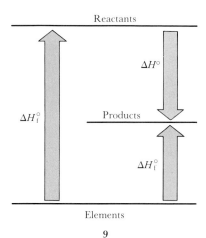

Reactants

ΔH°

ΔH_{f}°

Products

ΔH_{f}°

Elements

9

the standard reaction enthalpy is the difference between the standard reaction enthalpies for the formation of the products and the standard reaction enthalpies for the formation of the reactants (9). This is often written

$$\Delta H^{\circ} = \Delta H_{f}^{\circ}(\text{products}) - \Delta H_{f}^{\circ}(\text{reactants}) \qquad (4)$$

▼ **EXAMPLE 6.14 Using enthalpies of formation: I**

Use the information in Table 6.4 to calculate the standard enthalpy of combustion of benzene.

STRATEGY We can expect a negative value, since all combustions are exothermic. We always begin with the chemical equation, which in this case is

$$2C_6H_6(l) + 15O_2(g) \longrightarrow 12CO_2(g) + 6H_2O(l)$$

We can see that the standard reaction enthalpy is given by the difference between the enthalpies of formation of the two products (12 mol of CO_2 and 6 mol of H_2O) and the two reactants (2 mol of C_6H_6 and 15 mol of O_2). The enthalpy of formation of an element is zero; the other values can be taken from Table 6.4. To find the enthalpy of combustion, we must divide the standard reaction enthalpy by the number of moles of C_6H_6 taking part.

SOLUTION From Table 6.4, the total enthalpy change for the formation of the products is

$$\begin{aligned}\Delta H_{f}^{\circ}(\text{products}) &= 12 \text{ mol } CO_2 \times (-393.51 \text{ kJ/mol } CO_2) \\ &\quad + 6 \text{ mol } H_2O \times (-285.83 \text{ kJ/mol } H_2O) \\ &= -4722.12 \text{ kJ} + (-1714.98 \text{ kJ}) \\ &= -6437.10 \text{ kJ}\end{aligned}$$

The total enthalpy change for the formation of the reactants is

$$\begin{aligned}\Delta H_{f}^{\circ}(\text{reactants}) &= 2 \text{ mol } C_6H_6 \times (+49.0 \text{ kJ/mol } C_6H_6) \\ &= +98.0 \text{ kJ}\end{aligned}$$

The difference is

$$\Delta H^{\circ} = -6437.10 \text{ kJ} - (+98.0 \text{ kJ}) = -6535.1 \text{ kJ}$$

Since 2 mol of C_6H_6 molecules are burned in the reaction, the standard enthalpy of combustion of benzene is

$$\Delta H_{c}^{\circ} = \frac{-6535.1 \text{ kJ}}{2 \text{ mol } C_6H_6} = -3267.6 \text{ kJ/mol } C_6H_6$$

EXERCISE Calculate the standard enthalpy of combustion of glucose from the information in Table 6.4.

[*Answer*: -2808 kJ/mol $C_6H_{12}O_6$]

▲

▼ **EXAMPLE 6.15 Using enthalpies of formation: II**

Calculate the standard enthalpy of the reaction in which ammonium nitrate decomposes into dinitrogen oxide and water vapor.

STRATEGY We begin, as usual, by writing the chemical equation. Then we must calculate the standard enthalpies of formation of the products and of the reactants. Their difference is the standard reaction enthalpy we are asked to find. The data are in Table 6.4 and in Appendix 2A.

SOLUTION The reaction equation is

$$NH_4NO_3(s) \longrightarrow N_2O(g) + 2H_2O(g)$$

From Appendix 2A,

$$\Delta H_f^\circ(\text{products}) = 1 \text{ mol } N_2O \times (+82.05 \text{ kJ/mol } N_2O)$$
$$+ 2 \text{ mol } H_2O \times (-241.82 \text{ kJ/mol } H_2O)$$
$$= -401.59 \text{ kJ}$$

$$\Delta H_f^\circ(\text{reactants}) = 1 \text{ mol } NH_4NO_3 \times (-365.56 \text{ kJ/mol } NH_4NO_3)$$
$$= -365.56 \text{ kJ}$$

The difference is

$$\Delta H^\circ = -401.59 \text{ kJ} - (-365.56 \text{ kJ}) = -36.03 \text{ kJ}$$

EXERCISE Calculate the standard enthalpy of the reaction in which boron oxide and calcium fluoride give boron trifluoride and calcium oxide.

[*Answer*: +752 kJ]

▲

THE WORLD'S ENTHALPY RESOURCES

One of the most pressing questions facing civilization is, how will it continue to supply energy for manufacturing, transport, communication, and leisure? We may intuitively understand that we are using up our supplies of certain fuels, but the techniques developed in this chapter allow us to be more specific about the rate. A quantitative approach to any problem shows its magnitude more precisely and indicates when it may become critical. Since foods are fuels, they, too, can be discussed using the techniques we have developed in this chapter.

The sun is the origin of the energy stored in food and in most of the fuel we use (the exception is nuclear fuel, which we discuss in Chapter 22). Solar energy is captured by the chlorophyll molecules of green vegetation (Fig. 6.23) and used to drive the photosynthesis reaction. In this reaction, carbohydrates are built up from carbon dioxide in the air

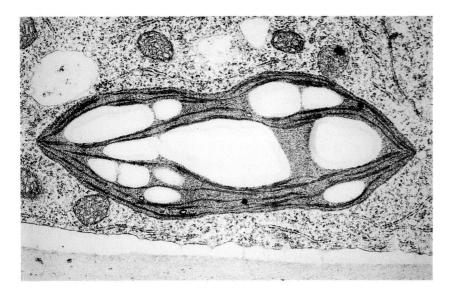

FIGURE 6.23 Chlorophyll, which gives leaves their green color, is contained in chloroplasts, like the one pictured here.

FIGURE 6.24 Highly concentrated energy being collected in the form of coal from an open-pit mine. There is no point in recovering the coal if more energy is needed than the recovered fuel will provide.

and water drawn from the soil. _Carbohydrates_ are organic compounds with empirical formulas that are often of the form $C_n(H_2O)_m$. One of the simplest is glucose ($C_6H_{12}O_6$), which acts as an energy supply in our blood. Another is cellulose, the structural material of plants; it consists of long chains of linked glucose molecules. Starch, the food reserve of plants, also consists of glucose molecules, but they are linked together in a branching structure.

The photosynthesis of glucose depends on a very complicated sequence of reactions, but the overall reaction is

$$6CO_2(g) + 6H_2O(l) \longrightarrow C_6H_{12}O_6(s) + 6O_2(g)$$

It would be very difficult to measure the reaction enthalpy of this synthesis directly. However, the synthesis is the reverse of the combustion

$$C_6H_{12}O_6(s) + 6O_2(g) \longrightarrow 6CO_2(g) + 6H_2O(l)$$

so the reaction enthalpy for the photosynthesis is the negative of the enthalpy of combustion of glucose. Since the latter is -2.8 MJ, the reaction enthalpy of the photosynthesis is $+2.8$ MJ. That is, to build 1 mol of glucose molecules (180 g of glucose), 2.8 MJ of energy must be supplied. The photosynthesis of 1 g of glucose (6 mmol of $C_6H_{12}O_6$) therefore requires 16 kJ of energy.

The energy content of all the solar radiation absorbed by vegetation on earth is enough to build about 6×10^{14} kg of glucose each year. Most of this glucose is converted to starch and cellulose. Hence, as long as its carbohydrates do not decay completely back to carbon dioxide and water, dead vegetation remains a store of energy. If losses through forest fires are ignored, this store increases by about 10^{19} kJ each year. This is about 30 times the current annual global industrial demand for energy, so the _amount_ of energy does not seem to present a problem. However, a problem becomes apparent when we attempt to _harvest_ the earth's energy: we need _high concentrations_ of energy-rich substances (Fig. 6.24). If these substances are not concentrated, the harvest may be uneconomical because fuel has to be used to extract, transport, and remove unwanted materials such as rocks. There is no point in mining or drilling for fuel if that process uses more fuel than is gained.

High concentrations of energy-rich compounds must be taken from where they have collected in the past, in deposits of *fossil fuels*—mainly coal, oil, and natural gas. Fossil fuels are the partially decomposed remains of vegetable and marine life that lived millions of years ago. Favorable rock formations or other conditions kept them out of contact with the atmosphere, so they have not been oxidized back to carbon dioxide and water. Coal, for example, originated as matter that collected at the bottoms of swamps, marshes, and bogs.

Of the 10^{19} kJ of energy stored annually by photosynthesis, it is believed that only about 10^{14} to 10^{15} kJ survive without being oxidized. However, only a fraction of that (estimated as 0.07%) forms deposits large enough to be economical sources of fuels. The rate at which energy is stored in a useful form is therefore only about 10^{11} kJ per year. We are using it at the rate of 3×10^{17} kJ per year, which is 3 million times more rapidly than it is being stored.

6.5 THE ENTHALPY OF FUELS

Many factors affect the decision as to which fuel is best for a particular job. They include its costs, the ease with which it burns, the pollution its use may cause, and the amount of fuel that must be supplied to achieve a particular effect. We consider only the last.

The *specific enthalpy* of a fuel is its enthalpy of combustion per gram (expressed without the minus sign). Its *enthalpy density* is the enthalpy of combustion per liter (similarly, without the sign). Fuels with high specific enthalpy release a lot of heat per gram when they burn. Hence, specific enthalpy is an important criterion when mass is of concern, as in airplanes and rockets. A fuel with a high enthalpy density releases a lot of heat per liter when it burns. Hence, enthalpy density is an important criterion when fuel storage volume must be considered. In some cases, both volume and mass are important; for example, increasing the size of an airplane to include more tanks for a less dense fuel may increase the retarding forces the airplane experiences to an uneconomical level.

Three readily available fuels are hydrogen, methane, and octane (Table 6.5). Methane is the major component of natural gas. Although octane is only one of a large number of components of gasoline, most of the molecules in a gallon are hydrocarbons with about eight carbon atoms, so octane is representative of the mixture. The data in Table 6.5

TABLE 6.5 Thermochemical properties of some fuels

Fuel	Combustion equation	ΔH_c°, kJ/mol	Specific enthalpy, kJ/g	Enthalpy density,* kJ/L
Hydrogen	$2H_2(g) + O_2(g) \longrightarrow 2H_2O(g)$	-286	142	13
Methane	$CH_4(g) + 2O_2(g) \longrightarrow CO_2(g) + 2H_2O(g)$	-890	55	40
Octane	$2C_8H_{18}(l) + 25O_2(g) \longrightarrow 16CO_2(g) + 18H_2O(g)$	-5471	48	3.8×10^4

*At atmospheric pressure and room temperature.

FIGURE 6.25 Calcium hydride releases hydrogen when it reacts with an acid.

show that one advantage of using gasoline in mobile engines arises from its high enthalpy density of 38 MJ/L. Because this value is high, the fuel tank need not be large to carry a large store of energy. Where low mass is important, as in a rocket, hydrogen may be an attractive fuel because of its high specific enthalpy. Mass is rarely a consideration in the supply of energy for domestic use. The higher enthalpy density of methane compared with hydrogen means that, for the same heating effect, a smaller volume of gas must be pumped through the pipes or supplied to the home in cylinders.

A great advantage of hydrogen over hydrocarbon fuels is that it produces no carbon dioxide when it burns. (Carbon dioxide is potentially harmful, since its presence in the atmosphere can lead to a worldwide rise in temperature through the insulating effect known as the "greenhouse effect.") The widespread adoption of hydrogen as a fuel—perhaps using solar-powered decomposition or electrolysis of water to produce it—would lead to what has been called a "hydrogen economy." However, one difficulty with hydrogen gas is its low enthalpy density. A solution might be to supply liquid hydrogen, which has a higher enthalpy density because the same volume contains many more molecules. Unfortunately, hydrogen condenses to a liquid only at low temperatures, and it is costly to refrigerate and store. Furthermore, even liquid hydrogen has such a low density (0.070 g/mL) that its enthalpy density is only 10 MJ/L, about a fourth that of gasoline, so that fuel tanks would have to be large.

Chemists are currently exploring the feasibility of synthesizing compounds that can be decomposed to release hydrogen as it is needed (Fig. 6.25). Candidates include the hydrides formed when some metals (including titanium, copper, and their alloys with other metals) are heated in hydrogen. These compounds carry hydrogen in a smaller volume than in liquid hydrogen itself, and they release it when heated or treated with acid. One example is the iron titanium hydride of approximate formula $FeTiH_2$. However, because of its iron and titanium content the compound is relatively dense, and its specific enthalpy is therefore small.

6.6 THE ENTHALPY OF FOODS

Over 50% by mass of our food is in the form of carbohydrates. One kind of carbohydrate is the cellulose of wood, stalks and leaves, the outer parts of seeds, and the walls of plant cells. Cellulose is ingested but not digested, and it helps move material through the intestines. The digestible carbohydrates are the starches and sugars. The first stage of their digestion is a breakdown into glucose, which is soluble in water and is carried through the body in the bloodstream. Glucose is therefore the carbohydrate that is actually used as fuel in the body.

When glucose is used in animal cells, it is oxidized to carbon dioxide and water; the overall reaction is thus the same as the combustion of glucose. The enthalpy of combustion is −2.8 MJ/mol, which corresponds to a specific enthalpy of 16 kJ/g: oxidizing 1 g of glucose produces 16 kJ of energy, enough to heat 1 L of water by about 4°C. The *average* specific enthalpy of all types of digestible carbohydrates, in-

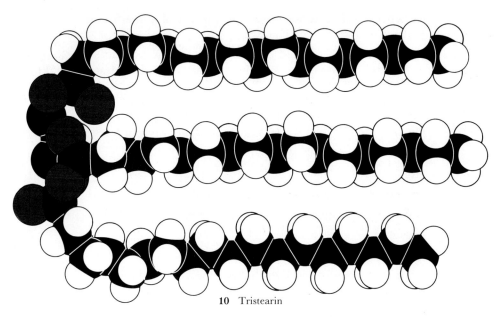

10 Tristearin

cluding starches, is about 17 kJ/g. Nutritional scientists often report this value in "food Calories" (Cal):

$$1 \text{ Calorie} = 4.184 \text{ kJ}$$

(Note the uppercase C, which distinguishes these Calories from calories.) The average specific enthalpy of carbohydrates is therefore about 4 Cal/g.

A second major energy source is *fats*, which are compounds with long $-CH_2-CH_2-CH_2-$ chains attached to a central unit. One example is tristearin, $C_{57}H_{110}O_6$ **(10).** Animals use fats as a long-term energy store. The enthalpy of combustion of tristearin is -75 MJ/mol, corresponding to a specific enthalpy of about 84 kJ/g or 20 Cal/g. This is significantly higher than the value for carbohydrates, and similar to the value for gasoline (compare octane in Table 6.5). The difference arises from the higher proportion of oxygen already present in carbohydrate molecules, as compared to fat molecules. Carbohydrate molecules are already partially oxidized, so a smaller enthalpy change occurs during complete oxidation to carbon dioxide and water. Fats are like gasoline in the sense that their molecules contain very little oxygen. That is, they are initially only very slightly oxidized and release a great deal of heat when complete oxidation finally takes place. It is interesting to note that our bodies store energy in compounds that resemble those found so convenient for storing energy in automobiles. Animals, airplanes, and automobiles all need to maximize their efficiency by storing energy in lightweight materials.

Another type of food is *proteins*, which are complex compounds of carbon, hydrogen, oxygen, and nitrogen made up of small units called "amino acids" (these are examined in Chapter 24). Proteins are large molecules, having molecular weights that sometimes reach 10^6 amu. They carry out many of the functions of a living cell and are too important to be used merely as fuel; nevertheless, they do become oxidized and, in mammals, form urea (CH_4ON_2; **11**). This change corresponds to a specific enthalpy of about 17 kJ/g, similar to that of carbohydrates.

11 Urea

TABLE 6.6 Thermochemical properties of some foods

| Food | Percentage composition | | | | Specific enthalpy, kJ/g |
	Water	Protein	Fat	Carbohydrate	
Apples	84.3	0.3	0	11.9	2.5
Beef	54.3	23.6	21.1	0	13.1
Bread	39.0	7.8	1.7	49.7	1.2
Cheese	37.0	26.0	33.5	0	17.0
Fish (cod)	76.6	21.4	1.2	0	3.1
Hamburger	40.9	15.8	14.2	29.1	17.3
Milk	87.6	3.3	3.8	4.7	2.6
Potatoes	80.5	1.4	0.1	17.7	3.5

The compositions and enthalpy resources of some typical foods are given in Table 6.6. The recommended consumption for 18- to 20-year-old males is 12 MJ/day; for females of the same age it is 9 MJ/day. That energy intake is not used merely to produce warmth; nor is it always simply stored as fat (the analog of the earth's deposits of fossil fuel). Food in animals, like fuel in engines, is used to do work. This work may be mechanical, like that required to move objects, including the animal itself. It may also be electrical or chemical, as in the propagation of nerve signals and in growth. We discuss this broader aspect of thermodynamics, in which work is considered as well as heat, in Chapters 16 and 17.

SUMMARY

6.1 Energy, the capacity to do work or provide heat, is stored in matter as **kinetic energy,** the energy of motion, and **potential energy,** the energy arising from position. Although one form of energy can be converted into the other, the **law of conservation of energy** implies that their sum, the **total energy,** is constant. The total energy of a **system,** or region of interest, is called its **internal energy.** Internal energy is a **state property,** one that is independent of how the substance or system was prepared; the change that occurs in the value of a state property due to a change from some initial state to a final state is independent of the path taken between these states. Internal energy may be changed by the transfer of energy as **heat.** When a system is isolated so that no energy transfers are possible, the **first law of thermodynamics** states that the internal energy remains constant. The heat transferred to a system at constant volume is equal to the increase in internal energy of the system, and vice versa.

6.2 The amount of heat transferred to or from a system at constant pressure is equal to the change of **enthalpy** H of the system. A change is **exothermic** if it releases heat ($\Delta H < 0$), and **endothermic** if heat is absorbed ($\Delta H > 0$). Enthalpy and internal-energy changes are measured by **calorimetry,** in which **heat capacity,** the proportionality between heat input and temperature rise, is used to interpret a temperature change as a heat transfer. Enthalpies of physical change include the **enthalpy of vaporization,** of **melting,** of **freezing,** and of **sublimation.** The enthalpy change for the reverse of any process is the negative of the enthalpy change for the forward process.

6.3 A **reaction enthalpy,** which is the difference between the enthalpies of the products and the reactants in a reaction, is usually reported for the substances in their **standard states**—their pure form at 1 atm pressure. Reaction enthalpies are used to determine the heat output of a reaction from the number of moles of reactants taking part. A special case of reaction enthalpy is the **enthalpy of combustion.** According to **Hess's law,** reaction enthalpies may be calculated from the most convenient **reaction sequence,** a series of (possibly hypothetical) reactions with the same net outcome as the reaction of interest.

6.4 Reaction enthalpies may also be calculated by combining **enthalpies of formation,** the enthalpy changes occurring when compounds are formed from their elements. Enthalpies of formation are often obtained from enthalpies of combustion.

6.5 and 6.6 The original source of the energy reserves in **fossil fuels** is the sun. Fuels are evaluated according to several criteria, including **specific enthalpy,** or the enthalpy of combustion per gram, and **enthalpy density,** or the enthalpy of combustion per liter.

EXERCISES

Kinetic and Potential Energy

In Exercises 6.1 and 6.2, calculate the kinetic energy of each object.

6.1 (a) A 140-g ball traveling at 40 m/s; (b) a 2000-kg automobile traveling at 55 mi/h.

6.2 (a) A 50-g bullet traveling at 1000 m/s; (b) a 100-lb sack of potatoes traveling at 3 ft/s.

The potential energy of an object of mass m at a height h above the surface of the earth is mgh, where $g = 9.81 \text{ m/s}^2$. In Exercises 6.3 and 6.4, calculate the potential energy (in joules or kilojoules) of each object.

6.3 (a) A 140-g ball at a height of 10 m; (b) a 2000-kg automobile, 10,000 ft up in the Rockies.

6.4 (a) A 1.0×10^5-kg airplane at a height of 10,000 m; (b) a 100-lb sack of potatoes on the back of a truck, 1.5 m above the ground.

6.5 How high could a 140-g ball reach (neglecting air resistance), if initially it had (a) a speed of 40 m/s; (b) a kinetic energy of 1 J, (c) a kinetic energy of 100 J?

6.6 How high could a 50-g bullet reach if (neglecting air resistance) the explosive charge gave it (a) a speed of 1000 m/s; (b) a kinetic energy of 1 J; (c) a kinetic energy of 100 J? Show that, for a given initial speed, the height a bullet can reach is independent of its mass.

Heat Capacity

6.7 The specific heat capacity of benzene is 1.05 J/(K · g) near room temperature. Calculate the heat needed to raise the temperature of 100.0 g of benzene from 10.0°C to 25.0°C.

6.8 The specific heat capacity of ethanol is 2.3 J/(K · g) near room temperature. Calculate the energy that must be removed as heat from 100.0 g of ethanol to cool it from 20.0°C to 10.0°C?

6.9 The specific heat capacity of water is 4.18 J/(K · g), and that of stainless steel is 0.51 J/(K · g). Calculate the heat that must be supplied to a 750.0-g stainless steel vessel containing 800.0 g of water to raise its temperature from 20.0°C to the boiling point of water. What percentage of the heat supplied is used to raise the temperature of the water?

6.10 The specific heat capacity of linseed oil is 1.84 J/(K · g), and that of copper is 0.38 J/(K · g). Calculate the heat that must be supplied to a 500-g copper vessel containing 200 g of linseed oil to raise its temperature from 20.0°C to 50.0°C. What percentage of the heat supplied is used to raise the temperature of the linseed oil?

6.11 How long would it take a 2.5-kilowatt (2.5-kW) electric kettle, constructed like the vessel in Exercise 6.9 and containing the same amount of water, to bring the water to boiling from 20°C?

6.12 How long would it take a 250-W electric heater to bring about the heating described in Exercise 6.10?

6.13 100.0 g of water at 62.5°C was poured into a calorimeter containing 100.0 g of water at 19.8°C, and the final temperature was 40.1°C. What is the heat capacity of (a) the calorimeter plus 100.0 g of water; (b) the calorimeter vessel alone?

6.14 50.0 g of water at 60.2°C was poured into a calorimeter containing 50.0 g of water at 18.7°C, and the final temperature was 35.0°C. What is the heat capacity of (a) the calorimeter plus 50.0 g of water; (b) the calorimeter vessel alone?

Enthalpy of Physical Change

In Exercises 6.15 and 6.16, calculate the enthalpy of vaporization from the data.

6.15 An electric heater was immersed in a flask of boiling ethanol (C_2H_6O), and 22.45 g of the ethanol was vaporized when 21.2 kJ of energy was supplied by the heater.

6.16 An experiment similar to that of Exercise 6.15 was carried out on liquid methane (CH_4) at its boiling point (-161°C), and 1.68 g was vaporized by 860 J of energy supplied as heat.

6.17 A sample of water was heated to boiling, and then a 50.0-W heater vaporized 22.3 g in 1020 s. What is the enthalpy of vaporization of water at its boiling point?

6.18 Benzene was vaporized under low pressure at 25°C, in an experiment in which a 75-W heater operated for 500 s led to the vaporization of 95 g of the liquid. Calculate the enthalpy of vaporization of benzene at its boiling point.

In Exercises 6.19 to 6.22, calculate the heat required and the time for which the heater should be operated in each case.

6.19 To vaporize 100.0 g of water at 100°C, using a 1.00-kW heater

6.20 To vaporize 100.0 g of ethanol at its normal boiling point (79°C), using a 1.00-kW heater

6.21 To melt 1.00 g of iron at its melting point, using a 1.00-kW heater ($\Delta H_{\text{melt}} = 13.8$ kJ/mol)

6.22 To melt 1.00 g of solid sodium at its melting point, using a 1.00-kW heater ($\Delta H_{\text{melt}} = 2.61$ kJ/mol)

Exercises 6.23 and 6.24 require the fact that strong sunshine delivers heat at about 1 kJ per square meter per second.

6.23 Calculate the mass of pure ethanol that can be vaporized in 10 min from a glass left in strong sunshine, supposing that the alcohol has a surface area of 30 cm² exposed to the sunshine and assuming that all heat goes into evaporation, not into an increase in temperature.

6.24 Calculate the mass of water that can be vaporized in 1 h from a swimming pool with an area of 50 m² exposed to strong sunshine, assuming that all heat goes into evaporation, not into an increase in temperature.

6.25 The standard enthalpies of combustion of graphite and diamond are −393.51 kJ/mol and −395.41 kJ/mol, respectively. Calculate the enthalpy of the graphite → diamond transition.

6.26 Elemental sulfur occurs in several forms; rhombic sulfur is the most stable under normal conditions, and monoclinic sulfur is somewhat less stable. The standard enthalpies of combustion of the two forms to sulfur dioxide are −296.83 kJ/mol and −297.16 kJ/mol, respectively. Calculate the enthalpy of the rhombic → monoclinic transition.

Calorimetry and Reaction Enthalpy

6.27 The heat capacity of a calorimeter was measured as 5.24 kJ/K. When a reaction was carried out in it, the temperature rose from 22.45 to 24.80°C. How much heat was produced by the reaction?

6.28 A calorimeter has a heat capacity of 3.57 kJ/K. When a reaction was carried out in it, the temperature rose from 23.55 to 26.88°C. How much heat was produced by the reaction?

6.29 50.0 mL of 0.500 M NaOH(aq) was poured into a simple calorimeter, and its temperature was measured. Then 50.0 mL of 0.500 M HNO$_3$(aq) at the same temperature was poured in, and the mixture was gently stirred. The temperature rose from 18.6°C to 21.3°C. The heat capacity of the calorimeter and contents was 525.0 J/K. Calculate (a) the enthalpy change observed, and (b) the enthalpy of the neutralization in kilojoules per mole of HNO$_3$ molecules.

6.30 25.00 mL of 0.700 M KOH(aq) and 25.00 mL of 0.700 M HCl(aq) were used in the same type of experiment as in Exercise 6.29. The temperature rose from 19.7°C to 21.8°C. Given that the heat capacity of the calorimeter and its contents was 488.1 J/K, calculate (a) the enthalpy change observed, and (b) the enthalpy of the neutralization per mole of HCl.

In Exercises 6.31 to 6.34, the calorimeter has heat capacity 6.27 kJ/K.

6.31 1.84 g of magnesium was oxidized to MgO in the calorimeter, and the temperature rose from 21.30°C to 28.56°C. Calculate the reaction enthalpy of the oxidation $2Mg(s) + O_2(g) \rightarrow 2MgO(s)$.

6.32 2.23 g of sulfur was oxidized to SO$_2$ in the calorimeter, and the temperature rose from 22.41°C to 25.70°C. Calculate the reaction enthalpy of $S_8(s) + 8O_2(g) \rightarrow 8SO_2(g)$.

6.33 Because of its impact on the analysis of biochemical processes, the enthalpy of combustion of glucose ($C_6H_{12}O_6$) is a very important quantity in thermochemistry. In an experiment to measure it, 1.22 g of glucose was burned in the calorimeter, and the temperature rose from 21.53°C to 24.56°C. Calculate the enthalpy of combustion of glucose.

6.34 Sucrose (cane sugar) is one of the carbohydrates we ingest, and its enthalpy of combustion is significant for that reason. In an experiment, 1.25 g of sucrose ($C_{12}H_{22}O_{11}$) was burned in the calorimeter, and the temperature rose from 23.24°C to 26.53°C. Calculate the enthalpy of combustion of sucrose. When you are backpacking, is it more efficient to carry glucose or sucrose?

In Exercises 6.35 and 6.36, classify each process as either exothermic or endothermic, and state which processes absorb (A) heat and which produce (P) heat when they occur.

6.35 (a) The formation of the gas acetylene, which is used for oxyacetylene welding:

$$2C(s) + H_2(g) \longrightarrow C_2H_2(g) \qquad \Delta H° = +227 \text{ kJ}$$

(b) The neutralization of hydrochloric acid with aqueous sodium hydroxide:

$$HCl(aq) + NaOH(aq) \longrightarrow NaCl(aq) + H_2O(l)$$
$$\Delta H° = -57.1 \text{ kJ}$$

(c) The dissolution of ammonium chloride in water:

$$NH_4Cl(s) \longrightarrow NH_4Cl(aq) \qquad \Delta H° = +15.2 \text{ kJ}$$

(d) The oxidation of nitrogen in the hot exhausts of jet engines and automobiles:

$$N_2(g) + O_2(g) \longrightarrow 2NO(g) \qquad \Delta H° = +180.6 \text{ kJ}$$

6.36 (a) The melting of ice:

$$H_2O(s) \longrightarrow H_2O(l) \qquad \Delta H° = +6.0 \text{ kJ}$$

(b) The industrial preparation of carbon disulfide (CS_2) from natural gas:

$$CH_4(g) + 4S(s) \longrightarrow CS_2(l) + 2H_2S(g)$$
$$\Delta H° = -106 \text{ kJ}$$

(c) The industrial preparation of carbon disulfide from coke and sulfur:

$$4C(s) + S_8(s) \longrightarrow 4CS_2(l) \qquad \Delta H° = +358.8 \text{ kJ}$$

(d) The dissolving of table salt in water:

$$NaCl(s) \longrightarrow NaCl(aq) \qquad \Delta H° = +3.9 \text{ kJ}$$

6.37 For each process in Exercise 6.35, state whether the temperature of the reaction mixture will rise or fall when the reaction is carried out in (i) an insulated container; (ii) a container that is not thermally insulated.
6.38 Repeat Exercise 6.37 for the processes in Exercise 6.36.

Enthalpy of Combustion

6.39 Ethanol can be produced by the fermentation of carbohydrates. There has been considerable interest in its use as a fuel, which would provide a means for using the energy of the sun without having to wait for the accumulation of fossil fuels. Calculate the heat produced when 100 g of ethanol is burned.
6.40 Methanol can also be used as a fuel. Calculate the heat produced when 100 g of methanol is burned.

6.41 Assume that coal, of density 1.5 g/cm³, is carbon graphite (it is in fact much more complicated, but this is a reasonable first approximation). Calculate the heat produced when a lump of coal of size 5 cm by 5 cm by 5 cm is burned. Estimate the mass of water that this will heat from 20°C to boiling.
6.42 The density of octane is 0.70 g/mL. Calculate the heat produced by burning 1.0 gal of octane, and estimate the mass of water it can heat from 20°C to boiling.

In Exercises 6.43 and 6.44, determine which of the two given substances can produce more heat when equal masses burn.

6.43 Aniline and phenol
6.44 Glucose and sucrose

Hess's Law

6.45 Using the standard enthalpy of combustion of methane when it forms carbon dioxide and liquid water, calculate its enthalpy of combustion when it forms water vapor.
6.46 Calculate the enthalpy of combustion of propane (C_3H_8) when it forms water vapor, using the value for its combustion to liquid water.

6.47 Use the standard enthalpy of combustion of methane to calculate the heat produced when 1 mol of CH_4 molecules burns in a limited supply of air and forms carbon monoxide and water.
6.48 Use the standard enthalpy of combustion of propane to calculate the heat produced when 1 mol of C_3H_8 molecules (propane) burns in a limited supply of air and forms carbon monoxide and water.

6.49 Use benzene's enthalpy of combustion to calculate the standard reaction enthalpy when that compound burns in a limited supply of air to form water and carbon monoxide.
6.50 Use toluene's enthalpy of combustion to calculate the standard reaction enthalpy when that compound burns in a limited supply of air to form water and carbon monoxide.

6.51 Two succeeding stages in the industrial manufacture of sulfuric acid are the combustion of sulfur and the oxidation of sulfur dioxide to the trioxide. From the standard reaction enthalpies

$$S_8(s) + 8O_2(g) \longrightarrow 8SO_2(g) \qquad \Delta H° = -2374.4 \text{ kJ}$$
$$2S(s) + 3O_2(g) \longrightarrow 2SO_3(g) \qquad \Delta H° = -791.4 \text{ kJ}$$

calculate the reaction enthalpy for the oxidation of sulfur dioxide to sulfur trioxide:

$$2SO_2(g) + O_2(g) \longrightarrow 2SO_3(g)$$

6.52 In the manufacture of nitric acid by the oxidation of ammonia, the first product is nitric oxide, which is then oxidized to nitrogen dioxide. From the standard reaction enthalpies

$$N_2(g) + O_2(g) \longrightarrow 2NO(g) \qquad \Delta H° = +180.6 \text{ kJ}$$
$$N_2(g) + 2O_2(g) \longrightarrow 2NO_2(g) \qquad \Delta H° = +66.4 \text{ kJ}$$

calculate the standard reaction enthalpy for the oxidation of nitric oxide to nitrogen dioxide:

$$2NO(g) + O_2(g) \longrightarrow 2NO_2(g)$$

6.53 Use the enthalpy of combustion of urea to calculate the standard reaction enthalpy for its further oxidation in which nitrogen dioxide is formed in place of elemental nitrogen:

$$2CO(NH_2)_2(s) + 7O_2(g) \longrightarrow$$
$$2CO_2(g) + 4H_2O(l) + 4NO_2(g)$$

6.54 Use the standard enthalpy of combustion of the amino acid glycine to calculate the standard reaction enthalpy for its oxidation in which nitrogen dioxide is formed in place of elemental nitrogen:

$$4NH_2CH_2CO_2H(s) + 13O_2(g) \longrightarrow$$
$$8CO_2(g) + 10H_2O(l) + 4NO_2(g)$$

Enthalpy of Formation

6.55 Calculate the enthalpy of formation of $PCl_5(s)$ from the enthalpy of formation of $PCl_3(l)$ and its chlorination reaction,

$$PCl_3(l) + Cl_2(g) \longrightarrow PCl_5(s) \qquad \Delta H° = -124 \text{ kJ}$$

6.56 Calculate the enthalpy of the oxidation of $SO_2(g)$ to $SO_3(g)$ from their enthalpies of formation.

*__6.57__ Calculate the enthalpy of formation of anhydrous aluminum chloride, given the following data in addition to those in Table 6.4:

$$2Al(s) + 6HCl(aq) \longrightarrow 2AlCl_3(aq) + 3H_2(g)$$
$$\Delta H° = -1007 \text{ kJ}$$
$$HCl(g) \longrightarrow HCl(aq) \qquad \Delta H° = -73 \text{ kJ}$$
$$AlCl_3(s) \longrightarrow AlCl_3(aq) \qquad \Delta H° = -323 \text{ kJ}$$

*__6.58__ Calculate the enthalpy of formation of $HCl(g)$, given the following information in addition to the information in Table 6.4 and Appendix 2A:

$$NH_3(aq) + HCl(aq) \longrightarrow NH_4Cl(aq) \qquad \Delta H° = -50.4 \text{ kJ}$$
$$NH_4Cl(s) \longrightarrow NH_4Cl(aq) \qquad \Delta H° = +16.4 \text{ kJ}$$
$$NH_3(g) \longrightarrow NH_3(aq) \qquad \Delta H° = -35.7 \text{ kJ}$$
$$HCl(g) \longrightarrow HCl(aq) \qquad \Delta H° = -73.5 \text{ kJ}$$

6.59 Calculate the enthalpy of the hydrogenation of ethylene in the reaction $C_2H_4(g) + H_2(g) \rightarrow C_2H_6(g)$ from the enthalpies of combustion of the two hydrocarbons.

6.60 Calculate the enthalpy of the hydrogenation of benzene in the reaction $C_6H_6(g) + 3H_2(g) \rightarrow C_6H_{12}(g)$ from the enthalpies of combustion of the two hydrocarbons.

6.61 The standard enthalpy of combustion of carbon is -393.5 kJ/mol C. What is the standard enthalpy of formation of carbon dioxide?

6.62 The standard enthalpy of combustion of sulfur to sulfur dioxide is -2374.4 kJ/mol S_8. What is the standard enthalpy of formation of sulfur dioxide?

In Exercises 6.63 and 6.64, calculate the standard enthalpy of formation of each compound from its standard enthalpy of combustion.

6.63 (a) Butane (C_4H_{10}); (b) glucose $(C_6H_{12}O_6)$; (c) glycine $(C_2H_5O_2N)$

6.64 (a) Pentane (C_5H_{12}); (b) sucrose $(C_{12}H_{22}O_{11})$; (c) urea $[(NH_2)_2CO]$

6.65 Use enthalpies of combustion to calculate the standard reaction enthalpy for the reaction in which phenol (C_6H_6O) reacts with ammonia to produce aniline (C_6H_7N) and water. The enthalpy of combustion of NH_3 to N_2 and H_2O is -383 kJ/mol.

6.66 Use enthalpies of combustion to calculate the standard reaction enthalpy for the hypothetical reaction in which acetic acid reacts with ammonia and oxygen to produce glycine $(C_2H_5O_2N)$ and water.

6.67 When 1.36 g of magnesium burns in oxygen to form magnesium oxide, 33.7 kJ of heat is evolved. Calculate the enthalpy of formation of magnesium oxide.

6.68 When 1.29 g of magnesium reacts with nitrogen to form magnesium nitride, 12.2 kJ of heat is evolved. Calculate the enthalpy of formation of the nitride.

6.69 The standard enthalpy of combustion of cyclopropane (C_3H_6) is -2091 kJ/mol. Calculate its standard enthalpy of formation.

6.70 The standard enthalpy of combustion of cyclohexane (C_6H_{12}) is -3920 kJ/mol. Calculate its standard enthalpy of formation.

In Exercises 6.71 to 6.74, calculate the standard reaction enthalpy for each reaction.

6.71 (a) Thermal decomposition of calcium carbonate:

$$CaCO_3(s) \longrightarrow CaO(s) + CO_2(g)$$

(b) The replacement of deuterium by ordinary hydrogen in heavy water:

$$H_2(g) + D_2O(l) \longrightarrow H_2O(l) + D_2(g)$$

(c) The decomposition of hydrogen peroxide to water and oxygen:

$$2H_2O_2(l) \longrightarrow 2H_2O(l) + O_2(g)$$

(d) The formation of carbon disulfide from natural gas:

$$2CH_4(g) + S_8(s) \longrightarrow 2CS_2(l) + 4H_2S(g)$$

6.72 (a) The thermal decomposition of magnesium carbonate:

$$MgCO_3(s) \longrightarrow MgO(s) + CO_2(g)$$

(b) The oxidation of water by fluorine to hydrogen fluoride and oxygen:

$$2F_2(g) + 2H_2O(l) \longrightarrow 4HF(aq) + O_2(g)$$

(c) The oxidation of ammonia:

$$4NH_3(g) + 5O_2(g) \longrightarrow 4NO(g) + 6H_2O(g)$$

(d) The redox reaction between hydrogen sulfide and sulfur dioxide:

$$16H_2S(g) + 8SO_2(g) \longrightarrow 3S_8(s) + 16H_2O(l)$$

6.73 (a) The final stage in the production of nitric acid, when nitrogen dioxide dissolves in and reacts with water:

$$3NO_2(g) + H_2O(l) \longrightarrow 2HNO_3(aq) + NO(g)$$

(b) The thermal decomposition of ammonium nitrate to nitrous oxide:

$$NH_4NO_3(s) \longrightarrow N_2O(g) + 2H_2O(g)$$

(c) The action of water on calcium carbide, a process still used for the production of acetylene:

$$CaC_2(s) + 2H_2O(l) \longrightarrow Ca(OH)_2(aq) + C_2H_2(g)$$

(d) The formation of boron trifluoride, which is widely used in the chemical industry:

$$B_2O_3(s) + 3CaF_2(s) \longrightarrow 2BF_3(g) + 3CaO(s)$$

6.74 (a) The disproportionation of potassium chlorate to a mixture of the perchlorate and the chloride:

$$4KClO_3(s) \longrightarrow 3KClO_4(s) + KCl(s)$$

(b) The formation of a sulfide by the action of hydrogen sulfide on an aqueous solution of a base:

$$H_2S(aq) + 2KOH(aq) \longrightarrow K_2S(aq) + 2H_2O(l)$$

(c) The final step in the industrial preparation of the fertilizer urea:

$$CO_2(g) + 2NH_3(g) \longrightarrow H_2O(g) + (NH_2)_2CO(s)$$

(d) The formation of phosphorous acid (H_3PO_3) by the action of water on phosphorus(III) chloride:

$$PCl_3(l) + 3H_2O(l) \longrightarrow H_3PO_3(aq) + 3HCl(g)$$

6.75 Calculate the standard enthalpy of formation of dinitrogen pentoxide from the data

$$2NO(g) + O_2(g) \longrightarrow 2NO_2(g) \qquad \Delta H° = -114.1 \text{ kJ}$$
$$4NO_2(g) + O_2(g) \longrightarrow 2N_2O_5(g) \qquad \Delta H° = -110.2 \text{ kJ}$$

and the standard enthalpy of formation of nitric oxide.

6.76 An important reaction that occurs in the atmosphere is $NO_2(g) \rightarrow NO(g) + O(g)$, which is brought about by sunlight. How much energy must be supplied by the sun to cause it? Calculate the standard enthalpy of the reaction from the following information:

$$NO(g) + O_3(g) \longrightarrow NO_2(g) + O_2(g)$$
$$\Delta H° = -183.0 \text{ kJ}$$
$$\Delta H_f°(\text{O atoms}) = +249.2 \text{ kJ/mol}$$
$$\Delta H_f°(O_3) = +142.7 \text{ kJ/mol}$$

Foods and Fuels

6.77 A minor component of gasoline is heptane (C_7H_{16}), which has a standard enthalpy of combustion of −4854 kJ/mol and a density of 0.68 g/mL. Calculate the specific enthalpy of heptane and its enthalpy density.

6.78 Another minor component of gasoline is toluene (C_7H_8), with a standard enthalpy of combustion of −3910 kJ/mol and a density of 0.867 g/mL. Calculate the specific enthalpy of toluene and its enthalpy density.

In Exercises 6.79 and 6.80, calculate the enthalpy density of each fuel.

6.79

Fuel	Formula	$\Delta H_c°$, kJ/mol	Density, g/mL
(a) Methane	$CH_4(g)$	−890	6.6×10^{-4}
(b) Propane	$C_3H_8(g)$	−2220	1.8×10^{-3}
(c) Benzene	$C_6H_6(l)$	−3268	0.88
(d) Methanol	$CH_4O(l)$	−726	0.79

6.80

Fuel	Formula	$\Delta H_c°$, kJ/mol	Density, g/mL
(a) Ethane	$C_2H_6(g)$	−1560	1.2×10^{-3}
(b) Acetylene	$C_2H_2(g)$	−1300	1.1×10^{-3}
(c) Anthracene	$C_{14}H_{10}(s)$	−7057	1.24
(d) Ethanol	$C_2H_6O(l)$	−1368	0.79

6.81 Calculate the specific enthalpy of magnesium from its enthalpy of combustion to magnesium oxide in oxygen. Would aluminum, which burns to aluminum oxide, be a better fuel if mass were the only consideration?

6.82 Calculate the specific enthalpy of phosphorus from its enthalpy of combustion in oxygen to P_4O_{10}. Would sulfur burned to either SO_2 or SO_3 be a more efficient fuel if mass were the only consideration?

6.83 One problem with fuels containing carbon is that they produce carbon dioxide when they burn, and so one consideration governing the selection of a fuel could be the heat per mole of CO_2 molecules produced. Calculate this quantity for (a) methane and (b) octane. If carbon dioxide production is an important consideration, should we go on eating? Calculate the heat produced per mole of CO_2 molecules from the combustion of glucose.

6.84 The booster rockets of the space shuttle use a mixture of powdered aluminum and ammonium perchlorate in the exothermic redox reaction

$$2Al(s) + 2NH_4ClO_4(s) \longrightarrow$$
$$Al_2O_3(s) + 2HCl(g) + 2NO(g) + 3H_2O(g)$$

Calculate the specific enthalpy of a stoichiometric mixture of aluminum and ammonium perchlorate. Would it be better to use magnesium in place of aluminum?

General

6.85 The enthalpy of formation of trinitrotoluene (TNT) is -67 kJ/mol. In principle it could be used as a rocket fuel, with the gases resulting from its combustion (nitrogen, carbon dioxide, and water vapor) streaming out the rear of the rocket to give the required thrust. In practice, of course, it would be extremely dangerous as a fuel because it is sensitive to shock. Explore its potential as a rocket fuel by calculating its enthalpy density. What other considerations are involved in the selection of a flight fuel? The density of TNT is 1.65 g/cm^3.

6.86 Calculate the enthalpy of vaporization of sodium chloride to a gas of ions from the following information and Table 6.4:

Atomization of sodium:

$$Na(s) \longrightarrow Na(g) \qquad \Delta H° = +108.4 \text{ kJ}$$

Ionization of sodium:

$$Na(g) \longrightarrow Na^+(g) + e^- \qquad \Delta H° = +495.8 \text{ kJ}$$

Dissociation of chlorine:

$$Cl_2(g) \longrightarrow 2Cl(g) \qquad \Delta H° = +242 \text{ kJ}$$

Electron attachment to chlorine:

$$Cl(g) + e^- \longrightarrow Cl^-(g) \qquad \Delta H° = -348.6 \text{ kJ}$$

6.87 Calculate the enthalpy of vaporization of potassium bromide to a gas of ions from the following information ($\Delta H_f° = -393$ kJ/mol):

Atomization of potassium:

$$K(s) \longrightarrow K(g) \qquad \Delta H° = +89.2 \text{ kJ}$$

Ionization of potassium:

$$K(g) \longrightarrow K^+(g) + e^- \qquad \Delta H° = +425.0 \text{ kJ}$$

Vaporization of bromine:

$$Br_2(l) \longrightarrow Br_2(g) \qquad \Delta H° = +30.9 \text{ kJ}$$

Dissociation of bromine:

$$Br_2(g) \longrightarrow 2Br(g) \qquad \Delta H° = +192.9 \text{ kJ}$$

Electron attachment to bromine:

$$Br(g) + e^- \longrightarrow Br^-(g) \qquad \Delta H° = -331.0 \text{ kJ}$$

6.88 Estimate the temperature of the flame obtained by burning methane in the stoichiometrically correct amount of oxygen, assuming that no heat escapes from the combustion chamber, that the enthalpy of combustion at 25°C may be used, and that the molar heat capacities of all the products of the combustion are approximately 40 J/(K · mol).

II

ATOMS, MOLECULES, AND IONS

This chapter introduces the main features of the structure of atoms and accounts for the form of the periodic table in terms of them. We begin to see how some of the properties of an element are related to its location in the table. The illustration shows the colors emitted by atoms of the alkali metals when their compounds are heated in a flame. These colors can be interpreted using the ideas described in this chapter.

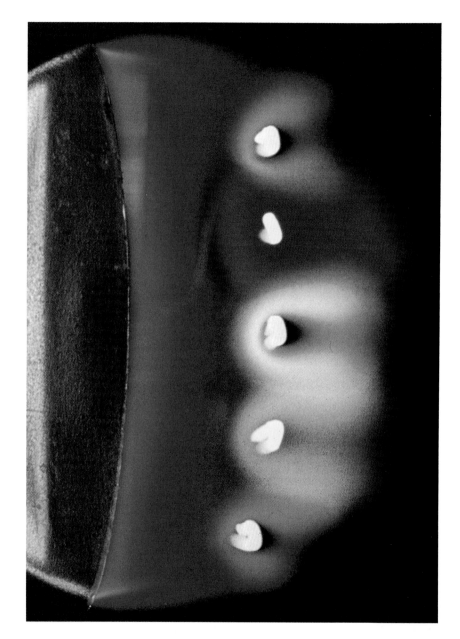

7

ATOMIC STRUCTURE AND THE PERIODIC TABLE

The structure of the hydrogen atom

7.1 Light
7.2 The spectrum of atomic hydrogen
7.3 Particles and waves

The structures of many-electron atoms

7.4 Orbital energies
7.5 The building-up principle

A survey of periodic properties

7.6 Blocks, periods, and groups
7.7 Periodicity of physical properties
7.8 Trends in chemical properties

The experiments described in Section 2.3 showed that an atom of atomic number Z consists of a central minute, massive, positively charged nucleus surrounded by Z electrons. When Ernest Rutherford proposed his model of the nuclear atom, he expected to be able to describe the motion of these electrons in terms of "classical mechanics," the laws of motion proposed by Isaac Newton in the seventeenth century. However, it quickly became clear that Newton's laws failed when they were applied to electrons in atoms. Their replacement by the laws of "quantum mechanics" caused an intellectual earthquake that shook science to its foundations.

We see a little of that earthquake in this chapter. We also see, in this and the next two chapters, the importance of *atomic structure*, the arrangement of electrons around the central nucleus of an atom. It is the key to understanding the properties of the elements, the compounds they can form, and the reactions they undergo.

THE STRUCTURE OF THE HYDROGEN ATOM

FIGURE 7.1 Flame tests are used to identify the elements in a compound. In particular, they provide an easy way of distinguishing the alkali metals. In each case except lithium, the colors come from energetically excited atoms. In lithium's case, LiOH molecules are responsible for the emission. (a) Lithium, (b) sodium, (c) potassium, and (d) rubidium.

Much of our current understanding of atomic structure has come from *spectroscopy*, the analysis of the light and other kinds of radiation emitted or absorbed by different substances. (The name comes from the Latin word for "appearance.") Some elements emit light of a distinctive color or mixture of colors when their compounds are heated in a flame (Fig. 7.1) or when their vapors are exposed to an electric discharge (a storm of electrons and ions passing between two electrodes). Sodium atoms emit the well-known yellow of some forms of highway and city lighting. Potassium atoms emit violet light. Rubidium gives a red flame (hence its name, from the Latin word for "red"), and cesium a blue one

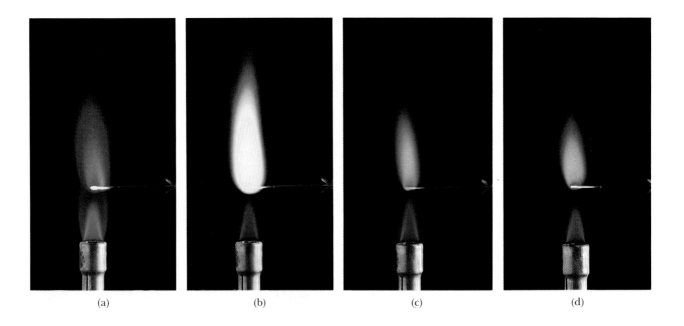

(a)　　　　　　(b)　　　　　　(c)　　　　　　(d)

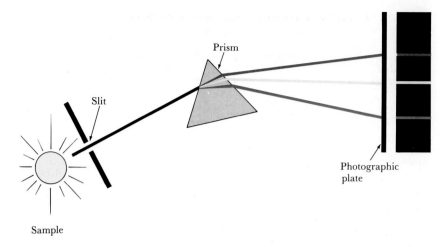

FIGURE 7.2 In a spectrometer, the light emitted by the sample is passed through a slit and then a prism. The latter separates the ray into different colors, which are recorded photographically. The spectral lines are the separate images of the slit.

(its name comes from the Latin word for "sky-blue"). The emitted light is in general a mixture of colors which can be separated by passing it through a prism, as sunlight is separated by raindrops to produce a rainbow. When a "spectrometer" is used, the separated colors are recorded photographically as a "spectrum" (Fig. 7.2). Since each ray of color gives an image of the slit the light passed through initially, the individual colors are recorded as a series of "spectral lines."

7.1 LIGHT

Light is *electromagnetic radiation*, a wave of oscillating electric and magnetic influences called *fields*. A magnetic field exerts a force on a moving charged particle. An electric field, the only field we consider here because it has the stronger effect on matter, exerts a force on any charged particle, moving or not. The "oscillation" of the electric field of light is such that its force acts first in one direction and then in the opposite direction.

Frequency and wavelength. One complete reversal of the direction of the field (the direction of the push) from an original direction to the opposite direction and back to the original direction, is called a "cycle." The number of cycles occurring per second is the *frequency* ν of the light (ν is the Greek letter nu). The unit of frequency is the "hertz" (Hz), defined as one cycle per second:

$$1 \text{ Hz} = 1 \text{ cycle per second}$$

The unit honors Heinrich Hertz, one of the pioneers of the study of electromagnetic radiation. In unit-analysis calculations we interpret 1 Hz as being 1/s, so 1 Hz × 1 s = 1.

TABLE 7.1 Color, frequency, and wavelength of light*

Color	Frequency, $\times 10^{14}$ Hz	Wavelength, nm	Energy per photon, $\times 10^{-19}$ J
X-rays and gamma rays	10^3 and above	3 and below	660 and above
Ultraviolet	10	300	6.6
Visible light			
Violet	7.1	420	4.7
Blue	6.4	470	4.2
Green	5.7	530	3.7
Yellow	5.2	580	3.4
Orange	4.8	620	3.2
Red	4.3	700	2.8
Infrared	3.0	1000	1.9
Microwaves and radio waves	3×10^{11} Hz and below	3×10^6 and above	2.0×10^{-22} J and below

*The values given are approximate but typical.

The frequency of light determines its color (see Table 7.1). The electric field of blue light, for example, oscillates at 6.4×10^{14} Hz. The light from a traffic signal changes frequency from 4.3×10^{14} Hz to 5.2×10^{14} Hz and then to 5.7×10^{14} Hz as it changes from red to yellow and then to green.

A wave is also characterized by its *wavelength* λ (the Greek letter lambda), or peak-to-peak distance (Fig. 7.3). Wavelength and frequency are related by

$$\lambda = \frac{c}{\nu} \tag{1}$$

where c is the speed of light, 3.00×10^8 m/s (more precisely 2.99792×10^8 m/s, corresponding to over 670 million mi/h); c is the speed at which *all* electromagnetic radiation travels through empty space. Equation 1 shows that a higher frequency means a shorter wavelength (Table 7.1). Higher-frequency blue light has a shorter wavelength (of about 470 nm) than lower-frequency red light (700 nm). The light from a traffic signal changes from 700 nm to 580 nm and then to 530 nm as it changes from red to yellow to green (Fig. 7.4).

Our eyes can detect electromagnetic radiation with wavelengths in the range 700 nm (red light) to 400 nm (violet light). The radiation in this range is called *visible light*. "White light," which includes sunlight, is a mixture of all frequencies of visible light. The entire span of electromagnetic radiation ranges from wavelengths of less than a picometer to

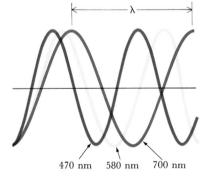

470 nm   580 nm   700 nm

FIGURE 7.3 A light wave is characterized by its wavelength λ (in nanometers) and its frequency ν, with red light having a longer wavelength than blue light. The crests of the wave move through space at a speed c, approximately 300,000 km/s. The frequency (in hertz) is related to the wavelength by $\nu = c/\lambda$.

wavelengths of more than several kilometers, but our eyes do not detect it all—either because radiation of some wavelengths is absorbed as it passes through the lens of the eye, or because the molecules in the retina do not respond to it. *Ultraviolet radiation* is radiation of higher frequency (shorter wavelength) than violet light (*ultra* is the Latin word for "beyond"). *Infrared radiation*, the radiation we experience as heat, is radiation of lower frequency than red light (*infra* is Latin for "below").

Photons. Quantum mechanics adds to our understanding of light as a wave by describing a ray of light as a stream of particles called "photons." The more intense the light, the greater the number of photons in the ray. Each photon is a packet or *quantum* of energy (from the Latin word for "amount"). We can feel the energy of the photons in the infrared radiation emitted by the sun as the warmth of sunlight.

The energy of a single photon is proportional to the frequency of the light. We write this proportionality as

$$E = h \times \nu \qquad (2)$$

where h is *Planck's constant*, a fundamental constant with the value 6.63×10^{-34} J/Hz. It is named for Max Planck, the German physicist who first introduced the idea that energy comes in packets. Blue light, for example, consists of a stream of photons, each having an energy

$$E = (6.63 \times 10^{-34} \text{ J/Hz}) \times (6.4 \times 10^{14} \text{ Hz})$$
$$= 4.2 \times 10^{-19} \text{ J}$$

A ray of red light also consists of a stream of photons, but because the frequency of the light is lower, each of its photons has less energy $(2.8 \times 10^{-19} \text{ J})$. The photon energies of light of various colors are given in Table 7.1.

The photoelectric effect. Evidence for the relation between the frequency and energy of photons comes from the *photoelectric effect*, the emission of electrons from the surface of metals when ultraviolet radiation strikes them. It is found that no electrons are emitted if the frequency of the light is below a certain threshold value characteristic of the metal. This is evidence for the existence of packets of energy, for if a photon has too low an energy, it cannot eject an electron.

By the law of conservation of energy, the kinetic energy of the *photoelectron*—the electron ejected by light—is equal to the energy of the incoming photon ($h\nu$) minus the energy needed to drive the electron out of the metal (a constant for a given metal):

$$E_K \quad = \quad h \times \nu \quad - \quad \text{a constant}$$

$$\underset{\text{of electron}}{\text{Kinetic energy}} = \underset{\text{photon}}{\text{energy of}} - \underset{\text{electron from metal}}{\text{energy to drive}}$$

Since this is the equation of a straight line,

$$E_K = -\text{constant} + h \times \nu$$
$$y = a + b \times x$$

we should expect a straight line of slope h when the kinetic energy of the emitted electrons is plotted against the frequency of the light. This

FIGURE 7.4 Each lamp in a traffic signal generates white light, a mixture of all colors of light, but the screens over the lamps allow only certain frequencies and wavelengths to pass.

FIGURE 7.5 When the kinetic energy of electrons ejected in the photoelectric effect is plotted against the frequency of the ultraviolet light, a straight line is obtained. The slope gives the value of Planck's constant. Note that below a certain minimum frequency (which depends on the metal) no photoelectrons are obtained.

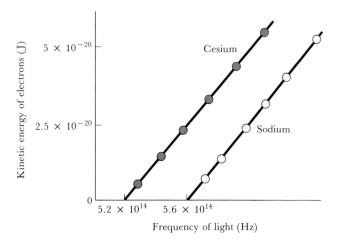

is found to be the case (Fig. 7.5). Because the slope of the line is equal to h, we can determine its value in the same experiment.

7.2 THE SPECTRUM OF ATOMIC HYDROGEN

In the visible region, the spectrum of atomic hydrogen consists of three lines. The most intense line (at 656 nm) is red, and the excited sample glows with this red light. Energetically excited hydrogen atoms also emit ultraviolet and infrared radiation, which can be detected electronically. The complete spectrum is shown diagramatically in Fig. 7.6.

Energy quantization. Although the complete spectrum of hydrogen looks like a jungle of lines, it actually follows a precise pattern. A part of the pattern was recognized by a Swiss schoolteacher, Joseph Balmer, in 1885. He noticed that the frequencies in and near the visible region fit the formula

$$\nu \propto \frac{1}{4} - \frac{1}{n^2}$$

with $n = 3, 4, \ldots$, and the symbol $\propto$ means "is proportional to." The lines this relation describes are now called the *Balmer series*. The full

FIGURE 7.6 The complete spectrum of atomic hydrogen. The lines belong to various groups called "series," two of which are shown.

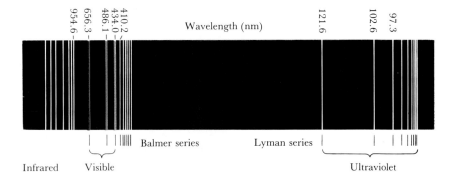

pattern of the spectrum was recognized by the Swedish spectroscopist Johannes Rydberg. He discovered that *all* the lines in the spectrum, including those in the infrared and ultraviolet regions, could be obtained by noting that the 4 in $\frac{1}{4}$ was 2^2 and then replacing it with 1^2, 3^2, 4^2, and so on. His formula is normally written

$$\nu = \mathfrak{R} \times \left(\frac{1}{n_f^{\,2}} - \frac{1}{n_i^{\,2}} \right) \qquad (3)$$

where the constant $\mathfrak{R}$, called the *Rydberg constant*, has the value 3.29×10^{15} Hz. Each series is obtained by designating $n_f = 1, 2$, and so on, in turn ($n_f = 2$ corresponds to the Balmer series). Each individual line in a series is obtained by designating $n_i = n_f + 1$, $n_f + 2$, and so on, in turn (so for the Balmer series, $n_i = 3, 4, \ldots$).

The key idea that connects the specific frequencies of the light emitted by any kind of atom with its structure is that each photon is emitted by one atom and carries energy away from the atom that emitted it. Heating or passing an electrical discharge through a sample provides energy that changes the structure of the atoms. As that distorted structure readjusts, all or some of the excess energy is lost, and the atom emits it as a photon of light (Fig. 7.7). If the energy of an atom decreases by ΔE, that amount of energy is carried away as a photon of light. Since the energy of a photon is $h\nu$, the frequency of the light generated by the atom is determined by the

Bohr frequency condition: $\Delta E = h\nu$ $\qquad (4)$

This relation is named in honor of the Danish scientist Niels Bohr (Fig. 7.8), who proposed it. It shows that if the decrease in the energy of the atom is large, then a photon of high-frequency light is generated. In that case we may detect blue or ultraviolet light coming from the sample. If the decrease in energy is small, then we may detect lower-frequency red or infrared radiation.

The scene is now set for a very striking conclusion. We have seen that the spectrum of atomic hydrogen consists of radiation with certain frequencies. We have also seen that each frequency represents a packet of energy carried away from the atom as a photon. We must then conclude that an electron in a hydrogen atom can have only certain energies; otherwise, the atom could emit all frequencies. This limitation is completely contrary to the predictions of classical mechanics, according to which an object can have *any* total energy. According to classical mechanics, a pendulum, for instance, can be given any initial energy simply by pushing it harder or more gently.

We summarize this discovery by saying that energy is "quantized":

The **quantization** of a property is the restriction of that property to certain values.

Macroscopic objects *seem* to us to be able to take on any energy. However, if we make very careful observations we find that they can accept energy only in discrete amounts. This is like pouring water into a bucket. Water *seems* to be a continuous fluid, and it seems that any

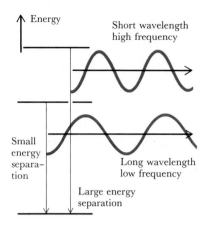

FIGURE 7.7 When an atom falls from a state of higher energy to one of lower energy, the energy difference is carried away as a photon of light. The greater the energy difference, the higher the frequency of the light.

FIGURE 7.8 Niels Bohr (1885–1962).

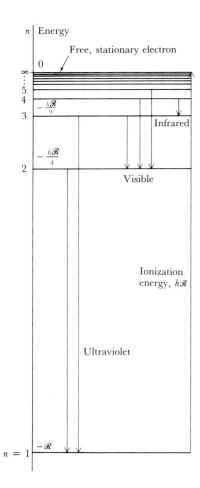

FIGURE 7.9 The energy levels predicted by Bohr, and some of the transitions responsible for the observed spectrum. The zero of energy corresponds to the widely separated nucleus and electron.

amount can be transferred. However, the smallest amount of water that can be added is one H_2O molecule, which could be thought of as one "quantum" of water. The amount of water that can be poured into the bucket is a multiple of this single "quantum" of water.

The Bohr model. Bohr attempted to account for the spectrum of atomic hydrogen by proposing what we now call the *Bohr model* of the atom. In this model it is supposed that the atom's single electron can travel only in certain circular orbits around the central nucleus, and that the energy of the atom is the sum of the electron's kinetic and potential energies in its orbit. Bohr deduced from his model that the energies of the allowed orbits are given by

$$E = -h \times \frac{\Re}{n^2} \qquad n = 1, 2, \ldots \qquad (5)$$

The minus sign means that the energy of the electron is taken to be zero when it is far from the nucleus and that the energy is lower when the electron is part of the atom. The energy levels predicted by Bohr's theory are shown in Fig. 7.9. The "quantum number" n is an integer (whole number) that labels the allowed orbits; $n = 1$ signifies the lowest-energy orbit, $n = 2$ the next higher, and so on. The orbit with $n = 1$ lies closest to the nucleus. When the electron is in this orbit, the atom is in its *ground state*, its state of lowest energy.

Bohr's formula gives the energy of each orbit. His final job was to calculate the decrease in energy when an electron falls from an initial high-energy orbit with quantum number n_i to a final orbit of lower energy and quantum number n_f. That energy decrease is

$$\Delta E = h \times \Re \left(\frac{1}{n_f^2} - \frac{1}{n_i^2} \right)$$

When Bohr set ΔE in this equation equal to the energy of the photon $h\nu$ and canceled h from both sides of the result, he obtained Eq. 3, which we know describes the spectrum. He also obtained a formula for $\Re$ in terms of Planck's constant and the mass and charge of the electron.* With the formula, he calculated that $\Re = 3.29 \times 10^{15}$ Hz, in excellent agreement with its experimental value. It is easy to imagine the thrill Bohr must have felt at this point in his calculations.

Bohr rounded off his work by calculating the radius of each of the allowed orbits. He found that the radius of the orbit with $n = 1$ is 53 pm, a value now called the *Bohr radius*. He also calculated the speed of the electron in its orbit, and for the ground state found that it was about 2200 km/s, or about 5 million miles per hour.

---

*Bohr's formula is

$$\Re = \frac{m_e e^4}{8h^3 \epsilon_0^2}$$

where m_e and e are the mass and charge of the electron, h is Planck's constant, and ϵ_0 is a fundamental constant with the value 8.85×10^{-12} C^2/(J · m).

▼ EXAMPLE 7.1 Calculating the wavelengths present in a spectrum

The ultraviolet lines in the atomic spectrum of hydrogen form the "Lyman series" (Fig. 7.6). These lines are generated as an electron falls from higher-energy orbits to the lowest-energy orbit, the one with $n = 1$. Calculate the wavelength of the ultraviolet line closest to the visible region.

STRATEGY Since ultraviolet radiation has a higher frequency than visible light, the lower its frequency the closer it lies to the visible region. We must therefore identify the transition that involves least change of energy but still gives rise to ultraviolet radiation. Since we are told that in an ultraviolet transition the electron ends up in the orbit with $n = 1$, it loses least energy if it falls from the next higher orbit, the one with $n = 2$. Therefore, we use the Rydberg formula with $n_i = 2$ and $n_f = 1$, and then convert the frequency to a wavelength.

SOLUTION From the Rydberg formula with $\mathcal{R} = 3.29 \times 10^{15}$ Hz,

$$\nu = \mathcal{R} \times \left(\frac{1}{1^2} - \frac{1}{2^2} \right) = \tfrac{3}{4}\mathcal{R}$$

$$= 2.47 \times 10^{15} \text{ Hz}$$

The wavelength corresponding to this frequency is obtained with Eq. 1 and the identity 1 Hz = 1/s:

$$\lambda = \frac{3.00 \times 10^8 \text{ m}}{1 \text{ s}} \times \frac{1 \text{ s}}{2.47 \times 10^{15}}$$

$$= 1.21 \times 10^{-7} \text{ m} = 121 \text{ nm}$$

EXERCISE Calculate the wavelength of the longest-wavelength line of the Balmer series, and identify its color.

[*Answer*: 656 nm; red]

▲

7.3 PARTICLES AND WAVES

Bohr's calculation was a spectacular numerical success. However, when all attempts to extend it to more complicated atoms failed, people began to suspect that it was unsound. That view was confirmed when further experiments were carried out on the behavior of matter.

The de Broglie relation. We have seen that a light ray, which in classical mechanics is treated as a wave, should in fact also be thought of as a stream of photons. The French scientist Louis de Broglie had the curious idea that the same *wave-particle duality*, or combined wavelike and particlelike character, should apply to matter too. In 1924 he suggested that we should also think of an electron as having the properties of a wave. He proposed that every particle has wavelike properties, including a wavelength that is related to its mass m and speed v by the

de Broglie relation: $\lambda = \dfrac{h}{m \times v}$ (6)

According to this relation, a heavy particle traveling at high speed has a small wavelength λ. A small particle traveling at low speed has a large wavelength.

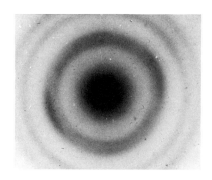

FIGURE 7.10 Davisson and Germer showed that electrons give a diffraction pattern when reflected from a crystal. G. P. Thomson, working in Aberdeen, Scotland, showed that they also give a diffraction pattern when they pass through a very thin gold foil. The latter is shown here. G. P. Thomson was the son of J. J. Thomson, who discovered the electron (Section 2.3). Both received Nobel Prizes—J. J. for showing that the electron is a particle, and G. P. for showing that it is a wave.

We cannot detect the wavelike character of ordinary objects because their wavelengths are so small. The wavelength of a 100-g tennis ball traveling at 65 km/h (40 mi/h) is less than 10^{-30} m, which is far smaller than the diameter of an atomic nucleus. The scientists who developed classical mechanics got excellent agreement with observation because the wavelike character of matter, of which they knew nothing, was completely undetectable for the objects they could observe (such as balls and planets). However, the wavelength of an electron moving at 2000 km/s is 360 pm, which is comparable to the diameter of an atom (about twice the Bohr radius, 106 pm). Hence, when trying to account for the properties of electrons in atoms, we must take their wave character into account.

One of the earliest experiments to prove the wavelike property of electrons was done by the American scientists Clinton Davisson and Lester Germer in 1927. They knew that when waves of any kind pass through a grid with a spacing comparable to their wavelength, they give rise to characteristic "diffraction patterns." Davisson and Germer showed that electrons give the expected pattern when they are reflected from a crystal, where the layers of atoms act as the grid. They also found that the pattern corresponds exactly to that expected of electrons with a wavelength given by the de Broglie relation (Fig. 7.10).

The uncertainty principle. Because of its wavelike character, we cannot say precisely where an electron is when it is traveling along a path. This is expressed numerically by the _uncertainty principle_ discovered by the German Werner Heisenberg in 1927. Heisenberg found that the more precisely we know the position of a particle, the less we can say about its speed, and vice versa (Fig. 7.11). More specifically, if we know the position of a particle of mass m to within a range Δx, then its speed must be uncertain by at least an amount Δv, where

$$\Delta x \times (m \times \Delta v) \geq \frac{h}{4\pi} \tag{7}$$

If Δx is zero (meaning we have perfect knowledge about the location), the only way this inequality can be satisfied is for Δv to be infinite

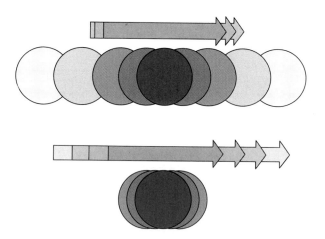

FIGURE 7.11 According to the uncertainty principle, the more precisely we know the position of a particle, the less precisely we can know its speed, and vice versa.

(meaning we are totally ignorant about the speed). Similarly, if the speed is certain (Δv is zero), the position must be completely uncertain (Δx is infinite). The uncertainty principle implies that Bohr's picture of electrons traveling in precise orbits cannot be valid, because an electron in such an orbit has a definite position and definite speed at every instant.

▼ **EXAMPLE 7.2** **Using the uncertainty principle**

Estimate the uncertainty of the speed of an electron in a hydrogen atom.

STRATEGY We must begin by deciding how precisely we know the position of the electron. That gives Δx, which we can substitute in Eq. 7 to find Δv. Since we know that the electron is *somewhere* in the atom, a rough measure of the uncertainty in position is the diameter of the atom (106 pm).

SOLUTION Since Δx is approximately 100 pm, the uncertainty in the speed must be at least

$$\Delta v = \frac{h}{4\pi \times m \times \Delta x}$$

$$= \frac{6.63 \times 10^{-34} \text{ kg} \cdot \text{m}^2/\text{s}}{4\pi \times (9.11 \times 10^{-31} \text{ kg}) \times (100 \times 10^{-12} \text{ m})}$$

$$= 6 \times 10^5 \text{ m/s}$$

(We have used the conversions 1 J/Hz = 1 J · s and 1 J = 1 kg · m²/s² on the second line of the calculation.) This uncertainty corresponds to nearly a million miles per hour!

EXERCISE How uncertain would the position of an electron be if we knew its speed to within 1 mm/s?

[*Answer*: 6 cm]

The atomic orbitals of hydrogen. According to quantum mechanics, an electron is spread out like a wave, and we can speak only of the *probability* of finding it at any point. The wave that summarizes how the electron is spread out through space is called its *wavefunction* ψ (the Greek letter psi). This wavefunction is large in some regions and small in others, just like a wave in water. According to the *Born interpretation* of the wavefunction, the probability of finding the electron at a given point is proportional to the *square* of ψ at that point. Therefore, if ψ is 0.1 at one point and −0.2 at another, there is four times the chance of finding the electron at the second point than at the first. In atoms, wavefunctions are called *atomic orbitals*, and we can think of them as defining a region of space in which there is a high probability of finding an electron. This name suggests something similar to, but less definite than, the orbits of the Bohr model. We shall see shortly that for most purposes it is possible to draw diagrams of atomic orbitals and to discuss their characteristic shapes without going into their mathematical details.

The equation that must be solved to find these orbitals was discovered by the German Erwin Schrödinger in 1926 (Fig. 7.12). When he solved it, he found that solutions of the equation exist only for certain energies. Hence, whereas Bohr had to *assume* that only certain orbits

FIGURE 7.12 Erwin Schrödinger (1887–1961).

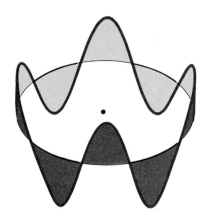

FIGURE 7.13 Only certain waves fit around a hydrogen atom. Since each wave has a characteristic energy, only certain energies are allowed.

existed, Schrödinger *deduced* from his equation that the energy of the atom is quantized. This can be understood by thinking of the electron as a wave that must fit around the nucleus (Fig. 7.13). Only certain wavelengths will fit, and each different wave corresponds to a certain energy. By a remarkable coincidence, the energy levels calculated from the Schrödinger equation turn out to be exactly the same as those obtained by Bohr (see Eq. 5). Since Bohr's calculations agree with the spectroscopic data for hydrogen, Schrödinger's model of the atom also agrees with experiment. However, it is more securely based in quantum mechanics than Bohr's model and can be extended to more complex atoms.

Schrödinger found that each atomic orbital could be specified by three numbers called "quantum numbers":

A **quantum number** is a number that labels the state of an electron and specifies the value of a property.

We have already seen an example: in the Bohr theory, each orbit is labeled by the quantum number n, and its value gives the energy of the orbit. Schrödinger's three quantum numbers (Tables 7.2 and 7.3; the fourth quantum number given there will be explained later) are the principal quantum number n, the azimuthal quantum number l, and the magnetic quantum number m_l.

The *principal quantum number* n specifies the energy of an electron in an atom in exactly the same way as in the Bohr model, and it can take the values 1, 2, 3, . . . up to infinity. In hydrogen, but not in atoms with more than one electron, all orbitals with $n = 2$ have the same energy, as do all orbitals with $n = 3$, and so on. All orbitals with the same value of n form a *shell* of the atom. Thus all orbitals with $n = 3$ (there are nine of them; the number is n^2 in general) form one shell. In the hydrogen atom, then, but not in atoms with more than one electron, all the orbitals of a shell have the same energy. The higher the value of n, the greater the average distance of the shell from the nucleus: an electron with $n = 1$ is usually close to the nucleus, one with $n = 2$ is usually more distant, and so on.

TABLE 7.2 Quantum numbers of the hydrogen atom

Name	Symbol	Values	Meaning
Principal	n	1, 2, . . .	Labels shell, specifies energy level
Azimuthal*	l	0, 1, . . . , $n - 1$	Labels subshell $l = 0, 1, 2, 3, 4, . . .$ $s\ \ p\ \ d\ \ f\ \ g\ \ . . .$
Magnetic	m_l	$l, l - 1, . . . , -l$	Labels orbitals of subshell
Spin-magnetic	m_s	$+\frac{1}{2}, -\frac{1}{2}$	Labels spin state

*Also called the *orbital angular momentum quantum number* because it specifies the angular momentum of the electron around the nucleus: angular momentum = $[l(l + 1)]^{1/2}\hbar/2\pi$.

TABLE 7.3 **Orbitals of the first four shells of the hydrogen atom**

Shell (n)	Subshell (l)	Orbital type	Number of orbitals, 2l + 1	Energy*
1	0	1s	1	-1
2	0	2s	1	$-\frac{1}{4}$
	1	2p	3	
3	0	3s	1	$-\frac{1}{9}$
	1	3p	3	
	2	3d	5	
4	0	4s	1	$-\frac{1}{16}$
	1	4p	3	
	2	4d	5	
	3	4f	7	

*As a multiple of $h \times \mathcal{R}$.

The orbitals belonging to a given shell are grouped into "subshells." A *subshell* of a given shell consists of all the orbitals with the same value of l. This *azimuthal quantum number* can take the values 0, 1, . . . up to $n - 1$, giving n values in all. This means that there is only one subshell of the $n = 1$ shell (the one with $l = 0$), there are two subshells of the shell with $n = 2$ ($l = 0$ and 1), and so on. It is common practice to refer to the subshells by letters rather than numbers, using the following correspondence:

$$l = 0 \quad 1 \quad 2 \quad 3$$
$$s \quad p \quad d \quad f$$

The shell for which $n = 3$ therefore consists of the three subshells labeled s, p, and d.

The different subshells correspond to the different speeds with which an electron can circulate around the nucleus. If the electron is in an s subshell, it does not circulate at all. If it is in a p subshell, it circulates at a particular rate. It circulates more quickly if it is in a d subshell, and so on.* In a hydrogen atom, but not in atoms with more than one electron, the energy of the electron (the sum of its kinetic and potential energies) is the same whatever subshell of a given shell it occupies.

Each subshell is made up of $2l + 1$ individual orbitals. These individual orbitals are labeled with the *magnetic quantum number* m_l, which, for a subshell of quantum number l, can take the $2l + 1$ values l, $l - 1$, $l - 2$, . . . down to $-l$. For example, since the orbitals in the p subshell of any shell have $l = 1$, there are three such orbitals and they have the magnetic quantum numbers $+1, 0$, and -1. It is often more convenient to use a different set of labels for these orbitals, and to name them p_x, p_y, and p_z. These alternative labels specify the shapes of the orbitals more directly, as we shall see.

*Specifically, the rate depends on the quantum number l as $[l(l + 1)]^{1/2}$, so the rotation rates of electrons in s, p, and d subshells are in the ratio 0 to 1.4 to 2.5.

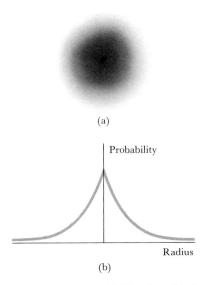

(a)

Probability

Radius

(b)

FIGURE 7.14 (a) The 1s orbital of hydrogen, drawn so that the density of shading represents the probability of finding the electron at any point. (b) The graph shows that the probability of finding the electron at any position along a given radius is greatest at the nucleus and decreases sharply with distance.

FIGURE 7.15 The simplest way of drawing an atomic orbital is as a boundary surface, a surface within which there is a high probability (typically 90%) of finding the electron. This is the boundary surface of an s orbital.

▼ **EXAMPLE 7.3** **Interpreting quantum numbers**

How many orbitals are there in the shell with $n = 4$?

STRATEGY We have to decide which subshells the $n = 4$ shell has, write down the number of orbitals in each one, and then add these numbers together. We know from the discussion above that l has values from 0 up to $n - 1$. We also know that the number of orbitals in a subshell of quantum number l is $2l + 1$. We must combine these two pieces of information.

SOLUTION Since the permitted values of l for $n = 4$ are $l = 0, 1, 2,$ and 3, the shell has $s, p, d,$ and f subshells. There are one s orbital, three p orbitals, five d orbitals, and seven f orbitals, giving 16 in all.

EXERCISE Calculate the total number of orbitals in the shell with $n = 6$. What is the general formula?

[*Answer*: 36; number of orbitals = n^2]

▲

The s orbitals. The lowest-energy orbital of the hydrogen atom is the *1s orbital* (the orbital with $n = 1$, $l = 0$, and $m_l = 0$). It is the only orbital permitted when $n = 1$. An electron that is in the region of space specified by a 1s orbital is said to "occupy" a 1s orbital and to be a "1s electron."

We can visualize the shape of the 1s orbital in several ways. Figure 7.14 uses shading to show the probability of finding the electron at any point in the region surrounding the nucleus. The shading is darkest close to the nucleus, since the electron is most likely to be found there. The shading becomes lighter with increasing distance from the nucleus, showing that there is a decreasing chance of finding the electron the further out we go from the nucleus. The shaded region is sometimes called the *charge cloud* of the electron. The illustration also shows a plot of the probability of the electron being found at any point along a given radius. Notice how the probability decays sharply with distance but never quite reaches zero.

All s orbitals are spherical. That is, the probability of finding the electron at a given distance from the nucleus is the same in every direction. The s orbital is therefore often drawn as a sphere called a *boundary surface* (Fig. 7.15): the surface of the sphere is the boundary within which there is about 90% probability of finding the electron. For a hydrogen atom the radius of the boundary surface is 140 pm.

The *2s orbital*, the s orbital belonging to the shell with $n = 2$, is similar to the 1s orbital but is spread over a larger volume. Its boundary surface, like that of the 1s orbital and all the other s orbitals, is spherical, so it too is normally drawn as a sphere.

The p orbitals. Three orbitals with $l = 1$ can occur for shells with n greater than 1. These orbitals, called *p orbitals*, have the shapes and orientations shown in Fig. 7.16. All three have the same shape, but each lies along one of three perpendicular axes. They are labeled with the names of those axes, which accounts for the notation p_x, p_y, and p_z. A p_x electron, an electron in a p_x orbital, is most likely to be found somewhere on the x axis.

A p orbital has a *nodal plane*, a plane on which a p electron will never be found, that extends through the nucleus. Whereas an s electron may be found at the nucleus, a p electron will never be found there.

The d and f orbitals. Five orbitals with $l = 2$, the *d orbitals*, can occur for shells with n greater than 2. Their boundary surfaces and labels are shown in Fig. 7.17. Four of the *d* orbitals have a double dumbbell shape; the d_{z^2} orbital is the one that is different.

For n greater than 3, *f orbitals* can occur. There are seven *f* orbitals for each value of $n > 3$, but their shapes are complicated and there is no need to try to remember them.

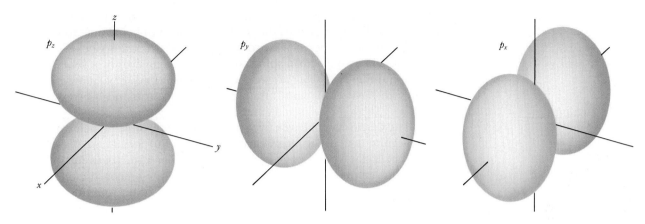

FIGURE 7.16 The boundary surfaces of the three *p* orbitals of a given subshell.

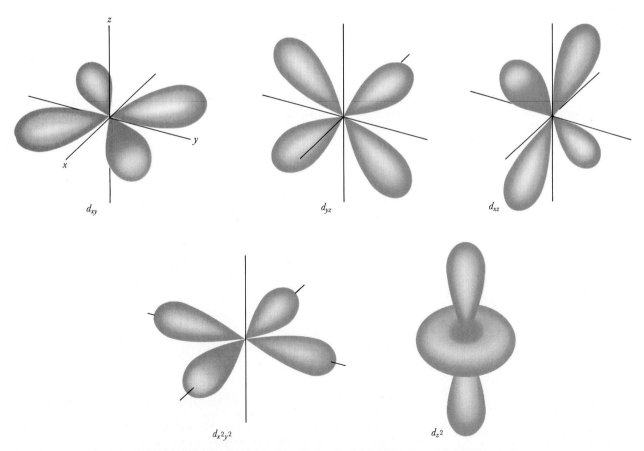

FIGURE 7.17 The boundary surfaces and labels of the five *d* orbitals of a shell with $n > 2$.

Electron spin. A careful analysis of the spectrum of atomic hydrogen shows that the lines do not have exactly the frequencies predicted by Schrödinger's calculations. The small differences were explained by a suggestion that had been made in 1925 by the Dutch-American physicists Samuel Goudsmit and George Uhlenbeck. They had proposed that an electron behaves in some respects like a sphere rotating on its axis, something like the way the earth rotates daily on its axis. This property is called the _spin_ of the electron.

An electron has only two spin states, represented by the arrows ↑ and ↓. A helpful *analogy* is to think of an electron as being able to spin at a constant rate either clockwise (the ↑ state) or counterclockwise (the ↓ state). The two electron spin states are distinguished by a fourth quantum number, the _spin-magnetic quantum number_ m_s. This quantum number can have only two values, $+\frac{1}{2}$ for a ↑ electron and $-\frac{1}{2}$ for a ↓ electron.

The discovery that only two spin states are possible helped explain an experiment that had been carried out by Otto Stern and Walter Gerlach in 1920. They had passed a beam of silver atoms between the poles of a powerful magnet that produced an inhomogeneous field (one that varies from place to place) and found that the beam split into two (Fig. 7.18). The explanation is based on the fact that a silver atom has an odd number of electrons ($Z = 47$) and in the Stern-Gerlach experiment behaves like a hydrogen atom with its one electron. As a result of the spin of the odd electron, the atom acts as a tiny magnet, and its path through the apparatus is bent by the laboratory magnet. The beam was split into two beams because atoms with an odd ↑ electron were pushed in one direction by the field, and those having an odd ↓ electron were pushed in the other direction.

The hydrogen atom: a summary. The electron in a hydrogen atom can occupy any one of the orbitals we have described. In the lowest state of

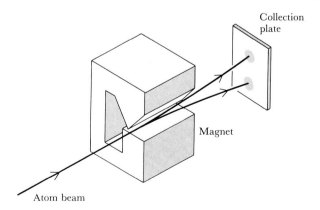

FIGURE 7.18 The quantization of electron spin is confirmed by the Stern-Gerlach experiment, in which a stream of atoms splits in two as it passes between the poles of a magnet. The atoms in one stream have an extra ↑ electron, and those in the other an extra ↓ electron.

the atom, the ground state, the electron occupies a $1s$ orbital and hence has the four quantum numbers

$$n = 1 \quad l = 0 \quad m_l = 0 \quad m_s = +\tfrac{1}{2} \text{ or } -\tfrac{1}{2}$$

(It can have either spin.) If it is given enough energy to reach the $n = 2$ shell, it may occupy any of the four orbitals in that shell (the single $2s$ or one of the three $2p$ orbitals), since these orbitals all have the same energy. A greater input of energy raises the electron into the shell with $n = 3$, where it can occupy any one of nine orbitals (one $3s$, three $3p$, and five $3d$ orbitals).

The electron is ejected from the atom when it is given enough energy to overcome the attraction of the nucleus. At this stage the atom has been "ionized." The energy needed to ionize a hydrogen atom from its ground state is the amount needed to raise the electron from the $n = 1$ orbital up to the zero of energy (the energy of the widely separated proton and electron). According to Fig. 7.9, this is equal to $h \times \mathfrak{R}$, or 2.18×10^{-18} J. To ionize a mole of hydrogen atoms therefore requires 6.02×10^{23} times this amount, or 1.31 MJ of energy.

THE STRUCTURES OF MANY-ELECTRON ATOMS

All neutral atoms other than hydrogen have more than one electron. The helium atom ($Z = 2$) has two, the lithium atom ($Z = 3$) has three, and in general an atom of an element with atomic number Z has Z electrons. All these are examples of *many-electron atoms*, atoms with more than one electron.

7.4 ORBITAL ENERGIES

A simple picture of a many-electron atom is one in which the electrons occupy orbitals like those of hydrogen, but with different energies. The nucleus of a many-electron atom is more highly charged than the hydrogen nucleus; it attracts the electrons more strongly, thus lowering their energy. However, there are also repulsions between the electrons, which raise their energy.

Effective nuclear charge. In the hydrogen atom, where there are no electron-electron repulsions, all the orbitals of a given shell have the same energy. In many-electron atoms, electron-electron repulsions cause the energy of an s orbital to be lower than that of a p orbital in the same shell, and that in turn is lower than the energy of a d orbital in that shell. The orbitals of a given *sub*shell are, however, equal in energy to one another. Each of the three $2p$ orbitals, for instance, has the same energy. This order of energy levels is shown in Fig. 7.19 and used for discussing atomic structure throughout the rest of the chapter.

The difference in energies of orbitals of the same shell can be traced to the shapes of the orbitals. There are two factors to take into account. First, as noted above, an s electron may be found very close to the

FIGURE 7.19 The relative energies of the orbitals in a many-electron atom. Each of the boxes represents one orbital and can hold up to two electrons. Note the change in the order of energies of the 3d orbitals after Z = 20.

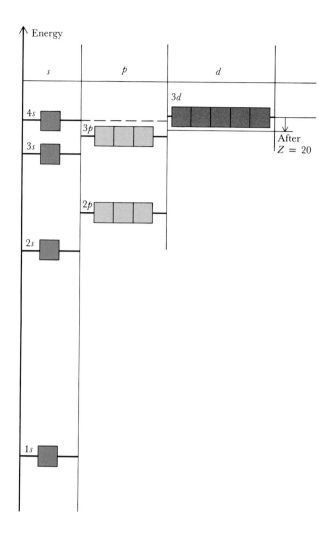

nucleus, but a *p* electron may not. We say that an *s* electron *penetrates* closer to the nucleus than a *p* electron. Second, each electron is repelled by the other electrons in the atom, and hence is less tightly bound to the nucleus than would be the case if those other electrons were absent. We say that each electron is *shielded* from the full attraction of the nucleus by the other electrons in the atom, and that the *effective* nuclear charge is lower than the actual charge (Fig. 7.20).

Now we bring the two factors together. Because an *s* electron penetrates closer to the nucleus than a *p* electron of the same shell, it is less strongly shielded from the nucleus by the atom's other electrons. It therefore feels a stronger effective nuclear charge than a *p* electron, and hence is bound more tightly. That is, an *s* electron has a slightly lower (more negative) energy than a *p* electron of the same shell. Similar differences in penetration and shielding apply between *p* and *d* orbitals, as *d* electrons are on average even farther from the nucleus than *p* electrons.

The effects of penetration and shielding can be large: a 4s orbital may be so much lower in energy than the 4p and 4d orbitals that it may also be lower in energy than the 3d orbital of the same atom. Whether or not this is in fact the case depends on the number of electrons in the atom. In some atoms the $4s$ orbital is lower in energy than the $3d$ orbitals, but in others it is higher in energy. We return to this point shortly, when we deal with the atomic structures of the transition metals.

The exclusion principle. The lowest total energy of an atom is not a configuration in which all its electrons occupy the $1s$ orbital. That arrangement is forbidden by a fundamental feature of nature, discovered by the Austrian Wolfgang Pauli in 1925:

Pauli exclusion principle: No more than two electrons may occupy any given orbital, and when two electrons do occupy one orbital their spins must be paired.

The spins of two electrons are *paired* if one is ↑ and the other is ↓. Paired spins are denoted ↑↓. In practice the exclusion principle means that no more than two electrons can enter each box of the energy-level diagram in Fig. 7.19. For $Z > 2$, the atom's electrons cannot all enter the $1s$ orbital, some must occupy higher-energy orbitals.

The configurations of hydrogen through lithium. The list of occupied orbitals in an atom is called its *electron configuration*. The hydrogen atom in its *ground state* or lowest-energy state has one electron in its $1s$ orbital. We can denote this by placing a single arrow in the box representing the $1s$ orbital in a diagram like that of Fig. 7.19, and we report the configuration of hydrogen as $1s^1$ ("one s one"). The lowest-energy configuration of helium ($Z = 2$) is that in which both electrons are in a $1s$ orbital; this configuration is reported as $1s^2$ ("one s two"). Instead of showing the whole diagram, it is simpler and just as informative to show only the occupied boxes in a "box diagram," as in

$$\text{He} \quad \boxed{\uparrow\downarrow}$$
$$1s^2$$

With this configuration, the $n = 1$ shell of helium is "closed":

A **closed shell** (or subshell) is a shell (or subshell) containing the maximum number of electrons allowed by the exclusion principle.

A closed shell is also described as being "complete."

▼ **EXAMPLE 7.4** **Counting electrons in closed shells**

How many electrons complete a closed shell with $n = 2$?

STRATEGY The exclusion principle limits the number of electrons in each orbital to a maximum of two. Therefore, we must count the number of orbitals in the shell and take each one to be doubly occupied. To find the number of orbitals with $n = 2$, we identify which subshells are included and the number of orbitals in each. The procedure is illustrated in Example 7.3.

SOLUTION The shell with $n = 2$ has two subshells; they contain the single $2s$ orbital and the three $2p$ orbitals, for four orbitals in all. The greatest number of electrons that can be accommodated in the shell is therefore eight.

EXERCISE How many electrons complete a closed d subshell?

[*Answer*: 10]

▲

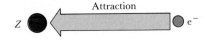

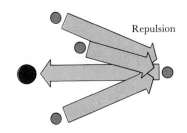

FIGURE 7.20 An electron near a bare nucleus experiences its full charge; in a many-electron atom, the electron is repelled by the other electrons and experiences a smaller net attraction. The overall effect is that of a smaller nuclear charge.

Lithium ($Z = 3$) has three electrons. Two can occupy the $1s$ orbital. The third electron occupies an orbital of the shell with $n = 2$. That shell has two subshells, with the $2s$ lower in energy than the $2p$. The third electron therefore enters the $2s$ orbital to give the configuration $1s^2 2s^1$:

Li —[⇅]—[↑]—
   $1s^2$ $2s^1$

We can think of the Li atoms as consisting of a *core* made up of inner closed shells (in this case, the single $n = 1$ shell) and, around that core, a single electron in a $2s$ orbital. Since they are so close to the nucleus, lithium's core electrons are much more tightly bound than the outer $2s$ electron, and they can be removed from the atom only if a lot of energy is supplied. In general, core electrons are much more "inert" than outer electrons and are not removed when the atom takes part in chemical reactions.

▼ EXAMPLE 7.5 Building up an electron configuration

Predict the lowest-energy electron configuration of boron.

STRATEGY Although we have not yet discussed all the rules for obtaining an electron configuration, we have enough information to deal with this atom. First we need to decide how many electrons are present (from the atomic number). Then we add arrows to the boxes in a diagram like that of Fig. 7.19, starting at the lowest orbital and filling each orbital before moving to the next higher orbital. Finally, we list the occupied orbitals as an electron configuration.

SOLUTION A boron atom has five electrons. Two enter the $1s$ orbital and fill the $n = 1$ shell. The $2s$ orbital is the next orbital to be occupied. It can accommodate two of the remaining electrons, so the fifth electron must occupy an orbital of the next available subshell, which Fig. 7.19 shows is the $2p$ subshell. The resulting configuration is $1s^2 2s^2 2p^1$:

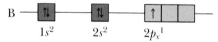

B —[⇅]—[⇅]—[↑][ ][ ]—
   $1s^2$ $2s^2$ $2p_x^{\,1}$

EXERCISE Predict the lowest-energy configuration of neon.

[*Answer*: $1s^2 2s^2 2p^6$]

▲

7.5 THE BUILDING-UP PRINCIPLE

The lowest-energy configuration of any element can be predicted with a generalization of the approach we used above for H, He, and Li. The procedure is called the *building-up principle* (some use its German name, the *Aufbau* principle). The principle leads to the configuration with the lowest *total* energy for the atom, taking into account the kinetic energies of the electrons, their attraction to the nucleus, and their repulsion of each other. It does this by specifying the order in which the orbitals are to be occupied as one electron after another is added, until all Z electrons are present. A clue to the order is given by the order of energy levels in Fig. 7.19:

$$1s < 2s < 2p < 3s < 3p < 4s < 3d$$

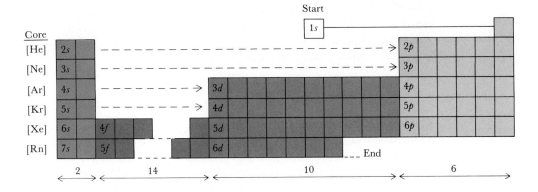

for $Z = 1$ through $Z = 20$, and

$$1s < 2s < 2p < 3s < 3p < 3d \lessgtr 4s < 4p < 5s < 4d$$

for $Z = 21$ through $Z = 38$. However, there is no need to memorize these orders, because we arrive at the same configuration if we add electrons in the order shown in Fig. 7.21, which matches the structure of the periodic table.

To assign a configuration to an element with atomic number Z, we proceed as follows:

1. Add Z electrons, one after the other, to the orbitals in the order shown in Fig. 7.21, but with no more than two electrons in any one orbital.

2. If more than one orbital in a subshell is available, add electrons with *parallel* spins to *different* orbitals of that subshell.

The second rule is called *Hund's rule*, for the German spectroscopist Fritz Hund who first proposed it. Electrons have "parallel spins" (denoted ⇈) if they spin in the same direction. The explanation of the first part of Hund's rule can be traced to electron repulsion. If electrons occupy different orbitals, they stay farther apart on average than if they are in the same orbital. As a result, they repel each other less, and the energy of the atom is lower. The explanation of why they should have parallel spins is complex and requires more advanced quantum-mechanics than we have covered here.

Lithium through sodium. The lithium atom, $1s^2 2s^1$, has been described above. Since it consists of a single $2s$ electron outside a heliumlike $1s^2$ closed-shell core, its configuration is more simply written $[He]2s^1$. It is often useful to think of an atom as a noble-gas core surrounded by electrons in the outermost shell. This outermost shell is called the *valence shell* of the atom, because its electrons are the ones mainly responsible for compound formation (and the number of bonds an atom can form is called its "valence"). All the atoms in a given period of the periodic table have the same core, as shown in Fig. 7.21. All the atoms of Period 2, for example, have the heliumlike $1s^2$ core, written [He], and all those of Period 3 have a neonlike $1s^2 2s^2 2p^6$ core, written [Ne].

The element with $Z = 4$ is beryllium. The first three electrons form $[He]2s^1$, as for Li. The fourth electron pairs with the $2s$ electron, giving

FIGURE 7.21 The order in which orbitals are filled according to the building-up principle. Each time an electron is added, move one place to the right until all the electrons are accommodated.

[He]$2s^2$. A Be atom therefore has a heliumlike core surrounded by a valence shell of two paired electrons. The next element, boron, has $Z = 5$. Its additional electron cannot enter the full $2s$ subshell; instead, it occupies one of three $2p$ orbitals, giving [He]$2s^2 2p^1$:

B

$1s^2 \quad 2s^2 \quad 2p_x{}^1$

The last electron can occupy any one of the three $2p$ orbitals, since they all have the same energy. Carbon, with $Z = 6$, has one more electron than boron and the configuration [He]$2s^2 2p^2$. According to Hund's rule, the sixth electron occupies a different $2p$ orbital from the one already occupied in boron and does so with a parallel spin:

C

$1s^2 \quad 2s^2 \quad 2p_x{}^1 2p_y{}^1$

Nitrogen has $Z = 7$ and one more electron than carbon, giving [He]$2s^2 2p^3$. Each p electron occupies a different orbital, and the three have parallel spins:

N

$1s^2 \quad 2s^2 \quad 2p_x{}^1 2p_y{}^1 2p_z{}^1$

Oxygen has $Z = 8$ and one more electron than nitrogen. The eighth electron must therefore pair with one already present, giving the configuration [He]$2s^2 2p^4$:

O

$1s^2 \quad 2s^2 \quad 2p_x{}^2 2p_y{}^1 2p_z{}^1$

Similarly, fluorine, with $Z = 9$ and one more electron than oxygen, has the configuration [He]$2s^2 2p^5$:

F

$1s^2 \quad 2s^2 \quad 2p_x{}^2 2p_y{}^2 2p_z{}^1$

The fluorine atom can be pictured as a heliumlike core surrounded by a valence shell which is complete except for one p electron. Neon, with $Z = 10$, has one more electron than fluorine. This electron completes the $2p$ subshell, giving [He]$2s^2 2p^6$:

Ne

$1s^2 \quad 2s^2 \quad 2p_x{}^2 2p_y{}^2 2p_z{}^2$

According to Fig. 7.21, the next electron enters the $3s$ orbital, the lowest-energy orbital of the next shell. The configuration of sodium is therefore [He]$2s^2 2p^6 3s^1$ or, more briefly, [Ne]$3s^1$, where [Ne] is the neonlike core.

We have seen that the configuration of the unreactive gas helium is a closed shell. So too is that of neon, another unreactive gas. We have also seen that the configuration of lithium, which is a reactive metal, is a single electron outside a core. So too is that of sodium, another reactive metal. Thus, with a few simple ideas about atomic structure, we have begun to account for the periodicity of the elements: the building-up principle leads periodically to analogous configurations and hence to similar chemical properties. The configurations of all the elements are listed in Appendix 2C.

▼ EXAMPLE 7.6 Deducing electron configurations

Predict the lowest-energy configuration of silicon. Which element already discussed has a similar configuration?

STRATEGY Since Si is in the same group as C, we can anticipate that the two elements have analogous ns^2np^2 configurations. The precise configuration is deduced with the building-up principle, beginning with the $1s$ orbital. Each time an electron is added, we move one place to the right in Fig. 7.21, noting which orbital is next to be filled. A short cut is to make use of the core configuration that corresponds to the period containing the element.

SOLUTION For silicon, $Z = 14$. The first 10 electrons give a neonlike $1s^22s^22p^6$ core, [Ne]. Now we can start at the left of the period in Fig. 7.21 that has [Ne] as core, and add four electrons. Two complete the $3s$ subshell. The remaining two occupy $3p$ orbitals. Hence, the lowest-energy configuration is $[Ne]3s^23p^2$. This is the analog of carbon, with $[He]2s^22p^2$.

EXERCISE Predict the lowest-energy configuration of the chlorine atom.

[*Answer*: $[Ne]3s^23p^5$]

The filling of d orbitals. The s and p subshells of the $n = 3$ shell are filled in an atom of argon, which is a colorless, odorless, and unreactive gas resembling neon. As a result of penetration and shielding, the $4s$ orbitals are lower in energy than the $3d$ orbitals and therefore are occupied next. Hence the next two configurations are $[Ar]4s^1$ for potassium and $[Ar]4s^2$ for calcium. Now the $3d$ orbitals come into line for occupation, and there is a change in the rhythm of the periodic table.

According to Fig. 7.21, the next 10 electrons (for scandium, with $Z = 21$, through zinc, with $Z = 30$) enter the $3d$ subshell. The configuration of scandium, for example, is $[Ar]3d^14s^2$, and that of its neighbor titanium is $[Ar]3d^24s^2$. Note that we write the $4s$ electrons *after* the $3d$ electrons. This reflects the change in order of the energies of the orbitals, with $3d$ below $4s$, which begins with scandium. The building-up principle leads to the correct overall configuration, even though the order in which electrons are added according to the principle is not precisely the same as the actual order of energies in the atom.

The configurations are now not quite as straightforward as before. This is because the half-complete subshell configuration d^5 and the complete subshell configuration d^{10} turn out to be more stable than simple theory suggests. In some cases, the neutral atom has a lower total energy if the $3d$ subshell is half completed (to d^5) or completed (to d^{10}) by the transfer of a $4s$ electron into it. For example, the configuration of chromium is $[Ar]3d^54s^1$, and that of copper is $[Ar]3d^{10}4s^1$. Other examples can be found in Appendix 2C.

After the $3d$ subshell is full, starting with gallium, the $4p$ orbitals are occupied. The configuration of arsenic (in Group V), for example, is obtained by adding three electrons to the $4p$ subshell outside the completed $3d$ subshell and is $[Ar]3d^{10}4s^24p^3$. This fourth period of the table contains 18 elements, since the $4s$ and $4p$ orbitals can accommodate 8 electrons and the $3d$ orbitals can accommodate 10. It is the first "long period" of the periodic table.

Next in line for occupation is the $5s$ orbital, followed by the $4d$ orbitals. As in Period 4, the energy of the $4d$ orbitals falls below that of the $5s$ orbital after two electrons have been accommodated in the $5s$ orbital. A

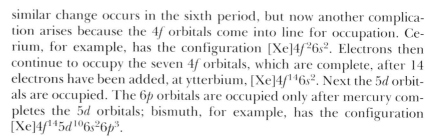

	III	IV	V
	Al	Si	P
Zn	Ga	Ge	As
Cd	In In$^+$	Sn Sn^{2+}	Sb Sb^{3+}
Hg	Tl Tl$^+$	Pb Pb^{2+}	Bi Bi^{3+}

FIGURE 7.22 The shaded boxes show elements that can lose s and p electrons in stages. The ions listed here are the ones that are formed when the s subshell of electrons is retained by the atom.

similar change occurs in the sixth period, but now another complication arises because the $4f$ orbitals come into line for occupation. Cerium, for example, has the configuration $[Xe]4f^26s^2$. Electrons then continue to occupy the seven $4f$ orbitals, which are complete, after 14 electrons have been added, at ytterbium, $[Xe]4f^{14}6s^2$. Next the $5d$ orbitals are occupied. The $6p$ orbitals are occupied only after mercury completes the $5d$ orbitals; bismuth, for example, has the configuration $[Xe]4f^{14}5d^{10}6s^26p^3$.

The configurations of ions. Cations are formed by the removal of electrons from the configuration predicted for the neutral atom. If the principal quantum number of the valence shell is n, we remove electrons in the order np first, then ns, and finally $(n-1)d$, until the appropriate number of electrons have been removed. Thus, for the Fe^{3-} ion, we work out the configuration of the Fe atom—$[Ar]3d^64s^2$—and remove three electrons from it. There are no $4p$ electrons, so the first two removed are $4s$ electrons. The third electron comes from the $3d$ subshell, giving $[Ar]3d^5$.

Anions are formed by adding enough electrons to the vacant orbitals of the valence shell to achieve the configuration of the next noble gas. Thus, to predict the electron configuration of the anion formed by a nitrogen atom, we first note that since nitrogen is in Group V (and hence has five valence electrons), three electrons are needed to reach a noble gas configuration, and therefore the ion will be N^{3-}. The configuration of the nitrogen atom is $[He]2s^22p^3$, with room for three more electrons in the $2p$ subshell. Thus, the N^{3-} ion has the configuration $[He]2s^22p^6$, the same as that of neon.

▼ **EXAMPLE 7.7** **Predicting the configurations of cations**

Predict the configurations of the In$^+$ and In^{3+} ions.

STRATEGY From the group and period of the element, we determine the configuration of its neutral atoms. Then one electron is removed to produce In$^+$, and two more to produce In^{3+}. We remove electrons from valence-shell p orbitals first, then from the s orbitals, and finally, if necessary, from the d orbitals.

SOLUTION Indium, in Group III, has three valence electrons. It is in Period 5 (so its valence shell has $n = 5$) and is preceded in Period 4 by the noble gas krypton (so its core is $[Kr]d^{10}$). Its ground state is therefore $[Kr]4d^{10}5s^25p^1$. One electron can be lost from the $5p$ orbital, giving In$^+$ as $[Kr]4d^{10}5s^2$. The next two electrons are lost from the $5s$ orbital, giving In^{3+} as $[Kr]4d^{10}$.

EXERCISE Give the configurations of copper(I) and copper(II) ions.
▲ [*Answer*: $[Ar]3d^{10}$; $[Ar]3d^9$]

A SURVEY OF PERIODIC PROPERTIES

The overall pattern of the periodic table was discovered by the Russian chemist Dmitri Mendeleev (Fig. 7.23) during a single day of furious thought, according to legend, on February 17, 1869. Mendeleev arranged the elements in groups and periods in order of increasing

FIGURE 7.23 Dmitri Ivanovich Mendeleev (1834–1907).

TABLE 7.4 Mendeleev's predictions for an unknown element "eka-silicon" compared with the actual properties of germanium

Property	Eka-silicon (E)	Germanium (Ge)
Molar mass	72 g/mol	72.6 g/mol
Density	5.5 g/cm^3	5.35 g/cm^3
Melting point	High	937°C
Appearance	Dark gray	Gray-white
Oxide	EO_2	GeO_2
Appearance	White solid	White solid
Acidic/basic	Amphoteric	Amphoteric
Density	4.7 g/cm^3	4.23 g/cm^3
Chloride	ECl_4	$GeCl_4$
Boiling point	Below 100°C	84°C
Density	1.9 g/cm^3	1.84 g/cm^3

atomic weight, not atomic number, for in his time the latter concept was unknown. However, because atomic weight almost always increases with atomic number, and hence with the number of electrons in the atom, Mendeleev was unwittingly arranging the atoms in order of increasing number of electrons. Since the periodicity of the elements reflects the periodicity of their configurations, he stumbled on the pattern of configurations without knowing anything about atomic structure.

Mendeleev's success also sprang from his chemical insight, which led him to leave gaps for elements that seemed to be needed to complete the pattern but were then unknown. He was even able to predict the properties of some of these missing elements. When later they were discovered, he turned out to be strikingly correct. For instance, his pattern required an element between gallium and arsenic and below silicon; the properties he predicted for this "eka-silicon" are shown in Table 7.4, where they are compared with the properties of germanium, which was discovered in 1886.

7.6 BLOCKS, PERIODS, AND GROUPS

The periodic table is divided into _blocks_ named for the last subshell that is occupied according to the building-up principle (Fig. 7.24). Sodium, [Ne]$3s^1$, and calcium, [Ar]$4s^2$, belong to the _s block_. Nitrogen, [He]$2s^2 2p^3$, and neon, [He]$2s^2 2p^6$, belong to the _p block_. Iron,

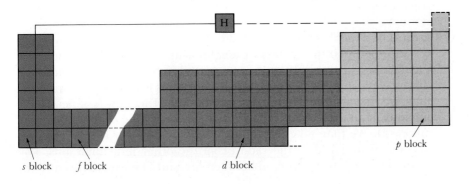

FIGURE 7.24 The block structure of the periodic table is based on the last subshell occupied by an element according to the building-up principle.

s block   *f* block   *d* block   *p* block

[Ar]$3d^6 4s^2$, belongs to the *d block* (the "transition elements"). Hydrogen is separated from the main table because, although it has an s^1 configuration, like lithium and sodium, it has quite different properties. The closed-shell noble gas helium is so similar to the noble gases neon and argon that it is treated as a member of the *p* block, even though it has no *p* electrons. The members of the *s* and *p* blocks are called the *main-group elements*. Calcium is a main-group element; iron is not.

The period number is the same as the principal quantum number of the valence shell of its atoms. Period 1 consists of the two elements (H and He) in which the single $1s$ orbital of the $n = 1$ shell is being occupied. Period 2 consists of the eight elements (Li through Ne) in which the $n = 2$ shell is being occupied. In Period 3 (Na through Ar), the $n = 3$ shell is being occupied, and so on.

With the exception of helium, the group number is equal to the sum of the numbers of *s* and *p* electrons in the valence shell. Lithium (Group I) has one valence-shell *s* electron, and bromine (Group VII) has seven valence-shell *s* and *p* electrons. All members of the same group have the same valence configuration (with different values of *n*).

▼ **EXAMPLE 7.8** **Writing an electron configuration from a group number**

Determine the lowest-energy configuration of a halogen atom.

STRATEGY We know that the group number of an element equals the number of *s* and *p* electrons in its valence shell. We can write the configuration by allowing two electrons to occupy an *s* orbital, leaving the rest to occupy *p* orbitals. The core configuration is that of the preceding noble gas.

SOLUTION The halogens are in Group VII, and hence have seven valence electrons. Two enter an *s* orbital, leaving five to enter *p* orbitals. Hence their characteristic electronic configuration is $s^2 p^5$. For Br (Period 4), as an example, the core is [Ar], and the configuration is [Ar]$3d^{10} 4s^2 4p^5$.

EXERCISE Give the valence-shell configuration of the atoms of the group that contains tin.

[*Answer:* $s^2 p^2$]

▲

7.7 PERIODICITY OF PHYSICAL PROPERTIES

The physical properties of the elements show a striking periodicity. This is particularly clear when we examine the variation of atomic sizes and the energies needed to remove electrons from atoms.

Atomic and ionic radii. Since the charge clouds of atoms and ions do not have sharp edges, it is necessary to define what we mean by their radii. The *metallic radius* of a metallic element is taken to be one-half the distance between the centers of neighboring atoms (**1**) in a solid sample. (We see how this distance is measured in Section 10.6.) Since that distance in copper is 270 pm, its metallic radius is 135 pm. To find *ionic radii*, we first assume that the distance between the centers of neighboring ions in an ionic solid is the sum of the ionic radii of the cation and anion (**2**). We then take the radius of the oxide ion (O^{2-}) as 140 pm, and use it to calculate the radii of all the other individual ions. For example, since the separation between Mg and O nuclei in magnesium

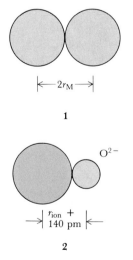

TABLE 7.5 Ionic radii of ions with noble-gas or pseudo-noble gas configurations, in picometers*

Li^+	Be^{2+}	B^{3+}	N^{3-}	O^{2-}	F^-
60	31	20	171	140	136
Na^+	Mg^{2+}	Al^{3+}	P^{3-}	S^{2-}	Cl^-
95	65	50	212	184	181
K^+	Ca^{2+}	Ga^{3+}	As^{3-}	Se^{2-}	Br^-
133	99	62	222	198	195
Rb^+	Sr^{2+}	In^{3+}		Te^{2-}	I^-
148	113	82		221	216
Cs^+	Ba^{2+}	Tl^{3+}			
169	135	95			

*The radius of an H^+ ion is the radius of a proton, about 0.001 pm (1 pm = 10^{-12} m).

oxide is 205 pm, the ionic radius of the Mg^{2+} ion is taken to be 205 pm − 140 pm = 65 pm. That value can then be used to deduce the ionic radius of Cl^- from measurements on magnesium chloride, and so on. The values of metallic and ionic radii obtained in this way are given in Tables 7.5 and 7.6, and some are illustrated in Fig. 7.25. Many metallic and cation radii are close to 100 pm. Anion radii are generally larger and often closer to 200 pm.

The tables show that metallic radius generally decreases from left to right across a period and increases down a group. The decrease across a period, as from Li to Ne, is a result of the increasing attraction between nucleus and electrons with increasing nuclear charge. The increase down a group, as from Li to Cs, is a result of the outermost electrons occupying shells that lie further from the nucleus. Figure 7.25 shows that all cations are smaller than their parent atoms. This is because a cation is formed when an atom loses one or more electrons, leaving its core exposed. The difference in radius can be large: the core is generally much smaller than the atom because the core electrons are so tightly bound to the nucleus. The metallic radius of Li is 145 pm, but the ionic radius of Li^+, the bare heliumlike $1s^2$ core, is only 60 pm. This is comparable to the difference in size between a cherry and its pit. Like metallic radii, and for the same reasons, cation radii decrease across a

TABLE 7.6 Metallic radii in picometers*

Li	Be				
145	105				
Na	Mg	Al			
180	150	125			
K	Ca	Ga	Ge		
220	180	130	125		
Rb	Sr	In	Sn	Sb	
235	200	155	145	145	
Cs	Ba	Tl	Pb	Bi	Po
266	215	190	180	160	190

*1 pm = 10^{-12} m.

100 pm

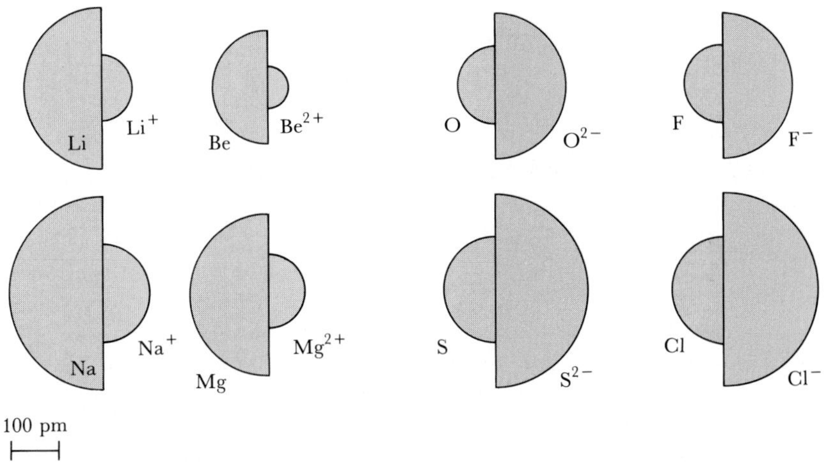

FIGURE 7.25 The metallic and ionic radii of some elements. Note that cations are smaller than their parent atoms, but anions are larger.

period and increase down each group. Figure 7.25 also shows that anions are larger than their parent atoms. The reason can be traced to the increased number of electrons in the anion and the repulsive effects they exert on each other. The variation in the radii of anions is the same as that for atoms and cations, with the smallest at the upper right of the periodic table, close to fluorine.

Ionization energies. The ease with which an electron can be removed from an atom is measured by its "ionization energy" I:

> The **ionization energy** of an atom is the minimum energy required to remove an electron from the ground state of a gaseous atom.

More precisely, the *first ionization energy* I_1 of an element E is the minimum energy needed to remove an electron from a neutral E atom:

$$E(g) \longrightarrow E^+(g) + e^-(g) \qquad \text{energy required} = I_1$$

The *second ionization energy* I_2 of the element is the minimum energy needed to remove an electron from the singly charged cation:

$$E^+(g) \longrightarrow E^{2+}(g) + e^-(g) \qquad \text{energy required} = I_2$$

The energy needed to ionize a gaseous sample of atoms can be supplied as heat. At constant pressure, the heat required per mole of atoms is equal to the *enthalpy of ionization* ΔH_{ion}:

$$E(g) \longrightarrow E^+(g) + e^-(g) \qquad \text{heat required} = \Delta H_{ion}$$

Ionization energies can be measured by analyzing atomic spectra. The Bohr frequency condition $\Delta E = h\nu$ shows that when an electron in

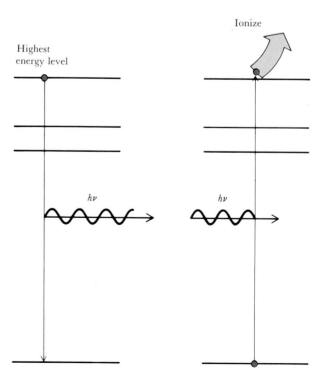

FIGURE 7.26 The shortest-wavelength (highest-frequency) light in an atom's spectrum (left) indicates the minimum amount of energy that is needed to eject an electron from the ground state (right).

the highest-energy orbital of an excited atom falls to the lowest-energy orbital available and re-forms the ground state of the atom, it emits the highest-frequency, shortest-wavelength light (Fig. 7.26). If we supply just slightly more than that energy difference ΔE to the ground-state atom, the electron is ejected. Therefore, the ionization energy of the atom is the energy corresponding to the shortest wavelength in its spectrum.

▼ **EXAMPLE 7.9** **Determining ionization energy**

The shortest-wavelength light emitted by the hydrogen atom has wavelength 91.1 nm. What is the ionization energy of hydrogen?

STRATEGY The 91.1-nm light is generated when an electron falls from its highest-energy orbital into the $1s$ orbital. Therefore, the energy corresponding to photons of 91.1-nm light is the maximum amount that can be supplied to an electron in a $1s$ orbital without ionization occurring. To find this energy, we convert wavelength to frequency and use the Bohr frequency condition.

SOLUTION Since $\nu = c/\lambda$, the frequency of 91.1-nm radiation is

$$\nu = 3.00 \times 10^8 \, \frac{m}{s} \times \frac{1}{91.1 \times 10^{-9} \, m}$$

$$= 3.29 \times 10^{15} \, \text{Hz}$$

This corresponds to an energy difference of

$$\Delta E = h\nu$$

$$= (6.63 \times 10^{-34} \, \frac{J}{Hz}) \times (3.29 \times 10^{15} \, \text{Hz})$$

$$= 2.18 \times 10^{-18} \, \text{J}$$

Hence, the ionization energy is 2.18×10^{-18} J. Ionization energies are normally reported for one mole of atoms, by multiplication by Avogadro's number. Hence, for 1 mol of hydrogen atoms,

$$I_1 = (6.022 \times 10^{23}) \times (2.18 \times 10^{-18} \, \text{J}) = 1.31 \, \text{MJ}$$

and the ionization energy per mole of atoms is 1.31 MJ/mol.

EXERCISE The shortest-wavelength light emitted by a sodium atom has wavelength 241 nm. What is the ionization energy per mole of atoms?

[*Answer*: 497 kJ/mol]

▲

The first and second ionization energies of the main-group elements are listed in Table 7.7. Most first ionization energies lie in the range 500 to 1000 kJ/mol. The second ionization energy of an element is always larger than its first, because more energy is required to remove an electron from a positively charged ion than from a neutral atom. In some cases (see the Group I elements) the second ionization energy is considerably larger than the first, but in others (see Group II) they are close in value. The energy needed to remove an electron from the core of an atom is always much larger than that needed to remove a valence electron. The core electrons have lower principal quantum numbers and are so close to the nucleus that they are strongly attracted to it.

TABLE 7.7 First and second (and some higher) ionization energies of the main-group elements, in kilojoules per mole

				H 1310			He 2370 5250
Li 519 7300	Be 900 1,760 14,800	B 799 2,420 3,660 25,000	C 1090 2350	N 1400 2860	O 1310 3390	F 1680 3370	Ne 2080 3950
Na 494 4560	Mg 736 1,450 7,740	Al 577 1,820 2,740 11,600	Si 786	P 1060	S 1000	Cl 1260	Ar 1520
K 418 3070	Ca 590 1,137 4,940	Ga 577	Ge 762	As 966	Se 941	Br 1140	Kr 1350
Rb 402 2650	Sr 548 1,060 4,120	In 556	Sn 707	Sb 833	Te 870	I 1010	Xe 1170
Cs 376 2420 3300	Ba 502 966 3,390	Tl 590	Pb 716	Bi 703	Po 812	At 920	Rn 1040

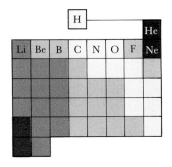

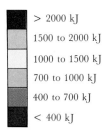

> 2000 kJ

1500 to 2000 kJ

1000 to 1500 kJ

700 to 1000 kJ

400 to 700 kJ

< 400 kJ

FIGURE 7.27 The variation of first ionization energy within the main groups of the periodic table. Red indicates highest, blue lowest.

Ionization energy varies periodically with atomic number. This is shown most clearly by the first ionization energies of the main-group elements (Fig. 7.27). With a few exceptions, the ionization energy rises from left to right across a period and then falls back to a lower value at the start of the succeeding period. The lowest values occur at the bottom left of the periodic table (near Cs), and the highest at the upper right (near He). In other words, less energy is needed to remove an electron from atoms of elements near Cs, and more energy is needed to remove one from atoms of elements near He. This is consistent with the fact that cesium is so highly reactive that it reduces water explosively.

The increase in first ionization energy across each period may be explained in terms of the trend in atomic radius. We have seen that atoms become smaller as the nuclear charge increases from left to right across the period. As a result of their smaller size, electrons of atoms on the right are closer to a more highly charged nucleus, are attracted more strongly to it, and hence are more difficult to remove. The decrease in ionization energy as we move down a group results from the outermost electron occupying an increasingly distant shell, with a larger value of the principal quantum number, and hence being bound less tightly. The small departures from these trends can usually be traced to the effects of repulsions between electrons, particularly electrons occupying the same orbital.

▼ EXAMPLE 7.10 Accounting for the variation of ionization energies

Suggest a reason for the small decrease in ionization energy between nitrogen (1400 kJ/mol) and oxygen (1310 kJ/mol).

STRATEGY Since the outermost electron is closer to the nucleus in O than in N, and the O nucleus is more strongly charged, we expect oxygen to have the higher (not lower) ionization energy. When faced with such a conflict between a prediction and an observation, we should look for an influence that has been ignored. Here, if there were a reason why the outermost electron would experience a strong repulsion from the other oxygen electrons, that might explain why less than the expected energy is needed to remove it. Therefore, we should examine the configurations of the atoms to see if they suggest a greater repulsion for the outermost electron in O than in N.

SOLUTION The outermost three electrons in N occupy three different $2p$ orbitals. When an additional electron is added to build O, it fills one of these already occupied orbitals. It therefore experiences a strong repulsion from the electron that is already present, and as a result less energy is needed to remove it. Hence, its ionization energy is lowered.

EXERCISE Account for the decrease in ionization energy between beryllium and boron.

[*Answer*: A new, higher-energy subshell being occupied.]

The low ionization energies of elements at the lower left of the periodic table account for the metallic character of those elements. A metal consists of a collection of cations surrounded by a sea of electrons that the atoms have lost. Only elements with low ionization energies can form such solids, since only they can lose their electrons easily. Because the elements that are far to the right of the periodic table have high ionization energies, they do not readily lose electrons and are not metals.

We noted in Section 2.7 that the common charge number of Group I cations is $+1$, and that of Group II cations is $+2$. This is easy to explain in terms of ionization energies. Although one electron can be removed quite easily from an alkali-metal atom (at a cost of 494 kJ/mol for sodium), 10 times as much energy (4560 kJ/mol) is needed to remove a second electron from the core. Hence, E^+ is the typical charge type of Group I cations. For the elements in Group II, the first two ionization energies are much less different (for magnesium they are 736 kJ/mol and 1450 kJ/mol), and it is energetically feasible to remove both electrons. However, a huge energy (7740 kJ/mol) would be needed to remove a third electron from magnesium, since it would have to come from the core. Hence, in Group II the expected charge type of the ions is E^{2+}.

Inert pairs. When we consider the ions that p-block elements can form, we need to take into account the difference in ionization energy between the s and p electrons in the valence shell. The p electrons are lost first, but as the s electrons have a substantially higher ionization energy, they might not be lost at all. It is found that the difference between the ionization energies for s and p valence electrons increases

TABLE 7.8 Electron-gain enthalpies (electron affinities) of the main-group elements, in kilojoules per mole*

				H −72			He +21
Li −60	Be +240	B −28	C −122	N +7	O −142 +844	F −328	Ne +29
Na −53	Mg +232	Al −44	Si −120	P −72	S −200 +532	Cl −349	Ar +35
K −48	Ca +156	Ga −29	Ge −117	As −77	Se −195	Br −325	Kr +39
Rb −47	Sr +52	In −29	Sn −121	Sb −101	Te −190	I −295	Xe +41

*Where two values are given, the first refers to the formation of the ion X^- from the neutral atom X; the second, to the formation of X^{2-} from X^-.

down a group. This is expressed by saying that the two valence-shell *s* electrons of a heavy element can act as an "inert pair":

> An **inert pair** is a pair of valence-shell *s* electrons that are tightly bound to the atom.

Thus, when Al forms an ion it gives up all its valence electrons and forms only Al^{3+} ions, but its heavier congener In loses three electrons in some compounds, to form In^{3+} ions, and only one electron in others, to form In^+ ions.

Electron affinity. Negative ions can be formed by adding one or more electrons to a gaseous atom or anion:

$$E(g) + e^-(g) \longrightarrow E^-(g)$$

$$E^-(g) + e^-(g) \longrightarrow E^{2-}(g)$$

The resulting enthalpy change is called the *electron-gain enthalpy* ΔH_{gain}, and it may be either positive (indicating an endothermic process) or negative (indicating an exothermic process). We speak of an element as having a high *electron affinity* if one of its atoms forms an anion exothermically, releasing energy as heat. That is, an element has a high electron affinity if the electron-gain enthalpy is negative. It has a low electron affinity if electron gain is endothermic or only weakly exothermic. Some values of ΔH_{gain} are given in Table 7.8.

Broadly speaking, electron gain gets more exothermic toward the upper right of the periodic table (Fig. 7.28) and is most exothermic close to fluorine. In these atoms the incoming electron occupies an orbital close to a highly charged nucleus. This position is energetically favorable; hence the transition is exothermic. Once an electron has entered the single vacancy in the valence shell of a Group VII atom, the shell is complete; any additional electron would have to begin a new shell. In that shell it would not only be farther from the nucleus, but would also feel the repulsion of the negative charge already present.

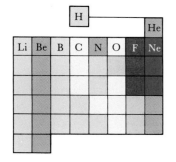

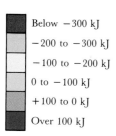

Below −300 kJ
−200 to −300 kJ
−100 to −200 kJ
0 to −100 kJ
+100 to 0 kJ
Over 100 kJ

FIGURE 7.28 Electron affinities of the main-group elements. Red indicates most exothermic, and blue most endothermic.

The addition of a second electron to a halide ion is therefore strongly endothermic. A consequence is that the ionic compounds of the halogens are built from singly charged ions such as F^-, and never from doubly charged ions such as F^{2-}.

A Group VI atom, such as O or S, has two vacancies in its valence shell and can accommodate two additional electrons. The first electron gain is exothermic. However, the second electron gain is endothermic because energy must be *supplied* to overcome repulsion by the negative charge already present. Since 142 kJ/mol of energy is released when the first electron is added to the neutral atom to form O^-, and 844 kJ/mol is needed to add a second electron to form O^{2-}, the net energy required is the difference, 702 kJ/mol.

▼ **EXAMPLE 7.11 Accounting for the variation in electron affinity**

Suggest a reason for the decrease in electron affinity between carbon and nitrogen.

STRATEGY The basic theory suggests that we should expect more energy to be released when an electron enters the N atom, because it is smaller than the C atom and its nucleus is more highly charged. That the opposite is observed suggests that we may have ignored the effect of electron repulsion in the resulting anions. We should therefore examine the configurations of the anions to see if the effect of electron repulsion is greater in N^- than in C^-. If it is, less energy will be released when the incoming electron attaches to the N atoms than when it attaches to the C atom.

SOLUTION On forming C^- from C, the incoming electron occupies an empty $2p$ orbital. On forming N^- from N, the incoming electron must occupy a $2p$ orbital that is already half full. Although the nuclear charge of nitrogen is greater than that of carbon, it is reduced by repulsion from the electron already in the $2p$ orbital; hence less energy is released when N^- is formed, and the electron affinity of nitrogen is lower than that of carbon.

EXERCISE Account for the large decrease in electron affinity between lithium and beryllium.
  [*Answer*: The additional electron enters $2s$ in Li, but $2p$ in Be, and a $2s$ electron is more tightly bound than a $2p$ electron.]
▲

Electronegativity. An atom is likely to form a cation if it has a low ionization energy and a low electron affinity. On the other hand, it is likely to form an anion if it has a high electron affinity and a high ionization energy. This suggests the following pattern of behavior:

Ionization energy	Electron affinity	Behavior
Low	Low	Form cation
High	High	Form anion

The table can be summarized neatly by introducing the concept of *electronegativity* χ (the Greek letter chi), defined as

$$\chi = \frac{\text{ionization energy} + \text{electron affinity}}{2}$$

TABLE 7.9 Electronegativities of the main-group elements

						H 2.1
Li 1.0	Be 1.5	B 2.0	C 2.5	N 3.0	O 3.5	F 4.0
Na 0.9	Mg 1.2	Al 1.5	Si 1.8	P 2.1	S 2.5	Cl 3.0
K 0.8	Ca 1.0	Ga 1.6	Ge 1.8	As 2.0	Se 2.4	Br 2.8
Rb 0.8	Sr 1.0	In 1.7	Sn 1.8	Sb 1.9	Te 2.1	I 2.5
Cs 0.7	Ba 0.9	Tl 1.8	Pb 1.8	Bi 1.9	Po 2.0	At 2.2

When both ionization energy and electron affinity are low, χ is low; when both are high, χ is high. This simplifies the table to

Electronegativity	Behavior
Low	Form cation
High	Form anion

An element with a low electronegativity is called "electropositive." The members of the *s* block, especially cesium, are the most electropositive elements.

Electronegativities are given in Table 7.9 for the main-group elements (for which the concept is most useful). The values in the table have been converted to a unitless scale devised originally by the American chemist Linus Pauling, who won a Nobel Prize for his work (and, for his other activities, the Nobel Peace Prize as well). The dependence of electronegativity on position in the periodic table is shown in Fig. 7.29. The trend to higher values toward the upper right follows the trend of ionization energies and electron affinities in that direction. Hence, fluorine and the Group VI and VII elements near it are the most electronegative, and cesium and the other alkali metals close to it are the most electropositive.

The high electronegativities of the halogens and high electropositivities of the alkali metals account for the formation of ionic compounds when they react together. Oxygen also has a very high electronegativity, so we would expect it to form the oxide ion when it reacts with the *s*-block elements. We shall see in more detail in Chapter 8 that electronegativity is indeed a useful guide to the type of compounds an element forms.

7.8 TRENDS IN CHEMICAL PROPERTIES

We can now illustrate how the various trends in atomic and ionic properties account for some of the chemical properties of the elements.

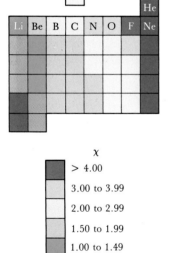

χ

> 4.00

3.00 to 3.99

2.00 to 2.99

1.50 to 1.99

1.00 to 1.49

0.75 to 0.99

< 0.75

FIGURE 7.29 The variation of electronegativity within the main-group elements of the periodic table. Red indicates most electronegative, and blue least electronegative (most electropositive).

TABLE 7.10 Characteristics of metals and nonmetals

Metals	Nonmetals
Physical properties	
Good conductors of electricity	Poor conductors of electricity
Malleable	Not malleable
Ductile	Not ductile
Lustrous	Not lustrous
They are usually:	
Solid	Solid, liquid, or gas
With high melting points	With low melting points
Of low volatility	Volatile
Good conductors of heat	Poor conductors of heat
Chemical properties	
React with acids	Do not react with acids
Form basic oxides (which react with acids)	Form acidic oxides (which react with bases)
Form cations	Form anions
Form ionic halides	Form covalent halides

(a)

(b)

The s-block elements. If an element belongs to the *s* block, its outermost electrons can be lost easily. The element (which we denote E) will be likely to form the ions E^+ (if it is in Group I) and E^{2+} (if it is in Group II). It is likely to be a reactive metal with all the features that the name "metal" implies (Table 7.10). Since ionization energies are lowest at the foot of each group, and the elements there lose outermost electrons most easily, those elements can be expected to react most vigorously.

The *s*-block elements show these expected properties. All are silver-gray metals that are too reactive to be found naturally in the "native" (uncombined) state (Fig. 7.30). All the Group I metals reduce water to hydrogen:

$$2K(s) + 2H_2O(l) \longrightarrow 2KOH(aq) + H_2(g)$$

The vigor of this reaction increases down the group. Lithium reacts gently, sodium moderately vigorously, and cesium with explosive violence. All the Group II metals except beryllium also reduce water:

$$Ca(s) + 2H_2O(l) \longrightarrow Ca(OH)_2(aq) + H_2(g)$$

However, the reaction is less violent than for the alkali metals, and magnesium reacts only with hot water.

All the *s*-block elements have basic oxides that react with water to form hydroxides:

$$CaO(s) + H_2O(l) \longrightarrow Ca(OH)_2(aq)$$

Beryllium, at the head of Group II, has the highest ionization energy of the block. It therefore loses its valence electrons less readily than its congeners do and has the least pronounced metallic character. The compounds of all the *s*-block elements (with the exception of beryllium) are ionic.

FIGURE 7.30 All the alkali metals are soft, silvery metals. Sodium is kept under paraffin oil to protect it from air, but a freshly cut surface (a) quickly becomes covered with oxides (b).

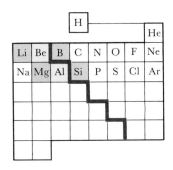

FIGURE 7.31 The pairs of elements represented by colored squares show a strong diagonal relationship to each other.

Diagonal relationships. Another feature of the periodic table becomes apparent when we compare neighbors on diagonal lines running from upper left to lower right, such as Li and Mg or Be and Al (Fig. 7.31). Such elements show a "diagonal relationship":

A **diagonal relationship** is a similarity between diagonal neighbors in the periodic table, especially for elements at the left of the table.

Diagonal relationships show up in many different ways and are very helpful for making predictions about the properties of elements and their compounds. The diagonal line dividing the metals from the non-metals is one example we have already met. A diagonal relationship shows up in the *s* block in the close similarity between the properties of Li and Mg. Lithium is the only member of Group I that burns in nitrogen and, like magnesium, forms a nitride:

$$6Li(s) + N_2(g) \longrightarrow 2Li_3N(s)$$

$$3Mg(s) + N_2(g) \longrightarrow Mg_3N_2(s)$$

The latter reaction is one reason why magnesium burns so brightly in air, for it combines exothermically with both the oxygen and the nitrogen.

The similarity between beryllium and its *p*-block diagonal neighbor aluminum is another example of a diagonal relationship. Like aluminum, beryllium is amphoteric. The two elements react with acids:

$$Be(s) + 2H^+(aq) \longrightarrow Be^{2+}(aq) + H_2(g)$$

$$2Al(s) + 6H^+(aq) \longrightarrow 2Al^{3+}(aq) + 3H_2(g)$$

(Aluminum seems not to react with nitric acid and, in fact, is often used to transport it. However, the acid actually oxidizes the surface of the metal to a protective film of aluminum oxide that prevents further reaction.) They also react with alkalis:

$$Be(s) + 2OH^-(aq) + 2H_2O(l) \longrightarrow [Be(OH)_4]^{2-}(aq) + H_2(g)$$

$$2Al(s) + 2OH^-(aq) + 6H_2O(l) \longrightarrow 2[Al(OH)_4]^-(aq) + 3H_2(g)$$

The reaction between aluminum and sodium hydroxide is partly responsible for the action of domestic drain cleaners, which are mixtures of sodium hydroxide and aluminum turnings. The reaction produces heat, which melts the fats and grease, and hydrogen, which helps to break up the mass. The sodium hydroxide also reacts with the fat, breaking it down into simpler soaplike substances that can be washed away. Beryllium is not used in drain cleaners because it is much more expensive than aluminum, and its salts are very poisonous.

The p-block elements. An element on the left of the *p* block has a low enough ionization energy to possess some of the metallic properties of the *s*-block elements, especially if it is in a late period. The dividing line between *p*-block metals and nonmetals—the red line in Fig. 7.31—runs diagonally across the table from aluminum to polonium. However, the ionization energies of all the *p*-block metals are quite high, and they are less reactive than the members of the *s* block.

In Group IV, lead and tin are metals. They are lustrous and malleable and conduct electricity, but they are not nearly as reactive as the s-block elements and many d-block elements. As a practical example, steel cans are tin-plated—covered in tin by dipping them into the molten metal—to protect them against corrosion, because tin is less reactive than iron. Lead was used for about 2000 years to make pipes for domestic water supplies, for it, too, is very resistant to corrosion. It is no longer used, because lead compounds are now known to be toxic. Lead and tin also show more than a hint of nonmetallic character, since they form amphoteric oxides:

$$SnO(s) + 2H^+(aq) \longrightarrow Sn^{2+}(aq) + H_2O(l)$$

$$SnO(s) + OH^-(aq) + H_2O(l) \longrightarrow [Sn(OH)_3]^-(aq)$$

The inert-pair effect is most important at the foot of a p-block group. We have seen that aluminum exists as Al^{3+} ions but indium, near the foot of the same group, forms both In^{3+} and In^+ ions. Tin, in Group IV, forms tin(IV) oxide when heated in air. In contrast, its heavier congener lead loses only its two p electrons and forms lead(II) oxide:

$$Sn(s) + O_2(g) \longrightarrow SnO_2(s)$$

$$2Pb(s) + O_2(g) \longrightarrow 2PbO(s)$$

Tin(II) oxide can be formed as a blue-black solid by adding alkali to a solution of tin(II) salt and then heating the precipitate in the absence of air. However, when it is heated to 300°C in air, it forms tin(IV) oxide (the inert pair now being lost) with the evolution of so much heat that the solid becomes incandescent (Fig. 7.32).

The characteristic property of elements well on the right of the p block is their ability to gain electrons and thus complete a closed-shell configuration. The elements at the upper right of the block, such as oxygen, sulfur, and the halogens, are therefore nonmetals and typically occur in ionic compounds as anions. Fluorine forms ionic or molecular compounds with every element except helium, neon, and argon.

The d-block elements. All d-block elements are metals. Elements that lie toward the left of the block, near scandium, are reactive and resemble the s-block metals. Those that lie to the right of the block, as copper and gold do, are less reactive and resemble the more chemically sluggish metals of the p block. The coinage metals can be used as such because they are so unreactive. The properties of the d-block elements are transitional between those of the s- and p-block elements; this is the origin of their alternative name, the "transition metals."

One very characteristic feature of the d-block elements is that many of them form a variety of cations with different charge types. Iron, we have seen, forms iron(II) and iron(III) ions, Fe^{2+} and Fe^{3+}. Copper forms copper(I) and copper(II), Cu^+ and Cu^{2+}. Potassium, an s-block metal, has a $4s^1$ valence configuration like copper but forms only one type of ion, K^+. The reason for the difference between copper and potassium can be seen by comparing their second ionization energies, which are 1960 kJ/mol and 3070 kJ/mol, respectively. To form Cu^{2+},

FIGURE 7.32 When tin(II) oxide is heated in air, it becomes incandescent as it reacts to form tin(IV) oxide. Even without being heated, it smolders and can ignite.

an electron is removed from the d subshell of $[Ar]3d^{10}$; but to form K^{2+}, an electron would have to be removed from potassium's argonlike core. Since such huge amounts of energy are not readily available in chemical reactions, potassium can lose only its $4s$ electron.

We are beginning to see why many chemists claim that the periodic table is the single most important discovery in their field. In the rest of this text we shall see that it is indeed an enormously powerful way of organizing knowledge.

SUMMARY

7.1 Light is **electromagnetic radiation;** its **frequency,** or number of oscillations per second, and **wavelength,** or peak-to-peak distance, are related and determine its color. According to **quantum mechanics,** light rays are also streams of **photons,** packets of energy of magnitude $h\nu$. Support for this view comes from the **photoelectric effect,** the ejection of electrons from the surfaces of metals by ultraviolet radiation.

7.2 When electrons in an atom change their spatial arrangement, the difference in energy is lost as a photon with a frequency given by the **Bohr frequency condition,** $\Delta E = h\nu$. The existence of **spectral lines** at discrete frequencies in atomic spectra indicates that the energy of atoms is **quantized,** or confined to certain values. The **Bohr model** was an attempt to explain the **Rydberg formula,** which summarized the spectrum of atomic hydrogen. In the model, the electrons are supposed to travel in definite orbits with a certain energy.

7.3 The Bohr model had to be rejected because electrons exhibit a **wave-particle duality,** a double wavelike and particlelike character, with a wavelength given by the **de Broglie relation,** $\lambda = h/mv$. The wave character of electrons is confirmed by the **diffraction pattern** formed in the **Davisson-Germer experiment.** Since they have wavelike properties, electrons do not travel in precisely determined paths; the precision with which the position and speed of an electron can be known is given by Heisenberg's **uncertainty principle,** $\Delta x(m\,\Delta v) \geq h/4\pi$. The wavelike character of electrons is taken into account by the **Schrödinger equation,** which gives the **orbital,** the region in which an electron is most likely to be found in an atom. The atomic orbitals of hydrogen are classified as s, p, d, and f, and they lie in a series of **shells** and **subshells.** Orbitals are labeled with the **quantum numbers** n, l, and m_l. In the hydrogen atom, all orbitals of a given shell have the same energy. The **ground state,** the state of lowest energy, of the hydrogen atom consists of an electron in the $1s$ orbital. The **Stern-Gerlach experi-**

ment, in which a beam of atoms is passed through a nonuniform magnetic field, shows that an electron may also have one of two **spin states** ($\uparrow$ and $\downarrow$).

7.4 The electronic structures of **many-electron atoms** are based on the orbitals deduced for the hydrogen atom. As a result of **penetration,** the approach of an electron to the nucleus through inner shells, and **shielding,** the repulsion by other electrons, the **effective nuclear charges** experienced by electrons in different subshells are different. Hence, the subshells have different energies. The electrons of an atom occupy the orbitals that result in the lowest total energy, subject to the requirements of Pauli's **exclusion principle,** that no more than two electrons can occupy a given orbital.

7.5 The ground-state **electron configuration,** or list of occupied orbitals, for any atom is obtained with the **building-up principle,** in which electrons are imagined as entering the orbitals of the atom one after the other in a certain order and in accordance with the exclusion principle. The availability of d orbitals in Period 4 and later periods accounts for the length of those periods.

7.6 The periodic table reflects the periodicity of the electron configurations of the elements. Its groups contain elements with analogous valence-electron configurations, and its periods correspond to the occupation of different shells. The **blocks** of the periodic table correspond to the occupation of subshells by the electrons added last according to the building-up principle.

7.7 **Metallic radii** and **ionic radii** vary periodically, with the largest to the lower left in the periodic table and the smallest at the upper right. **Ionization energy** and **electron affinity** (as measured by the **electron gain enthalpy**) also vary periodically, with smallest values at the lower left and greatest at the upper right. **Core electrons,** or electrons in completed inner shells, have much greater ionization energy than the **valence electrons,** the outermost electrons. The difference is less pronounced for a *d* subshell, so *d*-block electrons may be lost reasonably easily. The ability of elements in the *p* block to form cations with different charge numbers reflects the **inert-pair effect,** the relative inertness of a pair of valence-shell *s* electrons, which depends on differences between *s*- and *p*-electron ionization energies. **Electronegativity,** which indicates whether an atom is likely to lose or gain an electron in a compound, is greatest at the upper right of the periodic table.

7.8 As well as being closely related to their congeners in a group, a number of elements (especially those in Periods 2 and 3 toward the left of the table) show a strong **diagonal relationship** to each other.

EXERCISES

Frequency and Wavelength

In Exercises 7.1 and 7.2, calculate the wavelengths of the radiation.

7.1 (a) Radio waves corresponding to 98 MHz on the FM dial; (b) yellow light of frequency 5.3×10^{14} Hz; (c) the 2.0×10^{18} Hz x-rays given out when an electron beam strikes a block of copper.

7.2 (a) The UHF waves used for transmitting television that have frequency 700 MHz; (b) violet light of frequency 7.1×10^{14} Hz; (c) the 1420-MHz waves used by radioastronomers looking at insterstellar clouds of hydrogen atoms.

In Exercises 7.3 and 7.4, calculate the frequencies of the radiation.

7.3 (a) 530-nm green light; (b) 250-m AM radio waves; (c) the 149-pm x-rays generated from the particle accelerator called a "synchrotron."

7.4 (a) 1200-nm infrared radiation; (b) 3-cm radar waves; (c) the 86-pm "gamma radiation" emitted when the protons and neutrons inside the nucleus of an iron-57 atom adjust their positions.

7.5 The speed of sound in air is 331 m/s; calculate the wavelength of the sound waves corresponding to middle C on a piano (262 Hz).

7.6 The speed of sound in water at 20°C is 1482 m/s. Do you anticipate a shorter or a longer wavelength in water than in air? Calculate the wavelength of 262-Hz sound waves in water.

Photons

In Exercises 7.7 and 7.8, calculate the energy (i) of a single photon and (ii) per mole of photons for each type of electromagnetic radiation.

7.7 (a) The yellow 580-nm light emitted by a sodium highway lamp; (b) the red 640-nm light emitted by a neon sign; (c) the 154-pm x-rays emitted by copper when it is exposed to a high-energy electron beam.

7.8 (a) Violet 420-nm light; (b) the 470-nm light emitted by the mixture of argon and mercury vapor used in blue advertising signs; (c) the 1500-nm component of the infrared radiation emitted by a hot object.

7.9 Many fireworks mixtures depend on the highly exothermic combustion of magnesium to magnesium oxide, in which the heat causes the oxide to become incandescent and give out a bright white light. The color of the light can be changed by including nitrates and chlorides of elements that have spectra in the visible region (Fig. 7.33). Barium nitrate is often added to produce a yellow-green color. The excited barium ions generate 487-nm, 514-nm, 543-nm, 553-nm, and 578-nm light. Calculate the change in energy of the atom in each case, and the change in energy per mole of atoms.

7.10 Potassium salts are sometimes detected by the "flame test," in which they give a lilac-colored tint to a flame. The lilac light is a mixture of 405-nm and 767-nm radiation. What is the change in energy (a) of a potassium atom and (b) per mole of potassium atoms when these photons are generated?

FIGURE 7.33 The colors in this firework display are caused by excited atoms losing energy as radiation. (Fireworks by Grucci.)

7.11 A lamp rated at 100 W uses energy at the rate of 100 J/s. Suppose that 1.0% of this energy is used to produce yellow 580-nm light. Calculate the number of photons of yellow light generated each second. For how long must the lamp operate to generate 1 mol of yellow photons?

7.12 Suppose the lamp of Exercise 7.11 also uses 0.10% of the energy it consumes to produce blue 470-nm light. Calculate the number of photons of blue light generated each second, and the length of time the lamp must operate to produce 1 mol of blue photons.

7.13 The surface temperature of the sun is about 6000 K, and each square meter of its surface acts like a 0.64-W yellow lamp. How many yellow photons are produced per second per square meter of the sun's surface?

7.14 A high-power 100 MW carbon dioxide laser emits a pulse of 1.05-μm infrared radiation lasting 50 ns. How many photons does it generate?

7.15 Electrons are ejected by the photoelectric effect when the surface of the metal rubidium is illuminated with 300-nm ultraviolet radiation. The kinetic energy of each electron was measured and found to be 3.3×10^{-19} J. What is the maximum wavelength of light that can remove an electron from a sample of rubidium? What would be the kinetic energy of the electrons if 250-nm light were used?

7.16 The minimum energy needed to remove an electron from the surface of a sample of silver is 6.9×10^{-19} J. Could a photoelectric detector for ultraviolet 250-nm light be constructed using silver? If not, what is the maximum wavelength of light that could be detected photoelectrically by silver?

The Spectrum of Atomic Hydrogen

7.17 Calculate the wavelengths (in nanometers) of the first three lines of the Balmer series of atomic hydrogen.

7.18 Calculate the wavelengths (in nanometers) of the first three lines of the Lyman series of hydrogen.

7.19 Highly ionized atoms, in some cases having been stripped of all except one electron, occur in stars and contribute to their spectra. Given that the energy levels are proportional to $Z^2 \Re$ in ions with one electron, calculate the positions of the first three lines corresponding to the Balmer series in (a) He^+; (b) C^{5+}.

7.20 Repeat Exercise 7.19 for (a) Li^{2+}; (b) O^{7+}.

Particles and Waves

In Exercises 7.21 and 7.22, calculate the wavelength of each particle.

7.21 (a) An electron traveling at 2200 km/s; (b) an electron traveling at 1 cm/s; (c) a proton traveling at 1 cm/s; (d) a 150-g ball traveling at 1 cm/s.

7.22 (a) Yourself, running at 5 m/s; (b) yourself, running faster at 10 m/s; (c) an electron traveling at 10 m/s; (d) a hydrogen atom traveling at 10 m/s.

7.23 Who has the longer wavelength when traveling at the same speed, a woman or a (heavier) man?

7.24 At what speed would an electron have to travel to have a wavelength equal to the circumference of the first Bohr orbit in the hydrogen atom? What would then be its kinetic energy, as a multiple of $\Re$?

7.25 (a) If the position of an electron is known to within 100 pm, what is the minimum uncertainty in its speed? (b) Suppose instead that its speed is known to within 1.0 mm/s. What is the minimum uncertainty in its position?

7.26 (a) If the position of a proton is known to within 100 pm, what is the minimum uncertainty in its speed? (b) Suppose instead that its speed is known to within 1.0 mm/s. What is the minimum uncertainty in its position?

7.27 (a) Calculate the minimum uncertainty in the position of a speck of dust of mass 1.0 μg with a speed known to within 0.010 mm/s. (b) Suppose that you have located the speck to within 1.0 nm (about four atomic diameters). What is the minimum uncertainty in its speed?

7.28 (a) Calculate the minimum uncertainty in the speed of a speck of dust of mass 1.0 μg with a position known to within 1.0 μm. (b) Suppose that you have located the speck to within 0.10 nm (about one atomic diameter). What is the minimum uncertainty in its speed?

7.29 Suppose you were driving a 1000-kg vehicle, were caught by a speed trap, and were reported as traveling 1.0×10^7 meters per second. Does the uncertainty principle allow you to claim that you were in another vehicle over 10 meters away?

7.30 The new super collider will accelerate particles and direct them at a small target. If a particle of mass 1.7×10^{-32} kg were traveling at 3.1×10^3 meters per second, should the collider operators expect to repeatedly hit a 1.0×10^{-4}-m diameter target?

Atomic Orbitals

7.31 How many orbitals are there in the shells with principal quantum numbers (a) 1; (b) 2; (c) 3?

7.32 (a) How many orbitals are there in the shell with principal quantum number 5? (b) Identify the subshells of that shell (in the form 5s, . . .). (c) Suggest a general formula for the number of orbitals, given the value of n. (d) Test the formula with $n = 6$.

In Exercises 7.33 and 7.34, state the number of orbitals in each shell or subshell, and (except in the case of f orbitals) give their identities (in the form np_x, . . .).

7.33 (a) The $n = 1$ shell; (b) the $3d$ subshell; (c) the $4d$ subshell; (d) the $3s$ subshell.

7.34 (a) The $n = 2$ shell; (b) the $5f$ subshell; (c) the $4f$ subshell; (d) the $3f$ subshell.

In Exercises 7.35 and 7.36, give the value of the azimuthal quantum number for each subshell, and state how many orbitals each subshell contains.

7.35 (a) $3s$; (b) $4p$; (c) $5p$; (d) $3d$.

7.36 (a) $2s$; (b) $5s$; (c) $3p$; (d) $4f$.

In Exercises 7.37 and 7.38, decide how many orbitals are present in the shell or subshell.

7.37 (a) The $n = 1$ shell; (b) the $n = 2$ shell; (c) the $3d$ subshell; (d) the $4p$ subshell.

7.38 (a) The $n = 3$ shell; (b) the $n = 4$ shell; (c) the $3p$ subshell; (d) the $4s$ subshell.

In Exercises 7.39 and 7.40, list the allowed values of l and give the $s, p, d, \ldots$ designations of the subshells for each value of n.

7.39 (a) 1; (b) 2.

7.40 (a) 3; (b) 4.

In Exercises 7.41 and 7.42, determine which of the orbitals cannot occur.

7.41 (a) $2d$; (b) $5p$; (c) $3f$; (d) $10s$.

7.42 (a) $1p$; (b) $4f$; (c) $4g$ (the subshell with $l = 4$); (d) $4d$.

In Exercises 7.43 and 7.44, give the number of values (and the values themselves) that the magnetic quantum number m_l can have for each value of l that is given or implied by an orbital label.

7.43 (a) 0; (b) 4; (c) $2s$; (d) $3p$.

7.44 (a) 2; (b) 3; (c) $3d$; (d) $4f$.

In Exercises 7.45 and 7.46, specify the four quantum numbers in the order $\{n, l, m_l, m_s\}$ for each electron description. Where m_l cannot be determined, give all possible values.

7.45 (a) $3s(\uparrow)$; (b) $3s(\downarrow)$; (c) $3d(\downarrow)$; (d) $4f(\uparrow)$.

7.46 (a) $1s(\uparrow)$; (b) $2p(\downarrow)$; (c) $2p(\uparrow)$; (d) $4s(\uparrow)$.

In Exercises 7.47 and 7.48, several sets of quantum numbers are given in the order $\{n, l, m_l, m_s\}$. Determine which are unacceptable, and explain why.

7.47 (a) $\{1, 1, 1, \frac{1}{2}\}$; (b) $\{3, 0, 1, -\frac{1}{2}\}$; (c) $\{2, 2, 2, \frac{1}{2}\}$; (d) $\{2, 1, 0, \frac{1}{2}\}$.

7.48 (a) $\{0, 0, 0, \frac{1}{2}\}$; (b) $\{3, 2, 1, 0\}$; (c) $\{2, 1, -2, \frac{1}{2}\}$; (d) $\{1, 0, 0, \frac{1}{2}\}$.

7.49 Summarize the contributions of the following individuals to the development of modern quantum theory: Max Planck, Joseph Balmer, Johannes Rydberg, Niels Bohr, Louis de Broglie, Werner Heisenberg, Erwin Schrodinger, Otto Stern, and Walter Gerlach.

7.50 Name, give the symbol, and describe the four quantum numbers which are used to characterize electrons in atoms.

7.51 The $1s$ orbital of the hydrogen atom is a mathematical expression which we denote as psi (ψ). The probability of finding an electron at a given point depends upon the square of psi (ψ^2). If the values of ψ at 10 and 60 pm from the hydrogen nucleus are 0.84 and 0.32, respectively, then: (a) What is the probability of finding an electron at both 10 pm and 60 pm from the nucleus? Which is more probable? (b) As the nucleus is approached from 10 pm away would you expect the probability of finding an electron in a $1s$ orbital to increase or to decrease? (c) In a multielement atom would you expect the probability of locating an electron in a $2p$ to increase or to decrease as the nucleus is approached from a distance of about 60 pm? Explain.

7.52 As you will see in the next chapter, the overlap of atomic orbitals from different atoms can result in the formation of chemical bonds. The properties of the resulting bonds partially depends upon the shapes of the two overlapping atomic orbitals. Explain the fact that the overlap of two s atomic orbitals yields only one type of bond; yet, the overlap of two p atomic orbitals can yield two different types of bonds.

Penetration and Shielding

7.53 Atomic sodium emits light at 389 nm when an excited electron moves from a $4s$ orbital to a $3s$ orbital (this emission is, in fact, very weak), and at 330 nm when an electron moves from a $4p$ orbital to the same $3s$ orbital. What is the energy separation (in joules and kilojoules per mole) between the $4s$ and $4p$ orbitals?

7.54 Atomic potassium emits light at 365 nm when a $4d$ electron moves to a $4s$ orbital (this emission is very weak), and at 694 nm when a $4d$ electron moves to a $4p$ orbital (this emission is much more intense). What is the energy separation (in joules and kilojoules per mole) between the $4s$ and $4p$ orbitals? Why is the separation larger than for the same two orbitals in sodium (Exercise 7.53)?

The Building-Up Principle

In Exercises 7.55 and 7.56, draw a box diagram to represent each configuration, and indicate (with arrows, $\uparrow$ and $\downarrow$) the spin states of the electrons.

7.55 (a) $1s^2$; (b) $1s^1 2s^1$; (c) $2p^4$; (d) $1s^2 2s^2 2p^5$.

7.56 (a) $1s^1 3s^2$; (b) $4d^5$; (c) $3d^8$; (d) $2p_x^1 2p_y^2$.

In Exercises 7.57 to 7.60, predict the lowest-energy configuration of each atom.

7.57 (a) Li; (b) C; (c) N; (d) Ar.

7.58 (a) Cs; (b) B; (c) F; (d) Kr.

7.59 (a) Fe; (b) Bi; (c) Sn; (d) I.

7.60 (a) Sc; (b) Tl; (c) Pb; (d) Lu.

In Exercises 7.61 and 7.62, give the lowest-energy configuration of each ion.

7.61 (a) Na^+; (b) Cl^-; (c) Tl^+; (d) Pb^{2+}; (e) Pt^{2+}; (f) Sm^{3+}.
7.62 (a) Ca^{2+}; (b) Fe^{2+}; (c) Fe^{3+}; (d) O^{2-}; (e) Ho^{3+}; (f) Au^{3+}.

In Exercises 7.63 and 7.64, state which ions are likely to be formed.

7.63 (a) Ba^{2+}; (b) Al^{4+}; (c) Tl^+; (d) Tl^{3+}.
7.64 (a) Hg^{3+}; (b) Bi^{3+}; (c) At^{2-}; (c) Se^+.

7.65 Write the ground-state configurations of the elements of the second period of the periodic table. What would you predict as the lowest-energy configuration of the element in Group IV, Period 5?
7.66 Write the ground-state configurations of the elements of the third period of the periodic table. What would you predict as the ground-state configuration of the element in Group VI, Period 4?

7.67 Suggest ground-state electron configurations for the atoms of chromium and europium, and justify your results.
7.68 Suggest likely ground-state configurations for the new elements with atomic numbers 107 and 109.

The Periodic Table

In Exercises 7.69 and 7.70, identify the block of the periodic table to which each element belongs.

7.69 (a) Zirconium; (b) iron; (c) sulfur; (d) terbium.
7.70 (a) Cesium; (b) tin; (c) californium; (d) plutonium.

In Exercises 7.71 and 7.72, classify each element as a main-group element (MG) or transition metal (TM).

7.71 (a) Sulfur; (b) thallium; (c) iron; (d) platinum.
7.72 (a) Bromine; (b) silver; (c) gallium; (d) lead.

In Exercises 7.73 and 7.74, identify the group and block to which the element with the stated configuration belongs ([X] denotes a noble-gas configuration).

7.73 (a) $[Kr]4d^{10}5s^25p^1$; (b) $[Kr]4d^{10}5s^25p^5$;
(c) $[X]ns^2np^1$; (d) $[X]4d^55s^1$.
7.74 (a) $[Xe]6s^2$; (b) $[Xe]4f^{14}5d^{10}6s^26p^2$;
(c) $[X](n-1)d^2ns^2$; (d) $[X]4f^{14}5d^96s^1$.

In Exercises 7.75 and 7.76, use the location of each element in the periodic table to state the total number of electrons in the valence s and p subshells of its ground-state atoms.

7.75 (a) Rubidium; (b) radon; (c) oxygen; (d) phosphorus.
7.76 (a) Tellurium; (b) lead; (c) iodine; (d) sulfur.

7.77 What ions are likely to be formed by the elements listed in Exercise 7.75?

7.78 What ions are likely to be formed by the elements listed in Exercise 7.76?

Atomic and Ionic Radii

In Exercises 7.79 and 7.80, decide which statements are true, and explain why.

7.79 (a) The Na^+ ion is smaller than the K^+ ion.
(b) The Na^+ ion is smaller than the Na atom.
(c) F^- is smaller than the F atom.
(d) A sulfide ion is smaller than an oxide ion.
7.80 (a) Ions of the alkali metals are smaller than the ions of the alkaline-earth metals of the same period.
(b) Halogen atoms have smaller radii than the noble-gas atoms of the same period.
(c) A lithium cation is considerably larger than a fluoride anion.
(d) A nitride ion should be expected to be larger than an oxide ion.

7.81 The radius of the ammonium ion is 137 pm. If it were actually the ion of the "element" NH_4, where would NH_4 be located in the periodic table?
7.82 Phosphorus forms the phosphonium ion, PH_4^+, with radius 157 pm. Where would you expect the "element phosphonium" to lie in the periodic table?

Ionization Energy

In Exercises 7.83 and 7.84, determine which statements are true, and explain why.

7.83 (a) Ionization energies increase down a group.
(b) Ionization energies decrease from right to left across a period.
(c) All the halogens have higher first-ionization energies than all the noble gases.
(d) One reason why helium is unreactive is that its ionization energy is very high.
7.84 (a) The first ionization energy of cesium is greater than that of barium.
(b) The first ionization energy of He^+ is the same as the second ionization energy of the He atom.
(c) The second ionization energy of calcium is larger than that of magnesium.
(d) The first ionization energy of aluminum is smaller than that of magnesium.

7.85 Calculate the ionization energy of ground-state hydrogen from the Bohr formula for its energy levels. State the answer in joules and kilojoules per mole.
7.86 The energy levels of one-electron ions (such as He^+ and Li^{2+}) are given by the same expression as for hydrogen, but with $Z^2\mathcal{R}$ used in place of $\mathcal{R}$. Calculate the second ionization energy of helium, expressing the answer in joules and kilojoules per mole.

7.87 The first ionization energy of potassium is 418 kJ/mol. (a) Express that value as the energy needed to ionize one atom. (b) Calculate the maximum wavelength of light that could eject an electron from the K atom.

7.88 The first ionization energy of argon is 1520 kJ/mol. (a) Express that value as the energy needed to ionize one atom. (b) Calculate the maximum wavelength of light that could eject an electron from the Ar atom. (c) Why is there such a difference between the values of I_1 for argon and potassium (Exercise 7.87), even though their atomic numbers differ only by 1?

7.89 Light of wavelength 590 nm can be used to excite the $3s$ electron of sodium into a $3p$ orbital. Given that the ionization energy of sodium from its ground state is 494 kJ/mol, what would be the ionization energy of the atom from the excited configuration $[Ne]3p^1$?

7.90 Light of wavelength 770 nm can be used to excite the $4s$ electron of potassium into a $4p$ orbital, and 1240-nm light can then excite the $4p$ electron into a $5s$ orbital. The ionization energy of the ground-state atom is 418 kJ/mol. What is the ionization energy of the atom from the excited configuration $[Ar]5s^1$?

Electron Affinity

In Exercises 7.91 and 7.92, determine which statements are true, and explain why.

7.91 (a) More energy is released when an electron enters the $3p$ shell of chlorine than when it enters the $4p$ shell of bromine.

(b) The electron affinity of an ion is larger than that of the parent atom.

7.92 (a) The electron-gain enthalpy of an atom is equal to the ionization energy of the resulting ion.

(b) $F^-(g)$ has a lower energy than widely separated, stationary $F(g) + e^-(g)$.

7.93 The maximum wavelength of light that can eject an electron from a gaseous Li^- ion and lead to the formation of a neutral lithium atom is 2000 nm. Calculate the electron-gain enthalpy of lithium.

7.94 The maximum wavelength of light that can eject an electron from a gaseous C^- ion and lead to the formation of a neutral carbon atom is 980 nm. Calculate the electron-gain enthalpy of carbon.

Electronegativity

In Exercises 7.95 and 7.96, arrange the elements in order of increasing electronegativity.

7.95 B, Al, C, Si.
7.96 O, F, Cl, N.

In Exercises 7.97 and 7.98, arrange the elements in order of increasing electropositive character.

7.97 Na, K, Mg, Ca.
7.98 Al, Mg, Li, Be.

In Exercises 7.99 and 7.100, determine which one of each pair of elements will tend to acquire electrons from the other.

7.99 (a) H and C; (b) N and O; (c) O and F; (d) Cl and F.
7.100 (a) Al and Si; (b) H and S; (c) H and P; (d) Cl and O.

Trends in Chemical Properties

7.101 List three *physical* properties that are typical of metals.
7.102 State three *chemical* properties that are typical of nonmetals.

In Exercises 7.103 and 7.104, suggest whether each element is likely to be a metal (M) or nonmetal (NM).

7.103 (a) Beryllium; (b) cobalt; (c) gallium; (d) antimony.
7.104 (a) Francium; (b) einsteinium; (c) polonium; (d) element 109.

7.105 An element E has a steel-gray lustrous appearance but conducts electricity poorly; its oxide dissolves in water, but dissolves more readily in dilute aqueous sodium hydroxide; and with hydrochloric acid it gives a chloride of formula ECl_3. What type of element is E? Identify E.

7.106 An element E has a lustrous, silver-gray appearance but is brittle and is a poor conductor of electricity; the dioxide EO_2 dissolves in water, but more so when the water is either acidified or made alkaline. What type of element is E? Can it be identified from the information given?

7.107 Give three examples, with chemical equations, of the manner in which the inert-pair effect shows up in Group III.

7.108 Give three examples, with chemical equations, of the manner in which the inert-pair effect shows up in Group IV.

7.109 Illustrate, with chemical equations, some ways in which aluminum and beryllium resemble each other, both physically and chemically.
7.110 Illustrate, with chemical equations, some ways in which lithium and magnesium resemble each other, both physically and chemically.

7.111 In some forms of the periodic table, Cu, Ag, and Au are classified as belonging to a "Group IB," with the alkali metals in "Group IA." (a) What is the justification for this classification, in terms of electronic structure? (b) Do the chemical properties of the elements support it? (c) How are any differences between "Group IA" and "Group IB" to be explained?

7.112 In some forms of the periodic table, Zn, Cd, and Hg are classified as belonging to a "Group IIB," with the alkaline earths in "Group IIA." (a) What is the justification of this classification, in terms of electronic structure? (b) Do the chemical properties of the elements support it? (c) How are any differences between "Group IIA" and "Group IIB" to be explained? (d) Why are "A" metals more reactive than "B" metals?

General

*7.113 The electron in a Li^{2+} ion at the surface of a star fell from an orbital with $n = 3$ to an orbital with $n = 2$. (a) Can the resulting radiation from the Li^{2+} ionize a ground-state He^+ ion, given that the energy of a one-electron ion of atomic number Z is $-hZ^2\mathcal{R}/n^2$? (b) If not, what orbital must the He^+ electron be in initially if the He^+ is to be ionizable? (c) Show that *no* radiation from a He^+ ion can ionize a Li^{2+} ion in its ground state. (d) What is the principal quantum number of the lowest-energy orbital from which ionization can occur when Li^{2+} is exposed to any He^+ radiation? (e) Find a general expression for the principal quantum number n of the first orbital ionizable by He^+ $n = 2 \rightarrow n = 1$ radiation, in terms of the atomic number Z of a one-electron ion.

*7.114 Suppose that an electron could take up any of *three* spin orientations, and that *three* electrons could occupy a single orbital. (a) Write the electron configurations of the first ten elements in what would then be the periodic table. (b) Which element would begin Period 3, and which elements of Period 2 would it resemble? (c) Which elements would be the noble gases? (d) Which would be the first element of the *d* block?

*7.115 The atomic weight of iodine (126.9 amu) is less than that of tellurium (127.6 amu), yet iodine is located to the right of tellurium in the periodic table. (a) Account for this discrepancy. (b) Are there other, similar discrepancies in the table?

*7.116 (a) Suggest a value for the atomic number of the first element of the *g* block of the periodic table, on the assumption that the 5*g* orbitals are occupied after the 6*d* orbitals. (b) How many elements should there be in each series of *g*-block elements?

*7.117 The German physicist Lothar Mayer observed a periodicity in the physical properties of the elements at about the same time as Mendeleev was working on their chemical properties. Some of his observations can be reproduced by plotting the *molar volume*, the volume occupied per mole of atoms, for the solid form of the elements against atomic number. Do this for Periods 2 and 3, given the following densities for the elements in their solid forms:

Element	Li	Be	B	C	N	O
Density, g/cm^3	0.53	1.85	2.47	2.27	1.04	1.46

Element	F	Ne	Na	Mg	Al	Si
Density, g/cm^3	1.14	1.44	0.97	1.74	2.70	2.33

Element	P	S	Cl	Ar
Density, g/cm^3	1.82	2.09	2.03	1.66

Suggest a reason for the sharp change in molar volume between the *s*-block and *p*-block elements.

*7.118 Ionization energies are sometimes reported in electronvolts (eV), where 1 eV is the change in the energy of an electron produced by a potential difference of 1 V. Given that $1\,J = 1\,C \cdot V$, where C stands for the coulomb, the unit of charge, that the electron charge is 1.61×10^{-19} C, and that the energy change is equal to the charge multiplied by the potential difference, show that the conversion factor from kilojoules per mole to electronvolts is 1 eV = 96.49 kJ/mol. Express the ionization energies of the alkali metals in electronvolts.

*7.119 The radiation emitted and absorbed by atoms is often described in terms of its "wave number" $\tilde{\nu}$, the reciprocal of its wavelength ($\tilde{\nu} = 1/\lambda$). (a) Show that the relation between frequency ν and wave number is $\nu = \tilde{\nu}c$. (b) What is the minimum wave number of the radiation that can ionize a hydrogen atom in its ground state? (c) What is the conversion factor between wave numbers in inverse centimeters (cm^{-1}) and energy differences in kilojoules per mole? (d) Express the ionization energies of the alkali metals as a wavenumber in cm^{-1}.

*7.120 In the spectroscopic technique known as "photoelectron spectroscopy" (PES), ultraviolet light is directed at an atom or molecule. Electrons are ejected from the valence shell, and their kinetic energies are measured. Since the energy of the incoming ultraviolet photon is known and the kinetic energy of the outgoing electron is measured, the ionization energy can be deduced from the fact that the total energy is conserved. (a) Show that the speed (v) of the electron and frequency (ν) of the radiation are related by

$$h\nu = I + \tfrac{1}{2}m_e v^2$$

(b) Use this relation to calculate the ionization energy of a rubidium atom, given that light of wavelength 58.4 nm produces electrons with a speed of 2450 km/s.

This chapter explains, in terms of the electronic structures of atoms, how atoms bond together to form compounds. It shows how to predict the types of compounds an element can form, the number of bonds it can make with other atoms, and how the valence electrons are reorganized when a bond forms. As well as symbolizing bonding, the illustration shows a consequence of bonding: an adhesive forms chemical bonds with the two surfaces it joins.

Ionic bonds

8.1 The formation of ionic bonds

8.2 Variable valence

Covalent bonds

8.3 The electron-pair bond

8.4 Lewis structures of polyatomic molecules

8.5 Bond parameters

THE CHEMICAL BOND

Charge distributions in compounds

8.6 Assessing the charge distribution

8.7 Polarization

The concepts discussed in this chapter account for the formation and strength of chemical bonds in terms of the ideas about atomic structure developed in Chapter 7. In particular, we examine the role of the *valence electrons*, the outermost electrons of an atom, in bonding. The importance of these electrons to the strengths of bonds is the source of the name *valence*, which comes from the Latin word for "strength." "Vale!", "Be strong!", was what the Romans said on parting.

In the last chapter we dealt with the structures of atoms and ions—the simplest objects we need consider in chemistry. We are about to see how atoms group together into more complex structures, forming ionic and molecular compounds. By the end of this part of the book—in Chapter 11—our exploration of structures of increasing complexity will have brought us to bulk matter: to liquids, solids, and mixtures of the two.

IONIC BONDS

Ionic compounds are electrically neutral assemblies of cations and anions held together by the attraction between ions of opposite charge. This attraction is called an "ionic bond":

> An **ionic bond** is the attraction arising from the opposite charges of cations and anions.

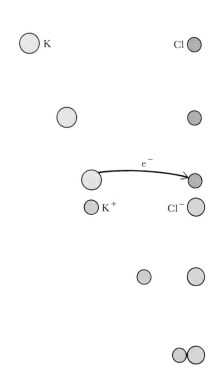

Since there are always other factors that contribute to the attraction between two ions, no bond is purely ionic. However, the ionic model is a good description of bonding in many compounds, particularly compounds containing elements from the *s* block, such as sodium chloride and magnesium oxide.

8.1 THE FORMATION OF IONIC BONDS

In this section we see how to use information about energy changes to judge whether or not an ionic bond is likely to form. We begin by considering a very simple case in which a single potassium atom and a single chlorine atom come together to form an *ion pair*, one cation next to one anion. This allows us to identify several contributions to the total energy change, even though the formation of a bond between one K atom and one Cl atom is unimportant in practice. We shall see that energy is released when a very electropositive atom (like K) transfers an electron to a very electronegative atom (like Cl) nearby. This energy release also occurs when a solid sample containing enormous numbers of ions is formed, so the ideas we develop concerning the formation of a single ion pair also pertain to real situations.

The formation of an ion pair. As a K atom and a Cl atom approach each other, at some point the K atom loses an electron, becoming a K^+ cation, and the Cl atom gains one, becoming a Cl^- anion (Fig. 8.1). The overall energy change accompanying this transfer is the sum of three contributions. One is the energy needed to form the K^+ cation; this is the ionization energy of potassium. The second is the energy released

FIGURE 8.1 As a potassium atom and a chlorine atom approach, there comes a point at which it is energetically favorable for an electron to move from the K atom to the Cl atom.

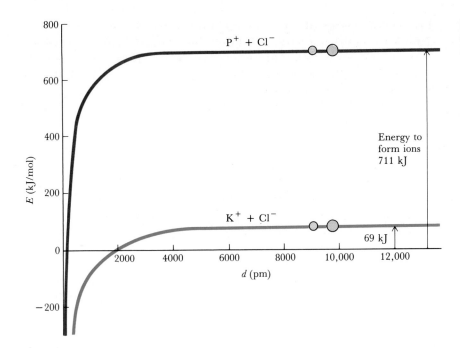

FIGURE 8.2 The energy of two ions relative to that of the parent atoms, which are taken as possessing zero energy. When the energy falls below zero, it is energetically advantageous for the ion pair to form. The P^+Cl^- case is discussed later.

when the Cl^- anion forms; this is the electron-gain energy of chlorine. The sum of the two contributions is

$$\begin{array}{lr} & \Delta H \\ K(g) \longrightarrow K^+(g) + e^- & +418 \text{ kJ} \\ \underline{Cl(g) + e^- \longrightarrow Cl^-(g)} & \underline{-349 \text{ kJ}} \\ K(g) + Cl(g) \longrightarrow K^+(g) + Cl^-(g) & +69 \text{ kJ} \end{array}$$

We see that 69 kJ of energy is *needed* to form 1 mol of each of the ions. So far, there seems to be no reason why the ions should form.

The third contribution, the potential energy arising from the attraction between the ions, is critical to bond formation. This potential energy depends on the separation d of the ions (Fig. 8.2), and with d in picometers is given by

$$E = \frac{z_A z_B}{d} \times (1.39 \times 10^5 \text{ kJ}) \qquad (1)$$

for 1 mol of the ion pairs; z_A and z_B are the charge numbers of the two ions (+1 and −1 for K^+ and Cl^- in our example).* The potential energy of K^+ and Cl^- is zero when the two ions are very far apart and cannot interact. It falls below −69 kJ when a pair of ions is separated by 2000 pm, about six ionic diameters. For instance, at $d = 1900$ pm

$$E = \frac{(+1) \times (-1)}{1900} \times (1.39 \times 10^5 \text{ kJ}) = -73 \text{ kJ}$$

---

*The potential energy of a charge q_1 at a distance d from another charge q_2 is given by $V = q_1 q_2 / 4\pi\epsilon_0 d$, where ϵ_0 is a fundamental constant with the value $8.85 \times 10^{-12} \text{ C}^2/\text{m} \cdot \text{J}$. The charge of a cation M^+ is 1.60×10^{-19} C, and that of an anion X^- is the negative of this value (the charge of an electron). Substitution of these values and multiplication by Avogadro's number produce the expression given.

The overall energy of a K^+ ion and a Cl^- ion is therefore lower than that of a K atom and a Cl atom when they are less than six diameters apart, and the formation of an ion pair K^+Cl^- is then energetically favored. When the two ions in each pair are in contact (at $d = 314$ pm), their potential energy is -443 kJ. The overall energy change in the formation of 1 mol of ion pairs from the atoms is then

$$\Delta E = \underset{\substack{\text{energy needed}\\\text{to form ions}\\\text{from atoms}}}{+69 \text{ kJ}} - \underset{\substack{\text{potential energy}\\\text{of attraction}\\\text{between ions}}}{443 \text{ kJ}}$$

$$= -374 \text{ kJ/mol}$$

This is a substantial decrease in energy, which suggests that ion-pair formation is favored.

The formation of an ionic solid. Now that we know when the formation of a certain ion pair is favored, our next step is to determine whether the energy changes favor the formation of a bulk ionic *solid* from the elements in their normal states. Is the energy of a sample of solid KCl lower than that of solid potassium and gaseous chlorine? We can answer this by imagining that the formation of KCl in the reaction

$$K(s) + \tfrac{1}{2}Cl_2(g) \longrightarrow KCl(s)$$

takes place in several steps. We find the enthalpy change for each step in Appendix 2A, and then add the individual changes to find the overall enthalpy change.

The steps are shown in Fig. 8.3. The diagram is called a *Born-Haber cycle* in recognition of the work of Max Born and Fritz Haber, who introduced it to analyze the enthalpy changes accompanying the formation of ionic solids. The first step is the formation of a gas of 1 mol of K atoms from solid potassium:

$$K(s) \longrightarrow K(g) \qquad \Delta H = +89 \text{ kJ}$$

The second is the "dissociation"—bond-breaking—of the gaseous chlorine molecules:

$$\tfrac{1}{2}Cl_2(g) \longrightarrow Cl(g) \qquad \Delta H = +122 \text{ kJ}$$

At this stage in the hypothetical process we have a gas of K and Cl atoms. Now we allow electron transfer from K atoms to Cl atoms; this forms a gas of widely separated ions. As we saw above,

$$K(g) + Cl(g) \longrightarrow K^+(g) + Cl^-(g) \qquad \Delta H = +69 \text{ kJ}$$

So far, all the enthalpy changes are endothermic, and their sum, $+280$ kJ, is the change of enthalpy needed to form a gas of ions from the starting materials.

Now the ions come together to form a solid. Just as in the formation of the single K^+Cl^- ion pair, this change is accompanied by a very large decrease in energy. The decrease is the net outcome of the attraction of every cation for every anion in the solid, less the repulsions that each cation has for the other cations and each anion for the other anions (Fig. 8.4). This sum is calculated by adding together a large number of

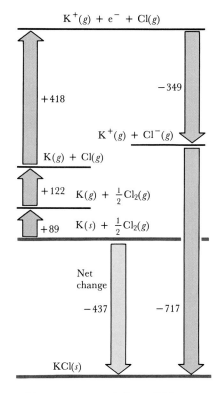

FIGURE 8.3 A Born-Haber cycle for the formation of KCl from potassium and chlorine. The sum of the enthalpy changes along the long (green) route is equal to the enthalpy change for the direct (pink) route.

terms like Eq. 1, each with the appropriate ion separation d in the denominator. The result is

$$K^+(g) + Cl^-(g) \longrightarrow KCl(s) \qquad \Delta H = -717 \text{ kJ}$$

The standard enthalpy change for the reverse of this process, the formation of the gas of K^+ and Cl^- ions from the solid KCl, is called the "lattice enthalpy" ΔH_L° of potassium chloride:

The **lattice enthalpy** of an ionic solid is the standard enthalpy change for the conversion of the solid to a gas of ions:

$$MX(s) \longrightarrow M^+(g) + X^-(g) \qquad \Delta H_L^\circ$$

All lattice enthalpies are endothermic because heat must be supplied to break up the solid and separate the ions. The lattice enthalpies for a number of solids are given in Table 8.1.

The overall enthalpy change for the formation of KCl(s) is the sum of two terms. One is the enthalpy change for the formation of the gas of ions from the starting materials. The second is the enthalpy change for the formation of solid KCl from the gas of ions:

$$
\begin{array}{lr}
 & \Delta H \\
K(s) + \tfrac{1}{2}Cl_2(g) \longrightarrow K^+(g) + Cl^-(g) & +280 \text{ kJ} \\
K^+(g) + Cl^-(g) \longrightarrow KCl(s) & -717 \text{ kJ} \\
\hline
K(s) + \tfrac{1}{2}Cl_2(g) \longrightarrow KCl(s) & -437 \text{ kJ}
\end{array}
$$

We see that there is a large reduction of enthalpy, $\Delta H = -437$ kJ, when 1 mol of solid KCl is formed from its elements.

The formation of KCl is exothermic overall because the transfer of an electron from $K(g)$ to $Cl(g)$ is not very endothermic ($\Delta H = +69$ kJ), and the energy invested in the electron transfer is more than recovered

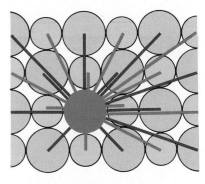

FIGURE 8.4 In an ionic solid, each ion interacts with all the others. Here we show some of the attractions (green lines) and repulsions (red lines) experienced by one anion. The net effect is attraction, partly because the anion is closest to ions of opposite charge.

TABLE 8.1   Lattice enthalpies at 25°C in kilojoules per mole

Halides

LiF 1046	LiCl 861	LiBr 818	LiI 759
NaF 929	NaCl 787	NaBr 751	NaI 700
KF 826	KCl 717	KBr 689	KI 645
AgF 971	AgCl 916	AgBr 903	AgI 887
BeCl$_2$ 3017	MgCl$_2$ 2524	CaCl$_2$ 2255	SrCl$_2$ 2153

Oxides

MgO 3850	CaO 3461	SrO 3283	BaO 3114

Sulfides

MgS 3406	CaS 3119	SrS 2974	BaS 2832

when the solid is formed. Now, though, let us consider the formation of the hypothetical solid KCl_2, in which the cation is K^{2+}. The energy required for the formation of K^{2+} is the sum of the first and second ionization energies:

$$\begin{array}{lr} & \Delta H \\ K(g) \longrightarrow K^+(g) + e^- & +418 \text{ kJ} \\ K^+(g) \longrightarrow K^{2+}(g) + e^- & +3070 \text{ kJ} \\ \hline K(g) \longrightarrow K^{2+}(g) + 2e^- & +3488 \text{ kJ} \end{array}$$

If we suppose that, because the cations have the same charges and about the same size, the lattice enthalpy of KCl_2 is about the same as that of $CaCl_2$ (2255 kJ), the Born-Haber cycle for the formation of KCl_2 (Fig. 8.5) results in an endothermic enthalpy change overall:

$$K(s) + Cl_2(g) \longrightarrow KCl_2(s) \qquad \Delta H = +868 \text{ kJ}$$

Hence, it is unlikely that KCl_2 will form.

▼ **EXAMPLE 8.1 Using a Born-Haber cycle to account for compound formation**

Use a Born-Haber cycle to show that $CaCl_2$ is a more likely compound of calcium and chlorine than CaCl.

STRATEGY One approach is to explore whether the disproportionation

$$2CaCl(s) \longrightarrow CaCl_2(s) + Ca(s)$$

is strongly exothermic, for if it is strongly exothermic, the reaction might be favorable. This involves calculating the enthalpy of formation of $CaCl_2$ and 2CaCl and taking their difference. In carrying out calculations on ionic solids, it is often a good idea to begin by drawing a Born-Haber cycle, to decide what information is needed. In this case we must draw two cycles, one leading from the elements to CaCl, and the other to $CaCl_2$. For the lattice enthalpy of the hypothetical solid CaCl we shall use the value for KCl. Enthalpy data are given in Appendix 2A.

SOLUTION The two Born-Haber cycles are shown in Fig. 8.6. The enthalpy of formation of CaCl(s) is given by the following sum:

$$\begin{array}{lr} & \Delta H \\ Ca(s) \longrightarrow Ca(g) & +178 \text{ kJ} \\ Ca(g) \longrightarrow Ca^+(g) + e^- & +590 \text{ kJ} \\ \tfrac{1}{2}Cl_2(g) \longrightarrow Cl(g) & +122 \text{ kJ} \\ Cl(g) + e^- \longrightarrow Cl^-(g) & -349 \text{ kJ} \\ Ca^+(g) + Cl^-(g) \longrightarrow CaCl(s) & -717 \text{ kJ} \\ \hline Ca(s) + \tfrac{1}{2}Cl_2(g) \longrightarrow CaCl(s) & -176 \text{ kJ} \end{array}$$

At this stage, it is plausible that CaCl could form since it releases energy. The enthalpy of formation of $CaCl_2(s)$ is given by the following sum:

$$\begin{array}{lr} & \Delta H \\ Ca(s) \longrightarrow Ca(g) & +178 \text{ kJ} \\ Ca(g) \longrightarrow Ca^+(g) + e^- & +590 \text{ kJ} \\ Ca^+(g) \longrightarrow Ca^{2+}(g) + e^- & +1137 \text{ kJ} \\ Cl_2(g) \longrightarrow 2Cl(g) & +244 \text{ kJ} \\ 2Cl(g) + 2e^- \longrightarrow 2Cl^-(g) & -698 \text{ kJ} \\ Ca^{2+}(g) + 2Cl^-(g) \longrightarrow CaCl_2(s) & -2255 \text{ kJ} \\ \hline Ca(s) + Cl_2(g) \longrightarrow CaCl_2(s) & -804 \text{ kJ} \end{array}$$

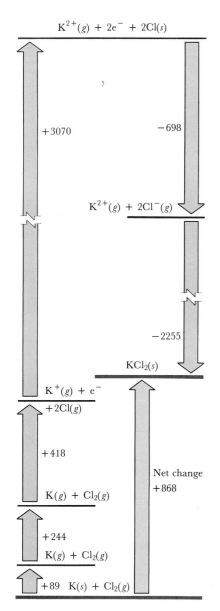

FIGURE 8.5 The Born-Haber cycle for the hypothetical compound KCl_2. Note that its formation is strongly endothermic.

Even though a considerable energy investment is required to ionize the calcium to Ca^{2+}, the lattice enthalpy is so much greater (on account of the strong attraction of the more highly charged Ca^{2+} ion) that overall the formation of $CaCl_2$ is much more exothermic than that of CaCl.

The enthalpy of the disproportionation is the difference of the two enthalpies of formation:

$$\Delta H = 1 \text{ mol } CaCl_2 \times \Delta H_f^\circ(CaCl_2) - 2 \text{ mol } CaCl \times \Delta H_f^\circ(CaCl)$$

$$= -804 \text{ kJ} - 2 \times (-176 \text{ kJ}) = -452 \text{ kJ}$$

Hence, the disproportionation is strongly exothermic, and $CaCl_2$ is likely to be the compound formed by calcium and chlorine.

EXERCISE Judge whether sodium and fluorine are likely to form NaF or NaF_2.

[*Answer*: NaF]

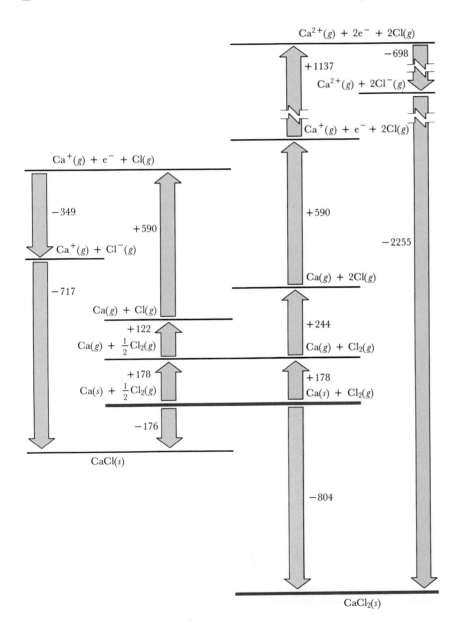

FIGURE 8.6 The Born-Haber cycles for the formation of CaCl and $CaCl_2$.

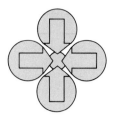

FIGURE 8.7 Small, highly charged ions attract ions of opposite charge very strongly.

Factors favoring ionic bonds. Ionic bonds can be expected to form only if the energy released in one hypothetical step, the formation of the solid ionic compound from the gas of "reactant" ions, exceeds the energy invested in the other hypothetical step, the formation of the reactant ions. The enthalpy of ion formation is quite small as long as the ionization energy of the cation-forming element is not very high and the electron affinity of the anion-forming element is not too low. These conditions effectively confine ionic bonds to compounds formed between strongly electropositive and strongly electronegative elements. The energy released in the hypothetical step in which the ionic solid forms from the gas of ions is the negative of the lattice enthalpy. Small, highly charged ions, which have strong interactions (Fig. 8.7), lead to high lattice enthalpies. The loss of all valence electrons from an *s*-block atom is therefore favored, for in that way the atom reaches the highest charge possible (such as Na^+ or Ca^{2+}) without breaking into the core. The completion of the valence shells of atoms to the right of the *p* block is also favored, for then they also reach the highest charge and hence the highest lattice enthalpy possible.

The formation of highly charged cations by breaking into a closed shell (to form K^{2+}, for instance) is not favored, because the large ionization energy needed to remove an electron from the core is not recovered from the resulting higher lattice enthalpy. The formation of highly charged anions by attaching electrons to closed-shell anions is also not favored, because even though a higher lattice enthalpy would result, that increased energy would not compensate for the extra energy needed to form the anion.

It is found, as we noted in Section 2.7, that ionic compounds are formed between elements toward the left of the periodic table (which provide the cations), and elements on the right of the table (which provide the anions). As shown in Fig. 8.8, *s*-block elements tend to form cations with charge number equal to their group number (for example,

I	II	III	IV	V	VI	VII	VIII
			H^+				
Li^+	Be^{2+}			N^{3-}	O^{2-}	F^-	
Na^+	Mg^{2+}	Al^{3+}		P^{3-}	S^{2-}	Cl^-	
K^+	Ca^{2+}	Ga^{3+}			Se^{2-}	Br^-	
Rb^+	Sr^{2+}	$In^+,$ In^{3+}	$Sn^{2+},$ Sn^{4+}	Sb^{3+}	Te^{2-}	I^-	
Cs^+	Ba^{2+}	$Tl^+,$ Tl^{3+}	$Pb^{2+},$ Pb^{4+}	Bi^{3+}		At^-	
Fr^+	Ra^{2+}						

FIGURE 8.8 The typical ions of the main-group elements.

Na$^+$ and Mg^{2+}). Elements to the lower left of the p block form cations with charge number equal to the group number or to the group number minus 2 (In^{3+} and In$^+$, Sn^{4+} and Sn^{2+}). Elements to the upper right of the p block form anions by acquiring enough electrons to complete the valence shell; they form anions with charge number equal to the group number minus 8 (as in O^{2-} and Cl$^-$).

▼ **EXAMPLE 8.2** **Predicting the compound formed by two elements**

Predict the formula of the compound likely to be formed by strontium and sulfur.

STRATEGY First we decide whether the compound is likely to be ionic by judging whether one element comes from the s block (or the lower left of the p block), and the other from toward the upper right of the p block. If they do, the s-block element forms a cation by losing all its valence electrons. The p-block element completes its valence shell. We then write a formula unit that is electrically neutral (as illustrated in Example 2.6).

SOLUTION Strontium (Group II) forms Sr^{2+} by loss of two electrons. Sulfur (Group VI) forms S^{2-} by gaining two electrons (not necessarily from the same Sr atom). The electrically neutral formula unit is SrS.

EXERCISE Predict the formula of the oxide formed by aluminum and oxygen.

[*Answer*: Al$_2$O$_3$]

▲

Pseudonoble-gas configurations. The elements near the left of the p block and the right of the d block include copper, zinc, and indium. Their electron configurations are

$$\text{Cu: [Ar]}3d^{10}4s^1 \quad \text{Zn: [Ar]}3d^{10}4s^2 \quad \text{In: [Kr]}4d^{10}5s^25p^1$$

Too much energy is required for these atoms to reach a noble-gas configuration by losing all their outer electrons. In fact, once the s and p electrons have been lost, the electrons in the complete d subshell have relatively high ionization energies, and the [Ar]$3d^{10}$ and [Kr]$4d^{10}$ configurations act like fragile closed shells. The configuration [noble gas]$3d^{10}$ is an example of a "pseudonoble-gas configuration":

A **pseudonoble-gas configuration** is a noble-gas core surrounded by a completed d subshell.

Elements that form cations with pseudonoble-gas configurations are shown in Fig. 8.9. Although the discarding of electrons to reach a pseudonoble-gas configuration accounts for the occurrence of some ions, that configuration is more easily broken up than a true noble-gas core.

Elements in the d block lose their s electrons first. However, most of the d-block atoms can also lose electrons from their d subshell and, as a result, can form a range of cations with different charges. This was mentioned in Section 7.8 and is discussed in more detail below.

Electron dot diagrams. A simple way of keeping track of the valence electrons when ionic bonds form is to write the "electron dot diagram" of each atom. Such a diagram consists of the chemical symbol and a dot

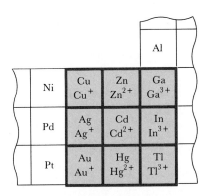

FIGURE 8.9 Elements that form ions with a pseudonoble-gas electron configuration.

for each valence electron. Some examples are

$$\text{H} \cdot \quad \text{He} : \quad : \overset{\cdot}{\underset{\cdot}{\text{N}}} \cdot \quad : \overset{\cdot}{\underset{\cdot\cdot}{\text{O}}} : \quad : \overset{\cdot\cdot}{\underset{\cdot\cdot}{\text{Cl}}} \cdot \quad \text{K} \cdot \quad \text{Mg} :$$

A single dot represents an electron alone in an orbital. The dots grouped in pairs represent electrons that occupy the same orbital. The dot diagram of nitrogen, for example, represents the electron configuration

N

$$[\text{He}] \; 2s^2 \quad 2p_x{}^1 2p_y{}^1 2p_z{}^1$$

Using electron dot diagrams, we can represent the formation of potassium chloride as the transfer of the single dot from a K atom into the single vacancy of a Cl atom:

$$\text{K} \cdot \; + \; : \overset{\cdot\cdot}{\underset{\cdot\cdot}{\text{Cl}}} \cdot \; \longrightarrow \; \text{K}^+ \; + \; : \overset{\cdot\cdot}{\underset{\cdot\cdot}{\text{Cl}}} :{}^-$$

The $s^2 p^6$ valence-electron configuration reached by each atom (the K atom has an [Ar] core) is called an *octet* of electrons. The dot diagram is a neat summary of the ability of some atoms to form ions by completing an octet, either by electron loss (for elements in the *s* block or lower left of the *p* block) or by electron gain (for elements on the right of the *p* block). We see later that dot diagrams also provide a powerful means for discussing the structures of molecular compounds.

▼ **EXAMPLE 8.3** **Interpreting the formulas of ionic compounds using electron dot diagrams**

Use the electron dot diagrams of aluminum and oxygen to interpret the formula for aluminum oxide (Al_2O_3) in terms of octet completion.

STRATEGY We know from charge-neutrality arguments that the oxide is $(Al^{3+})_2(O^{2-})_3$. We can write the electron dot diagrams of Al and O atoms based on their locations in the periodic table, by noting that each has the same number of valence electrons (and hence dots) as its group number. We must show that the valence electrons lost by two Al atoms fill the vacancies in the valence shells of three O atoms.

SOLUTION Al belongs to Group III and has three valence electrons ($: \overset{\cdot}{\text{Al}} \cdot$). Oxygen belongs to Group VI and has six valence electrons ($: \overset{\cdot}{\underset{\cdot}{\text{O}}} :$). The $2 \times 3 = 6$ electrons released by the two Al atoms can be accommodated in the $3 \times 2 = 6$ gaps of the three O atoms:

$$2 : \overset{\cdot}{\text{Al}} \cdot \; + \; 3 : \overset{\cdot}{\underset{\cdot}{\text{O}}} : \; \longrightarrow \; \underbrace{2\text{Al}^{3+} \; + \; 3 : \overset{\cdot\cdot}{\underset{\cdot\cdot}{\text{O}}} :{}^{2-}}_{\text{Al}_2\text{O}_3}$$

EXERCISE Justify the formulas of calcium nitride and sodium telluride using electron dot diagrams.

[*Answer*: $3\text{Ca}^{2+} + 2 : \overset{\cdot\cdot}{\underset{\cdot}{\text{N}}} :{}^{3-}, \; 2\text{Na}^+ + : \overset{\cdot\cdot}{\underset{\cdot\cdot}{\text{Te}}} :{}^{2-}$]

▲

8.2 VARIABLE VALENCE

We now turn to situations in which an element can form ions with different charge numbers. We saw some examples of these in Section 7.8, where tin was cited as forming Sn(II) and Sn(IV) compounds, and

(a)   (b)

FIGURE 8.10 (a) Fehling's solution (on the left) is an alkaline solution of Cu^{2+} ions and tartrate ions (from tartaric acid). It can be used as a test for mild reducing agents. The sample on the right is a solution of glucose. (b) When the glucose is added to Fehling's solution, it produces a red precipitate of copper(I) oxide, showing that it is a reducing agent.

indium as forming In(I) and In(III) compounds. Both *p*- and *d*-block elements show this *variable valence*, the ability to form ions with different charge numbers. We shall use the term oxidation number instead of charge number, so our discussion will apply to nonionic as well as ionic compounds.

Variable valence in the d block. Copper ($[Ar]3d^{10}4s^1$) is an example of a *d*-block element with variable valence. Not only does it form the $[Ar]3d^{10}$ Cu^+ ion by losing a $4s$ electron, but its relatively low second ionization energy (1960 kJ/mol) allows it to lose a second electron from the pseudonoble-gas core and form the $[Ar]3d^9$ Cu^{2+} ion as well. As a result, it can form both red copper(I) oxide (Cu_2O) and black copper(II) oxide (CuO):

$$2Cu^{2+}(aq) + 2OH^-(aq) + 2e^- \longrightarrow Cu_2O(s) + H_2O(l)$$

$$2Cu(NO_3)_2(s) \overset{\Delta}{\longrightarrow} 2CuO(s) + 4NO_2(g) + O_2(g)$$

The reduction of copper(II) ions can be carried out with SO_2 or some other mild reducing agent. In fact, "Fehling's solution" makes use of that reaction to test for mild reducing agents (Fig. 8.10); it is an alkaline solution of Cu^{2+} ions and tartrate ions ($C_4H_4O_6^{2-}$), the latter being included to prevent the precipitation of the copper(II) as the hydroxide in the basic solution.

The readiness with which copper loses a *d* electron to give copper(II) is shown by the ease with which copper(I) salts disproportionate in water. For example, when copper(I) oxide reacts with dilute sulfuric acid, it forms not copper(I) sulfate but copper(II) sulfate and copper metal (Fig. 8.11):

$$Cu_2O(s) + 2H^+(aq) \longrightarrow Cu(s) + Cu^{2+}(aq) + H_2O(l)$$

This reaction is the disproportionation

$$2Cu^+ \longrightarrow Cu + Cu^{2+}$$

FIGURE 8.11 When sulfuric acid is poured onto copper(I) oxide, a blue solution of copper(II) sulfate is formed, and copper metal is deposited. This is an illustration of a disproportionation reaction (Section 3.8).

Copper(I) salts can be prepared in aqueous solution, but only if they are so insoluble that they precipitate immediately, or if the solution is so reducing that the formation of copper(II) is suppressed or reversed. For example, adding iodide ions to copper(II) ions in solution leads to the formation of a white precipitate of insoluble copper(I) iodide:

$$2Cu^{2+}(aq) + 4I^-(aq) \longrightarrow 2CuI(s) + I_2(aq)$$

Similarly, copper(I) chloride can be prepared by reduction of copper(II) ions with sulfur dioxide:

$$2Cu^{2+}(aq) + 2Cl^-(aq) + SO_2(g) + 2H_2O(l) \longrightarrow$$
$$2CuCl(s) + SO_4^{2-}(aq) + 4H^+(aq)$$

Like the iodide, copper(I) chloride is kept from disproportionating by its insolubility.

The subtlety of the energy balances that are involved in breaking into the pseudonoble-gas core is shown by copper's congeners Ag and Au and its neighbor Zn, which give up d electrons to strikingly different extents even though they are close to each other in the periodic table. Zinc ($[Ar]3d^{10}4s^2$) forms no compounds in which the d subshell is used. All its compounds are colorless (incomplete d shells often lead to color, as in Cu^{2+}) and built from Zn^{2+} ions. Silver is much more stable as silver(I) than as silver(II), and its chemistry is dominated by silver(I) compounds. However, its d subshell can be broken into with a strong oxidizing agent, such as fluorine, which reacts with silver to form silver(II) fluoride. This brown solid is unstable and dissociates into silver(I) fluoride and fluorine when heated:

$$2AgF_2(s) \xrightarrow{\Delta} 2AgF(s) + F_2(g)$$

Gold, like copper, shows variable valence, but with oxidation numbers $+1$ and $+3$. In gold(III), two electrons have been lost from the d subshell. Gold also resembles copper in that the lower oxidation state readily disproportionates on contact with water:

$$3Au^+(aq) \longrightarrow 2Au(s) + Au^{3+}(aq)$$

The existence of iron(II) and iron(III) compounds is another sign that d electrons can be removed quite easily. However, in this case, they come from a subshell that is incomplete to begin with. The ground-state configuration of an iron atom is $[Ar]3d^64s^2$, and the loss of the two s electrons gives iron(II). This oxidation occurs, for example, when iron reacts with dilute hydrochloric or sulfuric acid:

$$Fe(s) + 2H^+(aq) \longrightarrow Fe^{2+}(aq) + H_2(g)$$

The pale green iron(II) sulfate heptahydrate, $FeSO_4 \cdot 7H_2O$, is a common laboratory chemical; it is manufactured by oxidizing the S_2^{2-} ion in the mineral pyrite ("fool's gold," FeS_2) with air and water:

$$2FeS_2(s) + 7O_2(g) + 2H_2O(l) \longrightarrow 2FeSO_4(aq) + 2H_2SO_4(aq)$$

A second electron can be removed from the d^6 subshell to produce iron(III). This oxidation occurs quite readily in aqueous solutions of iron(II) salts unless they are acidified. The Fe^{3+} ion has a pale violet

color in water. Amethyst, which has a similar color, is quartz (SiO_2) that is colored by some Fe^{3+} ions (Fig. 8.12). However, iron(III) solutions are usually yellowish brown because the violet is masked by the color of Fe^{3+} ions bonded to OH^- ions. Iron(III) is also formed when an oxidizing acid, such as hot concentrated sulfuric acid, reacts with iron:

$$2Fe(s) + 6H_2SO_4(aq, conc) \longrightarrow Fe_2(SO_4)_3(aq) + 3SO_2(g) + 6H_2O(l)$$

Variable valence arising from d electrons also shows itself in the existence of a very wide range of compounds that are built from *complexes*. These are clusters in which several (often six) ions or molecules are bonded to the same metal atom or ion to form a single large polyatomic molecule or ion. Examples of complexes are the ferrocyanide and ferricyanide ions, $[Fe(CN)_6]^{4-}$ and $[Fe(CN)_6]^{3-}$, respectively, in which a central Fe^{2+} or Fe^{3+} ion is bonded to six cyanide ions. Complex formation between Fe^{3+} and H_2O molecules, which occurs whenever a salt containing Fe^{3+} ions is dissolved in water, gives the "aquaion" $[Fe(H_2O)_6]^{3+}$. This complex is responsible for the purple color mentioned in the preceding paragraph; the replacement of one H_2O molecule by an OH^- ion, to give the complex anion $[Fe(OH)(H_2O)_5]^{2+}$, is responsible for the more common yellowish brown color of iron(III) solutions. The tartrate ions in Fehling's solution protect Cu^{2+} from precipitation by forming a complex with it.

Variable valence in the p block. Variable valence is also found among the elements close to the left edge of the p block (Fig. 8.13). As explained in Section 7.8, variable valence arises from the difference in energy between the s and p subshells and the extra energy needed to remove the two valence s electrons after the p electrons have been lost: in some reactions the electron pair is lost, in others it is not. We now discuss in more detail how this inert-pair effect helps to account for the chemical properties of the metallic elements in this part of the periodic table.

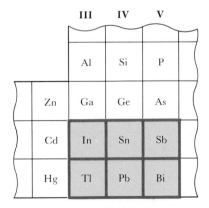

FIGURE 8.13 Variable valence in the p block. Atoms of the elements in red can lose all their valence electrons or all but an "inert pair" of valence s electrons.

FIGURE 8.14 The changes that occur when lead and its oxides are heated. Lead (gray) is oxidized to litharge, PbO (yellow). When that is heated to about 400°C, it is converted to Pb_3O_4 (red). Further heating converts it back to litharge.

The inert-pair effect becomes progressively more important on descending a group. In Group III, aluminum forms only aluminum(III), indium forms both indium(III) and indium(I), and the chemistry of thallium is dominated by thallium(I) compounds. In Group IV, oxidation number +2 becomes increasingly more important than +4 for elements lower in the group. Germanium(II) and tin(II) compounds have a strong tendency to lose the remaining two s electrons, and hence act as reducing agents. Conversely, lead(IV) tends to change to lead(II) by gaining two electrons, and hence is often a good oxidizing agent.

An illustration of the trend is that tin(II) oxide spontaneously ignites and smolders in air and forms the tin(IV) oxide:

$$2SnO(s) + O_2(g) \longrightarrow 2SnO_2(s)$$

On the other hand, the higher oxides of lead decompose to PbO when they are heated. We can see this tendency to favor Pb(II) over Pb(IV) in a number of ways. Heating lead in air (Fig. 8.14) results in litharge, a reddish yellow form of lead(II) oxide. Red lead, Pb_3O_4, which is used as a protective coating for iron and steel, is formed by heating lead(II) oxide to about 400°C. It is best regarded as $(Pb^{2+})_2(PbO_4)^{4-}$, in which two lead atoms are Pb(II) and the other is Pb(IV). However, the conversion of lead(II) to lead(IV) is short-lived, because at 470°C red lead changes to litharge again.

COVALENT BONDS

We introduced ionic bonds by showing that a considerable energy lowering results when a very electropositive atom (such as K) transfers an electron to a very electronegative atom (such as Cl) nearby. However, we shall now see that the same electron transfer does not lead to a lower energy when one of the elements is not electropositive. Since bonds between electronegative atoms *do* exist (as in PCl_3, SO_2, and ClO_3^-), this means that a second type of bond must account for the existence of molecules and polyatomic ions.

Suppose we carry through the ion-pair calculation for phosphorus and chlorine, a pair of elements that differ much less in electronegativity than K and Cl. Since the ionization energy of phosphorus is 1060 kJ/mol, we find

	ΔH
$P(g) \longrightarrow P^+(g) + e^-$	$+1060$ kJ
$Cl(g) + e^- \longrightarrow Cl^-(g)$	-349 kJ
$P(g) + Cl(g) \longrightarrow P^+(g) + Cl^-(g)$	$+711$ kJ

At least 711 kJ must be recovered from the energy supplied by the attraction between the P^+ and Cl^- ions before it would be energetically favorable for an ionic bond to form between them. That would be possible only if the centers of the ions could approach to within 200 pm of one another (see Fig. 8.2), but P^+ and Cl^- cannot come that close because 200 pm is less than the sum of their radii. Hence phosphorus and chlorine are unlikely to form an ionic bond. How, then, do we explain the existence of a molecule like PCl_3?

8.3   THE ELECTRON-PAIR BOND

An explanation of bonding in molecular compounds was proposed in 1916 by the American Gilbert Lewis (Fig. 8.15), one of the greatest of all chemists. With brilliant insight, and before anyone knew about quantum mechanics, Lewis identified the essential feature of the "covalent bond," the bond responsible for atoms forming molecules:

A **covalent bond** is a pair of electrons shared between two atoms.

In most cases, each atom contributes one electron to the pair it shares with its neighbor. However, in some cases one atom contributes both electrons. In either situation, the shared pair of electrons lies between the two neighboring atoms. A general rule for the occurrence of covalent bonds (and of molecular compounds) is:

Ionic bonding occurs whenever an *s*-block element is involved (with the exception of H and Be).
Covalent bonding occurs when both elements come from the *p* block.

In terms of electronegativity differences $\Delta\chi$:

Ionic bonding occurs when $\Delta\chi$ is larger than about 2.
Covalent bonding occurs when $\Delta\chi$ is smaller than about 1.

These are only very rough guides to the type of bonding that will occur, and when $\Delta\chi$ lies between about 1 and 2 the bonding is neither clearly ionic nor clearly covalent. We see later how to interpret this "intermediate" bonding.

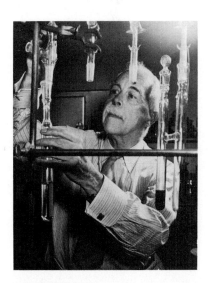

FIGURE 8.15 Gilbert Newton Lewis (1875–1946).

▼ **EXAMPLE 8.4** **Judging whether bonding is ionic or covalent**

Judge whether the bonding in phosphorus trichloride is likely to be ionic or covalent.

STRATEGY The distinction between ionic and covalent bonding depends on the electronegativity difference, as expressed above. Therefore, we should refer to Table 7.9 and calculate $\Delta\chi$.

SOLUTION Since $\chi(P) = 2.1$ and $\chi(Cl) = 3.0$, $\Delta\chi = 0.9$. This is less than 1, so we expect PCl_3 to be covalent.

EXERCISE Is the bonding in barium chloride likely to be ionic or covalent?

[*Answer*: Ionic]

▲

1

Lewis thought that, because the two electrons of a shared pair can attract the nuclei they lie between, they would pull the two atoms together (**1**). He had no way of knowing why it had to be a *pair* of electrons—two, and not some other number. The explanation came only with the development of quantum mechanics in the 1920s and the discovery of the Pauli exclusion principle.

The octet rule. We have already seen that when an ionic bond is formed, one atom loses electrons until it reaches a noble-gas octet and the other gains them until it has also reached one. Lewis suggested that atoms could also *share* electrons until they had reached a noble-gas configuration. He expressed this as the "octet rule":

> **Octet rule:** Atoms proceed as far as possible toward completing their octets by sharing electron pairs.

Nitrogen ($:\!\overset{\cdot}{\underset{\cdot}{N}}\!\cdot$) has five valence electrons and needs three more to complete its octet. Chlorine ($:\!\overset{\cdot\cdot}{\underset{\cdot\cdot}{Cl}}\!\cdot$) has seven valence electrons and needs to share one more to complete its octet. Argon ($:\!\overset{\cdot\cdot}{\underset{\cdot\cdot}{Ar}}\!:$) already has a complete octet and has no tendency to share any more electrons. Hydrogen ($H\cdot$) is an exception: it completes the heliumlike configuration $1s^2$, a "duplet" (rather than an octet) of electrons.

As long as no *d* orbitals are available for occupation, octet completion is the end of the line for any energy advantages that may come from sharing electron pairs. However, when *d* orbitals are available, more than eight electrons can be accommodated around an atom. Thus, although the octet rule works well for the Period 2 elements C, N, O, and F, it often fails for elements of later periods that have *d* orbitals available. The five bonds between P and Cl in PCl_5, for instance, require phosphorus to use its *d* orbitals to accommodate an additional pair of electrons beyond the octet. This molecule is said to have an *expanded octet*, a valence shell containing more than eight electrons. The octet rule also fails for boron, which often forms compounds in which it has only six valence electrons. Boron trifluoride, BF_3, is thus a molecule with an *incomplete octet*, one in which the valence shell has fewer than eight electrons.

At first sight it may appear that the Lewis octet rule has so many exceptions that it is useless. However, the existence of expanded and incomplete octets can be explained quite simply, and we can usually anticipate when they are likely to occur. We shall concentrate first on molecules that satisfy the octet rule, and then discuss these other cases.

The Lewis structures of diatomic molecules. The number of covalent bonds an atom can form depends on how many electrons it needs to complete its octet. This number is most easily found by using electron dot formulas.

Consider molecular hydrogen (H_2) as an example. The electron dot formula of each of the two hydrogen atoms is $H \cdot$, and each atom completes its heliumlike duplet by sharing its electron with the other:

$$H \cdot + \cdot H \text{ forms } H : H, \text{ denoted } H—H$$

H—H, in which the short horizontal line represents the shared electron pair, is the simplest example of a *Lewis structure*, a diagram showing how electron pairs are shared between atoms in a molecule.

A fluorine atom has seven valence electrons and needs one more to complete its octet. It can achieve this by sharing a pair, perhaps with another fluorine atom:

$$: \overset{..}{\underset{..}{F}} \cdot + \cdot \overset{..}{\underset{..}{F}} : \text{ forms } \left(: \overset{..}{F} : F :\right), \text{ denoted } : \overset{..}{\underset{..}{F}}—\overset{..}{\underset{..}{F}} :$$

The two circles have been drawn around the F atoms to show how each has completed an octet by sharing one electron pair. One difference between H_2 and F_2 is that the latter possesses "lone pairs" (or "non-bonding pairs") of electrons:

A **lone pair** of electrons is a pair of valence electrons that is not involved in bonding.

The lone pairs on neighboring F atoms repel each other and almost overcome the attractions that hold the F_2 molecule together. This is one of the reasons why fluorine is such a reactive gas: its atoms are bound together very weakly.

Multiple bonds. Two atoms can bond together by sharing more than one pair of electrons:

A **double bond** consists of two shared electron pairs.

A **triple bond** consists of three shared electron pairs.

We call the number of shared electron pairs the *bond order*. For instance, O_2 is formed when two oxygen atoms ($: \overset{.}{\underset{.}{O}} :$) share two pairs of electrons; the Lewis structure is obtained as follows:

$$: \overset{.}{\underset{.}{O}} : + : \overset{.}{\underset{.}{O}} : \text{ forms } \overset{.}{\underset{.}{O}} : \overset{.}{\underset{.}{O}} \text{ denoted } \overset{.}{\underset{.}{O}} = \overset{.}{\underset{.}{O}}$$

Each of the two atoms achieves a neonlike octet, and its bond order is 2. The Lewis structure of N_2 is obtained by drawing a diagram in which two atoms share three pairs of electrons, giving a bond order of 3.

$$: N \cdot + \cdot N : \text{ forms } : N : N : \text{ denoted } : N \equiv N :$$

The two atoms in N_2 are bound together strongly, because the three electron pairs lie between the nuclei, drawing them together tightly.

Writing Lewis structures. The Lewis structure of a diatomic molecule is obtained by writing the chemical symbols for the two atoms and deciding (from their locations in the periodic table) how many valence electrons each supplies. The sum of these numbers is the number of dots that must appear in the Lewis structure. In the case of carbon monoxide, for instance, the C atom has four valence electrons and the O atom has six, so 10 dots must appear in the Lewis structure. Moreover, the dots must be added to the structure in pairs so that each atom (other than hydrogen) ends up with an octet and all the dots are used. The same procedure is used to draw the Lewis structures of diatomic anions, but an additional electron is added for each negative charge.

If 16 electrons are available (as for two Ne atoms), each atom can have an octet without sharing, so no covalent bond is formed:

$$:\overset{..}{\underset{..}{X}}: \qquad :\overset{..}{\underset{..}{Y}}:$$

This is consistent with the monatomic character of the noble gases.

If 14 electrons are available (seven pairs of dots), each atom can have an octet if it has three lone pairs and one shared pair:

$$:\overset{..}{\underset{..}{X}}:\overset{..}{\underset{..}{Y}}: \qquad \text{or} \qquad :\overset{..}{\underset{..}{X}}{-}\overset{..}{\underset{..}{Y}}:$$

This is the case with the halogens, such as F_2 and I_2.

If 12 electrons are available, each atom can have two lone pairs and two bonding pairs:

$$\overset{..}{\underset{..}{X}}\colon\overset{..}{\underset{..}{Y}} \qquad \text{or} \qquad \overset{..}{\underset{.}{X}}{=}\overset{..}{\underset{.}{Y}}$$

This is the case with O_2. It is also the structure of the S_2 molecule, which occurs in the vapor of hot sulfur and is responsible for its violet hue.

If only 10 electrons are available, the atoms complete their octets by sharing three pairs and having one lone pair each:

$$:\overset{:}{\underset{:}{X}}\colon\overset{:}{\underset{:}{Y}}: \qquad \text{or} \qquad :X{\equiv}Y:$$

This is the case with N_2 and CO, both of which consist of atoms joined by a triple bond. We often find that *isoelectronic molecules*—molecules with the same number of atoms and the same number of valence electrons, such as N_2 and CO—have the same Lewis structures.

▼ **EXAMPLE 8.5** **Writing the Lewis structure of a diatomic ion**

Write the Lewis structure for the cyanide ion, CN^-.

STRATEGY In writing Lewis structures, it is sensible to be alert for isoelectronic molecules and ions, for they will have the same Lewis structures. The CN^- ion has $4 + 5 + 1 = 10$ electrons (four from the C atom, five from the N atom, and one for the additional negative charge), the same as N_2 and CO. We can therefore expect it to have the same Lewis structure as N_2 and CO and for the two atoms to be joined by a triple bond. A more systematic procedure is to count the number of electrons (here, 10) and then add them to the two atoms so that each atom has a complete octet.

SOLUTION The Lewis structure is formed as follows:

$$[\cdot \ddot{C} \cdot \ + \ \cdot \ddot{N} \ddot{\colon} \ + \ \cdot \,]^{-} \text{ forms } [\colon\!\ddot{C}\!\colon\!N\!\colon]^{-}, \text{ denoted } [\colon\!C\!\equiv\!N\!\colon]^{-}$$

As anticipated, there is a triple bond between the two atoms.

EXERCISE Write the Lewis structure for the BrO^+ ion.

[*Answer*: $[\ddot{Br}\!=\!\ddot{O}]^{+}$]

8.4 LEWIS STRUCTURES OF POLYATOMIC MOLECULES

Now we move on to polyatomic molecules and see that the same ideas apply to them as to diatomic molecules. Each atom of a polyatomic molecule completes its octet by sharing pairs of electrons with its immediate neighbors. Each pair of electrons shared by two neighbors is a covalent bond, just as in a diatomic molecule. A simple example is methane:

$$\begin{array}{ccccc} & H & & & H \\ & \cdot & & H & | \\ H\cdot\cdot C\cdot\cdot H & \text{ forms } & H\!\colon\!\ddot{C}\!\colon\!H, & \text{ denoted } & H\!-\!C\!-\!H \\ & \cdot & & H & | \\ & H & & & H \end{array}$$

This shows that the molecule consists of four carbon–hydrogen single bonds. What a Lewis structure does *not* show, however, is the three-dimensional arrangement of the atoms in space. (We shall see in the next chapter that a methane molecule consists of a carbon atom at the center of a regular tetrahedron of hydrogen atoms.) It is very important to remember that a Lewis structure is a two-dimensional diagram showing the links between atoms but not, except in special cases, the shape of the molecule.

Procedure for writing structures. The procedure for writing Lewis structures for polyatomic molecules is the same as for diatomic molecules. We use all the dots representing the valence electrons, and we arrange them so that each atom has an octet of electrons. The only complication is that we now need to know which atoms are linked to which other atoms. For instance, we need to know that the arrangement of atoms in carbon dioxide is OCO and not COO. As in this case, one clue is that the less electronegative atom is often the central atom. Another clue is that chemical formulas, especially simple ones, are often written with the central atom first, followed by the terminal atoms (the atoms attached to the central atom). An exception is the convention of writing acid formulas with hydrogen at the front. If the species is an oxoacid, those hydrogen atoms would be attached to oxygen atoms, which in turn are attached to the central atom. This is the case in H_2SO_4, which has the structure $(HO)_2SO_2$.

The procedure for writing Lewis structures for polyatomic molecules and ions can be summarized as follows:

Step 1. Calculate the total number of electron dots to be used by adding the numbers of valence electrons of all the atoms. Each hydrogen atom supplies one electron. Each main-group element supplies its group number of electrons. For a cation, subtract one dot for each positive charge. For an anion, add one dot for each negative charge. Divide this number by 2 to obtain the number of electron pairs.

Step 2. Arrange the chemical symbols for each atom in the formula so that the terminal atoms surround the central atom. Initial hydrogen atoms should be attached to any oxygen atoms present or, if no oxygen is present, to the central atom.

Step 3. Use electron pairs to form single bonds linking each atom to its neighbor. Then try to place any remaining electron pairs around the atoms so as to complete the necessary octet. If there are not enough electrons, use one or more of the lone pairs to form double or triple bonds to the central atom. Terminal halogen atoms always have four electron pairs and only form single bonds.

The third step may involve expansion of the octets of atoms other than those in Periods 1 and 2. We consider this possibility below.

▼ **EXAMPLE 8.6** **Writing Lewis structures for molecules**

Write the Lewis structure for acetic acid, CH_3COOH. In the —COOH group, both O atoms are attached to the C atom, and one of them is bonded to the final H atom.

STRATEGY We are able to anticipate that the CH_3— group, by analogy with methane, consists of a C atom joined to three H atoms by single bonds, and the fourth bond of that C atom links it to the other C atom. The full Lewis structure is obtained by working through the procedure set out above.

SOLUTION The total number of valence electrons is

$$3H \quad C \quad C \quad O \quad O \quad H$$
$$(3 \times 1) + 4 + 4 + 6 + 6 + 1 = 24$$

Twelve electron pairs must therefore be accommodated. The atomic arrangement suggested by the formula CH_3COOH is

$$
\begin{array}{ccc}
& H & O \\
H & C & C \\
& H & O \quad H
\end{array}
$$

We next draw dashes to link all bonded atoms with single bonds:

This accounts for seven pairs, leaving five pairs to be accommodated. We now arrange electron pairs so that each atom (except H) has an octet of electrons and forms its "usual" number of bonds, as given above:

$$
\begin{array}{c}
\text{H} \qquad\quad \overset{\cdot\cdot}{\text{O}}: \\
| \qquad\qquad \| \\
\text{H}-\text{C}-\text{C} \\
| \qquad\qquad \backslash \\
\text{H} \qquad\quad :\overset{\cdot\cdot}{\text{O}}-\text{H}
\end{array}
$$

EXERCISE Write a Lewis structure for the urea molecule, $(NH_2)_2CO$.

[*Answer*:
$$
\begin{array}{c}
\text{H} \quad \overset{\cdot\cdot}{\text{O}} \quad \text{H} \\
| \qquad \| \qquad | \\
\text{H}-\overset{\cdot\cdot}{\text{N}}-\text{C}-\overset{\cdot\cdot}{\text{N}}-\text{H}
\end{array}
$$
]

In the Lewis structure of an ionic compound containing a polyatomic ion, the cation and anion should be treated separately to show that they are individual ions and not linked by shared pairs. The structure of sodium sulfate (Na_2SO_4), for instance, is written

$$
2(\text{Na}^+) \left\{ \begin{array}{c}
:\overset{\cdot\cdot}{\text{O}}: \\
| \\
:\overset{\cdot\cdot}{\text{O}}-\text{S}-\overset{\cdot\cdot}{\text{O}}: \\
| \\
:\overset{\cdot\cdot}{\text{O}}:
\end{array} \right\}^{2-}
\qquad \text{rather than} \qquad
\begin{array}{c}
:\overset{\cdot\cdot}{\text{O}}: \\
| \\
\text{Na}-\overset{\cdot\cdot}{\text{O}}-\text{S}-\overset{\cdot\cdot}{\text{O}}-\text{Na} \\
| \\
:\overset{\cdot\cdot}{\text{O}}:
\end{array}
$$

Note that the charge belongs to the entire anion and is not localized on a particular atom. This is true of all polyatomic ions.

Resonance. In many cases it is possible to write several different Lewis structures for the same molecule. Three possible Lewis structures for the nitrate ion, NO_3^-, are

$$
\left\{ \begin{array}{c}
:\overset{\cdot\cdot}{\text{O}}: \\
| \\
:\overset{\cdot\cdot}{\text{O}}-\text{N}=\overset{\cdot\cdot}{\text{O}}:
\end{array} \right\}^{-}
\left\{ \begin{array}{c}
\overset{\cdot\cdot}{\text{O}} \\
\| \\
:\overset{\cdot\cdot}{\text{O}}-\text{N}-\overset{\cdot\cdot}{\text{O}}:
\end{array} \right\}^{-}
\left\{ \begin{array}{c}
:\overset{\cdot\cdot}{\text{O}}: \\
| \\
:\overset{\cdot\cdot}{\text{O}}=\text{N}-\overset{\cdot\cdot}{\text{O}}:
\end{array} \right\}^{-}
$$

The three structures differ only in the position of the double bond and therefore correspond to exactly the same energy. Since experiments have shown that all three O atoms and N—O bonds are identical (for instance, all three N—O bonds have the same length), the display of only one structure would give a false impression. The ion is taken to be a blend of all three Lewis structures, and this blending of structures, called *resonance*, is depicted as follows:

$$
\left\{ \begin{array}{c}
\text{O} \\
| \\
\text{O}-\text{N}=\text{O}
\end{array} \right\}^{-}
\longleftrightarrow
\left\{ \begin{array}{c}
\text{O} \\
\| \\
\text{O}-\text{N}-\text{O}
\end{array} \right\}^{-}
\longleftrightarrow
\left\{ \begin{array}{c}
\text{O} \\
| \\
\text{O}=\text{N}-\text{O}
\end{array} \right\}^{-}
$$

where, for clarity, we have omitted the oxygen lone pairs. The structure resulting from resonance is a *resonance hybrid* of the contributing Lewis structures. Resonance should not be thought of as the flickering of a molecule between Lewis structures, but as a *blend* of the individual structures (Fig. 8.16). This is a little like a mule being a mixture of a horse and a donkey, and not a creature that alternates between the two.

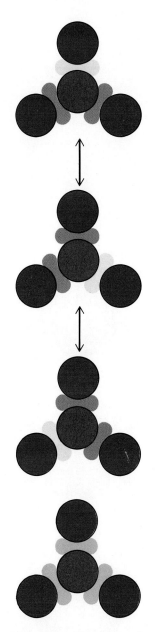

FIGURE 8.16 Resonance is a blending of structures to give an average of all the structures that contribute. In this diagram we represent single bonds with blue patches, double bonds with yellow patches, and the resonance blend with green patches.

▼ **EXAMPLE 8.7** **Writing resonance structures**

Suggest two resonance structures for the SO_2 molecule in which the S atom lies between two O atoms and the bond lengths are known to be the same.

STRATEGY Write a Lewis structure for the molecule, using the method of Example 8.6. Decide whether there is another equivalent structure that results from the interchange of a single and a double bond. Write the actual structure as a resonance hybrid of these Lewis structures.

SOLUTION Since all three atoms are members of Group VI, the total number of valence electrons in the molecule is $3 \times 6 = 18$. One structure is therefore $\ddot{\text{O}}{=}\ddot{\text{S}}{-}\ddot{\ddot{\text{O}}}{:}$. Interchanging the bonds gives $:\ddot{\ddot{\text{O}}}{-}\ddot{\text{S}}{=}\ddot{\text{O}}$, so the overall structure is the resonance hybrid

$$\ddot{\text{O}}{=}\ddot{\text{S}}{-}\ddot{\ddot{\text{O}}}{:} \longleftrightarrow :\ddot{\ddot{\text{O}}}{-}\ddot{\text{S}}{=}\ddot{\text{O}}$$

EXERCISE Write resonance hybrids for the acetate ion, $CH_3CO_2^-$. Recall that the structure of CH_3COOH was described in Example 8.6.

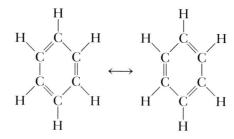

These are called *Kekulé structures* for the German chemist Friedrich Kekulé who first proposed (in 1865) that benzene had a "cyclic" structure with alternating single and double bonds. Compounds derived from benzene occur so widely in chemistry that the Kekulé structures are normally abbreviated to hexagons:

Benzene (C_6H_6) is another example of a molecule with a resonance-hybrid structure. A benzene molecule consists of a hexagonal ring of six carbon atoms, with a hydrogen atom attached to each C (**2**). Two structures that contribute to the resonance hybrid are

The circle-in-a-hexagon representation of the structure conveys the idea that all the carbon–carbon bonds are equivalent.

As well as equalizing bond lengths, resonance also stabilizes the molecule by lowering its total energy. This effect, which can be explained only in terms involving quantum mechanics, plays an important role in the chemical properties of benzene, which is more stable than expected for a molecule with three individual carbon–carbon double bonds.

Resonance results in the greatest stability when the contributing structures have equal energies. This is the case with all the structures

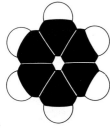

2 Benzene

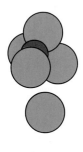

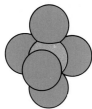

FIGURE 8.17 An N atom is too small to allow five Cl atoms to fit around it, but a P atom is big enough. Thus NCl_5 is unknown, but PCl_5 exists.

described so far, which have differed only in the positions of their double bonds. However, the best description of a molecule is obtained by considering its structure to be a blend of all possible Lewis structures, including those having different energies. In this case, the general rule is that the lowest-energy structures contribute most strongly to the overall structure. We discuss examples in the next few paragraphs.

Expanded octets. The octet rule is based on the idea that eight electrons fill a valence shell consisting of one s orbital and three p orbitals. However, if an atom has empty d orbitals available, it may be able to use them to accommodate more than eight electrons and hence to "expand its octet" to 10, 12, or even more electrons. That may allow the central atom of a molecule to form additional multiple bonds to the atoms attached to it, or to form bonds to more atoms. This is the case for the Period 3 elements of the p block, for which the $3d$ orbitals are only slightly higher in energy than the $3s$ and $3p$ orbitals; it is also true of the elements in later periods. However, because $2d$ orbitals do not exist, Period 2 elements cannot expand their octets. Another factor—and possibly the main factor—in determining whether more atoms can bond to a central atom than are allowed by the octet rule is the size of the central atom. A P atom is big enough for five Cl atoms to fit around it. An N atom is too small, so NCl_5 is unknown (Fig. 8.17).

Elements that can expand their octets show <u>*variable covalence*</u>, the ability to form different numbers of covalent bonds. Phosphorus is one example. It reacts directly with a limited supply of chlorine to form the colorless liquid phosphorus trichloride:

$$P_4(s) + 6Cl_2(g) \longrightarrow 4PCl_3(l)$$

The Lewis structure of the PCl_3 molecule is

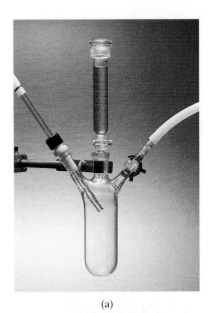

(a)

However, when excess chlorine is present or the trichloride reacts with more chlorine (Fig. 8.18), the pentachloride, a pale yellow crystalline solid, is produced:

$$P_4(s) + 10Cl_2(g) \longrightarrow 4PCl_5(s)$$

$$PCl_3(l) + Cl_2(g) \longrightarrow PCl_5(s)$$

Phosphorus pentachloride is an ionic solid consisting of PCl_4^+ cations and PCl_6^- anions that sublimes at 160°C to a gas of PCl_5 molecules. The Lewis structures of the polyatomic ions and the molecule are

(b)

FIGURE 8.18 Phosphorus trichloride is a liquid. As it drips into the flask (a), it reacts with chlorine gas to produce the very pale yellow solid phosphorus pentachloride (b), which is ionic and consists of PCl_4^+ cations and PCl_6^- anions.

Although PCl_4^+ is a polyatomic ion in which the P atom does not need to expand its octet, in PCl_6^- the P atom has expanded its octet to 12 with four extra electrons in two of its $3d$ orbitals. In PCl_5, the P atom has expanded its octet to 10 by using one $3d$ orbital.

▼ **EXAMPLE 8.8   Writing Lewis structures with expanded octets**

A fluoride of composition SF_4 is formed when fluorine diluted with nitrogen is passed over a film of sulfur at $-75°C$ in the absence of oxygen and moisture. Write the Lewis structure of sulfur tetrafluoride.

STRATEGY   Since a fluorine atom forms only single bonds, we anticipate that the Lewis structure consists of a shared pair between the central S atom and each of the four surrounding F atoms. However, if each F supplies 1 bonding electron and S already has 6 electrons in its valence shell, there are 10 electrons to accommodate. This octet expansion is possible because sulfur is in Period 3 and has empty $3d$ orbitals available. With that octet expansion in mind, we write the Lewis structure in the usual way, using the procedure given above.

SOLUTION   Sulfur ($:\!\overset{..}{S}\!:$) supplies six valence electrons, and each fluorine atom ($:\!\overset{..}{\underset{..}{F}}\!\cdot$) supplies seven. Hence there are $6 + 4 \times 7 = 34$ electrons to accommodate. Each F atom will have three lone pairs and a bonding pair shared with the central S atom. This suggests the structure

$$
\begin{array}{c}
:\!\overset{..}{\underset{..}{F}}\!: \\
:\!\overset{..}{\underset{..}{F}}\!\!-\!\!\overset{..}{S}\!: \\
:\!\overset{..}{\underset{..}{F}}\!:\quad:\!\overset{..}{\underset{..}{F}}\!:
\end{array}
$$

All 34 electrons are accounted for. Since the S atom has 10 electrons, which requires at least five orbitals, one $3d$ orbital must be used along with the four $3s$ and $3p$ orbitals.

EXERCISE   Write the Lewis structure for xenon tetrafluoride, XeF_4, and determine the number of electrons in the expanded octet.

[*Answer*: $:\!\overset{..}{\underset{..}{F}}\!\!-\!\!\overset{..}{Xe}\!\!-\!\!\overset{..}{\underset{..}{F}}\!:$; 12]

$$
\begin{array}{c}
:\!\overset{..}{\underset{..}{F}}\!:\quad:\!\overset{..}{\underset{..}{F}}\!:
\end{array}
$$

A molecule or ion may be a resonance hybrid of octet and expanded-octet Lewis structures. An example is the sulfate ion, SO_4^{2-}. Two of the numerous Lewis structures that may be written for the ion are

$$
\left\{
\begin{array}{c}
:\!\overset{..}{\underset{..}{O}}\!: \\
| \\
:\!\overset{..}{O}\!\!-\!\!S\!\!-\!\!\overset{..}{O}\!: \\
| \\
:\!\overset{..}{\underset{..}{O}}\!:
\end{array}
\right\}^{2-}
\qquad
\left\{
\begin{array}{c}
:\!\overset{..}{\underset{..}{O}}\!: \\
\| \\
:\!\overset{..}{O}\!\!=\!\!S\!\!=\!\!\overset{..}{O}\!: \\
| \\
:\!\overset{..}{\underset{..}{O}}\!:
\end{array}
\right\}^{2-}
$$

The sulfur octet is not expanded in the first, but is expanded (to 12 electrons) in the second. Since the two structures differ by more than the locations of double bonds, they can be expected to correspond to different energies. The actual structure of the ion is a resonance hybrid of these two structures and several similar structures in which the double bonds are in different locations. It turns out (by calculation based on the Schrödinger equation) that the expanded-octet structures have lower energy than the others. Hence, the best description of the molecule is as a resonance hybrid of all these structures, but with the expanded-octet structures most important.

Incomplete octets.   Some elements at the left of the p block, most notably boron, form compounds in which their atoms have incomplete octets. One example is the colorless gas boron trifluoride, BF_3, which has the Lewis structure

$$: \ddot{F} : \qquad\qquad : \ddot{F} :$$
$$: \ddot{F} : B : \ddot{F} : \qquad or \qquad : \ddot{F} - B - \ddot{F} :$$

The central boron atom has only six electrons. The octet could be completed if an F atom released two electrons and formed a B=F double bond. We would take this into account by expressing the structure as a resonance hybrid:

$$: \ddot{F} : \qquad : \ddot{F} : \qquad \ddot{F} \qquad : \ddot{F} :$$
$$: \ddot{F} - B - \ddot{F} : \longleftrightarrow : \ddot{F} - B = \ddot{F} \longleftrightarrow : \ddot{F} - B - \ddot{F} : \longleftrightarrow \ddot{F} = B - \ddot{F} :$$

Describing the molecule as a resonance hybrid of all four structures is better than describing it as the first structure alone. However, the first structure has a much lower energy than the other three (because the F atom has gained rather than lost electrons), so it alone is a fairly good description of the molecule. That is why we describe BF_3 as a molecule with an incomplete octet: the most important (lowest-energy) Lewis structure corresponds to a B atom with an incomplete octet.

Boron trifluoride is far from a laboratory curiosity, since it is produced industrially in large amounts (Fig. 8.19) by heating together boron oxide and calcium fluoride:

$$B_2O_3(s) + 3CaF_2(s) \xrightarrow{\Delta} 2BF_3(g) + 3CaO(s)$$

The boron oxide is obtained from the minerals borax and kernite, which are both forms of sodium borate, and the calcium fluoride is obtained from the mineral fluorspar. The reason so much BF_3 is manufactured is that the molecule can complete its octet by attaching to a lone pair of electrons on another molecule or ion and hence bring about a variety of reactions. For example, when the gas is passed over a metal fluoride, it forms the tetrafluoroborate anion BF_4^-:

$$: \ddot{F} : \qquad\qquad\qquad\qquad\quad \left[: \ddot{F} : \right]^-$$
$$: \ddot{F} - B \quad : \ddot{F} :^- \longrightarrow \left\{ : \ddot{F} - B - \ddot{F} : \right\}$$
$$: \ddot{F} : \qquad\qquad\qquad\qquad\qquad : \ddot{F} :$$

FIGURE 8.19 Boron trifluoride is used on a large scale in industry, especially in the manufacture of petrochemicals.

In this reaction the fluoride ion *coordinates*, or makes use of a lone pair to form a covalent bond, and the product is a complex ion. Another example of BF_3 coordination is its reaction with ammonia:

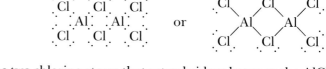

In this reaction the two gases combine to form a white solid. It is through analogous reactions, particularly with organic compounds, that boron trifluoride acts as an important industrial and laboratory "catalyst," a substance that speeds up reactions that otherwise would be too slow.

Boron trichloride is another molecule with an incomplete octet. It is made industrially (for use as a catalyst, like BF_3) by heating a mixture of boron oxide and carbon in chlorine:

$$B_2O_3(s) + 3C(s) + 3Cl_2(g) \longrightarrow 2BCl_3(g) + 3CO(g)$$

BCl_3 is a reactive gas. The trichloride of boron's congener aluminum, however, is not a gas but a volatile white solid that sublimes at 180°C to a gas of Al_2Cl_6 molecules. These molecules survive in the gas up to about 200°C but then fall apart into $AlCl_3$ molecules. The Al_2Cl_6 molecule exists because a Cl atom of one $AlCl_3$ molecule uses one of its lone pairs to coordinate to the Al atom of a neighboring $AlCl_3$ molecule:

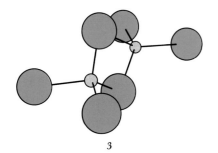

The two chlorine atoms that act as bridges between the $AlCl_3$ molecules lie above and below a plane defined by the other six atoms **(3)**. An interesting feature of the solid is that at its melting point (which is 192°C when it is kept under slightly greater than atmospheric pressure to prevent sublimation and allow the liquid to form), it nearly doubles in volume. This occurs because each Al^{3+} ion in the solid is surrounded by six Cl^- ions, but on melting, this compact arrangement breaks up as the liquid of Al_2Cl_6 molecules is formed.

3

Odd-electron molecules. Some molecules end up with an odd number of valence electrons. In such molecules, at least one atom cannot have an octet. This is the case with nitric oxide, which is prepared industrially by the oxidation of ammonia, as an intermediate in the production of nitric acid:

$$4NH_3(g) + 5O_2(g) \xrightarrow{\text{1000°C, Pt}} 4NO(g) + 6H_2O(g)$$

Since the N atom supplies 5 valence electrons and the O atom supplies 6, the NO molecule contains 11 valence electrons. One possible Lewis structure is

$$:\!N\!:\!\ddot{O}\!\cdot \quad \text{or} \quad \cdot N\!=\!\ddot{O}\!\cdot$$

with an unpaired electron on the nitrogen atom.

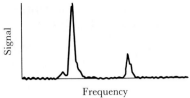

Other examples of molecules with unpaired electrons include the fragments, called "radicals," that are obtained when a molecule breaks up during a reaction:

A **radical** is a fragment of a molecule with at least one unpaired electron.

Examples include the methyl radical, formed by breaking the carbon–carbon bond in ethane (C_2H_6):

A methyl radical is denoted $\cdot CH_3$, the dot signifying the unpaired electron. Another example is the hydroxyl radical $\cdot \ddot{O} : H$ or, more simply, $\cdot OH$. This radical is formed briefly when a mixture of hydrogen and oxygen is ignited by a spark. It is also present in the upper atmosphere as a result of the action of the sun's radiation on water molecules.

Like most radicals, $\cdot CH_3$ and $\cdot OH$ are highly reactive and survive only for very short times under normal conditions. The reactivity of the hydroxyl radical is partly responsible for the explosive violence with which hydrogen and oxygen gases combine with each other. The methyl radical is involved in the explosion that occurs when ethane is ignited in air. Under abnormal conditions, though, a radical can survive indefinitely. Hydroxyl radicals have been detected in interstellar gas clouds, where they can survive for millions of years (Fig. 8.20). In

TABLE 8.2 Bond enthalpies of diatomic molecules

Molecule	B, kJ/mol
H_2	436
N_2	944
O_2	496
F_2	158
Cl_2	242
Br_2	193
I_2	151
HF	565
HCl	431
HBr	366
HI	299
CO	1074

that almost empty environment they collide very rarely with other molecules and so do not have the opportunity to react.

8.5 BOND PARAMETERS

A covalent bond between two specific atoms has characteristic features that, to some extent, are the same in all the molecules in which that bond occurs. The C—H bond, for example, has about the same strength and the same length in all organic compounds, and all O—N bonds are similar, no matter which compounds they occur in. These characteristic features are called *bond parameters*. We deal with two of them—bond strength and bond length—in this section.

Bond strength. The strength of a covalent bond is measured by the *bond enthalpy*, the enthalpy change that occurs when the bond dissociates. For example, the bond enthalpy of H_2 is the heat required at constant pressure to dissociate 1 mol of molecules:

$$H_2(g) \longrightarrow 2H(g) \qquad \Delta H° = +436 \text{ kJ}$$

All bond enthalpies are endothermic; they are normally denoted B and reported per mole of molecules, as in Table 8.2. They range from less than the 151 kJ/mol listed for I_2 up to 1074 kJ/mol for CO, the highest value for any diatomic molecule.

Average bond enthalpies. The bond enthalpy for a diatomic molecule is unambiguous: it indicates the strength of the bond between the only two atoms present. However, the bond enthalpy of any particular bond in a polyatomic molecule depends on the other atoms present. For instance, the values for the HO—H bond in water and the CH_3O—H bond in methanol are not quite the same; they are 492 kJ/mol and 435 kJ/mol, respectively. As in this example, variations in bond enthalpy are not usually very wide, so it is useful to list an average value, the *average bond enthalpy* (Table 8.3).

A multiple bond is stronger than a single bond between the same pair of elements. A C≡C triple bond (837 kJ/mol) is stronger than a C—C single bond (348 kJ/mol), and a C=C double bond is intermediate between the two (612 kJ/mol). The carbon-carbon bond enthalpy in benzene is 518 kJ/mol, which is greater than the average of the values for a single bond and a double bond (480 kJ/mol). This is a quantitative measure of how resonance lowers the energy of a molecule as a whole: it not only averages bonds but also strengthens them.

Average bond enthalpies can be used to explain chemical properties. For example, the enthalpies of E—H bonds (where E represents a Group IV element) decrease down the group from carbon (412 kJ/mol) to lead (205 kJ/mol), as shown in Fig. 8.21. This parallels a decrease in the stability of the hydrides of group members: methane (CH_4) can be kept indefinitely in air at room temperature, but stannane (SnH_4) decomposes to tin and hydrogen, and plumbane (PbH_4) has perhaps never been prepared except in trace amounts. Average bond enthalpies can also be used to estimate reaction enthalpies when accurate data are not available. This is shown in the next example.

TABLE 8.3 Average bond enthalpies in kilojoules per mole

C—H	412	N—H	388
C—C	348	N—N	163
C=C	612	N=N	409
C—C ↕ C=C	518 in benzene		
		O—H	463
		O—O	157
C≡C	837	F—H	565
C—O	360	Cl—H	431
C=O	743	Br—H	366
C—N	305	I—H	299

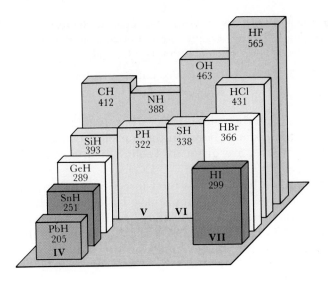

▼ **EXAMPLE 8.9** Estimating a reaction enthalpy from bond enthalpies

Estimate the enthalpy of combustion of ethanol vapor (ethanol is C_2H_5OH).

STRATEGY We expect a negative value because all combustions are exothermic. We begin by writing the chemical equation. To determine the overall change in enthalpy, we need to know the change in enthalpy that accompanies the breaking of all bonds in the reactants and the change that accompanies the formation of new bonds in the products (Fig. 8.22). We identify the bonds that are involved by writing the reaction in terms of Lewis structures. Then we use the average bond enthalpies of Table 8.3 to calculate the heat required to break all the bonds in the reactants, and the heat released when the products are formed from the atoms. The enthalpy of combustion is the difference between the two.

SOLUTION The reaction is

$$C_2H_5OH(g) + 3O_2(g) \longrightarrow 2CO_2(g) + 3H_2O(g)$$

In terms of Lewis structures, this is

$$
\begin{array}{cc}
\text{H} & \text{H} \\
| & | \\
\text{H—C—C—O—H} & + \ 3\text{O}{=}\text{O} \longrightarrow 2\text{O}{=}\text{C}{=}\text{O} + 3\text{H—O—H} \\
| & | \\
\text{H} & \text{H}
\end{array}
$$

The heat required to atomize the reactants is

$$5B(\text{C—H}) + B(\text{C—C}) + B(\text{C—O}) + B(\text{O—H}) + 3B(\text{O}{=}\text{O})$$
$$= 5 \times 412 \text{ kJ} + 348 \text{ kJ} + 360 \text{ kJ} + 463 \text{ kJ} + 3 \times 496 \text{ kJ}$$
$$= 4719 \text{ kJ}$$

The heat released when the atoms form the products is

$$2 \times 2B(\text{C}{=}\text{O}) + 3 \times 2B(\text{O—H}) = 2 \times 2 \times 743 \text{ kJ} + 3 \times 2 \times 463 \text{ kJ}$$
$$= 5750 \text{ kJ}$$

The difference is

$$\Delta H = 4719 \text{ kJ} - 5750 \text{ kJ} = -1031 \text{ kJ}$$

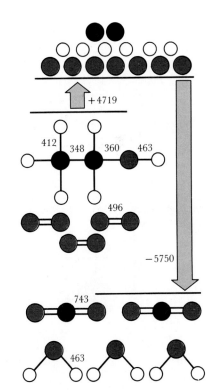

FIGURE 8.22 The enthalpy changes (in kilojoules per mole) involved in breaking the reactant molecules into atoms and then assembling products from the atoms.

TABLE 8.4 Average bond lengths

Bond	Bond length, pm
C—H	109
C—C	154
C=C	134
C≡C	120
C—C ↕ C=C } in benzene	139
C—O	143
C=O	122
O—H	96
N—H	101

Since 1 mol of C_2H_5OH molecules takes part in the reaction as written, we estimate the enthalpy of combustion of ethanol as -1031 kJ/mol. The experimental enthalpy of combustion is -1236 kJ/mol.

EXERCISE Estimate the standard enthalpy of formation of ammonia.

[*Answer*: -76 kJ/mol]

▲

Bond lengths. Another measurable property of covalent bonds is the *bond length*, the distance between the nuclei of two atoms joined by a bond. Some values are given in Table 8.4; a typical value is about 150 pm.

Bonds between heavy atoms tend to be longer than those between light atoms because heavier atoms have larger radii, which keep their nuclei farther apart. This is illustrated by the difference between F_2 (144 pm) and I_2 (267 pm). However, the strength of the bond also plays a role; stronger bonds result in shorter bond lengths:

Bond	Bond enthalpy, kJ/mol	Bond length, pm
C—C	348	154
C=C	612	134
C≡C	837	120

The length of the carbon–carbon bond in benzene (139 pm) is intermediate between those of C—C and C=C bonds. This is another case in which resonance blends the properties of the contributing structures.

A bond length is approximately the sum of two contributions called the *covalent radii* of the atoms (**4**). The O—H bond length in $CH_3\overline{CH_2OH}$, for example, is the sum of the covalent radius of H and that of O. Likewise, the length of the C—H bonds in the molecule is the sum of the covalent radii of H and C (**5**). Some covalent radii are given in Table 8.5. They are quite useful for estimating the sizes of molecules.

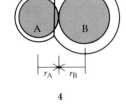

4

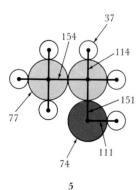

5

TABLE 8.5 Covalent radii* in picometers

		H 37			
Be 90	B 82	C 77 67 60	N 75 60 55	O 74 60	F 72
	Al 118	Si 111	P 120	S 102	Cl 98
	Ga 126	Ge 122	As 119	Se 117	Br 114
	In 144	Sn 141	Sb 138	Te 135	I 134

*Alternative radii refer to single, double, and triple bonds.

CHARGE DISTRIBUTIONS IN COMPOUNDS

In the extreme case of pure ionic bonding, electrons would be transferred completely from one atom to another. Likewise, in pure covalent bonding, each atom would share electrons equally with a neighbor. In actuality, no compound is either purely ionic or purely covalent, but rather is something intermediate between the two. No atom is infinitely electronegative or infinitely electropositive; hence no atom can be expected to gain complete control or lose complete control of the electrons in a bond. In this section we discuss the degree to which each atom controls the electrons in a molecule.

8.6 ASSESSING THE CHARGE DISTRIBUTION

When two atoms have similar electronegativities, as for those that bond covalently, little or no energy advantage is gained by transferring an electron from one atom to another. This suggests that if we can keep track of the rearrangement of electrons that occurs when various Lewis structures are formed, the structures involving the least electron migration between atoms are likely to be the most favored energetically.

The formal charge. A simple way of assigning ownership of electrons to the atoms in a molecule is to assume that each atom has an equal share in a bonding electron pair but owns its lone pairs completely. If this results in an atom having more electrons in the molecule than when it is a free, neutral atom, we say that the atom has a negative *formal charge* (FC) in the Lewis structure. If the assignment of electrons leaves an atom with fewer electrons than when it is free, we say that the atom has a positive formal charge. In general, we calculate the formal charge as

> FC = number of valence electrons in the free atom
> − number of lone-pair electrons
> − $\frac{1}{2}$ × number of shared electrons

The Lewis structures with formal charges closest to zero are likely to be the most favored energetically: the atoms in those structures have undergone the least redistribution of electrons relative to the free atoms, which is energetically desirable when there are no great differences of electronegativity between the atoms.

Since a C atom (Group IV) has four valence electrons, and an O atom (Group VI) has six, the formal charges in the Lewis structure $:C\equiv O:$ of carbon monoxide are

$$FC(C) = 4 - 2 - \frac{1}{2} \times 6 = -1$$

$$FC(O) = 6 - 2 - \frac{1}{2} \times 6 = +1$$

These results are normally expressed by writing +1 and −1 on the Lewis structure, as in $^{-1}C\equiv O^{+1}$. The calculation suggests that the Lewis structure corresponds to one negative charge being located on the C atom, and a positive charge on the O atom. However, it must never be forgotten that assigning one electron of a pair to one atom and another to the other atom is a very simplistic way of keeping track of the locations of electrons. We should not think of CO as actually having such a strongly pronounced distribution of charge.

The formal charges of the atoms in an electrically neutral molecule add up to zero. In an ion, the sum of the formal charges is the same as the charge number of the ion. (This is a useful way of checking the calculation of formal charge.) In CO, the C atom has formal charge -1, the O atom has formal charge $+1$, and the sum of the formal charges is zero. The formal charges of the Lewis structure $[:C{\equiv}N:]^-$ for the cyanide ion are written $^{-1}C{\equiv}N^0$, and they add up to -1, its charge number. They also suggest that the negative charge of the cyanide ion is centered on the carbon atom.

6a   **6b**

▼ **EXAMPLE 8.10   Calculating the formal charge**

Calculate the formal charges on the atoms in the two Lewis structures of a sulfate ion (**6a** and **6b**).

STRATEGY   We use the definition of formal charge given above, applying it to each atom in the two Lewis structures. Sulfur and oxygen both belong to Group VI and have six valence electrons. We must calculate the formal charge of each distinct type of O atom (that is, of O atoms that are bonded in different ways to the central S atom), but equivalent oxygen atoms have the same formal charge.

SOLUTION   For the S atom in (**6a**),

$$FC = 6 - 0 - \frac{1}{2} \times 8 = +2$$

and for that in (**6b**),

$$FC = 6 - 0 - \frac{1}{2} \times 12 = 0$$

All four O atoms in (**6a**) are equivalent, and for each one

$$FC = 6 - 6 - \frac{1}{2} \times 2 = -1$$

In (**6b**) the doubly bonded O atoms have formal charge

$$FC = 6 - 4 - \frac{1}{2} \times 4 = 0$$

and the singly bonded O atoms have FC $= -1$.

EXERCISE   Calculate the formal charges for the two Lewis structures of the phosphate ion (**7a** and **7b**).

[*Answer*: The formal charges are shown on the structures.]

7a   **7b**

Formal charge and plausible structures.   Now we apply the idea that the Lewis structures with the lowest formal charges are likely to be the most favored energetically. Because the formal charges of the SO_4^{2-} ion in (**6b**) are closer to zero than those in (**6a**), the former is likely to have lower energy and hence be the main contributor to the resonance hybrid. The formal charges on two resonance structures of BF_3 are shown in (**8a**) and (**8b**). We see that the lower values correspond to the

8a   **8b**

structure with the incomplete octet (**8a**); in the completed-octet structure (**8b**), one F atom has one unit of positive formal charge, which is energetically less favored.

The same idea also helps us to decide whether or not a particular arrangement of atoms is likely. For example, it suggests that the structure OCO (**9a**) is more likely for carbon dioxide than COO (**9b**), and that the structure NNO (**10a**) is more likely for dinitrogen oxide than NON (**10b**).

$$\overset{0\quad\ 0\quad\ 0}{O=C=O}\qquad\overset{0\quad +2\quad -2}{O=O=C}$$
9a   **9b**

$$\overset{-1\quad +1\quad\ 0}{N=N=O}\qquad\overset{-1\quad +2\quad -1}{N=O=N}$$
10a   **10b**

▼ **EXAMPLE 8.11** **Judging the plausibility of a structure**

Suggest a plausible chemical structure for the cyanate ion NCO^-, and write its Lewis structure.

STRATEGY Since O and N are both more electronegative than C, C is likely to be the central atom of this ion. Try to think of an isoelectronic molecule (one with the same number of valence electrons): that may provide a good clue to the structure of this ion. More systematically, we may calculate the formal charges on the three possible arrangements of atoms and select the one with formal charges closest to zero.

SOLUTION The NCO^- ion has the same number of atoms and valence electrons (16) as a CO_2 molecule, which we know to be OCO. This suggests that the atomic arrangement in the ion is NCO. The formal charges of the Lewis structures corresponding to the three possible arrangements are

$$\overset{-1\quad\ 0\quad\ 0}{:N=C=O:}\qquad\overset{-2\quad +1\quad\ 0}{:C=N=O:}\qquad\overset{-2\quad +2\quad -1}{:C=O=N:}$$

The formal charges are closest to zero in the structure at the left. That structure is therefore the likely one, as anticipated.

EXERCISE Suggest a plausible structure for ClF_3, and write its Lewis structure and formal charges.

[*Answer*: $\overset{\displaystyle :\overset{..}{\underset{..}{F}}:^0}{:\overset{..}{\underset{..}{F}}-\overset{0}{\underset{..}{Cl}}-\overset{..}{\underset{..}{F}}:}$]

▲

Oxidation number. When a covalent bond forms between two atoms with different electronegativities, the bonding pair of electrons lies slightly closer to the more electronegative atom. That is, one atom partially loses possession of the electron pair and, in a formal sense, undergoes oxidation (electron loss). The more electronegative atom gains partial control of the pair, increases its number of electrons, and hence in a formal sense undergoes reduction (electron gain). In HCl, for example, the Cl atom gains partial control of the bonding pair and gains some Cl^- character. The H partially loses control of the pair and hence partially resembles an H^+ ion. This situation is summarized by reporting that the oxidation number of Cl has decreased to -1 (the chlorine has been reduced) and the oxidation number of H has increased to $+1$ (it has been oxidized).

We can now see the origin of the rules of Section 3.9 (and particularly Table 3.4) for calculating the oxidation numbers of elements in compounds. The oxidation number is an *effective* ionic charge that is obtained by exaggerating the drift of electrons in a covalent bond and assuming that the transfer is complete. In the case of HCl, the oxidation numbers of the two elements are calculated by treating the chlorine atom as owning *both* electrons of the bond:

$$H:\overset{..}{\underset{..}{Cl}}: \quad \text{is exaggerated to} \quad H^+:\overset{..}{\underset{..}{Cl}}:^-$$

In the gaseous "interhalogen" compound chlorine monofluoride (ClF), the bonding pair is assigned to the F atom because it is more electronegative than the Cl atom:

$$:\overset{..}{\underset{..}{Cl}}:\overset{..}{\underset{..}{F}}: \quad \text{is exaggerated to} \quad :\overset{..}{\underset{..}{Cl}}^+:\overset{..}{\underset{..}{F}}:^-$$

The oxidation number of chlorine in this compound is therefore $+1$. The rules in Table 3.4 lead to the same values as those obtained by tracing the electron drifts that arise from electronegativity differences, but they are very much simpler to apply. That is especially so when several different types of atoms are present in the same molecule and in competition for the electrons.

8.7 POLARIZATION

We will now consider in greater detail the remark made previously that no bond is purely ionic or purely covalent. Consider a monatomic anion that is next to a cation. Because of the attraction of the cation's charge for the anion's electrons, the anion undergoes distortion from its normal spherical shape. We can think of this distortion as the tendency of an electron pair to move into the bonding region between the nuclei (**11**) and to form a covalent bond. Highly distorted ions are well on the way to becoming covalently bonded, and we can expect bonds to become more covalent as the distortion of the charge clouds increases.

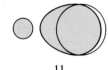

11

Polarizability and polarizing power. Atoms and ions that readily undergo large distortions are said to be highly *polarizable*. Ions that can cause large distortions are said to have high *polarizing power*. An anion can be expected to be highly polarizable if it is large, like the I^- ion. In such an ion, the nucleus exerts less control over its outermost electrons because they are farther away. As a result, their charge clouds are easily distorted. A cation has high polarizing power if it is small and highly charged, like an Al^{3+} ion.

Compounds composed of a small, highly charged cation and a large, polarizable anion will thus tend to have bonds with a significant covalent character. In Group II, the very small beryllium atom has strikingly different properties from its larger congeners. The Group II chlorides are largely ionic; however, beryllium chloride, which results from the action of carbon tetrachloride on beryllium oxide,

$$2BeO(s) + CCl_4(g) \xrightarrow{800°C} 2BeCl_2(g) + CO_2(g)$$

is not ionic. It consists of $BeCl_2$ molecules in the vapor, and of long chains of covalently bonded $BeCl_2$ units in the solid:

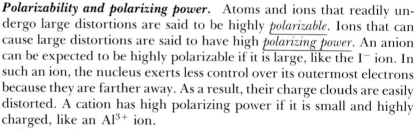

The covalent character of the silver halides increases from AgCl to AgI as the anion becomes more polarizable. One consequence is their decreasing solubility in water from AgCl to AgI. Another (for more complicated reasons) is their darkening of color from white for AgCl through pale yellow for AgBr to yellow for AgI (Fig. 8.23). Silver fluoride is freely soluble in water and is predominantly ionic.

Polarizability and polarizing power vary periodically through the periodic table and account for the diagonal relationships in it (see Section 7.8 and Fig. 7.31). Cations become smaller, more highly charged, and hence more strongly polarizing from left to right across a period. Thus, Be^{2+} is more strongly polarizing than Li^+, Mg^{2+} more than Na^+, and Al^{3+} more strongly still than Mg^{2+}. Cations become larger and hence less strongly polarizing down a group. Thus Na^+ is less strongly polarizing than Li^+, and K^+ is less polarizing still. Likewise, Mg^{2+} is less strongly polarizing than Be^{2+}. Since polarizing power increases from Li^+ to Be^{2+} but decreases from Be^{2+} to Mg^{2+}, it follows that the polarizing power of the diagonal neighbors Li^+ and Mg^{2+} should be roughly the same. Likewise, since B^{3+} is more strongly polarizing than Be^{2+} but Al^{3+} is less strongly polarizing than B^{3+}, the polarizing power of the diagonal neighbors Be^{2+} and Al^{3+} should also be roughly the same. Hence, we expect similarities in the properties of diagonally related neighbors, as we saw in Section 7.8.

Dipole moments. An oxidation number, we saw, is an exaggeration of the charge distribution in a compound. However, the *actual* charge distribution in a molecule has important consequences for the properties of the compound, including its volatility and its ability (if it is a liquid molecular compound) to act as a solvent.

In HCl the Cl atom has a small negative charge on account of its partial takeover of the bonding pair, and the H has a small positive charge. Such charges are called *partial charges* and are reported by writing $^{\delta+}H—Cl^{\delta-}$, where $\delta+$ indicates that there is a partial positive charge on the H atom, and $\delta-$ indicates a partial negative charge of equal magnitude on the Cl atom (δ, the Greek lowercase letter delta, is often used to indicate a small quantity). The molecule is said to have an "electric dipole":

An **electric dipole** is a positive charge next to an equal but opposite negative charge.

The dipole in the HCl molecule is denoted $\overset{+\longrightarrow}{H—Cl}$, where the + marks the location of the partial positive charge (Fig. 8.24).

The magnitude of an electric dipole is reported as the *electric dipole moment* μ (the Greek letter mu), in a unit called the *debye* (D). The unit is defined so that a single negative charge (an electron) separated by 100 pm from a single positive charge (a proton) has a dipole moment of 4.80 D.* The debye is named for the Dutch chemist Peter Debye,

*The 4.8 in the definition of the debye arises from an earlier system of units used for electric charge. The SI unit of dipole moment is $1\ C \cdot m$. It is the dipole moment of a charge of 1 C separated from a charge of -1 C by a distance of 1 m. The conversion is $1\ D = 3.336 \times 10^{-30}\ C \cdot m$.

FIGURE 8.23 The silver halides, which are formed by precipitation from silver nitrate and sodium halide solutions, become increasingly insoluble and darkly colored down the group from Cl to I.

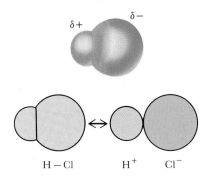

FIGURE 8.24 The HCl molecule has an electric dipole arising from the partial positive charge (red) on the H atom and the partial negative charge (green) on the Cl atom. The charges can be thought of as arising from resonance, in which a small proportion of an ionic structure blends into the pure covalent structure.

FIGURE 8.25 The ionic character of bonds increases with the difference in electronegativity between the two elements. Even in CsF (which has the largest electronegativity difference) the bond is only 95% ionic. No part of the curve is very precise, but the yellow area is particularly ambiguous.

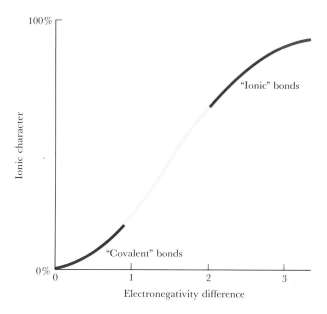

who made important studies of dipole moments. The dipole moment of HCl is 1.1 D, which is a typical value for a molecule.

A *polar bond* is a covalent bond between atoms carrying partial electric charges. One way of visualizing a polar bond (Fig. 8.24) is as a resonance hybrid of a pure covalent bond (in which the electron pair is exactly equally shared) and a pure ionic bond (in which an electron has been transferred completely):

$$H—\overset{..}{\underset{..}{Cl}}: \longleftrightarrow H^+ : \overset{..}{\underset{..}{Cl}} :^-$$

The ionic structure contributes more to the hybrid as the electronegativity difference between the two atoms increases (Fig. 8.25). As that difference increases, the partial charges $\delta+$ and $\delta-$ increase, so the dipole moment increases too.

An *approximate* relation between the electric dipole moment of a molecule AB and the electronegativities χ of the two atoms A and B is

$$\mu = \chi_A - \chi_B \qquad \text{in debye}$$

with the more electronegative atom forming the negative end of the dipole (since that atom tends to attract the electron pair). For example, since the electronegativities of H and Cl are 2.1 and 3.0, we predict

$$\mu = (3.0 - 2.1)\,D = 0.9\,D$$

in moderate agreement with the experimental value of 1.1 D. A C–H bond is predicted to be much less polar (only about 0.4 D), with the positive end of the dipole located on the H atom. This is so low a value that in most cases C–H bonds can be treated as almost nonpolar.

Even a diatomic molecule in which both atoms are identical can be regarded as a resonance hybrid of covalent and ionic structures:

$$: \overset{..}{\underset{..}{Cl}}{}^+ : \overset{..}{\underset{..}{Cl}} :^- \longleftrightarrow : \overset{..}{\underset{..}{Cl}}—\overset{..}{\underset{..}{Cl}} : \longleftrightarrow : \overset{..}{\underset{..}{Cl}} :^- \quad \overset{..}{\underset{..}{Cl}} :^+$$

However, because the Cl^+Cl^- and Cl^-Cl^+ forms contribute equally (as they have the same energy), there is no net electric dipole, and the bond is nonpolar (Fig. 8.26). We can think of the Cl_2 molecule as being held together by a bond that is largely covalent, but in which fluctuations in the positions of the bonding electron pair result in their sometimes being entirely on one Cl atom (giving Cl^+Cl^-) and for equal lengths of time being entirely on the other atom (giving Cl^-Cl^+).

These remarks can be summarized as follows. First, bonds between atoms of the same element in diatomic molecules (as in Cl_2) are nonpolar. Bonds between atoms of the same elements in polyatomic molecules, however, may be polar. In ozone (O_3), for instance, the central O atom is not identical to the outer two (it is joined to two atoms, whereas they are joined to one), and each bond is polar. Second, all bonds between atoms of different elements are polar, with the negative end of the dipole located on the more electronegative element. Third, in general, the greater the difference in electronegativities, the greater the polarity of the bond.

The extreme case of a polar bond would occur if one atom were so electronegative that it gained complete control of the bonding electron pair. Then only the ionic structure would contribute to the resonance, and the bond would be purely ionic. However, this extreme case is never completely reached, and so there is no such thing as a *purely* ionic bond. Even in CsF, where the electronegativity difference of 3.2 is the largest between any pair of elements, the fluorine atom wins only a 95% share of the electron pair.

The polarities of bonds affect the properties of molecules. However, the way in which the properties are affected depends on the arrangement of the bonds in space and, hence, on the shape of the molecule. So far we have said almost nothing about molecular shape, but in the next chapter we see how the Lewis theory can be adapted to deal with this feature too.

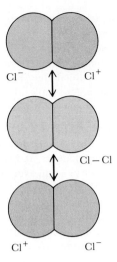

FIGURE 8.26 Ionic-covalent resonance occurs in Cl_2, but as the Cl^+Cl^- and Cl^-Cl^+ structures contribute equally, the molecule is nonpolar overall. Red indicates positive charge, green negative.

SUMMARY

8.1 An **ionic bond,** the attraction between oppositely charged ions, is formed between elements with widely different electronegativities. Whether or not an ionic bond can form depends on whether the energy needed to form the ions from the atoms is recovered from the strength of their interaction in the ionic solid. The latter is measured by the **lattice enthalpy,** the enthalpy change for the formation of a gas of ions from an ionic solid, which is large for small, highly charged ions. Ion formation generally proceeds as far as the completion of a noble-gas **octet.** Electron loss and gain are conveniently depicted in an **electron dot diagram,** in which valence electrons are shown as dots. In the p block, cations are formed by electron loss that proceeds as far as a **pseudonoble-gas configuration,** a closed core and a completed d subshell.

8.2 Variable valence may occur either because d electrons are used or, in the p block, because p electrons are lost more readily than s electrons.

8.3 Covalent bonds are formed between elements that do not differ greatly in electronegativity; each covalent bond consists of a **shared electron pair. Double bonds** and **triple bonds** consist of two and three shared pairs, respectively. A guide to the number of pairs that an atom can share, especially for Period 2 elements, is the **octet rule**—that an atom tends to complete its octet. The bonding arrangement in a molecule is shown by its **Lewis structure.**

8.4 When several Lewis structures can be written for a molecule, the actual structure is a **resonance hybrid,** a blend of them all, with the lowest-energy structures dominant. In benzene, the two most important resonance structures are the **Kekulé structures.** Resonance between single and double bonds equalizes bond lengths and strengths to intermediate values. Overall, resonance stabilizes molecules by lowering their total energy. Elements in Period 3 and later can form compounds with **expanded octets,** using their vacant d orbitals to accommodate more than eight electrons. This gives rise to **variable covalence,** the ability of elements to form different numbers of covalent bonds. Octet formation is not universal: some elements (notably boron) can form compounds in which they have an **incomplete octet,** and fragmentation of molecules gives rise to **radicals,** which have one or more unpaired electrons.

8.5 The strength of a covalent bond is measured by its **bond enthalpy,** the enthalpy change for the separation of the bonded atoms, and reaction enthalpies can be discussed in terms of an **average bond enthalpy** taken over a range of compounds. In general, the stronger the bond between a given pair of atoms, the shorter the **bond length,** the distance between the nuclei of the bonded atoms. Bond lengths may be expressed as sums of **covalent radii** typical of the atoms involved.

8.6 The reorganization of electrons that takes place on bond formation can be analyzed in terms of the **formal charge** of an atom, the charge it carries if it is assumed to

have half of each shared electron pair. The loss or gain of electrons that occurs when atoms form bonds is assessed from the **oxidation number** of each atom in the compound, its effective ionic charge.

8.7 Ions are distorted by **polarization** effects. Small, highly charged cations and anions are strongly polarizing, and large anions are highly polarizable. The distortion of ions represents a tendency toward covalent bonding. Covalent bonds between different atoms are **polar** in the sense that one atom (usually the more electronegative element) has a partial negative charge, and the other a partial positive charge: this gives rise to an electric **dipole,** a pair of equal and opposite partial charges, with a strength measured by the **dipole moment.** The periodicity of polarizing power is part of the explanation of diagonal relationships in the periodic table. There is no such thing as a *pure* ionic bond or a *pure* covalent bond.

EXERCISES

Formation of Ionic Bonds

In Exercises 8.1 to 8.4, state which ions each element is likely to form.

8.1 (a) Li; (b) Ca; (c) Ag; (d) Zn.
8.2 (a) Cs; (b) Al; (c) Cu; (d) Ti.

8.3 (a) F; (b) O; (c) N; (d) As.
8.4 (a) P; (b) Se; (c) I; (d) Te.

In Exercises 8.5 and 8.6, state which ions the elements of each group of the periodic table are likely to form.

8.5 (a) Group II; (b) Group V.
8.6 (a) Group III; (b) Group VI.

In Exercises 8.7 and 8.8, draw the electron dot formulas for the original atoms and the ions in each compound.

8.7 (a) Sodium chloride; (b) magnesium iodide; (c) calcium oxide; (d) strontium sulfide.
8.8 (a) Lithium fluoride; (b) lithium hydride; (c) aluminum oxide; (d) calcium nitride.

In Exercises 8.9 and 8.10, calculate the maximum distance at which it is energetically favorable for an ionic bond to form between atoms of each pair of elements (use the data in Tables 7.7 and 7.8).

8.9 (a) Sodium and chlorine; (b) chlorine and fluorine.
8.10 (a) Cesium and fluorine; (b) xenon and fluorine.

For Exercises 8.11 and 8.12, assume that two atoms of halogen X approach a single atom of metal M symmetrically from opposite sides. Calculate the distance at which it is energetically favorable for $X^-M^{2+}X^-$ to form (not

forgetting to take into account the repulsion between the anions) for M and X as given.

8.11 (a) Magnesium and fluorine; (b) sodium and chlorine.

8.12 (b) Calcium and chlorine; (b) potassium and bromine.

8.13 Explain why the lattice enthalpy of magnesium oxide (3850 kJ/mol) is greater than that of barium oxide (3114 kJ/mol).

8.14 Explain why the lattice enthalpy of magnesium oxide (3850 kJ/mol) is greater than that of magnesium sulfide (3406 kJ/mol).

8.15 Arrange the silver halides in order of increasing lattice enthalpy, without referring to tables of data.

8.16 Arrange the Group II oxides in order of increasing lattice enthalpy, without referring to tables of data.

8.17 Calculate the lattice enthalpy of silver fluoride from the data in Appendix 2A and the information:

Enthalpy of formation of Ag(g): +284 kJ/mol
Enthalpy of ionization of Ag(g): +731 kJ/mol
Enthalpy of formation of F(g): +79 kJ/mol
Enthalpy of formation of AgF(s): −205 kJ/mol

8.18 Calculate the lattice enthalpy of cesium chloride from the data in Appendix 2A and the following information:

Enthalpy of formation of Cs(g): +76 kJ/mol
Enthalpy of formation of Cl(g): +122 kJ/mol
Enthalpy of formation of CsCl(s): −443 kJ/mol

8.19 Calculate the lattice enthalpy of magnesium oxide from the data in the chapter and the information:

Enthalpy of formation of Mg(g): +148 kJ/mol
Enthalpy of formation of O(g): +249 kJ/mol
Enthalpy of formation of MgO(s): −602 kJ/mol

8.20 Calculate the lattice enthalpy of calcium sulfide from the data in the chapter and the information:

Enthalpy of formation of Ca(g): +178 kJ/mol
Enthalpy of formation of S(g): +279 kJ/mol
Enthalpy of formation of CaS(s): −482 kJ/mol

8.21 By assuming that the lattice enthalpy of $NaCl_2$ would be the same as that of $MgCl_2$, use enthalpy arguments to explain why $NaCl_2$ is an unlikely compound. The enthalpy of sublimation of sodium is +107 kJ/mol.

8.22 By assuming that MgCl would have the same lattice enthalpy as KCl, use enthalpy arguments to explain why MgCl is an unlikely compound.

Electron Dot Formulas

In Exercises 8.23 and 8.24, write the likely chemical formula and the electron dot formula for each of these ionic compounds.

8.23 (a) Potassium fluoride; (b) magnesium bromide; (c) aluminum sulfide; (d) calcium nitride.

8.24 (a) Cesium boride; (b) boron oxide; (c) aluminum nitride; (d) francium astatide.

In Exercises 8.25 and 8.26, decide which compounds are likely to be ionic, and explain why or why not.

8.25 (a) Magnesium oxide; (b) carbon dioxide; (c) nitrogen dioxide; (d) cesium chloride.

8.26 Aluminum chloride; (b) zinc chloride; (c) chlorine trifluoride; (d) boron carbide.

Variable Valence

In Exercises 8.27 and 8.28, give the ions commonly formed by each element.

8.27 (a) In; (b) Sn; (c) Ag; (d) Cu.
8.28 (a) Tl; (b) Pb; (c) Au; (d) Zn.

In Exercises 8.29 to 8.32, give the chemical equation for the reaction that could be used to prepare each compound.

8.29 (a) PCl_3; (b) PbO; (c) Cu_2O.
8.30 (a) PCl_5; (b) SnO; (c) CuO.
8.31 (a) SF_4; (b) SF_6; (c) CuCl; (d) $CuCl_2$.
8.32 (a) AgF_2; (b) AgCl; (c) CuI; (d) $CuBr_2$.

Lewis Structures

In Exercises 8.33 to 8.36, write the Lewis structure for each covalent compound.

8.33 (a) Hydrogen fluoride; (b) water; (c) ammonia; (d) methane.

8.34 (a) Iodine chloride, ICl; (b) hydrogen peroxide; (c) hydroxylamine, NH_2OH; (d) sulfur tetrachloride.

8.35 (a) Phosphorus pentafluoride; (b) sulfur hexafluoride; (c) boron trifluoride; (d) xenon tetrafluoride.

8.36 (a) Phosphorus pentachloride (as a solid and as a vapor); (b) iodine trichloride; (c) aluminum chloride (as a liquid and as a hot vapor); (d) xenon tetroxide.

In Exercises 8.37 and 8.38, write the Lewis structure for each ion.

8.37 (a) Phosphonium ion, PH_4^+; (b) hypochlorite ion, ClO^-; (c) cyanide ion, CN^-; (d) tetrahydridoaluminate ion, AlH_4^-.

8.38 (a) Nitronium ion, ONO^+; (b) chlorite ion, ClO_2^-; (c) superoxide ion, O_2^-; (d) tetrahydridoborate ion, BH_4^-.

In Exercises 8.39 and 8.40, give the Lewis structure for each salt.

8.39 (a) Sodium hypochlorite; (b) barium nitrate.
8.40 (a) Calcium nitrite; (c) potassium acetate.

In Exercises 8.41 and 8.42, write the Lewis structure for each organic compound.

8.41 (a) Methanol; (b) formaldehyde (H_2CO), which is used to preserve biological specimens as "formalin," its aqueous solution; (c) acetone (CH_3COCH_3), a solvent; (d) glycine [$CH_2(NH_2)CO_2H$], the simplest example of an amino acid, a building block of proteins.

8.42 (a) Ethanol; (b) methylamine (CH_3NH_2), a putrid-smelling substance formed when living matter, including fish flesh, decays; (c) formic acid (HCO_2H), a component of the venom injected by ants; (d) toluene ($CH_3C_6H_5$), a derivative of benzene.

In Exercises 8.43 and 8.44, write typical Lewis structures for each ion, allowing for resonance:

8.43 (a) Sulfate ion; (b) sulfite ion; (c) chlorate ion; (d) nitrate ion.

8.44 (a) Phosphate ion; (b) hydrogen sulfite ion; (c) perchlorate ion; (d) nitrite ion.

Bond Enthalpy

In Exercises 8.45 and 8.46, estimate the enthalpy change that occurs when each molecule is dissociated into its atoms.

8.45 (a) H_2O; (b) CO_2; (c) CH_3CO_2H; (d) CH_3NH_2.
8.46 (a) NH_3; (b) C_2H_4; (c) C_2H_5OH; (d) C_6H_6.

In Exercises 8.47 and 8.48, estimate the standard enthalpy of formation of each molecule in the gas phase.

8.47 (a) HCl; (b) H_2O_2.
8.48 (a) HBr; (b) NH_2OH.

In Exercises 8.49 and 8.50, use the information in Tables 8.3 and 6.2 to estimate the enthalpy of formation of each compound in the liquid state, given that the standard enthalpy of sublimation of carbon is +717 kJ/mol.

8.49 (a) H_2O; (b) CH_3OH; (c) C_6H_6.

8.50 (a) NH_3; (b) C_2H_5OH; (c) CH_3COCH_3 (acetone).

In Exercises 8.51 to 8.54, use average bond enthalpies to estimate the enthalpy of combustion of each gaseous compound.

8.51 (a) Acetylene; (b) octane; (c) methanol.

8.52 (a) Ethylene; (b) propanol ($CH_3CH_2CH_2OH$); (c) toluene.

8.53 The combustion of glycine (NH_2CH_2COOH) to (a) N_2; (b) NO_2. The mean bond enthalpy of NO_2 is 469 kJ/mol.

8.54 The combustion of ammonia to (a) N_2; (b) NO; (c) NO_2. The mean bond enthalpy of NO is 632 kJ/mol, and that of NO_2 is 469 kJ/mol.

Formal Charge

In Exercises 8.55 to 8.64, calculate the formal charges on the atoms in each Lewis structure.

8.55 (a) Water, H—O—H

(b) Ammonia, H—N—H

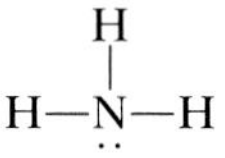

(c) Hydrazine,

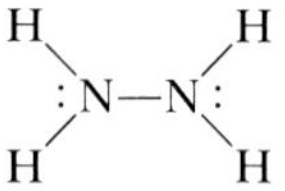

(d) Carbon dioxide, O=C=O

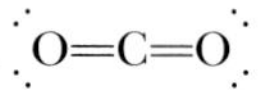

8.56 (a) Methane, H—C—H

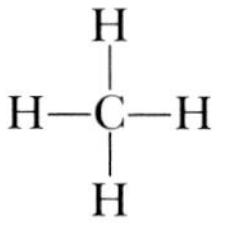

(b) Acetylene, H—C≡C—H

(c) Hydroxylamine,

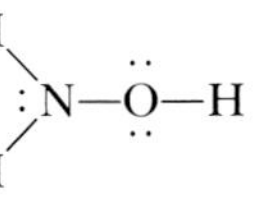

(d) A Kekulé structure of benzene

8.57 (a) 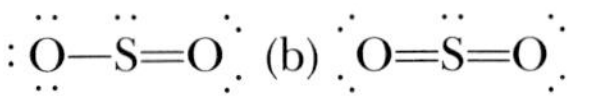 :O—S=O (b) O=S=O

8.58 (a), 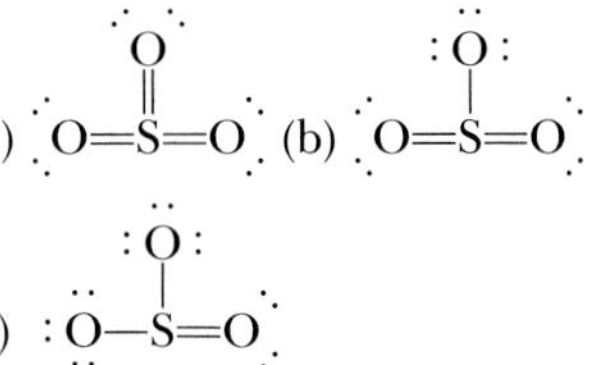 (b), (c)

8.59 (a), 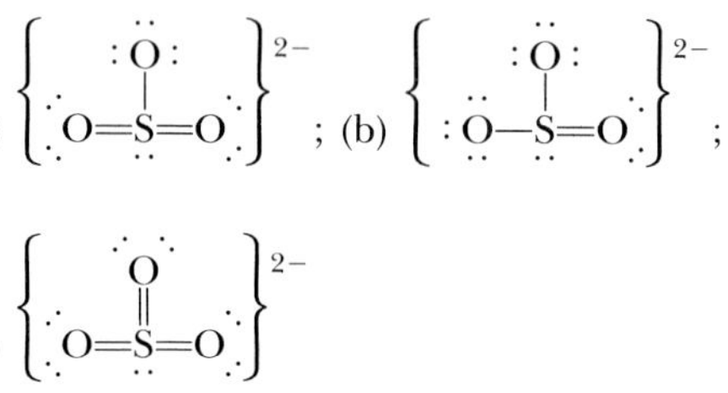 (b), (c)

8.60 (a), 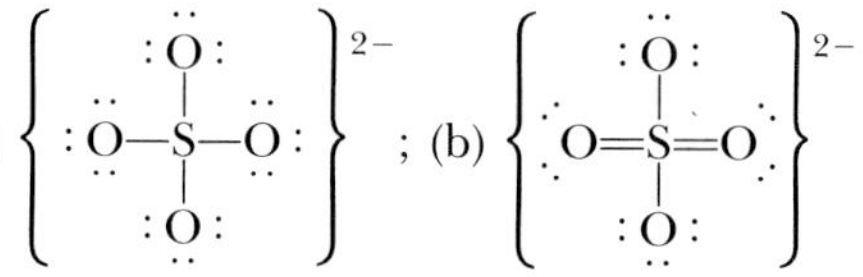 (b)

8.61 (a), 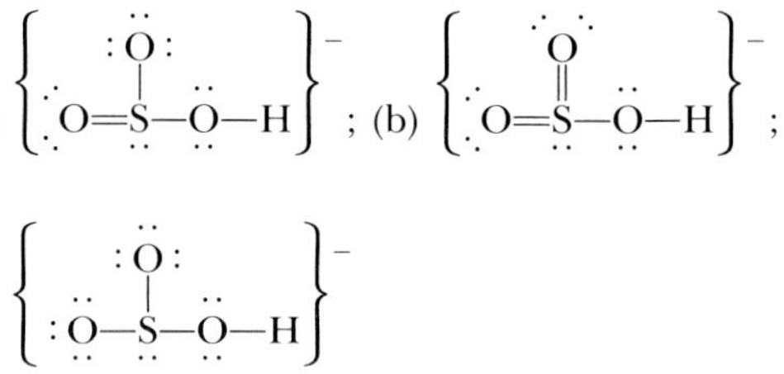 (b), (c)

8.62 (a), 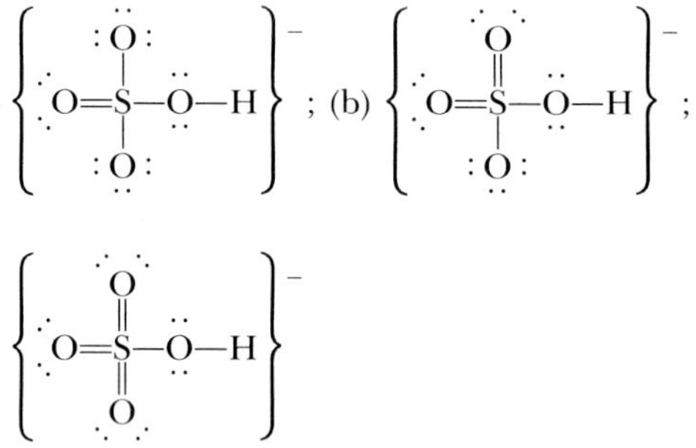 (b), (c)

8.63 (a) 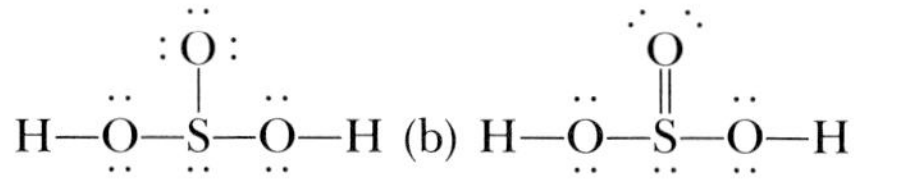 H—O—S—O—H (b) H—O—S—O—H

8.64 (a) 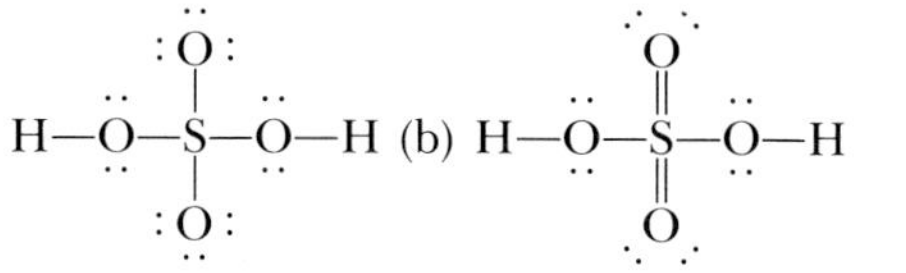 H—O—S—O—H (b) H—O—S—O—H

In Exercises 8.65 and 8.66, select from each pair of Lewis structures the one that is likely to make a dominant contribution to a resonance hybrid.

8.65 (a) 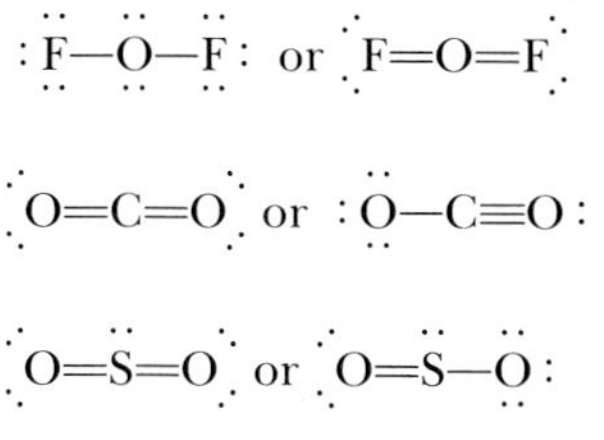 :F—O—F: or F=O=F

(b) O=C=O or :O—C≡O:

(c) O=S=O or O=S—O:

8.66 (a) 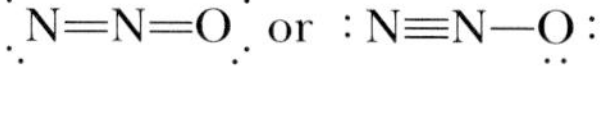 N=N=O or :N≡N—O:

(b) $\left\{ \begin{array}{c} \ddot{\text{O}} \\ \| \\ \ddot{\text{O}}=\text{N}-\ddot{\text{O}}\colon \end{array} \right\}^{-}$

or $\left\{ \begin{array}{c} \ddot{\text{O}}\colon \\ | \\ \colon\ddot{\text{O}}-\text{N}=\ddot{\text{O}} \end{array} \right\}^{-}$

(c) $\left\{ \begin{array}{c} \colon\ddot{\text{O}}\colon \\ | \\ \ddot{\text{O}}=\text{P}-\ddot{\text{O}}\colon \\ | \\ \colon\ddot{\text{O}}\colon \end{array} \right\}^{3-}$

or $\left\{ \begin{array}{c} \colon\ddot{\text{O}}\colon \\ | \\ \ddot{\text{O}}=\text{P}=\ddot{\text{O}} \\ | \\ \colon\ddot{\text{O}}\colon \end{array} \right\}^{3-}$

In Exercises 8.67 and 8.68, suggest a plausible chemical structure for each compound.

8.67 (a) SCl_4; (b) N_2H_4; (c) ICl_5; (d) XeF_4.

8.68 (a) N_2O_4; (b) N_2O; (c) $S_2O_3^{2-}$; (d) H_3O^+.

Polarization

In Exercises 8.69 and 8.70, arrange the given cations in order of increasing polarizing power, giving reasons for your decisions.

8.69 (a) Li^+, (b) Be^{2+}, (c) Sr^{2+}, and (d) H^+.

8.70 (a) K^+, (b) Mg^{2+}, (c) Al^{3+}, and (d) Cs^+.

In Exercises 8.71 and 8.72, arrange the given anions in order of increasing polarizability, giving reasons for your decisions.

8.71 (a) Cl^-, (b) Br^-, (c) N^{3-}, and (d) O^{2-}.

8.72 (a) N^{3-}, (b) P^{3-}, (c) I^-, and (d) At^-.

In Exercises 8.73 and 8.74, classify each compound as mainly ionic or significantly covalent.

8.73 (a) AgF; (b) AgI; (c) $AlCl_3$; (d) AlF_3.

8.74 (a) $BeCl_2$; (b) $CaCl_2$; (c) $FeCl_3$; (d) Fe_2O_3.

Polar Bonds

In Exercises 8.75 and 8.76, indicate with a dipole arrow ($\leftrightarrow$) the polarity of each bond, and estimate its dipole moment.

8.75 (a) O—H; (b) O—F; (c) F—Cl; (d) O—S.

8.76 (a) N—O; (b) C—O; (c) C—N; (d) N—H.

In Exercises 8.77 and 8.78, decide which of the molecules has nonpolar bonds.

8.77 (a) Br_2; (b) O_3; (c) CH_4; (d) H_2O_2.

8.78 (a) I_2; (b) S_8; (c) P_4; (d) C_6H_6.

General

8.79 Devise an argument to show that if the oxidation number of one element in a reaction is increased, then the oxidation number of at least one other element must be decreased. Go on to show that while the formal charges of elements in different resonance structures of a given molecular formula may differ, the oxidation numbers of the elements are the same.

8.80 Investigate whether the replacement of a carbon–carbon double bond by single bonds is energetically favored, by calculating the reaction enthalpy for the conversion of ethylene (C_2H_4) to ethane (C_2H_6).

8.81 Give chemical examples to justify the inclusion of the coinage metals in a single group of the periodic table. Explain any differences in their behavior.

8.82 Give examples of the similarities and differences between the coinage metals and the alkali metals. Account for the comparisons in terms of the electron configurations of the atoms.

This chapter explains how to account for and predict the shapes of molecules. It also shows how to bring bonding theory into line with quantum mechanics by describing molecular structure in terms of the orbitals occupied by electrons. The illustration shows the rods and cones in the light-sensitive retina of an eye. The initial process responsible for vision depends on a change of shape of the molecules inside these receptors.

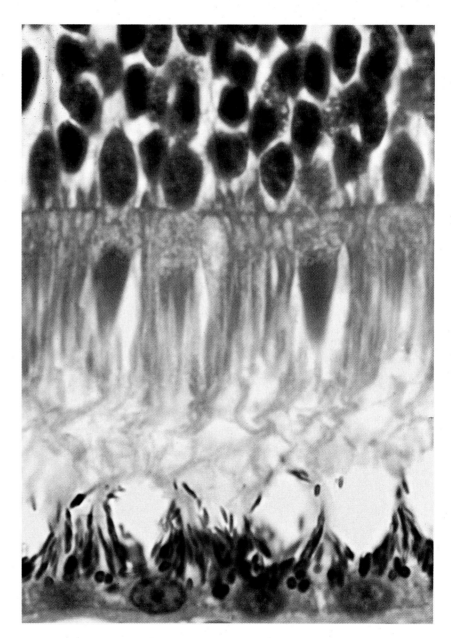

9

THE SHAPES OF MOLECULES

Electron-pair repulsions

9.1 The arrangement of
      electron pairs
9.2 Polar molecules

**The orbital model of
  bonding**

9.3 Hybridization
9.4 Bonding orbitals

Molecular orbital theory

9.5 Molecular orbitals
9.6 Bonding in Period 2
      diatomic molecules
9.7 Orbitals in polyatomic
      molecules
9.8 A perspective on
      chemical bonding

A molecule's shape is often an essential part of the explanation of its properties. We shall see, for instance, that the shape of the H₂O molecule is a factor in water's ability to act as a solvent for ionic compounds. In addition, the shapes of many organic molecules affect their odors, their tastes, and their action as drugs, as well as many of the reactions going on inside our bodies as we live, think, and work. Molecular

TABLE 9.1   The names of molecular shapes*

Number of atoms and (examples)	Shape		Polar?†
2 (H_2, HCl)	Linear		No if A_2 Yes if AB
3 (XY_2: OCO, FBeF, HCN)	Linear		No if BAB Yes if BAA or BAC
(XY_2E_2: HOH, HOF) (XY_2E: O_3, SO_2)	Angular		Yes
4 (XY_3: BF_3, CO_3^{2-}, NO_3^-)	Trigonal planar		No if AB_3
(XY_3E: NH_3, SO_3^{2-}, PCl_3)	Trigonal pyramidal		Yes
(XY_3E_2: ClF_3)	T-shaped		Yes
5 (XY_4E_2: XeF_4)	Square planar		No if AB_4
(XY_4E: SF_4)	Saw horse or distorted tetrahedron		Yes
(XY_4: CH_4, SO_4^{2-}, XeO_4)	Tetrahedral		No if AB_4

(continued)

*The conventional chemical formulas are written AB_n; when we want to specify the numbers of atoms and lone pairs on the central atom, we use XY_nE_m, where E denotes a lone pair. Thus, :NH_3 is an AB_3 molecule, and specifically XY_3E.
†Polar molecules are described in Section 9.2.

shapes can be accounted for in terms of the ideas we developed about atomic and molecular structure in Chapters 7 and 8.

In this chapter we deal mainly with the shapes of simple molecules—those containing no more than about a dozen atoms. Such molecules can be classified by shape using the names of geometric figures, as shown in Table 9.1. We recognize the shape of a molecule by noting the

Number of atoms and (examples)	Shape		Polar?†
6 (XY_5: PCl_5)	Trigonal bipyramidal		No if AB_5
(XY_5E, BrF_5)	Square pyramidal		Yes
7 (XY_6: SF_6, XeO_6^{4-})	Octahedral		No if AB_6
8 (XY_7: IF_7)	Pentagonal bipyramidal		No if AB_7

arrangement of the atoms in the molecule, and not the locations of any lone pairs that may be present. (Recall that lone pairs are pairs of valence electrons not taking part in bonding.) We classify the ammonia molecule NH_3 with its one lone pair, for example, as trigonal pyramidal, and the ammonium ion NH_4^+ as tetrahedral. Later we shall see that the four electron pairs in a water molecule are arranged tetrahedrally around the oxygen atom. However, as only two of the pairs form bonds between atoms, we classify the H_2O molecule as angular, not tetrahedral.

When we want to report shapes numerically, we report the *bond angles*. The bond angle in an A—B—C molecule or part of a molecule is the angle between the A—B and B—C bonds. The bond angle in H_2O, for instance, is the angle (104°) between the two O—H bonds.

ELECTRON-PAIR REPULSIONS

Our main task in this chapter is to explain the shapes of molecules and their bond angles in terms of the Lewis structures described in Chapter 8. To account for molecular shapes, we need the following simple addition to Lewis's ideas: *electron pairs repel each other*. As a result of this repulsion, the valence electron pairs on an atom (which include the lone pairs as well as the bonding pairs) take up positions as far from one another as possible, for then they repel each other least (Fig. 9.1). The positions of the atoms, and hence the shape of the molecule, are then dictated by the positions of the bonding electron pairs. This idea was first explored by the British chemists Nevil Sidgwick and Herbert Powell and is called the *valence-shell electron-pair repulsion theory*, or "VSEPR theory" for short.

9.1 THE ARRANGEMENT OF ELECTRON PAIRS

The VSEPR theory focuses on the positions taken up by a given number of electron pairs on the central atom of a molecule (the S atom in

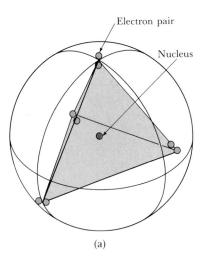 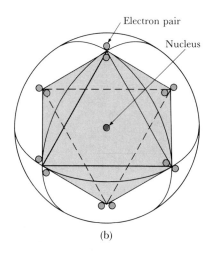

FIGURE 9.1 Electron pairs repel each other, and according to VSEPR theory, take up positions as far apart as possible. (a) Four electron pairs, (b) six.

(a)   (b)

SF_6 or the C atom in CO_2). These positions are predicted by assuming that all the pairs (the lone pairs as well as the bonding pairs) move as far apart as possible on the surface of an imaginary sphere surrounding the central atom. The "most-distant" arrangements of one through six electron pairs, which are found by using trigonometry, are shown in Table 9.2.

Lewis structures with single bonds. The first step in applying VSEPR theory to predict the shape of a molecule is to write down the Lewis

TABLE 9.2 Electron-pair arrangements according to VSEPR theory

Number of pairs	Arrangement of electron pairs*	Examples
2	Linear	
3	Trigonal planar	BF_3
4	Tetrahedral	CH_4, NH_3, H_2O
5	Trigonal bipyramidal	PCl_5, SF_4
6	Octahedral	SF_6, XeF_4

*Each line represents the position of one electron pair.

structure of the molecule. For the methane molecule (CH_4) and the sulfite ion (SO_3^{2-}), these are

$$H-\underset{\underset{H}{|}}{\overset{\overset{H}{|}}{C}}-H \qquad \left\{ \begin{array}{c} :\ddot{O}: \\ | \\ :\ddot{O}-S-\ddot{O}: \end{array} \right\}^{2-}$$

CH_4 is an example of an XY_4 molecule, and SO_3^{2-} an example of an XY_3E polyatomic ion, where X is the central atom, Y an attached atom, and E a lone pair. In both these structures, the central atom is surrounded by four electron pairs. (At this stage we do not distinguish between bonding pairs and lone pairs.) Table 9.2 indicates that the four electron pairs adopt a tetrahedral arrangement.

The next step is to decide which electrons are bonding pairs and which are lone pairs. Then, by noting the arrangement of the bonded atoms but ignoring the lone pairs, we identify the molecule's shape by referring to Table 9.1. Since in CH_4 each pair bonds an H atom, we predict that the molecule is tetrahedral with HCH angles of 109° **(1)**. Since in SO_3^{2-} three electron pairs bond O atoms, we predict that the ion is trigonal pyramidal with OSO angles of 109°. The methane molecule is known to be tetrahedral with HCH angles equal to 109°. However, although the sulfite ion is trigonal pyramidal, the observed bond angle is only 106°, and not the predicted 109°. Clearly, although VSEPR theory is broadly correct, it must be refined if it is to be used to make accurate predictions.

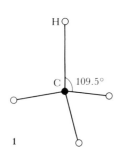

1

▼ **EXAMPLE 9.1** **Predicting the shape of a simple molecule**

Predict the shape of the BF_3 molecule.

STRATEGY The first step is to determine, from its Lewis structure, the number of electron pairs on the B atom. Then Table 9.2 gives the arrangement that minimizes their repulsions. We must then decide which pairs bond F atoms, and finally predict the shape of the molecule by matching the atomic arrangement to one of the shapes in Table 9.1.

SOLUTION The Lewis structure of BF_3 is

$$:\ddot{F}-\underset{}{\overset{\overset{\overset{..}{:}\ddot{F}:}{|}}{B}}-\ddot{F}:$$

The B atom has three electron pairs, which adopt a trigonal planar arrangement. Since each pair bonds an F atom, we predict that the molecule is trigonal planar with an FBF bond angle of 120°. That shape is confirmed by experiment.

EXERCISE Predict the shape of the PCl_5 molecule.

[*Answer*: Trigonal bipyramidal]

▲

The effect of lone pairs. The first refinement of VSEPR theory takes into account the greater repelling effect of a lone pair, as compared with a bonding pair. This difference in repulsion effects is sometimes explained by arguing that a lone pair is spread over a greater volume than a bonding pair (Fig. 9.2), since the latter is confined to the space between two atoms. Because a lone pair occupies more space, the argument continues, it repels other electron pairs from a greater region of space. Whatever the actual reason (no one really knows), VSEPR theory now includes the view that the strengths of electron-pair repulsions are ordered as follows:

Lone pair–lone pair repulsions
are stronger than
Lone pair–bonding pair repulsions
are stronger than
Bonding pair–bonding pair repulsions

More briefly,

(LP, LP) > (LP, BP) > (BP, BP)

It is therefore energetically most favorable for lone pairs to be as far from each other as possible. It is also energetically most favorable for bonding pairs to be far from lone pairs, even though that might bring them closer to other bonding pairs.

This additional feature accounts for the bond angle of the XY_3E sulfite ion. The four electron pairs adopt a tetrahedral arrangement. However, one of them is a lone pair **(2)** that repels the three bonding pairs, forcing them to move together slightly. Hence the OSO angle is reduced from the 109° of the regular tetrahedron toward the 106° observed experimentally. Note that although VSEPR theory can predict the *direction* of the distortion, it cannot predict its extent: we can predict that the angle will be less than 109° but cannot say it will be 3° less.

▼ EXAMPLE 9.2 Using VSEPR theory to predict shapes: I

Account for the shape of the water molecule (and of XY_2E_2 molecules in general).

STRATEGY According to VSEPR theory, we must count the number of electron pairs on the central atom and identify the arrangement they adopt, using Table 9.2. After that, we identify the bonding pairs, and hence identify the shape of the molecule with Table 9.1. Finally, we allow the lone pairs to move apart, at the expense of the bonding pairs moving together slightly.

SOLUTION The central atom has four pairs of electrons in a tetrahedral arrangement. In H_2O, the central oxygen atom has two lone pairs and two bonding pairs, giving an angular molecule **(3)**. Since the lone pairs move apart slightly and the bonding pairs move together to avoid them, we predict an HOH bond angle of less than 109°. This is in agreement with the experimental value of 104°.

EXERCISE Predict the shape of the NH_3 molecule (and of XY_3E molecules in general).

[*Answer*: Trigonal pyramidal **(4)**; YXY angle of less than 109°]

In trigonal bipyramidal (and pentagonal bipyramidal) molecules we must also distinguish between "axial" and "equatorial" electron pairs. An *axial* pair lies on the axis of the molecule **(5)**, but an *equatorial* pair lies in one of the positions around the "equator" of the molecule. A pair

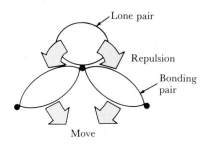

FIGURE 9.2 Lone pairs can be considered to repel electron pairs from a greater region of space than do bonding pairs, and so the latter move away from them.

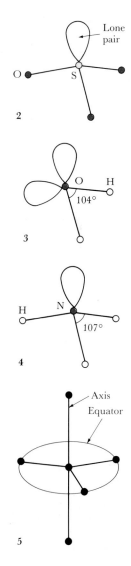

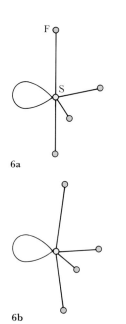

6a

6b

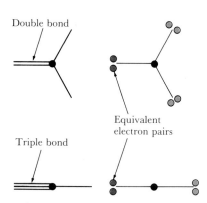

Double bond

Triple bond

Equivalent
electron pairs

FIGURE 9.3 For predicting the shapes of molecules, a double or triple bond is treated like a single pair of electrons (represented by red circles).

in an axial position repels three electron pairs strongly, but in the equatorial position it repels only two strongly. Therefore, if one of the pairs is a lone pair, it is energetically more favorable for it to be in an equatorial position.

▼ **EXAMPLE 9.3** Using VSEPR theory to predict shapes: II

Predict the shape of the sulfur tetrafluoride molecule.

STRATEGY We decide how many electron pairs are present by writing a Lewis structure of the molecule (as in Examples 8.4 and 8.5). Next, we determine the arrangement of the pairs by referring to Table 9.2. After that, we identify the bonding pairs and allow the molecule to distort so that lone pairs are as far from each other and from bonding pairs as possible. Finally, we identify the molecular shape using Table 9.1.

SOLUTION The Lewis structure of SF_4 was found in Example 8.8 to be

$$: \ddot{F} :$$
$$: \ddot{F} —\ddot{S} :$$
$$\ddot{F} \quad \ddot{F}$$

The S atom has five electron pairs, and the molecule is XY_4E. The pairs adopt a trigonal bipyramidal arrangement. The lone pair can be either axial or equatorial, but the latter arrangement **(6a)** is likely to have the lower energy. The energy is lowered still further if the four bonds bend away from the lone pair **(6b)**. This shape is the one found experimentally.

EXERCISE Predict the shape of the xenon tetrafluoride (XeF_4) molecule.
[*Answer*: Square planar]
▲

As this example shows, it is important to be alert to the presence of lone pairs in molecules, and not to conclude that every molecule with chemical formula AB_3 is trigonal planar or that every AB_4 molecule is tetrahedral. In each case, one or more lone pairs may be present and will affect the shape of the molecule. However, all XY_3 molecules (AB_3 molecules with no lone pairs) can be expected to have the same shape, as can all XY_3E molecules (AB_3 molecules with one lone pair) and all XY_3E_2 molecules.

Lewis structures with multiple bonds. The VSEPR theory includes rules about how to deal with multiple bonds. In a $Z{=}XY_2$ molecule in which the Z atom is joined to the central X atom by a double bond, the two electron pairs of the double bond hold the Z atom in nearly the same place as it would be in a $Z{-}XY_2$ molecule (Fig. 9.3). Similarly, in a $Z{\equiv}XY$ molecule, the three electron pairs of the triple bond hold the Z atom in the same place as a single electron pair would hold it. This suggests that, in using VSEPR theory to predict molecular shapes, we should treat a multiple bond as though it were a *single* electron pair.

Suppose we wanted to predict the shape of an ethylene molecule, $CH_2{=}CH_2$. The first step, as always, is to write its Lewis structure:

$$\begin{array}{ccc} H & & H \\ \diagdown & & \diagup \\ & C{=}C & \\ \diagup & & \diagdown \\ H & & H \end{array}$$

Each carbon is treated as having *three* electron pairs, since the two pairs of the double bond are treated as one. The arrangement around each carbon atom is therefore trigonal planar. We predict that the molecule will have the shape shown in (**7**), with the HCH and HCC angles both 120°. This is close to its actual shape.

A feature not brought out very clearly by VSEPR theory, however, is that the two CH_2 groups in ethylene lie in the same plane. Double bonds between carbon atoms always result in a planar arrangement of atoms, as explained in Section 9.4.

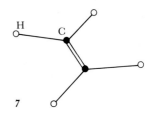

7

▼ **EXAMPLE 9.4** **Predicting the shape of a molecule with a multiple bond**

Suggest a shape for the acetylene molecule, HC≡CH.

STRATEGY The shape is given by VSEPR theory with the triple bond treated as a single electron pair. We write the Lewis structure, count the *effective* number of electron pairs around each C atom, and predict the arrangement of the atoms around each one.

SOLUTION The Lewis structure of the molecule is H—C≡C—H. Since the triple bond behaves as one electron pair, each C atom should be treated as having two electron pairs—one shared with the other C atom and the other bonding an H atom. Their arrangement on each atom is linear, with one pair diametrically opposite the other. The molecule is therefore also linear (**8**).

EXERCISE Predict the shape of a carbon suboxide molecule, C_3O_2, in which the atoms lie in the order OCCCO.

[*Answer*: Linear]

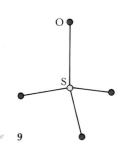

8

Treating multiple bonds as single pairs does away with any worries about which of several resonance structures to choose in determining shapes. This can be illustrated with the sulfate ion and the following two contributions to its resonance hybrid:

$$\left\{ \begin{array}{c} :\ddot{O}: \\ | \\ :\ddot{O}—S—\ddot{O}: \\ | \\ :\ddot{O}: \end{array} \right\}^{2-} \quad \left\{ \begin{array}{c} :\ddot{O}: \\ | \\ \ddot{O}=S=\ddot{O} \\ | \\ :\ddot{O}: \end{array} \right\}^{2-}$$

In the structure on the left the S atom has four bonding pairs, so we would predict a tetrahedral ion. In the other structure, the S atom is to be treated as though it had four electron pairs, so again the ion would be predicted to be tetrahedral (**9**).

The shapes of XY_aE_b molecules. Since the same shape is obtained for every resonance structure of a polyatomic molecule or ion, all we need to know to predict its shape is the number of lone pairs and atoms linked to the central atom. We do not need to know whether they are linked by single or multiple bonds. This simplification is the basis of a shortcut for predicting shapes.

First, count the total number of lone pairs and *atoms* (not bonds) attached to the central atom. Next, determine the arrangement of lone pairs and atoms using Table 9.2, and obtain the molecular shape from Table 9.1. Finally, take minor distortions into account by allowing for the greater repelling power of lone pairs.

▼ **EXAMPLE 9.5** **Predicting the shape of a molecule using the shortcut**

Predict the shape of an ozone molecule, O_3.

STRATEGY After writing the Lewis structure, we follow the steps outlined in the preceding paragraph.

SOLUTION The Lewis structure $:\!\ddot{O}\!=\!\ddot{O}\!-\!\ddot{O}\!:$ shows that the total number of lone pairs and atoms attached to the central atom is three: O_3 is an XY_2E molecule (in this special case, the X and Y atoms are all O atoms). Hence their arrangement is expected to be trigonal planar. Since two atoms are joined to the central atom, the molecule is angular (**10**). The two outer O atoms are bent away from the lone pair, giving an expected bond angle slightly smaller than 120°. Experimentally, the O_3 molecule is angular, and the O—O—O bond angle is 117°.

EXERCISE Predict the shape of a sulfur trioxide molecule.

[*Answer*: Trigonal planar]
▲

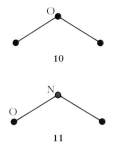

10

11

12

An even quicker way to predict the shape of a molecule is to find that it is isoelectronic with another molecule with a known shape, for their shapes will be the same. For instance, O_3 is isoelectronic with NO_2^- and SO_2. Because we know that O_3 is angular, we can predict that NO_2^- and SO_2 are angular too (**11** and **12**).

9.2 POLAR MOLECULES

The role of molecular shape will become increasingly clear in the remainder of this text. Here we introduce just one of many properties that depend on it: the *polarity* of a molecule, its possession of an electric dipole moment. (We discussed the concept of an electric dipole for a bond in Section 8.7.) A molecule may be either "polar" or "nonpolar":

A **polar molecule** is a molecule with a nonzero electric dipole moment.

It is very important to distinguish between a polar *molecule* and a polar *bond*. An individual bond is polar if the bonding electrons are shared unequally by the two bonded atoms. However, although each bond in a polyatomic molecule may be polar, the molecule as a whole might be nonpolar; this would be the case if its shape were such that the dipoles of the individual bonds pointed in opposite directions and canceled each other. An illustration of the importance of shape is the difference between *cis*-dichloroethylene (**13**) and *trans*-dichloroethylene (**14**): the bonds are the same in both molecules, but in the latter the dipoles (represented by the arrows) cancel and the compound is nonpolar. A simple test for polarity of molecules in a liquid is to see whether a stream of the liquid is deflected by an electrically charged rod (Fig. 9.4).

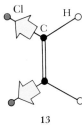

13

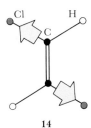

14

Tetrahedral XY_4 molecules with the same atom at each corner (as in CCl_4) are nonpolar because the dipoles of the four bonds cancel each other's effects. If one corner atom is replaced with a different atom (as in $CHCl_3$) or with a lone pair (as in NH_3 or any other XY_3E molecule), the cancellation does not occur and the molecule becomes polar. Similarly, we can think of an XY_2E_2 molecule (such as H_2O) as being derived from a nonpolar tetrahedral molecule by the replacement of two atoms with lone pairs. The bond dipoles no longer cancel, and the molecule is polar. The H_2O molecule is thus polar, and the O atom is the negative end of the dipole.

Whether or not a molecule with polar bonds is polar depends on the symmetry of the arrangement of its individual bonds: highly symmetrical arrangements (such as linear, planar triangular, and tetrahedral) are nonpolar. Table 9.1 shows which molecular shapes have canceling bond dipoles when all the atoms B joined to the central atom A are identical. Whether a given molecule is polar can often be decided by seeing whether it has one of these shapes or can be derived from one of them by loss or replacement of an atom.

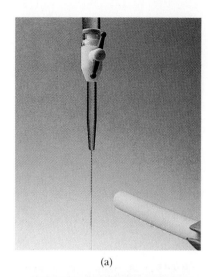

(a)

(b)

FIGURE 9.4 Whether or not a liquid is composed of polar molecules can be shown by running a stream of the liquid past a charged rod. A nonpolar liquid (such as *trans*-dichloroethylene) is undeflected (a), but a polar fluid (*cis*-dichloroethylene) is deflected (b).

▼ EXAMPLE 9.6 Predicting the polarity of a molecule

Predict whether (a) a boron trifluoride molecule or (b) an ozone molecule is polar.

STRATEGY In each case we must determine (using VSEPR theory) the shape of the molecule and whether it has any lone pairs on the central atom. Then, if the molecule has one of the shapes in Table 9.1 and the atoms joined to the central atom are identical, it is nonpolar. If the shape is not listed in the table, we must decide whether it can be obtained from one in the table by removal or replacement of an atom. If it can, and if the symmetry of the molecule is upset so that the bond dipoles do not cancel, then the molecule is polar.

SOLUTION (a) According to VSEPR theory, BF_3 is a trigonal planar XY_3 molecule. This is one of the shapes in Table 9.1. Because it has the same atom at each vertex, it is nonpolar **(15)**.
(b) According to VSEPR theory, the O_3 molecule is angular, with one lone pair on the central O atom **(10)**. If all three positions around that O atom were linked to O atoms, the molecule would be a nonpolar XY_3 molecule like BF_3. However, because one position is occupied by a lone pair, the dipole cancellation is upset and the molecule is polar.

EXERCISE Predict whether (a) SF_6 and (b) SF_4 are polar or nonpolar.
[*Answer*: (a) Nonpolar; (b) polar]

THE ORBITAL MODEL OF BONDING

The Lewis theory of the chemical bond and its extension, VSEPR theory, assume that each bonding electron pair is located between the two bonded atoms. However, we know from Chapter 7 that electrons are not located at precise positions, but are spread out over a region defined by the atomic orbitals they occupy. Our task in this section is to

15

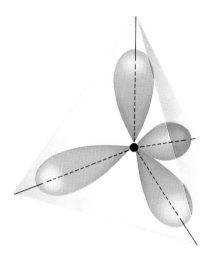

FIGURE 9.5 One s orbital and three p orbitals blend into four sp^3 hybrid orbitals pointing toward the corners of a tetrahedron.

see how to combine the Lewis approach to molecular structure with the description of a molecule in terms of orbitals. This extension also introduces concepts and a terminology that are used throughout chemistry.

9.3 HYBRIDIZATION

Suppose we were trying to express the locations of the four bonding electron pairs in CH_4 in terms of atomic orbitals on the C atom. At first sight this looks difficult because the valence shell of the C atom has one spherical $2s$ orbital and three dumbbell-shaped $2p$ orbitals that point along three perpendicular axes. This arrangement appears to suggest a structure with three of the C—H bonds at 90° to each other. According to VSEPR theory, though, and as confirmed by experiment, the four electron pairs are arranged as a tetrahedron with all four C—H bonds at 109° to each other.

s- and p-orbital hybridization. The way out of this difficulty lies in the fact that s and p orbitals are like waves centered on the nucleus. Like waves in water, the four orbitals produce patterns where they mix; the result of this mixing, in methane, is the four *hybrid orbitals*, or mixed orbitals, shown in Fig. 9.5. In other words, when the one s orbital and three p orbitals of the C atom in methane accommodate an octet of electrons, the combined charge cloud of all the electrons is the same as it would be if the electrons occupied the four hybrid orbitals. The description of the central atom in a molecule as having four pairs of electrons in s and p orbitals is *equivalent* to a description in which the four pairs occupy the four hybrid orbitals. As shown in Fig. 9.5, the four hybrid orbitals point toward the corners of a regular tetrahedron. This arrangement is in line with VSEPR theory and the electron-pair distribution in Table 9.2.

The formation of hybrid orbitals when atomic orbitals on an atom mix together is called *hybridization*. When one s orbital and three p orbitals mix, hybridization results in the four tetrahedral sp^3 *hybrids* of Fig. 9.5. (The notation sp^x means that the orbital is a mixture of one s orbital and x p orbitals.) Different patterns result from other mixtures. For example, mixing one s orbital and two p orbitals results in three sp^2 *hybrids* that point to the corners of an equilateral triangle (Fig. 9.6) and leaves one p orbital unchanged. The most important hybridizations are shown in Table 9.3; in every case, mixing N atomic orbitals gives N hybrids.

It is important to understand the relationship between hybridization and VSEPR theory. We use hybridization to describe the orbital arrangement of electron pairs once the molecule's shape has been predicted by VSEPR theory. If VSEPR theory predicts a tetrahedral arrangement of pairs, then we assume that one s and three p orbitals from the central atom hybridize into four sp^3 hybrids. If VSEPR theory predicts a trigonal planar arrangement, we assume one s and two p orbitals from the central atom hybridize into three sp^2 hybrids to produce it. Our analysis of hybridization *follows* from the shape; it is not a method for predicting the shape.

TABLE 9.3 Hybridization and molecular shape

Arrangement	Hybridization
Linear	sp
Trigonal planar	sp^2
Tetrahedral	sp^3
Square planar	sp^2d
Trigonal bipyramidal	sp^3d
Octahedral	sp^3d^2

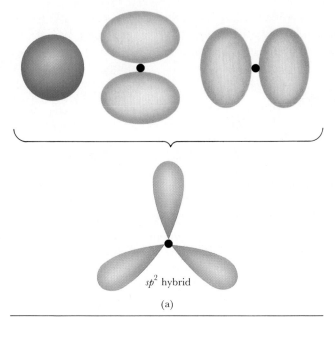

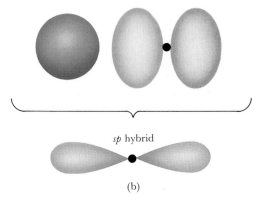

FIGURE 9.6 (a) One *s* orbital (blue) and two *p* orbitals (yellow) form three *sp*² hybrids (green) pointing to the corners of an equilateral triangle; (b) one *s* orbital and one *p* orbital form two linear *sp* hybrids.

*sp*² hybrid
(a)

sp hybrid
(b)

▼ EXAMPLE 9.7 Selecting a hybridization scheme

What is the hybridization of the B atom in BF_3?

STRATEGY We use VSEPR theory to predict the shape of the molecule, and then identify (in Table 9.3) the hybridization of the central atom that reproduces the predicted arrangement of electron pairs.

SOLUTION BF_3 is a trigonal planar molecule (see Example 9.1). We therefore take the B atom to be sp^2 hybridized, since three hybrid orbitals of this composition, and the three electron pairs they contain, point toward the corners of an equilateral triangle.

EXERCISE What is the hybridization of X in an XY_2 linear molecule?
[*Answer*: *sp* hybridized]

▲

Hybrids including d orbitals. When more than four electron pairs are present in the valence shell of the central atom, its *d* orbitals are needed to accommodate the expanded octet. If five pairs of valence electrons are present, so that five orbitals are needed, one *d* orbital must be

FIGURE 9.7 The sp^3d and sp^3d^2 hybrids, in which d orbitals are used as well as s and p orbitals.

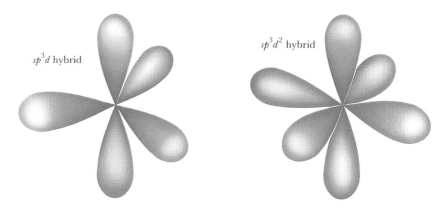

sp^3d hybrid

sp^3d^2 hybrid

occupied in addition to the four s and p orbitals. This gives five sp^3d hybrid orbitals, pointing to the corners of a trigonal bipyramid (Fig. 9.7). When each hybrid contains an electron pair, their trigonal bipyramidal arrangement reproduces the one predicted by VSEPR theory. Six electron pairs require two d orbitals in addition to the four s and p orbitals; in this case hybridization leads to six sp^3d^2 hybrids that point toward the corners of a regular octahedron (Fig. 9.7).

▼ **EXAMPLE 9.8 Describing a structure in terms of *d*-orbital hybridization**

Describe the structure of phosphorus pentafluoride in terms of the hybridization of the phosphorus atom.

STRATEGY Since phosphorus is a Period 3 element, we should be prepared for d-orbital involvement. We need to match the hybridization of the P atomic orbitals to the arrangement of electron pairs predicted by VSEPR theory. The first step is therefore to decide what that arrangement is; then we can refer to Table 9.3 for the corresponding hybridization.

SOLUTION The Lewis structure of the molecule is

and the central atom has five electron pairs. According to Table 9.2, these adopt a trigonal bipyramidal arrangement. From Table 9.3, this arrangement is matched by sp^3d hybridization. The five hybrid orbitals can accommodate the five electron pairs.

EXERCISE Account for the structure of sulfur hexafluoride in terms of the hybridization of the sulfur atom.

[*Answer*: sp^3d^2 octahedral hybridization]

9.4 BONDING ORBITALS

The final step we must take to bring the Lewis theory into line with quantum mechanics is to use hybrid orbitals to describe the bonds themselves. We know that we can think of electron pairs as occupying

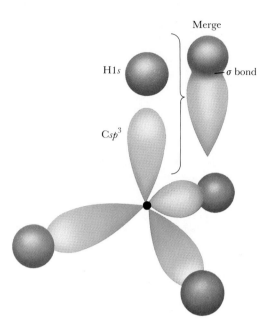

Merge

H1s

σ bond

Csp^3

FIGURE 9.8 Bonding orbitals in which hydrogen $1s$ orbitals have merged with four carbon sp^3 hybrid orbitals to give four σ bonds.

specific regions around the central atom. If a pair is bonding, we can also expect its two electrons to spread throughout the region between the central atom and the atom bonded to it.

σ bonds. We can picture bond formation as a merging of the orbitals of the two bonded atoms. A carbon sp^3 orbital can merge—the technical term for merging is *overlap*—with a $1s$ orbital on a neighboring hydrogen atom. This creates a charge cloud that spreads over and between the two nuclei and binds them together in a "smeared-out" version of Lewis's idea of a shared electron pair (Fig. 9.8). The merged charge cloud is called a σ *bond* ("sigma bond"). The letter σ is the Greek version of s and is used because, viewed along the bond, the charge cloud resembles an atomic s orbital. A σ bond is formed when two $1s$ orbitals merge, but it can also be formed when a p orbital merges with an s orbital or with another p orbital.

We can think of all the single bonds discussed so far as σ bonds formed by the overlap of orbitals on neighboring atoms. In H_2O, for instance, we can translate the Lewis picture of an O—H bond into a σ bond consisting of a smeared-out cloud of two electrons. All the bonds in the ethanol molecule (CH_3CH_2OH, Fig. 9.9) are σ bonds. The bond between the two C atoms is formed by the overlap of carbon sp^3 hybrids and is denoted (Csp^3, Csp^3). Using the same notation, we can write the bonds between the C and H atoms as (Csp^3, $H1s$), showing that they are formed by the overlap of carbon sp^3 hybrids with hydrogen $1s$ orbitals. The C—O bond is a (Csp^3, Osp^3) σ bond, formed where a carbon sp^3 hybrid merges with an oxygen sp^3 hybrid. The O—H bond is an (Osp^3, $H1s$) σ bond. The two lone pairs on the O atom occupy the two remaining sp^3 hybrids.

In the SF_6 molecule (in Fig. 9.9), each S—F bond is an (Ssp^3d^2, $F2p$) σ bond formed by the overlap of a $2p$ orbital on the F atom with one of the sp^3d^2 hybrid orbitals on the S atom. All this is much the same as in

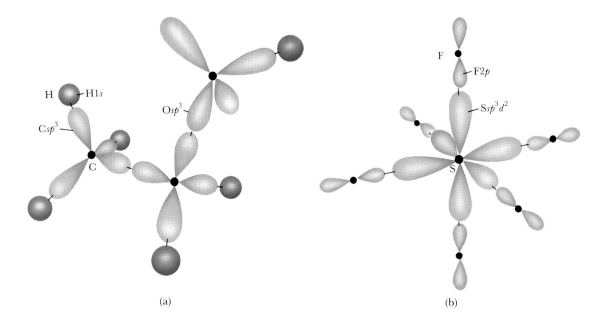

(a)

(b)

FIGURE 9.9 (a) The bonds in ethanol are all σ bonds, formed by various types of overlap. (b) In SF_6 all the σ bonds are formed by F2p atomic orbitals overlapping with S hybrid orbitals.

the Lewis and VSEPR pictures, except for the spreading of the electrons as a cloud of charge.

π bonds. We shift our attention now to double bonds. To picture the charge cloud of a double bond, we consider the VSEPR description of ethylene, C_2H_4 **(7)**, and its interpretation in terms of charge clouds.

All six atoms in ethylene lie in the same plane (Fig. 9.10), with HCH and CCH bond angles of 120°. This suggests that the C atoms are sp^2 hybridized and linked together by a (Csp^2, Csp^2) σ bond. The H atoms form (Csp^2, H1s) σ bonds with the remaining lobes of the hybrids. These five σ bonds account for 10 of the 12 valence electrons in the molecule. Since the C atoms are sp^2 hybridized, each C atom has an unused 2p orbital perpendicular to the plane of the σ bonds. These two 2p orbitals are so close together that they overlap to give a charge cloud that consists of one region above the plane of the molecule and one below (Fig. 9.11). This entire charge cloud, not either half separately, is called a π *bond* ("pi bond"). The Greek letter π, the equivalent of our p, is used because of the resemblance of the charge cloud to a p orbital when it is viewed along the bond. Therefore, the C=C double bond

FIGURE 9.10 An exploded view of the bonding structure in ethylene, showing the σ framework and the π bond formed by broadside overlap of carbon 2p orbitals.

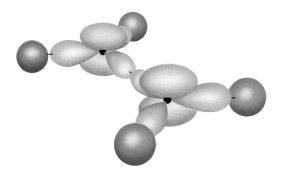

consists of an electron pair in a σ bond plus another electron pair in a π bond. This is the structure of all double bonds. In general:

A **single bond** is a σ bond.

A **double bond** is a σ bond plus a π bond.

A **triple bond** is a σ bond plus two π bonds.

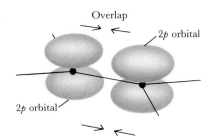

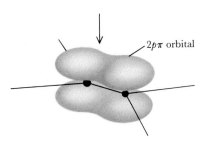

FIGURE 9.11 The structure of a π bond. Although it consists of two regions, the entire double charge cloud is one bond.

▼ **EXAMPLE 9.9** **Analyzing the structure of a molecule in terms of σ and π bonds**

Describe the structure of the acetic acid molecule (CH_3COOH) in terms of σ and π bonds and lone pairs in hybrid orbitals.

STRATEGY We must first write the Lewis structure of the molecule (this was done in Example 8.6) and then interpret each type of bond in terms of orbitals. We consider each single bond to be a σ bond formed by the overlap of neighboring orbitals, and each double bond as a σ bond plus a π bond. An atom with one double bond is sp^2 hybridized, with the spare p orbital used to form the π bond.

SOLUTION The Lewis structure is

$$\begin{array}{c} \text{H} \qquad \ddot{\text{O}}: \\ | \qquad \parallel \\ \text{H}-\text{C}-\text{C} \\ | \qquad \ddot{\text{O}}-\text{H} \\ \text{H} \qquad \ddot{} \end{array}$$

The CH_3 carbon atom is sp^3 hybridized and forms three (Csp^3, $H1s$) σ bonds by merging with the $1s$ orbitals of the three H atoms (Fig. 9.12). It uses the fourth sp^3 hybrid to form a σ bond with the second C atom. That atom is sp^2 hybridized (because it forms one double bond). The C—C single bond is therefore (Csp^3, Csp^2). The other two sp^2 hybrids on the second C atom are

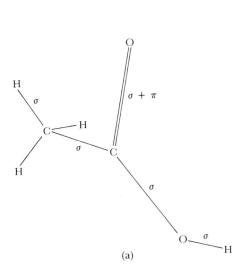

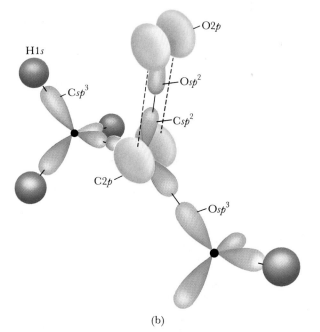

FIGURE 9.12 (a) A summary of the bonding in acetic acid and (b) an exploded view showing the various bonding orbitals in the molecule.

used to form σ bonds with the two O atoms. The O atom of the OH group is sp^3 hybridized, forming a (Csp^2, Osp^3) σ bond with the neighboring C atom and an (Osp^3, H1s) σ bond with the H atom. Its two lone pairs occupy the other two sp^3 hybrids. The other O atom can be pictured as being sp^2 hybridized. One of the hybrids is used to form the (Csp^2, Osp^2) σ bond. The other two carry the two lone pairs. The unhybridized 2p orbital overlaps with the unhybridized carbon 2p orbital and forms a (C2p, O2p) π bond. This structure is summarized in Fig. 9.12.

EXERCISE Account for the structure of benzene in terms of σ and π bonds.

▲   [*Answer*: See Fig. 9.13.]

The structure of a triple bond as one σ bond and two π bonds is best seen by considering acetylene. The Lewis structure is H—C≡C—H, and VSEPR theory indicates a linear structure. The C atoms are therefore sp hybridized, since that produces lobes at 180° to each other. Two of the lobes can merge to give a (Csp, Csp) σ bond, and the others can be used for the (Csp, H1s) σ bonds. The sp hybridization leaves two unhybridized 2p orbitals on each C atom. These merge to give *two* (C2p, C2p) π bonds at 90° to each other. Since there are 5 bonds in all, this accounts for all 10 valence electrons in the molecule.

Some properties of double bonds. The data in Section 8.5 show that a C=C double bond is stronger than a C—C single bond but weaker than the sum of two C—C single bonds. (This is not, however, true of all A=B and A—B bonds.) We also saw that a C≡C triple bond is weaker than the sum of three single bonds:

$$B(\text{C—C}) = 348 \text{ kJ/mol}$$

$$B(\text{C=C}) = 612 \text{ kJ/mol} \qquad B(\text{C≡C}) = \ 837 \text{ kJ/mol}$$

$$2 \times B(\text{C—C}) = 696 \text{ kJ/mol} \qquad 3 \times B(\text{C—C}) = 1044 \text{ kJ/mol}$$

FIGURE 9.13 An exploded view of the structure of benzene. The sp^2-hybridized carbon atoms form σ bonds with each other and with the hydrogen atoms. The remaining six 2p orbitals overlap to form π orbitals.

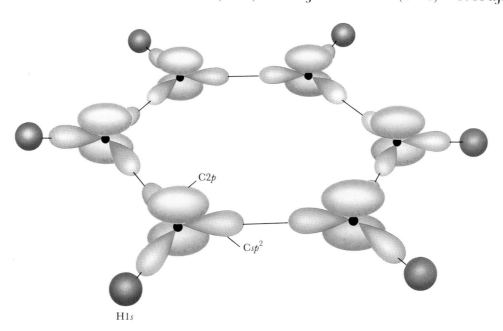

C2p

Csp^2

H1s

We can now see a part of the reason: whereas a single bond is a σ bond, the extra bonding in a multiple bond is a result of π bonding. In carbon (but not in general), π bonds are about 84 kJ/mol weaker than σ bonds because their electrons are in less favorable locations for pulling the atoms together.

We also remarked in Section 9.1 that a double bond prevents one part of a molecule from rotating relative to the other. This is summarized by saying that a double bond is *torsionally rigid* ("torsion" is twisting). The double bond of ethylene, for example, holds the molecule flat **(7)**. Again, the orbital theory of bonding provides an explanation. The drawing of the ethylene double bond in Fig. 9.11 shows that the two $2p$ orbitals overlap best if the sp^2 hybrids of the two —CH$_2$ groups lie in the same plane. If the —CH$_2$ groups are twisted, the overlap of the $2p$ orbitals lessens and the π orbital no longer forms. This removes the bonding effect of the two electrons in that orbital, and the energy of the molecule rises. In other words, the planar arrangement is rigid because energy must be supplied to change it by twisting one —CH$_2$ group relative to the other.

One consequence of the torsional rigidity of double bonds is that you can read these words. Vision depends on the shape of the molecule *retinal*, which is found in the image-receiving retina of the eye. In its normal form **(16)**, retinal is held rigid by its double bonds. When a photon of visible light enters the eye, it is absorbed by a retinal molecule. The photon's energy excites an electron out of the π orbital of the double bond marked by the arrow, reducing its torsional rigidity. Before the excited electron has time to fall back, the molecule is free to rotate around the remaining σ bond. When at last the double bond reforms, the molecule is trapped in its new shape **(17)**. This change of shape triggers a signal along the optic nerve and is interpreted by the brain as the sensation of vision.

A final feature clarified by the orbital picture of bonding is the rarity of multiple bonds between elements of periods higher than 2: although C=C and P=O bonds are common, Si=Si and P=S bonds are not. The reason can be traced to the very small π overlap that occurs between p orbitals on large atoms **(18)**. Period 3 atoms are too fat to come close enough for their p orbitals to overlap and give significant multiple bonding.

16

17

18

MOLECULAR ORBITAL THEORY

Lewis's original ideas were brilliant, and their extension in VSEPR theory made them even more powerful. However, they were little more than inspired guesswork. Lewis had no way of knowing why the electron pair, the essential feature of his approach, was so important. Another point to consider—something we have hidden until now—is that his approach fails for some molecules.

An example of the failure of the Lewis approach is the existence of the compound diborane (B$_2$H$_6$), a colorless gas that bursts into flame on contact with air. Diborane was prepared by the German chemist

Alfred Stock as a part of his pioneering work on boron chemistry, which he began in 1912. Stock produced the gas through the action of acid on magnesium boride (Mg_3B_2). The modern method is by the action of boron trifluoride on lithium tetrahydroborate ($LiBH_4$) dissolved in ether:

$$3LiBH_4 + 4BF_3 \longrightarrow 2B_2H_6 + 3LiBF_4$$

The problem with diborane from the point of view of Lewis structures is that it has only 12 valence electrons (three from each B atom and one from each H atom) but it needs at least 7 bonds—and therefore 14 electrons—to bind the eight atoms together! Diborane is therefore an example of an "electron-deficient compound":

> An **electron-deficient compound** is a compound with too few valence electrons for it to be assigned a Lewis structure.

Another puzzle of a different kind arose in connection with the Lewis description of O_2 as O=O, for oxygen is a magnetic substance. This is most clearly shown with liquid oxygen, a pale blue liquid that sticks to the poles of a magnet (Fig. 9.14). Magnetism (more precisely "paramagnetism" in this case, to distinguish it from the much more powerful "ferromagnetism" of iron) is a property of *unpaired* electrons, each of which acts as a tiny magnet and gives rise to a magnetic field. If all the electrons in a molecule are paired, any magnetic field generated by one electron is canceled by its partner, leaving the molecule nonmagnetic (technically, "diamagnetic"). The magnetism of O_2 therefore contradicts the Lewis structure for the molecule, which requires all its electrons to be paired.

The modern theory of chemical bonding, which was formulated when quantum theory was established in the late 1920s, overcomes all

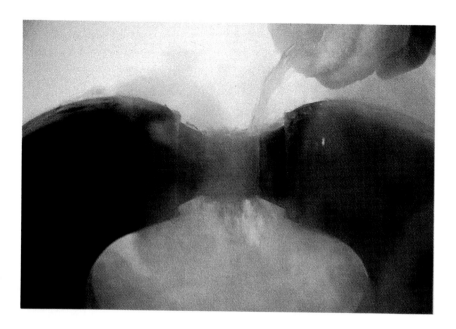

FIGURE 9.14 Oxygen is a paramagnetic substance; as a liquid it sticks to the pole of a magnet.

these difficulties. It shows very naturally why and when the electron pair is so important. It accommodates the boron hydrides just as naturally as it deals with methane and water. It also predicts that oxygen is paramagnetic. However, in doing all this, it also shows how close Lewis was to the truth and that his concepts can still be used with only minor refinement.

9.5 MOLECULAR ORBITALS

The σ and π bonds we have been discussing represent an intermediate stage between Lewis's concept of a localized electron pair and the modern theory of bonding. In the latter theory, electrons spread *throughout* the molecule in "molecular orbitals":

> A **molecular orbital** is a region in a molecule in which an electron is likely to be found.

In other words, whereas an atomic orbital represents the charge cloud of an electron belonging to an individual atom, an occupied molecular orbital is a charge cloud that spreads over all the atoms in a molecule. The description of chemical bonds in terms of molecular orbitals is called *molecular orbital theory*.

Molecular orbitals can be pictured as forming when the atoms of the molecule come together and their valence atomic orbitals merge. In the case of two widely separated hydrogen atoms, for example, we first picture their spherical 1s orbitals as two distinct orbitals (Fig. 9.15). As the atoms come closer, the two orbitals start to overlap; two molecular orbitals begin to form, and any electrons present spread over both atoms. According to molecular orbital theory, when two 1s orbitals overlap they form *two* molecular orbitals called *σ orbitals*. One, σ^*, has a nodal plane—a plane on which the electron is never found—midway between the two nuclei. Therefore, in one of the H_2 σ orbitals an electron may be found midway between the two nuclei, but in the orbital with a nodal plane it may not.

When N atomic orbitals merge, they form N molecular orbitals. Of these N molecular orbitals, one has no nodal planes between the atoms, and the remainder have successively greater numbers of nodal planes. As the number of nodal planes increases, the electrons that occupy the orbital are increasingly less likely to be found in the regions between the nuclei in the molecule.

Bonding and antibonding orbitals. Each molecular orbital has a certain energy. Detailed calculations (based on the Schrödinger equation) show that the greater the number of nodal planes in an orbital, the smaller the binding effect of the electrons that occupy it. In fact, electrons in orbitals with nodal planes can actually *destabilize*—raise the energy of—a molecule. This is because such electrons are more likely to be found outside the internuclear region, so that they tend to pull the nuclei apart (Fig. 9.15). The σ orbital without the nodal plane has

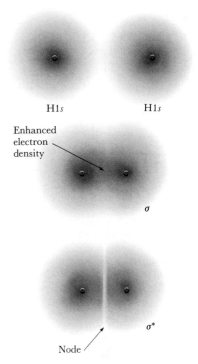

H1s   H1s

Enhanced electron density

σ

σ^*

Node

FIGURE 9.15 The formation of σ and σ^* orbitals when two hydrogen 1s orbitals overlap. Each one can hold up to two electrons.

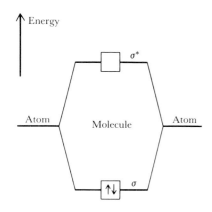

FIGURE 9.16 Two electrons in a bonding orbital (σ) lower the energy of a molecule. However, since electrons in an antibonding orbital are likely to be found outside the bonding region, they tend to break the bond by pulling the atoms apart. Therefore, the antibonding orbital (σ^*) corresponds to a higher energy for the molecule compared to the energy for the separate atoms.

low energy (Fig. 9.16), indicating that it has a stabilizing effect on the molecule. This orbital is known as a "bonding orbital":

> A **bonding orbital** is a molecular orbital which, if occupied, lowers the energy and so stabilizes a molecule.

The orbital with the nodal plane (σ^*) has high energy, representing a destabilizing effect. That orbital is therefore called an "antibonding orbital":

> An **antibonding orbital** is a molecular orbital which, if occupied, raises the energy and so destabilizes a molecule.

In simple cases, the destabilization caused by an antibonding orbital is about equal to or a little greater than the stabilization caused by the bonding orbital. Figure 9.16 is an example of a *molecular orbital energy-level diagram*, one that shows the relative energies of the orbitals in a molecule.

The role of the electron pair. We are about to make use of the building-up technique to fill molecular orbitals, just as we used it in Section 7.5 to fill atomic orbitals. That is, we shall accommodate electrons in the molecular orbitals, starting with the lowest-energy orbital and working toward higher energies. According to the Pauli exclusion principle, each orbital can accommodate up to two paired electrons. If two or more orbitals of the same energy are available, the electrons enter each one separately and adopt parallel spins.

We begin here with H_2, and deal with the other diatomic molecules in succeeding sections. In H_2, two atomic $1s$ orbitals (one on each atom) merge to form two molecular orbitals: we call the bonding orbital $1s\sigma$ and the antibonding orbital $1s\sigma^*$, the $1s$ showing the atomic orbitals from which the molecular orbitals are formed. Two electrons, one from each H atom, are available. Both occupy the bonding orbital (the lowest-energy orbital), resulting in the configuration $1s\sigma^2$. Since only the bonding orbital is occupied, the energy of the molecule is lower than that of the separate atoms, and H_2 exists as a molecule. At this point, one feature of the Lewis approach—the importance in bonding of the electron pair—emerges. The exclusion principle allows only *two* electrons to occupy any one molecular orbital, and their spins must be paired. Hence, two paired electrons form a single bond between two atoms.

▼ **EXAMPLE 9.10** **Accounting for the absence of bonding**

Use molecular orbital theory to suggest a reason why He_2 does not form.

STRATEGY The energy-level diagram for He_2 will resemble that for H_2, since both are built from atoms with their valence electrons in $1s$ orbitals. This suggests that He_2 can be discussed in terms of Fig. 9.16, but with four electrons instead of two. We need to use the building-up principle and then consider the net bonding effect of the resulting electron configuration. We should remember that an antibonding orbital destabilizes a molecule to about the same extent that a bonding orbital stabilizes it.

SOLUTION The first two electrons enter $1s\sigma$, which is then full. The remaining two must enter $1s\sigma^*$. The configuration of He_2 is therefore $1s\sigma^2 1s\sigma^{*2}$, and the bonding effect of $1s\sigma^2$ is canceled by the antibonding effect of $1s\sigma^{*2}$. Hence, there is no net bonding in He_2 and it is not a stable species.

EXERCISE Is He_2^+ a possible species?

[*Answer*: Yes]

▲

Potential-energy curves. The molecular orbital description of bonding can be taken further by exploring how the energy of a diatomic molecule varies as the bond length changes. This variation is given by the curve in Fig. 9.17, which is called a _molecular potential-energy curve_ because it shows how the energy of the molecule depends on the relative positions of its two atoms (potential energy, remember, is energy that depends on position). It resembles the curve in Fig. 5.15, which shows the intermolecular potential energy, but the minimum is much deeper there because forces that bind atoms in molecules are more than about 10 times as strong as intermolecular forces. At large separations, where the atoms are too far apart to interact, we set the potential energy equal to zero. Then the potential energy falls as the atoms approach one another and the low-energy bonding orbital forms and is occupied by the two electrons. However, the potential energy rises sharply when the separation is very small because the nuclei then repel each other strongly.

The molecular potential-energy curve enables us to predict the bond length in the molecule, something the Lewis theory could not do. Since a molecule adopts the atomic arrangement corresponding to the lowest energy, the experimental bond length is the separation corresponding to the minimum of the curve. When the curve is plotted for H_2 (by solving the Schrödinger equation), the minimum occurs at 74 pm, in agreement with the experimental value for the H_2 bond length.

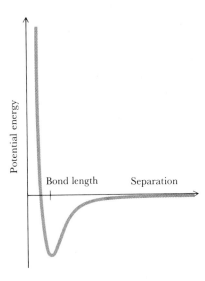

FIGURE 9.17 A molecular potential-energy curve shows how the energy of two atoms changes as they approach and form a bond. The minimum-energy point of the curve corresponds to the bond length of the molecule.

9.6 BONDING IN PERIOD 2 DIATOMIC MOLECULES

When we use molecular orbital theory to describe molecular structure, we begin with all the atoms at their actual positions in the molecule. (In more advanced treatments that is not necessary; the positions are found by calculating the arrangement that has the lowest energy.) Next, we form N molecular orbitals by merging all N valence-shell orbitals together. Finally, we use the building-up principle to accommodate the valence electrons supplied by the atoms.

Constructing the orbitals. The first step is to build up the molecular orbital energy-level diagram. At this stage we ignore the _electrons_ that are available and consider only the valence-shell atomic orbitals. Since Period 2 atoms have $2s$ and $2p$ orbitals in their valence shells, we form molecular orbitals from the overlap of these atomic orbitals. Because we use the _same_ atomic orbitals to form molecular orbitals for all the Period 2 diatomic molecules, they all have the _same_ energy-level dia-

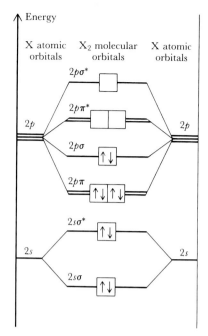

Energy

X atomic orbitals   X₂ molecular orbitals   X atomic orbitals

$2p\sigma*$

$2p\pi*$

$2p$ — $2p\sigma$ — $2p$

$2p\pi$

$2s\sigma*$

$2s$ — $2s\sigma$ — $2s$

FIGURE 9.18 The molecular-orbital energy-level diagram for the Period 2 diatomic molecules. Each box can hold up to two electrons; the configuration for N_2 is shown.

gram. That diagram is shown in Fig. 9.18, and the following discussion shows how it is constructed and used.

Two 2s orbitals overlap to form two sausage-shaped molecular orbitals, one bonding (the $2s\sigma$ orbital) and the other antibonding (the $2s\sigma*$ orbital). The three 2p orbitals on each neighboring atom can merge in two distinct ways. The two 2p orbitals that are directed toward each other along the internuclear axis form a bonding σ orbital ($2p\sigma$) and an antibonding $\sigma*$ orbital ($2p\sigma*$) where they overlap. (They are called σ orbitals for the same reason as those formed from s orbitals: when viewed along the bond, they look like s orbitals.)

Two 2p orbitals that are perpendicular to the internuclear axis overlap to form the bonding and antibonding _π orbitals_. (The bonding combination will turn out to be responsible for a π bond.) As there are two perpendicular 2p orbitals on each atom, two $2p\pi$ orbitals and two $2p\pi*$ orbitals are formed by their overlap. The resulting energy-level diagram is shown in Fig. 9.18. Pay special attention to the order of energies of the orbitals; it is critical for the discussion that follows.

Electron configurations and bond orders. Now that we know what molecular orbitals are available, we can obtain the ground-state electron configurations of the molecules using the building-up principle. Remember that each molecular orbital can hold up to two electrons. Two electrons in a bonding σ orbital form a σ bond, and two in a bonding π orbital form a π bond. Although a π orbital has a "double sausage" appearance, only two electrons can occupy it, and when they do they form _one_ π bond.

Consider N_2. Since nitrogen belongs to Group V, each atom supplies five valence electrons. A total of 10 electrons must therefore be assigned to the molecular orbitals. Two fill the $2s\sigma$ orbital. The next two fill the $2s\sigma*$ orbital. Next in line for occupation are the two $2p\pi$ orbitals, which can hold a total of four electrons. The last two electrons then enter the $2p\sigma$ orbital. The ground configuration is therefore

$$N_2:\quad (2s\sigma)^2(2s\sigma*)^2(2p\pi_x)^2(2p\pi_y)^2(2p\sigma)^2$$

or, more briefly,

$$N_2:\quad \sigma^2\sigma*^2\pi^4\sigma^2$$

The molecular orbital description of N_2 looks quite different from the Lewis description, N≡N. However, it is in fact very closely related. We can see this by calculating the _bond order_ BO, which is the net number of bonds after we allow for the cancellation of bonds by antibonds:

$$BO = \frac{1}{2} \times (\text{number of electrons in bonding molecular orbitals}$$
$$- \text{ number of electrons in antibonding molecular orbitals})$$

This definition is the generalization of the one given in Section 8.3, where we simply counted shared pairs. In N_2 there are eight electrons in bonding orbitals and two in antibonding orbitals, so

$$BO = \frac{1}{2} \times (8 - 2) = 3$$

Since its bond order is 3, N_2 is a _triply bonded_ molecule, just as the Lewis structure suggests.

▼ **EXAMPLE 9.11   Deducing the electron configurations of Period 2 diatomic molecules**

Deduce the electronic configuration of the fluorine molecule, and calculate its bond order.

STRATEGY   Since the Lewis structure of F_2 is $:\ddot{F}\!-\!\ddot{F}:$, we anticipate that the bond order is 1. To calculate it formally, we need to know the numbers of electrons in bonding and antibonding orbitals in the molecule. Therefore, we first use the building-up principle to accommodate the $2 \times 7 = 14$ valence electrons in the energy-level diagram. Then we calculate the bond order from the resulting configuration.

SOLUTION   The first 10 electrons repeat the N_2 configuration. The remaining four can enter the two antibonding $2p\pi^*$ orbitals. The configuration is therefore

$$F_2:\quad \sigma^2\sigma^{*2}\pi^4\sigma^2\pi^{*4}$$

The bond order is

$$BO = \frac{1}{2} \times [(2 + 4 + 2) - (2 + 4)] = 1$$

Hence, F_2 is a singly bonded molecule, in agreement with the Lewis structure F—F.

EXERCISE   Deduce the electronic configuration and bond order of the carbide ion (C_2^{2-}).

[*Answer:* $\sigma^2\sigma^{*2}\pi^4\sigma^2$, BO = 3]

▲

Earlier in this chapter we noted that Lewis theory cannot account for the paramagnetism of O_2. In molecular orbital theory the configuration of O_2 is obtained by feeding its 12 valence electrons into the molecular orbitals shown in Fig. 9.18. The first 10 repeat the N_2 configuration. The last two occupy the two separate $2p\pi^*$ orbitals with parallel spins. The configuration is therefore

$$O_2 \quad \boxed{\uparrow\downarrow}\!-\!\boxed{\uparrow\downarrow}\!-\!-\!\boxed{\uparrow\downarrow}\,\boxed{\uparrow\downarrow}\!-\!-\!-\!\boxed{\uparrow\downarrow}\!-\!-\!-\!-\!\boxed{\uparrow}\,\boxed{\uparrow}$$
$$(2s\sigma)^2\ (2s\sigma^*)^2\quad (2p\pi_x)^2(2p\pi_y)^2\quad (2p\sigma)^2\quad (2p_x\pi^*)^1(2p_y\pi^*)^1$$

This is a minor triumph: because the last two spins are not paired, their magnetic fields do not cancel, and the molecule is predicted to be paramagnetic—exactly as observed. The bond order of O_2 is

$$BO = \frac{1}{2} \times [(2 + 2 + 4) - (2 + 1 + 1)] = 2$$

This is consistent with the Lewis structure O=O. However, the Lewis structure conceals the fact that the double bond is actually a σ bond plus two *half* π bonds so that the molecule has two unpaired electrons.

9.7   ORBITALS IN POLYATOMIC MOLECULES

We will not go into the molecular orbital theory of polyatomic molecules except to say that it follows the same principles as for diatomic molecules. The only elaboration is that the molecular orbitals spread

FIGURE 9.19 Electrons in molecular orbitals spread thoughout the molecule and draw together all the atoms. The bonding effect of each electron pair that occupies an orbital is therefore shared throughout the molecule.

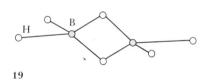

19

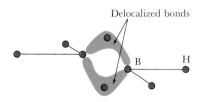

Delocalized bonds

FIGURE 9.20 The bonding in the electron-deficient molecule diborane. The two BH_2 groups are held together by two delocalized bonds formed by two electrons occupying a delocalized orbital.

over *all* the atoms in the molecule. Therefore, an electron pair that occupies a bonding orbital helps to bind together the *whole* molecule, not just an individual pair of atoms (Fig. 9.19). This picture of *delocalized orbitals*, molecular orbitals that spread over the entire molecule, accounts very neatly for both the existence of conventional molecules and the existence of electron-deficient compounds.

We can see how molecular orbital theory copes with electron-deficient molecules by examining diborane. Experimental information on the molecule shows that it consists of two BH_2 groups lying in a plane (like ethylene), but with two more H atoms located halfway between the two B atoms, one above the plane and one below **(19)**. This structure is consistent with the two B atoms being sp^3 hybridized and pinned together by two H atoms. Four hydrogen $1s$ orbitals overlap four of the sp^3 hybrids and lie in a plane. Two electrons occupy each orbital, so 8 of the 12 electrons are used to form B—H σ bonds in the normal way. The other two hybrids on each B atom overlap the orbitals of the two hydrogen atoms lying between the boron atoms. They form *delocalized* molecular orbitals (Fig. 9.20) that spread over the B, H, and B atoms in each case. (We are considering only a limited kind of delocalization; in a more general theory, we would construct orbitals that spread over all the atoms in the molecule.) The last four electrons occupy the two bonding orbitals created in this way, forming two delocalized B—H—B bonds, binding the two BH_2 units together.

The central feature of molecular orbital theory, which shows up most clearly in electron-deficient molecules, is that two electrons can form a delocalized bond and spread their bonding influence over several atoms. Hence, it is not always necessary to have two electrons present for every link in a molecule.

9.8 A PERSPECTIVE ON CHEMICAL BONDING

We have come a long way since the beginning of Chapter 8 and its introduction to chemical bonding. So many new concepts have been introduced that it may be helpful to pause and view them in perspective.

Chapter 8 introduced what seemed at first to be two different kinds of bonds. Then, through the concept of bond polarity, we saw that an ionic bond is a special case of a covalent bond. We also saw that Lewis's emphasis on the shared electron pair and the octet allowed us to account for the bonding in a wide variety of molecules and polyatomic ions. However, the Lewis approach is incomplete. It makes no predictions about the shapes of molecules, fails to explain the existence of some compounds, and leaves the importance of the electron pair unexplained.

This chapter has cleared up those points one by one. It showed how VSEPR theory uses Lewis structures to make predictions about the shapes of molecules. Moreover, we have seen that the locations of electron pairs predicted by VSEPR theory can be expressed in terms of orbitals by mixing s, p, and d orbitals together and then supposing that each electron pair occupies one of these hybrids. This brings Lewis's ideas more into line with the material in Chapter 7, where we saw that electrons are not localized in definite positions but are spread out over a region.

Finally, from the concept of electrons occupying atomic orbitals, we developed a model in which electrons occupy orbitals that spread through the entire molecule. These molecular orbitals provide a straightforward explanation of the role of the electron pair: only two electrons can occupy any given orbital, so a bond, an occupied bonding orbital, is complete as soon as two electrons are present. Molecular orbital theory not only echoes but also explains Lewis's theory. It easily accounts for the existence of electron-deficient compounds, since in any molecule a pair of electrons, being spread over several atoms, can bind groups of more than two atoms together simultaneously. Unlike the Lewis theory, molecular orbital theory explains how single electrons can help to bind atoms together. It also accounts for the paramagnetism of O_2.

Molecular orbital theory makes contact with the Lewis theory through the concept of bond order. The bond order shows that Lewis structures of the type A=B are really a shorthand for electron pairs in σ and π orbitals. This means that we can continue to use the Lewis theory in most cases, but now we have a deeper understanding of what the Lewis structures represent.

SUMMARY

9.1 The shapes of molecules are classified according to the relative locations of their atoms. **VSEPR theory** explains molecular shape in terms of the repulsions between electron pairs, which tend to move as far apart as possible. Repulsion strengths are in the order (LP, LP) > (LP, BP) > (BP, BP). A multiple bond is treated as equivalent to a single electron pair.

9.2 Depending on its shape and the polarity of its bonds, a molecule may be **polar,** that is, possess a nonzero dipole moment.

9.3 The locations of electron pairs can be expressed in terms of atomic orbitals by assuming that the latter **hybridize,** or mix together, to match the shape of the molecule. In general, N atomic orbitals give N hybrid orbitals.

9.4 Orbitals on one atom can **overlap** those of a neighbor to give either σ **bonds,** bonds with cylindrical symmetry about the axis, or π **bonds,** double-lobed electron distributions. A **single bond** is a σ bond, a **double bond** is a σ bond plus a π bond, and a **triple bond** is a σ bond plus two π bonds. Double bonds are **torsionally rigid,** or resistant to twisting.

9.5 Chemical bonds can be expressed in terms of **molecular orbitals,** which are orbitals that are formed by the overlap of atomic orbitals and spread throughout the molecule. They may be **bonding,** or energy-lowering, or they may be **antibonding,** or energy-raising. According to their shape, they are either σ **orbitals** or π **orbitals.** Electrons occupy the available orbitals in accord with the building-up principle. The importance of the electron pair in bond formation can be traced to the exclusion principle, which limits the occupation of a bonding orbital to two paired electrons. The dependence of a molecule's energy on its bond length is shown by a **molecular potential-energy curve.**

9.6 From the viewpoint of molecular orbital theory, a Lewis diagram is a representation of the **bond order,** the net number of bonds in a molecule, allowing for the cancellation of bonds by antibonds. Molecular orbital theory accounts for the paramagnetism of oxygen.

9.7 A molecular orbital is **delocalized:** the bonding influence of its electrons is spread over several atoms. This accounts for the existence of **electron-deficient** compounds, those for which Lewis structures cannot be written, in which two electrons bind groups of atoms together.

EXERCISES

VSEPR Theory

In Exercises 9.1 and 9.2, state the geometrical arrangement expected for each given number of electron pairs.

9.1 (a) Two; (b) three; (c) six.
9.2 (a) Four; (b) five; (c) eight.

In Exercises 9.3 and 9.4, state the number of electron pairs corresponding to each electron-pair arrangement.

9.3 (a) Tetrahedral; (b) linear; (c) octahedral.
9.4 (a) Trigonal planar; (b) trigonal bipyramidal.

In Exercises 9.5 and 9.6, state the shape of each molecule type and give two examples.

9.5 (a) XY_2E; (b) XY_2E_2; (c) XY_3E.
9.6 (a) XY_3E_2; (b) XY_4E_2; (c) XY_4E.

In Exercises 9.7 and 9.8, give as many examples as you can of molecules or ions that have each specified shape (look through this and later chapters for examples).

9.7 (a) Linear; (b) tetrahedral; (c) octahedral.
9.8 (a) Angular; (b) trigonal bipyramidal; (c) trigonal planar.

In Exercises 9.9 to 9.16, use VSEPR theory to predict the shape of each molecule or ion.

9.9 (a) H_2O; (b) CO_2; (c) SO_2; (d) O_3.
9.10 (a) H_2S; (b) CS_2; (c) XeF_2; (d) N_2O.
9.11 (a) H_3O^+; (b) ClO_2^-; (c) NO_2^-; (d) SO_3^{2-}.
9.12 (a) NH_4^+; (b) ClO_3^-; (c) ClO_4^-; (d) NO_3^-.
9.13 (a) SO_4^{2-}; (b) PO_4^{3-}; (c) I_3^-.
9.14 (a) SeO_3^{2-}; (b) IO_4^-; (c) I_3^+.
9.15 (a) SF_6; (b) PCl_5; (c) PF_3; (d) XeF_4.
9.16 (a) SCl_4; (b) XeO_4; (c) GeH_4; (d) ICl_3.

In Exercises 9.17 to 9.20, predict the shape about the indicated ion.

9.17 (a) $\overset{*}{C}H_3$—CH_3; (b) $\overset{*}{C}H_2$=CH_2; (c) $\overset{*}{C}H$≡CH.
9.18 (a) $\overset{*}{C}H_3$—CH_2—CH_3; (b) CH_3—$\overset{*}{C}H$=CH_2;
(c) HC≡$\overset{*}{C}$—CH=CH_2.

9.19 The amino acid glycine, $H_2\overset{*}{N}$—CH_2—$COOH$, one of the components from which proteins are built.
9.20 The amino acid serine, $CH_2(\overset{*}{O}H)$—$CH(NH_2)$—$COOH$, another component of proteins.

In Exercises 9.21 to 9.24, estimate the bond angles marked with arcs and lowercase letters.

9.21 (a) Ethylene:

(b) Acetic acid:

9.22 (a) Butane:

(b) Acetone:

9.23 (a) Acrolein, a pungent eye irritant in smoke:

(b) Azodicarbonamide, a compound that decomposes into N_2, CO, and NH_3 when it is heated, and thus is used to form the bubbles in polystyrene:

9.24 (a) Peroxyacetyl nitrate, an eye irritant in smog:

(b) Carvone, one of the components of oil of spearmint:

9.25 When BF_3 reacts with NH_3, a white solid of molecular formula F_3B—NH_3 is formed. Draw diagrams of the reactants and the product, and estimate the bond angles.

9.26 Hydrogen fluoride reacts with antimony pentafluoride to give a salt in which the cation is H_2F^+ and the anion is SbF_6^-. Draw diagrams of the reactants and the product, and estimate the bond angles.

Polar Molecules

In Exercises 9.27 to 9.32, predict whether each molecule is polar or nonpolar.

9.27 (a) CH_4; (b) CS_2; (c) PCl_5; (d) SF_4.
9.28 (a) CCl_4; (b) H_2S; (c) PCl_3; (d) XeF_4.

9.29 (a) C_6H_6; (b) C_2H_4.
9.30 (a) C_2H_2; (b) C_2H_6.

9.31 (a) CH_3OH; (b) $CH_3C_6H_5$ (toluene).
9.32 (a) CH_3CH_2OH; (b) HCHO (formaldehyde).

9.33 There are three forms of dichlorobenzene, $C_6H_4Cl_2$, which differ in the relative positions of Cl in the benzene ring **(a, b, c)**. (a) Which of the three forms are polar, and which are nonpolar? (b) Which has the largest dipole moment?

a   b   c

9.34 There are three forms of dichloroethylene, $C_2H_2Cl_2$, which differ in the locations of the chlorine atoms **(a, b, c)**. (a) Which of the forms are polar, and which are nonpolar? (b) Which has the largest dipole moment?

a   b   c

Hybridization

In Exercises 9.35 and 9.36, state the shape corresponding to each hybridization.

9.35 (a) sp^3; (b) sp; (c) sp^3d.
9.36 (a) sp^2; (b) sp^3d^2; (c) sp^2d.

In Exercises 9.37 and 9.38, state the hybridizations (there may be more than one) corresponding to each arrangement of electron pairs, and give an example of a molecule for each.

9.37 (a) Tetrahedral; (b) trigonal bipyramidal.
9.38 (a) Octahedral; (b) linear.

In Exercises 9.39 to 9.42, state the hybridization of each carbon atom.

9.39 (a) $CH_3CH_2CH_3$; (b) $CH_2{=}CH{-}CH_3$.
9.40 (a) $CH_3{-}C{\equiv}CH$; (b) $CH_2{=}C{=}C{=}CH_2$.

9.41 (a) C_6H_6 (benzene); (b) $HC{\equiv}C{-}C{\equiv}CH$.
9.42 (a) $CH_3C_6H_5$ (toluene); (b) HCHO (formaldehyde).

When a molecule does not have exactly one of the shapes shown in Table 9.1, the central atom is still hybridized, but the hybridization is only *approximately* one of the mixtures given in Table 9.3. In Exercises 9.43 and 9.44, state the approximate hybridization of the central atom in each molecule or ion, using VSEPR theory to predict the shape.

9.43 (a) NH_3; (b) NO_2^-; (c) H_2O.
9.44 (a) NH_3; (b) SF_4; (c) H_2S.

In Exercises 9.45 and 9.46, state the hybridization of each boldfaced atom.

9.45 (a) Acetic acid, $CH_3\mathbf{C}$

(b) Acetylsalicylic acid, the active component of aspirin

(c) Toluene

9.46 (a) Propylene, $CH_3{-}CH{=}CH_2$, the hydrocarbon used to make polypropylene
(b) β-Ionone, a molecule that is partly responsible for the odor of freshly cut hay

(c) 2,4,6-Trinitrotoluene, better known as TNT

$$CH_3$$

$$O_2N \qquad C \qquad NO_2$$

$$NO_2$$

Bonding Orbitals

In Exercises 9.47 and 9.48, state whether σ or π bonds or both are formed by the overlap of the given orbitals on neighboring atoms, where the z axis is the axis between them.

9.47 (a) (1s, 1s); (b) $(2p_z, 2p_z)$; (c) $(2p_x, 2p_x)$.
9.48 (a) (2s, 2s); (b) $(2s, 3p_z)$; (c) $(3d_{z^2}, 3d_{z^2})$.

In Exercises 9.49 to 9.52, give the compositions of the bonds—in the form $(Csp, H1s)$, as an example—in each molecule.

9.49 (a) C_2H_6; (b) C_2H_2; (c) HCN; (d) H_2O_2.
9.50 (a) C_2H_4; (b) CH_3OH; (c) CH_3CN; (d) HN_3.
9.51 (a) $H_2N—CH_2—COOH$ (glycine); (b) C_6H_5COOH (benzoic acid).
9.52 (a) $(NH_2)_2CO$ (urea); (b) the aspirin molecule in Exercise 9.45b.

Molecular Orbitals

In Exercises 9.53 to 9.56, deduce the ground-state electron configurations and bond orders of each molecule or ion.

9.53 (a) N_2; (b) F_2; (c) Be_2; (d) Ne_2.
9.54 (a) O_2; (b) C_2; (d) Li_2; (e) B_2.
9.55 (a) The carbide ion C_2^{2-}, as found in calcium carbide (CaC_2); (b) the peroxide ion O_2^{2-}.
9.56 (a) C_2^{2+}; (b) O_2^{2+}.

In Exercises 9.57 and 9.58, determine which of the given species are paramagnetic.

9.57 (a) O_2; (b) the superoxide ion O_2^-; (c) the ozonide ion O_3^-; (d) N_2^-.
9.58 (a) S_2; (b) the peroxide ion O_2^{2-}; (c) the azide ion N_3^-; (d) Br_2^+.

In Exercises 9.59 and 9.60, calculate the relevant bond orders and use them to predict which substance in each pair would be expected to have the stronger bond between atoms.

9.59 (a) N_2 or N_2^{2-}; (b) F_2 or F_2^{2-}; (c) Ne_2^{2+} or Ne_2.
9.60 (a) C_2 or C_2^{2+}; (b) C_2^{2-} or C_2; (c) O_2 or O_2^{2+}.

9.61 Which would you expect to have the longer bond, the oxygen molecule or the peroxide ion O_2^{2-}?

9.62 Which would you expect to have the shorter bond, the cyanogen radical CN or the cyanide ion CN^-?

In Exercises 9.63 and 9.64, write the configuration of each molecule and calculate its bond order.

9.63 (a) Cl_2; (b) S_2; (c) Ar_2.
9.64 (a) Br_2; (b) P_2; (c) Xe_2.

In Exercises 9.65 and 9.66, predict the configuration and bond order of each molecule or ion, using Fig. 9.18 as a rough guide to the configurations of these "heteronuclear" diatomic molecules (AB molecules in which A and B are different).

9.65 (a) NO^+; (b) CN^-; (c) NO; (d) ClO^-.
9.66 (a) CO; (b) BrO^-; (c) FCl; (d) ICl.

In Exercises 9.67 and 9.68, state the number of molecular orbitals that can be constructed from the given atomic orbitals.

9.67 (a) Three s orbitals; (b) two p orbitals; (c) six p orbitals.
9.68 (a) Ten p orbitals; (b) 1000 s orbitals, (c) two s orbitals.

In Exercises 9.69 and 9.70, state how many bonding and antibonding π orbitals can be expected in each molecule. How many are occupied?

9.69 (a) $CH_2{=}CH_2{-}CH{=}CH_2$; (b) $CH{\equiv}C{-}C{\equiv}CH$; (c) C_6H_6.
9.70 (a) $CH_2{=}CH{-}C{\equiv}C{-}CH{=}CH_2$;
(b) $CH{\equiv}C{-}C{\equiv}C{-}C{\equiv}CH$; (c) $C_{10}H_8$ (naphthalene).

General

9.71 The four molecules **(a)** to **(d)** are the nucleic acids involved in the genetic code. Identify the hybridizations of the atoms, the composition of the bonds, and the orbitals occupied by the lone pairs.

a

b

c

d

9.72 Nitrogen, phosphorus, oxygen, and sulfur exist as N_2, tetrahedral P_4, O_2, and ringlike S_8 molecules. Rationalize this fact in terms of the ability of the atoms of each element to form different types of bonds with each other. In each case, describe the bonding in terms of the hybridization of the atoms.

9.73 Xenon forms XeO_3, XeO_4^{2-} and XeO_6^{4-}, all of which are powerful oxidizing agents. Give their Lewis structures, and account for the bonding in terms of the hybridization of the xenon atom.

9.74 The halogens form compounds among themselves, called the "interhalogens." These have the formulas XX', XX_3', XX_5', and XX_7', where X is the heavier halogen. Predict their structures and bond angles. Which of them are polar?

*9.75 Confirm, using trigonometry, that the dipoles of the three bonds in a trigonal planar AB_3 molecule cancel, and hence that the molecule is nonpolar. Also show that the dipoles of the four bonds in a tetrahedral AB_4 molecule cancel so that the molecule is nonpolar.

9.76 The molecular configurations derived using Fig. 9.18 depend on the Pauli exclusion principle. Suppose we lived in a universe in which *three* electrons could occupy one orbital. What would be the electronic structure of O_2? Could Ar_2 exist? What rules for writing Lewis structures would you have proposed if you had been the G. N. Lewis who lived in that universe?

This chapter explains the origin of forces between molecules, and how those forces account for the condensation of gases to liquids and solids. We see that intermolecular forces account for the physical properties of liquids and the structures of solids, including metals, ionic solids, and molecular solids. The illustration shows the condensation of water on the interior of a flask: water vapor condensing to a liquid is a sign that there are forces acting between molecules.

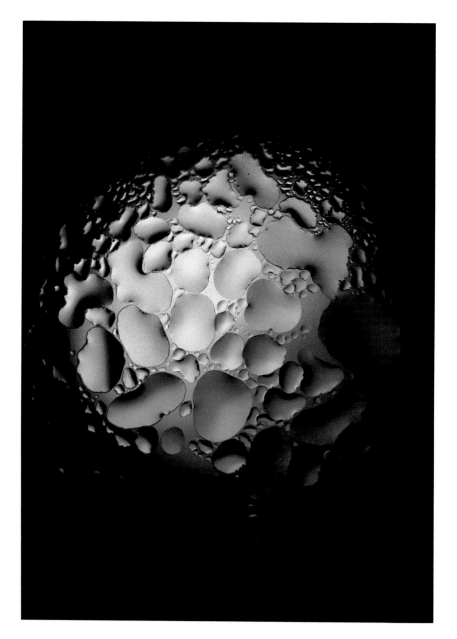

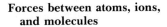

Forces between atoms, ions, and molecules

10.1 Ion and dipole forces
10.2 Hydrogen bonding

The properties of liquids

10.3 Viscosity and surface tension
10.4 Vapor pressure
10.5 Solidification

LIQUIDS AND SOLIDS

Solids

10.6 X-ray diffraction
10.7 Metals and semiconductors
10.8 Ionic solids
10.9 Other types of solids

Social revolutions have followed from the discovery of new liquids and solids. The discovery of natural sources of liquid hydrocarbons—petroleum—transformed society by providing fuel for light and mobile power sources in vehicles. The discovery of metals raised civilization out of the Stone Age. The development of steels gave materials strong enough to make possible the industrial revolution. The organic solids we call plastics, including polyethylene and nylon, have transformed people's lives during this century. We are still in the midst of the revolution that started with the discovery of semiconductors. We are at the start of the revolution that will follow from the discovery, in the late 1980s, of high-temperature superconductors—substances that conduct electricity without resistance at temperatures not far below room temperature.

In this chapter we consider some of the properties of the liquid and solid states of matter and the forces that bind them together. The only liquids we consider are those made up of molecules; these include water, ethanol, benzene, and the liquefied gases. We do not consider molten salts (which are liquid mixtures of ions) or molten metals. The solids we consider fall into four classes (Fig. 10.1):

Metals consist of cations held together by a sea of electrons.

Ionic solids are built from cations and anions.

Network solids consist of atoms linked together covalently throughout the extent of the solid.

Molecular solids are collections of individual molecules.

Some characteristics and examples of each type of solid are given in Table 10.1.

Almost all substances have only one liquid form (helium is the only exception), but many have several solid forms. Diamond and graphite, for instance, are two solid forms of carbon, and calcite and aragonite are two solid forms of calcium carbonate. The term used to distinguish

FIGURE 10.1 Examples of the four classes of solids considered in this chapter. Counterclockwise, from the upper left: an ionic solid, nickel(II) sulfate; a network solid, boron nitride; a molecular solid, carotene, the compound responsible for the color of carrots; and a metal, titanium.

TABLE 10.1 Examples and characteristics of solids

Class	Examples	Characteristics
Ionic	NaCl, KNO$_3$, CuSO$_4 \cdot 5H_2O$	Hard, rigid, brittle; high melting and boiling points; those soluble in water give conducting solutions
Network	B, C, black P, BN, SiO$_2$	Hard, rigid, brittle; high melting and boiling points; insoluble in water
Molecular	BeCl$_2$ S$_8$, P$_4$, I$_2$, ice, glucose, naphthalene	Soft; relatively low melting and boiling points; brittle if pure
Metals	s- and d-block elements	Malleable, ductile, lustrous; electrically and thermally conducting

the different physical states and their alternative forms is _phase_ (from the Greek word for "appearance"). Thus we can speak of the vapor, liquid, and solid phases of water. We can also speak of the diamond and graphite phases of solid carbon, and of the calcite and aragonite phases of calcium carbonate (Fig. 10.2).

FORCES BETWEEN ATOMS, IONS, AND MOLECULES

The bulk of this section is concerned with the origin of _intermolecular forces_, the interactions between neutral closed-shell atoms and molecules. These forces are also called _van der Waals forces_ after Johannes van der Waals, whose contribution to the study of the effects of intermolecular forces in real gases was described in Section 5.4.

FIGURE 10.2 The calcite (left) and aragonite (right) phases of calcium carbonate.

TABLE 10.2 Interionic and intermolecular forces

Type of interaction*	Distance dependence	Typical energy,† kJ/mol	Comment
Ion-ion	$1/d$	250	Only between ions
Ion-dipole	$1/d^2$	15	
Dipole-dipole	$1/d^3$	2	Between stationary polar molecules
	$1/d^6$	0.3	Between rotating polar molecules
London (dispersion)	$1/d^6$	2	Between all types of molecules
Hydrogen bonding	Contact	20	Between N, O, F; the link is a shared H atom

*Other intermolecular forces include the *ion-induced dipole interaction*, in which an ion polarizes a neighboring nonpolar molecule and interacts with the resulting dipole (typical strength: 3 kJ/mol), and the *dipole-induced dipole interaction*, in which a polar molecule causes the polarization (typical strength: 0.05 kJ/mol).
† Typical strengths are for distance $d = 500$ pm.

10.1 ION AND DIPOLE FORCES

The various types of intermolecular and ionic forces are summarized in Table 10.2. We have already dealt with the electrostatic forces between ions of opposite charge in an ionic solid (Section 8.1). Only slightly more complex is the force between an ion and the *partial* charges of the electric dipole of a polar molecule, or that between the partial charges of two dipoles. More problematic, though, is the origin of the interaction between nonpolar molecules.

Ion-ion interactions. As we saw in Section 8.1, the strength of the interaction between ions (more precisely, their potential energy, taking zero for their energy when they are far apart) is summarized by

$$E \propto \frac{-z_1 z_2}{d}$$

where z_1 and z_2 are the charge numbers of the ions, and d is their separation. The minus sign signifies that the energy becomes more negative—is further below zero—as the ions approach each other. The strength of the interaction decreases quite slowly with distance. For example, it falls to half its value when the separation doubles. As a result, it binds ions together over long distances, not just nearest neighbors.

The strong, long-range attraction between ions of opposite charge is responsible for the high lattice enthalpies and generally high melting and boiling points of ionic compounds (Table 10.3). Melting points and lattice enthalpies are particularly high for solids composed of small, highly charged ions, because the electrostatic forces can grip such ions together so strongly. This is shown by the fact that aluminum oxide melts only at 2015°C, and that magnesium oxide melts at 2800°C. The

TABLE 10.3 Normal melting and boiling points of ionic solids

Solid	Melting point, °C	Boiling point, °C
LiF	842	1676
LiCl	614	1382
NaCl	801	1413
KCl	776	1500s*
$MgCl_2$	708	1412
$CaCl_2$	782	2000
MgO	2800	3600
Al_2O_3	2015	2980

*The s indicates that this solid sublimes.

trend to lower melting points with increasing ion size is illustrated by the alkali metal chlorides (Fig. 10.3), for which the melting points fall from 801°C for NaCl to 645°C for CsCl.

Ion-dipole interactions. In an *ion-dipole interaction* a cation attracts the partial negative charge of an electric dipole **(1)**, or an anion attracts the partial positive charge **(2)**. The potential energy of this interaction decreases with the square of the distance between ion and dipole:

$$E \propto \frac{-z \times \mu}{d^2}$$

where z is the charge number of the ion, and μ is the dipole moment of the molecule. The strength of interaction decreases more rapidly with distance than in ion-ion interactions (as the inverse square of the separation, $1/d^2$ in place of $1/d$), because the partial charge at one end of the dipole tends to cancel the effect of the opposite partial charge at the other end: an electric dipole at a large distance seems to have no charge at all.

▼ **EXAMPLE 10.1** **Estimating the strength of an ion-dipole interaction**

Calculate the ratio of the potential energies of the interaction of a water molecule with Na^+ and with K^+ ions.

STRATEGY The potential energy of interaction is given by the expression above; the ratio of the values for the two ions can be calculated without knowing the constant of proportionality because it is canceled out when the ratio is formed. For the distance d, we take the radius of the cation, since we can assume that the partial negative charge of the H_2O molecule is on the surface of the O atom. The ionic radii are given in Table 7.6.

SOLUTION The ratio of energies is

$$\frac{E(K^+)}{E(Na^+)} = \frac{[d(Na^+)]^2}{[d(K^+)]^2} = \frac{(95 \text{ pm})^2}{(133 \text{ pm})^2} = 0.51$$

That is, the energy of interaction between an H_2O molecule and a K^+ ion is only about half that between H_2O and an Na^+ ion.

EXERCISE Calculate the ratio of the energies of the interaction of a water molecule with Na^+ and with Mg^{2+} ions.

[*Answer:* $E(Na^+)/E(Mg^{2+}) = 0.47$]

▲

The ion-dipole interaction is largely responsible for the *hydration* of cations in solution, the attachment of water molecules to a central ion (Fig. 10.4). The electric dipole of the H_2O molecule is such that the negative partial charge is on the O atom, and this atom is attracted to the positive charge of the cation. Hydration often persists in the solid and accounts for the existence of salt hydrates such as $Na_2CO_3 \cdot 10H_2O$ and $CuSO_4 \cdot 5H_2O$. Because the strength of the interaction increases as the separation of the ion and the dipole decreases, we expect small cations to be more strongly hydrated than large ones. This is consistent with the fact that lithium and sodium commonly form hydrated salts, whereas their larger congeners, potassium, rubidium, and cesium, do so only rarely. Ammonium salts are usually anhydrous for a similar

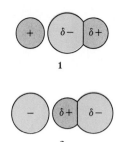

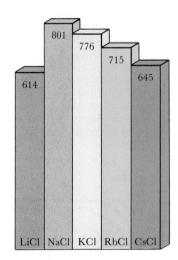

FIGURE 10.3 The melting points of the alkali metal halides show a variation that can be traced in large part to the strength of the interactions between the ions. The values given here are in degrees Celsius.

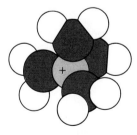

FIGURE 10.4 Ion-dipole interaction is an important contributor to the hydration of ions, particularly of cations. Here we see how several H_2O molecules attach to a central cation as a result of the interaction.

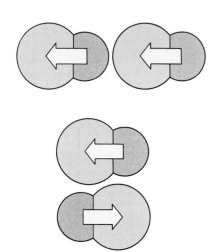

FIGURE 10.5 Polar molecules attract each other via the interaction between the partial charges of their electric dipoles (the yellow arrows). Either of these orientations is energetically favorable.

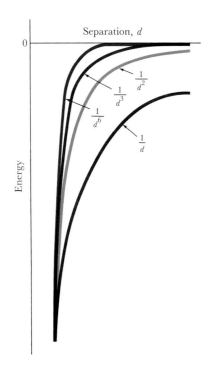

FIGURE 10.6 The dependence on distance of the interactions described in this chapter. Note that the ion-ion interaction dies to zero comparatively slowly (it is a "long-range" interaction); interactions proportional to $1/d^6$ die to zero rapidly (they are "short-range" interactions).

reason, since the NH_4^+ ion has about the same radius (143 pm) as a Rb^+ ion (148 pm). The effect of charge is shown by comparing barium and potassium. Their cations have similar radii (135 pm for Ba^{2+} and 133 pm for K^+), but barium salts are often hydrated because of their higher charge. Lanthanum, barium's neighbor, is both smaller (115 pm) and more highly charged (La^{3+}), and its salts include $La(NO_3)_3 \cdot 6H_2O$ and $La_2(SO_4)_3 \cdot 9H_2O$.

Dipole-dipole interactions. A *dipole-dipole interaction* is the attraction between the electric dipoles of polar molecules. When two neighboring polar molecules have either of the orientations shown in Fig. 10.5, the partial charges of opposite sign are closer together than those of like sign and there is a net attraction between the molecules. Since the dipole-dipole force arises from the interaction between *partial* charges, it is weaker than the attraction between oppositely charged ions. The relative strengths of dipole-dipole and ion-ion interactions can be seen by comparing the enthalpy of vaporization of the molecular solid hydrogen chloride (18 kJ/mol) with the lattice enthalpy of the ionic solid sodium chloride (787 kJ/mol). Nevertheless, the dipole-dipole interaction binds many polar molecules together strongly enough for heat to be released when they condense to a liquid.

The potential energy of the dipole-dipole interaction between molecules A and B with electric dipole moments μ_A and μ_B in a solid decreases with the cube of their separation:

$$E \propto \frac{-\mu_A \times \mu_B}{d^3}$$

Doubling the separation reduces the strength of the interaction by a factor of $2^3 = 8$ (Fig. 10.6). In gases and liquids, the interaction has a much smaller effective range because the molecules rotate, bringing like charges together some of the time. For rotating molecules, the potential energy of the interaction decreases as the sixth power of the separation:

$$E \propto \frac{-\mu_A^2 \times \mu_B^2}{d^6}$$

Doubling the separation of such molecules reduces the strength of the interaction by a factor of $2^6 = 64$, so their dipole-dipole interactions are unimportant unless the molecules are close together. The very short range of intermolecular forces between rotating molecules is one of the reasons why the ideal gas law is such a good description of the behavior of most real gases at normal pressures: in gases, molecules rotate and are, on average, far apart.

▼ EXAMPLE 10.2 Estimating the importance of dipole-dipole interactions in a gas

Estimate the relative importance of the dipole-dipole interactions in equal volumes of gaseous HCl and HBr at the same temperature and pressure, given that their dipole moments are 1.08 D and 0.80 D, respectively.

STRATEGY It follows from the last equation that the energy of this interaction is proportional to μ^4/d^6 (because in a pure gas A and B are the same type of molecule). We are given the dipole moments. For d, we know from Avogadro's principle that equal volumes of gas at the same temperature and pressure contain the same numbers of molecules. Therefore, the average separation of the molecules is the same in the two samples, and d^6 is canceled out when we form the ratio.

SOLUTION The ratio of the interaction energies is

$$\frac{E(\text{HCl})}{E(\text{HBr})} = \frac{\mu_{\text{HCl}}^4}{\mu_{\text{HBr}}^4} = \frac{(1.08\ \text{D})^4}{(0.80\ \text{D})^4} = 3.3$$

The interaction energy—and hence the departure from ideal-gas behavior—increases sharply as the dipole moment increases.

EXERCISE Calculate the ratio of potential energies arising from the dipole-dipole interactions in equal volumes of gaseous $CHCl_3$ ($\mu = 1.01$ D) and CH_3Cl (1.87 D) at the same temperature and pressure.

[*Answer*: 0.085]

▲

The London force. The *London force* can be traced to a continuous changing of the positions of the electrons in a molecule. As the electrons move about, some of their momentary locations may produce an instantaneous electric dipole in the molecule (Fig. 10.7). This is the case even if the *average* positions of the electrons are those of a nonpolar molecule; and even a noble gas atom can have an instantaneous dipole of this kind. The flickering instantaneous dipoles on pairs of neighboring molecules tend to align and come into step with each other, because the total energy is lower when the partial positive charge of one is near the partial negative charge of the other. As a result, the two instantaneous dipoles attract each other and the molecules stick together.

The London force (also called the "dispersion force") is named for the German-American physicist Fritz London, who first explained it. It acts between *all* types of molecules, polar as well as nonpolar, and is always attractive. It also acts between noble-gas atoms. This force is responsible for the existence of benzene as a liquid at normal temperatures, for the condensation of carbon dioxide to a solid, and for the condensation of oxygen, nitrogen, and the noble gases to liquids at low temperatures.

The London force increases with molecular weight. Larger molecules have more electrons, which produce greater fluctuations in

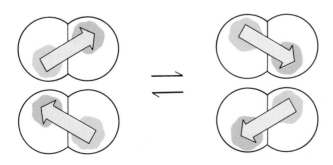

FIGURE 10.7 The London force arises from the attraction between two instantaneous dipoles (the yellow arrows). Although they are continuously changing direction, they remain in step in the most favorable orientation.

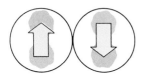

FIGURE 10.8 The instantaneous dipoles in two neighboring spherical molecules may be quite far apart and interact weakly. Those in two cylindrical molecules can be closer together and can interact more strongly.

3 Pentane

4 Neopentane

charge as they flicker between different positions, and these, in turn, lead to larger instantaneous dipoles. The strength of the interaction depends on the polarizability α (the ease of distortion, discussed in Section 8.7) of the molecules and is largest when both are highly polarizable. This fits the flickering-dipole picture of the interaction, because a highly polarizable molecule (α large) is one in which the nuclei have only weak control over their outer electrons. The potential energy of the London attraction decreases very rapidly (as the sixth power) with the separation of the two molecules:

$$E \propto \frac{-\alpha_A \times \alpha_B}{d^6}$$

This expression applies both to stationary molecules in a solid and to molecules that are free to rotate in a gas or liquid.

▼ **EXAMPLE 10.3** **Estimating the strength of the London interaction in a gas**

Calculate the relative change in the potential energy of the London interaction in a gas when its volume is doubled.

STRATEGY We know that the potential energy arising from the London force is inversely proportional to the sixth power of the separation of the molecules, so increasing the separation by a factor f decreases the interaction by a factor f^6. Therefore, we need to know how the separation changes when the volume of the sample is doubled. Since volume is given by (length)^3, doubling the volume increases the separation of the molecules by $2^{1/3}$ (because $2^{1/3} \times 2^{1/3} \times 2^{1/3} = 2$).

SOLUTION The average separation of the molecules increases by $2^{1/3} = 1.26$. Therefore, the potential energy of their interaction decreases by a factor of $(1.26)^6 = 4.00$. Doubling the volume of the gas thus reduces the importance of the interaction by a factor of 4.

EXERCISE Calculate the factor by which the potential energy due to the London force is increased when a gas is compressed to one-third its initial volume at constant temperature.

[*Answer*: 9]

▲

Since an H_2 molecule has only two electrons, its instantaneous dipoles are very small. As a result, it interacts so weakly with its neighbors that hydrogen gas condenses to a liquid only if the temperature is lowered to $-253°C$ (20 K). This is a general result: small, nonpolar molecules, with few electrons, produce a weak London force and hence have low melting and boiling points. However, it is very difficult to make precise predictions about boiling points, because the strength of the London force is also determined in part by the shape of the molecules and the way they pack together in the liquid. Both pentane **(3)** and neopentane **(4)**, for instance, have molecular formula C_5H_{12}, but they have markedly different boiling points (36°C and 10°C, respectively). Only a few of the atoms belonging to neighboring *spherical* molecules can approach each other closely, but atoms belonging to neighboring *cylindrical* molecules can get quite close to each other (Fig. 10.8). Hence the London interaction, which originates in the atoms, is stronger between cylindrical molecules than between spherical molecules.

Large differences in polarizability, however, can be more important than molecular shape. This is one of the reasons why hydrocarbons with up to 4 C atoms are gases, hydrocarbons with 5 through 17 C atoms (such as benzene and octane) are liquids, and the heavier hydrocarbons (those containing 18 or more C atoms) are waxy solids (Fig. 10.9). In each case, the compound with the greater molecular weight, and thus the greater number of electrons, experiences the stronger London interactions.

When heavier atoms are substituted for hydrogen atoms in a molecule, the resultant increase in the strength of the London force is striking. Methane boils at −162°C, but carbon tetrachloride, with its much more polarizable Cl atoms, is a liquid at room temperature and boils at 77°C. Carbon tetrabromide is even more polarizable. It is a solid at room temperature, melting at 90°C and boiling at 190°C. Covalent fluorides are generally more volatile than covalent chlorides, for fluorides have fewer electrons. Compare, for example, the values for CF_4 and CCl_4 in Table 10.4. The high nuclear charge of the fluorine nucleus and its firm control over the surrounding electrons also means that fluorocarbons are significantly less polarizable than hydrocarbons. Thus, although CF_4 has a greater molecular weight than benzene, its boiling point is 208°C lower (at −128°C, compared with 80°C for benzene).

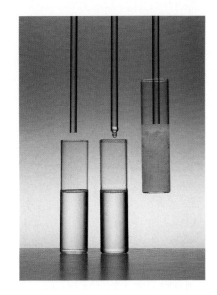

FIGURE 10.9 Hydrocarbons show how London forces increase in strength with molecular weight. Pentane is a mobile fluid (left), pentadecane ($C_{15}H_{32}$) a viscous liquid (center), and octadecane ($C_{18}H_{38}$) a waxy solid (right).

TABLE 10.4 Normal melting and boiling points of some molecular substances

Substance	Melting point, °C	Boiling point, °C	Substance	Melting point, °C	Boiling point, °C
Helium	−270 (3.5 K)	−269 (4.2 K)	HBr	−87	−67
			HI	−51	−35
Neon	−249	−246	H_2O	0	100
Argon	−189	−186	H_2S	−86	−60
Krypton	−157	−153	NH_3	−78	−33
Xenon	−112	−108	CO_2		−79s*
Hydrogen	−259	−253	SO_2	−76	−10
Nitrogen	−210	−196	CH_4	−183	−162
Oxygen	−218	−183	CF_4	−184	−128
Fluorine	−220	−188	CCl_4	−23	77
Chlorine	−101	−34	CH_3OH	−94	65
Bromine	−7	59	C_6H_6	6	80
Iodine	114	184	Glucose	142	d
HF	−83	20	Sucrose	184 d	
HCl	−114	−85			

*An s indicates that the solid sublimes, a d that it decomposes.

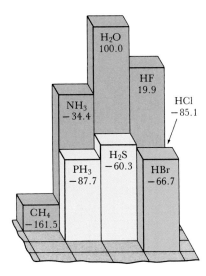

FIGURE 10.10 The boiling points of the hydrogen compounds of the *p*-block elements show a smooth variation, but those of nitrogen (NH_3), oxygen (H_2O), and fluorine (HF) are strikingly out of line.

FIGURE 10.11 The existence of the oceans is a result of hydrogen bonds between water molecules. If they were absent (as in H_2S), water would be a gas at room temperature.

The London interaction between polar molecules is usually stronger than their dipole-dipole interaction. The condensation of hydrogen chloride to a liquid is due more to the London force between the molecules than to their dipolar interaction.

▼ EXAMPLE 10.4 Accounting for a trend in boiling points

Explain the trend in the boiling points of the hydrogen halides, bearing in mind that the electronegativity differences *decrease* from HCl to HI:

$$HCl: -85°C \qquad HBr: -67°C \qquad HI: -35°C$$

STRATEGY We need to consider the strengths of the dipolar and London interactions and their likely effects on boiling points. Higher boiling points suggest stronger interactions. The size of the dipoles, and hence of the dipole-dipole interaction, increases with the difference in electronegativity between the hydrogen and halogen atoms. The strength of the London force increases with the number of electrons.

SOLUTION The electronegativity differences decrease from HCl to HI. Hence the molecules become less polar from HCl to HI, and their dipolar interactions decrease in strength, suggesting that the boiling points should decrease. This conflicts with the data. The number of electrons in the molecules increases from HCl to HI, so the London force is expected to increase too. Hence, the boiling points should increase from HCl to HI, in accord with the data. This suggests that the London force dominates the dipolar interactions for these molecules.

EXERCISE Account for the trend in the boiling points of the noble gases, which increase from He to Rn.

[*Answer:* London interactions increase in strength as atomic number increases.]

10.2 HYDROGEN BONDING

The existence of another type of intermolecular force becomes apparent when we examine trends in the boiling points of a range of compounds containing hydrogen (Fig. 10.10). Water boils at a much higher temperature (100°C) than hydrogen sulfide (−60°C). In fact, hydrogen sulfide is a gas at room temperature, even though an H_2S molecule has many more electrons—and hence stronger London forces—than an H_2O molecule. Ammonia and hydrogen fluoride also have higher boiling points than the analogous hydrogen compounds of their congeners. The unusually high boiling points of water and ammonia suggest that there are unusually strong forces between the molecules.

The strong intermolecular forces in NH_3, H_2O, and HF are the result of "hydrogen bonding":

A **hydrogen bond** is a link formed by a hydrogen atom lying between two strongly electronegative atoms.

The other intermolecular forces we have considered act between compounds built from any elements at all. Hydrogen bonding is exceptional because it requires the joint presence of hydrogen and two electronegative atoms; moreover, only N, O, and F are sufficiently electronegative to take part in hydrogen bonding. If we denote the hydrogen bond with three dots, as in A—H···B, where A and B are the two electronegative atoms, then the only important cases of hydrogen bonding are

5 O—H···O

N—H···N	O—H···N	F—H···N
N—H···O	O—H···O	F—H···O
N—H···F	O—H···F	F—H···F

The key to the formation of a hydrogen bond is the strongly polar nature of the A—H bond when A is strongly electronegative. The resulting partial positive charge on H can attract the lone-pair electrons of a neighboring atom B, especially if that atom is so electronegative that it also has a strong partial negative charge. Hydrogen is unique in being able to form this kind of bond because it is so small that the atom B can approach very closely to the partial charge of hydrogen and interact with it strongly.

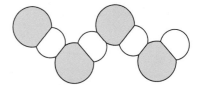

6 (HF)$_4$ chain

The strength of an O—H···O hydrogen bond (5) is around 20 kJ/mol. This is only a fraction of the strength of a normal O—H bond, which is 463 kJ/mol. However, when it exists it is strong enough to dominate all other types of intermolecular interaction. Hydrogen bonding is so strong that it can survive even in a vapor. Liquid hydrogen fluoride, for instance, contains zigzag chains of hydrogen-bonded HF molecules (6), and the vapor contains short fragments of these chains along with (HF)$_6$ rings. Acetic acid vapor contains pairs of molecules (7).

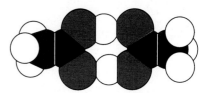

7 Acetic acid dimer

Hydrogen bonding is responsible for the hydration of oxoanions in solution and in salt hydrates: it contributes to the high degree of hydration in Ba(OH)$_2$·8H$_2$O, for example. We shall also see that it has a profound influence on the properties of water itself, including its existence as a liquid at room temperature (Fig. 10.11).

THE PROPERTIES OF LIQUIDS

A liquid consists of molecules that can move past their neighbors but not escape from them completely (Fig. 10.12). Three of the most characteristic properties of liquids are their ability to flow (which is shared by gases), their possession of a sharply defined surface (which distinguishes them from gases), and their tendency to vaporize into the space above them and to exert a vapor pressure. All three effects are related to the strengths of intermolecular forces.

10.3 VISCOSITY AND SURFACE TENSION

The ability of a liquid to flow is described by its _viscosity_, its resistance to flow: the higher the viscosity, the slower the flow. Viscosities are measured by observing how long it takes a given volume to flow through a

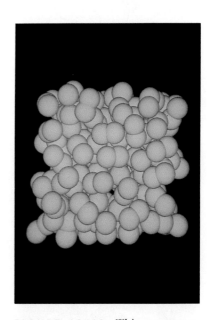

FIGURE 10.12 This computer-generated image of a liquid illustrates how the molecules, though close together, are able to move past each other more easily than in a solid.

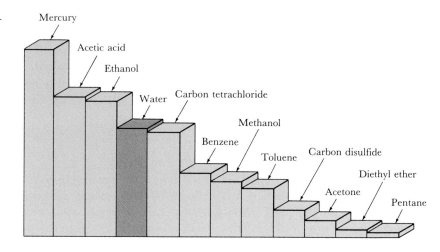

FIGURE 10.13 The relative viscosities of some common liquids at room temperature. Liquids are less viscous at higher temperatures.

Mercury
Acetic acid
Ethanol
Water Carbon tetrachloride
Methanol
Benzene
Toluene Carbon disulfide
Diethyl ether
Acetone
Pentane

8 Glycerol

FIGURE 10.14 The hydrocarbon molecules in oil are tangled together and flow past each other only slowly.

narrow tube; some values are compared with that of water in Fig. 10.13. A liquid with a high viscosity (like molasses at room temperature, or molten glass) is said to be "viscous."

Viscosity and intermolecular forces. The viscosity of a liquid arises from the forces between its molecules. The stronger the forces hindering the motion of the molecules, the greater the viscosity. Hydrogen bonding is particularly important in this respect because it can bind neighboring molecules together so strongly. This accounts for the fact that water has a greater viscosity than benzene, in which there is no hydrogen bonding. Phosphoric acid (H_3PO_4) and glycerol ($C_3H_7O_3$, **8**) are very viscous at room temperature on account of the numerous hydrogen bonds their molecules can form. Heavy hydrocarbon oils, which are not hydrogen-bonded, are also viscous. Their viscosity arises partly from the London force between molecules and partly because the long chainlike molecules get tangled together like a plate of cooked spaghetti (Fig. 10.14).

The viscosities of most liquids decrease with increasing temperature: the molecules have more energy at high temperatures and can move past their neighbors more readily. The viscosity of water at 100°C, for instance, is one-sixth its viscosity at 0°C, so the same amount flows through a tube six times more quickly. However, there are exceptions to this general rule. Rhombic sulfur, a molecular solid of S_8 rings, melts at 113°C and forms a mobile straw-colored liquid (Fig. 10.15). The viscosity of the liquid is low because the rings can move past each other readily. The viscosity increases if the liquid is heated further, because the S_8 rings break open into chains that become tangled. At still higher temperatures, the viscosity falls again as the color changes to a deep red-brown; now the S_8 chains are breaking up into smaller, more mobile, highly colored S_2 and S_3 molecules.

Surface tension and intermolecular forces. The effect of intermolecular forces is to draw molecules inward from the surface of a liquid, packing them together and forming a smooth surface. When water is poured from a beaker onto a table, it forms a puddle with a greater surface area than it had initially. As the puddle forms, H_2O molecules are being moved from the bulk water, where they are strongly attracted to each other, to the surface, where they are less strongly attracted (Fig. 10.16). In the absence of gravity, the water would form a sphere, for then the maximum number of molecules are surrounded by neighbors, and their total energy is lowest. However, in the presence of gravity, the lowest energy is achieved when the liquid is spread out in a nearly flat pool, even though fewer of the molecules are surrounded by neighbors. In a gravity-free environment, as in the space shuttle in orbit, the shape of liquid droplets is governed by surface tension alone. Tiny (0.01-mm diameter), perfect spheres that have been formed in space (Fig. 10.17) are now commercially available; they are used to calibrate particle sizes for powdered pharmaceuticals.

The energy needed to create new surface area by moving molecules from the bulk water to the surface is called the _surface tension_ γ (the Greek letter gamma) of the liquid. The greater the surface tension, the greater the energy needed to convert bulk molecules into surface molecules. The surface tension of water is about three times higher than that of most other common liquids; this is another consequence of its strong hydrogen bonds. The very high surface tension of mercury, more than six times that of water, can be traced to metallic bonding between its atoms.

Another observable effect of intermolecular forces is _capillary action_, the rise of liquids up very narrow tubes (the name comes from the Latin word for "hair"). The liquid rises up the tube because of attractive forces between its molecules and the tube's inner surface. These are forces of _adhesion_, forces that bind a substance to a surface; they are distinct from forces of _cohesion_, the forces that bind the molecules of a substance together to form a bulk material and that are responsible for condensation. However, the liquid can climb only so high, because its potential energy increases as it climbs. The final height h of the liquid is proportional to the surface tension and inversely proportional to its density ρ (the Greek letter rho) because more energy is needed to raise a dense liquid:

$$h = \frac{2\gamma}{\rho g r}$$

In this equation, r is the radius of the tube, and g is the acceleration due to gravity (9.81 m/s^2). The equation shows that narrower tubes lead to higher columns of liquid.

FIGURE 10.15 When sulfur is heated it melts, changes color, and undergoes changes in viscosity as the S_8 molecules break up.

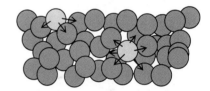

FIGURE 10.16 Surface tension arises from the attractive forces with which molecules in the bulk liquid pull on the molecules that form the surface. Since a molecule in the bulk liquid interacts with more neighbors than one on the surface, it has a lower energy.

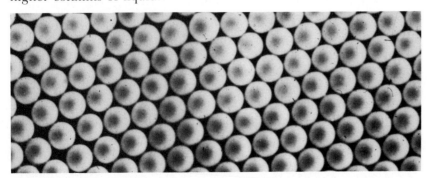

FIGURE 10.17 A sample of latex spheres produced on one of the space-shuttle missions. Each one is a perfect sphere of diameter 0.01 mm, formed by surface tension in the absence of gravity.

FIGURE 10.18 When the adhesive forces between a liquid and its glass container are stronger than the cohesive forces within the liquid, the liquid forms the meniscus shown here for water in glass (left). If the cohesive forces are stronger than the adhesive forces (as for mercury in glass), the meniscus is curved downward (right).

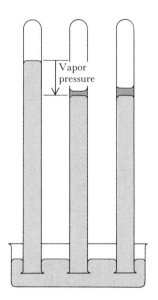

FIGURE 10.19 An apparatus for studying the pressure exerted by the vapor of a liquid. The vapor pressure (in Torr) exerted by the liquid above the mercury is equal to the distance by which the column of mercury is lowered. The vapor pressure is the same however much liquid water is present.

The _meniscus_ of a liquid (from the Greek word for "crescent") is the curved surface it forms in a narrow tube (Fig. 10.18). The meniscus of water in a glass capillary is curved upward (forming a ∪ shape) because the forces causing the water molecules to adhere to the glass are stronger than the forces of cohesion between water molecules. The water therefore tends to cover the greatest possible area of glass. The meniscus of mercury curves downward in glass (forming a ∩ shape). This is a sign that the cohesive forces between mercury atoms are stronger than the adhesive forces between mercury and glass, for in this case the liquid tends to reduce its contact with the glass.

10.4 VAPOR PRESSURE

We first encountered vapor pressure in Section 5.6, in connection with collecting gases over water; there we saw that it is a measure of the volatility of substances. In this section we discuss the concept of vapor pressure more precisely and examine its relation to the intermolecular forces.

The measurement of vapor pressure. Suppose we introduce a little water (or any other liquid) into the vacuum above the mercury in a barometer tube. The water immediately begins to evaporate and fill the space with vapor. As a result, the pressure rises, depressing the column of mercury a few millimeters. If we supply only a trace of water, it all evaporates, and the pressure exerted by the vapor depends on the amount of water added. However, if we add so much water that a little of it remains on the surface of the mercury in the tube (Fig. 10.19), the pressure exerted by the vapor becomes constant. That is, as long as some liquid water is present, at a fixed temperature its vapor exerts the same characteristic pressure. For example, at 40°C the mercury falls 55 mm, so the pressure exerted by the vapor is 55 Torr; this is so whether there is 0.1 mL or 1 mL of liquid water present. This characteristic pressure is the vapor pressure of the liquid. Some room-temperature values are given in Table 10.5.

A solid in a closed container also fills the space above it with its vapor, thus exerting a characteristic _sublimation vapor pressure_. However, the vapor pressures of most solids at room temperature are much smaller than those of liquids, and the rate at which a solid evaporates is often so low that its vapor never reaches its final pressure. We therefore concentrate on the vapor pressures of liquids.

Vapor pressure as a dynamic process. We can build up a precise definition of vapor pressure by thinking about what is happening at the upper surface of the water in Fig. 10.19. When water is first introduced into the vacuum above the mercury, some of its molecules leave the liquid and form a vapor. When molecules that have escaped from the liquid water strike the water's surface, they may be recaptured by the attractive intermolecular forces. As the number of molecules in the vapor increases, more of them strike the surface until eventually a stage is reached at which the number of molecules returning to the liquid exactly matches the number escaping from it. At this stage, the vapor is condensing as fast as the liquid is vaporizing. That is, the rate of vaporization, in moles of H_2O per second, is equal to the rate of condensation. The concentration of molecules in the vapor, and hence its pressure, then remain constant, and the liquid and vapor are in "dynamic equilibrium" (Fig. 10.20):

Dynamic equilibrium is a condition in which a forward process and its reverse are occurring simultaneously at equal rates.

"Dynamic" is a key word in this definition, and it implies continuous activity. Except at the beginning and end of the day, a busy store is often at dynamic equilibrium, with the number of customers arriving matching the number leaving. Dynamic equilibrium is quite unlike the "static equilibrium" of a ball at rest at the foot of a hill (where "static" indicates an absence of activity). A store is at static equilibrium before it opens for business.

In a closed container, a liquid and its vapor reach dynamic equilibrium when the pressure of the vapor has risen to a particular value that depends on the liquid and the temperature. This suggests the following definition:

The **vapor pressure** of a liquid is the pressure exerted by its vapor when the two phases are in dynamic equilibrium.

If vaporization is very slow, the pressure of the vapor need not be high for the condensation rate, which is proportional to the pressure, to match it. Hence low vapor pressure (like that of a solid or of cold water) is a sign that molecules leave the surface at a low rate. On the other hand, if vaporization is rapid, the pressure of the vapor must be high before condensation occurs at a matching rate. Hence high vapor pressure (like that of ether or hot water) is a sign that molecules leave the liquid surface at a great rate.

▼ **EXAMPLE 10.5** **Assessing the rate of vaporization**

The vapor pressure of mercury is 0.49 mTorr (0.49×10^{-3} Torr) when the weather is cool (10°C) but rises to 2.8 mTorr on a hot day in summer (30°C). What does this imply about the rates of vaporization of mercury at the two temperatures?

STRATEGY Since the vapor pressure is higher at the higher temperature, we know that the rate of vaporization is higher too. We can estimate how much higher because the rate of return to liquid from vapor is proportional to the vapor pressure; since the rate of escape is equal to the rate of return at equilibrium, it too is proportional to the vapor pressure.

SOLUTION The vapor pressures are in the ratio

$$\frac{2.8 \text{ mTorr}}{0.49 \text{ mTorr}} = 5.7$$

Therefore, the mercury vaporizes 5.7 times more rapidly at the higher temperature.

EXERCISE How much more rapidly does water evaporate at 30°C than at 10°C?

[*Answer*: 3.5 times]

TABLE 10.5 Vapor pressures at 25°C

Substance	Vapor pressure, Torr
Water*	23.8
Mercury	0.0017
Methanol	122.7
Ethanol	58.9
Benzene	94.6

*For values at other temperatures, see Table 5.8.

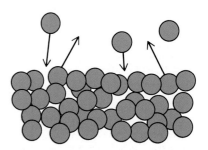

FIGURE 10.20 When a liquid and its vapor are in dynamic equilibrium, the rate at which molecules leave the liquid is equal to the rate at which they return.

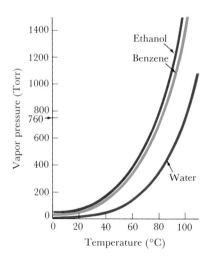

9 Dimethyl ether

10 Ethanol

FIGURE 10.21 The dependence of vapor pressure on temperature for ethanol (red), benzene (green), and water (blue). The normal boiling point is the temperature at which the vapor pressure is 1 atm (760 Torr).

In an open container such as a beaker of water, the vapor that is formed spreads away from the liquid. Little, if any, is recaptured at the surface of the liquid, the rate of condensation never increases to the point at which it matches the rate of vaporization, and a condition of dynamic equilibrium is never reached. Over time, the liquid *evaporates*, or vaporizes completely. Since the molecules of substances with high vapor pressures (such as gasoline and hot water) leave the liquid at a high rate, those substances evaporate more quickly than substances with low vapor pressures.

Vapor pressure increases with temperature (Fig. 10.21) because the molecules in the heated liquid move more energetically and can escape more readily from their neighbors. Practically all vaporization takes place from the surface of the liquid because the molecules there are least strongly bound and can escape to the vapor more easily than those in the bulk liquid.

Vapor pressure and molecular structure. A liquid has a low vapor pressure at room temperature if its intermolecular forces are strong. We therefore expect molecules capable of forming hydrogen bonds to exist as liquids that are much less volatile than others (Fig. 10.22). This is shown strikingly by dimethyl ether (**9**) and ethanol (**10**), which both have the molecular formula C_2H_6O and so might be expected to have similar vapor pressures. However, ethanol molecules each have an —OH group and can link together by forming hydrogen bonds. The ether molecules, which cannot form hydrogen bonds, interact only through London and dipole-dipole forces. As a result, ethanol is a liquid at room temperature while dimethyl ether is a gas.

Vapor pressure and boiling point. Now consider what happens as we heat water in an open container. When the temperature of the liquid is raised to the point at which the vapor pressure matches the pressure exerted by the atmosphere, vaporization can occur throughout the liquid because any vapor that is formed can drive back the atmosphere to make room for itself. Thus, bubbles of vapor form in the liquid and rise to the surface. Since vaporization no longer occurs only at the surface, it can proceed very rapidly. This is the condition we call "boiling"; the temperature at which it occurs is the "boiling point."

The boiling point increases with external pressure because a higher temperature must be reached before the vapor pressure matches the external pressure. Since we are normally interested in the properties of liquids under atmospheric pressure (specifically 1 atm or 760 Torr), it is convenient to report boiling points at that pressure:

The **normal boiling point** T_b of a liquid is the temperature at which its vapor pressure is equal to 1 atm.

Some boiling points are given in Table 10.4. By reporting the normal boiling point of water as 100°C (T_b = 373 K), we mean that its vapor pressure rises to 1 atm when it is heated to 100°C. Its boiling point is higher when the external pressure is increased, as in a pressure cooker or a steam boiler, and lower when the external pressure is lower. This phenomenon is familiar to mountain climbers and to inhabitants of high-altitude cities.

▼ EXAMPLE 10.6 Predicting the boiling point at a given pressure

Predict the boiling point of water at the atmospheric pressure of Denver, Colorado, which on a typical day is 650 Torr.

STRATEGY Because the atmospheric pressure (650 Torr) is lower than 760 Torr, we expect a boiling point that is lower than normal. To predict the boiling point, we can use Fig. 10.21 to find the temperature at which the vapor pressure of water reaches 650 Torr.

SOLUTION From Fig. 10.21, the vapor pressure of water is 650 Torr at 95°C. Hence its boiling point at that pressure is 95°C.

EXERCISE Predict the boiling point of ethanol on a day when the atmospheric pressure is 780 Torr.

[*Answer*: 79°C]

▲

Since boiling points are related to vapor pressure, they are related to the strengths of intermolecular forces. In general, a boiling point is high when the intermolecular forces are high, since a high temperature is then needed to raise the vapor pressure to 1 atm. This accounts for the anomalously high boiling point of water compared with that of hydrogen sulfide: hydrogen bonds are strong in water but absent in hydrogen sulfide.

A useful empirical observation known as *Trouton's rule* gives a relation between a liquid's enthalpy of vaporization ΔH°_{vap} and its boiling point T_b. The rule states that $\Delta H^\circ_{vap}/T_b$ is approximately equal to 85 J/(K · mol) for liquids in which hydrogen bonding is unimportant. Since $\Delta H^\circ_{vap}/T_b$ is a constant, a high boiling point goes with a high enthalpy of vaporization.

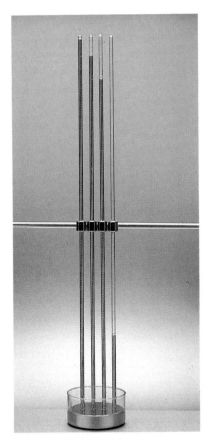

FIGURE 10.22 The height of the column of mercury (far left) is depressed by the vapor pressures of water, ethanol, and ether (left to right). Ether is so volatile because its molecules cannot form hydrogen bonds with each other.

▼ EXAMPLE 10.7 Using Trouton's rule

Estimate the enthalpy of vaporization of benzene from its boiling point, 80°C.

STRATEGY Since hydrogen bonding is unimportant in benzene, Trouton's rule may be used. We rearrange it to

$$\Delta H^\circ_{vap} = T_b \times 85 \text{ J/(K · mol)}$$

and substitute the data.

SOLUTION Since 80°C corresponds to 353 K, we have

$$\Delta H^\circ_{vap} = 353 \text{ K} \times 85 \text{ J/(K · mol)} = 30 \text{ kJ/mol}$$

The experimental value is 34 kJ/mol.

EXERCISE Estimate the enthalpy of vaporization of liquid bromine, which boils at 59°C.

[*Answer*: 28 kJ/mol (experimentally, 29.45 kJ/mol)]

▲

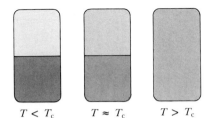

$T < T_c$ $T \approx T_c$ $T > T_c$

FIGURE 10.23 When the temperature of a liquid in a sealed container is raised, the pressure and density of the vapor increase. At and above the critical temperature T_c the density of the vapor is the same as the density of the liquid, and a single uniform phase fills the container.

The critical temperature. A liquid does not boil when it is heated in a closed, constant volume container. To see what happens instead, suppose initially that the tube shown in Fig. 10.23 contains liquid water

TABLE 10.6 Critical temperatures of some substances

Substance	Critical temperature, °C
Helium	−268 (5.2 K)
Neon	−229
Argon	−123
Krypton	−64
Xenon	17
Hydrogen	−240
Nitrogen	−147
Oxygen	−118
Chlorine	144
Bromine	311
HCl	52
HBr	90
HI	150
H_2O	374
H_2S	101
NH_3	132
CO_2	31
SO_2	158
SO_3	218
CH_4	−83
CF_4	−46
CCl_4	283
CH_3OH	240
C_6H_6	289

and water vapor at 24 Torr (its vapor pressure at 25°C). As the water plus vapor is heated, the vapor pressure increases, and at 100°C the water is in dynamic equilibrium with vapor at a pressure of 760 Torr. However, the water does not now boil because there is no space for the vapor to occupy. Instead the vapor pressure increases as the temperature increases. At 200°C the vapor pressure has risen to 11,700 Torr (15.4 atm); liquid and vapor are still in dynamic equilibrium, but now the vapor is very dense because it is at such a high pressure.

At 374°C, with the vapor pressure at 218 atm (so the container must be very strong!), the density of the vapor is so great that it is equal to that of the remaining liquid. Now there is no surface separating liquid from vapor, the two cannot be distinguished, and a single uniform substance fills the container. Because a substance that fills any container it occupies is by definition a gas, we conclude that we have reached a temperature above which the liquid phase does not exist. We have reached the "critical temperature" T_c of water:

The **critical temperature** of a substance is the temperature above which it cannot exist as a liquid.

The critical temperatures of some substances are listed in Table 10.6. Their values are important in practice, because it is useless to try to liquefy a gas by applying pressure if it is above its critical temperature. For example, since the critical temperature of carbon dioxide is 31°C, we know that it cannot be compressed to a liquid if its temperature is higher than 31°C. Since the critical temperature of helium is 5.2 K, the gas must be cooled almost to absolute zero before it can be liquefied by pressure.

A substance that is at a temperature above its critical temperature may be so dense that, although it is a gas, it can act as a solvent like a liquid. "Supercritical" carbon dioxide—carbon dioxide at high pressure but above its critical temperature—can dissolve organic compounds. It is used to remove caffeine from coffee beans and to extract perfumes from flowers. Supercritical hydrocarbons are used to dissolve coal and separate it from ash, and they have been proposed for extracting oil from oil-rich tar sands.

10.5 SOLIDIFICATION

A liquid solidifies when it is cooled to its freezing point. At this temperature and below, the molecules have such low energies that they are unable to escape from the forces that attract them to their neighbors. The molecules oscillate about their average positions but no longer migrate from site to site as they do in a liquid.

Freezing point and pressure. The "normal" freezing point T_f of a liquid is the temperature at which it freezes under 1 atm pressure. Under greater pressure, most liquids freeze at a higher temperature: pressure pushes the molecules together, reducing their separation, and they attract each other more strongly. The effect of pressure is usually quite small unless the pressure is very large. Iron, for example, melts at 1800 K under 1 atm pressure, and only a few degrees higher when the pressure is a thousand times as great. However, at the very center of the earth the pressure is high enough for iron to be solid despite the high temperatures there, and the earth's innermost core is solid.

Water is an exception to the general rule that freezing points increase with pressure. Water freezes at a *lower* temperature under pressure: at 1000 atm it freezes at −5°C. Like so many of water's properties, this too can be traced to hydrogen bonds, which hold the H_2O molecules in a more open structure in ice than in liquid water. This can be seen in the way the density of water varies with temperature (Fig. 10.24), which shows a maximum density at 4°C and a sharp density decrease on freezing. When pressure is applied to the sample, the water tends to remain liquid, for then it occupies a smaller volume. Conversely, ice tends to melt under pressure, for the more compact structure characteristic of the liquid is then favored.

The melting of ice under pressure is thought to contribute to the advance of glaciers. The weight of ice pressing on the edges of rocks deep under the glacier results in very high local pressures. The liquid forms despite the temperature being so low, and the glacier slides slowly downhill on a film of water.

Phase diagrams. A *phase diagram* is a summary in graphical form of the conditions of temperature and pressure at which the solid, liquid, and gaseous phases of a substance exist. The phase diagram for water is shown in Fig. 10.25. Any sample of water at a temperature and pressure that place it within the area labeled "ice" exists as ice. Similarly, a sample represented by any point in the "liquid" area is liquid. A sample at any point in the "vapor" region is a vapor. Using phase diagrams we can predict, almost at a glance, the phase of a substance at any given temperature and pressure.

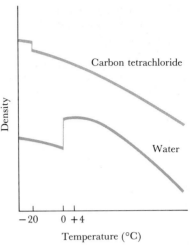

FIGURE 10.24 The variation of the densities of water and carbon tetrachloride with temperature. Note that ice is less dense than water at its freezing point and that the maximum density of water occurs at 4°C.

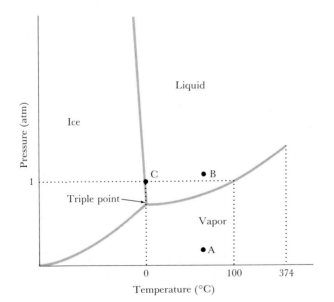

FIGURE 10.25 The phase diagram for water. The solid lines define the regions of pressure and temperature in which each phase is most stable. Note that the freezing point decreases slightly with increasing pressure.

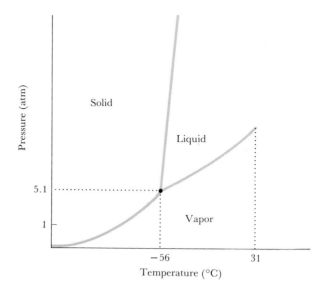

FIGURE 10.26 The phase diagram for carbon dioxide. Note the slope of the boundary between the solid and liquid phases. It shows that the freezing point rises as pressure is applied.

▼ **EXAMPLE 10.8** Interpreting a phase diagram

Use the phase diagram in Fig. 10.25 to predict the physical state of water at 60°C and pressures of (a) 0.2 atm and (b) 1.1 atm.

STRATEGY Locate the point corresponding to each set of conditions on the phase diagram. The region in which it lies shows the state of the sample under those conditions. If the point lies on one of the curves, both states are present in equilibrium.

SOLUTION (a) This is point A, and it lies in the vapor region.
(b) Increasing the pressure to 1.1 atm takes the system to point B, which lies in the liquid region. The sample is a liquid here since the applied pressure is greater than the vapor pressure and all the molecules have been pushed back into the condensed phase.

EXERCISE The phase diagram for carbon dioxide is shown in Fig. 10.26. What is the state of a sample at 10 atm and −15°C?

[*Answer*: Liquid]

▲

The lines separating the regions in a phase diagram are called *phase boundaries*. The points *on* a phase boundary (such as point C at 0°C and 1 atm on the phase diagram of water) show the conditions under which two phases (ice and liquid water in this case) are present in dynamic equilibrium. The phase boundary between liquid and vapor, which shows the pressures at which the vapor and liquid are in equilibrium, is therefore a plot of the vapor pressure of the liquid against the temperature.* The phase boundary between the solid and the vapor shows the

---

*The mathematical expression for the dependence of the vapor pressure on the temperature is called the "Clausius-Clapeyron equation":

$$\ln P' = \ln P + \frac{\Delta H^\circ_{vap}}{R}\left(\frac{1}{T} - \frac{1}{T'}\right)$$

where P is the vapor pressure at temperature T, and P' the vapor pressure at temperature T'. For the sublimation vapor pressure, ΔH°_{vap} is replaced by ΔH°_{sub}, the enthalpy of sublimation.

pressures at which the solid is in equilibrium with its vapor and, hence, is a plot of the sublimation vapor pressure of the solid.

The *triple point* is the point where the three phase boundaries meet. For water, it occurs at 4.6 Torr and 0.01°C. Under these conditions (and only these), all three phases (ice, liquid, and vapor) coexist in dynamic equilibrium. At the triple point, the vapor pressure of liquid water is equal to that of ice. Since the triple point of water is a characteristic, fixed property, it is a better choice for defining a temperature scale than the freezing or boiling point, both of which depend on the applied pressure. It is in fact used to define the size of the kelvin, and by definition there are exactly 273.16 kelvins between absolute zero and the triple point of water.

Since the triple point lies at the lowest point of the region labeled "liquid" in a phase diagram, it marks the lowest pressure at which the liquid can exist. For substances other than water, the triple point also marks the lowest *temperature* at which the liquid can exist. The phase diagram for carbon dioxide, for example, shows that the liquid phase exists only at temperatures higher than that of the triple point. It follows that liquid carbon dioxide can exist only if its pressure is greater than 5.1 atm and its temperature is greater than −56°C (but less than its critical temperature, +31°C). Hence, even though the temperature on Mars sometimes rises above −56°C, its carbon dioxide−rich atmosphere never rains liquid carbon dioxide because the atmospheric pressure there is only 5 Torr.

Liquid water can exist only if the pressure of the vapor is greater than 4.6 Torr. On a cold, dry morning on earth, the partial pressure of water in the air may be lower than 4.6 Torr, and frost may appear and disappear, like solid carbon dioxide on Mars, without liquid forming first.

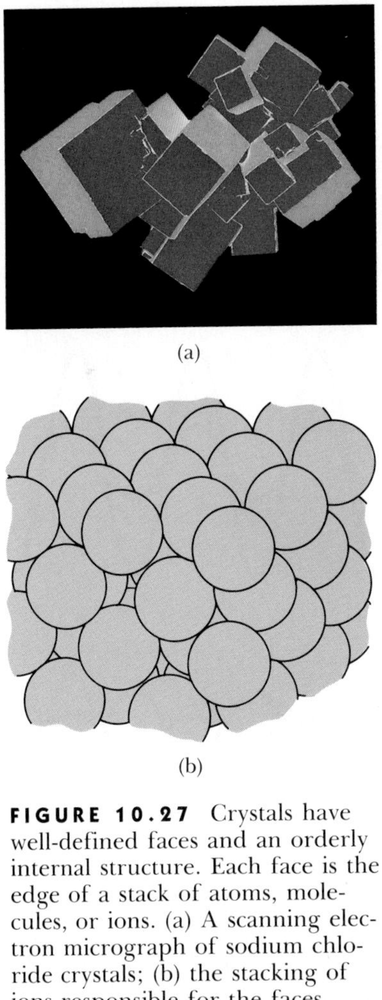

(a)

SOLIDS

A *crystalline solid* is a solid in which the atoms, ions, or molecules lie in an orderly array (Fig. 10.27). These solids usually have flat, well-defined surfaces, called *faces*, that make definite angles with their neighbors. These faces are the edges of orderly stacks of particles. An *amorphous solid* (from the Greek words for "without form") is one in which the atoms, ions, or molecules lie in a random jumble, as in butter and rubber. These solids do not usually have well-defined faces, unless they have been molded.

Many amorphous solids are mixtures of molecules that do not stack together very well. Butter, for instance, is a mixture of molecules with hydrocarbon chains of different lengths. Amorphous materials also include the various kinds of glass, such as the solid formed when silica (SiO_2) is melted and then allowed to cool quickly. Crystalline silica (quartz is one variety) is composed of an orderly, regular network of SiO_2 groups, but when the solid reforms after melting, the network is far more disorderly and the solid is no longer crystalline (Fig. 10.28).

(b)

FIGURE 10.27 Crystals have well-defined faces and an orderly internal structure. Each face is the edge of a stack of atoms, molecules, or ions. (a) A scanning electron micrograph of sodium chloride crystals; (b) the stacking of ions responsible for the faces.

FIGURE 10.28 Quartz is a crystalline form of silica with its atoms in an orderly network, as represented in two dimensions by (a). When molten silica solidifies it forms a glass in which the atoms form a disordered network (b).

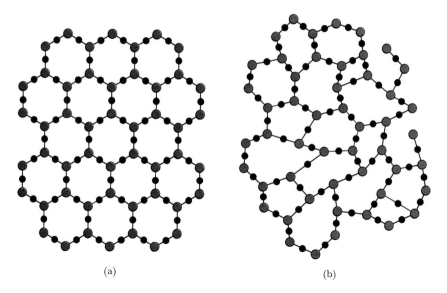

(a)   (b)

10.6 X-RAY DIFFRACTION

A very fruitful method for discovering the arrangement of atoms and ions in crystals makes use of *x-rays*, electromagnetic radiation with wavelengths from about 10 pm up to about 1000 pm (the exact limits are imprecise). The use of x-rays for this purpose was originally suggested by the German Max von Laue shortly after they were first discovered by another German, Wilhelm Röntgen, in 1895. The technique flourished in the hands of the Dutchman Peter Debye and the British father-and-son team of William and Lawrence Bragg and has become the most useful of all techniques for determining the structures of solids. Early workers in the field discovered the crystal structures of many simple solids. Through the work of Dorothy Hodgkin, Max Perutz, Rosalind Franklin, and numerous others, the structures of many complex biologically important molecules have also been determined.

Interference. To see why von Laue seized on x-rays as a means for exploring the interiors of solids, we need to know that "interference" may occur between waves. Imagine two waves of electromagnetic radiation in the same region of space. Where the peaks and troughs of the two waves coincide, they add together to give a wave of greater amplitude (Fig. 10.29). This enhancement of amplitude is called *constructive interference*. If the combined wave is detected photographically, a brighter spot is obtained than with either wave alone. However, if the peaks of one of the waves coincide with the troughs of the other, they partly cancel and give a wave of lower amplitude. This cancellation is called *destructive interference*. If this combined wave is detected photographically, a dimmer spot is obtained. No spot at all is detected if the peaks exactly match the troughs so that cancellation is complete.

The Bragg equation. An object standing in the path of waves of radiation can cause interference between them. This effect is called "diffraction":

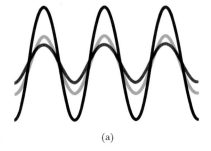

(a)

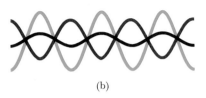

(b)

FIGURE 10.29 Interference between two waves (the red and green curves) may be either (a) constructive, to give a wave of greater amplitude, or (b) destructive, to give a wave of smaller amplitude.

Diffraction is interference between waves caused by an object in their path.

The resulting pattern of bright and dim spots is called a *diffraction pattern*.

A crystal can cause diffraction in a beam of x-rays, and when it is held at a certain angle to the beam, a bright spot of constructive interference is obtained. The angle θ (the Greek letter theta) at which this occurs is related to the wavelength λ of the x-rays and the separation d of the atoms in the crystal by the

Bragg equation: $\qquad 2d \sin \theta = \lambda \qquad\qquad$ (1)

Therefore, from the angle at which the spot is obtained and the wavelength of the radiation, the separation of the atoms can be calculated. Since $\sin \theta$ cannot be greater than 1, the smallest separation d that can be measured in this way is $\frac{1}{2}\lambda$. Von Laue realized that x-rays provide a way of exploring inside crystals because their wavelengths are so short: they can be used to measure distances comparable to the separation of atoms.

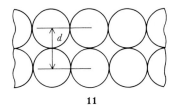

11

▼ EXAMPLE 10.9 Using the Bragg equation

When a crystal of pure copper was examined with x-rays of wavelength 154 pm, an intense spot from the layers marked in **(11)** was obtained for $\theta = 17.5°$. Calculate the metallic radius of copper.

STRATEGY The metallic radius (discussed in Section 7.7) is half the distance between the centers of neighboring atoms ($\frac{1}{2}d$ in **11**). We can find this distance from the value of θ by rearranging the Bragg equation to

$$d = \frac{\lambda}{2 \sin \theta}$$

and substituting the data.

SOLUTION Substituting $\lambda = 154$ pm and $\theta = 17.5°$ gives

$$d = \frac{154 \text{ pm}}{2 \times 0.301} = 256 \text{ pm}$$

The metallic radius of copper is half this distance, or 128 pm.

EXERCISE Calculate the metallic radius of silver, given that 70.8-pm x-rays produced an intense spot at $\theta = 7.10°$ from the same layers as in **(11)**.

[*Answer:* 143 pm]

▲

Figure 10.30 shows an *x-ray diffractometer*, the apparatus used for diffraction studies of single crystals. A beam of x-rays is generated by bombarding a metal with fast electrons. The beam passes through a slit in a screen and into the crystal of the substance being examined. Since there are many different atom-atom separations in a crystal, a large number of spots are found at many different angles, each one of which satisfies the Bragg equation. Buried inside this complicated pattern is the information about the positions of the atoms. The task of the experimenter is to extract and interpret that information in terms of the arrangement and sizes of the atoms.

FIGURE 10.30 An x-ray diffractometer like the one shown here is used to study the arrangements of atoms inside crystals. An x-ray source generates a narrow beam of rays. These rays are diffracted by the crystal, which can be rotated to sit at different angles relative to the incoming ray. The detector can also be rotated around the crystal. The apparatus is controlled by a computer, and thousands of readings are taken and then analyzed.

FIGURE 10.31 The stacks of oranges and other produce in a grocery store form faces with different slopes. They illustrate how atoms stack together in metals to give single crystals.

In modern instruments, the diffraction pattern is detected electronically, and the angles are analyzed on a computer linked directly to the diffractometer. The Bragg equation, or some refined version of it, is used in a series of complicated calculations to build up a complete three-dimensional picture of the interior of the crystal. Most of the structures we are about to discuss, as well as the metallic and ionic radii discussed in Chapter 7, were obtained in this way.

10.7 METALS AND SEMICONDUCTORS

The structures of metallic elements are quite easy to describe because all the atoms in a sample are identical. We can model a crystal of a pure metal, whether aluminum or gold or iron, by using identical spheres to represent the atoms and stacking them together like oranges in a grocery display. Indeed, if you look at a stack of oranges in a display, you can see how stacked spheres can form crystal planes (Fig. 10.31).

Close-packing. One kind of structure often found in metals is said to be "close packed":

A **close-packed structure** is one in which atoms occupy the smallest total volume with the least empty space.

There are two main ways of stacking identical atoms together into a close-packed structure. Both can be pictured as beginning with the two layers shown in Fig. 10.32a. The atoms of the second (upper) layer (layer B) lie in the dips of the first layer (layer A). A third layer can take up either of two arrangements, depending on the detailed electronic structure of the atoms.

One possibility is for the third layer of atoms to lie in the dips that are directly above the atoms of the first layer, as in Fig. 10.32b. The third layer then duplicates layer A, the next layer duplicates B, and so on. This results in an ABAB . . . pattern of layers. As can be seen from the illustration, each atom has three nearest neighbors in the plane below, six in its own plane, and three in the one above, giving twelve in all. This is reported by saying that the "coordination number" of each atom in the solid is 12:

The **coordination number** of an atom is the number of nearest neighbors it has in a solid.

FIGURE 10.32 A close-packed structure can be built up in stages. (a) The first layer (A) is laid down with minimum waste of space, and the second layer (B) lies in the dips, or depressions, of the first. Now there are two possibilities. (b) The third layer may lie above the spheres of the first layer, to give an hcp structure (ABABAB . . .). (c) Alternatively, the third layer can lie above the dips in the first layer to give a ccp structure (ABCABC . . .).

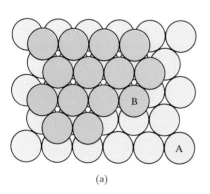

(a)

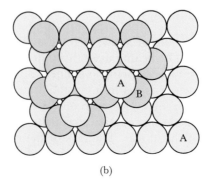

(b)

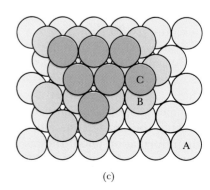

(c)

(For ionic solids, the coordination number of an ion is the number of nearest neighbors of opposite charge.) The structure just described is called the *hexagonally close-packed* (hcp) structure because it is possible to make out a hexagonal pattern in the arrangement of atoms (Fig. 10.33). Magnesium and zinc crystallize in this way.

The second possibility is for the atoms of the third layer to lie in the dips of the second layer that are not directly over the atoms of the first, as shown in Fig. 10.32c. If we call this layer C, the resulting structure has an ABCABC . . . pattern of layers. The coordination number here is also 12, since each atom has three nearest neighbors in the layer below, six in its own layer, and three in the layer above. This is called the *cubic close-packed* (ccp) structure because it exhibits a cubic pattern (Fig. 10.33). Aluminum, copper, silver, and gold crystallize in this way.

Some metals do not have close-packed structures. The electronic structures of their atoms are such that the solid has a lower total energy if the neighboring atoms lie in positions other than those corresponding to close-packing. One common form is shown in Fig. 10.34. Since in this structure an atom lies at the center of a cube formed by eight others, it is called the *body-centered cubic* (bcc) structure. Each atom has eight nearest neighbors, so the coordination number is 8. Iron, sodium, and potassium crystallize in this way.

Unit cells. The small units shown in Figs. 10.33 and 10.34 suggest the concept of a "unit cell":

> A **unit cell** is the smallest unit that, when stacked together repeatedly without gaps, can reproduce the entire crystal.

A unit cell is to a crystal as a brick is to a wall. Unit cells are sometimes drawn by representing each atom by a dot that marks the location of its center (Fig. 10.35). Notice that the ccp unit cell has a dot at the center of each face. For this reason it is also called the *face-centered cubic* (fcc) unit cell.

The number of atoms in a unit cell is counted by noting how atoms are shared between neighboring cells. For example, an atom at the

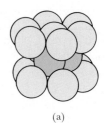

(a)

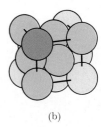

(b)

FIGURE 10.33 The fragments of (a) the hcp structure and (b) the ccp structure that give them their names.

FIGURE 10.34 The body-centered cubic (bcc) structure.

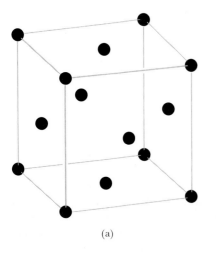

(a)

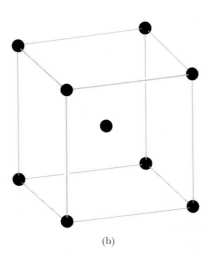

(b)

FIGURE 10.35 The unit cells of the (a) ccp and (b) bcc structures. The dots mark the locations of the centers of atoms.

center of a cell belongs entirely to that cell, but one at the center of a face is shared between two cells and counts as $\frac{1}{2}$ (of an atom). For an fcc structure, each of the eight corner atoms is shared by eight cells, so collectively they contribute 1 to the cell. Each atom at the center of each of the six faces contributes $\frac{1}{2}$, so jointly they contribute $6 \times \frac{1}{2} = 3$. The total number of atoms in the unit cell is therefore $1 + 3 = 4$, and the mass of the unit cell is four times the mass of one atom.

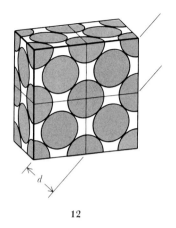

12

▼ EXAMPLE 10.10 Calculating the density of a metal

The metallic radius of copper is 128 pm, and it crystallizes with a fcc structure. Calculate the density of the metal.

STRATEGY We know that density is mass per unit volume and that it is an intensive property (Section 1.4). Since it is intensive, we can calculate the density for a sample of any size, including one as small as a unit cell. The volume of the unit cell (which is cubic) is the cube of its side. That can be obtained from the radius of the copper atom and the fact that the diagonal of one face of the cube is four times the radius (12). The mass of a unit cell is the sum of the masses of the atoms it contains, which is the number of atoms times the atomic mass of one of them. Bear in mind that typical metal densities are about 10 g/cm^3.

SOLUTION From (12), the Pythagorian theorem, and $r = 128$ pm, we have

$$\text{Side}^2 + \text{side}^2 = (4 \times 128 \text{ pm})^2$$

so

$$\text{Side}^2 = \frac{1}{2} \times (4 \times 128 \text{ pm})^2 = 1.31 \times 10^5 \text{ pm}^2$$

and

$$\text{Side} = 362 \text{ pm}$$

The volume of the unit cell is therefore

$$V = (362 \times 10^{-12} \text{ m})^3 = 4.75 \times 10^{-29} \text{ m}^3$$
$$= 4.75 \times 10^{-23} \text{ cm}^3$$

Each cell contains $8 \times \frac{1}{8} + 6 \times \frac{1}{2} = 4$ atoms. The mass of one atom is 63.54 amu, so the total mass of the unit cell is 4×63.54 amu $= 254.16$ amu. The conversion to grams (using Table 1.2) gives

$$\text{Mass} = 254.16 \text{ amu} \times \frac{1.6606 \times 10^{-24} \text{ g}}{1 \text{ amu}}$$
$$= 4.221 \times 10^{-22} \text{ g}$$

The density of the unit cell, and hence of the solid itself, is therefore

$$\text{Density} = \frac{4.221 \times 10^{-22} \text{ g}}{4.75 \times 10^{-23} \text{ cm}^3}$$
$$= 8.89 \text{ g/cm}^3$$

The experimental value is 8.92 g/cm^3.

EXERCISE Calculate the atomic radius of silver, given that its density is 10.5 g/cm^3 and it has a ccp unit cell.

[*Answer*: 144 pm]

Alloys. Alloys are homogeneous mixtures of two or more metals. Some common examples are listed in Table 10.7. Their structures are more complicated than those of pure metals because the two or more types of metal atoms have different radii. The packing problem is now akin to that of a storekeeper trying to stack oranges and grapefruit in the same pile. However, because the metallic radii of the d-block elements are all quite similar, for them the problem is not so acute as for some other metals. They form an extensive range of alloys with each other, simply because their atoms can replace one another with little distortion of the crystal structure.

Some alloys are stronger than their component metals alone. Copper containing 2% by mass of beryllium is much harder than pure copper, for the copper atoms are pinned together in the alloy by the small beryllium atoms lying between them. A useful feature of this alloy is that, because the high electrical conductivity of the copper allows charge to leak away rapidly, it does not produce sparks when struck. It is therefore used for tools in oil refineries, where a spark could be disastrous. In contrast, a nickel-barium alloy produces sparks easily, because barium has so many electrons that they help to form a conducting path through the air; it is used for the electrodes in spark plugs.

Some alloys are softer than their component metals. The presence of big bismuth atoms helps to soften a metal and lower its melting point, much as melons destabilize a stack of oranges. A low-melting-point alloy of lead, tin, and bismuth is used to control the water sprinklers in certain fire-extinguishing systems.

Electrical conduction. One of the most striking properties of a metal is its ability to conduct an electric current, or flow of electric charge. In _electronic conduction_, the type of conduction that occurs in metals, the charge is carried by electrons. In _ionic conduction_, the charge is carried by ions. This is the mechanism of conduction in a molten salt or an electrolyte solution. Because ions are too bulky to travel easily through most solids, the flow of charge through solids is almost always a result of electronic conduction.

The ability of a substance to conduct electricity is measured by its "resistance": the lower the resistance, the better it conducts.* Substances may be classified according to the resistance they show to the passage of a current:

An **insulator** is a substance that does not conduct electricity.

A **metallic conductor** is an electronic conductor with a resistance that increases as the temperature is raised.

A **semiconductor** is an electronic conductor with a resistance that decreases as the temperature is raised.

A **superconductor** is an electronic conductor that conducts electricity with zero resistance.

*Specifically, according to _Ohm's law_, the current I in amperes is related to the potential difference V in volts by $I = V/R$, where R is the resistance in ohms. The larger the resistance, the smaller the current for a given potential difference.

TABLE 10.7 Some common alloys

Alloy	Mass percentage composition
Brass	Up to 40% zinc in copper (yellow brass: 35% Zn)
Bronze	A metal other than zinc or nickel in copper (casting bronze: 10% Sn and 5% Pb)
Cupronickel	Nickel in copper (coinage cupronickel: 25% Ni)
Pewter	6% antimony and 1.5% copper in tin
Solder	Tin and lead
Stainless steel	Over 12% chromium in iron

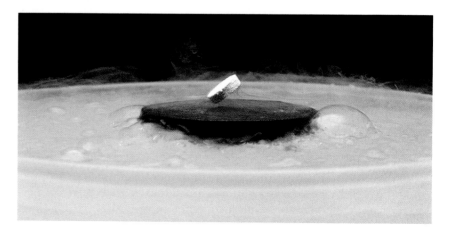

Insulators include gases, most solid ionic compounds, almost all organic compounds, and almost all molecular and covalent liquids and solids. Metallic conductors include all metals and some other solids. An example of a semiconductor is a crystal of pure silicon containing a tiny amount of arsenic or indium. Until 1987 most superconductors were metals (such as lead) or compounds cooled to close to absolute zero. However, in 1987 the first "high-temperature superconductors" were reported, which could be used at about 100 K (Fig. 10.36). These substances are important because liquid nitrogen (which boils at 77 K) can be used to cool them, in place of the very much more expensive liquid helium. They are complicated ionic oxides, often containing *f*-block ions.

Conduction by metals. Electrical conduction in metals can be explained in terms of orbitals that spread throughout the solid. As we saw in Section 9.7, when N atomic orbitals merge together in a molecule, they form N molecular orbitals. The same is true in a metallic solid, but now N is enormous (around 10^{23}). Instead of a few molecular orbitals with widely different energies, as is typical of small molecules, metals have a huge number of molecular orbitals with very similar energies. The energy levels are so close together that they form a nearly continuous band (Fig. 10.37).

FIGURE 10.37 When a large number of atomic orbitals overlap, they form the same large number of molecular orbitals; only the most strongly bonding and antibonding are shown here. Their energies lie in an almost continuous band. Each orbital can hold up to two electrons.

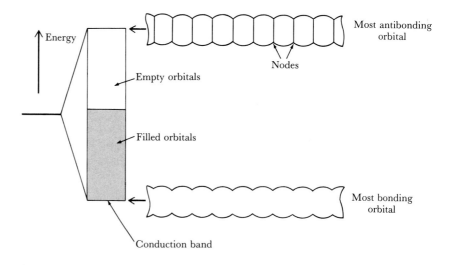

Consider a sample of an alkali metal, such as sodium. Each atom contributes one valence orbital (the 3s orbital in this case) and one valence electron. If there are N atoms in the sample, the N 3s orbitals merge to give a band of N molecular orbitals. Since each of the N atoms supplies one valence electron, a total of N electrons must be accommodated in the orbitals. These N electrons occupy the orbitals according to the building-up principle, filling the lower $\frac{1}{2}N$ of them (as two electrons can occupy each orbital), since that results in the lowest energy. An incompletely filled band of orbitals is called a _conduction band_ (Fig. 10.37). Since the orbitals in the band lie so close together, it takes very little energy to excite electrons from the topmost filled orbitals to the empty orbitals just above. These electrons can move freely through the solid and hence can carry an electric current.

The resistance of a metal increases with temperature because when a metal is heated, its atoms vibrate more vigorously. As they do so, they collide with any electrons traveling past them. This reduces the flow of electrons, and the solid is a poorer conductor at high temperatures than at low.

Insulators can also be described in terms of bands. In an insulator, electrons fill the entire band of orbitals. There is thus no conduction band and, moreover, there is a substantial _band gap_ before the next band of empty orbitals begins (Fig. 10.38). The electrons in the lower band can be excited into the upper band only by a large amount of energy. (The upper band is formed by overlap of higher atomic orbitals, but is analogous to the 3s band.) Hence the electrons of the lower band are effectively immobile, and the solid does not conduct electricity.

Semiconductors. One type of semiconductor consists of very pure silicon to which a minute amount of arsenic impurity has been added in a process called _doping_. The arsenic increases the number of electrons in the solid: an Si atom (Group IV) has four valence electrons, and an As atom (Group V) has five. The additional electrons enter the upper, empty band and hence enable the solid to conduct (Fig. 10.39). This type of material is called an _n-type semiconductor_, the n indicating the presence of excess _n_egatively charged electrons.

If silicon is doped with indium (Group III) instead of arsenic, the resulting solid has fewer electrons than the pure element, and the lower band is not completely full (Fig. 10.39). The solid is now electrically conducting because the uppermost occupied orbital has unoccupied neighbors. This type of material is called a _p-type semiconductor_; the p indicates that, because the bands contain fewer electrons they have less negative charge and, hence, in a sense, greater _p_ositive charge. Solid-state electronic devices (transistors and integrated circuits) are formed from "p-n junctions" in which a p-type semiconductor is in contact with an n-type semiconductor.

10.8 IONIC SOLIDS

As in the case of metals, the ions making up an ionic solid may be pictured as spheres, and the crystal structure as being built up by stacking them together. However, because cations and anions have different

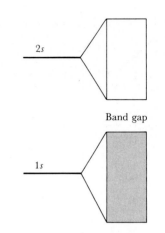

FIGURE 10.38 In an insulator, there is a substantial gap between the uppermost filled energy level and the first empty level, so the electrons are relatively immobile.

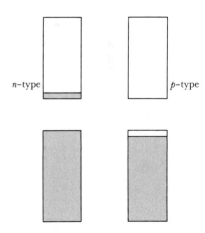

FIGURE 10.39 In an n-type semiconductor, the additional atoms provide extra electrons which occupy the upper band and carry the electric current. In a p-type semiconductor, the additional atoms remove electrons from the full band, and the electrons near the top of the filled orbitals are free to conduct.

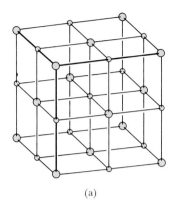

(a)

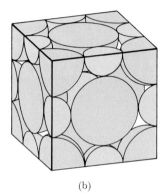

(b)

FIGURE 10.40 The arrangement of the ions in (a) the rock-salt structure, and (b) its unit cell. The pink dots show the centers of the cations, and the green dots the centers of the anions.

charges as well as different sizes, the packing problem is more complicated. Each unit cell is made up of ions with different radii and charges, yet it must be electrically neutral overall.

The rock-salt structure. One of the solutions to the packing problem is the *rock-salt structure*, the structure of the mineral form of sodium chloride and several other solids. In the rock-salt structure (Fig. 10.40), the anions (the green spheres, such as the Cl^- ion in NaCl itself) form a face-centered cubic (fcc) array. The smaller cations (the red spheres, such as the Na^+ in NaCl) also lie in an fcc pattern, one that fits snugly among the anions. The rock-salt structure therefore consists of two fcc arrays of ions fitting within each other. Each cation is surrounded by six anions, giving a coordination number of 6. Each anion is surrounded by six cations, so the anions also have coordination number 6. The structure is therefore described as having "(6, 6) coordination"; in this notation, the first number is the cation coordination number, and the second that of the anion.

▼ **EXAMPLE 10.11 Counting the ions in a unit cell**

How many ions are there in the unit cell shown in Fig. 10.40? Confirm that the cell is electrically neutral.

STRATEGY Since an ion at a corner or on a face or an edge is shared with neighboring unit cells, only a fraction of it (and its charge) belongs to the unit cell under discussion. An ion on a face is shared by two cells and counts as $\frac{1}{2}$ (of an ion). An ion on an edge is shared by four neighboring cells and so contributes $\frac{1}{4}$. An ion at a corner is shared by eight neighboring cells and contributes $\frac{1}{8}$ to each cell. To confirm electrical neutrality, we must count the numbers of cations and anions and make sure that their charges cancel.

SOLUTION Inspection of Fig. 10.40 gives the following count:

	Number of Ions	
Location	Na^+	Cl^-
Center of cell	1	0
On 6 faces	0	$6 \times \frac{1}{2} = 3$
On 12 edges	$12 \times \frac{1}{4} = 3$	0
At 8 corners	0	$8 \times \frac{1}{8} = 1$
Total	4	4

That is, the unit cell contains four cations and four anions. The total charge is zero, since the charge of $4Na^+$ is canceled by that of $4Cl^-$.

EXERCISE Repeat these calculations for the unit cell of the cesium chloride structure, shown in Fig. 10.41.

[*Answer*: One cation, one anion]

▲

As in the case of the rock-salt structure, crystal structures are often named for a particular substance although they apply to others as well. Several ionic compounds, including KCl, NaBr, MgO, and CaO, have the rock-salt structure (Table 10.8). So do the Group-I metal hydrides, such as sodium hydride (NaH), which are produced by heating the metals in hydrogen.

TABLE 10.8 Examples of crystal structures

Crystal structure	Examples*
Rock salt	**NaCl,** LiCl, KBr, RbI, MgO, CaO, AgCl
Cesium chloride	**CsCl,** CsI, TlSb
Fluorite	**CaF$_2$,** Na$_2$O, UO$_2$
Rutile	**TiO$_2$,** MnO$_2$, SnO$_2$, MgF$_2$
Perovskite	**CaTiO$_3$,** BaTiO$_3$, SrFeO$_3$, LaMnO$_3$

*The substance in bold type is the one that gives its name to the structure. For the fluorite, rutile, and perovskite structures, see Figs. 10.51 to 10.53.

The cesium chloride structure. The rock-salt structure is not the most favorable arrangement energetically when the radii of the cations and anions are very different. In this case a lower energy is obtained if the ions adopt the *cesium chloride structure* (Fig. 10.41). The bulky cation (the Cs$^+$ ion in CsCl) is surrounded by eight anions (Cl$^-$), and each anion is surrounded by eight cations, giving (8, 8) coordination. This structure is much less common than the rock-salt structure, but it illustrates how different structures accommodate ions of different relative sizes.

10.9 OTHER TYPES OF SOLIDS

The atoms in network solids are joined to their neighbors by covalent bonds, which form a network that extends throughout the crystal. Network solids show the strength of the covalent bonds that bind them by being very hard, rigid materials with high melting and boiling points. In contrast, molecular solids—solids composed of individual molecules—often have low melting points because only a small amount of thermal motion is needed to overcome the relatively weak intermolecular forces that hold the molecules in place. Titanium tetrachloride (TiCl$_4$), for instance, is a liquid (often yellow on account of iron impurities) that boils at 136°C; it freezes at −25°C to a molecular solid. The solid is composed of tetrahedral TiCl$_4$ units, held together largely by London forces between the electron-rich chlorine atoms. Not all molecular solids are soft: sucrose, a brittle solid, is a molecular solid in which the C$_{12}$H$_{22}$O$_{11}$ molecules are held together by hydrogen bonds between their numerous —OH groups. The bonding between sucrose molecules is so strong that by the time the melting point has been reached (at 184°C), the molecules themselves have started to decompose. The partially decomposed mixture is called caramel.

Network solids: diamond and graphite. Among the most important examples of network solids are diamond and graphite, two "allotropes" of carbon:

Allotropes are alternative forms of an element that differ in the way the atoms are linked.

Diamond is a crystalline form of elemental carbon that in nature is found embedded in a soft rock called kimberlite. This rock forms vol-

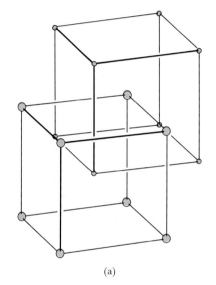

(a)

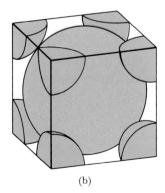

(b)

FIGURE 10.41 The structure of (a) cesium chloride and (b) its unit cell, with dots showing the locations of the ions.

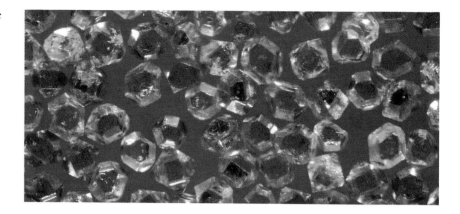

FIGURE 10.42 The crystals are tiny synthetic diamonds, 350 μm in diameter.

FIGURE 10.43 The structure of diamond. Each dot indicates the location of the center of a carbon atom. Each atom forms an sp^3 hybrid covalent bond with each of its four neighbors.

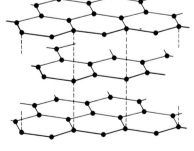

FIGURE 10.44 Graphite consists of layers of hexagonal rings of carbon atoms. The slipperiness of graphite is due to the ease with which the layers can slide over each other.

canic pipes that raise the diamonds from deep in the earth, where they have been formed under intense pressure. The erosion of kimberlite can result in diamonds being found in the sediments of some rivers. One method for making synthetic diamonds recreates the geological conditions that produce natural diamonds. This involves compressing graphite at over 80,000 atm and above 1500°C in the presence of small amounts of metals such as chromium and iron (Fig. 10.42). It is thought that the molten metals dissolve the graphite and then, as they cool, deposit crystals of diamond, which is less soluble in the molten metal than the graphite is.

Each C atom in diamond is covalently bonded to four neighbors through sp^3 hybrid σ bonds (Fig. 10.43). The structure resembles the steel framework of a large building, and its rigidity accounts for the great hardness of the solid. Diamond is one of the best conductors of heat and is used as a base for some integrated circuits so that they do not overheat. This thermal conductivity is also a result of the crystal structure, which rapidly transmits the vigorous vibration of an atom in a hot part of the crystal to distant, cooler parts.

Graphite is produced naturally as a result of changes that occur on ancient organic remains or on carbonates. Most commercial graphite is produced by heating coke rods to a high temperature in an electric furnace for several days. It is a black, lustrous, electrically conducting, slippery solid that sublimes at 3700°C. The solid consists of flat sheets of carbon atoms that are bonded covalently into hexagons (Fig. 10.44); the structure of each sheet resembles chicken wire. There are also covalent bonds between the sheets. However, because these bonds are so long and weak, the sheets can easily slide over each other. Electrons can readily move within each sheet, but much less readily from one sheet to another. Hence, graphite conducts electricity better through the length and breadth of the sheets than perpendicular to them.

Boron and boron carbide. Carbon's neighbor boron also forms network structures. The simplest allotrope of the element consists of B_{12} units linked together by covalent bonds. When boron oxide and coke are heated to a high temperature, they form boron carbide ($B_{12}C_3$), another network structure. Boron carbide, which is hard enough to be used as an abrasive, consists of B_{12} units that are pinned together by C atoms lying in the gaps between units. When boron trichloride and ammonia are heated together to a high temperature, they form boron nitride (BN), a white, fluffy, slippery, powder:

$$BCl_3(g) + NH_3(g) \xrightarrow{\Delta} BN(s) + 3HCl(g)$$

In this compound the average number of valence electrons per atom is four (three from B and five from N), the same as in carbon. It is therefore not surprising that its structure is like graphite's but with the flat planes of hexagons consisting of alternating B and N atoms. Unlike graphite, boron nitride is an insulator, because the electrons are trapped by the more electronegative nitrogen atoms and cannot move freely through the planes. Boron nitride occurs in various solid phases; under pressure at 1650°C it changes to a very hard, diamondlike structure known as *borazon*. This is almost as hard as diamond but resists oxidation better. It too is used as an abrasive.

A molecular solid: ice. Since molecules have such widely varying shapes, they stack together in solids in a wide variety of different ways. We consider only one example here—ice (Fig. 10.45). In ice, each O atom is surrounded by four H atoms. Two of these H atoms are linked to the O atom through σ bonds. The other two belong to neighboring H_2O molecules and are linked to the O atom by hydrogen bonds. As a result, the structure of ice is an open network of H_2O molecules held together by hydrogen bonds. Some of the hydrogen bonds break when the solid melts, and the molecules then pack less uniformly but more densely. The openness of the network in ice compared with that in the liquid explains why it has a lower density than liquid water (0.92 g/cm^3 and 1.00 g/cm^3, respectively, at 0°C). This feature was mentioned in Section 10.5, where we saw that it accounts for the unusual property that ice melts when subjected to pressure. Solid benzene and solid carbon dioxide, in contrast, have higher densities than their liquids. Their molecules are held together by London forces and pack together more closely in the solid than in the liquid (Fig. 10.46).

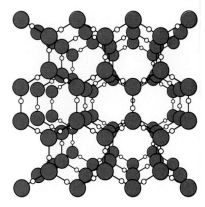

FIGURE 10.45 Ice is made up of water molecules that are held together by hydrogen bonds in a relatively open structure. Each O atom is surrounded tetrahedrally by four H atoms, two of which are σ-bonded to it and two of which are hydrogen-bonded to it.

FIGURE 10.46 Because of its open structure, ice is less dense than water and floats in it (left). Solid benzene, in contrast, is denser than liquid benzene, and "benzenebergs" sink in liquid benzene (right).

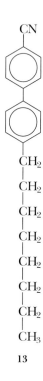

CN

CH₂
CH₂
CH₂
CH₂
CH₂
CH₂
CH₂
CH₃

13

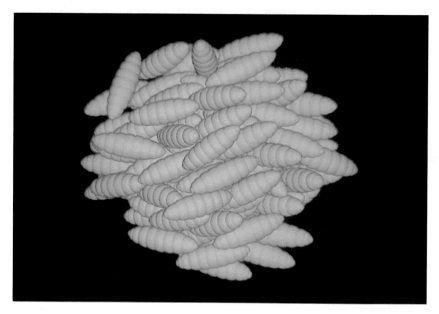

FIGURE 10.47 In liquid crystals the rodlike molecules are ordered in some directions, but not in all directions. This is the arrangement of molecules in the nematic phase.

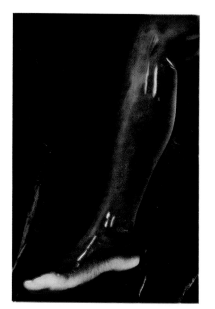

FIGURE 10.48 The temperature dependence of its structure accounts for changes in the optical properties of a cholesteric liquid crystal with temperature. When coated on a limb, it can be used to show the variation in skin temperature.

Liquid crystals. A special class of substances are the "liquid crystals" used in electronic displays. Liquid crystals flow like liquids, but their molecules lie in a moderately orderly array, as in a crystal. They are examples of substances in a "mesophase" (from the Greek words for "intermediate appearance"):

A **mesophase** is a state of matter showing some of the properties of both a liquid and a solid.

A typical liquid-crystal molecule (**13**) is long and rodlike. This shape allows the molecules to stack together like uncooked spaghetti: they lie parallel to each other but have freedom to move along their axes.

Depending on the particular substance and the temperature, a liquid crystal can exist in any of three classes of mesophases. The three phases differ in the arrangement of the molecules. In the *nematic phase* (from the Greek word for "threadlike") the molecules lie together as shown in Fig. 10.47. In the *smectic phase* (from the Greek word for "soapy"), the molecules form layers. In the *cholesteric phase* (from the Greek, unhelpfully, for "bile solid"), the molecules form smectic-like layers, but the molecules of neighboring layers are at different angles so that the liquid crystal has a helical arrangement of molecules.

The orientation of the molecules in liquid crystals changes in the presence of an electric field. It is this property that makes them useful in electronic displays, because the reorientation of the molecules also causes a change in their optical properties. Cholesteric liquid crystals are of interest because the repeat distance of the stacking (the pitch of the helix), and therefore the optical properties, depend on the temperature. This property is used in liquid-crystal thermometers (Fig. 10.48).

SUMMARY

10.1 The interactions that hold atoms, ions, and molecules together are called **cohesive forces. Metallic solids** are cations held together by electrons. Ions are held together in **ionic solids** by **ion-ion interactions.** In **network solids,** atoms are held together by covalent bonding that extends throughout the solid. In **molecular solids** the individual molecules are held together by **intermolecular forces.** Forces between ions and polar molecules include the **ion-dipole interaction,** which is responsible for the **hydration** of cations—the attachment of water molecules to an ion. Polar molecules attract each other through the **dipole-dipole interaction** between the partial charges of their dipoles. All molecules, polar or nonpolar, attract each other through the **London force,** the interaction between instantaneous electric dipoles.

10.2 The **hydrogen bond** is a link formed by a hydrogen atom lying between two strongly electronegative atoms (N, O, or F). It dominates the other interactions when it is present.

10.3 The mobility of molecules in liquids accounts for liquid flow; the stronger the intermolecular forces, the greater the **viscosity,** or resistance to flow. Hydrogen-bonded liquids are generally more viscous than others. The smooth surface of liquids is due to **surface tension,** the energy required to form their surfaces, which arises from the less favorable intermolecular forces at the surface. The **adhesion** of a liquid to the walls of its container accounts for **capillary action,** the tendency of a liquid to climb up inside a narrow tube, and the formation of a **meniscus,** a crescent-shaped surface.

10.4 A liquid heated in a closed container reaches **dynamic equilibrium** with its vapor, a condition in which the forward and reverse processes (in this case, of vaporization and condensation) continue at equal rates. Each liquid has a characteristic **vapor pressure,** the pressure of the vapor at equilibrium, that increases with temperature. A liquid in an open container boils when the vapor pressure is equal to the external pressure. A gas cannot be liquefied by the application of pressure if the temperature is above its **critical temperature.**

10.5 Freezing points depend weakly on pressure and normally increase as the pressure is raised. Water is unusual, since it melts at a lower temperature under pressure. The conditions under which solids, liquids, and gases are stable are shown on the **phase diagram** of the substance. The three phases are in dynamic equilibrium at the **triple point.**

10.6 A **crystalline solid** is an orderly array of atoms, ions, or molecules. The structures of crystals are determined by **x-ray diffraction,** the interference caused in a beam of x-rays by the layers of atoms in a solid.

10.7 The structures of metals can be described in terms of the stacking of spheres; in many cases a **close-packed structure,** a structure with the least waste of space, is adopted. The structure of a crystal can be summarized by drawing its **unit cell,** the unit that can reproduce the entire crystal when repeatedly stacked together. **Alloys,** which are homogeneous mixtures of metals, can be discussed similarly but as if they were composed of spheres of different radii. **Electric conduction** may be electronic or ionic. The temperature dependence of electronic conduction distinguishes **metallic conductors** from **semiconductors.** Electronic conduction may be discussed in terms of **bands** of orbitals. A metallic conductor is a substance in which the bands are incompletely filled. **Insulators** are substances in which the bands are completely filled. In an *n*-**type semiconductor,** the **doping** increases the number of valence electrons in the solid; in a *p*-**type semiconductor** it reduces them.

10.8 Two simple structures characteristic of several ionic compounds are the **rock-salt** and **cesium chloride** structures.

10.9 Network solids are often very hard. Diamond and boron nitride are two examples. However they may also be slippery (graphite is an example). Ice, a molecular solid, has an open structure, owing to the hydrogen bonds between its molecules. A **liquid crystal** is an example of a **mesophase,** a phase with characteristic features intermediate between liquid and solid. Its optical properties can be changed by electric fields and heat.

EXERCISES

Intermolecular Forces

In Exercises 10.1 to 10.4, list the kinds of intermolecular forces that might arise between pairs of the given molecules or atoms.

10.1 (a) Cl_2; (b) Ar; (c) HCl; (d) HF.

10.2 (a) N_2; (b) He; (c) H_2O; (d) O_3.

10.3 (a) C_6H_6; (b) C_6H_5Cl; (c) CH_4; (d) CCl_4.

10.4 (a) $CH_2{=}CH_2$; (b) $C_{10}H_8$ (naphthalene); (c) $CHCl{=}CHCl$; (d) $CCl_2{=}CCl_2$.

In Exercises 10.5 and 10.6, suggest, giving reasons, which substance in each pair is likely to have the higher normal boiling point.

10.5 (a) HCl and HBr; (b) HF and HCl; (c) CH_4 and SiH_4; (d) *cis*-CHCl=CHCl (**a**) and *trans*-CHCl=CHCl (**b**).

10.6 (a) H_2S and H_2Te; (b) NH_3 and PH_3; (c) CH_3Cl and CH_3Br; (d) *ortho*-dichlorobenzene (**a**) and *para*-dichlorobenzene (**b**).

In Exercises 10.7 to 10.10, determine which of the given substances are likely to form hydrogen bonds. Suggest the consequences in each case.

10.7 (a) HF; (b) NH_3; (c) CH_4; (d) CH_3OH (methanol).
10.8 (a) D_2O; (b) CH_3CH_2OH (ethanol); (c) CH_3CO_2H (acetic acid); (d) H_3PO_4 (phosphoric acid).
10.9 $C_4H_{10}O$ as (a) C_2H_5—O—C_2H_5 (ether); (b) $CH_3(CH_2)_3OH$ (butanol).
10.10 $C_4H_8O_2$ as (a) CH_3—CO—O—C_2H_5 (ethyl acetate); (b) $CH_3(CH_2)_2COOH$ (butanoic acid).

Vapor Pressure

In Exercises 10.11 and 10.12, when the given liquid was introduced into the "vacuum" above the mercury in a barometer, the mercury column was depressed as noted. Determine the vapor pressure of the compound in Torr.

10.11 A sample of chloroform ($CHCl_3$) at 19°C produced a depression of 151 mm.
10.12 A sample of bromoform ($CHBr_3$) at 29°C produced a depression of 7.6 mm.

10.13 Suppose you were to bubble 1.0 L of air slowly through water at 20°C, at which temperature the vapor pressure of water is 17.5 Torr. Estimate the loss of mass of water that would occur as a result of evaporation, assuming that the air becomes saturated.
10.14 Estimate the mass loss that occurs when 1.0 L of air is bubbled slowly through ethanol at 25°C, at which temperature the vapor pressure of ethanol is 60 Torr.

10.15 What mass of water can you expect to find in the air of a bathroom of dimensions 4 m by 3 m by 3 m when water has been left in the bathtub at 40°C, if its vapor pressure at that temperature is 7.4 kPa?
10.16 What mass of mercury vapor would you expect to find in 1.0 m^3 of the air of a laboratory if an unstoppered bottle of the liquid were left in it at 25°C, and the vapor pressure of mercury at that temperature is 0.224 Pa?

10.17 Use the vapor-pressure curve in Fig. 10.21 to estimate the boiling point of water when the pressure is (a) 700 Torr; (b) 770 Torr; (c) 100 Torr.
10.18 Use the vapor-pressure curve in Fig. 10.21 to estimate the boiling point of ethanol when the pressure is (a) 750 Torr; (b) 780 Torr; (c) 100 Torr.

In Exercises 10.19 and 10.20, use Trouton's rule to estimate the enthalpy of vaporization of each liquid (normal boiling points are given in parenthesis). Then comment on each set of results.

10.19 (a) Carbon tetrafluoride (−129°C); (b) carbon tetrachloride (77°C); (c) carbon tetrabromide (190°C).
10.20 (a) *ortho*-xylene (**a**; 144°C); (b) *meta*-xylene (**b**; 139°C); (c) *para*-xylene (**c**; 138°C).

Failure to satisfy Trouton's rule is often a sign that a strong molecular interaction, such as hydrogen bonding, is occurring. In Exercises 10.21 and 10.22, determine which of the liquids is abnormal in that sense, and suggest an explanation.

10.21 (a) Ethanol, which boils at 79°C and has $\Delta H°_{vap} = 44$ kJ/mol; (b) hydrogen selenide, which boils at −41°C and has $\Delta H°_{vap} = 20$ kJ/mol.

10.22 (a) Phosphorus trichloride, which boils at 74°C and has $\Delta H^\circ_{vap} = 31$ kJ/mol; (b) mercury, which boils at 357°C and has $\Delta H^\circ_{vap} = 58$ kJ/mol.

10.23 Suggest, with equations, methods of preparing the liquids in Exercise 10.21.

10.24 Suggest, with equations, methods of preparing the liquids in Exercise 10.22. For mercury, start with the mineral cinnabar (HgS).

Phase Diagrams

10.25 Using Fig. 10.25, the phase diagram for water, predict the state of a sample of water under the following sets of conditions: (a) 2 atm, 200°C; (b) 3 atm, 300°C; (c) 3 Torr, 0°C; (d) 218 atm, 374°C.

10.26 Using Fig. 10.26, the phase diagram for carbon dioxide, predict the state of a sample of carbon dioxide under the following sets of conditions: (a) 1 atm, −80°C; (b) 1 atm, −78°C; (c) 80 atm, 25°C; (d) 5.1 atm, −56°C.

10.27 The phase diagram for helium is shown in Fig. 10.49. Use it to answer the following questions:

(a) What is the maximum temperature at which superfluid helium-II can exist?

(b) What is the minimum pressure at which solid helium can exist?

(c) What is the maximum temperature at which liquid helium can exist?

(d) Can solid helium sublime?

10.28 The phase diagram for carbon, shown in Fig. 10.50, indicates the extreme conditions that are needed to form diamonds from graphite. Use the diagram to answer the following questions:

(a) At 2000 K, what is the minimum pressure needed to change graphite to diamond?

(b) What is the minimum temperature at which liquid carbon can exist at pressures below 10,000 atm?

(c) At what pressure do diamonds melt at room temperature?

(d) Are diamonds the expected form of carbon under normal conditions? If not, why can people wear them without having them compressed and heated?

The Classification of Solids

In Exercises 10.29 to 10.32, classify each solid according to its type of bonding.

10.29 (a) Sodium chloride; (b) solid nitrogen; (c) polyethylene; (d) sugar.

10.30 (a) Copper; (b) ice; (c) calcium carbonate; (d) nylon.

10.31 (a) Brass; (b) phosphorus; (c) wool; (d) cotton.

10.32 (a) Wood; (b) solder; (c) sulfur; (d) granite.

X-ray Diffraction

In Exercises 10.33 and 10.34, use the Bragg equation to calculate the distance between atomic layers if a bright spot was found at the given angle and wavelength.

10.33 At 27.2°, when 154-pm radiation is used.

10.34 At 17.1°, when 70.8-pm radiation is used.

In Exercises 10.35 and 10.36, calculate the metallic radii from each set of data, all of which relate to 154-pm x-rays.

10.35 (a) Opposite faces in a ccp nickel unit cell gave a diffraction spot at 12.7°.

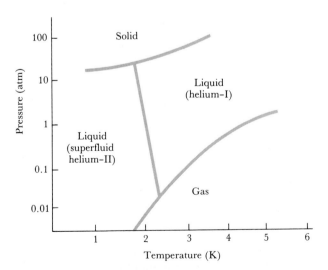

FIGURE 10.49 The phase diagram for helium.

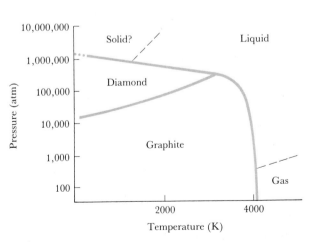

FIGURE 10.50 The phase diagram for carbon.

(b) Opposite faces in a bcc potassium unit cell gave a diffraction spot at 8.3°.

10.36 (a) Opposite faces in a ccp calcium unit cell gave a diffraction spot at 8.0°.

(b) Opposite faces in a bcc niobium unit cell gave a diffraction spot at 13.7°.

Close-Packed Structures

10.37 Aluminum crystallizes in a ccp structure. Its metallic radius is 125 pm.

(a) What is the length of a side of the unit cell?

(b) How many unit cells are there in 1 cm^3 of aluminum?

10.38 Calcium crystallizes in a ccp structure. Its metallic radius is 180 pm.

(a) What is the length of a side of the unit cell?

(b) How many unit cells are there in 1 cm^3 of calcium?

10.39 The metal polonium (which was named by Marie Curie after her native Poland) crystallizes in a *primitive cubic structure*, with an atom at each corner of a cubic unit cell. Calculate (a) the number of atoms per unit cell, (b) the coordination number, and (c) the length of a side of the unit cell, given that the metallic radius of polonium is 140 pm.

10.40 Potassium crystallizes in a bcc structure. Calculate (a) the number of atoms per unit cell, (b) the coordination number of the lattice, and (c) the length of a side of the unit cell, given that the metallic radius of potassium is 231 pm.

In Exercises 10.41 and 10.42, predict the density of each metal.

10.41 (a) Nickel (ccp structure); (b) rubidium (bcc structure). The metallic radii are 124 pm and 235 pm, respectively.

10.42 (a) Platinum (ccp structure); (b) cesium (bcc structure). The metallic radii are 138 pm and 266 pm, respectively.

In Exercises 10.43 to 10.46, calculate the metallic radius of the element from the information supplied:

10.43 The density of gold is 19.3 g/cm^3, and it crystallizes in a ccp structure.

10.44 The density of iridium, the densest element of all, is 22.5 g/cm^3, and it has a ccp structure.

10.45 The density of vanadium is 5.96 g/cm^3, and it has a bcc structure.

10.46 The density of molybdenum is 10.2 g/cm^3, and it has a bcc structure.

Alloys

In Exercises 10.47 and 10.48, calculate the relative number of atoms of each element contained in each alloy. The percentages are mass percentage compositions.

10.47 (a) 60% Na, 40% K; (b) yellow brass, containing 35% Zn in copper; (c) casting bronze, containing 10% Sn and 5% Pb in copper.

10.48 (a) 40% Na, 60% K; (b) coinage cupronickel, containing 25% Ni in copper; (c) pewter, containing 6% Sb and 1.5% Cu in tin.

In Exercises 10.49 and 10.50, calculate the mass percentage composition of an alloy with each stated atomic composition.

10.49 (a) One Cu atom for each Zn atom; (b) one Zn atom per 10 Cu atoms.

10.50 (a) One Sn atom for each Pb atom; (a) one Ni atom per 1000 Cu atoms.

Electrical Conduction

10.51 The electrical conductivity of graphite parallel to its planes is different from that perpendicular to the planes. Parallel to the planes the conductivity decreases as the temperature is raised, but perpendicular to them it rises. In what sense is graphite a metallic conductor or a semiconductor?

10.52 "Graphite bisulfates" are formed by heating graphite with a mixture of sulfuric and nitric acids. In the reaction the graphite planes are partially oxidized (so that, on average, about one positive charge is shared among 24 carbon atoms) and HSO_4^- anions are distributed between planes. What effects should this have on the electrical conductivity of graphite?

In Exercises 10.53 and 10.54, state whether each material is an *n*- or *p*-type semiconductor.

10.53 (a) Si doped with P; (b) Si doped with In; (c) Ge doped with Sb.

10.54 (a) Si doped with Al; (b) Ge doped with As; (c) Ge doped with Ga.

10.55 Silicon is a semiconductor even without being doped (diamond is too). Account for this property in terms of bands, using as a clue the idea that at absolute zero ($T = 0$) the conductivities of silicon and carbon are zero.

10.56 Zinc oxide is a semiconductor. Its conductivity increases when it is heated in a vacuum, but decreases when it is heated in oxygen. Account for these observations.

Ionic Solids

In Exercises 10.57 and 10.58, calculate the number of cations, anions, and formula units in each unit cell.

10.57 (a) The rock-salt unit cell shown in Fig. 10.40; (b) the fluorite (CaF_2) unit cell in Fig. 10.51. What are the coordination numbers of the ions in fluorite?

10.58 (a) The cesium chloride unit cell shown in Fig. 10.41; (b) the rutile (TiO_2) unit cell in Fig. 10.52. What are the coordination numbers of the ions in rutile?

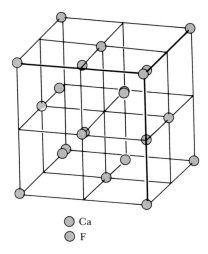

Ca
F

FIGURE 10.51 The fluorite unit cell.

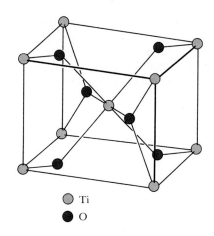

Ti
O

FIGURE 10.52 The rutile unit cell.

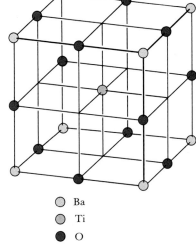

Ba
Ti
O

FIGURE 10.53 The perovskite unit cell.

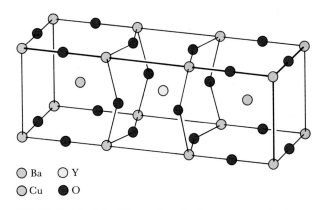

Ba   Y
Cu   O

FIGURE 10.54 The unit cell of a superconductor.

10.59 A unit cell of the mineral perovskite is shown in Fig. 10.53. What is its formula?

10.60 A unit cell of one of the new high-temperature superconductors is shown in Fig. 10.54. What is its formula?

*__10.61__ How many unit cells of the kind shown in Fig. 10.40 are there in a 1-mm³ grain of table salt? (Density of NaCl is 2.16 g/cm³.)

*__10.62__ How many unit cells of the kind shown in Fig. 10.40 are there in a 1-mm³ grain of potassium bromide? (Density of KBr is 2.75 g/cm³.)

In Exercises 10.63 and 10.64, calculate the density of each substance from the data in Table 7.6.

*__10.63__ (a) Sodium iodide (rock-salt structure); (b) cesium iodide (cesium chloride structure).

10.64 (a) Calcium oxide (rock-salt structure); (b) cesium bromide (cesium chloride structure).

Network Solids

10.65 Name two elements that, in at least one allotropic form, are network solids.

10.66 Name two compounds that exist as network solids.

10.67 The enthalpy of sublimation of diamond is 713 kJ/mol. What is the C—C bond enthalpy in the solid?

10.68 The enthalpy of formation of BN(*s*) is −254 kJ/ mol, and that of BN(*g*) is +647 kJ/mol. Given that the enthalpies of formation of gaseous B and N atoms are +563 kJ/mol and +473 kJ/mol, respectively, calculate the B—N bond enthalpy in the solid.

Molecular Solids

10.69 Name four elements that, in at least one allotropic form, exist as molecular solids at room temperature.

10.70 Name and give the formula of a solid molecular compound of each main-group element.

In Exercises 10.71 and 10.72, give the equation or equations for the formation of each compound. State whether each is liquid, solid, or gas at room temperature and pressure.

10.71 (a) PCl_3; (b) P_2O_5.
10.72 (a) PCl_5; (b) PBr_3.

In Exercises 10.73 and 10.74, identify the type of intermolecular forces that are responsible for the existence of each solid.

10.73 (a) Solid argon; (b) ice; (c) solid HCl; (d) glucose.
10.74 (a) Iodine; (b) solid benzene; (c) sucrose; (d) oxalic acid ($H_2C_2O_4$).

10.75 Figure 10.55a shows a unit cell of a molecular compound. What is its molecular formula?

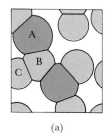

(a)

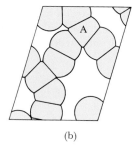

(b)

FIGURE 10.55 The unit cells for Exercises 10.75 and 10.76.

10.76 State the molecular formula of the compound with the unit cell shown in Figure 10.55b.

General

*10.77 Calculate the vapor pressure of water if, when 25.0 L of argon is bubbled through water at 80°C, 7.3 g of water vaporized.

*10.78 A small beaker of water was heated in a sealed room of volume 30 m³ that was filled with 2.0 kg of nitrogen at 20°C. At what temperature did the water boil?

*10.79 The normal boiling point of arsenic trichloride ($AsCl_3$) is 130.2°C, but under a pressure of 100 Torr it boils at 70.9°C.
(a) What mass of vapor would you expect to find in a room of volume 50 m³ if some of the liquid were left in an open beaker at 20°C?
(b) Would you expect a greater or smaller amount of arsenic trifluoride under the same conditions? Why?

10.80 Suggest a reason for the high surface tension of mercury.

10.81 Metals with bcc structures are not close-packed. Therefore, their densities would be greater if they were to shift to a ccp structure (under pressure, for instance). What would the density of tungsten be if it had a ccp structure rather than bcc? Its actual density is 19.4 g/cm³.

10.82 Draw a picture showing how cylinders may be close-packed in a single layer with their ends aligned. What is the coordination number of one layer of close-packed cylinders?

*10.83 Calculate the percentage of space occupied by the spheres in a cubic close-packed structure.

*10.84 The density of platinum is 21.450 g/cm³, and its unit cell has a ccp structure. Planes on opposite sides of the unit cell gave a reflection at 11.38° with 154.0-pm x-rays. Deduce a value for Avogadro's number from these data.

*10.85 All the noble gases except helium crystallize into ccp structures. Find an equation relating the atomic radius to the density of a ccp solid of atomic weight AW, and use it to deduce the atomic radii of the noble gases, given the following densities in grams per cubic centimeter: Ne, 1.20; Ar, 1.40; Kr, 2.16; Xe, 2.83; Rn, 4.4.

10.86 All the alkali metals crystallize into bcc structures.
(a) Find an equation relating the metallic radius to the density of a bcc solid element in terms of its atomic weight, and use it to deduce the metallic radii of the alkali metals, given the following densities in grams per cubic centimeter: Li, 0.53; Na, 0.97; K, 0.86; Rb, 1.53; Cs, 1.90.

Develop this concept further by combining your result here with that in Exercise 10.85.

(b) Find a factor for converting a bcc density to a ccp density for the same element.

(c) Determine what the densities of the alkali metals would be if they had ccp structures.

(d) Would any float on water?

10.87 (a) Suggest, in terms of energy bands, why the electrical conductivity of some substances increases when they are exposed to light.

(b) The band gap in amorphous selenium, which is used in the xerographic photocopying process, is 1.8 eV. What is the longest wavelength of light that can make it conduct?

(c) Can amorphous selenium be used in infrared burglar alarms?

In this chapter we explore the solubilities of substances in different solvents and under different conditions. We also see how the solute affects the physical properties of a solution. As the illustration shows, some substances can be dissolved by the action of a detergent. Here we see animal fats dissolving in a detergent solution.

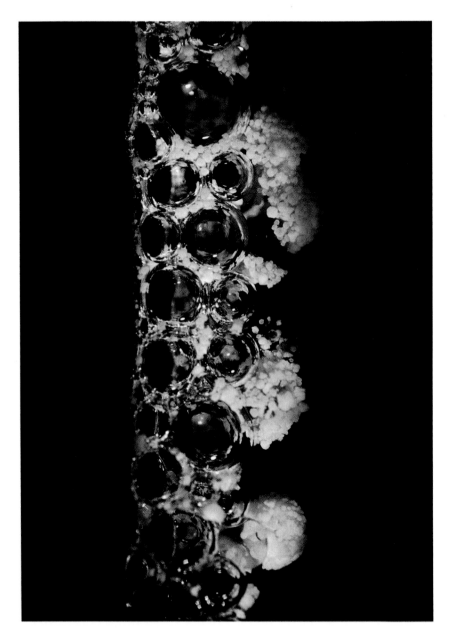

THE PROPERTIES OF SOLUTIONS

Measures of concentration

11.1 Emphasizing the
amount of solute
11.2 Emphasizing numbers
of molecules

Solubility

11.3 Saturation and solubility
11.4 The effect of pressure
on gas solubility
11.5 The effect of temperature
on solubility

Colligative properties

11.6 The lowering of vapor
pressure
11.7 Osmosis

Solutions are homogeneous mixtures of one substance—the solute—in another—the solvent. Aqueous solutions, those in which the solvent is water, are so important that we concentrate mainly on them in this chapter. Some aqueous solutions occur on a vast scale. The oceans, for instance, account for 1.4×10^{18} kg of the earth's surface water, which amounts to nearly 300,000 tons of water for each inhabitant. Although the main solutes in seawater are Na^+ and Cl^- ions (Table 11.1), it contains at least a trace of every naturally occurring element, and huge amounts of some. The rivers that feed the oceans are also solutions on a grand scale. Samples of the Columbia River taken at a point about 50 mi from its mouth show that each year about 1.8 trillion liters (1.8×10^{12} L) of water flow past and carry 83 million moles of PO_4^{3-} ions, 2.1 billion moles of NO_3^- ions, 27 billion moles of Si atoms (as various silicates), and 190 billion moles of HCO_3^- ions.

Chemists carry out many of their reactions in solution (usually on a much smaller scale), partly because the reactants are then mobile and can come together and react. We shall see numerous examples of the chemical aspects of solutions in Part III. In this chapter we discuss some of the physical properties of solutions in terms of the ions and molecules they contain.

MEASURES OF CONCENTRATION

When we do stoichiometric calculations for reactions taking place in solution, we need to know how many moles of solute there are in a given volume. For instance, to interpret an acid-base titration, we may need to know that 25.0 mL of a certain solution contains 0.100 mol of OH^- ions. To predict the effect of solutes on the physical properties of solutions, however, it turns out that we often need to know the *relative numbers* of solute and solvent molecules; the number of solute molecules per liter is less important (Fig. 11.1). For instance, if we know that one out of every hundred molecules in an aqueous solution is a solute molecule (the other 99 being water), then we can predict the boiling and freezing points of the solution. The different ways of measuring composition so as to emphasize either stoichiometry or relative numbers are summarized in Table 11.2 and described below.

11.1 EMPHASIZING THE AMOUNT OF SOLUTE

When we want to focus on the amount of solute present in a given volume of solution, we use either the molar concentration or the "mass concentration." We used the molar concentration in Section 4.4, but for convenience we review it here.

Molar concentration. By molar concentration, or molarity, we mean the number of moles of solute per liter of solution:

$$\text{Molar concentration} = \frac{\text{moles of solute}}{\text{liters of solution}}$$

TABLE 11.1 The principal components of seawater

Element	Principal form	Mass concentration, g/L
Cl	Cl^-	19.0
Na	Na^+	10.5
Mg	Mg^{2+}	1.35
S	SO_4^{2-}	0.89
Ca	Ca^{2+}	0.40
K	K^+	0.38
Br	Br^-	0.065
C	HCO_3^-, H_2CO_3, CO_3^{2-}, organic compounds	0.028

The unit of molar concentration is moles per liter (mol/L), written M. For low concentrations, it is often convenient to use $1\ mM = 10^{-3}\ M$.

We saw in Section 4.4 that the molar concentration acts as a conversion factor between volume of solution and number of moles of solute:

Number of moles of solute

$$= \text{volume of solution} \times \text{moles of solute per liter}$$

$$= \text{volume of solution} \times \text{molar concentration}$$

For example, if we have 25.00 mL of a $0.150\ M$ NaOH(aq) solution, then the number of moles of NaOH present is

$$\text{Moles of NaOH} = (25.00 \times 10^{-3}\ \text{L}) \times 0.150\ \frac{\text{mol NaOH}}{1\ \text{L}}$$

$$= 3.75 \times 10^{-3}\ \text{mol NaOH}$$

Because 1 mol/L is equal to 1 mmol/mL, the same conversion can be written more simply as

$$\text{Moles of NaOH} = 25.00\ \text{mL} \times 0.150\ \frac{\text{mmol NaOH}}{1\ \text{mL}}$$

$$= 3.75\ \text{mmol NaOH}$$

Since each NaOH unit gives one OH^- ion, we also know that the sample contains 3.75 mmol of OH^- ions.

Mass concentration. Closely related to molar concentration is the "mass concentration":

The **mass concentration** of a solution is the mass of solute per liter of solution:

$$\text{Mass concentration} = \frac{\text{mass of solute}}{\text{liters of solution}}$$

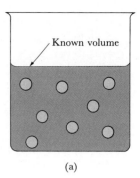

(a)

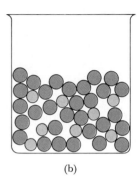

(b)

FIGURE 11.1 In some applications we need to emphasize the number of solute molecules or ions in a given volume (a); then we use the molarity or mass concentration. In others, we need to emphasize the relative numbers of solute and solvent molecules (b); then we use either the molality or the mole fraction.

TABLE 11.2 Concentration units

Measure	Notation	Units	Note
Molar concentration (molarity)	[X]	Moles per liter (M)	Moles of X per liter of solution
Mass concentration	c_X	Grams per liter	Grams of X per liter of solution
Molality		Moles per kilogram (m)	Moles of X per kilogram of solvent
Mole fraction	x		Number of moles as a fraction of the total number of moles of solute and solvent molecules: $x_A + x_B + \cdots = 1$
Parts per million	ppm		Number of molecules of the solute per million molecules of the solution
Mass percentage	%		Mass of a component, expressed as a percentage of the total mass

Dissolving 5.0 g of sodium chloride in enough water to make 500 mL of solution results in a mass concentration of

$$\text{Mass concentration} = \frac{5.0 \text{ g NaCl}}{0.500 \text{ L}} = 10 \text{ g NaCl/L}$$

Mass concentration is used mainly when the mass of solute in a given volume of solution is of interest. This would be the case, for instance, if we were judging whether it is economical to extract gold from seawater, for then we would want to know the mass of gold ions in a cubic kilometer of ocean (it is about 10 kg, since its mass concentration is about 10^{-8} g/L).

▼ **EXAMPLE 11.1** **Converting mass concentration to molar concentration**

Calculate the molar concentration of a 5.00-g/L aqueous solution of sodium chloride.

STRATEGY Mass concentration expresses the mass of solute per liter of solution; molar concentration is the number of moles of solute per liter. To obtain the latter from the former, we convert the 5.00 g of NaCl to moles of NaCl, using the molar mass of NaCl (58.44 g/mol) as the conversion factor.

SOLUTION Since 58.44 g NaCl = 1 mol NaCl, the conversion factor from grams to moles is

$$1 = \frac{1 \text{ mol NaCl}}{58.44 \text{ g NaCl}}$$

Therefore,

$$\text{Moles of NaCl/L} = \frac{5.00 \text{ g NaCl}}{1 \text{ L}} \times \frac{1 \text{ mol NaCl}}{58.44 \text{ g NaCl}}$$

$$= 0.0856 \text{ mol/L NaCl}$$

This concentration could be reported as 85.6 mM NaCl(aq).

EXERCISE Calculate the molar concentration of an aqueous sucrose ($C_{12}H_{22}O_{11}$) solution with a mass concentration of 10 g/L.

[*Answer*: 29 mM $C_{12}H_{22}O_{11}$(aq)]

▲

11.2 EMPHASIZING NUMBERS OF MOLECULES

The two principal measures of the relative numbers of solute and solvent molecules are the "mole fraction" and the "molality" of the solution. Another widely used measure is *parts per million* (ppm), the number of solute particles among 1 million solution molecules. Far away from sources of pollution, as in the South Pacific, the concentration of SO_2 in the atmosphere may be only 6×10^{-5} ppm, meaning that only 6 molecules out of 10^{11} are SO_2. However, in some large cities, SO_2 concentrations can reach 1.5 ppm, about 25,000 times higher (Fig. 11.2).

Mole fraction. The mole fraction is defined in Section 5.6 as the number of moles of molecules of a certain kind, expressed as a fraction of

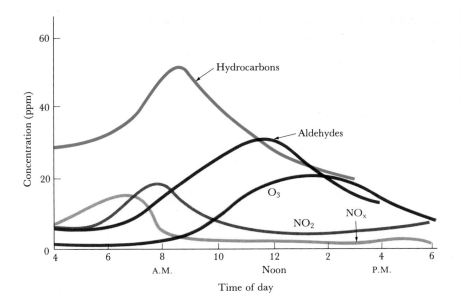

FIGURE 11.2 The concentrations (in parts per million) of some components of the atmosphere during the day in downtown Los Angeles in 1958, before emission controls were made very stringent.

the total number of moles of molecules present. For solute molecules in a nonelectrolyte solution, the mole fraction is

$$x_{\text{solute}} = \frac{n_{\text{solute}}}{n_{\text{solvent}} + n_{\text{solute}}}$$

where n_A is the number of moles of A that are present. A similar equation defines the mole fraction of the solvent molecules, x_{solvent}. A useful relation between the two is

$$x_{\text{solvent}} + x_{\text{solute}} = \frac{n_{\text{solvent}}}{n_{\text{solvent}} + n_{\text{solute}}} + \frac{n_{\text{solute}}}{n_{\text{solvent}} + n_{\text{solute}}}$$

$$= 1$$

A mole fraction must lie in the range $x_{\text{solute}} = 0$ (no solute) to $x_{\text{solute}} = 1$ (all solute). For an electrolyte solution, the mole fraction is calculated by treating the cations and anions as individual particles; hence

$$x_{\text{cations}} = \frac{n_{\text{cations}}}{n_{\text{cations}} + n_{\text{anions}} + n_{\text{solvent}}}$$

A similar expression may be written for the anions, and

$$x_{\text{solvent}} + x_{\text{cations}} + x_{\text{anions}} = 1$$

▼ EXAMPLE 11.2 Calculating a mole fraction

Calculate the mole fraction of sucrose ($C_{12}H_{22}O_{11}$) in a solution prepared by dissolving 5.00 g of sucrose in 100.0 g of water.

STRATEGY We need to know the numbers of moles of solute and solvent molecules in the solution. Therefore, we convert the masses of solute and solvent to moles, using molar masses, and then substitute the results into the definitions above. The answer can be checked by testing whether $x_{C_{12}H_{22}O_{11}} + x_{H_2O} = 1$.

SOLUTION From the molar masses of sucrose (342.3 g/mol) and water (18.02 g/mol), we have

$$342.3 \text{ g } C_{12}H_{22}O_{11} = 1 \text{ mol } C_{12}H_{22}O_{11}$$

$$18.02 \text{ g } H_2O = 1 \text{ mol } H_2O$$

These are used to convert the given masses to moles:

$$\text{Moles of } C_{12}H_{22}O_{11} = 5.00 \text{ g } C_{12}H_{22}O_{11} \times \frac{1 \text{ mol } C_{12}H_{22}O_{11}}{342.3 \text{ g } C_{12}H_{22}O_{11}}$$

$$= 0.0146 \text{ mol } C_{12}H_{22}O_{11}$$

$$\text{Moles of } H_2O = 100.0 \text{ g } H_2O \times \frac{1 \text{ mol } H_2O}{18.02 \text{ g } H_2O}$$

$$= 5.549 \text{ mol } H_2O$$

The total number of moles of molecules in the solution is

$$n_{total} = 0.0146 \text{ mol} + 5.549 \text{ mol} = 5.564 \text{ mol}$$

The two mole fractions are therefore

$$x_{C_{12}H_{22}O_{11}} = \frac{0.0146 \text{ mol}}{5.564 \text{ mol}} = 0.00262$$

$$x_{H_2O} = \frac{5.549 \text{ mol}}{5.564 \text{ mol}} = 0.9973$$

Note that $0.00262 + 0.9973 = 0.9999$, or virtually 1.

EXERCISE Calculate the mole fractions of H_2O and C_2H_5OH in a mixture of equal masses of water and ethanol.

[*Answer*: $x_{H_2O} = 0.719$, $x_{C_2H_5OH} = 0.281$]

Molality. The "molality" of a solution also emphasizes the relative numbers of solute and solvent molecules:

The **molality** of a solution is the number of moles of solute per kilogram of solvent:

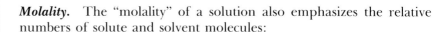

$$\text{Molality} = \frac{\text{moles of solute}}{\text{kilograms of solvent}}$$

The unit of molality (moles per kilogram of solvent) is often abbreviated to m and read "molal." The emphasis on *solvent* in the definition should be noted. It means that to prepare a 1 m NaCl(*aq*) solution, 1 mol of NaCl is dissolved in 1 kg of water (Fig. 11.3). Since 1 kg of solvent consists of a definite number of moles of molecules (55.5 mol of H_2O molecules for water), the higher the molality of a given solution, the higher the proportion of solute molecules.

The molality is a conversion factor from the mass of solvent in a sample to the number of moles of solute in it:

Number of moles of solute

= mass of solvent × moles of solute per kilogram of solvent

= mass of solvent × molality

FIGURE 11.3 A solution of given molality is prepared by dissolving a known mass of solute in a known mass of solvent.

▼ **EXAMPLE 11.3** **Preparing a solution of given molality**

Calculate the mass of potassium nitrate that should be added to 250 g of water to prepare a 0.200 m $KNO_3(aq)$ solution.

STRATEGY If we knew how many moles of KNO_3 were needed, we could convert that to grams, using the molar mass of KNO_3 (101.1 g/mol). We can find the required number of moles by using the molality as a conversion factor for the mass of solute, as set out above.

SOLUTION In a 0.200 m solution, the number of moles of KNO_3 in 0.250 kg of water is

$$\text{Moles of } KNO_3 = 0.250 \text{ kg } H_2O \times \frac{0.200 \text{ mol } KNO_3}{1 \text{ kg } H_2O}$$

$$= 0.0500 \text{ mol } KNO_3$$

Since 1 mol KNO_3 = 101.1 g KNO_3,

$$\text{Mass of } KNO_3 = 0.0500 \text{ mol } KNO_3 \times \frac{101.1 \text{ g } KNO_3}{1 \text{ mol } KNO_3}$$

$$= 5.06 \text{ g } KNO_3$$

EXERCISE Calculate the mass of potassium permanganate needed to prepare a 0.150 m $KMnO_4(aq)$ solution with 500 g of water.

[*Answer*: 11.6 g]

▲

The molality of a solution must be carefully distinguished from its molarity. The molality is the number of moles of solute per kilogram of *solvent*, emphasizing the relative numbers of solute and solvent particles. The molar concentration (molarity) is the number of moles of solute per liter of *solution;* it emphasizes the number of solute particles in a given volume of the solution.

▼ **EXAMPLE 11.4** **Converting between mole fraction and molality**

Express as a molality the composition of a solution of benzene (C_6H_6) in toluene ($CH_3C_6H_5$), reported as $x_{benzene} = 0.150$.

STRATEGY To work out a molality, we need to know the number of moles of solute (benzene) molecules and the mass of solvent. Suppose we consider a sample of the solution that consists of 1 mol of molecules (in total). The mole fraction tells us that the 1 mol of molecules consists of $x_{benzene}$ moles of benzene molecules and $(1 - x_{benzene})$ moles of toluene molecules. The number of moles of toluene molecules can be converted to a mass by using the molar mass of toluene (92.13 g/mol). Then we would know both the moles of solute and the mass of solvent.

SOLUTION The number of moles of benzene molecules in the sample is 0.150 mol, implying that the number of moles of toluene molecules is 0.850 mol. The total mass of toluene in the sample is then

$$\text{Mass of toluene} = 0.850 \text{ mol } CH_3C_6H_5 \times \frac{92.13 \text{ g } CH_3C_6H_5}{1 \text{ mol } CH_3C_6H_5}$$

$$= 78.3 \text{ g } CH_3C_6H_5$$

Therefore,

$$\text{Molality} = \frac{0.150 \text{ mol } C_6H_6}{78.3 \times 10^{-3} \text{ kg } CH_3C_6H_5} = 1.92 \text{ mol } C_6H_6/\text{kg } CH_3C_6H_5$$

EXERCISE Calculate the molality of a solution of toluene in benzene, given that the mole fraction of toluene is 0.150.

[*Answer*: 2.26 *m*]

SOLUBILITY

In this chapter we are focusing on aqueous solutions because they are so important, but many of our remarks apply equally to nonaqueous solutions. In the following discussion, remember that substances which dissolve to give solutions of ions that conduct electricity (e.g., sodium chloride and acetic acid) are called electrolytes (Section 2.7); those giving solutions that do not conduct electricity because the solute remains molecular (e.g., glucose and ethanol) are nonelectrolytes.

11.3 SATURATION AND SOLUBILITY

If we add 20 g of sucrose—cane sugar—to 100 mL of water at room temperature, all the sucrose dissolves. However, if we add 200 g, most dissolves but some does not (Fig. 11.4). When the solvent has dissolved all the solute it can and some undissolved solute remains, the solution is said to be "saturated."

The definition of solubility. If we could follow a single sucrose molecule in a saturated solution, we might find that at some instant it is part of the surface layer of a sucrose crystal (Fig. 11.5). Shortly after, the molecule might be found in solution. Still later, it might be buried more deeply in a crystal, under many layers of molecules that had settled on

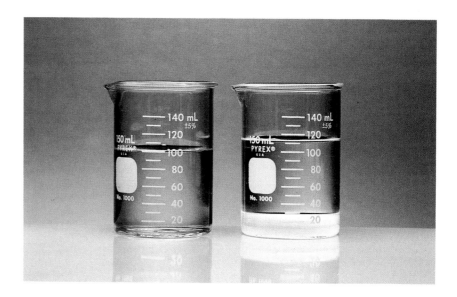

FIGURE 11.4 When a little sucrose is shaken with 100 mL of water, it all dissolves (left). However, when a large amount (more than 200 g) is added, some undissolved sucrose remains (right).

top of it. There it would remain until it became exposed again and was able to return to the solution. In other words, a saturated solution is another example of dynamic equilibrium (see Section 10.4), in which a forward process and its reverse occur at equal rates. In this case, the solute continues to dissolve, and it does so at a rate that exactly matches the rate of the reverse process, the return of solute from the solution. This suggests the following definition:

A **saturated solution** is a solution in which the dissolved and undissolved solute are in dynamic equilibrium.

Although we cannot follow a single molecule in a saturated solution, we can show experimentally that the equilibrium is dynamic and not static. One way to do so is to add solid silver iodide, containing some iodine-131 in place of the usual iodine-127, to a saturated solution of silver iodide. Iodine-131 is radioactive and can be detected with Geiger counters and other radioactivity-detection devices. After a time the solution becomes radioactive, but the total mass of dissolved solid remains unchanged. This shows that some I^- ions have dissolved and others have come out of solution, even though the solution was already saturated.

A saturated solution represents the limit of a solute's ability to dissolve in a given quantity of solvent. It is therefore a natural measure of the solute's "solubility" S:

The **solubility** of a substance in a solvent is the concentration of the saturated solution.

The solubilities of some substances are given in Table 11.3. They depend on the solvent, the temperature, and, for gases, the pressure.

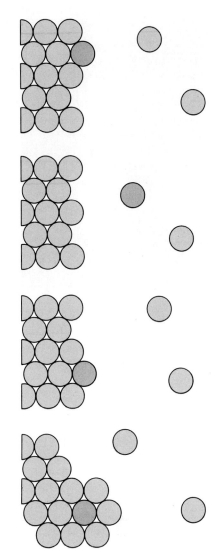

FIGURE 11.5 The solute in a saturated solution is in dynamic equilibrium with the undissolved solute. If we could follow a single solute particle (the red circle), we would sometimes find it in solution and sometimes in the solute.

TABLE 11.3   The solubilities of some substances

| Compound | Solubility, g solute/100 g solvent, in water at | | Other solvents |
	0°C	100°C or as specified	
NH_3	89.5	7.4	Organic solvents
NH_4NO_3	118	871	Alcohol, ammonia
$CaCl_2$	59.5	159	Alcohol
CaF_2	1.7×10^{-3}		
$CuSO_4 \cdot 5H_2O$	31.6	203.3	
HCl	82.3	56.1 at 60°C	Alcohol, benzene
MgO	6×10^{-4}	8×10^{-3} at 30°C	
AgF	182	205	
AgCl	7×10^{-5}	2×10^{-3}	

FIGURE 11.6 This Chile saltpeter has survived in the arid region where it is mined in Chile because there is too little groundwater to dissolve it and wash it away.

The dependence of solubility on the solute. Some substances are soluble in water, others sparingly (slightly) soluble, and others almost insoluble. We can know which behavior to expect by referring to the "solubility rules," which were given in Table 3.1. We used the rules in Chapter 3 to choose reagents for precipitation reactions; they are also of help in understanding the behavior of some everyday substances and the properties of minerals. Because of the solubility of most nitrates, for instance, they are rarely found in mineral deposits, for they are usually carried away by the water that trickles through the ground. An exception is the large deposit of sodium nitrate in the arid coastal region of Chile, where groundwater is absent. This "Chile saltpeter" (Fig. 11.6) was the main source of nitrates for fertilizers and explosives until the Haber process for ammonia was developed at the start of this century.

The low solubility of most phosphates is an advantage for skeletons, since bone consists largely of calcium phosphate (much of the rest is the protein collagen). However, this insolubility is inconvenient for agriculture, since it means that phosphorus, which is essential to the function of biological cells, is slow to circulate through the ecosystem. One of chemistry's achievements has been the development of manufacturing processes to speed phosphates on their way as fertilizers. The phosphates and hydrogen phosphates used for fertilizers are obtained from phosphate rocks (Fig. 11.7), principally the apatites—hydroxyapatite, $Ca_5(PO_4)_3OH$, and fluorapatite, $Ca_5(PO_4)_3F$—by treating them with concentrated sulfuric acid:

$$Ca_5(PO_4)_3OH(s) + 5H_2SO_4(aq) \longrightarrow$$
$$3H_3PO_4(aq) + 5CaSO_4(s) + H_2O(l)$$

The phosphate rocks themselves were once alive, for they are the crushed and compressed remains of the skeletons of prehistoric animals. Calcium hydrogen phosphate ($CaHPO_4$) is more soluble than calcium phosphate and is included in commercial phosphate fertilizers.

Just as hydrogen phosphates are more soluble than phosphates, so hydrogen carbonates (bicarbonates, HCO_3^-) are more soluble than carbonates. This difference is responsible for the behavior of *hard water*, water that contains dissolved calcium and magnesium salts. In particular, the difference accounts for the deposit of scale inside hot pipes and for the formation of a scum with soap in hard water. The

FIGURE 11.7 Mining of phosphate rock, the crushed remains of the skeletons of prehistoric animals.

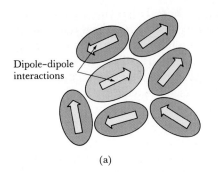

Dipole–dipole interactions

(a)

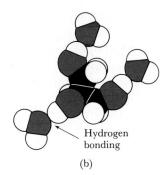

Hydrogen bonding

(b)

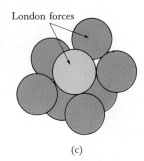

London forces

(c)

behavior of hard water begins with the fact that rainwater contains dissolved carbon dioxide, and hence some carbonic acid from the reaction

$$CO_2(g) + H_2O(l) \longrightarrow H_2CO_3(aq)$$

As the water runs along and through the ground, the carbonic acid reacts with the calcium carbonate of limestone or chalk and forms the more soluble hydrogen carbonate:

$$CaCO_3(s) + H_2CO_3(aq) \longrightarrow Ca(HCO_3)_2(aq)$$

These reactions are reversed when the water is heated in a kettle or furnace:

$$2HCO_3^-(aq) \xrightarrow{\Delta} CO_3^{2-}(aq) + CO_2(g) + H_2O(l)$$

The carbon dioxide is driven off, leaving carbonate ions in solution, and the almost insoluble calcium carbonate is deposited as scale.

The dependence of solubility on the solvent. In many instances, the dependence of the solubility of a substance on the identity of the solvent can be summarized by the rule that "like dissolves like." That is, a polar liquid, such as water, is generally a much better solvent than a nonpolar one (such as benzene) for ionic and polar compounds. Conversely, nonpolar liquids, including benzene and the tetrachloroethylene (C_2Cl_4) used for dry cleaning, are often better solvents for nonpolar compounds than for polar compounds (Fig. 11.8). The reason is that the energy of the solute molecules is similar in the solution to what it was in the original solid if the intermolecular forces in solution and solid are similar.

If the principal cohesive forces in a solute are hydrogen bonds, the "like dissolves like" rule implies that it is more likely to dissolve in a hydrogen-bonding solvent than in others. Sucrose, for example, dissolves readily in water but not in benzene. Similarly, if the principal cohesive forces are London forces, the best solvent is likely to be one held together by the same kind of forces. One example is carbon disulfide, which is a far better solvent for sulfur than is water (Fig. 11.9), because solid sulfur is a molecular solid of S_8 molecules held together by London forces.

Soaps and detergents. Modern soaps and detergents are a practical application of the principle of like dissolving like. Soaps are the sodium salts of organic acids with long hydrocarbon chains, including sodium

FIGURE 11.8 Like often dissolves like. (a) Intermolecular interactions help a polar solvent to dissolve other polar substances, (b) a hydrogen-bonding solvent to dissolve substances held together by hydrogen bonds, and (c) a solvent with strong London forces to dissolve nonpolar molecular solids.

FIGURE 11.9 The molecular solid sulfur does not dissolve in water (left) but does dissolve in carbon disulfide (right), with which its molecules have strongly favorable London interactions.

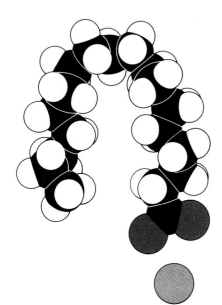

1 Sodium stearate

2 Polyphosphate ion

stearate **(1)**; we shall denote them NaA, where HA is the organic acid. The anions have a polar group (called the "head group") at one end of a long nonpolar group, the hydrocarbon chain. The anions (A^-) sink their nonpolar and thus *hydrophobic*, or water-repelling, hydrocarbon tails into a blob of grease. Their *hydrophilic*, or water-attracting, head groups remain on the surface of the grease blob, coating it with a skin of polar hydrogen-bonding groups (Fig. 11.10). The polar head groups enable the grease blob to dissolve in water and to be washed away.

A problem with soaps is that they form a scum in hard water. The scum is the product of a precipitation reaction that occurs because calcium salts are less soluble than sodium salts:

$$Ca^{2+}(aq) + 2A^-(aq) \longrightarrow CaA_2(s)$$

One way of avoiding the problem is to use another precipitation reaction to remove the Ca^{2+} ions from the water before the soap is used. This can be done by adding sodium carbonate ("washing soda") to the water and precipitating calcium carbonate:

$$Ca(HCO_3)_2(aq) + Na_2CO_3(aq) \longrightarrow CaCO_3(s) + 2NaHCO_3(aq)$$

Another way to avoid soap scum is to add polyphosphate ions to the water as a component of the detergent. Polyphosphate ions **(2)** are formed when phosphates are heated, and they consist of chains and rings of PO_4 groups. The first step in their formation is

$$HO-\overset{\overset{\displaystyle O}{\|}}{\underset{\underset{\displaystyle OH}{|}}{P}}-OH + HO-\overset{\overset{\displaystyle O}{\|}}{\underset{\underset{\displaystyle OH}{|}}{P}}-OH \overset{\Delta}{\longrightarrow} HO-\overset{\overset{\displaystyle O}{\|}}{\underset{\underset{\displaystyle OH}{|}}{P}}-O-\overset{\overset{\displaystyle O}{\|}}{\underset{\underset{\displaystyle OH}{|}}{P}}-OH + H_2O$$

Polyphosphate ions are big, and when they are added to hard water, they wrap around the calcium cations and hide them away from other anions with which they would normally precipitate. This wrapping up of one ion by another is called *sequestration* of the ion (from the Latin word for "hiding away"), and the polyphosphates are called "sequestering agents."

Modern commercial detergents are mixtures of compounds, the most important of which is the "surface-active agent," or "surfactant." Surfactant molecules are synthetic organic compounds that resemble the one shown below **(3)**. Like the stearate ion, they have a hydrophilic

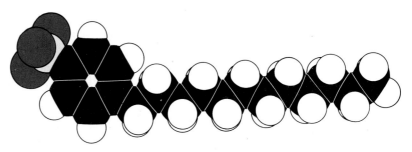

3 A typical surfactant molecule

head group and a hydrophobic tail, and they act similarly. Detergents also contain polyphosphates to sequester calcium ions and to adjust the acidity. Other additives in the mixture "fluoresce" (absorb ultraviolet light and then give out visible light) to give the impression of greater cleanliness.

11.4 THE EFFECT OF PRESSURE ON GAS SOLUBILITY

We have noted that solubility depends on the pressure the solution experiences. The strongest dependence is shown by gases, which are more soluble at higher pressures (Fig. 11.11). A practical application of this phenomenon is the production of soft drinks and champagne. In each case carbon dioxide is dissolved in the liquid under pressure (in champagne, as a result of fermentation that continues in the sealed bottle). When the bottle is opened the pressure is released, the solubility of the gas is greatly reduced, and the gas effervesces (bubbles out of solution) with a pop. A more serious consequence of the dependence of gas solubility on pressure is the additional nitrogen that dissolves in the blood of deep-sea divers. The dissolved nitrogen effervesces when the diver returns to the surface, resulting in the formation of numerous small bubbles in the bloodstream (Fig. 11.12). These bubbles can block the capillaries—the narrow vessels that distribute the blood—and starve the tissues of oxygen, causing the painful condition known as the "bends," which in serious cases can lead to death. The risk of the bends is reduced if helium is used instead of nitrogen to dilute the diver's oxygen supply, for helium is much less soluble than nitrogen.

Henry's law. The dependence of the solubility of a gas on its pressure was summarized in 1801 by the English chemist William Henry:

Henry's law: The solubility of a gas in a liquid is proportional to its partial pressure.

This law is normally written

$$S = k_H \times P$$

where P is the partial pressure of the gas, and k_H, which is called *Henry's constant*, depends on the gas, the solvent, and the temperature (see

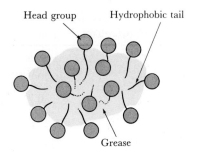

FIGURE 11.10 The hydrophobic tail of a soap or surfactant molecule enters the blob of grease, leaving the hydrophilic polar head group on the surface.

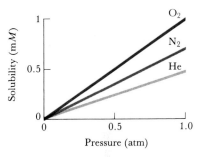

FIGURE 11.11 The variation of the solubilities of oxygen, nitrogen, and helium with the pressure. Note that the solubility of each gas is doubled when the pressure is doubled.

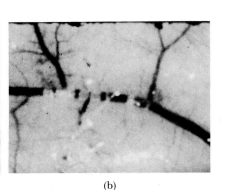

(a)   (b)

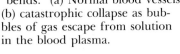

FIGURE 11.12 The small bubbles of air are responsible for the "bends." (a) Normal blood vessels; (b) catastrophic collapse as bubbles of gas escape from solution in the blood plasma.

TABLE 11.4 Henry's constant for gases in water at 20°C

Gas	k_H, mM/atm
Air	0.79
Carbon dioxide	23
Helium	0.37
Neon	0.5
Argon	1.5
Hydrogen	0.85
Nitrogen	0.7
Oxygen	1.3

Table 11.4). The law implies that, at constant temperature, doubling the partial pressure of a gas doubles its solubility.

▼ EXAMPLE 11.5 Using Henry's law

The lowest concentration of O_2 that can support aquatic life is about 0.13 mM (about 4 mg/L). Is the concentration in pond water at 20°C normally adequate?

STRATEGY To use Henry's law, we need the partial pressure of oxygen in air; that was calculated in Example 5.5 as 0.21 atm. We can take the value of k_H from Table 11.4.

SOLUTION For oxygen in water, $k_H = 1.3$ mM/atm; therefore

$$S = \frac{1.3 \text{ m}M}{1 \text{ atm}} \times 0.21 \text{ atm} = 0.27 \text{ m}M$$

This corresponds to 8.6 mg of oxygen per liter of water. Hence, under normal conditions, the concentration of oxygen in pond water is adequate.

EXERCISE Calculate the concentration of dissolved nitrogen under the same conditions.

[*Answer:* 0.5 mM]

▲

The increase in the solubility of a gas with pressure can be explained in terms of the dynamic equilibrium between gas molecules in the solution and those in the space above it (Fig. 11.13). When a solvent is saturated with dissolved gas, the rate at which gas molecules enter it matches the rate at which they leave it. If the pressure is increased, the rate at which molecules enter the solution increases because they hit its surface more often. As a result, the concentration of gas in the solution rises. However, the increased concentration of dissolved gas molecules means that more are available to escape back to the gas, so their rate of escape increases. A new equilibrium is reached when the two rates are equal, and this equilibrium corresponds to a higher concentration of gas in the solution than before, and hence to a higher solubility.

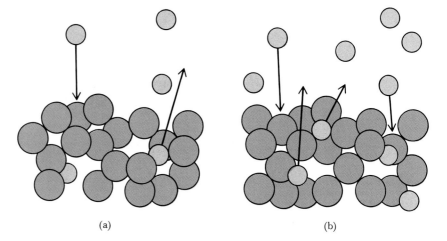

FIGURE 11.13 The effect of increased pressure, represented by the change from part a to part b, is to increase the rate at which gas molecules dissolve. The concentration of gas in the solution rises until the rate at which gas molecules leave the solution matches the new rate at which they arrive.

(a)   (b)

Le Chatelier's principle. The increase in gas solubility with pressure is an instance of a general characteristic of dynamic equilibria. This characteristic was first described by the French chemist Henri Le Chatelier in 1884:

> **Le Chatelier's principle:** A dynamic equilibrium tends to oppose any change in the conditions.

Accordingly, if the pressure of the gas in contact with the solution is increased, the dynamic equilibrium adjusts by increasing the concentration of gas in the solution. That tends to decrease the number of molecules of gas above the solution and hence to minimize the increase in pressure.

Le Chatelier's principle applies only to *dynamic* equilibria, not to static equilibria. A dynamic equilibrium is, in a sense, a living equilibrium: it can adjust because the forward and reverse processes are active and make the system responsive to changes. A pencil balanced on its point is in a condition of static equilibrium; it has no ability to recover its position when a force pushes it to the side. We shall see in this and the next few chapters that Le Chatelier's principle is very helpful for predicting the effects of changes in pressure, temperature, and composition.

11.5 THE EFFECT OF TEMPERATURE ON SOLUBILITY

In a discussion of the effect of temperature on solubility, we must distinguish between the *rate* at which a substance dissolves and the *concentration* it finally reaches. Although many substances dissolve more quickly at higher temperatures, their saturated solutions may have a lower concentration at the higher temperature.

All gases become *less* soluble as the temperature is raised. In contrast, most solids are *more* soluble in hot water than in cold (Fig. 11.14). The magnitude of the increase in solubility varies sharply from one substance to another. The solubility of sodium chloride in water increases by only about 10%, from 6.1 mol/kg to 6.7 mol/kg, when the temperature is raised from 0°C to 100°C, whereas that of silver nitrate increases by nearly 700%, from 7.2 mol/kg to 56.0 mol/kg, over the same temperature range. A few solids are less soluble at higher temperatures. The solubility of lithium sulfate, for instance, decreases by about 10% from 2.3 mol/kg at 0°C to 2.1 mol/kg at 100°C. Some compounds show a mixed behavior: the solubility of sodium sulfate increases up to 32°C but then decreases as the temperature is raised further.

The lower solubility of gases at higher temperatures is responsible for the tiny bubbles that appear when cool water from the faucet is left to stand in a warm room: the bubbles are air that dissolved when the water was cooler. A more important effect is *thermal pollution*, the damage caused to the environment by the waste heat of an industrial proc-

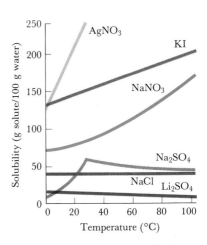

FIGURE 11.14 The variation with temperature of the solubilities of some substances in water.

FIGURE 11.15 This photograph of the outflow near a power-generating facility shows the temperature of the water as different colors. Oxygen is less soluble in the hot (red) regions than in ordinary river water.

ess (Fig. 11.15). One form this takes is a reduced oxygen concentration in rivers, resulting from the discharge of hot water from power stations. The problem is aggravated by the lower density of the warm water, which causes it to rise to the top; there, on account of the lower solubility of gases at high temperatures, it prevents oxygen from penetrating to the cooler water below.

Predicting the effect of temperature. According to Le Chatelier's principle, when a saturated solution is cooled or heated, the dynamic equilibrium adjusts by opposing the change in temperature. If heat is absorbed when the solute dissolves, we can expect more of the solute to dissolve if the temperature is raised. On the other hand, if dissolving gives out heat, we can expect the solubility to increase when the temperature is lowered. That is, a substance that dissolves endothermically becomes more soluble as the temperature is raised. Conversely, a substance that dissolves exothermically becomes less soluble as the temperature is raised.

Since dissolving normally occurs at constant pressure, the heat given out or absorbed is equivalent to an enthalpy change, called the *enthalpy of solution* ΔH_{sol}. The enthalpy of solution can be measured by calorimetry (although there are more accurate methods). Some values for very dilute solutions are given in Table 11.5; as in that table, enthalpies of solution are normally reported per mole of formula units or molecules.

TABLE 11.5 Enthalpies of solution ΔH_{sol} at 25°C in dilute aqueous solution, in kilojoules per mole*

Cation	Anion							
	Fluoride	**Chloride**	**Bromide**	**Iodide**	**Hydroxide**	**Carbonate**	**Nitrate**	**Sulfate**
Lithium	+4.9	−37.0	−48.8	−63.3	−23.6	−18.2	−2.7	−29.8
Sodium	+1.9	+3.9	−0.6	−7.5	−44.5	−26.7	+20.4	−2.4
Potassium	−17.7	+17.2	+19.9	+20.3	−57.1	−30.9	+34.9	+23.8
Ammonium	−1.2	+14.8	+16.0	+13.7			+25.7	+6.6
Silver	−22.5	+65.5	+84.4	+112.2		+41.8	+22.6	+17.8
Magnesium	−17.7	−160.0	−185.6	−213.2	+2.3	−25.3	−90.9	−91.2
Calcium	+11.5	−81.3	−103.1	−119.7	−16.7	−13.1	−19.2	−18.0
Aluminum	−27	−329	−368	−385				−350

*The value for silver iodide, for example, is the entry found where the row labeled Silver intersects the column labeled Iodide.

All gases dissolve exothermically and are, as we have noted, less soluble in hot solvent than in cold. As a result, some gases can be collected over hot water (as described in Section 5.6), even though they are too soluble in cold water for this to be feasible. Dinitrogen oxide is one gas that can be collected in this way.

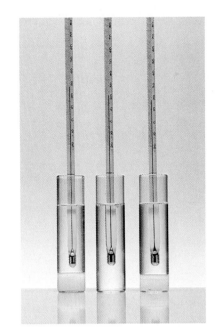

▼ **EXAMPLE 11.6   Predicting the temperature dependence of solubility**

Is silver bromide likely to be more or less soluble in hot water than in cold?

STRATEGY   The data in Table 11.5 are for very dilute solutions; they can be used in connection with Le Chatelier's principle only if the saturated solution is also very dilute. This is the case for silver bromide, which is a very sparingly soluble (that is, slightly soluble) salt. We have to determine whether the dissolution is endothermic or exothermic. If it is endothermic, the solubility will be greater in hot water than in cold. If it is exothermic, the solubility will be lower in hot water than in cold.

SOLUTION   The enthalpy of solution is +84 kJ/mol, an endothermic value. Heating therefore favors dissolving, and silver bromide should be more soluble in hot water. (It is.)

EXERCISE   Is magnesium fluoride likely to be more or less soluble in hot water than in cold?

[*Answer*: Less]

▲

FIGURE 11.16   The exothermic dissolving of lithium chloride (left) is shown by the rise in temperature above that of the original water (center); in contrast, ammonium nitrate dissolves endothermically (right).

Contributions to the enthalpy of solution.   Table 11.5 shows that some solids dissolve exothermically ($MgCl_2$ for example), and others endothermically (K_2SO_4). Lithium chloride and calcium chloride have quite strongly exothermic enthalpies of solution; when they are added to water the solution becomes noticeably warm (Fig. 11.16). The sign and magnitude of the enthalpy of solution can be understood by thinking of the overall dissolving process as taking place in two imaginary steps (Fig. 11.17). The first is the breakup of the solid, and the other the interaction of the separated molecules or ions with the solvent.

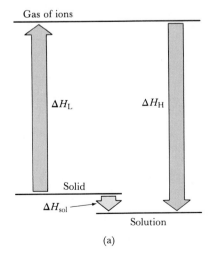

(a)

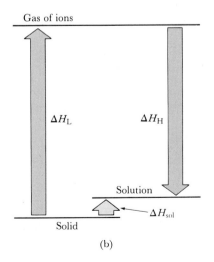

(b)

FIGURE 11.17   The enthalpy of solution is the sum of the enthalpy change required to separate the ions and the enthalpy change required for their hydration. The outcome is finely balanced. In some cases it is exothermic (a), and in others it is endothermic (b).

In the first step, the solid is imagined as being vaporized to a gas of ions or molecules. As we saw in Section 8.1, the standard change of enthalpy in this endothermic step is the lattice enthalpy ΔH_L of the solid. For NaCl it is the enthalpy change for the process

$$\text{NaCl}(s) \longrightarrow \text{Na}^+(g) + \text{Cl}^-(g) \qquad \Delta H_L$$

Some typical values are given in Table 8.1. As we noted, ionic compounds formed from small, highly charged ions (such as Mg^{2+} and O^{2-}) have high lattice enthalpies.

In the second step, the gas of ions dissolves in the solvent and heat is released as the ions are *solvated*, or surrounded by solvent molecules. Hydration is a special kind of solvation that occurs when the solvent is water. In aqueous solutions, water molecules hydrate cations as a result of ion-dipole interactions; they hydrate anions largely by hydrogen bonding. The change of enthalpy in this step is called the *enthalpy of hydration* ΔH_H. For sodium chloride, it is

$$\text{Na}^+(g) + \text{Cl}^-(g) \longrightarrow \text{Na}^+(aq) + \text{Cl}^-(aq) \qquad \Delta H_H$$

Hydration is always exothermic for ions (see Table 11.6). It is also exothermic for solutes that can form hydrogen bonds with the water. Such solutes include sucrose, glucose, acetone, and ethanol.

The enthalpy of solution is the sum of the lattice enthalpy and the enthalpy of hydration (Fig. 11.17). For NaCl we have

$$
\begin{aligned}
\text{NaCl}(s) &\longrightarrow \text{Na}^+(g) + \text{Cl}^-(g) & \Delta H_L &= +787 \text{ kJ} \\
\text{Na}^+(g) + \text{Cl}^-(g) &\longrightarrow \text{Na}^+(aq) + \text{Cl}^-(aq) & \Delta H_H &= -784 \text{ kJ} \\
\hline
\text{NaCl}(s) &\longrightarrow \text{Na}^+(aq) + \text{Cl}^-(aq) & \Delta H_{sol} &= +3 \text{ kJ}
\end{aligned}
$$

We see that the dissolving of NaCl is slightly endothermic. However, we also see that whether the overall process turns out to be endothermic or exothermic depends on a very delicate balance between the lattice enthalpy and the enthalpy of hydration. If sodium chloride had only a 0.5% smaller lattice enthalpy, +783 in place of +787 kJ/mol, it would dissolve exothermically instead of endothermically.

Predicting the enthalpy of solution from tabulated lattice enthalpies and enthalpies of hydration is very unreliable. It is like trying to calcu-

TABLE 11.6 Hydration enthalpies for some halides, in kilojoules per mole*

	Anion			
Cation	**F^-**	**Cl^-**	**Br^-**	**I^-**
H^+	-1613	-1470	-1439	-1426
Li^+	-1041	-898	-867	-854
Na^+	-927	-784	-753	-740
K^+	-844	-701	-670	-657
Ag^+	-993	-850	-819	-806

*The entry where the row labeled Na^+ intersects the column labeled Cl^-, for instance, is the enthalpy change when $\text{Na}^+(g) + \text{Cl}^-(g) \rightarrow \text{Na}^+(aq) + \text{Cl}^-(aq)$ and when the resulting solution is very dilute.

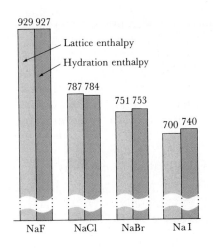

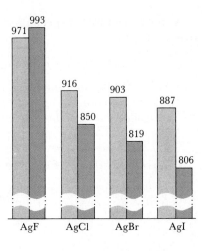

FIGURE 11.18 The lattice enthalpies and enthalpies of hydration of the sodium halides and the silver halides (in kJ/mol).

late the mass of the captain of a ship by measuring the mass of the ship with and without the captain aboard. However, it is reasonably safe to expect an endothermic value if the lattice enthalpy is very high. Similarly, we can predict an exothermic value if the lattice enthalpy is low and the enthalpy of hydration is large.

We can now explain why, with the exception of silver fluoride, the enthalpies of solution of the silver halides are more endothermic than those of the sodium halides. As we see in Fig. 11.18, the reason is that the lattice enthalpies of the silver halides are much more endothermic than their hydration enthalpies are exothermic, whereas this is not the case for the sodium halides. For instance, 129 kJ/mol more heat is needed to break up AgCl(s) than to break up NaCl(s), but only an extra 66 kJ/mol is released in the hydration step.

The enthalpies of hydration of NaF and AgF are far more exothermic than those of the other halides. This is probably due to the formation of strong hydrogen bonds between the F$^-$ ions and water molecules. The hydration of AgF releases more heat than is needed to break up solid silver fluoride, and the compound dissolves exothermically. However, the heat released when sodium fluoride is hydrated is not enough to overcome its large lattice enthalpy, and that compound dissolves endothermically.

Individual ion hydration enthalpies. The hydration enthalpies in Table 11.6 are the enthalpy changes that occur when widely separated cations and anions in a gas form a very dilute solution. It would be useful to know the values for the cations and anions separately, for then we could predict the value for NaCl(g) by adding together the values for Na$^+$(g) and Cl$^-$(g). A hint that values can be combined in this way comes from Fig. 11.18, which shows that the enthalpy of hydration of any silver halide (AgX) is more exothermic by about 66 kJ/mol than that of the corresponding sodium halide (NaX). Since the same halide ion (X$^-$) is present in each case, this suggests that the Ag$^+$ ion is more strongly hydrated, by 66 kJ/mol, than the Na$^+$ ion.

The enthalpies of hydration of cations cannot be measured without anions being present too. However, the value for the hydrogen ion has

TABLE 11.7 Ion hydration enthalpies ΔH_H at 25°C, in kilojoules per mole

Cations		Anions	
H^+	-1130	F^-	-483
Li^+	-558	Cl^-	-340
Na^+	-444	Br^-	-309
K^+	-361	I^-	-296
Rb^+	-335		
Cs^+	-303		
Ag^+	-510		
Mg^{2+}	-2003		
Ca^{2+}	-1657		
Sr^{2+}	-394		
Al^{3+}	-4797		

been measured with a special technique that makes use of a mass spectrometer. It is

$$H^+(g) \longrightarrow H^+(aq) \qquad \Delta H_H = -1130 \text{ kJ}$$

This value can be combined with the experimentally determined values in Table 11.6 to obtain the values for other ions. The resulting quantities are called *ion hydration enthalpies*. Some values are given in Table 11.7.

▼ **EXAMPLE 11.7** Calculation of ion hydration enthalpy

Calculate the hydration enthalpy of Cl^-, given that the enthalpy of hydration of a gas consisting of 1 mol of Ca^{2+} ions and 2 mol of Cl^- ions is -1387 kJ, and that the enthalpy of hydration of the Ca^{2+} ion is -707 kJ/mol.

STRATEGY We expect an exothermic value, but smaller than that for H^+ because the Cl^- ion is so much larger. We should write the equations for the processes that are involved and then see how we can use the information that is supplied.

SOLUTION We are given information regarding two processes:

$$Ca^{2+}(g) + 2Cl^-(g) \longrightarrow Ca^{2+}(aq) + 2Cl^-(aq) \qquad \Delta H_H = -1387 \text{ kJ} \quad \text{(a)}$$

$$Ca^{2+}(g) \longrightarrow Ca^{2+}(aq) \qquad \Delta H_H = -707 \text{ kJ} \quad \text{(b)}$$

Subtracting reaction b from reaction a gives

$$2Cl^-(g) \longrightarrow 2Cl^-(aq) \qquad \Delta H_H = -680 \text{ kJ}$$

The ion hydration energy of Cl^- is one-half this value, or -340 kJ/mol.

EXERCISE The enthalpy of hydration of a gas consisting of 1 mol of K^+ ions and 1 mol of Cl^- ions is -701 kJ. Calculate the hydration enthalpy of the K^+ ion from this information and the value for the Cl^- ion calculated in the example above.

[*Answer*: -361 kJ/mol]

▲

Ion hydration enthalpies show patterns. First, they are more strongly exothermic for ions with greater ionic charges:

Ion	Li^+	Be^{2+}	Al^{3+}
ΔH_H, kJ/mol	-558	-1435	-2537

The increase in ΔH_H reflects the much greater strength of the interaction between a more highly charged ion and the dipoles of the water molecules. Second, for ions of the same charge, hydration enthalpies are more strongly exothermic for ions with smaller radius:

	Cations			Anions		
	Li^+	Na^+	K^+	Cl^-	Br^-	I^-
Radius r, pm	60	95	133	181	195	216
ΔH_H, kJ/mol	-558	-444	-361	-340	-309	-296

The reason for this pattern is that a water molecule can approach a small ion more closely and hence interact with it more strongly. There are exceptions: Ag^+ is bigger than Na^+, but its hydration enthalpy is more exothermic. This may be because the Ag^+ ion can form covalent bonds with the hydrating water molecules.

The dependence of the hydration enthalpy on both ion charge and radius is expressed by the *Born equation*:

$$\Delta H_H = -69.7 \times 10^3 \text{ kJ/mol} \times \frac{z^2}{r}$$

where z is the charge number (for example, $z = +2$ for Mg^{2+}), and r is the radius of the ion in picometers. Experience has shown that the best results are obtained by using the ionic radii of anions as given in Table 7.6, but adding 85 pm to the radii given there for cations. Thus the hydration enthalpy of sodium ions is obtained by substituting $z^2 = 1$ and $r = 95 + 85 = 180$, which yields

$$\Delta H_H = -69.7 \times 10^3 \text{ kJ/mol} \times \frac{1}{180}$$

$$= -387 \text{ kJ/mol}$$

For Cl^-, on the other hand, we would take $r = 181$ from Table 7.6 directly, and obtain $\Delta H_H = -385$ kJ/mol.

▼ **EXAMPLE 11.8** Using the Born equation

Estimate the hydration enthalpy of Ca^{2+} from the value for Na^+.

STRATEGY The Born equation tells us that the hydration enthalpy is proportional to z^2/r; therefore, we can estimate the value for Ca^{2+} by allowing for the increase in charge number and the increase in radius from Na^+ to Ca^{2+}. Since both ions are cations, we must add 85 pm to the ionic radii in Table 7.6 before taking their ratio.

SOLUTION Since z is $+1$ for Na^+ and $+2$ for Ca^{2+}, the hydration enthalpy of Ca^{2+} will be $2^2 = 4$ times larger because of the increase in charge. However, the ionic radius of Ca^{2+} is 99 pm compared with 95 pm for Na^+, so the enthalpy will also be smaller by the factor

$$\frac{99 \text{ pm} + 85 \text{ pm}}{95 \text{ pm} + 85 \text{ pm}} = \frac{184 \text{ pm}}{180 \text{ pm}} = 1.02$$

Therefore, since the hydration enthalpy of Na^+ is -444 kJ/mol,

$$\Delta H_H(Ca^{2+}) = \frac{4}{1.02} \times (-444 \text{ kJ/mol}) = -1740 \text{ kJ/mol}$$

This compares well with -1657 kJ/mol in the table. The Born equation is often more reliable for *relative* values than for absolute values.

EXERCISE Predict the value of ΔH_H for Al^{3+} from the value for Mg^{2+}.
[*Answer*: -2185 kJ/mol]

▲

Enthalpy, solubility, and disorder. It may seem obvious that substances with exothermic enthalpies of solution should be soluble, because the solid loses energy as it dissolves. But why does ammonium nitrate dissolve in water, when in doing so it draws in heat from the surroundings and *gains* energy? In fact, a moment's thought shows that there is even

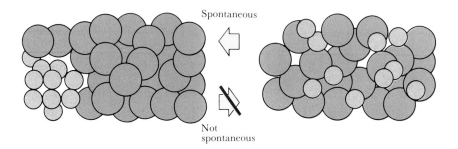

FIGURE 11.19 A universal tendency for energy and matter is to disperse in a disorderly way. Here we see energy (the red region) and matter (the blue) spreading out and becoming disordered. The reverse process is not spontaneous.

a problem with the idea of exothermic dissolution: because energy cannot be created or destroyed, the total energy of the system and its surroundings remains unchanged. Therefore, even for exothermic dissolving, although the energy of the solute-solvent system has gone down, that of its surroundings has gone up by the same amount. During endothermic dissolving the energy of the system goes up, but the energy of the surroundings decreases to the same extent. In each case there is no net change of energy. Why, then, does *either* process occur?

The answer lies in a very simple and natural idea: *energy and matter tend to disperse.* By "disperse" we mean spread out in a disorderly way (Fig. 11.19). Compared with a solid or a liquid, molecules are more dispersed if they are spread out as a gas or as the solute in a solution. Energy is more dispersed if it leaves a system and spreads through the surroundings as chaotic thermal motion (Fig. 11.20).

In this and the next five chapters, the idea that energy and matter tend to disperse will grow into a major principle of science and an explanation of many kinds of change. In particular, the idea accounts for the occurrence of *spontaneous changes*, changes that occur without us having to make them happen. Simple examples are the cooling of hot water to the temperature of its surroundings and the escape of a gas to fill the available space. We shall see that, as in these two examples, all spontaneous changes are accompanied by increased disorder. In contrast, changes that are not spontaneous must be driven if they are to take place at all. A beaker of water does not get hotter than its surroundings unless we immerse a heater in it or stir it vigorously. A gas does not collect into a smaller volume unless we compress it by mechanical means.

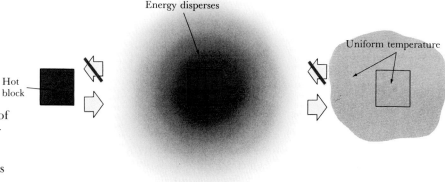

FIGURE 11.20 An example of a spontaneous and a nonspontaneous change. The unnatural change corresponds to the universe becoming less disorderly as energy collects in the block.

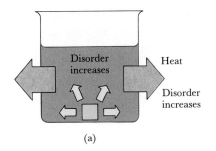

(a)

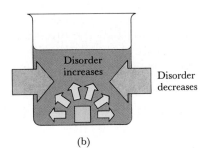

(b)

FIGURE 11.21 (a) When a solute dissolves exothermically, the system and the surroundings become more disordered, so the process is a natural one. (b) When a substance dissolves endothermically, energy must be concentrated into the system. Hence, there will be an overall increase in disorder and the process will be natural only if the disorder caused by the dispersing of the solute is great enough.

Here we consider a special case of the tendency to become disordered: the process of dissolving. If *overall* the solute-solvent system and its surroundings become more disordered when the solute is spread through the solvent, then dissolving is spontaneous and the substance is soluble. If *overall* the system and its surroundings are less disordered when the solute is spread through the solvent, then dissolving is not a spontaneous process and the substance is not soluble; on the contrary, if that substance is prepared in solution by mixing two reagents, it has a spontaneous tendency to precipitate out of solution.

When a substance dissolves, its ions or molecules become more disordered. When it dissolves exothermically, energy also spreads out into the surroundings as heat (Fig. 11.21a). Overall, the universe—the system and its surroundings—becomes more disorderly, so exothermic dissolving is a spontaneous process. However, it is spontaneous not because the *system* is losing energy, but because the universe is becoming more disorderly. The release of energy from the system simply adds to this overall increase in disorder.

When a substance dissolves endothermically, the ions or molecules also become more disorderly, and this again is a spontaneous process. However, because energy is drawn in, the surroundings become *less* disorderly (Fig. 11.21b). Whether or not the solid dissolves depends on the balance of these two effects. If the dissolving is only slightly endothermic (as for NaCl or even NH_4NO_3), the increased disorder of the solute has greater effect than the decreased disorder of the energy. The solute then has a tendency to dissolve because that corresponds to an overall increase in disorder. However, if the dissolving is very endothermic (as for MgO), the amount of energy that must collect in the solution is so great that the potentially greater disorder of the matter itself cannot compensate for it. Hence, a substance with a strongly endothermic enthalpy of solution does not have a tendency to dissolve because that would correspond to an overall increase in order.

We can now see why water is such a good solvent for ionic solids and for polar molecules, especially those able to form hydrogen bonds with H_2O molecules. Water molecules are so strongly polar that they hydrate cations and polar molecules very strongly. They also hydrate anions and hydrogen-bonding molecules strongly. As a result, the enthalpy of hydration is so great that it cancels, or nearly cancels, the lattice enthalpy. Hence, depending on the substance, either heat is released into the surroundings or only a small amount of heat must enter the solution from the surroundings. In either case, the total disorder of the system and its surroundings increases, and dissolving is a spontaneous process. Only if the lattice enthalpy is very large, or the enthalpy of

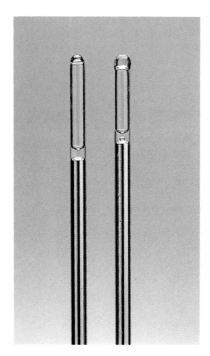

FIGURE 11.22 The vapor pressure is lowered by a nonvolatile solute. The barometer tube on the left has a small amount of pure water floating on the mercury. That on the right has a small amount of 0.1 m NaCl(aq) solution and has a lower vapor pressure.

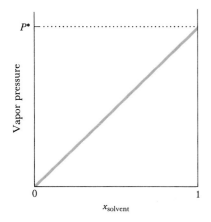

FIGURE 11.23 Raoult's law predicts that the vapor pressure of a solution is proportional to the mole fraction of the solvent molecules.

hydration very small, does so much energy need to be concentrated in the solution that the orderliness of the system and surroundings increases. In that case dissolving is not a spontaneous process, and the substance is insoluble.

COLLIGATIVE PROPERTIES

We now describe the effect of a solute on the physical properties of a solution. In particular, we consider four related changes. The first is the lowering of the vapor pressure of the solution below that of the pure solvent, which also gives rise to the second change: the raising of the normal boiling point of the solution above that of the pure solvent. Third, a solute lowers the freezing point of the solution below that of the pure solvent, and fourth it gives rise to an osmotic pressure (a property that is explained later). An important piece of experimental information is that all four effects depend only on the number of solute particles present in a solution, and not on their chemical composition. An aqueous glucose solution in which the glucose is present at a mole fraction of 0.01, with 1 glucose molecule per 100 molecules, has the same vapor pressure, boiling point, freezing point, and osmotic pressure as an aqueous sucrose solution in which the sucrose mole fraction is also 0.01. This is summarized by saying that all four properties are "colligative" properties:

A **colligative property** is a property that depends only on the number of solute particles present in a solution, and not on their chemical composition.

(*Colligative* means "depending on the collection.")

Since a colligative property depends only on the numbers of solute particles present, cations and anions in an electrolyte solution contribute separately to the total. Therefore, if 1 mol of NaCl is dissolved in a solvent, the solute consists of 2 mol of ions (1 mol of Na^+ ions and 1 mol of Cl^- ions). If 1 mol of $CaCl_2$ units is dissolved in a solvent, the total number of moles of solute is 3 mol of ions (1 mol of Ca^{2+} ions and 2 mol of Cl^- ions). On the other hand, in a nonelectrolyte solution, each solute molecule is present as a single unit. Therefore, when 1 mol of glucose is dissolved, the total amount of solute present is also 1 mol.

11.6 THE LOWERING OF VAPOR PRESSURE

A solution of a nonvolatile solute has a lower vapor pressure than the pure solvent (Fig. 11.22). For example, the vapor pressure of pure water at 40°C is 55 Torr, but that of a 0.1 M NaCl(aq) solution is only 44 Torr at the same temperature. Since a lower vapor pressure implies a higher boiling point, the presence of a nonvolatile solute raises the boiling point of the solution above that of the pure solvent.

Raoult's law. The French scientist François-Marie Raoult spent much of his life measuring vapor pressures. He found that the effect of a solute could be summarized as follows:

Raoult's law: The vapor pressure of a solution of a nonvolatile solute is proportional to the mole fraction of the *solvent* in the solution.

This is normally written

$$P = x_{\text{solvent}} \times P^*$$

where P^* is the vapor pressure of the pure solvent, x_{solvent} is the mole fraction of the solvent, and P is the vapor pressure of the solution (Fig. 11.23).

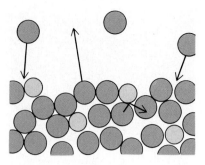

FIGURE 11.24 A nonvolatile solute particle (red) can block the escape of solvent particles, but has no effect on the return of the solvent from the vapor to the solution.

▼ **EXAMPLE 11.9   Using Raoult's law**

Calculate the vapor pressure at 100°C of a solution prepared by dissolving 5.00 g of sucrose in 100 g of water.

STRATEGY This becomes a straightforward application of Raoult's law once we know the mole fraction of the solvent (water) in the solution. (That mole fraction was calculated in Example 11.2.) We also need the vapor pressure of the pure solvent. We could get that from Table 5.8, but since the normal boiling point of water is 100°C, we know that its vapor pressure at that temperature is 760 Torr.

SOLUTION From Example 11.2, we have $x_{\text{H}_2\text{O}} = 0.997$. Therefore, since $P^* = 760$ Torr,

$$P = 0.997 \times 760 \text{ Torr} = 758 \text{ Torr}$$

EXERCISE Calculate the vapor pressure at 90°C of a solution prepared by dissolving 5.00 g of glucose in 100 g of water.

[*Answer*: 523 Torr]

The vapor pressure of a solvent is lowered by a nonvolatile solute because the solute blocks part of the surface and hence reduces the rate at which solvent molecules leave the solution (Fig. 11.24). However, the solute has no effect on the rate at which solvent molecules return, because a returning molecule can stick to any part of the surface (including solute particles). Since the rate of escape is reduced but the rate of return is unaffected, there is a net flow of molecules back to the solution. This reduces the pressure of the vapor to a new equilibrium value.

Modern experiments have shown that Raoult's law is reliable only at low concentrations, when there is a large separation between the solute molecules. However, many solutions behave approximately like an *ideal solution*, one that obeys Raoult's law at any concentration. Real solutions resemble ideal solutions more closely at lower concentrations; the agreement is quite good below about 10^{-1} M for nonelectrolyte solutions and 10^{-2} M for electrolyte solutions (in the latter, interactions between ions have a marked effect). We shall assume that all the solutions we meet are ideal.

The elevation of boiling point. Since the vapor pressure of a solution is lower than that of the pure solvent, the normal boiling point of a solution is higher than that of the pure solvent (Fig. 11.25). The increase, which is called the *elevation of boiling point*, is usually quite small. A 0.1 M aqueous sucrose solution, for instance, boils at 100.1°C. The ele-

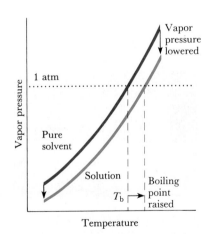

FIGURE 11.25 The lowering of the vapor pressure of a solution leads to an increase in its boiling point.

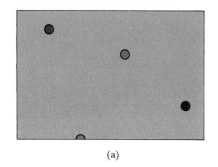

(a)

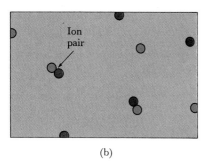

Ion pair

(b)

FIGURE 11.26 (a) In a very dilute solution, most solute ions move independently of each other and contribute separately to the molality. (b) As the ionic concentration increases, more ion pairs are formed, and the effective molality is different from the actual molality of ions. The drawing in b, for example, shows 10 actual ions but only 8 effective "particles."

TABLE 11.8   Boiling-point and freezing-point constants

Solvent	k_f, K/m	k_b, K/m	Solvent	k_f, K/m	k_b, K/m
Acetone	2.40	1.71	Cyclohexane	20.1	2.79
Benzene	5.12	2.53	Naphthalene	6.94	5.80
Camphor	39.7	5.61	Phenol	7.27	3.04
Carbon tetrachloride	29.8	4.95	Water	1.86	0.51

vation of boiling point is proportional to the molality m of the solution. For a molecular solute, the increase

$$\text{Elevation of boiling point} = k_b \times m$$

where k_b is the *boiling-point constant* of the solvent. Some values are given in Table 11.8.

When sodium chloride dissolves, each formula unit gives two ions. For very dilute solutions we may treat the cations and anions as contributing independently, so the total solute molality is twice the molality in terms of NaCl formula units. If the solution is not extremely dilute, however, ions of opposite charge do not move independently, but instead form ion pairs (Fig. 11.26). In this case, the effective molality of "particles"—individual ions or ion pairs—is different from the molality of NaCl treated as a collection of independent ions. The difference between the effective molality and the actual ionic molality is taken into account by introducing the "van't Hoff i factor" for electrolyte solutions:

$$\text{Elevation of boiling point} = i \times k_b \times m$$

In a very dilute solution, when all ions are independent, $i = 2$ for MX salts, $i = 3$ for MX_2 salts such as $CaCl_2$, and so on. The i factor is named for Johannes van't Hoff, a Dutch chemist who studied the properties of solutions and in 1901 was awarded the first Nobel Prize for chemistry.

Measurements of the boiling-point elevation caused by a known mass concentration of solute can be used to determine the molecular weights of solutes. However, the effect is so small and the results are so unreliable that the technique is rarely used.

The depression of freezing point.   A solute lowers the freezing point of a solution below that of the pure solvent. This effect is called the *depression of freezing point*. Seawater, an aqueous solution rich in Na^+ and Cl^- ions, freezes at a lower temperature than fresh water. In winter, salt is spread on highways in northern latitudes, since the salt lowers the freezing point of melted snow and prevents ice formation. Similarly, organic chemists can judge the purity of a compound by checking its melting point, which will be lower than normal if impurities are present.

This effect also accounts for the fact that mixtures freeze gradually, over a range of temperature, rather than solidifying at one precise

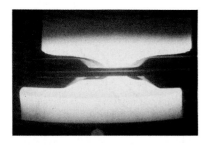

FIGURE 11.27   In the technique of zone refining, a molten zone is passed repeatedly from one end of a solid sample to the other. Impurities collect in the zone.

melting point. When a solution begins to freeze, only the solvent solidifies, and solute is left behind in the solution. As a result, the concentration of the remaining solution rises and its freezing point becomes even lower. The concentration of the solution continues to rise, and the freezing point continues to fall, until all the solvent has frozen. This phenomenon has been used to purify solvents. It is also the basis for a proposal to drag icebergs from polar regions as sources of fresh water, for the ice becomes largely salt-free as it forms. The technique called *zone refining*, which is used to prepare very pure samples of substances (such as silicon for use in semiconductors), also makes use of the effect. In zone refining, a cylindrical heater is moved several times from one end of a solid sample to the other, so that a molten zone is passed repeatedly through the solid (Fig. 11.27). Impurities dissolve in the molten zone as it passes, and the purer material freezes. The moving molten zone sweeps the impurities to one end of the sample, which can be discarded.

The effect of a solute on the freezing point of water is shown in Fig. 11.28: the solid-liquid boundary is shifted to lower temperatures as the solute concentration is increased. The figure also shows that the vapor pressure of the solution is lowered (as we have already seen).

The explanation of freezing-point depression lies in the effect of the solute on the rates at which solvent molecules form the solid and leave it to return to the liquid. At the freezing point of the pure solvent, these two rates are equal. When a solute is present, fewer solvent molecules from the liquid are in contact with the surface of the solid, because some of the places they occupied are now taken by solute particles (Fig. 11.29). As a result, they are slower to settle onto the surface and become part of the solid. However, the rate at which molecules leave the solid, which is pure solid solvent, is unchanged because a molecule can break away from the solid even though a solute particle is next to it in the solution. Therefore, when solute is added to the pure solvent there is a net flow of molecules away from the solid, and it melts. Only if the temperature is lowered will equilibrium be restored.

The depression of the freezing point of an ideal solution is proportional to the molality m. For a molecular solute,

$$\text{Depression of freezing point} = k_f \times m$$

where k_f is the *freezing-point constant* of the solvent (see Table 11.8). For an electrolyte solution we include the van't Hoff i factor on the right of the equation. The freezing-point constant is almost always larger than the boiling-point constant for the same solvent, implying that freezing-point depressions are larger than boiling-point elevations. A $0.1\ m$ aqueous sucrose solution, for instance, has

$$\text{Depression} = \frac{1.86\ \text{K}}{m} \times 0.1\ m = 0.2\ \text{K}$$

and hence freezes at $-0.2°C$.

The depression of freezing point can be used to determine the molecular weight of the solute. This is more reliable than using boiling-

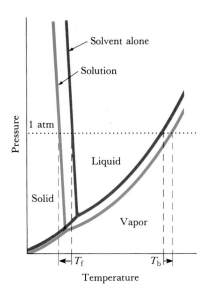

FIGURE 11.28 The effect of a solute on the freezing point of water. The effect on vapor pressure and boiling point is also shown.

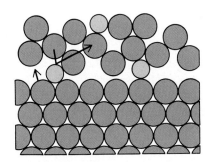

FIGURE 11.29 The rate at which solvent molecules leave the pure solid solvent is unaffected by the presence of solute particles nearby in the solution, but their rate of return to the solid is reduced.

point elevations because a freezing-point depression is generally larger and hence easier to measure. Moreover, heat-sensitive molecules are less likely to be damaged by freezing than by boiling, and the problems caused by decomposition products and evaporation are avoided. Since the organic compound camphor has a large freezing-point constant, solutes depress its freezing point significantly, and it is often used as a solvent in this kind of measurement.

▼ **EXAMPLE 11.10** **Using freezing-point depression to measure molar mass**

The addition of 0.24 g of sulfur to 100 g of carbon tetrachloride lowered its freezing point by 0.28 K. What is the molecular formula of solid sulfur?

STRATEGY Sulfur is a solid, unlike its diatomic neighbor chlorine, so we suspect that it is polyatomic (with a formula such as S_3 or S_4). The precise value of x in S_x can be found by measuring the molecular weight of sulfur and comparing it with the atomic weight. Since the freezing-point constant for carbon tetrachloride (in Table 11.8) can be used to convert a freezing-point depression to a molality, we can find the number of moles of sulfur molecules in the solution. We are given their total mass, so we can combine the two to obtain the molar mass. Then x can be found by dividing the molar mass of the molecules by the molar mass of sulfur atoms.

SOLUTION The molality of the solution is obtained by rearranging the freezing-point depression equation into

$$\text{Molality} = \frac{1}{k_f} \times \text{depression}$$

and inserting the data:

$$\text{Molality} = \frac{1}{29.8} \frac{m}{K} \times 0.28 \text{ K} = 0.0094 \, m$$

Therefore, the number of moles of S_x molecules in 100 g of solvent is

$$\text{Moles of } S_x = 0.100 \text{ kg} \times 0.0094 \frac{\text{mol } S_x}{1 \text{ kg}}$$

$$= 9.4 \times 10^{-4} \text{ mol } S_x$$

The total mass of sulfur present is 0.24 g. Therefore, the molar mass of sulfur is

$$\frac{0.24 \text{ g } S_x}{9.4 \times 10^{-4} \text{ mol } S_x} = 260 \text{ g/mol}$$

Since the molar mass of S atoms is 32.1 g/mol,

$$x = \frac{260 \text{ g/mol}}{32.1 \text{ g/mol}} = 8$$

Elemental sulfur is therefore composed of S_8 molecules.

EXERCISE The addition of 250 mg of eugenol, the compound responsible for the odor of oil of cloves, lowered the freezing point of 100 g of camphor by 0.62 K. Calculate the molar mass of eugenol molecules.
[*Answer*: 160 g/mol (actual: 164.2 g/mol)]

▲

11.7 OSMOSIS

Osmosis is the most important colligative property, both for life and in the laboratory. It is illustrated by the experiment shown in Fig. 11.30. A solution and a pure solvent are separated by a sheet of cellulose acetate (a material that is widely used as a transparent wrapper on candy boxes). Initially the heights of liquid inside and outside the tube are the same, but the solution height increases because pure solvent pushes through the membrane and into the solution, increasing the latter's volume. Equilibrium is reached when the pressure exerted by the additional height of solution can push solvent molecules back through the membrane at a matching rate.

The cellulose acetate membrane is *semipermeable*, because only certain types of molecules or ions can pass through it. Cellulose acetate, for example, allows water molecules to pass but not solute molecules or ions with their bulky coating of hydrating water molecules. The result, shown in Fig. 11.30, is called "osmosis":

> **Osmosis** is the passage of a solvent through a semipermeable membrane into a more concentrated solution.

(The name comes from the Greek word for "push.") The pressure needed to stop the flow of solvent is called the *osmotic pressure* Π (the Greek uppercase letter pi). The greater the osmotic pressure, the greater the height of the column of solution above that of the column of pure solvent.

Van't Hoff showed that the osmotic pressure is related to the molar concentration of the solution by

$$\Pi = i \times RT \times [\text{solute}]$$

where i is the van't Hoff i factor, R is the gas constant, T is the temperature in kelvins, and [solute] is the molar concentration of the solute in the solution. This is now known as the "van't Hoff equation." Since at 25°C, $RT = 24$ L · atm/mol (or 24 atm/M because 1 L/mol = 1/M), a handy form of the equation at this temperature is

$$\Pi = i \times [\text{solute}] \times 24 \text{ atm}/M \qquad \text{at } 25°C$$

The osmotic pressure of a 0.010 M solution of any nonelectrolyte is therefore 0.24 atm. This is enough to push a column of water to a height of over 2 m.

The explanation of osmosis can be found in the effect of the solute on the rates at which solvent molecules pass through the membrane from each side (Fig. 11.31). The rate is lower from the solution side because, although the same number of molecules jostle up to the membrane, only the solvent molecules can pass through. The two rates come into balance when the pressure has risen on the solution side so that more solvent molecules are pressed back through the membrane.

Osmometry. Like the other colligative properties, osmotic pressure can be used to determine molecular weights. The technique is called *osmometry*, and it involves measurement of the height reached by a column of solution of known mass concentration.

FIGURE 11.30 An experiment to illustrate osmosis. Initially the inverted tube contains sucrose solution, and the beaker contains pure water. At the stage shown here, water has passed into the solution by osmosis and its level has risen.

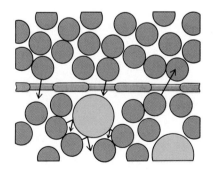

FIGURE 11.31 During osmosis, the presence of solute (green) on one side of the membrane hinders solvent molecules from passing through into the side containing pure solvent.

▼ EXAMPLE 11.11 Measuring molecular weight by osmometry

In an experiment to measure the molecular weight of polyethylene (PE, which consists of long —CH_2—CH_2— chains), 2.20 g of the plastic was dissolved in enough of the solvent toluene to produce 100 mL of solution. Its osmotic pressure at 25°C was measured as 1.10×10^{-2} atm (which corresponds to a 13-cm column of the toluene solution). Calculate the molecular weight of the polyethylene.

STRATEGY We know the mass of PE in the solution, so we can find its molecular weight if we know the number of moles of PE present. Since we know the volume of the solution, the latter can be found from the molar concentration [PE]. That in turn can be found from the osmotic pressure by using the van't Hoff equation.

SOLUTION From the van't Hoff equation with $i = 1$,

$$[PE] = \Pi \times \frac{1}{RT}$$

$$= 1.10 \times 10^{-2} \text{ atm} \times \frac{1}{24} \frac{M}{\text{atm}}$$

$$= 4.6 \times 10^{-4} \, M$$

The number of moles of PE in 100 mL of the solution is therefore

$$\text{Moles of PE} = 0.100 \text{ L} \times (4.6 \times 10^{-4} \text{ mol PE/L})$$

$$= 4.6 \times 10^{-5} \text{ mol PE}$$

Since the total mass of PE in the solution is 2.20 g,

$$\text{Molar mass of PE} = \frac{2.20 \text{ g PE}}{4.6 \times 10^{-5} \text{ mol PE}} = 4.8 \times 10^4 \text{ g/mol}$$

This molar mass corresponds to a molecular weight of 48,000 amu, and hence to a chain of 1700 —CH_2—CH_2— units.

EXERCISE The osmotic pressure of 3.0 g of polystyrene in enough benzene to produce 150 mL of solution was measured as 1.21 kPa at 25°C. Calculate the molar mass of polystyrene.

[*Answer*: 40 kg/mol]

The great advantage of osmometry over the other colligative methods is its very high sensitivity. Whereas a 0.01 M aqueous sucrose solution shows a boiling-point elevation of 0.005 K and a freezing-point depression of 0.02 K, it produces an osmotic pressure equivalent to a column of water 2 m high. The two small temperature changes are difficult to measure, but the large difference in height can be measured easily and accurately. This sensitivity is important when the solute has a very high molecular weight, as in the case of enzymes, proteins, and synthetic plastics. While the mass of these solutes in the solution may be appreciable, their molecular weights are so high (of the order of 10,000 amu and more) that the actual *number* of molecules present may be very small and the colligative properties correspondingly minute. The concentration of a solution of 0.10 g of hemoglobin (which has molecular weight 66,500 amu) in 100 mL of solution is only $1.5 \times 10^{-5} \, M$. It has an unmeasurably small freezing-point depression but an osmotic pressure equivalent to 4 mm of water, which is measurable.

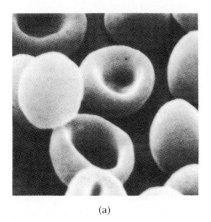

(a)

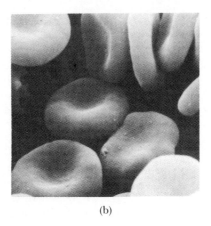

(b)

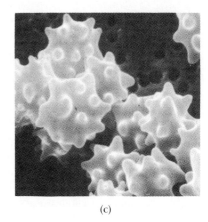

(c)

Some applications of osmosis. Biological cell walls act as semipermeable membranes that allow water, small molecules, and hydrated ions to pass while blocking the passage of the enzymes and proteins that have been synthesized within the cell. The higher concentration of solute inside a plant cell gives rise to osmosis, so water passes into the cell carrying nutrient molecules with it. This influx of water also keeps the cell turgid, or swollen. When the water supply is cut off, the turgidity is lost and the plant wilts. Salted meat is preserved from bacterial attack by osmosis. In this case the concentrated salt solution dehydrates bacteria by causing water to flow out of them.

Osmosis maintains the turgidity of the red corpuscles in blood (Fig. 11.32). The walls of these red blood cells are impermeable to sodium ions, so the presence of these ions on one side or the other affects the direction of osmotic flow. If the concentration of Na^+ ions in the surrounding blood plasma is too low, water flows into the corpuscles, and if the flow continues they burst. The walls of the blood capillaries are also semipermeable membranes. Since they are permeable to everything except big protein molecules, the latter's presence in blood plasma governs the direction of the flow of water through the capillary walls. If the concentration of protein in the blood plasma is reduced, as in cases of extreme hunger, water flows out of the plasma. This can be corrected, and a life saved, by intravenous injections of protein-enriched blood plasma.

Reverse osmosis. A modification of osmosis called *reverse osmosis* is used to remove salts from seawater so as to produce fresh water for drinking and irrigation. In reverse osmosis, a pressure greater than the osmotic pressure is applied to the solution side of the semipermeable membrane. This increases the rate at which water molecules leave the solution. The water is almost literally squeezed out of the salt solution through the membrane. The technological challenge is to make membranes that are strong enough to withstand the high pressures needed, yet permeable enough to permit a good flow of water without becoming clogged (Fig. 11.33). Commercial plants use cellulose acetate as the membrane, and pressures of up to 70 atm. The membrane is packed into containers, and each cubic meter of membrane can produce about 250,000 L of pure water a day.

FIGURE 11.32 (a) Red blood corpuscles need to be in solutions of the correct strength if they are to function properly. (b) If the solution is too dilute, water passes into them and they burst. (c) If it is too concentrated, water flows out of them and they shrivel up.

FIGURE 11.33 Water dripping out of a filter in the process of reverse osmosis.

SUMMARY

11.1 The measures of concentration that emphasize the number of solute molecules in a given volume of solution are **molar concentration** (molarity, or moles per liter of solution) and **mass concentration** (grams of solute per liter of solution).

11.2 The measures of concentration that emphasize the relative numbers of solute and solvent particles are **parts per million, mole fraction** (the number of moles of solute molecules as a fraction of the total), and **molality** (moles of solute per kilogram of solvent).

11.3 A **saturated solution** is one in which the dissolved and nondissolved solute are in dynamic equilibrium. This concept is used to define the **solubility** of a substance, which is its concentration in a saturated solution. Solubility depends on the identity of the solute and the solvent ("like dissolves like"), on the pressure, and on the temperature. Soaps and detergents contain ions and molecules with **hydrophobic** and **hydrophilic** parts that are attracted to oil and water, respectively.

11.4 The solubility of a gas increases with pressure, as described by **Henry's law** that the solubility is proportional to the partial pressure. This law is an example of **Le Chatelier's principle,** which states that a system at equilibrium tends to oppose a change in the conditions. Le Chatelier's principle applies to all dynamic equilibria.

11.5 The energy aspects of dissolving are discussed in terms of the **enthalpy of solution,** which is equal to the sum of the lattice enthalpy and the enthalpy of hydra-

tion for the solute. An enthalpy of hydration can be expressed as the sum of **ion hydration enthalpies** for each type of ion, based on the value for $H^+(aq)$ and estimated from the **Born equation.** The enthalpy of solution can be used to predict the effect of temperature on the solubility of a sparingly soluble compound. The direction of spontaneous change is the one that leads to an increased overall **dispersal** of matter and energy. Substances with strongly endothermic enthalpies of solution are unlikely to be soluble because, if they were to dissolve, too much energy would have to collect in the solution, decreasing the disorder of the universe.

11.6 Some of the physical properties of solutions are **colligative;** that is, they depend on the number of particles present and not their chemical identity. These include the **lowering of vapor pressure, elevation of boiling point,** and **depression of freezing point.** The lowering of vapor pressure is described by **Raoult's law,** which implies that the vapor pressure is proportional to the mole fraction of the solvent. Measurements of boiling-point elevation and freezing-point depression can be used to determine molecular weights.

11.7 Osmosis, the fourth colligative property, is the passage of a solvent through a **semipermeable membrane,** one that permits the passage of some types of particle but not all. **Osmometry,** the measurement of osmotic pressure, is used to determine molecular weights, especially those of high-molecular-weight compounds.

EXERCISES

Measures of Concentration

In Exercises 11.1 and 11.2, calculate the molar concentration of each solution.

11.1 (a) 5.0 g of sodium chloride in 250 mL of aqueous solution
(b) 10.0 g of glucose ($C_6H_{12}O_6$) in 250 mL of aqueous solution
11.2 (a) 5.0 g of anhydrous calcium chloride in 100 mL of aqueous solution
(b) 15 g of sucrose ($C_{12}H_{22}O_{11}$) in 500 mL of aqueous solution

The molar concentration of a pure solvent is the number of moles of solvent molecules present per liter of pure

solvent. In Exercises 11.3 and 11.4, calculate the molar concentration of the solvent under the given conditions.

11.3 Pure water (a) at 20°C, when its density is 0.998 g/mL; (b) at 100°C, when its density is 0.958 g/mL.
11.4 Pure ethanol (a) at 20°C, when its density is 0.789 g/mL; (b) at 30°C, when its density is 0.781°C.

In Exercises 11.5 and 11.6, calculate the mass of anhydrous solute needed to make 250 mL of each solution.

11.5 (a) 0.10 M NaCl(aq); (b) 0.10 M CaCl$_2$(aq); (c) 1.00 M C$_6$H$_{12}$O$_6$(aq).
11.6 (a) 0.10 M NaOH(aq), (b) 0.001 M Na$_2$SO$_4$(aq), (c) 0.001 M C$_{12}$H$_{22}$O$_{11}$(aq).

In Exercises 11.7 and 11.8, calculate the molality of each solution.

11.7 (a) 10 g of NaCl dissolved in 250 g of water; (b) 10 g of sucrose dissolved in 250 g of water; (c) 1.0 g of octane, C_8H_{18}, in 300 g of benzene; (d) 50 g of benzene in 50 g of toluene.

11.8 (a) 5.0 g of anhydrous calcium chloride dissolved in 50 g of water; (b) 5.0 g of $CaCl_2 \cdot 6H_2O$ dissolved in 50 g of water (calculate the molality of $CaCl_2$); (c) 50 g of water in 50 g of ethanol; (d) 50 g of ethanol in 50 g of water.

In Exercises 11.9 and 11.10, calculate the mass of anhydrous solute needed to prepare each solution using 250 g of solvent.

11.9 (a) 0.010 m NaOH(aq); (b) 0.10 m HCl(aq); (c) 1.0 m $CaCl_2$(aq); (d) 0.50 m $KMnO_4$(aq).

11.10 (a) 1.0 m H_2SO_4(aq); (b) 0.010 m $C_6H_{12}O_6$(aq); (c) 0.10 m $K_2Cr_2O_7$(aq); (d) 0.10 m Na_2CO_3(aq).

In Exercises 11.11 and 11.12, calculate the mass of hydrated solute needed to prepare the solution using 250 g of solvent.

11.11 Use $CaCl_2 \cdot 6H_2O$ to form a 1.0 m $CaCl_2$(aq) solution.

11.12 Use $Na_2CO_3 \cdot 10H_2O$, to form a 1.0 m Na_2CO_3(aq) solution. Then find a general expression for the mass of solute needed to prepare a solution of molality m_B when the hydrated solute has the formula $B \cdot xH_2O$.

In Exercises 11.13 and 11.14, calculate the mole fractions of the solutes and solvents in each solution.

11.13 (a) 50 g of water mixed with 50 g of ethanol; (b) a 0.10 m $C_6H_{12}O_6$(aq) solution.

11.14 (a) 1.0 g of benzene in 99 g of toluene, $CH_3C_6H_5$; (b) a 0.10 m $C_{12}H_{22}O_{11}$(aq) solution.

In Exercises 11.15 and 11.16, calculate the mole fractions of the cations, anions, and solvent molecules in each solution.

11.15 (a) A 0.10 m NaCl(aq) solution; (b) a 0.10 m Na_2CO_3(aq) solution.

11.16 (a) A 0.10 m $MgSO_4$(aq) solution; (b) a 0.10 m $Al_2(SO_4)_3$(aq) solution.

In Exercises 11.17 and 11.18, calculate the molality of each solution with the mass percentage composition as specified.

11.17 (a) A 5.0% NaCl(aq) solution; (b) a 10% aqueous sucrose solution.

11.18 (a) A 10.0% HCl(aq) solution; (b) a 50% aqueous ethanol solution (a "100 proof spirit").

Solubility

Solubilities are often expressed as the mass of solute that dissolves in 100 g of solvent. In Exercises 11.19 and 11.20, express the given solubilities in water in moles per liter.

11.19 (a) KCl: 28.1 g/100 g, density 1.15 g/mL; (b) NaCl: 35.7 g/100 g, density 1.20 g/mL; (c) AgCl: 7×10^{-5} g/100 g, density 1.00 g/mL; (d) NH_3: 28.1 g/100 g, density 0.917 g/mL.

11.20 (a) Sucrose ($C_{12}H_{22}O_{11}$): 200 g/100 g, density 1.33 g/mL; (b) $BaCl_2$: 30.5 g/100 g, density 1.24 g/mL; (c) MgF_2: 8 mg/100 g, density 1.00 g/mL; (d) $MgCl_2$: 53 g/100 g, density 1.27 g/mL.

Henry's Law

In Exercises 11.21 and 11.22, calculate the solubility in moles per liter M and milligrams per liter of each gas in water at 20°C and under the given pressure.

11.21 (a) O_2 at 1.0 atm; (b) O_2 at 0.21 atm; (c) CO_2 at 1.0 atm; (d) CO_2 at 0.1 atm.

11.22 (a) N_2 at 1.0 atm; (b) N_2 at 0.78 atm; (c) He at 1.0 atm; (d) He at 25 kPa.

11.23 The minimum mass concentration of oxygen required for the continuation of fish life is 4 mg/L.

(a) What is the minimum partial pressure of oxygen that would supply this concentration in water at 20°C?

(b) What is the minimum total atmospheric pressure that would give this partial pressure, assuming the normal atmospheric mass percentage composition?

11.24 A soft drink is made by dissolving carbon dioxide at 3 atm pressure in a waterlike liquid. What volume of carbon dioxide is released when a 250-mL can is opened? (Assume a temperature of 20°C throughout.)

Enthalpy of Solution

In Exercises 11.25 and 11.26, calculate the change in temperature that would be produced by dissolving 10.0 g of each compound in 100.0 mL of water, assuming that the enthalpies of solution given in Table 11.5 may be used in each case.

11.25 (a) NaCl; (b) NaBr; (c) $AlCl_3$; (d) NH_4NO_3.

11.26 (a) KCl; (b) $MgBr_2$; (c) KNO_3; (d) NaOH.

In Exercises 11.27 and 11.28, use the information in Table 11.5 to decide whether the solubility of each slightly soluble salt will increase (+) or decrease (−) as the temperature is raised.

11.27 (a) AgCl; (b) Li_2CO_3.

11.28 (a) $Ca(OH)_2$; (b) $BaSO_4$ ($\Delta H_{sol} = +19$ kJ/mol). (c) Why are the values in the table applicable only to slightly soluble salts?

11.29 The values in Table 11.7 depend on the accuracy of the hydration enthalpy calculated for the hydrogen ion. What would the value for (a) Cl^- and (b) Na^+ become if the value for the hydrogen ion were changed from -1130 kJ/mol to -1030 kJ/mol?

11.30 Suppose that the enthalpy of hydration of the fluoride ion had been calculated as -600 kJ/mol, and the values in Table 11.7 were based on that value. What would then be the values for (a) H^+ and (b) Cl^-?

In Exercises 11.31 to 11.34, estimate the enthalpy of solution of each compound from its lattice enthalpy and enthalpy of hydration, and suggest reasons for the trends in the estimated values.

11.31 (a) NaCl; (b) KCl.

11.32 (a) NaF; (b) NaCl; (c) NaBr; (d) NaI.

11.33 (a) MgF_2 lattice enthalpy $+2961$ kJ/mol; (b) $MgCl_2$.

11.34 (a) $CaCl_2$; (b) $SrCl_2$.

11.35 (a) Calculate the hydration enthalpy of Br^- from the enthalpy of solution of hydrogen bromide gas, -85 kJ/mol. Additional data is in Appendix 2.

(b) Use the value found in (a) to deduce the hydration enthalpy of Rb^+ from the enthalpy of solution of rubidium bromide, $+22$ kJ/mol, and its lattice enthalpy, 665.6 kJ/mol.

11.36 Calculate the hydration enthalpy of I^- from the enthalpy of solution of hydrogen iodide gas, -82 kJ/mol.

In Exercises 11.37 to 11.40, use the Born equation and the radii in Table 7.5 to estimate the hydration enthalpy of each ion.

11.37 (a) Na^+; (b) K^+; (c) Mg^{2+}; (d) Al^{3+}.

11.38 (a) Rb^+; (b) Cs^+; (c) Ca^{2+}; (d) Tl^{3+}.

11.39 (a) F^-; (b) Cl^-.

11.40 (a) Br^-; (b) I^-.

In Exercises 11.41 and 11.42, estimate the effective radius of each ion from its hydration enthalpy (in kJ/mol).

11.41 (a) NO_3^- (-389); (b) SO_4^{2-} (-1081).

11.42 (a) CO_3^{2-} (-1570); (b) ClO_4^- (-290).

Lowering of Vapor Pressure

In Exercises 11.43 to 11.48, calculate the vapor pressure (in Torr) of each aqueous solution (assumed to be ideal) at the stated temperature, taking the vapor pressure of pure water from Table 5.7.

11.43 (a) A solution at 100°C in which sucrose is present at a mole fraction 0.10; (b) the same solution with NaCl in place of sucrose.

11.44 (a) A solution at 80°C in which glucose is present at a mole fraction 0.050; (b) the same solution with $CaCl_2$ in place of glucose.

11.45 (a) A 1.0% by mass solution of urea, $CO(NH_2)_2$, at 40°C

(b) A saturated, aqueous solution of barium acetate at 25°C that has a molality of 2.5 m.

11.46 (a) A 2.0% by mass solution of the sugar fructose, $C_6H_{12}O_6$, at 50°C.

(b) A saturated solution of magnesium fluoride at 25°C, of concentration 1.2×10^{-3} M.

11.47 An aqueous solution at 80°C of (a) 1.0 m NaOH; (b) 1.0 m $MgCl_2$; (c) 1.0 m $C_6H_{12}O_6$.

11.48 A solution at 30°C of (a) 0.02 m KCl(aq); (b) 0.01 m $Fe(NO_3)_3$; (c) 0.01 m $C_{12}H_{22}O_{11}$.

In Exercises 11.49 and 11.50, calculate the molecular weight of the solute from the data.

11.49 The addition of 8.05 g of a compound to 100 g of benzene reduced the vapor pressure at 26°C from 100.0 Torr to 94.8 Torr.

11.50 The addition of 9.15 g of a compound to 100 g of ethanol reduced the vapor pressure at 78.4°C from 760 Torr to 740 Torr.

11.51 Find a relation between the percentage reduction of vapor pressure by a solute and the mole fraction of the solute in an ideal solution. Calculate its value for a solution of 50.0 g of sucrose in 250 g of water.

11.52 Find an approximate expression for the fractional reduction of vapor pressure by a solute in terms of the molality of an ideal solution and the molecular weight of the solvent. What fractional reduction should be observed for a 0.1 m aqueous solution of any nonvolatile, nonionic solute?

Boiling-Point Elevation

In Exercises 11.53 to 11.58, estimate the boiling-point elevation of each dilute solution, assuming it to be ideal.

11.53 (a) 0.10 m $C_6H_{12}O_6(aq)$; (b) 0.01 m NaCl(aq).

11.54 (a) 0.15 m $C_{12}H_{22}O_{11}(aq)$; (b) 0.001 m $CaCl_2(aq)$.

11.55 A saturated solution of lithium fluoride that has a solubility of 230 mg/100 g of water at 100°C.

11.56 A saturated solution of lithium carbonate, which has a solubility of 0.72 g/100 g of water at 100°C.

11.57 (a) An aqueous solution that has a vapor pressure of 751 Torr at 100°C

(b) An aqueous solution that freezes at -1.04°C

11.58 (a) A solution in benzene that has a vapor pressure of 740 Torr at 80.1°C, the normal boiling point of the pure solvent

(b) A solution in benzene that freezes at 2.0°C instead of 5.5°C, the normal melting point of the pure solvent

In Exercises 11.59 and 11.60, calculate the molecular weight of the solute from the given boiling-point data.

11.59 1.05 g of the substance dissolved in 100 g of carbon tetrachloride caused the boiling point to rise by 0.309 K.

11.60 2.20 g of the substance dissolved in 150 g of cyclohexane caused the boiling point to rise by 0.481 K.

11.61 Estimate the boiling point of seawater, given that it is approximately 0.5 M NaCl(aq). Assume that its density is 1 g/mL.

11.62 What mass of naphthalene should be added to 100 g of benzene to raise the latter's boiling point by 1%?

Freezing-Point Depression

In Exercises 11.63 to 11.66, estimate the freezing-point depression of the dilute solution, assuming it is ideal.

11.63 (a) 0.10 m $C_6H_{12}O_6$(aq); (b) 0.01 m NaCl(aq).

11.64 (a) 0.15 m $C_{12}H_{22}O_{11}$(aq); (b) 0.001 m CaCl$_2$(aq).

11.65 (a) A saturated solution of lithium fluoride that has a solubility of 120 mg/100 g of water at about 0°C

(b) An aqueous solution that has a vapor pressure of 751 Torr at 100°C

(c) An aqueous solution that boils at 101°C

11.66 (a) A saturated solution of lithium carbonate, of solubility 1.54 g/100 g of water at about 0°C

(b) A solution in benzene that has a vapor pressure of 740 Torr at 80.1°C, the normal boiling point of the pure solvent

(c) A solution in benzene that boils at 82.0°C instead of 80.1°C, the normal boiling point of the pure solvent

In Exercises 11.67 and 11.68, calculate the molecular weight of the solute from the data.

11.67 1.14 g of the substance in 100.0 g of camphor lowered the freezing point by 2.481 K.

11.68 2.11 g of the substance in 50.0 g of phenol lowered the freezing point by 1.753 K.

11.69 Estimate the freezing point of seawater, which may be taken to be 0.5 M NaCl(aq). Assume that its density is 1 g/mL.

11.70 What mass of naphthalene should be added to 100 g of benzene to lower the latter's melting point by 1°C?

Osmosis and Osmometry

In Exercises 11.71 to 11.74, estimate the osmotic pressure, in kilopascals and atmospheres, of each solution (at 20°C unless otherwise specified).

11.71 (a) 0.10 M $C_6H_{12}O_6$(aq); (b) 0.01 M NaCl(aq).

11.72 (a) 0.15 M $C_{12}H_{22}O_{11}$(aq); (b) 0.001 M CaCl$_2$(aq).

11.73 (a) A saturated solution of silver chloride, of solubility 0.07 mg/100 g of water at about 0°C

(b) An aqueous solution at 20°C that had a vapor pressure of 751 Torr at 100°C

(c) An aqueous solution at 25°C that boiled at 101°C

11.74 (a) A saturated solution of lithium carbonate, of solubility 1.54 g/100 g of water at about 0°C

(b) A solution in benzene at 25°C that had a vapor pressure of 740 Torr at 80.1°C, the normal boiling point of the pure solvent

(c) A solution in benzene at 20°C that boiled at 82.0°C instead of 80.1°C, the normal boiling point of the pure solvent

In Exercises 11.75 to 11.78, calculate the height to which osmosis will force each solution to rise in an arrangement like that shown in Fig. 11.30. In each case, take the density of the solution to be that of pure water at 20°C (0.998 g/mL); the density of Hg is 13.6 g/mL).

11.75 (a) 0.050 M $C_6H_{12}O_6$(aq); (b) 1.0 mM NaCl(aq).

11.76 (a) 3.0 mM $C_{12}H_{22}O_{11}$(aq); (b) 2.0 mM CaCl$_2$(aq).

11.77 A saturated solution of silver chloride, of solubility 0.07 mg/100 g of water at about 0°C

11.78 A saturated solution of lithium carbonate, of solubility 1.54 g/100 g of water at about 0°C.

In Exercises 11.79 to 11.82, calculate the height to which osmosis will force the solution to rise in an arrangement like that of Fig. 11.30 at 20°C, using the density of the solvent, toluene (0.867 g/mL), as that of the solution.

11.79 1.0 g of a polymer of molar mass 50 kg/mol in 150 g of toluene.

11.80 0.50 g of a polymer of molar mass 68 kg/mol in 200 mL of toluene.

*#**11.81** 150 mL of toluene containing 1.0 g of a polymer consisting of a mixture of molecules with the following molar masses and with mole fractions in the original solid shown in parentheses: 40 kg/mol (0.100), 45 kg/mol (0.300), 50 kg/mol (0.500), and 55 kg/mol (0.100)

*#**11.82** 50 mL of toluene containing 0.10 g of a polymer consisting of a mixture of molecules with the following molar masses and with mole fractions in the original solid shown in parenthesis: 20 kg/mol (0.050), 30 kg/mol (0.500), 40 kg/mol (0.050), 50 kg/mol (0.300), and 60 kg/mol (0.100)

In Exercises 11.83 and 11.84, calculate the molecular weight of the solute from the information given.

11.83 A solution of toluene (density 0.867 g/mL) containing 0.10 g of a polymer in 100 mL of solvent showed an 8.40-cm rise in an osmometer of the kind shown in Fig. 11.30 at 20°C.

11.84 A sample prepared by dissolving 0.010 g of a protein in 10 mL of water at 20°C showed a 5.22-cm rise in the apparatus of Fig. 11.30.

General

11.85 In terms of the tendency toward disorder, suggest why all gases should be less soluble at high temperatures than at low.

11.86 Deduce Raoult's law by assuming that the rate at which gas molecules return to a solution is proportional to the vapor pressure, and the rate at which they leave the solution is proportional to their mole fraction. Does the same kind of argument lead to Henry's law too?

11.87 Find an expression for the osmotic pressure of a solution in which the solute has mass m and is of general composition (that is, it is a mixture of several components J, of different molecular weights M_J and different mole fractions x_J).

III

RATES AND EQUILIBRIUM

One important feature of a chemical reaction is the rate at which it takes place. In this chapter we see how to define reaction rate and express it quantitatively. In doing so, we also see how to classify reactions according to their rates and how to use experimental information to predict the composition of a reaction mixture at any time after the reaction begins. We also see why most reactions go faster when they are heated or when catalysts are added to the mixture. The illustration shows the pattern produced by a complicated reaction. It can be understood in terms of the concepts we meet in this chapter, and people have speculated that a tiger's stripes arise in a similar way.

12

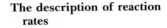

CHEMICAL KINETICS

The description of reaction rates

12.1 Reaction rates
12.2 Rate laws and reaction order

Reaction mechanisms

12.3 Elementary reactions
12.4 Chain reactions

Controlling rates of reactions

12.5 Temperature dependence of reaction rates
12.6 Catalysis

The study of the rates at which chemical reactions occur is called *chemical kinetics*. The subject is important in chemical industries because the designers of plants, processes, and equipment must know how the rates of the reactions are affected by changes of temperature, pressure, and concentration. It is important in biology and medicine, because health represents a balance between large numbers of reactions going on at once and responding to each other. Illness is often a sign that the rates of biologically important reactions have changed too much. Chemical kinetics is also important in the study of reactions in the atmosphere. Among other things, it deals with the reactions that maintain and destroy the ozone layer, and with the reactions that lead to the formation and removal of pollutants.

Chemical kinetics also provides a link between the topics of this part of the book and those treated in Part II. We saw in Part II that some of the physical properties of solutions, such as vapor pressure and osmosis, can be explained in terms of dynamic equilibria in which the rates of forward and reverse *physical* processes are balanced. In this chapter we see how to describe the rates of forward and reverse *chemical* processes. Once we have done that, we shall extend the discussion of dynamic equilibria to include the enormously important subject of chemical equilibrium, the main subject of Part III.

THE DESCRIPTION OF REACTION RATES

In this section we define the rate of a reaction and see how it is measured. We shall see that reaction rates show patterns that enable us to classify reactions into different types with characteristic features.

12.1 REACTION RATES

Rates in chemistry are defined like rates in other fields. A rate is the change in a property per unit time; it is calculated by dividing the total change by the time it takes for that change to occur. An everyday example is *speed*, the rate of change of position, which is distance traveled divided by the time taken. Likewise, the rate of a chemical reaction is the change in concentration of a substance divided by the time it takes for that change to occur. The concentration of TNT changes very rapidly during an explosion, so the rate of the reaction is very high. In a fermentation, the concentration of alcohol rises slowly, so the rate of that reaction is low.

The definition of rate. Although we shall soon see that we must refine this definition, initially we can write

$$\text{Rate} = \frac{\text{change in concentration}}{\text{time for the change to occur}}$$

If we let $\Delta[X]$ represent the change in molar concentration of a substance X, and Δt the time it takes for the change to occur, then we have

$$\text{Rate} = \frac{\Delta[X]}{\Delta t}$$

Suppose, for example, that we were studying the reaction

$$2HI(g) \longrightarrow H_2(g) + I_2(g)$$

and found that during an interval of 100 s the concentration of HI decreased by 0.50 M. Then the reaction rate would be

$$\text{Rate of decomposition of HI} = \frac{0.50 \text{ mol/L HI}}{100 \text{ s}}$$

$$= 5.0 \times 10^{-3} \text{ mol/(L} \cdot \text{s) HI}$$

As in this case, the unit of reaction rate is mol/(L $\cdot$ s), which is read as "moles per liter per second." More convenient values are sometimes obtained by using smaller units of molarity (millimoles per liter) or larger units of time (minutes or hours).

It is important to specify the substance for which the reaction rate is defined: in the above example, the rate of decomposition of HI is not the same as the rate of formation of H_2 or I_2, because only one H_2 molecule is formed for each pair of HI molecules that react. However, it is quite easy to use stoichiometric relations to convert a rate expressed in terms of one substance to a rate expressed in terms of another, as the following example shows.

▼ **EXAMPLE 12.1** **Expressing the reaction rate for different substances**

If the rate of decomposition of HI is 5.0 mmol/(L $\cdot$ s), what is the rate of formation of hydrogen in the same reaction?

STRATEGY We should expect a lower rate because only one molecule of H_2 is formed for every two HI molecules that decompose. Since we know the rate of change of the molar concentration of HI, we can calculate the rate of change of the concentration of H_2, using 2 mol HI = 1 mol H_2 to write a conversion factor.

SOLUTION Since 5.0 mmol/(L $\cdot$ s) HI corresponds to

$$5.0 \times 10^{-3} \frac{\text{mol HI}}{\text{L} \cdot \text{s}} \times \frac{1 \text{ mol H}_2}{2 \text{ mol HI}} = 2.5 \times 10^{-3} \text{ mol/(L} \cdot \text{s) H}_2$$

the H_2 formation rate is 2.5 mmol/(L $\cdot$ s), half the rate of decomposition of HI.

EXERCISE The rate of formation of ammonia from nitrogen and hydrogen is reported as 1.15 mol/(L $\cdot$ h) NH_3. What is the rate at which hydrogen is used?

[*Answer*: 1.72 mol/(L $\cdot$ h)]

▲

Instantaneous reaction rate. As the reactants are used up, the rates of most reactions change. We cannot speak of *the* rate of a reaction any more than we can speak of *the* speed of an automobile over an entire trip. To see how to take this changing rate into account, we consider the gas-phase decomposition of the volatile solid dinitrogen pentoxide, N_2O_5 (Fig. 12.1):

$$2N_2O_5(g) \longrightarrow 4NO_2(g) + O_2(g)$$

FIGURE 12.1 The darkening of color with time shows the formation of NO_2 as N_2O_5 decomposes at 65°C. The molar concentration of N_2O_5 at any stage can be calculated from the initial concentration, the measured concentration of NO_2, and the chemical relations between them.

The change in the concentration of N_2O_5 with time for a sample kept at 65°C is shown in Fig. 12.2. At any instant of time,

Rate of decomposition of N_2O_5

$$= \frac{\text{decrease in concentration of } N_2O_5}{\text{time for the change to occur}}$$

Suppose we wanted to find the rate at exactly 1000 s after the start of the reaction. By monitoring the pressure of the gas in the reaction vessel, we can measure the concentration of N_2O_5 at any instant. But how do we measure the *change* in the concentration at a specified instant?

One approach is to select an arbitrary interval—say 800 s—centered on the time of interest, and calculate the rate based on the decrease in concentration during that interval. This is represented by the red line in Fig. 12.2. We measure the N_2O_5 concentrations at 400 s before and at 400 s after the time of interest and calculate the difference to find the decrease in concentration over the 800-s interval. The reaction rate at the center of the interval (at 1000 s after the start of the reaction) is then *approximately*

$$\text{Rate} = \frac{1.69 \times 10^{-2} \text{ mol/L } N_2O_5}{800 \text{ s}} = 2.11 \times 10^{-5} \text{ mol/(L} \cdot \text{s) } N_2O_5$$

This result is only an approximation to the rate at 1000 s because the rate changes during that 800-s interval. We can make a better estimate by shortening the interval to 400 s, and basing the calculation on the change in concentration during that shorter interval (green in Fig. 12.2):

$$\text{Rate} = \frac{0.83 \times 10^{-2} \text{ mol/L } N_2O_5}{400 \text{ s}} = 2.08 \times 10^{-5} \text{ mol/(L} \cdot \text{s) } N_2O_5$$

This is still an approximation to the true rate at 1000 s, but it is a closer approximation.

An accurate value for the decomposition rate at 1000 s is obtained by reducing the interval centered on 1000 s until it is as short as possible. As explained in Appendix 1D, this is equivalent to drawing the *tangent* to the concentration curve at the time of interest and calculating its slope. The rate calculated in this way is called the *instantaneous rate* of the reaction at the specified time. In this case, the instantaneous rate at 1000 s is found to be 2.06×10^{-5} mol/(L · s) N_2O_5. In general, the instantaneous rate for any reaction at any time is obtained as the slope of the tangent to the concentration curve at that time. From now on, when we speak of a reaction rate, we shall always mean an *instantaneous* rate at a specified time.

Initial rates. Once a product has been formed in a reaction, it too may react—perhaps with the original reactants. Since this can make the analysis of reaction rates quite difficult, an important technique in chemical kinetics is to measure the reaction's *initial rate*, the rate at the start of the reaction when no products are present. Initial rates are measured just like other instantaneous rates, but with the tangent to the concentration curve drawn at the very start of the reaction (at time $t = 0$).

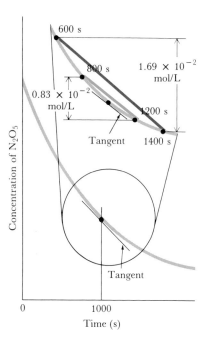

FIGURE 12.2 The time dependence of the concentration of N_2O_5 at 65°C as it decomposes into NO_2 and O_2. The reaction rate at any time after the start of the reaction is given by the slope of the tangent to the curve at that time. The red and green straight lines are approximations to the tangent.

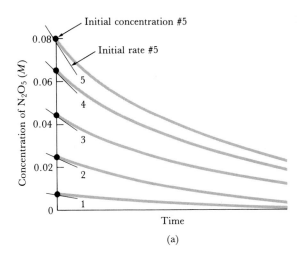

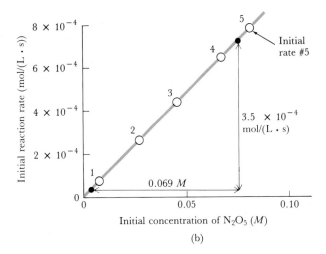

The initial rates of most reactions depend on the initial concentrations of the reactants. As an example, suppose we were to measure different masses of solid N_2O_5 into different flasks, immerse them all in a water bath at 65°C, and measure the initial rates at which the vapor in each flask decomposed. We would find greater initial rates of decomposition in the flasks with greater initial concentrations of N_2O_5 (Fig. 12.3a). The pattern in such data becomes clear when we look at actual values: the initial rate for an initial concentration of 0.08 M N_2O_5 vapor is twice the initial rate for an initial concentration of 0.04 M. This doubling of the rate when the concentration is doubled suggests that the initial rate is proportional to the initial molar concentration of vapor. That is,

$$\text{Initial rate} = k \times [N_2O_5]_0$$

where k is a constant and the subscript 0 specifies "initial value." This is the equation of a straight line with intercept 0 and slope k:

$$\text{Initial rate} = 0 + k \times [N_2O_5]_0$$
$$y \quad\quad = a + b \times \quad x$$

The assumed proportionality between initial rate and initial concentration can therefore be confirmed, and the value of k obtained, by plotting the initial rate against the initial concentration of N_2O_5. As Fig. 12.3b shows, a straight line is obtained, confirming the proportionality; the numerical value of k is the slope of the line:

$$k = \frac{3.6 \times 10^{-4} \text{ mol/(L} \cdot \text{s)}}{0.069 \text{ mol/L}} = 5.2 \times 10^{-3}/\text{s}$$

(Notice the unit of k: it is 1/s, "per second.") Once we know the value of k, we can use it to predict the initial rate of the same reaction at the same temperature for any initial concentration. For instance, if $[N_2O_5]_0 = 50$ mM and the temperature is 65°C, the initial rate of decomposition is

$$\text{Initial rate} = (5.2 \times 10^{-3}/\text{s}) \times (50 \times 10^{-3} \text{ mol/L } N_2O_5)$$
$$= 2.6 \times 10^{-4} \text{ mol } N_2O_5/(L \cdot s)$$

FIGURE 12.3 (a) The initial rate of decomposition of N_2O_5, as given by the slope of the tangent at time $t = 0$, is higher when the initial concentration of N_2O_5 is higher. (b) The initial rate is proportional to the initial concentration, as is shown by the straight line obtained when that rate is plotted against that concentration for the five samples of part a.

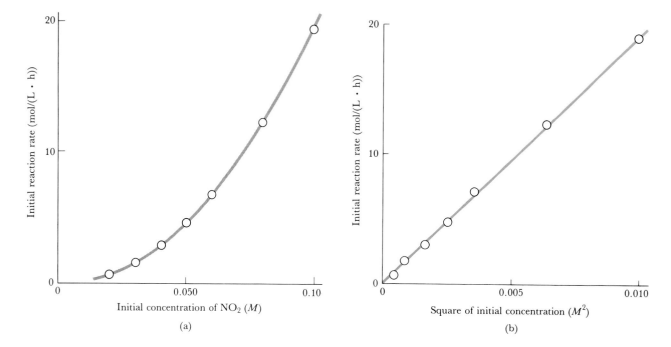

(a)

(b)

FIGURE 12.4 (a) When the initial rate of decomposition of NO_2 into NO and O_2 is plotted against its initial concentration, a straight line is not obtained. (b) A straight line is, however, obtained when that rate is plotted against the square of that concentration.

Not all reactions have initial rates that are proportional to the initial concentration of a reactant. When we plot the initial rate of the decomposition of nitrogen dioxide,

$$2NO_2(g) \longrightarrow 2NO(g) + O_2(g)$$

against the initial NO_2 concentration, we do not obtain a straight line (Fig. 12.4a). However, a plot of the initial rate against the *square* of the initial concentration, is a straight line (Fig. 12.4b). This shows that the initial rate of this reaction is proportional to $[NO_2]_0^2$, and therefore that

$$\text{Initial rate} = k \times [NO_2]_0^2$$

where k is a constant (not the same as that for N_2O_5, found above). The unit of this constant k is liters per mole per second, abbreviated L/(mol · s), as can be checked by dividing the unit for initial rate, mol/(L · s), by the square of the unit of concentration, $(mol/L)^2$. The numerical value of k is obtained from the slope of the graph and, for this reaction, is 0.54 L/(mol · s) at 300°C.

12.2 RATE LAWS AND REACTION ORDER

The expressions we have derived show how initial reaction rates depend on the initial concentration of reactant; because these are initial values, they imply the absence of any complicating effects arising from reactions of the products. The products of many reactions do not take part in other reactions to any significant extent, in which case the expression for the initial rate also gives the rate at later stages of the reaction. The only difference is that at each stage we must use the current value of the concentration. In the N_2O_5 decomposition, for example, when the N_2O_5 concentration has fallen to 10 mM from its initial 50 mM, the reaction rate is

$$\text{Rate} = k \times [N_2O_5]$$
$$= (5.2 \times 10^{-3}/\text{s}) \times (10 \times 10^{-3} \text{ mol/L } N_2O_5)$$
$$= 5.2 \times 10^{-5} \text{ mol/(L} \cdot \text{s)} N_2O_5$$

which is lower than the initial rate.

The proportionality of the reaction rate to $[N_2O_5]$ is confirmed by plotting the rate at different times after the start of the reaction against the concentration of N_2O_5 at each time (Fig. 12.5). This results in a straight line with the same slope as in Fig. 12.3. Hence, at each instant the remaining N_2O_5 reacts as if it were a smaller initial concentration. The rate of decomposition *at any stage of this reaction* is therefore described by

$$\text{Rate of decomposition} = k \times [N_2O_5]$$

where $[N_2O_5]$ is the concentration at the time of interest. This equation is an example of a "rate law":

A **rate law** is an equation expressing an instantaneous reaction rate in terms of the concentrations of the substances taking part in the reaction.

The constant k that appears in a rate law is called the *rate constant*. It is independent of the concentration of the reactants but depends on the reaction and the temperature. For the N_2O_5 decomposition at 65°C, as we have seen, $k = 5.2 \times 10^{-3}/\text{s}$.

The rate law for the NO_2 decomposition is

$$\text{Rate of decomposition} = k \times [NO_2]^2$$

with $k = 0.54$ L/(mol $\cdot$ s) at 300°C. This too is the same as the equation we found for the initial rate, except that the initial concentration is replaced by the value at the time of interest.

We shall see that the concentrations of substances taking part in reactions with similar rate laws change with time in a similar way. A useful first step in classifying reactions is therefore to group them according

FIGURE 12.5 (a) The instantaneous reaction rates of the N_2O_5 decomposition at different times during a reaction are obtained as the slopes of the tangents. (b) When these rates are plotted against the concentration of N_2O_5 remaining, a straight line is obtained.

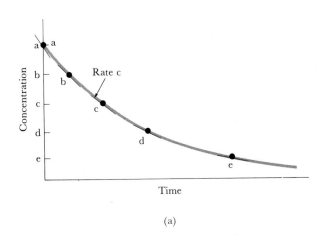

(a)

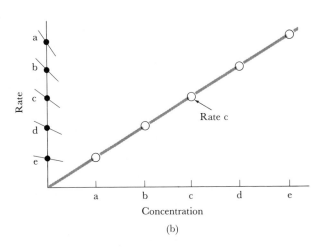

(b)

to their rate laws. The two rate laws we have seen so far differ in the power to which the concentration is raised. Both may be summarized by the equation

$$\text{Rate} = k \times [\text{reactant}]^a$$

with $a = 1$ for the N_2O_5 reaction, and $a = 2$ for the NO_2 reaction.

The N_2O_5 reaction is an example of a *first-order reaction*, a reaction for which the reaction rate is proportional to the concentration of a reactant (that is, $a = 1$). Doubling the concentration of the reactant in a first-order reaction doubles the reaction rate. The NO_2 decomposition is an example of a *second-order reaction*, one for which the reaction rate is proportional to the square of the concentration of a substance (that is, $a = 2$). Doubling the concentration of the reactant in a second-order reaction increases the reaction rate by a factor of $2^2 = 4$. A reaction has *zero order* if its rate is independent of the concentration. The name "zero order" comes from the mathematical fact that anything raised to the power zero is equal to one ($x^0 = 1$). In particular, $[\text{reactant}]^0 = 1$, so the rate law for a zero-order reaction is

$$\text{Rate} = k \times [\text{reactant}]^0 = k, \quad \text{a constant}$$

An example of a zero-order reaction is the decomposition of ammonia on a hot platinum wire:

$$2NH_3(g) \xrightarrow{\Delta,\ Pt} N_2(g) + 3H_2(g)$$

The ammonia decomposes at a constant rate until it has all disappeared. Then the reaction stops abruptly (Fig. 12.6).

Some reactions have more complicated rate laws than those discussed so far. The redox reaction between persulfate and iodide ions, for example,

$$S_2O_8{}^{2-}(aq) + 3I^-(aq) \longrightarrow 2SO_4{}^{2-}(aq) + I_3{}^-(aq)$$

obeys the rate law

$$\text{Rate} = k[S_2O_8{}^{2-}][I^-]$$

We say that this reaction is first-order with respect to (or "in") $S_2O_8{}^{2-}$ and first-order in I^-, where the *order* with respect to a substance is the power to which its concentration is raised in the rate law. Doubling either the $S_2O_8{}^{2-}$ concentration or the I^- concentration doubles the reaction rate. When, as in this example, the rate depends on the concentrations of more than one substance, we speak of its "overall order":

The **overall order** of a reaction is the sum of the powers to which the individual concentrations are raised in its rate law.

In general, if rate $= k[A]^a[B]^b \cdots$, then the overall order is $a + b + \cdots$. In the persulfate reaction, the overall order is two. The reduction of nitrogen dioxide by carbon monoxide,

$$NO_2(g) + CO(g) \longrightarrow NO(g) + CO_2(g)$$

obeys the rate law

$$\text{Rate} = k[NO_2]^2$$

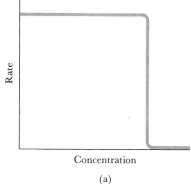

(a)

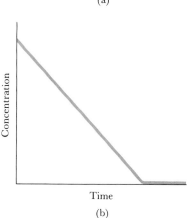

(b)

FIGURE 12.6 (a) The reaction rate of a zero-order reaction is independent of the concentration of the reactant. (b) In such a reaction, the concentration of the reactant falls at a constant rate until it is exhausted, at which time the reaction comes to an abrupt halt.

It is second-order in NO_2, zeroth-order in CO, and second-order overall. By "zeroth-order in CO" we mean that, although some CO must be present for the reaction to occur at all, the rate is independent of how much is present.

▼ **EXAMPLE 12.2 Classifying reactions by order**

The oxidation of sulfur dioxide to sulfur trioxide in the presence of platinum (Fig. 12.7) has the rate law

$$\text{Rate} = \frac{k[SO_2]}{[SO_3]^{1/2}}$$

What is the overall order of the reaction?

STRATEGY We have to identify the order with respect to each substance and then add these individual orders to find the overall order. Since $1/x^a = x^{-a}$, if a concentration occurs in a denominator, it has a negative order.

SOLUTION The rate law has the form

$$\text{Rate} = k[SO_2][SO_3]^{-1/2}$$

so it is first-order in SO_2 and of order minus one-half in SO_3. Since $1 + (-\frac{1}{2}) = \frac{1}{2}$, the reaction is half-order overall. Since the reaction's order in SO_3 is negative, the rate of the reaction *decreases* as the concentration of SO_3 increases.

EXERCISE The reduction of bromate ions (BrO_3^-) by bromide ions in acid solution,

$$BrO_3^-(aq) + 5Br^-(aq) + 6H^+(aq) \longrightarrow 3Br_2(aq) + 3H_2O(l)$$

has a reaction rate equal to $k[BrO_3^-][Br^-][H^+]^2$. What are the orders with respect to the reactants and the overall order?

[*Answer*: First-order in BrO_3^- and Br^-, second-order in H^+, fourth-order overall]

Some typical rate laws and rate constants are given in Table 12.1. Except in some special cases that are described later, a rate law cannot be deduced from a chemical equation; it must be determined experi-

FIGURE 12.7 When dried sulfur dioxide and oxygen are passed over hot platinum (left), they combine to form sulfur trioxide. This compound produces dense white fumes when it comes into contact with moisture in the atmosphere (right). The rate law for its formation, given in Example 12.2, shows that the rate of formation of SO_3 decreases as its concentration increases.

TABLE 12.1 Rate laws and rate constants

Reaction	Rate law*	Temperature, K	k
Gas phase			
$H_2 + I_2 \longrightarrow$ **2HI**	$k[H_2][I_2]$	500	4.3×10^{-4} mL/(mol · s)
		600	0.44
		700	63
		800	2.6×10^3
2HI $\longrightarrow H_2 + I_2$	$k[HI]^2$	500	6.4×10^{-6} mL/(mol · s)
		600	9.7×10^{-3}
		700	1.8
		800	92
$2N_2O_5 \longrightarrow 4NO_2 + O_2$	$k[N_2O_5]$	298	3.7×10^{-5}/s
		318	5.1×10^{-4}
		328	1.7×10^{-3}
		338	5.2×10^{-3}
$2N_2O \longrightarrow 2N_2 + O_2$	$k[N_2O]$	1000	0.76/s
		1050	3.4
$2NO_2 \longrightarrow 2NO + O_2$	$k[NO_2]^2$	573	0.54 L/(mol · s)
$C_2H_6 \longrightarrow 2CH_3$	$k[C_2H_6]$	973	5.5×10^{-4}/s
Cyclopropane $\longrightarrow$ propene	$k[cyclopropane]$	773	6.7×10^{-4}/s
Aqueous solution			
$H^+ + OH^- \longrightarrow H_2O$	$k[H^+][OH^-]$	298	1.5×10^{11} L/(mol · s)
$CH_3Br + OH^- \longrightarrow CH_3OH + Br^-$	$k[CH_3Br][OH^-]$	298	2.8×10^{-4} L/(mol · s)
$C_{12}H_{22}O_{11} + H_2O \xrightarrow{Acid} 2C_6H_{12}O_6$	$k[C_{12}H_{22}O_{11}][H^+]$	298	1.8×10^{-4} L/(mol · s)

*For the rate of formation or rate of reaction of the substance in bold type in the reaction column.

mentally. The N_2O_5 and NO_2 reactions both have equations of the form 2A → products, yet one is first-order and the other is second-order. Another example of a reaction with a rate law that shows little relation to its chemical equation is the reaction between iodine chloride and hydrogen,

$$2ICl(g) + H_2(g) \longrightarrow I_2(g) + 2HCl(g) \qquad rate = k[H_2][ICl]$$

The equation has two ICl molecules, but the rate law is proportional to $[ICl]$, not $[ICl]^2$. Another example is the iodination of acetone,

$$CH_3COCH_3(aq) + I_2(aq) \xrightarrow{Acid}$$
$$(CH_2I)COCH_3(aq) + HI(aq) \qquad rate = k[CH_3COCH_3][H^+]$$

in which $[I_2]$ does not appear in the rate law but $[H^+]$ does, even though H^+ does not appear in the chemical equation.

▼ EXAMPLE 12.3 Determining reaction order from experimental data

The data that follow show how the initial rate of the reaction

$$BrO_3^-(aq) + 5Br^-(aq) + 6H^+(aq) \longrightarrow 3Br_2(aq) + 3H_2O(l)$$

varies as the initial concentrations of the reactants are changed. What is the order of the reaction with respect to each reactant?

Initial concentration, M			Initial rate, mmol/(L · s) BrO_3^-
BrO_3^-	Br^-	H^+	
0.10	0.10	0.10	1.2
0.20	0.10	0.10	2.4
0.10	0.30	0.10	3.5
0.20	0.10	0.15	5.4

STRATEGY Suppose the concentration of a substance is increased by a certain factor f; if the reaction rate increases by f^a, then the reaction has order a in that substance. Therefore, we have to inspect the data to see how the rate changes when the concentration of each substance is changed. To isolate the effect of each substance, we should compare rows of data that differ in the concentration of only *one* substance at a time.

SOLUTION When the concentration of BrO_3^- is doubled from 0.10 M to 0.20 M (top two rows), the rate also doubles. Therefore, the reaction is first-order in BrO_3^-. When the concentration of Br^- is changed from 0.10 M to 0.30 M, by a factor of 3.0, the rate changes from 1.2 mmol/(L · s) to 3.5 mmol/(L · s), by a factor of 3.5/1.2 = 2.9. Therefore, allowing for experimental error, we can deduce that the reaction is also first-order in Br^-.

When the concentration of hydrogen ions is increased from 0.10 M to 0.15 M, by a factor of 1.5 determined from the second and fourth rows, the rate increases by a factor of 5.4/2.4 = 2.3. Thus we need to solve $1.5^a = 2.3$ for a. Taking logarithms of both sides gives

$$\log 1.5^a = \log 2.3 = 0.36$$

Then, since $\log 1.5^a = a \log 1.5$,

$$a = \frac{0.36}{\log 1.5} = \frac{0.36}{0.18} = 2.0$$

Hence, the reaction is second-order in H^+. The initial rate law (and perhaps the overall rate law) is therefore

$$\text{Rate} = k[BrO_3^-][Br^-][H^+]^2$$

and is fourth-order overall.

EXERCISE Repeat this example for the reaction between persulfate ions and iodide ions,

$$S_2O_8^{2-}(aq) + 2I^-(aq) \longrightarrow 2SO_4^{2-}(aq) + I_2(aq)$$

given the following data:

Initial concentration, M		Initial rate, mol/(L · s) $S_2O_8^{2-}$
$S_2O_8^{2-}$	I^-	
0.15	0.21	1.14
0.22	0.21	1.70
0.22	0.12	0.98

[*Answer*: $k[S_2O_8^{2-}][I^-]$]

An order need be neither a whole number nor positive, as we saw in Example 12.2. In the decomposition of ozone (O_3),

$$2O_3(g) \longrightarrow 3O_2(g) \qquad \text{rate} = \frac{k[O_3]^2}{[O_2]}$$

This is an example in which the concentration of a reaction product appears in the rate law and the reaction is of order -1 in O_2 (because $1/x = x^{-1}$). A negative order signifies that the reaction rate *decreases* as the concentration of the substance, in this case oxygen, increases.

▼ **EXAMPLE 12.4** Using a rate law with a negative power

By what factor does the rate of ozone decomposition change when the concentration of oxygen is increased by 20%?

STRATEGY Since the O_2 concentration occurs with a negative power, we expect the rate to decrease. We can use the rate law to calculate the ratio of the decomposition rates (a) with oxygen concentration of $[O_2]$ and (b) with that concentration increased by 20% to $1.2 \times [O_2]$. The ozone concentration and the rate constant will cancel out when we form the ratio.

SOLUTION The ratio of reaction rates is

$$\frac{\text{Rate}_2}{\text{Rate}_1} = \frac{k[O_3]^2/[O_2]_2}{k[O_3]^2/[O_2]_1} = \frac{[O_2]_1}{[O_2]_2}$$

$$= \frac{[O_2]_1}{1.2 \times [O_2]_1} = 0.83$$

The new rate is 83% of the original rate.

EXERCISE By how much should the oxygen concentration be reduced to increase the ozone decomposition rate by 50%?

[*Answer*: to 67% of the original]

▲

Some reactions do not have an overall order. The reaction

$$H_2(g) + Br_2(g) \longrightarrow 2HBr(g)$$

is found to obey the rate law

$$\text{Rate} = \frac{k[H_2][Br_2]^{3/2}}{[Br_2] + k'[HBr]}$$

The reaction is first-order in H_2, but because the rate law does not have the form rate $= k[H_2]^a[Br_2]^b[HBr]^c$, it does not have a specific order with respect to either Br_2 or HBr, and so it does not have an overall order.

The time dependence of first-order reactions. A major practical advantage of knowing the rate law and the rate constant is that they can be used to predict the concentrations of substances at any stage of a reaction. Conversely, knowledge of the time variation of the concentrations allows us to identify the pertinent rate law.

We begin with first-order reactions. Consider a reaction involving a substance A for which it has been found that

$$\text{Rate} = k[A]$$

An example is the rearrangement of the bonds in cyclopropane (C_3H_6; **1**) to form propene ($CH_3CH{=}CH_2$; **2**):

1 Cyclopropane

$$C_3H_6 \longrightarrow CH_3{-}CH{=}CH_2 \qquad \text{rate} = k[C_3H_6]$$

Another example is the decomposition of N_2O_5.

We often know the initial concentration $[A]_0$ of a reactant but want to know the concentration $[A]$ at some later time t. To find $[A]$, we need an expression for the concentration in terms of the time. Such an expression (which is obtained using calculus) is called an *integrated rate law*. For first-order reactions, the integrated rate law is

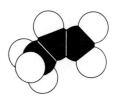

2 Propene

$$\ln \frac{[A]_0}{[A]} = kt \qquad (1)$$

The variation of $[A]$ with time as predicted by this equation is plotted in Fig. 12.8.* The graph, which is called an "exponential decay" curve, reproduces the initially steep but then slower rate of change of concentration observed experimentally (compare it with Fig. 12.2). Notice that a larger rate constant produces a more rapid initial decay.

▼ **EXAMPLE 12.5** **Calculating a concentration from a rate law**

Calculate the concentration of N_2O_5 remaining 600 s (10 min) after the start of its decomposition at 65°C if its initial concentration was 40 mM.

STRATEGY The concentration will be smaller than 40 mM, because some of the reactant has decomposed. Our aim is to use Eq. 1 to calculate the concentration of reactant at $t = 600$ s after the start of the reaction. To use it, we need the value of k in addition to the other data provided. We have already seen that $k = 5.2 \times 10^{-3}$/s for this reaction at 65°C (k is also given in Table 12.1). Equation 1 is an expression for $\ln([A]_0/[A])$, but we need $[A]$ itself. We can find it if we first take the natural antilogarithm of kt (in other words, enter e^x on your calculator, with x the value of kt).

SOLUTION Substituting for k and t in Eq. 1 gives

$$\ln \frac{[N_2O_5]_0}{[N_2O_5]} = \frac{5.2 \times 10^{-3}}{s} \times 600\ s = 3.1$$

The natural antilogarithm of 3.1 is 22 (that is, $e^{3.1} = 22$), so

$$\frac{[N_2O_5]_0}{[N_2O_5]} = 22$$

With $[N_2O_5]_0 = 40$ mM, this can be rearranged to

$$[N_2O_5] = \frac{40\ \text{m}M}{22} = 1.8\ \text{m}M$$

That is, after 600 s the concentration of N_2O_5 will have fallen from 40 mM to 1.8 mM.

*The dependence of the concentration on the time is obtained by taking the natural antilogarithm (e^x on a calculator) of both sides of the equation and rearranging the result to $[A] = [A]_0e^{-kt}$. This exponential dependence on the time is the origin of the name "exponential decay" for the shape of the curves in Fig. 12.8.

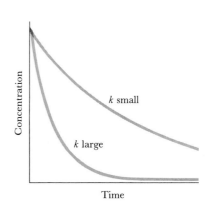

FIGURE 12.8 The characteristic shape of the time dependence of the concentration of a reactant in a first-order reaction is this "exponential decay." The larger the rate constant, the faster the decay from the same initial concentration.

Equation 1 is also used to test whether a reaction is first-order and, if it is, to determine its rate constant. Equation 1 is actually the equation of a straight line, since it can be rearranged as follows:

$$\ln [A] = \ln [A]_0 - k \times t$$
$$y \quad = \quad a \quad + b \times x$$

Therefore, if the reaction in question is first-order in A, a plot of $\ln [A]$ against t will give a straight line. If it does, the slope of the line is equal to $-k$.

▼ EXAMPLE 12.6 Measuring a rate constant

When cyclopropane **(1)** is heated to 500°C, it changes into propene **(2)**. The following data were obtained in one experiment:

t, min	0	5	10	15
[Cyclopropane], mM	1.5	1.24	1.00	0.83

Confirm that the reaction is first-order in C_3H_6 and calculate the rate constant.

STRATEGY As outlined above, we should plot $\ln [A]$ against t to see whether a straight line is obtained. If so, then the reaction is first-order. If you are more accustomed to working with base-10 logarithms,* use the relation

$$\log [A] = \log [A]_0 - \frac{k}{2.30} \times t$$

Plot $\log [A]$ against t, and equate the slope to $-k/2.30$.

SOLUTION We begin by drawing up a table of values for the plot:

t, min	0	5	10	15
log [cyclopropane]	0.18	0.093	0.00	-0.081

The points are plotted in Fig. 12.9: the graph is a straight line, confirming that the reaction is first-order in cyclopropane. The slope of the line is

$$\text{Slope} = \frac{\log [C_3H_6]_B - \log [C_3H_6]_A}{\text{time B} - \text{time A}}$$

$$= \frac{(-0.050) - (0.15)}{13.3 \text{ min} - 1.7 \text{ min}} = -0.017/\text{min}$$

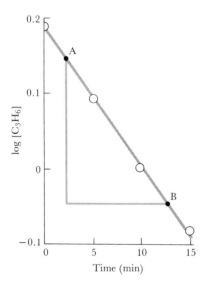

FIGURE 12.9 A test for a first-order reaction is to plot the logarithm of the reactant concentration against the time. This is simpler and more accurate than calculating the rate at each moment and plotting it against concentration. The slope of the line gives the rate constant.

*As is explained in Appendix 1B, a logarithm to the base 10 (log x) and a natural logarithm (ln x) are related by ln x = 2.303 log x.

Therefore, since $k = -2.30 \times$ slope,

$$k = -2.30 \times (-0.017/\text{min}) = 0.040/\text{min}$$

This is equivalent to $k = 6.7 \times 10^{-4}/\text{s}$, the value in Table 12.1.

EXERCISE Data on the decomposition of N_2O_5 at 25°C are as follows:

t, min	0	200	400	600	800	1000
$[N_2O_5]$, mM	15.0	9.6	6.2	4.0	2.5	1.6

Confirm that the reaction is first-order and find the value of k.

[*Answer*: $3.7 \times 10^{-5}/\text{s}$]

Equation 1 can also be rearranged to give the time needed to reach a specified concentration [A]:

$$t = \frac{1}{k} \ln \frac{[A]_0}{[A]} \tag{2}$$

▼ EXAMPLE 12.7 Using an integrated rate law

How long is needed for the concentration of N_2O_5 to decrease from 20 mM to 2.0 mM at 65°C?

STRATEGY This is a straightforward calculation; we need only substitute the data into the rearranged integrated rate law, Eq. 2. The rate constant has been given already, but, if necessary, we would use the appropriate value from Table 12.1.

SOLUTION Since $k = 19/\text{h}$ at this temperature,

$$t = \frac{1}{19/\text{h}} \times \ln \frac{20 \text{ m}M}{2.0 \text{ m}M}$$

$$= 0.0526 \text{ h} \times \ln 10 = 0.12 \text{ h}$$

which is about 7 min.

EXERCISE How long does it take for a concentration to decrease to one-hundredth of its initial value in a first-order reaction with $k = 1.0/\text{s}$?

[*Answer*: 4.6 s]

The half-life of first-order reactions. A widely used measure of reaction rates is the "half-life" $t_{1/2}$ of a substance:

The **half-life** of a substance is the time needed for its concentration to fall to one-half the initial value.

If the substance is denoted A, after one half-life the concentration of A falls from $[A]_0$ to $\frac{1}{2}[A]_0$. For a first-order reaction, from Eq. 2,

$$t_{1/2} = \frac{1}{k} \ln \frac{[A]_0}{\frac{1}{2}[A]_0}$$

$$= \frac{1}{k} \ln 2$$

FIGURE 12.10 The half-life of
a substance is the time needed for
its concentration to fall to half the
initial value. For first-order reactions, the half-life is the same
whatever the concentration at the
start of the period; that is, $t'_{1/2} = t_{1/2}$.

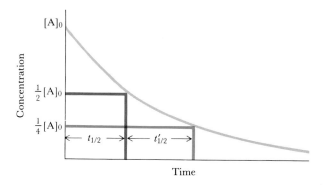

Since $\ln 2 = 0.69$, this is

$$t_{1/2} = \frac{0.69}{k} \tag{3}$$

The initial concentration does not appear in this expression, so the
half-life of a first-order reaction is independent of the initial concentration (Fig. 12.10). As we shall see, this is true only of first-order reactions.

For the N_2O_5 decomposition at 65°C,

$$t_{1/2} = \frac{0.69}{5.2 \times 10^{-3}/s} = 130 \text{ s}$$

For a sample with $[N_2O_5] = 10$ mM, the concentration decreases to
5 mM after 130 s. Since the half-life is independent of the concentration at the start of the period of interest, the concentration falls to
2.5 mM during the next 130 s, to half that again during the following
130 s, and so on.

▼ EXAMPLE 12.8 Using first-order half-lives

Calculate the time needed for the concentration of cyclopropane to fall to
(a) one-half and (b) one-fourth of its initial value at 500°C.

STRATEGY We first decide whether the reaction is first-order. If it is, the
time it takes for the concentration to fall to half its initial value is $t_{1/2}$, which
can be calculated with the rate constant and Eq. 3. We can find k in Table
12.1. The total time required for the concentration to fall to one-fourth its
initial value is the sum of two succeeding half-lives. For a first-order reaction
the two half-lives are equal, so the time required is $2t_{1/2}$.

SOLUTION Since $k = 6.7 \times 10^{-4}/s$,

$$t_{1/2} = \frac{0.69}{6.7 \times 10^{-4}/s} = 1.0 \times 10^3 \text{ s}$$

The time required for part (a) is therefore 1000 s, or 17 min. The time
required for part (b) is twice this, or 34 min.

EXERCISE Calculate the time needed for the concentration of N_2O to fall
to (a) one-half and (b) one-eighth of its initial value at 1000 K.

[*Answer*: (a) 0.91 s; (b) 2.7 s]

The half-life provides a quick way of detecting a first-order reaction. For any two different initial concentrations, we measure the time it takes each to fall by one half. If the two times are the same, the reaction is first-order.

The time dependence of second-order reactions. The integrated rate law for second-order reactions of the form

$$\text{Rate} = k[A]^2$$

is

$$\frac{1}{[A]} = \frac{1}{[A]_0} + kt \qquad (4)$$

The variation of concentration with time as predicted by this expression is shown in Fig. 12.11. The concentration decreases rapidly at first, but then changes more slowly than a first-order reaction with the same initial rate. Comparison of the two curves in Fig. 12.11 shows one difficulty with determining rate laws: the time dependences of the concentrations in first- and second-order reactions are not very different until well after the start of the reaction. Measurements must be made over several half-lives before the two orders can be distinguished.

We find the time needed for the concentration of A to change from some initial value $[A]_0$ to $[A]$ by rearranging the integrated rate law into

$$t = \frac{1}{k} \times \left(\frac{1}{[A]} - \frac{1}{[A]_0} \right) \qquad (5)$$

The half-life of A is found by setting $[A] = \frac{1}{2}[A]_0$:

$$t_{1/2} = \frac{1}{k} \times \left(\frac{2}{[A]_0} - \frac{1}{[A]_0} \right)$$

$$= \frac{1}{k[A]_0} \qquad (6)$$

In these second-order reactions the half-life *does* depend on the concentration. In fact, it is inversely proportional to the concentration. Thus, if it takes 10 s for a certain concentration to fall to half its initial value, it will take 20 s for that new concentration to fall to half its value. The doubling of the half-life after each half-life has passed is reflected in the flattening of the curves in Fig. 12.11.

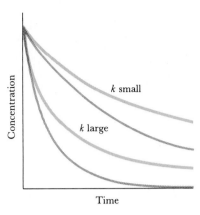

FIGURE 12.11 The characteristic time variation of the concentration of the reactant during a second-order reaction. The faint lines are the curves for first-order reactions with the same initial rates.

▼ **EXAMPLE 12.9** Using second-order half-lives

The second-order gas-phase decomposition of hydrogen iodide has $k = 1.8$ mL/(mol · s) at 700 K. How long will it take a 10-mM sample of hydrogen iodide to reach (a) one-half and (b) one-fourth of its initial concentration at that temperature?

STRATEGY Since the reaction is second-order and of the type rate = $k[A]^2$, Eq. 6 may be used. The total time required for the concentration to fall to one-fourth the initial concentration is the sum of the time it takes for the initial concentration $[HI]_0$ to fall to $\frac{1}{2}[HI]_0$ and the time for that concentration to fall to $\frac{1}{2} \times \frac{1}{2}[HI]_0$.

SOLUTION The first half-life, from $[HI]_0$ to $\frac{1}{2}[HI]_0$, is

$$t_{1/2} = \frac{1}{k[HI]_0}$$

Since

$$k[HI]_0 = 1.8 \times 10^{-3} \frac{L}{mol \cdot s} \times \left(10 \times 10^{-3} \frac{mol}{L}\right)$$

$$= 1.8 \times 10^{-5}/s$$

the value of $t_{1/2}$ is

$$t_{1/2} = \frac{1}{1.8 \times 10^{-5}/s} = 5.6 \times 10^4 \text{ s}$$

The second half-life is obtained by replacing $[HI]_0$ with its new starting value, $\frac{1}{2}[HI]_0$:

$$t_{1/2} = \frac{1}{k \times \frac{1}{2}[HI]_0} = \frac{2}{k[HI]_0}$$

This is twice the previous half-life, or 11.2×10^4 s. The total time required is therefore

$$t = 5.6 \times 10^4 \text{ s} + 11.2 \times 10^4 \text{ s} = 1.7 \times 10^5 \text{ s}$$

EXERCISE The rate constant for the second-order decomposition of NO_2 is $k = 0.54$ L/(mol · s) at 300°C. Calculate the time needed for an initial concentration $[NO_2] = 1.0$ mM to fall to (a) one-half and (b) one-eighth of that value.

[*Answer*: (a) 31 min; (b) 3 h 37 min]

▲

Some reactions of the form A + B → products have the overall second-order rate law

$$\text{Rate} = k[A][B]$$

This is the case, for example, for the reaction

$$CH_3Br(aq) + OH^-(aq) \longrightarrow CH_3OH(aq) + Br^-(aq)$$
$$\text{Rate} = k[CH_3Br][OH^-]$$

There are two cases in which a rate law of this kind can be simplified to a type we have already met. One is when a reactant has such a high concentration that it barely changes during the reaction. For instance, in the reaction given above, suppose the molar concentration of OH^- is 1.00 M, and the methyl bromide concentration is only 0.01 M; then at the end of the reaction, the OH^- concentration will have fallen to 1.00 M − 0.01 M = 0.99 M, or by only 1%. When this is so, the concentration $[OH^-]$ in this reaction, and the concentration [B] in general, can be treated as constant. Then the rate law can be simplified to

$$\text{Rate} = k'[A] \qquad \text{with} \quad k' = k \times [B]$$

A reaction such as this one, with a rate law that is *effectively* first-order because one substance has a constant concentration, is called a *pseudo-first-order reaction*. The time dependence of the concentration of A is given by Eq. 1 with a rate constant equal to $k[B]$. Thus, if $[OH^-] =$

1.0 M in the methyl bromide reaction, then

$$k' = 2.8 \times 10^{-4} \text{ L/(mol} \cdot \text{s)} \times 1.0 \text{ mol/L} = 2.8 \times 10^{-4}\text{/s}$$

and the half-life of the methyl bromide, as given by Eq. 3, is 2500 s.

A second simplifying case is that in which both reactants have the same concentration initially and react at the same rate. Since one CH_3Br molecule is consumed for each OH^- ion consumed, if their concentrations were equal at the start, their concentrations remain equal throughout the reaction. We can therefore set them equal in the rate law, which then gives

$$\text{Rate} = k[CH_3Br]^2 \quad \text{or} \quad \text{rate} = k[A]^2 \quad \text{in general}$$

In other words, the rate law now has the same second-order form as the one we have already discussed, and the concentration of either reactant can be calculated with Eq. 4.

REACTION MECHANISMS

Until now, we have used chemical equations to show the stoichiometry of reactions. For instance, the equation

$$H_2(g) + I_2(g) \longrightarrow 2HI(g)$$

is interpreted as meaning that 1 mol of H_2 reacts with 1 mol of I_2 to give 2 mol of HI: the equation does not mean that one particular H_2 molecule reacts with one particular I_2 molecule to produce two HI molecules. The question we now explore is how the rearrangement of atoms in the sample actually takes place. This type of question is of great importance, especially in organic chemistry, and many reactions can be understood, and their products predicted, once we know the details of events taking place between molecules.

12.3 ELEMENTARY REACTIONS

All but the simplest reactions are in fact the outcome of several, and sometimes many, steps called *elementary reactions*. In the formation of HI, these steps include the dissociation of the I_2 molecules into atoms and their attack on H_2 molecules. The probable identities of the elementary reactions involved in a reaction are discovered by identifying the rate law for the reaction and then trying to account for it in terms of a "mechanism":

A **reaction mechanism** is a series of elementary reactions that is proposed to account for the rate law of the overall reaction.

Elementary reactions are also written as chemical equations. However, these equations have a different significance: they show how *individual* atoms and molecules take part (Fig. 12.12) in different stages in a reaction. For example, one step in the formation of HI is the dissociation of an I_2 molecule. This is written in the usual way,

$$I_2(g) \longrightarrow 2I(g)$$

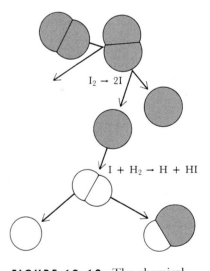

FIGURE 12.12 The chemical equations for elementary reactions show the *individual* events taking place between atoms and molecules that encounter each other. This figure illustrates two steps in the production of HI. In one ($I_2 \rightarrow 2I$), a collision results in the dissociation of an I_2 molecule. In the other, one of the I atoms so produced attacks an H_2 molecule ($I + H_2 \rightarrow HI + H$).

but since we are dealing with an elementary reaction, the equation is taken to mean that a particular I_2 molecule gives the two I atoms. This step may be followed by the attack of one of these I atoms on an H_2 molecule:

$$I(g) + H_2(g) \longrightarrow HI(g) + H(g)$$

Whether a chemical equation is expressing the behavior of individual atoms and molecules (the elementary reaction interpretation) or the overall stoichiometry is almost always clear from the context. The intended interpretation will be indicated whenever there is a risk of ambiguity.

The two elementary reactions discussed above differ in their "molecularity":

> The **molecularity** of an elementary reaction is the number of reactant molecules (or free atoms) taking part in it.

Once we know the molecularity of an elementary reaction, we can write its rate law from its equation. It must be stressed, however, that this is possible *only* for elementary reactions.

Unimolecular reactions. The elementary reaction $I_2 \rightarrow 2I$ is a *unimolecular reaction*, an elementary reaction in which a single reactant molecule (here I_2) changes into products. Another is the reaction in which an energetically excited ozone molecule vibrates so vigorously that it shakes off an oxygen atom:

$$O_3(g) \longrightarrow O_2(g) + O(g)$$

The unimolecular decomposition of ozone is one of the elementary reactions that contribute to the very complex reactions taking place in the upper atmosphere. Ozone molecules in the upper atmosphere are subjected to intense ultraviolet radiation from the sun. Upon absorbing the highly energetic photons, they dissociate into fragments that take part in further reactions.

In a unimolecular reaction, the excited molecules (for instance, the vigorously vibrating O_3 molecules in the upper atmosphere) fall apart at random. Since the number dissociating is proportional to the total number of molecules present, the reaction rate for the decomposition is proportional to the concentration. In other words, a unimolecular reaction has a first-order rate law:

$$A \longrightarrow \text{products} \qquad \text{rate} = k[A]$$

For the elementary ozone reaction, we can write

$$\text{Rate of } O_3 \text{ decomposition} = k[O_3]$$

Bimolecular reactions. The reaction between an iodine atom and a hydrogen molecule, $I + H_2 \rightarrow HI + H$, is an example of a *bimolecular reaction*, an elementary reaction in which two molecules or free atoms take part. In this reaction, an I atom collides with an H_2 molecule and carries off one of the H atoms. Another example is the reaction that occurs in the atmosphere when an O atom collides with an O_3 molecule. The force of the impact breaks one of the O—O bonds, and the incoming atom carries an O atom away from the molecule:

$$O(g) + O_3(g) \longrightarrow 2O_2(g)$$

A bimolecular reaction occurs only if two molecules or atoms meet. This suggests that the reaction rate should be proportional to the concentrations of both species, because the chance of two molecules meeting increases if the concentration of either is increased. In other words, a bimolecular reaction has an overall second-order rate law:

$$A + B \longrightarrow products \qquad rate = k[A][B]$$

For the elementary reaction between O and O_3,

$$Rate\ of\ O_2\ formation = k[O][O_3]$$

It is possible to think of a "termolecular reaction" in which three atoms or molecules collide simultaneously. However, the chance of this happening is very slight, and most reaction mechanisms ignore the possibility.

The overall reaction. Our task now is to see how a series of unimolecular and bimolecular elementary reactions combine to account for an overall reaction. We first propose a mechanism, writing down a series of elementary reactions that are likely to play a role in the overall reaction. (They are often suggested by additional experimental evidence.) The sum of the elementary reactions must be the overall reaction. Finally, we test the proposed mechanism by seeing if the rate laws for the elementary reactions combine to give the rate law for the overall reaction.

Some reactions are believed to be one-step processes in which one reactant molecule attacks another. This is the case for the reaction between methyl bromide and hydroxide ions, discussed earlier. The observed second-order rate law is consistent with a single bimolecular reaction in which an OH^- ion attacks a CH_3Br molecule. This case therefore involves the single elementary reaction

$$CH_3Br(aq) + OH^-(aq) \longrightarrow CH_3OH(aq) + Br^-(aq)$$

which is exactly the same as the overall reaction. Since the elementary step is a bimolecular reaction, its rate is proportional to the concentrations of the two reactants, and we can write

$$Rate = k[CH_3Br][OH^-]$$

This is in agreement with the observed second-order rate law for the overall reaction.

In building a reaction mechanism, we must take into account the possibility that the products of one step will be reactants in the next. Since the reverse reactions may also occur, we include them in the mechanism unless we know from the size of the rate constants that they are so slow that they can be ignored. These points can be illustrated with the reaction between nitrogen dioxide and carbon monoxide, which is second-order in NO_2:

$$NO_2(g) + CO(g) \longrightarrow NO(g) + CO_2(g) \qquad rate = k[NO_2]^2$$

The fact that CO does not appear in the rate law suggests that the reaction is not an elementary one, but rather has at least two steps. A mechanism has been proposed in which, in the first step, two NO_2 molecules collide to give an NO_3 molecule and the NO molecule:

Step 1. $2NO_2(g) \longrightarrow NO_3(g) + NO(g)$
Rate of formation of $NO_3 = k_1[NO_2]^2$

The NO_3 molecule does not appear in the overall reaction because it is not present at the beginning or at the end. However, it does take part in the elementary reactions that lead to the final product; it is therefore an example of a *reaction intermediate*, a substance that is produced and consumed during a reaction but does not appear in the overall equation. As in this case, a reaction intermediate is often a radical—a molecular fragment with one or more unpaired electrons. When a proposed mechanism involves a reaction intermediate, one test of the mechanism is to determine experimentally whether the intermediate is present while the reaction is in progress. This can often be done spectroscopically, since the intermediate is likely to absorb light of a characteristic wavelength.

In the second step, the NO_3 molecule collides with a CO molecule and gives up one of its O atoms:

Step 2. $NO_3(g) + CO(g) \longrightarrow NO_2(g) + CO_2(g)$
Rate of formation of $CO_2 = k_2[NO_3][CO]$

The overall reaction is the sum of these two steps:

$2NO_2(g) + NO_3(g) + CO(g) \longrightarrow$
$$NO_3(g) + NO_2(g) + NO(g) + CO_2(g)$$

which simplifies to

$$NO_2(g) + CO(g) \longrightarrow NO(g) + CO_2(g)$$

as required.

Rate-determining steps. In general, the overall reaction rate is determined by both steps. However, if the first step is much slower than the second, so that the products of the first step immediately take part in the second step, the overall rate of the reaction will be equal to the rate of the first step. In our example, as soon as an NO_3 radical is formed, it reacts with a CO molecule; hence, overall,

Rate of formation of CO_2 = rate of formation of NO_3
$$= k[NO_2]^2$$

in agreement with the observed rate law.

Step 1 in the reaction between NO_2 and CO is an example of a "rate-determining step":

The **rate-determining step** in a mechanism is the elementary reaction that is so much slower than the rest that it governs the rate of the overall reaction.

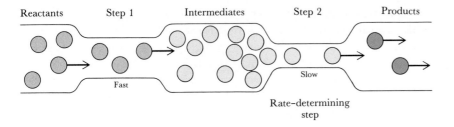

Reactants   Step 1   Intermediates   Step 2   Products

Fast

Slow

Rate-determining step

FIGURE 12.13 The rate-determining step in a reaction is the step that governs the rate at which products are formed in a multistep reaction.

A rate-determining step (Fig. 12.13) is like a slow ferry on the route between two cities. The rate at which the traffic arrives at its destination is governed by the rate at which it is ferried across the river, since this is much slower than the rate at which it travels along the roads to the ferry.

Another example of a mechanism with a rate-determining step is the one proposed for the decomposition of ozone in the upper atmosphere. The overall reaction is

$$2O_3(g) \longrightarrow 3O_2(g)$$

and the observed rate law, which was mentioned before, is

$$\text{Rate of decomposition of } O_3 = \frac{k[O_3]^2}{[O_2]}$$

If the rate law were simply $k[O_3]^2$, we might be tempted to suppose that the mechanism involves one bimolecular step:

$$O_3(g) + O_3(g) \longrightarrow 3O_2(g) \qquad \text{rate} = k[O_3]^2$$

However, the fact that O_2 appears in the rate law shows that a simple bimolecular elementary reaction between two O_3 molecules cannot be the mechanism. We need a more complex mechanism to account for the decrease in rate as the oxygen concentration increases. One possible mechanism is

Step 1. Unimolecular dissociation of excited O_3:

$$O_3(g) \longrightarrow O_2(g) + O(g) \qquad \text{rate} = k_1[O_3]$$

The O atoms are reaction intermediates. The presence of atomic oxygen in a sample of ozone irradiated with ultraviolet light is evidence for the occurrence of this step (Fig. 12.14). The reverse reaction may also occur:

$$O_2(g) + O(g) \longrightarrow O_3(g) \qquad \text{rate} = k_1'[O_2][O]$$

(We label each rate constant with the number of its step and use a prime to denote the reverse reaction.)

Step 2. Bimolecular attack of an O atom on an O_3 molecule:

$$O_3(g) + O(g) \longrightarrow 2O_2(g) \qquad \text{rate} = k_2[O_3][O]$$

The reverse reaction

$$2O_2(g) \longrightarrow O_3(g) + O(g) \qquad \text{rate} = k_2'[O_2]^2$$

FIGURE 12.14 The presence of O atoms in the upper atmosphere is partly responsible for the formation of auroras, the "northern lights" and "southern lights" that appear as colored bands of lights in northern and southern latitudes: excited oxygen atoms emit crimson and whitish-green light. Excited N_2 molecules give a pink light, and N_2^+ ions give out violet and blue light when they combine with electrons.

is so slow that it can be ignored. The overall forward reaction is the sum of the two forward elementary reactions:

$$O_3 \longrightarrow O_2 + O$$
$$\underline{O_3 + O \longrightarrow 2O_2}$$
$$2O_3 \longrightarrow 3O_2$$

Measurements of the rates of the elementary reactions show that the slowest one is that of step 2. We therefore select it as the rate-determining step and equate the overall rate to the rate of this step:

$$\text{Overall rate} = k_2[O_3][O]$$

Because the O atoms are intermediates, [O] must not appear in the expression for the overall rate. To eliminate it we must find a relation between the concentration of the intermediate O and the concentration of the reactants and products. For this, we consider only the relatively fast forward and reverse reactions in step 1 (which have the major effect on [O]) and ignore the much slower reaction in step 2. As we saw in Section 11.3, when forward and reverse processes both occur, a system reaches a state of dynamic equilibrium. There we were concerned with an equilibrium that produced a steady concentration of solute in a saturated solution. We are concerned with almost exactly the same thing here, the only difference being that we are now dealing with forward and reverse *chemical* reactions. These establish a dynamic *chemical* equilibrium in which the concentrations of the reactants and products remain constant even though the two reactions continue. This chemical equilibrium is reached when the rates of the forward and reverse reactions in step 1 are equal; that is, when

$$\text{Rate of forward reaction} = \text{rate of reverse reaction}$$

Remember, we are dealing only with step 1 here, because only that step has an appreciable reverse reaction. By substituting the two rate laws for that step, we find

$$k_1[O_3] = k_1'[O_2][O]$$

which we rearrange to

$$[O] = \frac{k_1[O_3]}{k_1'[O_2]}$$

When this equation is inserted into the expression we obtained for the overall rate, the result is

$$\text{Rate} = k_2[O_3][O] = k_2[O_3] \times \frac{k_1[O_3]}{k_1'[O_2]}$$

$$= \frac{k_1 k_2}{k_1'} \frac{[O_3]^2}{[O_2]}$$

This has exactly the same form as the observed rate law, in which we identify the observed overall rate constant k with the combination $k_1 k_2 / k_1'$ of the rate constants for the elementary steps.

The calculated and observed rate laws may be the same, but this does not *prove* that the proposed mechanism is correct: some other mechanism might also lead to the same rate law. Kinetic information can only *support* a proposed mechanism; it can never *prove* that a mechanism is the correct one. The acceptance of a suggested mechanism involves a process that is more like legal proof than mathematical proof, with evidence being assembled to give a convincing, consistent picture. In the case of the ozone decomposition, for instance, proof should include measurement of the concentration of oxygen atoms during the reaction, to see if it is proportional to $[O_3]/[O_2]$ as predicted by the equilibrium expression. The individual rate constants should also be measured to see if k is indeed equal to k_1k_2/k_1'.

▼ EXAMPLE 12.10 Deducing a rate law from a mechanism

The reaction $2NO(g) + O_2(g) \rightarrow 2NO_2(g)$ is thought to have the following mechanism:

Step 1. $2NO(g) \rightarrow N_2O_2(g)$ and its reverse (fast)

Step 2. $N_2O_2(g) + O_2(g) \rightarrow 2NO_2(g)$ (slow)

Find the rate law and the relation between the rate constant k for the overall reaction and the rate constants for the elementary reactions.

STRATEGY If we knew the rate law for the rate-determining step, we could identify the overall rate as equal to that rate. Therefore, we begin by identifying the rate-determining step (the slowest step) in the reaction mechanism. The rate law obtained at this stage often includes the concentrations of one or more of the reaction intermediates. These can be expressed in terms of the concentrations of reactants and products by assuming that the fast forward and reverse elementary reactions have reached equilibrium and their rates have become equal. We write first-order rate laws for unimolecular reactions, and second-order rate laws for bimolecular reactions.

SOLUTION The rate-determining elementary reaction is that of step 2, so we have

$$\text{Rate of formation of NO}_2 = k_2[N_2O_2][O_2]$$

We need the concentration of the intermediate N_2O_2. When the rates of the forward and reverse reactions in step 1 are equal,

$$k_1[NO]^2 = k_1'[N_2O_2]$$

This may be rearranged to

$$[N_2O_2] = \frac{k_1}{k_1'}[NO]^2$$

Substituting this into the overall rate law gives

$$\text{Rate of formation of NO}_2 = \frac{k_1k_2}{k_1'}[NO]^2[O_2]$$

which has the form

$$\text{Rate of formation of NO}_2 = k[NO]^2[O_2]$$

with $k = k_1k_2/k_1'$.

EXERCISE The proposed mechanism for a reaction is $AH + B \rightarrow BH^+ + A^-$ and its reverse, both of which are fast, followed by $A^- + AH \rightarrow$ products. Find the rate law for the reaction, with A^- treated as the intermediate.

[*Answer:* Rate = $(k_1k_2/k_1')[AH]^2[B]/[BH^+]$]

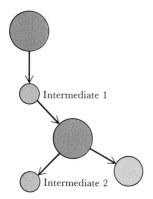

FIGURE 12.15 In a chain reaction, the product of one step (represented by a pink circle) is a reactant in another that in turn forms substances that can take part in more reactions.

12.4 CHAIN REACTIONS

The rate law for the reaction between hydrogen and bromine is quite complicated:

$$H_2(g) + Br_2(g) \longrightarrow 2HBr(g) \qquad \text{rate} = \frac{k[H_2][Br_2]^{3/2}}{[Br_2] + k'[HBr]}$$

This suggests that the reaction mechanism is also complicated. A "chain reaction" has been proposed (Fig. 12.15):

A **chain reaction** is a reaction in which an intermediate reacts to produce another intermediate.

In many chain reactions including the HBr synthesis, the reaction intermediate—which in this context is often called a *chain carrier*—is a radical, and the reaction is called a *radical chain reaction*. In a radical chain reaction, one radical reacts with a molecule to produce another radical, which goes on to attack another molecule. In the HBr synthesis, the chain carriers are the hydrogen atom ($H \cdot$) and the bromine atom ($Br \cdot$).

Radical chains. We can use the HBr synthesis to illustrate the characteristic steps of a chain reaction (Fig. 12.16). The first step in any chain reaction is *initiation*, the formation of radicals from a reactant:

$$Br_2 \xrightarrow{\Delta \text{ or light}} 2Br \cdot$$

Once radicals have been formed, the chain can *propagate*, that is, give rise to a series of reactions in which one radical reacts with a reactant molecule to produce another radical. The elementary reactions for the propagation of the chain are

$$Br \cdot + H_2 \longrightarrow HBr + H \cdot$$

$$H \cdot + Br_2 \longrightarrow HBr + Br \cdot$$

The radicals produced in these reactions can go on to attack other reactant (H_2 and Br_2) molecules, allowing the chain reaction to continue. The elementary reaction that ends the chain, called *termination*, occurs when radicals combine to form nonradical products:

$$2Br \cdot \longrightarrow Br_2$$

Some reactions can interrupt the propagation of the chain and result in *retardation*, the diversion of radicals from the formation of final products. For example, the elementary reaction

$$H \cdot + HBr \longrightarrow H_2 + Br \cdot$$

interferes with the net formation of the product HBr because it destroys some that has already been formed. However, it does not end the chain because the radical that is produced can go on to form more product.

The overall reaction may also be slowed by *inhibition*, in which radicals are removed by a reaction other than chain termination:

$$H \cdot + \text{other radicals} \longrightarrow \text{unreactive substances}$$

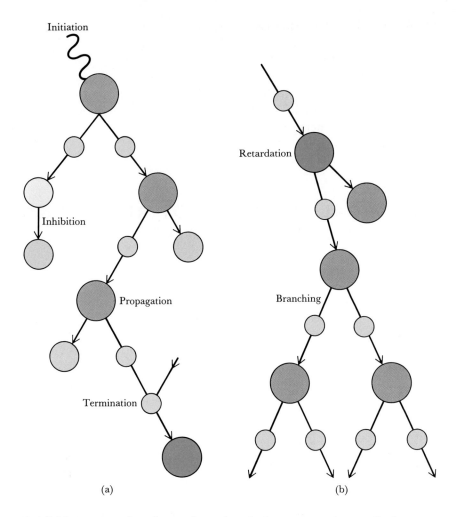

Initiation

Inhibition

Propagation

Termination

Retardation

Branching

(a)   (b)

FIGURE 12.16 (a) The characteristic steps of a chain reaction. The chain carriers, which are radicals in a radical chain reaction, are shown pink, the reactants blue, and the products green. (b) If the chain includes a branching step, the concentration of intermediates rises very rapidly and the reaction may become explosively fast.

Inhibition may take place when the chain-propagating radicals combine with impurities. Such "impurities" are sometimes added deliberately; this is the case with the food additives called "antioxidants," which are added to inhibit oxidation by removing chain-propagating radicals. Antioxidants are also added to plastics and rubber to prevent their degradation.

Explosions. In some cases a chain can propagate explosively. This is likely to happen when chain *branching* occurs, and more than one radical is formed in a propagation step. The characteristic pop that is heard when a mixture of hydrogen and oxygen is ignited is a consequence of chain branching. The two gases combine in a radical chain reaction in which the initiation step may be the formation of hydrogen atoms:

$$\textit{Initiation:} \quad H_2 \xrightarrow{\text{spark}} 2H\cdot$$

Two new radicals are formed when one hydrogen atom attacks an oxygen molecule:

$$\textit{Branching:} \quad H\cdot + O_2 \longrightarrow HO\cdot + \cdot O\cdot$$

Note that the oxygen atom, with valence-electron configuration $2s^2 2p_x^2 2p_y^1 2p_z^1$, has two electrons with unpaired spins (its complete Lewis

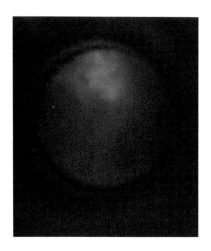

FIGURE 12.17 This flame front, inside the cylinder of an internal combustion engine, is a region where a radical chain reaction, in this case a combustion, is in progress.

FIGURE 12.18 A solid rod of iron can be heated in a flame, and it will not catch fire. However, a powder of very finely divided iron ignites in air because it presents a much greater surface area to attack by oxygen.

dot diagram is $\cdot \ddot{O} \cdot$). Two radicals are also produced when the oxygen atom attacks a hydrogen molecule:

$$\textit{Branching:} \quad \cdot O \cdot + H_2 \longrightarrow HO \cdot + H \cdot$$

As a result of these branching reactions, the chain produces a large number of radicals that can take part in even more branching steps. The reaction rate increases rapidly, and an explosion may occur (Fig. 12.17).

CONTROLLING RATES OF REACTIONS

The rates of chemical reactions are affected by concentrations, the exposed surface areas of the reactants, the temperature, and the presence of catalysts. We have already discussed the effect of concentration, which is taken into account in the rate laws. A striking illustration of the importance of the surface area exposed to a reagent, especially in reactions involving solids, is shown in Fig. 12.18: a dust of very finely divided iron particles ignites spontaneously in air. Dangerous explosions can occur in dusty environments, such as grain elevators and mines, for the minute particles have a large surface area exposed to the air and burn very rapidly.

12.5 TEMPERATURE DEPENDENCE OF REACTION RATES

Reaction rates almost always increase when the temperature is raised (Fig. 12.19), and an increase of 10°C from room temperature typically doubles the rate. (The factor by which the rate is increased usually lies in the range 1.5 to 4.) In this section we explore the reasons for this dependence of rate on temperature.

Arrhenius behavior. The fact that a reaction rate increases when the temperature is raised implies that the rate constant of the reaction has increased. The increase in k is often found to obey an equation suggested by Svante Arrhenius in 1889:

$$\ln k = \ln A - \frac{E_a}{RT} \tag{7}$$

The *Arrhenius parameters* A and E_a are constants that depend on the reaction, R is the gas constant, 8.31 J/(K · mol), and T is the temperature (in kelvins). The constant A is called the *preexponential factor*. The name comes from the expression obtained by taking the natural antilogarithms of both sides of the equation:

$$k = Ae^{-E_a/RT} \tag{8}$$

This shows that the rate constant depends exponentially on the temperature and that A is *pre*exponential in that it stands in front of the

FIGURE 12.19 Reaction rates almost always increase with temperature. The beaker on the left contains magnesium in cold water, and that on the right contains magnesium in hot water. An indicator has been added to show the presence of the alkaline solution that forms as the magnesium reacts.

exponential. The constant E_a is called the *activation energy* (we see its significance and the reason for this name shortly). When we compare the Arrhenius equation with the equation of a straight line,

$$\ln k = \ln A - \frac{E_a}{R} \times \frac{1}{T}$$

$$y \;=\; a \;+\; b \;\times\; x$$

we see that an *Arrhenius plot*, a graph of $\ln k$ against $1/T$, should be a straight line with intercept $\ln A$ at $1/T = 0$ and slope equal to $-E_a/R$ (Fig. 12.20). For the time being we treat E_a as an empirical parameter that characterizes the dependence of the rate on the temperature. As

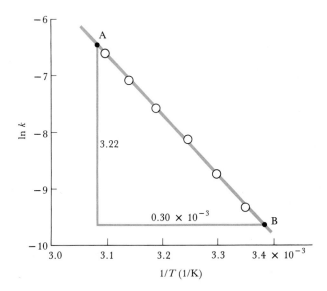

FIGURE 12.20 An Arrhenius plot is a graph of $\ln k$ against $1/T$. If, as here, the plot is a straight line, the reaction is said to show "Arrhenius behavior" in the temperature range studied, and the activation energy can be obtained from the slope.

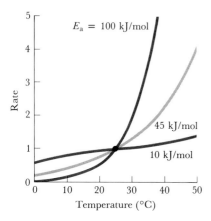

FIGURE 12.21 The curves show how the reaction rate depends on temperature for three values of the activation energy. The upper and lower curves represent large and small activation energies, and the middle curve a typical activation energy (45 kJ/mol). All values have been adjusted so that the reaction rate is 1 at 25°C.

Fig. 12.21 shows, the greater the activation energy E_a, the stronger the temperature dependence of the reaction rate. Reactions with small activation energies (about 10 kJ/mol, not a very steep Arrhenius plot) have rates that increase only slightly with temperature. Reactions with large activation energies (above about 60 kJ/mol, with a steep Arrhenius plot) have rates that depend strongly on the temperature.

▼ EXAMPLE 12.11 Measuring an activation energy

The second-order rate constant for the reaction between ethyl bromide and hydroxide ions in water,

$$C_2H_5Br(aq) + OH^-(aq) \longrightarrow C_2H_5OH(aq) + Br^-(aq)$$

was measured at several temperatures, with the following results:

Temperature, °C	25	30	35	40	45	50
$k, \times 10^{-5}$ L/(mol · s)	8.8	16	28	50	85	140

Find the activation energy of the reaction.

STRATEGY An activation energy is found by plotting $\ln k$ against $1/T$ and measuring the slope of the graph.

SOLUTION We begin by converting the temperatures to kelvins and drawing up a table of $\ln k$ and $1/T$:

Temperature, °C	25	30	35	40	45	50
T, K	298	303	308	313	318	323
$1/T, \times 10^{-3}$/K	3.36	3.30	3.25	3.19	3.14	3.10
$\ln k$	−9.34	−8.74	−8.18	−7.60	−7.07	−6.57

These data are plotted in Fig. 12.20. The slope of the line is

$$\text{Slope} = \frac{(\ln k)_B - (\ln k)_A}{(1/T)_B - (1/T)_A} = \frac{-3.22}{0.30 \times 10^{-3}/K}$$
$$= -1.07 \times 10^4 \text{ K}$$

Since the slope is equal to $-E_a/R$, we have $E_a = -R \times$ slope:

$$E_a = -8.31 \text{ J/(mol · K)} \times (-1.07 \times 10^4 \text{ K}) = 89 \text{ kJ/mol}$$

EXERCISE The rate constant for the second-order gas-phase reaction $HO(g) + H_2(g) \longrightarrow H_2O(g) + H(g)$ varies with the temperature as follows:

Temperature, °C	100	200	300	400
k, L/(mol · s)	1.1×10^{-9}	1.8×10^{-8}	1.2×10^{-7}	4.4×10^{-7}

Calculate the activation energy.

[*Answer*: 42 kJ/mol]

▲

Reactions that give a straight line when $\ln k$ is plotted against $1/T$ are said to show *Arrhenius behavior*. A wide range of reactions—not only those in the gas phase—show Arrhenius behavior. Even tropical fire-

TABLE 12.2 Arrhenius parameters for some reactions

Reaction	A	E_a, kJ/mol
First-order, gas phase		
Cyclopropane $\longrightarrow$ propene	2.0×10^{15}/s	272
CH_3—$CN \longrightarrow CH_3$—NC	4.0×10^{15}	160
$C_2H_6 \longrightarrow 2CH_3$	2.5×10^{17}	384
$N_2O \longrightarrow N_2 + O$	8.0×10^{11}	250
$2N_2O_5 \longrightarrow 2NO + O_2$	6.3×10^{14}	88
Second-order, gas phase		
$O + N_2 \longrightarrow NO + N$	1×10^{11} L/(mol · s)	315
$OH + H_2 \longrightarrow H_2O + H$	8×10^{10}	42
$2CH_3 \longrightarrow C_2H_6$	2×10^{10}	0
Second-order, in aqueous solution		
$C_2H_5Br + OH^- \longrightarrow C_2H_5OH + Br^-$	4.3×10^{11} L/(mol · s)	90
$CO_2 + OH^- \longrightarrow HCO_3^-$	1.5×10^{10}	38
Acid hydrolysis of sucrose	1.5×10^{15}	108

flies flash more quickly on hot nights than on cold, and their rate of flashing shows Arrhenius behavior over a small range of temperatures. Some Arrhenius parameters are listed in Table 12.2; a typical value is $E_a = 45$ kJ/mol. Once the Arrhenius parameters have been measured for a reaction, it is very easy to predict the rate constant of the reaction at any temperature that lies within the range of the original measurements (we cannot be sure that the reaction exhibits Arrhenius behavior outside that range).

▼ **EXAMPLE 12.12** **Using the Arrhenius equation**

Calculate the rate constant for the second-order conversion of sucrose to glucose and fructose in acid solution at 37°C (the temperature within the stomach).

STRATEGY This is a straightforward substitution of data into the Arrhenius equation. We take the parameters from Table 12.2 and express the temperature in kelvins. The units of k are the same as those of A.

SOLUTION At $T = 310$ K

$$RT = 8.31 \text{ J/(mol · K)} \times 310 \text{ K} = 2.58 \text{ kJ/mol}$$

Hence Eq. 7 becomes

$$\ln k = 34.9 - \frac{108 \text{ kJ/mol}}{2.58 \text{ kJ/mol}}$$

$$= 34.9 - 41.9 = -7.0$$

The natural antilogarithm of -7.0 is 9.1×10^{-4}. We conclude that $k = 9.1 \times 10^{-4}$ L/(mol · s), or 0.91 mL/(mol · s).

EXERCISE Calculate the rate constant for the decomposition of N_2O at 500°C.

[*Answer*: 1.0×10^{-5}/s]

▲

If a rate constant is already known to be k at some temperature T, its value k' at another temperature T' can be predicted from the activation energy alone. At the two temperatures

$$\ln k' = \ln A - \frac{E_a}{RT'}$$

$$\ln k = \ln A - \frac{E_a}{RT}$$

Subtraction of the second equation from the first gives

$$\ln k' - \ln k = \frac{-E_a}{RT'} - \left(\frac{-E_a}{RT}\right)$$

which can be rearranged to

$$\ln \frac{k'}{k} = \frac{E_a}{R}\left(\frac{1}{T} - \frac{1}{T'}\right) \tag{9}$$

▼ **EXAMPLE 12.13** **Predicting the rate constant at one temperature from its value at another temperature**

Calculate the rate constant at 35°C for the hydrolysis of sucrose, given that it is 0.91 mL/(mol · s) at 37°C.

STRATEGY Since the temperature is lower, we expect a smaller rate constant. We need the activation energy from Table 12.2; then we can simply substitute the data, using for k the value at 37°C. The temperatures must be expressed in kelvins.

SOLUTION Since $E_a = 108$ kJ/mol, Eq. 9 becomes

$$\ln \frac{k'}{k} = \frac{108 \text{ kJ/mol}}{8.31 \text{ J/mol}} \times \left(\frac{1}{310} - \frac{1}{308}\right)$$

$$= -0.272$$

(The K in the temperatures is canceled by the 1/K in the units of R.) Taking natural antilogarithms, we find $k'/k = 0.762$. Then, since $k = 0.91$ mL/(mol · s),

$$k' = 0.762 \times 0.91 \text{ mL/(mol · s)} = 0.69 \text{ mL/(mol · s)}$$

As we can see, the very high activation energy of the reaction makes its rate very sensitive to temperature.

EXERCISE The rate constant for the second-order reaction between CH_3Br and OH^- ions is 0.28 mL/(mol · s) at 25°C; what is it at 50°C?

▲ [*Answer*: 4.7 mL/(mol · s)]

Collision theory. So far, we have treated the Arrhenius parameters as quantities that summarize experimental observations on the temperature dependence of reaction rates. The *explanation* of Arrhenius behavior and an interpretation of these parameters come from the *collision theory* of elementary gas-phase bimolecular reactions. This theory supposes that molecules react only if they smash together with at least enough kinetic energy for bonds to be broken (Fig. 12.22). In this

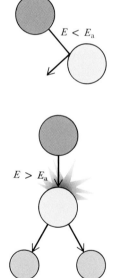

FIGURE 12.22 In the collision theory of chemical reactions, a reaction occurs only when two molecules collide with a certain minimum kinetic energy; otherwise they simply bounce apart. This figure and the next anticipate that E_{min} will turn out to be the activation energy E_a of the reaction.

respect, molecules behave like billiard balls: they bounce apart if they collide at low speed, but could smash into pieces if the impact were really powerful.

Consider a bimolecular reaction between substances A and B. Since the rate at which collisions occur between A and B molecules is proportional to their concentrations, we can write

$$\text{Rate of collisions} = \text{constant} \times [A][B]$$

If every collision were successful, we would conclude that this constant is equal to k. However, it is easy to see that not every collision can be successful. The kinetic theory of gases lets us calculate that, in a gas at 25°C and 1 atm pressure, each molecule collides with another within about 10^{-10} s. Therefore, if the reaction rate were equal to the rate at which molecules met, most gas-phase reactions would have half-lives of about 10^{-9} to 10^{-10} s. Since, as we have seen, some half-lives are minutes and even hours long, we must conclude that only a very tiny fraction of the collisions are successful.

In collision theory, it is assumed that a collision is successful only if the molecules collide with at least some minimum kinetic energy E_{min}. (Shortly, we shall identify this minimum energy with the activation energy.) Since the reaction rate is therefore only a fraction of the rate of collision, we may write

$$\text{Rate of reaction} = f \times \text{rate of collision}$$

$$= f \times \text{constant} \times [A][B]$$

where f is the fraction of the molecules with at least the minimum kinetic energy required for reaction. The fraction f can be calculated from the Maxwell distribution of speeds (Section 5.7 and Fig. 12.23). As shown by the shaded area under the blue curve in Fig. 12.23, at room temperature only a tiny fraction of the molecules have enough energy to react. At higher temperatures, as represented by the red curve and area, a much larger fraction of molecules can react.

The fraction of the molecules having a kinetic energy of at least E_{min} at a temperature T was calculated by the Austrian scientist Ludwig Boltzmann. He deduced that f is given by

$$\ln f = \frac{-E_{\text{min}}}{RT} \tag{10}$$

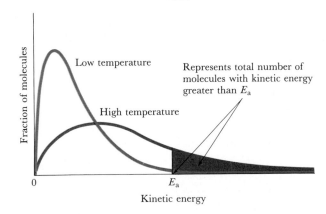

Low temperature

Represents total number of molecules with kinetic energy greater than E_a

High temperature

Fraction of molecules

0

E_a

Kinetic energy

FIGURE 12.23 The fraction of molecules that collide with at least kinetic energy E_a is given by the shaded area under each curve. This fraction increases greatly as the temperature is raised.

Substituting $E_{min} = 45$ kJ/mol into this expression gives $f = 1.3 \times 10^{-8}$ at 25°C. This shows that fewer than two collisions in a hundred million can be expected to lead to reaction.

If only a fraction f of the collisions are successful, the rate constant is reduced from k = constant to $k = f \times$ constant. Taking logarithms then yields

$$\ln k = \ln (f \times \text{constant})$$

$$= \ln (\text{constant}) + \ln f$$

When we substitute the Boltzmann expression for f, this becomes

$$\ln k = \ln (\text{constant}) - \frac{E_{min}}{RT}$$

If we identify the constant inside the logarithm as the preexponential factor A, and the minimum energy as the activation energy E_a, we obtain the Arrhenius equation (Eq. 7). Thus, according to collision theory, the preexponential factor A is the value the rate constant would have if every collision were successful. However, only collisions occurring with at least the activation energy E_a are successful, which reduces the rate constant from A to a fraction of its value.

Activation barriers. Collision theory can be extended to explain why the Arrhenius equation also applies to reactions in solution. The more general theory is called *activated-complex theory*. In this theory, two molecules approach and distort as they meet and form an *activated complex*, a combination of the two molecules that can either go on to form products or fall apart into the unchanged reactants. If we follow the potential energy of the two molecules (Fig. 12.24), we see that it rises as they merge to form the activated complex and then falls again as the product molecules form and separate. The entire curve in Fig. 12.24 is called the *reaction profile*, and the hump between reactants and products is the *activation barrier*.

Since only reactant molecules that have enough kinetic energy to cross the activation barrier can turn into products (the rest just bouncing apart unchanged), the height of the barrier is the activation energy E_a for the reaction. As in collision theory, the Boltzmann expression

FIGURE 12.24 In the activated-complex theory of chemical reactions, it is assumed that the potential energy of the reactant molecules rises as they approach one another, reaches a maximum as they form an "activated complex," and then decreases as the product molecules form and separate. Only molecules with enough energy to cross the activation barrier E_a can react when they meet.

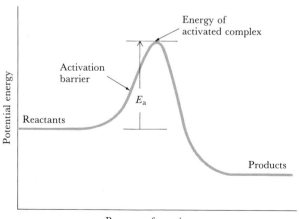

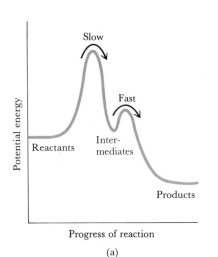

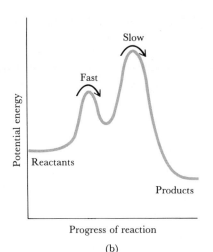

Slow

Fast

Potential energy

Reactants

Inter-
mediates

Products

Progress of reaction

(a)

Slow

Fast

Potential energy

Reactants

Products

Progress of reaction

(b)

FIGURE 12.25 In a multistep reaction, the reaction profile has several activation barriers; the highest corresponds to the rate-determining step. In (a) the first of two elementary reactions is rate-determining; in (b), the second.

gives the fraction of molecules that approach the barrier with sufficient energy to travel over it. Therefore, if A is identified as the rate constant in the absence of the barrier, the same argument as before leads to the Arrhenius equation. The difference between this and collision theory is that the reactants may be in solution, and meet as a result of the chaotic thermal motion happening to bring them together. In a gas, the reactants are in free flight until they collide.

In a multistep reaction, there is an activation barrier for each elementary reaction. The reaction profile then has several peaks, and the reaction rate is governed by the highest peak. The profile for the reaction between NO_2 and CO (Section 12.2) in which the first step is rate-determining, therefore looks like Fig. 12.25a. The profile for the reaction between NO and O_2 (Example 12.10), in which the second step is rate-determining, resembles Fig. 12.25b.

12.6 CATALYSIS

The rates of some reactions are increased by the addition of small amounts of certain substances, and these substances can often be recovered at the end of the reaction (Fig. 12.26). Such substances are called

(a)

(b)

FIGURE 12.26 A small amount of a catalyst, in this case potassium iodide in aqueous solution, can accelerate the decomposition of hydrogen peroxide to water and oxygen. This is shown by (a) the slow inflation of the balloon when no catalyst is present and by (b) its rapid inflation when a catalyst is present.

"catalysts" (from the Greek words meaning roughly "breaking down by coming together"):

A **catalyst** is a substance that increases the rate of a reaction without being consumed in the reaction.

The Chinese name for catalyst, "marriage broker," captures the sense quite well. An example of a reaction that can be catalyzed is the thermal decomposition of potassium chlorate:

$$2KClO_3(s) \xrightarrow{\Delta} 2KCl(s) + 3O_2(g)$$

The decomposition is very slow, even when the solid is heated to a high temperature. However, when a small amount of manganese dioxide is added to the potassium chlorate, oxygen is evolved briskly at the salt's melting point (356°C), and the manganese dioxide can be recovered at the end of the reaction. Another example is the use of finely divided iron to catalyze the synthesis of ammonia:

$$N_2(g) + 3H_2(g) \longrightarrow 2NH_3(g)$$

This catalyst was discovered by the German chemist Fritz Haber shortly before World War I, and it has led to the availability of ammonia on a huge scale. At first the ammonia was used largely for explosives, but now its principal destinations are fertilizers and plastics.

A catalyst speeds up a reaction by providing an alternative pathway from reactants to products. This new pathway has a rate-determining step with a lower activation energy than that of the original pathway (Fig. 12.27). At the same temperature, a greater fraction of reactant molecules can cross the barrier and turn into products. In terms of the highway-and-ferry analogy, the catalyst opens a new route with a more frequent ferry service.

Homogeneous and heterogeneous catalysis. A catalyst is *homogeneous* if it is present in the same phase as the reactants. For reactants that are gases, a homogeneous catalyst is also a gas. If the reactants are liquids, a homogeneous catalyst is a dissolved liquid or solid.

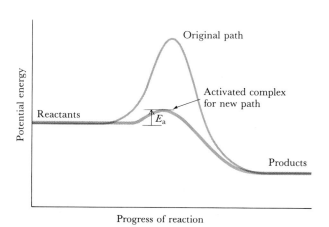

FIGURE 12.27 A catalyst lowers the activation energy of a reaction, allowing more reactant molecules to cross the barrier and form products.

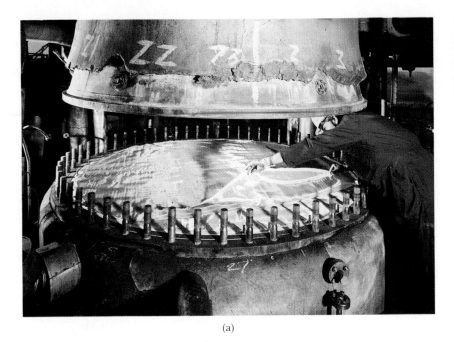

(a)

(b)

Dissolved bromine is a homogeneous liquid-phase catalyst for the decomposition of aqueous hydrogen peroxide:

$$2H_2O_2(aq) \xrightarrow{Br_2} 2H_2O(l) + O_2(g)$$

In the absence of bromine, a solution of hydrogen peroxide can be stored for a long time at room temperature; however, bubbles of oxygen form as soon as a drop of bromine is added. The mechanism of the bromine's action is believed to be its reduction to Br^- in a first step, followed by oxidation back to Br_2 in a second step:

$$Br_2(aq) + H_2O_2(aq) \longrightarrow 2Br^-(aq) + 2H^+(aq) + O_2(g)$$

$$2Br^-(aq) + H_2O_2(aq) + 2H^+(aq) \longrightarrow Br_2(aq) + 2H_2O(l)$$

When we add the two steps, the Br_2 and $2Br^-$ cancel, leaving the overall reaction as written above. Hence, although Br_2 molecules have participated in the reaction, they are not consumed.

A catalyst is *heterogeneous* if it is in a different phase from the reactants. The most common heterogeneous catalysts are the solids that are used in gas-phase or liquid-phase reactions. One example is the iron catalyst used in the Haber process. Another is finely divided vanadium pentoxide (V_2O_5), which is used as a catalyst in the contact process for the production of sulfuric acid:

$$2SO_2(g) + O_2(g) \xrightarrow{V_2O_5} 2SO_3(g)$$

The platinum catalyst used in the Ostwald process for the production of nitric acid from ammonia (Fig. 12.28) brings about a large increase in the oxidation number of nitrogen (from -3 to $+2$) and prevents the formation of molecular nitrogen, a less highly oxidized form:

$$4NH_3(g) + 5O_2(g) \xrightarrow{\Delta, Pt} 4NO(g) + 6H_2O(g)$$

FIGURE 12.28 (a) Installing a rhodium gauze catalyst for the production of nitric acid by the oxidation of ammonia, and (b) an enlarged view of the catalyst after use.

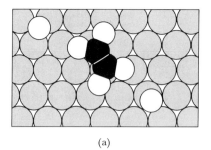

(a)

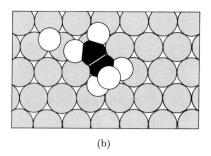

(b)

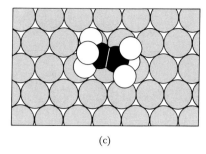

(c)

FIGURE 12.29 The reaction between ethylene, C_2H_4, and hydrogen on a metal surface. (a) The reactant molecules adsorb onto the surface, and the H_2 molecules dissociate into atoms. (b) A hydrogen atom moves across the surface, collides with a C_2H_4 molecule, and forms a C—H bond. (c) A second hydrogen atom collides with the C_2H_5— and forms another C—H bond, and the resulting C_2H_6 molecule leaves the surface.

The new reaction pathway opened by a heterogeneous catalyst generally depends on the catalyst's ability to *adsorb* the reactant, or attach the reactant to its surface. This adsorption often results in the dissociation of the reactant molecule—or at least the weakening of its bonds—and the reaction can then proceed more quickly (Fig. 12.29). The mechanism of the ammonia synthesis is not fully understood, but the rate-determining step may be the adsorption of N_2 molecules on the iron and the weakening of the strong N≡N triple bond.

Catalysts are used in the "catalytic converters" of automobiles to bring about the complete and rapid oxidation of unburned fuel. Some fuel always remains unburned in the exhaust gases leaving the engine, a mixture that includes carbon monoxide, the unburned hydrocarbon fuel, and the nitrogen oxides collectively referred to as NO_x. The pollution these substances can cause is reduced if the carbon compounds are oxidized to carbon dioxide and the NO_x reduced to nitrogen. However, although it is easy to find metal and metal oxide catalysts to speed up these reactions, selecting practical catalysts is very difficult. For instance, a catalyst that causes the reduction of NO_x to proceed beyond N_2 to NH_3 is undesirable, because ammonia will then be released into the atmosphere, where it may be oxidized back to NO_x. Furthermore, the catalyst must work even at the relatively low temperatures encountered when the engine is first started, for then the emission problem is worst. It is for this reason that cheap metals such as copper are less commonly used than the expensive noble metals, especially platinum: the latter are active at low temperatures. A further problem arises from the presence of sulfur in fuel, for the catalyst used to reduce nitrogen oxides may also catalyze the oxidation of SO_2 to SO_3, causing the vehicle to lay a trail of sulfuric acid in an urban, mobile version of the contact process.

Catalysts can be *poisoned*, or inactivated. A common cause of such poisoning is the adsorption of a molecule so tightly to the catalyst that it seals its surface against further reaction. Some heavy metals, especially lead, are very potent catalyst poisons, which is why lead-free gasoline must be used in engines fitted with catalytic converters. The elimina-

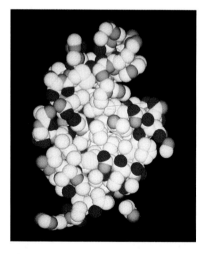

FIGURE 12.30 The lysozyme molecule shown here is a typical enzyme molecule. Lysozyme occurs throughout the body, including in tears and in the mucus of the nose. One of its functions is to attack the cell walls of bacteria and destroy them.

tion of lead has the further benefit of decreasing the amount of lead in the environment. This is important because lead is also a poison for people, who as living organisms depend on their own systems of catalysis to live and grow.

Enzymes. Biological catalysts are called *enzymes* (from the Greek words for "in yeast"). They are proteins with slotlike *active sites*, where reaction takes place (Fig. 12.30). The *substrate*, the molecule on which the enzyme acts, is recognized by its ability to fit into the slot as a key fits into a lock (Fig. 12.31). Once in the slot, the substrate undergoes reaction, which in many instances takes the form of an acid-base neutralization in which hydrogen ions are removed from one group and added to another. The modified substrate molecule is then released for use as a reactant in the next stage of the biochemical process, which is catalyzed by another enzyme, and the original enzyme molecule is free to receive the next substrate molecule.

One form of biological poisoning mirrors the effect of lead on a catalytic converter. The effectiveness of an enzyme is destroyed if an alien substrate attaches too strongly to the reactive site, for then the site is blocked and made unavailable to the true substrate (Fig. 12.32); as a result, the chain of biochemical reactions in the cell stops, and the cell dies. The action of nerve gases is believed to stem from their ability to block the enzyme-controlled reactions that allow impulses to travel through nerves. Arsenic, that favorite of fictional poisoners, acts in a similar way. After ingestion as As(V) in the form of arsenate ions (AsO_4^{3-}), it is reduced to As(III), which binds to the —SH groups of enzymes and inhibits their action.

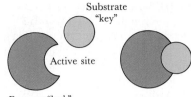

FIGURE 12.31 In the "lock-and-key" model of enzyme action, the correct substrate is recognized by its ability to fit into the active site.

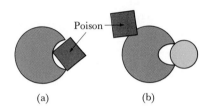

FIGURE 12.32 (a) An enzyme poison (represented by the orange square) can act by attaching so strongly to the active site that it blocks it, thereby taking the enzyme out of operation. (b) Alternatively, it may attach elsewhere, distorting the active site so that the substrate no longer fits.

SUMMARY

12.1 Chemical kinetics is the study of the rates of chemical reactions. The **initial rate** is the rate at the start of the reaction.

12.2 The **rate law** is the relation between the rate and the concentrations of substances present in the reaction mixture. The rate is reported in terms of the **rate constant** of the reaction, and reactions are classified by their **order,** the sum of the powers to which the concentrations are raised in the rate law. The rate law must be determined experimentally and cannot be deduced from the chemical equation for the reaction. Once the **integrated rate law,** the expression for the concentration in terms of time, is known, it may be used to predict the concentration at any stage of the reaction. The dependence of the **half-life,** the time needed for the concentration of a substance to fall to half its initial value, on concentration is a clue to the reaction order.

12.3 A **reaction mechanism** is a sequence of **elementary reactions,** or reaction steps involving individual molecules, that accounts for a rate law. Elementary reactions can be classified according to their **molecularity,** the number of molecules taking part. **Unimolecular reactions,** in which a single molecule changes to the products, obey first-order kinetics. **Bimolecular reactions,** elementary reactions between two molecules, obey second-order kinetics. A proposed reaction mechanism must agree with the observed rate law. The rate law predicted on the basis of a proposed mechanism is found by identifying the **rate-determining step,** the slowest elementary reaction, and finding expressions for the concentrations of any **reaction intermediates,** substances that are produced and consumed in the reaction and take part in it but are not present as initial reactants or final products.

12.4 Chain reactions are a class of reactions in which one intermediate reacts to produce another. In **radical chain reactions,** the intermediates, or **chain carriers,** are radicals. These reactions may include characteristic **initiation, propagation, retardation, inhibition,** and **termination** steps. Chain **branching,** in which more than one radical is produced in an elementary reaction, may lead to explosion.

12.5 The rate of a reaction is affected by the surface areas of the reactants, the presence of catalysts, the temperature, and the concentrations of the substances in the reaction mixture. The temperature dependence of the rate constant is expressed by the **Arrhenius equation** and two **Arrhenius parameters:** the **preexponential fac-**

tor A and the **activation energy** E_a. Their role is explained by **collision theory,** in which A is shown to be proportional to the frequency of collisions, and E_a to the minimum energy needed for reaction. This theory is generalized to **activated complex theory,** which supposes that an **activated complex,** an energetically excited combination of reactant molecules, is formed in the course of the reaction and that the reactants encounter an **activation barrier,** an energy barrier to reaction.

12.6 A **catalyst,** a substance that speeds a reaction but is not itself consumed, may be either **homogeneous** (in the same phase as the reactants) or **heterogeneous** (in a different phase). Catalysts act by providing a path with a lower activation energy. **Enzymes** are biological catalysts.

EXERCISES

The Definition of Rate

In Exercises 12.1 and 12.2, find the rate of each reaction in terms of the underlined substance.

12.1 (a) $2O_3(g) \rightarrow 3O_2(g)$, given that the rate of formation of O_2 is 1.5 mmol/(L · s)

(b) $H_2(g) + I_2(g) \rightarrow 2HI(g)$, given that the rate of reaction of I_2 is 2.6 mmol/(L · s)

(c) $3OCl^-(aq) \rightarrow ClO_3^-(aq) + 2Cl^-(aq)$, given that the rate of formation of Cl^- ions is 3.6 mol/(L · min)

(d) $2CrO_4^{2-}(aq) + 2H^+(aq) \rightarrow Cr_2O_7^{2-}(aq) + H_2O(l)$, given that the rate of formation of dichromate ions is 0.14 mol/(L · s)

12.2 (a) $2NO_2(g) \rightarrow N_2O_2(g) + O_2(g)$, given that the rate of reaction of NO_2 is 6.5 mmol/(L · s)

(b) $N_2(g) + 3H_2(g) \rightarrow 2NH_3(g)$, given that the rate of formation of NH_3 is 2.7 mmol/(L · s)

(c) $2ClO^-(aq) \rightarrow ClO_2^-(aq) + Cl^-(aq)$, given that the rate of formation of ClO_2^- is 1.6 mol/(L · min)

(d) $3MnO_4^{2-}(aq) + 4H^+(aq) \rightarrow 2MnO_4^-(aq) + MnO_2(s) + 2H_2O(l)$, given that the rate of oxidation of MnO_4^{2-} ions is 2.0 mol/(L · min)

12.3 What are the units for the rate constants of (a) zeroth-order, (b) second-order, and (c) third-order rate laws when the concentrations are expressed in moles per liter?

12.4 Since partial pressures are proportional to concentrations, rate laws for gas-phase reactions can also be expressed in terms of partial pressures, for instance as rate = kP_X for a first-order reaction. What are the units of the rate constants of (a) zeroth-order, (b) second-order, and (c) third-order gas-phase rate laws when the partial pressures are reported in Torr and the reaction rates in Torr per second? (d) If a rate constant is reported as 10/(Torr · s) at 298 K, what is its value in liters per mole per second?

Instantaneous Reaction Rate

The decomposition of gaseous dinitrogen pentoxide,

$$2N_2O_5(g) \longrightarrow 4NO_2(g) + O_2(g)$$

gave the data tabulated in Exercises 12.5 and 12.6. In each exercise, plot the concentration against time and estimate the N_2O_5 reaction rate at each of the times. Also plot, on the same graphs, the concentrations of the products of the reaction.

12.5 At 298 K:

t, s	0	4000	8000	12000	16000
$[N_2O_5]$, mM	2.15	1.88	1.64	1.43	1.25

12.6 At 308 K:

t, s	0	4000	8000	12000	16000
$[N_2O_5]$, mM	2.57	1.50	0.87	0.51	0.30

The synthesis of hydrogen iodide,

$$H_2(g) + I_2(g) \longrightarrow 2HI(g)$$

gave the results tabulated in Exercises 12.7 and 12.8. In each exercise, plot the concentrations against time, and report the instantaneous rate at each of the times. Plot the concentrations of the reactants, which are initially present with equal concentrations.

12.7 At 700 K:

t, s	0	1000	2000	3000	4000	5000
$[HI]$, mM	10	4.4	2.8	2.1	1.6	1.3

12.8 At 630 K:

t, s	0	2000	4000	6000	8000	10000
[HI], mM	30.0	26.1	23.0	20.6	18.7	17.1

Initial Rates

In Exercises 12.9 to 12.14, write the rate law for the reaction on the basis of the information supplied.

12.9 In the reaction $CH_3Br(aq) + OH^-(aq) \rightarrow CH_3OH(aq) + Br^-(aq)$, when the OH^- concentration was doubled the rate doubled, and when the CH_3Br concentration was increased by a factor of 1.2 the rate increased by the same factor.

12.10 In the reaction $2NO(g) + O_2(g) \rightarrow 2NO_2(g)$, when the NO concentration was doubled the rate increased by a factor of four, but when *both* the O_2 and NO concentrations were doubled the rate increased by a factor of eight.

12.11 The reaction $2ICl(g) + H_2(g) \rightarrow I_2(g) + 2HCl(g)$ produced the following data:

Concentration, mM		Rate of I_2 formation, mol/(L · s)
ICl	H_2	
1.5	1.5	3.7×10^{-7}
2.3	1.5	5.7×10^{-7}
2.3	3.7	14.0×10^{-7}

12.12 The reaction $NO_2(g) + O_3(g) \rightarrow NO_3(g) + O_2(g)$ produced the following data:

Concentration, mM		Rate of NO_3 formation, mol/(L · s)
NO_2	O_3	
0.38	0.38	6.3
0.21	0.70	6.3
0.21	1.39	12.5

12.13 For the reaction $A + B + C \rightarrow$ products, the following data were recorded:

Concentration, mM			Rate of formation of products, mmol/(L · s)
A	B	C	
1.25	1.25	1.25	8.7
1.97	1.25	1.25	13.7
1.25	3.02	1.25	51.0
1.97	3.02	2.01	129.0

*12.14 For the reaction $A + 2B + C \rightarrow$ products, the following data were recorded:

Concentration, mM			Rate of formation of products, mmol/(L · s)
A	B	C	
2.06	3.05	4.00	3.7
0.87	3.05	4.00	0.66
0.50	0.50	0.50	0.013
1.00	0.50	1.00	0.072

In Exercises 12.15 and 12.16, predict the initial rate of the reaction, using the rate constants in Table 12.1.

12.15 2.0 g of N_2O_5 vapor is confined to a 1.0-L container at 55°C and decomposes into NO_2 and O_2.

12.16 100 mg of ethane (C_2H_6), confined in a 250-mL container at 700°C, decomposes into methyl radicals in the first-order reaction $C_2H_6 \rightarrow 2CH_3$.

12.17 0.15 g of hydrogen and 0.32 g of iodine are confined to a 500-mL container at 700 K. By what factor would the initial reaction rate increase if (a) the mass of hydrogen were doubled; (b) 0.10 g of argon were added to the mixture?

12.18 100 mg of NO_2, confined to a 200-mL container, is heated to 300°C. It decomposes into NO and O_2. By what factor would the initial reaction rate increase if (a) the mass of NO_2 were doubled; (b) 100 mg of O_2 were added to the mixture?

Reaction Order

In each reaction of Exercises 12.19 and 12.20, identify the overall order and the order with respect to each substance. If an order cannot be designated, say so. Give the unit of the rate constant in each case, assuming the concentration is in moles per liter.

12.19 (a) $H_2(g) + I_2(g) \longrightarrow 2HI(g)$ rate $= k[H_2][I_2]$

(b) $2SO_2(g) + O_2(g) \xrightarrow{\text{Pt}} 2SO_3(g)$
$$\text{rate} = k[SO_2]/[SO_3]^{1/2}$$

(c) $BrO_3^-(aq) + 5Br^-(aq) + 6H^+(aq) \longrightarrow$
$3Br_2(aq) + 3H_2O(l)$ rate $= [BrO_3^-][Br^-][H^+]^2$

(d) $H_2O_2(aq) + 2I^-(aq) + 2H^+(aq) \longrightarrow$
$$I_2(aq) + 2H_2O(l) \quad \text{rate} = \frac{k[H_2O_2][I^-]}{1 + k'[H^+]}$$

12.20 (a) $S_2O_8^{2-}(aq) + 2I^-(aq) \longrightarrow$
$2SO_4^{2-}(aq) + I_2(aq)$ rate $= k[S_2O_8^{2-}][I^-]$

(b) $CH_3CHO(g) \longrightarrow CH_4(g) + CO(g)$
$$\text{rate} = [CH_3CHO]^{3/2}$$

(c) $NH_3(g) + D_2(g) \xrightarrow{\text{Fe}} NH_2D(g) + HD(g)$
$$\text{rate} = \frac{k[NH_3][D_2]^{1/2}}{(1 + k'[NH_3])^2}$$

(d) $2NH_3(g) \longrightarrow N_2(g) + 3H_2(g)$ rate $= k$

In Exercises 12.21 and 12.22, write the rate law for each reaction, using the experimental information provided. In each case give the units of the rate constant, assuming the concentrations are in moles per liter.

12.21 (a) The reaction $2A + B \rightarrow C + D$, which is found to be first-order in A, first-order in B, and second-order overall

(b) The reaction $A + B \rightarrow 2C$, which is found to be first-order in A, of order $-\frac{1}{2}$ in C, and half-order overall.

12.22 (a) The reaction $A + 2B \rightarrow C + D$, which is found to be second-order in A, first-order in B, and third-order overall

(b) The reaction $A + B \rightarrow C + D$, which is found to be first-order in A, half-order in B, of order $-\frac{3}{2}$ in D, and zeroth-order overall.

Half-Life

12.23 Calculate the time needed for a substance A, which decays in the first-order reaction $A \rightarrow B + C$, to fall to (a) one-quarter and (b) one-sixteenth of its initial concentration, given that its half-life is 200 s.

12.24 Calculate the time needed for a substance A, which decays in the first-order reaction $2A \rightarrow B + C$, to fall to (a) one-eighth and (b) one thirty-second of its initial concentration, given that its half-life is 1.0 min.

12.25 Calculate the time needed for a substance A, which decays in the second-order reaction $2A \rightarrow B + C$, to fall to (a) $\frac{1}{8}$ and (b) $\frac{1}{32}$ of its initial concentration of 0.20 M, given that it needed 100 s to fall to 0.10 M.

12.26 The decay of ^{14}C is first order, and the $t_{1/2}$ is 5800 years. While a plant or animal is living, it has a constant amount of ^{14}C tied up in its molecules. When the organism dies, the ^{14}C content decreases from radioactive decay, and the age of the ancient organism can be approximated if one knows the ^{14}C content of its remains. If the ^{14}C content of an ancient piece of wood was found to be $\frac{1}{4}$ that in living trees, then how old is the wood?

12.27 Find a general expression for the time needed for the concentration of A to fall to $1/2^n$ of its initial value, where $n = 1, 2, \ldots$, in terms of its half-life $t_{1/2}$, given that it decays in a first-order reaction.

12.28 The rate constant for the first order decomposition of N_2O_5 at 80°C is 0.15 s^{-1}. (a) Calculate the half life of this reaction at 80°C. (b) If one started with 10 g of N_2O_5, then how much would remain after two half lives?

Time Dependence of Concentrations

In Exercises 12.29 and 12.30, calculate the rate constant of each first-order reaction from the information provided.

12.29 (a) In the reaction $A \rightarrow B$, the concentration of A decreases to half its initial value in 1000 s.

(b) In the reaction $A \rightarrow B$ with an initial concentration [A] = 4.0 mM, the concentration of B rises to 3.0 mM in 100 s.

(c) In the reaction $2A \rightarrow B + C$, the concentration of B rises to 1.5 mM in 120 s, and after a very long time reaches 2.0 mM.

12.30 (a) In the reaction $2A \rightarrow B + C$, the concentration of A decreases to half its initial value in 125 s.

(b) In the reaction $2A \rightarrow B + C$ with an initial concentration [A] = 2.00 mM, the concentration of B rises to 0.75 mM in 1000 s.

(c) In the reaction $2A \rightarrow 3B + C$, the concentration of B rises to 4.5 mM in 1800 s, and after a very long time reaches 6.0 mM.

In Exercises 12.31 and 12.32, calculate the rate constant (for the decomposition of A) of each second-order reaction from the information provided.

***12.31** (a) In the reaction $2A \rightarrow B + C$, the concentration of A falls from 2.50 mM to 1.25 mM in 100 s.

(b) In the reaction $2A \rightarrow B$, the concentration of A falls from 2.0 M to 0.50 M in 200 s.

(c) In the reaction $2A \rightarrow B + C$ with an initial concentration [A] = 0.200 M, the concentration of B rises to 0.075 M in 155 s.

***12.32** (a) In the reaction $2A \rightarrow B + C + D$, the concentration of A falls from 0.10 mM to 0.05 mM in 12 min.

(b) In the reaction $2A \rightarrow B$, the concentration of A falls from 4.00 mM to 0.50 mM in 12 h.

(c) In the reaction $A \rightarrow 2B + C$ with an initial concentration [A] = 40 mM, the concentration of B rises to 70 mM in 15 h.

12.33 A substance A forms products B in the first-order reaction $A \rightarrow B$, in which the concentration of A falls to 20% of its initial value in 120 s. How long after the start of the reaction does the concentration of A fall to 10% of its initial value?

12.34 The reaction $2A \rightarrow B + C$ is first-order in A, and an initial concentration of 1.5 mM of A decreases to 0.25 mM in 180 s. How much more time is needed for the concentration of A to decrease to 0.010 mM?

12.35 In the reaction $A \rightarrow 2B$, which is first-order in A, it was observed that when the initial concentration of A was 2.0 mM, the concentration of B rose to 2.0 mM in 75 s. How much longer would be needed for it to rise to 3.0 mM?

12.36 In the reaction $2A \rightarrow B + C$, which is first-order in A, it was observed that when the initial concentration of A was 2.0 mM, the concentration of B rose to 0.50 mM in 175 min. From the beginning of the reaction, how much time is required for the concentration of B to rise to 0.75 mM?

In Exercises 12.37 and 12.38, calculate the time needed for each substance to reach the specified concentration, using information about the rate constant and reaction order from Table 12.1.

12.37 (a) For the concentration of 1.15 mmol of $N_2O_5(g)$ in a 250-mL container to fall to 0.55 mmol at 65°C

(b) Under the same conditions, for the O_2 concentration to rise to 0.80 mM

12.38 (a) For the concentration of 2.0 mmol of $C_2H_6(g)$ in a 500-mL container to fall to 0.50 mmol at 700°C

(b) Under the same conditions, for the CH_3 radical concentration to rise to 6.0 mM

In Exercises 12.39 to 12.42, confirm graphically that the reaction is first-order in the reactant, and find the rate constant.

12.39 $2N_2O_5(g) \rightarrow 4NO_2(g) + O_2(g)$ at 25°C:

t, s	0	4000	8000	12000	16000
$[N_2O_5]$, mM	2.15	1.88	1.64	1.43	1.25

12.40 $2N_2O_5(g) \rightarrow 4NO_2(g) + O_2(g)$ at 35°C:

t, s	0	4000	8000	12000	16000
$[N_2O_5]$, mM	2.57	1.50	0.87	0.51	0.30

12.41 $C_2H_6(g) \rightarrow 2CH_3(g)$ at 700°C:

t, s	0	1000	2000	3000	4000	5000
$[C_2H_6]$, mM	1.59	0.92	0.53	0.31	0.18	0.10

12.42 Cyclopropane $\rightarrow$ propene at 500°C:

t, s	0	1000	2000	3000	4000	5000
Cyclopropane, mM	4.57	2.34	1.19	0.61	0.31	0.16

In Exercises 12.43 to 12.46, confirm graphically that the reaction is second-order in the reactants, and evaluate the rate constant.

12.43 $2HI(g) \rightarrow H_2(g) + I_2(g)$ at 530 K:

t, s	0	4000	8000	12000	16000
$[HI]$, mM	45.5	44.5	43.6	42.7	41.8

12.44 $2HI(g) \rightarrow H_2(g) + I_2(g)$ at 580 K:

t, s	0	1000	2000	3000	4000
$[HI]$, mM	200	120	61	41	31

12.45 $H_2(g) + I_2(g) \rightarrow 2HI(g)$ at 780 K:

t, s	0	1	2	3	4
$[I_2]$, mM	1.00	0.43	0.27	0.20	0.16

12.46 $H_2(g) + I_2(g) \rightarrow 2HI(g)$ at 630 K:

t, s	0	100	300	600	1200	2400
$[I_2]$, mM	1.00	0.80	0.57	0.40	0.26	0.14

In Exercises 12.47 to 12.50, identify the order of the reaction and evaluate the rate constant.

12.47 $2A \rightarrow B$:

t, s	0	5	10	15	20
$[A]$, mM	100	14.1	7.8	5.3	4.0

12.48 $2A \rightarrow B$:

t, min	0	100	200	300	400	500
$[A]$, M	250	143	81	45	25	15

12.49 $A + B \rightarrow$ products, with $[A]_0 = [B]_0$:

t, s	0	20	40	60	80	100
$[A]$, M	2.04	0.30	0.16	0.11	0.08	0.07

12.50 $A + B \rightarrow$ products, with $[A]_0 = [B]_0$:

t, s	0	100	200	300	400	500
$[A]$, mM	250	61	33	24	18	15

Reaction Mechanisms

In Exercises 12.51 and 12.52, write the rate law for each elementary reaction, and classify it as unimolecular, bimolecular, or termolecular.

12.51 (a) $2NO(g) \longrightarrow N_2O_2(g)$
(b) $Cl_2(g) \longrightarrow 2Cl(g)$
(c) $2NO_2(g) \longrightarrow NO(g) + NO_3(g)$

12.52 (a) $CH_3Br(aq) + OH^-(aq) \longrightarrow$
$\qquad\qquad\qquad CH_3OH(aq) + Br^-(aq)$
(b) $C_2N_2(g) \longrightarrow 2CN \cdot (g)$
(c) $Ar(g) + 2O(g) \longrightarrow Ar(g) + O_2(g)$
(d) Suggest a reason why an argon atom is needed in reaction (c).

In Exercises 12.53 to 12.56, write the overall reaction for the given elementary reactions, and identify the reaction intermediates in the reaction mechanism proposed in each case, stating whether or not the intermediates are radicals. In the case of a chain reaction, identify the steps as initiation, propagation, and so on.

12.53 $ICl(g) + H_2(g) \longrightarrow HI(g) + HCl(g)$

$HI(g) + ICl(g) \longrightarrow HCl(g) + I_2(g)$

12.54 $Br_2(g) \longrightarrow 2Br(g)$

$Br(g) + H_2(g) \longrightarrow HBr(g) + H(g)$

$H(g) + Br_2(g) \longrightarrow HBr(g) + Br(g)$

$2Br(g) \longrightarrow Br_2(g)$

12.55 $Cl_2(g) \longrightarrow 2Cl(g)$

$Cl(g) + CO(g) \longrightarrow COCl(g)$

$COCl(g) + Cl_2(g) \longrightarrow COCl_2(g) + Cl(g)$

12.56 $N_2O_5(g) \longrightarrow NO_2(g) + NO_3(g)$

$NO_2(g) + NO_3(g) \longrightarrow NO_2(g) + NO(g) + O_2(g)$

$NO(g) + N_2O_5(g) \longrightarrow 3NO_2(g)$

In Exercises 12.57 to 12.60, deduce the rate law for the overall reaction from the proposed reaction mechanism.

12.57 The reaction between nitric oxide and bromine:

Step 1. $NO(g) + Br_2(g) \rightarrow NOBr_2(g)$ (slow)
Step 2. $NO(g) + NOBr_2(g) \rightarrow 2NOBr(g)$ (fast)

***12.58** The reaction between chlorine and chloroform (CHCl$_3$):

Step 1. $Cl_2(g) \rightarrow 2Cl(g)$ (fast, including reverse)
Step 2. $CHCl_3(g) + Cl(g) \rightarrow CCl_3(g) + HCl(g)$ (slow)
Step 3. $CCl_3(g) + Cl(g) \rightarrow CCl_4(g)$ (fast)

***12.59** The oxidation of iodide by hypochlorite:

Step 1. $OCl^-(aq) + H_2O(l) \rightarrow HOCl(aq) + OH^-(aq)$ (fast, including reverse)
Step 2. $I^-(aq) + HOCl(aq) \rightarrow HOI(aq) + Cl^-(aq)$ (slow)
Step 3. $HOI(aq) + OH^-(aq) \rightarrow IO^-(aq) + H_2O(l)$ (fast)

***12.60** The production of phosgene (COCl$_2$) from carbon monoxide and chlorine:

Step 1. $Cl_2(g) \rightarrow 2Cl(g)$ (fast, including reverse)
Step 2. $Cl(g) + CO(g) \rightarrow COCl(g)$ (fast, including reverse)
Step 3. $COCl(g) + Cl_2(g) \rightarrow COCl_2(g) + Cl(g)$ (slow)

In Exercises 12.61 and 12.62, suggest a mechanism for the reaction from its description.

12.61 Under certain conditions, the reaction $H_2(g) + Br_2(g) \rightarrow 2HBr(g)$ has the rate law $k[H_2][Br_2]^{1/2}$. However, the reaction hardly proceeds at all if another substance is added that removes hydrogen and bromine atoms very rapidly.

12.62 In the presence of iodine, *cis*-butene (**3**) changes into *trans*-butene (**4**) at a rate that is proportional to its concentration and to $[I]^{1/2}$. There is some evidence that an I atom attaches to the double bond of the butene molecule in the reaction.

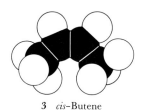

3 *cis*-Butene

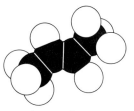

4 *trans*-Butene

The Temperature Dependence of Reaction Rates

In Exercises 12.63 and 12.64, calculate the rate constant for each reaction, using the given data and the activation energy in Table 12.2.

12.63 (a) The decomposition of N_2O_5 at 50°C, given that $k = 5.0 \times 10^{-4}$/s at 45°C

(b) The atmospheric reaction $O + N_2 \rightarrow NO + N$ at 700°C, given that $k = 9.7 \times 10^{10}$ L/(mol · s) at 800°C

12.64 (a) The dissociation of C_2H_6 into methyl radicals at 800°C, given that $k = 5.5 \times 10^{-4}$/s at 700°C

(b) The reaction between CO_2 and OH^- ions in water at blood temperature (37°C), given that $k = 1.5 \times 10^{10}$ L/(mol · s) at 25°C

In Exercises 12.65 and 12.66, calculate the activation energy of the reaction from the information provided.

12.65 For the first-order reaction $2N_2O(g) \rightarrow 2N_2(g) + O_2(g)$, the rate constant has the values 0.38/s at 1000 K and 0.87 s at 1030 K.

12.66 For the reaction $2HI(g) \rightarrow H_2(g) + I_2(g)$, the rate constant has the values 2.4×10^{-6} L/(mol · s) at 575 K and 6.0×10^{-5} L/(mol · s) at 630 K.

In Exercises 12.67 to 12.70, evaluate the activation energy of the reaction from an Arrhenius plot of the data:

***12.67** Cyclopropane $\longrightarrow$ propene:

T, K	750	800	850	900
k, 1/s	1.8×10^{-4}	2.7×10^{-3}	3.0×10^{-2}	0.26

***12.68** $C_2H_5I(g) \rightarrow C_2H_4(g) + HI(g)$:

T, K	660	680	720	760
k, 1/s	7.2×10^{-4}	2.2×10^{-3}	1.7×10^{-2}	0.11

***12.69** The acid hydrolysis of sucrose:

Temperature, °C	24	28	32	36	40
k, mL/(mol · s)	4.8	7.8	13	20	32

Predict the value of k at 37°C.

*12.70 The reaction between ethyl bromide (C_2H_5Br) and hydroxide ions in water:

Temperature, °C	24	28	32	36	40
k, mL/(mol · s)	1.3	2.0	3.0	4.4	6.4

Predict the value of k at 25°C.

Catalysis

*12.71 The activation energy of the reaction $H_2(g) + I_2(g) \rightarrow 2HI(g)$ is reduced from 184 kJ/mol to 59 kJ/mol in the presence of a platinum catalyst. By what factor will the reaction rate be increased by platinum at 600 K?

*12.72 The activation energy of the decomposition of ammonia to its elements is reduced from 350 kJ/mol to 162 kJ/mol in the presence of a tungsten catalyst. By what factor will the reaction rate be increased by tungsten at 700°C?

General

*12.73 The half-life of a substance taking part in a third-order reaction $A \rightarrow$ products is inversely proportional to the square of the initial concentration of A. How may this half-life be used to predict the time needed for the concentration to fall to (a) one-half, (b) one-fourth, and (c) one-sixteenth of its initial value?

*12.74 Could an activation energy ever be negative? In working out the answer, consider a reaction with a mechanism of more than one step.

12.75 An exothermic reaction with $\Delta H = -200$ kJ/mol has an activation energy of 100 kJ/mol. Estimate the activation energy of the reverse reaction.

*12.76 Use Le Chatelier's principle (Section 11.4) to predict the effect of temperature on the proportion of product in a mixture in which the forward and reverse reactions have reached dynamic equilibrium and the forward reaction is (a) exothermic and (b) endothermic. (c) Explain this effect in terms of the temperature dependence of the rate constants of the first-order forward and reverse reactions.

*12.77 The rate of a gas-phase reaction can sometimes be followed by monitoring the total pressure with a manometer. (This is possible when the number of gas-phase molecules changes, as in the N_2O_5 decomposition but not in the HI synthesis.) Treat the gases in the dinitrogen pentoxide decomposition as perfect, and plot the total pressure of the system against time, given that the initial pressure is 100 Torr and the data are as in Exercise 12.5.

12.78 At 329 K the total pressure of a dinitrogen pentoxide decomposition system varied with time as shown by the data that follow. Use the data to find the reaction rate in liters per mole per second at each time.

t, s	0	300	600	900	1200	1800	Long
P, kPa	12.3	29.3	39.5	46.0	50.2	54.5	57.5

12.79 When the rate of the reaction $2NO(g) + O_2(g) \rightarrow 2NO_2(g)$ was studied, it was found that the rate doubled when the O_2 concentration was doubled, but quadrupled when the NO concentration was doubled. Which of the following mechanisms accounts for these observations?

Mechanism 1

Step 1. $NO(g) + O_2(g) \rightarrow NO_3(g)$ (fast, including reverse)

Step 2. $NO(g) + NO_3(g) \rightarrow 2NO_2(g)$ (slow)

Mechanism 2

Step 1. $2NO(g) \rightarrow N_2O_2(g)$ (slow)
Step 2. $O_2(g) + N_2O_2(g) \rightarrow N_2O_4(g)$ (fast)
Step 3. $N_2O_4(g) \rightarrow 2NO_2(g)$ (fast)

*12.80 Suppose a calculation based on collision theory gave a value for a rate constant that was much less than the observed value, even though the activation energy had been taken into account. What improvement might be made to the theory?

In this chapter we see how to describe the dynamic equilibria reached by chemical reactions and how to predict the composition of the reaction mixture when equilibrium has been reached. We also see how reaction mixtures at equilibrium respond to changes in the pressure and the temperature at which the reaction is being carried out. The illustration shows the outcome of a precipitation reaction (in this case of organic materials in a nonaqueous solvent), which is one of the types of equilibria encountered in chemistry.

The description of chemical equilibrium

13.1 Reactions at equilibrium
13.2 The equilibrium constant
13.3 Heterogeneous equilibria

Equilibrium calculations

13.4 Specific initial concentrations
13.5 Arbitrary initial concentrations

CHEMICAL EQUILIBRIUM

The response of equilibria to the conditions

13.6 The effect of added reagents
13.7 The effect of pressure
13.8 The effect of temperature

We have seen several examples of physical processes—including vaporizing and dissolving—that reach dynamic equilibrium, the state in which the forward and reverse processes are continuing at equal rates. Now we shall see that exactly the same ideas apply to chemical reactions: they also approach a dynamic equilibrium in which the forward and reverse reactions continue at matching rates.

That chemical reactions tend toward (and often reach) dynamic equilibrium is of the greatest practical importance. It means that, like the vaporization of a liquid in a closed container, the change represented by the reaction does not always "go to completion," and some reactants may remain unchanged. However, by making use of the response of the equilibrium composition to changes in the pressure and temperature, we can often encourage the formation of a product and hence increase its yield.

THE DESCRIPTION OF CHEMICAL EQUILIBRIUM

The reaction between nitrogen and hydrogen in the Haber synthesis of ammonia provides a good example of chemical equilibrium. Nitrogen and hydrogen form ammonia in the reaction

$$N_2(g) + 3H_2(g) \longrightarrow 2NH_3(g)$$

However, the reverse reaction, the decomposition of ammonia, also occurs at the high temperatures used in the synthesis. At equilibrium, the forward and reverse reactions take place at the same rate, with the result that there may be only a small concentration of ammonia in the reaction mixture.

13.1 REACTIONS AT EQUILIBRIUM

Figure 13.1 shows how the molar concentrations of N_2, H_2, and NH_3 change with time in a mixture kept at 500°C. Fritz Haber (Section 12.6) made measurements like these and found that at first the concentration of NH_3 increases steadily, but that in due course it reaches a constant value. From that point on, the composition of the mixture remains the same even though some nitrogen and hydrogen are still present. This is exactly like the formation of a saturated solution, when some undissolved solute remains, and like the vaporization of a liquid in a closed container, when some liquid remains.

The final composition of the reaction mixture corresponds to a dynamic equilibrium in which the ammonia decomposes as fast as it is formed. We emphasize this dynamic character by writing the chemical equation of a reaction at equilibrium with a double-headed arrow:

$$N_2(g) + 3H_2(g) \rightleftharpoons 2NH_3(g)$$

This signifies that both the forward reaction and its reverse are continuing with equal rates.

To confirm that the equilibrium is dynamic, we could carry out two ammonia syntheses at 500°C with exactly the same starting conditions, except that in one of the syntheses we use D_2 in place of H_2 (Fig. 13.2). The two reaction mixtures reach equilibrium with the same composition, apart from the presence in one of them of D_2 and ND_3 in place of

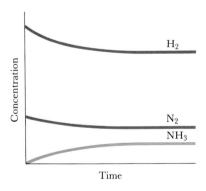

FIGURE 13.1 In the synthesis of ammonia, the concentrations of N_2, H_2, and NH_3 change with time until they finally settle into a state in which all three are present but there is no further net change in concentration.

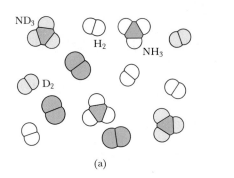
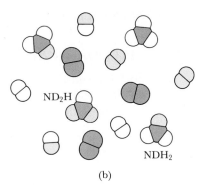

(a)   (b)

FIGURE 13.2 In an experiment showing that equilibrium is dynamic, two ammonia-synthesis reaction mixtures, one with H_2 and the other with D_2, are allowed to reach equilibrium. When they are mixed, as in (a), the concentrations of ammonia, nitrogen, and hydrogen do not change but in due course (b), NH_2D and NHD_2 appear in the products.

H_2 and NH_3. (We are ignoring the tiny differences that arise from the different masses of the molecules, which affect the equilibrium only very slightly.) The equilibrium mixtures are then combined and maintained at 500°C. Titration of a cooled sample shows that the concentration of ammonia remains constant in the mixture. However, if we analyze the sample with a mass spectrometer, we find that all possible isotopic species (H_2, HD, D_2, NH_3, NH_2D, NHD_2, and ND_3) are present. This scrambling of H and D must be a result of the forward and reverse reactions continuing in the mixture, for that allows the isotopes to exchange partners.

The steady, unchanging concentrations present when chemical equilibrium has been reached are different from those present when the substances in a mixture do not react at all. A mixture of hydrogen and oxygen can be stored unchanged indefinitely, but it is not in dynamic equilibrium because a single spark is enough to initiate the chain reaction that converts the mixture to water. Sometimes it is quite difficult to decide whether a mixture of reagents is actually at equilibrium or whether the composition of the mixture does not change simply because the reactions are very slow. One way to decide is to attempt to initiate a reaction with a spark, heat, or a flash of light.

13.2 THE EQUILIBRIUM CONSTANT

The key concept for discussing chemical equilibria quantitatively was identified by the Norwegians Cato Guldberg (a mathematician) and Peter Waage (a chemist) in 1864. They noticed that the concentrations of the reactants and products in a reaction mixture at equilibrium always satisfied a certain relation (which we specify shortly). This can be illustrated with the data in Table 13.1 for the reaction between acetic

TABLE 13.1 Esterification of acetic acid at 100°C

[Alcohol], M	[Acid], M	[Ester], M	[H_2O], M	Equilibrium constant K_c
0.10	7.29	1.71	1.71	4.0
0.32	7.07	2.93	2.93	3.8
0.75	5.86	4.14	4.14	3.9
3.33	3.33	6.67	6.67	4.0
13.8	1.42	8.58	8.58	3.8

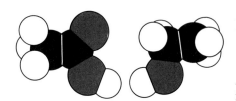

acid (CH_3COOH) and ethanol (C_2H_5OH), which produces ethyl acetate:

$$CH_3COOH + C_2H_5OH \rightleftharpoons CH_3COOC_2H_5 + H_2O$$

A compound formed by the reaction of an organic acid with an alcohol is called an *ester*, and this reaction is an example of an *esterification* (1). From now on we write the reaction

$$Acid + alcohol \rightleftharpoons ester + H_2O$$

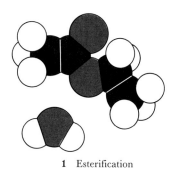

1 Esterification

The law of mass action. Suppose we prepare several different reaction mixtures with different initial compositions and allow them to reach equilibrium at 100°C. When we analyze the equilibrium mixtures, we will find that the reactants and products are present with different concentrations, as in the first four columns of Table 13.1 These values have a pattern that becomes clear when we calculate

$$K_c = \frac{[ester][H_2O]}{[acid][alcohol]}$$

for each equilibrium mixture. The values obtained are shown in the fifth column. The striking result is that, within experimental error, all the mixtures give the same value of K_c. This suggests that K_c, which is called the *equilibrium constant*, is a characteristic of the equilibrium composition of the reaction mixture. The subscript c denotes that the constant is being expressed in terms of molar concentrations. As we see in Table 13.1, equilibrium constants defined in terms of concentrations are not strictly constant but vary within a range, typically of about 10%. Hence the calculations described in the rest of this chapter are accurate to within about 10%. In more advanced work, equilibrium constants are defined in a manner that takes into account the influence of the molecules on each other, and their values are then strictly constant.

Guldberg and Waage studied a variety of reactions and found that, in every case, the equilibrium composition of a particular reaction could be expressed by an equilibrium constant. They proposed the "law of mass action" to summarize their conclusions:

Law of mass action: For a reaction of the form $aA + bB \rightleftharpoons cC + dD$, the concentrations at equilibrium satisfy

$$K_c = \frac{[C]^c[D]^d}{[A]^a[B]^b} \tag{1}$$

where K_c is a constant.

Notice that the products (C and D) occur in the numerator, and the reactants (A and B) in the denominator. Each concentration is raised to a power equal to the stoichiometric coefficient in the balanced equation. The molar concentrations [A], [B], and so on in the definition of K_c are the concentrations *at equilibrium*. The units of K_c depend on the stoichiometry of the reaction and are powers of M, such as M, M^2, and $1/M$. If all the concentrations cancel, K_c is unitless. We shall see examples shortly.

▼ EXAMPLE 13.1   Calculating an equilibrium constant

Nitrogen and hydrogen were confined inside a container at concentrations of 0.500 M and 0.800 M, respectively, and allowed to reach equilibrium. At equilibrium, the concentration of NH_3 had increased to 0.150 M. What is the value of the equilibrium constant for the reaction at the temperature of the experiment?

STRATEGY   Since we need to know the correct combination of equilibrium concentrations to calculate the value of K_c, and that combination follows from the stoichiometry, we begin by writing the chemical equation for the equilibrium and use the law of mass action to write the expression for K_c. In some cases we may realize that we need concentrations other than those given. If that is so, we can use the reaction stoichiometry to calculate them from the data. When all the concentrations have been assembled, we substitute them into the expression for K_c.

SOLUTION   At equilibrium we have

$$N_2(g) + 3H_2(g) \rightleftharpoons 2NH_3(g)$$

The law of mass action with $a = 1$, $b = 3$, $c = 2$, and $d = 0$ gives

$$K_c = \frac{[NH_3]^2}{[N_2][H_2]^3}$$

We know that $[NH_3] = 0.150\ M$. Since 1 mol of N_2 reacts to produce 2 mol of NH_3, the reduction in the concentration of N_2 resulting from the formation of NH_3 is

Reduction in N_2 concentration = increase in NH_3 concentration
$$\times \text{ moles of } N_2 \text{ per mole of } NH_3$$

$$= 0.150\ \frac{\text{mol } \cancel{NH_3}}{L} \times \frac{1 \text{ mol } N_2}{2 \text{ mol } \cancel{NH_3}}$$

$$= 0.075 \text{ mol/L } N_2$$

Since the initial concentration of N_2 is 0.500 M, at equilibrium the concentration is

$$[N_2] = 0.500\ M - 0.075\ M = 0.425\ M$$

A similar calculation gives, at equilibrium,

$$[H_2] = 0.800\ M - 0.225\ M = 0.575\ M$$

Substitution of all three equilibrium concentrations into the expression for K_c gives

$$K_c = \frac{(0.150\ M)^2}{(0.425\ M) \times (0.575\ M)^3} = 0.278/M^2$$

EXERCISE   A 5.00-g sample of dinitrogen tetroxide (N_2O_4) was allowed to vaporize in a 500-mL flask and reach equilibrium with its decomposition product, nitrogen dioxide. At equilibrium the sample contained 2.20 g of NO_2. Calculate the equilibrium constant for the decomposition.

[*Answer*: 0.150 M]

▲

Every reaction has its own characteristic equilibrium constant, with a value that depends only on the temperature. Whatever the initial composition of the reaction mixture, at a given temperature its equilibrium composition will always correspond (to within about 10%) to the value

TABLE 13.2 Equilibrium constants K_c for some reactions

Reaction	Temperature, K	K_c
$H_2(g) + Cl_2(g) \rightleftharpoons 2HCl(g)$	300	4.0×10^{31}
	500	4.0×10^{18}
	1000	5.1×10^{8}
$H_2(g) + Br_2(g) \rightleftharpoons 2HBr(g)$	300	1.9×10^{17}
	500	1.3×10^{10}
	1000	3.8×10^{4}
$H_2(g) + I_2(g) \rightleftharpoons 2HI(g)$	298	794
	500	160
	700	54
	763	46
$2BrCl(g) \rightleftharpoons Br_2(g) + Cl_2(g)$	300	377
	500	32
	1000	5
$2HD(g) \rightleftharpoons H_2(g) + D_2(g)$	100	0.52
	500	0.28
	1000	0.26
$F_2(g) \rightleftharpoons 2F(g)$	500	$7.3 \times 10^{-13}\ M$
	1000	$1.2 \times 10^{-4}\ M$
	1200	$2.7 \times 10^{-3}\ M$
$Cl_2(g) \rightleftharpoons 2Cl(g)$	1000	$1.2 \times 10^{-7}\ M$
	1200	$1.7 \times 10^{-5}\ M$
$Br_2(g) \rightleftharpoons 2Br(g)$	1000	$4.1 \times 10^{-7}\ M$
	1200	$1.7 \times 10^{-5}\ M$
$I_2(g) \rightleftharpoons 2I(g)$	800	$3.1 \times 10^{-5}\ M$
	1000	$3.1 \times 10^{-3}\ M$
	1200	$6.8 \times 10^{-2}\ M$

of K_c for that reaction. Therefore, to measure an equilibrium constant, we can take any convenient initial mixture of reagents, allow the reaction to reach equilibrium at the temperature of interest, measure the concentrations of the reactants and products, and use them in Eq. 1. The values in Table 13.2 were obtained in this way.

Equilibria favoring reactants or favoring products. Suppose we have a reaction in which the total number of reactant molecules in the chemical equation is equal to the total number of product molecules. The esterification reaction is such a reaction: one alcohol molecule and one acid molecule produce one ester molecule and one water molecule. Another example is the synthesis of hydrogen iodide, which reaches the equilibrium

$$H_2(g) + I_2(g) \rightleftharpoons 2HI(g) \qquad K_c = \frac{[HI]^2}{[H_2][I_2]}$$

The equilibrium mixture consists of hydrogen, iodine vapor, and hydrogen iodide. A third example is the redox equilibrium reached when aqueous solutions of iron(III) sulfate and cerium(III) sulfate are mixed:

$$Fe^{3+}(aq) + Ce^{3+}(aq) \rightleftharpoons Fe^{2+}(aq) + Ce^{4+}(aq) \qquad K_c = \frac{[Fe^{2+}][Ce^{4+}]}{[Fe^{3+}][Ce^{3+}]}$$

All three equilibria have the form

$$A + B \rightleftharpoons C + D \qquad K_c = \frac{[C][D]}{[A][B]}$$

In the HI synthesis, C and D both stand for HI. In each case the concentration units M cancel and K_c is a pure number.

For reactions of this kind, with the same number of molecules on each side, the value of K_c indicates whether the equilibrium lies in favor of the products (in the sense that they dominate the reaction mixture at equilibrium) or of the reactants (with them dominating). That is, it serves to indicate whether the reaction "goes" (lies in favor of the products) or "does not go" (lies in favor of the reactants). For reactions in which the numbers of molecules on the two sides of the equation are different, the conclusion is less obvious and depends on the starting concentrations. We deal with this class of reactions later.

Suppose that K_c for the first type of reaction is larger than 1 ($K_c > 1$). For the HI synthesis at 500 K, for example, $K_c = 160$. This value signifies that the numerator $[HI]^2$ is 160 times larger than the denominator $[H_2] \times [I_2]$ at equilibrium. Hence, this reaction and others for which $K_c > 1$ are reactions that favor the products. Reactions for which K_c is very much greater than 1 (by a factor of 10^3 or more) are often said to "go to completion," because the concentrations of the products at equilibrium are so much greater than the concentrations of the remaining reactants (Fig. 13.3).

Now suppose that K_c is less than 1 ($K_c < 1$); this is the case for the equilibrium between Fe^{3+} and Ce^{3+}, for which $K_c = 4 \times 10^{-4}$ at 25°C. This value signifies that the numerator $[Fe^{2+}] \times [Ce^{4+}]$ is smaller than the denominator $[Fe^{3+}] \times [Ce^{3+}]$, and the equilibrium concentrations of the products Fe^{2+} and Ce^{4+} are smaller than those of the reactants Fe^{3+} and Ce^{3+}. Hence, this reaction and others of this type for which $K_c < 1$ (particularly those for which K_c is very much smaller than 1) are reactions that favor the reactants. They "don't go" to an appreciable extent at the temperature in question (Fig. 13.3).

We should not conclude that a reaction for which $K_c < 1$ can never be carried out successfully. If we remove the product as soon as it is formed, its concentration will never reach the equilibrium value. This is one of the reasons why precipitation reactions are so effective, for the product is removed from solution the instant it is formed. The reactions used in industry rarely reach equilibrium, because yields are increased by removing the product before the reverse reaction becomes important. An example is the production of bromine by the oxidation of aqueous bromide with chlorine:

$$Cl_2(g) + 2Br^-(aq) \rightleftharpoons 2Cl^-(aq) + Br_2(aq)$$

Although the production of bromine is favored, the yield is improved by bubbling air through the solution; this encourages the newly formed bromine to vaporize. As Br_2 vapor is removed, more Br^- ion is oxidized in an attempt to reach equilibrium.

The direction of reaction. The equilibrium constant can be used to decide whether a reaction mixture of arbitrary composition will have a tendency to form more product or to decompose into reactants. Notice

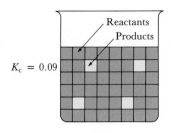

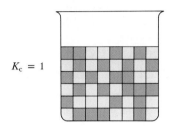

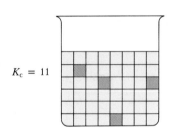

FIGURE 13.3 For reactions with the same number of reactant (blue) and product (yellow) molecules, the size of the equilibrium constant shows whether the reactants or the products are favored.

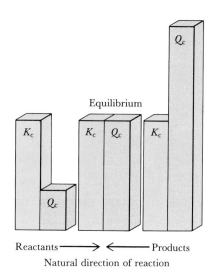

FIGURE 13.4 The relative sizes of the reaction quotient Q_c and the equilibrium constant K_c indicate the direction in which a reaction mixture tends to change. The progress toward equilibrium may, however, be unobservably slow.

that we say *tendency*, meaning "an inclination toward." It may be that the reaction is so slow in both directions that an equilibrium is not, in fact, ever reached. A mixture of hydrogen and oxygen has a tendency to form water; nevertheless, at room temperature and in the absence of an initiating spark, the reaction is very slow. The tendency, although present, is not realized, and the mixture of gases can be kept indefinitely. In dealing with equilibria, we have to bear in mind that the *rates* of the reactions may be too slow for our predictions to be fulfilled.

Now we turn to the general case, a reaction in which the numbers of product molecules might differ from the number of reactant molecules. The direction in which an arbitrary mixture tends to react (toward product or toward more reactants) is best predicted by comparing the equilibrium constant with the *reaction quotient* Q_c. This is defined like the equilibrium constant:

$$Q_c = \frac{[C]^c[D]^d}{[A]^a[B]^b} \tag{2}$$

However, the concentrations in Q_c are those of the *given mixture*, not necessarily the concentrations at equilibrium.

For an esterification reaction mixture,

$$Q_c = \frac{[\text{ester}][H_2O]}{[\text{acid}][\text{alcohol}]}$$

At the start of an esterification reaction, there are no products present, [ester] and $[H_2O]$ are both zero, and $Q_c = 0$. However, the reaction has a tendency to form products, and the value of Q_c has a tendency to grow toward K_c. Similarly, in a sample of pure ester and water, there are initially no "reactants" present, [acid] and [alcohol] are both zero, and Q_c for the reaction is infinite. In this case, there is a tendency for the reactants to form and for the value of Q_c to fall toward K_c. In each case (and in general), the mixture has a tendency to change toward its equilibrium composition, and Q_c tends to change toward K_c. This may be summarized as follows (Fig. 13.4):

If $Q_c > K_c$, reactants have a tendency to form.
If $Q_c = K_c$, the reaction is at equilibrium.
If $Q_c < K_c$, products have a tendency to form.

▼ **EXAMPLE 13.2** **Predicting the direction of reaction**

A mixture of hydrogen, iodine, and hydrogen iodide, each with a concentration of 0.0020 M, was introduced into a container heated to 490°C. At this temperature, $K_c = 46$ for the reaction $H_2(g) + I_2(g) \rightleftharpoons 2HI(g)$. Predict whether or not more HI will be formed.

STRATEGY Since the equilibrium constant is larger than 1, we expect the reaction to favor products. We anticipate that more HI will be produced in the reaction as it moves toward equilibrium. To confirm that this is so, we calculate the reaction quotient Q_c and compare it with the equilibrium constant.

SOLUTION The reaction quotient is

$$Q_c = \frac{[HI]^2}{[H_2][I_2]} = \frac{(0.002\ M)^2}{(0.002\ M) \times (0.002\ M)} = 1$$

The equilibrium constant is $K_c = 46$. Since $Q_c < K_c$, the reaction will tend to form more product.

EXERCISE A mixture of composition $[H_2] = 3$ mM, $[N_2] = 1$ mM, and $[NH_3] = 2$ mM was prepared and heated to 500°C, at which temperature $K_c = 0.11/M^2$. Calculate Q_c and predict whether ammonia will form or decompose.

[*Answer*: $1.5 \times 10^5/M^2$, decompose]

The relation of equilibrium composition to reaction rate. Although the concept of the equilibrium constant was discovered by analyzing experimental data, the reason for its existence is easy to understand. We know that chemical equilibria are dynamic and occur when the forward and reverse reactions in a reaction mixture have the same rates. Since reaction rates depend on (and change with) concentration, there will be a special set of reactant and product concentrations that correspond to equal forward and reverse rates. The equilibrium constant simply expresses the relationship between the concentrations that guarantees this equality of rates.

To see that this is so, consider the esterification reaction again. Experiments have shown that the rate laws for the forward and reverse reactions are both second-order overall:

$$\text{Acid} + \text{alcohol} \longrightarrow \text{ester} + H_2O \qquad \text{rate} = k[\text{acid}][\text{alcohol}]$$

$$\text{Ester} + H_2O \longrightarrow \text{acid} + \text{alcohol} \qquad \text{rate} = k'[\text{ester}][H_2O]$$

Setting the two rates equal, as they are at equilibrium, yields

$$k[\text{acid}][\text{alcohol}] = k'[\text{ester}][H_2O]$$

which we may rearrange into

$$\frac{[\text{Ester}][H_2O]}{[\text{Acid}][\text{alcohol}]} = \frac{k}{k'} = \text{constant}$$

This is exactly the form of the equilibrium constant. It also shows that the equilibrium constant is the ratio of the forward and reverse rate constants:

$$K_c = \frac{k}{k'}$$

If the rate constant k for the forward reaction is large compared with the constant k' for the reverse reaction, then the equilibrium constant is large and products are favored. If the rate constant for the reverse reaction is large compared with that for the forward reaction, the equilibrium constant is small and reactants are favored (Fig. 13.5).

Equilibria of multistep reactions. An apparently troublesome point arises when we interpret K_c as k/k'. Section 12.2 stressed that a rate law cannot in general be written simply by inspecting a chemical equation. However, the law of mass action *is* written from the chemical equation. Why is it that we can use the chemical equation to write the condition for the rates to be equal but cannot use it for the rates themselves?

The answer lies in the facts that each *individual* elementary step in a reaction mechanism is at equilibrium when the overall reaction is at

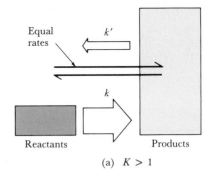

(a) $K > 1$

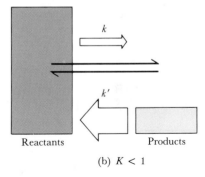

(b) $K < 1$

FIGURE 13.5 The equilibrium constant expresses the concentration condition under which the rates of the forward and reverse reactions are equal. (a) If the rate constant for the forward reaction is large compared with that for the reverse reaction, only a low concentration of reactants is needed to match the rate of decomposition of the products; then the equilibrium constant is large, and products are favored. (b) If the rate constant for the reverse reaction is large compared with that for the forward reaction, the equilibrium constant is small and reactants are favored.

equilibrium and each elementary step *can* be written from its chemical equation. (This point was explained in Section 12.3, where we did exactly that.) As an illustration of how this resolves the paradox, consider the ozone decomposition discussed in Section 12.3:

Step 1. $\quad\quad\quad\quad O_3(g) \longrightarrow O_2(g) + O(g) \quad\quad$ rate $= k_1[O_3]$

Reverse: $\quad O_2(g) + O(g) \longrightarrow O_3(g) \quad\quad\quad\quad$ rate $= k_1'[O_2][O]$

Step 2. $\quad\quad O_3(g) + O(g) \longrightarrow 2O_2(g) \quad\quad\quad$ rate $= k_2[O_3][O]$

Reverse: $\quad\quad\quad\quad 2O_2(g) \longrightarrow O_3(g) + O(g) \quad\quad$ rate $= k_2'[O_2]^2$

Overall: $\quad\quad\quad\quad 2O_3(g) \rightleftharpoons 3O_2(g) \quad\quad\quad K_c = \dfrac{[O_2]^3}{[O_3]^2}$

When the overall reaction is at equilibrium, all the individual steps are at equilibrium. The forward and reverse reactions in step 1 are in equilibrium when

$$k_1[O_3] = k_1'[O_2][O]$$

or when

$$\frac{k_1}{k_1'} = \frac{[O_2][O]}{[O_3]}$$

Likewise, the forward and reverse reactions in step 2 are in equilibrium when their rates are equal, or when

$$k_2[O_3][O] = k_2'[O_2]^2$$

This equation may be rearranged into

$$[O] = \frac{k_2'[O_2]^2}{k_2[O_3]}$$

which, when substituted into the expression for k_1/k_1', gives

$$\frac{k_1}{k_1'} = \frac{[O_2]}{[O_3]} \times \frac{k_2'[O_2]^2}{k_2[O_3]}$$

or

$$\frac{k_1}{k_1'} \times \frac{k_2}{k_2'} = \frac{[O_2]^3}{[O_3]^2}$$

This has exactly the same form as the equilibrium constant given by the law of mass action, with K_c the product of the individual rate constants:

$$K_c = \frac{k_1}{k_1'} \times \frac{k_2}{k_2'}$$

In other words, even though the overall rate law cannot be written from the overall equation, the expression for the equilibrium constant does follow from the overall equation. This is true for any mechanism for any reaction.

The equilibrium constant in terms of partial pressures. The total pressure of a gas-phase reaction mixture is readily monitored, and we shall see that it is sometimes convenient to express the equilibrium concentrations for such a reaction in terms of the partial pressures of the reactants and products. (Partial pressures were introduced in Section 5.6.) This is possible because partial pressure and molar concentration are proportional to each other, and either can be used in the definition of the equilibrium constant. The proportionality follows from the ideal gas law $PV = nRT$, which implies that the partial pressure P_X of a gas X is related to its molar concentration [X] by

$$P_X = \frac{n_X}{V} \times RT = [X] \times RT$$

When partial pressures are used, the equilibrium constant is written K_P. For a reaction of the form

$$aA(g) + bB(g) \rightleftharpoons cC(g) + dD(g)$$

K_P is defined as

$$K_P = \frac{(P_C)^c(P_D)^d}{(P_A)^a(P_B)^b} \tag{3}$$

where P_X is the partial pressure of gas X at equilibrium. The reaction quotient Q_P is defined similarly, but with the current partial pressures, which are not necessarily their values at equilibrium. In the ammonia synthesis, for example, we have

$$N_2(g) + 3H_2(g) \rightleftharpoons 2NH_3(g) \qquad K_P = \frac{(P_{NH_3})^2}{(P_{N_2})(P_{H_2})^3}$$

Suppose that, in a typical experiment at 500 K, it is found that $P_{NH_3} = 0.15$ atm, $P_{N_2} = 1.2$ atm, and $P_{H_2} = 0.81$ atm when the reaction reaches equilibrium. Then at that temperature,

$$K_P = \frac{(0.15 \text{ atm})^2}{(1.2 \text{ atm}) \times (0.81 \text{ atm})^3} = 0.035/\text{atm}^2$$

Some values of K_P for other reactions are given in Table 13.3. Note that if K_c has the units M^x or $1/M^x$, the corresponding value of K_P has the units $(\text{atm})^x$ or $1/\text{atm}^x$, respectively.

TABLE 13.3 Equilibrium constants K_P for some reactions

Reaction	Temperature, K	K_P
$N_2(g) + 3H_2(g) \rightleftharpoons 2NH_3(g)$	298	$6.8 \times 10^5/\text{atm}^2$
	400	$41/\text{atm}^2$
	500	$3.5 \times 10^{-2}/\text{atm}^2$
$H_2(g) + I_2(g) \rightleftharpoons 2HI(g)$	298	794
	500	160
	700	54
$2SO_2(g) + O_2(g) \rightleftharpoons 2SO_3(g)$	298	$4.0 \times 10^{24}/\text{atm}$
	500	$2.5 \times 10^{10}/\text{atm}$
	700	$3.0 \times 10^4/\text{atm}$
$N_2O_4(g) \rightleftharpoons 2NO_2(g)$	298	0.98 atm
	400	47.9 atm
	500	1700 atm

The numerical value of K_P and the corresponding value of K_c are in general different, but both describe the equilibrium composition. We can find the relation between their numerical values by expressing each partial pressure P_X as $[X]RT$:

$$K_P = \frac{([C]RT)^c \times ([D]RT)^d}{([A]RT)^a \times ([B]RT)^b}$$

$$= \frac{[C]^c[D]^d}{[A]^a[B]^b} \times \frac{(RT)^{c+d}}{(RT)^{a+b}}$$

$$= K_c \times (RT)^{(c+d)-(a+b)}$$

By letting $\Delta n = (c + d) - (a + b)$, which is the difference between the number of gas molecules on the right and that on the left of the chemical equation, we obtain

$$K_P = K_c \times (RT)^{\Delta n} \tag{4}$$

When the number of gas-phase molecules does not change during the reaction, as in

$$H_2(g) + I_2(g) \rightleftharpoons 2HI(g)$$

Δn is zero, $(RT)^0 = 1$, and K_P and K_c are numerically equal. Thus, for the HI synthesis at 490°C, $K_P = K_c = 46$.

▼ **EXAMPLE 13.3** Relating K_P and K_c

Calculate the value of K_P for the equilibrium $N_2O_4(g) \rightleftharpoons 2NO_2(g)$ at 25°C, given that $K_c = 0.040\ M$ at that temperature.

STRATEGY Since the units of K_c are moles per liter, we anticipate that the units of K_P are atmospheres because each molar concentration converts to a pressure. Equation 4 requires a value for Δn, which we find by writing the chemical equation and identifying the coefficients. At 25°C, $RT = 24.5$ L · atm/mol.

SOLUTION From the chemical equation we know that $\Delta n = 2 - 1 = 1$. Therefore,

$$K_P = K_c \times RT = 0.040\ \frac{\text{mol}}{\text{L}} \times 24.5\ \frac{\text{L} \cdot \text{atm}}{\text{mol}} = 0.98\ \text{atm}$$

EXERCISE Calculate the value of K_P for the synthesis of ammonia at 500°C, given that $K_c = 0.11/M^2$.

[*Answer*: $2.7 \times 10^{-5}/\text{atm}^2$]

▲

13.3 HETEROGENEOUS EQUILIBRIA

Chemical equilibria in which all the substances taking part are in the same phase are called *homogeneous equilibria*. All the equilibria described so far in this chapter (with the exception of precipitation reactions) are homogeneous. Many other reactions, however, lead to *heterogeneous equilibria*, equilibria in which at least one substance is in a different phase from the rest. We discussed examples of heterogeneous physical equilibria in earlier chapters. For instance, the equilibrium responsible for vapor pressure is heterogeneous, since it is an equilibrium between a gas and a liquid. Solubility equilibria are also heterogeneous, since the solution is a liquid and the undissolved solute is a solid or a gas.

A chemical example of a heterogeneous equilibrium is the equilibrium that results from the decomposition of calcium carbonate when it is heated in a sealed container:

$$CaCO_3(s) \xrightarrow{\Delta} CaO(s) + CO_2(g)$$

Experiments have shown that the concentration of CO_2 in equilibrium with the two solids depends only on the temperature and is independent of how much of each solid is present (as long as some is present).

Equilibrium constants for heterogeneous reactions. Equilibrium constants for heterogeneous reactions take an especially simple form. We can see this by writing the expression for the equilibrium constant for the $CaCO_3$ decomposition. According to the law of mass action, we should write

$$K'_c = \frac{[CO_2][CaO]}{[CaCO_3]}$$

where the prime on this equilibrium constant distinguishes it from the simpler version we are about to derive. To derive this simpler version, we need to think about the meaning of the "concentration" of a pure solid, in this case of CaO or $CaCO_3$. In particular, we need to understand that the molar concentration of a pure solid or liquid is independent of the amount present, and hence, although some may get used up or produced during a reaction, the concentration is a constant. To see that this is the case, we note that the molar concentration of a pure solid or liquid, the number of moles per unit volume, is proportional to its density, its mass per unit volume. Then, since density is an intensive property, it follows that molar concentration is intensive as well, and therefore that it is independent of the size of the sample. The following example illustrates this conclusion.

▼ **EXAMPLE 13.4** **Calculating the concentration of a pure solid substance**

Calculate the molar concentration of pure calcium oxide, given that its density is 3.3 g/cm³.

STRATEGY The molar concentration is the number of moles per liter. Therefore, we must express the density as grams per liter and then convert grams of CaO to moles using its molar mass, which is 56 g/mol.

SOLUTION Since 3.3 g/cm³ corresponds to 3.3×10^3 g/L, the molar concentration of pure CaO is

$$[CaO] = \text{moles of CaO per liter}$$

$$= \text{grams of CaO per liter} \times \text{moles of CaO per gram}$$

$$= 3.3 \times 10^3 \frac{\text{g CaO}}{L} \times \frac{1 \text{ mol CaO}}{56 \text{ g CaO}} = 59 \text{ mol/L CaO}$$

Note that the sample size does not enter this calculation. Hence, the same concentration, 59 M, is obtained whatever the size of the sample.

EXERCISE Calculate the concentration of H_2O in pure water at 25°C, at which temperature its density is 1.0 g/mL.

▲   [*Answer*: 55 M]

Both [CaO] and [CaCO₃] are constants in the expression for K_c'. Hence, if we rearrange the expression into

$$\frac{[CaCO_3]K_c'}{[CaO]} = [CO_2]$$

the term on the left is also a constant, which we may call K_c. Therefore, we have

$$K_c = [CO_2]$$

That is, when writing the expression for an equilibrium constant for a heterogeneous reaction, we can ignore the concentrations of any pure solids or liquids taking part in the reaction. The pure substances must be present for the equilibrium to exist, but they do not appear in the expression for K_c. The same applies to the expression for K_P:

$$K_P = P_{CO_2}$$

We now see that the equilibrium constant for the decomposition of calcium carbonate can be measured simply by measuring the pressure of the carbon dioxide that is present at equilibrium. Since at 800°C that pressure is 0.22 atm, $K_P = 0.22$ atm at 800°C.

Decomposition vapor pressure. The heterogeneous equilibrium between a solid and its gaseous decomposition products has a very close resemblance to the heterogeneous equilibrium responsible for the vapor pressure of a solid or liquid (Fig. 13.6). When calcium carbonate is heated to 800°C in a closed container, it decomposes until the partial pressure of carbon dioxide rises to 0.22 atm. At this pressure the reactant and products have reached dynamic equilibrium: the rate of decomposition of CaCO₃ is matched by the rate at which CO₂ molecules recombine with CaO. The only difference between this dynamic equilibrium and that responsible for the vapor pressure of a liquid is that here the condensed phase decomposes. Hence 0.22 atm is called the *decomposition vapor pressure* of calcium carbonate at 800°C. The ordinary vapor pressure thus can be considered to be a special kind of equilib-

FIGURE 13.6 (a) When calcium carbonate is heated, it decomposes and eventually reaches a state of dynamic equilibrium in which CaCO₃, CaO, and CO₂ are all present. This resembles (b) the vaporization of a liquid or solid, but in that process decomposition does not occur.

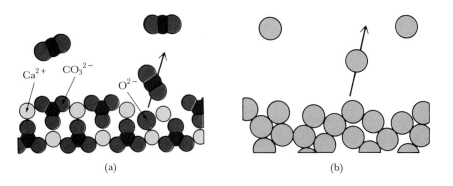

Ca²⁺   CO₃²⁻   O²⁻

(a)   (b)

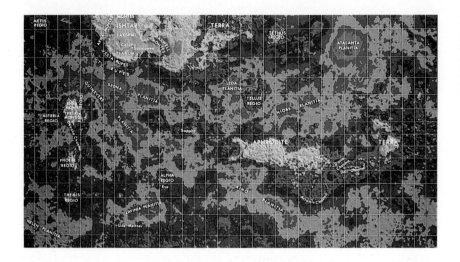

FIGURE 13.7 A radar image of the surface of Venus. Although the rocks are very hot, the partial pressure of carbon dioxide in the atmosphere is so great that carbonates may be abundant.

rium constant—one for a process in which the condensed phase, whether liquid or solid, does not decompose:

$$H_2O(l) \rightleftharpoons H_2O(g) \qquad K_P = P_{H_2O}$$

If calcium carbonate is heated to 800°C in an *open* container, as in a lime kiln or blast furnace, the gas escapes into the surroundings, the partial pressure of the carbon dioxide does not rise to 0.22 atm, equilibrium is never reached, and all the calcium carbonate decomposes. However, if $CaCO_3$ is heated in surroundings that are already so rich in carbon dioxide that its partial pressure exceeds 0.22 atm, then no net decomposition occurs; for every CO_2 molecule that is formed, one is quickly converted back to carbonate. This is probably what occurs on the very hot surface of Venus (Fig. 13.7), where the local partial pressure of carbon dioxide is 87 atm. This high pressure value has led to speculation that the planet's surface is rich in carbonates, in spite of its high temperature (about 500°C).

EQUILIBRIUM CALCULATIONS

The equilibrium constant of a reaction enables us to predict the composition of an equilibrium mixture of any given starting composition. It also allows us to make precise statements about how that composition will change if the conditions—the pressure, the temperature, and the proportions of the reagents—are changed. These applications are important throughout chemistry, for they are used to discuss solubilities; the behavior of acids, bases, and salts; and the outcome of redox reactions. An indication of their wide applicability is that the next four chapters are all based on the properties of equilibrium constants and the calculations related to them.

In this section we introduce the ideas that serve as a foundation for succeeding chapters. We see that there is a single unifying theme that, once grasped, accounts for a very wide range of behavior: at equilibrium, although the individual concentrations of substances in a reaction mixture may vary, collectively the concentrations must satisfy the law of mass action. Only when this condition is satisfied are all the substances taking part in the reaction in dynamic equilibrium with each other. Specifically, we show how to use an equilibrium constant to calculate the equilibrium composition of a reaction mixture. The calculations are sometimes quite tricky, and they are described in order of increasing difficulty.

13.4 SPECIFIC INITIAL CONCENTRATIONS

Some equilibrium calculations are quite straightforward because the concentrations of all the substances except one are known or can be calculated from the data.

One unknown concentration. Suppose we know the equilibrium concentrations of all but one of the substances taking part in a reaction. For example, we might know the concentrations of iodine vapor and hydrogen iodide at equilibrium and want to calculate the concentration of hydrogen. In such a case, if we know the equilibrium constant, we can simply rearrange the expression for K_c to give the unknown in terms of the concentrations that are known.

> ▼ **EXAMPLE 13.5** **Calculating one unknown concentration**
>
> A mixture of hydrogen and iodine was heated to 490°C, for which $K_c = 46$ in the reaction $H_2(g) + I_2(g) \rightleftharpoons 2HI(g)$. The equilibrium concentrations of I_2 and HI were measured spectroscopically as $[I_2] = 3.1$ mM and $[HI] = 2.7$ mM. Calculate the concentration of H_2 in the equilibrium mixture.
>
> **STRATEGY** Since one concentration is unknown, it can be calculated from K_c and the concentrations of the other substances. (If only K_P were available, it could be converted to K_c with Eq. 4.) We write the expression for the equilibrium constant, rearrange it to give the one unknown concentration, and then substitute the data.
>
> **SOLUTION** We first write the equilibrium constant for the reaction,
>
> $$K_c = \frac{[HI]^2}{[H_2][I_2]}$$
>
> and then rearrange it into an equation for the unknown:
>
> $$[H_2] = \frac{[HI]^2}{[I_2]} \times \frac{1}{K_c}$$
>
> Substituting the data then gives
>
> $$[H_2] = \frac{(2.7 \text{ m}M)^2}{3.1 \text{ m}M} \times \frac{1}{46} = 0.051 \text{ m}M$$
>
> **EXERCISE** The equilibrium constant at 500°C for the reaction $H_2(g) + CO_2(g) \rightleftharpoons CO(g) + H_2O(g)$ is $K_c = 0.18$. Calculate the concentration of H_2 in an equilibrium mixture in which $[CO_2] = 11$ mM, $[H_2O] = 3.0$ mM, and $[CO] = 41$ mM.
>
> ▲   [*Answer*: 62 mM]

Decomposition of a single substance. When hydrogen iodide is heated, it decomposes and reaches the equilibrium

$$2HI(g) \rightleftharpoons H_2(g) + I_2(g)$$

We might want to find the composition of the equilibrium mixture, based on a known initial concentration of HI. This can be calculated quite easily from the equilibrium constant, even though there are three unknowns [HI], [H_2], and [I_2], because the concentrations are linked by chemical relations between the substances. Since 1 mol of H_2 is produced for each mole of I_2 produced, the concentration of H_2 is always the same as that of I_2.

Suppose that the initial concentration of HI is C and that at equilibrium the concentrations of H_2 and I_2 are both x. From the relation 2 mol HI = 1 mol H_2, the number of moles of HI molecules that must have decomposed to give x mol of H_2 molecules is

$$\text{Moles of HI decomposed} = x \text{ mol } H_2 \times \frac{2 \text{ mol HI}}{1 \text{ mol } H_2}$$

$$= 2x \text{ mol HI}$$

Hence, the concentration of HI at equilibrium is $C - 2x$. These changes are best shown in a table:

	[HI]	[H_2]	[I_2]
Initial concentration	C	0	0
Change to reach equilibrium	$-2x$	$+x$	$+x$
Concentration at equilibrium	$C - 2x$	x	x

As is shown in the next example, the equilibrium constant can now be written in terms of the concentrations in the last line of the table and then solved for the unknown concentration x.

▼ EXAMPLE 13.6   Calculating the equilibrium composition for a decomposition reaction

Calculate the composition of the equilibrium mixture that results when pure hydrogen iodide is added to a container at a concentration of 2.1 mM and heated to 490°C, at which temperature $K_c = 0.022$.

STRATEGY   The reaction we are considering is

$$2HI(g) \rightleftharpoons H_2(g) + I_2(g) \qquad K_c = \frac{[H_2][I_2]}{[HI]^2} = 0.022$$

Since K_c is small, we expect only a small amount of HI to decompose. We need to substitute the entries in the bottom line of the table above into the expression for K_c, and use the resulting expression to find x. This and the data will give us all the equilibrium concentrations.

SOLUTION   Substituting the expressions for the equilibrium concentrations gives

$$K_c = \frac{x \times x}{(C - 2x)^2} = \left(\frac{x}{C - 2x} \right)^2$$

Taking the square root gives

$$\sqrt{K_c} = \frac{x}{C - 2x}$$

and multiplying through by $C - 2x$ gives

$$\sqrt{K_c}(C - 2x) = x$$

Collecting terms in x, we obtain

$$(1 + 2\sqrt{K_c})x = C\sqrt{K_c}$$

So

$$x = \frac{C\sqrt{K_c}}{1 + 2\sqrt{K_c}}$$

Substituting the data, with $\sqrt{K_c} = \sqrt{0.022} = 0.15$, finally yields

$$x = \frac{2.1 \text{ m}M \times 0.15}{1 + 0.30} = 0.24 \text{ m}M$$

Therefore, at equilibrium, $[H_2] = 0.24 \text{ m}M$, $[I_2] = 0.24 \text{ m}M$, and $[HI] = 2.1 \text{ m}M - 0.48 \text{ m}M = 1.6 \text{ m}M$.

EXERCISE The gas bromine monochloride (BrCl) decomposes into bromine and chlorine and reaches the equilibrium $2BrCl(g) \rightleftharpoons Br_2(g) + Cl_2(g)$, for which $K_c = 32$ at 500 K. If initially pure BrCl is present with $[BrCl] = 3.30 \text{ m}M$, what is its concentration in the mixture at equilibrium?

[*Answer*: 0.27 mM]

▲

Reactants in stoichiometric proportions. The equilibrium composition is also relatively easy to find when the reactants are mixed in *stoichiometric proportions*—that is, in the same proportions as their coefficients in the chemical equation. As an illustration, we consider the equilibrium between nitrogen, oxygen, and nitric oxide that occurs in a hot engine exhaust and contributes to the formation of nitrogen oxides in the environment:

$$N_2(g) + O_2(g) \rightleftharpoons 2NO(g) \qquad K_c = \frac{[NO]^2}{[N_2][O_2]}$$

The two reactants are present in stoichiometric proportions (1 mol of N_2 to 1 mol of O_2) if they both have the same concentration C initially. To determine the equilibrium composition, we assume that at equilibrium the concentration of N_2 has fallen by x. From the relations 1 mol $N_2 = 2$ mol NO and 1 mol $O_2 = 1$ mol N_2, we see that the O_2 concentration will also have fallen by x and the NO concentration will have risen from zero to $2x$:

	$[N_2]$	$[O_2]$	$[NO]$
Initial concentration	C	C	0
Change to reach equilibrium	$-x$	$-x$	$+2x$
Concentration at equilibrium	$C - x$	$C - x$	$2x$

Now we can express the equilibrium constant in terms of the equilibrium concentrations in the last line of the table:

$$K_c = \left(\frac{2x}{C - x}\right)^2$$

This equation can be solved for x, and the equilibrium concentrations found, just as in Example 13.6. Since a decomposition reaction forms products in stoichiometric proportions, it should not be surprising that we obtain similar results for these two cases.

▼ **EXAMPLE 13.7** **Calculating a composition starting from stoichiometric proportions**

Hydrogen and iodine, both in concentrations of 24 mM, are mixed and heated to 490°C in a container. Calculate the equilibrium composition of the reaction mixture.

STRATEGY The equation for the reaction is $H_2(g) + I_2(g) \rightleftharpoons 2HI(g)$, which has the same form as that for the nitric oxide synthesis above, so we can use the same table and the same equation for K_c. Again we rearrange the equation to find x, from which we can find the equilibrium concentrations. The value $K_c = 46$, given in Table 13.2, is larger than 1, so we expect significant HI formation.

SOLUTION Taking the square root of both sides of the equation for K_c gives

$$\sqrt{K_c} = \frac{2x}{C - x}$$

We rearrange this to

$$x = \frac{C\sqrt{K_c}}{2 + \sqrt{K_c}}$$

and substitute $C = 24$ mM and $\sqrt{K_c} = 6.8$ to obtain

$$x = \frac{24 \text{ m}M \times 6.8}{2 + 6.8} = 19 \text{ m}M$$

Therefore, the concentrations at equilibrium are

$$[HI] = 2x = 38 \text{ m}M$$

and

$$[H_2] = [I_2] = C - x = 24 \text{ m}M - 19 \text{ m}M = 5 \text{ m}M$$

EXERCISE The equilibrium constant for the reaction $N_2(g) + O_2(g) \rightleftharpoons 2NO(g)$ at 1200°C is $K_c = 1.00 \times 10^{-5}$. Calculate the equilibrium concentrations of NO, O_2, and N_2 in a container that initially held N_2 and O_2, each at a concentration of 0.114 M.

[*Answer*: $[NO] = 0.36$ mM, $[O_2] = [N_2] = 0.114$ M]

▲

13.5 ARBITRARY INITIAL CONCENTRATIONS

When the substances are not mixed in their stoichiometric proportions, it is still sometimes possible to solve the equilibrium-constant equation in a straightforward manner—but with a little more work.

As an illustration, we again consider the equilibrium between nitrogen, oxygen, and nitric oxide, but this time we assume that in a particular experiment nitrogen is initially present at the molar concentration $[N_2]_0$, and oxygen at $[O_2]_0$. Initially, as before, no nitric oxide is present. The table of changes is now

	$[N_2]$	$[O_2]$	$[NO]$
Initial concentration	$[N_2]_0$	$[O_2]_0$	0
Change to reach equilibrium	$-x$	$-x$	$+2x$
Concentration at equilibrium	$[N_2]_0 - x$	$[O_2]_0 - x$	$2x$

The concentrations in the last row of the table give the equilibrium constant as

$$K_c = \frac{(2x)^2}{([N_2]_0 - x) \times ([O_2]_0 - x)}$$

Direct solution of the equilibrium relation. In this example, and in many other cases that arise, it turns out that we need to solve a quadratic equation for x (Appendix 1C), an equation of the form

$$ax^2 + bx + c = 0 \tag{5}$$

Such an equation has two possible solutions:

$$x = \frac{-b + \sqrt{b^2 - 4ac}}{2a} \quad \text{and} \quad x = \frac{-b - \sqrt{b^2 - 4ac}}{2a} \tag{6}$$

We decide which of these two solutions to use by judging which of them is physically meaningful. For example, no concentration can be negative, so an x that gives a negative concentration can be rejected. Note, however, that x itself is a *change* in the concentration of a substance, so it may be either positive or negative.

▼ **EXAMPLE 13.8** **Calculating an equilibrium composition starting from a mixture**

Phosphorus pentachloride sublimes at 162°C and partially decomposes in the vapor phase to the trichloride. Calculate the concentrations of the reactants and products when the equilibrium

$$PCl_5(g) \rightleftharpoons PCl_3(g) + Cl_2(g) \qquad K_c = 0.800 \, M \text{ at } 340°C$$

is reached, given that initially all three substances are present at $0.120 \, M$.

STRATEGY Because in this case the equilibrium constant is not a pure number (meaning that the numbers of molecules are different on the two sides of the chemical equation), it is not obvious whether a large or small amount of decomposition will occur. We assume that the concentration of PCl_5 decreases by x in decomposing to reach equilibrium, and use the relations

$$1 \text{ mol } PCl_5 = 1 \text{ mol } PCl_3 \qquad 1 \text{ mol } PCl_3 = 1 \text{ mol } Cl_2$$

to construct a table of changes. We then proceed as in previous examples, expressing K_c in terms of x, solving the resulting equation for x, and writing the three equilibrium concentrations using the relations in the table.

SOLUTION The table of changes is

	[PCl₅]	**[PCl₃]**	**[Cl₂]**
Initial concentration	C	C	C
Change to reach equilibrium	$-x$	$+x$	$+x$
Concentration at equilibrium	$C - x$	$C + x$	$C + x$

where $C = 0.120\ M$, the initial concentration for all three substances. The last line gives the equilibrium constant as

$$K_c = \frac{[PCl_3][Cl_2]}{[PCl_5]} = \frac{(C + x) \times (C + x)}{C - x}$$

We can rearrange this expression into an equation for x by multiplying both sides by $C - x$ and expanding the numerator:

$$(C - x) \times K_c = C^2 + 2Cx + x^2$$

Further rearrangement gives the quadratic equation

$$x^2 + (2C + K_c)x + (C^2 - CK_c) = 0$$

and substitution of $C = 0.120\ M$ and $K_c = 0.800\ M$ gives

$$x^2 + (1.040\ M)x - 0.0816\ M^2 = 0$$

Comparison with Eq. 5 shows that the solutions are given by Eq. 6 with $a = 1$, $b = 1.040\ M$, and $c = -0.0816\ M^2$:

$$x = \frac{-1.040\ M + \sqrt{1.408\ M^2}}{2} = 0.0733\ M$$

$$x = \frac{-1.040\ M - \sqrt{1.408\ M^2}}{2} = -1.113\ M$$

The second solution gives the concentrations of PCl_3 and Cl_2 as $C + x = -0.993\ M$, which is impossible. Therefore, we take $x = 0.0733\ M$, which gives the equilibrium concentrations

$$[PCl_5] = C - x$$
$$= 0.120\ M - 0.0733\ M = 0.047\ M$$

$$[PCl_3] = [Cl_2] = C + x$$
$$= 0.120\ M + 0.0733\ M = 0.193\ M$$

EXERCISE Suppose the initial concentrations in this example are $[PCl_5] = 12$ mM, $[PCl_3] = 24$ mM, and $[Cl_2] = 36$ mM. Calculate the equilibrium concentrations at 340°C.
[*Answer*: $[PCl_5] = 2$ mM; $[PCl_3] = 34$ mM; $[Cl_2] = 46$ mM]

▲

Simplification by approximation. When the expression for the equilibrium concentration cannot be solved easily, it is possible to obtain an approximate solution. An example is the synthesis of ammonia:

$$N_2(g) + 3H_2(g) \rightleftharpoons 2NH_3(g) \qquad K_c = \frac{[NH_3]^2}{[N_2][H_2]^3}$$

Suppose the nitrogen and hydrogen had initial concentrations $[N_2]_0 = C$ and $[H_2]_0 = C'$. Then, using the relations 1 mol N_2 = 2 mol NH_3

and 1 mol N_2 = 3 mol H_2, we can construct the table of changes as follows:

	[N_2]	[H_2]	[NH_3]
Initial concentration	C	C'	0
Change to reach equilibrium	$-x$	$-3x$	$+2x$
Concentration at equilibrium	$C - x$	$C' - 3x$	$2x$

In terms of the equilibrium concentrations, K_c is then

$$K_c = \frac{(2x)^2}{(C - x) \times (C' - 3x)^3}$$

This is a very awkward equation to solve for x.

Now, however, suppose we know that the synthesis is being carried out under conditions that favor the reactants more than the products. Then we can assume that only a small amount of product will form, and x is likely to be small compared with C and C'. If that is so, then $C - x$ in the denominator can be approximated by C itself, and $C' - 3x$ can be approximated by C'. We are usually justified in making this approximation if x is no bigger than about 5% of the concentration from which it is being subtracted. For instance, if C were 0.10 M and x were 0.001 M, we could assume that $C - x$ is also 0.10 M, the same as C itself, within the number of significant figures in the data. Therefore, with x much smaller than C (which we write $x \ll C$), we would have

$$K_c = \frac{(2x)^2}{C \times (C')^3}$$

This can be solved for x by rearranging and taking square roots.

▼ **EXAMPLE 13.9 Finding an equilibrium concentration by approximation**

In an experiment at 700 K, at which temperature $K_c = 61/M^2$, a mixture of nitrogen and hydrogen was allowed to reach equilibrium. What was the concentration of ammonia at equilibrium if $[N_2]_0 = 0.010\ M$ and $[H_2]_0 = 0.0020\ M$?

STRATEGY Since the equilibrium constant has units, we cannot tell from its magnitude alone whether the products or the reactants are favored under the conditions of the experiment. One way to proceed is to assume that the proportion of product is small, do the calculations, and check the result to see if the assumption was valid.

SOLUTION We begin with the approximate expression for K_c, as derived above. Taking the square root of both sides gives

$$\sqrt{K_c} = \frac{2x}{\sqrt{C \times (C')^3}}$$

which we rearrange to

$$x = \frac{1}{2} \times \sqrt{K_c} \times \sqrt{C \times (C')^3}$$

Since $\sqrt{K_c} = 7.8/M$, $C = 0.010\ M$, and $C' = 0.0020\ M$, we have

$$x = \frac{1}{2} \times \frac{7.8}{M} \times \sqrt{8.0 \times 10^{-11}\ M^4} = 3.5 \times 10^{-5}\ M$$

Hence, the concentration of NH_3 in the equilibrium mixture is $2x = 7.0 \times 10^{-5}\ M$. Moreover, the assumption that $x \ll C'$ is justified because x is only 2% of C'.

EXERCISE In a similar experiment at 770 K, for which $K_c = 0.11/M^2$, the initial mixture consisted of $0.020\ M\ N_2$ and $0.010\ M\ H_2$. What is the equilibrium concentration of NH_3?

[*Answer:* $4.7 \times 10^{-5}\ M$]

THE RESPONSE OF EQUILIBRIA TO THE CONDITIONS

In the industrial production of chemicals such as NH_3, it is necessary to ensure that the equilibrium concentration of product is high enough for the process to be economical. How can that concentration be raised? By raising the temperature; by lowering it? By raising the pressure or lowering it? Or by what other means?

Chemical equilibria, being dynamic, are responsive to changes in the conditions under which the reaction takes place. If a change in the conditions increases the rate at which reactants change into products, then the equilibrium composition adjusts until the rate of the reverse reaction has risen to match the new forward rate. If a change reduces the rate of the forward reaction, then products decompose into reactants until the two rates are equal again. Since a catalyst affects the rates of both the forward and the reverse reactions equally, it has no effect on the composition of the mixture at equilibrium.

In Section 11.4 we saw that the response of dynamic equilibria to changes in the conditions can often be predicted using Le Chatelier's principle; in this section we apply the principle to chemical equilibria. It is important to realize, however, that Le Chatelier's principle only suggests a likely outcome; it does not provide an explanation or produce numerical values. For an explanation of the effect of changes on an equilibrium, we must examine their effect on the rates of the forward and reverse reactions. For a quantitative prediction of the effect, we have to perform a calculation based on the equilibrium constant.

13.6 THE EFFECT OF ADDED REAGENTS

Suppose we added water to an esterification mixture at equilibrium. According to Le Chatelier's principle, the equilibrium would tend to adjust so as to minimize the increase in the concentration of water (Fig. 13.8). This would be achieved if the reverse reaction formed more acid and alcohol:

$$\text{Ester} + H_2O \longrightarrow \text{acid} + \text{alcohol}$$

Conversely, if instead we had added more acetic acid, the equilibrium composition would have shifted in favor of the ester,

$$\text{Acid} + \text{alcohol} \longrightarrow \text{ester} + H_2O$$

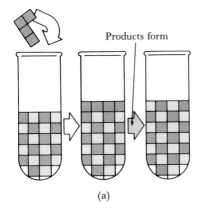

Products form

(a)

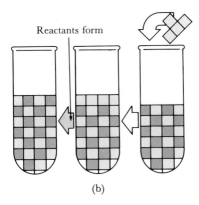

Reactants form

(b)

FIGURE 13.8 (a) When a reactant (blue) is added to a reaction mixture at equilibrium, the products have a tendency to form. (b) When a product (yellow) is added instead, the reactants tend to be formed. (For this reaction, $K_c = 1$.)

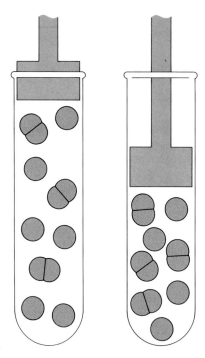

FIGURE 13.9 Le Chatelier's principle predicts that when pressure is applied to a reaction at equilibrium, the composition tends to shift in the direction that corresponds to the smaller number of gas-phase molecules.

The addition of water increases its concentration in the reaction mixture. This increases the rate of the reverse reaction, so more of the ester decomposes. This continues until so much acid and alcohol have formed that the forward rate rises until it matches the new reverse rate. Similarly, adding alcohol to the equilibrium mixture increases the forward rate. Then the concentrations of the products rise until the rate of the reverse reaction matches the new forward rate.

To calculate the effect of added reagent on an equilibrium, we make use of the fact that the equilibrium constant is independent of the individual concentrations. The equilibrium concentrations both before *and after* the extra reagent is added must satisfy the equilibrium expression, which for an esterification is

$$K_c = \frac{[\text{ester}][\text{H}_2\text{O}]}{[\text{acid}][\text{alcohol}]}$$

Since adding water increases the numerator [ester][H$_2$O], the denominator [acid][alcohol] must also increase to preserve the constancy of K_c at the given temperature. Hence, the equilibrium composition must shift in favor of the reactants until equilibrium is restored. We can calculate the increase in the reactant concentrations with one of the methods of the preceding section.

13.7 THE EFFECT OF PRESSURE

Equilibria respond to changes in pressure. We shall concentrate on gas-phase reactions here because they are most affected by pressure. Specifically, we shall examine the effect of pressure on an equilibrium in three different ways, using the predictions of Le Chatelier's principle, the explanation based on reaction rates, and numerical calculations using the equilibrium constant.

Predicting the effect of pressure from Le Chatelier's principle. According to Le Chatelier's principle, a gas-phase equilibrium responds to an increase in pressure by tending to shift in the direction that minimizes the increase. Since the formation of NH$_3$ from N$_2$ and H$_2$ decreases the number of gas molecules in the container, and hence the pressure the sample exerts, the equilibrium composition will tend to shift in favor of product. Therefore, to increase the yield of ammonia in the Haber process, the synthesis should be carried out at high pressure. The actual industrial process uses pressures of 250 atm and more. Some other equilibria respond similarly: when the pressure is increased, a reaction at equilibrium tends to adjust so as to reduce the number of molecules in the gas phase (Fig. 13.9).

▼ **EXAMPLE 13.10** **Predicting the effect of pressure**

Predict the effect of an increase in pressure on the equilibrium compositions for the reactions (a) N$_2$O$_4$(g) $\rightleftharpoons$ 2NO$_2$(g) and (b) H$_2$(g) + I$_2$(g) $\rightleftharpoons$ 2HI(g).

STRATEGY A glance at the chemical equation is normally enough to show which direction corresponds to a decrease in the number of gas-phase molecules. The composition of the equilibrium mixture will tend to shift in that direction.

SOLUTION (a) In the reverse reaction, two NO_2 molecules combine to form one N_2O_4 molecule. Hence, an increase in pressure favors the formation of N_2O_4. The effect is illustrated in Fig. 13.10.

(b) Since neither direction corresponds to a reduction of gas-phase molecules, increasing the pressure should have little effect on the composition of the equilibrium mixture.

EXERCISE Predict the effect of an increase in pressure on the equilibrium compositions for (a) $CH_4(g) + H_2O(g) \rightleftharpoons CO(g) + 3H_2(g)$ and (b) C(diamond) $\rightleftharpoons$ C(graphite). For the latter, consider the densities of the two solids, which are 3.5 g/cm³ for diamond and 2.0 g/cm³ for graphite.

[*Answer*: (a) Reactants favored; (b) diamond favored]

▲

Explaining the effect of pressure. To explain the effect of increased pressure on an equilibrium mixture, we look at the effect it has on the forward and reverse reaction rates. In the case of the $N_2O_4(g) \rightleftharpoons 2NO_2(g)$ equilibrium, the separation of N_2O_4 into two NO_2 molecules is a first-order process with a rate proportional to the pressure, whereas the recombination of the NO_2 is second-order, with a rate proportional to the square of the pressure. The higher-order reverse reaction rate depends more strongly on the pressure than the forward reaction rate and increases more rapidly as the pressure is increased. Thus the new equilibrium composition has a higher concentration of N_2O_4.

Predicting the effect of pressure. When the pressure on a sample is increased by pushing in a piston, the volume of the sample is decreased. This raises the concentrations of all the substances in the sample. We have already seen that the value of K_c is independent of the individual concentrations, so an increase in pressure leaves K_c unchanged. However, although the overall equilibrium constant is unchanged, the *individual* concentrations do in general adjust.

Consider the gas-phase equilibrium

$$N_2O_4(g) \rightleftharpoons 2NO_2(g) \qquad K_c = \frac{[NO_2]^2}{[N_2O_4]}$$

FIGURE 13.10 When pressure is applied to a gas-phase reaction, the equilibrium composition shifts in the direction that minimizes the increase in pressure. Note the lightening of the brown color of this equilibrium mixture (N_2O_4, colorless $\rightleftharpoons 2NO_2$, brown) as pressure is applied. Note also that immediately after the pressure is applied the gas is dark because the concentration of NO_2 has increased and has not yet had time to form N_2O_4.

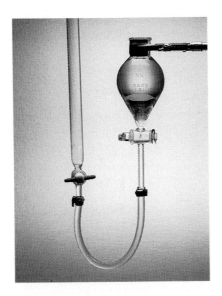 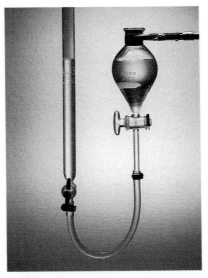 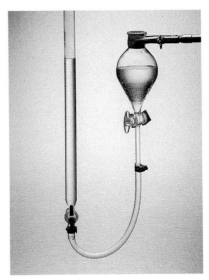

The molar concentration of each substance at equilibrium is the number of moles n of that substance present, divided by the total volume V:

$$[NO_2] = \frac{n_{NO_2}}{V} \qquad [N_2O_4] = \frac{n_{N_2O_4}}{V}$$

Hence,

$$K_c = \frac{n_{NO_2}{}^2 \times (1/V)^2}{n_{N_2O_4} \times 1/V} = \frac{n_{NO_2}{}^2}{n_{N_2O_4}} \times \frac{1}{V}$$

In order for this expression to remain constant when the volume is decreased and $1/V$ is increased, the factor multiplying $1/V$ must decrease. That is, when we confine the equilibrium mixture into a smaller volume, the number of moles of NO_2 must decrease and the number of moles of N_2O_4 must increase.

Suppose, however, that, at equilibrium, we increase the pressure within a reaction vessel by pumping in argon or some other inert gas. In this case the equilibrium composition is unaffected, because the reacting gases themselves continue to occupy the *same volume* and their individual concentrations remain unchanged.

13.8 THE EFFECT OF TEMPERATURE

Le Chatelier's principle predicts that a chemical equilibrium, like a solubility equilibrium, will respond to an increase in temperature by absorbing heat. That is, an increase in the temperature of a reaction mixture at equilibrium tends to shift the composition in the endothermic direction (Fig. 13.11).

▼ **EXAMPLE 13.11** Predicting the effect of temperature

Predict how the equilibrium composition of the ammonia synthesis mixture will tend to change when its temperature is raised.

STRATEGY The equilibrium composition will tend to shift in the direction that corresponds to endothermic reaction. We therefore need to know whether the formation of NH_3 or its decomposition is endothermic. We can use a table of standard enthalpies of formation (Table 6.4 or Appendix 2A) to help us decide.

SOLUTION From Table 6.4, we have

$$N_2(g) + 3H_2(g) \longrightarrow 2NH_3(g) \qquad \Delta H^\circ = -92 \text{ kJ}$$

Since the synthesis is exothermic, the reverse reaction is endothermic. Hence, heating the equilibrium mixture favors the decomposition of NH_3 to N_2 and H_2.

EXERCISE Predict the effect of raising the temperature on the equilibria (a) $N_2O_4(g) \rightleftharpoons 2NO_2(g)$ and (b) $O_2(g) + 2CO(g) \rightleftharpoons 2CO_2(g)$, using data from Table 6.4 or Appendix 2A.

▲   [*Answer*: (a) NO_2 favored; (b) CO and O_2 favored]

Explaining the effect of temperature. The explanation of the effect of temperature on equilibria is found in its effect on the forward and reverse reaction rates. As explained in Section 12.5, the higher the

FIGURE 13.11 When the equilibrium mixture for an endothermic reaction is heated, the equilibrium shifts toward products. The reaction here is the dehydration of cobalt(II) chloride, which changes color as it becomes anhydrous.

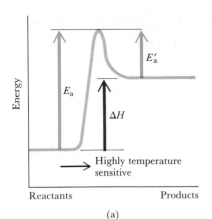

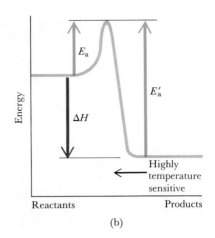

(a)   (b)

activation energy of a reaction, the more sensitive its rate is to temperature. Figure 13.12 shows that the activation energy of an endothermic reaction is larger than that of its reverse reaction. Therefore, the forward reaction rate increases more rapidly with temperature than the reverse reaction rate. As a result, when the temperature of the equilibrium mixture is raised, more reactants are converted to more products, until the concentration of products has risen enough for the reverse reaction rate to match the forward rate. The same argument applies to an exothermic reaction, but now the reverse reaction is more sensitive to temperature and generates more reactants when the temperature is raised.

The numerical calculation of the effect of temperature is quite straightforward: we simply use the value of K_c (or K_P) for the new temperature in calculations like those described already.* Values at various temperatures are given in Tables 13.2 and 13.3.

Haber's achievement. We can now comprehend Haber's problem, approach, and achievement in designing a process for the commercial production of ammonia. Since the synthesis of ammonia is exothermic, low temperatures favor the product. This is shown by the large decrease in the equilibrium constant: from $6.8 \times 10^5/atm^2$ at 25°C to $7.8 \times 10^{-5}/atm^2$ at 450°C, a change of 10 orders of magnitude. However, the rate at which nitrogen and hydrogen combine is virtually zero near room temperature, and at that temperature the reaction proceeds so slowly that it never reaches equilibrium. Haber was therefore faced with a dilemma. He had to use high temperatures to achieve an acceptable *rate* of conversion, but if he did, the *extent* of conversion would be very low.

---

*It is best to use the measured value of the equilibrium constant at the new temperature. However, it is also possible to estimate its value from an equation deduced by van't Hoff:

$$\ln \frac{K_P'}{K_P} = \frac{\Delta H°}{R} \times \left(\frac{1}{T} - \frac{1}{T'} \right)$$

where K_P' is the equilibrium constant at the temperature T', K_P the constant at T, and $\Delta H°$ the standard reaction enthalpy at the temperature T.

FIGURE 13.13 Fritz Haber
(1868–1934) (left) and Carl Bosch
(1874–1940) (right).

FIGURE 13.13 Fritz Haber
(1868–1934) (left) and Carl Bosch
(1874–1940) (right).

A part of the solution, as we saw in Section 12.7, was to use a catalyst. Haber found that iron oxide was reduced to porous iron in the hydrogen atmosphere within the reaction vessel, and that the iron lowered the activation energy for the reaction. As a result, the synthesis took place more quickly at moderately low temperatures. However, the presence of a catalyst speeds the reverse reaction as well as the forward reaction, leaving the equilibrium constant unchanged. In other words, although Haber achieved a higher rate of formation for ammonia, he simultaneously achieved a higher rate of decomposition! Although the equilibrium composition was reached more quickly, it contained very little ammonia.

Haber's solution to the equilibrium problem was to increase the pressure at which the synthesis took place. As we have seen, this shifts the equilibrium mixture in favor of the product and makes the process economical. This is where Carl Bosch (Fig. 13.13), Haber's chemical-

FIGURE 13.14 One of the high-pressure containers inside which the catalytic synthesis of ammonia takes place.

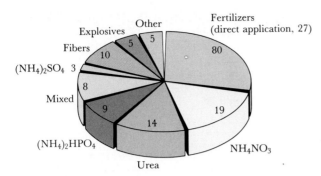

Fertilizers
(direct application, 27)

80

Other
5

Explosives
5

Fibers
10

(NH₄)₂SO₄ 3

8

Mixed

9

14

19

(NH₄)₂HPO₄

Urea

NH₄NO₃

FIGURE 13.15 The Haber process is still used to produce virtually all the ammonia manufactured in the world. This chart shows how the ammonia is used; most of it is used as fertilizer, either directly or after conversion to another compound.

engineer colleague, earned his Nobel Prize. He had to design the first high-temperature, high-pressure catalytic industrial process (Fig. 13.14), working at the limits of technology. He succeeded, and the Haber-Bosch process, which led to two Nobel Prizes, is still essentially the sole source of the ammonia produced worldwide (Fig. 13.15).

SUMMARY

13.1 Chemical equilibrium is a state of dynamic equilibrium in which the rate of formation of products is equal to their rate of decay back into reactants.

13.2 The composition of the reaction mixture at equilibrium is described by its **equilibrium constant** K_c. According to the **law of mass action,** for the general reaction $aA + bB \rightleftharpoons cC + dD$, the concentrations at equilibrium satisfy $K_c = [C]^c[D]^d/[A]^a[B]^b$. The equilibrium constant summarizes the combination of concentrations that guarantees that the rates of the overall forward and reverse reactions are equal. At equilibrium, although individual concentrations may vary, collectively they must satisfy the relation required by K_c (or K_P). In a multistep reaction, each step is at equilibrium when the overall reaction is at equilibrium. For a reaction with the same number of molecules on each side of the equation, the size of the equilibrium constant indicates whether the formation of products or reactants is favored. In general, we compare the **reaction quotient** with the equilibrium constant to determine in which direction a reaction will proceed.

13.3 Equilibria may be classified as **homogeneous** or **heterogeneous.** Equilibrium constants for the latter are written by ignoring the pure solids and liquids taking part in the reaction. Vapor pressures and solubilities, including **decomposition vapor pressures,** in which a substance decomposes as it vaporizes, are examples of heterogeneous equilibria.

13.4 and 13.5 Equilibrium calculations are carried out on the basis that the value of the equilibrium constant is fixed for a given temperature, and the equilibrium concentrations must combine to produce that value of K_c. A change in concentrations or pressure will produce a change in the equilibrium concentrations, but not in K_c. When the change results in a complicated equation, the computation can sometimes be simplified by assuming that a change in concentration is small compared with the concentration itself. This assumption must be checked at the end of the calculation. The calculation is also simplified if the substances are present in **stoichiometric proportions,** the same proportions as the stoichiometric coefficients in the chemical equation.

13.6 to 13.8 Equilibria are unaffected by the presence of a catalyst. They respond to pressure and temperature in accordance with Le Chatelier's principle. Reactions that reduce the number of gas-phase molecules are favored by an increase in pressure. Endothermic reactions are favored by an increase in temperature.

EXERCISES

Expressions for Equilibrium Constants

In Exercises 13.1 and 13.2, write the expression for the equilibrium constant K_c for each reaction, and state its units.

13.1 (a) $CO(g) + Cl_2(g) \rightleftharpoons COCl(g) + Cl(g)$
(b) $2SO_2(g) + O_2(g) \rightleftharpoons 2SO_3(g)$
(c) $H_2(g) + Br_2(g) \rightleftharpoons 2HBr(g)$
(d) $2O_3(g) \rightleftharpoons 3O_2(g)$

13.2 (a) $2NO(g) + O_2(g) \rightleftharpoons 2NO_2(g)$
(b) $SbCl_5(g) \rightleftharpoons SbCl_3(g) + Cl_2(g)$
(c) $2CH_3COOH(g) \rightleftharpoons (CH_3COOH)_2(g)$
(d) $N_2(g) + 2H_2(g) \rightleftharpoons N_2H_4(g)$

13.3 For the reaction $N_2(g) + 3H_2(g) \rightleftharpoons 2NH_3(g)$ at 400 K, $K_P = 41/atm^2$. Find the values (at the same temperature) of the equilibrium constants for the reactions
(a) $2NH_3(g) \rightleftharpoons N_2(g) + 3H_2(g)$
(b) $2N_2(g) + 6H_2(g) \rightleftharpoons 4NH_3(g)$
(c) $\frac{1}{2}N_2(g) + \frac{3}{2}H_2(g) \rightleftharpoons NH_3(g)$

13.4 The equilibrium constant for the reaction $2SO_2(g) + O_2(g) \rightleftharpoons 2SO_3(g)$ has the value $K_P = 2.5 \times 10^{10}/atm$ at 500 K. Find the values of the equilibrium constants for the following reactions at the same temperature:
(a) $SO_2(g) + \frac{1}{2}O_2 \rightleftharpoons SO_3(g)$
(b) $SO_3(g) \rightleftharpoons SO_2(g) + \frac{1}{2}O_2(g)$
(c) $3SO_2(g) + \frac{3}{2}O_2 \rightleftharpoons 3SO_3(g)$

The Calculation of Equilibrium Constants from Experimental Data

In Exercises 13.5 to 13.10, calculate the equilibrium constant K_c if concentration data are given or K_P if pressure data are given.

13.5 For $H_2(g) + I_2(g) \rightleftharpoons 2HI(g)$ at 460°C:

$[H_2(g)]$ at equilibrium, mM	$[I_2(g)]$ at equilibrium, mM	$[HI(g)]$ at equilibrium, mM
6.47	0.594	13.7
3.84	1.52	16.9
1.43	1.43	10.0

13.6 For $CH_3COOH + C_2H_5OH \rightleftharpoons CH_3COOC_2H_5 + H_2O$ at 100°C, with no water present initially:

$[CH_3COOH]$ initially, M	$[C_2H_5OH]$ initially, M	$[CH_3COOC_2H_5]$ at equilibrium, M
1.00	0.18	0.171
1.00	1.00	0.667
1.00	8.00	0.966

13.7 For $N_2O_4(g) \rightleftharpoons 2NO_2(g)$ at 25°C, with no NO_2 present initially, find K_c:

$[N_2O_4]$ initially, mM	Total pressure P, atm
6.28	0.212
13.6	0.425

13.8 For $2SO_2(g) + O_2(g) \rightleftharpoons 2SO_3(g)$ at 1000 K:

P_{SO_2} at equilibrium, atm	P_{O_2} at equilibrium, atm	P_{SO_3} at equilibrium, atm
0.309	0.353	0.338
0.456	0.180	0.364
0.564	0.102	0.333

13.9 For $NH_4HS(s) \rightleftharpoons NH_3(g) + H_2S(g)$ at 24°C:

P_{NH_3} at equilibrium, atm	P_{H_2S} at equilibrium, atm
0.309	0.309
0.364	0.258
0.539	0.174

13.10 For $N_2O_4 \rightleftharpoons 2NO_2$ in chloroform solution at 25°C:

$[N_2O_4]$ at equilibrium, mM	$[NO_2]$ at equilibrium, mM
0.129	1.17
0.227	1.61
0.405	2.13

13.11 When 2.20 g of hydrogen iodide is heated to 500 K in a sealed container, the resulting equilibrium mixture contains 1.90 g of hydrogen iodide. Calculate K_c for the decomposition reaction.

13.12 When 1.00 g of iodine is heated to 1000 K in a sealed container, the resulting equilibrium mixture contains 0.83 g of iodine molecules. Calculate K_c for the dissociation of $I_2(g)$ at this temperature.

13.13 When 25.0 g of ammonium carbamate ($NH_2CO_2NH_4$) was placed in an evacuated 250-mL flask and kept at 25°C, the pressure rose to 88 Torr. What is the value of K_P for the decomposition of this substance into ammonia and carbon dioxide?

13.14 Carbon monoxide and water vapor, each at 200 Torr pressure, were introduced into a 250-mL container. When the mixture had reached equilibrium at

700°C with its reaction products, CO_2 and H_2, the partial pressure of CO_2 was 88 Torr. Calculate the value of K_c for the reaction.

The Relation Between K_c and K_P

In Exercises 13.15 and 13.16, deduce the value of K_c for each reaction from the given value of K_P.

13.15 (a) For $N_2O_4(g) \rightleftharpoons 2NO_2(g)$, given that $K_P = 0.62$ atm at 45°C
(b) For $CaCO_3(s) \rightleftharpoons CaO(s) + CO_2(g)$, when $K_P = 167$ Torr at 1073 K
13.16 (a) For $2SO_2(g) + O_2(g) = 2SO_3(g)$, when $K_P = 3.4$/atm at 1000 K
(b) For $NH_4HS(s) \rightleftharpoons NH_3(g) + H_2S(g)$, when $K_P = 9.4 \times 10^{-2}$ atm² at 24°C

13.17 The equilibrium constant for the synthesis of ammonia is reported as 41/atm² at 400 K. What is its value in units of 1/Torr²?
13.18 The equilibrium constant for a reaction is reported as 1.5×10^{-5}/Torr³. What is its value in 1/atm³?

Homogeneous and Heterogeneous Equilibria

In Exercises 13.19 and 13.20, classify each equilibrium as homogeneous or heterogeneous.

13.19 (a) $Ca(OH)_2(aq) + CO_2(g) \rightleftharpoons CaCO_3(s) + H_2O(l)$
(b) $P(s, white) \rightleftharpoons P(s, red)$
(c) $H_2CO_3(aq) \rightleftharpoons H^+(aq) + HCO_3^-(aq)$
(d) $2KNO_3(s) \rightleftharpoons 2KNO_2(s) + O_2(g)$
13.20 (a) $H_2O(s) \rightleftharpoons H_2O(l)$
(b) $NH_4Cl(g) \rightleftharpoons NH_3(g) + HCl(g)$
(c) $4KClO_3(s) \rightleftharpoons 3KClO_4(s) + KCl(s)$
(d) $H_2O(l) \rightleftharpoons H^+(aq) + OH^-(aq)$

In Exercises 13.21 and 13.22, give three examples of heterogeneous equilibria involving *physical* changes of each substance.

13.21 (a) Water; (b) sulfur; (c) phosphorus.
13.22 (a) Ammonia; (b) carbon; (c) oxygen.

In Exercises 13.23 and 13.24, give an example of a homogeneous *chemical* equilibrium involving each group of substances and any others that are necessary to balance the equation.

13.23 (a) $N_2(g)$ and $NO(g)$; (b) $SO_2(g)$ and $SO_3(g)$; (c) $CH_3COOH(aq)$, $H^+(aq)$, and $CH_3CO_2^-(aq)$; (d) $H_2S(aq)$, $KOH(aq)$, $KHS(aq)$, and $H_2O(l)$.
13.24 (a) NH_3, NH_4^+, and OH^-; (b) CO_3^{2-} and HCO_3^-; (c) $NH_3(g)$, $O_2(g)$, and $NO(g)$; (d) $NO(g)$, $O_2(g)$, and $NO_2(g)$.

In Exercises 13.25 and 13.26, write the equation for K_c for each reaction.

13.25 (a) $MgCO_3(s) \rightleftharpoons MgO(s) + CO_2(g)$
(b) $2Br^-(aq) + Cl_2(g) \rightleftharpoons Br_2(g) + 2Cl^-(aq)$
(c) $Cu(s) + Cl_2(g) \rightleftharpoons CuCl_2(s)$
(d) $NH_4NO_3(s) \rightleftharpoons N_2O(g) + 2H_2O(g)$
13.26 (a) $NH_4Cl(s) \rightleftharpoons NH_3(g) + HCl(g)$
(b) $2KNO_3(s) \rightleftharpoons 2KNO_2(s) + O_2(g)$
(c) $AgCl(s) \rightleftharpoons Ag^+(aq) + Cl^-(aq)$
(d) $2Al(s) + 2OH^-(aq) + 6H_2O(l) \rightleftharpoons 2[Al(OH)_4]^-(aq) + 3H_2(g)$

In Exercises 13.27 and 13.28, calculate the molar concentration of each pure substance.

13.27 (a) $H_2O(s)$, of density 0.92 g/cm³ at 0°C
(b) $Cu(s)$, of density 8.94 g/cm³ at 25°C
(c) ethanol, of density 0.789 g/mL at 25°C
(d) $U(s)$, of density 18.95 g/cm³ at 25°C
13.28 (a) $H_2O(l)$, of density 0.9998 g/mL at 0°C
(b) methanol, of density 0.793 g/mL at 25°C
(c) $Fe(s)$, of density 7.86 g/cm³ at 25°C
(d) $H_2(l)$, of density 0.070 g/mL at −267°C.

In Exercises 13.29 and 13.30, express K_P for the equilibrium in terms of the *total* pressure of the system.

13.29 $NH_4HS(s) \rightleftharpoons NH_3(g) + H_2S(g)$, in the absence of an excess of either gas
13.30 $NH_2CO_2NH_4(s) \rightleftharpoons 2NH_3(g) + CO_2(g)$, in the absence of an excess of either gas

The Interpretation of the Equilibrium Constant

In Exercises 13.31 and 13.32, calculate the reaction quotient for each mixture from the data, and decide whether the mixture has a tendency to form reactants or products.

13.31 (a) A mixture consisting of 4.8 mM $H_2(g)$, 2.4 mM $I_2(g)$, and 2.4 mM $HI(g)$ at 460°C ($K_c = 49$)
(b) Equal concentrations of the three gases that appear in part (a)
13.32 A mixture that contains 1.0 M acetic acid, 2.0 M ethanol, 0.50 M ethyl acetate, and 5.0 M water, heated to 100°C ($K_c = 4.0$).

13.33 A 500-mL container was filled with 1.20 mmol of $SO_2(g)$, 0.50 mmol of $O_2(g)$, and 0.10 mmol of $SO_3(g)$ and heated to 700 K, at which temperature $K_P = 3.0 \times 10^4$/atm. Will more sulfur trioxide form?
13.34 Given that $K_P = 3.6 \times 10^{-2}$/atm² for the reaction $N_2(g) + 3H_2(g) \rightleftharpoons 2NH_3(g)$ at 500 K, calculate whether more ammonia will tend to form when a mixture of composition $[N_2] = 2.23$ mM, $[H_2] = 1.24$ mM, and $[NH_3] = 0.112$ mM is heated to 500 K.

Equilibrium Calculations

In Exercises 13.35 to 13.38, calculate the missing equilibrium concentration or its partial pressure, using data from Tables 13.2 and 13.3 as needed.

13.35 In a gas-phase equilibrium mixture of H_2, I_2, and HI at 490°C, [HI] = 2.21 mM and [I_2] = 1.46 mM.

13.36 In a gas-phase equilibrium mixture of H_2, Cl_2, and HCl at 500 K, [HCl] = 1.45 mM and [Cl_2] = 2.45 mM.

13.37 In an equilibrium mixture of PCl_5, PCl_3, and Cl_2 at 500 K, P_{PCl_5} = 0.15 atm and P_{Cl_2} = 0.20 atm. (K_P = 25 atm.)

13.38 In an equilibrium mixture of $SbCl_5$, $SbCl_3$, and Cl_2 at 500 K, P_{SbCl_5} = 0.15 atm and P_{SbCl_3} = 0.20 atm. (K_P = 3.5 × 10^{-4} atm.)

13.39 Calculate [OH$^-$] and [H$^+$] in pure water at 25°C, given that [H$^+$][OH$^-$] = 1.1 × 10^{-14} M^2 at equilibrium.

13.40 Calculate [Ag$^+$] and [S^{2-}] in a solution of silver sulfide in water at 25°C, given that [Ag$^+$]^2[S^{2-}] = 6.3 × 10^{-50} M^3 at equilibrium.

13.41 Calculate the solubility in g/L of silver chloride in (a) pure water and (b) 0.1 M NaCl(*aq*) at 25°C, given that [Ag$^+$][Cl$^-$] = 1.8 × 10^{-10} M^2 at equilibrium.

13.42 Calculate the solubility in g/L of lead(II) iodide in (a) pure water and (b) 0.1 M KI(*aq*) at 25°C, given that [Pb^{2+}][I$^-$]^2 = 1.0 × 10^{-9} M^3 at equilibrium.

13.43 An *isomerization* is a reaction in which the atoms of a molecule change their bonding arrangement; as no atoms are lost or gained, the equilibrium can be denoted A $\rightleftharpoons$ A′, where A and A′ are the two isomers. Find an expression for the equilibrium concentration of A in terms of its initial concentration [A]$_0$ and the equilibrium constant K_c.

13.44 Find an expression for the equilibrium concentration of A in the decomposition A $\rightleftharpoons$ C + D in terms of K_c and the initial concentration [A]$_0$.

In Exercises 13.45 to 13.50, calculate the equilibrium concentrations of the reactants and products for each decomposition reaction, using the data in Tables 13.2 and 13.3.

13.45 $Cl_2(g) \rightleftharpoons 2Cl(g)$ at (a) 1000 K and (b) 1200 K, starting with [Cl_2] = 1.0 mM

13.46 $F_2(g) \rightleftharpoons 2F(g)$ at (a) 500 K and (b) 1000 K, from the initial concentration [F_2] = 2.0 mM. Comment on the likely reactivities of chlorine and fluorine in light of this result and that in Exercise 13.45.

13.47 $2HBr(g) \rightleftharpoons H_2(g) + Br_2(g)$ at 500 K, starting with [HBr] = 1.2 mM

13.48 $2BrCl(g) \rightleftharpoons Br_2(g) + Cl_2(g)$ at 500 K, starting with [BrCl] = 1.4 mM

13.49 $PCl_5(g) \rightleftharpoons PCl_3(g) + Cl_2(g)$ at 400 K, for which K_P = 0.36 atm, starting with 1.0 g of phosphorus pentachloride in a 250-mL container.

13.50 $PCl_5(g) \rightleftharpoons PCl_3(g) + Cl_2(g)$ at 500 K, for which K_P = 25 atm, starting with 2.0 g of phosphorus pentachloride in a 300-mL container.

13.51 Let the equilibrium constants for the reactions $2H_2O(g) \rightleftharpoons 2H_2(g) + O_2(g)$ and $2CO_2(g) \rightleftharpoons 2CO(g) + O_2(g)$ be K_{P1} and K_{P2}, respectively. Show that the equilibrium constant for the reaction $CO_2(g) + H_2(g) \rightleftharpoons H_2O(g) + CO(g)$ is $K_{P3} = \sqrt{K_{P2}/K_{P1}}$, and evaluate it at 1565 K, at which temperature K_{P1} = 1.6 × 10^{-11} atm and K_{P2} = 1.3 × 10^{-10} atm.

13.52 Suppose that in the esterification reaction an amount A of acid and B of alcohol are mixed and heated to 100°C. Find an expression for the number of moles of ester that is present at equilibrium, in terms of A, B, and K_c. Evaluate the expression for A = 1.0 mol, B = 0.50 mol, and K_c = 3.5.

Response to Pressure

In Exercises 13.53 and 13.54, state whether reactants or products will be favored by an increase in the total pressure on each equilibrium.

13.53 (a) $2HD(g) \rightleftharpoons H_2(g) + D_2(g)$
(b) $Cl_2 \rightleftharpoons 2Cl(g)$
(c) $2O_3(g) \rightleftharpoons 3O_2(g)$
(d) $2NO(g) + O_2(g) \rightleftharpoons 2NO_2(g)$

13.54 (a) $Pb(NO_3)_2(s) \rightleftharpoons PbO(s) + NO_2(g) + O_2(g)$
(b) $2SO_2(g) + O_2(g) \rightleftharpoons 2SO_3(g)$
(c) $3NO_2(g) + H_2O(l) \rightleftharpoons 2HNO_3(aq) + NO(g)$
(d) $H_2O(g) + C(s) \rightleftharpoons H_2(g) + CO(g)$

13.55 The density of quartz (SiO_2) is greater than that of the glassy form of silica (also SiO_2). Would glass or quartz be favored under pressure?

13.56 The density of red phosphorus is 2.34 g/cm^3, and that of its white allotrope is 1.82 g/cm^3. Would you expect the red or the white variety to be favored under pressure?

*13.57 Let α be the fraction of PCl_5 molecules that have decomposed to PCl_3 and Cl_2, so that the amount of PCl_5 at equilibrium is $n(1 - \alpha)$, where the amount initially present is n. Write an equation for K_P in terms of α and the total pressure P, and solve it for α in terms of P. Calculate the fraction decomposed at 556 K, for which K_P = 4.96 atm, and pressures of (a) 0.5 atm and (b) 1.0 atm.

*13.58 Let α be the fraction of F_2 molecules that have dissociated (as in Exercise 13.57). Write an equation for K_P in terms of α and the total pressure P, and solve it for α in terms of P. Calculate the fraction decomposed at 800 K, for which K_P = 6.9 × 10^{-5} atm, and a pressure of (a) 0.5 atm and (b) 1.0 atm.

*13.59 Express the equilibrium $N_2O_4(g) \rightleftharpoons 2NO_2(g)$ in terms of the fraction α of N_2O_4 that has dissociated and the total pressure P of the reaction mixture, and show that when the extent of dissociation is small ($\alpha \ll 1$), α is inversely proportional to the square root of the total pressure ($\alpha \propto 1/\sqrt{P}$).

*__13.60__ Express K_P for the ammonia synthesis in terms of the total pressure P and the fraction α of nitrogen that has reacted in an initially stoichiometric mixture of nitrogen and hydrogen. Show that when α is small ($\alpha \ll 1$), it is proportional to the total pressure ($\alpha \propto P$).

The Response to Temperature

In Exercises 13.61 and 13.62, predict, using the information in Table 6.4 and Appendix 2A, whether each equilibrium will be shifted in favor of products by an increase in temperature.

__13.61__ (a) $N_2O_4(g) \rightleftharpoons 2NO_2(g)$
(b) $X_2(g) \rightleftharpoons 2X(g)$, where X is any element
(c) $CO_2(g) + 2NH_3(g) \rightleftharpoons$
$\qquad\qquad\qquad CO(NH_2)_2(g) + H_2O(g)$
(d) $Ni(s) + 4CO(g) \rightleftharpoons Ni(CO)_4(g)$

__13.62__ (a) Melting
(b) $CH_4(g) + H_2O(g) \rightleftharpoons CO(g) + 3H_2(g)$
(c) $CO(g) + H_2O(g) \rightleftharpoons CO_2(g) + H_2(g)$
(d) $2SO_2(g) + O_2(g) \rightleftharpoons 2SO_3(g)$

__13.63__ A mixture consisting of 2.23 mmol of N_2 and 6.69 mmol of H_2 in a 500-mL container was heated to 600 K (for which $K_P = 1.7 \times 10^{-3}$/atm^2) and allowed to reach equilibrium (with the product ammonia). Will more ammonia be formed if that equilibrium mixture is heated to 700 K, for which $K_P = 7.8 \times 10^{-5}$/atm^2?

__13.64__ A mixture consisting of 1.1 mmol of SO_2 and 2.2 mmol of O_2 in a 250-mL container was heated to 500 K (at which temperature $K_P = 2.5 \times 10^{10}$/atm) and

allowed to reach equilibrium (with the product SO_3). Will more sulfur trioxide form if that equilibrium mixture is cooled to 25°C, for which $K_P = 4.0 \times 10^{24}$/atm?

General

*__13.65__ When 1.00 g of phosphorus pentachloride was heated to 400 K in a closed 1.0-L container, analysis showed that 0.91 g remained at equilibrium. Calculate the value of K_c for the decomposition $PCl_5(g) \rightleftharpoons PCl_3(g) + Cl_2(g)$.

*__13.66__ The van't Hoff equation relates the equilibrium constant K'_P at temperature T' to its value K_P at T:

$$\ln \frac{K'_P}{K_P} = \frac{\Delta H°}{R} \times \left(\frac{1}{T} - \frac{1}{T'} \right)$$

where $\Delta H°$ is the standard enthalpy of the forward process. The temperature dependence of the equilibrium constant for the reaction $N_2(g) + O_2(g) \rightleftharpoons 2NO(g)$, which makes an important contribution to atmospheric nitrogen oxides, can be expressed as $\ln K_P = 2.5 - 181,000/T$. What is the standard reaction enthalpy of the forward reaction?

*__13.67__ The decomposition vapor pressure of ammonium hydrogen sulfide (NH_4HS) is 501 Torr at 298.3 K and 919 Torr at 308.8 K. Using the information in Exercise 13.66, estimate its enthalpy of dissociation and the temperature at which it would "boil" when the external pressure is 1 atm.

Two of the most important concepts in chemistry are those of "acid" and "base," which were first introduced in Chapter 3. Here we discuss more fully what these terms mean. We also see how to describe the behavior of acids and bases in terms of equilibrium constants, and we examine the molecular features that affect their strengths. The illustration shows a titration, in which the changing color of the indicator allows us to follow the progress of the reaction.

ACIDS AND BASES

The definitions of acids and bases

14.1 The Brønsted and Lewis definitions

14.2 Brønsted equilibria

Equilibria in solutions of acids and bases

14.3 Ionization constants

14.4 Strong and weak acids and bases

Hydrogen ion concentration and pH

14.5 The pH of solutions

14.6 Polyprotic acids

This chapter builds on the discussions of Chapter 3, our first encounter with acids, where they were described in terms of the Arrhenius definition: an acid is a compound that contains hydrogen and releases hydrogen ions in water. We soon saw, however, that a broader view of acids and bases is obtained if we adopt the Brønsted-Lowry definitions, in which an acid is any proton donor and a base is any proton acceptor.

Our understanding of acids and bases can be enriched considerably by combining the Brønsted-Lowry definitions with what we have learned about chemical equilibria. In particular, we shall see that we can set up a scale of strengths for acids and bases using equilibrium constants and that we can use that scale to make quantitative predictions about reactions involving acids and bases.

Table 14.1 summarizes some of the techniques for preparing and manufacturing common acids and bases.

THE DEFINITIONS OF ACIDS AND BASES

The Brønsted-Lowry definitions of acids and bases were introduced in Section 3.6. Those definitions focus attention on the transfer of H^+ from one molecule (the acid) to another (the base). Here we briefly review the central points of the theory based on those definitions and then extend it.

TABLE 14.1 **Preparation and manufacture of some common acids and bases**

Acid or base	Method of preparation or manufacture
Hydrohalic acids	1. Action of nonvolatile acid (H_2SO_4 or H_3PO_4) on metal halide:
	$$CaF_2(s) + 2H_2SO_4(l) \longrightarrow Ca(HSO_4)_2(s) + 2HF(g)$$
	(Use phosphoric acid if oxidation is a danger, as with HBr, HI.)
	2. Action of water on nonmetal halide:
	$$PCl_3(l) + 3H_2O(l) \longrightarrow H_3PO_3(aq) + 3HCl(aq)$$
Hypohalous acids	Action of halogen on water:
	$$Cl_2(g) + H_2O(l) \longrightarrow HOCl(aq) + HCl(aq)$$
Sulfuric acid	Contact process (catalytic oxidation of SO_2 followed by reaction with water)
Nitric acid	Ostwald process (catalytic oxidation of ammonia followed by reaction with water)
Acetic acid	Oxidation of ethanol
Sodium hydroxide	Electrolysis of brine
Calcium hydroxide	Thermal decomposition of limestone followed by reaction with water
Ammonia	Haber process (catalytic high-pressure synthesis)

14.1 THE BRØNSTED AND LEWIS DEFINITIONS

In the Brønsted-Lowry theory of acids and bases, *any* molecule or ion that acts as a proton donor is an acid, and *any* molecule or ion that acts as a proton acceptor is a base. We shall normally denote a Brønsted acid as HA, and a Brønsted base as B. Then proton donation from an acid to water, one of the most important cases of proton transfer, is

$$HA(aq) + H_2O(l) \longrightarrow H_3O^+(aq) + A^-(aq)$$

An H_2O molecule that has accepted a proton from a proton donor thus becomes a hydronium ion, H_3O^+. Similarly, proton donation from water to a base in aqueous solution is

$$H_2O(l) + B(aq) \longrightarrow BH^+(aq) + OH^-(aq)$$

Neutralization in the Brønsted-Lowry theory. The Brønsted-Lowry theory summarizes all acid-base neutralizations in aqueous solution with a single chemical equation:

$$H_3O^+(aq) + OH^-(aq) \longrightarrow 2H_2O(l)$$

This same net ionic equation (Fig. 14.1) applies to *all* neutralizations in water, because the acid molecule always produces H_3O^+ by donating a proton to an H_2O molecule, and the base molecule always produces OH^- by accepting a proton from another H_2O molecule.

Support for the view that all neutralizations in water are essentially the same comes from thermochemistry. When the reaction enthalpy of a neutralization is measured in very dilute solution (so that the ions do not interact with each other), approximately the same value, -57 kJ/mol, is obtained no matter what acid and base are used (Table 14.2). This suggests that all neutralization reactions are actually the same and that the enthalpy change is a result of the same proton-transfer reaction in every case.

Brønsted acids in nonaqueous solvents. The Brønsted-Lowry definitions specifically make no mention of the identity of the solvent and hence apply whatever it is. For instance, sodium amide ($NaNH_2$) dissolves in liquid ammonia to give a solution of Na^+ cations and NH_2^- anions. Since the amide ion can accept a proton, it is a Brønsted base:

$$NH_2^-(am) + H^+(am) \longrightarrow NH_3(l)$$

(Solution in ammonia is denoted *am*.) An ammonium ion in ammonia can donate a proton, so it is a Brønsted acid. It might, for example, donate a proton to an iodide ion in the solution:

$$NH_4^+(am) + I^-(am) \longrightarrow NH_3(l) + HI(am)$$

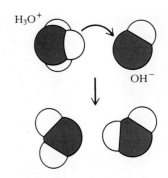

FIGURE 14.1 In the Brønsted-Lowry theory of acids and bases, all acid-base neutralizations amount to the transfer of protons from hydronium ions (H_3O^+) to hydroxide ions (OH^-).

TABLE 14.2 Standard enthalpies of neutralization at 25°C

Acid	Base	Salt	$\Delta H°$, kJ/mol
HCl	NaOH	NaCl	-57.1
HCl	KOH	KCl	-57.2
HNO_3	NaOH	$NaNO_3$	-57.3
2HCl	$Ba(OH)_2$	$BaCl_2$	$2 \times (-58.2)$

FIGURE 14.2 When gaseous HCl and NH_3 mix, they undergo a proton-transfer reaction with the formation of a cloud of NH_4Cl. According to the Brønsted-Lowry theory, the proton donor HCl is an acid, and the proton acceptor NH_3 is a base.

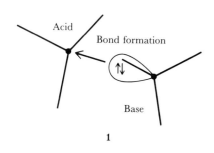

1

Neutralization in nonaqueous solvents follows the same pattern as in water. The solvated proton in liquid ammonia, the analog of the hydrated proton H_3O^+ in water, is the NH_4^+ ion, and the analog of the hydroxide ion is the amide ion, NH_2^-. All neutralizations in liquid ammonia are summarized as proton transfers between ammonium ions and amide ions:

$$NH_4^+(am) + NH_2^-(am) \longrightarrow 2NH_3(l)$$

The Brønsted-Lowry definitions also apply in the gas phase, where there is no solvent. Gaseous hydrogen chloride molecules can donate protons to gaseous ammonia molecules, with the production of a cloud of NH_4Cl particles (Fig. 14.2):

$$HCl(g) + NH_3(g) \longrightarrow \underbrace{Cl^-(s) + NH_4^+(s)}_{NH_4Cl(s)}$$

Hence, even in the absence of a solvent, HCl is a Brønsted acid and NH_3 is a Brønsted base.

Lewis acids and bases. Although the Brønsted-Lowry theory is much broader than the Arrhenius theory, it still excludes some reactions that resemble neutralizations but do not involve proton transfer. The reaction between boron trifluoride and ammonia, for instance,

$$BF_3(g) + NH_3(g) \longrightarrow BF_3NH_3(s)$$

is like a neutralization because one substance (an "acid") combines with another (a "base") to give a substance that is neither an acid nor a base (a "neutral" substance). However, since proton transfer does not occur, it is not a Brønsted acid-base reaction (Fig. 14.3).

Gilbert Lewis, whose great contribution to bonding theory was described in Chapter 8, suggested a way of defining acids and bases that includes all substances showing similar behavior:

A **Lewis acid** is an electron-pair acceptor.

A **Lewis base** is an electron-pair donor.

In his theory, Lewis pictured an acid and a base as sharing the electron pair provided by the base, and thus forming a covalent bond (**1**). The resulting combination is called a *complex*. Therefore, if the Lewis acid is denoted by A and the Lewis base by : B, then the fundamental reaction in the Lewis theory is

$$\underset{\text{acid}}{A} + \underset{\text{base}}{:B} \longrightarrow \underset{\text{complex}}{A—B}$$

We can easily verify that all Brønsted acids and bases are also Lewis acids and bases. We begin by noting that a proton can form a strong bond with a lone pair of electrons, as in the formation of H_3O^+:

$$\underset{\text{acid}}{H^+} + \underset{\text{base}}{:\overset{..}{O}H_2} \longrightarrow \underset{\text{complex}}{[H—\overset{..}{O}H_2]^+} \quad \text{or} \quad H_3O^+$$

Hence, the proton is an electron-pair acceptor and therefore a Lewis acid. We have already seen that anything that can donate a proton is a

Brønsted acid, so whatever acts as a Brønsted acid provides protons and hence acts as a Lewis acid too. A hydroxide ion, which has three lone pairs of electrons (2), is an electron-pair donor:

$$H^+ + :\overset{..}{\underset{..}{O}}H^- \longrightarrow H—\overset{..}{\underset{..}{O}}H \quad or \quad H_2O$$
$$\text{acid} \qquad \text{base} \qquad \text{complex}$$

The OH^- ion is therefore a Lewis base. The lone pair of NH_3, which the NH_3 molecule can use to combine with the Lewis acid H^+ to form NH_4^+, makes ammonia a Lewis base too. Since anything that accepts a proton is a Brønsted base, and accepting a proton always means sharing a lone pair of electrons with it, all Brønsted bases are Lewis bases too.

Lewis acids that are not Brønsted acids. Now we can go on to see that the Lewis definitions include a wider range of substances than the Brønsted-Lowry definitions. Consider BF_3 and NH_3, which, as previously noted, react together like an acid and a base but do not transfer a proton in the process. They are in fact a *Lewis* acid and a *Lewis* base. We can see this as soon as we write the Lewis structures of the two molecules and their product:

$$
\begin{matrix}
& F & & H & & F & & H \\
& | & & | & & | & & | \\
F—& B & & :N—H & & F—B—N—H \\
& | & & | & & | & & | \\
& F & & H & & F & & H
\end{matrix}
$$

The NH_3 molecule acts as an electron-pair donor, and hence it is a Lewis base. The B atom in BF_3 has an incomplete octet that it can complete by accepting an electron pair from a Lewis base; hence BF_3 is a Lewis acid. The formation of F_3BNH_3 is therefore the reaction of a Lewis acid (BF_3) with a Lewis base ($:NH_3$) to form the complex $F_3B—NH_3$.

The formation of a hydrated ion, such as $[Al(H_2O)_6]^{3+}$, is another example of a Lewis acid-base reaction. The metal cation is the Lewis acid, and the hydrating water molecules are the Lewis bases. Although H_2O has two lone pairs, only one is shared with the metal cation; the formation of one metal–oxygen bond causes the second lone pair to point in the wrong direction to be used for a second bond with the same atom (3). Therefore, when we are writing equations to show complex formation, we can express the water molecule as $:OH_2$ even though it has two lone pairs. To emphasize the fact that the hydration involves the formation of an Al—O bond, the hydrated ion is often written as the complex $[Al(OH_2)_6]^{3+}$.

▼ **EXAMPLE 14.1** **Identifying Lewis acids and bases**

Identify the Lewis acids and bases in (a) $Ni + 4CO \rightarrow Ni(CO)_4$; (b) $Ag^+ + 6H_2O \rightarrow [Ag(H_2O)_6]^+$; (c) $SO_3 + CuO \rightarrow CuSO_4$.

STRATEGY We can identify the electron-pair donor (the Lewis acid) and electron-pair acceptor (the Lewis base) by writing the Lewis structures (Section 8.3) for the reactants and products. In the case of ionic substances (c), it

FIGURE 14.3 When gaseous BF_3 and NH_3 mix, they react to form a cloud of BF_3NH_3. Although this is not an acid-base reaction according to the Brønsted-Lowry theory, it is one according to Lewis's more general theory. The electron-pair acceptor BF_3 is the Lewis acid, and the electron-pair donor NH_3 is the Lewis base.

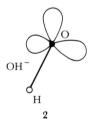

2

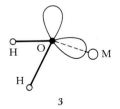

3

is clearer to consider the net ionic reaction, which in this case is $SO_3 + O^{2-} \rightarrow SO_4^{2-}$.

SOLUTION (a) $Ni + 4(:C\equiv O:) \longrightarrow$

$$
\underset{\text{acid}}{Ni} + 4(\underset{\text{base}}{:C\equiv O:}) \longrightarrow \; :O\equiv C - Ni - C\equiv O:
$$

(b) $\underset{\text{acid}}{Ag^+} + 6(\underset{\text{base}}{:OH_2}) \longrightarrow [Ag(OH_2)_6]^+$

(c) $\underset{\text{acid}}{O=S} + \underset{\text{base}}{:O:^{2-}} \longrightarrow \left\{ :O-S-O: \right\}^{2-}$

EXERCISE Identify the Lewis acids and bases in (a) $H^+ + Cl^- \rightarrow HCl$ and (b) $Al^{3+} + 6H_2O \rightarrow [Al(H_2O)_6]^{3+}$.

[*Answer*: (a) Acid H^+, base $:Cl:^-$; (b) acid Al^{3+}, base $H_2O:$]

The formation of a Brønsted acid by a Lewis acid. Another example of the wide applicability of the Lewis theory is its description of boric acid, $B(OH)_3$, a mild acid used as an antiseptic. Boric acid is so called because it dissolves in water to give an acidic solution (a solution with a higher hydronium ion concentration than that of pure water). However, the presence of hydronium ions in the solution is not a result of the ionization of $B(OH)_3$ itself, but of the reaction

$$B(OH)_3(s) + 2H_2O(l) \longrightarrow [B(OH)_4]^-(aq) + H_3O^+(aq)$$

This can be pictured as a Lewis acid-base reaction to form a complex that acts as a Brønsted acid:

$$
HO-B \; :O-H \; :O-H \longrightarrow \left\{ HO-B-OH \right\}^- + \left\{ H-O-H \right\}^+
$$

The $B(OH)_3$ acts as a Lewis acid and accepts a lone pair of the H_2O molecule (shown by a curved arrow). The complex $H_2O-B(OH)_3$ immediately acts as a Brønsted acid and donates a proton to another water molecule (shown by the second curved arrow).

The Lewis theory of acids and bases unifies a wide range of apparently different reactions. However, for our present purposes the theory is *too* broad; it shifts the focus away from the proton to the electron pair when, in fact, a large number of chemical reactions depend specifically on the properties of hydronium ions, protonated water molecules, in solution. Since the latter are so important, we use the Brønsted definitions in the remainder of this chapter. The broader usefulness of the Lewis theory will come to light later.

14.2 BRØNSTED EQUILIBRIA

Proton donating and proton accepting in solution are *dynamic* processes. For example, acetic acid molecules may donate protons to water molecules (or those of some other base):

$$CH_3COOH(aq) + H_2O(l) \longrightarrow CH_3CO_2^-(aq) + H_3O^+(aq)$$

At the same time, acetate ions may accept protons from hydronium ions and re-form acetic acid molecules:

$$CH_3CO_2^-(aq) + H_3O^+(aq) \longrightarrow CH_3COOH(aq) + H_2O(l)$$

When these two reactions occur at equal rates in solution, all the substances involved are in dynamic equilibrium:

$$CH_3COOH(aq) + H_2O(l) \rightleftharpoons CH_3CO_2^-(aq) + H_3O^+(aq)$$

The same is true of a solution of a base, where a proton-transfer equilibrium such as

$$NH_3(aq) + H_2O(aq) \rightleftharpoons NH_4^+(aq) + OH^-(aq)$$

is quickly established, with the forward and reverse reactions continuing at matching rates.

Conjugate acids and bases. In the forward reaction between NH_3 and H_2O, NH_3 accepts a proton and hence is a Brønsted base. In the reverse reaction, the ammonium ion donates a proton to the hydroxide ion. This makes NH_4^+ a Brønsted acid and, in particular, the "conjugate acid" of NH_3:

> The **conjugate acid** of a Brønsted base is the Brønsted acid formed when the base has accepted a proton.

("Conjugate" in this context means related.) Likewise, in the forward reaction, H_2O acts as a Brønsted acid and donates a proton to NH_3, becoming an OH^- ion in the process. This OH^- ion acts as a base in the reverse reaction by accepting the proton donated by the acid NH_4^+. Hence, OH^- is the "conjugate base" of H_2O:

> The **conjugate base** of a Brønsted acid is the Brønsted base formed when the acid has donated a proton.

Every Brønsted base has its conjugate acid, and every Brønsted acid its conjugate base (Table 14.3). Note that a conjugate acid is a Brønsted acid, just like any other acid, and a conjugate base is a Brønsted base. They are called "conjugate" to emphasize the fact that they are related to one another, not because they show any difference in behavior from other acids or bases.

All the equilibria discussed so far in this section have the form of the *Brønsted equilibrium*

$$Acid_1 + base_2 \rightleftharpoons base_1 + acid_2 \tag{1}$$

TABLE 14.3 Conjugate acids and bases

Acid	Base
HCl	Cl^-
HNO_3	NO_3^-
H_2SO_4	HSO_4^-
HSO_4^-	SO_4^{2-}
H_2CO_3	HCO_3^-
HCO_3^-	CO_3^{2-}
CH_3COOH	$CH_3CO_2^-$
H_2O	OH^-
OH^-	O^{2-}
H_3O^+	H_2O
H_2S	HS^-
HS^-	S^{2-}
NH_3	NH_2^-
NH_4^+	NH_3
$CH_3NH_3^+$	CH_3NH_2

and hence are described by the equilibrium constant

$$K_c = \frac{[\text{base}_1][\text{acid}_2]}{[\text{acid}_1][\text{base}_2]} \tag{2}$$

where base_1 is the conjugate base of acid_1, and acid_2 is the conjugate acid of base_2. For aqueous ammonia, we have

$$\underset{\text{acid}_1}{H_2O} + \underset{\text{base}_2}{NH_3} \rightleftharpoons \underset{\text{base}_1}{HO^-} + \underset{\text{acid}_2}{NH_4^+} \qquad K_c = \frac{[HO^-][NH_4^+]}{[H_2O][NH_3]}$$

For aqueous acetic acid,

$$\underset{\text{acid}_1}{CH_3COOH} + \underset{\text{base}_2}{H_2O} \rightleftharpoons \underset{\text{base}_1}{CH_3CO_2^-} + \underset{\text{acid}_2}{H_3O^+}$$

$$K_c = \frac{[CH_3CO_2^-][H_3O^+]}{[CH_3COOH][H_2O]}$$

This is the link we need with the ideas of Chapter 13: acids and bases in solution can be described with equilibrium constants that all have the same form, that of Eq. 2.

▼ EXAMPLE 14.2   Identifying conjugate acids and bases

Identify the acids and bases in (a) $H_2SO_4 + H_2O \rightleftharpoons HSO_4^- + H_3O^+$ in water and (b) $2NH_3 \rightleftharpoons NH_4^+ + NH_2^-$ in liquid ammonia.

STRATEGY We begin by looking to see if any familiar substances are present. If there are none, we identify as acid_1 the proton donor in the forward reaction; the other reactant, the proton acceptor in the forward reaction, is then base_2. The product differing from acid_1 by the loss of a proton is base_1, the conjugate base of acid_1. The other product, the result of base_2 accepting a proton, is acid_2, its conjugate acid.

SOLUTION (a) H_2SO_4 is familiar as sulfuric acid, and we identify it as the proton donor. This leaves H_2O as the base—the proton acceptor. On the right, the hydrogen sulfate ion HSO_4^- is the conjugate base of H_2SO_4. When H_2O gains a proton it becomes H_3O^+, so H_3O^+ is the conjugate acid. In summary, we have

$$\underset{\text{acid}_1}{H_2SO_4} + \underset{\text{base}_2}{H_2O} \rightleftharpoons \underset{\text{base}_1}{HSO_4^-} + \underset{\text{acid}_2}{H_3O^+}$$

(b) The equilibrium here is

$$\underset{\text{acid}_1}{NH_3(l)} + \underset{\text{base}_2}{NH_3(l)} \rightleftharpoons \underset{\text{acid}_2}{NH_4^+(am)} + \underset{\text{base}_1}{NH_2^-(am)}$$

with NH_3 acting both as an acid and as a base.

EXERCISE Identify the acids and bases in (a) $H_2CO_3 + H_2O \rightleftharpoons H_3O^+ + HCO_3^-$ in water and (b) $2CH_3COOH \rightleftharpoons CH_3CO_2^- + CH_3CO_2H_2^+$ in acetic acid.

[*Answer*: (a) Acid_1 H_2CO_3, base_2 H_2O, acid_2 H_3O^+, base_1 HCO_3^-; (b) acid_1 and base_2 CH_3COOH, base_1 $CH_3CO_2^-$, acid_2 $CH_3CO_2H_2^+$]

▲

It should be appreciated at this point that individual molecules and ions may be Brønsted acids and bases, and hence that solutions of salts may be acidic or basic. Since the acetate ion is a Brønsted base (because

FIGURE 14.4 Cations and anions are Brønsted acids and bases in water. This is shown by the colors of the indicators for four aqueous solutions: from left to right, NH_4^+, $CH_3CO_2^-$, CO_3^{2-}, PO_4^{3-}.

it can accept a proton), an aqueous solution that contains acetate ions is the solution of a base. We can therefore predict that aqueous solutions of acetate ions will be basic (so long as any other ions present are not stronger acids). Since the ammonium ion is a Brønsted acid (because it can donate protons), an aqueous solution of ammonium ions is a solution of an acid (Fig. 14.4). So long as the NH_4^+ ion is a stronger acid than the anion accompanying it in the solution is a base, we can predict that the solution of an ammonium salt will be acidic. Ammonium chloride, for example, gives an acidic solution because, as we shall see shortly, Cl^- is a much weaker base than NH_4^+ is an acid.

▼ EXAMPLE 14.3 Predicting the acidities of solutions

Predict whether solutions of (a) ammonium chloride and (b) sodium acetate are likely to be acidic or basic. Ignore the acid or base character of Cl^- ions and Na^+ ions (for reasons that are explained later).

STRATEGY All we need to know is whether the solutions contain ions that are Brønsted acids or bases. If the ions are bases, the solution will be basic; if they are acids, it will be acidic.

SOLUTION (a) NH_4^+ is an acid, so the solution will be acidic.
(b) $CH_3CO_2^-$ is a base, so the solution will be basic.

EXERCISE Predict the acidity or basicity of aqueous solutions of (a) NaCl and (b) NaOCl.

[*Answer*: (a) Neutral; (b) basic]

▲

Autoionization. The examples just below Eq. 2 show that water can act as both a Brønsted acid and a Brønsted base. It is therefore an *amphiprotic* substance, one that can both donate and accept protons. Since water is amphiprotic, one H_2O molecule can donate a proton to another:

$$H_2O + H_2O \rightleftharpoons OH^- + H_3O^+ \qquad K_c = \frac{[OH^-][H_3O^+]}{[H_2O]^2}$$

acid₁   base₂   base₁   acid₂

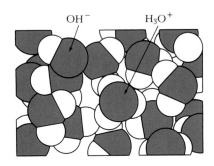

FIGURE 14.5 A sample of water is not a purely molecular liquid but also contains some hydronium ions and hydroxide ions that result from its autoionization.

This type of equilibrium, in which a substance ionizes itself, is called *autoionization* (Fig. 14.5). We will soon see that it plays a special role in the properties of acids and bases in water. Liquid ammonia is also amphiprotic and undergoes autoionization:

$$NH_3(l) + NH_3(l) \rightleftharpoons NH_2^-(am) + NH_4^+(am)$$

Pure acetic acid (denoted *ac* when it is the solvent) behaves similarly:

$$CH_3COOH(l) + CH_3COOH(l) \rightleftharpoons$$
$$CH_3CO_2^-(ac) + CH_3CO_2H_2^+(ac)$$

The water autoionization constant. The autoionization equilibrium of water is

$$H_2O(l) + H_2O(l) \rightleftharpoons OH^-(aq) + H_3O^+(aq) \qquad K_c = \frac{[OH^-][H_3O^+]}{[H_2O]^2}$$

In dilute aqueous solutions (the only ones we consider), the molar concentration of H_2O is almost constant at 56 M, which is the molar concentration of H_2O in pure water. Therefore, whenever we write an equilibrium constant for a reaction in aqueous solution, we can treat $[H_2O]$ as a constant and combine it with the equilibrium constant. For the autoionization equilibrium, we introduce the *water autoionization constant* (K_w) as $K_w = [H_2O]^2 \times K_c$, and hence write the autoionization equilibrium as

$$2H_2O(l) \rightleftharpoons OH^-(aq) + H_3O^+(aq) \qquad K_w = [OH^-][H_3O^+] \quad (3)$$

At equilibrium, the product of the concentrations of the H_3O^+ and OH^- ions must always be equal to K_w. This is true even if either one is modified by adding acid (which increases $[H_3O^+]$) or by adding base (which increases $[OH^-]$).

▼ EXAMPLE 14.4 Calculating the autoionization constant

At 25°C, the equilibrium concentration of hydronium ions in pure water is $1.0 \times 10^{-7}\ M$. Calculate the value of K_w.

STRATEGY We need to calculate $[OH^-] \times [H_3O^+]$ but are given only one of the ion concentrations. However, we know that in pure water, for each H_3O^+ formed by ionization, one OH^- ion is also formed. Therefore, the two concentrations are equal.

SOLUTION Since the concentrations are equal in pure water,

$$K_w = [H_3O^+]^2$$
$$= (1.0 \times 10^{-7}\ M)^2 = 1.0 \times 10^{-14}\ M^2$$

EXERCISE The concentration of NH_4^+ ions in liquid ammonia at −50°C is $3 \times 10^{-17}\ M$. Calculate the autoionization constant of ammonia at that temperature.

[*Answer:* $9 \times 10^{-34}\ M^2$]

The value of K_w calculated in Example 14.4, which is $1.0 \times 10^{-14}\ M^2$, applies to any aqueous solution at equilibrium at 25°C, whatever the individual concentrations of H_3O^+ and OH^-. Therefore, in any aqueous solution at 25°C,

$$[H_3O^+][OH^-] = 1.0 \times 10^{-14} \, M^2$$

If in *any* aqueous solution the concentration of OH^- ions is increased by adding base, the concentration of hydrogen ions will decrease to satisfy this equation. If the concentration of H_3O^+ is increased by adding acid, the concentration of OH^- ions will decrease similarly.

Calculations, discussions, and equations involving acids and bases are greatly simplified if we express the value of K_w (and several other quantities that we shall meet) in terms of their logarithms. We therefore use the quantity "pK_w," defined as

$$pK_w = -\log K_w$$

instead of K_w. At 25°C, when $K_w = 1.0 \times 10^{-14} \, M^2$,

$$pK_w = -\log (1.0 \times 10^{-14}) = 14.00$$

The value of pK_w is much easier to remember than K_w itself (and should be remembered). Notice that the power of 10 in K_w becomes the numeral to the left of the decimal point in pK_w (except for the minus sign), so the value of pK_w emphasizes the order of magnitude of K_w. This is the origin of the p (for power of 10) in pK_w.

EQUILIBRIA IN SOLUTIONS OF ACIDS AND BASES

The key feature of the Brønsted-Lowry theory is its emphasis on the equilibria that exist between conjugate acids and bases. Since proton transfer is so fast, as soon as a solution of an acid HA is prepared, the solution contains HA and its conjugate base A^- in their equilibrium concentrations. Similarly, a solution of a base B contains equilibrium concentrations of the base and its conjugate acid BH^+ virtually the instant it is prepared.

14.3 IONIZATION CONSTANTS

Many chemical reactions can be understood in terms of a competition between acids and bases of different strengths to donate or accept protons. Since so many reactions take place in water, an important aspect of this competition is the strength with which various acids and bases transfer protons to and from water.

Acid ionization constants. The Brønsted equilibrium for an acid HA in water is

$$HA(aq) + H_2O(l) \rightleftharpoons A^-(aq) + H_3O^+(aq) \qquad K_c = \frac{[A^-][H_3O^+]}{[HA][H_2O]}$$

As discussed above, we can treat the concentration of H_2O as a constant and combine it with K_c to obtain a new constant, called the *acid ionization constant* K_a and defined as $K_a = [H_2O] \times K_c$. Then

$$HA(aq) + H_2O(l) \rightleftharpoons A^-(aq) + H_3O^+(aq) \qquad K_a = \frac{[A^-][H_3O^+]}{[HA]} \quad (4)$$

HA might be acetic acid, in which case we would have

$$K_a = \frac{[CH_3CO_2^-][H_3O^+]}{[CH_3COOH]} = 1.8 \times 10^{-5}\ M$$

It is convenient to define the quantity

$$pK_a = -\log K_a \qquad (5)$$

Then, for acetic acid,

$$pK_a = -\log(1.8 \times 10^{-5}) = 4.74$$

An example of an acid with a less favorable ionization equilibrium (one for which less A^- and H_3O^+ forms) is hydrocyanic acid (HCN), which has $K_a = 4.9 \times 10^{-10}\ M$ and

$$pK_a = -\log(4.9 \times 10^{-10}) = 9.31$$

Note that the smaller K_a is, the larger pK_a is.

If the acid ionization equilibrium favors the formation of the conjugate base, then the acid HA is a strong proton donor. If it favors the reactants, so that the solution consists largely of nonionized HA molecules, then the acid is only a weak proton donor. Hence, a *smaller* pK_a (such as the 2.00 of chlorous acid, $HClO_2$) is a sign of a *stronger* acid, and a larger pK_a (such as the 7.53 of hypochlorous acid, HClO) is a sign of a weaker acid.

Base ionization constants. The Brønsted equilibrium for a base in aqueous solution is

$$H_2O(l) + B(aq) \rightleftharpoons OH^-(aq) + BH^+(aq) \qquad K_c = \frac{[OH^-][BH^+]}{[H_2O][B]}$$

Once again we treat $[H_2O]$ as a constant and combine it with K_c, this time obtaining the *base ionization constant* $K_b = [H_2O] \times K_c$. This gives us

$$H_2O(l) + B(aq) \rightleftharpoons OH^-(aq) + BH^+(aq) \qquad K_b = \frac{[OH^-][BH^+]}{[B]} \quad (6)$$

4 Methylamine

If B, for instance, is the organic base methylamine (CH_3NH_2; **4**), then BH^+ is the methylammonium ion $CH_3NH_3^+$, and the equilibrium is

$$H_2O(l) + CH_3NH_2(aq) \rightleftharpoons OH^-(aq) + CH_3NH_3^+(aq)$$

$$K_b = \frac{[OH^-][CH_3NH_3^+]}{[CH_3NH_2]}$$

Again we shall find it useful to define a logarithmic version of K_b:

$$pK_b = -\log K_b \qquad (7)$$

A base B is a strong proton acceptor if the equilibrium favors the formation of BH^+ (that is, if K_b is large). The base is a weak proton acceptor if the equilibrium favors the nonionized base B (that is, if K_b is small). Hence a base with a small pK_b is a stronger proton acceptor than a base with a larger pK_b. For instance, methylamine has $K_b = 3.6 \times 10^{-4}\ M$ and $pK_b = 3.44$; the weaker base aniline (**5**) has $K_b = 4.3 \times 10^{-10}\ M$ and $pK_b = 9.37$.

The relations between pK_a***,*** pK_b***, and*** pK_w***.*** A very useful relation exists between the pK_b of a base (such as NH_3) and the pK_a of its conjugate acid (here, NH_4^+). We can derive it by considering the equilibrium of the base NH_3 in water,

$$H_2O(l) + NH_3(aq) \rightleftharpoons OH^-(aq) + NH_4^+(aq)$$

$$K_b = \frac{[OH^-][NH_4^+]}{[NH_3]}$$

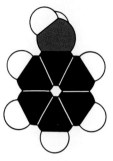

5 Aniline

and the equilibrium of its conjugate acid NH_4^+,

$$NH_4^+(aq) + H_2O(l) \rightleftharpoons NH_3(aq) + H_3O^+(aq)$$

$$K_a = \frac{[NH_3][H_3O^+]}{[NH_4^+]}$$

If we multiply the two ionization constants together, we get

$$K_a \times K_b = \frac{[NH_3][H_3O^+]}{[NH_4^+]} \times \frac{[OH^-][NH_4^+]}{[NH_3]}$$

$$= [H_3O^+][OH^-] = K_w$$

The same relation applies to all conjugate acids and bases:

$$K_a \times K_b = K_w \tag{8}$$

If K_b increases, then K_a must decrease to keep the product $K_a \times K_b$ equal to K_w. This shows that a stronger base must have a weaker conjugate acid, and, similarly, the stronger an acid, the weaker its conjugate base: HCN is a weak proton donor, and its conjugate base CN^- is a strong acceptor. CH_3COOH is a stronger proton donor, and its conjugate base $CH_3CO_2^-$ is a weaker proton acceptor. HCl is a very strong proton donor, and its conjugate base Cl^- is a very weak proton acceptor. Urea [$(NH_2)_2CO$; **6**] is a very weak proton acceptor, and its conjugate acid $NH_2CONH_3^+$ is a strong proton donor. Methylamine, CH_3NH_2, is a stronger proton acceptor, and its conjugate acid is a weaker proton donor. The oxide ion O^{2-} is a very strong proton acceptor, and its conjugate acid OH^- is a very weak proton donor.

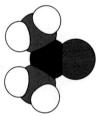

6 Urea

If we take the logarithms of both sides of Eq. 8, we obtain the equation

$$\log K_a + \log K_b = \log K_w$$

After a change of signs, this becomes

$$pK_a + pK_b = pK_w \tag{9}$$

As an example, since the pK_b of NH_3 is 4.75, the pK_a of NH_4^+ is

$$pK_a = pK_w - pK_b = 14.00 - 4.75 = 9.25$$

This shows that NH_4^+ is a weaker proton donor than acetic acid ($pK_a = 4.75$), but stronger than hypoiodous acid, HIO ($pK_a = 10.64$). The conjugate acid of urea, $NH_2CONH_3^+$, has $pK_a = 0.10$, showing that it is a stronger proton donor than any of the acids listed in Table 14.4.

TABLE 14.4 Acid and base ionization constants at 25°C

	Acid	K_a, M	pK_a
*Strongest weak acids**	Trichloroacetic acid, CCl_3COOH	3.0×10^{-1}	0.52
	Benzenesulfonic acid, $C_6H_5SO_3H$	2×10^{-1}	0.70
	Iodic acid, HIO_3	1.7×10^{-1}	0.77
	Sulfurous acid, H_2SO_3	1.6×10^{-2}	1.81
	Chlorous acid, $HClO_2$	1.0×10^{-2}	2.00
	Phosphoric acid, H_3PO_4	7.6×10^{-3}	2.12
	Chloroacetic acid, $CH_2ClCOOH$	1.4×10^{-3}	2.85
	Lactic acid, $CH_3CH(OH)COOH$	8.4×10^{-4}	3.08
	Nitrous acid, HNO_2	4.3×10^{-4}	3.37
	Hydrofluoric acid, HF	3.5×10^{-4}	3.45
	Formic acid, HCOOH	1.8×10^{-4}	3.75
	Benzoic acid, C_6H_5COOH	6.5×10^{-5}	4.19
	Acetic acid, CH_3COOH	1.8×10^{-5}	4.75
	Carbonic acid, H_2CO_3	4.3×10^{-7}	6.37
	Hypochlorous acid, HClO	3.0×10^{-8}	7.53
	Hypobromous acid, HBrO	2.0×10^{-9}	8.69
	Boric acid, $B(OH)_3$	7.2×10^{-10}	9.14
	Hydrocyanic acid, HCN	4.9×10^{-10}	9.31
Weakest weak acids	Phenol, C_6H_5OH	1.3×10^{-10}	9.89
	Hypoiodous acid, HIO	2.3×10^{-11}	10.64

	Base	K_b, M	pK_b
Weakest weak bases	Urea, $CO(NH_2)_2$	1.3×10^{-14}	13.90
	Aniline, $C_6H_5NH_2$	4.3×10^{-10}	9.37
	Pyridine, C_5H_5N	1.8×10^{-9}	8.75
	Hydroxylamine, NH_2OH	1.1×10^{-8}	7.97
	Nicotine, $C_{10}H_{14}N_2$	1.0×10^{-6}	5.98
	Morphine, $C_{17}H_{19}O_3N$	1.6×10^{-6}	5.79
	Hydrazine, NH_2NH_2	1.7×10^{-6}	5.77
	Ammonia, NH_3	1.8×10^{-5}	4.75
	Trimethylamine, $(CH_3)_3N$	6.5×10^{-5}	4.19
	Methylamine, CH_3NH_2	3.6×10^{-4}	3.44
	Dimethylamine, $(CH_3)_2NH$	5.4×10^{-4}	3.27
	Ethylamine, $C_2H_5NH_2$	6.5×10^{-4}	3.19
Strongest weak bases	Triethylamine, $(C_2H_5)_3N$	1.0×10^{-3}	2.99

*The terms *strong* and *weak* are explained later. Values for polyprotic acids—those capable of donating more than one proton—refer to the first ionization only.

14.4 STRONG AND WEAK ACIDS AND BASES

Once we know the equilibrium constants for the Brønsted equilibria of acids and bases (Table 14.4), we can classify acids and bases according to their strengths and seek reasons for their behavior. We continue to limit our discussion to acids and bases in water; their strengths in other solvents may be quite different.

Strong acids and bases. For a few acids (those listed in Table 14.5), the proton-transfer equilibrium favors proton loss so strongly that at ordinary concentrations (typically up to about 1 M) almost all the HA molecules present in aqueous solution are ionized. The solution contains H_3O^+ and A^- ions and almost no nonionized HA molecules. Acids of this kind are called *strong acids* (Fig. 14.6). Hydrogen chloride is a strong acid in water: a 0.1 M HCl(aq) solution consists of 0.1 M $H_3O^+(aq)$ ions and 0.1 M $Cl^-(aq)$ ions, with almost no HCl(aq) molecules remaining. Since hydrochloric acid is strong, its conjugate base (Cl^-) is very weak; the same is true of the conjugate base of any strong acid.

It is important to distinguish between a *concentrated acid* and a *strong acid*. Concentration refers to the number of moles of acid present per liter of solution; strength refers to the proton-donating power of the acid molecules. A strong acid, such as hydrochloric acid or perchloric acid, may be either concentrated (for example, at 10 M) or dilute 0.01 M); whichever the case, the acid will be almost fully ionized.

An aqueous solution of sodium hydroxide consists largely of $Na^+(aq)$ and $OH^-(aq)$ ions. It is an example of a *strong base*, one that consists largely of OH^- ions in water with few nonionized molecules present (Fig. 14.7). Some strong bases are listed in Table 14.6. The concentration of OH^- ions is known once we know the concentration of the base in solution and the number of OH^- ions each formula unit supplies. For example, since a formula unit of $Ba(OH)_2$ provides two OH^- ions, the hydroxide ion concentration in 0.020 M $Ba(OH)_2(aq)$ is 2 × 0.020 M = 0.040 M.

Weak acids and bases. Most acids are *weak acids* in the sense that $K_a \ll 1$. In a solution of a weak acid at ordinary concentrations (neither concentrated nor very dilute), the proton-transfer equilibrium

TABLE 14.5 The strong acids (in water)

*Weakest**	Hydrochloric acid, HCl
	Hydrobromic acid, HBr
	Hydroiodic acid, HI
	Sulfuric acid, H_2SO_4 (to HSO_4^-)
	Nitric acid, HNO_3
Strongest	Perchloric acid, $HClO_4$

*The listed acids are of equal strength in water; the grading from strongest to weakest refers to their behavior in other solvents.

TABLE 14.6 The strong bases (in water)

Group	Compounds	Examples
I	Hydroxides	NaOH, KOH
	Oxides*	Na_2O
II	Hydroxides (except Be)	$Sr(OH)_2$
	Oxides*	CaO

*In water, the oxide ion in the compound is converted to OH^-.

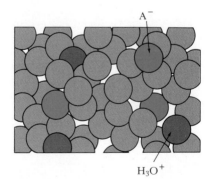

FIGURE 14.6 In a solution of a strong acid, nearly all the ionizable hydrogen atoms are present as hydronium ions, and the solution contains few nonionized acid molecules.

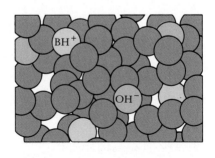

FIGURE 14.7 In a solution of a strong base, all the base molecules have accepted protons from water and are present as BH^+ ions.

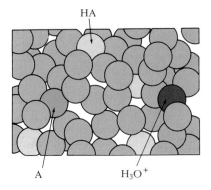

HA

A H_3O^+

FIGURE 14.8 In a solution of a weak acid, only some of the ionizable hydrogen atoms are present as hydronium ions, and the solution contains a high proportion of the original acid molecules (HA).

favors the nonionized HA (Fig. 14.8). All carboxylic acids, which are organic acids containing the $-C\overset{O}{\underset{OH}{\big<}}$ group, are weak acids in water. Acetic acid has a K_a of 1.8×10^{-5} M (pK_a = 4.75); although a 1 M $CH_3COOH(aq)$ solution contains some H_3O^+ and $CH_3CO_2^-$ ions, most of the solute is present as CH_3COOH. As with strong acids and concentrated acids, it is important to distinguish between a *weak acid* and a *dilute acid*: a weak acid may be concentrated or dilute. Weakness, like strength, refers to the proton-donating power of the acid molecules, not to the number of moles of acid per liter.

One practical difference between strong and weak acids is shown by dropping pieces of magnesium into 1 M HCl(aq) and into 1 M $CH_3COOH(aq)$. Hydrogen is evolved much more rapidly from the hydrochloric acid, which suggests that hydrogen ions are much more numerous in that acid than in acetic acid, even though the concentrations of the two acids are the same (Fig. 14.9). Another sign of the strength of HCl is its conductivity: 1 M HCl(aq) is a much better conductor of electricity than 1 M $CH_3COOH(aq)$. The concentration of ions (which carry electric current through the solution) is much greater in hydrochloric acid than in acetic acid (Fig. 14.10).

Ammonia in solution is largely present as nonionized NH_3 molecules. It is an example of a *weak base*, a base with $K_b \ll 1$. At ordinary concentrations, a weak base, like ammonia, is mainly present as nonionized molecules (Fig. 14.11). Other weak bases are listed in Table 14.4. All amines—substances containing the $-NH_2$ group—are weak bases in water: ammonia has $K_b = 1.8 \times 10^{-5}$ M (pK_b = 4.75), and only a small proportion exists as NH_4^+ ions in solution.

FIGURE 14.9 The effect of a strong acid (left) and a weak acid (right) on magnesium. Although the acids have the same concentrations, the rate of hydrogen evolution, which depends on the presence of hydrogen ions, is much greater in the strong acid.

(a)

(b)

(c)

Limiting strengths of acids and bases. Any acid HA that is a stronger proton donor than H_3O^+ will donate its proton to an H_2O molecule. Therefore, all such acids are strong and are completely ionized in water. The strongest acid that can actually exist in water is H_3O^+ itself, for anything stronger simply converts any H_2O molecules into H_3O^+. We say that all strong acids are _leveled_ to the same strength in water, and all behave as though they were solutions of H_3O^+ ions.

Weak acids are acids that are somewhat weaker proton donors than H_3O^+. In solutions of weak acids, both HA and H_3O^+ coexist, but if HA is a weaker proton donor than H_2O, the reaction

$$H_2O(l) + HA(aq) \longrightarrow OH^-(aq) + H_2A^+(aq)$$

is more important than

$$HA(aq) + H_2O(l) \longrightarrow A^-(aq) + H_3O^+(aq)$$

In this case HA acts as a proton acceptor, even though it contains hydrogen; hence it is a base. This is so for NH_3, which is a weaker proton donor than H_2O.

Any base that is a stronger proton acceptor than OH^- will remove a proton from an H_2O molecule and produce OH^- ions in aqueous solution. Therefore, all such bases are strong and are fully ionized in water. The strongest base that can actually exist in water is OH^- itself, for anything stronger simply converts any H_2O molecules to OH^- and behaves as a solution of OH^- ions. The oxide ion, for example, is a stronger base than OH^-, and in aqueous solution it exists only as OH^- as a result of the reaction

$$H_2O(l) + O^{2-}(aq) \longrightarrow 2OH^-(aq)$$

This accounts for the fact that we never encounter solutions of oxides: any dissolved oxide immediately reacts to form the hydroxide. An example is calcium oxide, which reacts vigorously with water to form calcium hydroxide.

FIGURE 14.10 For the same reason as in Fig. 14.9, hydrochloric acid is a much better electrical conductor than acetic acid with the same concentration. The samples are (a) deionized water, (b) 0.1 M $CH_3COOH(aq)$, and (c) 0.1 M $HCl(aq)$.

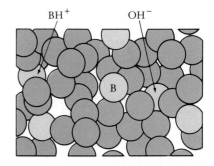

FIGURE 14.11 In a solution of a weak base, only a small proportion of the base molecules (B) have accepted protons from water.

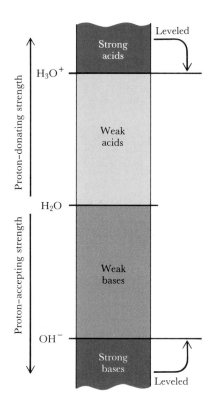

FIGURE 14.12 The proton-donating power of water is the dividing line between acids and bases in water. The strongest acid that can exist in water is the hydronium ion, and the strongest base is the hydroxide ion.

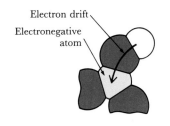

FIGURE 14.13 The greater the electronegativity of the central atom, the more the electrons in the O—H bond drift toward it, resulting in greater hydrogen bonding and a stronger acid.

The proton-donating and proton-accepting power of H_2O itself is thus the dividing line between strong and weak acids and bases in water (Fig. 14.12). Any proton donor that is a stronger donor than H_2O is a strong acid in water. Any proton donor that lies between H_3O^+ and H_2O in donating strength is a weak acid in water. Any substance that is a weaker proton donor than H_2O is a weak base in water. Any proton acceptor that is a stronger acceptor than OH^- is a strong base.

The relation of strength to structure. The pK_a values for acids (and pK_b values for bases) are very difficult to predict because they depend on a number of factors. These include the strength of the H—A bond, the strength of the H_2O—H bond in H_3O^+, and the extent to which the molecules and ions are hydrated. However, trends among series of compounds with similar structures can often be explained by focusing on the more important features.

A proton may be transferred reasonably easily from an HA to an H_2O molecule if, before it is transferred, it can form a hydrogen bond to the O atom in the H_2O molecule:

$$A—H\cdots O\diagdown^{H}_{H} \longrightarrow A^- + H_3O^+$$

Since hydrogen bonding is more likely when the A—H bond is polar, we expect that for acids with similar structures, *the more polar the H—A bond, the stronger the acid HA.* Since the polarity of the H—A bond increases as A becomes more electronegative, this implies that *the greater the electronegativity of A, the stronger the acid HA* (Fig. 14.13). An example is the difference between CH_4, which is not a Brønsted acid in water, and NH_4^+, which is. The difference between the electronegativities of N and H is 0.9 (see Table 7.9 and Fig. 14.14), compared with only 0.4 for C and H; hence the N—H bond is markedly more polar than the C—H bond. The high polarity of the O—H bond is one of the reasons why the —OH hydrogen atoms in many molecules can be donated. This is shown strikingly by phosphorous acid (H_3PO_3; **7**), which can donate the protons from its two —OH groups but not the proton attached directly to the phosphorus atom. Another example is CH_3COOH, in which the only ionizable hydrogen atom is the one attached to the oxygen atom.

Even though an H—A bond may be polar and able to form a hydrogen bond, it may be too strong to be broken. The polarity of the bond is like a handle on a weight; the weight can be grasped, but it may still be too heavy to lift. Since the H—A bond can be broken more readily if it is weak, we can expect that, for compounds with similar structures, *the weaker the H—A bond, the stronger the acid HA.* An example is the anomalous position of hydrofluoric acid among the hydrohalic acids. Even though the H—F bond is more polar than any other H—halogen bond, HF is a *weak* acid in water ($pK_a = 3.45$), whereas all the other hydrohalic acids are strong. A part of the reason is the strength of the H—F bond (562 kJ/mol, compared to 431 kJ/mol for HCl and 366 kJ/mol for HBr), which means that the proton is lost from HF only with difficulty. In water, many HF molecules form hydrogen bonds to water molecules (**8**), as though the tug-of-war for the proton had ended in a stalemate.

The strengths of oxoacids. The oxoacids form two structurally related series. In one, the number of O atoms attached to the central atom X is variable, as in the chlorine oxoacids HClO, $HClO_2$, $HClO_3$, and $HClO_4$. In the other series, the identity of the atom X is variable and the number of oxygen atoms is constant, as in the hypohalous acids HFO, HClO, HBrO, and HIO.

One trend observed within the first series is that *the greater the number of oxygen atoms attached to the central atom, the stronger the acid.* Since the oxidation number of the central atom increases as the number of O atoms increases, this implies that *the greater the oxidation number of the central atom, the stronger the acid.* The effect is demonstrated by the following acids:

	Acid			
	HClO	**HClO$_2$**	**HClO$_3$**	**HClO$_4$**
Oxidation number	+1	+3	+5	+7
pK_a	7.53	2.0	Strong	Strong

The effect of increasing the number of oxygen atoms is also shown very clearly by the difference between the acid strengths of alcohols and carboxylic acids. Alcohols are organic compounds that contain a hydroxyl group (—OH) covalently bonded to a carbon atom, as in ethanol (C_2H_5OH; **9**). They are extremely weak acids in water, partly because the oxygen atom is attached to carbon, which is not very electronegative. Ethanol, for instance, has p$K_a = 16$, and only a minute proportion of ethanol molecules ionize to their conjugate base, the ethoxide ion ($C_2H_5O^-$), in water. Carboxylic acids have an additional O atom next to the —OH group, as in acetic acid (**10**). Although carboxylic acids are weak acids, they are very much stronger than alcohols.

Additional oxygen atoms increase the strength of an acid through the stabilizing effect of the extra oxygens on the conjugate base. The second O atom of the carboxyl group, for instance, provides an additional electronegative atom over which the negative charge of the conjugate base can spread; this stabilizes the anion by resonance:

$$CH_3-C \overset{O^-}{\underset{O}{\big\langle}} \longleftrightarrow CH_3-C \overset{O}{\underset{O^-}{\big\langle}}$$

(The charges shown are formal charges.) Since the conjugate base has a lower energy as a result of this resonance, it has less of a tendency to accept a proton. With the base weakened in this way, the acid is stronger.

As the identity of the central atom in a series of oxoacids varies while the number of O atoms stays fixed, we observe that *the greater the electronegativity of the central atom, the stronger the oxoacid.* This is demonstrated by the following acids:

					He
B	C	N	O	F	Ne
−0.1	0.4	0.9	1.1	1.3	
Al	Si	P	S	Cl	Ar
−0.6	−0.3	0	0.4	0.9	
Ga	Ge	As	Se	Br	Kr
−0.5	−0.3	−0.1	0.3	0.7	
In	Sn	Sb	Te	I	Xe
−0.4	−0.3	−0.2	0	0.4	
Tl	Pb	Bi	Po	At	Rn
−0.3	−0.3	−0.2	−0.1	0.1	

FIGURE 14.14 The differences between the electronegativity of hydrogen (2.1) and those of the *p*-block elements. Note the high values for oxygen and fluorine. (Blue squares mark elements that are less electronegative than hydrogen.)

7 Phosphorous acid

8

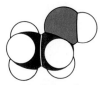

9 Ethanol

10 Acetic acid

	Acid			
	H_3AsO_4	H_3PO_4	HIO_4	$HClO_4$
Electronegativity	2.0	2.1	2.5	3.0
pK_a	2.30	2.12	1.64	Strong

A partial explanation is that electrons are withdrawn slightly from the O—H bond as the electronegativity of the central atom increases. As the electrons move toward the O atom, the O—H bond becomes more polar, and the more polar the bond, the more readily the O—H proton is lost, and hence the stronger the acid. In addition, the greater the electronegativity of the central atom, the greater the stability of the negatively charged conjugate base, and hence the greater the readiness of the acid to form it by donating a proton.

HYDROGEN ION CONCENTRATION AND pH

A knowledge of the molar concentration of hydrogen ions (more specifically, hydronium ions) in aqueous solutions is of the greatest importance in chemistry, biology, and medicine because it helps to explain many reactions. In this section we see how to calculate this concentration for solutions of acids or bases. In the next chapter we see how to do the calculation for mixtures of the two. All these calculations turn out to be applications of the techniques described in Chapter 13.

14.5 THE pH OF SOLUTIONS

Chemists usually report the hydrogen ion concentration of a solution in terms of its _pH_, defined as

$$pH = -\log [H_3O^+] \qquad (10)$$

where $[H_3O^+]$ is the numerical value of the molar concentration. The higher the pH, the lower the hydrogen ion concentration; an increase of one unit of pH corresponds to a factor-of-10 decrease in concentration. If $[H_3O^+]$ is greater than $1\,M$, the pH is negative, because $\log [H_3O^+]$ is then positive. The pH scale was introduced by the Danish chemist Søren Sørensen in 1909, in the course of his work on quality control in the brewing of beer. It is now used throughout chemistry, biochemistry, environmental chemistry, geology, industrial chemistry, medicine, and agriculture.

▼ **EXAMPLE 14.5** **Calculating a pH**

What is the pH of (a) human blood, in which the hydrogen ion concentration is $3.0 \times 10^{-7}\,M$, and (b) an acid in which the hydrogen ion concentration is $2.0\,M$?

STRATEGY In each case, we need only substitute the numerical value of the concentration into the definition, using logarithms to the base 10. (For advice about logarithms, see Appendix 1B.)

SOLUTION (a) The pH of blood is

$$pH = -\log(3.0 \times 10^{-7}) = 6.5$$

(b) The pH of 2.0 M acid is

$$pH = -\log 2.0 = -0.3$$

EXERCISE Calculate the pH of household ammonia, in which the hydrogen ion concentration is about $3 \times 10^{-12}\ M$.

[*Answer*: 11.5]

▲

The approximate pH of a solution can be judged with a strip of *universal indicator paper*, which turns different colors at different pH values. It is instructive to dip the paper into a selection of liquids and beverages to assess their relative acidities; some results are shown in Fig. 14.15. More precise measurements are made with a "pH meter"

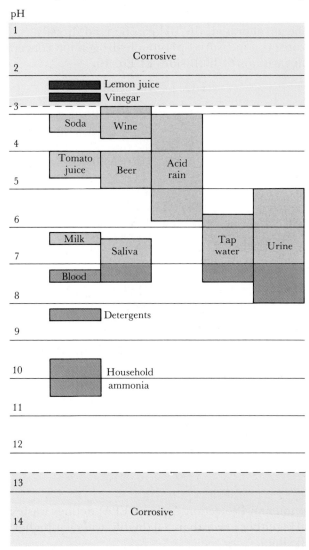

FIGURE 14.15 Typical pH values of some common aqueous solutions. The dotted lines indicate the pH beyond which substances are regarded as "corrosive."

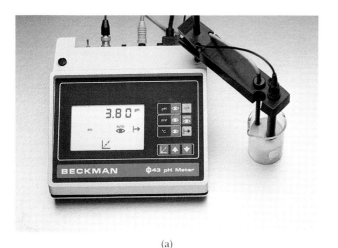

(a)

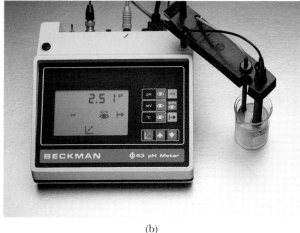

(b)

FIGURE 14.16 A pH meter is a voltmeter that measures the voltage between the two electrodes immersed in the solution. The display gives the pH directly. The samples are (a) orange juice and (b) lemon juice.

(Fig. 14.16). This consists of a voltmeter connected to two electrodes that dip into the solution. The voltage across the electrodes is proportional to the pH (as is explained in Chapter 17), so once the scale on the meter has been calibrated, the pH can be read directly.

We can convert a measured pH to a hydrogen ion concentration by taking antilogarithms of both sides of Eq. 10:

$$[H_3O^+] = 10^{-pH} \, M \tag{11}$$

The pH of fresh orange juice is about 3.5, so

$$[H_3O^+] = 10^{-3.5} \, M = 3 \times 10^{-4} \, M$$

Fresh lemon juice has a pH of 2.8, corresponding to a hydrogen ion concentration of $2 \times 10^{-3} \, M$, more than five times as high as that in orange juice. Lemon juice also tastes sharper because hydrogen ions stimulate the "sour" taste receptors on the surface of the tongue.

The pH of pure water is 7.00 at 25°C, which implies that

$$[H_3O^+] = 10^{-7.00} \, M = 1.0 \times 10^{-7} \, M$$

We conclude that of the 56 mol of H_2O molecules in 1 L of water, only 10^{-7} mol, or one molecule in 550 million, is ionized. This low concentration of ions explains why pure water is a very poor conductor of electricity. (Its reputation as a conductor stems from the ions present in impure water.)

We have seen that water is at the dividing line between acids and bases. Since its pH is 7, and pH increases as the hydrogen ion concentration decreases, we can use pH to distinguish between acidic and basic solutions:

pH	<7	7	>7
Solution	Acidic	Neutral	Basic

The Environmental Protection Agency (EPA) defines waste as "corrosive" if its pH is either lower than 3.0 (highly acidic) or higher than 12.5 (strongly alkaline).

The pH of a strong acid or base. Since a strong acid is essentially fully ionized in solution, its pH can be calculated directly from its concentration. A 0.10 *M* HCl(*aq*) solution, for instance, contains hydrogen ions at a concentration of 0.10 *M*, and its pH is

$$pH = -\log 0.10 = 1.0$$

The pH of a solution of a strong base can be found almost as easily, since it too is fully ionized. There is one additional step, because the concentration of the base gives the concentration of the OH^- ions, not of the hydronium ions. However, the water autoionization equilibrium adjusts to the presence of added OH^- to satisfy the equation

$$[H_3O^+][OH^-] = K_w$$

Hence, once we know $[OH^-]$, we can calculate $[H_3O^+]$ and then take its logarithm to find the pH. It is neater to combine the last two steps by dealing with logarithms directly. First, we take the logarithms of both sides of the equation, using the fact that $\log ab = \log a + \log b$, to obtain

$$\log [H_3O^+] + \log [OH^-] = \log K_w$$

The first term is $-pH$, and the third is $-pK_w$. The second can be expressed as a "pOH" value, where

$$pOH = -\log [OH^-]$$

Hence, changing signs throughout gives

$$pH + pOH = pK_w \qquad (12)$$

with $pK_w = 14.00$. Calculating the pH is now simple, for we can find the pOH directly from $[OH^-]$.

▼ **EXAMPLE 14.6** **Calculating the pH of a solution of a base**

Calculate the pH of a 0.020 *M* Ba(OH)$_2$(*aq*) solution.

STRATEGY We need to calculate the molar concentration of OH^- ions; once we know that, we can find pOH and use Eq. 12 to calculate pH. To find the concentration of OH^-, we note that each formula unit corresponds to two OH^- ions.

SOLUTION Since the concentration of OH^- ions is twice that of Ba(OH)$_2$, we have $[OH^-] = 0.040$ *M*. The pOH of the solution is therefore

$$pOH = -\log 0.040 = 1.40$$

and its pH is

$$pH = 14.00 - 1.40 = 12.60$$

EXERCISE Calculate the pH of a 1.5 m*M* Ca(OH)$_2$(*aq*) solution.

[*Answer*: 11.5]

▲

The pH of a weak acid or base. The pH of a solution of a weak acid is calculated using the techniques described in Chapter 13. For the "initial concentration" we use the molar concentration of acid molecules with none of them ionized. We assume that, to reach equilibrium, the acid ionizes until $[H_3O^+]$ has risen to some value x. We then draw up a table showing the changes that this implies, and use it to write an equation for the equilibrium constant (K_a in this instance) in terms of x. Finally, we solve the equation for x and then evaluate $-\log x$ to find the pH of the solution. It may be possible to simplify the resulting equations with approximations like those illustrated in Section 13.5, that is, by neglecting changes of less than about 5% of the initial concentration.

Calculations of pH are rarely reliable to more than about one decimal place (and even that may be overoptimistic). We shall therefore round the answers to one decimal place in such calculations.

▼ **EXAMPLE 14.7** Calculating the pH of a solution of a weak acid

Calculate the pH of $0.10\ M$ aqueous acetic acid.

STRATEGY Since acetic acid is weak, we expect a hydrogen ion concentration of much less than $0.10\ M$ and therefore a pH greater than 1 (since $-\log 0.1 = 1$). The value of K_a is found in Table 14.4. As long as x, the concentration of hydrogen ions, is small (which it generally is for weak acids), the equation relating x to K_a may be simplified by ignoring x in comparison with the initial concentration of acid. However, at the end of the calculation, the value of x should be checked to ensure that it is consistent with any approximation that has been made. If it is not, then the exact expression relating x and K_a must be solved (this often involves solution of a quadratic equation, Appendix 1C).

SOLUTION The following table shows the changes that occur in the solution as equilibrium is reached:

	[HA]	**[H₃O⁺]**	**[A⁻]**
Initial concentration	$0.10\ M$	0	0
Change to reach equilibrium	$-x$	$+x$	$+x$
Concentration at equilibrium	$0.10\ M - x$	x	x

In terms of x, then, at equilibrium,

$$K_a = \frac{[\text{A}^-][\text{H}_3\text{O}^+]}{[\text{HA}]} = \frac{x^2}{0.10\ M - x}$$

When x, the hydrogen ion concentration, is very small compared with $0.10\ M$ (less than about 5% of this value), this equation may be simplified to

$$K_a = \frac{x^2}{0.10\ M}$$

This rearranges to

$$x = \sqrt{0.10\ M \times K_a}$$

Since $x = [\text{H}_3\text{O}^+]$ and $K_a = 1.8 \times 10^{-5}\ M$, we have

$$[\text{H}_3\text{O}^+] = \sqrt{0.10\ M \times (1.8 \times 10^{-5}\ M)} = 1.3 \times 10^{-3}\ M$$

The approximation $x \ll 0.10\ M$ is valid because $1.3 \times 10^{-3}\ M \ll 0.10\ M$. Hence,

$$\text{pH} = -\log (1.3 \times 10^{-3}) = 2.9$$

EXERCISE Calculate the pH of $0.20\ M$ lactic acid.

[*Answer*: 1.9]

The calculation required to find the pH of a solution of a weak base is very similar. We first calculate $[OH^-]$; then we convert that to the corresponding value of $[H_3O^+]$.

▼ **EXAMPLE 14.8** **Calculating the pH of a solution of a weak base**

Calculate the pH of $0.10\ M\ NH_3(aq)$.

STRATEGY We expect a pH greater than 7 because the solution is basic. We can calculate the equilibrium concentration of OH^- ions using the technique explained in Example 14.7 (with K_b instead of K_a). Once we know $[OH^-]$ we can find pOH and convert it to pH using the equation pH + pOH = 14.00.

SOLUTION We assume that equilibrium is reached when $[OH^-]$ has risen from zero to x. Then the table of changes is as follows:

	$[NH_3]$	$[NH_4^+]$	$[OH^-]$
Initial concentration	$0.10\ M$	0	0
Change to reach equilibrium	$-x$	$+x$	$+x$
Concentration at equilibrium	$0.10\ M - x$	x	x

The equation for the base ionization constant is therefore

$$K_b = \frac{[NH_4^+][OH^-]}{[NH_3]} = \frac{x^2}{0.10\ M - x}$$

If $x \ll 0.10\ M$ (which we must verify later), $0.10\ M - x$ in the denominator can be replaced with $0.10\ M$. The solution of the simplified equation is

$$x = \sqrt{0.10\ M \times K_b}$$

From Table 14.4, $K_b = 1.8 \times 10^{-5}\ M$, so

$$x = \sqrt{0.10\ M \times (1.8 \times 10^{-5}\ M)} = 1.34 \times 10^{-3}\ M$$

This is much smaller than $0.10\ M$, so the approximation is valid.

Since $x = [OH^-]$, we have

$$pOH = -\log(1.34 \times 10^{-3}) = 2.87$$

Then, from Eq. 12, the pH of the solution is

$$pH = 14.00 - 2.87 = 11.1$$

EXERCISE Calculate the pH of $0.15\ M\ NH_2OH(aq)$.

[*Answer*: 9.6]

▲

One limitation of this approach to pH calculation is that it ignores the hydrogen ions that result from water autoionization. When the acid is in such low concentration that the calculation predicts a hydrogen ion concentration of less than $10^{-7}\ M$, we must not report its pH as being higher than 7: this is because the water autoionization itself provides hydrogen ions at a concentration of $10^{-7}\ M$. We can safely ignore the contribution of autoionization only when the calculated hydrogen ion concentration is substantially (about three times) higher than $10^{-7}\ M$.

14.6 POLYPROTIC ACIDS

Brønsted acids that can donate more than one proton are called *polyprotic acids*. Common examples are sulfuric acid and carbonic acid, each of which can donate two protons and hence is a "diprotic" acid. Another example is phosphoric acid (H_3PO_4), which can donate three protons. *Polyprotic bases* are Brønsted bases that can accept more than one proton. They include the CO_3^{2-} anion and the oxalate anion, $C_2O_4^{2-}$, both of which can accept one or two protons.

The equilibria of polyprotic acids in water are described by K_{a1} for the loss of the first proton, K_{a2} for the loss of the second, and so on. For carbonic acid,

$$H_2CO_3(aq) + H_2O(l) \rightleftharpoons HCO_3^-(aq) + H_3O^+(aq)$$

$$K_{a1} = \frac{[HCO_3^-][H_3O^+]}{[H_2CO_3]}$$

$$HCO_3^-(aq) + H_2O(l) \rightleftharpoons CO_3^{2-}(aq) + H_3O^+(aq)$$

$$K_{a2} = \frac{[CO_3^{2-}][H_3O^+]}{[HCO_3^-]}$$

In the second reaction HCO_3^-, the conjugate base of H_2CO_3, acts as an acid.

The equilibrium constant for the overall ionization

$$H_2CO_3(aq) + 2H_2O(l) \rightleftharpoons CO_3^{2-}(aq) + 2H_3O^+(aq)$$

$$K_a = \frac{[CO_3^{2-}][H_3O^+]^2}{[H_2CO_3]}$$

is the product of the two individual ionization constants:

$$K_a = K_{a1} \times K_{a2}$$

This is shown by multiplying the expressions for the two constants together and canceling $[HCO_3^-]$, which appears in both:

$$K_{a1} \times K_{a2} = \frac{[HCO_3^-][H_3O^+]}{[H_2CO_3]} \times \frac{[CO_3^{2-}][H_3O^+]}{[HCO_3^-]}$$

$$= \frac{[CO_3^{2-}][H_3O^+]^2}{[H_2CO_3]} = K_a$$

The strengths of polyprotic acids. The entries in Table 14.7 show that the strengths of polyprotic acids decrease as protons are lost. That is,

$$pK_{a1} < pK_{a2} < \cdots$$

The decrease is largely due to the greater difficulty of removing a positively charged proton from a negatively charged ion (such as HCO_3^-) than from the original neutral molecule, H_2CO_3. Sulfuric acid shows this decrease in strength very well. The H_2SO_4 molecule itself is a strong acid and loses its first proton to form its conjugate base, the hydrogen sulfate ion HSO_4^-; that ion, however, is a weak acid:

$$H_2SO_4(aq) + H_2O(l) \rightleftharpoons H_3O^+(aq) + HSO_4^{2-}(aq) \qquad \text{Strong}$$

$$HSO_4^-(aq) + H_2O(l) \rightleftharpoons H_3O^+(aq) + SO_4^{2-}(aq) \qquad pK_a = 1.92$$

TABLE 14.7 pK_a values of polyprotic acids

Acid	pK_{a1}	pK_{a2}	pK_{a3}
Sulfuric acid, H_2SO_4	Strong	1.92	
Oxalic acid, $(COOH)_2$	1.23	4.19	
Sulfurous acid, H_2SO_3	1.81	6.91	
Phosphorous acid, H_2PO_3	2.00	6.59	
Phosphoric acid, H_3PO_4	2.12	7.21	12.67
Tartaric acid, $C_4H_6O_6$	3.22	4.82	
Carbonic acid, H_2CO_3	6.37	10.25	
Hydrosulfuric acid, H_2S	6.88	14.15	

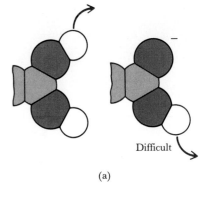

(a)

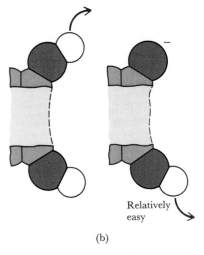

(b)

FIGURE 14.17 (a) The loss of a proton from a small molecule can have a strong influence on the loss of a second proton. (b) However, in a protein molecule, the second proton may come from a site far away from the first, and its loss is largely unaffected by the loss of the first.

Some proteins are complicated polyprotic acids that can donate dozens of protons in reactions. The decrease in acid strength is less pronounced for these proteins because the hydrogen atoms are usually lost from widely separated sites. The loss of a proton from one site has little influence on a distant site (Fig. 14.17).

▼ EXAMPLE 14.9 Calculating the pH of aqueous sulfuric acid

Calculate the pH of 0.100 M $H_2SO_4(aq)$ at 25°C, using the information in Table 14.7.

STRATEGY The pH depends on the total hydrogen ion concentration, with both ionization steps taken into account. The first ionization is complete, so it results in a hydrogen ion concentration equal to the concentration of the acid, 0.100 M. This corresponds to pH = 1.0. The second ionization adds to the hydrogen ion concentration slightly, so we should expect an overall pH slightly below 1.0. The contribution of the second ionization is obtained by treating the HSO_4^- formed in the first ionization as a weak acid of concentration 0.100 M and setting up a table of changes in the usual way. However, the initial concentration of hydrogen ions in the second ionization is not zero, but 0.100 M. Since $K_{a2} = 10^{-1.92} M = 0.012 M$ is not very small, it will be necessary to solve a quadratic equation without taking simplifying shortcuts.

SOLUTION The table of changes for the second ionization is as follows:

	$[HSO_4^-]$	$[H_3O^+]$	$[SO_4^{2-}]$
Initial concentration	0.100 M	0.100 M	0
Change to reach equilibrium	$-x$	$+x$	$+x$
Concentration at equilibrium	0.100 $M - x$	0.100 $M + x$	x

The equation for the acid ionization constant K_{a2} is therefore

$$K_{a2} = \frac{[SO_4^{2-}][H_3O^+]}{[HSO_4^-]} = \frac{x \times (0.100 \ M + x)}{0.100 \ M - x}$$

Substituting $K_{a2} = 0.012 M$ and rearranging gives

$$x^2 + (0.112 \ M)x - 0.0012 \ M^2 = 0$$

The solutions (Appendix 1C) are

$$x = \frac{-0.112 + \sqrt{(0.112)^2 + 4 \times 0.0012}}{2} = 0.010 \, M$$

$$x = \frac{-0.112 - \sqrt{(0.112)^2 + 4 \times 0.0012}}{2} = -0.122 \, M$$

Since the hydrogen ion concentration at equilibrium is $0.100 \, M + x$, these two solutions give

$$[H_3O^+] = 0.110 \, M \quad \text{or} \quad -0.022 \, M$$

The second value makes no sense and can be eliminated. Thus,

$$pH = -\log 0.110 = 0.96$$

This is slightly lower than 1.0, as anticipated.

EXERCISE Calculate the pH of $0.010 \, M \, H_2SO_4(aq)$ at 25°C.

[*Answer*: 1.8]

FIGURE 14.18 When carbon dioxide dissolves in water, it gives an acidic solution of H_2CO_3, as shown by the color change in the indicator. A significant proportion of the CO_2 molecules remain unchanged in the solution.

The anomalous behavior of carbonic acid. Carbonic acid solutions play an important role in environmental chemistry because they are formed when atmospheric carbon dioxide dissolves in rain and rivers (Fig. 14.18). Moreover, the acid is interesting because its pK_{a1} of 6.37 suggests that its first ionization is very weak. However, this value is based on the assumption that all the CO_2 molecules that dissolve form H_2CO_3. In fact, only about one molecule in 480 reacts to form the acid, while the great majority remain CO_2. The value of pK_{a1} reported in the table refers to

$$K_{a1} = \frac{[HCO_3^-][H_3O^+]}{[CO_2]}$$

where $[CO_2]$ is the total concentration of dissolved CO_2, whether as CO_2 or H_2CO_3. The true ionization constant is

$$K'_{a1} = \frac{[HCO_3^-][H_3O^+]}{[H_2CO_3]}$$

Because $[H_2CO_3]$ is much smaller than $[CO_2]$, the true ionization constant is much larger than that given in Table 14.4.

▼ EXAMPLE 14.10 Calculating a true ionization constant

Calculate the true pK_{a1} of H_2CO_3 from the information in Table 14.4 and the text.

STRATEGY We are asked to find the value of K'_{a1}, given the value of K_{a1} in Table 14.4. The expressions for the two constants differ in that the expression for K'_{a1} has $[H_2CO_3]$ in place of $[CO_2]$. Therefore, if we know the relation between these two concentrations, we can carry out the conversion. We are told in the text that the concentration of H_2CO_3 molecules is $\frac{1}{480}$ that of CO_2.

SOLUTION On substituting $[H_2CO_3] = \frac{1}{480} \times [CO_2]$ into the expression for the true ionization constant, we get

$$K'_{a1} = 480 \times \frac{[HCO_3^-][H_3O^+]}{[CO_2]}$$

$$= 480 \times K_{a1} = 2.1 \times 10^{-4} \, M$$

The true pK_{a1} is therefore 3.7, so carbonic acid is not nearly as weak as the value in the table suggests.

EXERCISE An acid HA formed when a substance dissolves has an apparent pK_a of 6.5, but it is known that only one in 800 molecules is in the form of HA. What is its true pK_a?

[*Answer*: 3.6]

▲

In calculating the pH of a polyprotic acid, it is normally sufficient (except at very low concentrations) to take only the first ionization into account: the second ionization is much less important than the first. In the case of H_2CO_3, for instance, $K_{a2} = 5.6 \times 10^{-11} \, M$, which is very much smaller than K_{a1}. That being so, we can ignore the contribution of the second ionization to the pH and treat the acid as a monoprotic (one-proton) acid with $pK_a = 6.37$. The pH of the solution is calculated exactly as in Example 14.7. Using the result

$$[H_3O^+] = \sqrt{\text{acid concentration} \times K_a}$$

which is derived in that example, we find that, for 0.020 M carbonic acid,

$$[H_3O^+] = \sqrt{0.020 \, M \times (4.3 \times 10^{-7} \, M)} = 9.3 \times 10^{-5} \, M$$

This corresponds to pH = 4.0. Note that we use the value of K_{a1} given in Table 14.4, not the true value, because 0.020 M is the *total* concentration of CO_2, not the actual concentration of molecules of H_2CO_3.

The second ionization of carbonic acid is important if we want to calculate the concentration of CO_3^{2-} ions at equilibrium, because the first ionization does not produce these ions directly.

▼ **EXAMPLE 14.11 Calculating the carbonate ion concentration in carbonic acid**

Calculate the carbonate ion concentration in 2.0 mM $H_2CO_3(aq)$ at 25°C.

STRATEGY The best procedure is to write the equilibrium that gives the required ion, and then inspect the expression for the equilibrium constant to see what additional information is needed. In this case we have

$$HCO_3^-(aq) + H_2O(l) \rightleftharpoons CO_3^{2-}(aq) + H_3O^+(aq) \quad K_{a2} = \frac{[CO_3^{2-}][H_3O^+]}{[HCO_3^-]}$$

When the ionization constant is rearranged to

$$[CO_3^{2-}] = K_{a2} \times \frac{[HCO_3^-]}{[H_3O^+]}$$

it shows that $[HCO_3^-]$ and $[H_3O^+]$ are needed. We assume (and later check) that because K_{a2} is so small, the second ionization barely affects the values calculated from the first ionization.

FIGURE 14.19 The effervescence that occurs when an antacid tablet is dissolved in water is the result of a disturbance that is passed along from one equilibrium to another.

SOLUTION Since from the first ionization we know that $[H_3O^+]$ = $[HCO_3^-]$, the two concentrations cancel in the expression above for $[CO_3^{2-}]$, and

$$[CO_3^{2-}] = K_{a2} = 5.6 \times 10^{-11} \, M$$

EXERCISE Calculate the concentrations of HS^- and S^{2-} in 0.010 M $H_2S(aq)$, and the pH of the solution.

[*Answer*: pH = 4.4; $[HS^-] = 3.6 \times 10^{-5} \, M$; $[S^{2-}] = 7.1 \times 10^{-15} \, M$]

The effervescence of carbon dioxide that occurs when an acid is added to a carbonate or a bicarbonate (Fig. 14.19) can be discussed in terms of Brønsted equilibria. The added hydrogen ions shift the equilibrium composition in the direction of the products in the reaction

$$H_3O^+(aq) + HCO_3^-(aq) \longrightarrow H_2O(l) + H_2CO_3(aq)$$

The carbonic acid in the solution is in equilibrium with dissolved CO_2 molecules according to

$$H_2CO_3(aq) \rightleftharpoons H_2O(l) + CO_2(aq)$$

However, when the H_2CO_3 concentration increases, the CO_2 concentration increases in response. The latter increase is so great that CO_2 bubbles out of solution with the familiar fizz. As in this everyday example, we often see chemical effects being transmitted along chains of equilibria, with the disturbance of one equilibrium affecting another equilibrium and perhaps still others. This is particularly important when acids and bases are both present in a single solution, as we see in the next chapter.

SUMMARY

14.1 A **Brønsted acid** is defined as a proton donor, and a **Brønsted base** as a proton acceptor. Wider definitions of acids and bases are provided by the **Lewis theory,** in which a **Lewis acid** is an electron-pair acceptor and a **Lewis base** is an electron-pair donor. A Lewis acid combines with a Lewis base to form a **complex.**

14.2 The key to the quantitative treatment of acids and bases is the rapid, dynamic **proton-transfer equilibrium** that exists in solution. This equilibrium can be expressed generally by introducing the concepts of **conjugate acids and bases,** the acids and bases formed by proton gain by a base and proton loss by an acid. We write all equilibria in the form

$$\text{Acid}_1 + \text{base}_2 \rightleftharpoons \text{base}_1 + \text{acid}_2$$

and express the equilibrium constants in the form

$$K_c = \frac{[\text{base}_1][\text{acid}_2]}{[\text{acid}_1][\text{base}_2]}$$

A special case is the **autoionization,** or self-ionization, of an **amphiprotic** solvent, one that can both donate and accept protons.

14.3 Since the concentration of water in a dilute solution is almost constant at 55 M, acid-base equilibria can all be discussed in terms of the **autoionization constant** of water K_w, the **acid and base ionization constants** K_a and K_b, and, for convenience, the corresponding pK values, where in each case p$K = -\log K$.

14.4 Ionization constants are measures of the strengths of acids and bases. A **strong acid** exists as the conjugate base in solution, with few nonionized molecules present. A **weak acid** is an acid with $K_a < 1$; a solution of a weak acid contains both the nonionized acid and the conjugate base. A **strong base** exists as the conjugate acid in solution, with few nonionized molecules present. A **weak base** is a base with $K_b < 1$; a solution of a weak base contains nonionized base molecules and the conjugate acid. The stronger the acid, the weaker its conjugate base.

The strongest acid in aqueous solution is the hydronium ion, and the strongest base is the OH^- ion. Stronger proton donors than H_2O are acids; stronger proton acceptors than H_2O are bases. The strengths of acids can (to some extent) be explained in terms of bond polarity, bond strength, electronegativity, oxidation number, and number of oxygen atoms.

14.5 The molar concentration of hydronium ions in a solution is often expressed in terms of its **pH,** with pH = $-\log [H_3O^+]$. This property is best measured with a pH meter, but it can be estimated with an indicator paper. The pH of pure water is 7; pH < 7 corresponds to an **acidic solution** (more H_3O^+ than OH^-), pH = 7 to a **neutral solution** (equal concentrations of H_3O^+ and OH^-), and pH > 7 to a **basic solution** (more OH^- than

H_3O^+). The pH of a weak acid is calculated using the equilibrium techniques described in Chapter 13. The pH of a weak base is calculated with an additional step that takes into account the base ionization equilibrium and water autoionization (by use of the equation pH + pOH = pK_w).

14.6 Special care must be taken with **polyprotic acids,** acids capable of donating more than one proton, and **polyprotic bases,** bases able to accept more than one proton; in principle, several ionization equilibria may contribute to the overall composition of a solution of such an acid or base. In most cases, however, the **first ionization constant** is much larger than the constants for subsequent ionizations, and it alone need be considered (unless it does not lead directly to the ion of interest).

EXERCISES

Brønsted Acids and Bases

In Exercises 14.1 and 14.2, identify the Brønsted acids and bases.

14.1 (a) HNO_3; (b) CH_3NH_2; (c) H_2S; (d) H_2O.
14.2 (a) Phenol, C_6H_5OH; (b) aniline, $C_6H_5NH_2$; (c) benzene, C_6H_6; (d) HCN.

In Exercises 14.3 and 14.4, give the conjugate acid or base for each substance.

14.3 (a) HNO_3; (b) HCO_3^-; (c) HCN; (d) CH_3COOH.
14.4 (a) $HClO_4$; (b) HPO_4^{2-}; (c) CH_3NH_2; (d) C_6H_5OH.

In Exercises 14.5 and 14.6, write the Brønsted equilibrium for each substance in water.

14.5 (a) HCOOH; (b) CH_3NH_2; (c) PO_4^{3-}; (d) CN^-.
14.6 (a) NH_2CH_2COOH; (b) Br^-; (c) H_3PO_2; (d) H_3PO_4.

In Exercises 14.7 to 14.10, state whether each oxide is likely to be acidic, amphoteric, or basic.

14.7 (a) CaO; (b) NO; (c) ClO; (d) K_2O.
14.8 (a) Rb_2O; (b) SeO_3; (c) BeO; (d) B_2O_3.
14.9 (a) Cl_2O_7; (b) GeO_2; (c) Ga_2O_3.
14.10 (a) As_2O_5; (b) TeO_3; (c) Tl_2O_3.

Lewis Acids and Bases

In Exercises 14.11 and 14.12, identify the Lewis acids and bases.

14.11 (a) NH_3; (b) Na^+; (c) Cl^-; (d) CO.
14.12 (a) H_2O; (b) Al^{3+}; (c) CN^-; (d) NO_2^-.

In Exercises 14.13 and 14.14, identify the reactions between Lewis acids and bases.

14.13 (a) A test for chlorides:
$BaCl_2(aq) + 2AgNO_3(aq) \rightarrow 2AgCl(s) + Ba(NO_3)_2(aq)$
(b) A neutralization reaction:
$2HNO_3(aq) + Ca(OH)_2(aq) \rightarrow Ca(NO_3)_2(aq) + 2H_2O(l)$
(c) The reaction of an alkali metal with water:
$2Na(s) + 2H_2O(l) \rightarrow 2NaOH(aq) + H_2(g)$
14.14 (a) The reaction of a metal oxide with water:
$Na_2O(s) + H_2O(l) \rightarrow 2NaOH(aq)$
(b) The laboratory preparation of a hydrogen halide: $PCl_3(l) + 3H_2O(l) \rightarrow H_3PO_3(aq) + 3HCl(aq)$
(c) The formation of uranium hexafluoride in one stage in the extraction of uranium: $UO_2 + 4HF \rightarrow UF_4 + 2H_2O$

Ionization Constants

In Exercises 14.15 and 14.16, express each ionization constant as a pK_a value.

14.15 (a) Formic acid, HCOOH: $1.77 \times 10^{-4}\,M$; (b) acetic acid, CH_3COOH: $1.75 \times 10^{-5}\,M$; (c) trichloroacetic acid, CCl_3COOH: $0.30\,M$; (d) benzoic acid, C_6H_5COOH: $6.46 \times 10^{-5}\,M$. (e) Suggest an explanation for the difference between the ionization constants for acetic acid and trichloroacetic acid.
14.16 (a) Hydrocyanic acid, HCN: $4.93 \times 10^{-10}\,M$; (b) nitrous acid, HNO_2: $4.3 \times 10^{-12}\,M$; (c) phenol, C_6H_5OH: $1.3 \times 10^{-10}\,M$; (d) 2,4,6-trichlorophenol **(11)**: $1 \times 10^{-6}\,M$. (e) Suggest an explanation of the difference between (c) and (d).

11

In Exercises 14.17 and 14.18, convert each given pK_a value to a K_a value, and arrange the acids in order of increasing strength.

14.17 (a) Phosphoric acid, H_3PO_4: 2.12; (b) phosphorous acid, H_3PO_3: 2.00; (c) selenic acid, H_2SeO_4: 1.92; (d) selenous acid, H_2SeO_3: 2.46. (e) Suggest a reason for the trends in values. How may (a) and (b) be prepared?
14.18 (a) Carbonic acid, H_2CO_3: 6.37; (b) germanic acid, H_2GeO_3: 8.59; (c) periodic acid, HIO_4: 1.64; (d) hypoiodous acid, HIO: 10.64. (e) Suggest a reason for the trends in values. How may (a) and (d) be prepared?

In Exercises 14.19 and 14.20, express each given K_b value as a pK_b value, and arrange the substances in order of increasing base strength.

14.19 (a) Ammonia: $1.8 \times 10^{-5}\,M$; (b) deuterated ammonia, ND_3: $1.1 \times 10^{-5}\,M$; (c) hydrazine, NH_2NH_2: $1.7 \times 10^{-6}\,M$; (d) hydroxylamine, NH_2OH: $1.07 \times 10^{-8}\,M$. (e) Suggest a reason for the differences between (a), (c), and (d).
14.20 (a) Methylamine, CH_3NH_2: $3.6 \times 10^{-4}\,M$; (b) dimethylamine, $(CH_3)_2NH$: $5.4 \times 10^{-4}\,M$; (c) trimethylamine, $(CH_3)_3N$: $6.5 \times 10^{-5}\,M$; (d) urea, $CO(NH_2)_2$: $1.3 \times 10^{-14}\,M$. (e) Suggest a reason for the weakness of urea as a base.

In Exercises 14.21 and 14.22, express each given pK_b value as a K_b value, and arrange the compounds in order of increasing base strength.

14.21 (a) Aniline (**12**, with X = H): 9.37; (b) 4-bromoaniline (**12**, with X = Br): 10.14; (c) 4-chloroaniline (**12**, with X = Cl): 9.85. Is there a trend that can be explained simply?
14.22 (a) 2-Fluoroaniline (**13**, with X = F): 10.80; (b) 3-fluoroaniline (**14**, with X = F): 10.50; (c) 4-fluoroaniline (**12**, with X = F): 9.35. (d) Is there a trend that can be explained simply?

14.23 Write the conjugate acid of each base in Exercise 14.21, and give its (1) K_a value and (2) pK_a value.
14.24 Write the conjugate acid of each base in Exercise 14.22, and give its (1) K_a value and (2) pK_a value.

In Exercises 14.25 and 14.26, arrange the bases in order of increasing strength, based on the given pK_a values of their conjugate values.

14.25 (a) Ammonia: 9.25; (b) methylamine, CH_3NH_2: 10.56; (c) ethylamine, $C_2H_5NH_2$: 10.72; (d) aniline, $C_6H_5NH_2$: 4.63. (e) Is there a simple pattern of strengths?
14.26 (a) Aniline: 4.63; (b) 2-hydroxyaniline (**13**, with X = OH): 4.72; (c) 3-hydroxyaniline (**14**, with X = OH): 4.17; (d) 4-hydroxyaniline (**12**, with X = OH): 5.47. (e) Is there a simple pattern of strengths?

In Exercises 14.27 and 14.28, calculate the percentage of molecules that are ionized in each solution.

14.27 (a) $0.20\,M$ $HCOOH(aq)$; (b) $0.20\,M$ $NH_3(aq)$.
14.28 (a) $0.15\,M$ $CH_3COOH(aq)$; (b) $0.10\,M$ NH_2NH_2, for which $pK_b = 5.77$. (c) What are the percentages when the concentrations are (1) reduced by half, (2) doubled?

Structure and Strength

In Exercises 14.29 to 14.32, arrange the acids in order of increasing strength. Refer to Table 14.4 to see whether your predictions are valid, and, if they are not, suggest an explanation.

14.29 (a) HF; (b) HCl; (c) HBr; (d) HI. (e) Where would HAt be expected to lie?
14.30 (a) HClO; (b) HBrO; (c) HIO. (d) Where would HAtO be expected to lie?

14.31 (a) HNO_2; (b) HNO_3. (c) How could each of these acids be prepared?
14.32 (a) H_2CO_3; (b) H_2GeO_3. (c) How could each of these acids be prepared?

pH of Strong Acids and Bases

In Exercises 14.33 and 14.34, express each given pH value as a hydrogen ion concentration.

14.33 (a) 1.0; (b) 7.00; (c) -0.50.
14.34 (a) 6.00; (b) 6.0; (c) 14.0.

In Exercises 14.35 and 14.36, express each given hydrogen ion concentration as a pH value.

14.35 (a) $2.0 \times 10^{-5}\,M$; (b) $1.0\,M$; (c) $5.0 \times 10^{-14}\,M$; (d) $5.0\,M$.
14.36 (a) $1.5 \times 10^{-7}\,M$; (b) $2.2\,mM$; (c) $10^{-7}\,M$; (d) $1.000 \times 10^{-6}\,M$.

In Exercises 14.37 and 14.38, calculate the pOH and the pH of each solution.

14.37 (a) $0.01\,M$ $KOH(aq)$; (b) $1\,mM$ $Ba(OH)_2(aq)$; (c) $0.01\,M$ $HNO_3(aq)$.

12

13

14

14.38 (a) 2 mM CsOH(aq); (b) 0.10 mM Sr(OH)$_2$(aq); (c) 1.0 M HClO$_4$(aq).

In Exercises 14.39 to 14.42, calculate the hydrogen ion concentration and pH of a solution prepared as described.

14.39 25 mL of 0.10 M HCl(aq) was added to 25 mL of 0.20 M NaOH(aq).

14.40 25 mL of 0.15 M HCl(aq) was added to 50 mL of 0.15 M KOH(aq).

14.41 14.0 g of sodium hydroxide was made up to 250 mL of solution, and 25.0 mL was pipeted into 50 mL of 0.20 HBr(aq).

14.42 0.150 g of barium hydroxide was made up to 50.0 mL of aqueous solution, and 25.0 mL was pipeted into 100 mL of 0.0010 M hydrochloric acid.

The pH of Weak Acids and Bases

In Exercises 14.43 and 14.44, calculate the pH of each solution at 25°C, using pK_a values from Table 14.4.

14.43 (a) 0.15 M CH$_3$COOH(aq); (b) 1.0 × 10^{-5} M CH$_3$COOH(aq); (c) 0.15 M CCl$_3$COOH(aq).

14.44 (a) 0.20 M CH$_3$CH(OH)COOH(aq) (lactic acid); (b) 1.5 × 10^{-5} M CH$_3$CH(OH)COOH; (c) 0.10 M C$_6$H$_5$SO$_3$H (benzenesulfonic acid).

In Exercises 14.45 and 14.46, calculate the hydroxide ion concentration, the pOH, and the pH of each aqueous solution of the given base at 25°C. When the pK_a is given, it refers to the conjugate acid.

14.45 0.15 M NH$_3$(aq), pK_b = 4.75.

14.46 (a) 0.10 M C$_6$H$_{11}$NH$_2$ (cyclohexylamine), pK_b = 3.36; (b) 0.20 M strychnine, for which the conjugate acid has pK_a = 8.26; (c) 0.015 M NH$_2$CH$_2$CH$_2$NH$_2$ (ethylenediamine), for which the conjugate acid, NH$_2$CH$_2$CH$_2$NH$_3^+$, has pK_a = 10.71.

14.47 When the pH of a 0.10 M solution of chlorous acid (HClO$_2$) was measured, it was found to be 1.2. What is the pK_a of this acid?

14.48 The pH of a 15 mM HNO$_2$(aq) solution was measured to be 2.63. What is the pK_a of nitrous acid?

14.49 A 25 mM aqueous solution of a base was found to have pH = 11.6. What are its pK_b and the pK_a of its conjugate acid?

14.50 When 150 mg of an organic base of molar mass 31.06 g/mol was dissolved in 50 mL of water, its pH was measured as 10.05. Calculate the pK_b of the base and the pK_a of its conjugate acid.

Polyprotic Acids

14.51 Calculate the pH of a 0.15 M H$_2$SO$_4$(aq) solution at 25°C.

14.52 Suppose a different source of data gave pK_{a2} = 2.00 for sulfuric acid. Calculate the pH of a 0.020 M H$_2$SO$_4$(aq) solution.

In Exercises 14.53 and 14.54, calculate the pH of each solution of a weak polyprotic acid at 25°C, ignoring second ionizations only when that approximation is justified.

14.53 (a) Aqueous hydrogen sulfide, 0.10 M H$_2$S(aq); (b) carbonic acid, 1.0 mM H$_2$CO$_3$(aq).

14.54 (a) Oxalic acid, 0.10 M (COOH)$_2$(aq); (b) tartaric acid, 0.15 mM C$_4$H$_6$O$_6$(aq).

In Exercises 14.55 and 14.56, calculate the concentration of the ions produced by each ionization of each acid, neglecting ionizations only when it is valid to do so.

14.55 (a) 0.10 M (COOH)$_2$, oxalic acid; (b) 0.10 mM B(OH)$_3$, boric acid, which acts as a monoprotic acid.

14.56 (a) 0.10 M H$_2$S(aq); (b) 15 mM H$_3$PO$_4$(aq).

General

14.57 Estimate the enthalpy of ionization of formic acid at 25°C, given that K_a is 1.765 × 10^{-4} M at 20°C and 1.768 × 10^{-4} M at 30°C.

14.58 Convert the van't Hoff equation for the temperature dependence of an equilibrium constant (stated in Exercise 13.66) to an expression for the temperature dependence of pK_w. Estimate the value of pK_w at blood temperature (37°C) from the enthalpy of the water ionization reaction. What is the pH of pure water at that temperature?

14.59 Estimate the enthalpy of the ionization of heavy water, D$_2$O, given the following data on its autoionization:

Temperature, °C	10	20	25	30	40	50
pK_w	15.44	15.05	14.87	14.70	14.39	14.10

What is the pD (pD = −log [D$_3$O$^+$]) of pure heavy water at 25°C? What equations in this chapter would be changed if fully deuterated compounds and heavy water were used throughout?

The main goal of this chapter is to explain how acids, bases, and their salts affect the pH of solutions. We also see how certain mixtures can be used to stabilize solutions against changes in pH, how to interpret changes in pH during acid-base titrations, and how to choose an indicator to detect an equivalence point accurately. The illustration shows a tiny section of an ocean: the pH of the water—which enables living organisms to survive and flourish—is maintained by some of the processes that we describe in this chapter.

15

ACIDS, BASES, AND SALTS IN WATER

Salts as acids and bases

15.1 Ions as acids and bases
15.2 The pH of mixed solutions

Titrations and pH curves

15.3 The variation of pH
       during a titration
15.4 Indicators and buffers

Solubility equilibria

15.5 The solubility product
15.6 Precipitation reactions
       and qualitative analysis
15.7 Complex formation

A striking illustration of the importance of pH is that we are likely to die if the pH of our blood plasma falls by more than 0.4 unit from its normal value of 7.4. This can happen as a result of disease or shock, both of which generate acidic conditions in the body. Death is also likely if the blood plasma becomes basic by more than 0.4 unit, as sometimes happens during the early stages of recovery from severe burns. To survive, the body must control its own pH, perhaps with the aid of physicians.

The pH of aqueous solutions—not only blood plasma, but also seawater, detergents, sap, and reaction mixtures—is controlled by the proton transfer that takes place between ions and water molecules. The presence of NH_4^+ ions, which are Brønsted acids, for example, makes an aqueous solution acidic; the presence of $CH_3CO_2^-$ ions, which are Brønsted bases, makes it basic. In this chapter we see how to calculate the pH of solutions that contain these ions. In doing so, we shall make use of two equations that were derived in Chapter 14. The first expresses the autoionization equilibrium:

$$pH + pOH = pK_w \qquad (1)$$

According to this equation, if the concentration of either H_3O^+ or OH^- increases, then the concentration of the other must fall to preserve the value of K_w. At 25°C (the only temperature we consider), $pK_w = 14.00$. The other equation expresses the relation between the strength of an acid and that of its conjugate base:

$$pK_a + pK_b = pK_w \qquad (2)$$

In this equation, pK_a and pK_b are the ionization constants of the conjugate acid and base, respectively. The equation implies that the stronger the acid (the smaller the pK_a), the weaker its conjugate base (the larger the pK_b).

Although we shall calculate pH values and pK values to the number of significant figures appropriate to the data, the answers may often be considerably less reliable than that. Thus, although we might calculate the pH of a solution as 8.82, in practice the result is unlikely to be reliable to more than one decimal place (pH = 8.8), and probably to no more than one significant figure (pH = 9). The reason is that we are ignoring a number of complicating effects, particularly interactions between the ions in solution.

SALTS AS ACIDS AND BASES

The experimental observation at the root of this chapter is that an aqueous solution of a salt derived from a weak acid or a weak base has a pH that is different from 7. We saw why in Section 14.2: the salt provides ions that act as acids or bases and affect the pH of the solution. This observation may not sound very interesting at first, but the effect of salts on the pH of solutions turns out to be fundamental to a lot of chemistry, and essential to much of biochemistry.

15.1 IONS AS ACIDS AND BASES

We can calculate the pH of an ammonium chloride solution (or of any other solution of a salt of a weak acid or weak base) by treating the calculation as a problem in chemical equilibrium. Initially the solution consists of the cations and anions of the salt in the solvent water. It quickly (in less than a billionth of a second) reaches equilibrium, since some of the cations donate protons to H_2O molecules:

$$NH_4^+(aq) + H_2O(l) \longrightarrow NH_3(aq) + H_3O^+(aq)$$

This increases the concentration of hydronium ions above their concentration in pure water, and the pH falls below 7. The actual concentration of H_3O^+ ions is calculated by finding the value that satisfies the ionization-constant equation for the proton-transfer equilibrium:

$$NH_4^+(aq) + H_2O(l) \rightleftharpoons NH_3(aq) + H_3O^+(aq) \qquad K_a = \frac{[NH_3][H_3O^+]}{[NH_4^+]}$$

▼ EXAMPLE 15.1 Calculating the pH of a salt solution: I

Calculate the pH of a 0.15 M $NH_4Cl(aq)$ solution.

STRATEGY We can suppose that the concentration of NH_4^+ ions decreases by x from its initial value, as the ions donate protons to H_2O to reach equilibrium. This increases the H_3O^+ ion concentration from 10^{-7} M, the value in pure water, to 10^{-7} $M + x$. If the salt concentration is not too low, the H_3O^+ ions produced in this way so outnumber those produced by autoionization that the original 10^{-7} M concentration can be ignored. However, we must check at the end of the calculation to ensure that the original 10^{-7} M is in fact not larger than about 5% of x.

To find x, we draw up a table of changes, substitute the equilibrium concentrations into the expression for the equilibrium constant (K_a in this instance), and solve for x. The pH of the solution is equal to $-\log x$. As in Chapter 13, we neglect changes that are no larger than about 5% of the starting concentration. We can find K_a for NH_4^+ from the equation $K_a \times K_b = K_w$ and the value of K_b for its conjugate base (NH_3) in Table 14.4.

SOLUTION If we ignore the original H_3O^+ concentration, the table of changes is as follows:

	$[NH_4^+]$	$[NH_3]$	$[H_3O^+]$
Initial concentration	0.15 M	0	0
Change to reach equilibrium	$-x$	$+x$	$+x$
Concentration at equilibrium	0.15 $M - x$	x	x

The expression for K_a then becomes

$$K_a = \frac{[NH_3][H_3O^+]}{[NH_4^+]} = \frac{x^2}{0.15\ M - x}$$

We note that x is likely to be much smaller than 0.15 M (even if the pH were as low as 3, x would be only 10^{-3} M). Thus it may be ignored in the denominator, so we have

$$K_a = \frac{x^2}{0.15\ M}$$

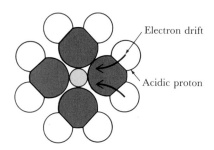

FIGURE 15.1 In water, metal cations exist as hydrated ions that can act as Brønsted acids. Although for clarity only four hydrating H_2O molecules are shown here, metal cations typically have six H_2O molecules bonded to them.

From this expression, we find that

$$x = \sqrt{0.15\,M \times K_a}$$

Since $K_b = 1.8 \times 10^{-5}\,M$ and $K_w = 1.0 \times 10^{-14}\,M^2$,

$$K_a = \frac{1.0 \times 10^{-14}\,M^2}{1.8 \times 10^{-5}\,M} = 5.6 \times 10^{-10}\,M$$

Therefore,

$$x = \sqrt{0.15\,M \times (5.6 \times 10^{-10}\,M)} = 9.2 \times 10^{-6}\,M$$

Hence,

$$\text{pH} = -\log\,(9.2 \times 10^{-6}) = 5.04$$

or, more realistically, pH = 5. (Note also that x is both much smaller than $0.15\,M$ and much larger than $10^{-7}\,M$, justifying the two approximations we used in the calculation.)

EXERCISE Calculate the pH of a 0.10 M aqueous solution of methylammonium chloride.

[*Answer*: 5.8]

▲

Hydrated cations. For a cation to be a Brønsted acid, it must have protons that it can donate. This would appear to rule out acid behavior in simple metal cations like Na^+ and Al^{3+}. However, metal cations are hydrated in solution, and the resulting hydrated ion can act as a Brønsted acid. This happens if an O—H bond in one of the hydrating H_2O molecules is weakened enough by the electron-withdrawing influence of the cation (Fig. 15.1). An example is Al^{3+}, which exists in solution as $[Al(H_2O)_6]^{3+}$. This complex is a Brønsted acid: the small, highly charged Al^{3+} ion polarizes the O—H bonds of the hydrating H_2O molecules and makes it possible for at least one or two protons to be lost:

$$[Al(H_2O)_6]^{3+}(aq) + H_2O(l) \rightleftharpoons [Al(H_2O)_5OH]^{2+}(aq) + H_3O^+(aq)$$

Some cations that act as acids in this way are listed in Table 15.1.

Acid character is typical of small, highly charged cations such as Al^{3+} and Fe^{3+}. Solutions of their salts are often about as acidic as dilute acetic acid (Fig. 15.2). An aqueous solution of titanium(III) sulfate, which contains the Brønsted acid $[Ti(H_2O)_6]^{3+}$, is so acidic that, like hydrochloric acid, it produces hydrogen sulfide when poured on sodium sulfide (Fig. 15.3). Hydrated Na^+ ions and other Group I cations are so weak as Brønsted acids that they are considered neutral (Table 15.1). The cations are too large to have an appreciable polarizing effect on the hydrating water molecules that surround them.

Anions as bases. The anion of a strong acid is a very weak Brønsted base in water. Therefore, if the cation is not an acid, the salt of strong acid gives a neutral solution in water (Table 15.2). The anion of a weak acid is a base, and, if the cation is not an acid, the salt of a weak acid gives rise to a basic solution in water. When formate ions (HCO_2^-), for example, are added to water as sodium formate (the salt of a weak acid), they remove protons from water molecules, leaving an excess of hydroxide ions in the solution:

$$HCO_2^-(aq) + H_2O(l) \longrightarrow HCOOH(aq) + OH^-(aq)$$

FIGURE 15.2 These four solutions show that hydrated cations can be significantly acidic. The test tubes contain (from left to right) deionized water, 0.1 M aluminum sulfate, 0.1 M titanium(III) sulfate, and 0.1 M acetic acid. All four test tubes contain the same universal indicator.

TABLE 15.1 The acid character and pK_a values of some cations in water

Neutral	Acidic
Li$^+$, Na$^+$, K$^+$, Mg^{2+}, Ca^{2+}	d-block ions, including Fe^{2+}, 5.10; Co^{2+}, 8.89; Ni^{2+}, 10.60; Cu^{2+}, 6.80; Fe^{3+}, 2.20
Basic	NH$_4^+$, 9.26; Pb^{2+}, 7.82;
None	Al^{3+}, 4.95

The water autoionization responds almost instantaneously to this increase: the hydronium ions that are present donate protons to the hydroxide ions until the autoionization equilibrium is reached again. The result, after only a billionth of a second or so, is a reduced concentration of hydronium ions and hence a solution with a pH higher than 7.

▼ **EXAMPLE 15.2** Calculating the pH of a salt solution: II

Calculate the pH of a 0.15 M NaCH$_3$CO$_2$(aq) solution.

STRATEGY Since a CH$_3$CO$_2^-$ anion is a weak base and the Na$^+$ cation is not acidic, we expect the solution to be basic (with pH > 7). We calculate the pH by assuming that, to reach the equilibrium

$$CH_3CO_2^-(aq) + H_2O(l) \rightleftharpoons CH_3COOH(aq) + OH^-(aq)$$

the OH$^-$ ion concentration must increase from its initial value in pure water (1.0×10^{-7} M) to x in the solution. We draw up a table of changes and use it to express the equilibrium constant K_b in terms of x; we then solve for x. At equilibrium, as long as x is no larger than about 5% of 0.15 M, we can approximate [CH$_3$CO$_2^-$] by its initial value, 0.15 M. The value of K_b for the base CH$_3$CO$_2^-$ is obtained from the relation $K_a \times K_b = K_w$ and the value of K_a for the conjugate acid (CH$_3$COOH) in Table 14.4.

SOLUTION The table of changes is

	[CH$_3$CO$_2^-$]	[CH$_3$COOH]	[OH$^-$]
Initial concentration	0.15 M	0	0
Change to reach equilibrium	$-x$	$+x$	$+x$
Concentration at equilibrium	0.15 $M - x$	x	x

Hence, in terms of x, the equilibrium constant is

$$K_b = \frac{[CH_3COOH][OH^-]}{[CH_3CO_2^-]} = \frac{x^2}{0.15\ M - x}$$

which we approximate as

$$K_b = \frac{x^2}{0.15\ M}$$

Table 14.4 shows that, for CH$_3$COOH, $K_a = 1.8 \times 10^{-5}$ M; hence, for its conjugate base CH$_3$CO$_2^-$ we have

$$K_b = \frac{1.0 \times 10^{-14}\ M^2}{1.8 \times 10^{-5}\ M} = 5.56 \times 10^{-10}\ M$$

Therefore,

$$x = \sqrt{0.15\ M \times (5.56 \times 10^{-10}\ M)} = 9.1 \times 10^{-6}\ M$$

FIGURE 15.3 A solution of titanium(III) sulfate is so acidic that it can release H$_2$S from some sulfides.

TABLE 15.2 The acid/base character of some anions in water*

Acidic

  HSO$_4^-$

Neutral

  Cl$^-$, Br$^-$, I$^-$

  NO$_3^-$, SO$_4^{2-}$, ClO$_4^-$

Basic

  F$^-$

  O^{2-}, OH$^-$, S^{2-}, HS$^-$

  CN$^-$

  CO$_3^{2-}$, HCO$_3^-$, PO$_4^{3-}$, HPO$_4^{2-}$, NO$_2^-$

  CH$_3$CO$_2^-$ and other carboxylates

*For pK_a and pK_b values, see Table 14.4.

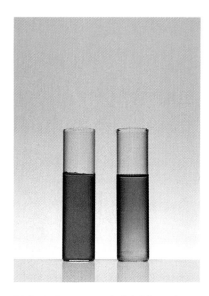

FIGURE 15.4 A 0.1 M acetic acid solution has a pH of about 3 (left). When acetate ions, which are bases, are added, the pH rises toward 7. The solution on the right is 0.1 M acetic acid plus 0.1 M sodium acetate. (The universal indicator was used.)

This is much smaller than 0.15 M, so the approximation $x \ll 0.15\,M$ is valid. It is also significantly larger than $10^{-7}\,M$, so we were justified in neglecting the concentration of OH^- in pure water. Continuing, we have

$$pOH = -\log\,(9.1 \times 10^{-6}) = 5.04$$

and the pH of the solution is

$$pH = 14.00 - 5.04 = 8.96$$

or, more realistically, pH = 9. This is on the basic side of neutral, as anticipated.

▲ **EXERCISE** Calculate the pH of a 0.10 M potassium benzoate solution.
[*Answer*: 8.6]

15.2 THE pH OF MIXED SOLUTIONS

Up until now, we have been considering a solution made up by adding a salt to pure water. We now consider the pH of a solution of a salt in which there is also some parent acid or base. This would be the case if we had added some ammonium chloride to ammonia, or some sodium acetate to acetic acid (Fig. 15.4). It would also be the case if we were partway through a titration and had not yet added enough hydrochloric acid to a solution of ammonia to reach the equivalence point. At an intermediate point, the solution would still contain unreacted NH_3 molecules as well as NH_4^+ and Cl^- ions.

When we add a base to an acid solution, we expect the pH of the solution to climb toward 7. This is true even if the base we add is the conjugate base (such as $CH_3CO_2^-$) of the acid already present (CH_3COOH). Similarly, when we add an acid to a basic solution, we expect the pH to fall toward 7. This is true even if the acid is the conjugate acid (such as NH_4^+) of the base (NH_3) already present. Adding sodium acetate to acetic acid should *increase* the pH of the solution (because we are adding base to an acid). Adding ammonium chloride to ammonia should *decrease* the pH (because we are adding acid to a base).

Both effects can be treated as the response of the equilibrium

$$Acid(aq) + H_2O(l) \rightleftharpoons base(aq) + H_3O^+(aq) \qquad K_a = \frac{[\text{base}][H_3O^+]}{[\text{acid}]}$$

to the addition of base ($CH_3CO_2^-$ if the acid is CH_3COOH) or acid (NH_4^+ if the base is NH_3). If base is added (as when sodium acetate is added to acetic acid), the equilibrium composition responds by shifting toward the reactants (the acid and water). This reduces the concentration of H_3O^+ ions, and the pH rises (Fig. 15.5a). If acid is added (as when we add ammonium chloride to aqueous ammonia), the equilibrium composition responds by shifting toward the products (H_3O^+ and base). This increases the concentration of hydrogen ions, and the pH falls (Fig. 15.5b).

The Henderson-Hasselbalch equation. We can calculate the pH of a mixed solution of the type described above by rearranging the equation for K_a to

$$[H_3O^+] = K_a \times \frac{[acid]}{[base]}$$

Taking the logarithms of both sides gives

$$\log [H_3O^+] = \log K_a + \log \frac{[acid]}{[base]}$$

and multiplying through by -1 yields

$$pH = pK_a - \log \frac{[acid]}{[base]} \qquad (3)$$

We can see how to simplify Eq. 3 by considering a typical case. Suppose we add sodium fluoride to hydrofluoric acid, a weak acid (Table 14.4). Then the concentration of the added base F^- is much greater than the concentration produced by the ionization of the weak acid. This is true in general, and a good approximation, therefore, is that [base] is the concentration of the added salt. Moreover, only a very small proportion of the acid molecules are ionized before the base is added, and even fewer are ionized after. Therefore, another good approximation is to take [acid] as the initial concentration of the acid. Similar approximations hold for the addition of, for example, ammonium chloride (or some other acid) to aqueous ammonia. Then [acid] is almost exactly the same as the concentration of the added salt, and [base] is very close to the initial concentration of ammonia.

Equation 3 is generally referred to as the *Henderson-Hasselbalch equation*. Since it is only the expression for K_a rewritten in terms of logarithms, it is usually just as quick to derive it as to try to remember it. As an example of its use, the pH of a solution containing $0.10\ M$ NaF(aq) and $0.20\ M$ HF(aq) is

$$pH = 3.45 - \log \frac{0.20}{0.10} = 3.15$$

or, more realistically, pH = 3. The pH of a mixed $0.10\ M$ NH_3 and $0.20\ M$ NH_4Cl aqueous solution is

$$pH = 9.25 - \log \frac{0.20}{0.10} = 8.95$$

or, more realistically, pH = 9. The pK_a value used in this second example is that of the acid NH_4^+, which is found by using Eq. 2 and the pK_b in Table 14.4.

Measuring pK_a. A special case of Eq. 3 arises when the concentrations of acid and base are equal. Suppose we prepare a solution of $0.10\ M$ HClO(aq) and $0.10\ M$ ClO$^-$(aq). Then for the solution,

$$pH = pK_a - \log \frac{0.10}{0.10} = pK_a$$

(because log 1 = 0). Since the ionization constant of hypochlorous acid is 7.53 (Table 14.4), the pH of the solution is also 7.53. Similarly, for *any* solution in which the acid and base have equal concentrations,

$$pH = pK_a$$

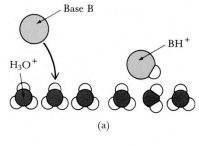

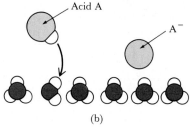

FIGURE 15.5 (a) When a base is added (as a salt) to a solution of an acid, the concentration of hydronium ions falls and the pH rises. (b) When an acid is added (as a salt), the pH falls.

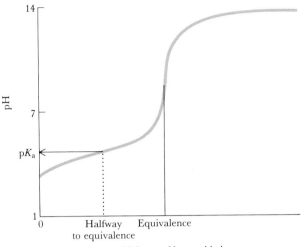

FIGURE 15.6 The pK_a of an acid can be measured by titrating a solution of the acid with a strong base, and noting the pH value halfway to the equivalence point.

Hence, we can predict the pH of solutions containing equal concentrations of acid and base simply by referring to Table 14.4.

This argument also shows that we can measure the pK_a of an acid or a base by making up a solution of that acid or base and an equal molar concentration of its conjugate base or acid (in the form of its salt), and then measuring the pH of the solution. In practice, pK_a is measured by recording the pH of the solution during a titration (Fig. 15.6) and noting its value halfway to the equivalence point, since then the concentrations of salt and acid are equal. Some of the pK_a and pK_b values listed in Table 14.4 were measured in this way.

TITRATIONS AND pH CURVES

The success of acid-base titration as an analytical technique depends on our ability to recognize the equivalence point, the point at which a chemically equivalent amount of acid has been added to a base (or vice versa). This is not always straightforward. Although pH = 7 signals neutrality (that is, $[H_3O^+]$ is the same as that in water), it does not signal the equivalence point if the salt is a Brønsted acid or base (because the ions of the salt are acids and bases). If we were titrating sodium hydroxide with formic acid, HCOOH, at the equivalence point the solution would contain equal concentrations of Na^+ ions and formate ions (HCO_2^-). However, a formate ion is a Brønsted base, so the solution would be basic, with pH greater than 7, at that point.

15.3   THE VARIATION OF pH DURING A TITRATION

If we are to detect the equivalence point accurately, we must know the pH values to expect during different types of acid-base titrations. Here we shall follow the pH through several kinds of acid-base titrations, to see what its value is just before, at, and just after the equivalence point. A graph of pH against volume of titrant added (like the one in Fig. 15.6), called a *pH curve*, is helpful in discussions of acid-base titrations.

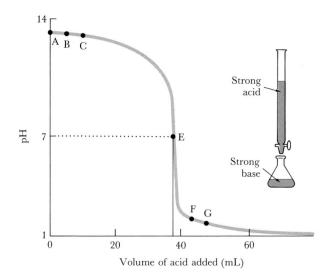

Titration of a strong base with a strong acid. As a first example, we consider the titration of a strong base (the analyte in the flask) with a strong acid (the titrant in the buret). To be specific, we consider the titration of aqueous sodium hydroxide with hydrochloric acid, but our discussion can be applied to a titration of any strong base with any strong acid.

▼ **EXAMPLE 15.3** **Calculating the pH before the equivalence point is reached in a strong acid–strong base titration**

Calculate the initial pH of 25.00 mL of 0.150 M NaOH(aq), and the pH after the addition of 5.00 mL of 0.100 M HCl(aq).

STRATEGY Since the sodium hydroxide is in excess in both cases, the solutions are basic; hence we expect pH > 7. Since OH$^-$ ions are in excess, a convenient approach is to calculate pOH first and then convert it to pH using the relation pH + pOH = pK_w. The concentration of OH$^-$ ions depends on the amount of OH$^-$ present and the total volume of the solution, both before and after the acid is added. The required volumes are given. We can find the number of moles of OH$^-$ ions in the combined base plus acid solution from the stoichiometry of the chemical reaction and the volume of acid added. The arithmetic is simpler if we express concentrations in millimoles per milliliter, recalling from Section 11.1 that a concentration in moles per liter is numerically the same as one in millimoles per milliliter.

SOLUTION The initial concentration of OH$^-$ ions is 0.150 M, corresponding to pOH = $-\log 0.150$ = 0.82. The initial pH is therefore

$$\text{pH} = 14.00 - 0.82 = 13.18$$

This is point A in Fig. 15.7. The original 25.00 mL of 0.150 M NaOH(aq) contains

$$\text{Moles of OH}^- = \text{volume of solution} \times \text{moles of OH}^- \text{ per liter}$$

$$= 25.00 \text{ mL} \times 0.150 \, \frac{\text{mmol OH}^-}{\text{mL}} = 3.75 \text{ mmol OH}^-$$

The added 5.00 mL of 0.100 M HCl(aq) contains

$$\text{Moles of } H_3O^+ = 5.00 \text{ mL} \times 0.100 \frac{\text{mmol } H_3O^+}{\text{mL}}$$

$$= 0.500 \text{ mmol } H_3O^+$$

We know that 1 mol of H_3O^+ reacts with 1 mol of OH^-, so this reacts with 0.500 mmol of OH^-; hence it reduces the amount of OH^- from 3.75 mmol to 3.25 mmol. At the same time, the volume of the solution increases from 25.00 mL to 30.00 mL. The concentration of OH^- ions therefore decreases to

$$[OH^-] = \frac{3.25 \text{ mmol } OH^-}{30.00 \text{ mL}} = 0.108 \text{ mol/L } OH^-$$

This corresponds to pOH = 0.97, so

$$\text{pH} = 14.00 - 0.97 = 13.03$$

This is point B in Fig. 15.7.

EXERCISE Calculate the pH of the solution after the addition of another 5.00 mL of acid.

[*Answer*: 12.90; point C in Fig. 15.7]

The equivalence point occurs after the addition of 37.5 mL of the base (this is easily verified by noting that 1 mol NaOH = 1 mol HCl and using the concentrations of the two solutions as conversion factors). At the equivalence point (point E in Fig. 15.7), the solution contains nonacidic Na^+ ions and a very weak base (Cl^-). Because neither ion affects the water autoionization, the solution is neutral, with pH = 7. Beyond the equivalence point, as more acid is added, all the additional H_3O^+ ions survive, and the pH falls below 7. All these hydronium ions contribute to the pH, but their concentration is reduced because they are contained in a larger volume of solution. The actual pH depends on the amount of excess acid added and the total volume of the solution.

▼ **EXAMPLE 15.4** **Calculating the pH after the equivalence point of a strong acid–strong base titration has been passed**

Calculate the pH of the solution of Example 15.3 if 5.00 mL of the acid titrant was added after the equivalence point was reached.

STRATEGY We need to find the concentration of H_3O^+ ions in the solution, given that all the ions survive that are supplied after the equivalence point has been passed, but they now occupy a larger volume of solution. The calculation therefore falls into two parts. In one, we calculate the number of moles of H_3O^+ ions in the solution. In the other, we calculate the total volume of solution. By dividing the number of moles by the total volume, we get $[H_3O^+]$ and hence the pH of the solution.

SOLUTION The number of moles of H_3O^+ ions in 5.00 mL of 0.100 M HCl(aq) is

$$\text{Moles of } H_3O^+ = \text{volume of solution} \times \text{moles of } H_3O^+ \text{ per liter}$$

$$= 5.00 \text{ mL} \times 0.100 \frac{\text{mmol } H_3O^+}{\text{mL}} = 0.500 \text{ mmol } H_3O^+$$

The total volume of solution is the sum of the initial volume (25.00 mL), the acid needed to neutralize the original base (37.5 mL), and the additional 5.00 mL of acid following equivalence, which is 67.5 mL. The hydronium ion concentration is therefore

$$[H_3O^+] = \frac{0.50 \text{ mmol } H_3O^+}{67.5 \text{ mL}} = 7.4 \times 10^{-3} \text{ mol/L } H_3O^+$$

and the pH of the solution is

$$pH = -\log (7.4 \times 10^{-3}) = 2.13$$

This is point F in Fig. 15.7.

EXERCISE Calculate the pH of the solution after the addition of another 5.00 mL of acid.

[*Answer*: 1.86; point G in Fig. 15.7]

▲

When a very large volume of acid has been added, the dilution caused by the original 25 mL of analyte becomes unimportant, and the concentration of H_3O^+ ions in the flask is negligibly different from that in the original acid titrant.

The shape of the pH curve in Fig. 15.7 is typical of all titrations in which a strong acid is added to a strong base. Initially the pH falls slowly. As the equivalence point is reached and passed, there is a sudden drop through pH = 7 as the hydrogen ion concentration surges up to a value typical of a diluted solution of the added acid. The pH then falls more slowly toward the pH value of the acid itself as the dilution caused by the original analyte solution becomes less and less important. The sudden decrease in pH thus marks the equivalence point; it may be detected with either a pH meter or an indicator (Fig. 15.8).

FIGURE 15.8 The end of a strong acid–strong base titration can be readily detected with an indicator. Three stages—(a) close to, (b) at, and (c) just after the equivalence point—are shown, with thymol blue as indicator.

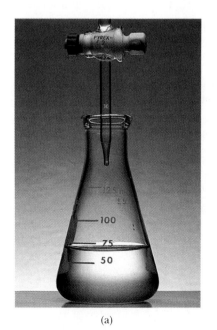

(a)

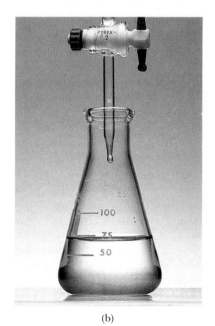

(b)

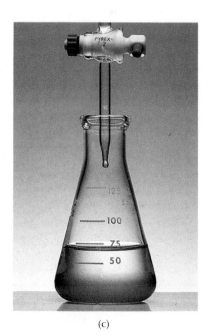

(c)

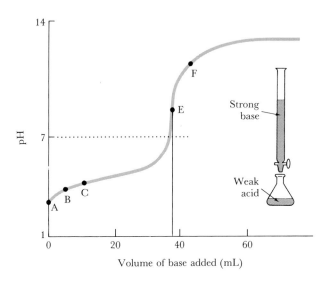

FIGURE 15.9 The variation of pH during the titration of a weak acid, 25 mL of 0.100 *M* HCOOH(*aq*), with a strong base, 0.150 *M* NaOH(*aq*). The equivalence point (point E) occurs on the basic side of pH = 7 because the salt anion HCO_2^- is a base.

Titration of a weak acid with a strong base. Suppose we were titrating a 25.00-mL sample of 0.100 *M* formic acid (HCOOH, the acid present in ant venom) with 0.150 *M* sodium hydroxide. At the equivalence point, which occurs after the addition of 16.7 mL of the base, the solution consists of sodium formate. Since an HCO_2^- ion is a Brønsted base, the solution is basic, with a pH greater than 7. But how basic? We need to know this if we are to identify the equivalence point accurately.

The initial pH of the solution of 0.100 *M* formic acid ($pK_a = 3.75$) is 2.37, point A in Fig. 15.9. (This is calculated by the procedure illustrated in Example 14.7.) At an intermediate stage of the titration, the solution consists of the remaining acid HCOOH and the base HCO_2^- formed by the addition of the strong base. An equation for its pH is found by using the same argument that led to the Henderson-Hasselbalch equation, Eq. 3.

▼ **EXAMPLE 15.5** Calculating the pH before the equivalence point is reached in a weak acid–strong base titration

Calculate the pH of a solution of 25.00 mL of 0.100 *M* HCOOH(*aq*) after the addition of 5.00 mL of 0.150 *M* NaOH(*aq*).

STRATEGY We expect the pH of the acid solution to rise when the base is added. Initially, the pH is 2.37. The pH of the mixed solution may be calculated by rearranging the equation for K_a in the equilibrium

$$HCOOH(aq) + H_2O(l) \rightleftharpoons HCO_2^-(aq) + H_3O^+(aq)$$

$$K_a = \frac{[HCO_2^-][H_3O^+]}{[HCOOH]}$$

to obtain

$$pH = pK_a - \log \frac{[HCOOH]}{[HCO_2^-]}$$

[HCOOH] is the concentration of acid remaining at this stage of the titration, and $[HCO_2^-]$ is the concentration of conjugate base that has been produced so far. Their volumes are equal (the HCOOH and HCO_2^- are present in the same solution in the same flask), so this equation may be written as

$$pH = pK_a - \log \frac{\text{moles of HCOOH}}{\text{moles of } HCO_2^-}$$

SOLUTION The number of moles of HCOOH present initially is

$$\text{Moles of HCOOH} = 25.00 \text{ mL} \times 0.100 \frac{\text{mmol HCOOH}}{\text{mL}}$$

$$= 2.50 \text{ mmol HCOOH}$$

The number of moles of base added is

$$\text{Moles of } OH^- = 5.00 \text{ mL} \times 0.150 \frac{\text{mmol } OH^-}{\text{mL}}$$

$$= 0.750 \text{ mmol } OH^-$$

Therefore, the amount of HCO_2^- present rises to 0.750 mmol HCO_2^-, and the amount of HCOOH present falls to 1.75 mmol HCOOH. Since the pK_a of formic acid (from Table 14.4) is 3.75, the pH is

$$pH = 3.75 - \log \frac{1.75}{0.750} = 3.38$$

This is point B in Fig. 15.9.

EXERCISE Calculate the pH of the solution after the addition of another 5.00 mL of base.

[*Answer*: 3.93; point C in Fig. 15.9]

▲

At the equivalence point, only the salt—sodium formate—is present. The HCO_2^- ion is a base, so the pH of the solution is greater than 7. To calculate its value, we use the technique of Example 15.2.

▼ **EXAMPLE 15.6 Calculating the pH at the equivalence point of a weak acid–strong base titration**

Calculate the pH of the solution of Example 15.5 at the equivalence point.

STRATEGY The solution contains the base HCO_2^-, so we expect pH > 7. Since the pH depends on the concentration of HCO_2^- in the solution, we calculate that first. Once that is known, the concentration of OH^- ions can be found from the equilibrium

$$HCO_2^-(aq) + H_2O(l) \rightleftharpoons HCOOH(aq) + OH^-(aq)$$

$$K_b = \frac{[HCOOH][OH^-]}{[HCO_2^-]}$$

as explained in Example 15.2. The value of K_b for HCO_2^- is related to the value of K_a for its conjugate acid HCOOH by the equation $K_a \times K_b = K_w$, and K_a can be found from Table 14.4.

SOLUTION The number of moles of HCO_2^- present at the equivalence point is equal to the initial number of moles of HCOOH, which from Example 15.5 is 2.50 mmol. Since the volume of base needed to reach equivalence is 16.7 mL, the total volume of the solution at equivalence is 25.00 mL + 16.7 mL = 41.7 mL. Therefore, the concentration of HCO_2^- ions is

$$[HCO_2^-] = \frac{2.50 \text{ mmol}}{41.7 \text{ mL}} = 0.0600 \ M$$

We assume that, in reaching the equilibrium shown above, a concentration x of OH^- ions is produced. Then by the reasoning of Example 15.2, and as long as x is no larger than about 5% of 0.06 M,

$$K_b = \frac{[HCOOH][OH^-]}{[HCO_2^-]} = \frac{x^2}{0.0600 \ M}$$

Since K_b works out to be $5.6 \times 10^{-11} \ M$, we have

$$x = \sqrt{0.0600 \ M \times (5.6 \times 10^{-11} \ M)} = 1.8 \times 10^{-6} \ M$$

(Since this is only 0.003% of 0.06 M, our approximation is justified.) It follows that

$$pOH = -\log(1.8 \times 10^{-6}) = 5.74$$

and therefore that

$$pH = 14.00 - 5.74 = 8.26$$

This is point E in Fig. 15.9.

EXERCISE Calculate the pH at the equivalence point of the titration of 25.00 mL of 0.010 M HClO(*aq*) with 0.020 M KOH(*aq*).

[*Answer:* 9.67]

▲

Once the equivalence point has been passed, the pH depends on the concentration of excess strong base. If another 5.00 mL of base (containing 0.750 mmol of OH^-) is added, the concentration of OH^- ions rises to

$$[OH^-] = \frac{0.750 \text{ mmol } OH^-}{46.7 \text{ mL}} = 0.0161 \ M$$

This corresponds to pOH = 1.79 and hence to pH = 12.21, which is point F in Fig. 15.9. The addition of more base results in the gradual increase of the pH toward 13.2, the pH of the 0.150 M NaOH(*aq*) solution itself.

The shape of the pH curve in Fig. 15.9 is typical of all titrations of a weak acid with a strong base. The pH of the weak acid solution initially rises slowly. It then passes rapidly through the pH of the equivalence point, which is *higher* than 7 because the salt anion is a base. Finally, it climbs slowly toward a strongly basic value.

Titration of a weak base with a strong acid. The titration of a weak base with hydrochloric acid gives a solution of the conjugate acid of the base. Thus, if ammonia is the base, the titration results in the production of

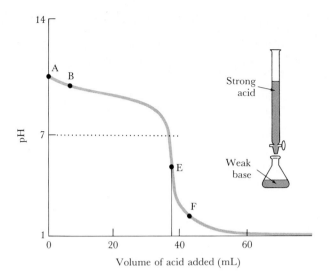

NH_4^+ ions. Because the NH_4^+ ion is a Brønsted acid, the solution is acidic and its pH is less than 7. The pH curve is shown in Fig. 15.10; its points are calculated in exactly the same way as for the titration of a weak acid. The pH begins above 7 and falls slowly at first, passes rapidly through the equivalence point on the acid side of neutrality, and then decreases slowly toward the pH of the strong acid titrant.

▼ **EXAMPLE 15.7** **Calculating the pH at the equivalence point of a weak base–strong acid titration**

Calculate the pH, at the equivalence point, of the titration of 25.00 mL of 0.150 M $NH_3(aq)$ with 0.100 M $HCl(aq)$.

STRATEGY We expect a pH < 7, since at the equivalence point the solution contains the acid NH_4^+. The number of moles of NH_4^+ present at the equivalence point is equal to the number of moles of NH_3 in the original solution. However, the concentration is lower, because titrant has been added to the initial solution so its volume is greater. This suggests that we should first calculate $[NH_4^+]$ from the total volume and then allow for the conversion of some to NH_3 in the equilibrium

$$NH_4^+(aq) + H_2O(l) \rightleftharpoons H_3O^+(aq) + NH_3(aq) \qquad K_a = \frac{[H_3O^+][NH_3]}{[NH_4^+]}$$

We can find the value of $[H_3O^+]$ in the same way as in Example 15.1: by assuming that it rises to x, expressing K_a in terms of x, and then solving the resulting equation for x. The value of K_a is obtained from the value of K_b for the conjugate base (NH_3) in Table 14.4.

SOLUTION The volume of acid needed to reach equivalence is 37.5 mL. (This is found in the usual way, by use of the conversion 1 mol NH_3 = 1 mol HCl and the concentrations of the two solutions.) The total volume of solution at equivalence is therefore 62.5 mL. Since the initial amount of NH_3 present is 3.75 mmol NH_3, the amount of NH_4^+ present at the equivalence point is 3.75 mmol NH_4^+. The concentration of NH_4^+ ions at equivalence is therefore

$$[NH_4^+] = \frac{3.75 \text{ mmol}}{62.5 \text{ mL}} = 0.0600 \; M$$

If the concentration of H_3O^+ rises to x to reach the equilibrium shown above, then by the procedure of Example 15.1, and assuming that $x \ll 0.06 \; M$, we have

$$K_a = \frac{[NH_3][H_3O^+]}{[NH_4^+]} = \frac{x^2}{0.0600 \; M}$$

Since $K_a = 5.6 \times 10^{-10} \; M$, this equation rearranges to

$$x = \sqrt{0.0600 \; M \times (5.6 \times 10^{-10} \; M)} = 5.8 \times 10^{-6} \; M$$

so

$$pH = -\log (5.8 \times 10^{-6}) = 5.24$$

This is point E in Fig. 15.10.

EXERCISE Calculate the pH at the equivalence point of the titration of 25.0 mL of 0.250 M $NH_3(aq)$ with 0.150 M $HCl(aq)$.

[*Answer*: 5.14]

15.4 INDICATORS AND BUFFERS

Automatic titrators (Fig. 15.11) monitor the pH of the titration solution. They detect the equivalence point by means of the characteristic rapid change in pH at equivalence. When automatic equipment is not available, equivalence points are detected by means of the change in color of an *indicator*, a water-soluble dye with a color that depends on the pH of the solution it is in. An indicator is a weak Brønsted acid that has one color when it is present in its acid form (HIn, where In stands for indicator) in an acidic solution, and another when it is present as its conjugate base (In^-) in a basic solution. One well-known naturally occurring indicator is litmus, which is red for pH < 6 and blue for pH > 8. Others are the organic dyes anthocyanidins (**1**), one of which is responsible for the colors both of red poppies and of blue cornflowers

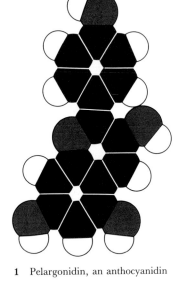

1 Pelargonidin, an anthocyanidin

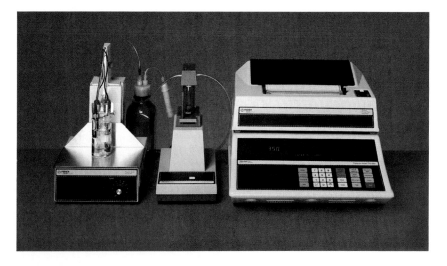

FIGURE 15.11 A modern acid–base titration apparatus, that detects the equivalence point automatically by monitoring the pH.

(a)   (b)

FIGURE 15.12 The same dye is responsible for the red of poppies (a) and the blue of cornflowers (b). The color difference is a consequence of the more acidic environment in poppies than in cornflowers.

(Fig. 15.12): in poppies the sap is acidic and in cornflowers it is basic, and the dye has the corresponding color in each.

The choice of indicator. Since litmus changes color close to pH = 7, it is suitable as an indicator in the titration of strong bases with strong acids, which have equivalence points close to pH = 7. However, because the pH of the solution sweeps through a wide range of values near the equivalence point, any indicator that changes color within about two units of pH = 7 can also be used, as long as the color change occurs over a narrow pH range (Fig. 15.13). Phenolphthalein is often used because its color change (at pH = 9) is sharp and easy to detect. In contrast, the first color change of thymol blue (from red to yellow) occurs in the pH range 1.2 to 2.8, which would give inaccurate results.

Phenolphthalein, which is an effective indicator in the pH range 8.2 to 10.0, can be used for titrations that reach equivalence near pH = 9; methyl orange, another indicator, should not be used for such titra-

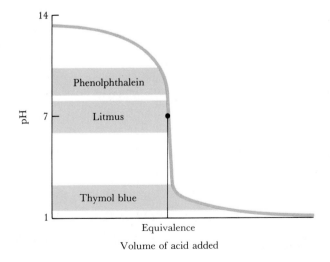

FIGURE 15.13 Ideally, an indicator should change color right at the equivalence point (pH = 7 for a strong acid–strong base titration). However, the change of pH is so abrupt that phenolphthalein can be used. Thymol blue would give a very misleading result. The shaded areas show the range of pH over which the color change occurs.

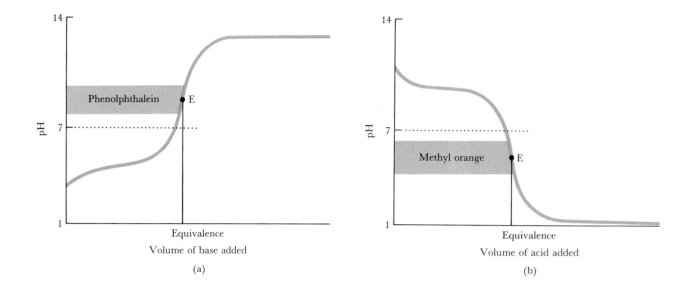

FIGURE 15.14 (a) Phenol-
phthalein (which changes color in
the shaded band) can be used to
detect the equivalence point of a
weak acid–strong base titration.
(b) Methyl orange can be used for
a strong acid–weak base titration.

tions because it changes color between pH 3.2 and pH 4.4 (Fig. 15.14).
However, in the titration of a strong acid with a weak base, the equivalence point occurs at pH < 7; hence methyl orange is suitable for this
titration, but phenolphthalein is not. Several indicators and their color
changes are shown in Fig. 15.15, and more are listed in Table 15.3.

Buffer action. One feature of the pH curves in Figs. 15.9 and 15.10 is
how slowly the pH changes halfway to the equivalence point, close to
pH = pK_a. This slow change means that when a small amount of acid
or base is added to a solution containing roughly equal concentrations

TABLE 15.3 Indicator color changes

Indicator	Acid color	pH range of color change	pK_a	Base color
Thymol blue	Red	1.2 to 2.8	1.7	Yellow
Methyl orange	Red	3.2 to 4.4	3.4	Yellow
Bromophenol blue	Yellow	3.0 to 4.6	3.9	Blue
Bromocresol green	Yellow	4.0 to 5.6	4.7	Blue
Methyl red	Red	4.8 to 6.0	5.0	Yellow
Bromothymol blue	Yellow	6.0 to 7.6	7.1	Blue
Litmus	Red	5.0 to 8.0	6.5	Blue
Phenol red	Yellow	6.6 to 8.0	7.9	Red
Cresol red	Yellow	7.2 to 8.8	8.2	Red
Thymol blue	Yellow	8.0 to 9.6	8.9	Blue
Phenolphthalein	Colorless	8.2 to 10.0	9.4	Pink
Alizarin yellow	Yellow	10.1 to 12.0	11.2	Red
Alizarin	Red	11.0 to 12.4	11.7	Purple

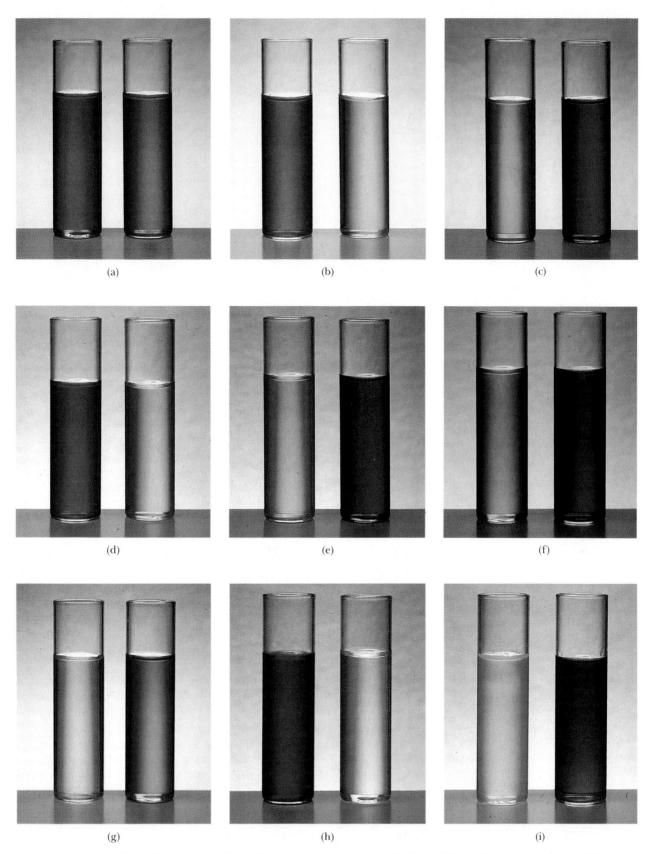

FIGURE 15.15 The color changes shown by common indicators: (a) thymol blue, (b) methyl orange, (c) bromo-cresol green, (d) methyl red, (e) bromothymol blue, (f) cresol red, (g) thymol blue, (h) phenolphthalein, and (i) alizarin.

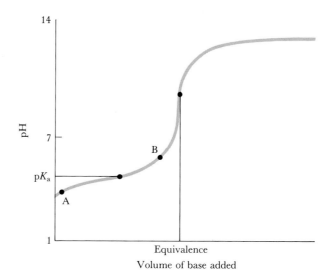

FIGURE 15.16 Although the ratio of salt to acid is increased considerably from A to B, the pH changes very little. In other words, the solution acts as a buffer in that region.

of salt and its parent weak acid or base, the pH changes much less than when the same amount is added to pure water (Fig. 15.16). Solutions that resist any change in pH when small amounts of acid or base are added are called *buffers*. Blood is maintained at pH = 7.4 by buffer action, as are other cell fluids. The oceans are maintained at about pH = 8.4 by a complex buffering process that depends on hydrogen carbonates and silicates.

Buffer action can be understood by considering the acetic acid equilibrium

$$CH_3COOH(aq) + H_2O(l) \rightleftharpoons CH_3CO_2^-(aq) + H_3O^+(aq)$$

(or any similar Brønsted equilibrium) and its response to the addition of acid or base. Suppose a solution contains large numbers of $CH_3CO_2^-$ ions and CH_3COOH molecules, in about equal concentrations. If a small amount of acid is added, the newly arrived H_3O^+ ions transfer protons to the $CH_3CO_2^-$ ions. This removes the hydronium ions from the solution and leaves its pH nearly unchanged. If a small amount of base is added instead, protons are removed from the CH_3COOH molecules. As a result, the OH^- ion concentration remains nearly unchanged, and consequently the H_3O^+ concentration (and the pH) is also left nearly unchanged. In other words, the buffer acid acts as a source of hydrogen ions, and its conjugate base acts as a sink for them (Fig. 15.17).

Figure 15.9 shows that the flattest part of the pH curve of a solution of a weak acid and its salt occurs where their concentrations are equal (halfway to the equivalence point). The pH is on the acid side of neutral and equal to the pK_a of the weak acid. The mixture is an *acid buffer*, one that stabilizes the pH at values close to pK_a. Similarly, Fig. 15.10 shows that a mixture of a weak base and its salt stabilizes the pH on the basic side of neutrality, close to pH = pK_a, where K_a is the ionization constant of the conjugate acid of the base. Mixtures in which the salt and acid (or base) have unequal concentrations also act as buffers, but they are less effective than those in which the concentrations are nearly equal.

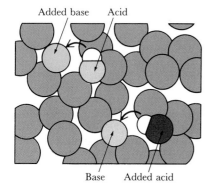

FIGURE 15.17 Buffer action depends on the acid molecules being able to supply protons when a base is added, and the base molecules being able to accept protons when an acid is added.

▼ EXAMPLE 15.8 Calculating the pH of a buffer solution

Calculate the pH of a buffer solution containing 0.040 M $Na_2HPO_4(aq)$ and 0.080 M $KH_2PO_4(aq)$ at 25°C.

STRATEGY Since buffer action depends on the simultaneous presence of acid and base, we should begin by identifying the weak acid and its conjugate base (the acid contains one more hydrogen atom than the base). Once they have been identified, we can write the equilibrium between them, rearrange the expression for K_a so it gives $[H_3O^+]$, and hence find the pH.

SOLUTION The acid is the dihydrogen phosphate ion $H_2PO_4^-$; its conjugate base is the hydrogen phosphate ion HPO_4^{2-}. The equilibrium to consider is

$$H_2PO_4^-(aq) + H_2O(l) \rightleftharpoons HPO_4^{2-}(aq) + H_3O^+(aq)$$

$$K_{a2} = \frac{[HPO_4^{2-}][H_3O^+]}{[H_2PO_4^-]}$$

where K_{a2} is the second ionization constant of phosphoric acid (from Table 14.6, $pK_{a2} = 7.21$). Rearranging the equation for K_{a2} gives

$$[H_3O^+] = K_{a2} \times \frac{[H_2PO_4^-]}{[HPO_4^{2-}]}$$

so

$$pH = pK_{a2} - \log \frac{[H_2PO_4^-]}{[HPO_4^{2-}]}$$

Substituting the given concentrations leads to

$$pH = 7.21 - \log \frac{0.080\ M}{0.040\ M} = 6.9$$

That is, the solution acts as a buffer close to pH = 7.

EXERCISE Calculate the pH of a buffer solution containing 0.040 M $NH_4Cl(aq)$ and 0.030 M $NH_3(aq)$ at 25°C.

[*Answer*: 9]

▲

Laboratory pH meters are often calibrated with a 0.025 M Na_2HPO_4 plus 0.025 M KH_2PO_4 aqueous solution, which has pH = 6.87 at 25°C. The computation demonstrated in Example 15.8 predicts pH = 7.2 for this solution; however, as noted at the start of the chapter, these calculations ignore ion-ion interactions, which modify the pH slightly.

Buffer capacity. The *capacity* of a buffer is the amount of acid or base that may be added before the buffer loses its ability to resist the change in pH. This loss may occur because much of the base (HPO_4^{2-} ions, in Example 15.8) has been converted to acid, or because most of the acid (the $H_2PO_4^-$ ions) has been converted to base.

The pH curve in Fig. 15.16 shows that the pH is held within a range of about one unit on either side of the optimum buffering value of pH = pK_a. A phosphate buffer can therefore be effective in the pH range 6.2 to 8.2, with the precise limits depending on the mixture. Since log 0.1 = −1 and log 10 = +1, Eq. 3 implies that the ratio of the acid concentration to the conjugate base concentration should not be less than 0.1 or more than 10, or else the pH of the solution will lie outside the moderately flat portion of the curve. The precise pH of the solution will vary slightly from $pK_a + 1$ to $pK_a - 1$ as acid or base is added.

▼ EXAMPLE 15.9 Estimating buffer capacity

Estimate the volume of $0.10\ M$ HCl(aq) that may be added to 25 mL of the phosphate solution of Example 15.8 before it stops acting as a buffer.

STRATEGY We assume that the buffer is viable until the ratio of the concentrations of acid and conjugate base has risen to about 10. We must therefore calculate the volume of added acid that will achieve this ratio. Each millimole of added HCl increases the amount of acid ($H_2PO_4^-$) by 1 mmol and decreases the amount of base (HPO_4^{2-}) by the same amount.

SOLUTION 25 mL of the buffer solution contains 1.0 mmol of HPO_4^{2-} and 2.0 mmol of $H_2PO_4^-$. If x mmol of HCl is added, these amounts change to $1.0 - x$ mmol of HPO_4^{2-} and $2.0 + x$ mmol of $H_2PO_4^-$. The ratio of concentrations rises to 10 when

$$\frac{2.0 + x}{1.0 - x} = 10$$

or when $x = 0.7$. That is, 0.7 mmol of HCl may be added before the capacity of the solution is reached. The volume of $0.1\ M$ acid that contains this amount is

Volume of acid = moles of HCl × liters of acid per mole of HCl

$$= 0.7\ \text{mmol HCl} \times \frac{1\ \text{mL HCl}(aq)}{0.1\ \text{mmol HCl}} = 7\ \text{mL HCl}(aq)$$

EXERCISE Calculate the buffer capacity of 25 mL of the ammonia plus ammonium chloride buffer of the exercise of Example 15.8 when $0.015\ M$ NaOH(aq) is added.

[*Answer*: 56 mL]

▲

The composition of blood plasma in which the concentration of HCO_3^- ions is about 20 times that of H_2CO_3, seems to be outside the range for optimal buffering. However, this is advantageous because the principal waste products of living cells are carboxylic acids, such as lactic acid (**2**). With its relatively high concentration of HCO_3^- ions, the plasma can absorb a significant surge of hydrogen ions from these carboxylic acids. Another consequence of the high proportion of HCO_3^- is that bodies are better able to withstand disturbances that lead to excess acid (disease and shock) than those leading to excess base (burns).

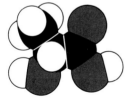

2 Lactic acid

SOLUBILITY EQUILIBRIA

We saw in Section 11.3 that in a saturated solution of a salt, the dissolved solute and undissolved solute are in dynamic equilibrium. We should therefore be able to express solubilities in terms of equilibrium constants. If so, we can use the techniques we have been developing to discuss their response to changes in the conditions. One such change is a modification of the pH of the solution by the addition of acid or base. Hence, the concepts developed since Chapter 11 allow us to predict how solubilities are affected by pH. The techniques in question are most useful for sparingly soluble salts, and we confine our attention to them.

15.5 THE SOLUBILITY PRODUCT

The equilibrium between solid silver chloride and its saturated solution is

$$AgCl(s) \rightleftharpoons Ag^+(aq) + Cl^-(aq)$$

The equilibrium constant for this heterogeneous equilibrium is called the _solubility product_ K_{sp} and is written

$$K_{sp} = [Ag^+][Cl^-]$$

You will recall that with heterogeneous equilibria (Section 13.3), the concentration of a pure solid does not appear in the equilibrium constant. A more complicated example is

$$Ag_3PO_4(s) \rightleftharpoons 3Ag^+(aq) + PO_4^{3-}(aq) \qquad K_{sp} = [Ag^+]^3[PO_4^{3-}]$$

The general formula for the solubility product of a salt that dissolves to give m cations M and n anions X is

$$K_{sp} = [M]^m[X]^n \tag{4}$$

We shall also find it convenient to use the quantity

$$pK_{sp} = -\log K_{sp} \tag{5}$$

Note that the higher the value of pK_{sp}, the lower the value of K_{sp} itself, and hence the less favorable the solubility equilibrium.

▼ EXAMPLE 15.10 Writing solubility products

Write the expressions for the solubility products of (a) silver sulfide and (b) mercury(I) chloride. Give the units of K_{sp} in each case.

STRATEGY To write the expression for a solubility constant, we proceed as for any equilibrium constant, beginning with the chemical equation for the equilibrium. The products (the ions) become the numerator of the expression; the concentration of the pure solid is considered to be constant, so it does not appear in the expression. We need to be careful with the formulas containing mercury(I), for which the cation is Hg_2^{2+}.

SOLUTION (a) The equilibrium equation is

$$Ag_2S(s) \rightleftharpoons 2Ag^+(aq) + S^{2-}(aq) \qquad K_{sp} = [Ag^+]^2[S^{2-}]$$

Since K_{sp} is the product of three concentrations, its units are (moles per liter)^3.
(b) Here the equilibrium is

$$Hg_2Cl_2(s) \rightleftharpoons Hg_2^{2+}(aq) + 2Cl^-(aq) \qquad K_{sp} = [Hg_2^{2+}][Cl^-]^2$$

Here, too, K_{sp} is the product of three concentrations, so its units are also (moles per liter)^3.

EXERCISE Write the expressions for the solubility products of (a) bismuth sulfide and (b) silver chromate. Give their units.

[_Answer_: (a) $[Bi^{3+}]^2[S^{2-}]^3$, M^5; (b) $[Ag^+]^2[CrO_4^{2-}]$, M^3]

▲

TABLE 15.4 Solubility products at 25°C

Compound	Formula	K_{sp}*	pK_{sp}
Aluminum hydroxide	$Al(OH)_3$	1.0×10^{-33}	33.00
Antimony sulfide	Sb_2S_3	1.7×10^{-93}	92.77
Barium carbonate	$BaCO_3$	8.1×10^{-9}	8.09
fluoride	BaF_2	1.7×10^{-6}	5.77
sulfate	$BaSO_4$	1.1×10^{-10}	9.96
Bismuth sulfide	Bi_2S_3	1.0×10^{-97}	97.00
Calcium carbonate	$CaCO_3$	8.7×10^{-9}	8.06
fluoride	CaF_2	4.0×10^{-11}	10.40
hydroxide	$Ca(OH)_2$	5.5×10^{-6}	5.26
sulfate	$CaSO_4$	2.4×10^{-5}	4.62
Copper(I) bromide	$CuBr$	4.2×10^{-8}	7.38
chloride	$CuCl$	1.0×10^{-6}	6.00
iodide	CuI	5.1×10^{-12}	11.29
sulfide	Cu_2S	2.0×10^{-47}	46.70
Copper(II) iodate	$Cu(IO_3)_2$	1.4×10^{-7}	6.85
oxalate	CuC_2O_4	2.9×10^{-8}	7.54
sulfide	CuS	8.5×10^{-45}	44.07
Iron(II) hydroxide	$Fe(OH)_2$	1.6×10^{-14}	13.80
sulfide	FeS	6.3×10^{-18}	17.20
Iron(III) hydroxide	$Fe(OH)_3$	2.0×10^{-39}	38.70
Lead(II) bromide	$PbBr_2$	7.9×10^{-5}	4.10
chloride	$PbCl_2$	1.6×10^{-5}	4.80
fluoride	PbF_2	3.7×10^{-8}	7.43
iodate	$Pb(IO_3)_2$	2.6×10^{-13}	12.59
iodide	PbI_2	1.4×10^{-8}	7.85

Solubility products can be obtained directly from solubilities. For example, the solubility of silver chloride in water at 25°C is $1.3 \times 10^{-5}\ M$. Since each AgCl unit gives one Ag^+ ion and one Cl^- ion, the concentrations of Ag^+ and Cl^- ions in the saturated solution are both also $1.3 \times 10^{-5}\ M$. The solubility product is therefore

$$K_{sp} = [Ag^+][Cl^-] = (1.3 \times 10^{-5}\ M)^2$$
$$= 1.7 \times 10^{-10}\ M^2$$

The value listed in Table 15.4 differs slightly from this computed value; it was found with a more accurate procedure that is discussed in Chapter 17.

We can also work in the opposite direction, and deduce the solubility S of a salt in water from its solubility product. For example, the value of K_{sp} for calcium fluoride refers to the equilibrium

$$CaF_2(s) \rightleftharpoons Ca^{2+}(aq) + 2F^-(aq)$$

Compound	Formula	K_{sp}*	pK_{sp}
sulfate	$PbSO_4$	1.6×10^{-8}	7.80
sulfide	PbS	1.3×10^{-28}	27.89
Magnesium ammonium phosphate	$MgNH_4PO_4$	2.5×10^{-13}	12.60
carbonate	$MgCO_3$	1.0×10^{-5}	5.00
fluoride	MgF_2	6.4×10^{-9}	8.19
hydroxide	$Mg(OH)_2$	1.1×10^{-11}	10.96
Manganese(II) sulfide	MnS	1.4×10^{-15}	14.85
Mercury(I) chloride	Hg_2Cl_2	1.3×10^{-18}	17.89
iodide	Hg_2I_2	1.2×10^{-28}	27.92
Mercury(II) sulfide	HgS (black)	1.6×10^{-52}	51.80
	HgS (red)	1.4×10^{-53}	52.85
Nickel(II) hydroxide	$Ni(OH)_2$	6.5×10^{-18}	17.19
Silver bromide	$AgBr$	7.7×10^{-13}	12.11
carbonate	Ag_2CO_3	6.2×10^{-12}	11.21
chloride	$AgCl$	1.6×10^{-10}	9.80
chromate	Ag_2CrO_4	9×10^{-12}	11.00
hydroxide	$AgOH$	1.5×10^{-8}	7.82
iodide	AgI	1.5×10^{-16}	15.82
phosphate	Ag_3PO_4	1.3×10^{-20}	19.89
sulfate	Ag_2SO_4	1.4×10^{-5}	4.85
sulfide	Ag_2S	6.3×10^{-51}	50.20
Strontium hydroxide	$Sr(OH)_2$	1.4×10^{-4}	3.85
sulfate	$SrSO_4$	3.2×10^{-7}	6.49
Zinc hydroxide	$Zn(OH)_2$	2.0×10^{-17}	16.70
sulfide	ZnS	1.6×10^{-24}	23.80

*The units of K_{sp} are (moles per liter)^n, where n is the number of ions in a formula unit.

At equilibrium (in the saturated solution), the concentration of Ca^{2+} ions must be S. If no other fluorides are present in the solution, the concentration of F^- ions must be $2 \times S$. Therefore,

$$K_{sp} = S \times (2 \times S)^2 = 4S^3$$

Since Table 15.4 shows $K_{sp} = 4.0 \times 10^{-11} M^3$, we have

$$S^3 = \frac{1}{4} \times (4.0 \times 10^{-11} M^3) = 1.0 \times 10^{-11} M^3$$

Taking the cube root* of this value gives

$$S = 2.2 \times 10^{-4} M$$

*A cube root can be taken either by using the y^x key on a calculator with $x = \frac{1}{3}$ or with logarithms (Appendix 1B).

FIGURE 15.18 In (a), the tube contains a saturated solution of zinc acetate. When acetate ions are added (as sodium acetate in the spoon), the solubility of the zinc acetate is significantly reduced, as shown in (b).

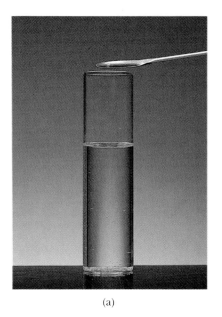

(a)　　　　　　　　　　　　　　　(b)

The common-ion effect. By a *common ion* we mean an ion that is common to two substances in the same mixture. The common ion in a solution of sodium acetate and zinc acetate, for example, is the $CH_3CO_2^-$ ion. When a soluble salt is added to a solution of a sparingly soluble salt that has an ion in common with it (as when sodium acetate is added to a zinc acetate solution, Fig. 15.18), the solubility of the sparingly soluble salt is reduced to much less than it would be in pure water. This reduction of the solubility of one salt by the addition of a common ion is called the *common-ion effect*.

The common-ion effect can be understood if we think about how the presence of a common ion affects the solubility equilibrium. In a saturated solution of silver chloride we have the equilibrium

$$AgCl(s) \rightleftharpoons Ag^+(aq) + Cl^-(aq) \qquad K_{sp} = [Ag^+][Cl^-]$$

and the solubility of silver chloride is

$$S = \sqrt{K_{sp}} = 1.3 \times 10^{-5} \, M$$

Adding sodium chloride to the solution increases the concentration of Cl^- ions. The system responds in such a way as to preserve the value of K_{sp}. This means that the concentration of Ag^+ ions must decrease, and it does so by precipitation of silver choride. The solubility of AgCl thus becomes lower than in pure water, as observed.

We can make a quantitative prediction of the effect of a common ion by using the fact that K_{sp} does not change when the additional ions are added. In any solution of silver and chloride ions, for example,

$$[Ag^+] = \frac{K_{sp}}{[Cl^-]}$$

where K_{sp} is a constant. Suppose that we try to dissolve silver chloride in 0.10 M NaCl(*aq*). The silver chloride dissolves until the concentra-

(a)   (b)

FIGURE 15.19 (a) $Al(OH)_3$ may be precipitated from an aluminum sulfate solution by ammonia. (b) It is amphoteric and dissolves in sodium hydroxide (left) or in hydrochloric acid (right). A universal indicator has been added to the solution to show the different pH values.

tion of Ag^+ ions reaches

$$[Ag^+] = \frac{1.6 \times 10^{-10} \, M^2}{0.10 \, M} = 1.6 \times 10^{-9} \, M$$

This is a 10,000-fold decrease from its solubility in pure water.

The effect of pH on the solubility of amphoteric oxides. Now we bring all the material of this chapter together and see how the solubilities of certain sparingly soluble compounds can be controlled by varying the pH. We begin by considering the amphoteric compound aluminum hydroxide. As explained in Section 3.5, an amphoteric compound is one that reacts with both acids and bases. We can therefore expect $Al(OH)_3$ to be soluble in both acids and bases, even if it is not soluble in water. That is, we expect its solubility to increase both as the pH of the solvent is raised by adding base and as the pH is lowered by adding acid (Fig. 15.19).

Aluminum hydroxide is, in fact, almost completely insoluble in water: the equilibrium

$$Al(OH)_3(s) \rightleftharpoons Al^{3+}(aq) + 3OH^-(aq) \qquad K_{sp} = [Al^{3+}][OH^-]^3$$

has $pK_{sp} = 33.0$. The value of K_{sp} corresponds to a solubility $S = 2.5 \times 10^{-9} \, M$. (This is obtained as explained above, but with $[OH^-] = 3 \times [Al^{3+}]$.) However, the compound is much more soluble in acid, since the additional hydrogen ions lower the OH^- ion concentration, causing more aluminum hydroxide to dissolve to preserve the value of K_{sp}.

The concentration of Al^{3+} ions in a saturated solution is related to the concentration of OH^- ions by

$$[Al^{3+}] = K_{sp} \times \frac{1}{[OH^-]^3}$$

This concentration is equal to the solubility S of the solid, because each

Al(OH)₃ unit that goes into solution gives one Al^{3+} ion. It can be expressed in terms of the hydrogen ion concentration and $K_w = [OH^-][H_3O^+]$:

$$S = [Al^{3+}] = K_{sp} \times \left(\frac{[H_3O^+]}{K_w}\right)^3$$

This equation can be used to find the solubility of Al(OH)₃ in acid solutions with various pH values. For example, we can calculate the solubility of the hydroxide in acid of pH = 3.5. Since this pH corresponds to $[H_3O^+] = 3 \times 10^{-4}\ M$, the solubility is

$$S = 1.0 \times 10^{-33}\ M^4 \times \left(\frac{3 \times 10^{-4}\ M}{1.0 \times 10^{-14}\ M^2}\right)^3$$

$$= 0.03\ M$$

This is an enormous increase (by a factor of about 10 million) over the solubility of aluminum hydroxide in pure water. The equation was used to calculate solubilities for the red part of the curve in Fig. 15.20; that of point A is the value just calculated.

Aluminum hydroxide is also soluble in alkali as a result of the reaction of the Lewis base OH^- with the Lewis acid Al(OH)₃, in which the aluminate ion is formed:

$$Al(OH)_3(s) + OH^-(aq) \rightleftharpoons Al(OH)_4^-(aq) \qquad K_c = \frac{[Al(OH)_4^-]}{[OH^-]}$$

At 25°C, $K_c = 40$. Here the solubility of Al(OH)₃ is equal to the equilibrium concentration of the aluminate ion, because each Al(OH)₃ unit that dissolves gives one $Al(OH)_4^-$ ion, and is therefore related to $[OH^-]$ by

$$S = [Al(OH)_4^-] = K_c \times [OH^-]$$

It is sometimes more convenient to express S in terms of the hydrogen ion concentration (and hence the pH). As usual, the connection between $[OH^-]$ and $[H_3O^+]$ is the water autoionization constant:

$$S = K_c \times \frac{K_w}{[H_3O^+]}$$

FIGURE 15.20 The dependence of the solubility of aluminum hydroxide on pH. The red part of the curve is the acid region, and the blue part the basic region. Points on the curve were calculated as described in the text.

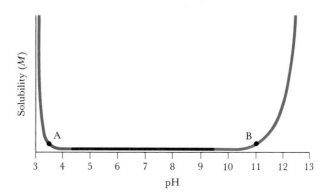

As an example, when the solution has pH = 11 (corresponding to $[H_3O^+] = 1 \times 10^{-11} M$), the solubility of $Al(OH)_3$ is

$$S = 40 \times \frac{1.0 \times 10^{-14} M^2}{1 \times 10^{-11} M} = 0.04 M$$

This is point B in Fig. 15.20. The pH dependence predicted by this equation is shown by the blue part of the curve.

The amphoterism of aluminum oxides and hydroxides is utilized in the commercial manufacture of aluminum from bauxite, which is largely alumina (Al_2O_3) but has silicate and Fe_2O_3 impurities. The crude bauxite is first treated with aqueous sodium hydroxide. The amphoteric alumina and acidic silicates dissolve, the former as aluminate ions, leaving the Fe_2O_3. The Fe_2O_3 would require a stronger base to dissolve, and can be removed by filtration. Next, aluminum ions are precipitated by making the solution acidic with dissolved carbon dioxide. The carbonic acid that is formed lowers the OH^- ion concentration, the aluminate ion decomposes in an attempt to preserve the equilibrium constant, and aluminum hydroxide precipitates. The precipitate is removed and heated to drive off water in the reaction

$$2Al(OH)_3(s) \xrightarrow{\Delta} Al_2O_3(s) + 3H_2O(g)$$

The reasonably pure oxide is now ready for the next stage, its electrolytic reduction to aluminum metal.

The effect of pH on the solubilities of salts. A compound need not be a hydroxide for its solubility to depend on pH: *any* Brønsted base responds to the addition of an acid. Suppose we have a saturated solution of a salt MA of the weak acid HA. Adding a strong acid such as HCl shifts the equilibrium

$$A^-(aq) + H_3O^+(aq) \rightleftharpoons HA(aq) + H_2O(l)$$

to the right, reducing the concentration of A^- ions in solution. As a result, the solubility equilibrium

$$MA(s) \rightleftharpoons M^+(aq) + A^-(aq)$$

also shifts to the right, in favor of dissolving. Anions that behave like this include OH^-, CO_3^{2-}, S^{2-}, $(CO_2)_2^{2-}$, and CrO_4^{2-}; in all cases their salts are more soluble in acid than in neutral water.

The effect is illustrated by calcium fluoride, which is very slightly soluble in water:

$$CaF_2(s) \rightleftharpoons Ca^{2+}(aq) + 2F^-(aq) \qquad K_{sp} = [Ca^{2+}][F^-]^2$$

At 25°C, $pK_{sp} = 10.40$. The fluoride ion is a base (Table 15.2), and it reacts as follows with any H_3O^+ ions that are added:

$$F^-(aq) + H_3O^+(aq) \rightleftharpoons HF(aq) + H_2O(l)$$

This reduces the concentration of F^- ions in the solution, so more CaF_2 dissolves.

15.6 PRECIPITATION REACTIONS AND QUALITATIVE ANALYSIS

Precipitation reactions can be discussed in terms of solubility products and used to analyze the compositions of some mixtures. The approach can be illustrated with the reaction that occurs when solutions of potassium iodide and lead nitrate are mixed:

$$Pb(NO_3)_2(aq) + 2KI(aq) \longrightarrow 2KNO_3(aq) + PbI_2(s)$$

The instant after the solutions are mixed, the concentrations of the Pb^{2+} and I^- ions are very high (typically about 1 M). However, the solubility product for the equilibrium

$$PbI_2(s) \rightleftharpoons Pb^{2+}(aq) + 2I^-(aq) \qquad K_{sp} = [Pb^{2+}][I^-]^2$$

is only 1.4×10^{-8} M^3 at 25°C, so solid lead iodide precipitates until the concentrations remaining in solution have been reduced to about 10^{-3} M. To bring about a precipitation reaction, we must contrive to produce ion concentrations that exceed the value of K_{sp} for the desired product.

The ion product and precipitation. A convenient way of discussing precipitation equilibrium is in terms of the *ion product* Q, the analog of the reaction quotient that we used in Section 13.2 to discuss the direction of reactions. It is defined like the solubility product,

$$Q = [M]^m[X]^n$$

but in place of the equilibrium molar concentrations of the ions we have their actual concentrations in a mixture that might or might not be at equilibrium. The ion product of lead(II) iodide, for example, is

$$Q = [Pb^{2+}][I^-]^2$$

Its value is obtained by substituting the actual ion concentrations that are present in a given solution. For example, the instant after equal volumes of 0.2 M $Pb(NO_3)_2$ and 0.4 M $KI(aq)$ solutions are mixed, the ion concentrations are

$$[Pb^{2+}] = 0.1\ M \qquad [I^-] = 0.2\ M$$

(The individual concentrations are reduced by half because the combined solution occupies twice the volume of either individual solution.) Hence,

$$Q = 0.1\ M \times (0.2\ M)^2 = 4 \times 10^{-3}\ M^3$$

We use Q exactly like the reaction quotient, to judge the direction in which a precipitation reaction will proceed—whether equilibrium will be reached by the formation of products or by the dissolution of a precipitate:

If $Q > K_{sp}$, the salt will precipitate.
If $Q = K_{sp}$, the solution is saturated.
If $Q < K_{sp}$, more salt will dissolve (if solid is present)

In our lead iodide example, $Q > K_{sp}$ when the solutions are first mixed, so lead iodide precipitates.

▼ **EXAMPLE 15.11** **Predicting whether a salt will precipitate**

Predict whether silver sulfate will precipitate when 1.0 mL of 1.0 mM $MgSO_4(aq)$ is poured into 100.0 mL of 0.50 mM $AgNO_3(aq)$.

STRATEGY The result can be predicted by determining whether Q exceeds K_{sp} for the silver sulfate (Ag_2SO_4). We begin by writing the expression for Q; then we can use the data to calculate the required ion concentrations to substitute into that expression. Concentrations must be based on the *total* volume of the combined solution. The value of K_{sp} is given in Table 15.4.

SOLUTION The ion product is

$$Q = [Ag^+]^2[SO_4{}^{2-}]$$

The total volume of the mixed solution is 101.0 mL. When the solutions are mixed, the volume occupied by $AgNO_3$ increases from 100.0 mL to 101.0 mL. Therefore, the concentration of the Ag^+ ions is decreased to

$$[Ag^+] = \frac{100.0 \text{ mL} \times 0.50 \text{ m}M}{101.0 \text{ mL}} = 0.50 \text{ m}M$$

The volume occupied by the $MgSO_4$ is increased from 1.0 mL to 101.0 mL, so the concentration of the $SO_4{}^{2-}$ ions is

$$[SO_4{}^{2-}] = \frac{1.0 \text{ mL} \times 1.0 \text{ m}M}{101.0 \text{ mL}} = 9.9 \times 10^{-3} \text{ m}M$$

Therefore, immediately after mixing,

$$Q = (0.50 \times 10^{-3} \, M)^2 \times (9.9 \times 10^{-6} \, M)$$

$$= 2.5 \times 10^{-12} \, M^3$$

Since $Q < K_{sp}$, no precipitation occurs.

EXERCISE Does barium sulfate precipitate when 1.0 mL of 1.0 mM $Na_2SO_4(aq)$ is poured into 250 mL of 0.1 M $BaCl_2(aq)$?

[*Answer:* $Q = 3.9 \times 10^{-7} \, M^2$; yes]

▲

The role of pH in qualitative analysis. The relationship between the ion product and solubility product explains some of the techniques that have been developed for *qualitative analysis*, the analysis of an unknown sample to discover what elements it contains. There are many different schemes of qualitative analysis, but all have features in common. One such feature is the formation of a precipitate when certain reagents are added to a solution of the sample, often after it has been made acidic or basic by the addition of acid or base. We shall work through a single procedure in the order actually adopted in the laboratory, explaining each step in terms of the solubility equilibrium involved. The scheme we shall adopt is only a fragment of the systematic procedure actually employed, but it is enough to illustrate the general idea.

(a)

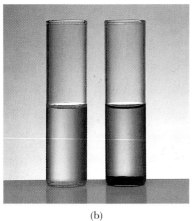

(b)

(c)

FIGURE 15.21 Sequence for the analysis of cations by selective precipitation. The original solution contains Ag^+, Cu^{2+}, and Zn^{2+} ions. (a) Addition of HCl precipitates AgCl, which can be removed by filtration. (b) Addition of H_2S to the remaining solution precipitates CuS, which can also be removed. (c) Making the solution basic precipitates ZnS.

The scheme is set out in Table 15.5 and illustrated in Figure 15.21. Certain chlorides (Table 15.5) are precipitated when hydrochloric acid is added to a solution of the sample. The idea behind this step is that silver, mercury(I), and lead(II) chlorides have such small values of K_{sp} that the addition of Cl^- ions to solutions of those metal ions raises Q above K_{sp}, with the result that the chlorides precipitate. The hydrogen ions provided by the acid play no role in this step; they simply accompany the chloride ions.

At a later stage in the typical analysis scheme, certain sulfides (Table 15.5) are precipitated by the addition of S^{2-} ions. The goal at this stage is to precipitate and identify some (but not all) of the metal cations that may still be in solution; that is, to achieve *selective precipitation* by making use of the widely different solubilities (and solubility products) of different compounds. Some metal sulfides (such as CuS, HgS, and Sb_2S_3) have extremely small solubility products ($pK_{sp} = 47$, 52, and 93, respectively) and are precipitated if there is the merest trace of S^{2-} ions in the solution. Such a very low concentration of S^{2-} ions is achieved by adding hydrogen sulfide to the acidified solution. The high hydrogen ion concentration ensures that almost all the H_2S is present in its nonionized form, and hence that very little is present as S^{2-}.

Next, ammonia is added to the solution. The base removes the hydrogen ions, and H_2S molecules are converted to S^{2-} ions. The high concentration of S^{2-} raises the ion product Q of any remaining metal sulfides above their K_{sp} values, and they precipitate. This step detects the presence of metals with sulfides having larger solubility products than those detected in the preceding step (Table 15.5), including zinc and manganese(II), with $pK_{sp} = 24$ and 15, respectively.

TABLE 15.5 Part of a qualitative analysis scheme

Step		Precipitate, pK_{sp}	
Add HCl(*aq*)	Sample → ○ → Precipitate may be	AgCl	9.8
		Hg_2Cl_2	17.9
		$PbCl_2$	4.8
Add H_2S(*aq*) (in acid solution)	→ ○ → Precipitate may be	Bi_2S_3	97.0
		CdS	28.4
		CuS	46.7
		HgS	51.8
		Sb_2S_3	92.8
H_2S(*aq*) (in basic solution)	→ ○ → Precipitate may be	FeS	17.2
		MnS	14.9
		NiS	23.9
		ZnS	23.8

15.7 COMPLEX FORMATION

When aqueous ammonia is added to a precipitate of silver chloride, the precipitate dissolves. This is a result of the formation of the complex $[Ag(NH_3)_2]^+$ and the favorable equilibrium

$$Ag^+(aq) + 2NH_3(aq) \rightleftharpoons [Ag(NH_3)_2]^+(aq)$$

which removes Ag^+ ions from the solution. Because the product $[Ag^+] \times [Cl^-]$ must remain equal to the value of K_{sp} for silver chloride, more silver chloride dissolves to make up for the loss of Ag^+ ions. Each new Ag^+ ion that goes into solution is immediately converted into $[Ag(NH_3)_2]^+$ by the ammonia present in the solution. If enough ammonia is present, almost all the precipitate dissolves. A similar process is used to remove the silver halide emulsion from exposed photographic film after it has been developed. In this case, the reagent used to form the complex is the thiosulfate ion, and the equilibrium is

$$Ag^+(aq) + 2S_2O_3{}^{2-}(aq) \rightleftharpoons [Ag(S_2O_3)_2]^{3-}(aq)$$

Formation constants. The Ag^+ ion is a Lewis acid (an electron-pair acceptor), and the NH_3 molecule is a Lewis base (an electron-pair donor). The reaction between them is a Lewis acid-base reaction, and the product is a complex. (The same is true of the reaction between Ag^+ ions and thiosulfate ions.) Here, then, is yet another kind of equilibrium that can affect the properties of a solution. The equilibrium constant for the formation of a complex is called a *formation constant* (or "stability constant") K_f and is defined in the usual way:

$$K_f = \frac{[\{Ag(NH_3)_2\}^+]}{[Ag^+][NH_3]^2}$$

For this equilibrium, $K_f = 1.6 \times 10^7/M^2$. Values for other equilibria are given in Table 15.6. (Note that we must be careful to distinguish the brackets that surround a complex, as in $[Ag(NH_3)_2]^+$, from the brackets that denote a molar concentration. There should be no problem: an ionic complex always has a charge written outside the brackets. However, to avoid ambiguity, here and on the next few pages we shall use braces { } to denote complex ions inside concentration brackets.)

TABLE 15.6   Formation constants in water at 25°C

Equilibrium	K_f
$Ag^+(aq) + 2Cl^-(aq) \rightleftharpoons [AgCl_2]^-(aq)$	$2.5 \times 10^5/M^2$
$Ag^+(aq) + 2Br^-(aq) \rightleftharpoons [AgBr_2]^-(aq)$	$1.3 \times 10^7/M^2$
$Ag^+(aq) + 2CN^-(aq) \rightleftharpoons [Ag(CN)_2]^-(aq)$	$5.6 \times 10^8/M^2$
$Ag^+(aq) + 2NH_3(aq) \rightleftharpoons [Ag(NH_3)_2]^+(aq)$	$1.6 \times 10^7/M^2$
$Au^+(aq) + 2CN^-(aq) \rightleftharpoons [Au(CN)_2]^-(aq)$	$2.0 \times 10^{38}/M^2$
$Cu^{2+}(aq) + 4NH_3(aq) \rightleftharpoons [Cu(NH_3)_4]^{2+}(aq)$	$1.2 \times 10^{13}/M^4$
$Hg^{2+}(aq) + 4Cl^-(aq) \rightleftharpoons [HgCl_4]^{2-}(aq)$	$1.2 \times 10^5/M^4$
$Fe^{2+}(aq) + 6CN^-(aq) \rightleftharpoons [Fe(CN)_6]^{4-}(aq)$	$7.7 \times 10^{36}/M^6$
$Ni^{2+}(aq) + 6NH_3(aq) \rightleftharpoons [Ni(NH_3)_6]^{2+}(aq)$	$5.6 \times 10^8/M^6$

We can calculate the solubility of a solid when complex formation occurs by ensuring that both the dissolving and the complex formation are at equilibrium. That is, in the case of AgCl in the presence of ammonia, we must ensure that the ion concentrations satisfy both the solubility equilibrium $[Ag^+] \times [Cl^-] = K_{sp}$ and the complex formation equilibrium K_f given above. In the only case we shall consider here, complex formation is so extensive that almost all the Ag^+ ions in the solution are converted to the complex; hence the solubility of the solid is equal to the concentration of the complex in the solution.

▼ EXAMPLE 15.12 Calculating solubility with complex formation taken into account

Calculate the solubility of silver chloride in $0.10\ M$ $NH_3(aq)$.

STRATEGY As noted above, complex formation in this solution leads to a much higher solubility than would be observed in pure water. The two equilibrium constants that must be satisfied are

$$AgCl(s) \rightleftharpoons Ag^+(aq) + Cl^-(aq) \qquad K_{sp} = [Ag^+][Cl^-]$$

$$Ag^+(aq) + 2NH_3(aq) \rightleftharpoons [Ag(NH_3)_2]^+(aq) \qquad K_f = \frac{[\{Ag(NH_3)_2\}^+]}{[Ag^+][NH_3]^2}$$

Since the formation constant is large, virtually all the Ag^+ ions will be present as the complex. Since the solubility S of AgCl is its molar concentration in the solution, and almost all the Ag^+ ions in solution will be present as the complex, we can assume that S is equal to the concentration of the complex. We can rearrange K_f to give the concentration of the complex and hence S. Then we can use K_{sp} and the fact that the electric charges in the solution balance—and hence that $[Cl^-] = [\{Ag(NH_3)_2\}^+]$—to find S in terms of $[NH_3]$.

SOLUTION The formation constant may be rearranged to

$$S = [\{Ag(NH_3)_2\}^+] = K_f \times [Ag^+][NH_3]^2$$

Since $[Ag^+]$ must satisfy the equation $K_{sp} = [Ag^+][Cl^-]$, we have $[Ag^+] = K_{sp}/[Cl^-]$ and therefore

$$S = K_f \times \frac{K_{sp}}{[Cl^-]} \times [NH_3]^2$$

However, $[Cl^-]$ is also equal to the solubility S (because no Ag^+ ion can go into solution without a Cl^- ion going too). Therefore, we can multiply both sides of the last equation by S, noting that $S = [Cl^-]$, to obtain

$$S^2 = K_f \times K_{sp} \times [NH_3]^2$$

and hence

$$S = \sqrt{K_f \times K_{sp}} \times [NH_3]$$

Now we can substitute the data:

$$S = \sqrt{(1.6 \times 10^7/M^2) \times (1.6 \times 10^{-10}\ M^2)} \times 0.10\ M$$

$$= 0.051 \times 0.10\ M = 5.1 \times 10^{-3}\ M$$

This is a 100-fold increase over the solubility in pure water, but it is still very low. The concentration of NH_3 needs to be high if much AgCl is to dissolve.

EXERCISE Calculate the solubility of silver bromide in $1.0\ M$ $NH_3(aq)$.

[*Answer*: 3.5 mM]

▲

Stepwise formation constants. Some Lewis bases can bond to certain cations in varying numbers. For these we can discuss the *stepwise formation constants* K_{f1}, K_{f2}, and so on, the formation constants for the successive addition of bases. An example is the series of complexes obtained when ammonia is added to a solution of copper(II) ions (Fig. 15.22):

$$Cu^{2+}(aq) + NH_3(aq) \rightleftharpoons [Cu(NH_3)]^{2+}(aq) \qquad K_{f1} = \frac{[\{Cu(NH_3)\}^{2+}]}{[Cu^{2+}][NH_3]}$$

$$[Cu(NH_3)]^{2+}(aq) + NH_3(aq) \rightleftharpoons [Cu(NH_3)_2]^{2+}(aq)$$

$$K_{f2} = \frac{[\{Cu(NH_3)_2\}^{2+}]}{[\{Cu(NH_3)\}^{2+}][NH_3]}$$

$$[Cu(NH_3)_2]^{2+}(aq) + NH_3(aq) \rightleftharpoons [Cu(NH_3)_3]^{2+}(aq)$$

$$K_{f3} = \frac{[\{Cu(NH_3)_3\}^{2+}}{[\{Cu(NH_3)_2\}^{2+}][NH_3]}$$

$$[Cu(NH_3)_3]^{2+}(aq) + NH_3(aq) \rightleftharpoons [Cu(NH_3)_4]^{2+}(aq)$$

$$K_{f4} = \frac{[\{Cu(NH_3)_4\}^{2+}]}{[\{Cu(NH_3)_3\}^{2+}][NH_3]}$$

The values of the stepwise constants are given in Table 15.7. The sum of the four stepwise equilibria is

$$Cu^{2+}(aq) + 4NH_3(aq) \rightleftharpoons [Cu(NH_3)_4]^{2+}(aq)$$

and the overall formation constant, the equilibrium constant for this reaction, is

$$K_f = K_{f1} \times K_{f2} \times K_{f3} \times K_{f4}$$
$$= 1.2 \times 10^{13}/M^4$$

FIGURE 15.22 Copper(II) ions act as Lewis acids and form a complex with the Lewis base ammonia. Intermediate ions are also present, but in excess ammonia most copper ions are present as the deep blue $[Cu(NH_3)_4]^{2+}$ ion.

▼ **EXAMPLE 15.13** **Calculating the concentration of a complex in solution**

Calculate the concentration of Cu^{2+} ions in a solution containing 0.020 M $Cu^{2+}(aq)$ plus 2.0 M $NH_3(aq)$.

STRATEGY We must remember that complexes other than the fully complexed ion, $[Cu(NH_3)_4]^{2+}$, are also present and must be taken into account. However, because the values of the stepwise equilibrium constants are so large and the ammonia is present in great excess, we can assume that the fully complexed ion is dominant in the solution. In that case, we can take the concentration of the complex to be 0.020 M. Moreover, we can take the concentration of NH_3 to be 2.0 M, because the amount used to form the complex is negligible compared with the total. We then need only substitute these values into the expression for K_f, rearranged to give $[Cu^{2+}]$.

SOLUTION We need the value of

$$[Cu^{2+}] = \frac{1}{K_f} \times \frac{[\{Cu(NH_3)_4\}^{2+}]}{[NH_3]^4}$$

If the complex concentration is approximately 0.020 M and the NH_3 concentration is approximately 2.0 M, then

TABLE 15.7 Stepwise formation constants for Cu^{2+} and NH_3

n	K_{fn}, $1/M$
1	1.8×10^4
2	4.1×10^3
3	9.5×10^2
4	1.7×10^2

$K = K_{f1} \times K_{f2} \times \cdots = 1.2 \times 10^{13}/M^4$

$$[Cu^{2+}] = \frac{M^4}{1.2 \times 10^{13}} \times \frac{0.020\ M}{(2.0\ M)^4}$$

$$= 1.0 \times 10^{-16}\ M$$

EXERCISE Calculate the concentration of Ni^{2+} ions in a solution of 0.10 M $Ni^{2+}(aq)$ plus 2.0 M $NH_3(aq)$, given that the formation constant of $[Ni(NH_3)_6]^{2+}(aq)$ is $5.6 \times 10^8/M^6$.

[*Answer*: $2.8 \times 10^{-12}\ M$]

Complex formation in the chemistry of gold. Complex formation is used in the extraction of gold from low-grade gold-containing rock. The formation constant for the complex $[Au(CN)_2]^-$ is very large, and as soon as any Au^+ ions are formed (through oxidation) in the presence of CN^- ions, the complex is produced. This removes the Au^+ from solution so that more of it will be formed. Therefore, even though gold is not oxidized by air normally, bubbling air through a suspension of gold-containing ore in the presence of cyanide ions leads to the formation of a solution of the complex:

$$4Au(s) + 8CN^-(aq) + O_2(g) + 2H_2O(l) \longrightarrow$$
$$4[Au(CN)_2]^-(aq) + 4OH^-(aq)$$

The unwanted material is removed by filtration, and zinc powder is added to the solution. This reduces the complex to metallic gold, which can then be removed by filtration.

Aqua regia ("kingly water"), a 3-to-1 mixture of concentrated hydrochloric acid and nitric acid, can also dissolve gold. In this process, the nitric acid is the oxidizing agent and the chloride ions form the complex $[AuCl_4]^-$ with Au^{3+} ions. As the formation of the complex removes Au^{3+} ions from the solution, the nitric acid oxidizes more metal until it has all reacted.

These examples demonstrate that ionic equilibria, particularly complex formation, can influence redox equilibria by removing a product of oxidation from the system. The next chapter prepares us for further discussion of redox equilibria, and Chapter 17 describes them in detail.

SUMMARY

15.1 The addition of any acid to water results in an acidic solution; the addition of any base results in a basic solution. A solution of a weak acid and one of its salts with a strong base is less acidic (with pH closer to 7) than a solution of the acid alone, because the anion is a base. A solution of a weak base and one of its salts with a strong acid is less basic (with pH closer to 7) than a solution of the base alone, because the cation is an acid.

Small, highly charged cations that are hydrated are often Brønsted acids.

15.2 The pH of a mixed solution (salt plus acid or base) is given by the **Henderson-Hasselbalch equation.** When the salt and the acid (or base) are present in equal concentrations, the pH of the solution is equal to the pK_a of the acid (or the conjugate acid of the base).

15.3 The equivalence point of the titration of a strong base with a strong acid is at pH = 7 because neither the cation nor the anion of the resulting salt is an acid or base. The pH at the equivalence point of the titration of a weak acid with a strong base is higher than 7, since the anion of the salt is a base. The pH at the equivalence point of the titration of a weak base with a strong acid is less than 7, since the cation of the salt is an acid. The variation of pH during a titration is shown by a **pH curve.**

15.4 At the equivalence point, the pH changes abruptly by several units; the change may be detected with an **indicator,** a Brønsted acid that shows different colors in its acid and conjugate base forms. The indicator must be chosen so that its color change occurs over a range that covers the pH of the equivalence point. The pH of a solution can be stabilized by a **buffer,** a mixture of a weak acid or base and its salt in approximately equal concentrations.

15.5 The solubility equilibrium of a sparingly soluble salt is described by an equilibrium constant called the **solubility product** K_{sp}. This equilibrium accounts for the **common-ion effect,** in which a sparingly soluble salt is less soluble in a solution that contains an ion in common with the salt. The solubility of a salt that behaves as an acid or a base (perhaps through hydrated-ion formation) depends on the pH of the solution.

15.6 Precipitation reactions may be discussed in terms of the **ion product** Q: precipitation can be expected if $Q > K_{sp}$. In one kind of **qualitative analysis,** a procedure for identifying the elements present in a sample, the size of Q is modified to bring about the **selective precipitation** of different compounds.

15.7 Some cations act as Lewis acids, forming **complexes** with anions and neutral molecules that act as Lewis bases. Such a complex is in equilibrium in solution, and the composition of the equilibrium mixture is described by the **formation constant.** In some cases, the Lewis base is best regarded as forming the complex in steps, with each step described by a **stepwise formation constant.** Complex formation can increase the solubility of a compound by removing cations from the solution and modifying the solubility equilibrium.

EXERCISES

Salts as Acids and Bases

In Exercises 15.1 and 15.2, state whether a solution of each salt is likely to be acidic, neutral, or basic.

15.1 (a) NH_4Br; (b) KF; (c) Na_2CO_3; (d) KBr.
15.2 (a) $K_2C_2O_4$; (b) CH_3NH_3Cl; (c) $FeCl_3$; (d) $Ca(NO_3)_2$.

In Exercises 15.3 to 15.6, calculate the pH of each aqueous solution at 25°C.

15.3 (a) 0.20 M $NaCH_3CO_2$; (b) 0.10 M NH_4Cl.
15.4 (a) 0.15 M CH_3NH_3Cl; (b) 0.20 M $NaHSO_3$.

15.5 (a) 250 mL of solution containing 10.0 g of potassium acetate; (b) 100 mL of solution containing 5.75 g of ammonium chloride; (c) 1.0 L of solution containing 10.0 g of potassium bromide.
15.6 (a) 50.0 mL of solution containing 1.0 g of sodium hydrogen sulfite; (b) 10 mL of solution containing 100 mg of silver nitrate; (c) 10 mL of solution containing 5.0 mg of methylammonium bromide.

The pH of Mixed Solutions

In Exercises 15.7 to 15.12, calculate the pH of each solution at 25°C.

15.7 (a) A solution that is 0.20 M HBrO(aq) and 0.10 M KBrO(aq)
(b) A solution that is 0.10 M HBrO(aq) and 0.20 M KBrO(aq)
15.8 (a) A solution that is 0.02 M NaCN(aq) and 0.05 M HCN(aq)
(b) A solution that is 0.05 M NaCN(aq) and 0.02 M HCN(aq)

15.9 A solution that is 0.05 M Na_2SO_4(aq) and 0.02 M $NaHSO_4$(aq)
15.10 A solution that is 0.10 M KH_2PO_4(aq) and 0.20 M Na_2HPO_4(aq)

15.11 (a) A solution that is 0.20 M NH_3(aq) and 0.10 M NH_4Cl(aq)
(b) A solution that is 0.10 M NH_3(aq) and 0.20 M NH_4Cl(aq)
15.12 (a) A solution that is 0.30 M CH_3NH_2(aq) and 0.10 M CH_3NH_3Br(aq)
(b) A solution that is 0.10 M CH_3NH_2(aq) and 0.30 M CH_3NH_3Br(aq)

15.13 A solution of equal concentrations of lactic acid and sodium lactate was found to have pH = 3.08. (a) What is the pK_a of lactic acid? (b) What would the pH of the solution be if the acid had twice the concentration of the salt?

15.14 A solution containing equal concentrations of saccharin and its sodium salt was found to have pH = 11.68. (a) What is the pK_a of saccharin? (b) What would the pH of the solution be if the salt had twice the concentration of the acid?

15.15 A solution containing equal concentrations of ammonia and ammonium chloride was found to have pH = 9.25. (a) What is the pK_b of ammonia? (b) What would be the pH of the solution if the salt had twice the concentration of the base?

15.16 A solution containing equal concentrations of ethylamine ($C_2H_5NH_2$) and ethylammonium bromide ($C_2H_5NH_3Br$) was found to have pH = 10.81. (a) What is the pK_b of ethylamine? (b) Suggest an explanation for the difference in strengths between ethylamine and ammonia. (c) What would be the pH of the solution if the base had twice the concentration as the salt?

15.17 Calculate the change in pH that occurs when 1.0 g of sodium fluoride is added to 25 mL of 0.10 M HF(aq). Ignore the small change in volume.

15.18 Calculate the change in pH that occurs when 10.0 g of anhydrous sodium carbonate is added to 250 mL of 0.50 M NaHCO$_3$(aq). Ignore the small change in volume.

15.19 Sodium hypochlorite is the active ingredient in many bleaches. Calculate the ratio of the concentrations of HOCl and OCl$^-$ in a solution of sodium hypochlorite buffered to pH = 6.50.

15.20 Aspirin is a derivative of salicylic acid, which has pK_a = 2.97. (a) Calculate the ratio of the concentrations of salicylic acid and its conjugate base in a solution buffered to pH = 4.50. (b) What pH would ensure equal concentrations?

15.21 The narcotic cocaine is a weak base with pK_b = 5.59. Calculate the ratio of the concentrations of cocaine and its conjugate acid in a solution of pH = 8.00.

15.22 Pyridine (C_5H_5N) is an organic base with a pK_b of 8.77. (a) Calculate the pH of a 0.10-M pyridine solution. (b) Calculate the ratio of the pyridinium salt ($C_5H_5NH^+$) to the pyridine at a pH of 5.00.

Strong Acid–Strong Base Titrations

In Exercises 15.23 to 15.26, calculate the volume of acid needed to reach equivalence; then use this volume to calculate the concentration of the salt at the equivalence point.

15.23 In 25.0 mL of 0.110 M NaOH(aq), being titrated with 0.150 M HCl(aq)

15.24 In 25.0 mL of 0.215 M KOH(aq), being titrated with 0.116 M HCl(aq)

15.25 In 25.0 mL of a solution containing 2.54 g of sodium hydroxide, being titrated with 0.150 M HCl(aq)

15.26 In 50.0 mL of a solution containing 2.88 g of potassium hydroxide, being titrated with 0.200 M HNO$_3$(aq)

In Exercises 15.27 to 15.30, calculate the pH at each specified stage.

15.27 For 25.0 mL of 0.110 M NaOH(aq) being titrated with 0.150 M HCl(aq), (a) initially; (b) after the addition of 5.0 mL of acid; (c) after the addition of another 5.0 mL of acid; (d) at the equivalence point; (e) after the addition of 5.0 mL of acid beyond the equivalence point; (f) after the addition of 10 mL of acid beyond equivalence.

15.28 For 25.0 mL of 0.215 M KOH(aq) being titrated with 0.116 M HCl(aq), (a) initially; (b) after the addition of 5.0 mL of acid; (c) after the addition of another 5.0 mL of acid; (d) at the equivalence point; (e) after the addition of 5.0 mL of acid beyond the equivalence point; (f) after the addition of 10 mL of acid beyond equivalence.

15.29 For the titration of Exercise 15.27, at (a) 0.1 mL and (b) 0.01 mL of titrant before equivalence; and (c) 0.01 mL and (d) 0.1 mL after equivalence.

15.30 For the titration of Exercise 15.28, at (a) 0.1 mL and (b) 0.01 mL of titrant before equivalence; and (c) 0.01 mL and (d) 0.1 mL after equivalence.

In Exercises 15.31 and 15.32, calculate the change in pH that would be observed on proceeding from 0.1 mL of titrant before the equivalence point is reached to 0.1 mL after it, thus demonstrating that in a strong acid–strong base titration the pH changes very rapidly at the equivalence point.

15.31 Using 0.100 M HCl to titrate 25.0 mL of 0.100 M NaOH

15.32 Using 0.200 M NaOH to titrate 25.0 mL of 0.200 M HCl

Weak Acid–Strong Base and Strong Acid–Weak Base Titrations

In Exercises 15.33 to 15.36, calculate the initial pH and the pH at the equivalence point.

15.33 25.0 mL of a 0.10 M CH$_3$COOH(aq) solution being titrated with 0.10 M NaOH(aq)

15.34 30 mL of a 0.20 M CH$_3$COOH(aq) solution being titrated with 0.20 M NaOH(aq)

15.35 15.0 mL of a 0.15 M NH$_3$(aq) solution being titrated with 0.10 M HCl(aq)

15.36 50.0 mL of a 0.25 M CH$_3$NH$_2$(aq) solution being titrated with 0.35 M HCl(aq)

In Exercises 15.37 to 15.40, calculate the pH of the solution at each specified stage.

15.37 For 25.0 mL of 0.110 M $NH_3(aq)$ being titrated with 0.150 M $HCl(aq)$, (a) initially; (b) after the addition of 5.0 mL of acid; (c) after the addition of another 5.0 mL of acid; (d) at the equivalence point; (e) after the addition of 5.0 mL of acid beyond the equivalence point; (f) after the addition of 10 mL of acid beyond equivalence.

15.38 For 25.0 mL of 0.215 M $CH_3NH_2(aq)$ being titrated with 0.116 M $HCl(aq)$, (a) initially; (b) after the addition of 5.0 mL of acid; (c) after the addition of another 5.0 mL of acid; (d) at the equivalence point; (e) after the addition of 5.0 mL of acid beyond the equivalence point; (f) after the addition of 10 mL of acid beyond equivalence.

15.39 For the titration of Exercise 15.37, at (a) 1.0 mL and (b) 0.1 mL of titrant before equivalence; and (c) 0.1 mL and (d) 1.0 mL after equivalence.

15.40 For the titration of Exercise 15.38, at (a) 1.0 mL and (b) 0.1 mL of titrant before equivalence; and (c) 0.1 mL and (d) 1.0 mL after equivalence.

Although the titrations of weak polyprotic acids are not described explicitly in the text, their treatment is straightforward. It involves combining the material in this chapter with that in Chapter 14. Begin in Exercises 15.41 and 15.42 by calculating the pH for each titration at each equivalence point.

15.41 (a) 0.20 M $H_2SO_4(aq)$, titrated with 0.20 M NaOH(aq)

(b) 0.20 M $(COOH)_2(aq)$, oxalic acid, titrated with 0.20 M NaOH(aq)

15.42 (a) 0.60 M $H_2SO_4(aq)$, titrated with 0.60 M NaOH(aq)

(b) 0.20 M $H_2CO_3(aq)$, titrated with 0.20 M NaOH(aq)

In Exercises 15.43 and 15.44, calculate the full pH curve for each titration.

***15.43** 0.30 M $H_2C_2O_4(aq)$, titrated with 0.20 M NaOH(aq)

***15.44** 0.10 M $H_2CO_3(aq)$, titrated with 0.10 M NaOH(aq)

Indicators

In Exercises 15.45 to 15.48, suggest a suitable indicator for each titration.

15.45 (a) Potassium hydroxide and nitric acid; (b) potassium hydroxide and nitrous acid; (c) sodium hydroxide and lactic acid; (d) sodium hydroxide and sulfuric acid.

15.46 (a) Sodium hydroxide and hypoiodous acid; (b) potassium hydroxide and hydrofluoric acid; (c) sodium hydroxide and oxalic acid; (d) potassium hydroxide and phosphoric acid.

15.47 (a) Hydrochloric acid and ammonia; (b) hydrochloric acid and ethylamine.

15.48 (a) Hydrochloric acid and ethylenediamine; (b) urea and hydrochloric acid.

In Exercises 15.49 and 15.50, for each indicator calculate the ratio of acid form to conjugate base form present in solution as the pH changes from 1 to 14 in unit steps.

15.49 (a) Methyl orange; (b) methyl red; (c) litmus; (d) phenolphthalein.

15.50 (a) Thymol blue; (b) bromophenol blue; (c) phenol red.

Buffers

In Exercises 15.51 and 15.52, calculate the pH change that occurs when 0.5 mL of 0.10 M $HCl(aq)$ is added to 10 mL of each solution.

15.51 (a) Pure water; (b) a solution that is 0.10 M $Na(CH_3CO_2)(aq)$ and 0.10 M $CH_3COOH(aq)$.

15.52 (a) 0.15 M $HCl(aq)$; (b) a solution that is 0.15 M $Na_2HPO_4(aq)$ and 0.10 M $KH_2PO_4(aq)$.

In Exercises 15.53 and 15.54, predict the pH region in which each buffer pair will be effective, assuming equal molar concentrations of acid and salt in each case.

15.53 (a) Sodium lactate and lactic acid; (b) potassium hydrogen phosphate and potassium phosphate; (c) potassium hydrogen phosphate and potassium dihydrogen phosphate.

15.54 (a) Sodium dihydrogen borate and boric acid; (b) sodium carbonate and sodium hydrogen carbonate; (c) ammonia and ammonium chloride.

In Exercises 15.55 and 15.56, use the pK_a values in Table 14.4 to suggest a buffer that would be effective at each stated pH.

15.55 (a) 2; (b) 3; (c) 7; (d) 12.
15.56 (a) 4; (b) 5; (c) 9; (d) 11.

In Exercises 15.57 and 15.58, estimate the volume of (a) 0.1 M $HCl(aq)$ and (b) 0.1 M $NaOH(aq)$ that may be added to the buffer solution before it ceases to be effective; in each case, state the buffer pH just before its capacity is exceeded.

15.57 25 mL of a solution containing 0.20 M $NaHCO_3(aq)$ and 0.20 M $Na_2CO_3(aq)$

15.58 100 mL of a solution containing 0.10 M $NH_3(aq)$ and 0.10 M $NH_4Cl(aq)$

***15.59** Tris is an organic buffer (B) that is commonly used in biochemistry experiments. In its acidic form (BH$^+$) it has a pK_a of 8.08. Calculate the pH of 1.00 L of solution that contains 0.050 moles of B and 0.100 moles of BH$^+$.

*15.60 Valine is an amino acid that acts like a diprotic acid (H_2A) in its acidic form. The $pK_{a1} = 2.29$ and the $pK_{a2} = 9.72$. (a) Calculate the two pH values at which valine solutions act as a strong buffer. (b) Calculate the pH of 1.00 L of solution containing 1.00 mol of H_2A and 1.00 mol of HA^-.

Solubility Products

In Exercises 15.61 and 15.62, write the expression for the solubility product of each compound.

15.61 (a) $AgBr$; (b) $Ca(OH)_2$; (c) Ag_2S; (d) Ag_2CrO_4.
15.62 (a) AgI; (b) Sb_2S_3; (c) $MgNH_4PO_4$; (d) $AgSCN$.

In Exercises 15.63 and 15.64, calculate K_{sp} and pK_{sp} from the given solubility.

15.63 (a) $AgBr$: $S = 8.8 \times 10^{-7}\ M$
      (b) $PbCrO_4$: $S = 1.3 \times 10^{-7}\ M$
      (c) $Ba(OH)_2$: $S = 0.11\ M$
      (d) $MgNH_4PO_4$: $S = 6.3 \times 10^{-5}\ M$
15.64 (a) AgI: $S = 9.1 \times 10^{-9}\ M$
      (b) $Ca(OH)_2$: $S = 0.011\ M$
      (c) Ag_3PO_4: $S = 2.7 \times 10^{-6}\ M$
      (d) Hg_2Cl_2: $S = 5.2 \times 10^{-7}\ M$

In Exercises 15.65 and 15.66, calculate the solubility at 25°C in moles per liter of each compound, using the data in Table 15.4.

15.65 (a) Ag_2S; (b) CuS; (c) $CaCO_3$; (d) $PbCl_2$.
15.66 (a) $PbSO_4$; (b) $Fe(OH)_2$; (c) Ag_2CrO_4; (d) $SrSO_4$.

15.67 (a) Fluoridation of city water supplies produces an $F^-(aq)$ concentration close to 0.05 mM. Is there a possibility that CaF_2 will precipitate in hard water in which the Ca^{2+} concentration is 0.2 mM?
      (b) Fluoride ions convert the hydroxyapatite, $Ca_5(PO_4)_3OH$, of teeth into fluorapatite, $Ca_5(PO_4)_3F$. The pK_{sp} values of the two compounds are 36 and 60, respectively. What are their solubilities?
      (c) Bacteria produce lactic acid, so the pH at the surface of a tooth falls to about 5. How does that affect tooth decay and its prevention?
15.68 The mass concentration of Mg^{2+} ions in seawater is about 1.3 μg/L. In the commercial recovery process, the magnesium is precipitated as the hydroxide. At what pH does the hydroxide precipitate?

The Common-Ion Effect

In Exercises 15.69 and 15.70, calculate the solubility of each substance.

15.69 (a) Silver chloride in 0.20 M $NaCl(aq)$; (b) mercury(I) chloride in 0.10 M $NaCl(aq)$; (c) lead(II) chloride in 0.10 M $CaCl_2(aq)$.
15.70 (a) Silver bromide in 1.0 mM $KBr(aq)$; (b) lead(II) bromide in 0.10 mM $KBr(aq)$; (c) magnesium ammonium phosphate in 0.10 M $NH_4Cl(aq)$.

The Effect of pH on Solubility

In Exercises 15.71 and 15.72, derive an expression for the solubility of each compound in terms of the pH, and then calculate the solubility at the given pH values.

15.71 (a) $Al(OH)_3$ at (1) pH = 7.0 and (2) pH = 4.5
      (b) $Zn(OH)_2$ at (1) pH = 7.0 and (2) pH = 6.0
15.72 (a) $Fe(OH)_3$ at (1) pH = 7.0 (2) pH = 3.0, and (3) pH = 11.0
      (b) $Fe(OH)_2$ at (1) pH = 7.0, (2) pH = 6.0, and (3) pH = 8.0

In Exercises 15.73 and 15.74, calculate the solubility of each sulfide in 0.1 M $H_2S(aq)$ at the stated pH values.

15.73 (a) ZnS at (1) pH = 1.0 and (2) pH = 9.0
      (b) Sb_2S_3 at (1) pH = 1.0 and (2) pH = 9.0
15.74 (a) MnS at (1) pH = 1.0 and (2) pH = 9.0
      (b) Bi_2S_3 at (1) pH = 1.0 and (2) pH = 9.0

In Exercises 15.75 and 15.76, find an expression for the pH dependence of the solubility of each compound, and then evaluate it at the specified pH values.

15.75 CaF_2 at (a) pH = 7.0 and (b) pH = 5.0
15.76 BaF_2 at (a) pH = 7.0 and (b) pH = 4.0

In Exercises 15.77 and 15.78, decide whether a precipitate will form when the given solutions are mixed.

15.77 (a) 25 mL of 0.1 M $Sr(NO_3)_2(aq)$ and 1.0 mL of 0.1 M $NaOH(aq)$
      (b) 1.0 mL of 1.0 mM $NaOH(aq)$ and 10 mL of 0.1 M $AlCl_3(aq)$
      (c) 0.1 mL of 1.0 mM $NaCl(aq)$ and 10.0 mL of 0.1 M $AgNO_3(aq)$
      (d) 1.0 mL of 1.0 M $K_2SO_4(aq)$ and 25 mL of 1.0 mM $CaCl_2(aq)$
15.78 (a) 25 mL of 0.1 M $CaCl_2(aq)$ and 0.1 mL of 0.1 M $Na_2CO_3(aq)$
      (b) 1.0 mL of 0.10 M $K_2CrO_4(aq)$ and 10 mL of 0.1 M $AgNO_3(aq)$
      (c) 0.1 mL of 0.10 M $NaOH(aq)$ and 50 mL of 0.1 M $FeCl_3(aq)$
      (d) 0.1 mL of 0.1 M $H_3PO_4(aq)$ and 10 mL of 0.1 M $AgNO_3(aq)$

15.79 A solution containing 0.10 M $Cu^{2+}(aq)$, 0.10 M $Fe^{2+}(aq)$, and 0.10 M $H_2S(aq)$ is acidified to pH = 0.5. Which sulfide precipitates?
15.80 A solution containing 0.20 M $Bi^{3+}(aq)$, 0.20 M $Mn^{2+}(aq)$, and 0.20 M $H_2S(aq)$ is acidified to pH = 0.3. Which sulfide precipitates?

15.81 The pK_{sp} values of the hydroxides of barium, calcium, copper(II), and manganese(II) are 2.3, 5.3, 19, and 13, respectively. In what order do they precipitate when potassium hydroxide solution is added to a solution that contains each one at a concentration 1 mM?
15.82 Suppose that two hydroxides, MOH and

$M'(OH)_2$, have $pK_{sp} = 12.0$, and that both are present in a solution at concentrations of 1 mM. Which precipitates first, and at what pH, when sodium hydroxide solution is added?

It is often useful to know whether two substances can be separated by selective precipitation from a solution. In Exercises 15.83 and 15.84, judge whether this is possible (to the extent of 99.99% purity) by adding the specified ions.

15.83 In a solution containing 0.010 M $Pb^{2+}(aq)$ and 0.010 M $Hg_2^{2+}(aq)$, by adding I^- ions

15.84 In a solution containing 50 mM $Ca^{2+}(aq)$ and 30 mM $Ag^+(aq)$, by adding SO_4^{2-} ions

Complex Formation

15.85 Calculate the concentration of $Cu^{2+}(aq)$ ions in a solution that was initially 0.10 M $Cu^{2+}(aq)$ and 1.0 M $NH_3(aq)$.

15.86 Calculate the concentration of $Ni^{2+}(aq)$ ions in a solution that was initially 0.25 M $Ni^{2+}(aq)$ and 2.5 M $NH_3(aq)$. The formation constant of $[Ni(NH_3)_6]^{2+}(aq)$ is $5.6 \times 10^8/M^6$.

15.87 Silver chloride is precipitated when a chloride solution is added to silver nitrate. The precipitate dissolves in aqueous ammonia. Suppose the initial concentration of the ammonia is 0.10 M NH_3. Calculate the final concentration of NH_3 in the solution and the solubility of AgCl in the solution.

15.88 Precipitated silver chloride also dissolves in hydrochloric acid as a result of the formation of $[AgCl_2]^-$ ions. Suppose that the concentration of acid is 1.0 M $HCl(aq)$. What is the solubility of AgCl in the acid?

In this chapter we develop the concept of entropy. We see how it accounts for the tendency of reaction mixtures to change until they reach equilibrium and how it can be used to predict equilibrium constants. We see, too, that reactants have certain properties that determine the composition of a reaction mixture at equilibrium. The illustration shows white-hot iron cooling: we shall see that the explanation of why iron cools can be adapted to explain why chemical reactions approach equilibrium.

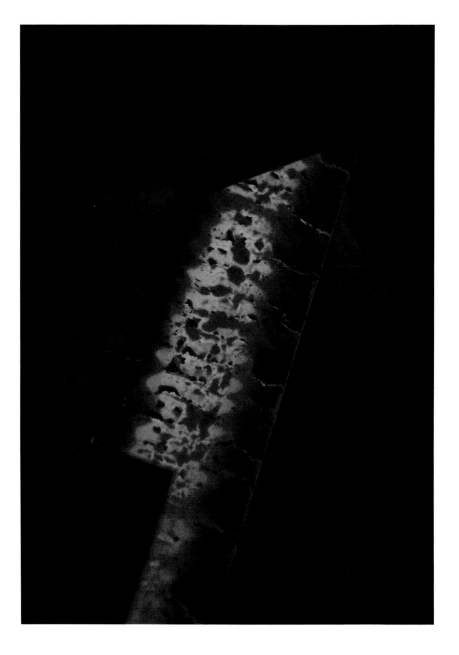

16

ENTROPY, FREE ENERGY, AND EQUILIBRIUM

The direction of spontaneous change

16.1 Entropy and spontaneous change
16.2 The entropy change in the surroundings
16.3 The second law

Free energy

16.4 Focusing on the system
16.5 Spontaneous reactions

Equilibria

16.6 Free energy and composition
16.7 The equilibrium constant

We have seen that physical and chemical changes tend toward equilibrium and that the composition of a system at equilibrium can be expressed in terms of an equilibrium constant. The big question is *why* this is so. What determines, for instance, whether or not a particular liquid has a tendency to vaporize? What determines whether a particular reaction mixture has a tendency to form more products or more reactants? We shall see in this chapter that the answer is embodied in a single unifying law—the second law of thermodynamics. The second law is of universal application throughout chemistry, for it underlies the tendency toward equilibrium of any type of change. It accounts for the tendency toward equilibrium in vaporizing, dissolving, and precipitating; it explains all the colligative properties; and it lies behind the equilibrium properties of Brønsted and Lewis acids and bases. We shall see in Chapter 17 that it also accounts for the tendency of redox reactions to approach equilibrium. It is not far from the truth to say that the second law summarizes chemistry.

The second law also has great practical importance: it can be used to *predict* the values of equilibrium constants. In particular, we shall see how to use tables of data like those at the back of this book to predict the value of the equilibrium constant for a wide range of reactions. We shall see how to predict solubility products, vapor pressures, pK_a values, and the values of the equilibrium constants for reactions in general. That is, we are about to see how to predict whether or not a reaction has a tendency to occur.

THE DIRECTION OF SPONTANEOUS CHANGE

In principle, the processes we have been discussing can run in either of two directions. For instance, depending on the pressure and temperature, water can either vaporize or condense. Section 11.5 introduced the idea that the actual direction of change—such as water's change from liquid to vapor if the pressure is 1 atm and the temperature is 100°C or more—is the direction that corresponds to the *dispersal* of energy and matter, their spreading in chaos. That is, when a change occurs, it leaves the universe (the system and its surroundings) in a more disordered state.

The tendency to disperse accounts for the expansion of a gas into a vacuum, for the gas becomes more disordered as its molecules spread. It also accounts for the tendency of substances to dissolve. We saw in Chapter 11 how the disorder of a sample increases as ions and molecules break away from a crystal and spread through a solvent. We shall now make these ideas quantitative, and express them in terms of properties that can be measured and calculated.

16.1 ENTROPY AND SPONTANEOUS CHANGE

A *spontaneous change* is a change that has a natural tendency to occur without having to be driven by some external influence. One simple example of a spontaneous physical change is the cooling of a block of

FIGURE 16.1 The direction of spontaneous change is for a hot block of metal (left) to cool to the temperature of its surroundings (right). A block at the same temperature as its surroundings does not spontaneously become hotter.

hot metal to the temperature of its surroundings (Fig. 16.1). The reverse change, a block of metal growing hotter than its surroundings, is not spontaneous. We can drive that reverse (nonspontaneous) change in a number of ways, for example, by forcing an electric current through the metal (which would heat it); but that reverse change will not occur unless it is driven. The expansion of a gas into a vacuum is also spontaneous (Fig. 16.2). The reverse change is not: a gas does not contract spontaneously into one part of a container. However, we can drive a gas into a smaller volume by pushing in a piston.

Although spontaneous changes are sometimes fast (as in the expansion of a gas), that is not always the case. A large block of metal cools spontaneously but slowly. Viscous oil has a spontaneous tendency to flow out of an upturned can, but that flow may be very slow. Throughout this chapter and the next we must be very careful not to assume that because a process *can* occur (because it generates disorder) it must take place at an observable rate. Thermodynamics tells us about *tendencies*; it is silent about rates.

Entropy as a measure of disorder. So far, we have mentioned only physical changes, in which no new substances are produced. Now we

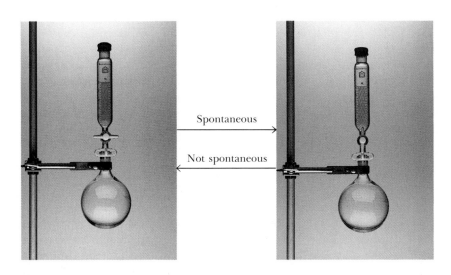

FIGURE 16.2 The direction of spontaneous change is for a gas to fill its container (right). A gas that already fills its container does not spontaneously collect in a small region of the container (left).

FIGURE 16.3 The vigorous combustion of magnesium is an example of a spontaneous chemical change. The reverse reaction, in which energy floods into magnesium oxide and converts it to the metal, is never observed.

FIGURE 16.4 Ludwig Boltzmann (1844–1906). His formula (using an earlier notation for natural logarithms) became his epitaph.

turn to chemical changes, in which new substances are formed, as in the formation of magnesium oxide when magnesium burns in oxygen (Fig. 16.3). Spontaneous chemical reactions are those in which there is an overall increase in disorder. However, before we can interpret a chemical reaction in those terms, we must know what it means for one substance to be "more disordered" than another. To decide this question, we calculate or measure their *entropy S*.

At one level, entropy is just another name for disorder; for a chemical change, we can often decide whether the entropy increases by deciding whether the disorder increases in the course of the reaction. Thus, entropy usually increases if a gas (a disorderly state of matter) is formed in a reaction and decreases if a gas is converted to a solid or a liquid (both of which are relatively orderly states—solids more so than liquids). When chlorine reacts with PCl_3 in the reaction

$$PCl_3(l) + Cl_2(g) \longrightarrow PCl_5(s)$$

a liquid and a gas are converted to a solid, so we can anticipate that the entropy of the reaction system decreases. (Do not be puzzled by the fact that this reaction proceeds in the direction of *decreasing* disorder—and hence of decreasing entropy—of the system: we shall see that the disorder of the system is only one contribution to the total disorder.) In the photosynthesis reaction

$$6CO_2(g) + 6H_2O(l) \longrightarrow C_6H_{12}O_6(s) + 6O_2(g)$$

there is a net reduction of disorder within the reaction system: simple reactant molecules come together to form highly organized glucose molecules. Hence, we can expect the entropy of the reaction system to decrease when this reaction occurs. Conversely, the entropy of the system increases when glucose burns in oxygen:

$$C_6H_{12}O_6(s) + 6O_2(g) \longrightarrow 6CO_2(g) + 6H_2O(l)$$

In this reverse reaction, the order of the large glucose molecules is lost as they break down into many small triatomic molecules.

A formula for calculating the entropy S of a substance was proposed by the Austrian Ludwig Boltzmann (Fig. 16.4):

$$S = k \ln W \tag{1}$$

The factor k, which is called *Boltzmann's constant*, is equal to R/N_A, where R is the gas constant and N_A is Avogadro's constant. Its value, $k = 1.38 \times 10^{-23}$ J/K, shows that the units of entropy are joules per kelvin (J/K). The logarithm (ln) is the natural logarithm. The disorder of the system is expressed by W, and is defined as the number of different ways in which the atoms in the sample can be arranged and yet have the same total energy. Before going any further, we must see what this means in practice.

The entropies of simple solids. Suppose we wanted to know the entropy of a solid made up of 20 heteronuclear diatomic molecules. (A real example might be a block of solid carbon monoxide or hydrogen chloride, but the number of molecules would then be closer to 10^{23} than to 20.) Suppose that the 20 molecules form a perfectly ordered crystal in which each molecule can lie in only one orientation (Fig. 16.5), and that the temperature is zero ($T = 0$, absolute zero) so that all motion has been quenched. We expect such a sample to have zero entropy because there is no disorder. This is confirmed by the Boltzmann formula: since there is only one way of arranging the molecules in the perfect crystal, $W = 1$ and (because $\ln 1 = 0$)

$$S = k \ln 1 = 0$$

Now suppose that each molecule can point in either of two directions in the solid (Fig. 16.6). Since each of the 20 molecules can have two orientations, the total number of ways of arranging them is

$$W = \underbrace{2 \times 2 \times \cdots \times 2}_{20 \text{ factors}} = 2^{20}$$

The entropy of this disorderly solid is therefore

$$S = k \ln 2^{20} = k \times (20 \ln 2)$$
$$= 1.9 \times 10^{-22} \text{ J/K}$$

(We have used $\ln x^a = a \ln x$, Appendix 1A.) This entropy is obviously higher than that of the perfectly ordered solid. If this solid contained 1 mol of CO molecules (that is, 6.02×10^{23} of them), its entropy would be

$$S = k \times (6.02 \times 10^{23} \ln 2) = 5.76 \text{ J/K}$$

When the entropy of 1 mol of CO is actually measured* near $T = 0$, the value obtained is 4.6 J/K. This is close enough to 5.8 J/K to suggest that in the crystal the molecules are indeed arranged nearly randomly, in one direction or the other, as in Fig. 16.6. The physical reason for this randomness is that the electric dipole moment of a CO molecule is very small, so there is little energy advantage in the molecules lying head to tail (as in Fig. 16.5); the molecules lie in either direction at random. For solid HCl, the same measurement gives $S = 0$, showing that at $T = 0$ the molecules are arranged in an orderly way; under the influence of their bigger electric dipole moment, the HCl molecules lie strictly head to tail at $T = 0$, as in Fig. 16.5.

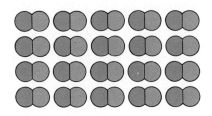

FIGURE 16.5 A sample of 20 heteronuclear diatomic molecules in a perfectly ordered arrangement at $T = 0$ has zero disorder and zero entropy ($S = 0$).

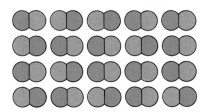

FIGURE 16.6 When each of the 20 molecules can take up either of two orientations without affecting the energy, there are $2 \times 2 \times \cdots = 2^{20}$ different possible arrangements. This illustration shows just one of them. The entropy of this sample is higher than that of the sample in Fig. 16.5.

▼ **EXAMPLE 16.1** Using the Boltzmann formula for entropy

The entropy of 1 mol of solid $FClO_3$ at absolute zero is 10.1 J/K. Suggest an interpretation.

STRATEGY The existence of nonzero entropy at $T = 0$ suggests that the molecules are not perfectly ordered. From the shape of the molecule (which can be obtained using VSEPR theory), we need to determine how many orientations W it is likely to be able to adopt in a crystal; then we can use the Boltzmann formula to see if that leads to the observed value of S.

*We shall not discuss how entropies are measured experimentally, other than to note that it involves the determination of heat capacities C down to very low temperatures and measurement of the area under the curve of a graph of C/T against T.

SOLUTION $FClO_3$ is a tetrahedral molecule, so we expect it to be able to take up any of four orientations in a crystal (Fig. 16.7). The total number of ways of arranging N of these molecules in a crystal is therefore

$$W = \underbrace{4 \times 4 \times \cdots \times 4}_{N \text{ factors}} = 4^N$$

The entropy is then

$$S = k \ln 4^N = k \times (N \times \ln 4)$$

Since in 1 mol of $FClO_3$ there are 6.02×10^{23} molecules,

$$S = (1.38 \times 10^{-23} \text{ J/K}) \times (6.02 \times 10^{23} \times \ln 4)$$

$$= 11.5 \text{ J/K}$$

This is reasonably close to the experimental value, which suggests that at absolute zero the molecules are oriented almost randomly.

EXERCISE Explain the observation that at absolute zero the entropy of 1 mol of solid N_2O is 6 J/K.

▲ [*Answer*: In the crystal, the orientations NNO and ONN are equally likely.]

At absolute zero in a perfect crystal, as in Fig. 16.5, there is perfect order and the entropy is zero. As the crystal is heated, the molecules start to move, and they become more disorderly as their thermal motion increases. The value of W, and hence of $\ln W$, increases, because although there is only one way of being orderly, there are many ways of being disorderly (think of the number of disorderly arrangements that are possible for a deck of cards). It follows that the entropy of a substance increases as it is heated and the sample becomes disorderly. Hence, we can expect the entropy of any substance at room temperature to be greater than zero, and to be large if the substance has a very disorderly arrangement of molecules and a lot of thermal motion.

Standard molar entropies. The entropies of many substances at different temperatures have been obtained either by calculation using Boltzmann's formula or by measurement of their heat capacities down to very low temperatures. Some "standard molar entropies" are given in Table 16.1, and a longer list can be found in Appendix 2A:

The **standard molar entropy** $S°$ of a substance is the entropy per mole of the pure substance at 1 atm pressure.

The unit of molar entropy is the joule per kelvin per mole, J/(K · mol). Table 16.1 gives standard molar entropies at 25°C; values for water at

FIGURE 16.7 Each $FClO_3$ molecule can take up one of four orientations at random at each site in the solid, so the entropy of solid $FClO_3$ is nonzero at $T = 0$.

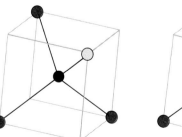 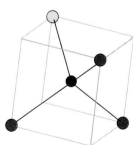 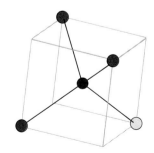 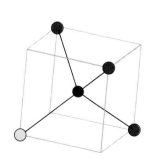

TABLE 16.1 Standard molar entropies of some substances at 25°C

Substance	$S°$, J/(K · mol)	Substance	$S°$, J/(K · mol)	Substance	$S°$, J/(K · mol)
Gases		*Liquids*		Graphite, C	5.7
Ammonia, NH_3	192.5	Benzene, C_6H_6	173.3	Lead, Pb	64.8
Carbon dioxide, CO_2	213.7	Ethanol, CH_3CH_2OH	160.7	Magnesium carbonate, $MgCO_3$	65.7
Helium, He	126.2	Water, H_2O	69.9	Magnesium oxide, MgO	26.9
Hydrogen, H_2	130.7	*Solids*		Sodium chloride, NaCl	72.1
Neon, Ne	146.3	Calcium carbonate, $CaCO_3$	92.9	Sucrose, $C_{12}H_{22}O_{11}$	360.2
Nitrogen, N_2	191.6	Calcium oxide, CaO	39.8	Tin	
Oxygen, O_2	205.1	Copper, Cu	33.2	Sn (white)	51.6
		Diamond, C	2.4	Sn (gray)	44.1

some other temperatures are shown in Table 16.2 for comparison. Note that all values are positive: a perfectly orderly state (at $T = 0$) has zero entropy, and all substances are more disordered at 25°C than at absolute zero.

The entropies of pure substances. The differences in the standard molar entropies of the substances in Tables 16.1 and 16.2 can be explained in terms of disorder. For example, the molar entropy of diamond, 2.4 J/(K · mol), is much lower than that of lead, 64.8 J/(K · mol). This is consistent with diamond having a much more rigid and orderly structure than lead. Also, the entropy of water increases as it is heated; this reflects the increasing disorder of the liquid as it gets hotter and its molecules undergo more vigorous thermal motion. The entropy of any substance increases as the temperature is raised.

The large entropy increase at the boiling point of water, from 87 J/(K · mol) for the liquid to 197 J/(K · mol) for the vapor, shows the large increase in disorder when a liquid changes to a (much more chaotic)

TABLE 16.2 The standard molar entropy of water at various temperatures

Phase	Temperature, °C	$S°$, J/(K · mol)
Solid	0	43.2
Liquid	0	65.2
	20	69.6
	50	75.3
	100	86.8
Vapor	100	196.9
	200	204.1

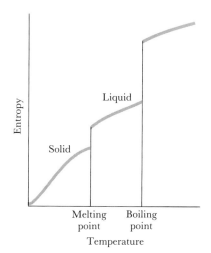

FIGURE 16.8 The entropy of a solid increases as the temperature is raised. It increases sharply when the solid melts to the more disordered liquid, and then increases steadily up to the boiling point. A second, even larger increase in entropy occurs when the liquid boils.

FIGURE 16.9 Gray tin and white tin are two allotropes of tin. The denser white metallic form is the more stable above 13°C. Here the powdery gray allotrope is forming on regions of the metallic white allotrope.

gas. A similar, but usually smaller, increase occurs when solids melt, because a liquid is more disordered than a solid (Fig. 16.8).

▼ EXAMPLE 16.2 Calculating the entropy of a physical change

Calculate the change in the entropy of 100 g of water when it freezes at 0°C in a refrigerator ice tray.

STRATEGY Since ice is a more orderly substance than liquid water, in which the molecules are free to move, we expect the ice to have the lower entropy; hence the change should be negative. We know the molar entropies of water and ice; the change in molar entropy is the difference between them. The entropy change for 100 g of water is the product of the change in molar entropy and the number of moles of H_2O molecules in 100 g of water, which we obtain from the molar mass of H_2O (18.02 g/mol).

SOLUTION The change in the standard molar entropy at 0°C is

$$\Delta S° = S°(ice) - S°(water)$$
$$= 43.2 \text{ J/(K} \cdot \text{mol)} - 65.2 \text{ J/(K} \cdot \text{mol)} = -22.0 \text{ J/(K} \cdot \text{mol)}$$

Since

Entropy change = moles of H_2O × entropy change per mole of H_2O

and

$$\text{Moles of } H_2O = 100 \text{ g } H_2O \times \frac{1 \text{ mol } H_2O}{18.02 \text{ g } H_2O} = 5.55 \text{ mol } H_2O$$

we find that

$$\text{Entropy change} = 5.55 \text{ mol } H_2O \times \frac{-22.0 \text{ J}}{1 \text{ K} \cdot \text{mol } H_2O} = -122 \text{ J/K}$$

The minus sign indicates that freezing decreases the entropy and that ice is less disordered than water at 0°C.

EXERCISE Use the information in Table 16.1 to calculate the change in molar entropy when 100 g of white tin (Fig. 16.9) changes to gray tin at 13.0° C. Which is the more ordered form?

[*Answer*: −6.3 J/K; gray]

▲

Standard reaction entropies. Example 16.2 shows that the entropies in Table 16.1 can be used to calculate the entropy change that occurs during a physical change. They can also be used to obtain the entropy change in a chemical reaction. This entropy change is called the "standard reaction entropy" $\Delta S°$ and is defined analogously to the standard reaction enthalpy (Section 6.3):

> The **standard reaction entropy** for any reaction is the difference between the entropy of the reactants in their standard states and the entropy of the products in their standard states:
>
> $$\Delta S° = S°(\text{products}) - S°(\text{reactants})$$

In this definition, $S°$(reactants) is the total standard entropy of the reactants, and $S°$(products) is that of the products.

▼ EXAMPLE 16.3   Calculating the standard reaction entropy

Calculate the standard reaction entropy for the synthesis of 2 mol of $NH_3(g)$ at 25°C in the reaction

$$N_2(g) + 3H_2(g) \longrightarrow 2NH_3(g)$$

STRATEGY We expect a decrease in entropy, because 4 mol of reactant gas molecules at 1 atm pressure occupy a larger volume than 2 mol of product gas molecules at the same pressure. To find the numerical value, we use the chemical equation to write an equation for $\Delta S°$, and substitute values from Table 16.1.

SOLUTION The reaction is

$$N_2(g) + 3H_2(g) \longrightarrow 2NH_3(g)$$

The standard reaction entropy is therefore

$$\Delta S° = 2 \text{ mol } NH_3 \times S°(NH_3, g)$$
$$- [1 \text{ mol } N_2 \times S°(N_2, g) + 3 \text{ mol } H_2 \times S°(H_2, g)]$$

$$= 2 \times 192.5 \text{ J/K} - (191.6 + 3 \times 130.7 \text{ J/K}) = -198.7 \text{ J/K}$$

As expected, the product is less disordered than the reactants.

EXERCISE Calculate the standard reaction entropy of the reaction $N_2O_4(g) \to 2NO_2(g)$ at 25°C.

[*Answer*: +175.8 J/K]

▲

16.2 THE ENTROPY CHANGE IN THE SURROUNDINGS

The entropy of water at 0°C is 22.0 J/(K · mol) higher than the entropy of ice at 0°C, which shows that liquid water is more disordered than ice. The freezing of water therefore corresponds to water becoming *less* disordered. Since we have asserted that a change to a less disordered state never occurs spontaneously, there must be an additional factor to account for the fact that water does freeze spontaneously below 0°C. Likewise, we know that nitrogen and hydrogen combine to form ammonia, yet we have just shown that NH_3 is *less* disordered than the reactants. Again, there must be another factor that explains why this reaction does occur spontaneously.

The additional factor is the change in the disorder of the surroundings. We have shown that the entropy of the *system*—the substance undergoing the physical change, or the reactants and products in a reaction—decreases when water freezes or ammonia forms. However, since freezing and the synthesis of ammonia are exothermic processes, heat passes into the surroundings when they take place. The heat released in each case stimulates chaotic thermal motion of the atoms in the surroundings and therefore increases their disorder (Fig. 16.10). If the increase in the disorder of the surroundings is greater than the decrease in the disorder of the system, there is an *overall* increase in the total disorder of the universe, and the change is spontaneous.

Calculating the entropy change from the enthalpy change. At first sight, calculating the entropy of the surroundings with Boltzmann's formula might look like an impossible task. How is it possible to calculate the number of ways of arranging all the atoms in the water bath, the labo-

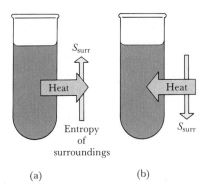

FIGURE 16.10 (a) In an exothermic process, heat escapes into the surroundings and increases their entropy. (b) In an endothermic process, the entropy of the surroundings decreases.

ratory, the country, and then the planet, all of which make up the surroundings? Fortunately, that calculation can be sidestepped by focusing on the entropy *changes* that occur when heat flows into the surroundings. When we deal only with changes, it turns out that we can use another formula for calculating the change in the entropy of the surroundings. Although the Boltzmann formula is still in the background (in the sense that the new formula can be derived from it), we do not need to use it directly. In what follows, we use the result obtained in Section 6.2 that a change in the enthalpy of a system is equal to the heat transferred to or from the system at constant pressure.

To derive a formula for ΔS_{surr}, the change in the entropy of the surroundings, we suppose that the process occurring in the system is accompanied by a change in enthalpy ΔH. This might be the exothermic enthalpy change of -6.00 kJ that occurs when 1 mol of H_2O freezes, or the -1202-kJ change when 2 mol of Mg forms 2 mol of MgO. When water freezes at constant pressure in an open container, 6.00 kJ of heat (in general, $-\Delta H$ if the enthalpy change is ΔH) flows into the surroundings and stirs up thermal motion there. We assume that the amount of disorder stirred up is proportional to the heat transferred:

$$\Delta S_{surr} \propto -\Delta H$$

Note that if the process is exothermic (if ΔH is negative because heat is released), then the entropy of the surroundings increases (ΔS_{surr} is positive). If the process is endothermic (ΔH is positive), then heat leaves the surroundings, reducing their disorder, and hence their entropy decreases (ΔS_{surr} is negative).

The change in entropy caused by a given amount of heat depends on the temperature. If the surroundings are hot, then they are already very chaotic, and a small inflow of heat from the system has relatively little impact (Fig. 16.11). If, on the other hand, the surroundings are cool, then they are relatively ordered and the same amount of heat can cause a great deal of disorder. The difference is like that between sneezing in a crowded street, which may pass unnoticed, and sneezing in a quiet library, which may not. This suggests that the entropy change caused by the transfer of a given amount of heat at constant pressure is inversely proportional to the temperature at which the transfer takes place, as shown in the equation at the top of the next page:

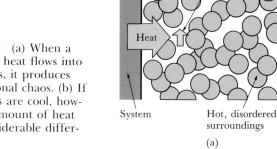

Small increase in entropy

Large increase in entropy

Heat

Heat

FIGURE 16.11 (a) When a given amount of heat flows into hot surroundings, it produces very little additional chaos. (b) If the surroundings are cool, however, the same amount of heat can make a considerable difference.

System

Hot, disordered surroundings

Cool, ordered surroundings

(a)

(b)

$$\Delta S_{surr} = \frac{-\Delta H}{T} \qquad (2)$$

This formula, which can be derived from the laws of thermodynamics, shows that the entropy of the surroundings increases if heat escapes into them, and that the increase is large when the temperature is low.

For the synthesis of 2 mol of NH_3 at 25°C, for which $\Delta H° = -92.2$ kJ, we find from Eq. 2 that

$$\Delta S_{surr} = \frac{-(-92.2 \times 10^3 \text{ J})}{298 \text{ K}} = +309 \text{ J/K}$$

This increase in the entropy of the surroundings is bigger than the decrease in the entropy of the system (in Example 16.3, we calculated this to be -199 J/K), so overall there is an increase in disorder when ammonia forms. Hence, the synthesis is spontaneous at 25°C.

The difference between the two entropy changes is much more pronounced for the combustion of magnesium in the reaction

$$2Mg(s) + O_2(g) \longrightarrow 2MgO(s) \qquad \Delta H° = -1202 \text{ kJ} \qquad \Delta S° = -217 \text{ J/K}$$

In this case, the change in the entropy of the surroundings at 25°C is $+4.03 \times 10^3$ J/K. This large increase overwhelms the relatively small decrease in the entropy of the system. Hence, as we know, the combustion of magnesium is spontaneous (once it is initiated).

Systems at equilibrium. Now consider the same calculation for the freezing of water at 0°C. We already know that the entropy of 1 mol of H_2O decreases by 22.0 J/K. Freezing is exothermic, and $\Delta H° = -6.00$ kJ; hence

$$\Delta S_{surr} = \frac{-(-6.00 \times 10^3 \text{ J})}{273 \text{ K}} = +22.0 \text{ J/K}$$

As if by magic, the increase in the entropy of the surroundings exactly cancels the decrease in the entropy of the system: overall, the entropy change is zero. Since spontaneous changes occur only when they are accompanied by an *increase* in entropy, we have to conclude that water does not freeze spontaneously at 0°C!

That is in fact the case. Water does *not* freeze spontaneously at 0°C: water and ice are *in equilibrium* at that temperature (Fig. 16.12b). Our

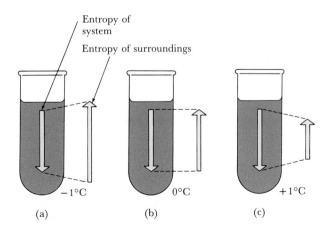

Entropy of system

Entropy of surroundings

(a)   −1°C

(b)   0°C

(c)   +1°C

FIGURE 16.12 All three diagrams show the entropy changes (the yellow arrows) when water freezes to ice. (a) At −1°C the increase in entropy of the surroundings is greater than the decrease in entropy of the system, and freezing is spontaneous. (b) The entropy changes are equal at 0°C. (c) The net change of entropy is negative at +1°C, and the reverse process, melting, is spontaneous.

calculation shows that, because the total entropy of the universe neither increases nor decreases when water becomes ice at 0°C, at that temperature water has no spontaneous tendency to freeze and ice has no spontaneous tendency to melt. In kinetic terms, at 0°C the rates at which ice melts and water freezes are exactly equal.

Suppose we raise the temperature to +1°C and calculate the change in entropy that accompanies the conversion of water to ice at that temperature. This hardly affects the change in the entropy of the system, which remains at −22.0 J/K for freezing. However, the accompanying change in the entropy of the surroundings is now

$$\Delta S_{surr} = \frac{-(-6.00 \times 10^3 \text{ J})}{274 \text{ K}} = +21.9 \text{ J/K}$$

The total entropy change ΔS_{total} is therefore

$$\Delta S_{total} = -22.0 \text{ J/K} + 21.9 \text{ J/K} = -0.1 \text{ J/K}$$

That is, the overall entropy decreases, showing that freezing is not a spontaneous process at +1°C. Its reverse, melting, is spontaneous, for it corresponds to an entropy change of +0.1 J/K.

Now suppose we lower the temperature to −1°C. We know from experience that freezing is now spontaneous, but it is interesting to see whether the entropy change shows that. By repeating the calculations at 272 K, we find that ΔS_{surr} is now +22.1 J/K, so the overall entropy change for freezing is +0.1 J/K, an increase. Hence freezing at −1°C corresponds to an overall increase in disorder and is indeed a spontaneous process.

In summary, we can now identify the freezing point or boiling point of a substance as the temperature at which the total entropy change accompanying the phase transition is zero. At higher temperatures, melting and vaporization lead to increased entropy and hence are spontaneous. At lower temperatures, melting and vaporization lead to lower total entropy, so the reverse processes—freezing and condensation—are spontaneous.

▼ **EXAMPLE 16.4** **Predicting the boiling point of a substance**

Predict the normal boiling point of liquid sodium, given that the entropy of 1 mol of Na(l) changes by +84.8 J/K when it forms a vapor at 1 atm pressure and that its enthalpy of vaporization is +98.0 kJ.

STRATEGY We know that at the normal boiling point the liquid and its vapor are in equilibrium at 1 atm pressure. That means that there is no change in total entropy when one phase changes to the other. We are given the entropy change for the system; we can find the entropy change for the surroundings from the enthalpy of vaporization and the temperature. Our task is therefore to find the temperature at which ΔS_{total}, the sum of the two entropy changes, is zero.

SOLUTION The overall entropy change is

$$\Delta S_{total} = +84.8 \text{ J/K} - \frac{98.0 \times 10^3 \text{ J}}{T}$$

This overall change is zero when $T = T_b$; that is, at the normal boiling point. Hence, we have

$$84.8 \text{ J/K} - \frac{98.0 \times 10^3 \text{ J}}{T_b} = 0$$

from which we find that

$$T_b = \frac{98.0 \times 10^3 \text{ K}}{84.8} = 1160 \text{ K} \quad \text{or} \quad 890°C$$

EXERCISE Calculate the melting point of chlorine, given that its enthalpy of melting is +6.41 kJ/mol and its entropy of melting is +37.3 J/(K · mol).
[*Answer*: 172 K]

▲

We can now clear up a point that was left unexplained in Section 10.4. We saw there that Trouton's rule is the observation that $\Delta H^\circ_{\text{vap}}/T_b$ is often quite close to 85 J/(K · mol) for liquids in which hydrogen bonding is unimportant. Since at the boiling point the total entropy change is zero, we may write (as we did in Example 16.4)

$$\Delta S + \Delta S_{\text{surr}} = \Delta S - \frac{\Delta H^\circ_{\text{vap}}}{T_b} = 0$$

This may be rearranged to yield

$$\Delta S = \frac{\Delta H^\circ_{\text{vap}}}{T_b}$$

Here the right-hand side is Trouton's constant, and the equation shows that that constant is, in fact, the entropy of vaporization of the liquid at its boiling point. Trouton's rule is explained if the entropy of vaporization at T_b is approximately the same for all liquids. That, in turn, is a reasonable assumption, since all liquids undergo approximately the same change in disorder when they vaporize. However, water and other hydrogen-bonded liquids are less disordered than most liquids, because of the hydrogen bonding between molecules. They therefore undergo a greater increase in disorder when they vaporize and hence have a higher entropy of vaporization (Fig. 16.13). The value for water, for example, is +109 J/(K · mol).

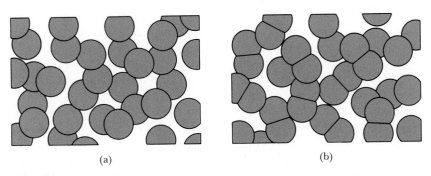

(a)   (b)

FIGURE 16.13 These diagrams represent the disorder in a liquid (a) without hydrogen bonds and (b) with hydrogen bonds that result in the pairing of molecules. There is more disorder in the former, so they undergo a smaller change in entropy on vaporization.

16.3 THE SECOND LAW

The calculations of the previous section show that when we want to express the direction of spontaneous change in terms of the entropy, we must consider the *total* entropy of the universe: the sum of the entropy changes in both the system and the surroundings. This is expressed by the "second law" of thermodynamics:

> **Second law of thermodynamics:** A spontaneous change is accompanied by an increase in the *total* entropy of the universe.

On one level this is just a restatement of the idea that spontaneous changes lead to an increase in overall disorder. However, on another level it goes much further, for soon we shall see that we can make numerical predictions: using the information in Table 16.1, we can calculate the entropy change in the system, and using Eq. 2, we can calculate it for the surroundings.

Exothermic reactions. The second law explains why exothermic reactions may occur spontaneously. These are reactions in which the heat released increases the disorder of the surroundings. Even though the disorder of the system may decrease (as in the combustion of magnesium, in which a gas is converted to a solid), the increased disorder of the surroundings is so great that overall the total entropy increases. In many exothermic reactions the disorder of the system also increases; this is so for the formation of hydrogen fluoride:

$$H_2(g) + F_2(g) \longrightarrow 2HF(g) \qquad \Delta S° = +14.1 \text{ J/K} \qquad \Delta H° = -546.4 \text{ kJ}$$

This reaction increases the disorder of the universe through both the change in the system and the change in its surroundings. As long as ΔH is reasonably large, a reaction with ΔS either positive *or* negative may be spontaneous (Fig. 16.14). In fact, exothermic reactions are common because entropy changes within the system are usually quite small compared with the accompanying increase in the entropy of the surroundings.

Endothermic reactions. The second law also shows why endothermic reactions can occur. These were a puzzle for the chemists who once believed that reactions ran only in the direction of decreasing enthalpy, for an endothermic reaction is one in which the products have a *higher* enthalpy than the reactants. One very striking example of an endothermic reaction, the reaction between barium hydroxide and ammonium thiocyanate, has already been mentioned (in Chapter 6), but there are plenty of others, including the decomposition of calcium carbonate at elevated temperatures.

As in the case of an exothermic reaction, the driving force of an endothermic reaction is the *total* entropy change that accompanies the reaction. Since heat flows in from the surroundings, their entropy decreases. However, there may still be an overall increase in entropy if the disorder within the system increases sufficiently. This is exactly the case for the reaction between barium hydroxide and ammonium thiocyanate. In that reaction, the two solids react to form a solution and a gas.

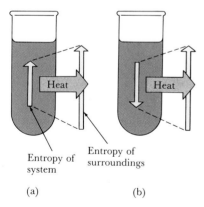

Entropy of system

Entropy of surroundings

(a)   (b)

FIGURE 16.14 In an exothermic reaction, (a) the overall entropy change is certainly positive when the entropy of the system increases; (b) the overall change may also be positive when the entropy of the system decreases.

As a result, there is a large increase in the entropy of the system, and hence an overall increase in entropy. This is also the case for the decomposition of calcium carbonate, which produces a gas.

Every endothermic reaction must be accompanied by increased disorder within the system (Fig. 16.15), but this is not so for exothermic reactions (Fig. 16.14). Hence, endothermic reactions are less common than exothermic reactions.

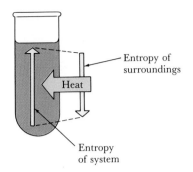

FIGURE 16.15 An endothermic reaction can occur only if the entropy of the system increases enough to overcome the decrease in the entropy of the surroundings.

FREE ENERGY

At this point, it seems that to predict the direction of a spontaneous change we have to carry out two separate entropy-change calculations—one for the system and one for the surroundings. However, there is a convenient way of combining the two calculations so that a *single* piece of information is enough.

16.4 FOCUSING ON THE SYSTEM

The total entropy change, the sum of the changes in the system and its surroundings, is

$$\Delta S_{\text{total}} = \Delta S + \Delta S_{\text{surr}}$$

ΔS is the change in the system, which we can calculate from the information in Table 16.1. If the pressure remains constant, the change in the entropy of the surroundings is given by Eq. 2, $\Delta S_{\text{surr}} = -\Delta H/T$. Therefore,

$$\Delta S_{\text{total}} = \Delta S - \frac{\Delta H}{T}$$

This equation tells us that we can calculate the total entropy change from information about the system alone (its temperature and the entropy and enthalpy changes it undergoes). That does not mean we are ignoring the surroundings. Their entropy change is simply expressed in terms of ΔH for the system.

It is common practice to rearrange the last equation into the form

$$-T\,\Delta S_{\text{total}} = \Delta H - T\,\Delta S$$

and to introduce a property called the *Gibbs free energy* G:

$$G = H - TS \qquad (3)$$

Josiah Gibbs (Fig. 16.16), for whom this state property is named, was a professor at Yale from 1869 until 1903 and was responsible for transforming thermodynamics from an abstract theory into a subject of great usefulness. Some insight into the origin of the name "*free* energy" can be obtained by thinking of G as the total enthalpy H of the system minus a quantity $T \times S$ that represents the part of the system's energy that is already disordered. Only the orderly part of the enthalpy, the

FIGURE 16.16 Josiah Willard Gibbs (1839–1903).

difference $H - TS$, is free to become more disordered and hence to cause change.

When a process takes place at constant temperature, the resulting changes in enthalpy and entropy produce a change in free energy of

$$\Delta G = \Delta H - T \, \Delta S \qquad (4)$$

By comparing this with the expression for ΔS_{total}, we see that

$$-T \, \Delta S_{\text{total}} = \Delta G \qquad (5)$$

To this point we have done nothing except reorganize the expression for the total entropy change. However, there are several important consequences.

Free energy and the direction of spontaneous change. The minus sign in Eq. 5 means that an increase in total entropy corresponds to a *decrease* in free energy. Therefore, in terms of G, the direction of spontaneous change (at constant pressure and temperature) is the direction of decreasing free energy.

If we plot the free energy of a reaction system against its changing composition as the reaction proceeds, we get a U-shaped curve (Fig. 16.17) with a minimum at some composition. The reaction will tend to proceed spontaneously toward the composition at the lowest point of the curve because that is the direction of decreasing free energy. The composition at the lowest point of the curve—the point of minimum free energy—thus corresponds to equilibrium. For a system at equilibrium, neither the forward change nor the reverse change is spontaneous, since both lead to an increase in free energy (a decrease in total entropy). Hence, if a reaction goes almost to completion, its free-energy minimum occurs very close to the composition corresponding to pure products. If a reaction does not go forward at all (in which case the reverse reaction goes to completion), its free-energy minimum occurs close to pure reactants.

Equation 4 shows very clearly the factors that govern the direction of spontaneous change. They are summarized in Table 16.3, which we can examine for factors that make ΔG negative. One condition that tends to do this is large negative ΔH, as for a strongly exothermic process such as combustion. Since large negative ΔH corresponds to a large increase in the entropy of the surroundings, once combustions

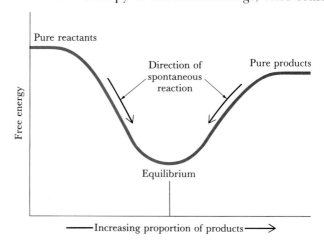

FIGURE 16.17 At constant temperature and pressure, the direction of spontaneous change is toward lower free energy. The equilibrium composition of a reaction mixture corresponds to the lowest point on the curve.

TABLE 16.3   Conditions that favor reaction

Enthalpy change	Entropy change	Spontaneous reaction?
Exothermic ($\Delta H < 0$)	Increase ($\Delta S > 0$)	Yes, $\Delta G < 0$
Exothermic ($\Delta H < 0$)	Decrease ($\Delta S < 0$)	If $\lvert T\,\Delta S\rvert < \lvert \Delta H\rvert$*
Endothermic ($\Delta H > 0$)	Increase ($\Delta S > 0$)	If $T\,\Delta S > \Delta H$
Endothermic ($\Delta H > 0$)	Decrease ($\Delta S < 0$)	No, $\Delta G > 0$

*$\lvert x\rvert$ means the absolute value of x (its value, disregarding its sign).

and other exothermic processes have been initiated, they are driven by the disorder they create in the surroundings.

However, ΔG can be negative even if ΔH is positive, provided that $T\,\Delta S$ is large and positive. This depends on two factors. One is the size and sign of ΔS, the change in the entropy of the system; the other is the temperature T. If disorder is created in the system (as in vaporization or the decomposition of calcium carbonate), then ΔS is positive and the $-T\,\Delta S$ term in Eq. 4 contributes a negative term to ΔG. However, the size of its contribution depends on the temperature, and even a large increase in disorder within the system can have a negligible influence if the temperature is low. The role of the entropy of the system becomes more important the higher the temperature. In other words, for reactions in which the entropy change is large (in either direction), the temperature may play an important role in determining whether or not the reaction is spontaneous.

▼ **EXAMPLE 16.5   Calculating the minimum decomposition temperature**

Assuming that the reaction enthalpy and reaction entropy are independent of temperature, estimate the temperature at which calcium carbonate is likely to decompose.

STRATEGY   Decomposition is likely to occur when the change in free energy for the reaction

$$CaCO_3(s) \longrightarrow CaO(s) + CO_2(g)$$

is negative. We can calculate the reaction enthalpy from the enthalpies of formation listed in Appendix 2A. We can calculate the entropy change from the entropies listed there. Finally, we can look for the temperature that makes $\Delta H° - T\,\Delta S°$ negative. The minimum temperature for decomposition is therefore $T = \Delta H°/\Delta S°$, for at all higher temperatures $T\,\Delta S°$ is larger than $\Delta H°$, and $\Delta H° - T\,\Delta S°$ is negative.

SOLUTION   From the enthalpies of formation in Appendix 2A, $\Delta H° = +178.3$ kJ. From Table 16.1, $\Delta S° = 39.8 + 213.7 - 92.9 = +160.6$ J/K. The minimum temperature at which decomposition can occur is therefore

$$T = \frac{\Delta H°}{\Delta S°} = \frac{178.3 \times 10^3\ \text{J}}{160.6\ \text{J/K}} = 1110\ \text{K}$$

That is, we expect the solid to decompose at temperatures higher than 1110 K, or higher than about 800°C.

EXERCISE Estimate the temperature at which magnesium carbonate can be expected to decompose when heated.

[*Answer*: 575 K]

▲

If a system happens to be at equilibrium, then there is no tendency for spontaneous change in either direction: $\Delta G = 0$ for such a system at constant temperature and pressure. This is the case for a liquid in equilibrium with its vapor, a solid in equilibrium with its liquid, or a solute in equilibrium with its saturated solution. It is also the case for a chemical reaction at equilibrium.

Standard free energies of formation and reaction. When nitric oxide reacts with oxygen to form nitrogen dioxide in the reaction

$$2NO(g) + O_2(g) \longrightarrow 2NO_2(g)$$

the entropy, enthalpy, and free energy of the system change. At 25°C, if 2 mol of pure NO at 1 atm pressure reacts with 1 mol of pure O_2 at 1 atm and produces 2 mol of pure NO_2 at the same pressure, the entropy changes by -147 J/K (-0.147 kJ/K) and the enthalpy changes by -114 kJ. The free energy therefore changes by

$$\Delta G° = \Delta H° - T \Delta S°$$

$$= -114 \text{ kJ} - 298 \text{ K} \times (-0.147 \text{ kJ/K}) = -70 \text{ kJ}$$

$\Delta G°$ is called the "standard reaction free energy" and is defined like the analogous standard reaction enthalpy and entropy:

The **standard reaction free energy** $\Delta G°$ for a reaction is the difference between the free energy of the products and that of the reactants in their standard states:

$$\Delta G° = G°(\text{products}) - G°(\text{reactants})$$

As before, the standard state of a substance is its pure form at 1 atm pressure.

In Section 6.4 we let the elements define a thermochemical "sea level," and we reported the enthalpy of a compound as the reaction enthalpy for its formation from its elements. The same can be done with free energies. We let the elements define the "sea level" of free energy (Fig. 16.18), and take the free energy of any compound to be the free energy of its formation from its elements—its height above "sea level," the "standard free energy of formation" $\Delta G_f°$.

The **standard free energy of formation** $\Delta G_f°$ of a compound is the standard reaction free energy per mole for its formation from its elements.

For example, the standard free energy of formation of hydrogen iodide gas at 25°C is obtained from the standard reaction free energy for

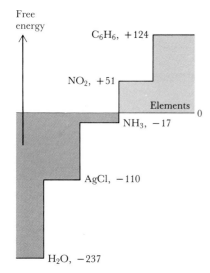

FIGURE 16.18 The standard free energy of formation of a compound is defined as the standard free energy change that accompanies its formation from its elements. Standard free energies of formation represent a "thermodynamic altitude" with respect to the elements at "sea level."

$$H_2(g) + I_2(s) \longrightarrow 2HI(g) \qquad \Delta G° = +3.40 \text{ kJ}$$

in which pure hydrogen at 1 atm pressure reacts with pure solid iodine to give pure hydrogen iodide gas at 1 atm pressure. It is expressed as free energy per mole of HI molecules:

$$\Delta G_f° = \frac{+3.40 \text{ kJ}}{2 \text{ mol HI}} = +1.70 \text{ kJ/mol HI}$$

Like the enthalpies of formation of elements, the standard free energies of formation of elements are zero.

Standard free energies of formation can be measured in a variety of ways, which include combining the enthalpy and entropy data in Tables 6.4 and 16.1. A list of values is given in Table 16.4, and a more extensive one in Appendix 2A.

▼ **EXAMPLE 16.6** **Calculating a standard free energy of formation**

Calculate the standard free energy of formation of HI(g) at 25°C from its standard entropy and its standard enthalpy of formation.

STRATEGY We can calculate $\Delta G°$ from the standard enthalpy and entropy of reaction by combining them according to Eq. 4:

$$\Delta G° = \Delta H° - T \Delta S°$$

The standard reaction enthalpy is found from the enthalpies of formation, using data from Appendix 2A. The standard reaction entropy is found as in Example 16.5, using data from Appendix 2A. The standard free energy of reaction must then be converted to a free energy per mole of HI.

SOLUTION The chemical equation is

$$H_2(g) + I_2(s) \longrightarrow 2HI(g)$$

From Appendix 2A, with $\Delta H_f° = 0$ for each of the elements,

$$\begin{aligned}\Delta H° =\ & 2 \text{ mol HI} \times \Delta H_f°(\text{HI}, g) - [1 \text{ mol } H_2 \times \Delta H_f°(H_2, g) \\ & + 1 \text{ mol } I_2 \times \Delta H_f°(I_2, s)]\end{aligned}$$

$$= 2 \times 26.48 \text{ kJ} - 0 = +52.96 \text{ kJ}$$

From Appendix 2A,

$$\begin{aligned}\Delta S° =\ & 2 \text{ mol HI} \times S°(\text{HI}, g) \\ & - [1 \text{ mol } H_2 \times S°(H_2, g) + 1 \text{ mol } I_2 \times S°(I_2, s)]\end{aligned}$$

$$= 2 \times 206.6 \text{ J/K} - (130.7 + 116.1) \text{ J/K} = +166.4 \text{ J/K}$$

Therefore, since $T = 298$ K,

$$\Delta G° = +52.96 \text{ kJ} - 298 \text{ K} \times 166.4 \text{ J/K} = +3.4 \text{ kJ}$$

The reaction produces 2 mol of HI(g), so

$$\Delta G_f° = \frac{+3.4 \text{ kJ}}{2 \text{ mol HI}} = +1.7 \text{ kJ/mol HI}$$

EXERCISE Calculate the standard free energy of formation of $NH_3(g)$ at 25°C.

[*Answer:* -16.5 kJ/mol NH_3]

TABLE 16.4 **Standard free energies of formation at 25°C***

Substance	$\Delta G_f°$, kJ/mol
Gases	
Ammonia, NH_3	-16.5
Carbon dioxide, CO_2	-394.4
Dinitrogen tetroxide, N_2O_4	$+97.9$
Nitrogen dioxide, NO_2	$+51.3$
Sulfur dioxide, SO_2	-300.2
Water, H_2O	-228.6
Liquids	
Benzene, C_6H_6	$+124.3$
Ethanol, CH_3CH_2OH	-174.8
Water, H_2O	-237.1
Solids	
Calcium carbonate, $CaCO_3$	-1128.8
Iron(III) oxide, Fe_2O_3	-742.2
Silver bromide, AgBr	-96.9
Silver chloride, AgCl	-109.8

*Additional values are given in Appendix 2A.

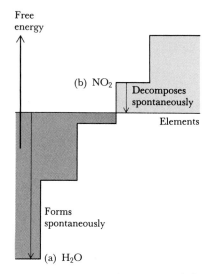

FIGURE 16.19 A compound, in this case H_2O, is thermodynamically stable with respect to its elements if its free energy of formation is negative. A compound is unstable, and tends to decompose into its elements, if its free energy of formation is positive. This is the case with NO_2.

16.5 SPONTANEOUS REACTIONS

The free energies of formation of compounds can be used to determine whether reactions involving those compounds are spontaneous. The simplest kind of reaction for a compound to take part in is decomposition into its elements (the reverse of its synthesis). We begin by considering that kind of reaction, and then go on to more complicated changes.

Thermodynamic stability. The standard free energy of formation of a compound is a measure of its stability as a compound, with respect to decomposition into its elements. If ΔG_f° is negative at a certain temperature, the elements have a spontaneous tendency to form the compound, and the compound is more stable than its elements at that temperature (Fig. 16.19a). If ΔG_f° is positive, the reverse reaction is spontaneous and the compound has a tendency to decompose into its elements (Fig. 16.19b). In the latter case there is no point in trying to synthesize the compound from its elements.

As an example, the standard free energy of formation of benzene is +124 kJ/mol at 25°C, which means that the synthesis

$$6C(s) + 3H_2(g) \longrightarrow C_6H_6(l) \qquad \Delta G^\circ = +124 \text{ kJ}$$

is not spontaneous. There is no point in trying to make benzene by exposing carbon to hydrogen gas at 25°C, even in the presence of a catalyst, as the product has no tendency to form. However, the reverse reaction

$$C_6H_6(l) \longrightarrow 6C(s) + 3H_2(g) \qquad \Delta G^\circ = -124 \text{ kJ}$$

is spontaneous at 25°C. The universe becomes more disorderly if benzene decomposes into carbon and hydrogen at 25°C. We say, therefore, that benzene is "thermodynamically unstable" with respect to its elements.

In general, a *thermodynamically unstable compound* is a compound with a positive standard free energy of formation; such a compound has a thermodynamic tendency to decompose into its elements. However, the *tendency* might not be realized in practice because the kinetics of the decomposition may be very slow. In the language of Section 12.6, the activation energy might be very high. Benzene can, in fact, be kept indefinitely without decomposing at all.

▼ **EXAMPLE 16.7** **Judging the stability of a compound**

Would it be worth looking for a catalyst to synthesize glucose from its elements at 25°C?

STRATEGY A synthesis reaction has a spontaneous tendency to proceed in the required direction only if the free energy of formation of the compound is negative. (This is true whether or not a catalyst is present, for a catalyst affects only the rate of a reaction, not its direction.) Therefore, we need only inspect a table of data (Table 16.4) to see whether ΔG_f° for glucose is negative.

SOLUTION From Appendix 2A, $\Delta G_f^\circ = -910$ kJ/mol. As this is negative, the synthesis has a spontaneous tendency to proceed in the required direction. Hence, on thermodynamic grounds at least, it would not be foolish to look for a catalyst that converts carbon, hydrogen, and oxygen directly into glucose.

EXERCISE Is it worth looking for a catalyst to prepare methylamine, CH_3NH_2, from its elements at 25°C?

[*Answer*: No]

▲

The spontaneous direction of reactions in general. We use standard free energies of formation like enthalpies of formation to calculate standard reaction free energies:

$$\Delta G^\circ = \Delta G_f^\circ(\text{products}) - \Delta G_f^\circ(\text{reactants})$$

That is, the standard reaction free energy is the difference between the total standard free energy for the formation of the products and that for the formation of the reactants. For the oxidation of ammonia,

$$4NH_3(g) + 5O_2(g) \longrightarrow 4NO(g) + 6H_2O(g)$$

we have

$$\Delta G^\circ = [4 \text{ mol NO} \times \Delta G_f^\circ(\text{NO}, g) + 6 \text{ mol } H_2O \times \Delta G_f^\circ(H_2O, g)]$$
$$-[4 \text{ mol } NH_3 \times \Delta G_f^\circ(NH_3, g) + 5 \text{ mol } O_2 \times \Delta G_f^\circ(O_2, g)]$$

$$= [4 \times 86.55 + 6 \times (-228.57)] \text{ kJ} - [4 \times (-16.45) + 0] \text{ kJ}$$

$$= -959.42 \text{ kJ}$$

The result shows that the oxidation of ammonia has a strong tendency to proceed. In the Ostwald process for producing nitric acid, this tendency is made a reality by passing the gases over a rhodium-platinum catalyst.

The same kind of calculation can be used to decide whether or not a compound has a tendency to decompose into simpler compounds (as distinct from its elements). The decomposition of hydrogen peroxide to water at 25°C is an example:

$$2H_2O_2(l) \longrightarrow 2H_2O(l) + O_2(g) \qquad \Delta G^\circ = -233.6 \text{ kJ at } 25°C$$

The negative ΔG° shows that pure liquid hydrogen peroxide has a tendency to decompose into water and oxygen under standard conditions at 25°C. This decomposition is rapid in the presence of MnO_2 as catalyst.

The calculation can also be used to judge whether or not two substances are likely to react to form a particular compound under standard conditions. It shows, for example, that there is no point in trying to produce liquid hydrogen peroxide from water and oxygen at 1 atm pressure and 25°C, because the reaction free energy is positive:

$$2H_2O(l) + O_2(g) \longrightarrow 2H_2O_2(l) \qquad \Delta G^\circ = +233.6 \text{ kJ at } 25°C$$

The reaction has no tendency to occur under those conditions.

▼ **EXAMPLE 16.8** **Estimating the temperature at which a reaction can occur**

Estimate the temperature at which it is thermodynamically possible for carbon to reduce iron(III) oxide to iron.

STRATEGY This is one reaction that takes place in a blast furnace, which we know (Fig. 3.18) operates in the region of 900°C. The reaction becomes feasible when the reaction free energy becomes negative. For an estimate of the temperature at which this occurs, we must find the temperature at which $\Delta H° - T \Delta S°$ becomes negative. We can calculate the reaction enthalpy and reaction entropy from the data in Appendix 2A, assuming that neither changes appreciably with temperature.

SOLUTION The reaction is

$$2Fe_2O_3(s) + 3C(s) \longrightarrow 4Fe(s) + 3CO_2(g)$$

From Appendix 2A,

$$\Delta H° = [4 \text{ mol Fe} \times \Delta H_f°(Fe, s) + 3 \text{ mol } CO_2 \times \Delta H_f°(CO_2, g)]$$
$$- [2 \text{ mol } Fe_2O_3 \times \Delta H_f°(Fe_2O_3, s) + 3 \text{ mol C} \times \Delta H_f°(C, s)]$$

$$= 3 \times (-395.5 \text{ kJ}) - 2 \times (-824.2 \text{ kJ}) = +461.9 \text{ kJ}$$

$$\Delta S° = [4 \text{ mol Fe} \times S°(Fe, s) + 3 \text{ mol } CO_2 \times S°(CO_2, g)]$$
$$- [2 \text{ mol } Fe_2O_3 \times S°(Fe_2O_3, s) + 3 \text{ mol C} \times S°(C, s)]$$

$$= 4 \times 27.3 \text{ J/K} + 3 \times 213.7 \text{ J/K} - (2 \times 87.4 + 3 \times 5.7) \text{ J/K}$$

$$= +558.4 \text{ J/K}$$

The temperature at which $\Delta G°$ changes from positive to negative is the solution of

$$461.9 \times 10^3 \text{ J} - T \times 558.4 \text{ J/K} = 0$$

which is

$$T = \frac{461.9 \times 10^3 \text{ J}}{558.4 \text{ J/K}} = 827 \text{ K}$$

The minimum temperature at which reduction occurs is therefore about 550°C.

EXERCISE What is the minimum temperature at which magnetite (Fe_3O_4) can be reduced to iron using carbon (to produce CO_2)?

[*Answer*: 930 K]

▲

EQUILIBRIA

Now we move to the climax of the chapter, the calculation of equilibrium constants from tabulated data. The first step is to find an equation relating the free energy to the composition of the system. That relation can be used to find the composition at which the free-energy change is zero, which, as we have seen, is the composition corresponding to equilibrium. In this way, we are able to calculate the equilibrium constant of almost any reaction, even those that have not been studied experimentally.

16.6 FREE ENERGY AND COMPOSITION

The reaction free energy is related to the composition of a reaction mixture by the formula

$$\Delta G = -RT \ln \frac{K}{Q} \qquad (6)$$

where Q is the reaction quotient (which depends on the composition), and K is the equilibrium constant of the reaction. At 25°C we use $RT = 2.48$ kJ. We shall not derive Eq. 6 here (the logarithm can in fact be traced back to the Boltzmann formula for the entropy), but the following remarks may help to show that it is plausible.

We have seen that spontaneous processes are processes for which $\Delta S_{total} > 0$. In Section 13.2, however, we identified spontaneous processes in terms of the relative sizes of the reaction quotient Q and the equilibrium constant K. Our most recent encounter with Q was in Section 15.6, where we saw that if $Q > K_{sp}$ then a solute precipitates, but if $Q < K_{sp}$ then more solid dissolves. This means that we need to reconcile two kinds of statements:

Statement	Equilibrium-constant language	Thermodynamic language
The reaction has a tendency to form products when	$Q < K$	$\Delta S_{total} > 0$
The reaction is at equilibrium when	$Q = K$	$\Delta S_{total} = 0$
The reaction has a tendency to form reactants when	$Q > K$	$\Delta S_{total} < 0$

In this table, K is K_{sp} for dissolving, K_c for a reaction in which we are concerned with concentrations, and K_P for a reaction in which we are concerned with pressures. For vaporization, K_P would be P, the vapor pressure itself. If we are interested in acid-base properties, K is the ionization constant K_a, K_b, or K_w.

One way of bringing the two languages into line is to suppose that

$$\Delta S_{total} \propto \ln \frac{K}{Q}$$

This works because the logarithm of a number greater than 1 (corresponding to $Q < K$) is positive (corresponding to $\Delta S_{total} > 0$); the logarithm of 1 (corresponding to $Q = K$) is zero (corresponding to $\Delta S_{total} = 0$); and the logarithm of a number less than 1 (corresponding to $Q > K$) is negative (corresponding to $\Delta S_{total} < 0$). Because $\Delta G = -T \Delta S_{total}$ (see Eq. 5), this relation may be written as

$$\Delta G \propto -T \ln \frac{K}{Q}$$

Except for the constant R, this is the same as Eq. 6.

16.7 THE EQUILIBRIUM CONSTANT

We are now very close to an equation for K in terms of free energy. Suppose we were considering a gas-phase equilibrium, such as the oxidation of nitric oxide, and were dealing with partial pressures. We would have

$$2NO(g) + O_2(g) \rightleftharpoons 2NO_2(g) \qquad Q_P = \frac{(P_{NO_2})^2}{(P_{NO})^2 P_{O_2}}$$

If each substance were pure and at 1 atm pressure, ΔG would have its *standard* value, $\Delta G°$. Under these conditions, $Q_P = 1$ since all the partial pressures are equal to 1 atm. Therefore, Eq. 6 would become (with $K = K_P$)

$$\Delta G° = -RT \ln K_P$$

This same equation applies for any equilibrium constant—for K_c, K_{sp}, K_a, K_b, or K_w. Therefore, for any kind of equilibrium,

$$\Delta G° = -RT \ln K \qquad (7)$$

Equation 7 is one of the most useful expressions in the whole of thermodynamics. It is the link between the equilibrium constant for *any* change—physical or chemical—and the standard reaction free energy. Since the latter can be found in tables such as those in Appendix 2A, Eq. 7 allows us to calculate K for any reaction for which free-energy data are available.

▼ **EXAMPLE 16.9** **Calculating an equilibrium constant**

Calculate K_P at 25°C for the equilibrium

$$N_2O_4(g) \rightleftharpoons 2NO_2(g) \qquad K_P = \frac{(P_{NO_2})^2}{P_{N_2O_4}}$$

STRATEGY The required value can be found by solving Eq. 7 for K_P. First, however, the standard reaction free energy must be calculated from the standard free energies of formation in Table 16.4. The units of K_P are found by inspecting its formula and noting that each partial pressure contributes an "atm."

SOLUTION From the data in Table 16.4:

$$\Delta G° = 2 \text{ mol } NO_2 \times \Delta G_f°(NO_2, g) - 1 \text{ mol } N_2O_4 \times \Delta G_f°(N_2O_4, g)$$

$$= 2 \times 51.3 \text{ kJ} - 97.9 \text{ kJ} = +4.7 \text{ kJ}$$

After rearranging Eq. 7 and substituting this value and $RT = 2.48$ kJ, we have

$$\ln K_P = \frac{-\Delta G°}{RT} = \frac{-4.7 \text{ kJ}}{2.48 \text{ kJ}} = -1.9$$

The natural antilogarithm of -1.9 is 0.15, so for this reaction $K_P = 0.15$ atm. If necessary, Eq. 4 of Chapter 13 could now be used to find K_c.

EXERCISE Calculate the equilibrium constant K_P for the reaction $N_2(g) + 3H_2(g) \rightleftharpoons 2NH_3(g)$ at 25°C.

[*Answer*: $6.0 \times 10^5/\text{atm}^2$]

▼ EXAMPLE 16.10 Calculating a vapor pressure

Calculate the vapor pressure of water at 25°C.

STRATEGY We can tackle this problem as soon as we realize that the vapor pressure is an equilibrium constant (this was explained in Section 13.3). The steps in the solution are the same as before: we write the equation for the vaporization and the corresponding equilibrium constant and then calculate the change in the standard free energy. The required data are found in Appendix 2A and again, at 25°C, $RT = 2.48$ kJ.

SOLUTION The vaporization equilibrium is

$$H_2O(l) \rightleftharpoons H_2O(g) \qquad K_P = P_{H_2O}$$

The standard free energy for this change is

$$\Delta G° = 1 \text{ mol } H_2O \times \Delta G_f°(H_2O, g) - 1 \text{ mol } H_2O \times \Delta G_f°(H_2O, l)$$

$$= -228.6 \text{ kJ} - (-237.1 \text{ kJ}) = +8.5 \text{ kJ}$$

Then, from Eq. 7,

$$\ln K_P = \frac{-8.5 \text{ kJ}}{2.48 \text{ kJ}} = -3.43$$

The natural antilogarithm of -3.43 is 0.032, so $K_P = 0.032$ atm at 25°C. This is equivalent to 24 Torr, which is close to the value listed in Table 10.5.

EXERCISE Calculate the vapor pressure of ethanol at 25°C from its standard free energies of formation, $\Delta G_f°(C_2H_5OH, l) = -174.8$ kJ/mol and $\Delta G_f°(C_2H_5OH, g) = -168.5$ kJ/mol.

[*Answer*: 60 Torr]

A feature of Chapters 14 and 15 on acids, bases, and salts can now be brought into focus. We saw there that pK is often more convenient to use than K itself. Now we can see why: since $\ln K$ is proportional to $\Delta G°$, pK is actually a disguised form of $\Delta G°$. In other words, all our pK manipulations were in fact calculations of the free energies of acid and base ionization reactions. A very similar replacement allows us to analyze the equilibria of redox reactions, but in those reactions $\Delta G°$ is expressed as a voltage. This is the subject we consider in the next chapter.

SUMMARY

16.1 A major question in chemistry is how to decide which reactions and processes are **spontaneous,** tending to occur without being driven. Changes are spontaneous if they correspond to an increase in disorder of the universe. This disorder can be expressed as **entropy:** the greater the disorder, the higher the entropy. An entropy may be calculated with the **Boltzmann formula,** $S = k \ln W$, or measured from the heat capacity. The entropy of any perfect crystal is zero at $T = 0$. Entropy increases with increasing temperature and when a substance melts

or vaporizes (the change on vaporization is larger than the change on melting). The change in entropy that accompanies the transformation of reactants into products, the **standard reaction entropy,** is calculated from the differences of their **standard molar entropies,** their entropies per mole under standard conditions.

16.2 The change in the entropy of the surroundings accompanying a reaction is calculated from the enthalpy change of the reaction system, with the formula $\Delta S_{surr} = -\Delta H/T$. Heat released to the surroundings increases their entropy; heat absorbed from the surroundings decreases their entropy. The total entropy change is zero for processes occurring at equilibrium.

16.3 According to the **second law of thermodynamics,** the direction of spontaneous change is that for which the total entropy of the universe increases. An exothermic reaction may be spontaneous because it releases heat to the surroundings and hence increases their entropy, even though the entropy of the reaction system might decrease. Endothermic reactions are less common. They are spontaneous only if the increase in the entropy of the system is great enough to offset the decrease in the entropy of the surroundings.

16.4 The total entropy change is normally expressed in terms of the **Gibbs free energy,** $G = H - TS$. The direction of spontaneous change (at constant pressure and temperature) is in the direction of decreasing free energy. For a system at equilibrium at constant temperature and pressure, $\Delta G = 0$. Whether or not a reaction is spontaneous can be judged from the **standard reaction free energy** at the temperature of interest: a reaction tends to occur spontaneously if the reaction free energy is negative. The standard reaction free energy can be calculated from **standard free energies of formation,** the standard reaction free energies for the formation of compounds from their elements.

16.5 The standard free energy of formation of a compound can be used to judge whether it is **thermodynamically unstable** with respect to decomposition into its elements or into other compounds. The temperature at which the reaction free energy becomes negative is the temperature above which a reaction can occur.

16.6 and 16.7 The reaction free energy depends on the composition of the reaction system. From the condition that at equilibrium the reaction free energy is zero, we can deduce that equilibrium constants are related to the standard reaction free energy by $\Delta G° = -RT \ln K$.

EXERCISES

Spontaneous Change

In Exercises 16.1 and 16.2, explain, in terms of an increase in disorder, why each change is spontaneous.

16.1 (a) Hot coffee cooling; (b) a pool ball rolling to a halt; (c) sugar dissolving in water.
16.2 (a) Water evaporating from damp clothes; (b) an automobile stopping when the brake is on; (c) a clock spring unwinding.

In Exercises 16.3 and 16.4, explain why each change is not spontaneous.

16.3 (a) A beaker of water becoming hotter than its surroundings; (b) a ball bouncing higher and higher.
16.4 (a) One half of a block of copper becoming hot at the expense of the other half; (b) a ball starting to roll across a horizontal table.

Entropy

In Exercises 16.5 to 16.8, use the Boltzmann formula to calculate the entropy of the system.

16.5 A stack of 10 pennies arranged (a) as heads up and (b) randomly heads up or down.
16.6 A pack of 52 cards (a) when new and (b) after thorough shuffling.

16.7 A sample of solid carbon monoxide containing 1 mol of CO molecules in which 90% are randomly oriented in either of two directions and 10% are perfectly ordered.

16.8 A sample of solid $FClO_3$ containing 1 mol of molecules in which 60% are randomly oriented in any of four directions and the remainder are perfectly aligned.

16.9 Nitric oxide forms rectangular $(NO)_2$ dimers when it condenses to a solid. Its molar entropy at absolute zero is close to 5 J/(K · mol). Suggest an interpretation of this entropy value.
16.10 The deuteration of methane produces CH_3D, and its molar entropy at absolute zero has been measured to be 12 J/(K · mol).
   (a) Suggest an interpretation of this value.
   (b) What do you predict as the entropy of CH_2D_2 at absolute zero?

In Exercises 16.11 and 16.12, calculate the entropy change for each physical change from the information in Table 6.2.

16.11 (a) 1 mol of H_2O freezing at 0°C; (b) 100 g of benzene melting at 6°C.
16.12 (a) 1 mol of H_2O evaporating at 100°C and 1 atm pressure; (b) 100 g of sodium melting at 371 K (its normal melting point at which ΔH_{melt} = 2.60 kJ/mol).

Standard Reaction Entropy

In Exercises 16.13 to 16.16, calculate the standard reaction entropy for each reaction at 25°C. In each case suggest a reason why the entropy change you calculated is reasonable.

16.13 (a) $2H_2(g) + O_2(g) \longrightarrow 2H_2O(l)$
(b) $6C(s) + 3H_2(g) \longrightarrow C_6H_6(l)$
(c) $2NO_2(g) \longrightarrow N_2O_4(g)$
(d) $H_2(g) + CO(g) \longrightarrow$
$\qquad\qquad\qquad H_2CO(g)$ (formaldehyde)

16.14 (a) $2CO(g) + O_2(g) \longrightarrow 2CO_2(g)$
(b) $C(s) + 2H_2(g) \longrightarrow CH_4(g)$
(c) $CaCO_3(s) \longrightarrow CaO(s) + CO_2(g)$
(d) $H_2(g) + D_2O(l) \longrightarrow D_2(g) + H_2O(l)$

16.15 (a) $2H_2O_2(l) \longrightarrow 2H_2O(l) + O_2(g)$
(b) $2CH_4(g) + S_8(s) \longrightarrow 2CS_2(l) + 4H_2S(g)$
(c) $2F_2(g) + 2H_2O(l) \longrightarrow 4HF(aq) + O_2(g)$
(d) $B_2O_3(s) + 3CaF_2(s) \longrightarrow 2BF_3(g) + 3CaO(s)$

16.16 (a) $CaC_2(s) + 2H_2O(l) \longrightarrow$
$\qquad\qquad\qquad Ca(OH)_2(s) + C_2H_2(g)$
(b) $CO_2(g) + 2NH_3(g) \longrightarrow$
$\qquad\qquad\qquad H_2O(l) + CH_4ON_2(s)$ (urea)
(c) $4NH_3(g) + 5O_2(g) \longrightarrow 4NO(g) + 6H_2O(g)$
(d) $4KClO_3(s) \longrightarrow 3KClO_4(s) + KCl(s)$

The Entropy Change in the Surroundings

In Exercises 16.17 and 16.18, calculate the change in the entropy of the surroundings when each given amount of heat is supplied to them.

16.17 (a) 100 kJ at 25°C; (b) 100 kJ at 1000 K; (c) 1 mJ at 10^{-9} K, the current world record low temperature.
16.18 (a) 1 J at 37°C, as in each heartbeat; (b) 100 kJ at 37°C; (c) 1 MJ at 20°C.

16.19 A human body generates heat at the rate of about 100 W (1 W = 1 J/s).
(a) At what rate do you generate entropy in surroundings assumed to be at 20°C?
(b) How much entropy do you generate each day?
16.20 An electric heater is rated at 2 kW.
(a) At what rate does it generate entropy in a room maintained at 28°C?
(b) How much entropy does it generate in the course of a day?
(c) Would more or less entropy be generated if the room were maintained at 25°C?

In Exercises 16.21 and 16.22, assume that the given substance surrounds a heater, and calculate the change in entropy in each case. So little heat is supplied that the temperature of the substance can be regarded as being constant.

16.21 10 J of heat is supplied to a block of copper at (a) 25°C; (b) 10 K; (c) 500 K.
16.22 1.0 kJ of heat is supplied to 1.0 L of water at (a) 0°C; (b) 25°C; (c) 99°C.

Entropy and Equilibrium

In Exercises 16.23 and 16.24, calculate (1) the overall entropy change and the entropy changes for (2) the sur-

roundings and (3) the substance when 1 mol of each substance evaporates at its normal boiling point.

16.23 (a) Argon; (b) methane; (c) ethanol; (d) benzene.

16.24 (a) Water; (b) mercury; (c) ammonia. Suggest a reason why these three substances have a significantly larger entropy of vaporization than those in Exercise 16.23.

In Exercises 16.25 and 16.26, calculate the molar entropy change that occurs when each substance melts at its normal melting point.

16.25 (a) Water; (b) ammonia; (c) argon.
16.26 (a) Benzene; (b) methane; (c) methanol. Is there a pattern in the values?

16.27 The entropy of vaporization of xenon is 76 J/(K · mol), and its enthalpy of vaporization is 12.6 kJ/mol. What is its boiling point?
16.28 The entropy of vaporization of liquid chlorine is 85.4 J/(K · mol), and its enthalpy of vaporization is 20.4 kJ/mol. What is its boiling point?

16.29 Estimate the boiling point of fluorine by assuming that the enthalpy of vaporization is similar to that of argon.
16.30 The enthalpy of vaporization of carbon disulfide is 26.7 kJ/mol. What is its boiling point?

16.31 Benzene boils at 80.0°C. Estimate its enthalpy of vaporization.
16.32 Carbon tetrachloride boils at 76.5°C. Estimate its enthalpy of vaporization.

Free Energy

16.33 Calculate the standard free energies for the reactions of Exercises 16.13 and 16.15 at 25°C from their enthalpies and entropies. Which reactions have equilibrium constants that favor the formation of products?
16.34 Calculate the standard free energies for the reactions in Exercises 16.14 and 16.16. Which reactions have equilibrium constants that favor the formation of products?

*__16.35__ Suppose a reaction reaches equilibrium at 200°C, and its standard reaction enthalpy is +100 kJ. Show, by considering the effect of temperature on the free energy, whether more products or more reactants are favored if the temperature is changed suddenly to (a) 205°C and (b) 195°C, assuming that the reaction enthalpy and entropy are unchanged.
*__16.36__ A reaction reaches equilibrium at 100°C and has a standard reaction enthalpy of −55 kJ. By calculating the free energy, decide whether more products or more reactants are favored when the temperature is changed suddenly to (a) 105°C and (b) 95°C, assuming that the reaction enthalpy and entropy are unchanged.

Standard Free Energy of Formation

In Exercises 16.37 to 16.40, calculate the standard free energy of formation for each compound at 25°C from enthalpies of formation and standard entropies.

16.37 (a) $NH_3(g)$; (b) $H_2O(l)$; (c) $CH_4(g)$; (d) $H_2O_2(l)$.
16.38 (a) $NaCl(s)$; (b) $H_2O(g)$; (c) $NO_2(g)$; (d) $N_2O_4(g)$.

16.39 (a) $CS_2(l)$; (b) $SO_2(g)$; (c) $CaCO_3(s)$; (d) $CO_2(g)$.
16.40 (a) $CO(g)$; (b) $SO_3(g)$; (c) $MgCO_3(s)$; (d) $C_6H_6(l)$.

In Exercises 16.41 and 16.42, use the standard free energies of formation in Appendix 2A to calculate the standard free energy for each reaction at 25°C.

16.41 (a) $H_2(g) + I_2(s) \longrightarrow 2HI(g)$
(b) $2SO_2(g) + O_2(g) \longrightarrow 2SO_3(g)$
(c) $CaCO_3(s) \longrightarrow CaO(s) + CO_2(g)$

16.42 (a) $2CH_3OH(l) + 3O_2(g) \longrightarrow$
$$2CO_2(g) + 4H_2O(l)$$
(b) $2CH_3OH(g) + 3O_2(g) \longrightarrow$
$$2CO_2(g) + 4H_2O(g)$$
(c) $NH_4Cl(s) \longrightarrow NH_3(g) + HCl(g)$
(d) $SbCl_5(g) \longrightarrow SbCl_3(g) + Cl_2(g)$

Thermodynamic Stability

In Exercises 16.43 and 16.44, determine which compounds are unstable with respect to decomposition into their elements at 25°C.

16.43 (a) CuO; (b) HCN; (c) O_3; (d) NO.
16.44 (a) PCl_5; (b) N_2H_4; (c) H_2O_2; (d) C_8H_{18} (octane).

16.45 Which compounds in Exercise 16.43 become more unstable as the temperature is increased?
16.46 Which compounds in Exercise 16.44 become more unstable as the temperature is increased?

16.47 Does potassium chlorate have a thermodynamic tendency to form potassium perchlorate and potassium chloride at 25°C? Might it do so at a higher temperature?
16.48 Does methanol have a thermodynamic tendency to decompose to carbon monoxide and hydrogen at 25°C? Is the tendency stronger at a higher temperature?

In Exercises 16.49 and 16.50, judge whether each oxide can be reduced by carbon to the metal at 1000 K (the number in parenthesis is the free energy of formation of the ore at 1000 K). In each case consider whether CO, CO_2, or both can be the products. The standard free energy of formation of CO at 1000 K is -200 kJ/mol, and that of CO_2 is -396 kJ/mol.

16.49 (a) Fe_2O_3 (-562 kJ/mol); (b) MnO_2 (-405 kJ/mol).
16.50 (a) TiO_2 (-762 kJ/mol); (b) Li_2O (-466 kJ/mol).

16.51 Calculate the standard reaction free energy of the water autoionization $2H_2O(l) \rightarrow H_3O^+(aq) + OH^-(aq)$ at 25°C.
16.52 Do $Cu^+(aq)$ ions have a thermodynamic tendency to form $Cu^{2+}(aq)$ ions and copper metal in water at 25°C?

Prediction of Equilibrium Constants

In Exercises 16.53 to 16.56, calculate the equilibrium constant at 25°C for each reaction.

16.53 (a) $H_2(g) + I_2(s) \longrightarrow 2HI(g)$
(b) $2SO_2(g) + O_2(g) \longrightarrow 2SO_3(g)$
(c) $CaCO_3(s) \longrightarrow CaO(s) + CO_2(g)$
16.54 (a) $2CH_3OH(l) + 3O_2(g) \longrightarrow$
$$2CO_2(g) + 4H_2O(l)$$
(b) $2CH_3OH(g) + 3O_2(g) \longrightarrow$
$$2CO_2(g) + 4H_2O(g)$$
(c) $NH_4Cl(s) \longrightarrow NH_3(g) + HCl(g)$
(d) $SbCl_5(g) \longrightarrow SbCl_3(g) + Cl_2(g)$

16.55 (a) $H_2(g) + 2O_2(g) \longrightarrow 2H_2O(l)$
(b) $2NO_2(g) \longrightarrow N_2O_4(g)$
(c) $H_2(g) + CO(g) \longrightarrow H_2CO(g)$ (formaldehyde)
16.56 (a) $2CO(g) + O_2(g) \longrightarrow 2CO_2(g)$
(b) $C(s) + 2H_2(g) \longrightarrow CH_4(g)$
(c) $H_2(g) + D_2O(l) \longrightarrow D_2(g) + H_2O(l)$

16.57 If $Q = 1.0$ atm for the following reaction at 25°C, will the reaction have a tendency to form products, to form reactants, or to be at equilibrium?

$$N_2O_4(g) \rightleftharpoons 2NO_2(g)$$

16.58 If $Q = 1 \times 10^{50}$ atm for the following reaction at 25°C, will the reaction have a tendency to form products, to form reactants, or to be at equilibrium?

$$C(s) + O_2(g) \rightleftharpoons CO_2(g)$$

In Exercises 16.59 and 16.60, use the information in Appendix 2A to calculate (1) the solubility constant and (2) the solubility of each sparingly soluble salt at 25°C.

16.59 (a) $AgCl$; (b) $AgBr$.
16.60 (a) AgI; (b) FeS; (c) $PbBr_2$; (d) $CaCO_3$.

In Exercises 16.61 and 16.62, calculate the standard reaction free energy and, where appropriate, the standard free energy of formation of the product, for each reaction at the given temperature.

16.61 (a) $N_2(g) + 3H_2(g) \rightleftharpoons 2NH_3(g)$, $K_P = 41$ atm^{-2}; at 400 K
(b) $2SO_2(g) + O_2(g) \rightleftharpoons 2SO_3(g)$, $K_P = 3.0 \times 10^4$ atm^{-1}; at 700 K

16.62 (a) $H_2(g) + I_2(g) \rightleftharpoons 2HI(g)$, $K_P = 160$; at 500 K
(b) $N_2O_4(g) \rightleftharpoons 2NO_2(g)$, $K_P = 47.9$ atm; at 400 K

In Exercises 16.63 and 16.64, find a relation between pK_{sp} and $\Delta G°$ for the dissolving equilibrium, and then calculate pK_{sp}, for each sparingly soluble compound at 25°C.

16.63 AgI.
16.64 AgCl.

General

16.65 Using only its boiling point ($-60.4°C$) as data, estimate how long it would take for 100 g of liquid hydrogen sulfide at its boiling point to evaporate when it is heated by a 10-W heater. (Recall that 1 W = 1 J/s.)

16.66 Calculate ΔG for the synthesis of water vapor from hydrogen and oxygen at 25°C when the partial pressures of the components are all (a) 0.1 atm and (b) 10^{-6} atm. Which reaction direction is spontaneous? What is the value of Q_P at equilibrium?

16.67 Deduce the van't Hoff equation for the temperature dependence of the equilibrium constant (Exercise 13.66),

$$\ln K' = \ln K - \frac{\Delta H°}{R} \times \left(\frac{1}{T'} - \frac{1}{T} \right)$$

Hint: Express K and K' in terms of $\Delta G°$ and assume that neither $\Delta H°$ nor $\Delta S°$ varies much with temperature. Go on to deduce an expression for the temperature dependence of the vapor pressure of a liquid.

16.68 The van't Hoff equation (Exercise 16.67) can be used to find the standard reaction enthalpy from the values of the equilibrium constant at different temperatures, and the resulting $\Delta H°$ can be combined with the standard reaction free energy from the equilibrium constant itself to obtain the standard reaction entropy. Hence all three thermodynamic properties of a reaction may be determined. Calculate $\Delta G°$, $\Delta H°$, and $\Delta S°$ for each of the following reactions at a temperature halfway between the given temperatures.
(a) $H_2(g) + I_2(g) \rightleftharpoons 2HI(g)$; $K_P = 160$ at 500 K, and $K_P = 54$ at 700 K
(b) $2SO_2(g) + O_2(g) \rightleftharpoons 2SO_3(g)$; $K_P = 2.5 \times 10^{10}$/atm at 500 K, and $K_P = 3.0 \times 10^4$/atm at 700 K

This chapter explains how chemical reactions are used to generate electric currents and how electric currents are used to bring about chemical reactions. It also extends the range of reactions that we can discuss in terms of equilibrium constants, and explains how a scale of oxidizing and reducing strengths is established and used. The illustration shows the interior of one cell of a lead-acid automobile battery while charging is in progress. The cell stores energy when it is charged and releases it when it is discharged.

ELECTROCHEMISTRY

Electrochemical cells

17.1 Cells and cell reactions
17.2 Practical cells

Thermodynamics and electrochemistry

17.3 Cell potential and reaction free energy
17.4 The electrochemical series
17.5 The dependence of cell potential on concentration

Electrolysis

17.6 The potential needed for electrolysis
17.7 The extent of electrolysis
17.8 Applications of electrolysis

17

Redox reactions were first introduced in Chapter 3. Our purpose in returning to them now is to discuss them using the powerful techniques we have been developing since that early introduction. As a reminder of the widespread occurrence and importance of these reactions, Tables 17.1 and 17.2 list some of the redox reactions used to produce elements and compounds from natural sources. Redox reactions are also used in the chemical generation and storage of electric power and in the reverse of this process—the use of electricity to bring about chemical change. This last application, electrolysis, was almost the first type of chemical change mentioned in Chapter 1, so our journey has come full circle. The branch of chemistry that describes the use of chemical reactions to produce electricity, the relative strengths of oxidizing and reducing agents, and the use of electricity to produce chemical change is called *electrochemistry*.

TABLE 17.1 Elements that are prepared by reduction

Element	Source	Process
*Easy**		
H_2	H_2O	Synthesis gas reaction:
		$CH_4(g) + H_2O(g) \xrightarrow{800°C, Ni} CO(g) + 3H_2(g)$
		Shift reaction:
		$CO(g) + H_2O(g) \xrightarrow{400°C, Fe/Cu} CO_2(g) + H_2(g)$
Cu	CuS	Copper smelting:
		$CuS(s) + O_2(g) \longrightarrow Cu(s) + SO_2(g)$
		Hydrometallurgy:
		$Cu^{2+}(aq) + H_2(g) \longrightarrow Cu(s) + 2H^+(aq)$
Moderately difficult		
P	PO_4^{3-}	Heat with carbon and sand in an electric furnace:
		$2Ca_3(PO_4)_2(l) + 6SiO_2(l) + 10C(s) \xrightarrow{1500°C} P_4(g) + 6CaSiO_3(l) + 10CO(g)$
Fe	Fe_2O_3	Blast furnace:
		$Fe_2O_3(s) + 3CO(g) \xrightarrow{900°C} 2Fe(l) + 3CO_2(g)$
Difficult		
Na	NaCl	Downs process:
		$2NaCl(l) \xrightarrow{\text{electrolysis at } 600°C} 2Na(l) + Cl_2(g)$
K	KCl	Reduction with sodium vapor:
		$KCl(l) + Na(g) \xrightarrow{700°C} K(l) + NaCl(s)$
Si	SiO_2	Reduction in electric furnace:
		$SiO_2(l) + 2C(s) \xrightarrow{1500°C} Si(l) + 2CO(g)$
Al	Al_2O_3	Hall process:
		$2Al_2O_3(\text{in cryolite}) + 3C(s) \xrightarrow{\text{electrolysis at } 900°C} 4Al(l) + 3CO_2(g)$
Ti	$TiCl_4$	Kroll process:
		$TiCl_4(g) + 2Mg(l) \xrightarrow{1000°C} Ti(s) + 2MgCl_2(l)$

*The methods are designated easy, moderately difficult, and difficult according to the strength of the reducing agent required. More difficult reductions require stronger agents.

TABLE 17.2 Elements that are prepared by oxidation

Element	Source	Process
*Easy**		
S	H_2S	Claus process:
		$2H_2S(g) + 3O_2(g) \longrightarrow 2SO_2(g) + 2H_2O(g)$
		$2H_2S(g) + SO_2(g) \xrightarrow{300°C, Fe_2O_3} 3S(g) + 2H_2O(g)$
Moderately difficult		
Cl_2	NaCl	Downs process (as for Na in Table 17.1)
Br_2, I_2	Br^-, I^- in brine	Oxidation and degassing:
		$Cl_2(g) + 2Br^-(aq) \longrightarrow 2Cl^-(aq) + Br_2(aq)$
Difficult		
F_2	F^-	Moissan's method:
		$HF(\text{with dissolved KF}) \xrightarrow{\text{electrolysis near 100°C}} F_2(g) + H_2(g)$
Au	Au	Cyanide process:
		$4Au(s) + 8CN^-(aq) + O_2(g) + 2H_2O(l) \longrightarrow 4[Au(CN)_2]^-(aq) + 4OH^-(aq)$
		$2[Au(CN)_2]^-(aq) + Zn(s) \longrightarrow 2Au(s) + Zn^{2+}(aq) + 4CN^-(aq)$

*The methods are designated easy, moderately difficult, and difficult according to the strength of the oxidizing agent required. More difficult oxidations require stronger agents.

The language of redox reactions was introduced in Section 3.4. In brief, oxidation is electron loss, as when H_2 is oxidized to H^+:

$$\textit{Oxidation:} \quad H_2(g) \longrightarrow 2H^+(aq) + 2e^-$$

The oxidation is brought about by an oxidizing agent, which accepts the electrons (and is reduced in the process). Reduction is electron gain, as when H^+ is reduced in the reaction

$$\textit{Reduction:} \quad 2H^+(aq) + 2e^- \longrightarrow H_2(g)$$

The reduction is brought about by a reducing agent, which donates the electrons (and is oxidized in the process). In the first of these two half-reactions, the reduced species H_2 (the substance with the relatively low oxidation number) is an electron donor and thus is acting as a reducing agent. In the second, the oxidized species H^+ (the substance with the relatively high oxidation number) is an electron acceptor and thus is acting as an oxidizing agent. The two half-reactions can be summarized as follows:

$$\underset{\text{oxidizing agent}}{\text{Oxidized form} + \text{electrons}} \underset{\text{oxidation}}{\overset{\text{reduction}}{\rightleftharpoons}} \underset{\text{reducing agent}}{\text{reduced form}}$$

ELECTROCHEMICAL CELLS

When we place a piece of zinc in an aqueous solution of copper sulfate, some of the zinc reacts, and as it does so, copper is deposited on the surface of the solid (Fig. 17.1). This shows that the redox reaction

$$Zn(s) + Cu^{2+}(aq) \longrightarrow Zn^{2+}(aq) + Cu(s)$$

FIGURE 17.1 When a lump of zinc is dropped into a beaker of copper sulfate solution, copper is deposited on the zinc and the copper sulfate is gradually replaced by the colorless zinc sulfate.

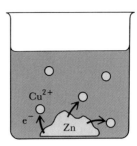

FIGURE 17.2 The reaction shown in Fig. 17.1 takes place at random all over the surface of the zinc as electrons are transferred to Cu^{2+} ions in the solution.

proceeds spontaneously in the direction of the arrow. As the reaction takes place, electrons are transferred from the zinc to Cu^{2+} ions nearby in the solution (Fig. 17.2). The electrons reduce the Cu^{2+} ions to Cu atoms, which stick to the surface of the zinc or form a finely divided solid deposit in the flask. The piece of zinc slowly reacts as its atoms give up electrons and form Zn^{2+} ions. The transfer of electrons from the zinc to the copper ions occurs at random all over the surface of the metal, and the reaction enthalpy is released as heat.

Now suppose that the zinc is separated from the copper solution in the arrangement shown in Fig. 17.3. This is called a *Daniell cell*, for the British chemist John Daniell. He invented the cell in 1836 when, in response to the growth in telegraphy, there was an urgent need for a steady source of electric current. The same redox reaction takes place as before, but now electrons can reach the Cu^{2+} ions only by traveling through the wire and light bulb. As Cu^{2+} ions are converted to neutral Cu atoms in one compartment, and Zn atoms are converted to Zn^{2+} ions in the other, ions—sulfate ions, if the solution is copper sulfate—must move between the two compartments (through the porous pot) to preserve electric neutrality and complete the electric circuit.

The reaction produces an orderly flow of electrons—an electric current—through the external wire. This is the essence of the generation of electric power by chemical reaction. Whenever we use a battery-driven device, we make use of a reaction in which reduction and oxidation take place in separate compartments while electrons flow between them through an external circuit.

17.1 CELLS AND CELL REACTIONS

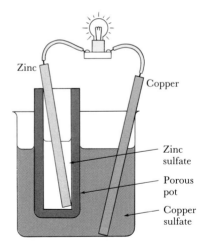

FIGURE 17.3 The Daniell cell consists of copper and zinc rods dipping into solutions of copper sulfate and zinc sulfate. The two solutions make contact through the porous pot, which allows ions to pass through to complete the electric circuit.

The device in Figure 17.3 is one kind of *electrochemical cell*, a device for producing electricity by means of a chemical reaction. An electrochemical cell consists of a container fitted with two electrodes and filled with an electrolyte (an ionic conductor). The electrodes are usually metal, but graphite is sometimes used. The electrolyte is usually an aqueous solution of ions, but it can be a molten salt or even a solid. The two electrodes share the same electrolyte in some cells; but if they are placed in different electrolytes, the two solutions are joined by a *salt bridge*. This is a tube containing a concentrated salt (potassium chloride or nitrate) in a jelly that acts as an electrolyte (Fig. 17.4) and allows ions to travel between the two compartments and complete the circuit. In practical cells such as those sold for use in the home, contact between the two solutions is often made through a membrane that allows ions to pass between compartments. A "battery" is really a collection of cells connected together, but the word is often used to mean a single cell.

Electrodes and couples. A feature of all electrochemical cells is that reduction occurs at one electrode and oxidation occurs at the other:

The **cathode** is the electrode at which reduction occurs.

The **anode** is the electrode at which oxidation occurs.

The cathode is marked "+" and the anode is marked "−" on the cells

FIGURE 17.4 A typical electro-chemical cell for use in the labo-ratory. The two electrodes are connected by a salt bridge, which completes the circuit within the cell.

and batteries sold in stores. Electrons leave the cell through the anode (hence the minus sign, because the electrons are lost from the cell there), travel through the external circuit, enter the cell again through the cathode (hence the plus sign, because the electrons are added to the cell there), and bring about the reduction (Fig. 17.5). In the Daniell cell, the source of electrons in the anode compartment is the oxidation

$$\textit{Anode reaction:} \quad \text{Zn}(s) \longrightarrow \text{Zn}^{2+}(aq) + 2\text{e}^-$$

The electrons are used in the cathode compartment for the reduction

$$\textit{Cathode reaction:} \quad \text{Cu}^{2+}(aq) + 2\text{e}^- \longrightarrow \text{Cu}(s)$$

The reduced and oxidized substances in each compartment form a redox *couple*. The two couples in the Daniell cell are Zn^{2+}/Zn and $\text{Cu}^{2+}/\overline{\text{Cu}}$. By convention, the oxidized form is written before the re-duced form in specifying a couple. This is the opposite of the conven-tion for specifying the electrode itself, in which the (reduced) metal is

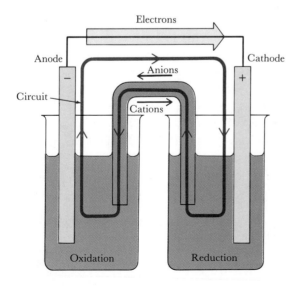

FIGURE 17.5 Electrons leave a cell at the anode ($-$), travel through the external circuit, and reenter the cell at the cathode ($+$). The source of the electrons is oxidation at the anode, and the electrons cause reduction at the cathode.

written first, as in $Zn(s)|Zn^{2+}(aq)$ and $Cu(s)|Cu^{2+}(aq)$:

Couple: oxidized form/reduced form

Electrode compartment: reduced form|oxidized form

The vertical line in the electrode notation represents the junction between the solid metal and the electrolyte solution and, in general, the junction between any two phases (such as gas and metal or undissolved salt and metal). It is normal practice to refer to the electrode compartment simply as the "electrode," and we shall do that from now on.

The *hydrogen electrode* $Pt|H_2(g)|H^+(aq)$ plays a special role in electrochemistry, as we shall see. In this electrode, hydrogen gas is bubbled over platinum dipping into a solution containing hydrogen ions (an acid). The platinum is present to make electrical contact with the solution; the redox couple is H^+/H_2. This notation shows that H^+ (more precisely, H_3O^+) is the oxidized species, and H_2 is the reduced species. When the hydrogen electrode acts as a cathode, the hydrogen ions are reduced:

Cathode reaction: $\quad 2H^+(aq) + 2e^- \longrightarrow H_2(g)$

When it acts as an anode, the hydrogen is oxidized:

Anode reaction: $\quad H_2(g) \longrightarrow 2H^+(aq) + 2e^-$

Whether an electrode acts as an anode or a cathode depends on its partner in the cell, as we shall see.

▼ **EXAMPLE 17.1** **Specifying an electrode**

One electrode that is commonly used in electrochemical cells is the silver–silver chloride electrode. It consists of a silver wire coated with silver chloride, dipping into a solution of chloride ions. When the electrode acts as a cathode, solid silver chloride is reduced to silver metal and chloride ions. Write the electrode half-reaction, the notation for the couple, and the notation for the electrode.

STRATEGY We need to identify the reduced and oxidized species, and then write out the couple and electrode as specified above. This means deciding how the electrode is constructed and using vertical lines to show the junctions between phases.

SOLUTION Since $AgCl(s)$ is reduced to $Ag(s)$ and $Cl^-(aq)$, the reaction at the cathode is

$$AgCl(s) + e^- \longrightarrow Ag(s) + Cl^-(aq)$$

Since the oxidized form is AgCl and the reduced forms are Ag and Cl^-, the couple is $AgCl/Ag, Cl^-$. The electrode consists of silver in contact with solid silver chloride, giving $Ag|AgCl(s)$, and the silver chloride is in contact with chloride ions, giving $Ag|AgCl(s)|Cl^-(aq)$.

EXERCISE The "calomel electrode" consists of mercury in contact with mercury(I) chloride (Hg_2Cl_2, or calomel), which is in contact with a solution of chloride ions. The mercury(I) chloride is reduced to mercury when the electrode acts as a cathode. Write the electrode reaction and the notation for the couple and the electrode.

[*Answer*: $Hg_2Cl_2(s) + 2e^- \longrightarrow 2Hg(l) + 2Cl^-(aq)$; $Hg_2Cl_2/Hg, Cl^-$;
$Hg|Hg_2Cl_2(s)|Cl^-(aq)$]

▲

Cell diagrams. The arrangement of electrodes in a cell is summarized by a *cell diagram* that makes use of the notation described above. The cathode is shown on the right, and the anode on the left. The cell diagram for a laboratory version of the Daniell cell is therefore

$$Zn(s)|Zn^{2+}(aq)|\ |Cu^{2+}(aq)|Cu(s)$$

$$\underset{\text{anode}}{\phantom{Zn(s)|Zn^{2+}(aq)}} \qquad \underset{\text{cathode}}{\phantom{Cu^{2+}(aq)|Cu(s)}}$$

The double vertical bar separating the two compartments represents the salt bridge. We know which electrode is the cathode because we know by observation of the changes taking place that the Cu^{2+} ions are being reduced. Shortly we shall see how to make an electrical measurement to decide which is the cathode; later we shall see how to predict it.

▼ **EXAMPLE 17.2** **Writing the cell diagram for a reaction**

Write the cell diagram for the reaction in which hydrogen is used to reduce $Fe^{3+}(aq)$ ions to $Fe^{2+}(aq)$ ions.

STRATEGY We should write the overall reaction, and then break it into oxidation and reduction half-reactions as explained in Section 3.10. We can then identify the electrode at which oxidation occurs as the anode; reduction identifies the cathode. If a redox couple does not include a solid metal, we should include platinum in the specification of the electrode (to provide electrical contact with the couple, as in the hydrogen electrode). The cell diagram is written anode $||$ cathode.

SOLUTION The overall reaction is

$$2Fe^{3+}(aq) + H_2(g) \longrightarrow 2Fe^{2+}(aq) + 2H^+(aq)$$

The half-reaction for the reduction of iron(III) is

$$Fe^{3+}(aq) + e^- \longrightarrow Fe^{2+}(aq)$$

so the redox couple is Fe^{3+}/Fe^{2+} and the electrode is written $Pt|Fe^{2+}(aq),Fe^{3+}(aq)$. The comma indicates that the solution contains both Fe^{2+} ions and Fe^{3+} ions without a junction separating them. Since the iron is reduced in the overall reaction, this electrode is the cathode. The half-reaction for the oxidation of H_2 is

$$H_2(g) \longrightarrow 2H^+(aq) + 2e^-$$

which occurs at the hydrogen electrode $Pt|H_2(g)|H^+(aq)$ acting as anode. The cell is therefore

$$Pt|H_2(g)|H^+(aq)|\ |Fe^{2+}(aq),Fe^{3+}(aq)|Pt$$

EXERCISE Devise a cell in which the reaction is the reduction of $Cu^{2+}(aq)$ to $Cu^+(aq)$ by magnesium.

[*Answer*: $Mg(s)|Mg^{2+}(aq)|\ |Cu^+(aq),Cu^{2+}(aq)|Pt$]

▲

Cell potential. The energy released by a redox reaction as it drives electrons through the external circuit can be used to heat the surroundings if the circuit includes a heater, or to do work (such as winding a cassette tape) if the circuit includes an electric motor. The energy available from a cell depends on the "electron-pushing" and "electron-pulling" power of the redox reaction. If the oxidation releases electrons readily and the reduction accepts them readily, electrons will be pushed and pulled through the circuit vigorously, and the reaction can provide a lot of energy.

FIGURE 17.6 Cell potential being measured with an electronic voltmeter, a device that draws negligible current so that the composition of the cell does not change during the measurement. Note that the display gives a positive potential when the + terminal of the meter is connected to the cathode.

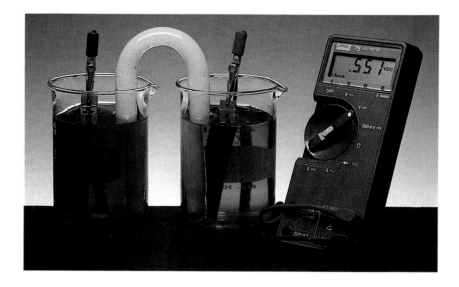

The pushing and pulling power of a cell reaction is measured as its *cell potential E*. (This is still often called the "electromotive force" of the cell or its "voltage.") The greater the cell potential, the greater the energy that a given number of electrons can release as they travel between the electrodes. That is, a high cell potential signifies that the cell reaction has a strong tendency to generate a current of electrons.

The SI unit of potential is the *volt* (V). It is defined so that one joule of energy is released when one coulomb of charge* travels between two electrodes that differ in potential by one volt:

$$1 \text{ J} = 1 \text{ C} \times 1 \text{ V}$$

Therefore, the energy released when a certain charge travels between electrodes differing in potential by a specified number of volts is

Energy in joules = charge in coulombs × potential in volts

The cell potential is by definition a positive quantity. Typical cells used in the home are rated at about 1.5 V, meaning that when 1.0 C of charge travels from one electrode to another, the energy released is about 1.0 C × 1.5 V = 1.5 J. The potential of a Daniell cell is about 1.1 V.

Cell potentials are measured with electronic voltmeters (Fig. 17.6). These are designed to give a positive reading (in volts) when the + and − terminals of the meter are connected to the + and − terminals, respectively, of the cell. Therefore, we can determine *experimentally* which electrode in a cell is the cathode (the + electrode) by finding the connection that results in a positive reading; the electrode that is joined to the + terminal of the voltmeter is then the cathode, and the other electrode is the anode.

A cell diagram is always written with the positive electrode (the cathode, the site of reduction) on the right (Fig. 17.5). We then always know

*A charge of 1 C corresponds to about 6.2×10^{18} electrons, or 1.0×10^{-5} mol of electrons, since each electron carries a charge of magnitude 1.602×10^{-19} C.

that electrons are tending to flow through the outside circuit from the electrode on the left of the diagram to the electrode on the right (in the direction we normally read). We therefore also know that the spontaneous direction of the cell reaction is oxidation in the left-hand or anode compartment, and reduction in the right-hand or cathode compartment (remember: r for right and reduction).

▼ **EXAMPLE 17.3 Deducing the reaction corresponding to a given cell**

The cell $Pt|H_2(g)|OH^-(aq)|O_2(g)|Pt$, which produces about 1.2 V at 25°C, has been used to produce electric power on some space missions. What is the cell reaction?

STRATEGY We should write the anode half-reaction as an oxidation of hydrogen, and the cathode half-reaction as a reduction of oxygen. As explained in Section 3.10, where we saw how to balance redox half-reactions, each half-reaction should be balanced using only H_2O, OH^- (the electrolyte is basic), and e^- in addition to H_2 and O_2. We can then add the two half-reactions (with electrons balanced) to obtain the equation of the overall reaction.

SOLUTION The anode half-reaction is

$$H_2(g) + 2OH^-(aq) \longrightarrow 2H_2O(l) + 2e^-$$

The cathode half-reaction is

$$O_2(g) + 2H_2O(l) + 4e^- \longrightarrow 4OH^-(aq)$$

To match electrons, we add two of the first half-reaction equations to the second, obtaining

$$2H_2(g) + O_2(g) \longrightarrow 2H_2O(l)$$

Hence the cell reaction is the formation of water from hydrogen and oxygen.

EXERCISE Write the cell reaction for the cell $Pt|O_2(g)|H^+(aq),H_2O_2(aq)|Pt$.
[*Answer*: $2H_2O_2(aq) \longrightarrow 2H_2O(l) + O_2(g)$]

▲

17.2 PRACTICAL CELLS

A *primary cell* is one that produces electricity from chemicals that are sealed into the cell when it is made. This type of cell cannot be recharged: once the cell reaction has reached equilibrium, the cell is discarded. A *secondary cell* is one that must be charged from some other energy supply before it can be used; this type is normally rechargeable (like an automobile battery). In the charging process, a nonequilibrium mixture of reactants is produced by an external source of electricity. When the cell is in use, it produces electricity as the reaction returns to equilibrium again. A *fuel cell* is a primary cell, but the reactants (the fuel) are supplied from outside while the cell is in use.

Primary cells. The workhorse of primary cells is the <u>*dry cell*</u> (Fig. 17.7) used to power portable electric equipment. It is also called the "Leclanché cell" for Georges Leclanché, the French engineer who invented it around 1866. The cell produces about 1.5 V initially, but with

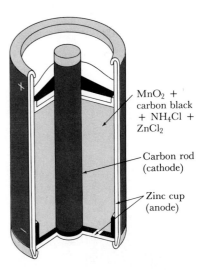

MnO_2 + carbon black + NH_4Cl + $ZnCl_2$

Carbon rod (cathode)

Zinc cup (anode)

FIGURE 17.7 A commercial dry cell consists of a graphite cathode in a zinc container that acts as the anode. The other components and the cell reaction are described in the text.

use this falls to about 0.8 V as reaction products accumulate inside it. If the cell is left unused for a day, its voltage may climb back to about 1.3 V as the products disperse through the electrolyte.

The zinc container of the cell is the anode. It supplies electrons to the outside circuit through the oxidation $Zn \rightarrow Zn^{2+} + 2e^-$. The electrolyte is a moist paste of ammonium chloride and zinc chloride in starch. The Zn^{2+} ions produced by the oxidation undergo a Lewis acid-base reaction with NH_3 produced from the NH_4^+ ions in the electrolyte:

$$Zn^{2+}(aq) + 2NH_4^+(aq) + 2OH^-(aq) \longrightarrow [Zn(NH_3)_2]^{2+}(aq) + 2H_2O(l)$$

This keeps the Zn^{2+} concentration low and so helps to maintain the voltage. At the end of the cell's life, so much $[Zn(NH_3)_2]^{2+}$ is present that its chloride salt crystallizes and decreases the ability of the electrolyte to conduct electricity. The salt can be encouraged to diffuse away from the anode by gently warming an exhausted cell for some hours; this partially restores the voltage. The cathode is a carbon rod surrounded by a mixture of manganese dioxide and carbon granules. Electrons arrive at this electrode through the external circuit and cause a complicated reduction reaction that can partly be represented as

$$MnO_2(s) + H_2O(l) + e^- \longrightarrow MnO(OH)(s) + OH^-(aq)$$

The OH^- ions migrate toward the zinc anode and take part in the Lewis acid-base reaction shown above.

Long-life, "high-power" *alkaline cells* are similar to the Leclanché cell, but the ammonium chloride is replaced by sodium or potassium hydroxide. They have a longer life because the zinc is not exposed to an acid environment like that caused by the NH_4^+ ions in the conventional cell. Because the ions move more easily through the electrolyte, they produce more power and a steadier current. Their higher cost arises largely from the difficulty of sealing the cells against hydroxide leakage.

A *mercury cell* (Fig. 17.8) also has a zinc anode, but the cathode is steel in contact with a mixture of mercury(II) oxide, potassium hydroxide, and zinc hydroxide. The cell's advantage is that it can be made very small but still produce a reasonably stable voltage (about 1.3 V) for long periods. The cathode reaction is the reduction $Hg^{2+}(s) + 2e^- \rightarrow Hg(l)$, and the overall cell reaction is

$$Zn(s) + HgO(s) \longrightarrow ZnO(s) + Hg(l)$$

Emergency batteries designed to activate automatically on contact with seawater (such as those in aircraft life vests) are manufactured without an electrolyte. Seawater floods in when they are immersed, and the redox reaction begins. One such cell uses a magnesium anode and a copper(I) chloride copper/cathode:

$$Anode: \quad Mg(s) \longrightarrow Mg^{2+}(aq) + 2e^-$$
$$Cathode: \quad CuCl(s) + e^- \longrightarrow Cu(s) + Cl^-(aq)$$

Secondary cells. The electrodes of a secondary cell must be chosen carefully. This is because both the reactants and the products of the cell reaction must be insoluble in the electrolyte, so that the electrodes retain their shape during charging and discharging.

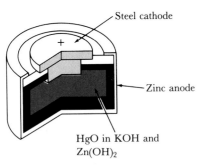

FIGURE 17.8 Cutaway view of a mercury cell.

Steel cathode

Zinc anode

HgO in KOH and $Zn(OH)_2$

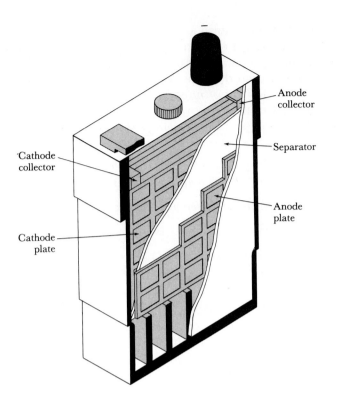

Anode collector

Separator

Cathode collector

Anode plate

Cathode plate

One of the most common secondary cells is the _lead-acid cell_ of the automobile battery. Each cell contains several plates that act as electrodes (Fig. 17.9). Since the total surface area of these plates is large, the battery can generate large currents on demand—at least for short periods, as for starting an engine. The plates are initially a hard lead-antimony alloy covered with a paste of lead(II) sulfate. The electrolyte is dilute sulfuric acid. During the first charging, some of the lead(II) sulfate is reduced to lead on one of the electrodes (the one that will act as the anode during discharge) and is oxidized to lead(IV) oxide on the other (the one that will act as the cathode).

The reaction at the anode of the lead-acid cell during discharge is

$$\text{Pb}(s) + \text{SO}_4{}^{2-}(aq) \longrightarrow \text{PbSO}_4(s) + 2e^-$$

In this reaction, metallic lead is oxidized to lead(II). The reaction at the cathode during discharge is

$$\text{PbO}_2(s) + \text{SO}_4{}^{2-}(aq) + 4\text{H}_3\text{O}^+(aq) + 2e^- \longrightarrow \text{PbSO}_4(s) + 6\text{H}_2\text{O}(l)$$

In this reaction, lead(IV) is reduced to lead(II). The equation for the overall cell reaction during discharge is therefore the sum of these two equations:

$$\text{Pb}(s) + \text{PbO}_2(s) + 2\text{H}_2\text{SO}_4(aq) \longrightarrow 2\text{PbSO}_4(s) + 2\text{H}_2\text{O}(l)$$

The cell reaction shows that sulfuric acid is used up during discharge. When the cell is recharged, the cell reaction is driven in reverse by the external supply of current, and sulfuric acid is produced. The state of charge of the cell can therefore be judged from the concentration of the sulfuric acid electrolyte, and that in turn can be judged from its density. If the battery is allowed to stand unused, a slow discharge

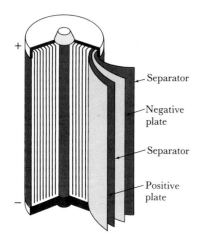

FIGURE 17.10 A rechargeable nickel-cadmium cell. The electrodes are assembled in a jelly-roll arrangement, separated by a layer soaked in moist sodium hydroxide or potassium hydroxide.

FIGURE 17.11 The electric eel *(Electrophorus electricus),* found in the Amazon River. The average potential difference it produces is about 350 V for a 3-ft-long eel.

occurs by the process known as "local action." In this process, impurities in the electrolyte are reduced by the electrodes, through reactions such as $Fe^{3+} + e^- \rightarrow Fe^{2+}$. Local action can discharge the cell by about a percent or two each day, but that is minimized if distilled water is used whenever the battery must be refilled, because fewer impurity ions are then introduced.

The rechargeable *nickel-cadmium cell* (Fig. 17.10) is widely used in portable electronic equipment. The electron supply is the cadmium oxidation

$$Cd(s) + 2OH^-(aq) \longrightarrow Cd(OH)_2(s) + 2e^-$$

The product electrons are used for the nickel reduction

$$Ni(OH)_3(s) + e^- \longrightarrow Ni(OH)_2(s) + OH^-(aq)$$

Since no gases are produced in either the charging or the discharging process, the cells can be sealed. This makes them ideal for use in transportable equipment.

Fuel cells. A simple fuel cell is one in which hydrogen gas—the fuel—is passed over one electrode, oxygen is passed over the other, and the electrolyte is aqueous potassium hydroxide. The electron supply is the oxidation of hydrogen at the anode, in the half-reaction $H_2(g) \rightarrow 2H^+(aq) + 2e^-$; the electron sink—the reaction in which the electrons are used—is the reduction of oxygen at the cathode: $O_2(g) + 2H_2O(l) + 4e^- \rightarrow 4OH^-(aq)$. As we saw in Example 17.3, the overall cell reaction is the formation of water. A version of this type of cell is used on the space shuttle (Fig. 6.8), one advantage being that the passengers can drink the product of the cell reaction.

Electric eels (Fig. 17.11) are mobile, natural fuel cells. They generate their electric charge in an "electric organ"—a battery of biological electrochemical cells fueled by food, with each cell providing about 0.15 V. It is an incidental feature of nature that whereas the eel's head is its cathode and its tail its anode, the electric catfish has the opposite polarity.

THERMODYNAMICS AND ELECTROCHEMISTRY

A cell with a nonzero potential is one in which the cell reaction has a spontaneous tendency to occur and to drive electrons through a circuit. Therefore, in the language of thermodynamics, a nonzero cell potential indicates that the cell reaction has a negative reaction free energy. This observation opens up a link between thermodynamics and electrochemistry.

17.3 CELL POTENTIAL AND REACTION FREE ENERGY

If ΔG for the cell reaction is large and negative, the reaction has a strong tendency to occur; hence we expect a large cell potential E (Fig. 17.12). If ΔG is negative but small, we expect only a small cell potential. If the reaction in the cell is at equilibrium, corresponding to $\Delta G = 0$, the reaction has no tendency to drive electrons through the circuit and the cell potential is zero. This suggests that the cell potential and the reaction free energy are proportional to each other:

$$\Delta G \propto -E$$

The negative sign is included on the right because a cell potential, which is always positive, indicates that the cell reaction is spontaneous and hence that ΔG is negative. If ΔG is positive for a certain reaction, then the *reverse* reaction is spontaneous and we must revise the cell diagram accordingly.

The precise relation between E and ΔG for a reaction in which n moles of electrons move from the anode to the cathode is*

$$\Delta G = -nFE \tag{1}$$

The constant F, the *Faraday constant*, is the charge per mole of electrons:

$$F = \text{charge of one electron} \times \text{number of electrons per mole}$$

$$= (1.602 \times 10^{-19}\ \text{C}) \times (6.022 \times 10^{23}/\text{mol}) = 96.5\ \text{kC/mol}$$

(A more precise value is 96.485 kC/mol.) F is named for the English scientist Michael Faraday (Fig. 17.13), who began life as the son of a blacksmith and came to be regarded as one of the greatest experimental scientists of the nineteenth century.

For the reaction in the Daniell cell, $n = 2$ mol because 2 mol of electrons migrate from 1 mol of Zn atoms to 1 mol of Cu atoms in the reaction

$$\text{Zn}(s) + \text{Cu}^{2+}(aq) \longrightarrow \text{Zn}^{2+}(aq) + \text{Cu}(s)$$

When the concentrations of the Cu^{2+} and Zn^{2+} ions are both $1\ M$, the cell potential is 1.1 V. Hence, for this reaction,

$$\Delta G = -(2\ \text{mol}) \times 96.5\ \text{kC/mol} \times 1.1\ \text{V} = -210\ \text{kJ}$$

Notice that if we write a reaction with all the stoichiometric coefficients multiplied by 2, as in

$$2\text{Zn}(s) + 2\text{Cu}^{2+}(aq) \longrightarrow 2\text{Zn}^{2+}(aq) + 2\text{Cu}(s)$$

then ΔG is also multiplied by 2, because now twice as much reactant is involved (e.g., 2 mol of Zn in place of 1 mol of Zn). However, n is also doubled because twice as many electrons are transferred. A factor of 2 thus appears on each side of Eq. 1, leaving E unchanged. It follows that if we multiply the equation for any cell reaction by any factor, the

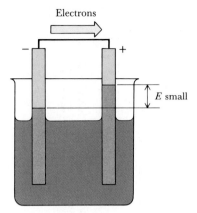

(a) ΔG small

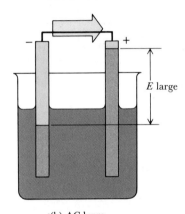

(b) ΔG large

FIGURE 17.12 The more negative the reaction free energy, the higher the cell potential.

*Equation 1 is obtained from thermodynamics. However, we can see that it is plausible by remembering that the free energy of a substance is its total enthalpy H less the energy stored in a disorderly way, $T \times S$. The difference $H - TS$, the free energy, is the energy stored in an *orderly* way, and only this energy can be used to drive a current in an orderly way through the circuit. (We cannot use the disorderly part of the energy to generate an orderly flow.) When it does so, the free energy changes by ΔG, and the energy of each electron in the current changes by $-e \times E$. If the current consists of n moles of electrons, with total charge $-nF$, their energy changes by $-nF \times E$. This change of energy is equal to the change in free energy in the system; equating $-nF \times E$ and ΔG gives Eq. 1.

FIGURE 17.13 Michael Faraday (1791–1867).

corresponding cell potential remains unchanged. That is, cell potential is an *intensive* property. This makes sense because we do not double the voltage of a cell by doubling its size. It is true that an automobile battery is large and gives 12 V, whereas a flashlight battery is small and gives 1.5 V; but this is so only because the larger battery consists of six 2-V cells joined together in series—it is not one single big cell.

We saw in Chapter 16 that a special role is played by the *standard reaction free energy* $\Delta G°$. We can measure this quantity for a cell simply by arranging for the reagents in the cell to be in their standard states, measuring the resulting *standard cell potential* $E°$, and making use of the following special case of Eq. 1:

$$\Delta G° = -nFE° \qquad (2)$$

We have seen that the standard states of solids and gases are their pure forms at a pressure of 1 atm. The standard state of an ion in solution is that in which its concentration is 1 M. Therefore,

> The **standard cell potential** of an electrochemical cell is the cell potential measured when the molar concentration of each type of ion taking part in the cell reaction is 1 M and all substances are at 1 atm pressure.

As an example, the standard potential of the Daniell cell at 25°C is its voltage when the concentrations of the Zn^{2+} ions and the Cu^{2+} ions in their two compartments are each 1 M. This is 1.1 V; so $\Delta G° = -210$ kJ as we calculated above.

There are thousands of possible cells and therefore thousands of possible standard cell potentials that could be tabulated. A great simplification is to consider each electrode of a cell as making a characteristic contribution to the cell potential. Then the cell potential is the sum of the contributions of its two electrodes (Fig. 17.14). For instance, the potential of the cell

$$Fe(s)|Fe^{2+}(aq)| \, |Ag^+(aq)|Ag(s) \qquad E° = 1.2 \text{ V}$$

can be thought of as consisting of one contribution from the silver electrode and another from the iron electrode. These individual contributions to the potential are called *electrode potentials*. When the cell is prepared in its standard state, they are called *standard electrode potentials* $E°$. The standard potential of a cell is the sum of its two standard electrode potentials:

$$E° = E°(\text{anode}) + E°(\text{cathode})$$

The standard hydrogen electrode. Since a voltmeter must be connected to two terminals, we cannot measure the potential of a single electrode. However, we can *assign* one electrode the value $E° = 0$. If we place that electrode in a cell with a second electrode, the second electrode can be held responsible for the entire standard cell potential, and we can measure its standard potential simply by measuring the standard cell potential. We can then form a second cell in which another electrode is combined with the second electrode. From the standard potential of the second electrode and the standard cell potential of this new cell, we

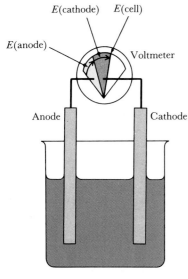

FIGURE 17.14 The cell potential can be thought of as the sum of the voltages of the anode and the cathode.

can work out the standard potential of the third electrode. This process can be continued until we have as many standard electrode potentials as we want.

We take the *standard hydrogen electrode* (SHE), which is the hydrogen electrode in its standard state (hydrogen ions at a concentration of 1 M and a hydrogen gas pressure of 1 atm) as the electrode assigned $E° = 0$. That is, we *define* the standard electrode potential of this electrode to be zero:

$$2H^+(aq) + 2e^- \longrightarrow H_2(g) \qquad E° = 0$$

Then, to measure the standard electrode potential of, for instance, the zinc electrode, we measure the standard potential of the zinc-hydrogen cell:

$$\underset{E°(\text{zinc})}{Zn(s)|Zn^{2+}(aq)|} \; \underset{E°(\text{SHE})}{|H^+(aq)|H_2(g)|Pt} \qquad E° = 0.76 \text{ V}$$

The fact that the hydrogen electrode is found to be the cathode shows that zinc is a stronger reducing agent than hydrogen and that it reduces hydrogen ions to hydrogen gas. Since, according to our convention, the zinc electrode contributes the entire 0.76 V to the standard cell potential, we write

$$Zn(s) \longrightarrow Zn^{2+}(aq) + 2e^- \qquad E° = +0.76 \text{ V}$$

For reasons that will shortly become clear, we always include the sign of an electrode potential. (That is unnecessary for cell potentials, since they are always positive.) If we form a cell with a copper electrode in place of the zinc, we find

$$\underset{E°(\text{SHE})}{Pt|H_2(g)|H^+(aq)|} \; \underset{E°(\text{copper})}{|Cu^{2+}(aq)|Cu(s)} \qquad E° = 0.34 \text{ V}$$

The fact that the copper electrode is now found to be the cathode shows that hydrogen is a stronger reducing agent than copper and that it reduces Cu^{2+} ions to copper metal. The copper electrode contributes the entire 0.34 V to the standard cell potential, so we write

$$Cu^{2+}(aq) + 2e^- \longrightarrow Cu(s) \qquad E° = +0.34 \text{ V}$$

Since zinc is a stronger reducing agent than hydrogen, and hydrogen is a stronger reducing agent than copper, we can now *predict* that zinc is a stronger reducing agent than copper. That is, when their ions are at 1 M concentrations, the spontaneous direction of reaction between zinc and copper is

$$Zn(s) + Cu^{2+}(aq) \longrightarrow Zn^{2+}(aq) + Cu(s)$$

and not the reverse. We can also use the two individual standard electrode potentials to predict the standard potential of the corresponding cell,

$$\underset{E°(\text{zinc})}{Zn(s)|Zn^{2+}(aq)|} \; \underset{E°(\text{copper})}{|Cu^{2+}(aq)|Cu(s)}$$

To do so, we simply add their standard electrode potentials:

Anode reaction: $\qquad Zn(s) \longrightarrow Zn^{2+}(aq) + 2e^- \quad . \quad E° = +0.76 \text{ V}$

$$\text{Cathode reaction:} \quad Cu^{2+}(aq) + 2e^- \longrightarrow Cu(s) \qquad E° = +0.34 \text{ V}$$

$$\text{Overall reaction:} \quad Cu^{2+}(aq) + Zn(s) \longrightarrow Cu(s) + Zn^{2+}(aq)$$
$$E° = 1.10 \text{ V}$$

Oxidation and reduction potentials. Because the +0.76-V potential of the zinc electrode arises through the oxidation of zinc, it is called a *standard oxidation potential*. Its value corresponds to a certain change in free energy, which we could work out from Eq. 2. Since reversing the reaction changes the sign of the reaction free energy and therefore that of the electrode potential, we can report the same information as

$$Zn^{2+}(aq) + 2e^- \longrightarrow Zn(s) \qquad E° = -0.76 \text{ V}$$

The −0.76 V here arises from a reduction and hence is called a *standard reduction potential*. The value +0.34 V for copper quoted above is also a standard reduction potential. If we wanted to express it as a standard oxidation potential we would simply reverse the half-reaction and change the sign of $E°$:

$$Cu(s) \longrightarrow Cu^{2+}(aq) + 2e^- \qquad E° = -0.34 \text{ V}$$

By convention, all standard electrode potentials are listed as reduction potentials. Hence, all the $E°$ values we quote from now on will refer to reduction half-reactions unless we state otherwise. A selection of values (at 25°C) is given in Table 17.3, and a longer table can be found in Appendix 2B. Note that the first column in the table gives the electron acceptor, the oxidized member of the couple. The third column gives the electron donor, the reduced member of the couple. The last column gives the standard reduction potential. Reduction potentials vary in a complicated way through the periodic table (Fig. 17.15); generally, however, the most negative are found toward the left, and the most positive toward the upper right.

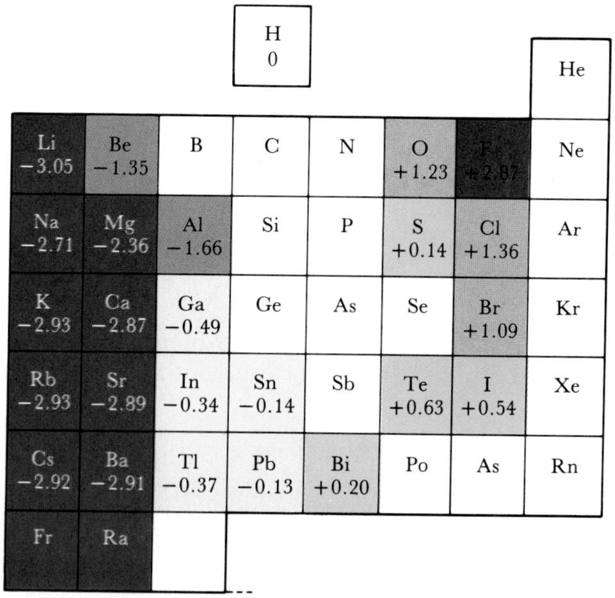

FIGURE 17.15 The variation of standard electrode potentials through the main groups of the periodic table.

TABLE 17.3   Standard electrode potentials at 25°C*

Oxidizing agent	Reducing agent	$E°$, V
Reduction half-reaction		
Strongly oxidizing		
F_2 + $2e^-$ ⟶ $2F^-$		+2.87
$S_2O_8^{2-}$ + $2e^-$ ⟶ $2SO_4^{2-}$		+2.05
Au^+ + e^- ⟶ Au		+1.69
Pb^{4+} + $2e^-$ ⟶ Pb^{2+}		+1.67
Ce^{4+} + e^- ⟶ Ce^{3+}		+1.61
MnO_4^- + $8H^+$ + $5e^-$ ⟶ Mn^{2+} + $4H_2O$		+1.51
Cl_2 + $2e^-$ ⟶ $2Cl^-$		+1.36
$Cr_2O_7^{2-}$ + $14H^+$ + $6e^-$ ⟶ $2Cr^{3+}$ + $7H_2O$		+1.33
O_2 + $4H^+$ + $4e^-$ ⟶ $2H_2O$		+1.23, +0.81 at pH = 7
Br_2 + $2e^-$ ⟶ $2Br^-$		+1.09
Ag^+ + e^- ⟶ Ag		+0.80
Hg_2^{2+} + $2e^-$ ⟶ $2Hg$		+0.79
Fe^{3+} + e^- ⟶ Fe^{2+}		+0.77
I_2 + $2e^-$ ⟶ $2I^-$		+0.54
O_2 + $2H_2O$ + $4e^-$ ⟶ $4OH^-$		+0.40, +0.81 at pH = 7
Cu^{2+} + $2e^-$ ⟶ Cu		+0.34
$AgCl$ + e^- ⟶ Ag + Cl^-		+0.22
$2H^+$ + $2e^-$ ⟶ H_2		0 (by definition)
Fe^{3+} + $3e^-$ ⟶ Fe		−0.04
O_2 + H_2O + $2e^-$ ⟶ HO_2^- + OH^-		−0.08
Pb^{2+} + $2e^-$ ⟶ Pb		−0.13
Sn^{2+} + $2e^-$ ⟶ Sn		−0.14
Fe^{2+} + $2e^-$ ⟶ Fe		−0.44
Zn^{2+} + $2e^-$ ⟶ Zn		−0.76
$2H_2O$ + $2e^-$ ⟶ H_2 + $2OH^-$		−0.83, −0.42 at pH = 7
Al^{3+} + $3e^-$ ⟶ Al		−1.66
Mg^{2+} + $2e^-$ ⟶ Mg		−2.36
Na^+ + e^- ⟶ Na		−2.71
Ca^{2+} + $2e^-$ ⟶ Ca		−2.87
K^+ + e^- ⟶ K		−2.93
Li^+ + e^- ⟶ Li		−3.05
	Strongly reducing	

*For a more extensive table, see Appendix 2B.

17.4 THE ELECTROCHEMICAL SERIES

An important application of standard electrode potentials is in ranking substances according to their reducing and oxidizing powers. That is, electrode potentials play the same role for redox reagents as pK values play for acids and bases. This should not be surprising: we saw at the end of Chapter 16 that pK values are standard free energies in disguise; Eq. 2 shows that standard electrode potentials are also free energies, expressed in volts.

Oxidizing and reducing strengths. We have seen already that zinc (with $E° = -0.76$ V) is a stronger reducing agent than either hydrogen ($E° = 0$) or copper ($E° = +0.34$ V). This suggests that the *more negative* the standard reduction potential of a couple, the greater the reducing power of the reduced species in the couple (the Zn in Zn^{2+}/Zn, for instance). Stated differently, the lower the position of a couple in Table 17.3, the greater the reducing power of the substance in the third column. It follows that when two couples form a cell, the couple lower down in the table forms the *anode* of the cell (because the reduced species in the couple gets oxidized as the reaction proceeds), and the couple higher up forms the *cathode* (because the oxidized member of the couple is reduced). To write the cell reaction, we would write the half-reaction for the lower couple as an oxidation (reversing it and changing the sign of its electrode potential), and that for the upper couple as a reduction (as in the table). This may be summarized as follows:

Lower couple   Upper couple

Anode compartment| |Cathode compartment

Red $\longrightarrow$ ox + e⁻   Ox + e⁻ $\longrightarrow$ red

This arrangement guarantees that the cell potential is positive and therefore that the reaction has a negative free energy; that is, the reaction has a spontaneous tendency to occur.

Table 17.3 is arranged with the most strongly oxidizing substances at the top of the leftmost column, and the most strongly reducing substances at the bottom of the third column. In this form, it is called the *electrochemical series*. The oxidized species in a couple in the list has a tendency to oxidize the reduced species in any couple that lies below it: the free energy for that reaction is negative, and the reaction is spontaneous. The reduced species in a couple has a tendency to reduce the oxidized species in any couple that lies above it. Hence, we can see at a glance the direction in which a particular redox reaction will tend to run (when the ion concentrations are 1 M and any gases are at 1 atm pressure).

▼ **EXAMPLE 17.4** Writing a cell diagram using the electrochemical series

Write the diagram for a cell in which one electrode is $Pt|Cl_2(g)|Cl^-(aq)$ and the other is $Pt|Br_2(l)|Br^-(aq)$. Give the cell reaction.

STRATEGY From its position in the periodic table we would expect Cl_2 to be a stronger oxidizing agent than Br_2 and therefore anticipate that Cl_2 will form the cathode (because it will be reduced). More formally, the strategy in such problems is to find the couples in the electrochemical series and then write the cell diagram with the electrode for the upper couple on the right and that of the lower couple on the left. To write the cell reaction, reverse the lower half-reaction (converting a reduction to an oxidation) and add it to the upper half-reaction.

SOLUTION Since the Cl_2/Cl^- couple lies above the Br_2/Br^- couple, the cell diagram is

$$Pt|Br_2(l)|Br^-(aq)| |Cl^-(aq)|Cl_2(g)|Pt$$

The two half-reactions are

$$\text{Cathode half-reaction:} \quad Cl_2(g) + 2e^- \longrightarrow 2Cl^-(aq)$$

$$\text{Anode half-reaction:} \quad 2Br^-(aq) \longrightarrow Br_2(l) + 2e^-$$

Therefore, the overall reaction is

$$Cl_2(g) + 2Br^-(aq) \longrightarrow 2Cl^-(aq) + Br_2(l)$$

EXERCISE Write the cell diagram and the equation for the cell reaction for the redox couples Ag^+/Ag and Cu^{2+}/Cu^+.
[*Answer*: $Pt|Cu^+(aq),Cu^{2+}(aq)| |Ag^+(aq)|Ag(s)$;
$Ag(s) + Cu^{2+}(aq) \longrightarrow Ag^+(aq) + Cu^+(aq)$]

(a)

▼ EXAMPLE 17.5 Predicting oxidizing power using the electrochemical series

Can aqueous potassium permanganate be used to oxidize iron(II) to iron(III) under standard conditions in acid solution?

STRATEGY We need to look for the relative positions of the two couples in the electrochemical series. One couple is Fe^{3+}/Fe^{2+}; the other is $MnO_4^-,H^+/Mn^{2+},H_2O$. If the latter lies above the Fe^{3+}/Fe^{2+} couple, its oxidized species (MnO_4^-) can oxidize the reduced species (Fe^{2+}) of that couple under standard conditions.

SOLUTION In Table 17.4 we see that the Fe^{3+}/Fe^{2+} couple lies at +0.77 V, which is below the permanganate couple (at +1.51 V). Therefore, permanganate can oxidize Fe^{2+} to Fe^{3+} in acid solution (Fig. 17.16).

EXERCISE Can mercury displace zinc from aqueous zinc sulfate under standard conditions?

[*Answer*: No]

▼ EXAMPLE 17.6 Predicting the direction of a reaction with the electrochemical series

Write the equation for the chemical reaction that occurs when tin is placed in a solution of $Fe^{2+}(aq)$ and $Fe^{3+}(aq)$ ions. Write the diagram for a cell that could make use of the reaction.

STRATEGY Again we need to inspect the electrochemical series for the two redox couples (Sn^{2+}/Sn and Fe^{3+}/Fe^{2+}). The reduced species of the couple that is lower in the series has a tendency to reduce the oxidized species of the higher couple. The cell is formed with the lower, reducing couple on the left (as the anode).

SOLUTION Since the Sn^{2+}/Sn couple lies below the Fe^{3+}/Fe^{2+} couple in Table 17.4, the tin reduces $Fe^{3+}(aq)$ to $Fe^{2+}(aq)$. The half-reactions are

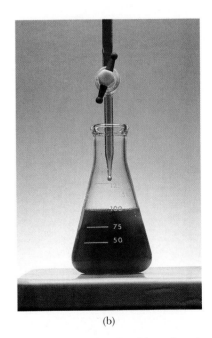

(b)

FIGURE 17.16 In this redox titration, the purple potassium permanganate is being used to oxidize the pale green Fe^{2+} solution (a). The equivalence point is detected by noting when the purple color of the permanganate persists (b).

Anode:	$Sn(s) \longrightarrow Sn^{2+}(aq) + 2e^-$	$E° = +0.14$ V
Cathode:	$Fe^{3+}(aq) + e^- \longrightarrow Fe^{2+}(aq)$	$E° = +0.77$ V
Overall:	$Sn(s) + 2Fe^{3+}(aq) \longrightarrow Sn^{2+}(aq) + 2Fe^{2+}(aq)$	$E° = +0.91$ V

The cell corresponding to this reaction is

$$Sn(s)|Sn^{2+}(aq)|\,|Fe^{2+}(aq),Fe^{3+}(aq)|Pt \qquad E° = 0.91 \text{ V}$$

EXERCISE Write the equation for the reaction that can occur when copper is added to a solution of $Co^{2+}(aq)$ and $Co^{3+}(aq)$ ions.

[*Answer*: $Cu(s) + 2Co^{3+}(aq) \longrightarrow Cu^{2+}(aq) + 2Co^{2+}(aq)$]

To judge the oxidizing strength of an oxidized substance, we note its location in the first column of the electrochemical series. The higher it is, the greater its oxidizing power. This is illustrated by the oxidation of Br^- ions by chlorine, described in Example 17.4. As is shown there, Cl_2 is a stronger oxidizing agent than Br_2, so the oxidation of Br^- to Br_2 is spontaneous. It is, in fact, used to produce bromine from brine that contains bromide ions.

▼ **EXAMPLE 17.7** Predicting relative oxidizing strengths

Is an acidified permanganate solution a more powerful oxidizing agent than an acidified dichromate solution under standard conditions?

STRATEGY We inspect the electrochemical series to see if the permanganate ion MnO_4^- lies above the dichromate ion $Cr_2O_7^{2-}$. We confirm the result by calculating the cell potential.

SOLUTION Since the $MnO_4^-,H^+/Mn^{2+},H_2O$ couple (+1.51 V) lies above the $Cr_2O_7^{2-},H^+/Cr^{3+},H_2O$ couple (+1.33 V), MnO_4^- is the stronger oxidizing agent. We confirm this as follows:

Reduction: $MnO_4^-(aq) + 8H^+(aq) + 5e^-(aq) \longrightarrow Mn^{2+}(aq) + 4H_2O(l)$
$$E° = +1.51 \text{ V}$$

Oxidation: $2Cr^{3+}(aq) + 7H_2O(l) \longrightarrow Cr_2O_7^{2-}(aq) + 14H^+(aq) + 6e^-$
$$E° = -1.33 \text{ V}$$

The equation for the overall reaction is obtained by adding 6 times the reduction half-reaction to 5 times the oxidation half-reaction (so that the numbers of electrons supplied and accepted will match). The cell potential is the sum of the two electrode potentials, which is 0.18 V. This indicates that the reaction free energy is negative (because $E° \propto -\Delta G°$) and hence that the reaction is spontaneous. That is, permanganate ions can oxidize dichromate ions in acid solution.

EXERCISE Which is the stronger reducing agent, Zn or Ni?

[*Answer*: Zn]

The reactions of metals with acids. The production of hydrogen by the action of an acid on a metal is a redox reaction in which the hydrogen ions of the acid are reduced to H_2. Since the standard electrode potential of the H^+/H_2 couple is zero, only substances with negative standard reduction potentials (those below hydrogen in the electrochemical series) can bring this reduction about. We can therefore predict that Mg, Fe, Ca, Sn, and Pb all have a thermodynamic tendency to produce hydrogen. (However, as always, thermodynamics is silent about rates, and although the tendency might exist, the rate of the process might be very slow.) On the other hand, since metals with positive standard electrode potentials (those above hydrogen) cannot reduce hydrogen ions, we know that they do not produce hydrogen when acted on by dilute acid. This is the case for Cu and the noble metals Ag, Pt, and Au. These metals may, however, be able to reduce the anions of oxoacids, which are often more strongly oxidizing than the hydrogen ion.

▼ **EXAMPLE 17.8 Judging whether a metal can reduce an oxoacid**

Show that copper can reduce nitric acid to nitric oxide but not to hydrogen.

STRATEGY We begin by inspecting the electrochemical series (using the longer one in Appendix 2B): copper can reduce the oxidized species in any couple lying above it. To confirm this conclusion, we should show that the resulting redox reaction has a negative free energy.

SOLUTION The Cu^{2+}/Cu couple (+0.34 V) lies above H^+/H_2 (zero), but below the $NO_3^-,H^+/NO,H_2O$ couple (+0.96 V). Therefore, copper cannot reduce hydrogen ions to hydrogen but can reduce NO_3^- ions to NO in an acid medium. We confirm this conclusion as follows:

Reduction:
$$2NO_3^-(aq) + 8H^+(aq) + 6e^- \longrightarrow 2NO(g) + 4H_2O(l)$$
$$E^\circ = +0.96 \text{ V}$$

Oxidation:
$$3Cu(s) \longrightarrow 3Cu^{2+}(aq) + 6e^-$$
$$E^\circ = -0.34 \text{ V}$$

Overall: $3Cu(s) + 2NO_3^-(aq) + 8H^+(aq) \longrightarrow$
$$3Cu^{2+}(aq) + 2NO(g) + 4H_2O(l) \qquad E^\circ = 0.62 \text{ V}$$

The cell potential corresponds to a negative reaction free energy and hence to a spontaneous reaction.

EXERCISE Can mercury reduce nitric acid to (a) hydrogen and (b) nitric oxide?

[*Answer:* (a) No; (b) yes]

▲

Passivation. When a thermodynamic tendency is not realized in practice, there is a good chance that the reason is kinetic, that is, related to the rates of reactions. A glance at the standard electrode potential of aluminum (−1.66 V) suggests that, like magnesium, it should give hydrogen with hydrochloric acid. However, it does not. Aluminum does not react with dilute acid because any Al^{3+} ions that are produced immediately form a hard, unreactive, almost impenetrable layer of oxide on the surface of the metal (Fig. 17.17). This prevents further reaction: we say that the metal has been *passivated*, or protected from further reaction by a surface film. The passivation of aluminum is of great commercial importance, because it enables the metal to be used for, among many other things, airplanes and window frames in buildings. Aluminum containers are used to transport nitric acid, since once the surface is passivated no further reaction occurs.

Corrosion. The electrochemical series gives us some insight into that depressing feature of everyday life, corrosion. Any element lower than the $H_2O/H_2,OH^-$ couple in the electrochemical series has a tendency

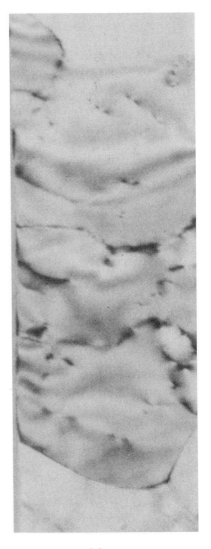

0.2 μm

FIGURE 17.17 This micrograph of a slice through the surface of aluminum shows the passivating layer of oxide that protects it from further attack.

FIGURE 17.18 Iron nails stored in oxygen-free water (left) do not rust because the water is too weakly oxidizing. However, when oxygen is present (right), the oxidation is favored and rust soon forms.

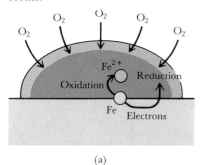

(a)

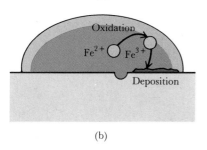

(b)

FIGURE 17.19 The mechanism of rust formation. (a) Oxidation of the iron occurs at a point out of contact with the oxygen of the air, where the metal surface behaves as an anode in a tiny electrochemical cell. (b) Further oxidation of Fe^{2+} to Fe^{3+} results in the deposition of rust on the surface.

to be oxidized by water as a result of the half-reaction

$$2H_2O(l) + 2e^- \longrightarrow H_2(g) + 2OH^-(aq) \qquad E° = -0.83 \text{ V}$$

This standard electrode potential is measured at an OH^- concentration of $1\ M$, which corresponds to pH = 14, a strongly basic solution. At pH = 7, this couple has the potential $E = -0.42$ V. Since the Fe^{2+}/Fe couple has almost the same potential ($E° = -0.44$ V), iron has only a very slight tendency to be oxidized by pure water. For this reason, iron pipes can be used in water-supply systems, and iron can be stored in oxygen-free water without rusting (Fig. 17.18). However, if the iron is exposed to damp air, with both oxygen and water present, the half-reaction

$$O_2(g) + 4H^+(aq) + 4e^- \longrightarrow 2H_2O(l) \qquad E° = +1.23 \text{ V}$$

must be taken into account. The electrode potential of this reaction at pH = 7 is +0.81 V, well above the value for the Fe^{2+}/Fe couple. Hence, oxygen and water can jointly oxidize iron and form rust.

The mechanism of rusting is interesting. A drop of water on the surface of iron acts as the electrolyte in a tiny electrochemical cell (Fig. 17.19). At the edge of the drop, where there is dissolved oxygen close to the metal, the oxygen has a tendency to oxidize the iron by means of the reaction given above. However, the electrons withdrawn from the metal by this oxidation can be replaced by electrons from another part of the conducting metal—in particular, from iron lying beneath the oxygen-poor region of the drop. Iron atoms there give up their electrons, form Fe^{2+} ions, and drift away into the surrounding water. There they meet more oxygen and become oxidized to Fe^{3+} by the oxidizing agent in any couple above Fe^{3+}/Fe^{2+} in the electrochemical series, including oxygen. As they do so, they precipitate as a hydrated iron(III) oxide, $Fe_2O_3 \cdot H_2O$, the brown, insoluble substance we call rust. If the water carries dissolved ions it is more highly conducting, and rusting can occur more rapidly. That is one reason why salt in the air of coastal cities (and in that of inland cities after salt has been used on highways) can be so damaging.

The prevention of corrosion. The simplest way to prevent corrosion is to ensure that the surface of the metal is not exposed to air and water. This can be achieved by painting. A more sophisticated method for iron is to _galvanize_ the metal, or coat it with an unbroken film of zinc, either by dipping it into molten zinc (as is done for automobiles; Fig. 17.20) or by electroplating (a process that is described later). Zinc lies below iron in the electrochemical series, so if a scratch exposes the iron underneath the coating, the more strongly reducing zinc can release electrons to the iron. Hence the zinc, not the iron, is oxidized. The zinc itself survives exposure on the unbroken surface because, like aluminum, it is passivated by a protective oxide.

Zinc plating is better than tin plating, because the Sn^{2+}/Sn couple is more positive than the iron couple. As soon as a tin-plated can is scratched, the more strongly reducing iron supplies electrons to the tin and is oxidized rapidly (Fig. 17.21). This is why tin-plated steel cans, which were common until aluminum began to replace them, corrode so rapidly once they have been damaged.

FIGURE 17.20 Pieces of metal being galvanized by immersion in a bath of molten zinc.

It is not possible to galvanize structures as large as ships, pipelines, and bridges, but a similar measure called *cathodic protection* can be used instead. Instead of the entire surface being covered with a more strongly reducing metal, a block of the active metal (such as magnesium or zinc) is connected to the structure that is to be protected. Then the block—and not the structure—will supply electrons at the demand of the oxygen. The block, which is called a *sacrificial anode*, becomes oxidized but is cheaper to replace than the structure it protects. For similar reasons, automobiles generally have "negative ground systems" as part of their electric circuitry. That is, the body of the car is connected to the anode of the battery. The decay of the anode in the battery is the sacrifice that helps preserve the vehicle itself.

17.5 THE DEPENDENCE OF CELL POTENTIAL ON CONCENTRATION

So far, we have used electrode potentials and the electrochemical series to discuss redox reactions qualitatively. We can now take the discussion one step further and obtain a very useful quantitative result: standard electrode potentials may be used to calculate the equilibrium constant of any reaction that can be expressed as the sum of an oxidation half-reaction and a reduction half-reaction.

Calculating the equilibrium constant. We saw in Section 16.7 that the standard free energy of a reaction is related to its equilibrium constant by

$$\Delta G^\circ = -RT \ln K$$

(This is Eq. 6 of Chapter 16.) We have also seen that the standard reaction free energy is related to the standard cell potential by Eq. 2 of this chapter ($\Delta G^\circ = -nFE^\circ$). Combining the two equations gives

$$\ln K = \frac{nF}{RT} \times E^\circ \qquad (3)$$

(a)

(b)

FIGURE 17.21 The Sn^{2+}/Sn couple lies above the Fe^{2+}/Fe couple in the electrochemical series. When a tin-plated iron can is scratched (a), the iron is rapidly oxidized by Sn^{2+} ions present in the damp environment (b).

The combination RT/F occurs frequently in electrochemistry; at 25°C it has the value

$$\frac{RT}{F} = \frac{8.314 \text{ J}}{\text{K} \cdot \text{mol}} \times 298.15 \text{ K} \times \frac{1}{9.649 \times 10^4} \frac{\text{mol}}{\text{C}}$$

$$= 0.02569 \text{ J/C} = 0.02569 \text{ V}$$

Converting from natural logarithms to common logarithms (using $\ln x = 2.303 \log x$) and substituting into Eq. 3 yields

$$\log K = \frac{n \times E°}{2.303 \times 0.02569 \text{ V}}$$

$$= \frac{n \times E°}{0.0592 \text{ V}} \tag{4}$$

Since we can calculate $E°$ from standard electrode potentials, we can now also calculate equilibrium constants. For example, since the cell potential for the overall reaction

$$\text{Zn}(s) + \text{Cu}^{2+}(aq) \rightleftharpoons \text{Zn}^{2+}(aq) + \text{Cu}(s) \qquad K_c = \frac{[\text{Zn}^{2+}]}{[\text{Cu}^{2+}]}$$

is 1.10 V and $n = 2$ for the reaction as written, we have

$$\log K_c = \frac{2 \times 1.10 \text{ V}}{0.0592 \text{ V}} = 37.2$$

Taking the antilogarithm gives $K_c = 1.6 \times 10^{37}$. Now we know not only that the reaction is spontaneous as written, but also that equilibrium is reached only when the concentration of Zn^{2+} ions is over 10^{37} times that of Cu^{2+} ions. For all practical purposes, the reaction goes to completion.

▼ EXAMPLE 17.9 Calculating an equilibrium constant

Calculate the equilibrium constant for the reaction $\text{AgCl}(s) \rightleftharpoons \text{Ag}^+(aq) + \text{Cl}^-(aq)$ at 25°C. The reaction at the silver–silver chloride electrode was determined in Example 17.1.

STRATEGY We should recognize the equilibrium constant for this reaction as the solubility product for silver chloride in water. Its value can be found from the standard potential of the cell for which it is the overall reaction, using Eq. 4. We thus need to write the cell diagram corresponding to the reaction, obtain its standard potential from the data in Table 17.3, and substitute that into Eq. 4. The value of n may be obtained by inspection of the half-reactions in the two half-cells.

SOLUTION The cell diagram corresponding to the reaction is

$$\text{Ag}(s)|\text{Ag}^+(aq)| \, |\text{Cl}^-(aq)|\text{AgCl}(s)|\text{Ag}(s)$$

The half-reactions and their electrode potentials are

Anode:	$\text{Ag}(s) \longrightarrow \text{Ag}^+(aq) + \text{e}^-$	$E° = -0.80 \text{ V}$
Cathode:	$\text{AgCl}(s) + \text{e}^- \longrightarrow \text{Ag}(s) + \text{Cl}^-(aq)$	$E° = 0.22 \text{ V}$

These reactions show that $n = 1$. The sum of the two electrode potentials is -0.58 V, so, from Eq. 4,

$$\log K = \frac{1 \times (-0.58 \text{ V})}{0.0592 \text{ V}} = -9.80$$

The antilogarithm of -9.80 is 1.6×10^{-10}. Since the units of K_{sp} for AgCl are M^2, we conclude that $K_{sp} = 1.6 \times 10^{-10} \, M^2$, as in Table 15.4.

EXERCISE Calculate the solubility product of mercury(I) chloride.

[*Answer*: $1.3 \times 10^{-18} \, M^3$]

▲

Example 17.9 shows that we now have an electrochemical method for finding the solubility product, and hence the solubility, of a sparingly soluble salt. This is a much more accurate method than trying to measure the minute amount of solid that dissolves in a liter of water.

The Nernst equation. To see how cell potentials depend on concentration and pressure, we make use of the equation relating ΔG to composition, which we obtained in Section 16.6:

$$\Delta G = -RT \ln \frac{K}{Q}$$

Q is the reaction quotient. For the copper-zinc reaction,

$$Zn(s) + Cu^{2+}(aq) \longrightarrow Zn^{2+}(aq) + Cu(s) \qquad Q_c = \frac{[Zn^{2+}]}{[Cu^{2+}]}$$

The molar concentrations in the expression for Q are the actual concentrations in a given electrolyte solution, not necessarily those at equilibrium. In a cell in which $[Zn^{2+}] = 0.10 \, M$ and $[Cu^{2+}] = 0.0010 \, M$, for instance, $Q = 100$. It will be more convenient to use the relation in the form

$$\Delta G = -RT \ln K + RT \ln Q$$
$$= \Delta G° + RT \ln Q$$

Since ΔG is related to the cell potential by Eq. 1, and $\Delta G°$ is related to the *standard* cell potential by Eq. 2,

$$-nFE = -nFE° + RT \ln Q$$

This can be rearranged to the *Nernst equation,*

$$E = E° - \frac{RT}{nF} \ln Q \qquad (5)$$

This equation is named for Walther Nernst, the German chemist who first derived it. At 25°C, and with the natural logarithm converted to a common logarithm, we have

$$E = E° - \frac{0.0592 \text{ V}}{n} \times \log Q \qquad (6)$$

▼ **EXAMPLE 17.10** **Using the Nernst equation**

Calculate the potential at 25°C of a Daniell cell in which the molar concentration of Zn^{2+} ions is $0.10 \, M$ and that of the Cu^{2+} ions is $0.0010 \, M$.

STRATEGY All we have to do is substitute the concentrations or pressures into the expression for Q and then use Eq. 6. If necessary, the standard cell potential may be calculated as the difference between the two standard reduction potentials.

SOLUTION Since the standard potential of the Daniell cell is 1.10 V, its potential when the Zn^{2+} concentration is 0.10 M and the Cu^{2+} concentration is 0.0010 M is

$$E = 1.10 \text{ V} - \frac{0.0592 \text{ V}}{2} \log \frac{0.10}{0.0010} = 1.04 \text{ V}$$

EXERCISE Calculate the potential of the cell $Zn|Zn^{2+}(aq)||Fe^{2+}(aq)|Fe$ when the Fe^{2+} concentration is 0.10 M and the Zn^{2+} concentration is 1.50 M.

▲ [*Answer*: 0.32 V]

Electrical measurement of pH. One important application of the Nernst equation is to the measurement of pH (and, through pH, of the pK_a and pK_b of acids and bases, as explained in Section 15.2). This application makes use of a cell in which one electrode is sensitive to the hydrogen ion concentration. A hydrogen electrode in combination with a calomel electrode, $Hg(l)|Hg_2Cl_2(s)|Cl^-(aq)$, connected through a salt bridge, is one possible cell. The half-reaction for the calomel electrode (which is described in the exercise for Example 17.1) is

$$Hg_2Cl_2(s) + 2e^- \longrightarrow 2Hg(l) + 2Cl^-(aq) \qquad E° = +0.27 \text{ V}$$

Suppose we set up a cell of the form

$$Pt|H_2(g)|H^+(aq)| \, |Cl^-(aq)|Hg_2Cl_2(s)|Hg(l) \qquad E° = +0.27 \text{ V}$$

Then the cell reaction and reaction quotient when the pressure of the hydrogen is 1 atm are

$$H_2(g) + Hg_2Cl_2(s) \longrightarrow 2H^+(aq) + 2Cl^-(aq) + 2Hg(l)$$
$$Q = [H^+]^2[Cl^-]^2$$

(The concentrations of the H^+ and Cl^- ions are unrelated, since they are contained in different electrode compartments and the Cl^- concentration is fixed for a given electrode.) We want to know the concentration of hydrogen ions in the anode compartment. Substituting Q into the Nernst equation gives

$$E = E° - \frac{0.0592 \text{ V}}{2} \times \log [H^+]^2[Cl^-]^2$$

$$= 0.27 \text{ V} - \frac{0.0592 \text{ V}}{2} \times \log [Cl^-]^2 - \frac{0.0592 \text{ V}}{2} \log [H^+]^2$$

$$= 0.27 \text{ V} - 0.0592 \text{ V} \log [Cl^-] - 0.0592 \text{ V} \log [H^+]$$

The first two terms on the right are constants that we can combine into a single constant E'. Then

$$E = E' - 0.0592 \text{ V} \times \log [H^+]$$

$$= E' + 0.0592 \text{ V} \times pH$$

That is, the cell potential is proportional to the pH, and by measuring it we can determine the pH.

In practice the hydrogen electrode is difficult to use because it is awkward to set up, and it settles down to give a stable reading only very sluggishly. The *glass electrode*, a thin-walled glass bulb containing an electrolyte (Fig. 17.22), is very much easier to use and has a potential

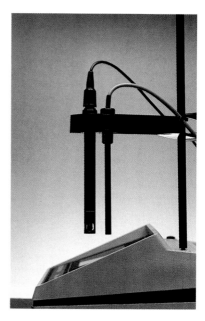

FIGURE 17.22 A glass electrode used to measure pH. This electrode is usually used in conjunction with a calomel electrode.

that is proportional to the pH. It is normally used in conjunction with a calomel electrode that makes contact with the test solution through a salt bridge. The cell potential is measured, but the pH of the solution is displayed directly. The meter is calibrated with a phosphate buffer of known pH, such as that described in Section 15.4.

Since other commercially available electrodes are sensitive to other ions, "pX meters" are available that can be used to measure the concentrations of various ions X. These electrodes are useful for monitoring industrial processes and in pollution control.

ELECTROLYSIS

We now turn to the use of electricity to bring about a chemical change—the reverse of the generation of electricity by chemical reaction. Much of our discussion of electrochemical cells applies to electrolysis, but there are two major differences. First, an electrolysis cell always has a single electrolyte, and the two electrodes share the same compartment. Second, the conditions are usually far from standard: gas pressures are rarely 1 atm, and ion concentrations are generally quite different from 1 M.

A simple electrolysis cell is shown in Fig. 17.23. It has the same construction as the electrochemical cells we have discussed already, with two electrodes dipping into an electrolyte, but it is used differently: current is forced through the cell by connecting it to an external supply. In the case of a cell containing copper ions, electrons enter the cell through the cathode, causing reduction there:

$$Cu^{2+}(aq) + 2e^- \longrightarrow Cu(s)$$

Copper metal is deposited on the cathode (which need not be made of copper). Electrons are removed from the cell through the anode, the site of oxidation. If that electrode is made of copper, the reaction is

$$Cu(s) \longrightarrow Cu^{2+}(aq) + 2e^-$$

and copper ions go into solution. The net outcome of the process in this case is the transfer of copper from the anode to the cathode.

In order for electrons to enter the electrolytic cell at the cathode, that electrode must be connected to the negative electrode—the anode—of another cell (or of some other source of electricity). The anode is where electrons leave a cell that is producing electricity, and hence the anode of the other cell provides the supply of electrons that enters the electrolytic cell. Similarly, the electrons that leave the electrolytic cell at that cell's anode enter the other electron-supplying cell through its positive electrode—its cathode.

17.6 THE POTENTIAL NEEDED FOR ELECTROLYSIS

We mentioned the electrolysis of water in Chapter 2 and are now in a position to discuss it more fully. The general idea behind electrolysis is to use an electric current to drive a reaction in the reverse of its sponta-

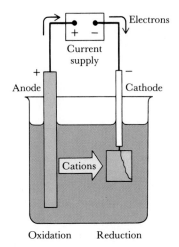

FIGURE 17.23 A simple electrolysis cell. The current enters the cell at the cathode, which is connected to the negative (−) terminal of the external supply. It leaves through the anode, which is connected to the positive (+) terminal.

neous direction. This can be done by connecting the electrodes of the electrolysis cell to a supply with a greater potential than the electrolysis cell would have if it were being used to produce electricity. Then current is forced to flow in the reverse of its spontaneous direction, and the cell reaction is driven in reverse. The electrolysis reaction

$$2H_2O(l) \longrightarrow 2H_2(g) + O_2(g)$$

illustrates this point. It is the reverse of the cell reaction discussed in Example 17.3:

$$2H_2(g) + O_2(g) \longrightarrow 2H_2O(l)$$

$$Pt|H_2(g)|OH^-(aq)|O_2(g)|Pt \qquad E = 1.23 \text{ V at pH} = 7$$

To reverse this cell reaction, we must supply at least 1.23 V to the electrolysis cell, with the negative terminal of the external supply connected to the hydrogen electrode (so that the H_2O is reduced to H_2 there). This voltage requirement can be expressed in terms of the potentials needed for the two electrodes. In neutral water, where pH = 7, the two half-reactions we want to drive by electrolysis are

$$2H_2O(l) + 2e^- \longrightarrow H_2(g) + 2OH^-(aq) \qquad E = -0.42 \text{ V}$$

$$4OH^-(aq) \longrightarrow O_2(g) + 2H_2O(l) + 4e^- \qquad E = -0.81 \text{ V}$$

That is, we must supply at least 1.23 V from an outside source: 0.42 V to achieve the production of hydrogen, and 0.81 V to achieve the production of oxygen.

Overpotential. In practice, appreciable product formation is obtained by electrolysis only if the applied potential is significantly greater than the cell potential. The additional potential that must be applied beyond the cell potential is called the _overpotential_. The overpotential for the production of hydrogen and oxygen using platinum electrodes is about 0.6 V, so about 1.8 V (and not 1.23 V) must be applied to the electrolytic cell before hydrogen and oxygen are evolved at an appreciable rate. Of this 0.6 V of cell overpotential, 0.5 V is needed to drive the oxygen evolution, and only 0.1 V is needed for the hydrogen evolution. The overpotentials required when other electrode materials are used may be quite different. Hydrogen evolution using a lead electrode requires an overpotential of 0.6 V, whereas oxygen evolution requires 0.3 V.

Competing reductions. If ions are present in water that is being electrolyzed, it is possible that they, rather than the water, will be reduced at the cathode. The standard reduction potential of Na^+ ions, for example, is

$$Na^+(aq) + e^- \longrightarrow Na(s) \qquad E° = -2.71 \text{ V}$$

This is much more negative than the potential for the reduction of water to hydrogen, so the reduction of Na^+ ions in water is unimportant. However, if Cu^{2+} ions are present instead, we have to consider the possibility that the reduction

$$Cu^{2+}(aq) + 2e^- \longrightarrow Cu(s) \qquad E° = +0.34 \text{ V}$$

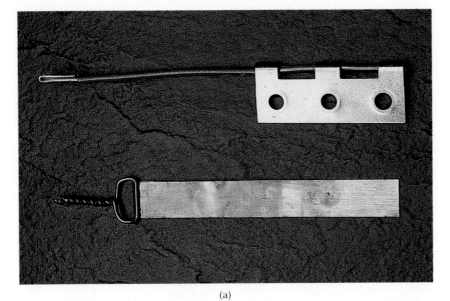

(a)

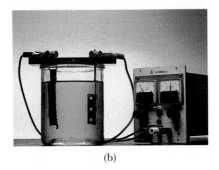

(b)

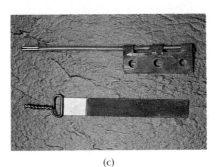

(c)

will occur, for Cu^{2+} ions are thermodynamically much easier to reduce than water. Moreover, since the overpotential for metal deposition is generally quite small, we would expect to find copper deposited rather than hydrogen evolved when copper ions are present in aqueous solu-tion. This is in fact the process that is used to refine copper into the pure metal (Fig. 17.24).

The reduction potential for neutral water, -0.42 V, serves as the dividing line between couples that can be reduced electrolytically in aqueous solution and those that cannot. Any oxidized species with a standard reduction potential greater than about -0.42 V can be reduced electrolytically in water. However, if we attempt to reduce an oxidized species with a potential below about -0.42 V, hydrogen will be produced instead. Aluminum ions ($E° = -1.66$ V), for exam-ple, cannot be reduced to the metal in water; however, silver ions ($E° = +0.80$ V) can be reduced to silver.

Competing oxidations. An alternative possibility is that ions that are present may be oxidized by the anode. For instance, when Cl^- ions are present in water that is being electrolyzed, it is possible that they, and not the water, will be oxidized. In neutral water the two oxidation potentials are

$$4OH^-(aq) \longrightarrow O_2(g) + 2H_2O(l) + 4e^- \qquad E = -0.81 \text{ V}$$

$$2Cl^-(aq) \longrightarrow Cl_2(g) + 2e^- \qquad E° = -1.36 \text{ V}$$

Since only 0.81 V is needed to drive the first reaction, but 1.36 V is needed for the second, it appears that oxygen should be the product. However, the overpotential for oxygen production is very high, and in practice chlorine may be produced instead.

17.7 THE EXTENT OF ELECTROLYSIS

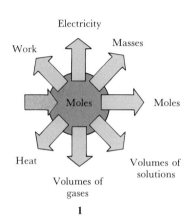

1

We now focus on the amount of substance that can be electrolyzed by a given amount of electricity. This is where we add the final arrow to the mole diagram (**1**). The data for our calculation might be the amount of electricity supplied, and the desired outcome of the calculation the number of moles, grams, or liters of product that the electrolysis reaction forms.

Faraday's law. If we were interested in the electrolytic deposition of copper, we could write the reduction equation

$$Cu^{2+}(aq) + 2e^- \longrightarrow Cu(s)$$

and from it obtain the relation

$$2 \text{ mol e}^- = 1 \text{ mol Cu}$$

We could then use this relation to form a conversion factor for predicting the number of moles of Cu atoms produced when a given number of moles of electrons is supplied through the cathode. For example, if 4.0 mol of electrons is supplied, then

$$\text{Moles of Cu} = \text{moles of electrons}$$
$$\times \text{ moles of Cu per mole of electrons}$$

$$= 4.0 \text{ mol e}^- \times \frac{1 \text{ mol Cu}}{2 \text{ mol e}^-} = 2.0 \text{ mol Cu}$$

This simple calculation is summarized by the following statement, which is a combined, modern version of Faraday's law:

Faraday's law of electrolysis: The number of moles of product formed in an electrolysis cell by an electric current is chemically equivalent to the number of moles of electrons supplied.

From the number of moles of product formed, we can calculate the masses of the products or, if they are gases, their volumes.

▼ **EXAMPLE 17.11** **Calculating the volume of gas produced in an electrolysis**

Calculate the volume of hydrogen produced at STP by the electrolysis of water when 0.050 mol of electrons is supplied.

STRATEGY We begin by writing the reduction half-reaction and finding a relation between moles of H_2 and moles of electrons taking part in the reduction. This can be used to form a factor for converting from moles of electrons supplied to moles of H_2 produced. However, we are asked for the volume of gas produced, not the number of moles. The volume can be calculated by multiplying the number of moles of H_2 by its molar volume at STP (22.4 L/mol).

SOLUTION The reduction half-reaction is

$$2H_2O(l) + 2e^- \longrightarrow H_2(g) + 2OH^-(aq)$$

so 2 mol e^- = 1 mol H_2. Therefore,

Moles of H_2 = moles of e^- × moles of H_2 per mole of e^-

$$= 0.050 \text{ mol } e^- \times \frac{1 \text{ mol } H_2}{2 \text{ mol } e^-} = 0.025 \text{ mol } H_2$$

Since the molar volume of H_2 at STP is 22.4 L/mol, this amount of hydrogen corresponds to

$$\text{Volume of hydrogen} = 0.025 \text{ mol } H_2 \times \frac{22.4 \text{ L}}{\text{mol}} = 0.56 \text{ L } H_2$$

EXERCISE Calculate the volume of oxygen produced in a similar electrolysis when 0.020 mol of electrons is supplied.

[*Answer*: 0.11 L O_2]

▲

Measuring the amount of electricity. The final link between the amount of product formed and the quantity of electricity used is the determination of the number of moles of electrons supplied during the electrolysis. The Faraday constant, F = 96.5 kC/mol, gives the charge per mole of electrons; hence we can use the relation

$$1 \text{ mol } e^- = 96.5 \text{ kC}$$

to write a conversion factor for converting coulombs supplied to moles of electrons supplied. That is, once we know the number of coulombs supplied, we can calculate

Moles of electrons = coulombs supplied × moles per coulomb

$$= \text{coulombs supplied} \times \frac{1 \text{ mol } e^-}{96.5 \text{ kC}}$$

To find the charge that has been supplied, we need to know that an electric current of one ampere supplies one coulomb of charge per second. That is, 1 A = 1 C/s, and so

Coulombs supplied = current in amperes × time in seconds

For instance, the total charge passing through a television set that uses a 2.0-A current for 1 h is

Coulombs supplied = 2.0 A × 3600 s = 7200 C   or   7.2 kC

Hence, by measuring the current and the time for which it flows, we can calculate the number of moles of electrons supplied. The number of moles of electrons passing through the television set, for instance, is

$$\text{Moles of electrons} = 7.2 \text{ kC} \times \frac{1 \text{ mol e}^-}{96.5 \text{ kC}} = 0.075 \text{ mol e}^-$$

Note that even appreciable currents, flowing for long periods, supply very few moles of electrons.

▼ **EXAMPLE 17.12** **Predicting the mass of an element produced by electrolysis**

Aluminum is produced by the electrolysis of aluminum oxide dissolved in molten cryolite (Na_3AlF_6). Calculate the mass of aluminum that can be produced in one day in an electrolysis cell operating continuously at 100,000 A.

STRATEGY The focus of calculations like this is the relation between the number of moles of electrons supplied and the number of moles of Al atoms produced. We work from the data toward the answer. The charge supplied (in coulombs) is the product of the current and the time for which it flows. Thus charge is converted to the number of moles of electrons supplied using the Faraday constant. The number of moles of electrons is converted to a number of moles of Al atoms. Finally, the number of moles of Al atoms is converted to grams using the molar mass of Al (26.98 g/mol).

SOLUTION Since there are 86,400 s in a day,

$$\text{Charge in coulombs} = 100,000 \text{ A} \times 86,400 \text{ s} = 8.64 \times 10^9 \text{ C}$$

The number of moles of electrons supplied is

$$\text{Number of moles of e}^- = 8.64 \times 10^9 \text{ C} \times \frac{1 \text{ mol e}^-}{9.65 \times 10^4 \text{ C}}$$

$$= 8.95 \times 10^4 \text{ mol e}^-$$

Since $Al^{3+} + 3e^- \rightarrow Al$, we know that 3 mol e$^-$ = 1 mol Al; hence

$$\text{Number of moles of Al} = 8.95 \times 10^4 \text{ mol e}^- \times \frac{1 \text{ mol Al}}{3 \text{ mol e}^-}$$

$$= 2.98 \times 10^4 \text{ mol Al}$$

Finally, since 1 mol Al = 26.98 g Al,

$$\text{Mass of Al} = 2.98 \times 10^4 \text{ mol Al} \times \frac{26.98 \text{ g Al}}{1 \text{ mol Al}}$$

$$= 8.04 \times 10^5 \text{ g Al} \quad \text{or} \quad 804 \text{ kg Al}$$

The fact that the production of each Al atom requires three electrons accounts for the very high consumption of electricity that is characteristic of aluminum production plants.

EXERCISE Calculate the mass of magnesium produced by the electrolysis of magnesium chloride in a cell operating all day at 50,000 A.

[*Answer*: 544 kg]

▲

17.8 APPLICATIONS OF ELECTROLYSIS

A cathode acts as a very powerful reducing agent. Increasing the voltage between the electrodes forces electrons into the electrolyte and can bring about a reduction such as $Cu^{2+}(aq) + 2e^- \rightarrow Cu(s)$. An anode acts as a very powerful oxidizing agent (Fig. 17.25). When the applied potential drags a current out of the cell, the electrons are stripped from compounds and ions near the anode, as in the oxidation $2F^-(aq) \rightarrow F_2(g) + 2e^-$. The oxidizing power of an anode can be made so great (by increasing the voltage) that even fluoride ions, which cannot be oxidized by any chemical reagent, can be oxidized electrolytically to the element. This is how Henri Moisson first isolated fluorine (in 1886) from an anhydrous molten mixture of potassium fluoride and hydrogen fluoride. It is still the way that fluorine is prepared commercially.

Sodium ion electrolysis. Another industrial application of electrolysis is the production of sodium metal by the *Downs process*. The electrolyte is molten sodium chloride, mined as rock salt, with a little calcium chloride added to lower its melting point of 830°C to a more economical 600°C. The principal design requirement for the cell (Fig. 17.26) is to keep the sodium and chlorine produced by the electrolysis out of contact with each other and with air.

We have already seen that if aqueous rather than molten sodium chloride is used as the electrolyte, then hydrogen is produced at the cathode, not sodium. The reduction of the electrolyte solution produces OH^- ions, so the product formed in the cathode chamber is sodium hydroxide. Because the overpotential of oxygen is so high, chlorine is the oxidation product in the anode compartment. However, if the chlorine is allowed to mix with the sodium hydroxide produced at the cathode, it is oxidized to hypochlorite:

$$Cl_2(g) + OH^-(aq) \longrightarrow ClO^-(aq) + Cl^-(aq) + H^+(aq)$$

Sodium hypochlorite is widely used as household bleach. Bleaches act by oxidizing colored compounds, and they kill bacteria by oxidation as

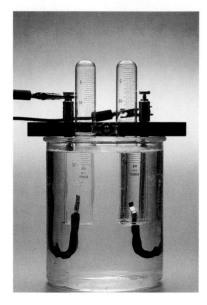

FIGURE 17.25 The electrochemical production of chlorine. An anode can be made a strong enough oxidizing agent to remove electrons from chloride ions. This is the only method available for producing fluorine from fluoride ions.

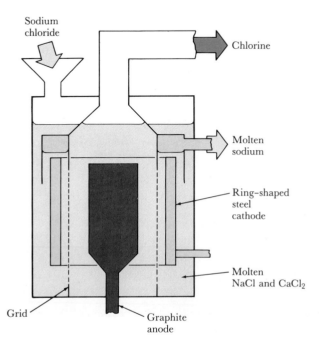

FIGURE 17.26 In the Downs process, molten sodium chloride is electrolyzed with a graphite anode (which oxidizes the Cl^- ions to chlorine) and a steel cathode (which reduces the Na^+ ions to sodium). The sodium and chlorine are kept apart by the hoods surrounding the electrodes.

well. This process has been suggested as a means of treating sewage in coastal areas: the sewage would be mixed with seawater (brine) and electrolyzed; the hypochlorite ions so produced would oxidize the organic matter, making it harmless.

Electroplating. A common industrial application of electrolysis is <u>*electroplating*</u>, the deposition of a thin film of metal on an object by electrolysis. The object to be plated is made the cathode. Metal is deposited on the cathode from ions supplied to the electrolyte solution by the oxidation of the anode, which is made of the metal that is to be deposited. The plated object is usually metal, but plastic objects can also be plated if they are first coated with graphite. The electrolyte is an aqueous solution of a salt of the plating metal; however, to eliminate side reactions and control the rate at which metal atoms are deposited on the cathode (so that a smooth surface is obtained), the salt must be chosen very carefully. In silver plating, for example, it has been found that good results are obtained if the silver ions are present as $[Ag(CN)_2]^-$.

One of the most familiar examples of electroplating is "chrome plating," in which a thin layer of chromium is deposited on another metal (Fig. 17.27). The electrolyte is prepared by dissolving CrO_3 in dilute sulfuric acid. The electrolysis reduces chromium(VI) first to chromium(III) and then to chromium metal, the overall reduction being

$$CrO_3(aq) + 6H^+(aq) + 6e^- \longrightarrow Cr(s) + 3H_2O(l)$$

The chromium is deposited on the cathode as a hard protective film. Because *six* electrons must be supplied for each atom of chromium deposited, large amounts of electricity are needed. Attempts to cut the cost of electricity by half by starting from a chromium(III) salt have failed because the hydrated ion $[Cr(H_2O)_6]^{3+}$ is so stable that it cannot

FIGURE 17.27 Electroplating is the process of depositing a thin layer of one metal on another. Here, chromium is being deposited on steel components.

be reduced. When chromium(VI) is used, the chromium(III) produced in the first reduction stage is already bound to the cathode surface, and it can be reduced quite easily.

We began our study of chemistry, in Chapter 1, with an elementary account of electrolysis. In the intervening chapters, we progressed to a much deeper understanding of that phenomenon, and we can now make both qualitative and quantitative predictions about it. We have encountered other types of reaction along the way, and have achieved a similar depth of understanding and capability with them. We have, in fact, built up a considerable store of knowledge about the methods and principles of chemistry. In Part IV, we shall put them to use to explain the characteristics of individual elements.

SUMMARY

17.1 Electrochemistry is the branch of chemistry concerned with the chemical generation of electricity and the use of electricity to cause chemical change. It provides information about redox reactions and reactions that can be expressed as sums of redox reactions. Electrochemical measurements are made on **electrochemical cells,** which consist of two electrodes dipping into an electrolyte. Reduction occurs at the **cathode,** and oxidation at the **anode.** The half-reaction at either electrode defines its redox **couple,** the oxidized and reduced forms of the substances taking part in the reaction. A cell is specified by its **cell diagram,** in which the cathode appears on the right. The voltage produced by a cell is reported as the (positive) **cell potential.** A positive cell potential shows that the reducing agent of the left-hand couple (in the cell diagram) has a natural tendency to reduce the oxidizing agent of the right-hand couple.

17.2 A **primary cell** is one that produces electricity from reagents built into it when it is manufactured. A **secondary cell** is a cell that must be charged. A **fuel cell** produces electricity from reagents that are supplied to it while it is operating. An example of a primary cell is the Leclanché **dry cell.** An example of a collection of secondary cells is the **lead-acid battery** used in automobiles.

17.3 The potential of a cell is proportional to the Gibbs free energy of the cell reaction. The reporting of cell properties is simplified by the use of the **standard cell potential,** the potential (in volts) of a cell when all the ions taking part in the redox reaction are at 1 M concentration and any gases are at 1 atm pressure. The standard cell potential is the sum of the contributions of the two electrodes: these **standard electrode potentials** are defined in relation to the **standard hydrogen electrode,** which by convention is taken to contribute zero volts; they may be reported as either **oxidation potentials** or **reduction potentials.**

17.4 The **electrochemical series** is a list of redox couples arranged in order of their reduction potentials. If the couple ox/red lies below another couple in the series, then red is the stronger reducing agent, and ox the weaker oxidizing agent. The lower of the two couples forms the anode of a cell, and the half-reaction it undergoes is red → ox + e$^-$. Only metals with negative reduction potentials can reduce acids to hydrogen. However, kinetic factors may interfere, and a strongly reducing metal is oxidized very slowly if it is **passivated** by an inert oxide film. Electrochemical processes include **corrosion** and such corrosion-prevention methods as **galvanizing** and the use of a **sacrificial anode.**

17.5 The electrode potentials for the two couples in a reaction give the equilibrium constant for the cell reaction. The relation between the cell potential and the concentrations or pressures of the reagents and products of a cell reaction is given by the **Nernst equation.** Of particular importance is the Nernst equation for a cell containing a hydrogen electrode, for the cell potential is proportional to the pH. A **glass electrode** is a more conveniently used version of the hydrogen electrode.

17.6 Electrolysis is the use of an electric current to bring about a chemical reaction. The applied voltage must overcome the cell potential by an amount equal to the **overpotential.**

17.7 and 17.8 The amount of product formed by a given current is described by **Faraday's law of electrolysis** in terms of moles of electrons and substances. The number of moles of electrons supplied is found by noting the time for which a known current flows through the cell. In an electrolytic cell, the cathode is connected to the negative terminal of a supply of electric current. A cathode is a very powerful reducing agent, and an anode a very powerful oxidizing agent. Electrolysis is used commercially for producing aluminum, sodium, chlorine, sodium hydroxide, sodium hypochlorite, and fluorine (among other substances). It is also used for **electroplating,** the electrochemical deposition of a metal on an object.

EXERCISES

Assume a temperature of 25°C (298 K) for the following exercises, unless instructed otherwise.

Cells and Cell Reactions

In Exercises 17.1 and 17.2, write the diagram for each electrode and give the reduction half-reaction.

17.1 (a) Zinc metal in contact with zinc cations; (b) platinum dipping into a solution of iron(II) and iron(III) salts; (c) chlorine gas in contact with chloride ions; (d) the calomel electrode.

17.2 (a) Silver in contact with silver ions; (b) platinum in contact with cerium(III) and cerium(IV) ions; (c) oxygen gas in contact with hydroxide ions; (d) the silver–silver iodide electrode.

In Exercises 17.3 and 17.4, write the electrode at which each half-reaction would occur.

17.3 (a) $Cd^{2+}(aq) + 2e^- \longrightarrow Cd(s)$
(b) $S_2O_8^{2-}(aq) + 2e^- \longrightarrow 2SO_4^{2-}(aq)$
(c) $Hg_2Cl_2(s) + 2e^- \longrightarrow 2Hg(l) + 2Cl^-(aq)$
17.4 (a) $U^{4+}(aq) + e^- \longrightarrow U^{3+}(aq)$
(b) $2Hg^{2+}(aq) + 2e^- \longrightarrow Hg_2^{2+}(aq)$
(c) $O_2(g) + H_2O(l) + 2e^- \longrightarrow$
$$HO_2^-(aq) + OH^-(aq)$$

In Exercises 17.5 to 17.8, write the electrode that you would set up to study each given couple. All ions are in aqueous solution.

17.5 (a) Cu^{2+}/Cu; (b) Cl_2/Cl^-; (c) $O_2,H_2O/OH^-$; (d) Pt^{2+}/Pt.
17.6 (a) Pb^{2+}/Pb; (b) Pb^{4+}/Pb^{2+}; (c) $AgBr/Ag,Br^-$; (d) $O_2,H^+/H_2O$.

17.7 (a) $Fe(OH)_2/Fe,OH^-$; (b) $MnO_2,H^+/Mn^{2+},H_2O$.
17.8 (a) $Cd(OH)_2/Cd,OH^-$;
(b) $Cr_2O_7^{2-},H^+/Cr^{3+},H_2O$.

In Exercises 17.9 and 17.10, write the diagram for each cell.

17.9 (a) A hydrogen electrode in combination with a silver electrode

(b) A chlorine electrode in combination with a silver-silver chloride electrode
(c) A calomel electrode in combination with a hydrogen electrode
(d) A copper-copper(II) electrode in combination with an iron(III)-iron(II) electrode
17.10 (a) An acid oxygen electrode in combination with a hydrogen electrode
(b) A manganese(III)-manganese(II) electrode in combination with a chromium(III)-chromium(II) electrode
(c) A silver-silver iodide electrode in combination with an iodine-iodide electrode
(d) Two hydrogen electrodes at different pressures

In Exercises 17.11 to 17.16, write the half-reactions and the cell reaction for each cell.

17.11 (a) $Ag(s)|Ag^+(aq)| |Fe^{3+}(aq),Fe^{2+}(aq)|Pt$
(b) $Pt|H_2(g, P_L)|HCl(aq)|H_2(g, P_R)|Pt$
(c) $U(s)|U^{3+}(aq)| |V^{2+}(aq)|V(s)$
17.12 (a) $Cu(s)|Cu^{2+}(aq)| |Cu^+(aq)|Cu(s)$
(b) $Pt|I_3^-(aq),I^-(aq)| |Cl^-(aq)|Cl_2(g)|Pt$
(c) $Ag(s)|AgCl(s)|Cl^-(aq)| |I^-(aq)|AgI(s)|Ag(s)$
17.13 (a) $Hg(l)|Hg_2Cl_2(s)|Cl^-(aq)| |Cl^-(aq)|AgCl(s)|Ag(s)$
(b) $Sn(s)|Sn^{2+}(aq)| |Sn^{4+}(aq),Sn^{2+}(aq)|Pt$
(c) $Pt|Sn^{4+}(aq),Sn^{2+}(aq)| |Sn^{2+}(aq)|Sn$
(d) $Au(s)|Au^+(aq)| |Au^{3+}(aq)|Au(s)$
17.14 (a) $Hg(l)|Hg_2^{2+}(aq)| |Cl^-(aq)|Hg_2Cl_2(s)|Hg(l)$
(b) $Pt|Sn^{4+}(aq),Sn^{2+}(aq)| |Pb^{4+}(aq),Pb^{2+}(aq)|Pt$
(c) $Pt|Pb^{4+}(aq),Pb^{2+}(aq)| |Sn^{4+}(aq),Sn^{2+}(aq)|Pt$
(d) $Pt|O_2(g)|H^+(aq)| |OH^-(aq)|O_2(g)|Pt$

In Exercises 17.15 to 17.18, devise a cell for each given cell reaction.

17.15 (a) $Cr(s) + Zn^{2+}(aq) \longrightarrow Cr^{2+}(aq) + Zn(s)$
(b) $Cr^{3+}(aq) + Zn(s) \longrightarrow Cr(s) + Zn^{2+}(aq)$

(c) $H_2(g) + Cl_2(g) \longrightarrow 2HCl(aq)$
(d) $3Au^+(aq) \longrightarrow 2Au(s) + Au^{3+}(aq)$

17.16 (a) $Mn(s) + Ti^{2+}(aq) \longrightarrow Mn^{2+}(aq) + Ti(s)$
(b) $Mn^{2+}(aq) + Ti(s) \longrightarrow Mn(s) + Ti^{2+}(aq)$
(c) $Cr^{2+}(aq) + Ce^{4+}(aq) \longrightarrow Cr^{3+}(aq) + Ce^{3+}(aq)$
(d) $2Cu^+(aq) \longrightarrow Cu(s) + Cu^{2+}(aq)$

17.17 (a) $AgBr(s) \longrightarrow Ag^+(aq) + Br^-(aq)$, a dissolution

(b) $H_3O^+(aq) + OH^-(aq) \longrightarrow$
    $2H_2O(l)$, a Brønsted-Lowry neutralization
(c) $Cd(s) + 2Ni(OH)_3(s) \longrightarrow$
    $Cd(OH)_2(s) + 2Ni(OH)_2(s)$,
the reaction in the nickel-cadmium cell

17.18 (a) $AgNO_3(aq) + KI(aq) \longrightarrow AgI(s) + KNO_3(aq)$,
a precipitation
(b) $H_3O^+(aq, conc_1) \longrightarrow H_3O^+(aq, conc_2)$, a dilution
(c) $Zn(s) + 2MnO_2(s) + 2NH_4Cl(aq) \longrightarrow$
    $[Zn(NH_3)_2]Cl_2(aq) + 2MnO(OH)(s)$,
the reaction in a dry cell

Sometimes you are given only sketchy information about the cell reaction, but it is still possible to think of an appropriate cell. In Exercises 17.19 and 17.20, write the cell corresponding to the reaction.

17.19 Acidified potassium permanganate solution used to oxidize iron(II) chloride
17.20 Acidified sodium dichromate used to oxidize mercury(I) to mercury(II)

Standard Electrode Potentials

In Exercises 17.21 to 17.24, give the standard electrode potential of each couple.

17.21 (a) Cu^{2+}/Cu; (b) Fe^{3+}/Fe^{2+}; (c) H^+/H_2.
17.22 (a) Ce^{4+}/Ce^{3+}; (b) Cu^+/Cu; (c) Cl_2/Cl^-.
17.23 (a) I_3^-/I^-; (b) $Hg_2Cl_2/Hg,Cl^-$; (c) $AgBr/Ag,Br^-$.
17.24 (a) $O_2,H_2O/HO_2^-,OH^-$;
(b) $MnO_2,H^+/Mn^{2+},H_2O$; (c) $S_2O_8^{2-}/SO_4^{2-}$.

In Exercises 17.25 to 17.28, predict the voltage that would be produced under standard conditions by each combination of electrodes. State which terminal would be the cathode, and which the anode.

17.25 (a) $Cu(s)|Cu^{2+}(aq)$ and $Cu(s)|Cu^+(aq)$
(b) $Ag(s)|AgI(s)|I^-(aq)$ and $Ag(s)|AgCl(s)|Cl^-(aq)$
17.26 (a) $Ag(s)|Ag^+(aq)$ and $Pt|Fe^{2+}(aq),Fe^{3+}(aq)$
(b) $U(s)|U^{3+}(aq)$ and $V(s)|V^{2+}(aq)$
17.27 (a) $Hg(l)|Hg_2^{2+}(aq)$ and $Hg(l)|Hg_2Cl_2(s)|Cl^-(aq)$
(b) $Pt|Sn^{2+}(aq),Sn^{4+}(aq)$ and $Pt|Pb^{2+}(aq),Pb^{4+}(aq)$
17.28 (a) $Sn(s)|Sn^{2+}(aq)$ and $Pt|Sn^{2+}(aq),Sn^{4+}(aq)$
(b) $Au(s)|Au^+(aq)$ and $Au(s)|Au^{3+}(aq)$

The Electrochemical Series

In Exercises 17.29 to 17.34, arrange the given elements in order of increasing reducing character.

17.29 Cu, Zn, Cr, Fe
17.30 Li, Na, K, Mg

17.31 U, V, Ti, Al
17.32 Ni, Sn, Au, Ag

17.33 S_8, H_2, O_2, Cl_2, F_2
17.34 Pb, Sn, Br_2, H_2, Hg, I_2

In Exercises 17.35 and 17.36, state which reducing agent in each pair of couples will reduce the oxidizing agent in the other couple (under standard conditions).

17.35 (a) K^+/K and Na^+/Na
(b) Cl_2/Cl^- and Br_2/Br^-
(c) In^{3+}/In^{2+} and Sn^{4+}/Sn^{2+}
(d) V^{3+}/V^{2+} and Ti^{3+}/Ti^{2+}
17.36 (a) La^{3+}/La and Na^+/Na
(b) Cu^{2+}/Cu and I_3^-/I^-
(c) F_2/F^- and $S_2O_8^{2-}/SO_4^{2-}$
(d) U^{4+}/U^{3+} and Fe^{3+}/Fe^{2+}

In Exercises 17.37 and 17.38, determine which reactions are spontaneous in the forward direction under standard conditions.

17.37 (a) $Cl_2(g) + 2Br^-(aq) \longrightarrow 2Cl^-(aq) + Br_2(l)$
(b) $3Cd(s) + 2Bi^{3+}(aq) \longrightarrow 3Cd^{2+}(aq) + 2Bi(s)$
*(c) $5Ag(s) + KMnO_4(aq) + 8HCl(aq) \longrightarrow$
    $5AgCl(s) + MnCl_2(aq) + KCl(aq) + 4H_2O(l)$
17.38 (a) $Pb^{2+}(aq) + Cu^{2+}(aq) \longrightarrow Pb^{4+}(aq) + Cu(s)$
(b) $2Fe^{3+}(aq) + 2I^-(aq) \longrightarrow 2Fe^{2+}(aq) + I_2(s)$
*(c) $3K_2S_2O_8(aq) + 2Cr(NO_3)_3(aq) + 7H_2O(l) \longrightarrow$
$2K_2SO_4(aq) + K_2Cr_2O_7(aq) + 6HNO_3(aq) + 4H_2SO_4(aq)$

17.39 Explain how electrode potentials allow us to predict that copper will react with dilute nitric acid, forming copper(II) nitrate, but not with dilute hydrochloric acid.
17.40 A chemist working with nuclear waste needs to oxidize Pu^{3+} to Pu^{4+}. Suggest a suitable reagent.

17.41 Can acidified aqueous potassium permanganate be used to oxidize iodide ions to iodine?
*17.42** A chemist is interested in the compounds formed by the *d*-block element manganese, and wants to find a way of preparing $Mn^{3+}(aq)$ from $Mn^{2+}(aq)$. Would acidified sodium dichromate be suitable?

17.43 Can chromium displace titanium from titanium(II) solutions?
17.44 Chlorine is used to displace bromine from brine. Could oxygen be used instead? If so, why isn't it used?

17.45 Suggest two metals that could be used for the cathodic protection of a titanium pipeline. What factors other than relative electropositive character need to be considered in practice?

17.46 A chromium-plated steel bicycle handlebar is damaged. Will rusting be encouraged or retarded by the chromium?

17.47 Which oxoanions, ions, or elemental form of manganese can

    (a) reduce hydrogen cations to H_2 in aqueous solution

    (b) oxidize peroxide ions to O_2 in basic aqueous solution

    (c) be oxidized by water

    (d) be reduced by water

17.48 Which oxoanions, ions, or elemental form of chromium can

    (a) reduce hydrogen cations to H_2 in aqueous solution

    (b) oxidize peroxide ions to O_2 in basic aqueous solution

    (c) be oxidized by water

    (d) be reduced by water

Free Energies and Electrode Potentials

In Exercises 17.49 and 17.50, predict the standard potential of the cell corresponding to each cell reaction.

17.49 (a) $Zn(s) + Cu^{2+}(aq) \longrightarrow Zn^{2+}(aq) + Cu(s)$
    (b) $2H_2(g) + O_2(g) \longrightarrow 2H_2O(l)$, in acid
    (c) $Ag^+(aq) + Cl^-(aq) \longrightarrow AgCl(s)$

17.50 (a) $Fe^{2+}(aq) + Ce^{4+}(aq) \longrightarrow$
$$Fe^{3+}(aq) + Ce^{3+}(aq)$$
    (b) $H_3O^+(aq) + OH^-(aq) \longrightarrow 2H_2O(l)$
    (c) $Ag^+(aq) + I^-(aq) \longrightarrow AgI(s)$

17.51 Use standard electrode potentials to calculate the standard reaction free energy of the disproportionation $3Au^+(aq) \rightarrow 2Au(s) + Au^{3+}(aq)$.

17.52 Use standard electrode potentials to calculate the standard reaction free energy of the disproportionation $2Cu^+(aq) \rightarrow Cu(s) + Cu^{2+}(aq)$.

The Nernst Equation

In Exercises 17.53 and 17.54, give the expression for the potential of each cell in terms of the concentrations of the ions and in terms of the pressures of the gases taking part in the cell reaction.

17.53 (a) $Ag(s)|AgCl(s)|HCl(aq)| \; |HCl(aq)|H_2(g)|Pt$
    (b) $Pt|H_2(g)|HCl(aq)|AgCl(s)|Ag(s)$

17.54 (a) $Ag(s)|AgI(s)|KI(aq)| \; |AgNO_3(aq)|Ag(s)$
    (b) $Pt|H_2|HCl(aq)| \; |NaOH(aq)|O_2(g)|Pt$

In Exercises 17.55 and 17.56, calculate the potential of each cell. Pressure units are summarized in Table 5.4.

17.55 (a) $Pt|H_2(g, 1.0 \text{ atm})|HCl(aq, 1.0 \, M)| \; |HCl(aq, 2.0 \, M)|H_2(g, 1.0 \text{ atm})|Pt$

    (b) $Pt|H_2(g, 10 \text{ atm})|HCl(aq, 0.10 \, M)|H_2(g, 0.1 \text{ atm})|Pt$

17.56 (a) $Pt|Cl_2(g, 100 \text{ Torr})|HCl(aq, 0.10 \, M)| \; |HCl(aq, 0.001 \, M)|Cl_2(g, 800 \text{ atm})|Pt$

    (b) $Pt|Cl_2(g, 1.0 \text{ mbar})|HCl(aq, 0.01 \, M)|Cl_2(g, 1.0 \text{ kbar})|Pt$

In Exercises 17.57 and 17.58, calculate the value of Q for the cell reaction, given the measured values of the cell potential.

17.57 $Pt|Sn^{4+}(aq),Sn^{2+}(aq)| \; |Pb^{4+}(aq),Pb^{2+}(aq)|Pt$; $E = 1.33 \text{ V}$

17.58 $Pt|O_2(g)|H_3O^+(aq)| \; |Cr_2O_7^{2-}(aq),Cr^{3+}(aq), H_3O^+(aq)|Pt$; $E = 0.10 \text{ V}$

17.59 Could an electrochemical cell be used as a pressure gauge? Discuss the possibility in terms of the cell SHE| |HE, where HE is a hydrogen electrode and SHE is a standard hydrogen electrode.

17.60 Show how a silver–silver chloride electrode and a hydrogen electrode can be used to measure (a) pH; (b) pOH.

17.61 Explain how electrochemical cells can be used to measure the pK_a of lactic acid and the pK_b of ammonia.

17.62 In a neuron (a nerve cell), the concentration of K^+ ions inside the cell is about 20 to 30 times that outside. What potential difference between the inside and the outside of the cell would you expect to measure if the difference is due only to the imbalance of potassium ions? Which would be more positive, the inside or the outside?

Equilibrium Constants

In Exercises 17.63 and 17.64, calculate the equilibrium constant for each reaction. Report an acid or base ionization constant as a pK_a, pK_b, or pK_w value.

17.63 (a) $AgBr(s) \rightleftharpoons Ag^+(aq) + Br^-(aq)$
    (b) $Sn^{2+}(aq) + Pb^{4+}(aq) \rightleftharpoons Sn^{4+}(aq) + Pb^{2+}(aq)$
    (c) $Cr(s) + Zn^{2+}(aq) \rightleftharpoons Cr^{2+}(aq) + Zn(s)$
    (d) $2H_2O(l) \rightleftharpoons H_3O^+(aq) + OH^-(aq)$

17.64 (a) $Hg_2Cl_2(s) \rightleftharpoons Hg_2^{2+}(aq) + 2Cl^-(aq)$
    (b) $AgI(s) \rightleftharpoons Ag^+(aq) + I^-(aq)$
    (c) $In^{3+}(aq) + U^{3+}(aq) \rightleftharpoons In^{2+}(aq) + U^{4+}(aq)$
    (d) $Fe(s) + 3Ti^{4+}(aq) \rightleftharpoons Fe^{3+}(aq) + 3Ti^{3+}(aq)$

17.65 A chemist wants to make a range of silver(II) compounds. Can aqueous sodium persulfate be used to oxidize silver(I) compound to silver(II)? If so, what will be the equilibrium constant for the reaction?

17.66 A chemist suspects that manganese(III) may be involved in an unusual biochemical reaction and wants to prepare some of its compounds. Can aqueous potassium permanganate be used to oxidize manganese(II) to manganese(III)? If so, what will be the equilibrium constant of the reaction?

Electrolysis

In Exercises 17.67 and 17.68, determine whether or not each metal may be deposited electrolytically from aqueous solution. Base your answer only on the potentials listed in Table 17.3. Do not consider other factors such as passivation or overpotential.

17.67 (a) Mn; (b) Al; (c) Ni; (d) Au.
17.68 (a) Cr; (b) Pt; (c) Cu; (d) U.

In Exercises 17.69 and 17.70, write the half-reactions for each electrolysis.

17.69 (a) The deposition of copper; (b) the evolution of hydrogen; (c) the production of sodium in the Downs process; (d) the production of chlorine in the Downs process.
17.70 (a) The deposition of silver; (b) the evolution of oxygen from acidified water; (c) the production of hypochlorite ion from brine.

In Exercises 17.71 and 17.72, determine the number of moles of each italicized substance that is produced by electrolysis when 1 mol of electrons is supplied.

17.71 (a) *Copper* from aqueous copper sulfate; (b) *aluminum* from aluminum oxide dissolved in molten cryolite; (c) *hydrogen* from acidified water; (d) *oxygen* from acidified water.
17.72 (a) *Silver* from aqueous silver nitrate; (b) *chromium* from acidified chromium(VI) oxide solution; (c) *chlorine* from molten sodium chloride; (d) *sodium hypochlorite* from brine.

In Exercises 17.73 and 17.74, calculate the number of moles of electrons supplied by each current, flowing for the given time.

17.73 (a) 1 A for 1 s; (b) 5 A for 1 min; (c) 100 A for 1 h.
17.74 (a) 1 mA for 1 day; (b) 2 A for 3 h; (c) 100,000 A for 1 week.

17.75 An electric heater rated at 2 kW uses a current of 18 A for 1 h. How many moles of electrons pass through it in that time?
17.76 A portable cassette player uses 150 mA of current. Calculate the number of moles of electrons its batteries generate when it is on for 1 h.

In Exercises 17.77 and 17.78, calculate the mass of each italicized substance that is produced during electrolysis with a 0.50-A current for 24 h.

17.77 (a) *Nickel* from aqueous nickel(II) nitrate; (b) *zinc* from aqueous zinc sulfate; (c) *fluorine* from a mixture of potassium and hydrogen fluorides; (d) *sodium hydroxide* from the electrolysis of brine, but without allowing the chlorine to mix with the solution.
17.78 (a) *Aluminum* from aluminum oxide; (b) *chromium* from chromium(VI) oxide; (c) *oxygen* from acidified water; (d) *calcium hypochlorite* from the electrolysis of aqueous calcium chloride.

In Exercises 17.79 and 17.80, calculate the volume of each italicized (and assumed ideal) gas that is produced by electrolysis with a 1.00-A current for 3600 s when the product is collected under the stated conditions.

17.79 (a) *Hydrogen* from water at STP; (b) *oxygen* from acidified water at 25°C and 1.00 atm; (c) *fluorine* from hydrogen fluoride and potassium fluoride at −10°C and 750 Torr.
17.80 (a) *Chlorine* from sodium chloride at STP; (b) *bromine* from molten potassium bromide at 200°C and 0.10 atm; (c) *mercury* from molten mercury(I) fluoride at 1000°C and 100 Torr.

In Exercises 17.81 and 17.82, calculate the time for which the electrolysis must continue to form 1.0 g of each italicized product.

17.81 (a) *Aluminum* from aluminum oxide using 1.0 A; (b) *chromium* from chromium(VI) oxide using 0.50 A; (c) *oxygen* from acidified water using 100 mA; (d) *calcium hypochlorite* from the electrolysis of aqueous calcium chloride using 10 A.
17.82 (a) *Nickel* from aqueous nickel(II) nitrate using 1.0 A; (b) *zinc* from aqueous zinc sulfate using 0.50 A; (c) *fluorine* from potassium and hydrogen fluorides using 10.0 A; (d) *sodium hydroxide* from the electrolysis of brine using 25 A, but without allowing the chlorine to mix with the solution.

In Exercises 17.83 and 17.84, determine the charge number of the cation in solution from the information given.

17.83 When a titanium chloride solution was electrolyzed for 500 s with a 120-mA current, 15 mg of titanium was deposited.
17.84 When a mercury nitrate solution was electrolyzed for 1200 s with a 210-mA current, 0.26 g of mercury was produced.

*****17.85** Thomas Edison was faced with the problem of measuring the electricity that each of his customers had used. His first solution was to use a *zinc coulometer*, an electrolysis cell in which the quantity of electricity is measured by weighing the mass of zinc deposited. However, only some of the current used by the customer passed through the coulometer. What mass of zinc would be deposited in one month (of 31 days) if 1 mA of current passed through the cell continuously?
*****17.86** An alternative solution to the problem described in Exercise 17.85 is to collect the hydrogen produced by electrolysis and measure its volume. What volume would be collected at STP under the given use conditions?

General

*17.87 Calculate the standard electrode potential of (a) the Cu^+/Cu couple from those of the Cu^{2+}/Cu^+ and Cu^{2+}/Cu couples; (b) the Fe^{2+}/Fe couple from the values for the Fe^{3+}/Fe^{2+} and Fe^{3+}/Fe couples. Base your calculation on the relation $\Delta G^\circ = -nFE^\circ$ and the fact that ΔG° of an overall reaction is the sum of the ΔG° for each reaction into which it may be divided.

*17.88 Use standard electrode-potential data to calculate the solubility of (a) $AgCl(s)$; (b) $Hg_2Cl_2(s)$; (c) $PbSO_4(s)$.

*17.89 Dental amalgam, a solid solution of silver and tin in mercury, is used for filling tooth cavities. Two of the reduction half-reactions that the filling can undergo are

$$3Hg_2{}^{2+}(aq) + 4Ag(s) + 6e^- \longrightarrow 2Ag_2Hg_3(s)$$
$$E^\circ = +0.85 \text{ V}$$

$$Sn^{2+}(aq) + 3Ag(s) + 2e^- \longrightarrow Ag_3Sn(s)$$
$$E^\circ = -0.05 \text{ V}$$

Suggest a reason why, when you accidentally bite on a piece of aluminum foil with filled teeth, you may feel pain.

*17.90 The reaction that powers our bodies is the oxidation of glucose:

$$C_6H_{12}O_6(aq) + 6O_2(g) \longrightarrow 6CO_2(g) + 6H_2O(l)$$

During a day's normal activities, a person uses the equivalent of about 10 MJ of energy. Estimate the average current through the body over the course of a day, assuming that all the energy we use arises from the reduction of O_2 in the glucose oxidation reaction. Also estimate your power in watts.

IV

THE ELEMENTS

In Part IV, we survey the properties of individual elements and families of elements and see how they illustrate the concepts developed in previous chapters. We begin by discussing the properties of hydrogen, the alkali metals, and the alkaline earth metals. The illustration shows potassium dissolving in a mixture of organic solvents, losing its outermost electrons, and turning the solution blue.

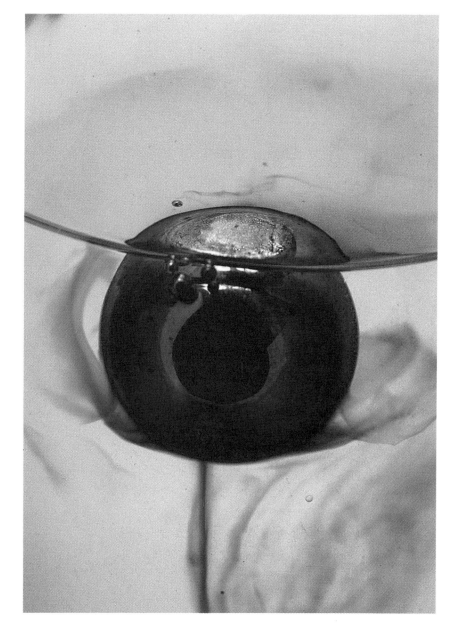

18

HYDROGEN AND THE s-BLOCK ELEMENTS

Hydrogen

18.1 The element hydrogen

18.2 Some important hydrogen compounds

Group I: the alkali metals

18.3 The Group I elements

18.4 Some important Group I compounds

Group II: the alkaline earth metals

18.5 The Group II elements

18.6 Some important Group II compounds

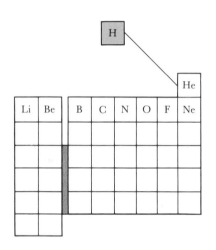

This chapter is the first of four chapters in which we work systematically but selectively through the periodic table, illustrating earlier statements about how the table helps to organize the properties of the elements and their compounds. Two points should be kept in mind as we do so. The first is that, mainly as a result of the small size of its atoms, the element at the head of each group (e.g., Li in Group I and Be in Group II) differs significantly from its congeners. Hydrogen, at the head of the entire periodic table, is particularly distinctive; for that reason we treat it separately and not as a member of a group or even a block. The second point is the importance of diagonal relationships, particularly between Period 2 and Period 3 elements toward the left of the periodic table. Thus we can expect Li to resemble its diagonal neighbor Mg, and Be to resemble its p-block diagonal neighbor Al.

HYDROGEN

There are three isotopes of hydrogen. The most abundant is ^{1}H, with a single proton for its nucleus. Deuterium (^{2}H or D) has one neutron in addition to the proton in the nucleus. Since the natural abundance of deuterium is 0.02%, there are likely to be only two ^{2}H atoms in any 10,000 atoms of hydrogen. The presence of the additional neutron doubles the mass of the atom without changing its electron configuration, and for no other element is there such a significant difference between isotope masses. This difference makes deuterium very valuable for keeping track of hydrogen atoms in a series of reactions: we can synthesize a molecule with a higher than normal proportion of deuterium, and then use mass spectrometry to identify the product molecules that contain the extra deuterium.

The third isotope, tritium (^{3}H or T), is radioactive. We can follow its progress through a complex series of reactions by noting which products are radioactive.

Hydrogen is the most abundant element in the universe. However, there is little free hydrogen on earth because H_2 molecules, being so light, move at high average speeds and escape from the atmosphere. The molecular hydrogen that does survive was originally formed by the action of bacteria on ancient vegetable and animal remains, and then trapped beneath rock formations.

18.1 THE ELEMENT HYDROGEN

FIGURE 18.1 When the runny oil (top) is hydrogenated, it is converted to a solid—a fat (bottom). At the molecular level, hydrogen converts carbon-carbon double bonds to single bonds and hence increases the flexibility of the molecules. This enables them to pack together more closely.

About 3×10^8 kg of hydrogen is used in the United States each year. Although hydrogen's most spectacular use is as a rocket fuel for the space shuttle, that accounts for less than 1% of the total (about 3×10^5 kg per flight). About 50% of the hydrogen used in industry is converted to ammonia by the Haber process (the high-pressure, high-temperature catalytic synthesis of ammonia from nitrogen and hydrogen). Through the reactions of ammonia, hydrogen finds its way into the numerous important products for which ammonia is the starting point. Hydrogen also enters the economy through the production of methanol:

$$2H_2(g) + CO(g) \xrightarrow{\Delta,\ \text{pressure, catalyst}} CH_3OH(l)$$

This process takes place at about 300°C and 250 atm, over a catalyst of zinc, chromium, manganese, and aluminum oxides. Through methanol, hydrogen finds its way into formaldehyde (HCHO), which is used in the manufacture of plastics, and into acetic acid, which is used in the manufacture of synthetic fibers. Methanol is also used as an additive in gasoline, where it promotes smooth combustion and hence helps to improve the octane rating of the fuel.

Hydrogen is also used in the *hydrometallurgical extraction* of copper—its extraction from ores by reduction in aqueous solution:

$$Cu^{2+}(aq) + H_2(g) \longrightarrow Cu(s) + 2H^+(aq)$$

In this process, ores containing copper oxide and sulfide are dissolved in sulfuric acid, and then hydrogen is bubbled through the solution. About a third of the hydrogen that is manufactured is used for this and other hydrometallurgical reductions.

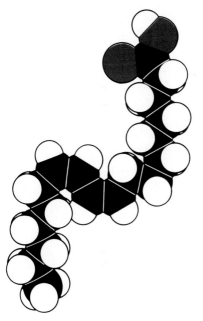

1   Linoleic acid

▼ **EXAMPLE 18.1   Judging the feasibility of reduction with hydrogen**

Is the hydrometallurgical extraction of copper with hydrogen thermodynamically favorable under standard conditions at 25°C?

STRATEGY   We can judge whether H_2 can reduce Cu^{2+} by checking to see whether the H^+/H_2 couple lies below the Cu^{2+}/Cu couple in the electrochemical series (in Appendix 2B).

SOLUTION H^+/H_2 ($E° = 0$) does lie below Cu^{2+}/Cu ($E° = +0.34$ V) in the electrochemical series. Hence the reduction is feasible under standard conditions.

EXERCISE   Can aluminum be extracted by hydrometallurgical reduction of Al^{3+} ions with hydrogen?

[*Answer*: No]

▲

Hydrogen is used in the food industry for converting vegetable oils to shortening (Fig. 18.1). There it is added to carbon-carbon double bonds in the reaction called *hydrogenation*:

$$H_2(g) + \ldots C{=}C \ldots \xrightarrow{220°C,\ 2\ \text{to}\ 10\ \text{atm, Ni}} \ldots CH{-}CH \ldots$$

This converts a double bond to a single bond. Oil and fat molecules have long hydrocarbon chains; however, the molecules from which oils are derived **(1)** have more C=C double bonds than the molecules from which fats are derived **(2)**. Because double bonds resist twisting, chains that contain them are stiff and do not pack together well, so the result is a liquid. When they are hydrogenated, the double bonds become single bonds, the chains become much more flexible, and the molecules pack together and form a solid.

2   Stearic acid

FIGURE 18.2 When water is added to calcium hydride, it oxidizes the hydride ion to molecular hydrogen, which forms the bubbles we see here.

Manufacture. Natural supplies of hydrogen gas are far too small to satisfy the needs of industry, so most commercial hydrogen is obtained from low-molecular-weight hydrocarbons. The process that is used depends on two catalyzed reactions. The first is a *reforming reaction*, a reaction in which the hydrocarbon and steam are converted to carbon monoxide and hydrogen over a nickel catalyst:

$$CH_4(g) + H_2O(g) \xrightarrow{800°C, \text{ Ni}} CO(g) + 3H_2(g)$$

The mixture of products, which is called *synthesis gas*, is the starting point for the manufacture of numerous compounds, including methanol. In a second reaction, the carbon monoxide in the synthesis gas reacts with more water in the *shift reaction*:

$$CO(g) + H_2O(g) \xrightarrow{400°C, \text{ Fe/Cu}} CO_2(g) + H_2(g)$$

where Fe/Cu denotes a catalyst made of iron and copper.

Hydrogen is also produced by the electrolysis of water, but that process is economical only where electricity is cheap. Currently, chemists are seeking ways of using sunlight to drive the *water-splitting reaction*, the photochemical decomposition of water into its elements:

$$2H_2O(l) \xrightarrow{\text{light}} 2H_2(g) + O_2(g) \qquad \Delta G° = +474 \text{ kJ}$$

The decomposition is not spontaneous, but it can be driven in the desired direction if the light supplies enough energy. The principal (and largely unsolved) problem is to find a suitable catalyst.

Hydrogen is prepared in the laboratory by the reduction of hydrogen ions with a metal that has a negative reduction potential, as in

$$Zn(s) + 2HCl(aq) \longrightarrow ZnCl_2(aq) + H_2(g)$$

Another useful source is the oxidation of a hydride with water (Fig. 18.2):

$$CaH_2(s) + 2H_2O(l) \longrightarrow Ca(OH)_2(aq) + 2H_2(g)$$

TABLE 18.1   Properties of hydrogen

Valence configuration	$1s^1$
Atomic number	1
Symbol	H
Isotopes*	^{1}H (99.98% abundance, 1.008 amu)
Deuterium	^{2}H or D (0.02% abundance, 2.014 amu)
Tritium	^{3}H or T (radioactive, 3.016 amu)
Atomic weight	1.008 amu
Normal form†	Colorless, odorless gas
Boiling point	−253°C (20 K)
Melting point	−259°C (14 K)

*The isotope ^{1}H is sometimes called protium.
†"Normal form" means the state and appearance of the element at 25°C and 1 atm pressure.

This reaction is a convenient source of hydrogen for filling weather balloons.

Physical properties. The physical properties of hydrogen are summarized in Table 18.1. It is a colorless, odorless, tasteless gas, almost insoluble in water, that condenses to a colorless liquid at 20 K. Its low boiling point and insolubility stem from its very weak intermolecular interactions. Being nonpolar, H_2 molecules can attract each other only by London forces. However, even these forces are weak, because each molecule has only two electrons and hence only a very small instantaneous electric dipole. One striking physical property of the liquid is its very low density (0.09 g/cm³), less than one-tenth that of water (Fig. 18.3). This accounts for the fact that hydrogen occupies more than two-thirds of the total volume of the space shuttle's main fuel tank (Fig. 18.4). In the language of Section 6.5, liquid hydrogen has a low enthalpy density.

The special characteristics of hydrogen. The smallness of hydrogen atoms allows them to take part in one of the most important types of intermolecular interaction—hydrogen bonding. As we have seen, hydrogen bonding has far-reaching consequences, among them the low vapor pressure of water, ammonia, and hydrogen fluoride. It also accounts for the expansion of water when it freezes, and hence the fact that ice floats on water: in the solid, hydrogen bonds hold the molecules apart in an open structure. Hydrogen bonding is responsible for the solubility in water of many organic molecules that contain —OH and —NH₂ groups and for the rigidity of cellulose, where it forms cross-links between neighboring chains of glucoselike carbohydrate molecules. In addition, hydrogen bonding plays a crucial part in controlling the shapes of protein and DNA molecules: it is strong enough to give rise to recognizable molecular structures but weak enough for those structures to be easily modified.

Another reason for hydrogen's special position in chemistry is the exceptionally small size of its cation. The latter, the proton, has $\frac{1}{10,000}$ the diameter of the neutral atom, and is far smaller than any other cation. As a result, it can attract electrons very strongly, making the hydrogen

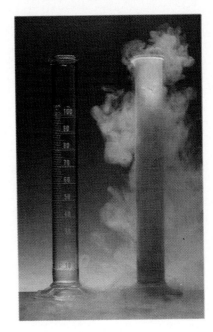

FIGURE 18.3 The two measuring cylinders contain the same mass of liquid: the liquid on the left is water, that on the right is liquid hydrogen, which is one-tenth as dense.

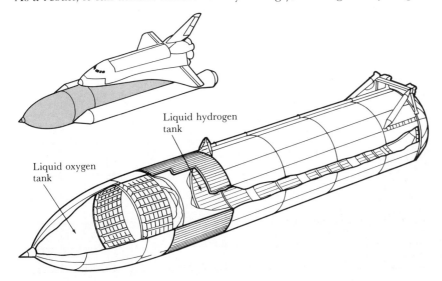

Liquid hydrogen tank

Liquid oxygen tank

FIGURE 18.4 The arrangement of fuel tanks in the space shuttle. Note the very large size of the hydrogen tank, compared with that of the oxygen tank. Liquid hydrogen has a very low enthalpy density.

ion the strongest Lewis acid. We see that ability in the formation of H_3O^+ and NH_4^+ with, respectively, H_2O and NH_3.

In many cases, particularly when the hydrogen atom is attached to an O atom, a halogen atom, or a positively charged N atom (as in NH_4^+), it can be transferred (as a proton) to another molecule. This proton transfer leads to the rich chemistry of the proton donors that we call Brønsted acids and the proton acceptors that we call Brønsted bases. Because proton transfer is very fast, acid-base equilibria respond immediately to changes in the conditions, particularly changes in the concentrations of other acids and bases. We explored that area of chemistry in Chapters 14 and 15.

18.2 SOME IMPORTANT HYDROGEN COMPOUNDS

Hydrogen forms a huge number of compounds. However, most of them are best treated as compounds of the other elements they contain. Sulfuric acid, for instance, is best regarded as a compound of sulfur or as a representative of the oxoacids. All the main-group elements, with the exception of the noble gases and (possibly) indium and thallium, form compounds with hydrogen. So do most of the elements on the left and right of the *d* block (but not those at the center of the block; there is no iron hydride, for instance). Here we consider only the binary compounds, compounds of hydrogen that contain one other element.

The binary compounds of hydrogen are traditionally divided into three classes called "saline" (saltlike), "molecular," and "metallic" (Fig. 18.5). However, like most classifications in chemistry, this one is by no means rigid, and the characteristics and structures of the compounds blend from one class to another.

Saline hydrides. The *saline hydrides* are compounds of hydrogen and a strongly electropositive metal (any member of the *s* block, with the exception of Be). They are formed by heating the metal in hydrogen:

$$2K(s) + H_2(g) \xrightarrow{\Delta} 2KH(s)$$

These compounds are white, high-melting-point solids that contain the hydride ion H^-. Their crystal structures are analogous to those of the

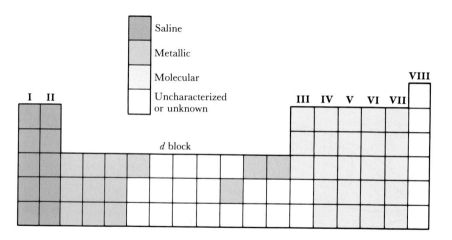

FIGURE 18.5 The different classes of binary hydrogen compounds in the periodic table.

metal halides; the alkali metal hydrides, for instance, have the rock-salt structure.

The single positive charge of the hydrogen atomic nucleus has only weak control over the two electrons in the H^- ion. The ionic radius of H^- depends on the cations present, but lies between the values for F^- (136 pm) and Cl^- (181 pm), as indicated in (3). The readiness with which one of the electrons can be lost means that ionic hydrides are powerful reducing agents:

$$H_2(g) + 2e^- \longrightarrow 2H^-(aq) \qquad E° = -2.25 \text{ V}$$

The reducing power of the hydride ion is similar to that of sodium metal, for the H_2/H^- couple is not very far above Na^+/Na in the electrochemical series.

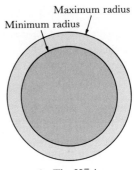

Maximum radius
Minimum radius

3   The H^- ion

▼ **EXAMPLE 18.2**   **Judging the reducing power of hydride ions**

Determine whether hydride ions can form hydrogen when a saline hydride reacts with water.

STRATEGY   We know that sodium reduces water to hydrogen, and have just seen that hydride ions have a similar reducing power; hence we should expect hydride ions to reduce water to hydrogen. More formally, we should determine whether the H_2/H^- couple lies below the $H_2O/OH^-,H_2$ couple for the reduction of water in the electrochemical series. The series is given in Appendix 2B, but we need the potential at pH = 7, which was given in Table 17.3.

SOLUTION   The reduction half-reaction is

$$2H_2O(l) + 2e^- \longrightarrow 2OH^-(aq) + H_2(g) \qquad E = -0.42 \text{ V}$$

The H_2/H^- couple ($E° = -2.25$ V) lies well below this couple. Hence, H^-
▲ ions can reduce water.

Saline hydrides reduce water as soon as they come into contact with it:

$$NaH(s) + H_2O(l) \longrightarrow NaOH(aq) + H_2(g)$$

Because this reaction produces H_2, and because the saline hydrides are produced by direct reaction, they are potentially useful as transportable sources of hydrogen.

Molecular compounds of hydrogen.   The molecular compounds of hydrogen consist of discrete molecules. They are formed by the nonmetals, and in most cases they are gases such as ammonia and the hydrocarbons methane, ethylene, and acetylene, or liquids such as water, ethanol, and benzene. This class of hydrogen compounds includes the electron-deficient boron hydrides, but these are best treated as compounds of boron.

The molecular compounds of hydrogen are often prepared by direct action of the elements, as in the Haber process for producing ammonia or in the synthesis of HCl:

$$N_2(g) + 3H_2(g) \xrightarrow{\text{400 to 600°C, 150 to 600 atm, Fe}} 2NH_3(g)$$

$$H_2(g) + Cl_2(g) \xrightarrow{\text{light}} 2HCl(g)$$

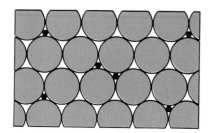

FIGURE 18.6 In a metallic hydride, the small hydrogen atoms occupy gaps—called "interstices"—between the larger metal atoms.

The vigor of these and similar reactions varies widely: the synthesis of ammonia requires a catalyst, but the reaction of hydrogen and chlorine is violent once it has been initiated. Hydrogen reacts explosively with fluorine as soon as the two gases are mixed.

The binary molecular compounds of hydrogen may be prepared in a less dramatic way by _protonation_, or proton transfer to a Brønsted base such as S^{2-}:

$$FeS(s) + 2HCl(aq) \longrightarrow FeCl_2(aq) + H_2S(g)$$

Another example is the protonation of the carbide ion C_2^{2-} to produce acetylene,

$$:C{\equiv}C:^{2-} + 2H_2O \longrightarrow H{-}C{\equiv}C{-}H + 2OH^-$$

which occurs when water acts on calcium carbide. Volatile acids can be prepared in this way, by using a less volatile acid as the proton donor:

$$CaF_2(s) + H_2SO_4(l) \longrightarrow CaSO_4(s) + 2HF(g)$$

The reaction proceeds to the right as the volatile product is removed. When oxidation of the product by sulfuric acid is a danger, an acid that is less strongly oxidizing than sulfuric acid should be used, as in the production of HBr:

$$KBr(s) + H_3PO_4(aq) \longrightarrow KH_2PO_4(aq) + HBr(g)$$

A third route to the binary molecular compounds of hydrogen is _deprotonation_, or the removal of a proton from a Brønsted acid, as in the production of phosphine (PH_3):

$$PH_4^+(aq) + OH^-(aq) \longrightarrow PH_3(g) + H_2O(l)$$

Metallic hydrides. The _metallic hydrides_ are so called because they are electrically conducting. They also have variable composition, with a variable number of hydrogen atoms occupying gaps (interstices) between metal atoms in the solid (Fig. 18.6). These black, powdery solids are prepared by heating certain d-block metals in hydrogen; for example,

$$2Cu(s) + H_2(g) \xrightarrow{\Delta} 2CuH(s)$$

Since the metallic hydrides release their hydrogen (as H_2) when heated or treated with acid, they are also currently being investigated as a means for storing and transporting hydrogen intended for use as a fuel.

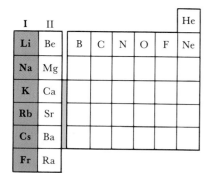

GROUP I: THE ALKALI METALS

The members of Group I, the "alkali metals," and some of their properties are listed in Table 18.2. The valence electron configurations are ns^1, where n is the period number. The physical and chemical properties of the element are dominated by the ease with which the single valence electron can be removed.

(a)

(b)

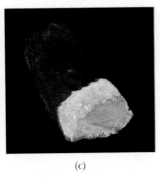
(c)

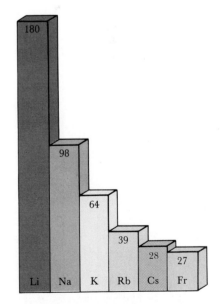
(d)

FIGURE 18.7 The alkali metals of Group I: (a) lithium, (b) sodium, (c) potassium, and (d) rubidium (left), and cesium (right). Francium has never been isolated in visible quantities. The first three corrode rapidly in moist air; rubidium and cesium are even more reactive and have had to be photographed sealed away from contact with the atmosphere.

18.3 THE GROUP I ELEMENTS

All the Group I elements are too easily oxidized to be found in the uncombined state, and very strong reducing agents are needed to extract them from their compounds. The pure metals are obtained by electrolysis of their molten salts or, in the case of potassium, by exposing molten potassium chloride to sodium vapor:

$$KCl(l) + Na(g) \xrightarrow{750°C} NaCl(s) + K(g)$$

Although the equilibrium constant for this reaction is not particularly favorable, the reaction proceeds to the right because potassium is more volatile than sodium and is driven off by the heat.

Physical properties. All the Group I elements are soft, silver-gray metals (Fig. 18.7). Lithium is the hardest, but even so it is softer than lead. The melting points decrease down the group (Fig. 18.8). Cesium,

TABLE 18.2 The Group I elements: the alkali metals

Valence configuration: ns^1
Normal form*: Soft silver-gray metals

Z	Name	Symbol	Atomic weight, amu	Melting point, °C	Boiling point, °C	Density, g/cm³
3	Lithium	Li	6.94	180	1360	0.53
11	Sodium	Na	22.99	98	900	0.97
19	Potassium	K	39.10	63	777	0.86
37	Rubidium	Rb	85.47	39	705	1.53
55	Cesium	Cs	132.91	28	686	1.90
87	Francium	Fr	223	27	677	

*At 25°C and 1 atm.

FIGURE 18.8 The melting points of the alkali metals decrease smoothly down the group. (Values are given in degrees Celsius.)

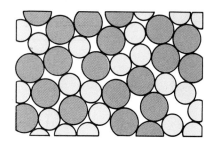

FIGURE 18.9 The atomic radii of sodium and potassium are such that their atoms pack together poorly, and an alloy of the two is a liquid at room temperature.

which melts at 28°C, is barely a solid at room temperature. Some alloys of sodium and potassium are liquid at room temperature because their atoms pack together poorly and hence produce a fluid structure (Fig. 18.9). They are used as coolants in some nuclear reactors since, being metals, they conduct heat very well and are not decomposed by radiation. Lithium was an element with few applications until thermonuclear weapons (which use lithium-6 to generate the explosion, Section 22.9) were developed after World War II.

Trends in reduction potentials. Since the first ionization energies of the alkali metals are so low (Table 7.7), only a small energy investment is needed to produce singly charged cations. Consequently, almost all their compounds are predominantly ionic, with the metal present as an M^+ ion.

The ease with which the valence electron is lost in solution results in the Group I elements having couples with strongly negative reduction potentials. Although the general trend is for reduction potentials to become more negative down the group, lithium is anomalous (Fig. 18.10); it is one of the most strongly reducing metals in the periodic table. Period 2 anomalies such as this one can often be traced to the small size of the atoms. The origin of the strong reducing power of lithium is the strongly exothermic hydration of the small Li^+ ions, which favors the ionization of the element.

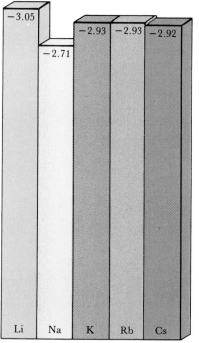

FIGURE 18.10 The standard reduction potentials of the alkali metals. Note the anomalous value for Li. (Values are given in volts.)

▼ **EXAMPLE 18.3** Accounting for an electrode potential

Account for the anomalous value of the standard electrode potential of the Li^+/Li couple by setting up a thermodynamic cycle and comparing it with Na^+/Na.

STRATEGY We can analyze $E°$ by setting up the cycles shown in Fig. 18.11 and using them to find a reason why the half-reaction $Li(s) \rightarrow Li^+(aq) + e^-$ is so favorable thermodynamically in comparison with $Na(s) \rightarrow Na^+(aq) + e^-$. Although we should express the cycles in terms of free energies (because $-nFE° = \Delta G°$), we can find a qualitative explanation by considering only enthalpy changes (which dominate the entropy changes in these processes). Numerical data for expressing the cycles quantitatively can be found in Tables 7.7, 7.8, and 11.6 and in Appendix 2A.

SOLUTION The diagram shows that $\Delta H°$ for the half-reaction $Li(s) \rightarrow Li^+(aq) + e^-$ is the sum of the sublimation and ionization enthalpies of lithium less the hydration enthalpy of Li^+ ions. Since the Li^+ cation is so small, the hydration is strongly exothermic. In the analogous process for sodium, although the formation of a gas-phase Na^+ ion is less endothermic, its larger ionic radius results in a smaller hydration enthalpy. The net result is that the formation of Na^+ in water is more endothermic than the formation of Li^+, so its formation is less favorable.

EXERCISE Predict, using a similar cycle, whether potassium will lie lower or higher than sodium in the electrochemical series.

[*Answer*: Lower; it is more reducing]

Reaction with water and ammonia. As we have seen, a strong thermodynamic tendency to occur does not necessarily mean that a reaction is

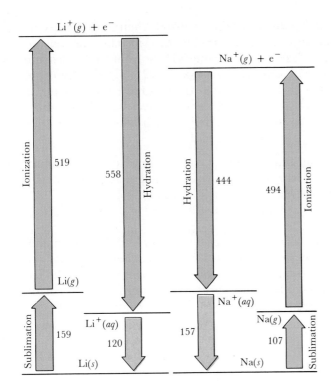

FIGURE 18.11 The thermodynamic cycles used to analyze the standard reduction potentials of the Li^+/Li and Na^+/Na couples in water (values are in kJ/mol). Note that the half-reaction $M^+(aq) + e^- \longrightarrow M(s)$ is less exothermic for lithium than for sodium, indicating that the reverse process is more favorable for lithium and hence that lithium has a more negative reduction potential.

vigorous (or even that it occurs at all). Thus, all the Group I elements reduce water, as in the reaction

$$2Na(s) + 2H_2O(l) \longrightarrow 2NaOH(aq) + H_2(g)$$

and do so with a vigor that increases uniformly down the group (this is illustrated in Fig. 2.4). However, the reaction between water and lithium is gentle, even though the reduction potential of lithium is strongly negative. The reaction is vigorous enough with potassium to ignite the hydrogen produced, and dangerously explosive with rubidium and cesium. The danger with these last two elements stems from their being denser than water, so they sink and react beneath the surface; the rapidly evolved hydrogen gas then forms a shock wave that can shatter the vessel.

▼ EXAMPLE 18.4 Accounting for a reaction rate

Suggest a reason for the slow reaction between lithium and water.

STRATEGY We saw in Chapter 12 that the rates of reactions depend on the activation energy, the energy the reactants need in order to cross the activation barrier and form products. Therefore, we should look for a step in the reaction that requires a lot of energy before it can occur.

SOLUTION In the course of the reaction, lithium atoms must be removed from a tightly bonded solid and ionized. Both steps require a large amount of energy, so the activation energy can be expected to be high and the reaction slow.

EXERCISE Predict, using a similar argument, whether sodium or potassium will react more vigorously with water.

[*Answer*: K]

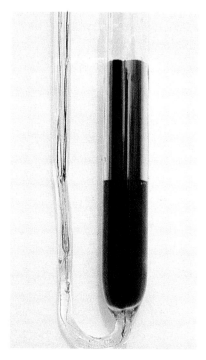

FIGURE 18.12 Sodium dissolves in liquid ammonia to form the deep blue solution in the lower half of the tube. At higher concentrations, the metal-ammonia solution becomes bronze in color, as in the top half of the tube.

The alkali metals are also oxidized when they dissolve in liquid ammonia. However, unlike their behavior in water, the released electrons occupy cavities formed by the NH_3 solvent molecules and give dark blue _metal-ammonia solutions_ (Fig. 18.12). These solutions of electrons (and cations of the metal) are excellent reducing reagents. As the metal concentration is increased, the blue gives way to bronze and the solution begins to conduct electricity like a liquid metal.

Reaction with nonmetals. All alkali metals react directly with most nonmetals (other than the noble gases). However, only lithium reacts with nitrogen, in which it burns (like its diagonal neighbor magnesium) to form the nitride:

$$6Li(s) + N_2(g) \xrightarrow{\Delta} 2Li_3N(s)$$

When the alkali metals react with oxygen, the principal product varies down the group (Fig. 18.13). The smaller cations form compounds with small anions (the O^{2-} ion) predominantly, and the larger cations form compounds with progressively larger anions (peroxide O_2^{2-}, and superoxide O_2^-). Lithium, for example, forms mainly the oxide:

$$4Li(s) + O_2(g) \xrightarrow{\Delta} 2Li_2O(s)$$

The oxides of the other elements are produced by decomposition of their carbonates, rather than by direct reaction with oxygen; for example,

$$K_2CO_3(s) \xrightarrow{\Delta} K_2O(s) + CO_2(g)$$

The O^{2-} ion is such a strong Brønsted base that when a Group I oxide dissolves in water it immediately forms the hydroxide:

$$O^{2-}(aq) + H_2O(l) \longrightarrow 2OH^-(aq)$$

FIGURE 18.13 Although the alkali metals give a mixture of products when they burn in oxygen, lithium gives mainly the oxide (left), sodium the very pale yellow peroxide (center), and potassium the yellow superoxide (right).

Sodium reacts with oxygen to form mainly the pale yellow peroxide:

$$2\text{Na}(s) + \text{O}_2(g) \xrightarrow{\Delta} \text{Na}_2\text{O}_2(s)$$

The remaining metals form mainly the superoxide when they react with O_2:

$$\text{K}(s) + \text{O}_2(g) \xrightarrow{\Delta} \text{KO}_2(s)$$

Potassium superoxide, a paramagnetic yellow solid, is used to improve the quality of air in enclosed spaces by the reaction

$$4\text{KO}_2(s) + 2\text{CO}_2(g) \longrightarrow 2\text{K}_2\text{CO}_3(s) + 3\text{O}_2(g)$$

Much of the metallic potassium currently produced is used to manufacture potassium superoxide for this purpose. A similar reaction occurs with the Group I peroxides; for sodium, for example,

$$2\text{Na}_2\text{O}_2(s) + 2\text{CO}_2(g) \longrightarrow 2\text{Na}_2\text{CO}_3(s) + \text{O}_2(g)$$

Lithium peroxide reacts similarly. Since it has a lower formula weight, a smaller mass of Li_2O_2 can increase the oxygen content of a given amount of air, so it is favored for use in spacecraft.

18.4 SOME IMPORTANT GROUP I COMPOUNDS

The properties of the principal compounds of the Group I elements are summarized in Table 18.3. The presence of these elements in compounds can be detected quite easily by the characteristic colors they give to flames. As we saw at the beginning of Chapter 7, lithium gives deep red, sodium bright yellow, potassium lilac, rubidium red, and cesium blue. The color that is characteristic of lithium is emitted by LiOH molecules formed briefly in the flame, not by Li atoms themselves.

TABLE 18.3 Properties of Group I compounds

Compounds	Formula*	Comment
Oxides	M_2O	Formed by decomposition of carbonates; strong bases; react with water to form hydroxides
Hydroxides	MOH	Formed by reduction of water with the metal or from the oxide; strong bases
Carbonates	M_2CO_3	Soluble in water; weak bases in water; most decompose into oxides when strongly heated
Hydrogen carbonates	$MHCO_3$	Weak bases in water; can be obtained as solids
Nitrates	MNO_3	Decompose to nitrite and evolve oxygen when strongly heated
Salts of strong acids	MA	Usually soluble in water to give neutral solutions
Salts of weak acids	MA	Give basic solutions

*M = Group I metal.

Lithium compounds. Lithium is typical of an element at the head of its group, in that it differs significantly from its congeners and has a diagonal relationship with a neighbor—in this case, magnesium. The differences stem in part from the small size of the Li^+ cation. This gives it a high polarizing power, and hence a tendency toward covalence in its bonding. A further consequence of the small size of the Li^+ ion is strong ion-dipole interactions, with the result that lithium salts are often hydrated.

Commercial availability of lithium has increased during the past few decades, and so has the variety of the metal's applications. Lithium compounds are used in batteries, ceramics, and lubricants. They are also used in medicine, where small daily doses of lithium carbonate have been found to be an effective treatment for the mental disorder known as manic depression; the mode of action is still not understood, despite a great deal of research.

Sodium chloride. The annual tonnage of sodium chloride used by industry actually exceeds that of the traditionally chart-topping sulfuric acid. However, sodium chloride is not normally included in the top fifty chemicals because it is available naturally and does not have to be manufactured. Sodium chloride is mined as rock salt (Fig. 18.14) or obtained by evaporation from brine. The vastness of both resources raises the interesting question of the origin of the saltiness of the oceans. Why are the oceans nearly 30 times as rich in Na^+ ions as in K^+ ions (Table 11.1), even though these two elements occur with similar overall abundance on earth?

Three factors contribute to the predominance of sodium in the oceans. One is that K^+ ions, being larger than Na^+ ions, have weaker

FIGURE 18.14 Mining salt in a salt mine.

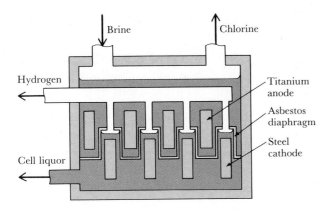

FIGURE 18.15 A diaphragm cell for the electrolytic preparation of sodium hydroxide from brine. The diaphragm prevents the chlorine from mixing with the hydrogen and the sodium hydroxide. The liquid is drawn off, and the water is partially evaporated. The unconverted sodium chloride crystallizes, leaving the sodium hydroxide in solution.

ion-dipole interactions with H_2O molecules and therefore its salts are slightly less soluble. Another is that Na^+ ions are less strongly bound in minerals than K^+ ions, so they are washed out of the minerals (and, eventually, into the oceans) more easily. Third, K^+ ions are more essential to plant growth than Na^+ ions, and plant roots have efficient methods for collecting the scarcer K^+ ions from the water that trickles past them in the soil. In effect, plants selectively filter out the K^+ ions, leaving the abundant Na^+ ions to flow on to the sea. The major deposits of potassium chloride (as the mineral sylvite) were probably left by bodies of water that formed and died in arid environments devoid of vegetation.

Sodium hydroxide. Sodium chloride is used in large quantities in the electrolytic production of chlorine and sodium hydroxide ("caustic soda") from brine. The hydroxide is a soft, waxy, white, corrosive solid. It is an industrial chemical of considerable importance, since it provides an inexpensive base for the production of other sodium salts.

Brine electrolysis is second only to aluminum extraction in the use of electricity for chemical production. Most modern electrolysis methods use a "diaphragm cell" (Fig. 18.15), in which compartments containing steel and titanium electrodes are separated by porous asbestos diaphragms. In the electrolysis

$$2NaCl(aq) + 2H_2O(l) \longrightarrow 2NaOH(aq) + Cl_2(g) + H_2(g)$$

1 mol of Cl_2 is produced for each 2 mol of NaOH; hence the industry must coordinate its hydroxide production with the demand for chlorine. This can be difficult, since a rise in the demand for one product is not always accompanied by a rise in demand for the other.

Sodium sulfate. Anhydrous sodium sulfate is called "salt cake," and the decahydrate ($Na_2SO_4 \cdot 10H_2O$) is called "Glauber's salt." Much of the sodium sulfate used in industry (for papermaking and as a substitute for phosphates in detergents) now comes from natural sources, particularly the sulfate-rich underground brines found in Texas. In countries lacking natural supplies, sodium sulfate is produced by the action of concentrated sulfuric acid on sodium chloride:

$$H_2SO_4(l) + 2NaCl(s) \xrightarrow{\Delta} Na_2SO_4(s) + 2HCl(g)$$

This is another reaction that proceeds in the desired direction because a volatile product escapes.

Sodium carbonate. Sodium carbonate decahydrate ($Na_2CO_3 \cdot 10H_2O$) was widely used earlier in this century as "washing soda." It was added to water to precipitate Mg^{2+} and Ca^{2+} ions as carbonates, as in

$$Ca^{2+}(aq) + CO_3^{2-}(aq) \longrightarrow CaCO_3(s)$$

and to provide an alkaline environment (CO_3^{2-} is a Brønsted base). Anhydrous sodium carbonate, or "soda-ash," is used in large amounts in the glass industry as a source of sodium oxide, into which it decomposes when heated. In this reaction, the carbonate ion decomposes into the Lewis acid CO_2 and the Lewis base O^{2-}:

$$CO_3^{2-}(s) \xrightarrow{\Delta} O^{2-}(s) + CO_2(g)$$

In more detail, this equation may be written

Sodium carbonate has been prepared by the *Solvay process* since the beginning of this century. However, in the 1940s a large deposit of the mineral trona, $Na_2CO_3 \cdot NaHCO_3 \cdot 2H_2O$, was found about 500 m beneath the surface in Wyoming. (The formula $Na_2CO_3 \cdot NaHCO_3 \cdot 2H_2O$ should be interpreted as signifying that the solid contains equal numbers of CO_3^{2-} and HCO_3^- ions, the requisite number of Na^+ cations, and two H_2O molecules for each formula unit.) Except for the cost of transportation to distant parts of the country, mining this mineral is more economical than using the Solvay process. That process is, however, still used in one major plant in the northeastern United States and in other parts of the world.

The Solvay process is simple in principle; overall, it is described by the equation

$$2NaCl(aq) + CaCO_3(s) \longrightarrow Na_2CO_3(s) + CaCl_2(s)$$

However, to bring about this reaction inexpensively requires chemical cunning; the flow chart for the process is shown in Fig. 18.16. First, carbonate is converted to hydrogen carbonate in a process that starts with the thermal decomposition of calcium carbonate:

$$CaCO_3(s) \xrightarrow{\Delta} CaO(s) + CO_2(g)$$

The carbon dioxide resulting from this reaction is converted to carbonic acid by dissolving it in water:

$$CO_2(g) + H_2O(l) \longrightarrow H_2CO_3(aq)$$

In practice, since this stage occurs in aqueous ammonia, it is accompanied by the Brønsted acid-base reaction

$$H_2CO_3(aq) + NH_3(aq) \longrightarrow NH_4^+(aq) + HCO_3^-(aq)$$

which yields the hydrogen carbonate.

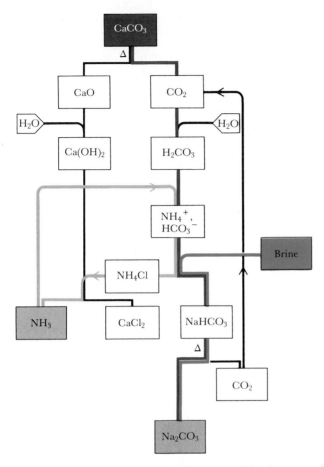

FIGURE 18.16 The flow of materials in the Solvay process for the production of sodium carbonate. The origin of the carbonate anion is indicated by the red track, and that of the sodium ion by the green track. The closed ammonia cycle is shown in orange.

The next (and last) stage begins with a precipitation reaction with brine:

$$NaCl(aq) + NH_4HCO_3(aq) \longrightarrow NaHCO_3(s) + NH_4Cl(aq)$$

Although sodium salts are soluble in water, this reaction proceeds because $NaHCO_3$ is not very soluble in the concentrated reaction mixture. The solid product is removed and converted to anhydrous sodium carbonate by heating:

$$2NaHCO_3(s) \xrightarrow{\Delta} Na_2CO_3(s) + CO_2(g) + H_2O(g)$$

The carbon dioxide is recirculated.

Two consecutive Brønsted acid-base reactions are used to recover the ammonia. First the lime produced in the calcium carbonate decomposition is slaked:

$$CaO(s) + H_2O(l) \longrightarrow Ca(OH)_2(aq)$$

(In this reaction water acts as a Brønsted acid, and O^{2-} as a base.) Then the OH^- ion from the $Ca(OH)_2$ is used as a base with NH_4^+ from the NH_4Cl solution:

$$NH_4^+(aq) + OH^-(aq) \longrightarrow NH_3(aq) + H_2O(l)$$

Overall, therefore, the ammonia recovery reaction is

$$2NH_4Cl(aq) + CaO(s) \longrightarrow 2NH_3(aq) + CaCl_2(aq) + H_2O(l)$$

The calcium chloride can be used to de-ice highways (as explained during the discussion of colligative properties in Section 11.6).

Potassium compounds. The principal mineral sources of potassium are sylvite (KCl) and carnallite (KCl · $MgCl_2$ · $6H_2O$). Potassium chloride is incorporated directly into fertilizers as a source of essential potassium, but for crops such as potato and tobacco, which cannot tolerate high Cl^- ion concentrations, potassium nitrate is used instead. The latter is formed by crystallization from a mixed solution of potassium chloride and sodium nitrate:

$$NaNO_3(aq) + KCl(aq) \longrightarrow NaCl(s) + KNO_3(aq)$$

The success of this process depends on the low solubility of sodium chloride, compared with that of potassium nitrate.

Potassium compounds are generally more expensive than the corresponding sodium compounds, but sometimes their advantages outweigh the expense. Potassium nitrate releases oxygen when heated, in the reaction

$$2KNO_3(s) \xrightarrow{\Delta} 2KNO_2(s) + O_2(g)$$

and is used to facilitate the ignition of matches. It is more expensive but less hygroscopic (water-absorbing) than the corresponding sodium compound, because the K^+ cation is larger and has weaker ion-dipole interactions with H_2O molecules. For the same reason, potassium nitrate is also the oxidizing agent in old-fashioned black gunpowder, a mixture of the nitrate, charcoal, and sulfur. It reacts as follows:

$$2KNO_3(s) + 4C(s) \longrightarrow K_2CO_3(s) + 3CO(g) + N_2(g)$$

$$2KNO_3(s) + 2S(s) \longrightarrow K_2SO_4(s) + SO_2(g) + N_2(g)$$

The two reactions result in the rapid formation of 6 mol of gas molecules from 4 mol of KNO_3, causing a sudden explosive expansion. Potassium carbonate is used as a source of potassium oxide in glassmaking and in the manufacture of soft (liquid) soaps; sodium produces hard (solid) soaps.

GROUP II: THE ALKALINE EARTH METALS

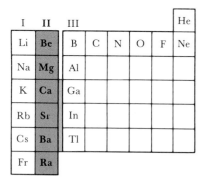

The members of Group II are listed with some of their properties in Table 18.4. Calcium, strontium, and barium are called the "alkaline earth metals," since their "earths"—the old name for oxides—are basic. The term "alkaline earths" is often extended loosely to all the members of the group.

18.5 THE GROUP II ELEMENTS

All the elements of the group are metals (Fig. 18.17) and too reactive to occur in the uncombined state. Beryllium occurs mainly as beryl ($3BeO$ · Al_2O_3 · $6SiO_2$), sometimes in crystals so big that they weigh sev-

TABLE 18.4 The Group II elements

Valence configuration: ns^2
Normal form: Soft silver-gray metals

Z	Name	Symbol	Atomic weight, amu	Melting point, °C	Boiling point, °C	Density, g/cm³
4	Beryllium	Be	9.01	1285	2470	1.85
12	Magnesium	Mg	24.31	650	1100	1.74
18	Calcium	Ca	40.08	840	1490	1.53
38	Strontium	Sr	87.62	770	1380	2.58
56	Barium	Ba	137.34	710	1640	3.59
88	Radium	Ra	226.03	700	1500	5.00

(a)

(b)

(c)

(d)

(e)

FIGURE 18.17 The elements of Group II: (a) beryllium, (b) magnesium, (c) calcium, (d) strontium, and (e) barium. All four central elements of the group (magnesium through barium) were discovered by Humphry Davy in a single year, 1808. The two outer elements were discovered later: beryllium in 1828 (by Friedrich Wöhler) and radium in 1898 (by Pierre and Marie Curie).

FIGURE 18.18 An emerald is a crystal of beryl with some Cr^{3+} ions, which are responsible for its color.

eral tons. Emerald is a form of beryl; its green color is caused by Cr^{3+} ions that are present as impurities (Fig. 18.18). Magnesium occurs in seawater and as the mineral dolomite ($CaCO_3 \cdot MgCO_3$).

Physical properties and preparation. Beryllium has grown in importance in recent years, both because of its low density, which makes it suitable for the construction of missiles and satellites, and for its useful nuclear properties (discussed in Chapter 22). Since Be atoms have so few electrons, thin sheets of the metal are transparent to x-rays and can be used as windows in x-ray tubes (to allow the rays to escape). The metal is obtained by electrolytic reduction of molten beryllium chloride. Much of the metal that is produced is added in small amounts to copper; the small Be atoms pin the Cu atoms together in a structure that is more rigid than pure copper, but they do not destroy copper's excellent ability to conduct electricity. The hard electrically conducting alloy is formed into nonsparking tools for use in oil refineries.

Magnesium is a silver-white metal that is protected in air by a film of white oxide and hence looks dull gray. It has a low density (two-thirds that of aluminum) and is widely used as a component of alloys in applications where lightness and toughness are needed, as for airplanes. The unalloyed metal burns vigorously in air (it reacts with the nitrogen and the carbon dioxide as well as the oxygen, especially when it is sprayed with water); hence magnesium is used in fireworks and incendiary devices.

Metallic magnesium is produced by reduction, either electrolytically or with a chemical reducing agent. In the chemical reduction of magnesium oxide (obtained from the decomposition of dolomite), "ferrosilicon," an alloy of iron and silicon, is used as the reducing agent at about 1200°C. At this temperature the magnesium produced is immediately vaporized, so even though the equilibrium does not favor the reduction, the process continues because magnesium is removed as soon as it is formed.

The electrolytic method of magnesium production uses seawater as the principal raw material (Fig. 18.19). The first stage is the precipita-

FIGURE 18.19 This facility is a magnesium extraction plant near Salt Lake City, Utah. Magnesium ions are also present in the remains of former seas.

tion of magnesium hydroxide ($K_{sp} = 1.1 \times 10^{-11}\ M^3$) using lime (calcium hydroxide):

$$Mg^{2+}(aq) + 2OH^-(aq) \longrightarrow Mg(OH)_2(s)$$

(The lime is produced by the thermal decomposition of calcium carbonate in shells dredged up from the ocean floor.) The precipitated magnesium hydroxide is then filtered off and treated with hydrochloric acid:

$$Mg(OH)_2(s) + 2HCl(aq) \longrightarrow MgCl_2(aq) + 2H_2O(l)$$

The acid is obtained by the action of chlorine (a byproduct of the electrolysis) on natural gas:

$$CH_4(g) + Cl_2(g) \longrightarrow CH_3Cl(g) + HCl(g)$$

Finally, the magnesium chloride is dried and transferred to an electrolysis cell. In the electrolysis, magnesium is produced at the cathode, and chlorine at the anode:

$$MgCl_2(l) \xrightarrow{\text{electrolysis}} Mg(l) + Cl_2(g)$$

The true alkaline earth metals—calcium, strontium, and barium—are obtained either by electrolysis or by reduction with aluminum in a version of the thermite process; for example, for barium,

$$3BaO(s) + 2Al(l) \xrightarrow{\Delta} Al_2O_3(s) + 3Ba(l)$$

All three metals are partially passivated in air by a protective surface layer of oxide. Barium reacts particularly quickly with oxygen and may ignite in moist air. Although this is a difficult problem for the barium industry, like most problems it can be turned to advantage in particular contexts. Barium is in fact used as a "getter" in vacuum tubes, since it combines with any oxygen and air that remain after manufacture and leaves a better vacuum. Calcium is used on a larger scale for much the same type of reaction: it is added to some steels to remove remaining traces of oxides.

Trends in properties. The valence electron configuration of the atoms of the Group II elements is ns^2, where n is the period number. The second ionization energy is low enough that the energy required to form the doubly charged cation can be recovered from the increased lattice enthalpy. Hence, the Group II elements occur with oxidation number +2 in all their compounds, generally as the cation M^{2+}. All are low in the electrochemical series, with strongly negative reduction potentials. Their reducing power increases down the group, Be having the least and Ra the greatest.

▼ EXAMPLE 18.5 Predicting the properties of an element

Before reading any further, suggest some of the physical and chemical properties you would expect beryllium to possess.

STRATEGY The element's group and period are a clue to its properties. We should also decide whether it is likely to have a diagonal relationship with a neighbor with known properties. Pay attention to whether the element is a metal, a nonmetal, or on the border between the two.

SOLUTION Beryllium is at the head of Group II and can be expected to resemble its diagonal neighbor aluminum. Although all the Group II elements are metals, Be is likely to be the least metallic. We can expect it to have an amphoteric oxide, and its compounds to be more covalent than those of its congeners. Since Be is a small atom, we might expect it to form bonds with no more than four other atoms and predict that a covalently bonded BeX_4 group will be apparent in a number of its compounds.

EXERCISE Which element from another group will magnesium resemble in its physical and chemical properties?

[*Answer*: Li]

All the Group II elements except beryllium reduce water; for example,

$$Ca(s) + 2H_2O(l) \longrightarrow Ca(OH)_2(aq) + H_2(g)$$

Beryllium does not react with water even when red hot: it is passivated (like aluminum) by an oxide film that survives even at high temperatures. Magnesium reacts when heated, and calcium reacts even with cold water. The metals all reduce hydrogen ions to hydrogen, but beryllium does not react with nitric acid—again because it becomes passivated by a film of oxide. Beryllium shows its amphoteric character by being the only member of the group that dissolves (like aluminum) in aqueous sodium hydroxide. In doing so, it forms a beryllate ion through a Lewis acid-base reaction:

$$Be(s) + 2OH^-(aq) + 2H_2O(l) \longrightarrow [Be(OH)_4]^{2-}(aq) + H_2(g)$$

This formation of a tetrahedral BeX_4 covalent unit (**4**) is a typical feature of beryllium's chemistry.

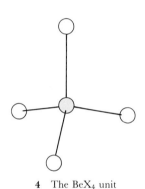

4 The BeX_4 unit

18.6 SOME IMPORTANT GROUP II COMPOUNDS

The properties of some compounds of the Group II elements are summarized in Table 18.5. Like the alkali metals, the compounds of these elements are best discussed in terms of their anions. However, certain features are worth emphasizing here. Apart from a hint of nonmetallic character in beryllium, these elements have all the chemical characteristics that are typical of metals, such as forming M^{2+} cations and having basic oxides and hydroxides.

The alkaline earth metals can be detected in compounds by the colors they give to flames. Calcium gives orange-red, strontium crimson, and barium yellow-green (Fig. 18.20). Fireworks often contain their salts (typically nitrates and chlorates, for the anions then act as an additional supply of oxygen) together with magnesium powder. Sparklers are made of aluminum powder in place of magnesium, so that they burn more slowly.

Beryllium compounds. Beryllium compounds are very toxic and must be handled with great caution. Their properties are dominated by the highly polarizing character of the Be^{2+} ion and its small size. The strong polarizing power results in moderately covalent compounds, and the small size limits to four the number of groups that can attach to the ion. These two features together are responsible for the promi-

TABLE 18.5 Properties of some Group II compounds

Compounds	Formula*	Comment
Oxides	MO	Formed by decomposition of carbonates; give strongly basic solutions when they react with water (BeO is amphoteric); withstand high temperatures
Hydroxides	$M(OH)_2$	Formed by action of water on oxides or precipitation from salt solution; sparingly soluble in water (except Ba); strong bases (except Be, which is amphoteric)
Carbonates	MCO_3	Most decompose on heating; only very slightly soluble in water
Hydrogen carbonates	$M(HCO_3)_2$	Unstable as solids; more soluble in water than carbonates
Nitrates	$M(NO_3)_2$	Decompose into NO_2 and O_2 when heated; soluble in water
Salts of strong acids	MA_2	When soluble in water, give neutral solutions except for Be, which forms an acidic hydrated ion
Salts of weak acids	MA_2	When soluble in water, generally give basic solutions

*M = Group II metal.

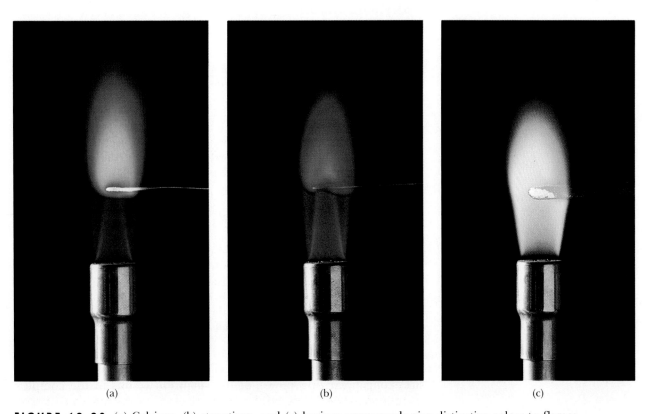

    (a)   (b)   (c)

FIGURE 18.20 (a) Calcium, (b) strontium, and (c) barium compounds give distinctive colors to flames.

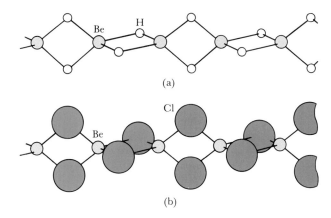

FIGURE 18.21 The structures of (a) BeH_2 and (b) $BeCl_2$.

nence of the tetrahedral BeX_4 unit, as mentioned in Example 18.5. This unit is found not only in the oxoanions of beryllium, but also in the chloride and the hydride (Fig. 18.21). The chloride is produced by the action of chlorine on the oxide in the presence of carbon:

$$BeO(s) + C(s) + Cl_2(g) \xrightarrow{\text{600 to 800°C}} BeCl_2(g) + CO(g)$$

In this compound, the Be atoms act as Lewis acids and accept electron pairs from the Cl atoms of neighboring $BeCl_2$ groups, thus forming a chain of tetrahedral $BeCl_4$ units.

▼ EXAMPLE 18.6 Accounting for the bonding in beryllium hydride

Show, in terms of delocalized orbitals, that beryllium hydride can be expected to be stable.

STRATEGY We need to show that all the valence electrons can be accommodated in orbitals that have a predominantly bonding character, for then the compound will have a lower energy than its separated atoms. Recall from Chapter 9 that given N atomic orbitals, we can form N molecular orbitals, of which about half will be bonding. Therefore, we begin by counting the atomic orbitals, then we count the electrons, and finally we determine whether there are enough bonding orbitals to accommodate the electrons without using too many antibonding orbitals.

SOLUTION In a chain of N BeH_2 units, there are N $2s$ orbitals on the Be atoms and $2N$ $1s$ orbitals on the H atoms, making $3N$ atomic orbitals in all. In each chain there are therefore $3N$ delocalized molecular orbitals. Of these $3N$ orbitals, about $\frac{3}{2}N$ are bonding and $\frac{3}{2}N$ are antibonding. Each Be atom contributes two electrons, and each H atom one electron, so overall there are $2N + N + N = 4N$ electrons to accommodate. Since two electrons can occupy each orbital, these electrons require $2N$ orbitals. Therefore, the bonding orbitals are filled, and only half the antibonding orbitals need be used. This suggests that, because there are more bonding electrons than
▲ antibonding electrons, the chain will be stable.

The Be^{2+} ion is hydrated to $[Be(H_2O)_4]^{2+}$ in aqueous solution, and the strong polarizing power of the small ion results in the complex being acidic (like $[Al(H_2O)_6]^{3+}$):

$$[Be(H_2O)_4]^{2+}(aq) + H_2O(l) \rightleftharpoons [Be(H_2O)_3OH]^+(aq) + H_3O^+(aq)$$

Ions more complex than this are also formed, and precipitation of a solid hydroxide begins when enough base has been added to bring the average number of OH^- groups per Be atom to one. As in the case of the analogously amphoteric aluminum hydroxide, the beryllium hydroxide precipitate dissolves when more alkali is added.

Magnesium compounds. Magnesium oxide is formed by thermal decomposition of the hydroxide or the carbonate (burning the metal in air gives the nitride as well as the oxide). It dissolves only very slowly and slightly in water. One of its most striking properties is its ability to withstand high temperatures, for it melts only at 2800°C. This high stability (and its corresponding very high lattice enthalpy of 3850 kJ/mol) can be traced to the small ionic radii of the Mg^{2+} and O^{2-} ions and hence to their very strong electrostatic interaction with each other. The oxide has two other characteristics that make it useful: it conducts heat very well, and it conducts electricity poorly. All three properties lead to its use as an insulator in electric heaters.

Magnesium hydroxide is a mild base. It is only very slightly soluble in water, but instead forms a white colloidal "suspension"—a mist of small particles dispersed through a liquid—that is known as "milk of magnesia" and used as a stomach antacid. Because this base is relatively insoluble, it is not absorbed from the stomach but remains to act on whatever acids are present. Hydroxides have the advantage over bicarbonates (which are also used as antacids) that their neutralization does not lead to the formation of carbon dioxide and its consequence, belching. The disadvantage of milk of magnesia is that the product of neutralization is magnesium chloride, which can act as a purgative. Magnesium sulfate ("Epsom salts") is also a popular purgative. Its action, and that of the chloride, appears to be to inhibit the reabsorption of water from the intestine. The resulting increased flow of water triggers the mechanism that results in defecation.

FIGURE 18.22 The green of vegetation is caused by the absorption of red and blue light by the chlorophyll molecule in leaves; the green light is not absorbed and is reflected, giving the leaves their green color.

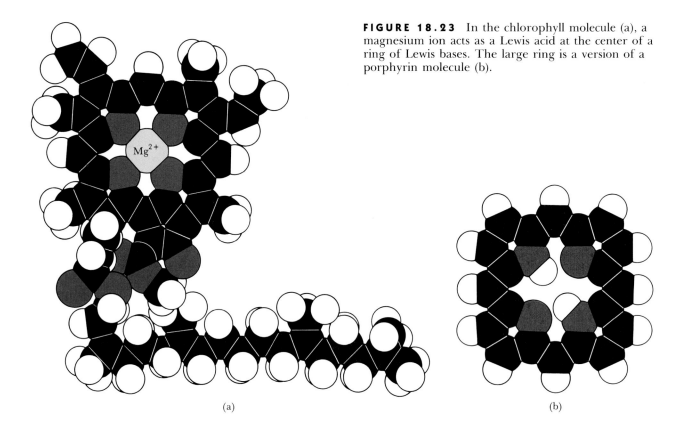

FIGURE 18.23 In the chlorophyll molecule (a), a magnesium ion acts as a Lewis acid at the center of a ring of Lewis bases. The large ring is a version of a porphyrin molecule (b).

(a)   (b)

FIGURE 18.24 Marble is a dense form of calcium carbonate, and is often colored by ionic impurities.

Probably the most important compound of magnesium is chlorophyll (Fig. 18.22), the name of which comes from the Greek words for "green leaf." This green compound captures light from the sun and channels its energy into photosynthesis:

$$6CO_2(g) + 6H_2O(l) \xrightarrow{\text{light, chlorophyll}} C_6H_{12}O_6(aq) + 6O_2(g)$$

In the chlorophyll molecule (Fig. 18.23), the Mg^{2+} acts as a Lewis acid, and the N atoms of the large ringlike "porphyrin" organic group act as Lewis bases. One function of the Mg^{2+} ion (which lies just above the plane of the ring) appears to be to keep the porphyrin ring rigid. This helps to ensure that the energy captured from incoming photons does not spread away as thermal motion before it has been used to bring about a chemical reaction.

Calcium carbonate. Calcium occurs naturally as chalk (calcium carbonate, $CaCO_3$) and limestone (a harder form of the carbonate). Marble is a dense form of calcium carbonate that can be given a high polish; it is often colored by impurities, most commonly iron ions (Fig. 18.24). The two most common forms of pure calcium carbonate are calcite and aragonite (see Fig. 10.2). All these carbonates are the fossilized remains of marine life. As explained in Section 11.3, the "hardness" of hard water is due to the presence of calcium and magnesium salts (particularly their hydrogen carbonates), dissolved out of the rocks through which the water has trickled.

▼ EXAMPLE 18.7 Accounting for the effect of heat on hard water

Suggest a reason why calcium carbonate is precipitated when hard water, which contains Ca^{2+} and HCO_3^- ions and dissolved CO_2, is boiled.

STRATEGY We should look for a chain of equilibria that connects the heating to the precipitation, for then the effect can be transmitted along the chain. We can draw on the fact that gases are less soluble in hot water than in cold, which suggests that carbon dioxide will be driven out of solution by the heating. We would then consider the effect of the decreased CO_2 concentration on other equilibria.

SOLUTION The removal of CO_2 from the solution by heating encourages the Lewis acid-base decomposition

$$H_2CO_3(aq) \longrightarrow H_2O(l) + CO_2(g)$$

to proceed as written. The resulting reduction in the H_2CO_3 concentration promotes the Brønsted acid-base reaction

$$\underset{\text{acid}_1}{HCO_3^-(aq)} + \underset{\text{base}_2}{HCO_3^-(aq)} \longrightarrow \underset{\text{base}_1}{CO_3^{2-}(aq)} + \underset{\text{acid}_2}{H_2CO_3(aq)}$$

Since this increases the CO_3^{2-} ion concentration, it causes the precipitation

$$Ca^{2+}(aq) + CO_3^{2-}(aq) \longrightarrow CaCO_3(s)$$

▲ The net result is the precipitation of calcium carbonate.

Calcium oxide and hydroxide. Calcium carbonate decomposes to the oxide (quicklime) when heated:

$$CaCO_3(s) \xrightarrow{\Delta} CaO(s) + CO_2(g)$$

This decomposition requires a higher temperature (about 800°C) than that of $MgCO_3$. The difference can be explained by recognizing that the larger Ca^{2+} ion is a weaker Lewis acid than the small Mg^{2+} ion and hence is less effective at removing an O^{2-} ion from a neighboring CO_3^{2-} ion in the solid. The addition of water to quicklime, CaO, produces slaked lime, $Ca(OH)_2$, and a considerable amount of heat. Slaked lime is the form in which lime is normally sold, for quicklime can set fire to moist wood and paper.

A solution of calcium hydroxide, which is sparingly soluble in water, is called "lime water." It is used in a test for carbon dioxide, which reacts with it to form a suspension of the even less soluble calcium carbonate:

$$Ca(OH)_2(aq) + CO_2(g) \longrightarrow CaCO_3(s) + H_2O(l)$$

Lime is produced in enormous quantities throughout the world, being sixth in annual tonnage. About 40% of this output is used in metallurgy. In ironmaking it is used as a Lewis base; its O^{2-} ion reacts in the blast furnace with the silica impurities in the ore to form a liquid slag:

$$CaO(s) + SiO_2(s) \longrightarrow CaSiO_3(l)$$

SiO_2 is a Lewis acid in this reaction. About 50 kg of lime are needed to produce a ton of iron.

FIGURE 18.25 A highly magnified view of the surface of mortar.

Lime is used as an inexpensive base in industry, as well as to adjust the pH of soils in agriculture. Perhaps surprisingly, it is also used to *remove* Ca^{2+} ions from water. Its role here is to convert HCO_3^- to CO_3^{2-} by providing OH^- ions:

$$HCO_3^-(aq) + OH^-(aq) \longrightarrow CO_3^{2-}(aq) + H_2O(l)$$

The increase in the concentration of CO_3^{2-} ions encourages the precipitation of Ca^{2+} ions in the reaction

$$Ca^{2+}(aq) + CO_3^{2-}(aq) \longrightarrow CaCO_3(s)$$

This removes both the Ca^{2+} ions that were present initially and the Ca^{2+} ions that were added as lime, and overall the Ca^{2+} ion concentration is reduced.

The reaction that converts calcium oxide to calcium carbide,

$$CaO(s) + 3C(s) \xrightarrow{2200°C} CaC_2(s) + CO(g)$$

is a bridge between inorganic and organic compounds. The carbide ion ($:C\equiv C:^{2-}$) is a Brønsted base and is readily protonated by water, resulting in the formation of the organic compound acetylene:

$$C_2^{2-}(s) + 2H_2O(l) \longrightarrow C_2H_2(g) + 2OH^-(aq)$$

Structural calcium. Calcium compounds are often used in (or as) structural components; their rigidity stems from the strength with which the small, highly charged Ca^{2+} cation interacts with its neighbors. "Mortar" consists of about one part slaked lime and three parts sand (largely silica, SiO_2). It sets to a hard mass as the lime reacts with carbon dioxide from the air to form the carbonate (Fig. 18.25). Calcium is also found in the rigid structural components of living things, either as the calcium carbonate of shells or as the calcium phosphate of bone. About a kilogram of calcium is present in an adult human body—mostly in the form of insoluble calcium phosphate, but also as Ca^{2+} ions in fluids within our cells. The calcium in newly formed bone is in dynamic equilibrium with the calcium ions in the body fluids, and about 20 g of calcium is exchanged between the two each day.

Tooth enamel is a hydroxyapatite, a mineral of composition $Ca_5(PO_4)_3OH$. Tooth decay can occur when acids attack the enamel:

$$Ca_5(PO_4)_3OH(s) + 4H_3O^+(aq) \longrightarrow$$
$$5Ca^{2+}(aq) + 3HPO_4^{2-}(aq) + 5H_2O(l)$$

The principal agents of tooth decay are the carboxylic acids (acids with a —COOH group) produced when bacteria act on the remains of food. A more resistant tooth coating is obtained when the OH^- ions in the apatite are replaced by F^- ions to give a fluorapatite:

$$Ca_5(PO_4)_3OH(s) + F^-(aq) \longrightarrow Ca_5(PO_4)_3F(s) + OH^-(aq)$$

The addition of fluoride ions to domestic water supplies in the form of NaF is now a widespread practice. Another strategy for strengthening the protective coating of teeth is the use of fluoridated toothpastes, containing either tin(II) fluoride or sodium monofluorophosphate (Na_2FPO_3).

Calcium sulfate is a component of several rigid materials that are utilized in all kinds of solid structures. Alabaster is a compact form of gypsum ($CaSO_4 \cdot 2H_2O$); it resembles marble but is softer, and is used for ornamental work. When gypsum is heated, it loses some of the water of hydration and becomes "plaster of Paris," $2CaSO_4 \cdot H_2O$. This is so called because it was originally obtained from the gypsum mined in the Montmartre district of Paris. Plaster of Paris sets to a rigid mass when water is added and the hydration is restored. Since rehydration causes the solid to swell, the plaster takes a good impression from a mold. Calcium sulfate is also a component of cements and concrete, which are described in the next chapter.

SUMMARY

18.1 Hydrogen is the most abundant element in the universe. It is manufactured by the **reforming reaction,** which produces **synthesis gas,** a mixture of H_2 and CO, and by a subsequent **shift reaction.** Alternative procedures are electrolysis and the reduction of $H^+(aq)$ ions by metals. The principal uses of hydrogen are in the synthesis of ammonia and methanol, the reduction of ores, and the hydrogenation of oils. It is also used as a rocket fuel. Hydrogen can take part in hydrogen bonding between electronegative elements. The transfer of the hydrogen ion (the proton) is the central feature of Brønsted acid-base behavior.

18.2 Hydrogen forms three classes of binary compounds with other elements. The **saline hydrides,** which contain the **hydride ion** H^-, are formed by s-block elements and are oxidized readily by water. The **molecular compounds** are formed by elements in the p block, and most are gases or volatile liquids. Some of the d-block elements form **metallic hydrides,** in which the hydrogen atoms occupy gaps between the metal atoms.

18.3 The elements of Group I are the **alkali metals.** They are all strongly reducing metals that react with water to form hydroxides. Lithium is diagonally related to magnesium and forms compounds that are more markedly covalent than those of its congeners. All the elements react directly with the nonmetals (other than

the noble gases) and burn in air to form an oxide, peroxide, or superoxide.

18.4 **Sodium chloride** is a major industrial chemical. It is electrolyzed to chlorine and **sodium hydroxide** in a diaphragm cell, converted to **sodium sulfate** with sulfuric acid, and used in the **Solvay process** for the production of **sodium carbonate.** Potassium compounds tend to be less hygroscopic than sodium compounds.

18.5 The elements of Group II are the **alkaline earth metals.** A rich source of magnesium is seawater, from which the metal is extracted by precipitation and then electrolysis. Metallic character increases down the group. Beryllium has a strong diagonal resemblance to aluminum, forming amphoteric oxides and compounds that are markedly covalent. Structurally, its compounds are dominated by the tetrahedral BeX_4 group. All the Group II metals other than beryllium reduce water with a vigor that increases down the group and form compounds that contain the M^{2+} cation.

18.6 **Calcium carbonate** is a major source of **lime** and **slaked lime;** the latter is widely used as an inexpensive base. Calcium appears in a number of structurally rigid compounds, including the **calcium phosphate** of bone and the **calcium sulfate** of alabaster, gypsum, and plaster of Paris. **Calcium carbide** is an important link between inorganic and organic compounds.

EXERCISES

Hydrogen

In Exercises 18.1 and 18.2, write the balanced chemical equation for each reaction.

18.1 (a) The reaction between lithium and water
   (b) The reaction between red hot magnesium and water

   (c) The formation of synthesis gas
   (d) The formation of sodium hydrogen carbonate by bubbling carbon dioxide through aqueous sodium carbonate
18.2 (a) The reaction between hydrogen and sodium
   (b) The shift reaction

(c) The overall reaction in which the formation of synthesis gas is followed by the shift reaction

(d) The reactions that occur when magnesium burns in damp air.

In Exercises 18.3 and 18.4, suggest, with balanced equations, a method for preparing each compound from starting materials of your choice.

18.3 (a) LiH; (b) H_2S; (c) HBr; (d) PH_3.
18.4 (a) RbH; (b) HF; (c) HI; (d) SbH_3.

In Exercises 18.5 and 18.6, classify each compound as saline, molecular, or metallic.

18.5 (a) Sodium hydride; (b) water; (c) titanium hydride.
18.6 (a) Lithium hydride; (b) ammonia; (c) palladium hydride.

***18.7** Draw the crystal structure of sodium hydride and calculate its density.
***18.8** Draw the crystal structure of potassium hydride and calculate its density.

18.9 Describe two processes for the industrial preparation of hydrogen. What are the economic considerations involved in the selection of a process to use in a plant?

18.10 What are the advantages and disadvantages of hydrogen as a fuel in vehicles? In terms of the heating value per unit volume, is methane or hydrogen more economical to pump through pipelines? Relative to volume and mass, would it be advantageous to replace the liquid hydrogen used in the space shuttle with liquid methane (of density 0.42 g/mL)?

18.11 Estimate the standard reaction enthalpy of (a) the synthesis gas reaction, (b) the shift reaction, and (c) the overall reaction, all at 25°C.

18.12 Calcium hydride is used as a portable source of hydrogen. What volume of the gas (at STP) can be produced from 1.0 kg of the hydride? What mass of water should be supplied for the reaction?

18.13 Calculate the number of moles of H_2 molecules used each year in the United States, given that around 3×10^8 kg of the gas is used. What fraction of this production is required to make 15×10^9 kg of ammonia (the approximate annual production)?

18.14 Calculate the volume that the annual United States production of hydrogen (3×10^8 kg) would occupy if it were stored as (a) a gas at 25°C and 1 atm; (b) a liquid of density 0.09 g/cm^3.

18.15 Describe the evidence for the remark that hydrogen can act as both a reducing agent and an oxidizing agent, giving chemical equations in each case.

18.16 Is there any chemical support for the view that hydrogen should be classified as a member of Group I?

18.17 What information about the character of an element can be deduced from its reaction with hydrogen?
18.18 Under what conditions does hydrogen react with (a) chlorine, (b) sodium, and (c) ethylene. What are the products in each case?

In Exercises 18.19 and 18.20, suggest, with chemical equations, a method for the preparation of each compound.

18.19 (a) Ammonia; (b) hydrogen sulfide; (c) hydrogen chloride.
18.20 (a) Water; (b) calcium hydride; (c) hydrogen iodide.

18.21 Estimate the density of heavy water, D_2O, from the atomic weight of deuterium (2.0 amu) and the density of ordinary water (1.0 g/mL at 25°C).

18.22 Fluorine forms one binary hydride, oxygen forms two, and nitrogen forms three. Identify them, and write their Lewis structures. State and explain the differences between the properties of the two oxygen compounds.

Group I

18.23 Name the minerals that are the principal sources of sodium and potassium. Describe how the elements are obtained in each case.
18.24 Name the sources of sodium chloride, sodium hydroxide, and sodium carbonate, and give two industrial uses for each compound.

18.25 Give the chemical names and formulas of the minerals (a) rock salt; (b) carnallite; (c) sylvite; (d) trona.
18.26 Give the chemical names of (a) caustic soda; (b) salt cake; (c) soda ash; (d) washing soda.

18.27 Write the chemical equations for the Solvay process for manufacturing sodium carbonate. Draw a flow diagram showing how the products of intermediate reactions are recirculated.

18.28 Describe, with equations, how lithium differs from its congeners.

In Exercises 18.29 and 18.30, construct a Born-Haber cycle to demonstrate that the disproportionation is strongly exothermic, and explain why alkali metal halides of formula MX_2 are hence never found.

18.29 $NaCl_2(s) \longrightarrow NaCl(s) + Na(s)$
18.30 $KBr_2(s) \longrightarrow KBr(s) + K(s)$

18.31 Explain why the sea is salty with sodium chloride and not with potassium chloride. Suggest reasons why salts of sodium and potassium are more widely used in industry than salts of any other element.

18.32 What justification is there for regarding the ammonium ion NH_4^+ as an analog of a Group I metal cation? Consider solubility, charge, size, etc.

18.33 Sodium carbonate is often supplied as the decahydrate $Na_2CO_3 \cdot 10H_2O$. What mass of this solid should be used to prepare 250 mL of a 0.100 M $Na_2CO_3(aq)$ solution?

18.34 Explain how the following compounds may be prepared starting from sodium chloride and any acid: (a) $NaOH$; (b) $NaNO_3$; (c) Na_2SO_4; (d) HBr.

Group II

18.35 Name the mineral forms used as sources of beryllium and magnesium. Show, with equations, how the elements are extracted.

18.36 Give the chemical names and formulas of (a) limestone; (b) quicklime; (c) calcite; (d) dolomite; (e) gypsum; (f) Epsom salts.

18.37 Describe, with equations, how magnesium is extracted from seawater. Draw a flow diagram showing how the reagents and by-products are recirculated. If 1 L of seawater contains 1.35 g of Mg^{2+}, what volume of seawater must be processed to obtain 1 kg of magnesium metal? For how long must a current of 10 A be passed through an electrolytic cell to obtain that mass of magnesium?

18.38 Explain, with equations, how aqueous sodium hydroxide reacts with (a) beryllium; (b) aluminum.

18.39 Give examples, with equations, of how beryllium differs from its congeners.

18.40 How do the properties of calcium confirm its position near the top of the electrochemical series?

18.41 List the sources of calcium carbonate, quicklime, slaked lime, and calcium phosphate, and name two uses for each compound. Suggest reasons why calcium forms rigid compounds.

18.42 Explain, with equations, how the following compounds may be prepared from calcium carbonate: (a) $Ca(OH)_2$; (b) $Ca(NO_3)_2$; (c) C_2H_2.

18.43 Explain what is meant by "hard water" and (with equations) how it may be softened.

18.44 How may beryllium be distinguished from magnesium (a) physically; (b) chemically?

In Exercises 18.45 and 18.46, construct a Born-Haber cycle to demonstrate that the disproportionation is strongly exothermic, and explain why alkaline earth metal halides of formula MX are hence never observed.

18.45 $2MgCl(s) \longrightarrow Mg(s) + MgCl_2(s)$
18.46 $2CaBr(s) \longrightarrow Ca(s) + CaBr_2(s)$

18.47 Explain why calcium fluoride is more soluble in acidic solution than in water.

18.48 Estimate the standard reaction enthalpy, free energy, and equilibrium constant at 25°C of each of the reactions (a) $CaCO_3(s) \longrightarrow CaO(s) + CO_2(g)$ and (b) $MgCO_3(s) \longrightarrow MgO(s) + CO_2(g)$. By assuming that the entropy changes are the same in both cases, estimate the decomposition temperatures of the two compounds.

Groups I and II: Comparative Chemistry

18.49 The valence electron configurations of the Group I and Group II elements are ns^1 and ns^2, respectively. Show how these configurations account for the similarities and differences between the two groups.

18.50 Arrange hydrogen and the Group I and Group II elements in electrochemical order, starting with the most strongly reducing element. Suggest reasons for the order.

18.51 Give examples showing that lithium resembles magnesium and beryllium resembles aluminum.

18.52 How may solutions of barium chloride and sodium chloride be distinguished chemically?

18.53 Suggest what compounds could be included with magnesium powder in fireworks to give red, orange, yellow, green, blue, and lilac colors. What anions would it be reasonable to use?

****18.54** How does the position of an element in the electrochemical series affect the method that is employed to extract it? Discuss thermodynamically the possibility of using carbon to extract calcium from limestone.

18.55 Is there any evidence in this chapter to support the view that metal character decreases from left to right across a period?

18.56 It has been proposed that the ability of a cation to polarize anions is proportional to its charge divided by its radius. Use this criterion to arrange the s-block elements in order of increasing polarizing power. Do the resulting values support the diagonal relationships within the block?

18.57 Find three redox reactions that are given in the chapter. Identify the reducing and oxidizing agents, and write the corresponding reduction and oxidizing half-reactions.

18.58 Find three Lewis acid-base reactions in the chapter, and one precipitation reaction. Identify the Lewis acids and bases.

18.59 Classify each step in the Solvay process by reaction type (Lewis acid-base, Brønsted acid-base, and so on). Then determine the mass of ammonia needed to produce 1.0 kg of soda ash.

This chapter is the first of two
that describe the p-block elements
and their rich chemical properties.
In it we see the transition from the
predominantly metallic character
of the Group III elements to the
predominantly nonmetallic
character of the Group V
elements. The illustration shows a
section through the mineral
bauxite: one of the problems we
discuss in this chapter is that of
purifying the mineral before it is
used as a source of aluminum.

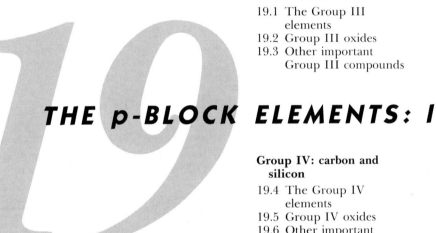

Group III: boron and aluminum

19.1 The Group III elements
19.2 Group III oxides
19.3 Other important Group III compounds

THE p-BLOCK ELEMENTS: I

Group IV: carbon and silicon

19.4 The Group IV elements
19.5 Group IV oxides
19.6 Other important Group IV compounds

Group V: nitrogen and phosphorus

19.7 The Group V elements
19.8 Hydrogen and halogen compounds
19.9 Group V oxides and oxoacids

The p block of the periodic table consists of Groups III through VIII. The characteristics of its members range from almost typical metal behavior (as shown by aluminum) to typical nonmetal behavior (as we see in the lighter halogens and the noble gases). The p block includes carbon and oxygen, the elements central to life and natural intelligence, as well as silicon and germanium, the elements central to modern technology and artificial intelligence.

Since the p-block elements form such a wide and important range of compounds (including the oxides, the oxoacids, the halides and the millions of organic compounds), we shall discuss them in three stages. This chapter describes some of the elements of Groups III through V (from boron to nitrogen and phosphorus). Chapter 20 deals with Groups VI through VIII (from oxygen to the noble gases). In Part V we return to carbon and describe the compounds that traditionally are treated as "organic."

GROUP III: BORON AND ALUMINUM

The elements in Group III and some of their properties are listed in Table 19.1. We concentrate on the two most important members of the group, boron and aluminum.

19.1 THE GROUP III ELEMENTS

Boron, at the head of Group III, shows the characteristics of a nonmetal in most of its chemical reactions. It has acidic oxides and hydroxides and forms an interesting and extensive range of binary molecular compounds with hydrogen. However, metallic character increases down the group, and even boron's immediate neighbor aluminum is a

TABLE 19.1 The Group III elements

Valence configuration: ns^2np^1

Z	Name	Symbol	Atomic weight, amu	Melting point, °C	Boiling point, °C	Density, g/cm³	Normal form
5	Boron	B	10.81	2030	3700	2.47	Brown nonmetallic powder
13	Aluminum	Al	26.98	660	2350	2.70	Silver-white metal
31	Gallium	Ga	69.72	30	2070	5.91	Silver metal
49	Indium	In	114.82	157	2050	7.29	Silver-white metal
81	Thallium	Tl	204.37	304	1460	11.87	Soft metal

metal. Nonetheless, aluminum is sufficiently far to the right in the periodic table to show a hint of nonmetal character. Thus it is amphoteric; it reacts both with nonoxidizing acids to form aluminum ions,

$$2Al(s) + 6H^+(aq) \longrightarrow 2Al^{3+}(aq) + 3H_2(g)$$

and with hot aqueous alkali to form aluminate ions,

$$2Al(s) + 2OH^-(aq) + 6H_2O(l) \longrightarrow 2[Al(OH)_4]^-(aq) + 3H_2(g)$$

▼ EXAMPLE 19.1 Predicting group trends

Predict the trend in oxidation numbers that can be expected for the elements in Group III.

STRATEGY The group number tells us the maximum oxidation number, but we should also consider the influence of the inert-pair effect (Section 7.7) and how it is likely to vary through the group.

SOLUTION In Group III we expect a maximum oxidation number of +3. The inert-pair effect is unimportant at the top of the group, so we expect the oxidation states of B and Al to be +3 in all their compounds. (This is true, except for some complex solids that are mentioned later.) The oxidation number +1 becomes increasingly important as we move down the group, and we can expect it to be common in thallium. In fact, Tl(I) compounds are as common as Tl(III) compounds.
▲

Boron. Boron is mined as borax and kernite ($Na_2B_4O_7 \cdot xH_2O$, where x is 10 and 7, respectively). Large deposits are found in volcanic regions, such as in the Mojave Desert region of California. In the extraction process for boron, the minerals are converted to boron oxide with acid and then reduced with magnesium:

$$B_2O_3(s) + 3Mg(l) \xrightarrow{\Delta} 2B(s) + 3MgO(s)$$

A purer product is obtained by reducing a volatile compound with hydrogen on a heated filament (tantalum is used, as it has a very high melting point):

$$2BBr_3(g) + 3H_2(g) \longrightarrow 2B(s) + 6HBr(g)$$

Boron production remains quite low despite the element's desirable hardness and lightness. There is an opportunity for a new, young Charles Hall (the inventor of the process used to extract aluminum) to transform the picture for boron as Hall did for aluminum.

Elemental boron has several allotropes. It is typically either a gray-black, nonmetallic, high-melting-point solid or a dark brown powder with a structure built from clusters of twelve atoms (Fig. 19.1). Boron fibers are incorporated into plastics to produce a tough material that is stiffer than steel yet lighter than aluminum and of interest for use in aircraft, missiles, and body armor. The element is very inert and is attacked by only the strongest oxidizing agents.

Aluminum. Aluminum is the most abundant metallic element in the earth's crust. However, the aluminum content of most minerals is low, and the commercial source of aluminum, bauxite, is a hydrated, im-

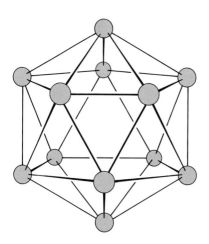

FIGURE 19.1 The structure of boron is based on linked twelve-atom units like the one shown here. Since the unit has twenty faces, it is called an "icosahedron" (from the Greek words for "twenty-faced").

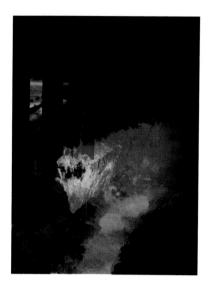

FIGURE 19.2 In the Hall process, aluminum oxide is dissolved in molten cryolite. Here the crust on the molten cryolite in the electrolysis pot is being broken.

pure oxide. The element is obtained by the *Hall process*, in which purified alumina (Al_2O_3), obtained from bauxite, is dissolved in molten cryolite (Na_3AlF_6) and then electrolyzed using carbon electrodes (Fig. 19.2). The overall reaction is

$$2Al_2O_3 + 3C \longrightarrow 4Al + 3CO_2$$

and Al^{3+} ions are reduced to Al atoms at the cathode. From the reduction half-reaction, $Al^{3+} + 3e^- \rightarrow Al$, we can deduce that a current of 1 A must flow for 80 h to produce 1 mol (27 g) of aluminum—about enough for one soft-drink can. The very high electricity consumption (three electrons must be supplied for every Al atom that the industry produces) means that the cost of electricity is of overriding importance in aluminum production. Thus, aluminum recycling is economically desirable: it is cheaper to recycle the metal than to manufacture it (Fig. 19.3).

Aluminum is a light, strong metal and an excellent electrical conductor. Although in principle it is susceptible to oxidation ($E° = -1.66$ V for the Al^{3+}/Al couple in water), it is resistant to corrosion because it is passivated by a stable oxide film that forms on its surface. The thickness of the oxide layer can be increased by making aluminum the anode of an electrolytic cell; the result is called "anodized aluminum." Dyes may be added to the dilute sulfuric acid electrolyte, to produce surface layers with different colors. Brown and bronze anodized aluminum is widely used in modern architecture.

Aluminum's low density, wide availability, and corrosion resistance make it ideal for construction. For use in airplanes it is usually alloyed with copper and silicon. Its lightness and good electrical conductivity lead to its use in overhead power lines. Another of its applications depends on its highly exothermic *thermite reaction* with oxides:

$$Fe_2O_3(s) + 2Al(s) \xrightarrow{\Delta} 2Fe(l) + Al_2O_3(s) \qquad \Delta H° = -852 \text{ kJ}$$

This reaction is used to weld iron rails or large pipes and to reduce oxide ores for which carbon is too weak a reducing agent. Two examples are

$$2Al(l) + Cr_2O_3(s) \longrightarrow Al_2O_3(s) + 2Cr(l)$$

$$2Al(l) + 3BaO(s) \longrightarrow Al_2O_3(s) + 3Ba(l)$$

FIGURE 19.3 Aluminum extraction is so energy-intensive (and hence expensive) that it is economical to recycle used aluminum cans and other objects.

19.2 GROUP III OXIDES

Boron, a nonmetal, has acidic oxides. Aluminum, its metallic neighbor, has amphoteric oxides. The oxides of both elements are important in their own right, as sources of the elements and as the starting point for the manufacture of other compounds.

Boron oxide and boric acid. Boric acid, $B(OH)_3$, is obtained commercially by the action of sulfuric acid on borax; it crystallizes as a white solid that melts at 171°C. It is toxic to bacteria and many insects and has long been used as a mild antiseptic. As explained in Section 14.1, the $B(OH)_3$ molecule acts as a Lewis acid by accepting a lone pair of electrons from an H_2O molecule. The complex so formed is a Brønsted acid:

$$(OH)_3B + :OH_2 \longrightarrow (OH)_3B\text{---}OH_2$$
$$\underset{\text{acid}}{\text{Lewis}} \qquad\qquad\qquad \underset{\text{acid}}{\text{Brønsted}}$$

$$(OH)_3B\text{---}OH_2(aq) + H_2O(l) \rightleftharpoons [B(OH)_4]^-(aq) + H_3O^+(aq)$$

Boric acid also retards the spread of flames in cellulosic materials, particularly paper. The scrap paper used to manufacture home insulation contains about 5% boric acid, to reduce the risk of fire.

Some of the chemical properties of the *p*-block elements can be understood by noting whether an oxide is the *anhydride* of an oxoacid—an oxide that forms the acid when it reacts with water. It is also helpful to note whether an oxide is the true anhydride or only the *formal anhydride*—an oxide that has the formula of an oxoacid less the elements of water but does not react with water to produce the acid: an example is CO, the formal anhydride of formic acid, HCOOH. A number of anhydrides can be formed by simply heating the oxoacid. This is the case with boric acid, which dehydrates on heating to its anhydride, boron oxide (B_2O_3):

$$2B(OH)_3(s) \xrightarrow{\Delta} B_2O_3(s) + 3H_2O(g)$$

Because it melts (at 450°C) to a liquid that dissolves many metal oxides, boric acid is used as a flux to clean metals before they are soldered or welded.

Alumina. The purified alumina (Al_2O_3) used in the Hall process is extracted from bauxite by the *Bayer process*. The crude bauxite is first treated with aqueous sodium hydroxide, which dissolves the amphoteric alumina and the acidic silica, leaving behind the iron oxide impurities as "red mud." Then carbon dioxide is bubbled through the dissolved alumina to make it acidic. The resulting decrease in the OH^- ion concentration encourages the decomposition of aluminate ions:

$$[Al(OH)_4]^-(aq) \longrightarrow Al(OH)_3(s) + OH^-(aq)$$

The neutral hydroxide precipitates, and after it is removed, it is dehydrated to the oxide by heating to 1200°C:

$$2Al(OH)_3(s) \xrightarrow{\Delta} Al_2O_3(s) + 3H_2O(g)$$

FIGURE 19.4 Some of the impure forms of alumina that are prized as gems: (a) alumina with Cr^{3+} ion impurities is ruby; (b) with Fe^{3+} and Ti^{4+} ions, it is sapphire; and (c) with Fe^{3+} ions, it is topaz.

Alumina exists in a variety of solid phases with different crystal structures. As α-alumina, it is the very hard substance corundum; impure microcrystalline corundum is the purple-black abrasive known as emery. A less dense and more reactive form of the oxide is γ-alumina. This form adsorbs water and is used as a support in chromatography. Some impure forms of alumina are highly prized (Fig. 19.4).

▼ EXAMPLE 19.2 Predicting the properties of alumina

What chemical properties can be predicted for alumina? What properties can be expected for aqueous solutions obtained by the action of acids on alumina?

STRATEGY We can decide from the location of aluminum in the periodic table whether alumina is likely to be acidic, amphoteric, or basic. A diagonal relationship might also give us a clue. An acid acting on alumina is likely to give a salt of Al^{3+} ions; this is a small, highly charged cation, and the properties of such ions in water were discussed in Section 15.1.

SOLUTION Aluminum is close to the border between metals and nonmetals and is a diagonal neighbor of the amphoteric element beryllium. Hence, we expect its oxide to be amphoteric too (it is). We expect the oxide to react with acids to give solutions of Al^{3+} ions:

$$Al_2O_3(s) + 6H_3O^+(aq) \longrightarrow 2Al^{3+}(aq) + 9H_2O(l)$$

The resulting cation is actually $[Al(H_2O)_6]^{3+}$, with six Lewis-base H_2O molecules linked to the Al^{3+} ion acting as a strong Lewis acid. We would expect the strong polarizing power of the small, highly charged Al^{3+} ion to result in the $Al^{3+}(aq)$ ion being a Brønsted acid, and solutions of aluminum salts to be acidic (they are):

$$[Al(H_2O)_6]^{3+}(aq) + H_2O(l) \rightleftharpoons [Al(H_2O)_5OH]^{2+}(aq) + H_3O^+(aq)$$

EXERCISE What reaction can be expected between alumina and an alkali?
[*Answer*: The formation of aluminate]

▲

Alums and aluminates. One of the most important aluminum salts resulting from the action of an acid on alumina is aluminum sulfate:

$$Al_2O_3(s) + 3H_2SO_4(aq) \longrightarrow Al_2(SO_4)_3(aq) + 3H_2O(l)$$

Aluminum sulfate is "papermaker's alum," which is used in the paper industry to coagulate cellulose fibers into a hard, nonabsorbent surface. True alums (from which aluminum takes its name) are mixed sulfates of formula $M^+(M')^{3+}(SO_4^{2-})_2 \cdot 12H_2O$; they include potassium alum, $KAl(SO_4)_2 \cdot 12H_2O$, and ammonium alum, $NH_4Al(SO_4)_2 \cdot 12H_2O$.

γ-Alumina dissolves in alkalis to give solutions of the aluminate ion:

$$Al_2O_3(s) + 2OH^-(aq) + 3H_2O(l) \longrightarrow 2[Al(OH)_4]^-(aq)$$

A principal use for sodium aluminate in conjunction with aluminum sulfate is in water purification. The acidic aluminum cation (from the sulfate) precipitates aluminum hydroxide when it is mixed with the aluminate ion:

$$2Al^{3+}(aq) + 6[Al(OH)_4]^-(aq) \longrightarrow 8Al(OH)_3(s)$$

The aluminum hydroxide precipitates as a "floc"—a fluffy, extensive network—that captures impurities in the water and can be filtered off. Using aluminum for both the cation and the anion gives the greatest possible bulk of impurity-collecting alumina because the reaction has no by-products.

19.3 OTHER IMPORTANT GROUP III COMPOUNDS

Both boron and aluminum form a range of compounds that are useful either because they have unusual physical properties or because they are chemically unique.

Boron carbides and boron nitride. When boron is heated to high temperatures with carbon, it forms boron carbide $B_{12}C_3$, an extremely hard, high-melting solid:

$$12B(s) + 3C(s) \xrightarrow{\Delta} B_{12}C_3(s)$$

The solid consists of B_{12} groups that are pinned together by C atoms occupying the gaps between them.

When boron is heated to white heat in ammonia, boron nitride (BN) is formed as a fluffy, slippery powder:

$$2B(s) + 2NH_3(g) \xrightarrow{\Delta} 2BN(s) + 3H_2(g)$$

Boron nitride has a graphitelike structure in which flat planes of carbon hexagons are replaced by planes of hexagons of alternating B and

FIGURE 19.5 The structure of boron nitride resembles that of graphite, since it consists of flat planes of hexagons. However, the hexagons consist of alternating B and N atoms (in place of C atoms) and are stacked differently. (Compare with Fig. 10.44.)

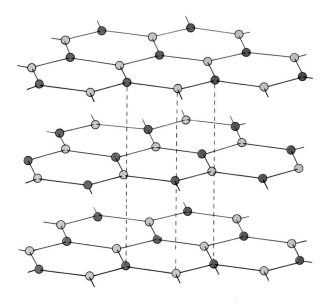

N atoms (Fig. 19.5). Unlike graphite, BN is white and does not conduct electricity. Under high pressure it is converted to a very hard, diamondlike crystal form called borazon.

Boron halides. The boron halides are produced either by direct reaction of the elements at high temperature or from the oxide. The most important is boron trifluoride BF_3, a widely used industrial catalyst produced by the reaction between boric oxide, calcium fluoride, and sulfuric acid:

$$B_2O_3(s) + 3CaF_2(s) + 3H_2SO_4(l) \longrightarrow$$
$$2BF_3(g) + 3CaSO_4(s) + 3H_2O(l)$$

Boron trichloride is produced commercially by the action of the halogen on the oxide in the presence of carbon:

$$B_2O_3(s) + 3C(s) + 3Cl_2(g) \xrightarrow{500^\circ C} 2BCl_3(g) + 3CO(g)$$

▼ **EXAMPLE 19.3** **Predicting the properties of the boron halides**

Suggest how the melting and boiling points of the boron halides will vary as the halogen is changed from fluorine down to iodine.

STRATEGY We know that in the absence of hydrogen bonding, London forces generally dominate the intermolecular interactions. Therefore, we must decide how the London forces will vary as the halogen is changed.

SOLUTION The number of electrons in the molecules increases as the halogen is changed from fluorine down to iodine, so the London forces become stronger. We therefore expect melting and boiling points to increase from BF_3 to BI_3. In fact, BF_3 and BCl_3 are gases, BBr_3 is a liquid, and BI_3 is a solid at room temperature.

EXERCISE What chemical feature is suggested by the incomplete-octet electronic structure of the boron halides?

[*Answer:* Their ability to act as Lewis acids]

In all its halides, the B atom has an incomplete octet; the compounds consist of trigonal planar molecules with an empty $2p$ orbital perpendicular to the molecular plane. The empty orbital allows the boron halides to act as strong Lewis acids, and the fluoride and the chloride are widely used as catalysts on this account.

Aluminum halides. Aluminum chloride, another major industrial catalyst, is prepared by the action of chlorine on aluminum or alumina in the presence of carbon:

$$2Al(s) + 3Cl_2(g) \longrightarrow 2AlCl_3(g)$$

$$Al_2O_3(s) + 3C(s) + 3Cl_2(g) \longrightarrow 2AlCl_3(g) + 3CO(g)$$

Aluminum chloride is an ionic solid in which each Al^{3+} ion is surrounded by six Cl^- ions, but it sublimes at 192°C to a vapor of Al_2Cl_6 molecules. The Al_2Cl_6 molecule **(1)** is an example of a *dimer*, the union of two identical molecules. The formation of Al_2Cl_6 from two $AlCl_3$ molecules is an example of Lewis acid-base complex formation between identical molecules: the Al atoms are the acids, and a Cl atom on the neighboring Al atom provides a lone pair and acts as a Lewis base:

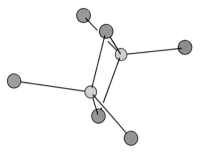

1 Aluminum chloride dimer

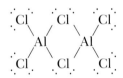

Because the halogen atoms of aluminum bromide and iodide are more polarizable than the Cl atoms of Al_2Cl_6, the Al_2Br_6 and Al_2I_6 molecules survive in the solid as well as in the vapor state.

The aluminum halides react with water with a considerable evolution of heat. When the chloride is exposed to moist air, it produces fumes of hydrochloric acid (Fig. 19.6). The ionic aluminum chloride hexahydrate ($AlCl_3 \cdot 6H_2O$), the chloride salt of $[Al(H_2O)_6]^{3+}$, is prepared by dissolving alumina in concentrated hydrochloric acid. It is used as a deodorant and antiperspirant, its role being to kill the bacteria that feed on perspiration and convert it into unpleasant-smelling compounds. The hexahydrate does not form the anhydrous chloride when it is heated, because the Al—O bonds are too strong to be replaced by Al—Cl bonds. Instead, heating produces the oxide:

$$2[Al(H_2O)_6]Cl_3(s) \longrightarrow Al_2O_3(s) + 6HCl(g) + 9H_2O(g)$$

Boranes and borohydrides. Boron forms a remarkable series of binary compounds with hydrogen—the *boranes*. These include diborane, B_2H_6, and more complex compounds such as decaborane, $B_{10}H_{14}$. Anionic versions of these compounds, the *borohydrides*, are also known; the most important is BH_4^- as sodium borohydride, $NaBH_4$.

Sodium borohydride is a white crystalline solid produced by the reaction between sodium hydride and boron chloride in a nonaqueous solvent. The net ionic reaction is

$$4H^- + BCl_3 \longrightarrow BH_4^- + 3Cl^-$$

with the Na^+ present as spectator ions. The compound is prepared industrially by chemical onslaught on borax using sodium metal and

FIGURE 19.6 When anhydrous aluminum chloride is left exposed to moist air, it reacts to form hydrochloric acid. Here we see white fumes of ammonium chloride forming where the hydrochloric acid reacts with ammonia.

hydrogen, in the presence of silica at 500°C and under pressure. The product is extracted from the mixture by dissolving it in liquid ammonia. It is a very useful and quite widely used reducing agent.

At pH = 14 (strongly alkaline conditions) the potential of the half reaction

$$H_2BO_3^-(aq) + 5H_2O(l) + 8e^- \longrightarrow BH_4^-(aq) + 8OH^-(aq)$$

is 1.24 V. Since this is well below the potential of the Ni^{2+}/Ni couple (-0.23 V), the half-reaction can be used to reduce Ni^{2+} ions to metallic nickel. This is the basis of the *chemical plating* of nickel using BH_4^- as the reducing agent. The advantage of chemical plating over electroplating is that the object being plated does not have to be an electrical conductor.

The starting point for the production of the boranes is the reaction (in organic solvent) of sodium borohydride with boron trifluoride:

$$4BF_3 + 3BH_4^- \longrightarrow 3BF_4^- + 2B_2H_6$$

The product B_2H_6 is diborane (**2**), a colorless gas that bursts into flame in air. On contact with water, it is immediately oxidized to boric acid and hydrogen:

$$B_2H_6(g) + 6H_2O(l) \longrightarrow 2B(OH)_3(aq) + 6H_2(g)$$

When diborane is heated to high temperatures, it decomposes into hydrogen and pure boron:

$$B_2H_6(g) \longrightarrow 2B(s) + 3H_2(g)$$

This is a useful route to the pure element, but when the heating is less severe, more complex boranes are formed. When diborane is heated to 100°C, for instance, it forms decaborane, $B_{10}H_{14}$, a solid that melts at 100°C. Decaborane is stable in air, is oxidized by water only slowly, and is an example of the general rule that boranes of higher molecular weights are less flammable.

The boranes are electron-deficient compounds. As mentioned in Section 9.7, these are compounds for which Lewis structures cannot be written because too few electrons are available. However, as we also saw, there is no difficulty in accounting for their structure in terms of a delocalized bond in which the bonding power of one electron pair is shared by more than two atoms (**2**).

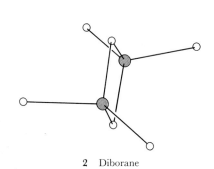

2   Diborane

Aluminum hydrides. The most important compound of aluminum and hydrogen is lithium aluminum hydride ($LiAlH_4$). This white solid is prepared by the reaction of lithium hydride with aluminum chloride in an organic solvent (diethyl ether):

$$4LiH + AlCl_3 \longrightarrow LiAlH_4 + 3LiCl$$

Its importance lies in its use as a reducing agent, especially in organic chemistry. Lithium aluminum hydride reacts with aluminum chloride to form aluminum hydride AlH_3, which is best regarded as an ionic hydride:

$$3AlH_4^- + AlCl_3 \longrightarrow 4AlH_3 + 3Cl^-$$

The aluminum analogs of the higher boranes are unknown.

GROUP IV: CARBON AND SILICON

The elements of Group IV and some of their properties are listed in Table 19.2. Carbon forms a very wide range of compounds and gives rise to a special branch of chemistry, organic chemistry, which we introduce in Part V. However, the oxides and oxoanions of carbon are typically "inorganic" and are described here.

▼ **EXAMPLE 19.4**   **Predicting the properties of a group**

Indicate the general properties that can be expected for the elements of Group IV.

STRATEGY   To make predictions about these elements, we need to consider their location in the periodic table, their classifications as metals or nonmetals, and the consequences of that character. The role of the inert pair should also be considered.

SOLUTION   The elements can be expected to show a trend in properties from nonmetallic to metallic down the group. In fact, carbon and silicon are nonmetals, whereas germanium, tin, and lead are metals (with some nonmetallic characteristics). Carbon can be expected to be neither malleable nor ductile, to have acidic oxides, and to form compounds that are largely covalent. Tin and lead can be expected to be malleable and ductile and have a

		III	**IV**	V			He
Li	Be	B	**C**	N	O	F	Ne
		Al	**Si**	P			
		Ga	**Ge**	As			
		In	**Sn**	Sb			
		Tl	**Pb**	Bi			

TABLE 19.2   The Group IV elements

Valence configuration: ns^2np^2

Z	Name	Symbol	Atomic weight, amu	Melting point, °C	Boiling point, °C	Density, g/cm^3	Normal form
6	Carbon	C	12.01	3700s*		1.9 to 2.3	Black nonmetal (graphite)
						3.2 to 3.5	Transparent nonmetal (diamond)
14	Silicon	Si	28.09	1410	2620	2.33	Gray, lustrous, nonmetal
32	Germanium	Ge	72.59	959	2850	5.32	Gray-white lustrous metal
50	Tin	Sn	118.69	232	2720	7.29	White lustrous metal
82	Lead	Pb	207.19	328	1760	11.34	Blue-white lustrous metal

*The symbol s denotes that the element sublimes.

lustrous appearance. They can also be expected to show a hint of nonmetallic character and hence to be amphoteric. We can expect increasing levels of inertness for the valence-shell *s* electrons as we move down the group, with the electron pair fully active in carbon and almost completely inert in lead.

EXERCISE Suggest equations for the reaction of tin with (a) hot concentrated hydrochloric acid and (b) hot alkali.

$$[\textit{Answer}: \text{(a) } Sn(s) + 2HCl(aq) \longrightarrow SnCl_2(aq) + H_2(g);$$
$$\text{(b) } Sn(s) + 2OH^-(aq) + 2H_2O(l) \longrightarrow [Sn(OH)_4]^{2-}(aq) + H_2(g)]$$

▲

19.4 THE GROUP IV ELEMENTS

We expect carbon, because it is at the head of its group, to be different from its congeners. It is, and the differences are more pronounced in Group IV than anywhere else in the periodic table. Some of these differences stem from the wide occurrence of C=C and C=O double bonds, compared with the relative rarity of Si=Si and Si=O double bonds. Carbon dioxide, which consists of discrete O=C=O molecules, is a gas that we can exhale. Silicon dioxide (silica), which consists of networks of —O—Si—O— groups, is a mineral we can stand on.

The valence-electron configuration is s^2p^2 for all members of the group. All four electrons are approximately equally available for bonding in the lighter elements, and carbon and silicon are characterized by their ability to form four covalent bonds. However, toward the bottom of the group the energy separation between the *s* and *p* orbitals increases, and the *s* electrons become progressively less available for bonding.

Carbon. Solid carbon exists as graphite and as diamond, with graphite the thermodynamically more stable allotrope under normal conditions. Graphite is produced by passing a heavy electric current through coke rods for several days (the coke is obtained by heating coal in the absence of air). Synthetic diamonds are usually made under high pressure at high temperature (Section 10.9). A more recently developed technique makes use of the thermal decomposition of methane on a hot wire. The carbon atoms that are formed settle on a cooler surface as both graphite and diamond; however, the hydrogen atoms that are also produced react more quickly with the graphite than with the diamond to form volatile hydrocarbons, so more diamond than graphite survives.

Soot and carbon black consist of very small crystals of graphite. Carbon black, which is produced by heating gaseous hydrocarbons to nearly 1000°C in the absence of air, is used in pigments and printing inks and to reinforce rubber. "Activated carbon" consists of granules of microcrystalline carbon that are obtained by heating waste organic matter in the absence of air and then processed to increase their porosity. Its very large surface area (up to about 2000 m²/g) enables it to absorb impurities from liquids and gases. It is used in air purifiers, gas masks, aquarium water filters, and, on a much larger scale, in the emission-control canisters of automobiles.

The properties and uses of the allotropes of carbon reflect their crystal structures, which are described in Section 10.9. Graphite consists of

FIGURE 19.7 Three of the common forms of silica (SiO$_2$): quartz (top left), quartzite (bottom left), and cristobalite (right) in which the black parts are obsidian, a volcanic rock that also contains silica.

flat planes of carbon atoms in a hexagonal network. Since electrons are free to move through the planes, graphite is a black, lustrous, electrically conducting solid. One of its uses is for the electrodes of electrochemical cells (as in the Hall process). Its slipperiness, which results from the ease with which the flat planes move past each other when impurities are present, leads to its use as a lubricant. It is also used as the "lead" in pencils; layers of graphite are rubbed off by friction as the pencil is moved across the paper. In diamond, each carbon atom is linked tetrahedrally to its four neighbors, with all electrons localized in C—C σ bonds. Diamond is a rigid, transparent, electrically insulating solid. It is the hardest substance known and the best conductor of heat (about five times better than copper). These last two properties make it an ideal abrasive, for it can wear down all other substances yet the heat generated by friction is conducted rapidly away.

Silicon. Silicon occurs widely as silicates in rocks and as the silica (SiO$_2$) of sand (Fig. 19.7). It is obtained from quartzite, a granular form of quartz, by reduction with high-purity carbon in an electric arc furnace:

$$SiO_2(l) + 2C(s) \longrightarrow Si(l) + 2CO(g)$$

The impure product is then exposed to chlorine to form silicon tetrachloride:

$$Si(s) + 2Cl_2(g) \longrightarrow SiCl_4(l)$$

This is distilled and then reduced to a purer form of the element with hydrogen:

$$SiCl_4(l) + 2H_2(g) \longrightarrow Si(s) + 4HCl(g)$$

For use in semiconductors, more purification may be needed. In one process, a large single crystal is grown by pulling a solid rod of the

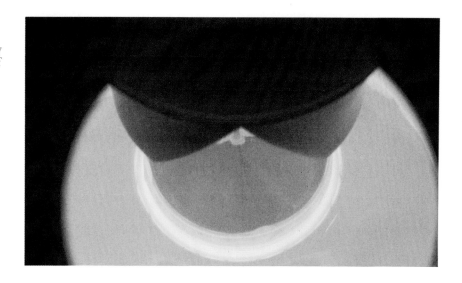

FIGURE 19.8 Pulling a single crystal of silicon from molten silicon. The crystal is withdrawn very slowly from the melt. (Courtesy of Siltec Silicon.)

element slowly from the melt (Fig. 19.8). Then the crystals are zone refined to an impurity concentration of less than one atom per billion Si atoms. Alternative techniques include the thermal decomposition of silane (SiH_4), which produces a highly crystalline form of silicon when the atoms settle onto a cool surface, and the decomposition of silane by an electric discharge. The latter method produces an amorphous form of silicon with a significant hydrogen content. Amorphous silicon is used in photovoltaic devices, which produce electricity from sunlight.

Germanium, tin, and lead. Germanium is recovered from the flue dust of plants processing zinc ores (in which it occurs as an impurity). It is used mainly—and increasingly—in the semiconductor industry. Tin and lead are obtained very easily from their ores and have been known since antiquity. Tin occurs mainly as the mineral cassiterite (SnO_2) and is obtained from it by reduction with carbon at 1200°C:

$$SnO_2(s) + C(s) \xrightarrow{1200°C} Sn(l) + CO_2(g)$$

The main problem with this essentially simple process is that the iron that is present as an impurity must be reoxidized without oxidizing the tin as well. This delicate balance is achieved by passing oxygen through the vigorously stirred molten tin. The principal lead ore is galena (PbS). It is "roasted" in air, which converts it to PbO, and then reduced with coke.

Tin is expensive and not very strong, but it is resistant to corrosion. Its main use is in tin-plating, which accounts for about 40% of its consumption. However, as the Sn^{2+}/Sn couple ($E° = -0.14$ V) lies above the Fe^{2+}/Fe couple ($E° = -0.44$ V) in the electrochemical series, the tin plate promotes the oxidation of the iron underneath if the tin surface is damaged. Tin is also used in some alloys (Table 10.6).

One very important property of lead is its durability (its chemical inertness), which makes it useful in the construction industry. The inertness of lead under normal conditions can be traced to the passivation of its surface by oxides, chlorides, and sulfates. This passivation allows lead to be used for transporting hot concentrated sulfuric acid but not nitric acid, for lead nitrate is soluble. Another important property of lead is its density, which makes it useful not only as a radiation shield, because its numerous electrons absorb high-energy radiation, but also for soundproofing buildings, because its heavy atoms move sluggishly and transmit sound waves poorly. Nowadays, the main use of lead is for the electrodes of storage batteries.

19.5 GROUP IV OXIDES

Carbon forms several oxides, but the most important are carbon monoxide and carbon dioxide. In striking contrast to these two carbon oxides, which are gases, silicon dioxide is a solid. Derivatives of the oxides and oxoacids of both elements are very important components of landscapes and are a rich source of raw materials.

Carbon monoxide. Carbon monoxide is produced when carbon or an organic compound burns in a limited supply of air, as in automobile engines and cigarette smoking. Commercially it is produced as synthesis gas in the reaction (Section 18.1)

$$CH_4(g) + H_2O(g) \longrightarrow CO(g) + 3H_2(g)$$

▼ **EXAMPLE 19.5 Finding a method for preparing an oxide**

Suggest a method for preparing carbon monoxide, making use of the fact that it is the formal anhydride of an oxoacid.

STRATEGY We should identify the oxoacid of which CO is the formal anhydride by adding one or more H_2O units to its formula. The gas may then be prepared from the acid by dehydration.

SOLUTION Adding CO and H_2O gives H_2CO_2, the formula of formic acid. This suggests that dehydration of formic acid is a source of carbon monoxide. In the laboratory, dehydration is carried out by heating it to 150°C with concentrated sulfuric acid, a powerful dehydrating agent:

$$HCOOH(l) \xrightarrow{150°C, H_2SO_4} CO(g) + H_2O(l)$$

EXERCISE Suggest a method for preparing carbon suboxide, O=C=C=C=O.
[*Answer*: Dehydration of HOOC—CH₂—COOH (malonic acid)]
▲

Carbon monoxide is a colorless, flammable, almost water-insoluble, and very toxic gas that condenses to a colorless liquid at −90°C. It is not very reactive, partly on account of its high bond enthalpy (1074 kJ/mol, the highest for any molecule). However, it is a Lewis base, and the lone pair on the carbon atom forms covalent bonds with *d*-block atoms and ions. An example of this type of complex formation is its reaction with nickel to give the intensely poisonous liquid nickel carbonyl:

$$Ni(s) + 4CO(g) \xrightarrow{50°C, 1\ atm} Ni(CO)_4(l)$$

Complex formation is also responsible for carbon monoxide's toxicity: it attaches more strongly than oxygen to the iron in hemoglobin; as the blood is then unable to transport sufficient oxygen, the victim suffocates.

Carbon monoxide is a reducing agent. This is its role in the production of a number of metals, most notably iron in a blast furnace (see Fig. 3.17):

$$Fe_2O_3(s) + 3CO(g) \xrightarrow{800°C} 2Fe(l) + 3CO_2(g)$$

Carbon dioxide and the carbonates. The colorless gas carbon dioxide is formed whenever carbon or organic compounds burn in a free supply of air, in fermentation (the conversion of carbohydrates to ethanol by yeast), and by the action of acids on carbonates. Since its triple point is at 5 atm, it cannot exist as a liquid at ordinary atmospheric pressures. Solid carbon dioxide, which is sold as "dry ice," therefore sublimes directly to the gas—a property that makes it convenient for use as a refrigerant.

Carbon dioxide is a weak Lewis acid. It dissolves in water to form carbonic acid, a weak diprotic Brønsted acid, in a Lewis acid-base reaction in which an H_2O molecule is the Lewis base:

$$CO_2(aq) + H_2O(l) \rightleftharpoons H_2CO_3(aq)$$

Carbonic acid is the parent of the carbonates and the hydrogen carbonates (the bicarbonates). The Lewis acid-base reaction that forms carbonates is reversed when a carbonate is heated, and most carbonates decompose into the corresponding oxide and carbon dioxide:

$$CO_3{}^{2-}(s) \xrightarrow{\Delta} O^{2-}(s) + CO_2(g)$$

Solid hydrogen carbonates can be isolated for the alkali metals only; others decompose to carbon dioxide and the carbonate when the water is removed.

Carbonates and hydrogen carbonates evolve carbon dioxide when they are treated with acid: the hydrogen ion, a very strong Lewis acid, separates the Lewis base O^{2-} from the Lewis acid CO_2:

$$\left\{ \begin{matrix} O \\ \diagdown \\ C-O \\ \diagup \\ O \end{matrix} \right\}^{2-} + H^+(aq) \longrightarrow \begin{matrix} O \\ \| \\ C \\ \| \\ O \end{matrix} + O-H^-(aq)$$

This property is responsible for the action of "baking powder," a mixture of sodium hydrogen carbonate and an acid, which may be either sodium alum (in which the hydrated aluminum cation is the acid) or calcium dihydrogen phosphate, $Ca(H_2PO_4)_2$. For the latter,

$$2NaHCO_3(s) + Ca(H_2PO_4)_2(s) \xrightarrow{\Delta}$$
$$Na_2HPO_4(s) + CaHPO_4(s) + 2CO_2(g) + 2H_2O(g)$$

The carbon dioxide is trapped as bubbles in the dough, causing it to rise. However, to produce a good dough, the CO_2 should be released in two stages (hence "double-acting" baking powder). In the first stage, at about room temperature, a small release makes small cavities (Fig. 19.9). In the second, at baking temperature, these cavities are expanded by a general release of the gas. The first release is achieved

FIGURE 19.9 Double-acting baking powder first forms small cavities in dough when it is moistened. These are later expanded by a second release of carbon dioxide. In this illustration, a similar process takes place when yeast is used, but its action is biochemical and involves enzymes.

FIGURE 19.10 Some impure forms of silica: amethyst (in which the color is due to Fe^{3+} impurities), onyx, and agate.

by including the organic acid tartaric acid, which acts on the HCO_3^- ions as soon as the dough is moistened.

Silica and glass. Silica occurs naturally as quartz and as sand, an impure form of quartz colored golden brown by iron oxide impurities. Some precious and semiprecious stones are impure silica (Fig. 19.10); flint is silica colored black by carbon impurities.

Some sense can be made of the bewildering array of structures exhibited by silica and the silicates by picturing them as arrangements of SiO_4 tetrahedra. In silica itself, each corner O atom is shared by two Si atoms. Hence, each tetrahedron contributes one silicon atom and $4 \times \frac{1}{2} = 2$ oxygen atoms to the solid. The structure of quartz is complicated, for it is built from helical chains of SiO_4 units wound around each other. When it is heated to about 1500°C, it changes to another arrangement, that of the mineral cristobalite (Fig. 19.11). This structure is simpler to describe: its Si atoms are arranged like the C atoms in diamond, but with an O atom between each pair of neighboring Si atoms.

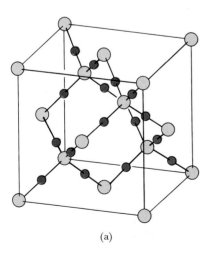

(a)

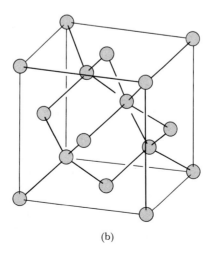

(b)

FIGURE 19.11 The structure of cristobalite (a) resembles that of diamond (b), but an O atom (red) lies between each pair of Si atoms (gray).

FIGURE 19.12 The reaction between hydrofluoric acid and glass is used to etch glass. The surface of the glass is covered with wax, a design is scratched in the wax, and acid is poured over it. This etched glass bowl was designed by the artist Frederick Carder in the 1920s.

Silica resists attack by Brønsted acids but gives way to attack by the strong Lewis base F^- from hydrofluoric acid, with which it forms fluorosilicate ions (Fig. 19.12):

$$SiO_2(s) + 6HF(aq) \longrightarrow SiF_6^{2-}(aq) + 2H_3O^+(aq)$$

It is also attacked by the Lewis base OH^- in hot molten sodium hydroxide and by the Lewis base O^{2-} in the carbonate anion of molten sodium carbonate:

$$SiO_2(s) + 2Na_2CO_3(l) \xrightarrow{1400°C} Na_4SiO_4(s) + 2CO_2(g)$$

The sodium silicate produced by this reaction is soluble, and it is sold commercially as "water glass." It is used in detergents, partly as an alkaline buffer (SiO_4^{4-} is a Brønsted base) and partly to keep dirt from settling back onto the fabric. The SiO_4^{4-} ions act by attaching to dirt particles and hence giving them a negative charge, which prevents them from merging into larger, insoluble particles (Fig. 19.13).

When a solution of sodium silicate is acidified, a gelatinous precipitate of silica is produced:

$$4H_3O^+(aq) + SiO_4^{4-}(aq) \longrightarrow SiO_2(s) + 6H_2O(l)$$

After it is washed and dried, this "silica gel" has a very high surface area (about 700 m^2/g) and is useful as a drying agent, a support for catalysts, a packing for chromatography columns, and a thermal insulator.

Many of the Si—O bonds break when silica (usually in the form of sand) is heated to around 1600°C. The orderly structure of the crystals is not regained when the Si—O bonds re-form as the melt cools, and an amorphous, glassy material, called fused silica, results. The presence of metal oxides in the melt results in *glass* itself. In the process the silica again shows its Lewis-acid character, forming bonds with the Lewis-base O^{2-} ions to give —Si—O$^-$M$^+$ groups in place of some of the Si—O—Si links of pure silica.

The addition of sodium oxide alone to silica does not give a very durable glass. Greater durability is obtained by including Ca^{2+} ions too. Almost 90% of the glass now manufactured is *soda-lime glass*. This glass, which is used for windows and bottles, contains about 12% Na_2O, prepared by the action of heat on sodium carbonate (the "soda"), and 12% CaO (the "lime"), prepared by heating calcium carbonate. When the proportions of soda and lime are reduced and 16% B_2O_3 is added, a *borosilicate glass*, such as "Pyrex," is produced. Because borosilicate glasses do not expand much when heated, they survive rapid heating and cooling and are used for ovenware.

Colored glass (Fig. 19.14) is produced by adding other substances; cadmium sulfide and selenide, for instance, give ruby glass. Ordinary soda-lime glass is often very pale green because of the iron impurities it contains. Cobalt blue glass is colored by Co^{2+} ions. Brown beer-bottle glass is colored by iron sulfides. "Photochromic" sunglasses, which darken in sunlight, contain silver and copper ions. As in the photographic process (Section 3.10), the energy of sunlight causes the Ag^+ ions to be reduced to silver metal:

$$Ag^+(s) + Cu^+(s) \xrightarrow{light} Ag(s) + Cu^{2+}(s)$$

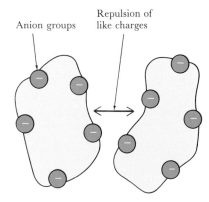

FIGURE 19.13 When ions attach to dirt particles, neighboring particles repel each other and do not collect into larger, insoluble particles.

In sunglasses, however, the oxidized reducing agent (in this case Cu^{2+}) cannot escape, and as soon as the light is removed it oxidizes the silver again.

The silicates. Silicon is the most abundant element in the earth's crust after oxygen, and most of it is found in silicates. As noted above, their structures can be pictured in terms of SiO_4 tetrahedra with various negative charges. For some purposes it is convenient to think of an SiO_4^{4-} unit as an Si^{4+} ion in the center of four surrounding O^{2-} ions, but in reality each Si—O bond has considerable covalent character. The differences between the various silicates arise from the different numbers of negative charges on the tetrahedra, the different numbers of corner O atoms shared, and the manner in which chains and sheets of the linked tetrahedra lie together.

▼ EXAMPLE 19.6 Identifying the structure of a silicate

The pyroxenes consist of chains of SiO_4 units, with two corner O atoms of each tetrahedron shared by neighbors (Fig. 19.15a) and two negative charges on each unit. What is the formula of each repeated unit?

STRATEGY We need to count the net number of atoms of each element in the unit, with shared atoms counting as $\frac{1}{2}$.

SOLUTION Each unit has one Si atom and $2 + 2 \times \frac{1}{2} = 3$ O atoms. Since each unit has a double negative charge, the repeated unit has formula SiO_3^{2-}.

EXERCISE Some silicates consist of the units shown in Fig. 19.15b. What is the repeating unit?

[*Answer*: $Si_4O_{11}^{6-}$]

▲

FIGURE 19.14 Colored glass. The colors are produced in a variety of ways: rich red hues are produced by cadmium sulfide and selenide or by finely divided gold particles. Blues are often produced by cobalt ions. This leaded glass window was produced by Tiffany Studios, ca. 1900.

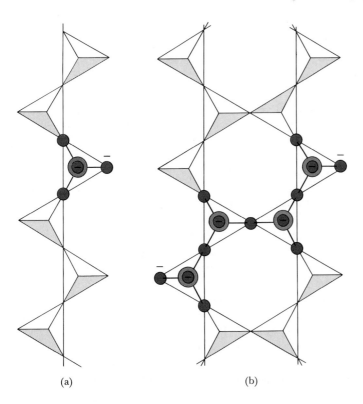

(a)   (b)

FIGURE 19.15 (a) The basic structural feature of the pyroxenes is this unit. (b) The amphiboles have a ladderlike structure. Each tetrahedron represents an SiO_4 unit, and a shared corner represents a shared O atom.

FIGURE 19.16 The aluminosilicate mica cleaves into thin transparent sheets. It has been used for windows in furnaces.

The simplest silicates, the *orthosilicates*, are built from SiO_4^{4-} ions. They are not very common, but include the sodium silicate we have already considered and the mineral zircon ($ZrSiO_4$), which is used as a substitute for diamond in jewelry. The *pyroxenes* consist of chains of SiO_4 units in which two corner O atoms are shared by neighboring units (Fig. 19.15a). As is shown in Example 19.6, the repeating unit has formula SiO_3^{2-}. Electrical neutrality is provided by cations spaced out along the chain. The pyroxenes include jade, $NaAl(SiO_3)_2$.

The chains of units can link together to form the ladderlike *amphiboles* (Fig. 19.15b), which include tremolite, $Ca_2Mg_5(Si_4O_{11})_2(OH)_2$. Tremolite is one of several minerals called "asbestos," which are characterized by a fibrous structure and an ability to withstand heat. Their fibrous quality reflects the way the ladders lie together but can easily be torn apart. The SiO_4 tetrahedra can also link together to form sheets. An example is talc, a hydrated magnesium silicate, $Mg_3(Si_2O_5)_2(OH)_2$. Talc is soft and slippery because the silicate sheets slide over each other.

Aluminosilicates and ceramics. More complex (and more widely occurring) structures are obtained when some of the notional Si^{4+} ions in silicates are replaced by Al^{3+} ions to form the *aluminosilicates*. The difference of one positive charge per ion is made up by the presence of extra cations. These cations account for the difference in properties between the silicate talc and the aluminosilicate mica (Fig. 19.16), of which one form is $KMg_3(Si_3AlO_{10})(OH)_2$. In the latter, the sheets of tetrahedra are held together by extra K^+ ions. Although it cleaves neatly into layers when the sheets are torn apart, mica is not slippery like talc.

The *feldspars* are aluminosilicates in which more than half the Si^{4+} ions have been replaced by Al^{3+} ions. They are the most abundant silicate materials on earth; granite is a compressed mixture of mica, quartz, and feldspar. When some of the cations between the crystal layers are washed away as these rocks weather, the structure crumbles to clay. *Ceramics* are created by heating clays so as to drive out the water between the sheets of tetrahedra. What is left is a rigid mass of tiny interlocking crystals, bound together by glassy silica (Fig. 19.17). China clay, which is used to make porcelain and china, is a form of aluminum aluminosilicate; it is, however, reasonably free of the iron impurities that make many clays look reddish brown. It is used in large amounts to coat paper (including this page) to give a smooth, nonabsorbent surface.

When aluminosilicates are melted and then allowed to solidify, various *cements* are obtained. The most widely used is Portland cement, which is made by heating a mixture of silica, clay, and limestone to about 1500°C. The cooled mass is then ground very fine, and some gypsum ($CaSO_4 \cdot 2H_2O$) is added. The main components of the complex mixture are various calcium silicates and aluminates. When water is added, complex reactions occur and the mass sets to a solid. "High-alumina cement," which is resistant to corrosion in harsh environments, is made by fusing alumina, lime, and silica.

Silicones. The *silicones* consist of long —O—Si—O— chains in which the remaining two bonding positions on each silicon atom are occupied

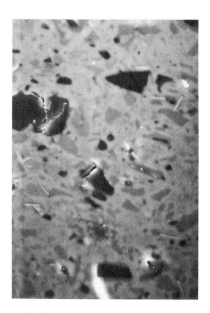

FIGURE 19.17 A photomicrograph of fired porcelain, showing that the minute crystals are bound together by amorphous silica.

by organic groups such as the methyl group, —CH$_3$ (Fig. 19.18). Silicones are used to waterproof fabrics because the oxygen atoms of the silicones attach to the fabric, leaving the hydrophobic (water-repelling) organic groups to act like tiny inside-out umbrellas sticking up out of the fabric's surface. For similar reasons, methyl silicones are biologically inert and survive intact when exposed to body fluids. Since they do not coagulate blood or stick to body tissues, they are used for surgical and cosmetic implants.

Tin and lead oxides. Tin(II) oxide is obtained by adding alkali to tin(II) aqueous solutions and then heating the product in the absence of air:

$$Sn^{2+}(aq) + 4OH^-(aq) \longrightarrow [Sn(OH)_4]^{2-}(aq)$$

$$[Sn(OH)_4]^{2-}(aq) \xrightarrow{\Delta} SnO(s) + 2OH^-(aq) + H_2O(l)$$

This blue-black solid smolders in air, and at 300°C it becomes incandescent as it oxidizes to tin(IV) oxide. The tin(II) and tin(IV) oxides are amphoteric and dissolve in acids and alkalis.

Like tin(II) oxide, lead(II) oxide is amphoteric. The dioxide, lead(IV) oxide, is a maroon solid and a strong oxidizing agent. Thus there is a marked difference in stability between the +2 and +4 oxidation states of tin, which is generally more stable as +4, and those of lead, which is generally more stable as +2. As discussed in Section 7.7, this is a sign of the greater importance of the inert pair in lead. When lead(IV) oxide reacts with warm aqueous acids, it oxidizes the water and the lead(II) salt is obtained, not the lead(IV) salt:

$$2PbO_2(s) + 4HCl(aq) \xrightarrow{warm} 2PbCl_2(aq) + 2H_2O(l) + O_2(g)$$

The relative ease with which lead(II) and lead(IV) are converted from one to the other is utilized in the lead-acid storage battery (Section 17.2).

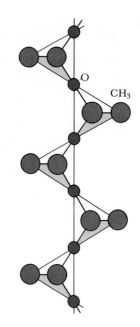

FIGURE 19.18 A typical silicone structure. The hydrocarbon groups give the substance its water-repelling quality.

19.6 OTHER IMPORTANT GROUP IV COMPOUNDS

In this section we discuss a few of the numerous compounds, other than oxides, formed by the Group IV elements, concentrating on those that are of commercial importance.

Carbides. There are three classes of carbides: *saline* (or saltlike), *covalent*, and *interstitial*. The Group I and II metals form saline carbides when their oxides are heated with carbon. One of the most important examples is calcium carbide (CaC$_2$). The anion is C$_2^{2-}$, and therefore more properly regarded as the "acetylide ion"—the conjugate base of the very weak Brønsted acid acetylene.

The covalent carbides include silicon carbide (SiC), which is sold as carborundum. It is prepared by heating silica with carbon:

$$SiO_2(s) + 3C(s) \xrightarrow{2000°C} SiC(s) + 2CO(g)$$

The pure material is colorless, but iron impurities normally make it almost black. Because it is very hard, and its crystals fracture in a way that leaves them with sharp edges (Fig. 19.19), carborundum is used as an abrasive.

FIGURE 19.19 Carborundum crystals, showing the sharp fractured edges that give the substance its abrasive power.

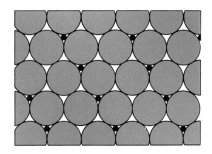

FIGURE 19.20 The structure of an interstitial carbide, in which carbon atoms lie between metal atoms, producing a rigid structure.

The interstitial carbides are formed by the direct reaction of a *d*-block metal and carbon at temperatures above 2000°C. In these compounds, C atoms occupy the gaps between the metal atoms (Fig. 19.20), like the H atoms in metallic hydrides. Here, however, the C atoms pin the metal atoms together into a rigid structure. This results in extremely hard substances with melting points often well above 3000°C. Tungsten carbide (W_2C) is used for the cutting surfaces of drills, and iron carbide (Fe_3C) is an important component of steel.

Halides. All the elements in Group IV form liquid molecular tetrachlorides. The least stable is $PbCl_4$, which decomposes to $PbCl_2$, a solid, when it is warmed to about 50°C. Carbon tetrachloride was once widely used as an industrial solvent; now, however, because it is so toxic, it is used primarily as the starting point for the manufacture of chlorofluorocarbons. It is formed by the action of chlorine on methane:

$$CH_4(g) + 4Cl_2(g) \xrightarrow{650°C} CCl_4(g) + 4HCl(g)$$

Chlorofluorocarbons are manufactured from CCl_4 by successive replacement of Cl atoms by F atoms, using hydrogen fluoride in the presence of the catalyst SbF_5:

$$CCl_4(l) + HF(l) \longrightarrow CFCl_3(g) + HCl(g)$$

Chlorofluorocarbons, which are sold as "Freons," are used as refrigerants and, in many countries, are still used as aerosol propellants. The damage they cause to the ozone layer of the atmosphere when the aerosol is used is the result of a series of reactions. When the

FIGURE 19.21 The disappearance of ozone from a region above Antarctica is thought to be a result of its destruction by chlorofluorocarbons. The colors show the concentration of ozone as measured in September 1986. The ozone "hole" is the oval of gray and violet colors over Antarctica.

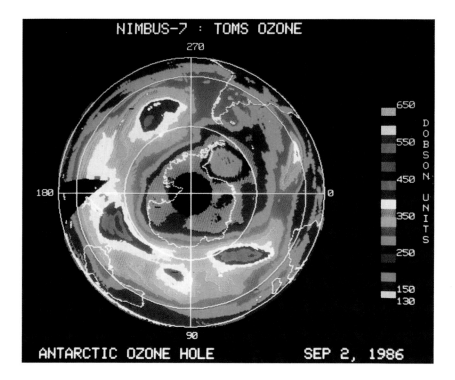

chlorofluorocarbons first reach the stratosphere, they are broken down by radiation, and Cl atoms are produced. These atoms attack ozone:

$$Cl(g) + O_3(g) \longrightarrow ClO(g) + O_2(g)$$

and are then regenerated by reaction with oxygen atoms:

$$ClO(g) + O(g) \longrightarrow Cl(g) + O_2(g)$$

Because of this regeneration, a single atom of chlorine survives in the stratosphere for between four and ten years; during that time it can destroy countless ozone molecules (Fig. 19.21).

Silicon reacts directly with chlorine to form silicon tetrachloride (we saw this earlier, as one step in the purification of silicon). This compound differs strikingly from CCl_4 in that it reacts readily with water:

$$SiCl_4(l) + 2H_2O(l) \longrightarrow SiO_2(s) + 4HCl(aq)$$

▼ **EXAMPLE 19.7** **Accounting for a difference between carbon and silicon**

Suggest a reason for the difference in reactivity with water between CCl_4 and $SiCl_4$.

STRATEGY Since C is a Period 2 element and Si is a Period 3 element, we should consider the differences between these two types of element. Another clue is that water often acts as a Lewis base and attacks a molecule that can act as a Lewis acid.

SOLUTION Since Si belongs to Period 3, it is bigger than a C atom and can expand its octet by using its d orbitals to accommodate the lone pair of the attacking Lewis base H_2O. The C atom in CCl_4 cannot do this, because it is too small and has no d orbitals available; hence it cannot act as a Lewis acid.

EXERCISE Would you expect SiH_4 to react with OH^- ions in water?
[*Answer*: Yes, for a similar reason]
▲

Cyanides. The cyanide ion is CN^-. The parent acid from which the cyanides are derived is hydrogen cyanide, HCN, which is made by heating ammonia, methane, and air in the presence of a platinum catalyst:

$$2CH_4(g) + 2NH_3(g) + 3O_2(g) \xrightarrow{1100°C,\ Pt} 2HCN(g) + 6H_2O(g)$$

Cyanides are strong Lewis bases that form a range of complexes with metals. They are also famous as poisons. When they are ingested, they combine with enzymes that regulate the transfer of oxygen to cellular tissues, and death results from suffocation.

Hydrides. The hydrocarbons are so numerous that they have a whole chapter—Chapter 23—to themselves. Silicon forms a much smaller number of compounds with hydrogen; the simplest is silane (SiH_4), the analog of methane. Silane is formed by the action of lithium aluminum hydride on silicon halides in ether (water cannot be used as solvent since it reacts):

$$SiCl_4 + LiAlH_4 \longrightarrow SiH_4 + LiCl + AlCl_3$$

Silane is much more reactive than methane and bursts into flame on contact with air. Although it is stable in pure water, it forms SiO_2 when a trace of alkali is present:

$$SiH_4(g) + 2H_2O(l) \xrightarrow{\text{OH}^-} SiO_2(s) + 4H_2(g)$$

The more complicated silanes, such as SiH_3—SiH_2—SiH_3 (the analog of propane), decompose on standing.

Sulfides.   The sulfur analog of carbon dioxide is carbon disulfide (CS_2), a volatile and very poisonous liquid. Its reputation for having a vile smell is undeserved, because the odor is in fact due not to CS_2 but to impurities. It is prepared in a reaction between methane and sulfur:

$$CH_4(g) + 4S(g) \xrightarrow{600°C,\ Al_2O_3} CS_2(g) + 2H_2S(g)$$

In the laboratory it is sometimes used as a solvent for sulfur; as explained in Section 11.3, its ability to dissolve sulfur stems from the strong London forces it exerts on S_8 molecules.

GROUP V: NITROGEN AND PHOSPHORUS

The elements in Group V and some of their properties are listed in Table 19.3. The most important members of the group are nitrogen and phosphorus, and we concentrate on them.

TABLE 19.3   The Group V elements

Valence configuration: ns^2np^3

Z	Name	Symbol	Atomic weight, amu	Melting point, °C	Boiling point, °C	Density, g/cm³	Normal form
7	Nitrogen	N	14.01	−210	−196		Colorless gas
15	Phosphorus	P	30.97	44	280	1.82	White or red nonmetal
33	Arsenic	As	74.92	613s*		5.78	Gray nonmetal
51	Antimony	Sb	121.75	630	1640	6.69	Blue-white lustrous metal
83	Bismuth	Bi	208.98	271	1650	8.90	White-pink metal

*The symbol s denotes that the element sublimes.

19.7 THE GROUP V ELEMENTS

The Group V elements range in character from the nonmetals nitrogen and phosphorus to the largely metallic bismuth. This range is reflected in their chemical properties: all the oxides of nitrogen and phosphorus are acidic, and bismuth's oxide is basic.

Nitrogen. Nitrogen, the principal component of air, is a remarkable element. The considerable bond strength of the N_2 molecule makes it almost as inert as the noble gases. However, in combination with other elements it has a rich and interesting chemistry. Nitrogen is obtained by the fractional distillation of liquid air; any remaining oxygen is removed by passing the gas over hot copper.

Nitrogen's main use is as the raw material for the Haber synthesis of ammonia. This is the principal initial means of *fixing* the element, that is, of combining it with other elements (as in NH_3). Once fixed, nitrogen can be converted to other compounds, such as nitric acid, fertilizers, explosives, and plastics, and can be used by plants for their metabolism. Natural fixation reactions can be initiated by lightning, which converts some atmospheric nitrogen to its oxides that are then washed into the soil by rain. Nitrogen is also fixed by bacteria that are found in nodules on the roots of leguminous plants, including clover, beans, peas, and alfalfa (Fig. 19.22). An intensely active field of research is the search for catalysts that can mimic these bacteria and fix nitrogen at normal temperatures.

Nitrogen differs sharply from its congeners. It is highly electronegative ($\chi = 3.0$, the same as for chlorine) and, as its atoms are small, can form multiple bonds by using its p orbitals. Its d orbitals are too high in energy to be available for bonding. Nitrogen is also unusual in having one of the widest ranges of oxidation numbers of any element: nitrogen compounds are known for each whole-number oxidation number

FIGURE 19.22 The bacteria that inhabit these nodules on the root of a pea plant are responsible for fixing atmospheric nitrogen and making it available to the plant.

FIGURE 19.23 The white (top) and red (bottom dish) allotropes of phosphorus. White phosphorus is stored under water because it ignites in air.

from −3 (as in NH_3) to +5 (as in nitric acid and the nitrates) and for some fractional oxidation numbers as well (such as the $-\frac{1}{3}$ in the azide ion, N_3^-).

Phosphorus. Phosphorus occurs widely as various kinds of phosphate rocks—particularly the apatites, which are forms of calcium phosphate, $Ca_3(PO_4)_2$. The element is obtained by heating the rocks with carbon in an electric furnace in the presence of sand. The latter removes the calcium by forming a molten slag of calcium silicate:

$$2Ca_3(PO_4)_2(s) + 6SiO_2(s) + 10C(s) \xrightarrow{\Delta}$$
$$P_4(g) + 6CaSiO_3(l) + 10CO(g)$$

The phosphorus vapor is condensed to white phosphorus, a soft, white, reactive, poisonous molecular solid consisting of tetrahedral P_4 molecules. The reactivity of this allotrope is partly due to the strain associated with the acute angles between the bonds. White phosphorus (the name comes from the Greek words for "the light bringer") ignites on contact with air and is normally stored under water (Fig. 19.23).

When it is heated in the absence of air, white phosphorus changes into red phosphorus. This denser allotrope is less reactive, but it can be ignited by friction and so is used in matches. Its structure is uncertain but may consist of chains of linked P_4 tetrahedra.

19.8 HYDROGEN AND HALOGEN COMPOUNDS

The Group V elements have negative oxidation numbers in their binary compounds with hydrogen (−3 for N in NH_3, for instance). Hence these compounds are formed by reduction of the elements, and they tend to be reducing agents themselves. By far the most important is ammonia.

Ammonia. The *Haber-Bosch process* for the synthesis of ammonia (Section 13.8) stands at the head of the industrial nitrogen chain:

$$N_2(g) + 3H_2(g) \xrightarrow{\text{400 to 600°C, 150 to 600 atm, Fe}} 2NH_3(g)$$

Nitrogen for the synthesis is obtained from the atmosphere, and hydrogen from the reforming and shift reactions described in Section 18.1.

Ammonia is a pungent, toxic gas that condenses to a colorless liquid at −33°C. The liquid resembles water in its physical properties, including its ability to act as a solvent. Gaseous ammonia is very soluble in water because NH_3 molecules can form hydrogen bonds to H_2O molecules. Although the aqueous solution is sometimes called "ammonium hydroxide," the compound NH_4OH has not been isolated. Ammonia is a weak Brønsted base ($pK_b = 4.75$) in water and a reasonably strong Lewis base, particularly toward *d*-block elements. An example of its ability to form complexes with the latter is its reaction with $Cu^{2+}(aq)$ ions to give a deep blue complex (Fig. 19.24):

$$Cu^{2+}(aq) + 4NH_3(aq) \longrightarrow [Cu(NH_3)_4]^{2+}(aq)$$

We shall see many of these complexes in Chapter 21.

The neutralization of aqueous ammonia with a Brønsted acid gives the corresponding ammonium salt, in which the cation is the ammonium ion NH_4^+. Solutions of ammonium salts are acidic, since NH_4^+ is a weak Brønsted acid. Ammonium salts decompose when heated:

$$NH_4Cl(s) \xrightarrow{\Delta} NH_3(g) + HCl(g)$$

The pungent smell of decomposing ammonium carbonate was the reason for its use in "smelling salts," and the pungency of heated ammonium chloride (*sal ammoniac*) was known to the Ammonians, the worshippers of the Egyptian god Amun. If the anion of an ammonium salt is an oxidizing ion, the ammonium cation may be oxidized; as in

$$NH_4NO_3(s) \xrightarrow{170°C} N_2O(g) + 2H_2O(g)$$

$$2NH_4NO_3(s) \xrightarrow{300°C} 2N_2(g) + O_2(g) + 4H_2O(g)$$

This second reaction can be explosively violent, which accounts for the use of ammonium nitrate as a component of dynamite. Ammonium nitrate has a high nitrogen content (35% by mass) and is very soluble in water; these two characteristics make it attractive as a fertilizer, and that is now its principal use.

Hydrazine. Hydrazine (NH_2NH_2) is an oily, colorless liquid in which nitrogen has the oxidation number -2 (compared with -3 in ammonia). It is produced by the gentle oxidation of ammonia by alkaline hypochlorite solution:

$$2NH_3(aq) + OCl^-(aq) \longrightarrow N_2H_4(aq) + Cl^-(aq) + H_2O(l)$$

Its physical properties are very similar to those of water (its melting point is 1.5°C, its boiling point 113°C). However, it is dangerously explosive and, for this reason, is normally used in aqueous solution. The exothermicity of its combustion makes it a valuable rocket fuel: the Apollo lunar missions used a mixture of liquid N_2O_4 and a derivative of hydrazine (methylhydrazine, CH_3NHNH_2) to land on and leave the moon. These two liquids, with the nitrogen in one having oxidation number -2 and in the other $+4$, ignite as soon as they mix, producing a large volume of gas:

$$4CH_3NHNH_2(l) + 5N_2O_4(l) \longrightarrow 9N_2(g) + 12H_2O(g) + 4CO_2(g)$$

A valuable application of hydrazine back on earth is to eliminate dissolved oxygen from the water used in high-pressure, high-temperature steam furnaces (to avoid corrosion):

$$N_2H_4(aq) + O_2(aq) \longrightarrow N_2(g) + 2H_2O(l)$$

Phosphine. The hydrogen compounds of other members of Group V are much less stable than ammonia, and they decrease in stability down the group.

Phosphine is a poisonous gas that smells slightly of garlic and bursts into flames in air if it is slightly impure. It is much less soluble than

FIGURE 19.24 When aqueous ammonia is poured into aqueous copper sulfate, the dark blue complex is formed by a Lewis acid-base reaction.

ammonia in water, a fact that again points to the importance of hydrogen bonding: NH_3 can form hydrogen bonds with water, but PH_3 cannot. Aqueous solutions of phosphine are neutral, for PH_3 has only an extremely weak tendency to accept a proton ($pK_b = 27.4$).

▼ **EXAMPLE 19.8** **Suggesting a preparation of phosphine**

Suggest a method of preparing phosphine based on the fact that PH_3 is the conjugate acid of the phosphide ion P^{3-}.

STRATEGY Since PH_3 is the (extremely weak) conjugate acid of the Brønsted base P^{3-}, we can expect to produce it by protonation with a Brønsted acid. Even a very weak acid should be adequate, so long as kinetic factors are not unfavorable.

SOLUTION We can expect to form phosphine by proton transfer from H_2O acting as a weak acid. The net ionic reaction is

$$2P^{3-}(s) + 6H_2O(l) \longrightarrow 2PH_3(g) + 6OH^-(aq)$$

It can be achieved in practice with calcium phosphide, a readily available phosphide. The overall reaction is then

$$Ca_3P_2(s) + 6H_2O(l) \longrightarrow 2PH_3(g) + 3Ca(OH)_2(aq)$$

EXERCISE Suggest a method for preparing phosphine based on the fact that PH_3 is the conjugate base of PH_4^+.

[*Answer*: Deprotonation, as in
$$PH_4I(aq) + KOH(aq) \longrightarrow PH_3(g) + KI(aq) + H_2O(l)]$$

▲

Phosphorus halides. Phosphorus trichloride (PCl_3) and phosphorus pentachloride (PCl_5) are the two most important halides of phosphorus. The trichloride is prepared by direct chlorination of phosphorus. The liquid product must be distilled away before it can react further; it fumes in moist air and reacts readily with water to form phosphorous acid:

$$PCl_3(l) + 3H_2O(l) \longrightarrow H_3PO_3(s) + 3HCl(g)$$

This reaction illustrates a common feature of the reaction of nonmetal halides: they react with water to give an oxoacid without undergoing a change of oxidation number.

The trichloride also takes part in many reactions with other substances. It is a major intermediate for the production of phosphorus compounds for use in pesticides, oil additives, and flame retardants. Phosphorus pentachloride is made by allowing the trichloride to react with more chlorine.

▼ **EXAMPLE 19.9** **Predicting the result of a nonmetal halide reaction**

What products will form when phosphorus pentachloride reacts with water?

STRATEGY We have just seen that a nonmetal halide typically reacts with water without undergoing a change of oxidation number. Hence, we expect the reaction to produce the oxoacid of the same oxidation number as the starting material.

SOLUTION The oxidation number of P in PCl_5 is +5. Hence we expect the oxoacid product to be phosphoric acid, H_3PO_4, which also has phosphorus with oxidation number +5. The reaction is in fact

$$PCl_5(s) + 4H_2O(l) \longrightarrow H_3PO_4(l) + 5HCl(g)$$

and is violent.

EXERCISE What products can be expected when arsenic tribromide reacts with water?

[*Answer*: H_3AsO_3 and HBr]

▲

An interesting feature of phosphorus pentachloride is that it is an *ionic* solid of tetrahedral PCl_4^+ cations and octahedral PCl_6^- anions, but it vaporizes to trigonal bipyramidal PCl_5 molecules. Phosphorus pentabromide is also molecular in the vapor but ionic as the solid, but in the solid the anions are simply Br^- ions, presumably because of the difficulty of fitting six bulky Br atoms around a central P atom. The same difficulty arises, but in a much more severe form, with the smaller N atom (**3**); even the trihalides of nitrogen (with the exception of NF_3) are explosively unstable (Fig. 19.25), and pentahalides of nitrogen are unknown.

3 Nitrogen triiodide

19.9 GROUP V OXIDES AND OXOACIDS

Although the nitrogen oxides can be confusing at first sight, we shall see that their properties and interconversions can be understood by keeping track of their oxidation numbers, which range from +1 to +5. All nitrogen oxides are acidic, and in discussing them we shall also meet the nitrogen oxoacids (Table 19.4). In atmospheric chemistry, where the nitrogen oxides play an important role in both maintaining and polluting the atmosphere, they are referred to generally as NO_x.

Phosphorus compounds appear in the top 50 industrial chemicals by virtue of phosphoric acid, which stands ninth in rank. Its principal use

TABLE 19.4 The oxides and oxoacids of nitrogen

| Oxidation number | Oxide | | Corresponding acid | |
	Formula	Name	Formula	Name
+5	N_2O_5	Dinitrogen pentoxide	HNO_3	Nitric acid
+4	NO_2*	Nitrogen dioxide		
	N_2O_4	Dinitrogen tetroxide		
+3	N_2O_3	Dinitrogen trioxide	HNO_2	Nitrous acid
+2	NO	Nitric oxide; nitrogen oxide		
+1	N_2O	Nitrous oxide; dinitrogen oxide	$H_2N_2O_2$	Hyponitrous acid

*$2NO_2 \rightleftharpoons N_2O_4$.

FIGURE 19.25 Nitrogen triiodide (NI_3) is so unstable—partly because three bulky I atoms are attached to a small N atom—that it explodes when it is lightly touched. The explosive compound is $NI_3 \cdot NH_3$, since NI_3 itself has not been isolated.

is in the manufacture of fertilizers, which accounts for 85% of its production. Phosphate (fertilizer) production also takes two-thirds of the production of sulfuric acid, for sulfuric acid is used in its manufacture. A structural theme in the oxides of phosphorus is the tetrahedral PO_4 unit, similar to the unit that occurs in the oxides of its neighbor silicon.

Dinitrogen oxide. Dinitrogen oxide (or nitrous oxide, N_2O) is the oxide of nitrogen with the lowest oxidation number (+1). It is formed by gently heating ammonium nitrate:

$$NH_4NO_3(s) \xrightarrow{170°C} N_2O(g) + 2H_2O(g)$$

N_2O is a colorless, very soluble gas with a slight odor that is used as an anesthetic. Since it is tasteless, is nontoxic in small amounts, and dissolves readily in fats, it is used as a foaming agent and propellant for whipped cream. Dinitrogen oxide is quite unreactive; in particular, it does not react at ordinary temperatures with the halogens, the alkali metals, or ozone. However, organic matter burns in it after being ignited.

Nitric oxide. Nitric oxide (NO) is prepared industrially by the catalytic oxidation of ammonia:

$$4NH_3(g) + 5O_2(g) \xrightarrow{850°C, \text{ Pt}} 4NO(g) + 6H_2O(g)$$

Its formation from the elements is endothermic,

$$N_2(g) + O_2(g) \longrightarrow 2NO(g) \qquad \Delta H° = +181 \text{ kJ}$$

partly as a result of the high dissociation enthalpy of nitrogen. Hence, by Le Chatelier's principle, the equilibrium composition shifts in favor of NO as the temperature is raised. As a result, it is formed in the hot exhausts of airplane and automobile engines.

Nitric oxide contributes to the problem of acid rain and the formation of smog. It also contributes, like the chlorofluorocarbons, to the destruction of the ozone layer:

$$NO(g) + O_3(g) \longrightarrow NO_2(g) + O_2(g)$$
$$NO_2(g) + O(g) \longrightarrow NO(g) + O_2(g)$$

The net result of these two reactions is the destruction of an ozone molecule and the regeneration of a nitric oxide molecule, which is then free to destroy more ozone. Because NO molecules are regenerated in this way, a small amount of nitric oxide can eliminate a large amount of ozone.

In the laboratory, nitric oxide can be prepared by reducing a nitrite with a mild reducing agent:

$$2NO_2^-(aq) + 2I^-(aq) + 4H^+(aq) \longrightarrow 2NO(g) + I_2(aq) + 2H_2O(l)$$

It is a colorless gas that is oxidized rapidly to nitrogen dioxide on exposure to air (Fig. 19.26):

$$2NO(g) + O_2(g) \longrightarrow 2NO_2(g)$$

Both NO and NO_2 are odd-electron, paramagnetic molecules.

FIGURE 19.26 Nitric oxide may be prepared by the reduction of nitrite ions with iodide ions in acid solution. The product is colorless, but when it mixes with the oxygen of the air it forms the brown gas NO_2.

Nitrogen dioxide. Nitrogen dioxide is a choking, poisonous, brown gas that contributes to the color of smog. In the gas phase it exists in equilibrium with its colorless dimer N_2O_4; in the solid only the dimer exists. When it dissolves in water, NO_2 disproportionates and forms nitric acid:

$$3NO_2(g) + H_2O(l) \longrightarrow 2HNO_3(aq) + NO(g)$$

Nitrogen dioxide undergoes the same reaction in the atmosphere, thus contributing to the formation of acid rain (Section 20.3). Industrially the dioxide is made by oxidizing nitric oxide obtained from the oxidation of ammonia. In the laboratory it is prepared by heating lead nitrate:

$$2Pb(NO_3)_2(l) \xrightarrow{400°C} 4NO_2(g) + 2PbO(s) + O_2(g)$$

Nitrous acid. When a mixture of the odd-electron molecules NO and NO_2 is cooled to −20°C, they combine to form dinitrogen trioxide, N_2O_3 **(4)**. This compound, in which the oxidation number of the nitrogen is +3, is unstable as a gas but can be collected as a dark blue liquid (Fig. 19.27). It is the anhydride of nitrous acid (HNO_2), and it forms that acid when it dissolves in water:

4   Dinitrogen trioxide

$$N_2O_3(g) + H_2O(l) \longrightarrow 2HNO_2(aq)$$

Nitrous acid has not been isolated in the pure form but is quite widely used in aqueous solution.

Nitrous acid is a weak acid ($pK_a = 3.37$) and the parent of the nitrites, compounds that contain the angular nitrite ion (NO_2^-). However, in practice nitrites are more easily produced by the reduction of nitrates with hot metal; for example,

$$KNO_3(l) + Pb(s) \xrightarrow{350°C} KNO_2(s) + PbO(s)$$

Most nitrites are soluble in water and mildly toxic. Despite their toxicity, nitrites have been used for the cosmetic treatment of meat because they inhibit the oxidation of the blood (which turns the meat brown) and form a pink complex with the hemoglobin. This complex is responsible for the pink color of ham and other cured meats.

Nitric acid. Nitric acid (HNO_3), one of the most widely used industrial acids, is produced by the three-stage *Ostwald process*:

1. Oxidation of ammonia, from oxidation number −3 to +2:

$$4NH_3(g) + 5O_2(g) \xrightarrow{850°C, \, 5\, atm, \, Pt/Rh} 4NO(g) + 6H_2O(g)$$

2. Oxidation of nitric oxide, from oxidation number +2 to +4:

$$2NO(g) + O_2(g) \longrightarrow 2NO_2(g)$$

3. Disproportionation in water, from oxidation number +4 to +5 and +2:

$$3NO_2(g) + H_2O(l) \longrightarrow 2HNO_3(aq) + NO(g)$$

FIGURE 19.27 Dinitrogen trioxide (N_2O_3) condenses to a deep blue liquid that freezes at −100°C to a pale blue solid as shown above; on standing, it turns green as a result of partial decomposition into nitrogen dioxide.

Nitric acid is a colorless liquid that boils at 86°C and is normally used in aqueous solution. Concentrated nitric acid is often pale yellow as a result of the partial decomposition of the acid to NO_2.

▼ **EXAMPLE 19.10 Suggesting a preparation of a nitrogen oxide**

Suggest a method for preparing dinitrogen pentoxide, N_2O_5.

STRATEGY We know that the oxides of nitrogen are acidic, so we expect N_2O_5 to be the anhydride of an oxoacid. If we knew which oxoacid, we could consider producing the oxide by dehydration of the acid. Therefore, we have to identify the acid of which N_2O_5 is the anhydride (by adding one or more H_2O units to N_2O_5).

SOLUTION Adding H_2O to N_2O_5 gives $H_2N_2O_6$, which is two HNO_3 units. Hence, we can expect to produce the oxide by dehydrating nitric acid. The reaction that is used in practice is

$$2HNO_3(l) \xrightarrow{\Delta, \ P_4O_{10}} N_2O_5(s) + H_2O(l)$$

EXERCISE What property other than Brønsted acidity can be anticipated for nitric acid?

[*Answer*: Oxidizing agent]

▲

Since nitrogen attains its highest oxidation number (+5) in HNO_3, nitric acid is an oxidizing agent as well as a Brønsted acid. Some metals (particularly aluminum) become protected by an oxidized layer when they are attacked by nitric acid; the noble metals gold, platinum, iridium, and rhodium are not attacked at all. However, they will be attacked if Cl^- ions are also present, as in aqua regia (Section 15.7), as a result of the formation of the complex $[AuCl_4]^-$.

Nitric acid is the parent of the nitrates, compounds that contain the planar triangular nitrate ion (NO_3^-). Most nitrates are soluble in water. When heated, nitrates of heavy metals decompose (much like the carbonates) into nitrogen dioxide and the metal oxide, as in

$$2Pb(NO_3)_2(l) \xrightarrow{\Delta} 2PbO(s) + 4NO_2(g) + O_2(g)$$

The nitrates of the lighter metals decompose into the nitrite and oxygen, as in

$$2KNO_3(l) \xrightarrow{\Delta} 2KNO_2(s) + O_2(g)$$

This reaction makes nitrates useful as an additional supply of oxygen in matches, flares, and explosives.

Phosphorus(III) oxide and phosphorous acid. White phosphorus burns in a limited supply of air to form phosphorus(III) oxide, P_4O_6 **(5)**. In structure, its molecules are closely related to the P_4 molecules of phosphorus, but an O atom lies between each pair of P atoms. Phosphorus(III) oxide is the anhydride of phosphorous acid, H_3PO_3 **(6)**, and is converted to that acid by cold water. Although its formula suggests that it should be triprotic, H_3PO_3 is in fact *di*protic since one of the H atoms is attached directly to the P atom.

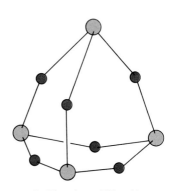

5 Phosphorus(III) oxide

6 Phosphorous acid

Phosphorus(V) oxide and phosphoric acid. When phosphorus burns in a plentiful supply of air (Fig. 19.28), it is oxidized to phosphorus(V) oxide, P_4O_{10} (7). This compound is the anhydride of phosphoric acid, H_3PO_4 (8), a triprotic Brønsted acid. A P_4O_{10} molecule is a P_4O_6 molecule with an O atom attached to each of its P atoms. This oxide has a great affinity for water, and so it is widely used in the laboratory as a drying agent (to maintain a moisture-free atmosphere inside a container) and as a dehydrating agent (to remove the elements of water from compounds, as in the preparation of acid anhydrides).

Pure phosphoric acid, H_3PO_4, is a colorless solid of melting point 42°C, but in the laboratory it is normally a syrupy liquid on account of the water it has absorbed. Its high viscosity can be traced to extensive hydrogen bonding between neighboring H_3PO_4 molecules. Although phosphorus has a high oxidation number (+5), the acid shows appreciable oxidizing power only at temperatures above about 350°C; hence it may be used where nitric acid and sulfuric acid would be too oxidizing.

Most phosphoric acid is manufactured by the action of sulfuric acid on phosphate rock:

$$Ca_3(PO_4)_2(s) + 3H_2SO_4(l) \longrightarrow 2H_3PO_4(l) + 3CaSO_4(s)$$

Since fluorine is also commonly present in the rock (as fluorapatite), a more realistic version of the process is

$$Ca_5F(PO_4)_3(s) + 5H_2SO_4(l) + 10H_2O(l) \longrightarrow$$
$$3H_3PO_4(l) + 5CaSO_4 \cdot 2H_2O(s) + HF(g)$$

The HF is pumped off and absorbed in silicon tetrafluoride, and the resulting H_2SiF_6 is used to fluoridate water supplies.

Phosphoric acid is the parent of the phosphates, which contain the tetrahedral PO_4^{3-} anion and are of great biological and commercial importance. Phosphate rock is mined in huge quantities, especially in Florida and Morocco. After being crushed, it is treated with sulfuric acid to produce a mixture of sulfates and phosphates called "superphosphate," a major fertilizer. The phosphate rock may be treated with phosphoric acid rather than sulfuric acid to produce a mixture with a higher phosphate content; the resulting mixture of calcium phosphates is sold as "triple superphosphate." We have already described (in Section 18.6) the role of calcium phosphate in skeletons and teeth.

Polyphosphates. When undiluted phosphoric acid is heated, it undergoes a reaction in which two molecules combine and H_2O is eliminated:

$$\underset{\substack{| \\ OH}}{\overset{\substack{O \\ \|}}{HO-P-O-H}} + \underset{\substack{| \\ OH}}{\overset{\substack{O \\ \|}}{HO-P-OH}} \xrightarrow{\Delta} \underset{\substack{| \\ OH}}{\overset{\substack{O \\ \|}}{HO-P-O}}\underset{\substack{| \\ OH}}{\overset{\substack{O \\ \|}}{-P-OH}} + H_2O$$

The product, $H_4P_2O_7$, is pyrophosphoric acid. Further heating gives even more complicated products that contain chains and rings of PO_4 groups and are called *polyphosphoric acids*. In this regard, phosphorus shows a resemblance to its glass-forming Group IV neighbor silicon.

Polyphosphate ions, as salts of the polyphosphoric acids, are added to detergents to sequester the cations responsible for hardness and

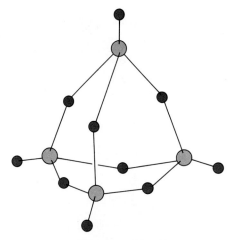

7 Phosphorus(V) oxide

8 Phosphoric acid

FIGURE 19.28 Phosphorus smolders in air to produce phosphorus(V) oxide, P_4O_{10}. When ignited, it produces dense white clouds of the oxide.

prevent their precipitation (see Section 11.3). A second function of these ions is to attach to the surfaces of dirt particles, giving them a negative charge. Earlier we saw what happens when sodium silicate is used in this way: neighboring dirt particles then repel each other and disperse into the surrounding water. Finally, being Brønsted bases, the polyphosphate ions buffer the solution to a pH at which the detergent molecules can act most effectively.

The most important polyphosphate is adenosine triphosphate, ATP (**9**), a compound that is found in every living cell. The triphosphate part of this molecule is a chain of three phosphate groups. Its conversion to the diphosphate ADP

$$\underset{\substack{|\\O}}{\overset{\substack{O\\||}}{-\!O\!-\!P\!-\!O}}\!-\!\underset{\substack{|\\O}}{\overset{\substack{O\\||}}{P}}\!-\!O\!-\!\underset{\substack{|\\O}}{\overset{\substack{O\\||}}{P}}\!-\!O + H_2O \longrightarrow$$

$$\underset{\substack{|\\O}}{\overset{\substack{O\\||}}{-\!O\!-\!P\!-\!O}}\!-\!\underset{\substack{|\\O}}{\overset{\substack{O\\||}}{P}}\!-\!O + HPO_4{}^{2-}(aq) + H^+(aq)$$

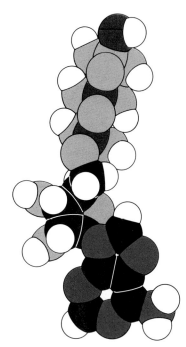

9 Adenosine triphosphate

provides the energy that fuels life, for this reaction enables plants and animals to persist and grow and the ones with brains to think. The death of the organism releases the phosphate to the ecosystem, where it may lie as phosphate rock until nature or industry sends it on its way again.

SUMMARY

19.1 The oxidation number +3 is dominant in the lighter elements of Group III, and metallic character increases down the group. Boron is a hard nonmetal with several allotropic forms; it shows a diagonal relationship to silicon. Aluminum, obtained by the electrolytic **Hall process,** is a metal but has an amphoteric oxide. Although low in the electrochemical series, it is passivated by a tough layer of oxide that protects it from further reaction. Its reducing power is shown in the **thermite reaction.**

19.2 Boron has an acidic oxide—a Lewis acid that in water forms boric acid, a Brønsted acid. Alumina is obtained from bauxite by the **Bayer process.** It reacts with acids to form the acidic $[Al(H_2O)_6]^{3+}$ ion, which occurs in the alums. It also reacts with alkalis to give **aluminates.**

19.3 The boron and aluminum **halides** are Lewis acids, with the aluminum halides forming Lewis acid-base dimers (the chloride only in the vapor state). Boron forms an extensive series of electron-deficient **boranes** with structures that depend on delocalized bonds. The **borohydride anion** $BH_4{}^-$ is an important reducing agent that is used in the **chemical plating** of nickel. The most important hydride of aluminum is $LiAlH_4$, which is a useful reducing agent.

19.4 Metallic character also increases down Group IV but is less pronounced than in Group III; the inert-pair effect plays a more important role in this group, particularly in tin and lead. Carbon and silicon are nonmetals. Germanium, tin, and lead are metals but show amphoteric properties.

19.5 The oxides of carbon are acidic, carbon dioxide being the anhydride of **carbonic acid,** the parent of the **carbonates.** Carbon monoxide is a Lewis base and an important reducing agent. Silicon also has an acid oxide, silica, and it forms an extensive and important collection of **silicates** and **aluminosilicates,** in which Al atoms take the place of some Si atoms. Their structures can be understood in terms of linked SiO_4 tetrahedra. Molten silica with added s-block oxides cools to a **glass.** Tin and lead form a much more limited (and amphoteric) set of oxides, with the oxidation number +4 dominant in tin and +2 in lead.

19.6 Carbides are classified as **saline, covalent,** or **interstitial.** The halides of Group IV elements are progressively less covalent down the group and show the trend from nonmetallic to metallic character in the elements. The **silanes** (such as SiH_4) are much less stable than the hydrocarbons. Part of this instability rises from silicon's ability to expand its octet and accommodate the lone pair of an attacking Lewis base.

19.7 The elements in Group V are mainly nonmetals,

with metallic character becoming pronounced only at the bottom of the group (in Sb and Bi). Nitrogen exists in a wide range of oxidation numbers (from -3 to $+5$).

19.8 Industrial **fixation** of nitrogen (its combining with other elements), is achieved by the Haber process for the synthesis of ammonia. Ammonia is the most important of the binary hydrides of the group; it is more stable and a much stronger base than **phosphine,** PH_3.

19.9 Ammonia is the industrial source of the oxoacids and oxides of nitrogen, particularly by its oxidation to **nitric acid,** the parent of the soluble **nitrates,** in the **Ostwald process.** The oxides of nitrogen span the oxidation numbers +1 to +5. The oxides of phosphorus are also acidic, being the anhydrides of **phosphorous acid** and the much more important **phosphoric acid.** The structural theme of the oxoanions of phosphorus is the PO_4 tetrahedron, as SiO_4 is the theme in the silicates. When heated, phosphoric acid forms various **polyphosphates,** of which the most significant biologically is the molecule ATP, adenosine triphosphate.

EXERCISES

In Exercises 19.1 and 19.2, identify the group and period to which each p-block element belongs.

19.1 (a) Al; (b) P; (c) Si; (d) Ga; (e) Ge; (f) In.
19.2 (a) C; (b) N; (c) B; (d) Sn; (e) Bi; (f) Tl.

In Exercises 19.3 and 19.4, write the ground state valence electron configuration of each atom.

19.3 (a) B; (b) Si; (c) Ge; (d) N; (e) Bi; (f) In.
19.4 (a) Al; (b) C; (c) Ga; (d) P; (e) As; (f) Tl.

Group III

19.5 Outline the sources, preparation, properties, and uses of boron and aluminum.

19.6 List and explain (with equations where appropriate) how alumina is used.

19.7 Explain, with equations, the chemical principles involved in the Bayer process for the purification of alumina.

19.8 Explain, with equations, how you could prepare aluminum hydride from naturally available substances.

19.9 Explain, with equations, how you could prepare decaborane from naturally available substances.

19.10 Account for the difference between the first ionization energies of boron and aluminum. Why does boron have a greater electronegativity than aluminum?

*__19.11__ Plot the first ionization energies and electronegativities of the Group III elements, and predict their values for the next member of the group that might one day be synthesized. What would be its atomic number?

In Exercises 19.12 and 19.13, balance and classify each reaction.

19.12 (a) $B_2H_6 + O_2 \longrightarrow B_2O_3 + H_2O$
(b) $B_2H_6 + H_2O \longrightarrow B(OH)_3 + H_2$
(c) $LiH + AlCl_3 \longrightarrow LiAlH_4 + LiCl$
19.13 (a) $B_2H_6 + NaBH_4 \longrightarrow Na_2B_{12}H_{12} + H_2$
(b) $B_2O_3 + C + Cl_2 \longrightarrow BCl_3 + CO$
(c) $LiAlH_4 + H_2SO_4 \longrightarrow AlH_3 + Li_2SO_4 + H_2$

19.14 What properties account for the rarity of aluminum metal until the end of the nineteenth century? What makes it such a useful metal? How may magnesium and aluminum be distinguished?

19.15 List the properties (with equations where appropriate) that support the view that beryllium and aluminum have a diagonal relationship. How can the relationship be explained?

19.16 In what ways is boron different from its congeners? What accounts for the differences? In what ways is boron similar to its congeners?

19.17 Calculate the minimum mass of aluminum required to reduce 100 kg of chromium(III) oxide to chromium. What mass of chromium is obtained?

19.18 Calculate the maximum mass of aluminum that would be obtained in a day by the Hall process in a plant using a current of 100,000 A.

19.19 Suppose a heater could be devised to use the oxidation of either iron or aluminum as a source of energy. Which metal has the more favorable specific enthalpy?

19.20 Identify the cathode and the electrode in each of the following cells, and calculate the standard cell potential. Show the natural direction of the reaction in each case.
(a) $Fe(s)|Fe^{3+}(aq)$ in combination with $Al(s)|Al^{3+}(aq)$
(b) $Cr(s)|Cr^{3+}(aq)$ in combination with $Al(s)|Al^{3+}(aq)$
(c) $Al(s)|Al^{3+}(aq)$ in combination with $Zn(s)|Zn^{2+}(aq)$
(d) $Al(s)|Al^{3+}(aq)$ in combination with $Mg(s)|Mg^{2+}(aq)$

19.21 In what sense is boric acid an acid? Write the Lewis structure of boric acid, and predict the shapes of the $B(OH)_3$ molecule and the $[B(OH)_4]^-$ ion.

19.22 Explain why an aqueous solution of Al^{3+} is acidic. What changes occur as sodium hydroxide is added to such a solution? Give the chemical equations for the reactions.

19.23 The first ionization energies of the Group II elements decrease smoothly down the group, but in Group III the values for gallium and thallium are both higher than for aluminum. Suggest a reason for this difference.

*\ **19.24** The standard enthalpies of formation of $AlBr_3(g)$ and its dimer are -411 kJ/mol and -1021 kJ/mol, respectively, and the enthalpies of formation of $Al(g)$ and $Br(g)$ are $+326$ kJ/mol and $+111.9$ kJ/mol, respectively. Use these values to calculate the mean bond enthalpies of the Al—Br bonds in each case, and to estimate the bond enthalpies of the Al—Br and Al—Br—Al bonds. Which bonds would you expect to be the longer?

*\ **19.25** The standard enthalpies of formation of $BH_3(g)$ and diborane are $+100$ kJ/mol and $+36$ kJ/mol, respectively, and the enthalpies of formation of $B(g)$ and $H(g)$ are $+563$ kJ/mol and $+218$ kJ/mol, respectively. Use these values to calculate the mean bond enthalpies of the B—H bonds in each case, and to estimate the bond enthalpies of the B—H and B—H—B bonds. Which bonds would you expect to be the longer?

19.26 The standard free energy of formation of $Tl^{3+}(aq)$ is $+215$ kJ/mol at 25°C. Calculate the standard reduction potential of the Tl^{3+}/Tl couple.

19.27 The standard reduction potential of the Al^{3+}/Al couple in aqueous solution at 25°C is -1.66 V. Calculate the standard free energy of formation of $Al^{3+}(aq)$. What factors account for the difference from the value for $Tl^{3+}(aq)$ in the previous example?

Group IV

19.28 Outline the sources, preparation, properties, and uses of carbon and silicon.

19.29 Suggest, with equations, how a pure sample of sodium silicate could be prepared from naturally occurring substances.

In Exercises 19.30 and 19.31, balance and classify each reaction.

19.30 (a) $CH_4 + S \longrightarrow CS_2 + H_2S$
(b) $Mg_2Si + H_2SO_4 \longrightarrow MgSO_4 + SiH_4$
(c) $Sn + KOH + H_2O \longrightarrow K_2Sn(OH)_6 + H_2$
(d) $Pb(NO_3)_2 \longrightarrow PbO + NO_2 + O_2$

19.31 (a) $CS_2 + S_2Cl_2 \longrightarrow CCl_4 + S$
(b) $Si_2Cl_6 \longrightarrow Si_6Cl_{14} + SiCl_4$
(c) $GeCl_4 + SO_3 \longrightarrow Ge(SO_4)_2 + S_2O_5Cl_2$
(d) $Na_4Pb_9 + CH_3CH_2Cl \longrightarrow$
$(CH_3CH_2)_4Pb + Pb + NaCl$

19.32 What properties support the placing of carbon and silicon in the same group? In what ways does carbon show differences from silicon, and how may they be explained?

19.33 Is there any evidence for diagonal relationships between members of Group IV and members of Groups III and V?

19.34 Summarize the evidence supporting the statement that metallic character increases down Group IV and is less pronounced there than in Group III.

19.35 List the characteristic properties of the carbonates of Group I and Group II elements. Devise a convenient chemical test for a carbonate.

19.36 Describe the structures of silicate materials, including the aluminosilicates, and indicate how their structures correlate with their physical properties.

19.37 What is the empirical formula of a potassium silicate in which the silicate tetrahedra share (a) one oxygen; (b) two oxygens and form a chain? (In each case, there are single negative charges on the unshared oxygens.) Suggest a reason why carbon does not form such a series of carbonates.

19.38 The standard reduction potential of the Pb^{2+}/Pb couple is -0.13 V. How does this value, in combination with other influences including kinetic effects, affect the way in which metallic lead can be used?

19.39 Suggest a reason for the observation that methane is stable when bubbled through aqueous sodium hydroxide, but silane reacts rapidly. Write a chemical equation for the silane reaction.

19.40 When the mineral dolomite, $CaCO_3 \cdot MgCO_3$, is heated, it gives off carbon dioxide and forms a mixture of a metal oxide and a metal carbonate. Which oxide is formed, CaO or MgO? Which carbonate, $CaCO_3$ or $MgCO_3$?

19.41 Suppose that the stability of carbonates when they are heated depends on the ability of the metal cation to polarize carbonate ion and remove an oxide ion from it, thus releasing carbon dioxide. On that basis, predict the order of thermal stability of the Group I and Group II metal carbonates. Comment on the likely stability of aluminum carbonate.

19.42 The $C{=}O$ bond enthalpy in carbon dioxide is 805 kJ/mol, whereas the $C{-}O$ bond enthalpy is 360 kJ/mol. Do these data support the existence of discrete CO_2 molecules rather than a silicalike CO_2 structure? How is your answer affected by taking entropy into account?

19.43 Write the Lewis structure for the silicate anion SiO_4^{4-}, giving the formal charges and oxidation numbers of the atoms. Use VSEPR theory to predict the shape of the ion.

19.44 The free energy of the reaction $C(s) + CO_2(g) \rightarrow 2CO(g)$ is negative at 25°C, but changes sign above 980 K. What is the equilibrium composition of the reaction mixture at 980 K?

19.45 The reduction of tin(IV) oxide by carbon will proceed at moderately low temperatures, but it is normally carried out at over 980 K, in the presence of abundant carbon monoxide. Write the two chemical equations, one for C and the other for CO as reducing agent. What is the *kinetic* advantage of using carbon monoxide as the reducing agent instead of carbon itself?

*19.46 The enthalpy of formation of $SiO_2(g)$ molecules is about −322 kJ/mol at 25°C. Estimate the $Si{=}O$ bond enthalpy. The $Si{-}O$ bond enthalpy is 466 kJ/mol. Use the two bond-enthalpy values to account for the occurrence of silica as a solid and the occurrence of complex silicates.

19.47 Summarize the uses of carbonates, and justify these uses in terms of the thermodynamic and acid/base properties of the carbonates. Suggest an explanation for the greater solubility of hydrogen carbonates (bicarbonates) than of carbonates.

19.48 Explain what a *glass* is, and how it is distinguished from a crystal. Describe how glass is prepared. "Lead crystal" is prepared by including lead(II) oxide: what physical properties might it be expected to have?

*19.49 Sketch a graph showing the fractions of solute present as H_2CO_3, HCO_3^-, and CO_3^{2-} as the pH of carbonic acid is changed from 0 to 14 by the addition of aqueous sodium hydroxide.

19.50 Clean rain has pH = 6, and very acid rain can reach pH = 3. What are the main forms of carbonate ion present in each case? The pH of seawater is around 8.3. What is the principal form of the dissolved carbon dioxide?

Group V

Name each compound in Exercises 19.51 to 19.54.

19.51 (a) HNO_2; (b) NO; (c) H_3PO_3; (d) N_2O_3.
19.52 (a) HNO_3; (b) N_2O; (c) H_3PO_4; (d) N_2O_5.

19.53 (a) $H_2N_2O_2$; (b) $Na_4P_2O_7$; (c) $Na_2N_2O_3$; (d) NaN_3.
19.54 (a) NH_4NO_2; (b) $NaKH_2P_2O_7$; (c) $Pb(N_3)_2$; (d) N_2O_4.

In Exercises 19.55 to 19.58, write the chemical formula for each compound.

19.55 (a) Nitrous oxide; (b) nitric acid; (c) phosphorus pentoxide; (d) sodium azide.
19.56 (a) Nitric oxide; (b) phosphorous acid; (c) phosphine; (d) hyponitrite ion.

19.57 (a) Ammonium nitrate; (b) magnesium nitride; (c) calcium phosphide; (d) hydrazine.
19.58 (a) Dinitrogen trioxide; (b) ammonium dihydrogen phosphate; (c) disodium dihydrogen pyrophosphate; (d) phosphorus(III) oxide.

19.59 Give the chemical formulas of three compounds discussed in this chapter in which (a) nitrogen and (b) phosphorus have the oxidation number +3. Do the same for oxidation number +5.

In Exercises 19.60 to 19.63, balance each equation.

19.60 (a) $NH_3 + CuO \longrightarrow N_2 + Cu + H_2O$
(b) $NH_3 + Br_2 \longrightarrow N_2 + NH_4Br$
(c) $Al_2O_3 + C + N_2 \longrightarrow AlN + CO$
(d) $NH_3 + O_2 \longrightarrow NO + H_2O$
19.61 (a) $NH_2OH \longrightarrow N_2 + NH_3 + H_2O$
(b) $NH_3 + F_2 \longrightarrow NF_3 + NH_4F$
(c) $HNO_3 + H_2O + As_2O_3 \longrightarrow N_2O_3 + H_3AsO_4$
(d) $NH_3 + O_2 \longrightarrow N_2 + H_2O$
19.62 (a) $Ca_3(PO_4)_2 + C \longrightarrow Ca_3P_2 + CO$
(b) $P_4 + KOH + H_2O \longrightarrow PH_3 + KH_2PO_2$
(c) $OPBr_3 + Mg \longrightarrow PO + MgBr_2$
(d) $Ca_3(PO_4)_2 + SiO_2 + C$
$\longrightarrow CaSiO_3 + CO + P_4$
19.63 (a) $H_3PO_4 + PbO_2 \longrightarrow PbP_2O_7 + H_2O$
(b) $P_4 + H_2O + e^- \longrightarrow PH_3 + OH^-$
(c) $NH_3 + P_4 \longrightarrow PN_2H + H_2$
(d) $Ca_5(PO_4)_3F + H_2SO_4 + H_2O \longrightarrow$
$H_3PO_4 + CaSO_4 \cdot 2H_2O + HF$

19.64 Arsenic(III) sulfide is oxidized by acidic hydrogen peroxide solution to the arsenate ion AsO_4^{3-}. Write the chemical equation. Give the reduction and oxidation half-reactions for the reaction, and confirm that the production of arsenate is thermodynamically favored under standard conditions.

19.65 Describe the evidence that supports the view that metallic character increases down Group V. What evidence is there to support the statement that the group as a whole is more nonmetallic than Group IV?

19.66 Account for the variation of first ionization energies, electronegativities, and atomic and ionic radii down Group V. How does this variation correlate with the properties of the elements?

19.67 Suggest reasons why nitrogen occurs as N_2 molecules but white phosphorus occurs as P_4 molecules (in white phosphorus and its vapor). What is the angle between neighboring P—P bonds in the P_4 molecule? Is this value relevant to the chemical properties of white phosphorus?

19.68 Calculate the oxidation number of nitrogen in each of the following substances, and suggest (with equations) methods for transforming them from one to another: (a) NO; (b) N_2O; (c) HNO_2; (d) N_3^-.

19.69 Calculate the oxidation number of phosphorus in each of the following compounds, and suggest (with equations) methods for transforming them from one to another: (a) P_4; (b) PH_3; (c) H_3PO_4; (d) P_4O_6.

19.70 Explain the problems associated with the industrial synthesis of ammonia, and describe the chemical principles that were used to overcome them.

19.71 What problems might arise if phosphine became of such industrial importance that it had to be manufactured on a large scale? What would be a possible synthetic route to phosphine, starting from readily available sources? What could the by-products be used for?

19.72 Urea, $(NH_2)_2CO$, reacts with water to form ammonium carbonate. Write the balanced equation, and calculate the number of kilograms of ammonium carbonate that are obtained from 10 kg of urea.

19.73 Nitrous acid reacts with hydrazine in acidic solution to form hydrazoic acid (HN_3). Write the balanced chemical equation, and indicate how sodium azide may be prepared.

19.74 Lead azide, $Pb(N_3)_2$, is used as a detonator. What volume of nitrogen does 1.0 g of the solid produce when it decomposes at STP? Would 1.0 g of mercury(II) azide, which is also used as a detonator, produce a larger or smaller volume?

19.75 Sodium azide is used to inflate protective bags in automobiles. What mass of the solid is needed to provide 100 L of nitrogen at 1.5 atm pressure and 20°C?

19.76 List the oxides of nitrogen and any oxoacids of which they are the anhydrides. List the acids in order of strengths. Show with equations how, starting from ammonia, all the acids may be prepared.

19.77 Solid dinitrogen pentoxide exists as $NO_2^+NO_3^-$. Write the Lewis structures of the two ions, and predict their shapes from VSEPR theory.

19.78 Estimate the enthalpies of formation of $NF_3(g)$ and $NCl_3(g)$ from the relevant bond enthalpies, and give the main reasons why NF_3 is an exothermic compound whereas NCl_3 is endothermic.

*19.79 Demonstrate, on the basis of the reaction enthalpy for the dissociation of ammonium hydride into ammonia and hydrogen, that ammonium hydride is unlikely to be stable. For the lattice enthalpy of NH_4H, use the value for RbH (673 kJ/mol), which has ions of approximately the same size. Would entropy considerations work against the conclusion that NH_4H is unstable?

*19.80 Estimate the enthalpy of formation of solid ammonium hydroxide, using the lattice enthalpy for RbOH. Show that NH_4OH is unlikely to be stable with respect to decomposition into ammonia and water.

19.81 Zinc can reduce nitrate ions to ammonia in basic aqueous solution. Write the balanced equation. The ammonia can be passed into excess hydrochloric acid, and the remaining hydrochloric acid titrated with aqueous sodium hydroxide, so this reaction can be used to determine the concentration of nitrate ions in a solution. Calculate the nitrate concentration in a solution from the following information: volume of solution, 25.00 mL; volume of 0.250 M HCl(aq) used, 50.00 mL; volume of 0.150 M NaOH(aq) required for neutralization, 28.22 mL.

19.82 The standard free energy of formation of nitric oxide at 25°C is positive (+86.6 kJ/mol), yet (a) it can be made and (b) it does not decompose into its elements. Calculate the equilibrium constant for the decomposition, and account for these observations.

*19.83 Use bond enthalpies to estimate the enthalpy of formation of a hypothetical N_4O_6 molecule with the same structure as P_4O_6. Show that its decomposition into two N_2O_3 molecules is strongly exothermic.

19.84 Describe, with equations, the stages in the Ostwald process for the manufacture of nitric acid, and explain the chemical principles involved in each stage.

*19.85 The hydrolysis of lithium nitride with heavy water produces ND_3. Suggest a method for the production of deuterated nitric acid, $DNO_3(l)$, starting from elemental reagents, and write the chemical equations for each stage. How could you verify the identity of the product and distinguish it from $HNO_3(l)$?

19.86 Phosphorus pentabromide exists as PBr_5 molecules in the vapor, but as the ionic solid $PBr_4^+Br^-$. Write the Lewis structure for the cation, and predict its shape from VSEPR theory.

Group Comparisons

In Exercises 19.87 and 19.88, arrange the elements in order of (a) increasing electropositive character and (b) increasing reducing power.

19.87 Aluminum, gallium, indium, thallium, tin, germanium

19.88 Arsenic, antimony, bismuth, gallium, tin, indium

In Exercises 19.89 and 19.90, arrange the elements in order of increasing electronegativity.

19.89 Carbon, silicon, hydrogen, nitrogen, phosphorus

19.90 Carbon, boron, nitrogen, phosphorus, arsenic

*19.91 What evidence is there for diagonal relationships between elements in Groups III, IV, and V? Where appropriate, give equations to illustrate your answer.

*19.92 What evidence is there to support the view that the nonmetallic character of the elements increases with the group number? Give equations to illustrate your answer.

*19.93 What clues suggest the presence of the intervening d-block elements when the characteristics of the elements in Group II and Group III are compared?

*19.94 Show, with chemical equations, how differences between the properties of B, C, and N and the properties of their congeners can be explained in terms of the small size of their atoms.

This chapter continues the description of the p-block elements. Here we meet the elements of Group VI, including oxygen, the most abundant element on earth and one that occurs in numerous compounds, and the vigorously reactive halogens of Group VII. We also discuss the noble gases, a group of elements that until a few years ago were thought to form no compounds at all. One source of the halogen fluorine is the mineral fluorite, and the illustration shows the rich color that fluorite crystals emit—as fluorescence—when they are exposed to ultraviolet radiation.

THE p-BLOCK ELEMENTS: II

Group VI: oxygen and sulfur

20.1 The Group VI elements

20.2 Compounds with hydrogen

20.3 Some important compounds of sulfur

Group VII: the halogens

20.4 The Group VII elements

20.5 Halides

20.6 Halogen oxides and oxoacids

Group VIII: the noble gases

20.7 The Group VIII elements

20.8 Compounds of the noble gases

The elements in Groups VI through VIII of the *p* block are almost all nonmetals. The metallic character increases down each group, but we are now so far to the right of the periodic table that only one true metal—polonium—is found among them. A notable feature of this half of the *p* block is the sharp difference between the reactive halogens in Group VII and the nearly inert noble gases of Group VIII. The latter are almost chemically dead, and with them we seem to reach the end of chemistry (but not of this text).

GROUP VI: OXYGEN AND SULFUR

The elements of Group VI and some of their properties are listed in Table 20.1. The members of this group are collectively called the *chalcogens*, a name derived from the Greek words for "brass giver," because they are found in copper ores, and copper is a component of brass.

20.1 THE GROUP VI ELEMENTS

The two most important members of the group are oxygen and sulfur, and we deal mainly with them.

Oxygen. Oxygen is the most abundant element in the earth's crust and accounts for 23% by mass of the atmosphere. Some of our present atmospheric oxygen has been produced by the photochemical action of sunlight on water that steams up from the interior of the hot earth. Most of the rest has been produced by photosynthesis. Nearly 2×10^{10} kg of liquid oxygen is obtained each year in the United States (about 80 kg per inhabitant) by fractional distillation of liquid air. The biggest consumer is the steel industry, which needs about 1 ton of oxygen to produce 1 ton of steel. The gas is used in much smaller quantities in medicine, where it is administered to relieve strain on the heart and to act as a stimulant, and in oxyacetylene welding.

Oxygen is a colorless, tasteless, odorless gas of O_2 molecules that condenses to a pale blue liquid at $-183°C$. Although O_2 has an even number of electrons, two of them are unpaired and the molecule is

TABLE 20.1 **The Group VI elements (the chalcogens)**

Valence configuration: ns^2np^4

Z	Name	Symbol	Atomic weight, amu	Melting point, °C	Boiling point, °C	Density, g/cm³	Normal form
8	Oxygen	O	16.00	−219	−183		Colorless paramagnetic gas (O_2)
				−192	−112		Blue gas (ozone, O_3)
16	Sulfur	S	32.06	115	445	2.09	Yellow nonmetallic solid
34	Selenium	Se	78.96	220	685	4.81	Gray nonmetallic solid
52	Tellurium	Te	127.60	450	1390	6.25	Silver-white nonmetallic solid
84	Polonium	Po	210	254	960	9.40	Gray metal

paramagnetic. Molecular-orbital theory readily accounts for this unusual feature by showing that the last two electrons occupy two different π orbitals, which they can do without pairing their spins. This paramagnetism has a practical application, for the oxygen content of incubators and other life-support systems can be monitored by measuring the magnetism of the gas they contain.

Elemental oxygen also occurs as the allotrope ozone (O_3) in the "ozone layer," high in the atmosphere, where it is formed by the effect of solar radiation. Its total abundance in the atmosphere is equivalent to a layer that, at normal temperature and pressure, would cover the earth to a thickness of 3 mm. Ozone can be made in the laboratory by passing an electric discharge through oxygen. It is a blue gas that condenses at $-112°C$ to an explosive liquid resembling dark blue ink (Fig. 20.1). Its pungent smell (its name comes from the Greek word for smell) can often be detected near electrical equipment or after lightning.

Sulfur. Sulfur is widely distributed as sulfide ores, which include galena (PbS), argentite (Ag_2S), and cinnabar (HgS). It is also found as deposits of the native element (brimstone), formed by bacterial action on H_2S. Native sulfur is mined by the ingenious *Frasch process*, which makes use of sulfur's low melting point and low density. In this process (Fig. 20.2), water at about 165°C and under pressure is pumped down the outermost of three concentric pipes to melt the sulfur trapped beneath deep rock layers. Compressed air is passed down the innermost pipe to force a frothy mixture of sulfur, air, and hot water up the middle pipe.

Since sulfur is a by-product of a number of metallurgical processes (especially the extraction of copper from its sulfide ores) and must be removed from sulfur-rich petroleum, the recovery of sulfur from these sources is replacing the mining of the native element. A major source of recovered sulfur is the *Claus process*, in which some of the H_2S that occurs in oil and natural-gas wells is first oxidized to sulfur dioxide:

$$2H_2S(g) + 3O_2(g) \longrightarrow 2SO_2(g) + 2H_2O(l)$$

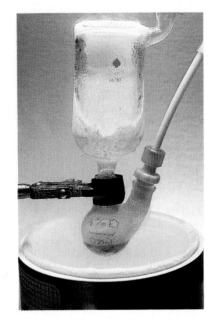

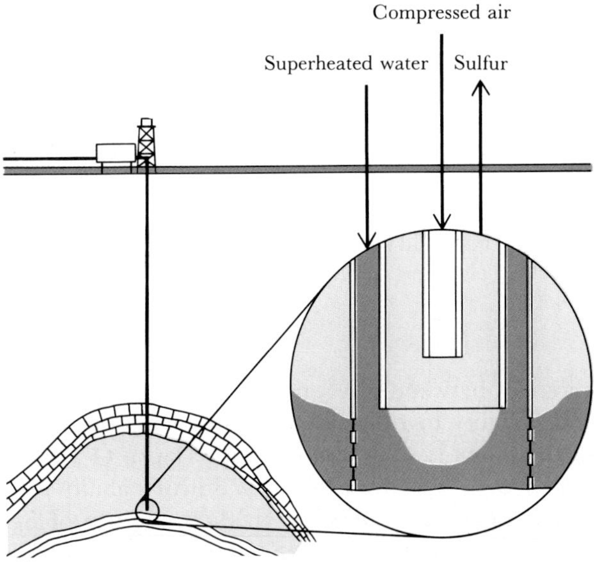

FIGURE 20.1 Liquid ozone is a dark blue, highly unstable liquid.

Compressed air

Superheated water | Sulfur

FIGURE 20.2 In the Frasch process, superheated water—water under pressure and heated to above 100°C—is pumped down the outermost of three concentric pipes; compressed air is pumped down the innermost pipe to force a mixture of sulfur, air, and hot water up the middle pipe.

FIGURE 20.3 One of the two most common forms of sulfur is this blocklike rhombic sulfur. It differs from the needlelike monoclinic sulfur in the manner in which the S_8 rings stack together.

Then the remainder of the hydrogen sulfide is oxidized by reaction with this SO_2:

$$2H_2S(g) + SO_2(g) \xrightarrow{\text{300°C, Al}_2\text{O}_3} 3S(s) + 2H_2O(g)$$

Sulfur is of major industrial importance. Most of the sulfur that is produced is used to make sulfuric acid, but an appreciable amount is used to vulcanize rubber.

Elemental sulfur is a yellow, tasteless, almost odorless, water-insoluble, nonmetallic molecular solid of crownlike S_8 rings. The most stable allotrope under normal conditions is rhombic sulfur (Fig. 20.3). The changes that sulfur undergoes as it is heated are discussed in Section 10.3 and illustrated in Fig. 10.15. Sulfur vapor has a blue tint from the S_2 molecules present in it. The latter are paramagnetic, like O_2.

Selenium and tellurium. Selenium and tellurium occur in sulfide ores; they are also recovered from the anode sludge that is formed during the electrolytic refining of copper. Both elements have several allotropes, the most stable consisting of long, zigzag chains of atoms. Although these allotropes look like silver-white metals, they are poor electrical conductors. Since the conductivity of selenium is increased in the presence of light, it is used in photoelectric devices and in photocopying machines. Selenium also occurs as a deep red solid composed of Se_8 molecules.

Trends and differences. Oxygen, sulfur, and selenium are typical nonmetals; tellurium is a nonmetal with some metallic character, and polonium is a metal. Electron affinities and electronegativities decrease down the group (Fig. 20.4), and covalent and ionic radii increase (Fig. 20.5).

▼ **EXAMPLE 20.1** Predicting the characteristics of oxygen

What can be expected as the principal chemical characteristics of oxygen?

STRATEGY We should be able to predict oxygen's main characteristics from its location in the periodic table and its position at the head of a group.

SOLUTION The O atom has a small radius and has no energetically available d orbitals. Since it is close to the top right of the periodic table, it can be expected to have a high electronegativity (it is actually the most electronegative element after fluorine). Since it is in Group VI and is electronegative, we can expect that in almost all its compounds it will have the oxidation number $6 - 8 = -2$ (the exceptions include compounds with fluorine, and the O_2^-, O_2^{2-}, and O_3^- ions). As a consequence of its small size and high electronegativity, it should have the ability to draw out the higher oxidation numbers of other elements, such as the $+6$ of sulfur in SO_4^{2-} and the $+8$ of ▲ xenon in XeO_4.

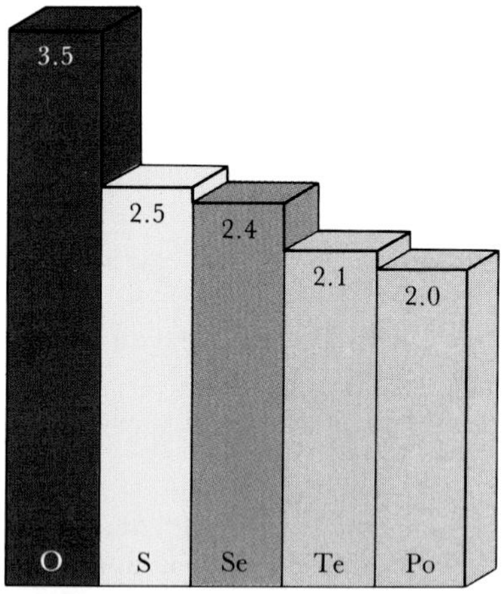

The differences between oxygen and sulfur are emphasized by the latter's striking ability to *catenate*, or form chains of atoms. Oxygen's ability to form chains is very limited, with H_2O_2, O_3, and the anions O_2^-, O_2^{2-}, and O_3^- the only examples. Sulfur's ability is much more developed. It appears, for instance, in the existence of S_8 rings and their fragments. The existence of —S—S— links in proteins is another

FIGURE 20.4 The electronegativities of the Group VI elements.

O 3.5
S 2.5
Se 2.4
Te 2.1
Po 2.0

example of catenation and an important contribution to their shapes.

Oxygen combines directly, often vigorously, and usually exothermically with all but a few elements (principally the noble gases and the noble metals). It forms compounds with all elements except the lightest noble gases (He, Ne, Ar, and Kr). The properties of oxides provide one of the chemical criteria for distinguishing between metals (which form basic oxides) and nonmetals (which form acidic oxides). Sulfur is similarly reactive, and it combines with almost all the elements except the noble gases and noble metals. A notable difference from oxygen is that sulfur does not react directly with nitrogen, but sulfur-nitrogen compounds are known.

20.2 COMPOUNDS WITH HYDROGEN

By far the most important compound of oxygen and hydrogen is water (H_2O). Two other compounds of hydrogen and a Group VI element that play an important role in chemistry and the economy are hydrogen peroxide (H_2O_2) and hydrogen sulfide (H_2S).

Water. Water is available on a huge scale worldwide, but in various states of purity. Municipal water supplies normally undergo several stages of purification. The reservoirs in which they are stored usually contain a low concentration of copper ions (added as copper sulfate) to restrict the growth of algae. The first steps in treating water taken from a reservoir are normally to pass it over activated carbon and to add chlorine to kill bacteria. Then aluminum sulfate and aluminate ions are added so that other impurities will be captured in the floc of aluminum hydroxide formed by their reaction (see Section 19.2). Solid particles are removed by filtration; if the water is to be fluoridated to prevent dental decay, H_2SiF_6 is added. Swimming-pool water is maintained in good condition by the addition of hypochlorite ions (as the sodium or calcium salt) or organic bactericides, and then kept at a pH that is comfortable for the swimmers yet optimal for the action of the additives. A pH of about 7.5 is typical. The acid HSO_4^- (added as sodium hydrogen sulfate) is used to lower the pH if necessary, and the base CO_3^{2-} (as sodium carbonate) is used to raise it.

High-purity water for special applications is obtained by distillation or *ion exchange*, the exchange of one type of ion in a solution by another. In the latter process, the water is passed through a "zeolite," an aluminosilicate with a very open structure that can capture ions such as Mg^{2+} and Ca^{2+}, often in exchange for H^+ ions. Very pure water is needed industrially for use in high-pressure boilers, where even tiny amounts of impurity can lead to the formation of sediment. Industrial boiler water is probably the purest high-tonnage chemical available, for it must have no more than 0.02 ppm of impurity (approximately 2 g of impurity in 100,000 kg of water); this is not far short of the purity of zone-refined silicon. The oxygen concentration of the water is reduced to a very low level by reduction with hydrazine, as is mentioned in Section 19.8.

By extrapolation from the boiling points of the other Group VI H_2X compounds, we would predict water to be a gas condensing at around $-100°C$, not a liquid boiling at $+100°C$ (Fig. 20.6). The anomalously

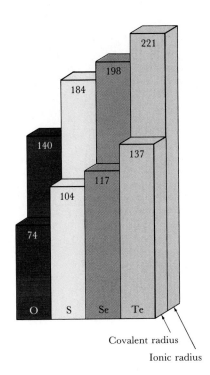

FIGURE 20.5 The covalent radii of the Group VI elements (their contribution to the lengths of covalent bonds) increase down the group, as do their ionic radii. The values shown are in picometers.

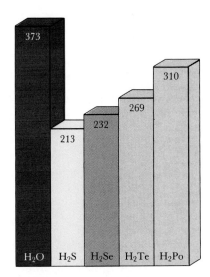

FIGURE 20.6 The normal boiling points (in kelvins) of the H_2X compounds of Group VI. Note the anomalously high value for water.

high boiling point of water (or, equivalently, its anomalously low vapor pressure) is a result of extensive hydrogen bonding between H_2O molecules. Hydrogen bonding, and the open structure that it gives in the solid, is also responsible for the unusual property that the liquid is less dense than the solid.

The important chemical properties of water include its high polarity, which together with its ability to form hydrogen bonds to anions gives it a unique ability to act as a solvent for ionic compounds. Water is also amphiprotic—able both to donate and to accept protons—and hence can act both as a Brønsted acid and as a Brønsted base:

$$CH_3COOH(aq) + H_2O(l) \rightleftharpoons CH_3CO_2^-(aq) + H_3O^+(aq)$$

$$H_2O(l) + NH_3(aq) \rightleftharpoons OH^-(aq) + NH_4^+(aq)$$

The strengths with which H_2O donates and accepts protons make it the dividing line between weak acids and weak bases.

Water is an oxidizing agent:

$$2H_2O(l) + 2e^- \longrightarrow 2OH^-(aq) + H_2(g) \qquad E = -0.42 \text{ V at pH} = 7$$

A simple example is its reaction with the alkali metals, as in

$$2Na(s) + 2H_2O(l) \longrightarrow 2NaOH(aq) + H_2(g)$$

However, unless the other reagent is a strong reducing agent, water acts as an oxidizing agent only at high temperatures, as in the reforming reaction,

$$CH_4(g) + H_2O(g) \xrightarrow{\Delta} CO(g) + 3H_2(g)$$

Water is also a very mild reducing agent, the half-reaction being

$$4H^+(aq) + O_2(g) + 4e^- \longrightarrow 2H_2O(l) \qquad E = +0.81 \text{ V at pH} = 7$$

Few substances are strong enough as oxidizing agents to remove electrons from water and bring this reaction about, but one that can is fluorine:

$$2H_2O(l) + 2F_2(g) \longrightarrow 4HF(aq) + O_2(g)$$

Water is a Lewis base, for an H_2O molecule can donate one of its lone pairs to a Lewis acid, as in the formation of complexes such as $[Fe(H_2O)_6]^{3+}$. This same property is the source of water's ability to *hydrolyze* other substances, or react with them to form a new oxygen-element bond; for example,

$$PCl_5(s) + 4H_2O(l) \longrightarrow H_3PO_4(aq) + 5HCl(aq)$$

We noted in Section 19.9 that a characteristic reaction of nonmetal halides is their reaction with water to form an oxoacid without change of oxidation number. We now see that these reactions are hydrolyses.

Hydrolyses can also occur with change of oxidation number. Such is the case in the reaction between chlorine and water, which can be pictured as a hydrolysis in which the water drives one of the Cl atoms out of the molecule:

$$Cl-Cl(g) + H_2O(l) \longrightarrow Cl-OH(aq) + HCl(aq)$$

This is a disproportionation reaction, since the oxidation number of Cl changes from 0 to +1 (in HOCl) and −1 (in HCl).

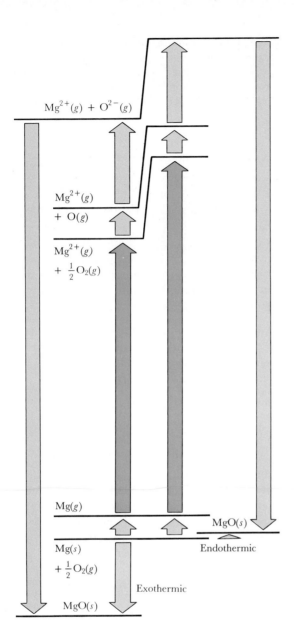

$Mg^{2+}(g) + O^{2-}(g)$

$Mg^{2+}(g)$
$+ O(g)$

$Mg^{2+}(g)$
$+ \frac{1}{2}O_2(g)$

$Mg(g)$

MgO(s)

$Mg(s)$
$+ \frac{1}{2}O_2(g)$

Endothermic

Exothermic

MgO(s)

FIGURE 20.7 The Born-Haber cycle for the formation of MgO: if the ionization energy were only a little larger (as shown on the right), the compound would be endothermic and less likely to form.

▼ EXAMPLE 20.2 Accounting for a property of oxides

Suggest a reason why ionic oxides are formed only by elements located to the left in the periodic table.

STRATEGY Thermodynamic processes are often best discussed and analyzed with the help of a Born-Haber cycle. Not only does this show the various factors that contribute to the explanation, but it also helps to identify the major contribution. Therefore, we shall construct a Born-Haber cycle for the formation of an oxide (using data from Appendix 2A) and use it to identify (if possible) the most important contributions.

SOLUTION The Born-Haber cycle for the formation of MgO is shown (on the left) in Fig. 20.7. The lattice enthalpy is large, owing to the small size

1 Hydrogen peroxide

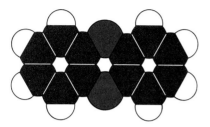

2 Anthraquinone

and high charge of the O^{2-} ion. Nevertheless, since the formation of O^{2-} is endothermic, the ionization enthalpy of the metal ion cannot be too large, for otherwise the overall energy investment required to form both the cation and the oxide anion would not be recoverable. Hence, only elements toward the left of the table (those with low ionization energies) can be expected to form ionic oxides. ▲

The conjugate base of H_2O is the hydroxide ion OH^-; the latter's conjugate base is the oxide ion O^{2-}. Since O^{2-} is such a strong Brønsted base, soluble oxides are immediately and fully protonated in water in the reaction

$$O^{2-}(aq) + H_2O(l) \longrightarrow 2OH^-(aq)$$

This proton-transfer reaction, together with the formation of ionic oxides by electropositive elements, accounts for the general observation that the oxides of metals are basic.

Hydrogen peroxide. Hydrogen peroxide, H_2O_2 (**1**), was once manufactured by the electrolysis of dilute sulfuric acid. It is now produced in a process in which anthraquinone (**2**; abbreviated A) in a hydrocarbon solvent is first made to react with hydrogen,

$$A + H_2(g) \xrightarrow{\text{Ni}} AH_2$$

and then air is passed through the product:

$$AH_2 + O_2(g) \longrightarrow A + H_2O_2$$

Hydrogen peroxide is a very pale blue liquid that is appreciably denser than water (1.44 g/mL at 25°C). In its other physical properties it is quite similar to water (its melting point is −0.4°C, and its boiling point is 152°C). Chemically, though, hydrogen peroxide and water differ greatly. The presence of the second oxygen atom makes H_2O_2 a very weak acid ($pK_{a1} = 11.75$). It can act as an oxidizing agent in both acidic and basic solutions:

$$2Fe^{2+}(aq) + H_2O_2(aq) + 2H^+(aq) \longrightarrow 2Fe^{3+}(aq) + 2H_2O(l)$$

$$2Mn^{2+}(aq) + 4H_2O_2(aq) + 4OH^-(aq) \longrightarrow$$
$$2MnO_2(s) + 6H_2O(l) + O_2(g)$$

It can also act as a reducing agent:

$$2MnO_4^-(aq) + 5H_2O_2(aq) + 6H^+(aq) \longrightarrow$$
$$2Mn^{2+}(aq) + 8H_2O(l) + 5O_2(g)$$

$$Cl_2(g) + H_2O_2(aq) + 2OH^-(aq) \longrightarrow 2Cl^-(aq) + 2H_2O(l) + O_2(g)$$

An interesting feature of the oxidation of H_2O_2 is that the O_2 so produced is sometimes in an excited state and emits light as it discards its excess energy. This process is an example of *chemiluminescence*, the emission of light by products formed in energetically excited states (Fig. 20.8) when reagents are mixed.

Hydrogen peroxide is normally sold commercially as a 30% (by mass) aqueous solution. When used as a hair bleach (in a 6% solution), it acts by oxidizing the pigments in the hair. Since it oxidizes unpleasant efflu-

FIGURE 20.8 Chemiluminescence—the emission of light as a result of a chemical reaction—occurs when hydrogen peroxide is added to a solution of the organic compound perylene. Although hydrogen peroxide itself can fluoresce, in this instance the fluorescence is emitted by the organic compound.

ents without producing any harmful by-products, H_2O_2 is becoming increasingly widely used for the control of pollution.

The conjugate base of H_2O_2 is the hydroperoxide ion HO_2^-, and that ion's conjugate base is the peroxide ion O_2^{2-}. The latter ion occurs in sodium peroxide, the product of the combustion of sodium in air. Organic peroxides are oxidizing molecular compounds that contain the link —O—O—. They include peroxyacetyl nitrate (PAN; **3**), the eye irritant in smog.

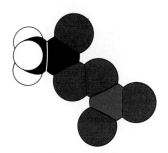

3   Peroxyacetyl nitrate

Hydrogen sulfide. The gas hydrogen sulfide (H_2S) is formed either by protonation of the sulfide ion, a Brønsted base, in a reaction such as

$$FeS(s) + 2HCl(aq) \longrightarrow FeCl_2(aq) + H_2S(g)$$

or by direct reaction of the elements at 600°C. When it is being used to precipitate sulfides in qualitative analysis, a safer and more convenient method of production is the hydrolysis of thioacetamide:

$$CH_3-\overset{\displaystyle S}{\underset{\displaystyle NH_2}{C}} + 2H_2O(l) \longrightarrow CH_3-\overset{\displaystyle O}{\underset{\displaystyle O^-}{C}} + NH_4^+(aq) + H_2S(aq)$$

All the Group VI H_2X compounds other than water are toxic gases with offensive odors. They are insidious poisons because they paralyze the olfactory nerve and soon cannot be smelled any more. Rotten eggs smell of hydrogen sulfide because egg proteins contain sulfur and give off the gas when they decompose. Another sign of the formation of sulfides in eggs is the pale green discoloration sometimes seen in cooked eggs where the white meets the yolk: this is a deposit of iron(II) sulfide.

Hydrogen sulfide dissolves in water to give a solution that, as a result of its oxidation by dissolved air, slowly becomes cloudy with small particles of sulfur. This is an example of a *colloidal dispersion*, one in which tiny particles—though large enough to contain thousands of atoms— are dispersed through the liquid. Hydrogen sulfide is a weak diprotic acid ($pK_{a1} = 6.88$, $pK_{a2} = 14.15$) and is the parent of the hydrogen sulfides (which contain the HS^- ion) and the sulfides (which contain the S^{2-} ion). The sulfides of the *s*-block elements are moderately soluble, whereas the sulfides of the heavy *p*- and *d*-block metals are generally very insoluble.

A sulfur analog of hydrogen peroxide also exists and is an example of a *polysulfane*, a molecular compound of composition $HS-S_n-SH$, where *n* can take on any of the values 0, 1, 2, . . . , 6. In these substances, sulfur shows its ability to catenate. The polysulfide ions obtained from the polysulfanes include two ions found in lapis lazuli (Fig. 20.9).

20.3   SOME IMPORTANT COMPOUNDS OF SULFUR

The common oxides and oxoacids of sulfur are summarized in Table 20.2. The most important are the dioxide and the trioxide and the corresponding sulfurous and sulfuric acids.

FIGURE 20.9 The blue stones in this ancient Egyptian ornament are lapis lazuli. This semiprecious stone is an aluminosilicate colored by S_2^- and S_3^- impurities. The blue color is due to S_3^-, and the hint of green to S_2^-.

4 Sulfur dioxide

TABLE 20.2 Some sulfur oxides and oxoacids

Oxide	Oxidation number of sulfur	Name	Oxoacid	Name
SO_3	+6	Sulfur trioxide	H_2SO_4	Sulfuric acid
	+5		$H_2S_2O_6$	Dithionic acid*
SO_2	+4	Sulfur dioxide	H_2SO_3	Sulfurous acid
	+3		$H_2S_2O_4$	Dithionous acid*
	+2†		$H_2S_2O_3$	Thiosulfuric acid

*Only the salts of these acids (dithionates and dithionites) are known.
†The average oxidation number of the two sulfur atoms.

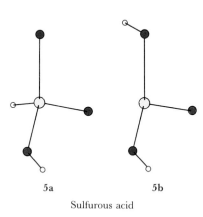

5a   5b

Sulfurous acid

Sulfur dioxide and the sulfites. Sulfur burns in air to form sulfur dioxide, SO_2 (**4**), a colorless, choking, poisonous gas. Sulfur dioxide occurs in the atmosphere as a result of volcanic activity, the combustion of sulfur-rich fuels, and the industrial roasting of sulfide ores.

Sulfur oxides in the atmosphere are referred to as SO_x (read "sox"). About 70 billion kilograms of the dioxide result from the decomposition of vegetation and from volcanic emission. In addition, 100 billion kilograms of hydrogen sulfide are formed naturally and then produce the dioxide through oxidation by atmospheric oxygen:

$$2H_2S(g) + 3O_2(g) \longrightarrow 2SO_2(g) + 2H_2O(g)$$

Industry and vehicles contribute another 150 billion kilograms of the dioxide of which about 70% comes from oil and coal combustion—mainly in electricity-generating plants. The average atmospheric concentration of SO_2 in rural areas in the northern hemisphere is about 1×10^{-6} mol/L.

Sulfur dioxide is an acidic oxide; it is the anhydride of sulfurous acid (H_2SO_3), which is the parent of the hydrogen sulfites (or bisulfites) and the sulfites:

$$SO_2(g) + H_2O(l) \longrightarrow H_2SO_3(aq)$$

The H_2SO_3 molecule is an equilibrium mixture of two molecules (**5a** and **5b**); in the former it resembles phosphorous acid, since one of the H atoms is attached directly to the S atom. These sulfurous acid molecules are also in equilibrium with SO_2 molecules, each of which is surrounded by a cage of water molecules. The evidence for this is that when the solution is cooled, crystals of composition $SO_2 \cdot xH_2O$, with x about 7, are obtained. Substances like this, in which a molecule sits in a cage, are called *clathrates* (from the Greek word for "cage"). Clathrates are also formed by other gases, including methane, carbon dioxide, and the noble gases.

In sulfur dioxide and the sulfites, sulfur has the oxidation number +4, which is intermediate in its range of −2 to +6. Hence these compounds can act as either reducing or oxidizing agents (Fig. 20.10). For instance, HSO_3^- ions oxidize HS^- ions in a reaction that produces thiosulfate ions ($S_2O_3^{2-}$):

$$4HSO_3^-(aq) + 2HS^-(aq) \longrightarrow 3S_2O_3^{2-}(aq) + 3H_2O(l)$$

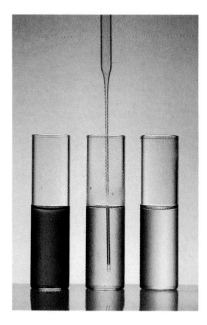

FIGURE 20.10 When SO_2 is bubbled (center) through an aqueous solution of bromine (left), it reduces the Br_2 to colorless bromide ions (right).

Sodium thiosulfate, or photographer's "hypo," can also be made by boiling aqueous sodium sulfite with sulfur, which is oxidized:

$$SO_3^{2-}(aq) + S(s) \longrightarrow S_2O_3^{2-}(aq)$$

In Chapter 3 we saw that thiosulfate ions are used to remove any residual silver halide from photographic-film emulsions by forming a soluble complex of silver.

6 Sulfur trioxide

Sulfur trioxide and the sulfates. By far the most important reaction of sulfur dioxide is its oxidation to sulfur trioxide, SO_3 (**6**):

$$2SO_2(g) + O_2(g) \longrightarrow 2SO_3(g)$$

The direct reaction is very slow, so an SO_2 molecule in the atmosphere survives for a few days before it is oxidized to SO_3. The oxidation occurs much more rapidly on the walls of buildings and in the presence of metal cations like Fe^{3+}, which may be dissolved in droplets of water. Other routes from SO_2 to SO_3 include reaction with OH, H_2O_2, or the O_3 formed by the effect of sunlight on air and water vapor. The sulfur trioxide produced by these processes reacts with atmospheric water to form dilute sulfuric acid:

7 Sulfur trioxide trimer

$$SO_3(g) + H_2O(l) \longrightarrow H_2SO_4(aq)$$

This, together with the nitric acid formed by dissolved NO_x, falls as "acid rain."

At normal temperatures sulfur trioxide is a volatile liquid (its boiling point is 45°C), composed of triangular planar SO_3 molecules. In the solid, and to some extent in the liquid, these molecules form trimers (unions of three molecules) of composition S_3O_9 (**7**), and larger combinations, as well.

Sulfuric acid is produced in the *contact process*. Sulfur is first burned in oxygen, and the resulting SO_2 is oxidized to SO_3 over V_2O_5:

$$S(g) + O_2(g) \xrightarrow{1000°C} SO_2(g)$$

$$2SO_2(g) + O_2(g) \xrightarrow{500°C,\ V_2O_5} 2SO_3(g)$$

The sulfur trioxide is then absorbed in 98% concentrated sulfuric acid to give a dense, oily liquid called "oleum." It is converted to the acid by diluting the oleum with water. Sulfuric acid is a colorless, corrosive, oily liquid that boils (and decomposes) at about 300°C. It has three chemically important properties: it is a Brønsted acid, a dehydrating agent, and an oxidizing agent (Fig. 20.11).

Sulfuric acid is a strong acid in the sense that its first ionization is almost complete in dilute aqueous solution. However, its conjugate base HSO_4^- is only a weak acid:

$$H_2SO_4(aq) + H_2O(l) \longrightarrow HSO_4^-(aq) + H_3O^+(aq)$$

$$HSO_4^-(aq) + H_2O(l) \rightleftharpoons SO_4^{2-}(aq) + H_3O^+(aq) \qquad pK_{a2} = 1.92$$

The cheapness of sulfuric acid results in its being widely used in industry, particularly for the production of fertilizers, petrochemicals, dyestuffs, and detergents. About two-thirds of the H_2SO_4 produced is used in the manufacture of phosphate fertilizers.

FIGURE 20.11 Sulfuric acid is an oxidizing agent. When some concentrated acid is poured onto solid sodium bromide, the Br^- ions are oxidized to bromine, giving a brown coloration. (This is effectively the reverse of the reaction in Fig. 20.10.)

FIGURE 20.12 When concentrated sulfuric acid is poured onto sucrose, the sucrose is dehydrated, leaving a frothy black mass of carbon.

Sulfuric acid is a powerful dehydrating agent. In a spectacular demonstration of this property, a little of the concentrated acid is poured on sucrose ($C_{12}H_{22}O_{11}$). A black, frothy mass of carbon forms (Fig. 20.12) as a result of the reaction.

$$C_{12}H_{22}O_{11}(s) \xrightarrow{\text{conc } H_2SO_4} 12C(s) + 11H_2O(l)$$

The froth is caused by CO and CO_2 that are also formed. A similar reaction is the production of carbon monoxide from formic acid:

$$HCOOH(l) \longrightarrow CO(g) + H_2O(l)$$

Sulfuric acid and water mix exothermically, and if water is added to the acid, the mixture spits liquid as the droplets vaporize explosively with the heat that is released. The proper technique is always to add the acid to the water, and never the reverse.

▼ **EXAMPLE 20.3** Predicting a reaction of sulfuric acid

What reaction can be expected between copper and concentrated sulfuric acid?

STRATEGY In general, either the hydrogen ion or an oxoacid's anion may act as an oxidizing agent. We should therefore decide (from the electrochemical series) whether copper can reduce hydrogen ions to hydrogen. If not, we should also consider whether the copper can reduce the anion of the acid.

SOLUTION Copper lies above hydrogen in the electrochemical series (see Appendix 2B), so it cannot reduce H^+ to H_2. However, in H_2SO_4 the sulfur has its highest oxidation number (+6), which suggests that the anion might be reduced. This is so, and copper is oxidized to copper(II) while the acid is reduced to sulfur dioxide:

▲ $$Cu(s) + 2H_2SO_4(aq, conc) \xrightarrow{\Delta} CuSO_4(aq) + SO_2(g) + 2H_2O(l)$$

Sulfur halides. Sulfur reacts directly with all the halogens except iodine. It ignites in fluorine and burns brightly to give sulfur hexafluoride, SF_6 **(8)**, a colorless, tasteless, odorless, nontoxic, thermally stable, and water-insoluble gas. Its stability is, in essence, due to the F atoms that surround the central S atom and protect it from attack. The ionization energy of SF_6 is very high, because any electron that is removed from the molecule must come from the highly electronegative F atoms. Even strong electric fields cannot strip off these electrons, so SF_6 is a good gas-phase electrical insulator. It is, in fact, a much better insulator than air, and it is used in switches on high-voltage power lines.

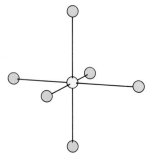

8 Sulfur hexafluoride

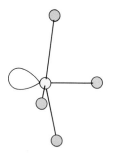

9 Sulfur tetrafluoride

▼ **EXAMPLE 20.4** Predicting the properties of a sulfur halide

A second product of the reaction of fluorine and sulfur is sulfur tetrafluoride, SF_4 **(9)**, another gas. Predict some of its properties.

STRATEGY We should consider what we know about atoms like sulfur that can expand their octet, and the consequences of that. Another clue is the general behavior of nonmetal halides (Section 19.8).

SOLUTION Because the S atom is less protected in SF_4 than in SF_6, we should anticipate that it is open to attack by a Lewis base (such as water). We know that nonmetal halides are often hydrolyzed without change of oxidation number (+4 in this case), and we should expect that behavior here. We therefore predict that SF_4 is much more reactive than the hexafluoride, and that it is hydrolyzed rapidly to sulfur dioxide and hydrogen fluoride on contact with water:

▲

$$SF_4(g) + 2H_2O(l) \longrightarrow SO_2(aq) + 4HF(aq)$$

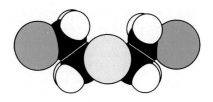

10   Mustard gas

Sulfur reacts directly with chlorine. One product is disulfur dichloride, S_2Cl_2, a yellow liquid with a disgusting smell; it is used to provide sulfur in the vulcanization of rubber. When disulfur dichloride reacts with more chlorine in the presence of iron(III) chloride as a catalyst, sulfur dichloride (SCl_2), an evil-smelling red liquid, is produced. Sulfur dichloride reacts with ethylene to give "mustard gas" **(10)**, which has been used in chemical warfare. Mustard gas causes blisters, discharge from the nose, and vomiting; it also destroys the cornea of the eye. All in all, it is easy to see why ancient civilizations associated sulfur with the underworld.

GROUP VII: THE HALOGENS

The elements of Group VII are the _halogens_; some of their properties are summarized in Table 20.3.

20.4   THE GROUP VII ELEMENTS

The rich chemical properties of the halogens can be traced to the fact that their atoms, which have the configuration ns^2np^5, need only one more electron to reach a closed-shell configuration.

TABLE 20.3   **The Group VII elements (the halogens)**

Valence configuration: ns^2np^5

Z	Name	Symbol	Atomic weight, amu	Melting point, °C	Boiling point, °C	Density, g/cm³	Normal form
9	Fluorine	F	19.00	−220	−188		Almost colorless gas
17	Chlorine	Cl	35.45	−101	−34		Yellow-green gas
35	Bromine	Br	79.91	−7	59	3.12	Red-brown liquid
53	Iodine	I	126.90	114	184	4.95	Purple-black nonmetallic solid
85	Astatine	At*	210	300	350		

*Radioactive.

FIGURE 20.13 Fluorine is produced in a large-scale adaptation of the electrolytic method that was used to isolate it originally. This is the interior of a preparation plant, showing the electrolysis cells.

Fluorine. Fluorine occurs widely in many minerals, including fluorspar, CaF_2; cryolite, Na_3AlF_6; and the fluorapatites, $Ca_5F(PO_4)_3$. It is so strong an oxidizing agent that only an anode can be made more strongly oxidizing (by increasing its positive charge); hence only in an electrolysis cell can fluorine be driven out of its compounds by removing an electron from F^- ions. The normal procedure for producing elemental fluorine is to electrolyze an anhydrous mixture of potassium fluoride and hydrogen fluoride at about 75°C, with a carbon anode.

Fluorine is a reactive, almost colorless gas of F_2 molecules. It was little used before the development of the nuclear power industry but now is produced on a large scale—at the rate of about 5 million kilograms a year in the United States (Fig. 20.13). Most of this fluorine is used to make the volatile solid UF_6, as part of the procedure for processing nuclear fuel. Much of the rest is used in the production of SF_6 for electrical equipment and in the production of fluorinated hydrocarbons such as "Teflon."

Chlorine. Chlorine is widely available as sodium chloride and is obtained by electrolysis, either of molten rock salt or of brine (Fig. 17.26). It is a pale yellow-green gas of Cl_2 molecules that condenses at −34°C. Chlorine is used in a number of industrial processes, including the manufacture of plastics, solvents, and organic chemicals in general. It is used as a bleach in the paper and textile industries and as a disinfectant in water treatment—an application that has made large-scale community living possible. The "chlorine" smell of chlorinated water comes largely from the amines (organic compounds containing the —NH_2 group) that have become chlorinated (converted to —NHCl).

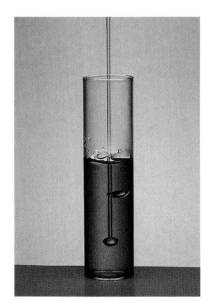

FIGURE 20.14 When chlorine is bubbled through a solution of bromide ions, it oxidizes them to bromine, which colors the solution brown. The yield is improved by bubbling air through the solution, which removes the bromine as vapor and allows the reaction to continue.

Bromine and iodine. Chlorine is also used to produce bromine from brine wells through the oxidation of Br^- ions (Fig. 20.14):

$$2Br^-(aq) + Cl_2(g) \longrightarrow Br_2(l) + 2Cl^-(aq)$$

Air is bubbled through the solution to vaporize the bromine and drive it out.

Bromine is a corrosive, red-brown, fuming liquid of Br_2 molecules with a penetrating odor (its name comes from the Greek word for "stench"). It is increasingly used in industrial chemistry because of the ease with which it can be added to and removed from organic molecules. Organic bromides are incorporated into textiles as fire retardants and are used as pesticides; inorganic bromides, particularly AgBr, are used as photographic emulsions. An application of a different kind takes advantage of the very high density of some aqueous bromide solutions. Saturated aqueous zinc bromide, for example, has a density of 2.3 g/mL and is used in the oil industry to control the escape of oil from deep wells. The tall column of solution that results when this solution is poured down a well exerts a very high pressure at its base.

Iodine occurs as iodide ions in natural brines, and as an impurity in Chile saltpeter. It was once obtained from seaweed, which contains high concentrations accumulated from seawater: 2000 kg of seaweed produce about 1 kg of iodine. The best modern source is the brine from oil wells, since the oil itself was produced by the decay of marine organisms that had accumulated the iodine while they were alive. The element is produced by oxidation of I^- ions with chlorine and is a blue-black lustrous solid of I_2 molecules. Its vapor is purple.

Iodine is only slightly soluble in water, in which it gives a brown solution; in other solvents it dissolves to give a wide variety of colors that arise from the different interactions between the I_2 molecules and the solvent (Fig. 20.15). It is much more soluble in water if I^- ions are present, because the triiodide ion I_3^- is formed:

$$I_2(aq) + I^-(aq) \longrightarrow I_3^-(aq)$$

Table 4.2 summarizes how this solution is used in redox titrations. The element itself has few direct uses, but a solution in alcohol is used as a mild oxidizing antiseptic. Iodine is an essential element—in minute quantities—for living systems; a deficiency in humans leads to goiter, a swelling of the thyroid gland in the neck. Iodides are added to table salt (to produce "iodized salt") to avoid this deficiency.

The interhalogens. The halogens form compounds among themselves. These *interhalogens* have the formulas XX', XX'_3, XX'_5, and XX'_7, where X is the heavier (and larger) of the two halogens X and X'. Only some of the possible combinations have been prepared (Table 20.4), partly

FIGURE 20.15 The color of an iodine solution depends strongly on the identity of the solvent: from left to right, water, carbon tetrachloride, benzene, ethanol.

TABLE 20.4 The known interhalogens

XF_n				XCl_n		XBr_n
ClF Colorless gas	ClF$_3$ Colorless gas	ClF$_5$ Colorless gas				
BrF Pale brown gas	BrF$_3$ Pale yellow liquid	BrF$_5$ Colorless liquid		BrCl Red-brown gas		
IF Unstable	IF$_3$ Yellow solid	IF$_5$ Colorless liquid	IF$_7$ Colorless gas	ICl Red solid	I$_2$Cl$_6$ Yellow solid	IBr Black solid

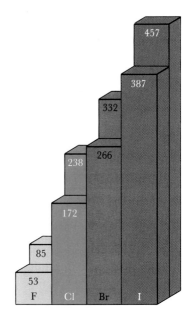

FIGURE 20.16 The melting and boiling points of the halogens. The values shown are in kelvins.

for steric reasons. Thus, whereas seven F atoms can fit around an I atom to give IF_7, seven Cl atoms cannot; hence ICl_7 is unknown.

All the interhalogens are prepared by the direct reaction of two halogens, the product formed being determined by the proportions of reactants used. In general, the interhalogens have physical properties that are intermediate between those of their parent halogens. Their chemical properties are dominated by the decreasing X—X′ bond enthalpy as X′ becomes heavier. The fluorides of the heavier halogens are all very reactive. Bromine trifluoride is so reactive that even asbestos burns in it.

Trends and differences. The halogens are a closely related family of elements that (apart from some characteristics of fluorine) show a smooth variation of properties. Their melting and boiling points and enthalpies of melting and vaporization all increase steadily down the group (Fig. 20.16), as we would expect when London forces between molecules are dominant. Electronegativities decrease down the group (Fig. 20.17), and covalent and ionic radii increase smoothly (Fig. 20.18).

▼ **EXAMPLE 20.5** **Predicting the relative oxidizing abilities of the halogens**

Predict the trend in oxidizing character of the halogens in aqueous solution.

STRATEGY Trends in oxidizing ability can be predicted by examining the standard reduction potentials of the X_2/X^- couples, where X is a halogen. The more positive the value of $E°$, the more strongly oxidizing will be the X_2. Data are given in Appendix 2B.

SOLUTION The standard reduction potentials of the X_2/X^- couples change from the highly positive oxidizing value for fluorine (+2.87 V) to the only slightly positive value for iodine (+0.54 V). Hence, whereas fluorine, chlorine, and bromine are good oxidizing agents, iodine is only weakly oxidizing.

EXERCISE Which halide ion is the strongest reducing agent in aqueous solution?

[*Answer*: I^-]

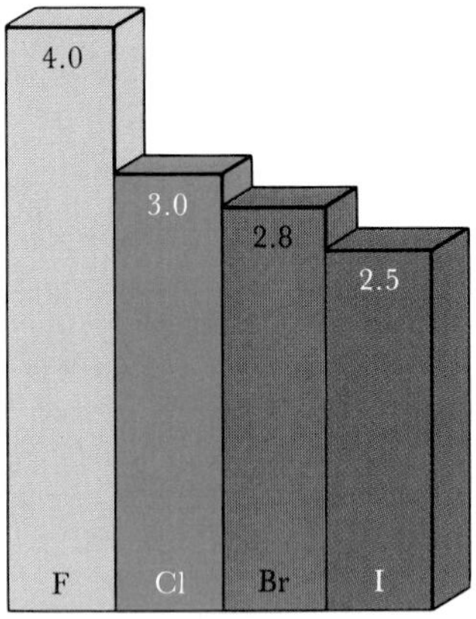

▲

A pattern is also exhibited by the compounds of the halogens. All the halogens form hydrogen halides that are Brønsted acids: all of these compounds except HF are strong acids in water, and the strengths increase from HCl to HI largely as a result of the weakening H—X bond and of decreasing polarity down the group. They form an extensive series of oxoacids with strengths that increase with the number of oxygen atoms and with the electronegativity of the halogen atom (as we saw in Section 14.4). Perchloric acid, for example, with four oxygen atoms and a highly electronegative halogen atom, is a very strong Brønsted acid in water. Moreover, all the halogens except fluorine exhibit a variety of oxidation numbers, typically (but not only) −1, 0, 1, 3, 5, and 7.

Fluorine, the first member of the group, has a number of peculiarities that stem from its high electronegativity, small size, and lack of

FIGURE 20.17 The electronegativities of the halogens.

available *d* orbitals. It is the most electronegative element in the periodic table and has a negative oxidation number (−1) in all its compounds. Its high electronegativity and small size help it to bring out high oxidation numbers in other elements. (Its smallness helps in this, for it allows several F atoms to fit around a central atom, as in IF_7.) Oxygen, however, is sometimes better in this role than fluorine, because an O atom increases the oxidation number of the other element by 2 rather than by 1; hence fewer O atoms are needed to produce the same oxidation number. In XeO_4, for example, Xe has the oxidation number +8, but XeF_8 is unknown.

Because the fluoride ion is so small, the lattice enthalpies of its ionic compounds tend to be high (Table 8.1). One result is the lower solubility of fluorides as compared with other halides (although there are some exceptions, including AgF). This is one of the reasons why the oceans are salty with chlorides rather than fluorides, even though fluorine is more abundant than chlorine in the earth's crust.

Fluorine also shows thermodynamic and kinetic differences from its congeners. These can be traced to the very low bond enthalpy of F_2 and the considerable strengths of the bonds fluorine forms with other elements. A large change in enthalpy occurs when a weak F—F bond is replaced by a strong bond between fluorine and another element; as a result, fluorine's reactions are strongly exothermic, so there is a strong thermodynamic tendency for them to proceed. Moreover, the small amount of energy needed to produce F atoms from F_2 molecules results in low activation energies, so the reactions are also fast.

Covalent fluorides are usually more volatile than the corresponding covalent chlorides because the former have fewer electrons, and so the London forces between them are weaker. However, there are some striking exceptions; as an example, hydrogen fluoride is less volatile than hydrogen chloride. This is explained by fluorine's ability to take part in hydrogen bonding, with the result that the HF molecules stick together strongly in the liquid.

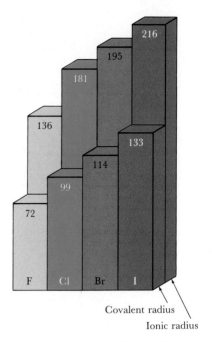

FIGURE 20.18 The covalent and ionic radii of the halogens. The values shown are in picometers.

20.5 HALIDES

The hydrogen halides HX can be prepared by direct reaction of the elements:

$$H_2(g) + X_2(g) \longrightarrow 2HX(g)$$

Fluorine reacts explosively in a radical chain reaction as soon as the gases are mixed. A mixture of hydrogen and chlorine explodes when exposed to light. Bromine and iodine react with hydrogen much more slowly. A less hazardous laboratory source of the hydrogen halides is the action of a nonvolatile acid on a metal halide, as in

$$CaF_2(s) + 2H_2SO_4(aq,\ conc) \longrightarrow Ca(HSO_4)_2(aq) + 2HF(g)$$

after which the product is distilled. Since Br^- and I^- are oxidized by sulfuric acid, phosphoric acid is used in the preparation of HBr and HI:

$$KI(s) + H_3PO_4(aq) \xrightarrow{\Delta} KH_2PO_4(aq) + HI(g)$$

TABLE 20.5 The hydrogen halides

	HF	HCl	HBr	HI
Molecular weight, amu	20.01	36.46	80.92	127.91
Melting point, °C	−83	−114	−87	−51
Normal boiling point, °C	19	−85	−67	−35
pK_a at 25°C	3.45	Strong	Strong	Strong
Bond enthalpy, kJ/mol	574	428	363	295
Bond length, pm	92	127	141	161

All the hydrogen halides are colorless, pungent gases (Table 20.5), but hydrogen fluoride is a liquid at temperatures below 19°C. Its lower volatility is a sign of extensive hydrogen bonding, and short zigzag chains of hydrogen-bonded molecules, up to about $(HF)_5$, survive to some extent in the vapor. All the hydrogen halides dissolve in water to give acidic solutions. All except hydrofluoric acid are strong acids, with hydroiodic acid the strongest.

Hydrofluoric acid has the distinctive property of reacting with glass and silica. The reaction is the attack of the strong Lewis base F^- on the Si atoms, replacing the O atoms:

$$6HF(aq) + SiO_2(s) \longrightarrow H_2SiF_6(aq) + 2H_2O(l)$$

This reaction is used in glass etching (Fig. 19.12). The interiors of lamp bulbs are frosted with a solution of hydrofluoric acid and ammonium fluoride in a similar reaction (Fig. 20.19).

Anhydrous metal halides may be formed by direct reaction:

$$Cu(s) + Cl_2(g) \longrightarrow CuCl_2(s)$$

They are also formed when the halogen reacts with an oxide of the metal in the presence of a reducing agent, as in

FIGURE 20.19 When a mixture of sulfuric acid and sodium fluoride is swirled inside a flask (a), the reaction with the silica in the glass frosts its surface (b).

(a)   (b)

$$Cr_2O_3(s) + 3C(s) + 3Cl_2(g) \xrightarrow{\Delta} 2CrCl_3(s) + 3CO(g)$$

Halides—not only anhydrous halides—are also formed by neutralization and by precipitation. An example of the latter is

$$BaCl_2(aq) + 2KF(aq) \longrightarrow BaF_2(s) + 2KCl(aq)$$

20.6 HALOGEN OXIDES AND OXOACIDS

The halogen oxoacids are listed in Table 20.6. As in the case of the oxides of nitrogen, it is helpful to keep track of the oxidation number of the halogen in these compounds. By doing so we can make informed estimates of the relative strengths of the acids as proton donors and, since they are *oxo*acids, of their oxidizing strengths.

▼ **EXAMPLE 20.6** **Predicting trends in oxoacid strength**

Predict the relative strengths of the oxoacids of chlorine in water.

STRATEGY The relevant guideline was discussed in Section 14.4: the higher the oxidation number, the stronger the acid.

SOLUTION The oxidation numbers of chlorine in the oxoacids are as follows:

$$\overset{+1}{HClO} \quad \overset{+3}{HClO_2} \quad \overset{+5}{HClO_3} \quad \overset{+7}{HClO_4}$$

Hence, we predict (and observe) the order of strength

$$HClO < HClO_2 < HClO_3 < HClO_4$$

EXERCISE What is likely to be the typical reaction for a halogen oxoacid with an oxidation number that is intermediate in the halogen's range? Write the equation for the reaction of ClO^- ions.

▲ [*Answer*: Disproportionation; $3ClO^-(aq) \longrightarrow ClO_3^-(aq) + 2Cl^-(aq)$]

TABLE 20.6 **The halogen oxoacids**

Oxidation number	General formula*	General acid name	General ion formula*	General ion name	Known examples	pK_a
7	HXO_4	Perhalic acid	XO_4^-	Perhalate	$HClO_4$ $HBrO_4$ HIO_4	Strong Strong 1.64
5	HXO_3	Halic acid	XO_3^-	Halate	$HClO_3$ $HBrO_3$ HIO_3	Strong Strong 0.77
3	HXO_2	Halous acid	XO_2^-	Halite	$HClO_2$ $HBrO_2$	2.00 Unstable
1	HXO	Hypohalous acid	XO^-	Hypohalite	HFO $HClO$ $HBrO$ HIO	Unstable 7.53 8.69 10.64

*X stands for the halogen atom.

Hypohalous acids. The hypohalous acids HOX are prepared by direct reaction of the halogen on water:

$$Cl_2(g) + H_2O(aq) \longrightarrow HOCl(aq) + HCl(aq)$$

Hypofluorous acid is so unstable that it survives only at the freezing point of water; at higher temperatures, fluorine oxidizes water to oxygen:

$$2F_2(g) + 2H_2O(l) \longrightarrow 4HF(aq) + O_2(g)$$

Hypohalite ions (OX^-) are formed when the aqueous solution of a base is used in place of water, as in the preparation of sodium hypochlorite by the electrolysis of brine:

$$Cl^-(aq) + 2OH^-(aq) \longrightarrow ClO^-(aq) + H_2O(l) + 2e^-$$

Hypochlorites are useful for their ability to oxidize organic material. This is the reason for the use of calcium hypochlorite to purify swimming-pool water, and of aqueous alkaline sodium hypochlorite as household bleach. The ClO^- ion is a weak base:

$$ClO^-(aq) + H_2O(l) \rightleftharpoons HOCl(aq) + OH^-(aq)$$

Chlorates. Chlorate ions (ClO_3^-), in which chlorine has the high oxidation number of +5, are formed when chlorine reacts with hot aqueous alkali:

$$3Cl_2(g) + 6OH^-(aq) \xrightarrow{\Delta} ClO_3^-(aq) + 5Cl^-(aq) + 3H_2O(l)$$

They decompose when heated, to an extent that depends on whether or not a catalyst is present:

$$4KClO_3(l) \xrightarrow{\Delta} 3KClO_4(s) + KCl(s)$$

$$2KClO_3(l) \xrightarrow{\Delta,\ MnO_2} 2KCl(s) + 3O_2(g)$$

The latter reaction is a convenient laboratory source of oxygen.

Chlorates have a number of uses that stem from their oxidizing ability. Potassium chlorate is used as an oxygen supply in fireworks and in safety matches. The heads of matches consist of a paste of potassium chlorate, antimony sulfide, sulfur, and powdered glass; the striking strip contains red phosphorus. Potassium chlorate is used, rather than the cheaper sodium chlorate, because it is less hygroscopic (water-absorbing). Sodium chlorate is widely used as a weed killer and defoliant in agriculture. It has also been used in warfare. The principal use of sodium chlorate, however, is as a source of chlorine dioxide (ClO_2). The chlorine in ClO_2 has oxidation number +4, so a chlorate must be reduced to form it. Sulfur dioxide is a convenient reducing agent:

$$2NaClO_3(aq) + SO_2(g) + H_2SO_4(aq,\ dilute) \longrightarrow$$
$$2NaHSO_4(aq) + 2ClO_2(g)$$

Chlorine dioxide has an odd number of electrons and is a paramagnetic yellow-green gas. It is used to bleach paper pulp; it does so by oxidizing various pigments in the pulp without degrading the wood fibers.

FIGURE 20.20 The booster rockets of the space shuttle are fueled by a mixture of ammonium perchlorate and aluminum powder. Here they are being filled with the mixture.

Perchlorates. Perchlorate ions are prepared by electrolytic oxidation of aqueous solution of chlorates:

$$ClO_3^-(aq) + H_2O(l) \longrightarrow ClO_4^-(aq) + 2H^+(aq) + 2e^-$$

Perchloric acid, $HClO_4$, is prepared by the action of concentrated hydrochloric acid on sodium perchlorate, followed by distillation. It is a colorless liquid and the strongest of all common acids. Since chlorine has its highest oxidation number of $+7$ in the perchlorates, they are powerful oxidizing agents; contact between perchloric acid and even a small amount of organic material can result in a dangerous explosion. One spectacular example of their oxidizing ability under controlled conditions is the use of a mixture of ammonium perchlorate and aluminum powder in the booster rockets of the space shuttle (Fig. 20.20). The ammonium ions and the aluminum are the "fuel," and the perchlorate ions are the source of oxygen.

GROUP VIII: THE NOBLE GASES

The elements of Group VIII, the noble gases, are listed with some of their properties in Table 20.7. Until their first compounds were prepared in 1962, their closed-shell electron configurations were taken to indicate that these elements are chemically inert; they are still sometimes incorrectly called the "inert gases."

20.7 THE GROUP VIII ELEMENTS

All the Group VIII elements occur in the atmosphere as monatomic gases; the most common is argon, which is more abundant than carbon dioxide. The noble gases other than helium and radon (for which there are richer sources) are obtained by the fractional distillation of liquid air.

TABLE 20.7 The Group VIII elements (the noble gases)

Valence configuration: ns^2np^6
Normal form: colorless monatomic gases

Z	Name	Symbol	Atomic weight, amu	Melting point, °C	Boiling point, °C
2	Helium	He	4.00		−269 (4.2 K)
10	Neon	Ne	20.18	−249	−246
18	Argon	Ar	39.95	−189	−186
36	Krypton	Kr	83.80	−157	−153
54	Xenon	Xe	131.30	−112	−108
86	Radon*	Rn	222	−71	−62

*Radon is radioactive.

Helium. Helium, the most abundant element in the universe after hydrogen, is rare on earth because its atoms are so light that they reach high average speeds and escape from the atmosphere. It is found mixed with natural gases trapped under rock formations in some locations (notably Texas), where it has collected as a result of the emission of α particles by radioactive elements. An α particle is a helium nucleus (He^{2+}), and an atom of the element forms when an α particle picks up two electrons from its surroundings:

$$He^{2+} + 2e^- \longrightarrow He$$

Helium gas is twice as dense as hydrogen under the same conditions; however, because it is nonflammable it is used to provide buoyancy in airships. The element has the lowest boiling point of any substance (4 K), and it does not freeze to a solid at any temperature unless pressure is applied to hold the light, mobile atoms together. These properties make helium useful in the study of matter at very low temperatures—the field known as *cryogenics* (from the Greek word for "frost"). Below 2 K, liquid helium shows the remarkable property of *superfluidity*, the ability to flow without viscosity. The phase diagram in Fig. 20.21 shows the regions of pressure and temperature at which the different phases are stable. Helium is the only substance known to have more than one liquid phase.

The other noble gases. Neon, which emits a red glow when an electric current flows through it, is widely used in advertising signs (Fig. 20.22). Argon is used to provide an inert atmosphere for welding (to prevent oxidation) and to fill some types of light bulbs, where its function is to conduct heat away from the filament. Krypton gives an intense white

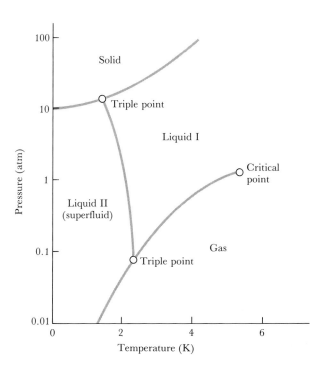

FIGURE 20.21 The phase diagram for helium.

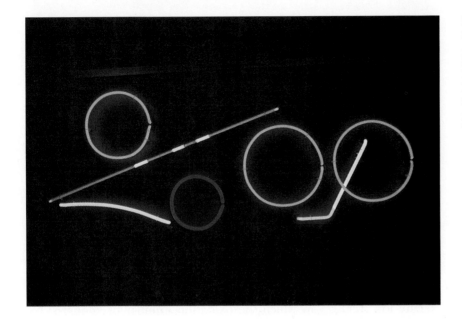

light when a current is passed through it, and so is used in airport runway lighting. As it is produced by nuclear fission, its atmospheric abundance is one measure of worldwide nuclear activity. Xenon is used in photographic flashbulbs because an electric discharge through it—a miniature lightning flash—gives an intense white light; it is also being investigated as a nonpoisonous anesthetic.

The radioactive gas radon seeps out of the ground as a product of radioactive processes deep in the earth. There is now some concern that its accumulation in buildings can lead to dangerously high levels of radiation.

20.8 COMPOUNDS OF THE NOBLE GASES

The ionization energies of the noble gases are relatively high, but they decrease down the group; krypton's is low enough for electrons to be lost to very electronegative elements, especially fluorine. No compounds of helium, neon, and argon are known (except under very special conditions). Radon is known to react with fluorine, but its radioactivity makes it so dangerous to handle that chemists have not been inclined to study its compounds. Since krypton forms only one known neutral molecule, KrF_2, this leaves xenon as the noble gas with the richest chemistry. It forms several compounds with fluorine and oxygen, and compounds with Xe—N and Xe—C bonds have been reported. In 1988 the first compound with a Kr—N bond was reported, but it is stable only below $-50°C$.

Xenon fluorides. The starting point for the preparation of xenon compounds is the production of xenon difluoride (XeF_2) and xenon tetrafluoride (XeF_4) by heating a mixture of the elements to 300°C to 400°C. At higher pressures, fluorination produces xenon hexafluoride (XeF_6). All three fluorides are crystalline solids. In the gas phase, all are covalent molecular compounds. Solid xenon hexafluoride, however, is ionic, its complex structure consisting of XeF_5^+ cations bridged by F^- anions.

The xenon fluorides can be used as powerful *fluorinating agents*, reagents for attaching fluorine atoms to other substances. The tetrafluoride will even fluorinate platinum:

$$Pt(s) + XeF_4(s) \longrightarrow Xe(g) + PtF_4(s)$$

Xenon oxides. The xenon fluorides are used to prepare xenon oxides and oxoacids and, in a series of disproportionations, to bring the oxidation number of xenon up to +8. First, xenon tetrafluoride is hydrolyzed to xenon trioxide (XeO_3):

$$6XeF_4(s) + 12H_2O(l) \longrightarrow$$
$$2XeO_3(aq) + 4Xe(g) + 3O_2(g) + 24HF(aq)$$

The trioxide is the anhydride of xenic acid (H_2XeO_4). Xenon trioxide reacts with aqueous alkali to form the hydrogen xenate ion $HXeO_4^-$, which slowly disproportionates into xenon and the octahedral perxenate ion XeO_6^{4-}, in which the oxidation number of xenon is +8:

$$2HXeO_4^-(aq) + 2OH^-(aq) \longrightarrow$$
$$XeO_6^{4-}(aq) + Xe(g) + O_2(g) + 2H_2O(l)$$

Perxenate solutions are yellow and, as expected for compounds of elements with such a high oxidation number, very powerful oxidizing agents.

When barium perxenate is treated with sulfuric acid, it is dehydrated to its anhydride, xenon tetroxide (XeO_4). With this compound, which is an explosively unstable gas, our journey through the p block of the periodic table comes to an end with a bang.

SUMMARY

20.1 The members of Group VI of the periodic table are called the **chalcogens.** Oxygen occurs mainly as O_2 but also as its allotrope **ozone.** It is a highly electronegative element that brings out the high oxidation numbers of elements. Sulfur is mined by the **Frasch process** or recovered by the **Claus process,** the oxidation of H_2S by SO_2. It is less electronegative than oxygen but more able to **catenate,** or form chains. Oxygen and sulfur react directly with almost all other elements, but sulfur does not react with nitrogen.

20.2 The most important compound of oxygen is **water,** which is amphiprotic (both a Brønsted acid and a Brønsted base), a Lewis base, an oxidizing agent, and a weak reducing agent. It is the parent of the **oxides** and **hydroxides.** Oxides of metals are basic; those of non-metals are acidic. **Hydrogen peroxide,** a very weak acid,

is also an oxidizing agent and a reducing agent and the parent of the **peroxides.** The sulfur analog of water is **hydrogen sulfide,** the parent acid of the **sulfides.** Sulfur forms a series of **polysulfanes,** of formula H_2S_n.

20.3 Important oxides of sulfur include **sulfur dioxide** and **sulfur trioxide,** the anhydrides of **sulfurous acid** and **sulfuric acid,** respectively. Sulfuric acid is made by the **contact process** and is a Brønsted acid, a dehydrating agent, and an oxidizing agent. Sulfur reacts with the halogens to form the **sulfur halides,** which include the very stable **sulfur hexafluoride.**

20.4 The members of Group VII are called the **halogens.** They include **fluorine,** the most electronegative element of all. The halogens are characterized by their smoothly varying properties. The character of an **inter-**

halogen, a binary compound of two different halogens, is generally intermediate between that of its component elements. Elements higher in the group have high electronegativities and are strongly oxidizing. The iodide ion is a reducing agent. Fluorine is highly reactive, largely because of its low bond enthalpy and the strengths of the bonds it forms with other elements.

20.5 and 20.6 All the halogens form **hydrogen halides** that dissolve in water to give the **hydrohalic acids;** except for hydrofluoric acid, these acids are all strong acids in water. The halogens form an important series of oxi-dizing oxoacids and oxoanions, of which the most important are **hypochlorous, chloric,** and **perchloric acids.**

20.7 and 20.8 The **noble gases** of Group VIII all have closed-shell electron configurations. Only xenon forms an extensive series of compounds, of which the most important are the fluorides, oxides, and oxoacids. The fluorides of xenon are strong fluorinating agents. They are hydrolyzed to the oxides, which are strong oxidizing agents.

EXERCISES

In Exercises 20.1 to 20.4, identify the group and period to which each element belongs.

20.1 (a) O; (b) Br; (c) Sb; (d) Kr.
20.2 (a) F; (b) Te; (c) At; (d) Po.

20.3 (a) Ga; (b) As; (c) I; (d) Rn.
20.4 (a) Tl; (b) Se; (c) Xe; (d) Bi.

In Exercises 20.5 to 20.8, write the ground-state electron configuration of the atoms of each element.

20.5 (a) He; (b) O; (c) S; (d) Br.
20.6 (a) F; (b) Ne; (c) Cl; (d) Se.

20.7 (a) Xe; (b) Po.
20.8 (a) Te; (b) Ar.

In Exercises 20.9 and 20.10, arrange the elements in the order of increasing oxidizing power in water.

20.9 Iodine, bromine, selenium
20.10 Chlorine, bromine, tellurium

In Exercises 20.11 to 20.14, arrange the elements in the order of increasing electronegativity.

20.11 Chlorine, oxygen, bromine
20.12 Sulfur, bromine, polonium

20.13 Hydrogen, arsenic, sulfur, fluorine
20.14 Hydrogen, carbon, phosphorus, selenium

In Exercises 20.15 and 20.16, outline the source and method of production of each element.

20.15 (a) Oxygen; (b) chlorine; (c) argon.
20.16 (a) Sulfur; (b) fluorine; (c) helium.

In Exercises 20.17 to 20.20, balance each equation and classify its reaction as Lewis, Brønsted, or redox.

20.17 (a) $KClO_3 \longrightarrow KCl + O_2$
(b) $O_3 + H^+ + e^- \longrightarrow O_2 + H_2O$
(c) $OF_2 + OH^- \longrightarrow O_2 + F^- + H_2O$
(d) $H_2S + O_2 \longrightarrow SO_2 + H_2O$
20.18 (a) $O_3 + KOH \longrightarrow KO_3 + O_2 + H_2O$
(b) $O_3 + H_2O + e^- \longrightarrow O_2 + OH^-$
(c) $F_2 + NaOH \longrightarrow OF_2 + NaF + H_2O$
(d) $H_2S + O_2F_2 \longrightarrow SF_6 + HF + O_2$
20.19 (a) $S_2F_2 + H_2O \longrightarrow S_8 + HF + H_2S_2O_6$
(b) $S_2O_3^{2-} + Cl_2 + H_2O \longrightarrow HSO_4^- + H^+ + Cl^-$
(c) $I_4O_9 \longrightarrow I_2O_5 + I_2 + O_2$
(d) $I_2O_5 + BrF_5 \longrightarrow FBrO_2 + IF_5$
20.20 (a) $SCl_2 + NaF \longrightarrow S_2Cl_2 + SF_4 + NaCl$
(b) $SO_3^{2-} + HCO_2^- \longrightarrow$
$$S_2O_3^{2-} + C_2O_4^{2-} + OH^- + H_2O$$
(c) $CO + I_2O_5 \longrightarrow I_2 + CO_2$
(d) $XeF_6 + OH^- \longrightarrow$
$$XeO_6^{4-} + Xe + O_2 + F^- + H_2O$$

20.21 What properties support the statement that nonmetallic character increases toward the top right of the periodic table?

20.22 What features of the p-block elements support the view that the first member of each group is distinct from its congeners? Is there much evidence for diagonal relationships in the second (right) half of the p block?

Group VI

20.23 Give the equations for two methods for producing oxygen in the laboratory.

20.24 What atomic characteristics account for the fact that oxygen is a component of the atmosphere but sulfur is found underground?

In Exercises 20.25 to 20.28, give the chemical formula of each substance.

20.25 (a) Sulfuric acid; (b) calcium sulfite; (c) ozone; (d) barium peroxide.

20.26 (a) Sulfurous acid; (b) sodium hydrogen sulfite; (c) the superoxide ion; (d) hydrogen telluride.

20.27 (a) Sodium thiosulfate; (b) sulfur dichloride; (c) sodium tellurate; (d) galena.

20.28 (a) Sodium selenate; (b) disulfur dichloride; (c) sulfur hexafluoride; (d) cinnabar.

20.29 In what ways is oxygen different from its congeners, both physically and chemically? In what ways does oxygen resemble sulfur?

20.30 Describe and explain the changes that occur when (a) oxygen is cooled and (b) sulfur is heated.

20.31 Compare and contrast the hydrogen compounds of the Group VI elements. To what extent do they reflect differences in the electronegativities of the elements?

20.32 How would you establish that the formula of water is H_2O and that it is a molecular compound? Summarize its chemical properties, giving two examples (with equations) of each aspect of its character.

20.33 Summarize, in the form of a labeled periodic table, the acid and base character of the oxides of the elements. Why are the oxides of metals basic, but the oxides of nonmetals acidic?

20.34 Calculate the enthalpy change accompanying the formation of an oxide ion in the half-reaction $O(g) + 2e^- \rightarrow O^{2-}(g)$, given the following data regarding barium oxide:

  Enthalpy of atomization of Ba(s): +180 kJ/mol
  Ionization enthalpies of Ba(g): +689, +971 kJ/mol
  Dissociation enthalpy of $O_2(g)$: +249 kJ/mol
  Lattice enthalpy of BaO(s): +3152 kJ/mol
  Formation enthalpy of BaO(s): −554 kJ/mol

20.35 Calculate the enthalpy of formation of a peroxide ion in the half-reaction $O_2(g) + 2e^- \rightarrow O_2^{2-}(g)$. Use the data given in Exercise 20.34 and the following data for barium peroxide: enthalpy of formation −634 kJ/mol; lattice enthalpy = 3027 kJ/mol.

20.36 Would you expect the formation of ozone from oxygen to be favored by high temperatures or low temperatures? State, giving reasons, whether the reaction entropy is likely to be positive or negative.

20.37 When lead(II) sulfide is treated with hydrogen peroxide, the possible products are either (a) lead(IV) oxide and sulfur dioxide or (b) lead(II) sulfate. In terms of the reaction free energy, which product or products are more likely?

20.38 Thermodynamic arguments are useful for discussing whether oxide and sulfide ores can be reduced to the metal with coke. The graphs in Fig. 20.23 show

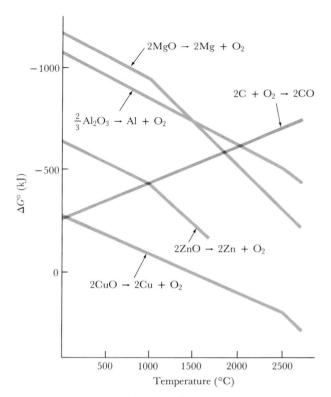

FIGURE 20.23 This "Ellingham diagram" shows how the free energy of individual redox reactions varies with temperature.

the variation of the free energies of formation of four metal oxides and that of carbon monoxide. Given that the reaction $MO + C \rightarrow M + CO$ (where M means metal) is spontaneous if its free energy is negative, decide whether carbon can be used to reduce (a) ZnO; (b) Al_2O_3; (c) HgO; (d) MgO. Where it can be so used, state the minimum temperature at which the reduction can take place.

20.39 When the enthalpy of vaporization of water is divided by its boiling point (on the Kelvin scale), the result is 110 J/K · mol. For hydrogen sulfide the same calculation gives 88 J/K · mol. What is the significance of these values, and why is one larger than the other?

20.40 The O=O and S=S bond enthalpies are 498 kJ/mol and 429 kJ/mol, respectively; the O—O and S—S bond enthalpies are 144 kJ/mol and 266 kJ/mol, respectively. Show that these values indicate that chains of sulfur atoms are thermodynamically stable (ignoring entropy effects) with respect to decomposition into S_2 molecules, but that chains of oxygen atoms are unstable.

20.41 Describe the production of the sulfur oxides and oxoacids, and outline their contribution to the problem of acid rain.

20.42 If you were titrating 0.10 M $H_2S(aq)$ with 0.10 M NaOH(aq), at what pH would you have a 0.10 M NaHS(aq) solution?

20.43 Describe and explain the variation in the acid strengths of H_2S, H_2SO_3, and H_2SO_4.

20.44 Outline the methods for the preparation of sulfides, and describe the properties of these compounds. What are the consequences and applications of the very low solubilities of heavy-metal sulfides?

20.45 Do the properties of sulfides give any clue to the metallic or nonmetallic character of elements?

20.46 "Fool's gold" is FeS_2 and is so called because it looks like gold. What physical and chemical tests would distinguish it from gold?

In Exercises 20.47 to 20.50, write the Lewis structure for, and predict the shape of, each molecule or polyatomic ion.

20.47 (a) O_3; (b) SO_3; (c) SF_4.
20.48 (a) SO_2; (b) SO_4^{2-}; (c) SeF_6.
20.49 (a) H_2Te; (b) SeO_3^{2-}.
20.50 (a) H_2Se; (b) TeO_4^{2-}.

20.51 Explain how you could distinguish a solution of sodium sulfite from a solution of sodium sulfate.

20.52 Calculate the volume of pure sulfuric acid (density 1.84 g/mL) that can be produced from 10 kg of sulfur.

Group VII

In Exercises 20.53 to 20.56, name each compound.

20.53 (a) HBr; (b) IBr; (c) ClO_2; (c) $NaIO_3$.
20.54 (a) HI; (b) IF_3; (c) HClO; (d) $NaClO_2$.
20.55 (a) KI_3; (b) Cl_2O_7; (c) $HClO_2$; (d) $HClO_4$.
20.56 (a) BrF_4^-; (b) $HBrO_3$; (c) NH_4ClO_4; (d) OF_2.

In Exercises 20.57 to 20.60, write the chemical formula for each compound.

20.57 (a) Hydroiodic acid; (b) sodium triiodide; (c) potassium perchlorate; (d) chloric acid.
20.58 (a) Hypoiodous acid; (b) perchloric acid; (c) sodium chlorate; (d) iodine heptafluoride.
20.59 (a) Sodium chlorite; (b) cryolite; (c) hypobromous acid.
20.60 (a) Sodium periodate; (b) fluorspar; (c) chloric acid.

In Exercises 20.61 and 20.62, give the chemical formulas and names of three compounds of the element in which the halogen oxidation number is (a) +1; (b) +5; (c) −1.

20.61 Chlorine.
20.62 Iodine.

In Exercises 20.63 and 20.64, identify the oxidation number of the halogen in each substance or ion.

20.63 (a) ClO_2; (b) IO^-; (c) BrF_3; (d) $NaIO_3$.
20.64 (a) I_3^-, (b) IF_7; (c) $NaIO_4$; (d) BrO_2^-.

20.65 Summarize the properties of the halogens and the main trends in their properties. Is there any indication of increasing metallic character down the group?

20.66 Summarize and account for the differences between fluorine and its congeners. What do we mean when we say that fluorine is a reactive element?

20.67 Describe how the halogens are obtained from their natural sources, and explain why those particular techniques are used.

20.68 Summarize the uses of the Group VII elements and their compounds, including uses that have been mentioned in previous chapters.

In Exercises 20.69 and 20.70, write the Lewis structure for each molecule or ion, and predict its shape.

20.69 (a) ClO_2^-; (b) IO_3^-; (c) BrF_3.
20.70 (a) IO_4^-; (b) ClF_5; (c) I_3^-.

20.71 Explain why iodine is much more soluble in a solution of potassium iodide than in pure water.

20.72 Predict the physical and chemical properties of astatine, including its standard reduction potential, its electronegativity, and the strengths of its oxoacids.

20.73 Describe the variation in the strengths of the halogen oxoacids, and explain that variation in terms of oxidation numbers and electronegativities.

20.74 Arrange the halogen oxoacids in the order of increasing oxidizing strengths (as judged from the standard reduction potentials of couples involving their ions), and suggest an interpretation of that order. Give examples of the use of the halogen oxoacids and their salts to cause oxidation.

20.75 The booster rockets on the space shuttle are fueled by a mixture of aluminum and ammonium perchlorate. Write the chemical equation for the reaction, assuming that the products are nitrogen, water, hydrochloric acid, and aluminum oxide. Calculate the mass of ammonium perchlorate that should be carried for each kilogram of aluminum.

20.76 Sodium iodate is an impurity in Chile saltpeter, which was once the major source of iodine. The element was obtained by the reduction of a solution of the iodate with sodium hydrogen sulfite. Write the chemical equation for the reduction, and calculate the number of kilograms of sodium hydrogen sulfite needed to produce 1.0 kg of iodine.

20.77 The electron affinity of F(g) is lower than that of Cl(g), yet the standard reduction potential of the F_2/F^- couple is more negative than that of the Cl_2/Cl^- couple. Comment on this observation, and suggest an explanation.

20.78 Anhydrous hydrogen fluoride is made by the action of concentrated sulfuric acid on fluorspar. Write the chemical equation for the process. Explain, with equations, why silica is an undesirable impurity in this process.

20.79 The standard free energy of formation of hydrogen iodide is +1.70 kJ/mol at 25°C. Calculate the equilibrium composition of the gas at 1.0 atm pressure. (In practice, decomposition is slow unless a catalyst is present.)

20.80 Set up a thermodynamic cycle to identify the origin of the energy difference between hydrogen chloride, which has an exothermic enthalpy of formation ($\Delta H_f^\circ = -92.3$ kJ/mol at 25°C), and hydrogen iodide, which has an endothermic enthalpy of formation ($\Delta H_f^\circ = +26.5$ kJ/mol at 25°C).

20.81 Account for the observation that melting and boiling points generally *decrease* from fluoride to iodide for ionic halides, but *increase* from fluoride to iodide for molecular halides.

20.82 Account for the observation that solubility in water generally *increases* from fluoride to iodide for ionic halides, but *decreases* from fluoride to iodide for less ionic halides. Illustrate the observation with halides of Group I elements.

20.83 The interhalogen BrF_5 has a nonzero dipole moment. Is that consistent with the structure predicted by VSEPR theory?

20.84 Concentrations of Cl^- ions can be measured gravimetrically by precipitating silver chloride, with silver nitrate used as the precipitating reagent in the presence of dilute nitric acid. The white precipitate is filtered out and weighed. Calculate the Cl^- ion concentration in 25.00 mL of a solution that gave a silver chloride precipitate of mass 3.050 g. Why is the method inappropriate for measuring fluoride ion concentrations?

20.85 Concentrations of F^- ions can be measured by adding lead(II) chloride solution and weighing the lead(II) chlorofluoride (PbClF) precipitate. Calculate the molar concentration of F^- ions in 25.00 mL of a solution that gave a lead chlorofluoride precipitate of mass 0.765 g.

20.86 Suppose 25.00 mL of an aqueous solution of iodine was titrated with 0.025 M $Na_2S_2O_3(aq)$, using starch as the indicator. The blue color of the starch-iodine complex disappeared when 28.45 mL of the thiosulfate solution had been added. What was the molar concentration of I_2 in the original solution?

*20.87 The concentration of hypochlorite ions in a solution can be determined by adding a known volume to a solution containing excess I^- ions, which are oxidized to iodine, and then measuring the resulting iodine concentration by titration with sodium thiosulfate (see Exercise 20.86). In an experiment, 10.00 mL of hypochlorite solution was added to a potassium iodide solution, which in turn required 28.34 mL of 0.110 M $Na_2S_2O_3(aq)$ to reach the equivalence point. Calculate the molar concentration of ClO^- ions in the original solution.

Group VIII

20.88 Describe the trends in the physical and chemical properties of the noble gases, and account for them in terms of the noble gas atoms.

In Exercises 20.89 and 20.90, give the oxidation number of the noble gas atom in each compound or ion.

20.89 (a) KrF_2; (b) XeF_2; (c) XeO_3; (d) XeF_6.
20.90 (a) KrF_4; (b) XeO_4^{2-}; (c) XeO_6^{4-}.

20.91 Summarize (with equations) the means of preparing xenon compounds, including xenon fluorides, oxides, oxoacids, and oxosalts.

20.92 Suggest reasons why the noble gas compounds were unknown until the 1960s.

20.93 Predict the order of the strengths of the xenon oxoacids.

20.94 Predict some aspects of the chemistry of radon.

The Main-Group Elements

20.95 List the evidence that shows that metallic character decreases from the lower left of the periodic table to the upper right.

20.96 Summarize the properties of the hydrogen compounds of the main-group elements, and show how they illustrate trends that occur throughout the periodic table.

20.97 Summarize the properties of the oxides, oxoacids, and oxosalts of the main-group elements, and show how they illustrate trends that occur throughout the periodic table.

20.98 To what extent are gas-phase ionization-energy and electron-affinity values a poor guide to the behavior of substances in aqueous solution?

20.99 Summarize the evidence for diagonal relationships throughout the main-group elements, and suggest reasons why they are less important on the right side of the periodic table.

20.100 Identify one interesting compound of each Period 2 and Period 3 element that would result in the world being a much poorer place if that element did not exist. Give reasons for each choice.

20.101 Summarize the properties of the elements in Groups VI through VIII. How do they support the view that the least metallic elements are found near the top right of the periodic table?

20.102 Summarize the structures and properties of the oxoacids of the elements in Groups VI through VIII. Do they continue the trend observed in Groups III through V of the p block?

20.103 Discuss the usefulness (or lack of usefulness) of the concept of oxidation number in explaining the properties of the p-block elements.

20.104 Summarize, with examples and explanations, the principal differences between the elements in the s and p blocks.

20.105 Chapters 18 to 20 have been organized by group. Discuss whether it would be helpful to organize the elements according to period, and give examples of trends that could be displayed more effectively in that way.

The elements of the d block share some fascinating properties. In particular, they form an important series of compounds, the coordination compounds, that have colors and magnetic properties that can be explained by a single unified theory. One of the elements in the d block is iron. The illustration shows oxygen being blown through molten iron to oxidize the impurities it contains. This is one of the steps in the manufacture of steels, which are alloys of iron with other d-block elements.

The *d*-block elements and their compounds

21.1 Trends in properties
21.2 The elements scandium through nickel
21.3 The elements copper through mercury

THE d-BLOCK ELEMENTS

Complexes of the *d*-block elements

21.4 The preparation and stability of complexes
21.5 The structures of complexes
21.6 Isomerism

Crystal field theory

21.7 The effects of ligands on *d* electrons
21.8 The electronic structures of many-electron complexes

The members of the *d* block, widely known as the *transition metals*, are the workhorse elements of the periodic table (Fig. 21.1). They include the constructional metals—notably iron and copper—that helped civilization rise from the Stone Age and that are still very widely used in the modern world. They include the metals of the new technologies, such as titanium for the aerospace industry and vanadium for catalysts in the petrochemical industry. They also include the precious metals silver, platinum, and gold, which are prized as much for their beauty, rarity, and durability as for their usefulness.

The *d* block is so called because its members have atoms with ground-state electron configurations in which, according to the building-up principle, *d* orbitals are being occupied. However, as explained in Section 7.5, the building-up principle is only a formal procedure for arriving at an atom's configuration; in fact, in *d*-block atoms the *d* electrons are more strongly bound than the valence *s* electrons. Hence, when a *d*-block atom becomes a cation, the valence *s* electrons are lost first, after which a variable number of *d* electrons may be lost. The configuration of iron, for example, is $[Ar]3d^6 4s^2$, and its two most common ions are Fe^{2+} with configuration $[Ar]3d^6$ and Fe^{3+} with configuration $[Ar]3d^5$.

Since there are five *d* orbitals in a given shell, and since each orbital can accommodate up to two electrons, there are 10 elements in each row of the *d* block. However, the seven 4*f* orbitals begin to be occupied at lanthanum, and the 14 elements of the first row of the *f* block (the lanthanides) delay the completion of the *d* block. Although we shall not discuss the *f* block in detail, we should be aware of its presence, because it affects the properties of elements that follow it in Period 6.

THE *d*-BLOCK ELEMENTS AND THEIR COMPOUNDS

All the *d*-block elements are metals (and we shall often refer to them as the "*d* metals"). Most are good electrical conductors, silver being the best of all elements at room temperature. They are usually malleable,

FIGURE 21.1 The elements in the *d* block of the periodic table. Note that the *f* block intervenes at the beginning of the block in Periods 6 and 7.

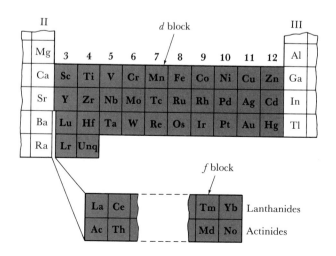

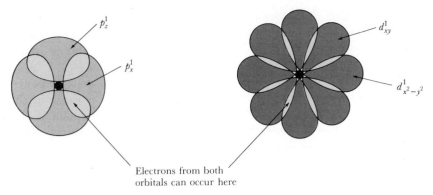

FIGURE 21.2 Two *d* electrons are well separated in space—more so than two *p* electrons—and hence interact with each other relatively weakly. The regions of space that the two electrons can share in each case are shown in tan.

ductile, lustrous, and silver-white in color, and they generally have higher melting and boiling points than the main-group elements. However, there are some notable exceptions: mercury has so low a melting point that it is a liquid at room temperature; copper is red-brown in color, and gold is yellow.

21.1 TRENDS IN PROPERTIES

The properties of the *d* metals can be traced in large measure to the directional character of the lobes of the *d* orbitals. Because of this directionality, electrons in different *d* orbitals occupy markedly different regions of space. As a result, their repulsions are weak, and two *d* electrons interact less than two *s* electrons. They also interact less than two *p* electrons, for even when the latter occupy different *p* orbitals, they are not so well separated as two *d* electrons (Fig. 21.2). Another consequence of the very directional character of *d* orbitals is that the nucleus is only poorly shielded by a *d* electron: in effect, the nuclear charge is exposed through the gaps between the lobes of the *d* orbital.

Trends in metallic radii. Both the atomic number and the number of *d* electrons increase from left to right along each period. However, because *d* electrons repel each other weakly, the increasing charge of the nucleus succeeds in drawing them inward. This has two consequences.

First, the increasing nuclear charge reduces the radii of the atoms. Thus, a $3d^6 4s^2$ iron atom (with metallic radius 126 pm) is smaller than a $3d^1 4s^2$ scandium atom (160 pm), even though the iron atom has more *d* electrons (Fig. 21.3). Nevertheless, the range of *d*-metal radii is not very great, and some of the atoms of one *d* metal can be replaced by atoms of another without causing too much strain in a solid. The *d* metals can therefore be mixed together to form a wide range of alloys, of which the numerous steel alloys are important examples.

Second, as the nuclear charge increases from left to right across the period, the first ionization energies also increase—from 632 kJ/mol for

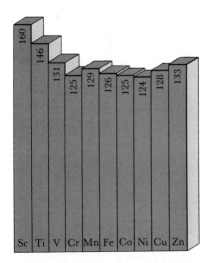

FIGURE 21.3 The metallic radii of the elements of the first row of the *d* block, in picometers. The radii decrease from Sc to Ni, on account of the increasing nuclear charge and the poor shielding of the *d* electrons.

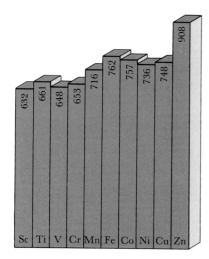

FIGURE 21.4 The first ionization energies of the first row of the *d* metals increase up to Fe. After that, electron–electron repulsions become important, since the *d* orbitals become doubly occupied. Values are given in kilojoules per mole.

scandium to 762 kJ/mol for iron in the first row of the block (Fig. 21.4). The decrease in ionization energy after Fe and the increase in radius (Fig. 21.3) are results of the double occupation of *d* orbitals at this stage, which results in stronger electron–electron repulsions.

The elements in the second and third rows of the block show the same general trends in radii and ionization energies as those in the first. However, although there is a general increase in metallic radius from the first row to the second (because a new quantum shell is being occupied), the metallic radii of the elements in the third row are about the same as in the second row. This failure of the third-row metallic radii to increase as expected is called the *lanthanide contraction*. It is so called because it results from the presence of the *f* block (and in particular the lanthanides). The *f* electrons that are present in the lanthanides are even poorer shields than *d* electrons. As a result, there is a marked decrease in metallic radius along the *f* block, as the nuclear charge increases and pulls the electrons inward. When the *d* block resumes (at lutetium), the metallic radius has fallen from 188 pm to 157 pm. Hence, all the elements that follow have smaller radii than might be expected.

One effect of the lanthanide contraction is the high density of the Period 6 elements (Fig. 21.5). These elements are so dense because their metallic radii are comparable to those of the elements in Period 5, while their atomic weights are about twice as large. The atomic weight of iridium, for example, is nearly twice that of rhodium in the row above (192 amu compared with 103 amu), but the metallic radii of Ir and Rh are almost identical. (Iridium is actually one of the two densest elements; its neighbor osmium is the other.) Another effect of the lanthanide contraction is the low reactivity of gold and platinum—their "nobility." This can be traced to the fact that their valence electrons are relatively close to the nucleus and hence tightly bound.

FIGURE 21.5 The densities (in grams per cubic centimeter) of the *d* metals at 25°C. The lanthanide contraction has a pronounced effect on the densities of the elements in Period 6 (front row), which are among the densest of all elements.

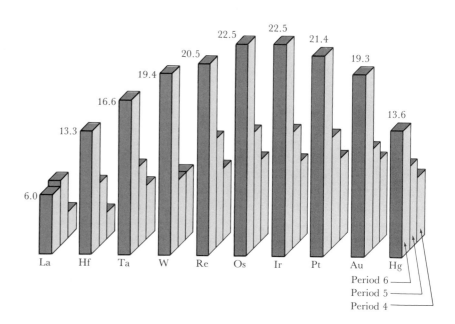

Trends in chemical properties. The principal chemical characteristics of the *d* metals are their ability to form numerous complexes (which is described in Section 21.4), their action as catalysts, and their extensive range of oxidation numbers.

The wide range of *d*-metal oxidation numbers (Table 21.1) might seem daunting at first sight. However, it is helpful to keep in mind that, with the exception of Hg, the elements at the ends of each row of the *d* block have only one oxidation number other than zero. Scandium, for example, occurs only with oxidation number +3, and zinc only with oxidation number +2. All the other elements of the row can have at least two oxidation numbers. The most common oxidation numbers of

TABLE 21.1 **The oxidation numbers of the *d*-block elements***

Sc	Ti	V	Cr	Mn	Fe	Co	Ni	Cu	Zn
	1	1	1	1		1	1	**1**	
	2	2	2	**2**	**2**	**2**	**2**	**2**	**2**
3	3	3	**3**	**3**	**3**	**3**	3	3	
	4	**4**	4	4	4	4	4		
		5	5	5	5				
			6	6	6				
				7					

Y	Zr	Nb	Mo	Tc	Ru	Rh	Pd	Ag	Cd
	1	1	1		1	1		**1**	
	2	2	2	2	**2**	2	**2**	2	**2**
3	3	3	3	3	**3**	**3**	3	3	
	4	4	**4**	**4**	**4**	4	**4**		
		5	**5**	5	5	5			
			6	6	6	6			
				7	7	7			
					8	8			

Lu	Hf	Ta	W	Re	Os	Ir	Pt	Au	Hg
	1	1		1	1	1		**1**	**1**
		2	2	2	2	2	**2**	2	**2**
3	3	3	3	3	**3**	**3**		**3**	3
	4	4	4	**4**	**4**	4	4		
		5	5	5	5	5	5		
			6	6	6	6	6		
				7	7				
					8				

*Important oxidation numbers are shown in bold type. This table should be considered as only provisional: chemists are challenged by an absence and they often strive to show that an unlisted oxidation number of an element can in fact be prepared.

copper, for example, are +1 (as in CuCl) and +2 (as in $CuCl_2$). Elements close to the center of each row have the widest range of oxidation numbers; this is shown strikingly by Mn, with its seven known oxidation numbers. Another helpful fact is that elements toward the left of the block and low down in a group are more frequently found with their higher oxidation numbers.

As with main-group elements, oxidation numbers help to explain the properties of d-metal compounds. A compound in which an element has an oxidation number high in its range tends to be a good oxidizing agent. An example is the oxidizing power of the permanganate ion MnO_4^- (in which Mn has oxidation number +7) in acidic solution:

$$MnO_4^-(aq) + 8H^+(aq) + 5e^- \longrightarrow Mn^{2+}(aq) + 4H_2O(l)$$
$$E° = +1.51 \text{ V}$$

Compounds in which the oxidation number is intermediate in the element's range often disproportionate, as copper(I) compounds do in water:

$$2Cu^+(aq) \longrightarrow Cu(s) + Cu^{2+}(aq)$$

Compounds in which the element has an oxidation number low in its range, such as compounds containing Fe^{2+}, are often good reducing agents:

$$Fe^{3+}(aq) + e^- \longrightarrow Fe^{2+}(aq) \qquad E° = -0.44 \text{ V}$$

In addition, although most d-metal oxides are basic, the oxides of a given element show a shift toward acidic character with increasing oxidation number. A good example is provided by the family of chromium oxides:

Oxide	Oxidation number	Property
CrO	+2	Basic
Cr_2O_3	+3	Amphoteric
CrO_3	+6	Acidic

The last oxide, CrO_3, is the anhydride of chromic acid, H_2CrO_4, the parent of the chromates.

Among other similarities, the elements on the left side of the d block resemble the s-block metals by being much more difficult to extract from their ores than the metals on the right side. Indeed, as we move across the d block from left to right, we encounter the elements in roughly the reverse of the order in which they became available for exploitation. On the far right are copper and zinc, which jointly were responsible for the Bronze Age. That age was succeeded by the Iron Age, as higher temperatures became attainable and iron-ore reduction feasible. Finally, on the left of the block, we have metals that require such extreme conditions for their extraction—including utilization of the thermite process and electrolysis—that they have become widely available only in this century (Fig. 21.6).

| (a) | (b) | (c) |

FIGURE 21.6 These three artifacts represent the progress that has been made in the extraction of the *d* metals. (a) An ancient bronze chariot axle cap from China made from a metal that was easy to extract. (b) An early iron steam engine made from a metal that was moderately easy to extract once high temperatures could be achieved. (c) A twentieth-century airplane engine with titanium components which had to await high temperatures and advanced technology before the element became available.

21.2 THE ELEMENTS SCANDIUM THROUGH NICKEL

Some of the physical properties of the elements in the first row of the *d* block, from scandium to nickel, are summarized in Table 21.2. Here we consider only their most important chemical properties.

Scandium and titanium. Scandium was first isolated in 1937. It is a reactive metal (reacting with water about as vigorously as calcium and tarnishing in air) with few uses. The small, highly charged Sc^{3+} ion is

TABLE 21.2 Properties of the *d*-block elements scandium through nickel

Z	Name	Symbol	Valence electron configuration	Atomic weight, amu	Melting point, °C	Boiling point, °C	Density, g/cm³
21	Scandium	Sc	$3d^1 4s^2$	44.96	1540	2800	2.99
22	Titanium	Ti	$3d^2 4s^2$	47.90	1670	3300	4.51
23	Vanadium	V	$3d^3 4s^2$	50.94	1920	3400	6.09
24	Chromium	Cr	$3d^5 4s^1$	52.00	1860	2600	7.19
25	Manganese	Mn	$3d^5 4s^2$	54.94	1250	2120	7.47
26	Iron	Fe	$3d^6 4s^2$	55.85	1540	2760	7.87
27	Cobalt	Co	$3d^7 4s^2$	58.93	1494	2900	8.80
28	Nickel	Ni	$3d^8 4s^2$	58.71	1455	2150	8.91

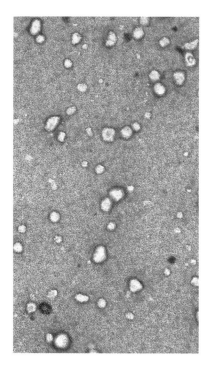

FIGURE 21.7 This microphotograph of the surface of white paint shows the presence of small crystals—"crystallites"—of titanium dioxide, which are responsible for its brilliant whiteness.

strongly hydrated in water (like Al^{3+}), and the resulting $[Sc(H_2O)_6]^{3+}$ complex is about as strong a Brønsted acid as acetic acid.

Titanium is a light, strong metal, and it is used primarily where these properties are vital—as in airplanes. It is resistant to corrosion because a skin of oxide forms on its surface. The principal sources of the metal are the ores ilmenite ($FeTiO_3$) and rutile (TiO_2).

Titanium requires vigorous reducing conditions for extraction from its ores. It was not exploited commercially until quite recently, when the demand from the aerospace industry began to grow. The metal is obtained by treating the oxide with chlorine in the presence of coke:

$$TiO_2(s) + 2C(s) + 2Cl_2(g) \xrightarrow{1000°C} TiCl_4(g) + 2CO(g)$$

The chloride is then reduced with liquid magnesium:

$$TiCl_4(g) + 2Mg(l) \longrightarrow Ti(s) + 2MgCl_2(s)$$

The most stable oxidation number of titanium is +4. Its most important compound is the dioxide (TiO_2), a brilliantly white, nontoxic, stable compound that is used as the white pigment in paints (Fig. 21.7). Titanium tetrachloride ($TiCl_4$) is a liquid that boils at 136°C and fumes in moist air because it is hydrolyzed to the dioxide:

$$TiCl_4(l) + 2H_2O(l) \longrightarrow TiO_2(s) + 4HCl(aq)$$

This reaction is used for skywriting, both on the grand scale of military smokescreens and in advertising.

▼ **EXAMPLE 21.1 Predicting the properties of titanium solutions**

Suggest what might happen when an aqueous solution of titanium(III) sulfate is poured on sodium sulfide.

STRATEGY We must determine the consequences of the fact that Ti^{3+} is a small, highly charged ion; Al^{3+} is a helpful analogy here. Such ions were discussed in Section 15.1. The fact that the sulfide ion is a Brønsted base is another clue.

SOLUTION The small Ti^{3+} ion is strongly hydrated, and, by analogy with Al^{3+}, we expect it to be a Brønsted acid. Since the S^{2-} ion is a Brønsted base, we expect an acid-base (proton-transfer) reaction, with the formation of H_2S:

$$[Ti(H_2O)_6]^{3+}(aq) + H_2O(l) \rightleftharpoons [Ti(H_2O)_5OH]^{2+}(aq) + H_3O^+(aq)$$

$$2H_3O^+(aq) + S^{2-}(aq) \longrightarrow H_2S(g) + 2H_2O(l)$$

Titanium(III) sulfate is in fact sufficiently acidic for this reaction to occur at
▲ an appreciable rate.

Titanium forms a series of oxides called *titanates*, which are prepared by heating TiO_2 with a chemically equivalent amount of the oxide or carbonate of a second metal. Barium titanate ($BaTiO_3$) is *piezoelectric*, which means that it becomes electrically charged when it is mechanically distorted. This property leads to its use in phonograph heads and for underwater sound detection, applications in which a mechanical vibration must be converted into an electrical signal.

Vanadium, chromium, and manganese.   Pure vanadium, which is a soft silver-gray metal, is produced by reducing vanadium pentoxide (V_2O_5) with calcium or reducing vanadium chloride with magnesium. When other metals are added to iron in steelmaking, it is usually not the pure metal that is added, as this would be uneconomical, but a *ferroalloy* of the metal, an alloy with iron and carbon that is less expensive to produce. This is the case with vanadium, which is normally added to iron as "ferrovanadium," a mixture of about 85% V, 12% C, and 3% Fe by mass. It is prepared by reducing vanadium pentoxide with aluminum in the presence of iron.

Vanadium pentoxide is the most important compound of vanadium. This orange-yellow solid is used as an oxidizing agent and as an oxidizing catalyst in the contact process for making sulfuric acid. The wide range of colors of vanadium compounds, including the blue of vanadyl ion VO^{2+}, accounts for their use as glazes in the ceramics industry.

Chromium is a bright, lustrous, corrosion-resistant metal that got its name from its colorful compounds (its name comes from the Greek word for "color"). It is obtained from chromite ore ($FeCr_2O_4$) by reduction with carbon in an electric arc furnace. The main product of this reduction is ferrochromium, which is used in the production of stainless steels (they typically contain about 15% chromium by mass). Chromium(IV) oxide (CrO_2) is used to coat "chrome" recording tapes because it responds better to high-frequency magnetic fields than do conventional "ferric" (Fe_2O_3) tapes. Another major use of chromium is in chrome plating (Section 17.8).

Sodium chromate (Na_2CrO_4), a yellow solid that is one of the most important chromium compounds, is made commercially by heating a mixture of chromite ore and sodium carbonate in the presence of oxygen:

$$4FeCr_2O_4(s) + 8Na_2CO_3(s) + 7O_2(g) \xrightarrow{\Delta} 8Na_2CrO_4(s) + 2Fe_2O_3(s) + 8CO_2(g)$$

The chromate is converted into the orange dichromate ion ($Cr_2O_7^{2-}$) in the presence of acid:

$$2CrO_4^{2-}(aq) + 2H^+(aq) \longrightarrow Cr_2O_7^{2-}(aq) + H_2O(l)$$

In the laboratory, acidified solutions of dichromates, in which the oxidation number of Cr is +6, are useful oxidizing agents:

$$Cr_2O_7^{2-}(aq) + 14H^+(aq) + 6e^- \longrightarrow 2Cr^{3+}(aq) + 7H_2O(l)$$
$$E° = +1.33 \text{ V}$$

Sodium chromate and sodium dichromate are the starting points for the production of a number of pigments, corrosion inhibitors, fungicides, and ceramics.

▼ EXAMPLE 21.2 Suggesting a means of synthesizing a chromate

Suggest a method for preparing lead(II) chromate, the pigment "chrome yellow" ($PbCrO_4$), which is widely used for marking highways.

STRATEGY A major clue to a solution is that a pigment for use on highways must be insoluble in water; hence a precipitation reaction might be a feasible route. We can check in the solubility rules (Table 3.1) that lead(II) chromate is insoluble, and then look for reagents suitable for a double replacement reaction.

SOLUTION Lead(II) nitrate is a soluble lead salt. With sodium chromate, it will give a precipitate of lead(II) chromate, which can be separated from the remaining sodium nitrate solution by filtration:

$$Pb^{2+}(aq) + CrO_4^{2-}(aq) \longrightarrow PbCrO_4(s)$$

▲

Sulfur dioxide reduces sodium chromate to $Cr(OH)SO_4$, a compound used in the tanning of leather. The compound attaches to the collagen in the animal skins, making it insoluble and protecting it from biological degradation, but leaving it flexible.

Manganese is a gray metal that resembles iron. It is much less resistant to corrosion than chromium and becomes coated with a thin brown layer of oxide when exposed to air. The metal is rarely used alone but is an important component of alloys. In steel, where it is added as ferromanganese, it helps to remove sulfur by forming a sulfide. It also increases iron's hardness, toughness, and resistance to abrasion. Manganese is also alloyed with various nonferrous metals. One such alloy is manganese bronze (39% zinc, 1% manganese, some iron and aluminum, and the rest copper), which is very resistant to corrosion and is used for marine propellors. The increasing use of aluminum cans for beverages has helped to increase the consumption of manganese, since it is alloyed with the aluminum to increase the stiffness of the metal container.

A rich supply of manganese lies in the "manganese nodules" that litter the ocean floors (Fig. 21.8). These nodules range in diameter from millimeters to meters and are lumps of the oxides of iron, manganese, and other elements. However, since this source is technically difficult to exploit, manganese is currently obtained by the thermite process from pyrolusite, a mineral form of manganese dioxide (MnO_2). Ferromanganese, which contains about 80% Mn and 20% Fe by mass, is obtained by heating a mixture of pyrolusite, coke, and limestone in an electric furnace.

Manganese lies at the center of its row and occurs with a wide variety of oxidation numbers. The most stable is +2, but +4 and +7 also play important roles in manganese compounds. Its most important compound is manganese dioxide, a black solid used in dry cells, as a decolorizer to conceal the green tint of glass, and as the starting point for the production of other manganese compounds. As examples of the last use, the green manganate ion (MnO_4^{2-}) and the purple per-

FIGURE 21.8 Manganese "nodules" litter the ocean floor and are potentially a valuable source of the element.

manganate ion (MnO_4^-) may be produced from manganese dioxide by oxidation and disproportionation in acid solution:

$$2MnO_2(s) + 4KOH(s) + O_2(g) \xrightarrow{\Delta} 2K_2MnO_4(s) + 2H_2O(l)$$

$$3MnO_4^{2-}(aq) + 4H^+(aq) \longrightarrow 2MnO_4^-(aq) + MnO_2(s) + 2H_2O(l)$$

However, this is not a good commercial route to permanganate because the disproportionation also produces manganese dioxide, the original starting material. Electrolytic oxidation of the manganate ion is used instead:

$$MnO_4^{2-}(aq) \longrightarrow MnO_4^-(aq) + e^-$$

Potassium permanganate, in which the manganese has its highest oxidation number ($+7$), is useful as an oxidizing agent in acid solution:

$$MnO_4^-(aq) + 8H^+(aq) + 5e^- \longrightarrow Mn^{2+}(aq) + 4H_2O(l)$$
$$E° = +1.51\ V$$

FIGURE 21.9 Pyrite (FeS_2), or "fool's gold," is so called because it looks a little like gold.

Its usefulness stems not only from its thermodynamic tendency to oxidize other substances, but also from its ability to act through a variety of alternative mechanisms and hence to discover a reaction path with low activation energy. It is used for oxidations in organic chemistry and as a mild disinfectant.

Iron. Iron, the most widely used of all the d metals, is the second most abundant metal in the earth's crust (aluminum is first). Its principal ores are the oxides hematite (Fe_2O_3) and magnetite (Fe_3O_4). The sulfide ore pyrite (FeS_2), which is also called "fool's gold" because of its misleading appearance (Fig. 21.9), is also widely available; it is not used in steelmaking because the sulfur is difficult to remove.

Iron ore is reduced to iron in a blast furnace, as is illustrated in Fig. 3.17. The furnace is filled with a "charge" of ore, limestone, and coke. The calcium oxide produced in the furnace by the thermal decomposition of the limestone helps to remove the impurities in the ore:

$$CaO(s) + SiO_2(s) \xrightarrow{\Delta} CaSiO_3(l)$$

$$CaO(s) + Al_2O_3(s) \xrightarrow{\Delta} Ca(AlO_2)_2(l)$$

$$6CaO(s) + P_4O_{10}(s) \xrightarrow{\Delta} 2Ca_3(PO_4)_2(l)$$

The mixture of products is liquid at the temperature of the furnace and floats on the denser molten iron. It is drawn off and used to make rocklike materials for the construction industry. The liquid iron is run off as "pig iron," an impure product that contains about 5% carbon.

The first stage in steelmaking is to lower the carbon content of the pig iron and to remove the remaining impurities, which include silicon, phosphorus, and sulfur. The modern approach uses a version of the early *Bessemer converter*, which is essentially a big metal pot lined with basic material—typically dolomite, $CaMg(CO_3)_2$—that decomposes to oxides at the operating temperature. Air is blown through the molten impure iron in the converter, the impurities react with the basic lining to form a molten slag, and the carbon is oxidized and removed as the monoxide. In the modern version, called the *basic oxygen process*, oxygen

TABLE 21.3 The compositions of different steels

Ingredient added to iron	Typical amount	Effect
Manganese	0.5 to 1.0%	Increases strength and hardness but lowers ductility
	13%	Increases wear resistance
Nickel	<5%	Increases strength and shock resistance
	>5%	Enhances corrosion resistance (stainless steel) and heat resistance
Chromium		Increases hardness and wear resistance
	>12%	Enhances corrosion resistance (stainless steel, with nickel)
Vanadium		Increases hardness
Tungsten	<20%	Increases hardness, especially at high temperatures

FIGURE 21.10 In the basic oxygen process, a blast of oxygen and powdered limestone is used to purify the molten iron by, respectively, oxidizing and combining with the impurities in it.

and powdered limestone are forced through the molten metal (Fig. 21.10). In the second stage, steels are produced by adding the appropriate metals, often as ferroalloys, to the molten iron (Table 21.3).

Iron is not a very hard metal. Its physical properties are improved by partial reaction with carbon, for the carbides so formed strengthen the solid. Its corrosion resistance is greatly improved by alloying to form steels. Iron has the property of *ferromagnetism*, the ability to be permanently magnetized. Ferromagnetism results when a large number of electrons in the metal spin in the same direction simultaneously (Fig. 21.11), and hence generate a strong magnetic field. The spins are aligned by exposing the metal to another magnetic field, such as that

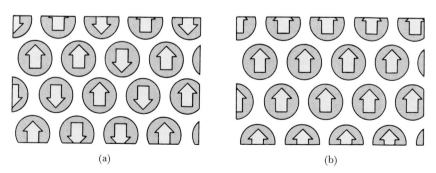

(a)   (b)

FIGURE 21.11 A ferromagnetic material, such as iron, magnetite, or cobalt, consists of atoms in which many electrons spin in the same direction and give rise to a strong magnetic field: (a) before magnetization; (b) after magnetization. The magnetization persists when the magnetizing influence has been removed. The yellow arrows represent the spins of each atom.

FIGURE 21.12 These solutions show the typical colors of iron ions in aqueous solution. From left to right: $FeSO_4$, $FeCl_2$, $Fe_2(SO_4)_3$, $FeCl_3$.

caused by an electric current, and the alignment persists when the field is removed. Some oxides of iron (including magnetite) are also ferromagnetic.

Iron is quite reactive and corrodes in moist air. It reacts with acids that have nonoxidizing anions, evolving hydrogen and forming iron(II) salts. The colors of these salts vary from pale yellow to a dark green-brown; $[Fe(H_2O)_6]^{2+}$ itself is pale green (Fig. 21.12). Iron(II) salts are quite readily oxidized to iron(III) salts. The oxidation is slow in acidic solution but rapid in basic solution, where insoluble iron(III) hydroxide, $Fe(OH)_3$, is precipitated. Although $[Fe(H_2O)_6]^{3+}$ ions are pale purple, the colors of aqueous solutions of iron(III) salts are dominated by the yellow $[Fe(H_2O)_5OH]^{2+}$ ion, which is the conjugate base of the acidic hydrated ion:

$$[Fe(H_2O)_6]^{3+}(aq) + H_2O(l) \rightleftharpoons H_3O^+(aq) + [Fe(H_2O)_5OH]^{2+}(aq)$$

Solutions of Fe^{3+} are sufficiently acidic to produce carbon dioxide from sodium carbonate. To preserve iron(III) salts in water, it is necessary to shift the equilibrium in favor of the hydrated ion by adding acid and maintaining a pH close to zero. If the pH rises above 2, a red-brown hydrated iron(III) oxide (a version of rust) precipitates, often as a colloid.

Compounds in which iron has oxidation number 0 are also known. When iron is heated in carbon monoxide, it reacts to form iron penta-carbonyl, $Fe(CO)_5$, a yellow molecular liquid with boiling point 103°C. The CO group is not very electronegative, so the number of electrons

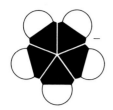

1 Cyclopentadienide ion

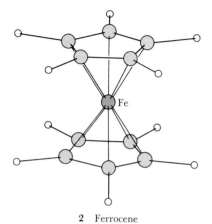

2 Ferrocene

on the Fe atom is barely changed from the number on the free atom. Hence the iron is ascribed oxidation number 0 in this and similar compounds. Iron(II) chloride reacts with the cyclopentadienide ion $C_5H_5^-$ **(1)** to form the molecular compound ferrocene, $Fe(C_5H_5)_2$, a wittily named *sandwich compound* **(2)** and one of a number of similar compounds that are more formally known as *metallocenes* (Fig. 21.13), compounds in which a metal atom lies between two organic rings. This orange solid (of melting point 174°C) is insoluble in water and thermally stable up to 500°C. Other *d* metals can be used to produce sandwiches with different fillings.

Iron is essential to the diet, largely because of its role as the site at which oxygen molecules attach to the oxygen-carrying protein hemoglobin. A healthy adult human body contains around 3 g of iron, mostly as hemoglobin. Around 1 mg is lost daily in sweat, feces, and hair (and about 20 mg is lost during menstruation), so about 1 mg must be ingested daily to maintain a proper balance. Iron deficiency, or anemia, results in reduced transport of oxygen to the brain and muscles; an early symptom is chronic tiredness.

Cobalt and nickel. Economically workable deposits of cobalt are found in association with copper sulfide. Miners called it *kobold* (which means "mischievous spirit" in German) because it interfered with the production of the copper. Cobalt is a silver-gray metal that is used mainly for alloying with iron. Most interest in cobalt arises from its ability to form complexes, as described in Section 21.4. Cobalt is another essential element, for it is a component of vitamin B_{12}.

About 70% of the western world's nickel (named for "Old Nick," for much the same reason as the one underlying "cobalt") comes from iron and nickel sulfide ores that were brought close to the surface by the impact of a meteor in Sudbury, Ontario. The first step in obtaining the metal from its ore (which is in fact a complex mixture of sulfides) is to roast it in air to form the oxide:

$$2NiS(s) + 3O_2(g) \longrightarrow 2NiO(s) + 2SO_2(g)$$

The oxide is then either reduced with carbon and refined electrolytically or reduced and purified by the *Mond process*. In the latter, the reduction is carried out with hydrogen:

$$NiO(s) + H_2(g) \xrightarrow{\Delta} Ni(s) + H_2O(g)$$

Then the impure Ni is exposed to carbon monoxide, when it forms nickel carbonyl, $Ni(CO)_4$, a volatile, poisonous liquid of boiling point 43°C:

$$Ni(s) + 4CO(g) \xrightarrow{50°C} Ni(CO)_4(g)$$

The carbonyl is decomposed to pure nickel by heating it to about 200°C.

Nickel is a hard, silver-white metal used mainly for the production of stainless steel and for alloying with copper to produce cupronickels, the alloys used for coins. Cupronickels are slightly yellow but are whitened

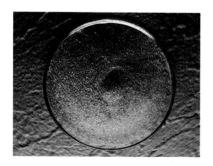

FIGURE 21.13 A sample of ferrocene, the first and one of the most stable "sandwich" compounds in which a metal atom lies between two planar organic rings.

TABLE 21.4 Properties of the *d*-block elements copper through mercury

Z	Name	Symbol	Valence electron configuration	Atomic weight, amu	Melting point, °C	Boiling point, °C	Density, g/cm³
29	Copper	Cu	$3d^{10}4s^1$	63.54	1085	2580	8.93
47	Silver	Ag	$4d^{10}5s^1$	107.87	962	2160	10.50
79	Gold	Au	$5d^{10}6s^1$	196.97	1064	2850	19.28
30	Zinc	Zn	$3d^{10}4s^2$	65.37	420	913	7.14
48	Cadmium	Cd	$4d^{10}5s^2$	112.40	321	770	8.65
80	Mercury	Hg	$5d^{10}6s^2$	200.59	−39	357	13.55

by the addition of small amounts of cobalt. Nickel is also used as a catalyst, especially for the addition of hydrogen to organic compounds, as in the manufacture of edible fats from vegetable oils. Its most stable oxidation number is +2, and $[Ni(H_2O)_6]^{2+}$ ions are green. Nickel(III) is reduced to nickel(II) by cadmium at the cathode during the discharge of rechargeable nickel-cadmium cells:

Cathode: $Ni(OH)_3(s) + e^- \longrightarrow Ni(OH)_2(s) + OH^-(aq)$

Anode: $Cd(s) + 2OH^-(aq) \longrightarrow Cd(OH)_2(s) + 2e^-$

21.3 THE ELEMENTS COPPER THROUGH MERCURY

The six elements in the two groups at the far right of the *d* block are the *coinage metals* copper, silver, and gold, all of which have $d^{10}s^1$ electron configurations, and their neighbors zinc, cadmium, and mercury, with $d^{10}s^2$ configurations. Some of their properties are summarized in Table 21.4

▼ **EXAMPLE 21.3** Accounting for the low reactivity of the coinage metals

Explain why the coinage metals, particularly gold, are not very reactive.

STRATEGY We need to decide why the coinage metals do not lose electrons readily. A clue is that the elements stand close to the end of the *d* block and that gold, the least reactive, is also preceded by *f*-block elements earlier in its period.

SOLUTION The low reactivity of the coinage metals is partly due to the poor shielding of the *d* electrons and hence the tight grip that the nucleus can exert on the outermost electrons. This effect is enhanced in Period 6 by
▲ the lanthanide contraction.

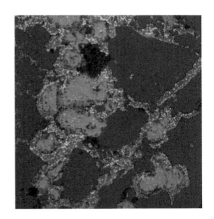

FIGURE 21.14 A color-enhanced x-ray photograph showing copper (red) filling the fissures of a rock; the metal can be recovered from such sources by mining.

Copper. Copper is unreactive enough for some to be found native (Fig. 21.14), but most is produced from its sulfides—particularly the ore chalcopyrite ($CuFeS_2$). The crushed and ground ore is separated from excess rock by *froth flotation*. In this process (Fig. 21.15), the powdered ore is combined with oil, water, and detergents. Then air is blown through the mixtures; the oil-coated ore floats to the surface with the detergent froth, and the unwanted copper-poor residue—the "gangue"—sinks to the bottom.

Processes for extracting metals from their ores are generally classified as *pyrometallurgical*, if high temperatures are used, or *hydrometallurgical*, if aqueous solutions are used. Copper is extracted by both methods. In the pyrometallurgical process for the extraction of copper, the enriched ore is first roasted in air:

$$4CuFeS_2(s) + 7O_2(g) \longrightarrow 4CuS(s) + 2Fe_2O_3(s) + 4SO_2(g)$$

(This and the next step can contribute an alarming amount of SO_2 to the atmosphere unless precautions are taken to contain it.) The CuS is then *smelted*, or reduced by melting with a reducing agent. In this step, air is blown through the molten mixture to remove the sulfur as SO_2:

$$CuS(s) + O_2(g) \longrightarrow Cu(l) + SO_2(g)$$

Although oxygen appears to be acting as a reducing agent in this reaction, it and the Cu^{2+} ions in the sulfide are in fact jointly oxidizing the S^{2-} ions to SO_2; hence S^{2-} is the actual reducing agent. Limestone and sand, which are added to the mixture, form a molten slag that removes many of the impurities.

In hydrometallurgical processes for extracting copper, Cu^{2+} ions are first formed by the action of sulfuric acid on the ores. Then the metal is obtained by reducing these ions in aqueous solution. In principle, the reduction can be performed with any reducing agent that has a more negative electrode potential than copper. Cheap, and therefore economically viable, reducing agents include hydrogen,

$$Cu^{2+}(aq) + H_2(g) \longrightarrow Cu(s) + 2H^+(aq)$$

and scrap iron,

$$Cu^{2+}(aq) + Fe(s) \longrightarrow Cu(s) + Fe^{2+}(aq)$$

The reduction may also be carried out electrolytically. The resulting crude copper is refined electrolytically: it is made into anodes and plated onto cathodes of pure copper. The rare metals obtained from the anode sludge are of substantial help in paying for the electricity used in the electrolysis.

Copper is an excellent electrical conductor when pure and is widely used in the electrical industry. It is also alloyed with zinc to form brass, with tin to form bronze, and with nickel to form cupronickel (Table 21.5). We saw in Chapter 17 that as copper lies above hydrogen in the electrochemical series, it cannot displace hydrogen from acids; however, it can be oxidized by oxidizing acids, as in the reaction

$$3Cu(s) + 8H^+(aq) + 2NO_3^-(aq) \longrightarrow 3Cu^{2+}(aq) + 2NO(g) + 4H_2O(l)$$

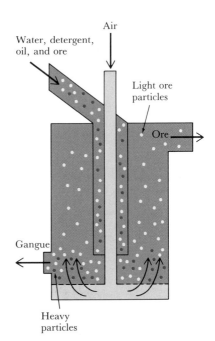

FIGURE 21.15 The froth flotation process, which is used to separate higher-quality copper ore from unwanted impurities.

TABLE 21.5   Nonferrous alloys

Alloy	Composition
Brass	Up to 40% zinc in copper (yellow brass: 35% zinc)
Bronze	A metal other than zinc or nickel in copper (casting bronze: 10% tin, 5% lead)
Cupronickel	Nickel in copper (coinage cupronickel: 25% nickel)
Pewter	6% antimony and 1.5% copper in tin

FIGURE 21.16 Copper corrodes in air to form a pale green layer of "basic copper carbonate." This patina, or incrustation, passivates the surface.

Copper corrodes in moist air as a result of oxidation caused by a mixture of water, oxygen, and carbon dioxide:

$$2Cu(s) + H_2O(l) + O_2(g) + CO_2(g) \longrightarrow Cu_2(OH)_2CO_3(s)$$

The pale green product is called "basic copper carbonate" and is responsible for the green patina of copper and bronze objects (Fig. 21.16). It adheres to the surface, protects the metal, and gives a pleasing appearance.

Copper, in common with the other coinage metals, forms compounds with oxidation number $+1$; however, in water, copper(I) salts disproportionate into metallic copper and copper(II) ions. The latter exist as pale blue $[Cu(H_2O)_6]^{2+}$ ions in water. This is the color of copper(II) sulfate solutions; the deeper blue of the solid pentahydrate, $CuSO_4 \cdot 5H_2O$, is caused by $[Cu(H_2O)_4]^{2+}$ ions (the fifth H_2O links this ion to the sulfate anion). Copper(II) salts, particularly the acetate, chloride, carbonate, and hydroxide, are used as pesticides because they are toxic to fungi, algae, and bacteria.

Silver and gold. Very little silver occurs native: most is obtained as a by-product of the refining of copper and lead, and a considerable amount is recycled through the photographic industry. Silver lies above hydrogen in the electrochemical series and so does not reduce $H^+(aq)$ to hydrogen. However, silver is oxidized by nitric acid in a reaction resembling copper's, but leading to the production of Ag^+ ions:

$$3Ag(s) + 4H^+(aq) + NO_3^-(aq) \longrightarrow 3Ag^+(aq) + NO(g) + 2H_2O(l)$$

Silver(I) does not disproportionate in aqueous solution, and almost all silver compounds have oxidation number $+1$. Apart from silver nitrate and silver fluoride, silver salts are generally only very slightly soluble in water. Silver nitrate ($AgNO_3$) is the most important compound of silver and the starting point for the manufacture of silver halides for use in photography.

Gold is so inert that most of it is found native. Pure gold is "24-karat gold," and its alloys with silver and copper, which differ in hardness and hue, are classified according to the proportion of gold they contain (Fig. 21.17). For example, 10- and 14-karat golds contain, respectively, $\frac{10}{24}$ and $\frac{14}{24}$ parts by mass of gold.

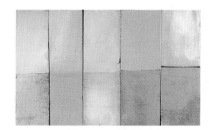

FIGURE 21.17 The color of commercial gold depends on its composition: from left to right, 8-karat gold, white gold, 18-karat gold, and 24-karat gold. This white gold is 6 parts Au and 18 parts Ag.

▼ EXAMPLE 21.4 Predicting the reactions of gold

Should we expect gold to react with acids?

STRATEGY We need to consider (by inspection of the location of gold in the electrochemical series, Appendix 2B) whether it can be oxidized by H^+ ions or by oxoanions. Even if neither reaction is viable, it may still be possible for a reaction to occur if oxidized gold is effectively removed from solution by complex formation (a point discussed in Section 15.7).

SOLUTION Gold lies well above hydrogen in the electrochemical series and is too noble to react even with nitric acid. Both the following gold couples lie above H^+/H_2 and above NO_3^-, H^+/NO, H_2O:

$$Au^+(aq) + e^- \longrightarrow Au(s) \qquad\qquad E° = +1.69 \text{ V}$$

$$Au^{3+}(aq) + 3e^- \longrightarrow Au(s) \qquad\qquad E° = +1.40 \text{ V}$$

$$NO_3^-(aq) + 4H^+(aq) + 3e^- \longrightarrow NO(g) + 2H_2O(l) \qquad E° = +0.96 \text{ V}$$

However, we saw in Section 15.7 that gold does react with aqua regia (a mixture of concentrated nitric and hydrochloric acids) because the Au^{3+} ions form the complex ion $[AuCl_4]^-$ with Cl^- ions:

$$Au(s) + 6H^+(aq) + 3NO_3^-(aq) + 4Cl^-(aq) \longrightarrow$$
$$[AuCl_4]^-(aq) + 3NO_2(g) + 3H_2O(l)$$

Even though the equilibrium constant for the oxidation of Au to Au^{3+} is very unfavorable, the reaction proceeds because Au^{3+} ions are effectively ▲ removed from solution.

The most common oxidation number for gold in its compounds is +3. Gold(I) compounds are also known, but in aqueous solution they tend to disproportionate into metallic gold and gold(III) unless they are stabilized by complex formation. This means of stabilization accounts for the ability of air to oxidize gold to gold(I) if cyanide ions are present (Section 15.7).

Zinc, cadmium, and mercury. Zinc is found mainly as the sulfide ZnS in sphalerite, often in association with lead ores. The ore is concentrated by froth flotation, and the metal is extracted by smelting with coke, when the zinc distills off. Cadmium is obtained similarly. Zinc and cadmium are silvery, reactive metals. Zinc is used mainly for galvanizing iron; like copper, it is protected by a hard film of basic carbonate, $Zn_2(OH)_2CO_3$, that forms on contact with air.

Zinc and cadmium are similar to each other but differ sharply from mercury. Zinc is amphoteric (like its main-group neighbor aluminum). It reacts with acids to form Zn^{2+} ions and with alkalis to form the zincate ion $[Zn(OH)_4]^{2-}$:

$$Zn(s) + 2OH^-(aq) + 2H_2O(l) \longrightarrow [Zn(OH)_4]^{2-}(aq) + H_2(g)$$

Galvanized containers should therefore not be used for transporting alkalis. For that purpose cadmium, which is lower down the group, is more metallic, and hence has a more basic oxide, can be used instead. However, this is hazardous, as cadmium salts are toxic. Zinc and cadmium have the oxidation number +2 in all their compounds.

Mercury occurs mainly as the mineral cinnabar (HgS), from which it is separated by froth flotation and then roasting in air:

$$HgS(s) + O_2(g) \xrightarrow{\Delta} Hg(g) + SO_2(g)$$

The volatile metal is distilled off and condensed. Mercury is unique in being the only metallic element that is liquid at room temperature, and with the long range of temperatures over which it is liquid (from its melting point of $-39°C$ to its boiling point of $357°C$) it is well suited for its use in thermometers.

Mercury lies above hydrogen in the electrochemical series, so it is not oxidized by hydrogen ions. However, it does react with nitric acid:

$$3Hg(l) + 8H^+(aq) + 2NO_3^-(aq) \longrightarrow$$
$$3Hg^{2+}(aq) + 2NO(g) + 4H_2O(l)$$

In compounds it has the oxidation number $+1$ or $+2$. Compounds with the former are unusual in that the mercury(I) cation is the covalently bonded diatomic ion $(Hg\text{-}Hg)^{2+}$, written Hg_2^{2+}. The zinc and cadmium analogs of this cation, Zn_2^{2+} and Cd_2^{2+}, have recently been found in their molten salts.

Mercury(I) chloride is insoluble, and like silver chloride it, too, precipitates when chloride solution is added to a sample undergoing qualitative analysis. However, unlike silver chloride, which dissolves when ammonia is added, mercury(I) chloride disproportionates to metallic mercury and mercury(II) ions:

$$Hg_2^{2+}(aq) \longrightarrow Hg(l) + Hg^{2+}(aq)$$

Since the mercury turns the white precipitate a blackish gray (Fig. 21.18), it can be distinguished from silver chloride. Hence, the presence of mercury can be distinguished from the presence of silver.

FIGURE 21.18 When ammonia is added to a silver chloride precipitate, the precipitate dissolves. However, when ammonia is added to a precipitate of mercury(I) chloride, mercury metal is formed by disproportionation, and the mass turns gray.

COMPLEXES OF THE *d*-BLOCK ELEMENTS

One of the most outstanding properties of the *d* metals is their ability to act as Lewis acids and to form numerous complexes. One example is the formation of the ferrocyanide ion $[Fe(CN)_6]^{4-}$, in which the Lewis acid Fe^{2+} forms bonds by sharing electron pairs provided by the CN^- ions. Another is the formation of the neutral complex $Ni(CO)_4$, in which Ni acts as a Lewis acid and the CO molecules act as Lewis bases. Main-group elements also form complexes, as in $[Al(H_2O)_6]^{3+}$, but the range of compounds is much wider for the *d* block than for the *s* and *p* blocks.

A great deal of modern chemical research focuses on the structure, properties, and uses of complexes, partly because they take part in many biological reactions. Hemoglobin and vitamin B_{12}, for example, are both complexes—one of iron and the other of cobalt. Complexes are often brightly colored and magnetic, and they are used in analysis, color science, and catalysis. They are now being examined for use in solar-energy conversion, in atmospheric nitrogen fixation, and as pharmaceuticals.

21.4 THE PREPARATION AND STABILITY OF COMPLEXES

Many complexes are prepared simply by mixing solutions of a *d*-metal ion with the appropriate Lewis base (Fig. 21.19); for example,

$$Fe^{2+}(aq) + 6CN^-(aq) \longrightarrow [Fe(CN)_6]^{4-}(aq)$$

Complexes are formed by *substitution reactions*, in which one Lewis base expels another and takes its place. That is the process occurring in the reaction above, since the CN^- ions actually drive H_2O molecules out of the $[Fe(H_2O)_6]^{2+}$ complex. A less complete replacement occurs when Cl^- ions are added to an iron(II) solution,

$$[Fe(H_2O)_6]^{2+}(aq) + Cl^-(aq) \longrightarrow [Fe(H_2O)_5Cl]^+(aq) + H_2O(l)$$

or when ammonia is added to a copper(II) solution:

$$[Cu(H_2O)_6]^{2+}(aq) + 4NH_3(aq) \longrightarrow$$
$$[Cu(NH_3)_4(H_2O)_2]^{2+}(aq) + 4H_2O(l)$$

The impressive changes of color that often accompany these substitution reactions (Fig. 21.20) are one of the features we explore here.

We must be careful, as always, to distinguish *thermodynamic* stability from *kinetic* stability. Although a complex may be thermodynamically unstable, there may be such a high activation barrier to reaction that the complex survives for long periods. A complex that survives only for short periods is called *labile*. An example of the distinction between thermodynamic and kinetic stability is found in the different behaviors of $[Co(NH_3)_6]^{2+}$ and $[Co(NH_3)_6]^{3+}$. Both undergo substitution reactions in acidified water:

$$[Co(NH_3)_6]^{2+}(aq) + 6H_3O^+(aq) \longrightarrow [Co(H_2O)_6]^{2+}(aq) + 6NH_4^+(aq)$$

$$[Co(NH_3)_6]^{3+}(aq) + 6H_3O^+(aq) \longrightarrow [Co(H_2O)_6]^{3+}(aq) + 6NH_4^+(aq)$$

Both reactions have negative reaction free energy, so both ammonia complexes are thermodynamically unstable. However, whereas the first reaction reaches equilibrium in a few seconds at room tempera-

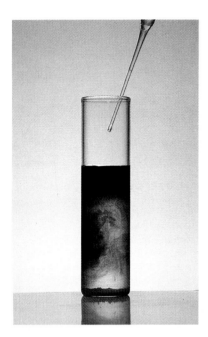

FIGURE 21.19 When cyanide ions (in the form of potassium cyanide) are added to a solution of iron(II) sulfate, they displace the H_2O molecules from the Fe^{2+} ion and produce a new complex, the more strongly colored ferrocyanide ion.

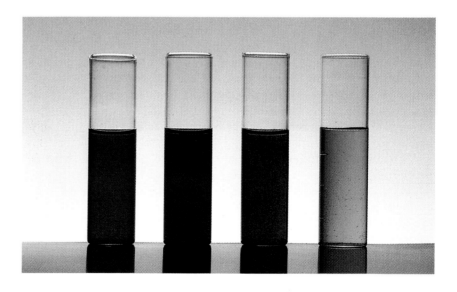

FIGURE 21.20 Some of the highly colored compounds that result when complexes are formed. Clockwise, starting at the red solution: $FeSCN^{2+}$, $[Co(SCN)_4]^{2-}$, $[Cu(NH_3)_4]^{2+}$, $K_2[CuBr_4]$.

ture, the second requires weeks. That is, $[Co(NH_3)_6]^{2+}$ is labile as well as unstable; $[Co(NH_3)_6]^{3+}$ is unstable but nonlabile.

21.5 THE STRUCTURES OF COMPLEXES

The central metal ion in a complex is a Lewis acid, and the groups attached to it are Lewis bases. These groups are called *ligands* (from the Latin word for "bound"). Some common examples are listed in Table 21.6. We say that the ligands "coordinate" to the metal when they form the complex. This is the origin of the term *coordination compound*, which means either a neutral molecular complex, such as $Ni(CO)_4$, or an ionic compound in which at least one of the ions is a complex, as in potassium ferrocyanide, $K_4[Fe(CN)_6]$. The ligands directly attached to the central ion in a complex (and which are enclosed between brackets in the chemical formula) make up the *coordination sphere* of the central ion.

Types of complexes. Many important complexes have either six or four ligands in their coordination spheres and are called, respectively, "six-

TABLE 21.6 The names of common ligands

Anionic ligands		Neutral ligands	
Formula	Name	Formula	Name
F^-	Fluoro	H_2O	Aqua
Cl^-	Chloro	NH_3	Ammine
Br^-	Bromo	NO	Nitrosyl
I^-	Iodo	CO	Carbonyl
OH^-	Hydroxo	$NH_2CH_2CH_2NH_2$	Ethylenediamine (en)
O^{2-}	Oxo		
CN^-	Cyano (as M-CN)		
CN^-	Isocyano (as M-NC)		
SCN^-	Thiocyanato (as M-SCN)		
SCN^-	Isothiocyanato (as M-NCS)		
NO_2^-	Nitrito (as M-ONO)		
NO_2^-	Nitro (as M-NO$_2$)		
CO_3^{2-}	Carbonato		
SO_4^{2-}	Sulfato		
$C_2O_4^{2-}$	Oxalato (ox)		

Ethylenediaminetetraaceto (EDTA)

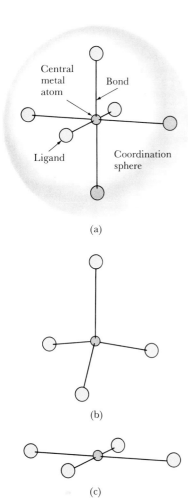

(a)

(b)

(c)

FIGURE 21.21 The various components of a complex, which may be either neutral or charged. (a) Almost all six-coordinate complexes are octahedral. Four-coordinate complexes are either (b) tetrahedral or (c) square planar.

coordinate" and "four-coordinate" complexes (Fig. 21.21). The ligands of six-coordinate complexes, such as the CN^- in $[Fe(CN)_6]^{4-}$, are almost always arranged at the corners of an octahedron: such complexes are called "octahedral complexes." It is conventional to represent these octahedral complexes with the simplified diagram **(3)**. The ligands of four-coordinate complexes are usually found either at the corners of a tetrahedron or (most notably for Pt^{2+} and Au^{3+} complexes) at the corners of a square. The two types are called "tetrahedral complexes" and "square-planar complexes," respectively.

Many ligands, including H_2O, NH_3, and CN^-, occupy only one site in a coordination sphere and are called *monodentate ligands* (for "one-toothed"). Some ligands can simultaneously occupy more than one site and are called *polydentate ligands* ("many-toothed"). Ethylenediamine (NH_2—CH_2—CH_2—NH_2; **4**) has a tooth (a nitrogen lone pair) at each end, and is therefore a *bidentate ligand*. This ligand occurs widely in coordination chemistry and is abbreviated to "en," as in $[Co(en)_3]^{3+}$ **(5)**.

Chelates. The metal atom in $[Co(en)_3]^{3+}$ lies at the center of the three ligands as though being pinched by three molecular claws. It is an example of a *chelate*, a complex containing at least one polydentate ligand forming a ring of atoms that includes the central metal atom. (The name comes from the Greek word for "claw.") An example of a hexadentate chelating ligand, one that can occupy all six octahedral sites around an ion and grip it in a single six-pronged claw, is ethylenediaminetetraacetic acid, EDTA (Fig. 21.22). EDTA demonstrates the fact that main-group elements can form complexes, for it forms a complex with Ca^{2+} and is used to soften water and remove scale (which is $CaCO_3$). It also forms a complex with Pb^{2+} and thus acts as an antidote to lead poisoning.

Chelating ligands are quite common in nature. Mosses and lichens secrete chelating ligands to capture essential metal ions from the rocks they inhabit. Chelate formation also lies behind the body's strategy of running a fever when infected by bacteria. The higher temperature kills bacteria by reducing their ability to synthesize a particular iron-chelating ligand.

The nomenclature of complexes. The names of some common ligands and their abbreviations for use in formulas are given in Table 21.6. The names of all anionic ligands end in -o (as in cyano for CN^- as a ligand), the rules being that -ide, -ite, and -ate suffixes change to -o,

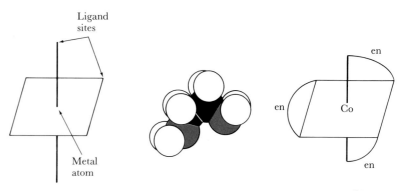

3 Simplified octahedral complex

4 Ethylenediamine

5 $[Co(en)_3]^{3+}$

-ito, and -ato, respectively. The most important neutral ligands are H_2O (which is called aqua when it is a ligand), NH_3 (ammine), and CO (carbonyl). The metals are named with their oxidation numbers in parentheses, as in cobalt(II) and cobalt(III).

If several different ligands are present in a complex, they are named in alphabetical order. The number of each is shown with the same prefix as is used for naming simple compounds (Table 2.8):

$[Fe(H_2O)_5Cl]^+$: pentaaquachloroiron(II) ion
$[Cr(NH_3)_4Cl_2]^+$: tetraamminedichlorochromium(III) ion

If the ligand is complicated—for example, if it already contains prefixes in its name (as in ethylenediamine)—then the prefix denoting the number of ligands is changed to bis- or tris- (in place of di- and tri-, respectively). For example,

$[Co(en)_3]^{3+}$: tris(ethylenediamine)cobalt(III) ion

If the complex itself is an anion, the suffix -ate is added to the stem of the metal's name (sometimes in its Latin form as listed in Table 2.1: ferr-, for example, for iron). Hence we have

$[Fe(CN)_6]^{4-}$: hexacyanoferrate(II) ion

We have already used the old name for this complex: the ferrocyanide ion.

The names of coordination compounds are built in the same way as those of simple compounds, with the cation named before the anion:

$NH_4[Pt(NH_3)Cl_3]$: ammonium amminetrichloroplatinate(II)

Some of these names can become extremely long; nevertheless, they can always be picked apart to determine precisely what compound is intended.

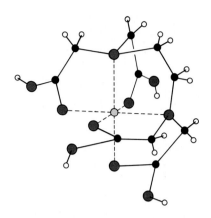

FIGURE 21.22 EDTA is a hexadentate chelating ligand. Its two N atoms and four of its O atoms can occupy all six octahedral bonding positions around a central metal atom.

▼ **EXAMPLE 21.5** **Naming a coordination compound**

Name the compound $[Co(NH_3)_3(H_2O)_3][Fe(CN)_6]$.

STRATEGY We need to identify the cation and the anion, name each separately, and then combine them. For each, we should first note the name and oxidation number of the metal; then identify the ligands; and finally, string the names together in alphabetical order with the appropriate prefixes, ending with the name of the metal.

SOLUTION The anion might be either $[Fe(CN)_6]^{3-}$ or $[Fe(CN)_6]^{4-}$. If it were the former, the cation would be $[Co(NH_3)_3(H_2O)_3]^{3+}$, implying cobalt(III); if it were the latter, the cation would be a complex of cobalt(IV), which is unlikely (see Table 21.1). The anion must therefore be $[Fe(CN)_6]^{3-}$, a complex of iron(III), because each CN^- ligand contributes one negative charge, and Fe^{3+} is required to give an overall charge of -3. This anion is therefore hexacyanoferrate(III), also commonly called the ferricyanide ion. The cation has three ammine ligands and three aqua ligands; it is therefore triamminetriaquacobalt(III). The name of the compound is therefore triamminetriaquacobalt(III) hexacyanoferrate(III).

EXERCISE Name the compound $[Fe(H_2O)_5OH][Cr(H_2O)_2(ox)_2]$.
[*Answer*: Pentaaquahydroxoiron(III) diaquabisoxalatochromate(II)]
▲

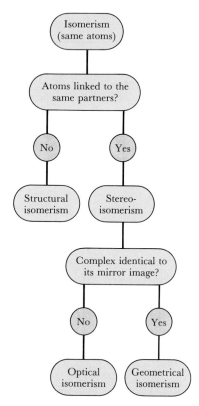

FIGURE 21.23 This chart shows the different types of isomerism. We shall encounter some of them again in our discussions of organic chemistry.

FIGURE 21.24 Solutions of the two coordination compounds [Co(NH₃)₅Br]SO₄ (left) and [Co(NH₃)₅SO₄]Br (right). Although they are built of exactly the same atoms, they are different compounds with their own characteristic properties.

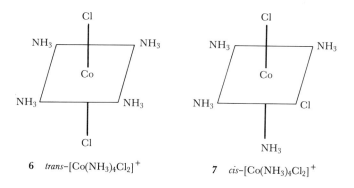

6 *trans*-[Co(NH₃)₄Cl₂]⁺   **7** *cis*-[Co(NH₃)₄Cl₂]⁺

21.6 ISOMERISM

A characteristic feature of complexes and coordination compounds is the existence of "isomers":

Isomers are compounds that contain the same numbers of the same atoms in different arrangements.

(The name comes from the Greek words for "equal parts.") Isomers are different compounds built from the same set of parts (**6** and **7**). These ions differ only in the positions of the Cl^- ligands, but they are distinct substances, for they have different physical and chemical properties.

Isomerism—the existence of isomers—can take a number of forms (Fig. 21.23). The two major classes of isomers are *structural isomers*, in which the atoms have different partners, and *stereoisomers*, in which the atoms have the same partners but are arranged differently in space. For instance, $[Co(NH_3)_5Br]SO_4$ and $[Co(NH_3)_5SO_4]Br$ are structural isomers because the Br^- ion is a ligand of the cobalt in the former but is an accompanying anion in the latter. The two compounds are physically quite different: the bromo complex is red, for example, and the sulfato complex is violet (Fig. 21.24). The two complexes shown in diagrams **6** and **7** are stereoisomers. Both have the formula $[Co(NH_3)_4Cl_2]^+$, so all the atoms have the same partners in both; however, the arrangement of these atoms in space differs from one isomer to the other.

Geometrical and optical isomerism.   The two $[Co(NH_3)_4Cl_2]^{2+}$ stereoisomers are *geometrical isomers*, compounds in which the coordination spheres are identical except for the locations of the ligands. The isomer with the Cl^- ligands on opposite sides of the central atom (**6**) is called the *trans* isomer (from the Latin word for "across"), and that with the ligands on the same side (**7**) is called the *cis* isomer (from the Latin word for "on this side").

A more subtle type of stereoisomerism is *optical isomerism*. This occurs when a molecule (or ion) and its mirror image are not identical, as for the two complexes shown in Fig. 21.25. A molecule that is distinct from its own mirror image, as a left hand is distinct from a right hand, is said to be *chiral* (from the Greek word for "hand"). Pairs of optical isomers,

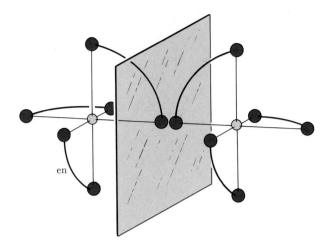

like pairs of hands, are called *enantiomers* (from the Greek words for "opposite parts").

▼ **EXAMPLE 21.6** **Identifying optical isomerism**

Which of the following complexes are chiral, and which form enantiomeric pairs:

(a)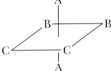

(b)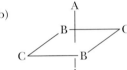

(c)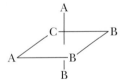

(d)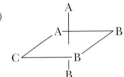

STRATEGY By drawing the mirror image of each complex, we should be able to judge whether any rotation of the original molecule will cause it to match its image. If not, then the complex is chiral. We can determine which form enantiomeric pairs by finding pairs that are the mirror images of each other.

SOLUTION In the pairs of complexes below, the original is on the left, and its mirror image is on the right.

(a)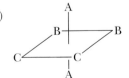

The image is identical to the original.

(b)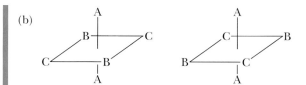

If we rotate the image about A-A, we obtain

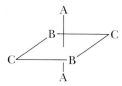

which is identical to the original.

(c)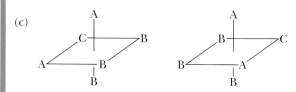

The original complex is chiral.

(d)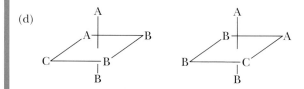

The original complex is chiral.

Complexes (c) and (d) are chiral because no rotation can make either match its mirror image. However, when the image of (c) is rotated by 180° around the vertical A-B axis, it becomes identical to (d); hence (c) and (d) form an enantiomeric pair.

EXERCISE Repeat this example for the following complexes:

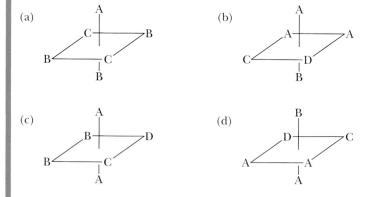

[*Answer*: (a) and (c) are not chiral; (b) and (d) are chiral and enantiomeric]

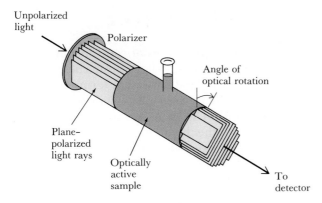

FIGURE 21.26 In an experiment to demonstrate optical activity, light is passed through a polarizer (such as a sheet of Polaroid) so that the wave oscillates in a single plane. After the light has passed through an optically active sample, its plane of polarization is found to be at a different angle. The planes show the polarization of each ray in the beam.

Optical activity. Enantiomers have identical physical and chemical properties, with two exceptions: they can react differently with other chiral compounds, and they are *optically active*. That is, they have different effects on plane-polarized light. *Plane-polarized light* is light in which the wave motion occurs in a single plane. This is different from ordinary unpolarized light, such as the light from an ordinary electric lamp, in which the wave motion occurs at all angles around the direction of travel. Light is polarized by passing it through a polarizing filter, which eliminates all wave motion but that in a single plane. When it is then passed through a solution of optically active molecules, the angle of the plane of polarization changes (Fig. 21.26). Separate solutions of enantiomers at the same concentration rotate the plane of polarization in equal and opposite directions. The enantiomer that rotates the plane clockwise (as seen by the observer looking along the beam toward the approaching light) is called the *(+)-enantiomer*. The one that rotates it counterclockwise is called the *(−)-enantiomer*.

When a chiral compound is synthesized from reagents that are not themselves chiral, a *racemic mixture*—a mixture containing equal amounts of the two enantiomers—is generally formed (as when a factory produces pairs of gloves). This is the case when $[Co(en)_3]^{3+}$ is prepared from ethylenediamine, which is not chiral itself, and $[Co(H_2O)_6]^{3+}$, which is also not chiral. A racemic mixture is optically inactive because the clockwise rotation caused by one enantiomer is canceled by the counterclockwise rotation caused by the other. Since enantiomers form crystals that are the mirror images of each other, racemic mixtures can occasionally be separated by first crystallizing the sample and then sorting the crystals by hand. Biological syntheses of chiral molecules often produce only one enantiomer, so many molecules in our bodies have a definite "handedness." This topic is taken up in Section 24.9.

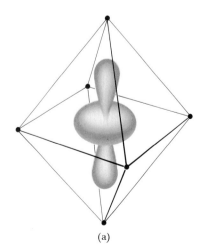
(a)

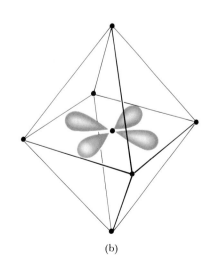
(b)

FIGURE 21.27 In the crystal field theory of complexes, the lone pairs of electrons on the ligands (a) are treated as negative point charges (b).

CRYSTAL FIELD THEORY

The most notable physical properties of coordination compounds are their colors and their magnetism. These properties, and some differences in their stabilities, can be discussed in terms of _crystal field theory_. This theory was originally devised to explain the colors of solids— particularly ruby, in which Cr^{3+} ions are responsible for the color. A more complete version of the theory is called "ligand field theory," but only the simpler version is described here.

21.7 THE EFFECTS OF LIGANDS ON *d* ELECTRONS

In crystal field theory, each ligand lone pair is represented by a negative point charge (Fig. 21.27). The electronic structure of the complex is then expressed in terms of the electrostatic interactions—the "field"— between these charges and the electrons and nucleus of the central metal ion. We begin by considering a complex containing a single *d* electron, such as $[Ti(H_2O)_6]^{3+}$, in which the configuration of Ti^{3+} is $[Ar]3d^1$; we then treat complexes with several *d* electrons.

The ligand-field splitting. Since the central ion of a complex is positively charged, the negative charges representing the ligand lone pairs are attracted to it. This results in an overall lowering of energy and is a major factor in the stability of the complex. The stability of the $[Ti(H_2O)_6]^{3+}$ ion, for instance, can be ascribed largely to the strong attraction between the Ti^{3+} ion and a negative charge representing the lone pair on each of the six H_2O ligands. However, when we examine the structure of the complex in more detail, we have to consider the fact that its single valence electron interacts differently with the ligand charges when it occupies different *d* orbitals (Fig. 21.28).

In an octahedral complex, three of the five *d* orbitals (d_{xy}, d_{yz}, and d_{zx}), which are called _t orbitals_, point between the ligand positions. An electron occupying any of these orbitals avoids the repulsion of the ligands slightly and so has a lower energy than if it were uniformly spread

FIGURE 21.28 In an octahedral complex with a central *d*-metal ion, (a) a d_{z^2} orbital points directly toward two ligands, and an electron that occupies it has a relatively high energy; (b) a d_{xy} orbital points between the ligands, and an electron that occupies it has a relatively low energy.

(a)

(b)

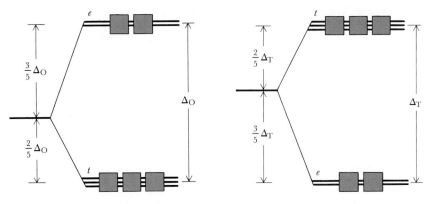

FIGURE 21.29 The energy levels of the d orbitals in (a) an octahedral complex with the ligand-field splitting Δ_O and (b) in a tetrahedral complex with the ligand-field splitting Δ_T. Each box can hold two electrons.

around the nucleus. The remaining two d orbitals (d_{z^2} and $d_{x^2-y^2}$), which are called *e orbitals*, point directly at the ligands. If an electron occupies either of them, it experiences a stronger repulsion than if it were spread around the metal nucleus uniformly. That is, a t electron has a lower energy than if it were spread spherically around the central ion, and an e electron has a higher energy.

These ideas are summarized in the energy-level diagram of Fig. 21.29(a). The energy separation between the two sets of orbitals is called the *ligand-field splitting* Δ_O (the O denotes octahedral) and typically amounts to about 10% of the total energy of interaction of the central ion with the ligands. The t orbitals lie at an energy that is $\frac{2}{5}\Delta_O$ below the average energy (the energy that an electron would have if the directional properties of the orbitals were ignored), and the e orbitals lie at an energy that is $\frac{3}{5}\Delta_O$ above the average. Since the t orbitals have the lower energy, we can predict that in the ground state of the $[\text{Ti}(\text{H}_2\text{O})_6]^{3+}$ complex, the single d electron occupies one of them in preference to an e orbital, and hence that the electron configuration of the complex is t^1. This can be represented by the box diagram

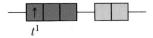

t^1

In a tetrahedral complex, the three t orbitals point more directly at the ligands than do the two e orbitals. As a result, the t orbitals here have a *higher* energy than the e orbitals (Fig. 21.29b). The ligand-field splitting Δ_T (where the T denotes tetrahedral) is generally smaller than in octahedral complexes because the d orbitals do not point as directly at the ligands and there are fewer repelling ligands.

Light absorption by d^1 complexes. The t electron of the $[\text{Ti}(\text{H}_2\text{O})_6]^{3+}$ complex can be excited into one of the e orbitals if it absorbs a photon of energy equal to Δ_O (Fig. 21.30). Since a photon carries an energy $h\nu$, where h is Planck's constant and ν is its frequency, it can be absorbed if its frequency satisfies

$$h\nu = \Delta_O$$

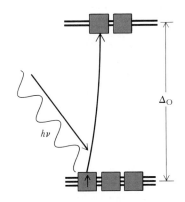

FIGURE 21.30 When a complex is exposed to light of the correct frequency, an electron can be excited to a higher-energy orbital (from t to e in this octahedral complex) as the light is absorbed.

The frequency and wavelength λ of light are related by $\lambda = c/\nu$, where c is the speed of light (Section 7.1); hence the wavelength of light absorbed and the ligand-field splitting are related by

$$\Delta_O = \frac{hc}{\lambda} \tag{1}$$

That is, the greater the splitting, the shorter the wavelength of the light that is absorbed by the complex.

▼ **EXAMPLE 21.7** **Calculating Δ_O from spectroscopy**

The $[Ti(H_2O)_6]^{3+}$ complex absorbs light of wavelength 510 nm. What is the value of Δ_O for 1 mol of this complex?

STRATEGY We can use Eq. 1 to calculate Δ_O for a single complex, and then multiply the result by Avogadro's number to find the value for 1 mol of complex.

SOLUTION Inserting the data into Eq. 1 and using 1 Hz = 1/s gives

$$\Delta_O = \frac{(6.626 \times 10^{-34}\ \text{J/Hz}) \times (2.998 \times 10^8\ \text{m/s})}{510 \times 10^{-9}\ \text{m}}$$

$$= 3.895 \times 10^{-19}\ \text{J}$$

On multiplying by Avogadro's number and rounding, we obtain

$$\Delta_O = (6.022 \times 10^{23}) \times (3.895 \times 10^{-19}\ \text{J}) = 235\ \text{kJ}$$

This is about 10% of the total interaction energy between the Ti^{3+} ion and its six H_2O ligands.

EXERCISE A different complex absorbed light of wavelength 422 nm; calculate the value of Δ_O for 1 mol of this complex.

[*Answer*: 283 kJ]

▲

The spectrochemical series. The technique illustrated in the example can be used to measure the ligand-field splittings in a range of different complexes. It is found that a particular ligand always produces a greater or smaller splitting than another ligand, regardless of the identity of the central ion. Hence ligands can be arranged in a *spectrochemical series* according to the Δ_O they produce:

Strong-field ligands

CN^-, CO
NO_2^-
NH_3

H_2O
OH^-
F^-
SCN^-, Cl^-
Br^-
I^-

Weak-field ligands

It is convenient to classify ligands below the horizontal line above as *weak-field ligands*, and those above it as *strong-field ligands*. We see from

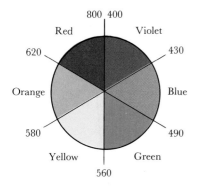

FIGURE 21.31 The perceived color of a complex in white light is the complementary color of the light it absorbs. In this color wheel, complementary colors are diametrically opposite each other. The numbers are approximate wavelengths in nanometers.

the series that any complex of a particular metal ion has a smaller Δ_O value if it contains weak-field ligands than if it contains strong-field ligands. We can therefore go on to say that the complex absorbs longer-wavelength light if it has weak-field ligands than if it has strong-field ligands.

The effect of ligands on color. We saw in Section 7.1 that white light is a mixture of all wavelengths of electromagnetic radiation from about 420 nm (violet) to 700 nm (red). When some of these wavelengths are removed from a beam of white light (by passing the light through a sample that absorbs certain wavelengths), the emerging light is no longer white. For example, if red light is absorbed from white light, the light that remains is green. If green is removed, the light appears red. We say that red and green are each other's *complementary colors*—each is the color that white light becomes when the other is removed. Complementary colors are shown on the "color wheel" in Fig. 21.31.

We now see that if a substance looks blue (as does copper sulfate(II) solution, for instance), then it is absorbing orange (620-nm) light. Conversely, if we know the wavelength (and therefore the color) of the light that a substance absorbs, then we can predict the color of the substance by noting the complementary color on the color wheel. Thus, since $[Ti(H_2O)_6]^{3+}$ absorbs 510-nm light, which is green light, the complex looks purple (Fig. 21.32). However, it is important to realize that the color of a compound is actually a combination of very subtle effects, and such simple predictions can be misleading. One problem is that compounds absorb light over a range of wavelengths and may absorb in several regions of the spectrum. Chlorophyll, for example, absorbs both red and blue light, leaving only the wavelengths near green to be reflected from vegetation. The following discussion is a very simplified version of the real situation.

Since weak-field ligands give small splittings, the complexes they form can be expected to absorb long-wavelength light. The long wavelengths correspond to red light, so these complexes can be expected (as a first approximation) to exhibit colors near green. Since strong-field ligands give large splittings, the complexes they form should absorb short wavelengths. Because short-wavelength light corresponds to the violet end of the visible spectrum, such complexes can be expected to have colors near orange and yellow (Fig. 21.33). This is part of the reason for the color change that occurs when ammonia is added to aqueous copper(II) sulfate: strong-field ammine (NH_3) ligands replace four of the weak-field aqua (H_2O) ligands of the $[Cu(H_2O)_6]^{2+}$ ion. The absorption shifts to higher energies and shorter wavelengths, from orange light to yellow light, and the perceived color shifts from blue toward violet.

21.8 THE ELECTRONIC STRUCTURES OF MANY-ELECTRON COMPLEXES

The electron configurations of d^n complexes—complexes containing n electrons in the d orbitals of the central metal—are obtained by following the rules of the building-up principle. There are three t orbitals; therefore, according to the Pauli exclusion principle, there can be up to

FIGURE 21.32 Since $[Ti(H_2O)_6]^{3+}$ absorbs green 510-nm light, it looks purple in white light.

FIGURE 21.33 The effect of changing the ligands in octahedral cobalt(III) complexes. Weak-field ligands are on the left, strong-field ligands are on the right.

six t electrons in a complex (a maximum of two electrons in each orbital). There are two e orbitals, and hence there can be up to four e electrons.

We add the n electrons to the t and e orbitals of the complex in the arrangement that gives the lowest overall energy (but with no more than two electrons in any one orbital). For this, we use the orbital energy-level diagram in Fig. 21.30 for octahedral complexes, and the diagram in Fig. 21.31 for tetrahedral complexes. The other important type of complex, square planar, presents a more complicated case that we shall not deal with here.

High- and low-spin complexes. The lowest-energy electron configurations for d^1 through d^3 octahedral complexes can be written without difficulty. There are three t orbitals, and because all three have the same energy, each electron can occupy a separate t orbital. According to Hund's rule (Section 7.5) these electrons will have parallel spins:

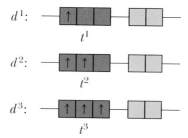

A conflict arises at $n = 4$ for octahedral complexes. The fourth electron can enter a t orbital, resulting in a t^4 configuration. However, to do so it must enter an orbital that is already half occupied, and hence experience a strong repulsion from the electron already there:

Alternatively, it can avoid that strong repulsion by occupying an empty e orbital to give a t^3e^1 configuration:

However, it will then experience a strong repulsion from the ligands. The configuration that is actually adopted is the one that leads to the lower overall energy. If Δ_O is large (as it is for strong-field ligands), signifying strong ligand repulsion of an e electron, then t^4 will give the lower energy. If Δ_O is small (for weak-field ligands), then t^3e^1 will be the lower-energy configuration and hence the one adopted.

▼ **EXAMPLE 21.8** **Predicting the electron configuration of a complex**

Predict the electron configuration of an octahedral d^5 complex with (a) strong-field ligands and (b) weak-field ligands, and give the number of unpaired electrons in each case.

STRATEGY We have to decide whether the lowest energy is reached with all the electrons in the lowest set of orbitals, in which case there will be strong electron-electron repulsions, or with some electrons occupying the upper set of orbitals too. If the splitting Δ_O is large, the lowest overall energy may be obtained by occupying the lower orbitals despite the strong electron-electron repulsions that will occur. If Δ_O is small, electrons are likely to occupy upper orbitals.

SOLUTION   (a) In the strong-field case, all five electrons enter the t orbitals, and to do so some of them must pair:

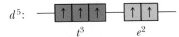

There is one unpaired electron in this configuration.
(b) In the weak-field case, the five electrons occupy all five orbitals without pairing:

$$d^5: \quad \boxed{\uparrow}\boxed{\uparrow}\boxed{\uparrow} \quad \boxed{\uparrow}\boxed{\uparrow}$$
$$\qquad\qquad t^3 \qquad\quad e^2$$

There are now five unpaired electrons.

EXERCISE   Predict the electron configuration and the number of unpaired electrons of an octahedral d^6 complex with (a) strong-field ligands and (b) weak-field ligands.

[*Answer*: (a) t^6 (0); (b) t^4e^2 (4)]

▲

The configurations predicted for complexes with $n = 1$ through 10 are listed in Table 21.7. Note that alternative configurations occur for $n = 4$ through $n = 7$ for octahedral complexes. In tetrahedral complexes the ligand field is always too weak to correspond to anything other than the weak-field case, so there is no need to trouble about the alternative configurations.

TABLE 21.7   Electronic configurations of d^n complexes

Number of d electrons	Octahedral complexes		Tetrahedral complexes
0	—		
1	t^1		e^1
2	t^2		e^2
3	t^3		e^2t^1
	Low-spin	*High-spin*	
4	t^4	t^3e^1	e^2t^2
5	t^5	t^3e^2	e^2t^3
6	t^6	t^4e^2	e^3t^3
7	t^6e^1	t^5e^2	e^4t^3
8	t^6e^2		e^4t^4
9	t^6e^3		e^4t^5
10	t^6e^4		e^4t^6

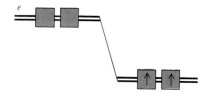

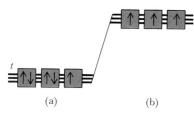

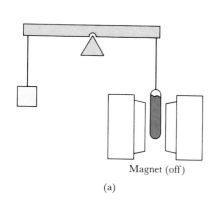

FIGURE 21.34 A strong-field ligand is likely to lead to a low-spin complex (a); a weak-field ligand is likely to lead to a high-spin complex (b).

A d^n complex with the maximum possible number of unpaired spins is called a *high-spin complex*. A d^n complex with the minimum possible number of unpaired spins is called a *low-spin complex*. Tetrahedral complexes are almost always high-spin complexes. For octahedral complexes, when the two alternatives exist we can predict whether a complex is likely to be a high-spin or low-spin complex by noting where the ligands lie in the spectrochemical series. If they are strong-field ligands, we expect a low-spin complex; if they are weak-field ligands, we expect a high-spin complex. This rule of thumb is summarized in Fig. 21.34.

The magnetic properties of complexes. As we saw in Section 9.6 (in connection with O_2) a compound with unpaired electrons is *paramagnetic*; such a compound is pulled into a magnetic field. A substance without unpaired electrons is *diamagnetic* and is pushed out of a magnetic field. The two types of substances can be distinguished experimentally with the apparatus shown in Fig. 21.35: a sample is hung from a balance so that it lies between the poles of an electromagnet. When the magnet is turned on, a paramagnetic substance is pulled into the field and appears to weigh more when the magnet is turned on than when it is off. A diamagnetic substance is pushed out of the field and appears to weigh less.

Many d-metal complexes have unpaired d electrons and are therefore paramagnetic. We have just seen that a high-spin d^n complex has more unpaired electrons than a low-spin d^n complex. The former is therefore more strongly paramagnetic and is drawn more strongly into a magnetic field. Moreover, whether a complex is a high-spin or low-spin complex depends on the ligands present: strong-field ligands tend to result in low-spin and thus weakly paramagnetic substances, whereas weak-field ligands tend to result in high-spin, strongly paramagnetic substances. This suggests that we should be able to modify the magnetic properties of a complex by changing its ligands.

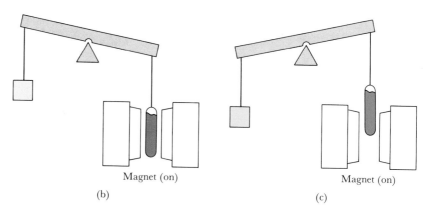

FIGURE 21.35 The magnetic character of a complex can be studied with the arrangement shown here. (a) A sample is hung from a balance so that it lies between the poles of an electromagnet. (b) When the magnet is turned on, a paramagnetic sample is drawn into the magnetic field, making the sample appear to weigh more. (c) In contrast, a diamagnetic sample is pushed out of the field when the magnet is turned on.

▼ EXAMPLE 21.9 Predicting the magnetic properties of a complex

What magnetic properties can be expected when the hydrated Fe^{2+} ion is complexed with cyanide ions?

STRATEGY We should decide from their positions in the spectrochemical series whether the complex has weak-field or strong-field ligands. Then we should judge whether the reaction changes the complex from a high-spin to a low-spin complex or vice versa. In either case we can expect a change in its magnetic properties.

SOLUTION The Fe^{2+} ion in water is the d^6 $[Fe(H_2O)_6]^{2+}$ ion. Since H_2O is a weak-field ligand, we predict a high-spin configuration with four unpaired electrons. The ion is therefore predicted (and found) to be paramagnetic. When cyanide ions are added to the solution, they form the $[Fe(CN)_6]^{4-}$ ion. Now the ligands are strong-field ligands, and the resulting complex is a low-spin t^6 complex; it has no unpaired electrons, and it is not paramagnetic. In other words, the ligand substitution reaction has had the effect of quenching the paramagnetism.

EXERCISE What magnetic properties can be expected when NO_2^- ligands in an octahedral complex are replaced by Cl^- ligands in (a) a d^6 complex; (b) a d^3 complex?

[*Answer*: (a) The complex becomes paramagnetic; (b) no change in magnetic properties]

As we have seen, the spectroscopic and magnetic properties of d-metal complexes are closely interrelated. In more advanced work, we could add their thermodynamic stability (and, to some extent, their labilities) to this list. Each of these properties is related to the magnitude of the ligand-field splitting, and they are all brought together in a single explanation by a single, unified theory.

SUMMARY

21.1 All the d-block elements are metals. Their metallic radii are all quite similar, permitting extensive alloy formation. The **lanthanide contraction** results in the third-row elements (those in Period 6) having radii very close to those of the second-row elements and hence being very dense. The principal chemical characteristics of the d-block elements are their ability to form numerous complexes, their action as catalysts, and their ability to occur with a wide range of oxidation numbers. With the exception of Hg, the elements at the ends of each row of the block have only one oxidation number other than zero. Elements close to the center of each row have the widest range of oxidation numbers. Elements toward the left of the block and lower down a group tend to be found with their higher oxidation numbers. Although most d-block oxides are basic, the oxides of a given metal show a shift toward acidic character with increasing oxidation number.

21.2 Many of the d metals are prepared as **ferroalloys** with iron, and used in steelmaking. The ease of extraction of the metals from their ores generally increases from left to right across the d block. The metals on the left are extracted by versions of the thermite process. Iron is extracted by reduction with carbon monoxide in a blast furnace.

21.3 The metals on the right of the d block are extracted by either **pyrometallurgical** or **hydrometallurgical** processes; the former include **roasting,** or heating in air, and **smelting,** or melting with a reducing agent. The ores are sometimes concentrated by **froth flotation,** in which the ore floats upward in a froth of detergent. The metals are often purified either by electrolysis or by the formation of a volatile compound that is subsequently decomposed.

21.4 and 21.5 A characteristic property of the d metals is the ability of their atoms and ions to act as Lewis acids and form **complexes** with Lewis bases that become **ligands.** Complexes with six ligands in the **coordination sphere,** which is made up of the ligands directly attached to the metal, are usually octahedral; those with four ligands are either square planar or tetrahedral. Ligands may be either **monodentate** or **polydentate;** the latter can form **chelates,** compounds in which a ring surrounds a central metal atom.

21.6 Ligands may occur in different arrangements, leading to **isomerism,** the existence of different compounds built from the same atoms. The classes of isomerism are summarized in Fig. 21.23. **Chiral** molecules, those that cannot be superimposed on their mirror image, are **optically active** in that they rotate the plane

of polarization of polarized light. **Enantiomeric pairs** of isomers, which are the mirror images of each other, produce equal but opposite rotations of this plane of polarization.

21.7 In the simple version of **ligand field theory** known as **crystal field theory,** the effect of each ligand is represented by a point negative charge. The charges split the five d orbitals into two groups: three t **orbitals** and two e **orbitals** (in octahedral and teterahedral complexes), with energy separation Δ (Δ_O in octahedral complexes, and Δ_T in tetrahedral complexes). This separation is called the **ligand-field splitting;** it can be estimated from the wavelength of light absorbed by a complex. Each complex has the **complementary color** of the color of the light it absorbs. The size of the splitting increases along the **spectrochemical series** of the ligands, from **weak-field ligands** to **strong-field ligands.**

21.8 Octahedral d^4 to d^7 complexes may have either a **high-spin** or a **low-spin** configuration. The configuration that is adopted is the one that produces the lowest overall energy. High-spin complexes are more strongly paramagnetic than low-spin complexes.

EXERCISES

The Elements and Their Configurations

Name each element in Exercises 21.1 and 21.2.

21.1 (a) Mn; (b) Rh; (c) Re; (d) W.
21.2 (a) Sc; (b) Zr; (c) Ta; (d) Y.

In Exercises 21.3 and 21.4, give the chemical symbol of each element.

21.3 (a) Copper; (b) osmium; (c) mercury; (d) technetium.
21.4 (a) Gold; (b) silver; (c) niobium; (d) ruthenium.

For Exercises 21.5 to 21.8, draw the d block as a 10×3 grid of squares. Then locate each element on it.

21.5 (a) Scandium; (b) chromium; (c) iron; (d) nickel.
21.6 (a) Zinc; (b) manganese; (c) cobalt; (d) mercury.
21.7 (a) Zirconium; (b) lanthanum; (c) silver; (d) molybdenum.
21.8 (a) Rhenium; (b) gold; (c) niobium; (d) hafnium.

For Exercises 21.9 and 21.10, take the columns of the d block as being numbered 3 through 12, and the rows as 1 through 3. Then, without referring to a printed periodic table, identify (by symbol and name) the element in each specified [row, column] location.

21.9 (a) [1, 3]; (b) [2, 12]; (c) [1, 11]; (d) [3, 10].
21.10 (a) [1, 5]; (b) [1, 8]; (c) [3, 9]; (d) [3, 3].

In Exercises 21.11 and 21.12, write the ground-state electron configuration of the atoms of each element.

21.11 (a) Sc; (b) V; (c) Cu; (d) Au.
21.12 (a) Ti; (b) Cr; (c) Cd; (d) Hg.

Trends in Properties

21.13 Account for the trends in the following metallic radii:

(a) Sc (160 pm), Ti (146 pm), and V (131 pm)
(b) Cu (128 pm), Ag (144 pm), and Au (144 pm)

21.14 Account for the trends in the following densities (in grams per cubic centimeter):

(a) Zn (7.14), Cd (8.64), and Hg (13.6)
(b) Fe (7.86), Co (8.90), and Ni (8.90)

21.15 Explain what is meant by the "lanthanide contraction," and give examples of its effect. Suggest reasons why the p-block metals have less negative reduction potentials than the s-block metals. Is there evidence for a "d-block contraction" analogous to the lanthanide contraction?

21.16 Explain how the oxidation numbers of the *d*-block elements help to explain their chemical properties. Give examples to illustrate the points you make.

In Exercises 21.17 and 21.18, outline the process, with chemical equations where possible, by which each element is prepared in the pure state.

21.17 (a) Titanium; (b) copper; (c) mercury.
21.18 (a) Nickel; (b) vanadium; (c) zinc.

Scandium through Nickel

21.19 Explain how sodium dichromate can be prepared from chromite ore. Suggest, with equations, how chromium(VI) oxide might be obtained from chromite ore.

21.20 Decide whether or not acidified aqueous sodium dichromate can be used to bring about each of the following oxidations:

(a) Bromide ions to bromine
(b) Manganese(II) ions to manganese dioxide
(c) Chloride ions to chlorine
(d) Silver(I) ions to silver(II) ions

21.21 Give the systematic names and chemical formulas of the following substances: (a) rutile; (b) hematite; (c) pyrolusite; (d) magnetite; (e) pyrite; (f) ilmenite.

21.22 Explain how the concentrations of chromate ions and dichromate ions may be calculated from the pH of a solution containing them both. If 250 mL of an aqueous solution contains 4.0 g of Na_2CrO_4, what are the concentrations of CrO_4^{2-} and $Cr_2O_7^{2-}$ ions at a pH of (a) 5; (b) 7; (c) 9?

21.23 The standard reduction potentials of the couples Cr^{3+}/Cr^{2+} and Cr^{2+}/Cr are -0.41 V and -0.91 V, respectively. Calculate the standard electrode potential of the Cr^{3+}/Cr couple.

21.24 Express the acidic disproportionation of the manganate ion into the permanganate ion and manganese dioxide in terms of oxidation and reduction half-reactions.

21.25 Write the chemical equation for the electrolytic oxidation of manganate ions to permanganate. What is the *minimum* potential needed to achieve oxidation?

21.26 Describe, with chemical equations, the processes that occur in a blast furnace during the production of iron from its ores. What role does the limestone play? Discuss the processes thermodynamically, and classify the reactions by type (redox, Lewis, and so on).

21.27 By considering electron configurations, suggest a reason why iron(III) compounds are readily prepared from iron(II), but the conversion of nickel(II) or cobalt(II) to nickel(III) or cobalt(III) is much more difficult.

21.28 Iron(III) hydroxide reacts with aqueous hypochlorite ions in basic solution to form the ferrate ion, FeO_4^{2-}. Write the chemical equation for the reaction, and suggest what properties might be expected for ferrates.

21.29 Iron, cobalt, and nickel are often grouped together as the *iron triad*. What similarities do they exhibit that justifies this grouping? How may these similarities be explained?

Copper through Mercury

21.30 Explain, with equations, how (a) the coinage metals and (b) zinc, cadmium, and mercury are obtained from their ores.

21.31 Describe briefly how the properties of the *d* metals vary down a group, illustrating your answer with examples drawn from the coinage metals and zinc, cadmium, and mercury.

21.32 Describe the chemical evidence for treating the coinage metals as a single group.

21.33 Describe the chemical evidence for treating zinc, cadmium, and mercury as a single group.

21.34 Compare the chemical properties of copper and zinc, and account for the differences in terms of their electron configurations.

21.35 What chemical tests would distinguish brass from bronze?

21.36 Explain as fully as possible, with diagrams of the structures of the hydrated ions, the changes that occur when (a) anhydrous copper(II) sulfate is moistened, (b) it is dissolved in water, and (c) aqueous ammonia is added to the resulting solution.

***21.37** Calculate the standard reduction potential of the Cu^+/Cu couple, given that the values for Cu^{2+}/Cu and Cu^{2+}/Cu^+ are $+0.34$ V and $+0.15$ V, respectively.

***21.38** Calculate the standard reduction potential of the Au^{3+}/Au^+ couple, given that the values for Au^{3+}/Au and Au^+/Au are $+1.40$ V and $+1.69$ V, respectively.

21.39 Calculate the equilibrium constant for the disproportionation of copper(I) ions in aqueous solution at 25°C from the standard reduction potentials in Appendix 2B. Explain why the result implies that although a soluble copper(I) salt is unstable in water, a slightly soluble copper(I) salt might survive.

21.40 Calculate the equilibrium constant for the disproportionation of gold(I) ions in aqueous solution at 25°C from the standard reduction potentials in Appendix 2B.

21.41 Summarize, giving equations and suggesting explanations, how (a) the coinage metals, (b) cadmium, and (c) mercury react with acids and alkalis.

21.42 Would it be wise to try to protect an iron object by plating it with copper instead of zinc?

Complexes

Name each complex in Exercises 21.43 and 21.44.

21.43 (a) $[Fe(CN)_6]^{4-}$; (b) $[Co(NH_3)_6]^{3+}$; (c) $[Co(H_2O)(CN)_5]^{2+}$; (d) $[Co(NH_3)_5SO_4]^+$.

21.44 (a) $[Fe(CN)_6]^{3-}$; (b) $[Fe(H_2O)_5OH]^{2+}$; (c) $[Co(NH_3)_4(H_2O)Cl]^{2+}$; (d) $[Ir(en)_3]^{3+}$.

Name each compound in Exercises 21.45 and 21.46.

21.45 (a) $[Cr(NH_3)_5SO_4]Cl$; (b) $[Cr(en)_3][Cr(ox)_3]$; (c) $[Pt(NH_3)_2Cl_2]$; (d) $K_3[Co(NO_2)_6]$.

21.46 (a) $[Pt(en)_2Cl_2]Cl_2$; (b) $[Co(en)_3][Fe(CN)_6]$; (c) $K_3[Fe(ox)_3]$; (d) $Na[Cr(EDTA)]$.

In Exercises 21.47 and 21.48, write the formula for each compound.

21.47 (a) Potassium hexacyanochromate(III)
(b) Pentaamminesulfatocobalt(III) chloride
(c) Tetraamminediaquacobalt(III) bromide
(d) Sodium diaquabis(oxalato)ferrate(III)

21.48 (a) Triammineaquadihydroxochromium(III) chloride
(b) Potassium tetrachloroplatinate(III)
(c) Tetraaquadichloronickel(II)
(d) Potassium tris(oxalato)rhodium(III) chlorohydroxybis(oxalato)rhodium(III) octahydrate

Isomerism

In Exercises 21.49 and 21.50, state the type of isomerism shown by the given complexes and compounds.

21.49 (a) $[Cr(NH_3)_5Cl]Br$ and $[Cr(NH_3)_5Br]Cl$
(b) *cis*- and *trans*-$[Pt(NH_3)_2Cl_2]$
(c) $[Co(en)_3]^{3+}$

21.50 (a) $[Pt(NH_3)_4SO_4](OH)_2$ and $[Pt(NH_3)_4(OH)_2]SO_4$
(b) $[Pt(NH_3)_4][PtCl_6]$ and $[Pt(NH_3)_4Cl_2][PtCl_4]$
(c) $[Co(NH_3)_5NO_2]Br_2$ and $[Co(NH_3)_5(ONO)]Br_2$

21.51 Can a tetrahedral complex show (a) stereoisomerism; (b) geometrical isomerism; (c) optical isomerism?

21.52 Draw the *cis*-diammine-*cis*-diaqua-*cis*-dichlorochromium(III) ion, and comment on its isomerism. What can be said about the *trans*-diammine isomer?

In Exercises 21.53 and 21.54, determine which of the given complexes are chiral and which form enantiomeric pairs.

*21.53 (a)

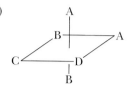

(b)

21.54 (a)

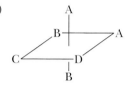

(b)

21.55 Suggest a chemical test for distinguishing between
(a) $[Ni(en)_2(NO_2)_2]Cl_2$ and $[Ni(en)_2Cl_2](NO_2)_2$
(b) $[Ni(en)_2I_2]Cl_2$ and $[Ni(en)_2Cl_2]I_2$

Crystal Field Theory

21.56 Explain how crystal field theory connects colors with the magnetic properties of complexes.

21.57 Would you expect the *p* orbitals of a main-group atom to be split by either an octahedral or a tetrahedral ligand field?

*21.58 Suggest the form that the orbital energy-level diagram would take for a square-planar complex, and discuss how the building-up principle applies.

In Exercises 21.59 and 21.60, draw an orbital energy-level diagram showing how the electrons are accommodated in each complex ion.

21.59 (a) $[Co(NH_3)_6]^{3+}$; (b) $[NiCl_4]^{2-}$; (c) $[Fe(H_2O)_6]^{3+}$; (d) $[Fe(CN)_6]^{3-}$.

21.60 (a) $[Zn(H_2O)_6]^{2+}$; (b) $[CoCl_4]^{2-}$; (c) $[Co(CN)_6]^{3-}$; (d) $[CoF_6]^{3-}$.

21.61 Explain the difference between a weak-field ligand and a strong-field ligand. What measurements can be used to classify them as such?

21.62 Describe the changes that occur in a compound's properties when weak-field ligands are replaced by strong-field ligands.

In Exercises 21.63 and 21.64, state the color of a sample that absorbs light of the given wavelength.

21.63 (a) 410 nm; (b) 650 nm; (c) 480 nm.
21.64 (a) 570 nm; (b) 580 nm; (c) 590 nm.

In Exercises 21.65 and 21.66, suggest the wavelength that a compound will absorb most strongly if it has the stated color in white light.

21.65 (a) Yellow; (b) green; (c) blue.

21.66 (a) Purple; (b) red; (c) orange.

21.67 Suggest a reason why $Zn^{2+}(aq)$ ions are colorless. Would you expect zinc compounds to be paramagnetic?
21.68 Suggest a reason why copper(II) compounds are often colored but copper(I) compounds are colorless. Which oxidation number gives paramagnetic copper compounds?

In Exercises 21.69 and 21.70, calculate the ligand-field splitting for each complex and arrange the ligands in a spectrochemical series, knowing that each complex absorbs light of the given wavelength.

21.69 $[CrCl_6]^{3-}$ (740 nm); $[Cr(H_2O)_6]^{3+}$ (575 nm); $[Cr(NH_3)_6]^{3+}$ (460 nm)
21.70 $[Co(H_2O)_6]^{3+}$ (540 nm); $[Co(NH_3)_6]^{3+}$ (435 nm); $[Co(CN)_6]^{3-}$ (295 nm)

▼ ▼ ▼ ▼

In this chapter we examine some of the events that accompany changes in the structures of atomic nuclei. These include radioactivity, the conversion of one element into another, and the release of energy for the generation of nuclear power. Nuclear power, in particular, presents chemists with several demanding problems, and we see some of their attempts to solve them. The illustration shows part of the fuel-rod assembly for an advanced gas-cooled nuclear reactor.

▲ ▲ ▲ ▲

NUCLEAR CHEMISTRY

Nuclear stability

22.1 Nuclear structure and nuclear radiation
22.2 The identities of daughter nuclides
22.3 The pattern of nuclear stability
22.4 Nucleosynthesis

Radioactivity

22.5 Measuring radioactivity
22.6 The rate of nuclear disintegration

Nuclear power

22.7 Nuclear fission
22.8 Nuclear fusion
22.9 Chemical aspects of nuclear power

So far, we have taken the atomic nucleus to be little more than an unchanging passenger in chemical reactions. However, nuclei can change. *Nuclear chemistry* is the study of the structure of atomic nuclei, the changes this structure undergoes, and the consequences of those changes for chemistry. Nuclei that change their structure spontaneously and emit radiation are called *radioactive*. Many such unstable nuclei occur naturally. For instance, all nuclei with $Z > 83$ (from polonium onward in the periodic table) are unstable and radioactive, and all the lighter elements have isotopes that are unstable to some degree.

Nuclear chemistry is central to the development of nuclear power because techniques must be found for the purification and recycling of nuclear fuels and for the disposal of hazardous radioactive waste. The numerous other applications of nuclear chemistry include the investigation of reaction mechanisms, the dating of archeological objects, and the development of methods for fighting cancer.

NUCLEAR STABILITY

We introduced some features of the structure of atomic nuclei in Section 2.3. The key points and terms are summarized below, after which we shall examine the different types of nuclear radiation and the changes in nuclear structure responsible for them.

22.1 NUCLEAR STRUCTURE AND NUCLEAR RADIATION

A simple, but for our purposes adequate, picture of the nucleus is as a tightly bound collection of *nucleons*, the general name for protons p and neutrons n. These two types of nucleons have almost exactly the same mass (Table 2.2), but a proton has a single positive charge and a neutron is electrically neutral. Nucleons are bound together within the nucleus by an attractive force that is so strong over short distances that it overcomes the electrostatic repulsion between protons (Fig. 22.1). Provided the number of protons is not too large, therefore, the nucleus may be stable. However, in a nucleus containing more than 83 protons, these charged particles repel each other so strongly that the nucleus is unstable and hence radioactive.

Nuclides. The number of protons in a nucleus is the atomic number Z of the element. The total number of nucleons (protons plus neutrons) in a nucleus is the mass number A. The number of neutrons that can accompany a given number of protons in a nucleus is variable over a small range. This variability gives rise to *isotopes*, atoms with the same atomic number (the same number of protons) but different mass numbers (because they have different numbers of neutrons); two examples are carbon-12 and carbon-14. The isotopes of an element are chemically almost identical, because their atoms have the same number of electrons and hence the same electron configuration.

Each distinct isotope of an element is called a *nuclide*. Thus, carbon-12 is one nuclide, carbon-14 is another, and uranium-235 is still an-

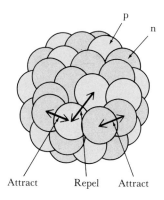

FIGURE 22.1 A nucleus can be pictured as a collection of protons (pink) and neutrons (gray). The protons repel each other electrostatically, but a strong force that acts between all the particles holds the nucleus together.

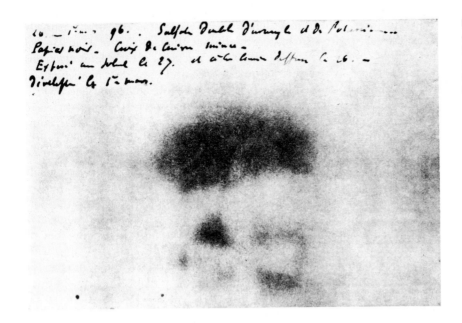

FIGURE 22.2 Becquerel discovered radioactivity when he noticed that an unexposed photographic plate left near some uranium oxide became fogged. This is one of his original plates.

other. A nuclide is denoted by the symbol $_Z^A$E, where E is the chemical symbol of the element, Z its atomic number, and A its mass number. All nuclides with the same value of Z are isotopes of the element E, but no two distinct nuclides have the same values of both Z and A. Strictly speaking, Z is redundant (because the symbol E implies the value of Z, as He implies $Z = 2$); but when we are keeping track of changes in nuclei, it is useful to display it. We shall sometimes denote an individual proton ($A = 1$, $Z = 1$) by $_1^1$p, and an individual neutron ($A = 1$, $Z = 0$) by $_0^1$n when we want to be specific about their masses and charges.

Radiation from nuclei. Radioactivity was discovered by the French scientist Henri Becquerel in 1896 (Fig. 22.2). His discovery was taken up by Ernest Rutherford, who identified three different types of radioactivity by observing the effect of electric fields on the paths of the rays (Fig. 22.3). He called the three types α (alpha), β (beta), and γ (gamma) radiation. As we shall see, α radiation is more typical of the heavy elements, and β radiation is more typical of the lighter ones; γ radiation often accompanies α and β radiation but may occur alone.

Rutherford found that α rays are repelled from a positively charged electrode. This led him to propose that they are a stream of positively charged particles, which he called α *particles*. He was able to deduce the ratio of the charge of an α particle to its mass from the size of the deflection caused by electric and magnetic fields, and hence to identify α particles as the nuclei of helium atoms: $_2^4$He^{2+}. This identification was confirmed when helium gas was detected near a substance that emitted α radiation, for a helium atom forms when an α particle captures two electrons during collisions with other atoms:

$$_2^4\text{He}^{2+} + 2\text{e}^- \longrightarrow {_2^4}\text{He}$$

An α particle is denoted $_2^4\alpha$, or simply α. We can think of it as a tightly bound cluster of two protons and two neutrons.

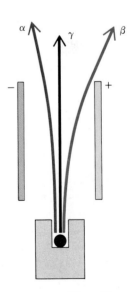

FIGURE 22.3 The effects of an electric field on nuclear radiation. The deflections identify α rays as positively charged, β rays as negatively charged, and γ rays as uncharged.

Rutherford also found that β rays are attracted to a positively charged electrode, which suggested that they consist of a stream of negatively charged particles. That the particles are actually electrons was confirmed by measuring their charge and mass. Electrons emitted by nuclei (but not electrons from outside the nucleus) are called β _particles_ and denoted β. The important characteristics of the β particle are its single negative charge and the fact that its mass is negligible in comparison with the masses of protons and neutrons. (We saw in Chapter 2 that the mass of an electron is only about $\frac{1}{1840}$ that of a nucleon.) Since a β particle has no protons or neutrons, its mass number is zero. Since its charge is -1, we sometimes (as when balancing the equations that we meet later) denote it $_{-1}^{0}e$, but the subscript is not a true atomic number.

Finally, Rutherford found that γ radiation is unaffected by electric fields. He identified it as electromagnetic radiation like light, but of much higher frequencies—greater than about 10^{20} Hz—and with wavelengths below about 1 pm (the wavelength of visible light is about 500,000 times greater). Since electromagnetic radiation consists of photons with energy proportional to their frequency, γ radiation can be regarded as a stream of photons with very high energies; each photon is emitted by a single nucleus as that nucleus discards energy. A γ-ray photon is denoted γ. Like all photons, γ-ray photons are massless and uncharged.

Nuclear disintegration. The emission of an α particle or a β particle from a nucleus occurs as a result of _nuclear disintegration_, the partial breakup of a nucleus. The disintegration is only partial, in the sense that, after it takes place, a less massive nucleus is left behind (Fig. 22.4). This less massive nucleus, called the _daughter nucleus_, may be the nucleus of some other element. For example, when a radon-222 nucleus emits an α particle, a polonium-218 nucleus is formed. This process (and any other disintegration) can be described in the form of a chemical equation:

$$^{222}_{86}\text{Rn} \longrightarrow {}^{218}_{84}\text{Po} + \alpha$$

When a carbon-14 nucleus emits a β particle, a nitrogen-14 nucleus is formed:

$$^{14}_{6}\text{C} \longrightarrow {}^{14}_{7}\text{N} + \beta$$

In each case, _nuclear transmutation_, or the conversion of one element into another, has taken place. Nuclear transmutation, particularly of lead into gold, was the dream of the alchemists and the root of modern chemistry. However, only in this century has it been recognized in nature and achieved in the laboratory.

It is common for α and β radiation to be accompanied by γ radiation, because the ejection of an α or β particle often leaves the nucleons of the daughter nucleus in a high-energy arrangement (Fig. 22.5); a γ-ray photon is emitted as the nucleus rearranges to a state of lower energy. As in the case of excited atoms, the photon has a frequency ν given by the Bohr frequency condition $\Delta E = h\nu$ (Section 7.2). Gamma rays have very high frequency because the energy difference between the excited and ground nuclear states is very large.

Experiments done since Rutherford's time have shown that other particles can be emitted when nuclei disintegrate (Table 22.1). One of

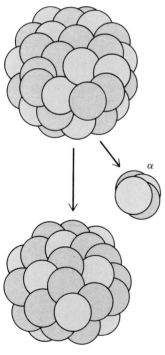

FIGURE 22.4 In the process of nuclear disintegration, a nucleus ejects a particle (in this case an α particle) and becomes a daughter nucleus.

the most important is the _positron_ $^{0}_{1}e$ (or more simply β^{+}), which has the same mass as an electron but a single positive charge.

22.2 THE IDENTITIES OF DAUGHTER NUCLIDES

We can predict the identity of a daughter nuclide by noting how the atomic number and mass number change when a particle is ejected from the parent nucleus.

α disintegration. When an α particle is ejected from a nucleus, it carries away two units of positive charge and a mass equivalent to four nucleons. The loss of two units of positive charge (corresponding to the loss of two protons from the nucleus) reduces the atomic number by 2. Thus, when a radium-226 nucleus, with $Z = 88$, undergoes α decay, the fragment remaining is a nucleus of atomic number 86, and hence the nucleus of an atom of the noble gas radon. Since the α particle carries away two neutrons as well as two protons, the mass number is reduced by 4. Hence α decay of radium-226 results in the formation of radon-222. This transmutation—a "nuclear reaction"—is summarized by the equation

$$^{226}_{88}\text{Ra} \longrightarrow {}^{222}_{86}\text{Rn} + {}^{4}_{2}\alpha$$

or, more simply,

$$^{226}\text{Ra} \longrightarrow {}^{222}\text{Rn} + \alpha$$

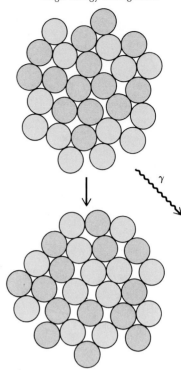

High–energy arrangement

Low–energy arrangement

FIGURE 22.5 Nuclear disintegration may result in the formation of a daughter nucleus with nucleons in a high-energy arrangement; as they adjust into a lower-energy arrangement, the excess energy is released as a γ-ray photon.

TABLE 22.1 Nuclear radiation

Radiation	Comments	Particle*	Example
α	Not penetrating but damaging Speed: $<10\%$ of c†	Helium-4 nucleus $^{4}_{2}\text{He}^{2+}$, $^{4}_{2}\alpha$, α	$^{226}_{88}\text{Ra} \longrightarrow {}^{222}_{86}\text{Rn} + \alpha$
β	Moderately penetrating Speed: $<90\%$ of c	Electron $_{-1}^{0}e$, β^{-}, β	$^{3}_{1}\text{H} \longrightarrow {}^{3}_{2}\text{He} + \beta$
γ	Very penetrating; often accompanies other radiation Speed: c	Photon	$^{60}\text{Co}‡ \longrightarrow {}^{60}\text{Co} + \gamma$
β^{+}	Moderately penetrating Speed: $<90\%$ of c	Positron $^{0}_{1}e$, β^{+}	$^{22}_{11}\text{Na} \longrightarrow {}^{22}_{10}\text{Ne} + \beta^{+}$
p	Moderate to low penetration Speed: $<10\%$ of c	Proton $^{1}_{1}\text{H}^{+}$, $^{1}_{1}\text{p}$, p	$^{53}_{27}\text{Co} \longrightarrow {}^{52}_{26}\text{Fe} + \text{p}$
n	Very penetrating Speed: $<10\%$ of c	Neutron $^{1}_{0}\text{n}$, n	$^{137}_{53}\text{I} \longrightarrow {}^{136}_{53}\text{I} + \text{n}$

*Alternative symbols are given for each particle; often it is sufficient to use the simplest (the one on the right).
†c is the speed of light.
‡An energetically excited state of a nucleus.

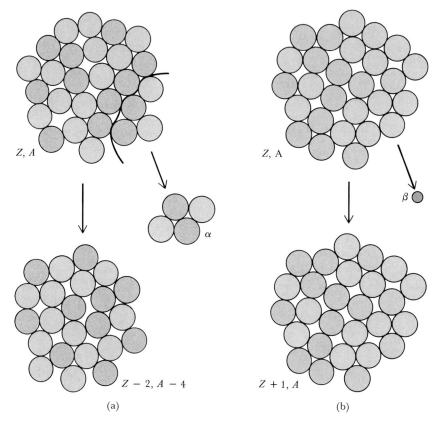

FIGURE 22.6 (a) When a nucleus ejects an α particle, the atomic number of the atom decreases by 2 and the mass number decreases by 4. (b) When a β particle is ejected, the atomic number increases by 1 and the mass number remains unchanged. Pink circles represent protons, and gray circles represent neutrons.

In general, when a nuclide undergoes α decay, the mass number decreases by 4 and the atomic number decreases by 2, because the total numbers of protons and of neutrons are unchanged (Fig. 22.6a).

▼ EXAMPLE 22.1 Identifying an element formed by α decay

Identify the element produced by α decay of cesium-114.

STRATEGY We need to ensure that the mass numbers and atomic numbers balance in the equation for the nuclear reaction. First, we should identify the initial nuclide and write its chemical symbol. Then we write the nuclear reaction, with the mass number and atomic number of the daughter nuclide written as A and Z, respectively. We can find the values of A and Z from the fact that the sum of the mass numbers and the sum of the atomic numbers are both unchanged in the decay. Once we know A and Z, we can identify the daughter nuclide.

SOLUTION The initial nuclide is $^{114}_{55}\text{Cs}$, so the reaction is

$$^{114}_{55}\text{Cs} \longrightarrow {}^{A}_{Z}\text{E} + {}^{4}_{2}\alpha$$

We need $114 = A + 4$ and $55 = Z + 2$; hence $A = 110$ and $Z = 53$. The element with $Z = 53$ is I, so the nuclide that is produced is iodine-110; the reaction is

$$^{114}\text{Cs} \longrightarrow {}^{110}\text{I} + \alpha$$

EXERCISE Identify the nuclide produced by α decay of uranium-235.

[*Answer*: thorium-231]

▲

β disintegration. When a β particle is ejected from a nucleus, it carries away one unit of negative charge. The loss of one unit of negative charge can be interpreted as the conversion of a neutron into a proton within the nucleus:*

$$^1_0\text{n} \longrightarrow {}^1_1\text{p} + {}^{0}_{-1}\text{e} \qquad \text{or} \qquad \text{n} \longrightarrow \text{p} + \beta$$

This implies that the atomic number of the daughter nuclide is greater by 1 than before, because an additional proton is now present. The mass number is unchanged because the total number of protons and neutrons in the nucleus is unchanged (Fig. 22.6b). For example, when sodium-24, with atomic number 11, undergoes β decay, the daughter nuclide is an atom of the element of atomic number 12—magnesium— with the same mass number:

$$^{24}_{11}\text{Na} \longrightarrow {}^{24}_{12}\text{Mg} + {}^{0}_{-1}\text{e}$$

Note that, as required, the mass number on the left (24) is equal to the sum of the mass numbers on the right; similarly, the atomic number on the left (11) is equal to the sum of the atomic numbers on the right. Once we know the equation is balanced (for A and Z), we can write it more simply as

$$^{24}\text{Na} \longrightarrow {}^{24}\text{Mg} + \beta$$

▼ **EXAMPLE 22.2** **Identifying an element formed by β decay**

Identify the element produced by β decay of lithium-11.

STRATEGY We should proceed as in Example 22.1, but·for a nuclear reaction in which an electron is ejected from the nucleus.

SOLUTION The initial nuclide is $^{11}_3\text{Li}$, and the nuclear reaction is

$$^{11}_3\text{Li} \longrightarrow {}^A_Z\text{E} + {}^{0}_{-1}\text{e}$$

We need $11 = A + 0$ and $3 = Z - 1$; hence $A = 11$ and $Z = 4$. The element with $Z = 4$ is Be, so the nuclide produced is beryllium-11. The nuclear reaction is therefore

$$^{11}\text{Li} \longrightarrow {}^{11}\text{Be} + \beta$$

EXERCISE Identify the nuclide produced by β decay of tritium.

[*Answer*: helium-3]

▲

---

*Modern research has shown that another particle, the *neutrino ν*, is also emitted in some nuclear reactions, including the one shown here, which is better written $\text{n} \rightarrow \text{p} + \beta + \nu$. However, as the neutrino has a mass much smaller than the electron (and is almost certainly massless) and is electrically neutral, its emission has no effect on the identity of the daughter nuclide. For our purposes, it can be ignored.

FIGURE 22.7 (a) In the process of electron capture, a nucleus captures one of the surrounding electrons. The effect is to convert a proton into a neutron, so Z decreases by 1 but the mass number remains unchanged. (b) The same outcome is obtained in positron emission, since, in effect, a proton discards its positive charge and becomes a neutron.

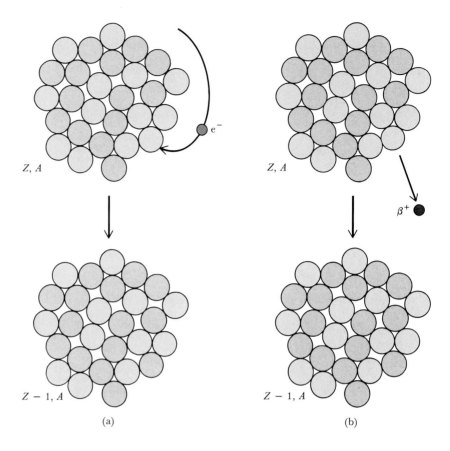

(a) (b)

Other modes of disintegration. Numerous other modes of nuclear transmutation have been observed. In *electron capture* (Fig. 22.7a), a nucleus captures one of its own atom's *s* electrons and the atomic number is reduced by 1. (We saw in Chapter 7 that an *s* electron can be found right at the nucleus, so it runs the risk of being captured by the attractive nuclear forces that act at very short distances there.) An example is the nuclear reaction

$$\ce{^{44}_{22}Ti} + \ce{^{0}_{-1}e} \longrightarrow \ce{^{44}_{21}Sc}$$

In *positron emission*, a positron ($\ce{^{0}_{1}e}$ or β^{+}) is ejected; this also reduces the atomic number by 1 without change of mass number (Fig. 22.7b):

$$\ce{^{43}_{22}Ti} \longrightarrow \ce{^{43}_{21}Sc} + \beta^{+}$$

▼ EXAMPLE 22.3 Identifying the products of other kinds of nuclear transmutations

Identify the nuclides that result from (a) electron capture by potassium-40 and (b) positron emission by sodium-22.

STRATEGY Again we can proceed as in Example 22.1. We write the skeletal (unbalanced) equation for the nuclear reaction, balance the atomic numbers and mass numbers, and identify the product element from its atomic number.

SOLUTION The skeletal equations are

(a) $\ce{^{40}_{19}K} + \ce{^{0}_{-1}e} \longrightarrow \ce{^{A}_{Z}E}$ (b) $\ce{^{22}_{11}Na} \longrightarrow \ce{^{A}_{Z}E} + \ce{^{0}_{1}e}$

These balance when we write

$$\text{(a)} \quad {}^{40}_{19}\text{K} + {}^{0}_{-1}\text{e} \longrightarrow {}^{40}_{18}\text{E} \qquad \text{(b)} \quad {}^{22}_{11}\text{Na} \longrightarrow {}^{22}_{10}\text{E} + {}^{0}_{1}\text{e}$$

(a) The element with $Z = 18$ is Ar, so the reaction produces argon-40.
(b) The element with $Z = 10$ is Ne, so the reaction produces neon-22. The nuclear reactions are therefore

$$\text{(a)} \quad {}^{40}\text{K} + \text{e}^- \longrightarrow {}^{40}\text{Ar} \qquad \text{(b)} \quad {}^{22}\text{Na} \longrightarrow {}^{22}\text{Ne} + \beta^+$$

EXERCISE Identify the nuclides that result from (a) proton emission from cobalt-53 and (b) electron capture by copper-64.

[*Answer*: (a) iron-52; (b) nickel-64]

22.3 THE PATTERN OF NUCLEAR STABILITY

Observations on the stability of nuclides and the modes of radioactive decay for unstable nuclei are summarized in Fig. 22.8. The points (that is, values of A and Z) representing nonradioactive, stable nuclei form a narrow band, the so-called "band of stability." This is surrounded by the "sea of instability," made up of nuclides that are unstable and decay with the emission of radiation. For low atomic numbers (up to about 20), the stable nuclides are those having approximately equal numbers

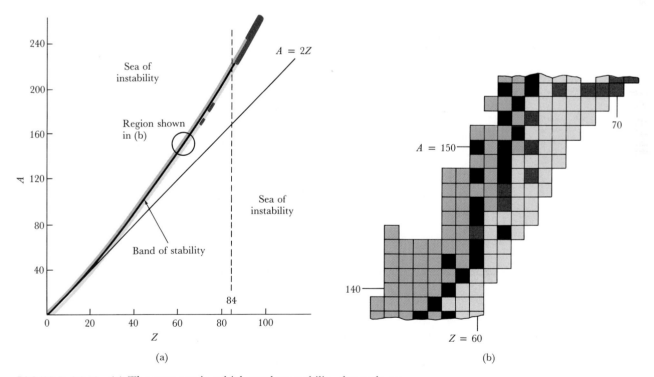

FIGURE 22.8 (a) The manner in which nuclear stability depends on atomic number and mass number. Nuclides along the narrow black band (the band of stability) are generally stable. Nuclides in the blue region are likely to emit a β particle, and those in the red region are likely to emit an α particle. Nuclei in the pink region are likely either to emit positrons or to undergo electron capture. (b) A magnified view of the diagram near $Z = 60$, showing the structure of the band of stability.

of neutrons and protons (those for which A is close to $2Z$). For higher atomic numbers, all known nuclides—both stable and unstable—have more neutrons than protons (so $A > 2Z$).

Predicting the type of disintegration. Figure 22.8 gives us a clue as to the type of disintegration a given nuclide is likely to undergo. Unstable isotopes that lie above the band of stability (such as $^{24}_{11}Na$) have a high proportion of neutrons (are "neutron-rich"); they can reach the band by ejecting a β particle, since this corresponds to the conversion of a neutron into a proton:

$$^{24}_{11}Na \longrightarrow {}^{24}_{12}Mg + \beta$$

Unstable isotopes of elements that lie below the band of stability (such as $^{7}_{4}Be$) have a low proportion of neutrons (are "proton-rich"); they can move toward the band by ejecting a positron, as in

$$^{29}_{15}P \longrightarrow {}^{29}_{14}Si + \beta^{+}$$

or by electron capture, as in

$$^{7}_{4}Be + {}_{-1}^{0}e \longrightarrow {}^{7}_{3}Li$$

As we have noted, all nuclei with $Z > 83$ are unstable and radioactive; they disintegrate mainly by α-particle emission. (On the other hand, very few nuclides with $Z < 60$ emit α particles.) To reach stability, nuclides with $Z > 83$ must discard protons (to reduce their atomic number), and they generally need to lose neutrons as well. Their disintegration often proceeds in a stepwise fashion: first one α particle is ejected, then another α particle or a β particle is ejected, then perhaps one or more α or β particles, until a stable nucleus—usually an isotope of lead (the element with the "safe" atomic number 82)—is formed. Such a stepwise decay path is called a *radioactive series*. Three radioactive series begin from naturally occurring nuclides:

Uranium-238 series. Starts at uranium-238 and ends at lead-206
Uranium-235 series. Starts at uranium-235 and ends at lead-207
Thorium-232 series. Starts at thorium-232 and ends at lead-208

The first of these series is shown in Fig. 22.9. A knowledge of radioactive series was important when radioactivity was first studied, because the only source of radioactive materials was the decay of heavy elements (such as uranium) through a series of products. Their importance now lies in the fact that they summarize the behavior of the components of nuclear fuels.

▼ EXAMPLE 22.4 Identifying the number of α and β emissions

How many α and β particles must be emitted for uranium-238 to transmute into lead-206?

STRATEGY We know that the loss of one α particle reduces the atomic number by 2 and the mass number by 4, so we can find the total reduction in A and Z when N_α α particles are lost. Similarly, we know that the loss of one β particle increases the atomic number by 1 and leaves the mass number unchanged, so we know that the loss of N_β β particles increases the atomic number by N_β. Given the original nuclide and the end product, we can identify the total change in atomic number and in mass number required for transmutation; hence we can set up equations for N_α and N_β.

SOLUTION The total required change in atomic number is $82 - 92 = -10$. The change when N_α α particles and N_β β particles are ejected is $N_\beta - 2N_\alpha$. Therefore, one equation is

$$N_\beta - 2N_\alpha = -10$$

The total required change in mass number is $206 - 238 = -32$. Since β particles have no effect on mass number, the second equation is

$$4N_\alpha = 32$$

Hence, $N_\alpha = 8$ and $N_\beta = 6$. That is, 8 α particles and 6 β particles must be lost.

EXERCISE How many α and β particles must be emitted for thorium-232 to change to lead-208?

[*Answer*: 6α, 4β]

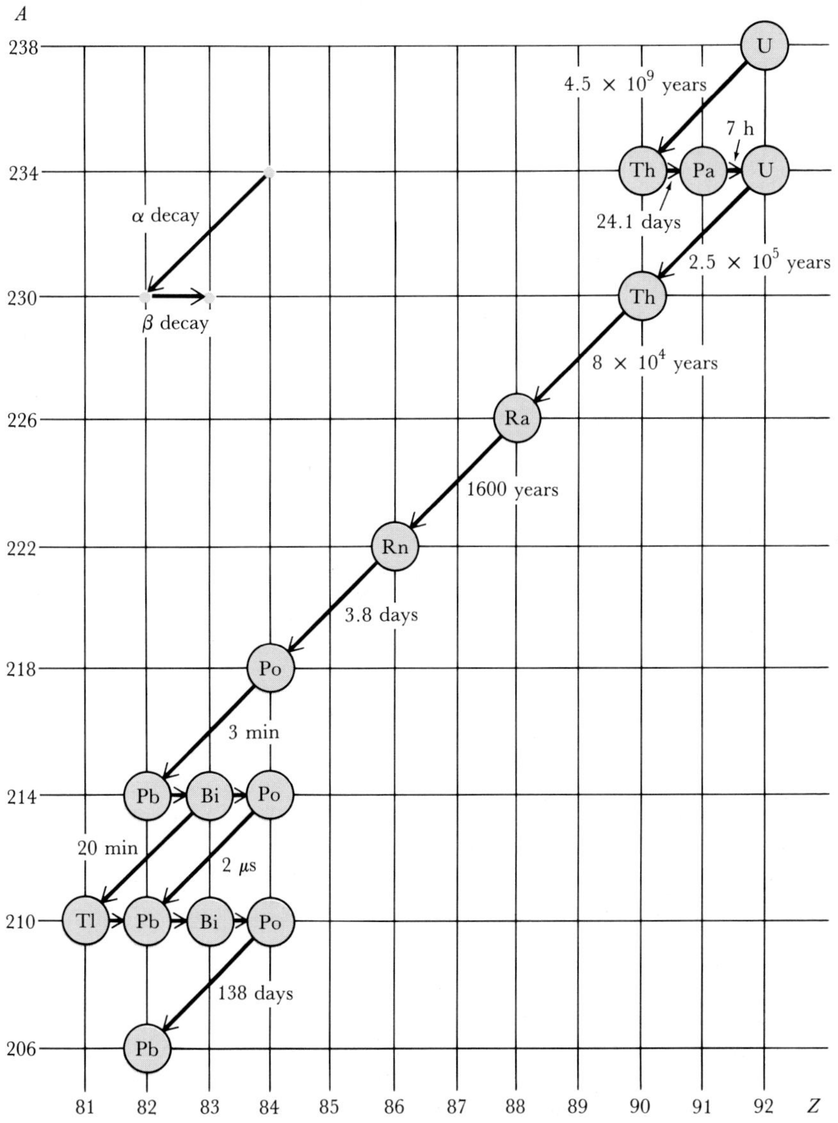

FIGURE 22.9 The uranium-238 series. The times are the half-lives of the nuclides (as explained in Section 22.6).

22.4 NUCLEOSYNTHESIS

The formation of elements is called *nucleosynthesis*. This process occurs naturally in the interiors of stars and is responsible for the formation of many of the elements now found on earth; in this sense, human flesh is stardust. Nucleosynthesis can be brought about artificially on earth by bombarding nuclei with elementary particles or other nuclei that have been accelerated to high speeds in a particle accelerator (Fig. 22.10). High speed is essential if the projectile particles are charged, because they must overcome the electrostatic repulsion of the target nucleus to approach it closely. Heating a substance to a very high temperature—millions of degrees, as in the interior of a star—is another way of achieving high kinetic energy and hence closeness of approach.

FIGURE 22.10 The Stanford Linear Accelerator is one of the devices used to accelerate charged particles to such high kinetic energies that they can penetrate into the nucleus. One outcome is the formation of elements that do not occur naturally.

Neutron-induced transmutation. It is quite easy for a neutron to get very close to a target nucleus, because it is not repelled electrostatically by the nuclear charge. An example of neutron-induced transmutation is the formation of iron-59 from iron-58:

$$\ce{^{58}_{26}Fe} + n \longrightarrow \ce{^{59}_{26}Fe}$$

This is the first step in the preparation of the intensely radioactive nuclide cobalt-60, which is used in some forms of radiation therapy. The second step is β decay of the iron-59 to cobalt-59:

$$\ce{^{59}_{26}Fe} \longrightarrow \ce{^{59}_{27}Co} + \beta$$

In the final step, the cobalt-59 absorbs another neutron from the incident beam and is transformed to cobalt-60:

$$\ce{^{59}_{27}Co} + n \longrightarrow \ce{^{60}_{27}Co}$$

The equation for the overall reaction is the sum of the equations for the elementary reactions:

$$\ce{^{58}_{26}Fe} + 2n \longrightarrow \ce{^{60}_{27}Co} + \beta$$

A nucleosynthesis reaction in which element E′ is formed from element E is often shown in the compact form

$$E(\text{particles in, particles out})E'$$

For this case the compact form is

$$\ce{^{58}_{26}Fe}(2n, \beta)\ce{^{60}_{27}Co}$$

Transmutation with charged particles. If it has enough kinetic energy, a proton, α particle, or another nucleus can approach the nucleus of a target atom so closely that the attractive force between nucleons overcomes the electrostatic repulsion. A process of this kind that occurs in stars leads to the formation of oxygen-16 from carbon-12:

$$\ce{^{12}_{6}C} + \alpha \longrightarrow \ce{^{16}_{8}O} + \gamma$$

One that has been achieved in an accelerator on earth is the synthesis of the artificial element with atomic number 103, lawrencium, by bombarding californium with boron nuclei.

▼ EXAMPLE 22.5 Writing equations for nucleosynthesis reactions

Complete the following equations, which illustrate the synthesis of three artificial elements.
(a) $^{238}_{92}U + {}^2_1H \longrightarrow ? + 2n$
(b) $^{241}_{95}Am + ? \longrightarrow {}^{243}_{97}Bk + 2n$
(c) $^{246}_{96}Cm + ? \longrightarrow {}^{254}_{102}No + 4n$

STRATEGY As in the earlier examples, we have to select particles or nuclides that balance the mass numbers and atomic numbers in the equations. (a) We can identify the new element from its atomic number by referring to the periodic table. (b and c) We need to identify the particle from its charge and mass number.

SOLUTION (a) The mass numbers are balanced if

$$238 + 2 = A + 2 \times 1$$

from which $A = 238$. The atomic numbers are balanced if

$$92 + 1 = Z + 2 \times 0$$

from which $Z = 93$. Element 93 is Np, so the nuclide produced is neptunium-238.
(b) From the mass-number balance we have $241 + A = 243 + 2 \times 1$, from which $A = 4$. The atomic-number balance gives us $95 + Z = 97 + 2 \times 0$, so $Z = 2$. Hence the incoming particle is the α particle, $^4_2\alpha$.
(c) Here the mass-number balance gives us $246 + A = 254 + 4 \times 1$, so $A = 12$. The atomic-number balance yields $96 + Z = 102 + 4 \times 0$, from which $Z = 6$. A particle with atomic number 6 and mass number 12 is the nucleus of carbon-12.

The complete nuclear reactions are therefore

$$\text{(a)} \qquad ^{238}_{92}U + {}^2_1H \longrightarrow {}^{238}_{93}Np + 2n$$

$$\text{(b)} \qquad ^{241}_{95}Am + {}^4_2\alpha \longrightarrow {}^{243}_{97}Bk + 2n$$

$$\text{(c)} \qquad ^{246}_{96}Cm + {}^{12}_6C \longrightarrow {}^{254}_{102}No + 4n$$

EXERCISE Complete the following nuclear reactions:
(a) $? + \alpha \longrightarrow {}^{243}_{96}Cm + n$
(b) $^{242}_{96}Cm + \alpha \longrightarrow {}^{245}_{98}Cf + ?$
(c) $^{249}_{98}Cf + ? \longrightarrow {}^{263}_{106}Unh + 4n$

[*Answer:* (a) $^{240}_{94}Pu$; (b) n; (c) $^{18}_8O$]

RADIOACTIVITY

One of the most important social aspects of radioactivity is the damage it can do to biological tissues. The amount of damage depends on the strength of the source, the type of radiation, and the length of exposure. Another important aspect of radioactivity is its persistence: how long do we need to store nuclear waste before it may be considered safe? Such issues can be examined quantitatively by drawing on some of the material discussed earlier in the text, particularly the techniques of chemical kinetics (Chapter 12).

22.5 MEASURING RADIOACTIVITY

Becquerel detected and measured radioactivity by exposing photographic film to a source of radiation. He gauged the intensity of the radiation by the degree of blackening of the developed film. The blackening results from the same redox processes as in ordinary photography (Section 3.10), except that the initial oxidation of the halide ions is caused by nuclear radiation. Becquerel's technique is still used to monitor the exposure of workers to radiation. Two other devices are also commonly used to detect and measure radiation.

A *Geiger counter* (Fig. 22.11) is essentially a cylinder containing a low-pressure gas and two electrodes. Radiation ionizes the atoms in the cylinder and hence allows a brief flow of current between the electrodes. One type of Geiger counter converts this current into an audible click from a loudspeaker, the rapidity of the clicks indicating the intensity of the radiation. A *scintillation counter* makes use of the fact that certain substances (notably zinc sulfide) give a flash of light—a "scintillation"—when exposed to radiation. The intensity of the radiation is measured by counting the scintillations electronically.

Penetrating power. The three principal types of nuclear radiation penetrate matter to different extents. The least penetrating is α radiation. The massive, highly charged α particles interact so strongly with matter that they slow down, capture electrons from surrounding material, and change into bulky He atoms before traveling very far. However, even though α particles do not penetrate far into matter, they are very damaging, because the energy of their impact can knock atoms out of molecules and displace ions from their sites in crystals. When they impinge on biological matter, the damage caused by α particles can lead to illness. If DNA and the enzymes that interpret its protein-building messages are damaged, the result may be cancer. Most α radiation is absorbed by the surface layer of the skin, but inhaled and ingested particles can cause internal damage.

Next in penetrating power is β radiation. The fast electrons of these rays can penetrate about 1 cm of flesh before electrostatic interactions with nuclei and other electrons bring them to a standstill.

FIGURE 22.11 The detector in a Geiger counter consists of a gas (typically argon together with a little ethanol vapor, or neon and some bromine vapor) in a container, with a high potential difference (500 to 1200 V) applied between the container walls and a central wire. An electric current passes between the electrodes (walls and wire) when radiation ionizes the gas.

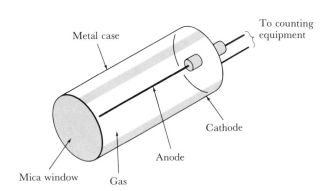

Metal case

To counting equipment

Cathode

Anode

Mica window

Gas

TABLE 22.2 **Radiation units**

| Property | Unit | | Definition |
	Name	Symbol	
Activity	Curie	Ci	3.7×10^{10} disintegrations per second
	Becquerel	Bq	1 disintegration per second ($1 \text{ Ci} = 3.7 \times 10^{10}$ Bq)
Absorbed dose	Radiation absorbed dose	rad	10^{-2} J/kg
	Gray	Gy	1 J/kg (1 Gy = 100 rad)
Dose equivalent	Roentgen equivalent man	rem	$Q \times$ rad*
	Sievert	Sv	100 rem

*Q is the relative biological effectiveness of the radiation. Normally $Q = 1$ for x, γ, and β radiation, and $Q = 20$ for α radiation and fast neutrons. A further factor of 5 (that is, $5 \times Q$) is used for bone under certain circumstances.

Most penetrating of all is γ radiation. The uncharged, high-energy γ-ray photons can pass right through the body and cause damage by ionizing molecules in their path. Protein molecules and DNA that have been damaged in this way cannot carry out their functions, and the results can be radiation sickness and cancer. Sources of γ rays must be surrounded by walls built from lead bricks or thick concrete to shield people from this penetrating radiation.

The activity of radioactive sources. The intensity of the radiation from a radioactive element depends on the element's _activity_, the number of nuclear disintegrations that occur per second. Disintegrations can be counted by recording the number of clicks of a Geiger counter or the number of flashes of a scintillation counter; each click or flash indicates that one disintegration has occurred. Various units have been introduced to report the intensity of radiation sources and the extent of their effects on matter (particularly living tissue), and they are summarized in Table 22.2. We shall discuss only the most commonly used units.

Measurements (using a Geiger counter or a scintillation counter) have shown that 3.7×10^{10} nuclear disintegrations occur each second in 1 g of radium-226. This rate of disintegration is used as a unit of activity, the _curie_ (Ci), named for (and defined by) Marie Curie, the Polish-French chemist who discovered radium and polonium (Fig. 22.12):

$$1 \text{ Ci} = 3.7 \times 10^{10} \text{ disintegrations per second}$$

The more active the source, the greater the number of nuclear disintegrations per second and hence the greater its activity in curies. A 1-g

FIGURE 22.12 Marie Sklodowska Curie, 1867–1934.

sample of radium-226 is a "1-Ci source" of radioactivity. If a certain cobalt-60 source undergoes 4.2×10^{13} disintegrations per second, its activity in curies is

$$\text{Activity} = 4.2 \times 10^{13} \text{ disintegrations per second}$$
$$\times \frac{1 \text{ Ci}}{3.7 \times 10^{10} \text{ disintegrations per second}}$$
$$= 1.1 \times 10^3 \text{ Ci} \quad \text{or} \quad 1.1 \text{ kCi}$$

The principal natural source of radioactivity in the human body is potassium-40, which occurs in about 0.01% abundance among the potassium ions found throughout the body and which emits β particles and positrons. Since the body of an adult human is a 0.1-μCi source, about 37,000 potassium-40 nuclei disintegrated in your body in the time it took you to read this sentence (about 10 s). Our own bodies contribute about 20% of the total radiation we receive from natural sources.

Doses of radioactivity. The *dose* of radiation is the energy deposited in a sample (in particular, the human body) that has been exposed to radiation. The unit for reporting dose is the *rad,* which stands for *rad*iation *a*bsorbed *d*ose:

> 1 **rad** is the amount of radiation that deposits 10^{-2} J of energy per kilogram of tissue.

A dose of 1 rad would correspond, if the radiation were absorbed uniformly through the body, to a 65-kg person absorbing a total of 0.65 J of energy from the incident radiation. This may seem a minute amount, for it is the same as the amount of energy transferred as heat when a 2-mg droplet of boiling water touches the skin. However, the energy of the particles of radiation is highly localized—like the impact of a bullet, but on a much smaller scale—and not spread over a region the size of even a tiny water droplet. As a result, the incoming particles can break individual atomic bonds as they collide with molecules in their path.

The extent of radiation damage to living tissue depends on the type of radiation absorbed and the type of tissue absorbing it. A dose of 1 rad of γ radiation causes about the same amount of damage as 1 rad of β radiation, but α particles are about 20 times more damaging (even though they are the least penetrating). A factor called the *relative biological effectiveness* Q must therefore be included in comparisons of the damage that a given dose of each type of radiation may cause. For β and γ radiation Q is set arbitrarily at about 1, so for α radiation Q is close to 20. The precise figure depends on the total dose, the dose rate, and the type of tissue, but these values are typical.

The *dose equivalent* is the actual dose modified to take into account the different destructive powers of the various types of radiation in combination with various types of tissue. It is obtained by multiplying the actual dose (in rads) by the value of Q for the radiation type. The result is expressed in a unit called the *roentgen equivalent man* (rem):

$$\text{Dose equivalent in rem} = Q \times \text{absorbed dose in rad}$$

TABLE 22.3   Health hazards of radiation

Dose equivalent, rem	Effect
0 to 25	None observable
25 to 50	Decrease of white-blood-cell count
100 to 200	Nausea; marked decrease in white-blood-cell count
500	Fifty percent likelihood of death within 30 days

Wilhelm Roentgen was the discoverer of x-rays, penetrating electromagnetic radiation with longer wavelengths than γ rays (typically about 100 pm).

The damage caused by various dose equivalents of radiation is summarized in Table 22.3. A dose of 30 rad of γ radiation corresponds to a dose equivalent of 30 rem, enough to cause a reduction in the number of white blood cells (the cells that fight infection), but 30 rad of α radiation corresponds to 600 rem, which is enough to cause death. We each typically receive an average annual dose equivalent from natural sources of about 0.2 rem, but this figure varies, depending on our lifestyle and where we live. As noted above, about 20% of this is radiation from our own bodies. About 30% comes from the cosmic rays (a mix of γ rays and high-energy elementary particles from outer space) that continuously bombard the earth, and 40% comes from radon seeping out of the ground. The remaining 10% is a result largely of medical diagnostic techniques. Emissions from nuclear facilities contribute about 0.1%.

22.6   THE RATE OF NUCLEAR DISINTEGRATION

We can discuss the rate at which radioactive nuclei disintegrate—and hence the activity of sources and the persistence of radioactivity—using the techniques of Chapter 12 for describing the rates of chemical reactions.

The law of radioactive decay and the nuclear half-life.   The disintegration of a nucleus,

$$\text{Parent nucleus} \longrightarrow \text{daughter nucleus} + \text{radiation}$$

is a nuclear version of unimolecular decay, with an unstable nucleus taking the place of an excited molecule. As in the case of unimolecular chemical reactions (Section 12.4), the rate law for the decay is first-order; that is, the relation between the rate of decay and the number N of radioactive nuclei present may be written

$$\text{Rate of decay} = k \times N$$

where k (the rate constant in chemical kinetics) is called the *decay constant* in this context. The rate law is called the *law of radioactive decay*. It

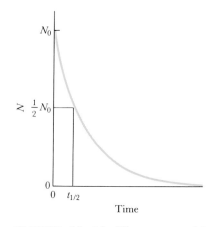

FIGURE 22.13 The exponential decay of the number of radioactive nuclei in a sample implies that the activity of the sample also decays exponentially with time. The curve is characterized by the half-life $t_{1/2}$.

implies that the more radioactive nuclei there are in a sample, the greater the overall rate of disintegration and the more active the sample.

As in the case of a first-order chemical reaction, we can calculate the number N of nuclei remaining after a time t with the formula

$$\ln \frac{N}{N_0} = -kt \tag{1}$$

where N_0 is the number present initially (at $t = 0$). As explained in Section 12.3, this is the equation of an exponential decay. Hence, the activity of a radioactive sample decays exponentially toward zero (Fig. 22.13).

▼ **EXAMPLE 22.6 Using the law of radioactive decay**

If a 1.0-g sample of tritium is stored for 5.0 years, what mass of that isotope will remain? The decay constant is 0.0564/year.

STRATEGY The number of tritium nuclei in the sample is proportional to the total mass m of tritium, so Eq. 1 can be expressed as

$$\ln \frac{m}{m_0} = -kt$$

where m_0 is the initial mass of tritium. The question can now be answered by substituting the data and taking antilogarithms.

SOLUTION Substitution of the data gives

$$\ln \frac{m}{m_0} = -0.0564/\text{year} \times 5.0 \text{ years} = -0.282$$

Taking the natural antilogarithm gives

$$\frac{m}{m_0} = 0.754$$

Since $m_0 = 1.0$ g, the mass remaining after 5.0 years will be 0.75 g.

EXERCISE The decay constant for fermium-244 is 210/s. What mass of the isotope will remain after a 1.0-μg sample of fermium-244 is kept for 10 ms?

[*Answer*: 0.12 μg]

▲

Radioactive decay is normally discussed in terms of the half-life $t_{1/2}$ of the nuclide:

The **half-life** of a radioactive nuclide is the time required for half the initial number of nuclei to disintegrate.

The relation between $t_{1/2}$ and k is found by setting $N = \frac{1}{2}N_0$ in Eq. 1. This gives

$$\ln \tfrac{1}{2} = -kt_{1/2}$$

Substituting $\ln \frac{1}{2} = -0.69$ then yields

$$t_{1/2} = \frac{0.69}{k}$$

The larger the value of k, the shorter the half-life of the nuclide.

Half the nuclei originally present in a radioactive sample disintegrate in a time equal to its half-life. The same time is then required for half the remaining nuclei to disintegrate. Therefore, after $2 \times t_{1/2}$ the number of nuclei will have fallen to $\frac{1}{2} \times \frac{1}{2} = \frac{1}{4}$ of the original number. After another $t_{1/2}$, the number remaining will be $\frac{1}{2} \times \frac{1}{2} \times \frac{1}{2} = \frac{1}{8}$ the original number, and so on.

Some half-lives are given in Table 22.4. The table shows that the values span a very wide range, with some nuclides having a fleeting existence and others surviving for billions of years. As an example, consider strontium-90, for which the half-life is 28 years. Strontium-90 is one of the nuclides that occurs in the fallout from nuclear explosions. Because it is chemically very similar to calcium, it may accompany calcium wherever that element occurs in the environment. Thus, it can be incorporated into bones, and once there, it continues to generate radiation for many years. Even after three half-lives (84 years), one-eighth of the original strontium-90 still survives. As a general guide, it takes about 10 half-lives (for strontium-90, 280 years) for the radioactivity of a sample to become negligible (at that time the sample retains $1/2^{10}$, or about $\frac{1}{1000}$, of its original activity).

The rate at which nuclear disintegration occurs is independent of the temperature, the physical state of the material containing the nuclide, and whether or not the atoms are part of a compound. This is because the forces that bind, and occasionally eject, nucleons in a nucleus are so strong that the relatively feeble energies associated with thermal motion and chemical bonding are completely negligible in comparison. The half-lives of nuclei are fixed once and for all by the forces acting within each nucleus, and nuclear disintegration cannot be accelerated by heating or slowed by cooling.

Isotopic dating. The fact that the half-life of a nuclide is unchangeable is put to practical use in *isotopic dating*, the determination of the ages of archeological artifacts by measuring the activity of the radioactive iso-

TABLE 22.4 **Radioactive half-lives and activities**

Nuclide	Half-life $t_{1/2}$	Activity, Ci/g
Tritium	12.3 years	9.7×10^3
Carbon-14	5.73×10^3 years	4.4
Carbon-15	2.4 s	3.0×10^{11}
Potassium-40	1.26×10^9 years	7.1×10^{-6}
Cobalt-60	5.26 years	1.1×10^3
Strontium-90	28.1 years	1.4×10^2
Iodine-131	8.05 days	1.2×10^5
Radium-226	1.60×10^3 years	1.00
Uranium-235	7.1×10^8 years	2.1×10^{-6}
Uranium-238	4.5×10^9 years	3.5×10^{-7}
Fermium-244	3.3 ms	1.4×10^{13}

topes they contain. The most important dating technique is *radiocarbon dating*, in which the β decay of carbon-14 is utilized.

Radiocarbon dating depends on three characteristics of carbon-14. First, carbon-14 is a naturally occurring isotope of carbon and is present in all living things. Second, it is radioactive and has a half-life of 5730 years. Third, the supply of carbon-14 atoms in the environment is nearly constant, since they are produced when nitrogen nuclei in the atmosphere are bombarded by neutrons:

$$^{14}_{7}\text{N} + \text{n} \longrightarrow {}^{14}_{6}\text{C} + \text{p} \qquad \text{or} \qquad {}^{14}_{7}\text{N}(\text{n, p}){}^{14}_{6}\text{C}$$

The neutrons originate from the collisions of cosmic rays with other nuclei.

The carbon-14 atoms produced in the atmosphere enter living organisms through photosynthesis and digestion. They leave them by the normal processes of excretion and respiration, and by transmutation, in which their nuclei decay at a steady rate. As a result of this continuous intake and loss of carbon-14, all living things have a constant proportion of the isotope among their very much more numerous carbon-12 atoms. In other words, there is a fixed ratio (about $1/10^{12}$) of carbon-14 atoms to carbon-12 atoms in living tissues.

When the organism dies, it no longer exchanges carbon with its surroundings. However, the carbon-14 nuclei already inside it continue to disintegrate with a constant half-life. Hence the ratio of carbon-14 to carbon-12 decreases steadily after death, and the ratio observed in a sample of dead tissue can be inserted into Eq. 1 and used to determine the time since death. In the modern version of the technique (Fig. 22.14), which requires only milligrams of sample, the carbon

FIGURE 22.14 In the modern version of the carbon-14 dating technique, a mass spectrometer (inside the cylinder) is used to determine the proportion of carbon-14 atoms in a sample.

atoms are converted to C^- ions by bombardment of the sample with cesium atoms, the C^- ions are then accelerated with electric fields, and the carbon isotopes are finally separated and counted with a mass spectrometer. In a simpler version of the technique, similar to the original method developed by Willard Libby in Chicago in the years following World War II, larger samples are used and β radiation from the sample is measured. As is illustrated in the next example, the time since death is calculated by comparing the activity of the ancient sample with the activity of a modern one.

▼ **EXAMPLE 22.7** **Interpreting radiocarbon dating**

A 1.00-g sample of carbon from wood found in an archeological site gave 7900 carbon-14 disintegrations in a 20-hour period. In the same period, 1.00 g of carbon from a modern source underwent 18,400 disintegrations. Calculate the age of the sample.

STRATEGY To use Eq. 1, we need the values of k, N, and N_0. The value of k is obtained from the half-life, of carbon-14, which is known (Table 22.4). The number of disintegrations during the 20-hour period is proportional to the number of carbon-14 nuclei present in the two samples. If we suppose that the proportion of carbon-14 in the atmosphere was the same when the ancient sample was alive as it is now, then we can take the original proportion of carbon-14 in the ancient sample to have been the same as that found in the modern sample. We can therefore set N/N_0 equal to the ratio of the numbers of disintegrations in the two samples.

SOLUTION The decay constant is

$$k = \frac{0.69}{t_{1/2}} = \frac{0.69}{5730 \text{ years}} = 1.2 \times 10^{-4}/\text{year}$$

Then, from Eq. 1, the age of the ancient sample is

$$t = -\frac{1}{k} \times \ln \frac{N}{N_0}$$

$$= -\frac{1 \text{ year}}{1.2 \times 10^{-4}} \times \ln \frac{7900}{18,400}$$

$$= 7.0 \times 10^3 \text{ years}$$

EXERCISE A 250-mg sample of carbon from wood underwent 15,300 carbon-14 disintegrations in 36 hours. Estimate the time since the death of the sample.

[*Answer*: 6.4×10^3 years]

▲

The reliability of radiocarbon dating depends on the constancy of the proportion of carbon-14 in the environment. The proportion does vary slightly because the intensity of cosmic radiation varies over the centuries, and that must be taken into account in precise determinations of age. This can be done in a number of ways. For example, radiocarbon dating measurements can be calibrated against a sample of carbon from an ancient tree, the age of which has been determined by counting growth rings.

NUCLEAR POWER

One of the most significant contributions to both the creation and the solution of significant social and economic problems has been the discovery that large amounts of energy are released in certain types of nuclear reactions. This section outlines some of the principles involved. It also explores some of the contributions being made by chemistry to the solution of the very pressing problems arising from the use of nuclear power.

22.7 NUCLEAR FISSION

So far, the nuclear disintegrations we have described have been processes in which a nucleus has ejected one or two subatomic particles. However, an important type of nuclear disintegration is *fission*, the breaking of a nucleus into two smaller nuclei of similar mass. This is the kind of disintegration we examine here.

Types of nuclear fission. Fission that takes place without being initiated by the impact of other particles is called *spontaneous nuclear fission*. It occurs when the oscillation of a heavy nucleus causes about half the protons and neutrons to break away (Fig. 22.15). An example is the disintegration of americium-244 into iodine and molybdenum:

$$^{244}_{95}\text{Am} \longrightarrow ^{134}_{53}\text{I} + ^{107}_{42}\text{Mo} + 3\text{n}$$

Fission does not occur in precisely the same way in every instance: the fission products may also include isotopes of these elements as well as other elements.

Induced nuclear fission is fission caused—often artificially—by bombarding a heavy nucleus with neutrons (Fig. 22.16). Nuclei that can undergo induced fission are called *fissionable*. For most nuclides, fission is induced only if the impinging neutrons are moving so rapidly that they smash into the nucleus and drive it apart; uranium-238 is one such nuclide. Some nuclides, however, can be nudged into breaking apart even if the incoming neutrons are slow. Such nuclides are called *fissile*; they include uranium-235, uranium-233, and plutonium-239, the fuels of nuclear power.

Induced nuclear fission can be *self-sustaining*; that is, once it is initiated, it can continue even if the supply of neutrons from outside is discontinued. This is the case when more neutrons are produced by a fission event than are needed to induce it initially. Self-sustaining fission occurs with uranium-235, which undergoes numerous fission processes, including

$$^{235}_{92}\text{U} + \text{n} \longrightarrow ^{142}_{56}\text{Ba} + ^{92}_{36}\text{Kr} + 2\text{n}$$

If the two product neutrons strike two other fissile nuclei, then after the next round of fission there will be four free neutrons, which can induce fission in four more nuclei. In the language of Section 12.5, neutrons are carriers in a branched chain reaction (Fig. 22.17).

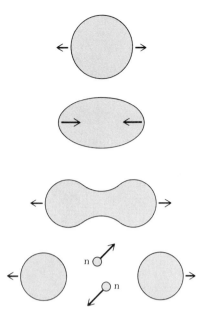

FIGURE 22.15 In spontaneous nuclear fission, the oscillations of a heavy nucleus in effect tear the nucleus apart, forming two smaller nuclei of similar mass along with some neutrons.

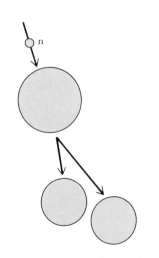

FIGURE 22.16 In induced nuclear fission, an incoming neutron causes the nucleus to break apart.

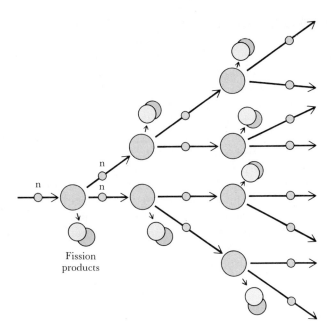

Fission
products

Nuclear explosions. When a nuclear branched chain reaction is allowed to run freely, the cascade of released neutrons can result in the fissioning of all the available uranium-235 in only a fraction of a second. This is what happens in a nuclear explosion.

Some of the neutrons produced in a chain reaction inevitably escape into the surroundings. Enough neutrons are captured to sustain the chain reaction only if a large enough number of uranium nuclei are present in the sample. That is, there is a *critical mass*, a mass of fissionable material below which so many neutrons escape from the material that the fission chain reaction is not sustained. If a sample is *supercritical*, with a mass in excess of the critical mass, then enough neutrons will induce fission for the chain reaction to result in an explosion. The critical mass for a solid sphere of pure plutonium of normal density is about 15 kg, which makes a sphere about the size of a grapefruit. The critical mass is smaller if the metal is compressed, or imploded. In nuclear weapons this is done by detonating a conventional explosive that surrounds a sphere of plutonium. Then the nuclei are pressed closer together so that they are more effective at blocking the escape of neutrons. The critical mass can be as low as 5 kg for highly compressed plutonium.

The technically difficult step in designing a nuclear weapon is to convert a given mass of fissile material into a supercritical mass so rapidly that the chain reaction occurs uniformly throughout the metal. This can be done by shooting two blocks of material toward each other (as was done in the bomb that fell on Hiroshima), or by the implosion of a single subcritical mass (as in the bomb that destroyed Nagasaki). A strong neutron emitter, typically polonium, is also included to initiate the chain reaction. The products of the fission process, which consist of dozens of different nuclides (most of them radioactive), are the *fallout* of a nuclear explosion.

Controlled fission. An alternative to explosive fission is a controlled chain reaction sustained by a limited supply of neutrons. This is achieved in a *nuclear reactor* (Fig. 22.18), where control rods made of neutron-absorbing elements, such as boron and cadmium, are inserted between the fuel elements—rods containing the fissile material (such as ^{235}U). The fuel elements are surrounded by a *moderator*, a substance such as graphite or heavy water (D_2O) that slows the neutrons. Slow neutrons are absorbed more effectively than fast neutrons by the fissile uranium-235 and also allow the control rods to act more efficiently.

One of the many problems connected with nuclear power is its dependence on the availability of fissile material: uranium-235 reserves are only about 1% of those of the nonfissile uranium-238. One solution is to synthesize fissile nuclides from other elements. In a *breeder reactor*, a reactor that is used to create nuclear fuel, the neutrons are not moderated. Their high speeds result not only in the fission of, and hence energy release from, the uranium-238, but also in the formation of some fissile plutonium-239. The chain of nuclear reactions is

$$^{238}_{92}U + n \longrightarrow {}^{239}_{92}U$$

$$^{239}_{92}U \xrightarrow{23.5\ min} {}^{239}_{93}Np + \beta$$

$$^{239}_{93}Np \xrightarrow{2.4\ days} {}^{239}_{94}Pu + \beta$$

The plutonium can be separated and used as fuel (or for warheads).

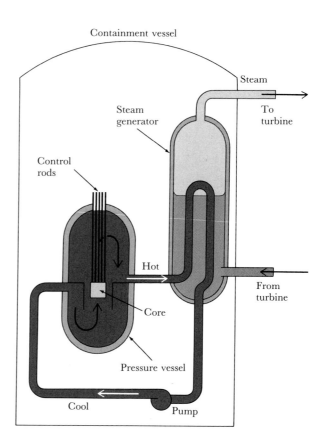

FIGURE 22.18 A schematic diagram of one type of nuclear reactor. This is a pressurized-water reactor, in which the coolant is water under pressure.

Binding energy. Energy is released during fission because the nucleons that were originally in a single large nucleus can interact more strongly in the resulting smaller fragments. Since the forces between nucleons are so strong, even a small change in their relative positions results in a significant change in energy, and the difference is released as heat.

The exothermicity of nuclear fission is calculated by comparing the masses of the nuclei before and after fission. This procedure is founded on Albert Einstein's theory of relativity, which implies that the mass of an object is an indication of its energy content. The greater the mass of an object, the greater its energy. Specifically, the total energy E and the mass m are related by Einstein's famous equation,

$$E = mc^2 \qquad (2)$$

where c is the speed of light (3.00×10^8 m/s). Loss of energy is always accompanied by a loss of mass. Mass loss accompanies the exothermic energy changes considered in earlier chapters, but the energies involved are so small that the mass losses are undetectably small. For example, when 100 g of water cools from 100°C to 20°C, it loses 33 kJ of energy as heat; this corresponds to a mass loss of only 3.7×10^{-10} g. Even in a strongly exothermic chemical reaction, such as one that releases 1000 kJ of energy, the mass of the products is only 10^{-8} g less than that of the reactants; this is outside the range of all but extremely precise measuring instruments. However, in a nuclear reaction the energy changes are very large, the corresponding mass loss is measurable, and the observed change in mass can be used to calculate the amount of energy released. The energy loss is a result of the rearrangement of the protons and neutrons when new nuclei are formed from old.

Changes in nuclear mass and energy are normally discussed in terms of "nuclear binding energy" E_{bind}:

> The **nuclear binding energy** of an element is the energy released when Z protons and $A - Z$ neutrons come together to form a nucleus.

This energy is calculated from the change in mass that occurs when the nucleus forms from the separated nucleons and they interact more strongly. For example, the binding energy of iron-56 (with 26 protons and 30 neutrons) is calculated from

$$\Delta m = \text{total mass of 26 protons and 30 neutrons}$$
$$- \text{mass of } {}^{56}_{26}\text{Fe nucleus}$$

Because the energy arising from the rearrangement of atomic electrons is insignificant in comparison with the energy arising from the rearrangement of nucleons, we generally calculate the binding energy as the energy released when Z hydrogen atoms (in place of protons) combine with $A - Z$ neutrons to form a neutral atom (rather than the nucleus itself). For the formation of iron-56, then, we would have

$$\Delta m = \text{mass of 26 hydrogen atoms and 30 neutrons}$$
$$- \text{mass of } {}^{56}_{26}\text{Fe atom}$$

We would substitute this expression in Einstein's formula to obtain

$$E_{\text{bind}} = \Delta m \times c^2$$

▼ EXAMPLE 22.8 Calculating the nuclear binding energy

Calculate the nuclear binding energy of 1 mol of helium-4, given the following masses: ^{4}He, 4.0026 amu; ^{1}H, 1.0078 amu; n, 1.0087 amu.

STRATEGY The nuclear binding energy is the energy released in the nuclear reaction

$$2\,^1\text{H} + 2\text{n} \longrightarrow\ ^4\text{He}$$

We begin by calculating the difference between the masses of products and reactants; then we convert the result from atomic mass units to kilograms, using the factor from inside the back cover; and finally we use the Einstein relation to calculate the energy corresponding to this loss of mass.

SOLUTION The mass loss is

$$\Delta m = 2 \times 1.0078\ \text{amu} + 2 \times 1.0087\ \text{amu} - 4.0026\ \text{amu}$$
$$= 0.0304\ \text{amu}$$

In kilograms, this is

$$\Delta m = 0.0304\ \text{amu} \times \frac{1.6605 \times 10^{-27}\ \text{kg}}{1\ \text{amu}}$$
$$= 5.05 \times 10^{-29}\ \text{kg}$$

Then, from Eq. 2,

$$E_{\text{bind}} = \Delta m \times c^2$$
$$= (5.05 \times 10^{-29}\ \text{kg}) \times (3.00 \times 10^8\ \text{m/s})^2$$
$$= 4.55 \times 10^{-12}\ \text{kg} \cdot \text{m}^2/\text{s}^2 = 4.55 \times 10^{-12}\ \text{J}$$

Multiplying by Avogadro's number, we find that the binding energy of a mole of helium-4 nuclei is 2.7×10^9 kJ, which is 10 million times larger than the energy of a typical chemical bond.

EXERCISE Calculate the binding energy of 1 mol of carbon-12 nuclei.

[*Answer*: 8.9×10^9 kJ]

▲

The heat released during fission. The nuclear binding energies of the stable isotopes of the elements are plotted in Fig. 22.19. The vertical axis shows the binding energy divided by the mass number of the nuclide (E_{bind}/A), or the binding energy per nucleon. The actual binding energy of uranium-235, for instance, is 235 times the value shown in the graph. The greater the binding energy per nucleon, the more stable the nucleus is (and the lower its total energy), relative to smaller nuclei into which it might fragment or from which it can be synthesized.

The graph shows that nucleons are bonded together most strongly in iron and in nuclei with atomic numbers close to that of iron. This is one of the reasons for the high abundance of iron in the universe. The binding energy per nucleon is less for elements heavier than iron;

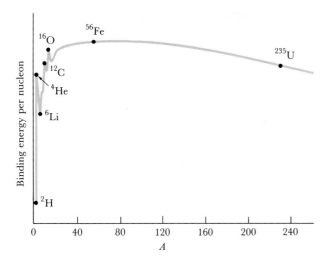

FIGURE 22.19 The variation of nuclear binding energy per nucleon among the elements. The maximum binding energy occurs at iron-56, showing that this nucleus has the lowest total energy of all since its nucleons are most tightly bound.

hence, when a U atom disintegrates into two smaller nuclei, the binding energy per nucleon increases and energy is released. The energy released when a nucleus undergoes fission could be calculated from the binding energies shown in the graph, but it is just as simple to calculate it by subtracting the mass of the original nucleus from the total mass of the product particles.

▼ **EXAMPLE 22.9** **Calculating the energy released during fission**

Calculate the energy (in joules) released when 1.0 g of uranium-235 undergoes fission and forms barium-142 and krypton-92 in the nuclear reaction

$$^{235}_{92}U + n \longrightarrow {}^{142}_{56}Ba + {}^{92}_{36}Kr + 2n$$

The masses of the particles are ^{235}U, 235.04 amu; ^{142}Ba, 141.92 amu; ^{92}Kr, 91.92 amu; n, 1.0087 amu.

STRATEGY If we know the mass loss, we can interpret it as an energy, using Einstein's equation. Therefore, we must calculate the total mass of the particles on each side of the equation and then substitute the difference into Eq. 2.

SOLUTION

$$\text{Mass of products} = \text{mass(Ba)} + \text{mass(Kr)} + 2 \times \text{mass(n)}$$
$$= 141.92 \text{ amu} + 91.92 \text{ amu} + 2 \times 1.0087 \text{ amu}$$
$$= 235.86 \text{ amu}$$
$$\text{Original mass} = \text{mass(U)} + \text{mass(n)}$$
$$= 235.04 \text{ amu} + 1.0087 \text{ amu} = 236.05 \text{ amu}$$

Hence the mass loss per uranium nucleus is 0.19 amu. This is equivalent to 3.1×10^{-28} kg and to an energy of

$$E = (3.1 \times 10^{-28} \text{ kg}) \times (3.00 \times 10^8 \text{ m/s})^2$$
$$= 2.8 \times 10^{-11} \text{ J}$$

Since the number of ^{235}U atoms in 1.0 g of uranium-235 (with molar mass 235 g/mol) is 2.6×10^{21}, the total energy released when all have undergone fission is

$$E = (2.6 \times 10^{21}) \times (2.8 \times 10^{-11} \text{ J})$$

$$= 7.3 \times 10^7 \text{ kJ}$$

Since a power output of 1 kW for 1 h (called "one kilowatt-hour" and abbreviated 1 kW · h) is equivalent to 3600 kJ, the 1 g of uranium can supply 2.0×10^4 kW · h of energy.

EXERCISE Another way in which uranium-235 can undergo fission is

$$^{235}_{92}\text{U} + \text{n} \longrightarrow {}^{135}_{52}\text{Te} + {}^{100}_{40}\text{Zr} + \text{n}$$

Calculate the energy (in kilowatt-hours) released when 1.0 g of uranium-235 undergoes fission in this way. The additional masses needed are Te, 134.92 amu; Zr, 99.92 amu.

[*Answer*: 2.2×10^4 kW · h]

▲

22.8 NUCLEAR FUSION

We see from Fig. 22.19 that there is a large increase in nuclear binding energy (and hence a lowering in total energy) on going from hydrogen to its heavier isotopes and then to the first few light elements. This suggests that large amounts of energy should be released if H nuclei could be made to fuse together to form He or Li.

The principal difficulty in achieving fusion arises from the strong electrostatic repulsion between protons when they approach each other closely enough to fuse together. The heavier isotopes of hydrogen are therefore used in fusion reactions; their nuclei fuse together more readily, because the attraction arising from their additional neutrons helps to overcome the repulsion between approaching protons. The high kinetic energies needed for successful collisions are achieved by raising the temperature to millions of degrees.

One fusion method uses deuterium and tritium in the following sequence of nuclear reactions:

$$\text{D} + \text{D} \longrightarrow {}^3\text{He} + \text{n}$$

$$\text{D} + \text{D} \longrightarrow \text{T} + \text{p}$$

$$\text{D} + \text{T} \longrightarrow {}^4\text{He} + \text{n}$$

$$\underline{\text{D} + {}^3\text{He} \longrightarrow {}^4\text{He} + \text{p}}$$

Overall reaction: $\quad 6\text{D} \longrightarrow 2{}^4\text{He} + 2\text{p} + 2\text{n}$

The method of Example 22.9 can be used to determine that the overall reaction releases 3×10^8 kJ from each gram of deuterium (corresponding to 10^9 kW · h, the energy generated when the Hoover dam operates at full capacity for about an hour). Additional tritium is supplied to the reaction chamber to facilitate the processes. Since it has a very low natural abundance and is radioactive, the tritium is generated by bombarding lithium-6 with neutrons in the immediate surroundings of the reaction zone:

$$^6\text{Li} + \text{n} \longrightarrow \text{T} + {}^4\text{He}$$

It is on account of this reaction (and the same reaction in thermonuclear bombs) that lithium-6 is being extracted from stocks of lithium, with the result that the atomic weight of the remaining, commercially available element is slowly rising as the proportion of lithium-7 grows.

Nuclear fusion is very difficult to achieve in practice on account of the vigor with which the charged nuclei must be hurled at each other. One way of accelerating them to high enough speeds is to heat them with a fission explosion: this method is used in producing a *thermonuclear explosion*, in which a fission bomb (containing uranium or plutonium) is exploded to ignite a lithium-6 fusion bomb. A more constructive and controlled approach is to heat a *plasma*, or ionized gas, by passing an electric current through it. The very hot plasma, and its very fast ions, is kept away from the walls of the container with magnetic fields. This method of initiating fusion is the subject of intense research and is beginning to show signs of success (Fig. 22.20).

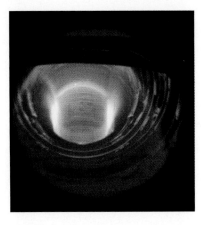

FIGURE 22.20 Research into controlled nuclear fusion is being carried out in several countries. Here we see the hot ionized gas inside the Tokomak fusion test reactor at the Princeton Plasma Physics Laboratory.

22.9 CHEMICAL ASPECTS OF NUCLEAR POWER

Chemists have important contributions to make to the development of nuclear power and to the elimination of the legitimate worries associated with its use. Their three principal areas of involvement are the preparation of the fuel itself, the recovery of important fission products, and the safe disposal (or utilization) of nuclear waste.

Nuclear fuel. Uranium is mined as several minerals, the most important being pitchblende, UO_2 (Fig. 22.21). The aim of uranium refining processes is in general not only to reduce the ore to the metal but also to *enrich* it; that is, to increase the abundance of a specific isotope, in this case uranium-235. The natural abundance of uranium-235 is 0.7%, and the goal of enrichment is to increase that to about 3%.

One extraction scheme begins with the oxidation of the ore by nitric acid:

$$3UO_2(s) + 8HNO_3(aq) \longrightarrow 3UO_2(NO_3)_2(aq) + 2NO(g) + 4H_2O(l)$$

The uranium(VI) nitrate is extracted with solvents, purified, and then heated to 300°C, at which temperature it decomposes to uranyl oxide:

$$UO_2(NO_3)_2(s) \xrightarrow{300°C} UO_3(s) + NO(g) + NO_2(g) + O_2(g)$$

The nitrogen oxides are collected, dissolved in water, and recycled as the source of nitric acid for the first stage of the process.

After this oxidation and purification, the reduction sequence begins. First, the uranium(VI) is reduced to uranium(IV) with hydrogen:

$$UO_3(s) + H_2(g) \xrightarrow{700°C} UO_2(s) + H_2O(g)$$

FIGURE 22.21 A common uranium ore—pitchblende, a variety of uraninite, UO_2.

(a)   (b)

FIGURE 22.22 Two fluorides of uranium used in the extraction of uranium. (a) UF_4. (b) Small white flecks of solid UF_6 on the glass porthole of a freezer-sublimer in a uranium enrichment plant.

The pure uranium oxide is then treated with hydrofluoric acid to produce uranium tetrafluoride, a green solid (Fig. 22.22):

$$UO_2(s) + 4HF(aq) \longrightarrow UF_4(s) + 2H_2O(l)$$

One possible route at this stage is the reduction of uranium tetrafluoride with magnesium:

$$UF_4(s) + 2Mg(s) \longrightarrow U(s) + 2MgF_2(s)$$

The uranium metal is then fabricated into fuel rods about 3 cm in diameter and 1 m long. An alternative route is the oxidation of the tetrafluoride to the hexafluoride, a volatile yellowish white solid:

$$UF_4(s) + F_2(g) \xrightarrow{450°C} UF_6(s)$$

The hexafluoride route is used if the uranium is to be enriched.

The enrichment process makes use of the small mass difference between uranium-235 and uranium-238. One of the first procedures to be developed (as a part of the Manhattan Project to build the first atomic bomb) made use of the different diffusion rates that result from this mass difference: it follows from Graham's law (Section 5.7) that the rates of diffusion of $^{235}UF_6$ (molecular weight 349 amu) and $^{238}UF_6$ (molecular weight 352 amu) should be in the ratio

$$\frac{\text{Rate of diffusion of } ^{235}UF_6}{\text{Rate of diffusion of } ^{238}UF_6} = \sqrt{\frac{352}{349}}$$

$$= 1.004$$

Since this ratio is so close to 1, the uranium hexafluoride vapor must be allowed to diffuse repeatedly through the porous barriers designed for the purpose, which are screens with large numbers of minute holes. In practice it is allowed to do so thousands of times (some of the stages were shown in Fig. 5.31).

Since the diffusion process is technically demanding and uses a lot of energy, scientists and engineers continue to look for alternative enrichment procedures. One of these utilizes a centrifuge that rotates samples of uranium hexafluoride vapor at very high speed. This causes the heavier $^{238}UF_6$ molecules to be thrown outward and collected as a solid on the outer parts of the rotor, leaving a higher proportion of $^{235}UF_6$ closer to the axis of the rotor, from where it can be removed.

Once the uranium hexafluoride has been enriched to the required degree, it is reduced to UO_2 by reaction with hydrogen and water:

$$UF_6(s) + H_2(g) + 2H_2O(g) \xrightarrow{600°C} UO_2(s) + 6HF(g)$$

The hydrogen fluoride is recycled for use earlier in the process. The enriched oxide can now be reduced to the metal with magnesium, but modern reactors generally use the oxide directly because it is less reactive and has a higher melting point than the metal.

Spent-fuel processing. The processing of spent nuclear fuel is a much more complex task than the fuel's initial preparation. There are three problem areas: the recovery of any remaining uranium-235, the extraction of any plutonium produced, and the disposal of the highly radioactive and largely useless fission products.

The uranium and plutonium are separated by _solvent extraction_, a separation process that makes use of their differing solubilities in various solvents. The "purex process" (from _p_lutonium-_u_ranium _r_eduction _ex_traction) utilizes the differing solubilities of plutonium and uranium compounds in a mixture of 80% kerosene and 20% of the organic compound tri-tertiary-butyl phosphate. When mixed with water, the plutonium(IV) and uranium(VI) oxides dissolve preferentially in the organic solvent, and most of the other fission products dissolve preferentially in the water. The organic solution containing the uranium and plutonium is then removed, and the plutonium(IV) in it is reduced to plutonium(III). When the solution is again mixed with water, the uranium(VI) remains in the organic solvent while the plutonium(III) dissolves preferentially in the water. Now the two elements may be separated by separating the two solutions. Further purification and reaction stages produce pure UO_2 and PuO_2. One hazard is the possibility that the plutonium will accumulate in a critical mass, which in concentrated aqueous solution is only about 500 g.

Nuclear waste. The highly radioactive fission (HRF) products in used nuclear fuel rods must be stored for about ten half-lives before their level of radioactivity is no longer dangerous. The current approach to storage is to incorporate the HRF products into a glass—a solid, complex network of silicon and oxygen atoms (Fig. 22.23). One of the reasons why glass appears to be suitable is that most of the fission products are oxides that are themselves "network formers"; that is, they promote the formation of a relatively disorderly Si—O network rather than crystallization into a regular array of atoms. Crystallization is dangerous because the cracks between crystalline regions leave the incorporated

FIGURE 22.23 Molten glass for storing nuclear waste is being poured from a platinum crucible into a steel bar mold.

radioactive material exposed to moisture, which might dissolve them and carry them away from the storage area.

Experiments are currently in progress to see how storage conditions affect the rate of _leaching_, the dissolution of substances in any water that happens to percolate through the solid. It is known that the leaching rate is very sensitive to temperature and that it increases about 100-fold between 25°C and 100°C. Storage in cool caverns might therefore seem to be an answer. However, the radioactive decay of the fission products is a source of heat (on a larger scale, radioactive decay is responsible for the heat in the interior of the earth), and the temperature of the glass blocks can rise even if their surroundings are cool.

Another problem is that radiation can damage the structure of the block and perhaps open up channels that will allow water to spread into its interior. The most important type of radiation in this respect is α decay, because the heavy, energetic α particles displace the atoms they strike and open up cavities in the solid. There has also been a fear (which so far seems unfounded) that the α particles will collect as bubbles of helium gas and break up the glass from within as a result of the pressure they exert. The two effects may be self-canceling, and the channels opened up by α particles may allow the helium to escape. In current experiments, intense α-particle emitters such as plutonium-238 and californium-244 are incorporated into the glass, subjecting it to a dose in a few years that actual HRF glasses receive only after centuries, yet the glass seems to exhibit very little breakdown.

SUMMARY

22.1 Radioactivity, the emission of radiation by nuclei, gives rise to α **radiation** (consisting of α particles, which are helium-4 nuclei), β **radiation** (consisting of β particles, which are fast electrons emitted from nuclei), and γ **radiation** (very-high-frequency, short-wavelength, electromagnetic radiation). Of the three, α radiation is the least penetrating but the most damaging; γ radiation is the most penetrating.

22.2 A new nuclide is produced in the process of **nuclear transmutation** when a nucleus undergoes a **nuclear reaction,** the simplest forms of which are α decay and β decay. The **daughter nuclide,** the new nuclide that is produced in a nuclear reaction, is identified by balancing mass numbers and atomic numbers.

22.3 Heavy nuclides decay in a stepwise **radioactive series,** a sequence of α and β emissions, until a stable nuclide (often an isotope of lead) is formed.

22.4 Nucleosynthesis is the synthesis of elements; it occurs when particles are captured by a nucleus. Nucleo-synthesis occurs naturally in stars and was the source of the elements currently on earth. It may also be brought about artificially by accelerating particles to speeds that are high enough to overcome electrostatic repulsions.

22.5 The **intensity** of a radiation source depends on the number of nuclear disintegrations occurring per second and is expressed in **curies.** The **dose** of radiation received by a sample is expressed in **rads,** and the **dose equivalent,** which takes into account the greater damaging effect of α rays is expressed in **rems.**

22.6 The number of radioactive nuclei in a sample decreases exponentially in time, in accordance with the **law of radioactive decay,** a first-order rate law. Half the nuclei in any sample of a nuclide decay in a time equal to the **radioactive half-life** of the nuclide. Half-lives are independent of temperature and the state of chemical combination of the radioactive atom. One application of radioactive decay is **radiocarbon dating,** which is based on the measurement of the proportion of carbon-14 in a sample.

22.7 and 22.8 Nuclear **fission** is the breaking of a nucleus into two nuclei of similar size; it may be either **spontaneous** or **induced** (by neutrons). A **fissionable** nucleus is one that can undergo induced fission, and a **fissile** nucleus is one for which fission can be induced by slow neutrons. **Sustained fission** is a **nuclear chain reaction** in which neutrons are the chain carriers; it is achieved if each fission event results in the production of several neutrons and the mass of the fissionable sample is **supercritical** in the sense of exceeding a certain critical value. Energy is released during fission because nucleons become rearranged to states of lower energy; the release of energy can be discussed in terms of **binding energies** and calculated from the mass differences between reactants and products using **Einstein's formula,** $E = mc^2$. Nuclear **fusion,** the formation of a larger nucleus from smaller ones, also releases energy.

22.9 The chemical problems connected with nuclear power include the production of the fuel—and especially the **enrichment** of natural uranium to increase the proportion of uranium-235 in the fuel. Enrichment is carried out either by diffusion or with a centrifuge. **Fuel processing** includes the recovery of plutonium from used fuel by the **purex process.** The disposal of **nuclear waste** is problematic because of the long-lived, intense radioactivity; one possible solution is to incorporate it in glass.

EXERCISES

Nuclear Structure and Nuclear Radiation

In Exercises 22.1 to 22.4, determine the numbers of protons, neutrons, and nucleons in each nuclide.

22.1 (a) ^2_1H; (b) $^{24}_{12}Mg$; (c) $^{109}_{47}Ag$; (d) $^{244}_{96}Cm$.
22.2 (a) $^{12}_6C$; (b) $^{90}_{38}Sr$; (c) $^{197}_{79}Au$; (d) $^{258}_{101}Md$.

22.3 (a) T; (b) ^{32}S; (c) ^{128}I; (d) ^{238}Pu.
22.4 (a) D; (b) ^{16}O; (c) ^{57}Fe; (d) ^{241}Am.

In Exercises 22.5 and 22.6, determine the numbers of protons, neutrons, and nucleons in each nuclide, and write its chemical symbol.

22.5 (a) Bromine-81; (b) krypton-90; (c) berkelium-247.
22.6 (a) Chlorine-35; (b) cobalt-60; (c) unnilpentium-262.

In Exercises 22.7 and 22.8, calculate the wavelength and energy of photons of γ radiation with the given frequency.

22.7 (a) 2×10^{19} Hz; (b) 5×10^{21} Hz.
22.8 (a) 1.0×10^{20} Hz; (b) 1.0×10^{22} Hz.

An excited nucleus emits a γ photon as its nucleons adjust into a more stable arrangement. In Exercises 22.9 and 22.10, calculate the frequency and wavelength of the γ radiation emitted as a result of the given change of energy.

22.9 (a) 4.1×10^{-14} J; (b) 1.6×10^{-13} J.
22.10 (a) 1.3×10^{-13} J; (b) 3.2×10^{-13} J.

Nuclear energy changes are often expressed in *millions of electronvolts* (MeV), where 1 MeV is the kinetic energy gained by an electron when it falls through a potential difference of 1 MV; that is, 1 MeV = 1.60×10^{-13} J. In Exercises 22.11 and 22.12, calculate the frequency and wavelength of the γ radiation emitted by a nuclear transition with the given energy.

22.11 (a) Cobalt-60: 1.33 MeV; (b) iron-59: 1.10 MeV.
22.12 (a) Thorium-233: 0.087 MeV; (b) uranium-235: 0.187 MeV.

Nuclear Transmutation

In Exercises 22.13 to 22.16, identify the daughter nuclides formed when the specified decay occurs.

22.13 (a) β decay of tritium
(b) β decay of carbon-14
(c) α decay of radium-226
(d) α decay of americium-241
22.14 (a) β decay of lithium-8
(b) α decay of magnesium-27
(c) α decay of gold-179
(d) α decay of uranium-233

22.15 (a) β^+ decay of boron-8
(b) β^+ decay of cobalt-56
(c) Electron capture by cobalt-56
(d) Proton emission from cobalt-53
22.16 (a) β^+ decay of argon-35
(b) β^+ decay of iridium-188
(c) Electron capture of iridium-188
(d) β^+ decay accompanied by proton emission from cesium-144

In Exercises 22.17 and 22.18, determine what particle is emitted when the specified change occurs, and write the nuclear reaction.

22.17 (a) Sodium-24 to magnesium-24
(b) Tin-128 to antimony-128
(c) Lanthanum-130 to barium-130
(d) Thorium-228 to radium-224

22.18 (a) Cesium-142 to barium-142
(b) Gold-188 to platinum-188
(c) Uranium-229 to thorium-225
(d) Mercury-192 to gold-192

Patterns of Nuclear Stability

In Exercises 22.19 to 22.22, suggest which type of decay each nuclide is likely to undergo, and name the daughter nuclide.

22.19 (a) Beryllium-10; (b) nitrogen-12; (c) copper-68; (d) copper-60.

22.20 (a) Bromine-87; (b) bromine-74; (c) cadmium-103; (d) xenon-140.

22.21 (a) Plutonium-232; (b) plutonium-246; (c) berkelium-243; (d) americium-246.

22.22 (a) Uranium-227; (b) uranium-240; (c) unnilpentium-260; (d) neptunium-240.

In Exercises 22.23 to 22.26, identify the nuclides in the radioactive series, if the string of particles is given in the order of emission.

22.23 Uranium-235 series: $\alpha,\beta,\alpha,\beta,\alpha,\alpha,\alpha,\beta,\alpha,\beta,\alpha$

22.24 Neptunium-237 series: $\alpha,\beta,\alpha,\alpha,\beta,\alpha,\alpha,\alpha,\beta,\alpha,\beta$

22.25 Curium-242 series: $\alpha,\alpha,\alpha,\alpha,\alpha,\alpha,\alpha,\beta,\beta,\alpha,\beta, \ldots$

22.26 Uranium-233 series: $\alpha,\alpha,\beta,\alpha,\alpha,\alpha,\beta,\alpha,\beta$

Nucleosynthesis

In Exercises 22.27 and 22.28, complete the equations of each nuclear reaction, and write the completed equation in the form E(in,out)E'.

22.27 (a) $^{14}N + ? \longrightarrow {}^{17}O + p$
(b) $^{13}C + n \longrightarrow ? + \gamma$
(c) $? + n \longrightarrow {}^{249}Bk + \beta^-$
(d) $^{243}Am + n \longrightarrow {}^{244}Cm + ? + \gamma$

22.28 (a) $? + p \longrightarrow {}^{21}Na + \gamma$
(b) $^{1}H + p \longrightarrow {}^{2}H + ?$
(c) $^{15}N + p \longrightarrow {}^{12}C + ?$
(d) $^{20}Ne + ? \longrightarrow {}^{24}Mg + \gamma$

In Exercises 22.29 and 22.30, complete each nuclear equation and write it as a "chemical equation."

22.29 (a) $^{20}Ne(\alpha,?)^{16}O$
(b) $^{20}Ne(^{20}Ne,^{16}O)?$
(c) $^{44}Ca(?,\gamma)^{48}Ti$
(d) $^{27}Al(^{2}H,?)^{28}Al$

22.30 (a) $?(\gamma,\beta)^{20}Ne$
(b) $^{44}Ti(e^-,\beta^+)?$
(c) $^{241}Am(?,4n)^{248}Fm$
(d) $?(n,\beta^-)^{244}Cm$

In Exercises 22.31 and 22.32, complete each fission-reaction equation.

22.31 (a) $^{244}Am \longrightarrow {}^{134}I + {}^{107}Mo + 3?$
(b) $^{235}U + n \longrightarrow ? + {}^{138}Te + 2n$
(c) $^{235}U + n \longrightarrow {}^{101}Mo + {}^{132}Sn + ?$

22.32 (a) $^{239}Pu + n \longrightarrow {}^{98}Mo + {}^{138}Te + ?$
(b) $^{239}Pu + n \longrightarrow {}^{100}Tc + ? + 4n$
(c) $^{239}Pu + n \longrightarrow ? + {}^{133}In + 3n$

The Measurement of Radioactivity

In Exercises 22.33 and 22.34, express the activity of the source in curies.

22.33 A source in which there are 3.7×10^6 disintegrations per second

22.34 A source in which there are 3.7×10^{11} disintegrations per second

In Exercises 22.35 and 22.36, calculate the number of disintegrations that occur in the given time period for a source of the given activity.

22.35 (a) A 1.0-Ci source in 1.0 s; (b) a 1.0-mCi source in 1 ms.

22.36 (a) A 5.0-mCi source in 1.0 s; (b) a 10^{-8}-Ci source in 1.0 day.

A certain Geiger counter is known to respond to $\frac{1}{1000}$ of the radiation emitted by a sample. In Exercises 22.37 and 22.38, calculate the activity of the source (in curies), given the number of clicks in the stated period.

22.37 (a) 370 clicks in a 10-s period; (b) 1000 clicks in a 100-s period.

22.38 (a) 1.4×10^5 clicks in 1.0 h; (b) 1 click per second.

In Exercises 22.39 and 22.40, calculate the dose in rads and the dose equivalent in rems.

22.39 A 1.0-kg sample absorbs an energy of 1.0 J as a result of exposure to β radiation.

22.40 A 1.0-g sample absorbs an energy of 1.0 J as a result of exposure to α radiation.

22.41 Suppose someone is exposed to a source of β radiation that results in a dose rate of 1.0 rad/day. Given that nausea begins after a dose equivalent of about 100 rem, how soon will the symptoms of radiation sickness be apparent?

22.42 Suppose someone is exposed to a source of α radiation that results in a dose rate of 2.0 mrad/day. Given that nausea begins after a dose equivalent of about 100 rem, how soon will the symptoms of radiation sickness be apparent?

The Rate of Nuclear Disintegration

In Exercises 22.43 and 22.44, calculate the decay constant of each radioactive nuclide from the given half-life.

22.43 (a) Tritium, 12.3 years; (b) carbon-14, 5730 years; (c) potassium-40, 1.3 billion years; (d) einsteinium-243, 20 s.

22.44 (a) Lithium-8, 0.84 s; (b) nitrogen-13, 10.0 min; (c) cobalt-60, 5.26 years; (d) nobelium-255, 180 s.

In Exercises 22.45 and 22.46, use Table 22.4 to calculate the time required for the activity of the given sample of radioactive nuclei to decrease to the specified extent.

22.45 (a) A 1.0-Ci radium-226 source decreasing to 0.50 Ci

(b) A 1.0-Ci cobalt-60 source decreasing to 0.25 Ci

(c) A 0.1-μCi potassium-40 source decreasing to $\frac{1}{8}$ that activity

22.46 (a) A 1.0-mCi iron-59 source (for which $t_{1/2}$ = 45.1 d) decreasing to 0.50 mCi

(b) A 1.0-Ci strontium-90 source decreasing to $\frac{1}{8}$ that activity

(c) A 1.0-mCi iodine-131 source decreasing to $\frac{1}{64}$ of that activity

22.47 Estimate the activity of a 1-mCi radon-222 source (half-life, 3.82 days) after two weeks have passed.

22.48 Estimate the activity of a 1-Ci cobalt-60 source (half-life, 5.26 years) after a century has passed.

In Exercises 22.49 and 22.50, calculate the fraction of the original sample remaining after the stated time.

22.49 (a) Carbon-14 after 1000 years

(b) Tritium after 1.0 year

(c) Strontium-90 after 19 years

(d) Iodine-131 after 24 h

22.50 (a) Carbon-14 after 10,000 years

(b) Potassium-40 after 4.5 billion years (the age of the earth)

(c) Uranium-235 after 18 billion years (thought to be the age of the universe)

(d) Strontium-90 after 75 years

In Exercises 22.51 and 22.52, calculate the time needed for the activity of each source to change as stated.

22.51 (a) A 1.0-Ci radium-226 source to decay to 0.10 Ci

(b) A 1.0-μCi potassium-40 source to decay to 10 nCi

(c) A 10-Ci cobalt-60 source to decay to 8 Ci

22.52 (a) A 1.0-mCi thorium-234 source (half-life, 24.1 days) to decay to 0.10 mCi.

(b) A 0.010-mCi strontium-90 source to decay to 0.0010 mCi

(c) A 1.0-Ci iodine-131 source to decay to 1.0 mCi

In Exercises 22.53 and 22.54, calculate the activity of each substance from its half-life starting from the law of radioactive decay and the relation between $t_{1/2}$ and k.

22.53 (a) 1.0 mg of pure radium-226; (b) 1.0 g of pure $^{235}UO_2$

22.54 (a) 1.0 g of pure natural carbon, with carbon-14 in its natural steady-state abundance of 1 in 10^{12} atoms; (b) 1.0 g of cobalt containing 1.0% cobalt-60.

In Exercises 22.55 to 22.58, calculate the age (since death) of the sample, given that 1.00 g of carbon from a modern source undergoes 920 disintegrations per hour.

22.55 A 1.00-g sample of carbon from wood that gave 5500 disintegrations in a 24-h period

22.56 A 2.50-g sample of carbon from wood that gave 4000 disintegrations in a 20-h period

22.57 A 250-mg sample of carbon from a textile that gave 1900 disintegrations in a 10-h period

22.58 A 500-mg sample of carbon from a wooden artifact that gave 900 disintegrations in a 15-h period.

Nuclear Masses and Binding Energies

In Exercises 22.59 and 22.60, calculate the mass loss or gain during each process.

22.59 (a) 1.0 kg of copper (specific heat, 0.39 J/g · K) is heated from 25°C to 1000°C.

(b) 1.0 mol of $H_2O(l)$ freezes to ice at 0°C.

(c) 1.0 mol of $H_2O(l)$ is formed from its elements at 25°C.

22.60 (a) 1000 kg of iron (specific heat, 0.45 J/g · K) is heated from 25°C to 1000°C.

(b) 1.0 mol of $H_2O(l)$ vaporizes at 100°C.

(c) 1.0 mol of $C_6H_6(l)$ is formed from its elements at 25°C.

22.61 Calculate the energy in joules that is equivalent to (a) 1 g of matter; (b) 1 amu of matter. What are the equivalent energies in millions of electronvolts (1 MeV = 1.60×10^{-13} J).

22.62 The sun emits radiant energy at the rate of 3.9×10^{26} J/s. To what rate of loss of mass does that correspond?

In Exercises 22.63 and 22.64, calculate the binding energy per nucleon for each nuclide (with mass as given), expressing the result in joules per nucleon and millions of electronvolts per nucleon (1 MeV = 1.60×10^{-13} J).

22.63 (a) ^{4}He, 4.0026 amu; (b) ^{12}C, 12 amu; (c) ^{56}Fe, 55.9349 amu; (d) ^{235}U, 235.0439 amu.

22.64 (a) ^{2}H, 2.0141 amu; (b) ^{16}O, 15.9949 amu; (c) ^{58}Ni, 57.9353 amu; (d) ^{239}Pu, 239.0522 amu.

In Exercises 22.65 and 22.66, calculate the energy released per gram of starting material in each nuclear reaction, with masses as given.

22.65 Some of the reactions involved in nuclear fusion:

(a) D + D $\longrightarrow$ ^{3}He + n (D, 2.0141 amu; ^{3}He, 3.0160 amu)

(b) $^3He + D \longrightarrow {}^4He + p$ (^4He, 4.0026 amu)

(c) $^7Li + p \longrightarrow 2\,{}^4He$ (^7Li, 7.0160 amu)

(d) $D + T \longrightarrow {}^4He + p$ (T, 3.0160 amu)

22.66 The *Bethe chain* of nuclear reactions that release energy in stars:

(a) $^{12}C + p \longrightarrow {}^{13}N$ (^{13}N, 13.0057 amu)

(b) $^{13}N \longrightarrow {}^{13}C + \beta^+$ (^{13}C, 13.0034 amu)

(c) $^{13}C + p \longrightarrow {}^{14}N$ (^{14}N, 14.0031 amu)

(d) $^{14}N + p \longrightarrow {}^{15}O$ (^{15}O, 15.0031 amu)

(e) $^{15}O \longrightarrow {}^{15}N + \beta^+$ (^{15}N, 15.0001 amu)

(f) $^{15}N + p \longrightarrow {}^{12}C + {}^4He$

Deduce the overall reaction for the Bethe chain. How many moles of H atoms would have to be converted to helium per second if this were the only source of the energy output of the sun (3.9×10^{26} J/s).

Nuclear Fuel

22.67 What is meant by the *processing* of nuclear fuel? Outline the procedure used for separating plutonium from uranium in a used fuel rod. Classify the reactions involved.

ORGANIC CHEMISTRY

This chapter introduces some of the principles of organic chemistry by describing the structure, synthesis, and properties of the hydrocarbons—the binary compounds of carbon and hydrogen. These compounds are the basic structures from which the formulas of all organic compounds are derived. One very important hydrocarbon—polyethylene—is shown in the illustration. This is a magnified view of part of an expanded-polyethylene sheet.

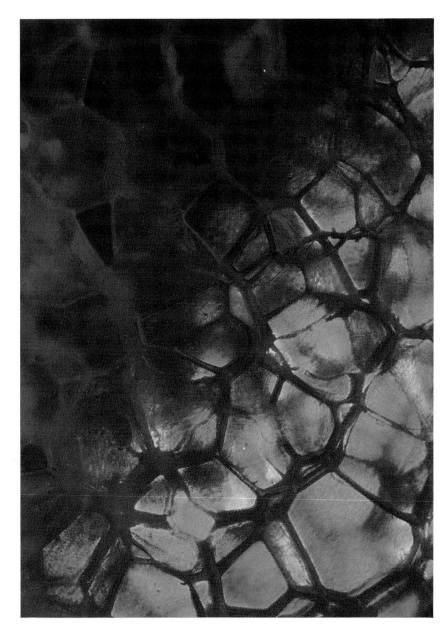

THE HYDROCARBONS

The alkanes

23.1 Isomerism
23.2 Alkane nomenclature
23.3 The properties of
     alkanes

**The alkenes and the
   alkynes**

23.4 Alkene nomenclature
23.5 The carbon–carbon
     double bond
23.6 Alkene polymerization
23.7 Alkynes

Aromatic hydrocarbons

23.8 Arene nomenclature
23.9 Reactions of aromatic
     hydrocarbons

Carbon forms a vast number of compounds; around 6 million are already known, and many new ones are being reported every day. The branch of chemistry that deals with these compounds is called *organic chemistry*, and the compounds themselves (with a few exceptions, such as the carbonates) are called *organic compounds*. The name reflects the erroneous view once held that organic compounds could be formed only by living organisms. Modern organic chemistry deals with both naturally occurring and synthetic compounds, including plastics, pharmaceuticals, petrochemicals, fuels, and foods. Since it provides a link between the properties of atoms and the functioning of living organisms, through organic chemistry we come to biochemistry and hence to the chemical foundations of life itself.

Because the compounds of carbon are so numerous, it is convenient to organize them into families of compounds that exhibit structural similarities. One of the simplest families consists of the *hydrocarbons*, the binary compounds of carbon and hydrogen. Their molecules are made up of chains, rings, and networks of carbon atoms, with the remaining bonding positions occupied by hydrogen atoms. In Chapter 24 we shall meet compounds that can be regarded as being derived from the hydrocarbons through the replacement of hydrogen atoms with atoms of other elements.

It became clear in the early days of organic chemistry that the hydrocarbon family could itself be subdivided into two families. It was found (by means that included the combustion analysis technique described in Section 4.3) that some hydrocarbons have a lower ratio of H atoms to C atoms than others. For example, ethane (C_2H_6) has three H atoms for each C atom, but ethylene (C_2H_4) has only two. This led to a distinction between compounds like ethane that are "saturated" and compounds like ethylene that are "unsaturated." It is now known that the supposedly "unused" carbon valences in unsaturated compounds are in fact used in multiple carbon–carbon bonds, so the two families are now defined as follows:

A **saturated** hydrocarbon is one with no carbon–carbon multiple bonds.

An **unsaturated** hydrocarbon is one with at least one carbon–carbon multiple bond.

THE ALKANES

The two simplest saturated hydrocarbons are methane (CH_4; **1**) and ethane (C_2H_6; **2**). The methane molecule is tetrahedral, as predicted by VSEPR theory. Each of its C—H bonds can be pictured as formed by the overlap of an sp^3 hybrid orbital on the C atom and a $1s$ orbital on the H atom. We can think of an ethane molecule as having been formed from a CH_4 molecule by inserting a *methylene group*, —CH_2—, between the C atom and one of the H atoms. We can emphasize this by writing its formula as CH_3—CH_2—H, or more briefly as CH_3CH_3. Both C atoms in ethane are sp^3 hybridized, and the C—C bond is formed by the overlap of two sp^3 orbitals on neighboring atoms (Fig.

1 Methane

2 Ethane

23.1). The sp^3 hybridization reflects the shape of the molecule, which is tetrahedral around each C atom, with bond angles close to 109°. We shall normally write only the two-dimensional Lewis structures,

H
|
H—C—H   and
|
H

H H
| |
H—C—C—H
| |
H H

but must not forget that the actual molecules are three-dimensional.

The way we picture the formula for ethane as the result of inserting a methylene group into the formula for methane suggests the concept of a "homologous series" of compounds:

A **homologous series** is a family of compounds with molecular formulas that are obtained by repeatedly inserting a given group (most commonly —CH_2—) into a parent structure.

The homologous series that starts with the formula for methane is the family of hydrocarbons called the "alkanes":

The **alkanes** are a homologous series of hydrocarbons with formulas that are derived from CH_4 by inserting —CH_2— groups, and having molecular formulas C_nH_{2n+2}.

They include methane itself and its homologs ethane (C_2H_6), propane (C_3H_8), and butane (C_4H_{10}):

H H H
| | |
H—C—C—C—H
| | |
H H H

Propane

H H H H
| | | |
H—C—C—C—C—H
| | | |
H H H H

Butane

The first few alkanes are found in varying proportions in natural gas and as components of petroleum; their names and some of their physical properties are given in Table 23.1. The alkanes were once called the

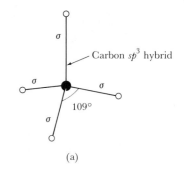

(a)

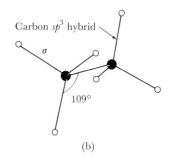

(b)

FIGURE 23.1 The structure of the bonds in (a) methane and (b) ethane. In each case the carbon atom is sp^3 hybridized and forms σ bonds with its neighbors. All bond angles are approximately 109°.

TABLE 23.1 **The properties of some alkanes**

Formula	Name	Melting point, °C	Boiling point, °C	Density,* g/mL
CH_4	Methane	−182	−162	
C_2H_6	Ethane	−183	−89	
C_3H_8	Propane	−188	−42	
C_4H_{10}	Butane	−138	−1	
C_5H_{12}	Pentane	−130	36	0.626
C_6H_{14}	Hexane	−95	69	0.659

*At 25°C.

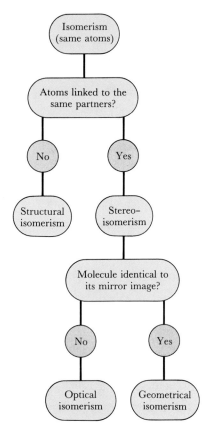

FIGURE 23.2 Classification of the isomers of organic compounds.

"paraffins," from the Latin words for "little affinity." As we shall see, this is a good summary of their chemical properties, for they are not very reactive.

23.1 ISOMERISM

The fourth alkane, C_4H_{10}, introduces a very important feature of organic chemistry—the possibility of isomerism. As we saw in Section 21.6 in connection with d-metal complexes, isomers are compounds that are built from the same atoms, but in different arrangements. In alkanes, the same number of C and H atoms may be linked into different arrangements. The chart first shown as Fig. 21.24, in which the various kinds of isomerism are classified, is shown again in Fig. 23.2, with modifications that make it more relevant to organic compounds.

The alkanes show *structural isomerism*—isomerism in which the same atoms are linked to different neighbors. It is possible, for example, to insert a —CH_2— group into the C_3H_8 molecule in two different ways:

$$
\begin{array}{ccc}
\text{H—C—C—C—H} & \longrightarrow & \text{H—C—C—C—C—H} \quad \text{or} \quad \text{H—C—C—C—H} \\
\text{CH}_3\text{CH}_2\text{CH}_2\text{CH}_3 & & (\text{CH}_3)_3\text{CH} \\
\text{Butane} & & \text{Methylpropane}
\end{array}
$$

(Although the —CH_2— group can be inserted in other places, the resulting molecules can all be twisted into one or other of these two isomers.) Both compounds are gases, but butane condenses at $-1°C$ and methylpropane condenses at $-12°C$.

Instead of drawing the complete structural formula, it is often sufficient to give only the shortened form, as in $CH_3CH_2CH_2CH_3$ for butane and $(CH_3)_3CH$ for methylpropane, showing how the atoms are grouped together in these two C_4H_{10} isomers. When a group of atoms is attached to a "main-chain" carbon, we usually place it in parenthesis at the *right* of the group to which it is attached, as in —$CH(CH_3)$— in the formula $CH_3CH_2CH_2CH(CH_3)CH_2CH_3$. However, the shortest possible formula is sometimes obtained by writing an attached group in front of the carbon atom to which it is attached, particularly at the beginning of a chain. Thus we write $(CH_3)_3CH$, not $CH_3CH(CH_3)CH_3$, because the latter is longer. When writing a short formula, we look for the shortest unambiguous way to describe the compound. When several —CH_2— groups are repeated, we may collect them together; hence we could write $CH_3(CH_2)_2CH_3$ for butane.

▼ **EXAMPLE 23.1 Writing the formulas of isomeric molecules**

Draw Lewis structures for all the alkanes of formula C_5H_{12}, and write the short formula for each.

STRATEGY We have to decide which *distinct* molecules can be constructed by inserting one —CH_2— group into the formula for each of the two C_4H_8 molecules given above. Distinct molecules are those that cannot be changed into each other simply by rotating either the entire formula or parts of the formula on the page. One approach is to insert —CH_2— groups into different parts of the molecule, and then to discard formulas that repeat those already obtained. The short formulas summarize the structures in the briefest unambiguous way by showing how the atoms are grouped and linked together.

SOLUTION From butane we can form

(a) $CH_3CH_2CH_2CH_2CH_3$ (b) $CH_3CH_2CH(CH_3)_2$

From methylpropane we can form

(c) $CH_3CH_2CH(CH_3)_2$ (d) $C(CH_3)_4$ (e) $(CH_3)_2CHCH_2CH_3$

Molecules (b) and (c) are obviously the same, while the atoms of molecule (e) are joined together in the same arrangement as (b) and (c), even though they look different as drawn on the page; hence (b), (c), and (e) are all the same (see Fig. 23.3). There are therefore only *three* distinct isomers with formula C_5H_{12}: (a), (b), and (d).

EXERCISE Write the short formulas for the five isomeric alkanes of formula C_6H_{14}.

[*Answer*: $CH_3(CH_2)_4CH_3$, $CH_3(CH_2)_2CH(CH_3)_2$, $CH_3CH_2CH(CH_3)CH_2CH_3$, $CH_3CH_2C(CH_3)_3$, $(CH_3)_2CHCH(CH_3)_2$]

23.2 ALKANE NOMENCLATURE

The "IUPAC* rules" for systematically naming organic compounds begin with the names given to the alkanes. However, as the rules sometimes result in very cumbersome names for common compounds, alternative names are also widely used.

*International Union of Pure and Applied Chemistry.

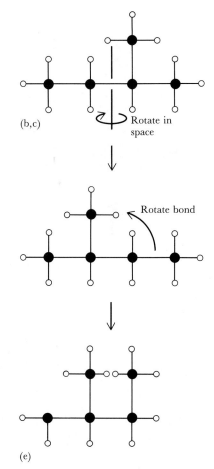

FIGURE 23.3 The rotations around carbon–carbon bonds and the overall rotations in space that show that the structures (b), (c), and (e) in Example 23.1 all represent the same molecule.

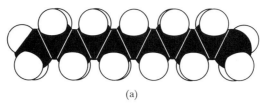

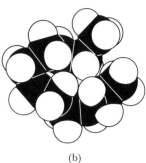

(a)

(b)

FIGURE 23.4 A "straight-chain" hydrocarbon is in fact a zigzag chain of carbon atoms. It may be found (a) fully extended, (b) rolled up into a ball, or in some intermediate shape.

3 *n*–Octane

Unbranched alkanes. The first step in naming an alkane is to distinguish "unbranched" from "branched" compounds. In an *unbranched alkane* (or "straight-chain" alkane), all the carbon atoms are located on a string of carbon atoms —C—C—C—C (as in butane). A *branched alkane* also has one or more carbon "side chains" (as in methylpropane). Although an unbranched alkane is usually written with the carbon atoms in a straight line, they actually take a zigzag shape. Moreover, because neighboring —CH_2— groups can rotate around the bonds that join them to each other, a "straight-chain" molecule may even be found rolled up into a ball (Fig. 23.4).

The systematic names of the first few unbranched alkanes are given in Table 23.2. They all end in -ane. The names of alkane molecules with one through four carbon atoms (methane, ethane, propane, and butane) have historical origins; the remainder are derived from the Greek numbers. The common names of the unbranched alkanes are the same as the systematic names, but they carry the prefix *n*-, standing for *normal* (and signifying an unbranched chain), as in *n*-octane **(3)**.

▼ EXAMPLE 23.2 Naming unbranched alkanes

Give the systematic names of (a) $CH_3(CH_2)_4CH_3$ and (b) $CH_3(CH_2)_{14}CH_3$, and (c) write the formula for decane.

STRATEGY We can identify the alkane by counting the number of C atoms and referring to Table 23.2. In the last part we should work backward, identifying the number of C atoms from the name.

SOLUTION (a) The number of carbon atoms is 6; hence the compound is hexane. (b) The molecule has 16 carbon atoms, so its name is hexadecane. (c) We note that the name decane signifies an unbranched alkane with 10 carbon atoms. Its formula is therefore $CH_3(CH_2)_8CH_3$.

EXERCISE Give the systematic names of (a) $CH_3CH_2CH_2CH_3$, (b) $CH_3(CH_2)_6CH_3$, and (c) $CH_3(CH_2)_{16}CH_3$, and (d) write the formula for heptane.

▲ [*Answer*: (a) Butane; (b) octane; (c) octadecane; (d) $CH_3(CH_2)_5CH_3$]

Branched alkanes. For the purpose of naming the branched alkanes, we treat the side chains as *substituents*, atoms or groups that have been substituted for hydrogen atoms and are attached to an unbranched backbone like ribs on a spine. First, we identify the longest unbranched chain of carbon atoms in the molecule as the backbone, and give it the name of the corresponding alkane. Thus,

$$CH_3-\overset{\overset{\displaystyle CH_3}{|}}{\underset{\underset{\displaystyle H}{|}}{C}}-CH_3$$

is a substituted propane because its longest unbranched chain (in bold type) contains three carbon atoms; also,

$$CH_3-\overset{\overset{\displaystyle CH_3}{|}}{\underset{\underset{\displaystyle CH_3}{|}}{C}}-CH_2-\overset{\overset{\displaystyle CH_3}{|}}{CH}-CH_3$$

TABLE 23.2 Alkane nomenclature

Number of C atoms	Formula	Name of alkane	Name of group
1	CH_3-	Methane	Methyl
2	CH_3-CH_2-	Ethane	Ethyl
3	$CH_3-CH_2-CH_2-$	Propane	Propyl
	$CH_3 \diagdown CH-, \; (CH_3)_2CH- \atop CH_3 \diagup$		Isopropyl*
4	$CH_3-CH_2-CH_2-CH_2-$	Butane	Butyl
	$CH_3 \diagdown CH-CH_2-, \; (CH_3)_2CHCH_2- \atop CH_3 \diagup$	Isobutane* (methylpropane)	Isobutyl*
	$CH_3-\overset{\overset{\displaystyle CH_3}{|}}{\underset{\underset{\displaystyle CH_3}{|}}{C}}-, \; (CH_3)_3C-$	Methylpropane	*tert*-Butyl*,†
5	$CH_3-CH_2-CH_2-CH_2-CH_2-$ or $CH_3(CH_2)_4-$	Pentane	Pentyl
6	$CH_3(CH_2)_5-$	Hexane	Hexyl
7	$CH_3(CH_2)_6-$	Heptane	Heptyl
8	$CH_3(CH_2)_7-$	Octane	Octyl
9	$CH_3(CH_2)_8-$	Nonane	Nonyl
10	$CH_3(CH_2)_9-$	Decane	Decyl
11	$CH_3(CH_2)_{10}-$	Undecane‡	Undecyl

*This name is not formed systematically but is widely used.
†*tert*- denotes "tertiary."
‡Names of alkanes with more carbon atoms are formed similarly, as in pentadecane for $C_{15}H_{32}$ and hexadecane for $C_{16}H_{34}$.

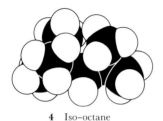

4 Iso–octane

is a substituted pentane because its longest unbranched chain has five C atoms. Since the latter's molecular formula is C_8H_{18}, it is an isomer of octane (its common name is isooctane; **4**); nevertheless, according to the IUPAC rules, it is named as a derivative of pentane.

Next, we identify and name the substituents. We call CH_3— a *methyl group*, and CH_3CH_2— an *ethyl group*, as noted in Table 23.2. In general, groups derived from alkanes are called *alkyl groups* and denoted by the letter R— (so R— may stand for CH_3— or C_2H_5—, and so on). Their names are obtained from the alkane names by changing the ending -ane to -yl. We then combine the names of the substituents with the name of the backbone.

$$CH_3-\underset{\underset{H}{|}}{\overset{\overset{CH_3}{|}}{C}}-CH_3 \qquad \text{methylpropane}$$

If there are several substituents, we name them in alphabetical order:

$$CH_3-CH_2-\underset{\underset{CH_3}{|}}{CH}-\underset{\underset{CH_2CH_3}{|}}{CH}-CH_2-CH_2-CH_2-CH_3 \qquad \text{an ethylmethyloctane}$$

If there are two or three substituents of the same kind, we use the prefix di- (for 2) or tri- (for 3):

$$CH_3-\underset{\underset{CH_3}{|}}{\overset{\overset{CH_3}{|}}{C}}-CH_2-\underset{\underset{CH_3}{|}}{CH}-CH_3 \qquad \text{a trimethylpentane}$$

The substituents are alphabetized by their name, not any prefix they have, as in 5-ethyl-2,3-dimethyloctane. In Chapter 24 we shall meet many other (nonhydrocarbon) substituents. Ones that will be mentioned again in this chapter are halogen atoms. Their presence turns an alkane into a *haloalkane* (or an "alkyl halide"), such as chloroethane, CH_3CH_2Cl. In haloalkane names, the halogen atom is denoted by the prefix halo; chloro- and bromo- are two such prefixes.

Specifying locations. Some compounds differ only in the locations at which the substituents are attached to the backbone. To specify these locations, we number the C atoms of the backbone in order, starting at the end of the molecule that results in the substituents having the lower numbered locations. Then the numbers of the C atoms to which the substituents are attached are given as prefixes:

$$\underset{1}{CH_3}-\underset{2}{\underset{\underset{CH_3}{|}}{\overset{\overset{CH_3}{|}}{C}}}-\underset{3}{CH_2}-\underset{4}{\underset{\underset{}{|}}{\overset{\overset{CH_3}{|}}{CH}}}-\underset{5}{CH_3} \qquad \text{2,2,4-trimethylpentane}$$

Numbering the chain from the right would result in the name 2,4,4-trimethylpentane. This is rejected because it gives one of the substituents a higher number.

▼ EXAMPLE 23.3 Naming branched alkanes

Name the alkanes (b) and (d) in Example 23.1, and write the formula for 5-ethyl-2,3-dimethyloctane.

STRATEGY We need to identify and name the longest chains and then number their C atoms so that the lowest numbers for the locations are obtained. We can then build the names by specifying the substituents and their locations. To deduce a formula from a name, we follow the procedure in reverse: we determine the number of C atoms in the backbone, write them as an unbranched chain, and then attach the substituents to the specified locations. All other bonding positions are occupied by hydrogen atoms.

SOLUTION Molecules (b) and (d) of Example 23.1 are

$$
\text{(b)} \quad H-\overset{\overset{\displaystyle H}{|}}{\underset{\underset{\displaystyle H}{|}}{C}}-\overset{\overset{\displaystyle H}{|}}{\underset{\underset{\displaystyle H}{|}}{C}}-\overset{\overset{\displaystyle CH_3}{|}}{\underset{\underset{\displaystyle H}{|}}{C}}-\overset{\overset{\displaystyle H}{|}}{\underset{\underset{\displaystyle H}{|}}{C}}-H, \qquad \text{(d)} \quad H-\overset{\overset{\displaystyle H}{|}}{\underset{\underset{\displaystyle H}{|}}{C}}-\overset{\overset{\displaystyle CH_3}{|}}{\underset{\underset{\displaystyle CH_3}{|}}{C}}-\overset{\overset{\displaystyle H}{|}}{\underset{\underset{\displaystyle H}{|}}{C}}-H
$$

The backbones are (b) a butane and (d) a propane. Naming the substituents then gives (b) 2-methylbutane and (d) 2,2-dimethylpropane.

The compound 5-ethyl-2,3-dimethyloctane is a substituted octane, so its backbone is

$$-C-C-C-C-C-C-C-C-$$

The prefix 5-ethyl indicates an ethyl group (CH_3CH_2-) at carbon atom 5. The prefix 2,3-dimethyl indicates that there are two methyl (CH_3-) groups, one at carbon atom 2 and the other at carbon atom 3. The molecule is

$$
H-\overset{\overset{\displaystyle H}{|}}{\underset{\underset{\displaystyle H}{|}}{C}}-\overset{\overset{\displaystyle CH_3}{|}}{\underset{\underset{\displaystyle H}{|}}{C}}-\overset{\overset{\displaystyle H}{|}}{\underset{\underset{\displaystyle CH_3}{|}}{C}}-\overset{\overset{\displaystyle H}{|}}{\underset{\underset{\displaystyle H}{|}}{C}}-\overset{\overset{\displaystyle CH_2CH_3}{|}}{\underset{\underset{\displaystyle H}{|}}{C}}-\overset{\overset{\displaystyle H}{|}}{\underset{\underset{\displaystyle H}{|}}{C}}-\overset{\overset{\displaystyle H}{|}}{\underset{\underset{\displaystyle H}{|}}{C}}-\overset{\overset{\displaystyle H}{|}}{\underset{\underset{\displaystyle H}{|}}{C}}-H
$$

with short (!) form $(CH_3)_2CHCH(CH_3)CH_2CH(CH_2CH_3)CH_2CH_2CH_3$. This alkane is an isomer of $C_{12}H_{26}$.

EXERCISE Name the alkanes (a) $CH_3CH_2C(CH_3)_2CH_2CH_3$ and (b) $(CH_3)_2CHCH_2CH(CH_2CH_3)_2$, and (c) write the formula for 3,3,5-triethylheptane.

[*Answer*: (a) 3,3-dimethylpentane; (b) 4-ethyl-2-methylhexane; (c) $(CH_3CH_2)_3CCH_2CH(CH_2CH_3)_2$]

▲

Cycloalkanes. Cyclohexane, C_6H_{12}, is an example of a *cycloalkane*, an alkanelike hydrocarbon in which the carbon atoms form a ring. All the C atoms in cyclohexane are approximately sp^3 hybridized, and the ring

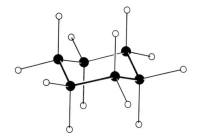

5 Chair cyclohexane

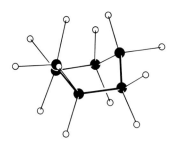

6 Boat cyclohexane

they form is nonplanar. In equations, the molecule is often represented by the simple hexagon

Cyclohexane

The chemical properties of most of the cycloalkanes are very similar to those of the alkanes themselves (cyclopropane is an exception), and we do not need to treat them separately. However, cyclohexane is an interesting molecule partly because it can adopt several *conformations*—shapes that can be interchanged by rotation about bonds, without breaking and re-forming the bonds. The two principal conformations are called the "chair" (**5**) and "boat" (**6**) forms. A sample of cyclohexane consists of molecules that are continually changing from one conformation to the other. The chair form has the lower energy (by about 25 kJ/mol), and at room temperature a molecule spends about 90% of its time in this shape.

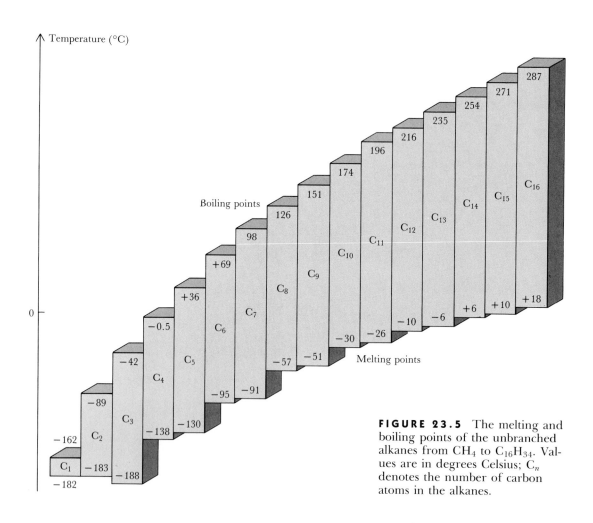

FIGURE 23.5 The melting and boiling points of the unbranched alkanes from CH_4 to $C_{16}H_{34}$. Values are in degrees Celsius; C_n denotes the number of carbon atoms in the alkanes.

TABLE 23.3 **Hydrocarbon constituents of petroleum**

Hydrocarbons	Boiling range, °C	Name
C_1 to C_4	−160 to 0	Gas
C_6 to C_{11}	30 to 200	Gasoline
C_{10} to C_{16}	180 to 400	Kerosene, fuel oil
C_{16} and up	350 and above	Lubricants
C_{20} and up	Low-melting-point solids	Paraffin wax
C_{35} and up	Soft solids	Asphalt

23.3 THE PROPERTIES OF ALKANES

The alkanes show a smooth variation of physical properties. They become less volatile with increasing molecular weight (Fig. 23.5), as we would expect for compounds in which London forces are dominant. The lightest members, methane through butane, are gases at room temperature. Pentane is a volatile liquid, and hexane through undecane are moderately volatile liquids that are present in gasoline (Table 23.3). Kerosene, a fuel used in jet engines and diesel engines, is a mixture of hydrocarbons. It contains a number of alkanes in the range C_{10} to C_{16} (that is, having 10 to 16 carbon atoms). Lubricating oils are mixtures in the range C_{16} to C_{22}, most of the molecules being branched. The heavier members of the series include the paraffin waxes and asphalt. All alkanes are insoluble in water.

Unbranched alkanes tend to have higher melting points, boiling points, and enthalpies of vaporization than their branched isomers. This difference can be traced to the weaker London forces that exist between branched molecules: the atoms in neighboring molecules cannot lie as close together as they do in their unbranched isomers (Fig. 23.6).

Chemical properties The alkanes—or "paraffins"—are unaffected by concentrated sulfuric acid, boiling nitric acid, potassium permanganate, or boiling aqueous sodium hydroxide. One reason for their low chemical reactivity is thermodynamic, for the C—C and C—H bonds

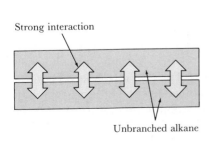
Strong interaction

Unbranched alkane

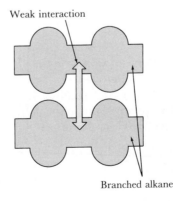
Weak interaction

Branched alkane

FIGURE 23.6 The atoms in neighboring straight-chain alkanes can lie close together. However, the branches of neighboring substituted alkane molecules prevent the close approach of all the atoms, so the London forces are weaker and branched-chain alkanes are more volatile.

TABLE 23.4 The strengths of some C—X bonds

Bond	Bond enthalpy, kJ/mol	Bond	Bond enthalpy, kJ/mol
C—C	348	C—N	305
C=C	612	C=N	613
C≡C	837	C≡N	890
C—C ↕ C=C } in benzene	518	C—F	484
		C—Cl	338
C—H	412	C—Br	276
C—O	360	C—I	238
C=O	743		

are strong (Table 23.4); there is little energy advantage in replacing them with other bonds, except most notably for C=O, C—OH, and C—F bonds. Another reason is related to the fact that the carbon atom has a complete octet of electrons, is small, and has no d orbitals available. Hence it cannot readily accommodate the electron pairs of an attacking Lewis base (such as OH^-).

Alkanes are not completely inert, and one of their commercially most important reactions is oxidation, especially their complete combustion to carbon dioxide and water:

$$CH_4(g) + 2O_2(g) \longrightarrow CO_2(g) + 2H_2O(g) \qquad \Delta H° = -890 \text{ kJ}$$

This exothermic reaction (which proceeds by a radical chain mechanism) and similar reactions involving the higher homologs are the basis of the use of alkanes as fuels. For reasons related to the details of the reaction mechanism, branched-chain hydrocarbons burn more smoothly in engines than unbranched alkanes; the latter explode prematurely, causing the noise called "knocking." The standard of good antiknock behavior (but not the best behavior possible) is 2,2,4-trimethylpentane, which, as we have seen, is also called isooctane. The standard of poor antiknock behavior is the straight-chain hydrocarbon heptane, C_7H_{16}. Every automobile fuel is given an "octane rating" that is the percentage (by volume) of isooctane that must be blended with heptane to match the performance of that fuel. A gasoline rated as "100 octane" has the same antiknock characteristics as pure isooctane.

Isomerization reactions. One of the aims of an oil refinery is to convert unbranched alkanes into the more smoothly burning branched alkanes, using a catalytic *reforming* process (Fig. 23.7). In one version of this process, the unbranched alkane is passed over a Lewis-acid catalyst (typically an aluminum halide) at about room temperature:

FIGURE 23.7 The tower at the center of this oil refinery is where the catalytic reforming reaction takes place; that reaction increases the proportion of branched-chain isomers in a mixture of hydrocarbons.

$$CH_3—CH_2—CH_2—CH_3 \xrightarrow{\text{AlCl}_3, \text{ room temperature}} CH_3—\overset{\overset{\displaystyle CH_3}{|}}{CH}—CH_3$$

This reaction is an example of an *isomerization*, a reaction in which a compound is converted into one of its isomers. In practice, a mixture of products is formed. Fragmentation ("cracking") into alkanes with lower molecular weights also occurs as the temperature is raised.

Substitution reactions. Alkanes also undergo "substitution reactions":

> A **substitution reaction** is a reaction in which an atom or group of atoms is substituted for an atom in the reactant molecule.

For an alkane, the displaced atom is a hydrogen atom **(7)**. An example is the reaction between methane and chlorine. A mixture of these two substances is stable in the dark, but in sunlight, when exposed to ultra-violet radiation, or when they are heated, they react:

$$CH_4(g) + Cl_2(g) \longrightarrow CH_3Cl(g) + HCl(g)$$

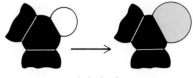

7 Substitution

The reaction does not produce only chloromethane, CH_3Cl, but instead leads to a mixture that also contains dichloromethane (CH_2Cl_2), trichloromethane ($CHCl_3$), and tetrachloromethane (CCl_4). Trichloromethane is better known as chloroform and was one of the early anesthetics. Tetrachloromethane, which is commonly called carbon tetrachloride, has been used as a solvent and in fire extinguishers; however, the realization that it is toxic has limited its use.

THE ALKENES AND THE ALKYNES

8 Ethylene

The simplest unsaturated hydrocarbon is ethylene, C_2H_4 or $CH_2{=}CH_2$ **(8)**, the parent of a homologous series called the "alkenes":

> The **alkenes** are a homologous series of hydrocarbons with formulas derived from ethylene by insertion of —CH_2— groups, and having the molecular formulas C_nH_{2n}.

The next homolog of the series is propylene, $CH_3CH{=}CH_2$. The double bond makes alkenes much more reactive than alkanes. One consequence is that they are very rare in oil wells.

Because alkenes are important as the starting points for a number of manufacturing processes, one of the first steps in petrochemical processing is to convert the abundant alkanes into alkenes. This is achieved by removing hydrogen atoms from neighboring carbon atoms by catalytic *dehydrogenation*:

$$CH_3CH_3(g) \xrightarrow{Cr_2O_3,\ 500°C} CH_2{=}CH_2(g) + H_2(g)$$

Alkenes are also formed in the laboratory when HBr is removed from a bromoalkane by heating it in alcohol with sodium hydroxide:

$$CH_3{-}\overset{\overset{\displaystyle Br}{|}}{C}H{-}\overset{\overset{\displaystyle H}{|}}{C}H_2 + OH^- \xrightarrow{\text{organic solvent, 55°C}}$$

$$CH_3{-}CH{=}CH_2 + Br^- + H_2O$$

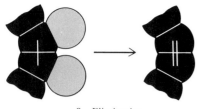

9 Elimination

This reaction is an example of a *dehydrohalogenation*, in this case a dehydrobromination. More generally, it is an example of an "elimination reaction" **(9)**:

An **elimination reaction** is a reaction in which two groups or atoms on neighboring carbon atoms are removed from a molecule, leaving a multiple bond between the carbon atoms.

23.4 ALKENE NOMENCLATURE

The systematic name of an alkene is obtained from the name of the corresponding alkane by changing the ending to -ene. The systematic name of ethylene is therefore ethene, and that of propylene (C_3H_6) is propene.

In all cases except ethylene and propylene it is necessary to specify the location of the double bond. We do this by numbering the C atoms in the backbone and then reporting the *lower* of the numbers of the two atoms joined by the double bond. The numbering starts at the end of the backbone that results in the lowest number for the double-bond position:

$$\overset{4}{C}H_3 - \overset{3}{C}H_2 - \overset{2}{C}H = \overset{1}{C}H_2 \qquad \overset{1}{C}H_3 - \overset{2}{C}H = \overset{3}{C}H - \overset{4}{C}H_3$$

1-Butene   2-Butene

The naming of substituted alkenes is illustrated in the following example.

▼ **EXAMPLE 23.4** **Naming alkenes**

Name   (a) $CH_3CH_2CH_2CH=CH_2$,   (b) $CH_2=CHCH_2CH_2CH_3$,   and
(c) $(CH_3)_2CHCH=CH_2$.

STRATEGY We identify the corresponding alkane by counting the number of atoms in the longest chain that contains the double bond; then we form the name by changing the ending to -ene. We use a numerical prefix to denote the position of the double bond, and name and specify the location of any other substituent.

SOLUTION (a) $CH_3CH_2CH_2CH=CH_2$ has a C_5 chain and is therefore a pentene. Since the double bond joins atoms 1 and 2, it is 1-pentene.
(b) $CH_2=CHCH_2CH_2CH_3$ is the same molecule written in the reverse order, so it is also 1-pentene.
(c) $(CH_3)_2CHCH=CH_2$, which can be written as $CH_3CH(CH_3)CH=CH_2$, has a C_4 unbranched chain containing the double bond, it is a butene. The correct numbering is

$$\overset{4}{C}H_3 - \overset{3}{C}H(CH_3) - \overset{2}{C}H = \overset{1}{C}H_2$$

which makes this molecule a substituted 1-butene. A methyl substituent is attached to carbon atom 3. It is therefore 3-methyl-1-butene.

EXERCISE Name (a) $CH_3CH=CH_2$; (b) $(CH_3CH_2)_2CHCH=CHCH_3$;
(c) $(CH_3)_2CHCH_2CH_2CH=CH_2$.
▲   [*Answer*: (a) Propene; (b) 4-ethyl-2-hexene; (c) 5-methyl-1-hexene]

23.5 THE CARBON–CARBON DOUBLE BOND

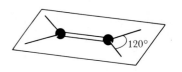

10 The double bond

The C=C double bond of alkenes consists of a σ bond and π bond. Its presence in alkenes makes them significantly different from alkanes in shape, in the types of isomers they form, and in their reactions.

The effect on shape. A double bond resists being twisted. All four atoms attached to the C=C group lie in the same plane and are locked into that arrangement by the π bond (**10**). As a result, alkene molecules are less flexible than the corresponding alkanes and cannot roll up into so compact a ball. Consequently, alkene molecules do not pack together as closely as alkanes and hence have lower melting points (Fig. 23.8).

The effect of double bonds on melting points is shown in the difference between fatty oils and soft solids such as shortening, as we saw in Section 18.1. One step in the manufacture of shortening from vegetable oils is the addition of hydrogen to the unsaturated hydrocarbon chains, to convert a runny oil into a useful fat. In practice this is done by passing the vegetable oil and hydrogen under pressure (2 to 10 atm) over a nickel catalyst at about 200°C. Fish oils are highly unsaturated: this helps them to remain fluid in cold aquatic environments.

Geometrical isomerism. The isomerism characteristic of double bonds becomes apparent in 2-butene (C_4H_8). As the C=C bond is resistant to twisting, the two molecules

$$
\begin{array}{ccc}
CH_3 \quad CH_3 & & CH_3 \quad H \\
\diagdown \quad \diagup & & \diagdown \quad \diagup \\
C{=}C & \text{and} & C{=}C \\
\diagup \quad \diagdown & & \diagup \quad \diagdown \\
H \quad\quad H & & H \quad\quad CH_3 \\
\textit{cis}\text{-2-Butene} & & \textit{trans}\text{-2-Butene}
\end{array}
$$

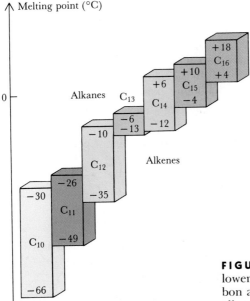

Melting point (°C)

Alkanes

Alkenes

							+18 C_{16} +4
					+10 C_{15} −4	
				+6 C_{14} −12		
			C_{13} −6 −13			
		−10 C_{12} −35				
	−26 C_{11} −49					
−30 C_{10} −66						

FIGURE 23.8 The melting point of an alkene is usually lower than that of the alkane with the same number of carbon atoms. The values shown here are for straight-chain alkanes and 1-alkenes, in degrees Celsius.

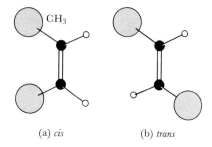

(a) *cis*   (b) *trans*

11   2–Butene

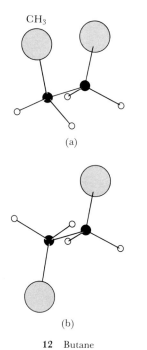

(a)

(b)

12   Butane

are distinct substances (**11**): both are gases, but the *cis* isomer condenses at 4°C and the *trans* isomer condenses at 1°C. This distinctiveness of character is not exhibited by the two forms of butane (**12**), which change from one to the other as one end of the molecule rotates relative to the other end. A sample of butane is a *single* substance containing continually interchanging conformations of C_4H_{10} molecules.

The relation between *cis*-2-butene and *trans*-2-butene is an example of geometrical isomerism, in which atoms are bonded to the same neighbors but have different arrangements in space (Fig. 23.2). As we have indicated here, the two geometrical isomers of 2-butene (and, by analogy, the isomers of other alkenes) are distinguished by the prefixes *cis* ("this side") and *trans* ("across").

▼ **EXAMPLE 23.5**   **Naming geometrical isomers**

Name the following two pairs of geometrical isomers:

(a) $\begin{array}{c} Cl \\ \diagdown \\ H \end{array} C{=}C \begin{array}{c} Cl \\ \diagup \\ H \end{array}$ and $\begin{array}{c} Cl \\ \diagdown \\ H \end{array} C{=}C \begin{array}{c} H \\ \diagup \\ Cl \end{array}$

(b) $\begin{array}{c} Cl \\ \diagdown \\ H \end{array} C{=}C \begin{array}{c} H \\ \diagup \\ CH_3 \end{array}$ and $\begin{array}{c} Cl \\ \diagdown \\ H \end{array} C{=}C \begin{array}{c} CH_3 \\ \diagup \\ H \end{array}$

STRATEGY   We should name the molecules systematically, and then add the prefix *cis*- or *trans*-, depending on whether the substituents are on the same side or on opposite sides of the double bond. As pointed out earlier, a chlorine atom is denoted by the prefix chloro-.

SOLUTION   (a) Both molecules are dichloro-substituted ethenes. The substituents are on atoms 1 and 2; hence they are both 1,2-dichloroethenes. The left-hand molecule has both chlorine atoms on the same side of the double bond, so it is *cis*-1,2-dichloroethene (there is only one possible location for the double bond); the other is *trans*-1,2-dichloroethene.
(b) Both molecules are substituted C_3 alkenes, derivatives of propene. In both a Cl atom is present on carbon atom 1, so both are geometrical isomers of 1-chloropropene. In the left-hand molecule the Cl substituent is on the opposite side of the double bond from the methyl group of the propene backbone, so the molecule is *trans*-1-chloropropene; the other is *cis*-1-chloropropene.

EXERCISE   Name the following pairs of geometrical isomers:

(a) $\begin{array}{c} CH_3 \\ \diagdown \\ H \end{array} C{=}C \begin{array}{c} H \\ \diagup \\ CH_2CH_3 \end{array}$ and $\begin{array}{c} CH_3 \\ \diagdown \\ H \end{array} C{=}C \begin{array}{c} CH_2CH_3 \\ \diagup \\ H \end{array}$

(b) $\begin{array}{c} Cl \\ \diagdown \\ H \end{array} C{=}C \begin{array}{c} H \\ \diagup \\ CH_2CH_2CH_3 \end{array}$ and $\begin{array}{c} Cl \\ \diagdown \\ H \end{array} C{=}C \begin{array}{c} CH_2CH_2CH_3 \\ \diagup \\ H \end{array}$

[*Answer*: (a) *trans*-2-pentene and *cis*-2-pentene;
(b) *trans*-1-chloro-1-pentene and *cis*-1-chloro-1-pentene]

▲

The conversion from one geometrical isomer into another is called *cis-trans isomerization*; it can occur only if the double bond is broken (as by the absorption of a photon of light), leaving the molecule free to rotate around the remaining single σ bond. We have already seen an example: in Section 9.4, we noted that the *cis-trans* isomerization of retinal is responsible for the initiation of vision.

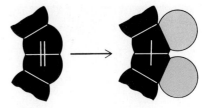

13 Addition

Addition reactions of alkenes. The most characteristic reaction of a C=C double bond is "addition" **(13)**, the reverse of elimination:

> An **addition reaction** of an unsaturated compound is a reaction in which a reactant is added to the two atoms joined by a multiple bond.

An example is the *hydrogenation* of an alkene—the addition of hydrogen atoms to the alkene and its conversion to an alkane:

$$CH_3CH{=}CHCH_3 + H_2 \xrightarrow{\text{Ni, 500°C}} CH_3CH_2CH_2CH_3$$

The states of the reactants and products are rarely given for organic reactions. This is because the solvent is often some other organic compound or the reaction takes place at a surface, as in this example.

The occurrence of addition reactions can be traced to the weakness of the C=C double bond, as compared to the strengths of the single bonds in the addition product. In the hydrogenation reaction, the reaction enthalpy is the difference between the energy invested in breaking the C=C and H—H bonds of the reactants and the energy returned by the formation of two C—H bonds and a C—C bond. In terms of the bond enthalpies B (Section 8.5)

$$\Delta H = B(C{=}C) + B(H{-}H) - [2 \times B(C{-}H) + B(C{-}C)]$$

$$= 612 \text{ kJ} + 436 \text{ kJ} - (2 \times 412 \text{ kJ} + 348 \text{ kJ})$$

$$= -124 \text{ kJ}$$

That is, the reaction is exothermic, releasing 124 kJ of energy for each mole of alkene that reacts. Similar calculations for other addition reactions lead to similar exothermic reaction enthalpies; for example,

$$Cl_2 + \overset{\diagdown}{}C{=}C\overset{\diagup}{} \longrightarrow \overset{\overset{\text{Cl Cl}}{|\;\;|}}{-\overset{|}{C}-\overset{|}{C}-} \qquad \Delta H = -170 \text{ kJ}$$

$$Br_2 + \overset{\diagdown}{}C{=}C\overset{\diagup}{} \longrightarrow \overset{\overset{\text{Br Br}}{|\;\;|}}{-\overset{|}{C}-\overset{|}{C}-} \qquad \Delta H = -95 \text{ kJ}$$

$$HCl + \overset{\diagdown}{}C{=}C\overset{\diagup}{} \longrightarrow \overset{\overset{\text{H Cl}}{|\;\;|}}{-\overset{|}{C}-\overset{|}{C}-} \qquad \Delta H = -55 \text{ kJ}$$

The incorporation of a halogen into a compound is called *halogenation*. Alkenes provide one method by which this can be done, because

FIGURE 23.9 When bromine water (brown) is mixed with an alkene (colorless), the bromine atoms add to the molecule at the double bond, giving a colorless product.

halogens undergo addition reactions with alkenes, as in the formation of 1,2-dichloroethane:

$$CH_2{=}CH_2 + Cl_2 \longrightarrow CH_2Cl{-}CH_2Cl$$

Halogenation, and in particular bromination with bromine water (an aqueous solution of bromine), is used as a simple test for alkenes (Fig. 23.9).

The addition of a hydrogen halide to an alkene with the formation of a haloalkane is called *hydrohalogenation* (it is the reverse of the dehydrohalogenation by which alkenes can be made):

$$CH_2{=}CH_2 + HBr \xrightarrow{0^\circ C} CH_3{-}CH_2Br$$

In a typical method, the gaseous hydrogen halide is bubbled through the alkene or a solution of the alkene in an organic solvent, such as acetic acid. A low temperature is needed to reduce the rate of the reverse reaction.

23.6 ALKENE POLYMERIZATION

Alkenes react with themselves in a process called *addition polymerization*. Thus, an ethylene molecule may form a bond to another ethylene molecule, another ethylene molecule may add to that, and so on, until a long hydrocarbon chain has grown. The initial alkene, such as ethylene, is called the *monomer* (from the Greek words for "one part"); the product, the chain of covalently linked monomers, is called a *polymer*

TABLE 23.5 Some addition polymers

Name	Monomer Formula	Polymer	Typical polymer names
Ethylene	$CH_2{=}CH_2$	$-(CH_2{-}CH_2)_n-$	Polyethylene
Vinyl chloride	$CH{=}CH_2$ | Cl	$-(CH{-}CH_2)_n-$ | Cl	Polyvinyl chloride, PVC
Styrene	$CH{=}CH_2$ (benzene ring)	$-(CH{-}CH_2)_n-$ (benzene ring)	Polystyrene
Acrylonitrile	$CH{=}CH_2$ | CN	$-(CH{-}CH_2)_n-$ | CN	Orlon
Propylene	$CH{=}CH_2$ | CH_3	$-(CH{-}CH_2)_n-$ | CH_3	Polypropylene
Methyl methacrylate	$O{=}C{-}O{-}CH_3$ | $C{=}CH_2$ | CH_3	$-\left\{ \begin{array}{c} O{=}C{-}O{-}CH_3 \\ \| \\ {-}C{-}CH_2{-} \\ \| \\ CH_3 \end{array} \right\}_n-$	Plexiglas, Lucite
Tetrafluoroethylene	$CF_2{=}CF_2$	$-(CF_2{-}CF_2)_n-$	Teflon, PTFE

("many parts"). The simplest addition polymer is polyethylene, $-(CH_2CH_2)_n-$, which consists of long chains of $-CH_2CH_2-$ units. Actual polymer molecules include a number of branches, which are new chains that have grown from intermediate points in the original chain.

The plastics industry has developed polymers from a number of monomers of the form $CHX=CH_2$, where X is either an atom (such as Cl in vinyl chloride, $CHCl=CH_2$) or a group of atoms (such as CH_3- in propylene). These give polymers of formula $-(CHXCH_2)_n-$, such as polyvinyl chloride (PVC), $-(CHClCH_2)_n-$, and polypropylene, $-[CH(CH_3)CH_2]_n-$. The polymers, some of which are listed in Table 23.5 and illustrated in Fig. 23.10, differ in appearance, rigidity, transparency, and resistance to weathering. One of the more recent exten-

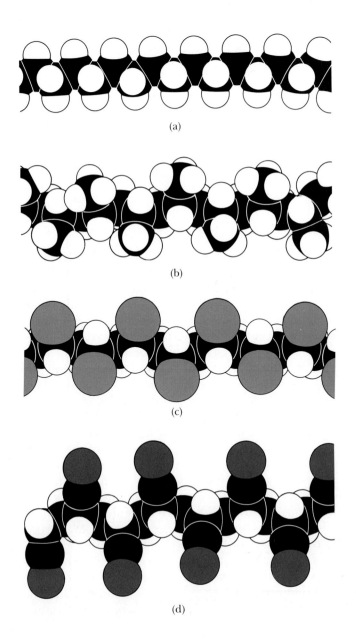

(a)

(b)

(c)

(d)

FIGURE 23.10 Four of the polymers produced by the addition polymerization of alkenes: (a) polyethylene; (b) polypropylene; (c) polyvinyl chloride; (d) polyacrylonitrile, which is sold as "Orlon." Polymers often exhibit chain branching.

sions of the range of polymers is polytetrafluoroethylene (PTFE), which is sold as Teflon; the monomer is fully fluorinated ethylene (CF_2=CF_2), and the polymer consists of —$(CF_2CF_2)_n$— chains. Teflon is very resistant to chemical attack, partly because of the great strength of the C—F bond (Table 23.4), and survives under harsh conditions.

Polymerization processes. A widely used polymerization procedure is *radical polymerization*, in which the chain is propagated by radicals. In a typical process, a monomer (such as ethylene) is compressed to about 1000 atm and heated to 100°C in the presence of a small amount of an organic peroxide (a compound of formula R—O—O—R, where R may be an alkyl group). The chain is initiated by dissociation of the O—O bond, giving two radicals:

$$R—O—O—R \longrightarrow R—O \cdot + \cdot O—R$$

These radicals attack monomer molecules CHX—CH_2 (with X = H for ethylene itself) and form a new radical:

$$R—O \cdot + CH_2{=}\underset{\underset{X}{|}}{\overset{\overset{H}{|}}{C}} \longrightarrow R—O—CH_2{-}\underset{\underset{X}{|}}{\overset{\overset{H}{|}}{C}} \cdot$$

The chain now propagates as this radical attacks other monomer molecules:

$$R—O—CH_2{-}\underset{\underset{X}{|}}{\overset{\overset{H}{|}}{C}} \cdot + CH_2{=}\underset{\underset{X}{|}}{\overset{\overset{H}{|}}{C}} \longrightarrow R—O—CH_2{-}\underset{\underset{X}{|}}{\overset{\overset{H}{|}}{C}}{-}CH_2{-}\underset{\underset{X}{|}}{\overset{\overset{H}{|}}{C}} \cdot$$

Propagation continues until all the monomer has been used up or until two chains link together. The product consists of long chains of formula —$(CHXCH_2)_n$— in which n can reach many thousands.

Another very important polymerization procedure utilizes *Ziegler-Natta catalysts*. These catalysts, named for the German chemist Karl Ziegler and the Italian chemist Giulio Natta, are typically made from titanium tetrachloride and an alkylaluminum (a compound in which an alkyl group is bonded to an aluminum atom) such as triethylaluminum, $Al(CH_2CH_3)_3$. Ziegler-Natta catalysts bring about polymerization at low temperatures and pressures and lead to the creation of high-density materials (Fig. 23.11). The mechanism of their action is still uncertain, but it is believed to involve the formation of a complex between the titanium (a Lewis acid) and the alkene (a Lewis base).

The high density of Ziegler-Natta polymers arises from the character of the addition process, which leads to a regular arrangement of substituents along the polymer chain. Ziegler-Natta polymerization produces either an *isotactic polymer*, in which the substituents are all on the same side of the chain, or a *syndiotactic polymer*, in which the groups alternate. As a result of this "stereoregularity," the chains can pack together well and form a dense, highly crystalline material. In contrast, radical polymerization gives a poorly packing *atactic* product, one in

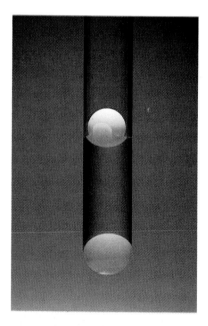

FIGURE 23.11 The two samples of polyethylene in the test tube were produced by different processes. The one that floats was produced by high-pressure (radical) polymerization but the one that has sunk to the bottom was produced with a Ziegler-Natta catalyst. The latter has a higher density, since the chains pack together better.

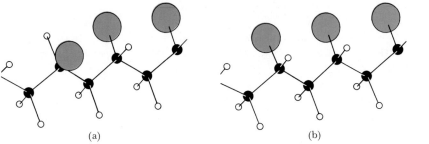
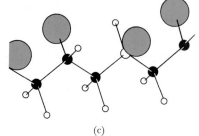

(a)   (b)   (c)

FIGURE 23.12 (a) In an atactic polymer, the substituents lie randomly along the sides of the chain. The stereoregular polymers produced by Ziegler-Natta catalysts may be (b) isotactic (all on one side) or (c) syndiotactic (alternating).

which the groups are attached randomly, on one side or the other, along the chain (Fig. 23.12).

Rubber. Rubber is a natural polymer formed, at least conceptually, from isoprene monomers:

$$H_3C \quad\quad H$$
$$C-C$$
$$H_2C \quad\quad CH_2$$

Isoprene

Natural rubber is obtained from the bark of the rubber tree as a milky white liquid, called "latex" (Fig. 23.13), that consists of a suspension of rubber particles in water. (Latex, from the Latin word for "liquid," is also the white fluid inside dandelion and milkweed stalks.) The rubber itself is a soft white solid that becomes even softer when warm. It is used for pencil erasers and has been used as crepe for the soles of shoes.

The softening that occurs when natural rubber becomes warm is greatly reduced by vulcanization, a process invented by Charles Goodyear in 1839. In this process, rubber is heated with sulfur. The sulfur atoms form cross-links between polyisoprene chains and produce a three-dimensional network of atoms. Since the chains are pinned together, vulcanized rubber does not become sticky as the temperature is raised. High concentrations of sulfur, leading to very extensive cross-linking, result in the hard material called "ebonite." In the commercial production of rubber products such as tires, finely divided carbon black is mixed into the rubber during vulcanization; the carbon helps to strengthen the material, and by coloring it black also helps to protect it from damage by sunlight. Another important ingredient is an "antioxidant," an additive that combats oxidation by trapping radicals formed by air and sunlight before they can attack the double bonds of the polymer.

Chemists were unable to synthesize rubber for a long time, even though they knew it was a polymer of isoprene. The enzymes in the

FIGURE 23.13 Collecting latex from a rubber tree in Malaysia, one of its principal producers.

FIGURE 23.14 (a) In natural
rubber, the isoprene units are
polymerized to be all *cis*. (b) The
harder material gutta-percha is
the all-*trans* polymer.

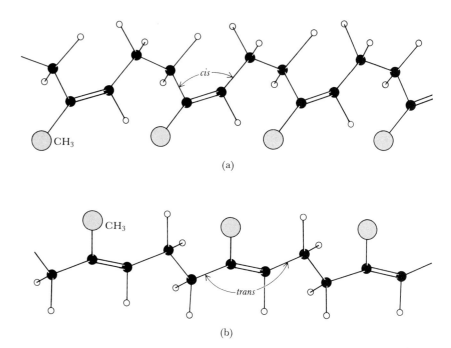

(a)

(b)

rubber tree produce a stereoregular polymer in which all the links
between monomers are in a *cis* arrangement (Fig. 23.14); straightforward radical polymerization, however, produces a random mixture of
cis and *trans* links and a sticky, useless product. The stereoregular polymer was finally achieved with a Ziegler-Natta catalyst, and almost pure,
rubbery *cis*-polyisoprene can now be produced. *trans*-Polyisoprene, in
which all the links are *trans*, is also known and produced naturally: it is
the hard material "gutta-percha" that was once used for golf balls.

Chemists have modified the isoprene monomer and have developed
a variety of synthetic rubbers, or *elastomers* ("elastic polymers"). Some of
these elastomers are *copolymers*, polymers formed from a mixture of
monomers. The advantage of butyl rubber, a copolymer of a little isoprene with isobutylene, $(CH_3)_2C{=}CH_2$, over natural rubber stems
from the fact that isobutylene has only one double bond; because this
polymer has fewer double bonds in the chain than occur in natural
rubber, it is less likely to be attacked and degraded.

23.7 ALKYNES

A third family of hydrocarbons consists of molecules that are more
unsaturated than the alkenes. They contain a carbon–carbon triple
bond:

The **alkynes** are a homologous series of hydrocarbons with formulas
that are derived from acetylene, $HC{\equiv}CH$, by insertion of $-CH_2-$
groups, and having molecular formulas C_nH_{2n-2}.

The names of the alkynes are formed like those of the alkenes but with
the suffix -yne, as in ethyne and propyne. The systematic name of the

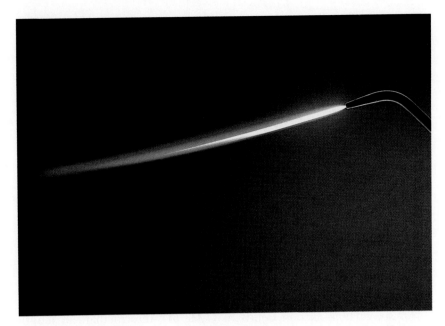

FIGURE 23.15 Oxyacetylene welding makes use of the very hot flame produced when acetylene burns in oxygen. The flame can be used under water, since the oxygen is supplied as well as the acetylene.

parent compound HC≡CH is ethyne, but it is almost always known as acetylene. The next homolog is propyne, $CH_3C≡CH$.

Acetylene burns with an intensely hot flame that can reach 3000°C (Fig. 23.15). As we saw in Section 18.6, it is formed by the action of water on calcium carbide, in which the reaction is the protonation of the Brønsted base C_2^{2-}:

$$:C≡C:^{2-} + 2H_2O \longrightarrow H—C≡C—H + 2OH^-$$

In general, alkynes are prepared like alkenes. Among the preparation methods is the elimination of 2HBr from dibromoalkanes; however, a higher temperature is needed to remove the second HBr:

$$\underset{\overset{|}{Br}\;\;\overset{|}{Br}}{CH_3CH—CH_2} + OH^- \xrightarrow{\text{organic solvent, 80°C}} \underset{\overset{|}{Br}}{CH_3C}=CH_2 + Br^- + H_2O$$

$$\underset{\overset{|}{Br}}{CH_3C}=CH_2 + OH^- \xrightarrow{\text{organic solvent, 120°C}} CH_3C≡CH_2 + Br^- + H_2O$$

The reactions of alkynes resemble those of the alkenes; the principal reaction is addition across the triple bond. Alkynes can be hydrogenated, halogenated, hydrohalogenated, and hydrated:

$$CH_3—C≡CH + 2HX \longrightarrow CH_3—CX_2—CH_3$$

The —C≡C— unit is linear, and alkynes are rodlike in the vicinity of the triple bond.

AROMATIC HYDROCARBONS

Chemists were once puzzled by a group of hydrocarbons that, although known to be unsaturated, are much less reactive than the alkenes. One sign of this lower reactivity is that the compounds, though unsaturated, do not decolorize bromine water. The simplest member of the class is benzene, C_6H_6, a pungent, colorless liquid that freezes at 5.5°C. (Benzene is also very dangerous, since exposure to it may lead to cancer.)

The pungency of benzene gave rise to the name *aromatic* for all the members of the class. This name is still widely used in chemistry to denote compounds derived from benzene and compounds closely related to it, even though many are odorless. The modern tendency, however, is to refer to such hydrocarbons as *arenes* (from *aromatic* and *-ene,* the latter denoting double bonds as in the alkenes). Compounds that are not aromatic, which include alkanes, alkenes, and alkynes, are called *aliphatic* (from the Greek word for "fat").

14 Benzene

All six carbon–carbon bonds in C_6H_6 are intermediate in length (140 pm) between the C—C single bond (153 pm) and the C=C double bond (133 pm). As we saw in Section 8.4, one representation of the benzene ring **(14)** emphasizes the equivalence of the benzene bonds and their intermediate character between single and double bonds. In Lewis terms, the circle represents the resonance between two energetically equivalent Kekulé structures. In molecular-orbital terms, it represents the six electrons that occupy delocalized π orbitals. The σ framework of the molecule is formed by the overlap of sp^2 hybrid orbitals on the C atoms, either with each other or (for the C—H bonds) with $1s$ orbitals on the H atoms; this was illustrated in Fig. 9.13. This hybridization leaves a $2p$ orbital on each carbon atom free to overlap similar orbitals on its two neighbors, thus forming the π molecular orbitals that spread around the ring.

The aromatic compounds also include *polycyclic*, or many-ringed, analogs of benzene. Two of the most important polycyclic aromatics are naphthalene ($C_{10}H_8$) and anthracene ($C_{14}H_{10}$):

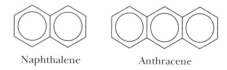

Naphthalene   Anthracene

Both compounds can be obtained by the distillation of coal. Indeed, coal itself, which is a very complex mixture of complicated molecules, consists of very large networks containing regions in which aromatic rings can be identified (Fig. 23.16). When coal is "destructively" distilled—heated so that it decomposes and vaporizes—the sheetlike molecules break up, and the fragments include aromatic hydrocarbons and their derivatives. Naphthalene and anthracene are examples of homologs of benzene, with formulas derived by inserting —C_4H_2— groups into the formula for benzene.

FIGURE 23.16 A schematic representation of the structure of coal. When coal is heated in the absence of air, the structure breaks up and a complex mixture of products—many of them aromatic—is obtained.

23.8 ARENE NOMENCLATURE

Many aromatic compounds have common names that indicate their natural origins. Thus toluene, $CH_3C_6H_5$ **(15)**, was originally obtained from *Tolu balsam*, the aromatic resin of a South American tree. Its systematic name is methylbenzene; the methyl group (CH_3—) is treated as a substituent of the benzene ring.

Isomers can result when more than one substituent is present, as in the three dimethylbenzenes:

15 Toluene

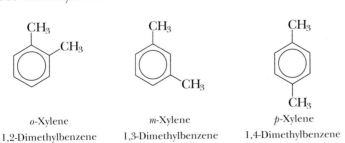

o-Xylene	*m*-Xylene	*p*-Xylene
1,2-Dimethylbenzene	1,3-Dimethylbenzene	1,4-Dimethylbenzene

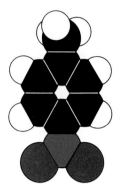

16 *p*-Nitrotoluene

17 Location numbering

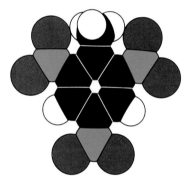

18 2,4,6-Trinitrotoluene

All three isomers have the common name xylene (they were originally obtained from the distillation of wood, for which the Greek word is *xulon;* they are now obtained from coal and petroleum). In the common name, the three relative positions of the substituents are denoted by the prefixes *ortho-* (abbreviated *o-*), *meta-* (*m-*), and *para-* (*p-*). The three prefixes come for the Greek words meaning "directly," "after," and "beyond," and hence suggest the sequence of positions around the ring. The same prefixes are used to name isomers when the two substituents are different from each other, as in *p*-nitrotoluene (**16**), which contains the nitro group (—NO$_2$).

As we have seen with the xylenes, systematic names are obtained by regarding the compounds as substituted benzenes. The numbers used to specify the locations of the substituents are shown in the margin (**17**). All three xylenes are dimethylbenzenes (a benzene ring with two methyl substituents). As shown above, they are 1,2-dimethylbenzene, 1,3-dimethylbenzene, and 1,4-dimethylbenzene. This nomenclature can readily be extended to more complicated molecules with more than two substituents. For example, the systematic name of trinitrotoluene (**18**, the explosive TNT) is 1-methyl-2,4,6-trinitrobenzene.

▼ **EXAMPLE 23.6** **Naming derivatives of benzene**

Give the systematic names of the following substances:

(a) CH$_2$CH$_3$ / CH$_2$CH$_3$ (b) CH$_2$CH$_3$ / CH$_3$ (c) CH$_3$ / CH$_3$ CH$_3$

STRATEGY We need to identify the substituents and their locations, listing them in alphabetical order as prefixes to "benzene."

SOLUTION (a) Two (di-) —CH$_2$CH$_3$ (ethyl) groups are located in positions 1 and 2; the compound is 1,2-diethylbenzene.
(b) An ethyl group is located in position 1, and a methyl group in position 3; the compound is 1-ethyl-3-methylbenzene.
(c) Three (tri-) methyl groups are located at positions 1, 3, and 5; the compound is 1,3,5-trimethylbenzene.

EXERCISE Name the following compounds:

(a) CH$_2$CH$_2$CH$_3$ / CH$_2$CH$_2$CH$_3$ (b) CH$_2$CH$_3$ / CH$_2$CH$_2$CH$_3$ (c) CH$_2$CH$_3$ / CH$_3$ CH$_3$

[*Answer*: (a) 1,4-dipropylbenzene; (b) 1-ethyl-3-propylbenzene; (c) 1-ethyl-3,5-dimethylbenzene]

When a benzene ring is a substituent in another backbone, as in styrene **(19)**, the C_6H_5— group is called "phenyl" ("phene" is an old name for benzene). The systematic name of styrene, the monomer used for the production of polystyrene, is therefore phenylethene.

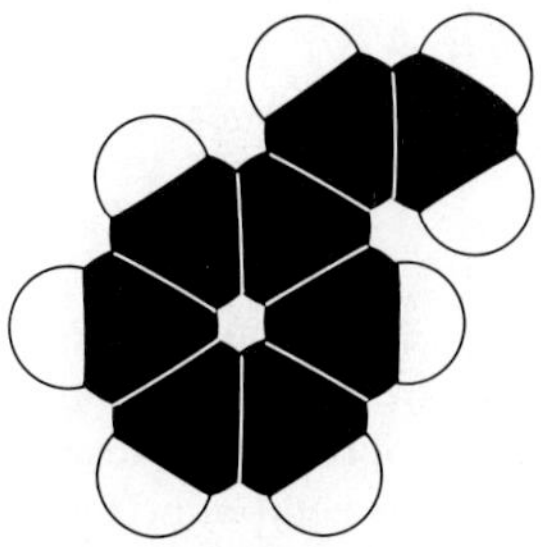

19 Styrene

23.9 REACTIONS OF AROMATIC HYDROCARBONS

Although both are unsaturated hydrocarbons, arenes are markedly different from alkenes. In the reactions of arenes the aromatic ring is usually left unchanged, but in the reactions of alkenes the double bond is usually lost. In fact, the benzene ring is so unreactive that it usually needs to be coaxed into reaction by a catalyst. We shall see in Chapter 24 that the reactions of aromatic compounds are much more extensive once the ring has undergone substitution.

Substitution reactions. In contrast to alkenes, for which addition is the dominant reaction type, arenes predominantly undergo substitution. For example, bromine immediately adds to a double bond of an alkene; however, it reacts with benzene only in the presence of a catalyst—typically iron(III) bromide—and the result is a substitution in the ring, not addition across a double bond:

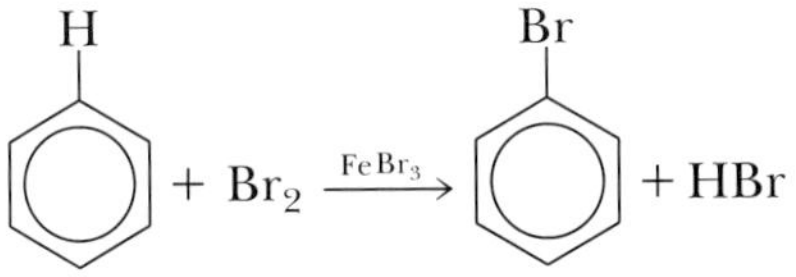

The fluorination of benzene with F_2 is explosively violent; chlorination with Cl_2 requires a catalyst ($AlCl_3$ or $FeCl_3$). Iodination with I_2 of benzene cannot be carried out directly.

Friedel-Crafts alkylation. A part of the organic chemist's expertise is the ability to build larger molecules by using substitution reactions to attach groups to aromatic rings. Alkyl groups can be attached to benzene, to build a more extensive network of carbon atoms, using *Friedel-Crafts alkylation*. This reaction, which is named for the French chemist Charles Friedel and his American coworker James Crafts, uses aluminum chloride as a catalyst:

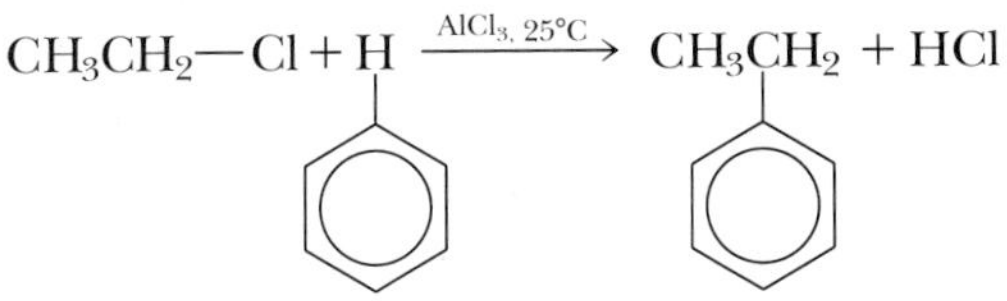

The reaction takes place quite vigorously when a small amount of the catalyst is added to a mixture of the two liquid reagents. It is important because it produces a new C—C bond and hence can be used to build up a complex molecule.

SUMMARY

23.1 The **hydrocarbons** are binary compounds of hydrogen and carbon that may be classified as shown in Fig. 23.17. They form several **homologous series,** each consisting of compounds with formulas that are related by the insertion of a particular group (which is often —CH_2—). The **alkanes** have the general formula C_nH_{2n+2} and form a homologous series based on methane. The alkanes include numerous structural **isomers,** molecules with the same atoms in a different arrangement, and their classification is summarized in Fig. 23.2.

23.2 The names of the alkanes are the basis of the IUPAC systematic nomenclature of organic compounds. In naming organic compounds, it is important to differentiate between **branched** and **unbranched** carbon chains. Ringlike saturated hydrocarbons are known as **cycloalkanes.** One example of this class, cyclohexane, exists in several different **conformations,** or forms that differ only due to twisting around a single bond.

23.3 The alkanes are **saturated compounds**—compounds having no multiple carbon–carbon bonds—with low reactivity. Their typical reactions are **substitution reactions,** the replacement of one atom or group by another, which generally occur by a radical mechanism. **Isomerization** reactions, conversions of a compound into one of its isomers, are brought about by Lewis-acid catalysts.

23.4 The **alkenes** have the general formula C_nH_{2n} and contain a carbon–carbon double bond. They are formed by an **elimination reaction,** in which atoms attached to neighboring carbon atoms are removed, typically **dehydrogenation** of alkanes, the removal of H_2, or **dehydrohalogenation** of haloalkanes, the removal of a hydrogen halide.

23.5 A double bond is torsionally rigid and leads to the existence among the alkenes of **geometrical isomers,** isomers that differ in the spatial arrangement of the atoms. The characteristic reaction of alkenes is **addition** to the double bond.

23.6 Alkene **monomers,** individual small molecules, can react with themselves and undergo **addition polymerization** by **radical chain reactions.** These mechanisms lead to **atactic** polymers, in which there is an irregular arrangement of substituents along the polymer chain. **Stereoregular** polymers, with a regular arrangement of substituents along the polymer chain, may be **isotactic** or **syndiotactic** and are produced if **Ziegler-Natta catalysts** are used. The stereoregular addition polymerization of isoprene leads to **rubber,** and other monomers can be used to produce other **elastomers,** or elastic polymers.

23.7 The **alkynes** have the general formula C_nH_{2n-2} and contain a carbon–carbon triple bond. Their reactions resemble those of the alkenes. The —C≡C— group is linear.

23.8 and 23.9 Aromatic hydrocarbons, or **arenes,** are hydrocarbons based on the benzene ring. Arenes are

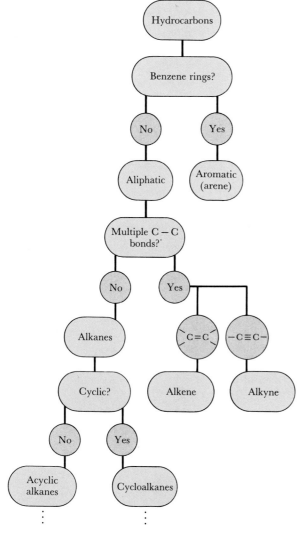

FIGURE 23.17 The classification of hydrocarbons. The dotted lines indicate that the classification continues into cyclic and acyclic compounds as for alkanes.

much less reactive than alkenes. The typical reactions of arenes are **ring substitution reactions.** Examples include **halogenation,** the incorporation of a halogen atom and **Friedel-Crafts alkylation,** the attachment of an alkyl group, using aluminum chloride as catalyst.

EXERCISES

Nomenclature: Aliphatic Compounds

In Exercises 23.1 and 23.2, state whether each compound is an alkane, alkene, or alkyne, given that no rings are involved.

23.1 (a) CH_4; (b) C_3H_8; (c) $C_{10}H_{18}$; (d) C_3H_6.
23.2 (a) C_2H_2; (b) $C_{16}H_{34}$; (c) $C_{10}H_{20}$; (d) C_7H_{16}.

In Exercises 23.3 and 23.4, name each unbranched alkane.

23.3 (a) C_2H_6; (b) C_4H_{10}; (c) $C_{10}H_{22}$; (d) $C_{17}H_{36}$.
23.4 (a) C_3H_8; (b) C_5H_{12}; (c) C_8H_{18}; (d) $C_{12}H_{26}$.

In Exercises 23.5 and 23.6, write the molecular formula (in the form C_nH_{2n+2}) of each alkane.

23.5 (a) Ethane; (b) propane; (c) decane; (d) hexadecane.
23.6 (a) Butane; (b) pentane; (c) heptadecane; (d) heptane.

23.7 (a) Explain the meaning of the term "homologous series," and give three examples. (b) Write the general formula for a cycloalkane consisting of n carbon atoms.
23.8 Write the general formula for a "diene," an unsaturated compound containing two carbon–carbon double bonds. What would the formula be if the molecule were cyclic?

23.9 Identify the alkane responsible for ions of molar mass (in grams per mole) (a) 16; (b) 58.1; (c) 72.2.
23.10 Identify the alkene responsible for ions of molar mass (in grams per mole) (a) 56.1; (b) 98.2; (c) 168.3.

In Exercises 23.11 to 23.14, give the systematic name of each compound.

23.11 (a) $CH_3CH_2CH_3$; (b) $CH_3(CH_2)_6CH_3$; (c) $CH_2=CH(CH_2)_4CH_3$; (d) $(CH_3)_2CHCH_2CH(CH_3)_2$.
23.12 (a) $CH_3CH_2CH_2CH_3$; (b) $CH\equiv C(CH_2)_2CH_3$; (c) $CH_2(CH_2CH_3)_2$; (d) $CH_3(CH_2)_3CH(CH_2CH_3)-(CH_2)_4CH_3$.
23.13 (a) Isooctane; (b) ethylene; (c) propylene.
23.14 (a) n-Heptane; (b) acetylene; (c) butylene.

In Exercises 23.15 to 23.18, write the structural formula for each compound.

23.15 (a) 3-Methylpentane; (b) 4-ethyl-3,3-dimethylheptane; (c) 5,5-dimethyl-1-hexene.
23.16 (a) 4,4-Dimethylnonane; (b) 4-butyl-5,5-diethyl-1-decyne; (c) 5-ethyl-1,3,6-heptatriene.
23.17 (a) Methylbutane; (b) 2-methylpentane; (c) 2,7,8-trimethyldecane; (d) 5-methyl-4-propylnonane.
23.18 (a) Methylpropane; (b) 4-ethyl-3,3-dimethylheptane; (c) 2,3,5-trimethyl-4-propylheptane.

Isomerism

23.19 Write and name all the isomers of the alkanes (a) C_4H_{10}; (b) C_5H_{12}.
23.20 Write and name all the isomers (including geometrical isomers) of the alkenes (a) C_4H_8; (b) C_5H_{10}.

23.21 Are any additional isomers of the compounds in Exercise 23.19 possible if cycloalkanes are included?
23.22 Are any of the compounds in Exercises 23.19 and 23.20 optically active? If so, which ones?

In Exercises 23.23 and 23.24, use bond enthalpies to decide which isomer of the pair has the lower energy.

23.23 2-Pentene and cyclopentane
23.24 2-Hexene and cyclohexane

Molecular Formulas

23.25 A 2.14-g sample of a hydrocarbon formed 3.32 g of water and 6.48 g of carbon dioxide in a combustion experiment. Deduce its empirical formula, and state whether it is an alkane, an alkene, or an alkyne.
23.26 In another combustion experiment, a 3.69-g sample of a hydrocarbon formed 4.50 g of water and 11.7 g of carbon dioxide. Deduce its empirical formula, and state whether it is an alkane, an alkene, or an alkyne.

In Exercises 23.27 and 23.28, deduce the molecular structure of the hydrocarbon reactant from the information given.

23.27 With chlorine in the presence of light, C_4H_{10} gives two isomers of formula C_4H_9Cl.
23.28 With chlorine in the presence of light, C_6H_{14} gives two isomers of formula $C_6H_{13}Cl$.

23.29 The alkane C_6H_{14} reacts with chlorine in the presence of light to give two isomers of formula $C_6H_{13}Cl$. When one of these isomers is treated with potassium hydroxide in an organic solvent, it produces one alkene, C_6H_{12}; however, the other isomer produces two isomeric alkenes. Propose formulas for all the compounds.

23.30 A hydrocarbon contains 88% by mass of carbon and 12% by mass of hydrogen and has molar mass 82 g/mol. It decolorizes bromine water, and 1.5 g of the hydrocarbon reacts with 400 mL of hydrogen in the presence of a nickel catalyst. Identify the hydrocarbon, and write its structural formula.

Reactions of Alkanes and Alkenes

23.31 Use the bond enthalpies in Table 8.3 to calculate the enthalpy of combustion of (a) alkanes; (b) alkenes; (c) alkynes containing n carbon atoms.

23.32 Use bond enthalpies to estimate the heat output per gram of (a) alkanes; (b) alkenes; (c) alkynes containing n carbon atoms.

In Exercises 23.33 to 23.40, predict the products of, classify, and write the chemical equation for each reaction.

23.33 Sodium hydroxide is heated with (a) 2-bromopropane; (b) 3-bromopentane in an organic solvent.

23.34 Hydrogen bromide is mixed with (a) ethylene; (b) propene.

23.35 Chlorine is mixed with (a) ethane; (b) propane and exposed to light.

23.36 2-Methylpropene and hydrogen are passed over a platinum catalyst.

23.37 $CH_3(CH_2)_3CH(OH)CH_3$ is heated with concentrated sulfuric acid to 120°C.

23.38 $CH_3(CH_2)_2CH(OH)CH_3$ is heated with concentrated phosphoric acid to 80°C.

23.39 1-Iodoheptane is heated with potassium hydroxide in an organic solvent.

23.40 Hydrogen bromide reacts with (a) 2-methyl-2-butene; (b) 1-pentene.

23.41 Suggest a series of reactions that will convert cyclopentane into cyclopentene.

23.42 Cyclohexane and 1-hexene are isomers. How may they be distinguished chemically?

Addition Polymers

In Exercises 23.43 and 23.44, name the polymer that is formed when each monomer is polymerized.

23.43 (a) Ethene; (b) propene; (c) isoprene.
23.44 (a) Chloroethene; (b) phenylethene.

The molecular weights of polymers are measured by osmometry (Section 11.7). In Exercises 23.45 and 23.46, calculate the molecular weight of the polymer from the given data (at 25°C), and estimate the number of monomer units in the polymer chain.

23.45 2.0 g of polyethylene in 250 mL of toluene gave rise to an osmotic pressure of 0.71 Pa.

23.46 2.84 g of polystyrene in 150 mL of toluene gave rise to an osmotic pressure of 0.18 Pa.

Nomenclature: Aromatic Compounds

In Exercises 23.47 and 23.48, name each aromatic compound.

23.47

(a)

(b)

(c)

(d)

23.48

(a)

(b)

(c)

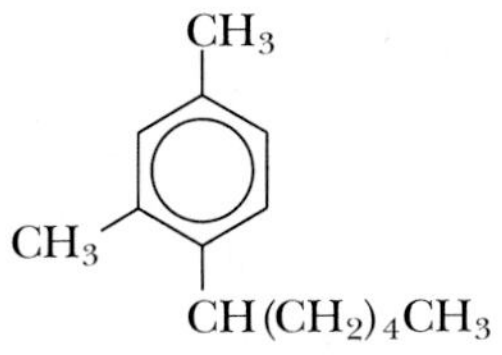

(d)

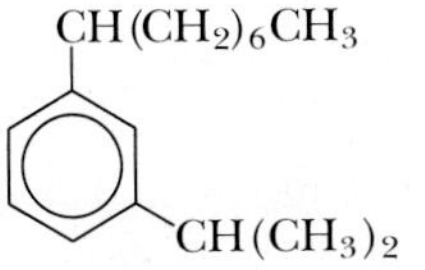

In Exercises 23.49 and 23.50, write the structural formula of each compound.

23.49 (a) Toluene; (b) *p*-chlorotoluene; (c) 4-chloromethylbenzene; (d) 2,4,6-trimethylnitrobenzene.

23.50 (a) *m*-Xylene; (b) 1,2,-dichlorobenzene; (c) 1-ethyl-2,5-dichlorobenzene; (d) 3-phenylpropene.

Reactions of Aromatic Compounds

23.51 Suggest a method for preparing cyclohexene from benzene.

23.52 Suggest a method for preparing 2-phenylpropane from propene and benzene.

23.53 Which of the three possible dichlorobenzenes are polar compounds?

23.54 Which of the possible trichlorobenzenes are polar compounds?

The presence of particular groups of atoms in a molecule give it characteristic, predictable properties. In this chapter we meet half a dozen or so of these groups and see how they are used to build new materials and contribute to the molecules that participate in the processes of life. Many modern plastics, for instance, consist of molecules that have been linked together by reactions between their functional groups. The illustration shows one of them being used in a computer that makes use of fluid flow through complex molded channels rather than the flow of electric current through a circuit.

24

FUNCTIONAL GROUPS AND BIOMOLECULES

The hydroxyl group

24.1 Alcohols and ethers
24.2 Phenols

The carbonyl group

24.3 Aldehydes and ketones
24.4 Carbohydrates

The carboxyl group

24.5 Carboxylic acids
24.6 Esters

**Functional groups
   containing nitrogen**

24.7 Amines
24.8 Amino acids
24.9 DNA and RNA

As we noted in Chapter 23, we can regard many organic compounds as having hydrocarbon backbones to which are attached groups of atoms called *functional groups*, groups that undergo characteristic reactions. Functional groups, such as —OH and —Br, bring organic compounds to life. We have in fact already seen how a functional group can enliven the hydrocarbons, for the presence of a C=C double bond, which is one example of such a group, gives rise to the characteristic properties of the alkenes. The genetic molecule DNA, which we explore at the end of this chapter, is very rich in functional groups, and it is correspondingly rich in chemical properties.

Some common functional groups are listed in Table 24.1. Chemists use them to build new pharmaceuticals, plastics, and biologically active

TABLE 24.1 Common functional groups

Structural formula	Short form	Prefix or suffix in compounds	Class of compound	Typical example
C=C	C=C	-ene	Alkene	$CH_2=CH_2$
—C≡C—	C≡C	-yne	Alkyne	$CH≡CH$
—O—H	OH	Hydroxy- -ol	Alcohol	CH_3OH
C=O	CO	-al	Aldehyde	CH_3CHO
		-one	Ketone	CH_3COCH_3
—C(=O)O—H	COOH	Carboxy- -oic acid	Carboxylic acid	CH_3COOH
—N(H)(H)	NH₂	Amino- -amine	Amine	CH_3NH_2
—N(=O)(O)	NO₂	Nitro-	Nitro compound	CH_3NO_2
—F, —Cl, —Br, —I	F, Cl, Br, I	Halo- (fluoro-, chloro-, bromo-, iodo-)	Haloalkane	CH_3Cl
—C≡N	CN	Cyano-	Nitrile	CH_3CN
—N≡C	NC	Isocyano-	Isocyanide	C_6H_5NC
—O—	O	R-oxy-	Ether	CH_3OCH_3
—S(=O)(=O)—O—H	SO₃H	Sulfo- -sulfonic acid	Sulfonic acid	$C_6H_5SO_3H$
—S—H	SH	Mercapto- -thiol	Thiol	CH_3SH

materials. Similarly, living cells use them to form proteins and to bring about the biochemical processes of life. The functional groups are like a kit of parts from which the whole range of organic molecules, and ultimately people, are constructed.

THE HYDROXYL GROUP

The *hydroxyl group* —OH is an —O—H group covalently bonded to another atom. It must distinguished from the hydroxide ion (OH^-) of inorganic hydroxides, which is a separate diatomic ion. The —OH group occurs in several classes of organic compounds, including alcohols, phenols, and carboxylic acids. As we shall see, these compounds are distinguished by the other groups that are attached to the carbon atom carrying the —OH group. We shall also see that the ability of the hydroxyl group to take part in hydrogen bonding is responsible for some of the distinctive physical properties of alcohols.

24.1 ALCOHOLS AND ETHERS

One of the best-known organic compounds is ethanol, or ethyl alcohol, CH_3CH_2OH. In chemistry it is common for a single compound to inspire the name of an entire class of related compounds, and that is the case here: *alcohol* is now taken to mean any compound that contains the hydroxyl group not connected directly to an aromatic ring or to a C=O group. The name "alcohol" comes from the Arabic words for "fine powder." The term gradually came to mean the "essence" of a thing— and in particular of the liquid obtained by distilling wine.

A major source of ethanol is the fermentation of carbohydrates using the enzymes in yeast, but it is also obtained by the acid-catalyzed hydration of ethylene:

$$CH_2{=}CH_2 + H_2O \xrightarrow{\ H^+\ } CH_3CH_2OH$$

Another alcohol, methanol (or methyl alcohol) CH_3OH, was originally obtained from the distillation of wood and hence was known as "wood alcohol." It is now obtained either by the oxidation of hydrocarbon gases or from carbon monoxide and hydrogen:

$$CO(g) + 2H_2(g) \xrightarrow{\ 400°C,\ 150\ atm,\ ZnO\ } CH_3OH(g)$$

Methyl alcohol is a volatile, colorless liquid that causes blindness and death when ingested.

Nomenclature of alcohols. Alcohols are divided into three classes according to the number of hydrogen atoms attached to the carbon atom to which the —OH group is bonded:

A **primary alcohol** has the structure R—C—OH, with H atoms above and below the carbon.

A **secondary alcohol** has the structure R—C—OH.

$$\begin{matrix} & H & \\ & | & \\ R & - C - & OH \\ & | & \\ & R & \end{matrix}$$

A **tertiary alcohol** has the structure R—C—OH.

$$\begin{matrix} & R & \\ & | & \\ R & - C - & OH \\ & | & \\ & R & \end{matrix}$$

The organic groups R need not be the same; they may be either aliphatic (such as CH_3—) or aromatic (such as C_6H_5—). Ethanol and benzyl alcohol are both primary alcohols; isopropyl alcohol is a secondary alcohol:

Benzyl alcohol   Isopropyl alcohol

The systematic name of an alcohol is formed by changing the final -e of the parent alkane's name to -ol. Thus, CH_3OH is derived from methane and hence is methanol. If, to avoid ambiguity, we need to state the locations of substituents, we number the carbon atoms as explained in Section 23.2, starting at the carbon atom that gives the location of the —OH group the lower number. The systematic name of isopropyl alcohol, for example, is 2-propanol.

▼ **EXAMPLE 24.1** **Classifying and naming alcohols**

Classify and name (a) $CH_3CH_2CH_2CH_2OH$, (b) $CH_3CH(OH)CH_2CH_3$, and (c) $(CH_3)_2CHCH_2CH_2OH$.

STRATEGY For the classification, we have to decide how many groups other than H are attached to the C—OH group. For the name, we must identify the longest unbranched alkane chain, use Table 23.2 to find the name of the corresponding substituted alkane, and change the ending -e to -ol.

SOLUTION (a) The molecule has a —CH_2—OH group and so is a primary alcohol. A C_4 alkane is butane, so the alcohol is a butanol. There is more than one butanol because the hydroxyl group can be attached to different carbon atoms; therefore, the group's location on carbon atom 1 must be specified, giving the name 1-butanol. (b) Since the molecule has a ⟩CH—OH group, it is a secondary alcohol. It too is a butanol, and as the hydroxyl group is attached to carbon atom 2, it is 2-butanol. (c) The —CH_2—OH group signifies a primary alcohol. Writing out the structural formula and numbering the backbone C atoms,

$$\begin{matrix} & CH_3 & & & \\ & | & & & \\ CH_3 - & CH - & CH_2 - & CH_2 - & OH \\ 4 & 3 & 2 & 1 & \end{matrix}$$

we see that it is 3-methyl-1-butanol. The alternative numbering, which results in the name 2-methyl-4-butanol, is rejected because it gives —OH a higher-numbered location.

EXERCISE Classify and name the following alcohols: (a) $CH_3(CH_2)_4$-CH_2OH, (b) $(CH_3CH_2)_2CHOH$, and (c) $CH_3CH_2C(CH_3)_2OH$.

[*Answer*: (a) primary, 1-hexanol; (b) secondary, 3-pentanol; (c) tertiary, 2-methyl-2-butanol]

1 Ethylene glycol

Ethylene glycol, 1,2-ethanediol, $HOCH_2CH_2OH$ (**1**), is an example of a *diol*, a compound with two hydroxyl groups. It is a component of radiator antifreeze solutions and is used in the manufacture of some synthetic fibers. Its action as an antifreeze stems from its high solubility (on account of its hydroxyl groups, which can form hydrogen bonds with water), coupled with the disruptive effect its short hydrocarbon backbone has on the structure of ice: the presence of ethylene glycol tends to prevent the formation of hydrogen bonds between water molecules. Ethylene glycol is not corrosive and will not damage the radiator. It is also not very volatile, so it does not boil away readily.

Glycerol, $HOCH_2CH(OH)CH_2OH$ (**2**), has three hydroxyl groups and is thus a *triol*. It is a syrupy, viscous liquid that dissolves readily in water and is widely used as a skin softener in cosmetic preparations and as an antidrying medium and sweetening agent in toothpaste and packaged coconut. Its characteristic physical properties—its high viscosity, its high boiling point, its solubility in water, and its water-retaining power—all result from its ability to form hydrogen bonds with itself and with water.

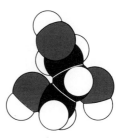

2 Glycerol

Properties and reactions of alcohols. Alcohols with low molecular weights are liquids. This is a sign of the importance of hydrogen bonding, for alcohols have much lower vapor pressures than the parent hydrocarbons. Their ability to form hydrogen bonds also accounts for alcohols being much more soluble than hydrocarbons in water. Both methanol and ethanol mix with water in all proportions.

Alcohols are amphiprotic. They are very weak Brønsted acids, as we can see with ethanol:

$$CH_3CH_2OH(aq) + H_2O(l) \rightleftharpoons$$
$$CH_3CH_2O^-(aq) + H_3O^+(aq) \qquad pK_a = 16$$

Ethanol is actually so weak an acid that it cannot turn litmus red. The conjugate base $CH_3CH_2O^-$, the *ethoxide ion*, is also formed when sodium metal reacts with ethanol, which it does more gently than with water (Fig. 24.1):

$$2Na(s) + 2CH_3CH_2OH(l) \longrightarrow 2Na(CH_3CH_2O)(aq) + H_2(g)$$

Alcohols can also act as Brønsted bases, because the lone pairs of electrons on the O atom of the —OH group can accept protons from other acids:

$$CH_3CH_2OH(aq) + H_3O^+(aq) \rightleftharpoons CH_3CH_2OH_2^+(aq) + H_2O(l)$$

The conjugate acid $CH_3CH_2OH_2^+$ is an example of an *oxonium ion*, an ion of the form ROH_2^+ (the hydronium ion, H_3O^+, is a special case of

FIGURE 24.1 Sodium reacts with ethanol much less vigorously than it reacts with water. The products are hydrogen and a solution of sodium ethoxide, an electrolyte.

an oxonium ion, one with R = H). The formation of an oxonium ion is the first step in various acid-catalyzed reactions of alcohols, including their dehydration to alkenes:

$$CH_3CH_2OH \xrightarrow{\text{conc } H_2SO_4,\ 170°C} CH_2{=}CH_2 + H_2O$$

Dehydration is a potential route from the ethanol produced by fermentation to the polyethylene normally obtained from petrochemicals.

Alcohols can be used as reagents in halogenation reactions (reactions in which a halogen atom is introduced into a molecule). For example, concentrated hydrobromic acid converts alcohols into bromoalkanes:

$$CH_3CH_2{-}OH + HBr \xrightarrow{\text{warm}} CH_3CH_2{-}Br + H_2O$$

Such brominations are important because the products can be used to synthesize other compounds.

Oxidizing agents, including the oxygen of the air and acidified sodium dichromate solution, produce the compounds known as "aldehydes" from primary alcohols (we deal with aldehydes more fully in Section 24.3). Acetaldehyde, CH_3CHO, for example, is produced from ethanol; we can express this schematically as:

$$CH_3{-}\underset{\underset{H}{|}}{\overset{\overset{OH}{|}}{C}}{-}H \xrightarrow{\text{oxidation removes 2H}} CH_3{-}\overset{O}{\underset{H}{C}}$$

As we shall see, aldehydes are readily oxidized; hence the oxidation of a primary alcohol often proceeds still further, to produce a "carboxylic acid" (such as acetic acid, CH_3COOH):

$$CH_3{-}\overset{O}{\underset{H}{C}} \xrightarrow{\text{oxidation adds oxygen}} CH_3{-}\overset{O}{\underset{O{-}H}{C}}$$

These last two schemes show the sequence of events that cause wine to sour as its ethanol is oxidized by air to acetic acid.

The oxidation of secondary alcohols produces the compounds called "ketones" (such as acetone, CH_3COCH_3, Section 24.3):

$$CH_3{-}\underset{\underset{CH_3}{|}}{\overset{\overset{OH}{|}}{C}}{-}H \xrightarrow{\text{oxidation removes 2H}} CH_3{-}\overset{O}{\underset{CH_3}{C}}$$

Ketones are much more difficult to oxidize than aldehydes, and the oxidation does not go any further (unless it occurs very vigorously, as in a combustion).

Ethers. We can think of an alcohol as being derived from H_2O by replacing one H atom with an alkyl group. The product of replacing the second H atom with an alkyl group is an *ether*, a compound of the form R—O—R.

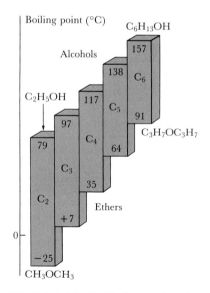

FIGURE 24.2 Hydrogen bonding occurs in alcohols but not in ethers. As a result, the boiling points of ethers are lower than those of the isomeric alcohols. Values are in degrees Celsius. All the alkyl groups here are straight chains.

$$\text{H—O—H} \qquad \text{CH}_3\text{CH}_2\text{—O—H} \qquad \text{CH}_3\text{CH}_2\text{—O—CH}_2\text{CH}_3$$

Water   Ethanol   Diethyl ether

(Diethyl ether is the "ether" that was once used as an anesthetic.) Ethers are more volatile than the alcohols of the same molecular weight because they are not hydrogen-bonded (Fig. 24.2).

As ethers are not very reactive toward many reagents, they are useful solvents for other organic compounds. However, ethers are flammable: diethyl ether is so easily ignited that it can cause dangerous fires. Like alcohols, ethers can undergo acid-catalyzed substitution reactions.

▼ **EXAMPLE 24.2** **Predicting the product of a substitution reaction**

Predict the outcome of the reaction between Br^- ions and diethyl ether in the presence of a strong acid.

STRATEGY We have seen that in the halogenation of alcohols, —OH is driven out of R—OH, leaving R—Br as the product; we may suspect that —OR will be driven out of R—OR, leaving R—Br as the product.

SOLUTION Building on this analogy, we can write the reaction as

$$\text{CH}_3\text{CH}_2\text{—O—CH}_2\text{CH}_3 + \text{HBr} \xrightarrow{\text{H}^+} \text{CH}_3\text{CH}_2\text{OH} + \text{CH}_3\text{CH}_2\text{Br}$$

The product $\text{CH}_3\text{CH}_2\text{OH}$ may go on to produce more of the bromoethane in a continuation of the reaction.

EXERCISE Predict the final products of the acid-catalyzed reaction between $\text{CH}_3\text{CH}_2\text{OCH}_2\text{CH}_2\text{CH}_3$ and Br^- ions.

[*Answer*: $\text{CH}_3\text{CH}_2\text{Br}$ and $\text{CH}_3\text{CH}_2\text{CH}_2\text{Br}$]

▲

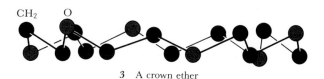

3 A crown ether

Cyclic ethers (ethers with molecules in the form of a ring) with alternating —CH_2CH_2— units and O atoms are called *crown ethers*. The name reflects the crownlike shape the molecules adopt (**3**). The importance of crown ethers stems from their ability to bind very strongly to metal cations (such as K^+), effectively encasing them in a hydrocarbon shell and making them soluble in nonpolar solvents. Potassium permanganate, for example, dissolves readily in benzene when a crown ether with six O atoms is added to it (Fig. 24.3). The resulting solution can be used to carry out oxidations in organic solvents.

24.2 PHENOLS

The compound obtained when a hydroxyl group is attached directly to an aromatic ring is a *phenol*, Ar—OH, not an alcohol. Phenol itself, a white, low-melting-point, crystalline molecular solid, is benzenol,

FIGURE 24.3 Both flasks contain potassium permanganate and benzene, in which the permanganate is almost insoluble. A crown ether has been added to the flask on the right, and as the purple color shows, the potassium permanganate has dissolved.

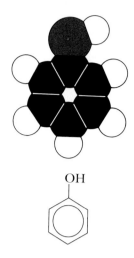

OH

4 Phenol

OH

CH(CH$_3$)$_2$

H$_3$C

5 Thymol

OH

O—CH=CH$_2$

CH$_2$CH=CH$_2$

6 Eugenol

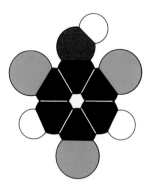

7 2,4,6-Trichlorophenol

C$_6$H$_5$OH (**4**). At one time it was obtained from the distillation of coal tar, but now it is mainly produced synthetically from benzene. Many phenols occur naturally, and some are partly responsible for the fragrances of plants. They are often components of *essential oils*, the oils that can be distilled from flowers and leaves. Thus thymol (**5**) is the active ingredient of oil of thyme, and eugenol (**6**) that of oil of cloves.

Properties and reactions of phenols. Although phenols resemble alcohols in that they both contain C—OH groups, the location of the C atom in an aromatic ring results in markedly different chemical properties. In particular, although phenols are weak acids, they are much stronger than the alcohols:

$$OH + H_2O \rightleftharpoons O^- + H_3O^+ \qquad pK_a = 9.89$$

Phenol was in fact once known as "carbolic acid." The conjugate base of phenol, C$_6$H$_5$O$^-$, is the *phenoxide ion*. The greater acidity of phenols as compared with alcohols stems from the ability of the negative charge of the phenoxide ion to spread over the benzene ring, which stabilizes the conjugate base.

Although the benzene ring itself is not particularly reactive, substituents (including the —OH group) can often sensitize it to attack. The reaction between chlorine and aqueous phenol, for instance, continues as far as the antibacterial agent 2,4,6-trichlorophenol (**7**), even without a catalyst:

$$OH + 3Cl_2 \longrightarrow Cl \cdots OH \cdots Cl + 3HCl$$

Phenolic resins. The phenoxide ion has an even higher accumulation of negative charge in the ring than phenol, for the O$^-$ atom is a better supplier of electrons than the —OH group. We therefore would expect that when phenol is in basic solution (where it is present largely as phenoxide ions), it is even more reactive than phenol itself. This feature is employed in the manufacture of *phenolic resins* (or "phenol-formaldehyde resins"). In essence, formaldehyde (HCHO) reacts with phenol in basic solution to give a mixture of ortho and para products. Schematically, we can represent this as the unbalanced equation

$$O^- + HCHO \longrightarrow O^- CH_2OH \quad and \quad O^- \cdots CH_2OH$$

These products lose water when heated, as in the reaction

The products of this step can react with more phenoxide ions, and the result is an extensively cross-linked polymer of the form

Phenolic resins are used as plywood adhesives, as binders for fiberglass insulation, and in molded electrical components.

THE CARBONYL GROUP

The *carbonyl group* $\ce{C=O}$ occurs in several classes of compounds and has distinctive properties. As we saw in Section 9.4, the double bond in the carbonyl group consists of a σ bond formed from a carbon sp^2 orbital and a π bond formed from one of its $2p$ orbitals. In this section we discuss two closely related families of compounds that contain the group; these are the aldehydes and the ketones, already mentioned as products of the oxidation of alcohols. Then we consider the carbohydrates, which contain the carbonyl group in a disguised form.

24.3 ALDEHYDES AND KETONES

Aldehydes and ketones both contain the carbonyl group but differ in the number of H atoms directly attached to it:

Aldehydes are compounds of the form $\ce{R-\overset{\displaystyle O}{\underset{\displaystyle \|}{C}}-H}$.

Ketones are compounds of the form $\ce{R-\overset{\displaystyle O}{\underset{\displaystyle \|}{C}}-R}$.

The R— groups may be either aliphatic or aromatic. The $\ce{-C\overset{O}{\underset{H}{}}}$

TABLE 24.2 **Aldehydes and ketones**

Formula	Name	Melting point, °C	Boiling point, °C
Aldehydes			
$H-CHO$	Formaldehyde,* methanal	-92	-21
CH_3-CHO	Acetaldehyde,* ethanal	-124	21
CH_3-CH_2-CHO	Propanal	-81	49
C_6H_5-CHO	Benzaldehyde*	-26	179
Ketones			
$CH_3-CO-CH_3$	Acetone,* propanone	-95	56
$CH_3CH_2-CO-CH_3$	Butanone	-87	80
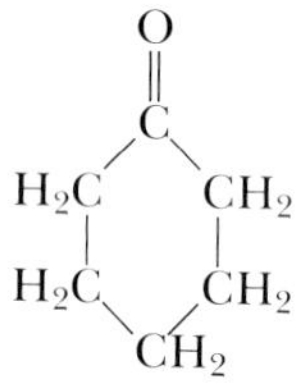 (cyclohexanone structure)	Cyclohexanone	-16	156

*Common name.

group is normally written —CHO, as in HCHO (formaldehyde), the simplest aldehyde. The liquid "formalin" used to preserve biological specimens is an aqueous solution of formaldehyde. Wood smoke contains formaldehyde, and formaldehyde's destructive effect on bacteria is one of the reasons why the smoking of food helps to preserve it. Some properties of common aldehydes and ketones are given in Table 24.2.

The formulas of aldehydes differ from those of *primary* alcohols by two H atoms:

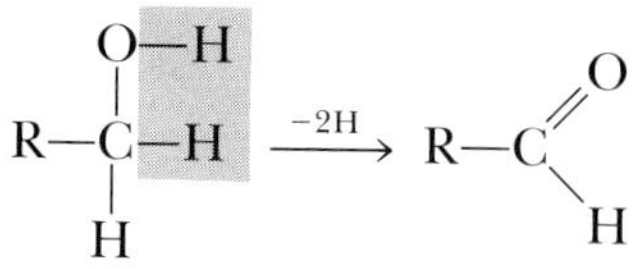

This difference is reflected in the name "aldehyde," which is derived from *al*cohol-*dehyd*rogenated. Acetaldehyde is in fact a product of ethanol oxidation in the liver, and its accumulation in the blood is one of the causes of hangovers.

The formulas of ketones differ by two H atoms from those of *secondary* alcohols:

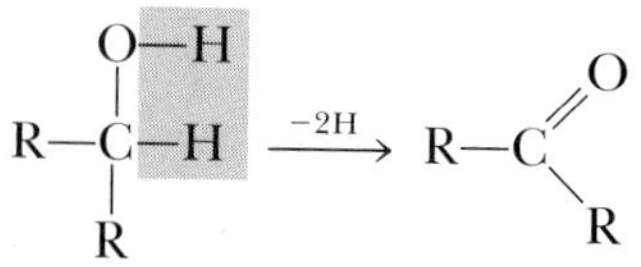

The name "ketone" is derived from the German name (*Keton*) for the simplest member, acetone (CH₃—CO—CH₃).

Aldehydes occur naturally in essential oils and are partly responsible for the flavors of many fruits and the odors of plants. Benzaldehyde, C_6H_5CHO **(8)**, has the characteristic aroma of cherries and almonds. Cinnamaldehyde **(9)** occurs in oil of cinnamon, and vanillin **(10)** in vanilla beans. Ketones can be fragrant, as for example carvone **(11)**, a component of the essential oil of spearmint, which is used in chewing gum.

Nomenclature. The systematic name of an aldehyde is formed by identifying the parent alkane (with the —CHO carbon atom included in the count of C atoms) and changing the final -e of the name to -al. Thus, acetaldehyde (CH_3CHO) is treated as a derivative of ethane, and its systematic name is ethanal. When the carbon atoms must be numbered to specify the locations of substituents, we start with 1 at the —CHO group.

Ketones are named systematically by changing the -e in the name of the parent alkanes to -one. Thus, as acetone (CH₃—CO—CH₃) is a derivative of propane, its systematic name is propanone. When numbering is necessary, it begins at the end of the molecule that results in the lower number for the location of the carbonyl group; hence CH₃CH₂CH₂—CO—CH₃ is 2-pentanone.

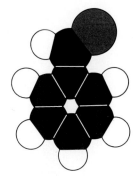

8 Benzaldehyde

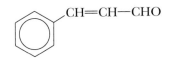

9 Cinnamaldehyde

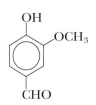

10 Vanillin

▼ **EXAMPLE 24.3** **Naming aldehydes and ketones**

Name the compounds (a) CH₃CH₂CH₂CHO; (b) CH₃CH(CHO)CH₂CH₃; (c) CH₃CH₂COCH₃.

STRATEGY We simply need to follow the rules as given above for aldehydes (a, b) and ketones (c). The first step in every case is to identify the longest chain (by writing it out if necessary), including the carbonyl-group carbon atom. We can then identify the corresponding alkane by referring to Table 23.2.

SOLUTION (a) This aldehyde is a derivative of butane, so its name is butanal.
(b) We write the structure out, so as to identify the longest chain:

$$
\begin{array}{c}
\text{CH}_3 \\
| \\
\underset{4}{\text{CH}_3}-\underset{3}{\text{CH}_2}-\underset{2}{\text{CH}}-\underset{1}{\text{CHO}}
\end{array}
$$

This is the derivative of a substituted C₄ alkane and therefore, like (a), is a butanal. Since the carbon atom of the —CHO group is numbered 1 in aldehydes, the methyl substituent is at carbon atom 2 and the compound is 2-methylbutanal.
(c) This ketone is also a derivative of butane, and hence it is a butanone. There is no need to write it 2-butanone (although that is often done), because the only alternative position for the carbonyl group would result in an aldehyde, not a ketone.

EXERCISE Name the following compounds: (a) CH₃(CH₂)₃CHO; (b) CH₃CH₂CH(CHO)(CH₂)₃CH₃; (c) CH₃(CH₂)₂COCH₂CH₃.
[*Answer*: (a) pentanal; (b) 2-ethylhexanal; (c) 3-hexanone]

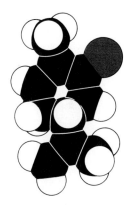

11 Carvone

FIGURE 24.4 An aldehyde (left) produces a brick-red precipitate of copper(I) oxide with Fehling's solution, but a ketone (right) does not.

Preparation. Since aldehydes and ketones differ from primary and secondary alcohols by the loss of two H atoms, both can be prepared by oxidation of the corresponding alcohol. Schematically,

Aldehydes:

$$CH_3CH_2-\overset{\overset{\displaystyle O-H}{|}}{\underset{\underset{\displaystyle H}{|}}{C}}-H \xrightarrow{\text{oxidation}} CH_3CH_2-\overset{\overset{\displaystyle O}{\parallel}}{C}{\diagdown}_H$$

Ketones:

$$CH_3-\overset{\overset{\displaystyle O-H}{|}}{\underset{\underset{\displaystyle CH_3}{|}}{C}}-H \xrightarrow{\text{oxidation}} CH_3-\overset{\overset{\displaystyle O}{\parallel}}{C}{\diagdown}_{CH_3}$$

In oxidizing a primary alcohol to an aldehyde, oxidation of the product aldehyde is avoided by using a mild oxidizing agent, such as acidified sodium dichromate. Schematically,

$$CH_3CH_2CH_2OH \xrightarrow{\text{Na}_2\text{Cr}_2\text{O}_7(aq),\ \text{H}_2\text{SO}_4(aq)} CH_3CH_2CHO$$

Aldehydes are also prepared by use of organic solvents, since this reduces the risk of further oxidation. Formaldehyde is prepared industrially (for the manufacture of phenolic resins) by the catalytic oxidation of methanol:

$$2CH_3OH(g) + O_2(g) \xrightarrow{600°C,\ Ag} 2HCHO(g) + 2H_2O(g)$$

There is less risk of overoxidation for ketones than for aldehydes, since a C—C bond has to be broken. Dichromate oxidation of secondary alcohols produces the ketone in good yield with little additional oxidation. Schematically,

$$CH_3CH_2CH(OH)CH_3 \xrightarrow{\text{Na}_2\text{Cr}_2\text{O}_7(aq),\ \text{H}_2\text{SO}_4(aq)} CH_3CH_2COCH_3$$

Reactions. One sign of the difference between the ease of oxidation of aldehydes and that of ketones is the ability of aldehydes to reduce Fehling's solution (a basic solution of Cu^{2+} and tartrate ions), producing a brick-red precipitate of Cu_2O; ketones do not react with Fehling's solution (Fig. 24.4). Schematically,

Aldehydes: $CH_3CH_2CHO \xrightarrow{\text{Fehling's solution}} CH_3CH_2COOH + Cu_2O(s)$

Ketones: $CH_3COCH_3 \xrightarrow{\text{Fehling's solution}}$ no reaction

Nor do ketones produce a silver mirror—a coating of silver on the test tube—with "Tollens' reagent," a solution of Ag^+ ions in aqueous ammonia (Fig. 24.5):

Aldehydes: $CH_3CH_2CHO \xrightarrow{\text{Tollens' reagent}} CH_3CH_2COOH + Ag(s)$

Ketones: $CH_3COCH_3 \xrightarrow{\text{Tollens' reagent}}$ no reaction

The carbonyl group reacts with a cyanide ion to give a *cyanohydrin*, a compound containing both an —OH group and a —C≡N group:

$$CH_3-\overset{\overset{\displaystyle O}{\parallel}}{C}{\diagdown}_{CH_3} + HCN \longrightarrow CH_3-\overset{\overset{\displaystyle OH}{|}}{\underset{\underset{\displaystyle CH_3}{|}}{C}}-C{\equiv}N$$

FIGURE 24.5 An aldehyde (left) produces a silver mirror with Tollens' reagent, but a ketone (right) does not.

In practice, a compound containing the carbonyl group is treated with sodium cyanide and dilute sulfuric acid; the reaction succeeds with all aldehydes, but with only a few ketones (acetone among them). Because the reaction results in a new C—C bond, it is useful for building up molecules containing more carbon atoms. For example, the cyanohydrins so formed are readily hydrolyzed to carboxylic acids:

$$
\begin{array}{c} OH \\ | \\ CH_3\!-\!C\!-\!C\!\equiv\!N \\ | \\ CH_3 \end{array}
+ 2H_2O \longrightarrow
\begin{array}{c} OH \quad O \\ | \quad \diagup \\ CH_3\!-\!C\!-\!C \\ | \quad \diagdown \\ CH_3 \quad OH \end{array}
+ NH_3
$$

(The fact that the hydrolysis produces a carboxylic acid with an additional —OH group need not concern us: that extra functional group can be very useful in organic syntheses.)

Aldehydes combine with alcohols in the presence of dry hydrogen chloride to form *hemiacetals*, compounds with both an —O— ether link and an —OH group. For example,

$$
\begin{array}{c} O \\ \diagup\!\!\diagup \\ CH_3\!-\!C \\ \diagdown \\ H \end{array}
+ CH_3CH_2OH \xrightarrow{H^+}
\begin{array}{c} OH \\ | \\ CH_3\!-\!C\!-\!O\!-\!CH_2CH_3 \\ | \\ H \end{array}
$$

The hemiacetal (the prefix *hemi* is the Greek word for "half") usually cannot be isolated but readily reacts with more alcohol to form the corresponding *acetal*, a type of double ether that contains the —O—C—O— group:

$$
\begin{array}{c} OH \\ | \\ CH_3\!-\!C\!-\!O\!-\!CH_2CH_3 \\ | \\ H \end{array}
+ CH_3CH_2OH \xrightarrow{H^+}
$$

$$
\begin{array}{c} O\!-\!CH_2CH_3 \\ | \\ CH_3\!-\!C\!-\!O\!-\!CH_2CH_3 \\ | \\ H \end{array}
+ H_2O
$$

If the original aldehyde contains a hydroxyl group, hemiacetal formation may occur *intramolecularly*, or within the molecule itself:

$$
\begin{array}{c} O \\ \diagup\!\!\diagup \\ HO\!-\!CH_2CH_2CH_2CH_2\!-\!C \\ \diagdown \\ H \end{array}
\xrightarrow{H^+}
\begin{array}{c} O \\ \| \\ C\!-\!H \\ \diagup \quad\\ CH_2 \qquad O\!-\!H \\ | \qquad | \\ CH_2 \quad CH_2 \\ \diagdown \quad \diagup \\ CH_2 \end{array}
\longrightarrow
\begin{array}{c} OH \quad H \\ \diagdown\diagup \\ C \\ \diagup \quad \diagdown \\ CH_2 \quad O \\ | \qquad | \\ CH_2 \quad CH_2 \\ \diagdown \quad \diagup \\ CH_2 \end{array}
$$

The resulting compound is a cyclic hemiacetal that is stabilized by ring formation. We shall see in a moment that this reaction is of immense

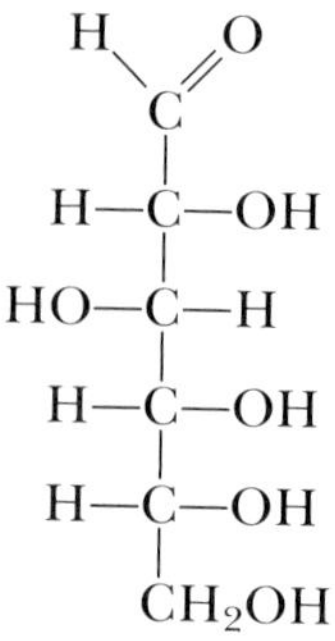

12 Glucose

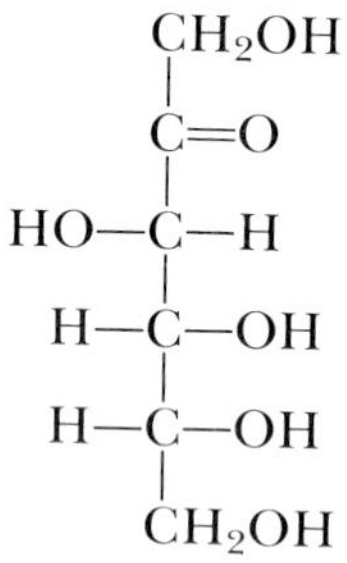

13 Fructose

14 A pyran ring

15 Furan

importance in understanding the structure of cellulose, the most abundant organic chemical in the world.

Finally, in the reverse of the reactions by which they are made, aldehydes and ketones can be reduced to alcohols. The reducing agents used are usually sodium borohydride ($NaBH_4$) in ethanol and lithium aluminum hydride ($LiAlH_4$) in ether.

24.4 CARBOHYDRATES

The *carbohydrates* are so called because they often have the empirical formula $CH_{2n}O_n$, or $C(H_2O)_n$, which suggests a hydrate of carbon. They include glucose ($C_6H_{12}O_6$; **12**), which is an aldehyde, and fructose (also $C_6H_{12}O_6$; **13**), the sugar that occurs widely in fruit, which is a ketone. They also include cellulose and starch.

Pyranose and furanose forms. Glucose, being both an alcohol and an aldehyde and having a reasonably long and flexible backbone, can form a cyclic hemiacetal:

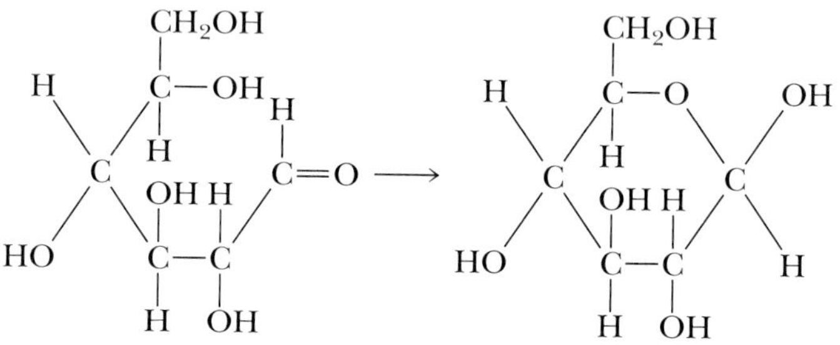

The six-membered ring is called the *pyranose* form of glucose because it resembles the cyclic ether pyran (**14**), and it is specifically named a glucopyranose. Fructose is also present as a cyclic hemiacetal in solution, mainly as a pyranose six-membered ring but also (to the extent of about 30%) as a five-membered ring. The latter is called its *furanose* form, specifically a fructofuranose, since it resembles furan (**15**), another cyclic ether.

Glucopyranose is normally represented by the line diagram (**16**), and fructofuranose by (**17**). However, it must be remembered that in solution the cyclic hemiacetal forms are in equilibrium with their aldehyde and ketone open-chain forms (**12**, **13**). Evidence for the presence of

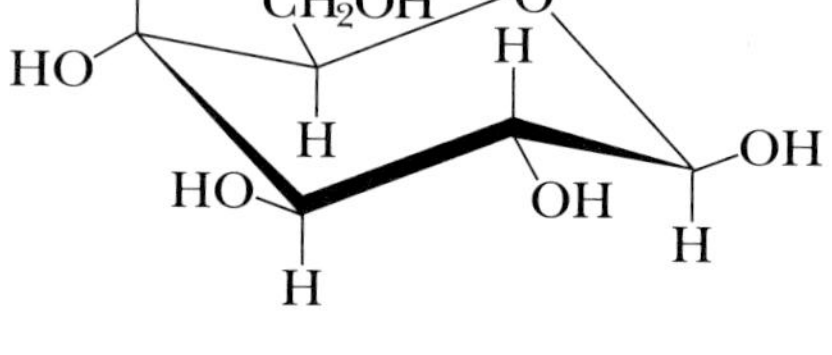

16 Glucopyranose

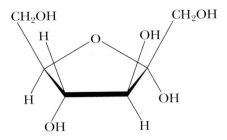

17   Fructofuranose

FIGURE 24.6 Glucose is a reducing sugar, and it produces a precipitate with Fehling's solution (left). Sucrose is not a reducing sugar, and so it does not react with Fehling's solution (right).

these forms is the fact that glucose undergoes a typical reaction of an aldehyde—oxidation. With Fehling's solution, glucose gives a brick-red precipitate of copper(I) oxide. If we show only the —CHO group of the molecule explicitly, we can express the reaction as

$$\ldots -CHO \xrightarrow{\text{Fehling's solution}} \ldots -COOH + Cu_2O(s)$$

Fructose is also oxidized, but in this case the site of oxidation is the —OH group on the carbon atom next to the carbonyl group:

$$\ldots -\overset{\overset{\displaystyle O}{\|}}{C}-\overset{\overset{\displaystyle OH}{|}}{\underset{\underset{\displaystyle H}{|}}{C}}- \xrightarrow{\text{Fehling's solution}} \ldots -\overset{\overset{\displaystyle O}{\|}}{C}-\overset{\overset{\displaystyle O}{\|}}{C}- + Cu_2O(s)$$

Sugars that are oxidized by Fehling's solution and Tollens' reagent are called *reducing sugars*. Both glucose and fructose are reducing sugars (Fig. 24.6).

Oligosaccharides.   Carbohydrates that are more complex than glucose and fructose can usually be regarded as being built up of these smaller units. The most familiar example is sucrose, or cane sugar, $C_{12}H_{22}O_{11}$ **(18)**, which consists of a glucopyranose unit bonded to a fructofuranose unit. Since it consists of two C_6 carbohydrate units, it is called a *disaccharide*; each individual unit is a *monosaccharide*. A string of several saccha-

18   Sucrose

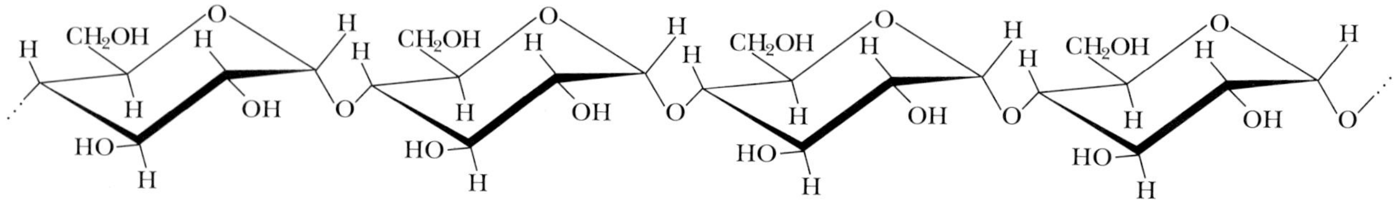

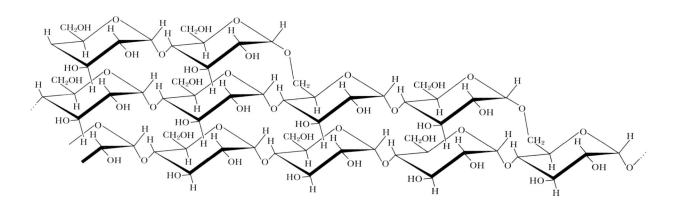

FIGURE 24.7 The amylose molecule, one component of starch, is a polysaccharide; it consists of glucose units linked together to give a moderately branched structure.

ride units is called an *oligosaccharide* (the prefix is from the Greek word for "a few").

The hemiacetal rings of sucrose are not in equilibrium with their open-chain forms, because the oxygen atom that would become the carbonyl oxygen now serves as a bridge between linked rings. Therefore, sucrose does not show the typical reactions of an aldehyde or ketone and, in particular, is not a reducing sugar. However, it is very soluble in water (because of its numerous hydroxyl groups), and it is cut (hydrolyzed) into its component sugars by enzymes in the digestive system.

Polysaccharides. The *polysaccharides* consist of chains of many glucose units linked together. This category includes starch, which we can digest (hydrolyze), and cellulose, which we cannot.

Starch is made up of two components: amylose and amylopectin. Amylose, which accounts for about 20 to 25% of most starches, is composed of chains consisting of several thousand glucose units linked together, as shown in Fig. 24.7. Amylopectin is also composed of glucose chains (Fig. 24.8), but its chains are linked into a branched structure and its molecules are much larger, having typically about a million glucose units.

FIGURE 24.8 The amylopectin molecule is another component of starch; it has a more highly branched structure than amylose.

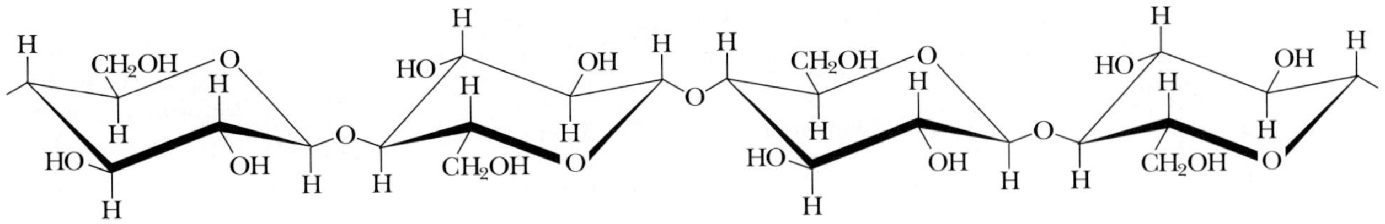

FIGURE 24.9 Cellulose is yet another polysaccharide that is constructed from glucose units. However, here the linking between units results in long, flat ribbons that can produce a fibrous material through hydrogen bonding.

Cellulose is the structural material of plants. Like starch, it is a polymer of glucose, but the units link differently, forming flat, ribbonlike strands (Fig. 24.9); these strands can lock together through hydrogen bonds into a rigid structure that for us (but not for the microbes inside termites) is indigestible. The difference between cellulose and starch shows nature at its most economical and elegant, for only a small modification of the linking between glucose units results on the one hand in an important foodstuff and on the other in a versatile construction material.

THE CARBOXYL GROUP

We now turn to a functional group that combines two we have already met: the *carboxyl group* $-\overset{\overset{\text{O}}{\|}}{\text{C}}-\text{OH}$ is a combination of the carbonyl and hydroxyl groups. Although it is normally abbreviated to $-COOH$, we must remember that the group does not contain an $O-O$ bond. Loss of a hydrogen ion from the hydroxyl group results in the *carboxylate group*, $-CO_2{}^-$. The two oxygen atoms in the latter are equivalent as a result of resonance,

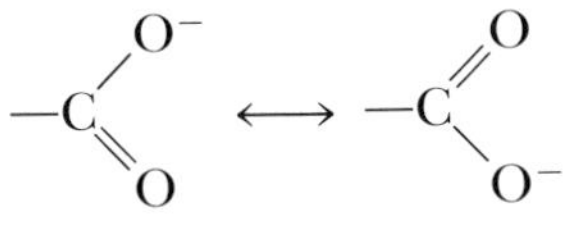

$$-\overset{\overset{\text{O}^-}{\diagup}}{\underset{\diagdown}{\text{C}}}_{\text{O}} \quad \longleftrightarrow \quad -\overset{\overset{\text{O}}{\diagup}}{\underset{\diagdown}{\text{C}}}_{\text{O}^-}$$

which is why we write $-CO_2{}^-$ and not $-COO^-$.

24.5 CARBOXYLIC ACIDS

Compounds containing the carboxyl group are called *carboxylic acids*. The properties of some of them are summarized in Table 24.3.

Nomenclature of carboxylic acids. Many carboxylic acids are known by their common names. The simplest is formic acid, HCOOH, the acid injected by ants. The next higher homolog is acetic acid, CH_3COOH

TABLE 24.3 Carboxylic acids

Formula	Name	pK_{a1}	pK_{a2}	Melting point, °C	Boiling point, °C
H—COOH	Formic acid,* methanoic acid	3.75		8	101
CH₃—COOH	Acetic acid,* ethanoic acid	4.76		17	118
CH₂Cl—COOH	Chloroacetic acid*	2.86		63	189
CH₃CH₂—COOH	Propanoic acid	4.87		−21	141
C₆H₅—COOH	Benzoic acid	4.20		122	249
COOH | COOH	Oxalic acid,* ethanedioic acid	1.23	4.28	190d†	
COOH | CH₂ | COOH	Malonic acid,* propanedioic acid	2.83	5.69	136d	

*Common name.
†d signifies that the acid decomposes.

(the acid of vinegar), which is formed when the ethanol in wine is oxidized by air:

$$CH_3—CH_2—OH(aq) + O_2(g) \longrightarrow CH_3—\overset{\overset{\displaystyle O}{\|}}{C}—OH(aq) + H_2O(l)$$

The systematic names of carboxylic acids are obtained by changing the -e at the end of the parent alkane's name to -oic acid. Thus, acetic acid is regarded as a derivative of ethane and named ethanoic acid.

▼ **EXAMPLE 24.4 Naming carboxylic acids**

Name (a) $CH_3CH_2CH_2COOH$ and (b) $(COOH)_2$.

STRATEGY We should proceed as outlined above, first identifying the parent alkane. If more than one functional group is present, we use the prefixes di-, tri-, and so on.

SOLUTION (a) The parent alkane of $CH_3CH_2CH_2COOH$ is butane, so the acid is butanoic acid.
(b) The parent alkane of $(COOH)_2$ is ethane, and it has two carboxyl groups; hence its systematic name is ethanedioic acid (the final -e of ethane is retained here to make the name easier to read). Its common name is oxalic acid.

EXERCISE Name the acids (a) $CH_3(CH_2)_6COOH$ and (b) $CH_2(COOH)_2$.
▲ [*Answer*: (a) octanoic acid; (b) propanedioic acid]

Preparation and reactions. Carboxylic acids can be prepared by oxidizing primary alcohols and aldehydes, using either acidified aqueous potassium permanganate or air with an appropriate catalyst. The latter is much cheaper and is used industrially to produce acetic acid (with cobalt acetate as the catalyst). In some cases, alkyl groups can be oxidized directly to carboxyl groups. This process is very important industrially, and among other applications it is used for the oxidation of p-xylene:

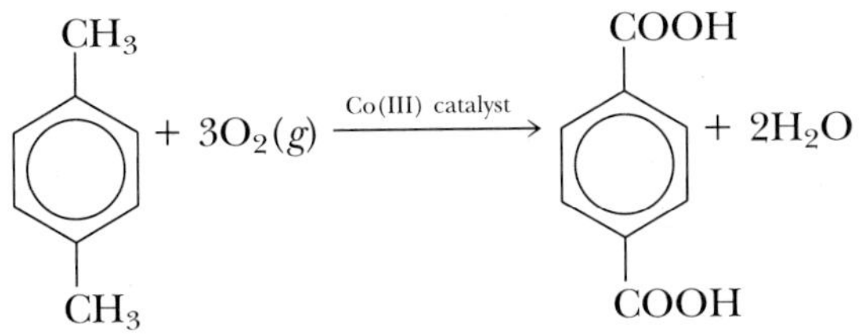

The product, terephthalic acid, is used in the production of artificial fibers, as we shall see.

In a carboxyl group, the electronegative carbonyl oxygen atom partially withdraws electrons from the neighboring hydroxyl O—H bond, weakening it; hence carboxylic acids are typical weak Brønsted acids. Their acid strengths differ strikingly from those of alcohols: for ethanol, $pK_a = 16$; however, when CH_3CH_2OH is converted to CH_3COOH the pK_a falls to 4.8, representing a change of 11 orders of magnitude in ionization constant. The acid character of carboxylic acids (which typically have pK_a values in the range 4 to 5) is shown by their formation of salts with bases. Since the anions are Brønsted bases, solutions of these salts are basic (see Section 15.1).

Carboxylic acids with long hydrocarbon chains occur in oils and fats (as we see below) and are called *fatty acids*. Their sodium salts are used as soap; a typical example is sodium stearate, the sodium salt of stearic acid, $CH_3(CH_2)_{16}COOH$. The ability of soaps to act as surfactants is explained in Section 11.3.

24.6 ESTERS

The organic product of the reaction between a carboxylic acid and an alcohol is called an *ester*. Acetic acid and ethanol, for example, react readily when heated to about 100°C, especially if the mixture is acidified with a strong acid (to act as catalyst). The product of this *esterification* is ethyl acetate, a fragrant liquid:

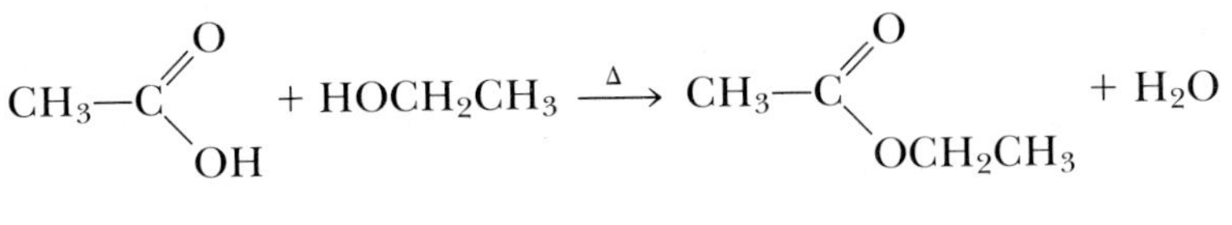

The occurrence of esters. Many esters have pleasant odors and contribute to the flavors of fruits (Table 24.4). Flavor—the joint impact of taste and odor—actually arises from a very complicated mixture of ingredients. The way a substance tastes is the product of a particular blend of sweetness due to sugars and sourness due to carboxylic acids. Its odor is produced by the mixture of all the volatile chemicals present. These include esters, as well as aldehydes and ketones. An apple

TABLE 24.4   The odors of some esters

Formula	Odor*
$CH_3{-}CH_2{-}CH_2{-}C\overset{O}{\underset{O-CH_2-CH_2-CH(CH_3)_2}{\big\backslash}}$	Pear
$CH_3{-}C\overset{O}{\underset{O-CH_2-CH_2-CH_2-CH_2-CH_3}{\big\backslash}}$	Banana
$CH_3{-}CH_2{-}CH_2{-}C\overset{O}{\underset{O-CH_2-CH_2-CH_2-CH_2-CH_3}{\big\backslash}}$	Banana
$CH_3{-}C\overset{O}{\underset{O-CH_2-(CH_2)_6-CH_3}{\big\backslash}}$	Orange
$\underset{CH_3}{\overset{CH_3}{\big\backslash}}CH{-}CH_2{-}CH_2{-}C\overset{O}{\underset{O-CH_2-CH_2-CH\underset{CH_3}{\overset{CH_3}{\big\langle}}}{\big\backslash}}$	Apple
$CH_3{-}CH_2{-}CH_2{-}C\overset{O}{\underset{O-CH_2-CH_2-CH_2-CH_2-CH_3}{\big\backslash}}$	Apricot

*The odors of the esters are reminiscent of these fruits; the fruits do not necessarily contain the esters.

contains about 20 carboxylic acids (with alkyl chains up to about 19 atoms long), nearly 30 alcohols (with a similar range of chain lengths), about 70 esters produced from the most abundant alcohols present, and about three dozen other components. Food chemists often try to recreate flavors with smaller numbers of components (Table 24.5), which is why synthetic flavors are often not as subtle as the real thing.

The naturally occurring esters include fats and oils. These are two members of a wide class of substances called *lipids*, which are naturally occurring compounds that dissolve in nonpolar solvents but not in water. The vegetable oil triolein and the animal fat tristearin (**10** in Section 6.6), for instance, are esters formed from glycerol and oleic acid or stearic acid. Other oils and fats have similar structures.

Esterification. Ester formation is an example of a "condensation reaction" (**19**):

> A **condensation reaction** is a reaction in which two molecules combine to form a larger one, and a small molecule is eliminated.

In an esterification, the eliminated molecule is H_2O. Esterification may *look* like an acid-base neutralization, with the alcohol seeming to play the role of a base, and the product, the ester, being named like a salt (as

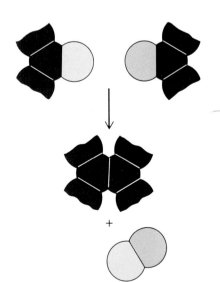

19   Condensation

TABLE 24.5 Components of synthetic pineapple flavor

Formula	Proportion, %
$CH_3-CH_2-CH_2-CH_2-C\overset{O}{\underset{O-CH_2-CH_3}{}}$	66
$CH_3-CH_2-CH_2-CH_2-C\overset{O}{\underset{O-CH_2-CH_2-CH(CH_3)_2}{}}$	22
$CH_3-CH_2-CH_2-CH_2-C\overset{O}{\underset{O-CH_2-CH=CH_2}{}}$	5
$CH_3-C\overset{O}{\underset{O-CH_2-CH_3}{}}$	5
$CH_2(OH)-CH(OH)-CH_2(OH)$, glycerol	1
Lemon oil	1

in "ethyl acetate," like sodium acetate). However, it is very important to understand that this is a misleading analogy. Experiments—using oxygen-18 and mass spectrometry to keep track of the fate of the ^{18}O atoms—have shown that the oxygen atom in the H_2O molecule produced by an esterification comes from the carboxylic acid, and not from the hydroxyl group of the alcohol:

$$CH_3-C\overset{O}{\underset{OH}{}} + HO-CH_2CH_3 \longrightarrow CH_3-C\overset{O}{\underset{O-CH_2CH_3}{}} + H_2O$$

Polyesters and condensation polymerization. Polymers formed by linking together carboxylic acids and alcohols are called *polyesters* and are widely used for making artificial fibers. Polyesters are examples of *condensation polymers*, polymers formed by a chain of condensation reactions.

A typical example of a polyester is the material sold as Dacron, a polymer produced by the esterification of terephthalic acid with ethylene glycol. The first condensation is

$$HOOC-\langle\bigcirc\rangle-COOH + HO-CH_2-CH_2-OH \longrightarrow$$
$$HOOC-\langle\bigcirc\rangle-CO-O-CH_2-CH_2-OH$$

Another ethylene glycol unit can now condense with the carboxyl group on the left of the product, and another terephthalic acid unit can condense with the hydroxyl group on the right. As a result the polymer can grow at both ends and in due course becomes

$$HOOC-\langle\bigcirc\rangle-CO-\left\{OCH_2CH_2OOC-\langle\bigcirc\rangle-CO\right\}_n-OCH_2CH_2OH$$

In free-radical polymerization, side chains can start to grow out from the main chain; however, in condensation polymerization, growth can occur only at the functional groups, so chain branching is much less likely. As a result, polyester molecules can be made to lie side by side by forcing them through a small hole, and so can be spun into yarn (Fig. 24.10).

FUNCTIONAL GROUPS CONTAINING NITROGEN

Nitrogen brings to organic compounds a rich variety of properties. In this section we consider the amines and compounds related to them.

24.7 AMINES

One of the most important families in which nitrogen occurs in a functional group are the *amines*, compounds derived from ammonia by replacing one or more H atoms:

$$\underset{\text{Ammonia}}{H-\overset{\overset{\displaystyle H}{|}}{N}-H} \qquad \underset{\text{Methylamine}}{CH_3-\overset{\overset{\displaystyle H}{|}}{N}-H} \qquad \underset{\text{Dimethylamine}}{CH_3-\overset{\overset{\displaystyle H}{|}}{N}-CH_3} \qquad \underset{\text{Trimethylamine}}{CH_3-\overset{\overset{\displaystyle CH_3}{|}}{N}-CH_3}$$

In each case the molecule is trigonal pyramidal, with a lone pair of electrons on the nitrogen atom and σ bonds formed from its approximately sp^3 hybrid orbitals.

Amines occur widely, both as natural products and in synthetic organic materials. Many have a pungent, often unpleasant odor. Because proteins are organic compounds of nitrogen, amines are often the end products of the decomposition of living matter and, together with sulfur compounds, responsible for its stench. The names of the two *diamines* (amines with two amino groups), putrescine, $NH_2(CH_2)_4NH_2$, and cadaverine, $NH_2(CH_2)_5NH_2$, speak for themselves.

Nomenclature of amines. Amines fall into three classes, depending on the number of carbon atoms that are directly attached to the nitrogen atom:

A **primary amine** has the structure $R-\overset{\overset{\displaystyle H}{|}}{N}-H$.

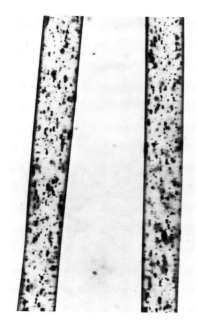

FIGURE 24.10 Dacron, a polyester, being spun. In the process, the long polyester chains are forced to lie parallel to each other, giving the material a fibrous texture. This is a section along a fiber, magnified 1000 times.

A **secondary amine** has the structure

$$R\!-\!\overset{\displaystyle H}{\underset{\displaystyle |}{N}}\!-\!R.$$

A **tertiary amine** has the structure

$$R\!-\!\overset{\displaystyle R}{\underset{\displaystyle |}{N}}\!-\!R.$$

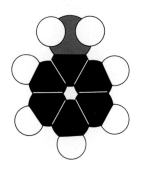

NH_2

20 Aniline

(Note that this classification scheme is different from that for alcohols.) A _quaternary ammonium ion_ is a tetrahedral ion of the form R_4N^+, such as the tetramethylammonium ion, $(CH_3)_4N^+$. The R groups, which may be either aliphatic or aromatic, need not all be the same. Methylamine is a primary amine, dimethylamine is a secondary amine, and trimethylamine is a tertiary amine. The $-NH_2$ group itself is called an _amino group_.

Amines are named by specifying the groups attached to the nitrogen in alphabetical order. Thus $CH_3CH_2NH_2$, a primary amine, is ethylamine; $(CH_3CH_2)_2NCH_2CH_2CH_3$, a tertiary amine, is diethylpropylamine. Many amines have common names. For instance, phenylamine, $C_6H_5NH_2$ (**20**), is aniline.

▼ **EXAMPLE 24.5** **Naming amines**

Classify and name the amines (a) $CH_3CH_2CH_2NH_2$, (b) $CH_3NHC_6H_5$, and (c) $(C_6H_5)_3N$.

STRATEGY To classify an amine, we simply count the number of carbon atoms attached directly to the nitrogen atom. To name it, we identify the substituents on the nitrogen atom and list them in alphabetical order.

SOLUTION (a) This amine has the form RNH_2, so it is a primary amine. Since $CH_3CH_2CH_2-$ is propyl, the compound is propylamine.
(b) This compound has the form R_2NH, so it is a secondary amine. Since CH_3- is methyl and C_6H_5- is phenyl-, it is methylphenylamine.
(c) This molecule has the form R_3N and is a tertiary amine. Since C_6H_5- is phenyl, the compound is triphenylamine.

EXERCISE Classify and name the amines (a) $CH_3(CH_2)_3CH_2NH_2$, (b) $(CH_3CH_2)_2NH$, and (c) $(CH_3)_3N$.
[_Answer_: (a) primary, pentylamine; (b) secondary, diethylamine; (c) tertiary, trimethylamine]

▲

Preparation of amines. Amines are prepared by several methods. The simplest is by the reaction of a haloalkane with ammonia:

$$CH_3CH_2\!-\!Br + NH_3 \longrightarrow$$
$$CH_3CH_2\!-\!NH_2 + HBr \qquad \text{as } CH_3CH_2NH_3{}^+Br^-$$

The direct action of ammonia on haloalkanes has the disadvantage that it leads to a mixture of products. That is, the haloalkane can react with the primary amine product to give the secondary amine:

$$CH_3CH_2\!-\!Br + CH_3CH_2\!-\!NH_2 \longrightarrow$$
$$(CH_3CH_2)_2NH + HBr \text{ as } (CH_3CH_2)_2NH_2{}^+Br^-$$

The free amine derived from this reaction can react further to give the tertiary amine. Other syntheses have been developed to avoid this difficulty.

▼ **EXAMPLE 24.6** **Suggesting a synthesis of an amine**

Suggest a means for synthesizing ethylamine that begins with the reaction between bromomethane and CN^- ions.

STRATEGY Like OH^-, we can expect the CN^- ion to substitute for the bromine atom in the bromoalkane. The problem then is to convert the resulting nitrile into an amine. Since the cyano group is $-C\equiv N$ and a primary amine is CH_2-NH_2, we should expect reduction to be appropriate.

SOLUTION The first stage in the reaction is the substitution

$$CH_3-Br + CN^- \longrightarrow CH_3-C\equiv N + Br^-$$

Reduction then produces the amine:

$$CH_3-C\equiv N \xrightarrow{\text{addition of 4H}} CH_3-CH_2-NH_2$$

In practice, $LiAlH_4$ is found to be a suitable reducing agent.

EXERCISE What bromoalkane should you select for the synthesis of isobutylamine by this procedure?

▲
[*Answer*: 2-bromopropane]

Reactions of amines. An important and characteristic chemical feature of amines is that, like their parent ammonia, they are Lewis and Brønsted bases. The Lewis-base character of amines allows them to form complexes with metal ions acting as Lewis acids, as they do in chlorophyll and the numerous coordination compounds of the *d*-block elements.

In general the alkylamines (amines formed from alkanes, such as methylamine) are stronger Brønsted bases than ammonia. This is so primarily because the alkyl groups can supply electrons to the nitrogen and help stabilize the conjugate acid:

$$CH_3NH_2 + H_2O \rightleftharpoons CH_3NH_3^+ + OH^- \qquad pK_b = 3.38$$

In contrast, the arylamines (amines formed from aromatic compounds, such as aniline, for which $pK_b = 9.37$) are weaker. Here the lone pair is partially delocalized over the benzene ring, and the *base* is stabilized:

$$C_6H_5NH_2 + H_2O \rightleftharpoons C_6H_5NH_3^+ + OH^- \qquad pK_b = 9.37$$

Amides and polyamides. Primary and secondary amines behave in some respects like alcohols in that they condense with carboxylic acids; for example,

21

The major product is an *amide*, a compound formed by the reaction of an amine and a carboxylic acid. Natural and synthetic amides are common. The pain-relieving drug sold as Tylenol (**21**) is an amide.

In the laboratory, amides are formed by the action of ammonia or an amine on an "acyl chloride," a compound of formula R—CO—Cl:

$$CH_3-C\underset{Cl}{\overset{O}{\diagup}} + NH_2CH_3 \longrightarrow CH_3-C\underset{NH-CH_3}{\overset{O}{\diagup}} + HCl$$

An alternative synthesis (and the great importance of this reaction will become clear in a moment) is simply to heat the ammonium salt of a carboxylic acid:

$$CH_3C\underset{O^-}{\overset{O}{\diagup}} + NH_4^+ \xrightarrow{\Delta} CH_3C\underset{NH_2}{\overset{O}{\diagup}} + H_2O$$

The product of this reaction, CH_3CONH_2, is acetamide.

Condensation polymerization by amide formation leads to the *polyamides*, substances more commonly known as "nylons." A typical polyamide is nylon-6,6, which is a copolymer of 1,6-hexanediamine, $H_2N(CH_2)_6NH_2$, and adipic acid, $HOOC(CH_2)_4COOH$. The 6,6 in the name indicates the numbers of C atoms in the monomers.

For a condensation polymerization, it is necessary to mix chemically equivalent amounts of the two reactants. This can be accomplished by mixing nearly equal proportions of the acid and base that are to be used. In the case of polyamide formation, the starting materials form "nylon salt" by proton transfer, as in

$$HOOC(CH_2)_4COOH + H_2N(CH_2)_6NH_2 \longrightarrow$$
$$H_3N^+(CH_2)_6NH_3^+ + {}^-O_2C(CH_2)_4CO_2^-$$

At this point, the excess acid or amine can be removed. Then, when the nylon salt—an ammonium salt of a carboxylic acid—is heated, the condensation begins. The first step is

$$\underset{O}{\overset{^-O}{\diagdown}}C-(CH_2)_4-C\underset{O^-}{\overset{O}{\diagup}} + H_3\overset{+}{N}-(CH_2)_6-\overset{+}{N}H_3 \longrightarrow$$

$$\underset{O}{\overset{^-O}{\diagdown}}C-(CH_2)_4-C\underset{NH-(CH_2)_6-\overset{+}{N}H_3}{\overset{O}{\diagup}} + H_2O$$

The amide grows at both ends through further condensations, and the final product is

$$\underset{HO}{\overset{O}{\diagdown}}C(CH_2)_4\overset{O}{\overset{\|}{C}}\left[NH(CH_2)_6NH\overset{O}{\overset{\|}{C}}(CH_2)_4\overset{O}{\overset{\|}{C}}\right]_n NH(CH_2)_6NH_2$$

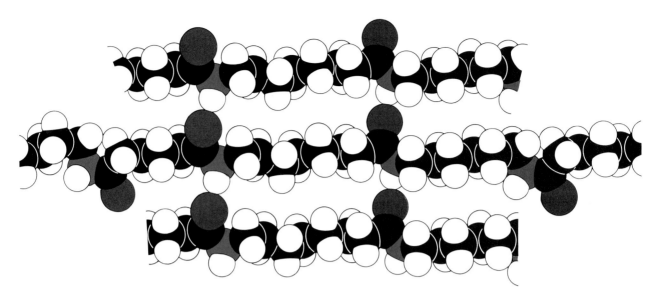

FIGURE 24.11 The strength of nylon fibers is yet another sign of the presence of hydrogen bonds, this time between neighboring polyamide chains.

The long polyamide (nylon) chains can be spun into fibers (like polyester molecules) or molded. Some of the strength of nylon fibers arises from the N—H···O=C hydrogen bonding that can occur between neighboring chains (Fig. 24.11). However, this ability to form hydrogen bonds also accounts for nylon's tendency to absorb moisture, since H_2O molecules can become hydrogen-bonded to the chains and worm their way in among them. The ability of an —NH— group to become charged, forming —NH_2^+—, for example, accounts for the buildup of electrostatic charge on nylon fabrics and carpets that are subject to friction.

24.8 AMINO ACIDS

An *amino acid* is a carboxylic acid that also contains an amino group. An example is glycine, NH_2CH_2COOH, the amino acid derived from acetic acid by substituting an amino group for a methyl hydrogen atom.

Optical activity. We saw in Section 21.6 (in the context of complexes) that a chiral molecule is one that is not identical to its mirror image. We also saw that a physical property stemming from chirality (and one way of detecting it) is optical activity, the ability to rotate the plane of polarization of light. Chirality and optical activity occur in organic compounds, particularly among amino acids. The simplest amino acid, glycine, is not chiral; it is identical to its mirror image (Fig. 24.12). However, when a CH_3— group is substituted for one of the two —CH_2— hydrogen atoms, we get alanine, $CH_3CH(NH_2)COOH$, which *is* chiral.

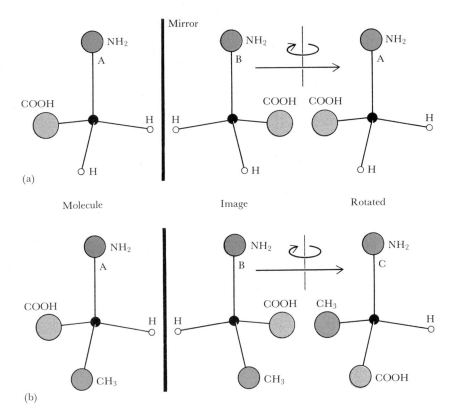

(a)

Molecule   Image   Rotated

(b)

FIGURE 24.12 (a) A glycine molecule can be superimposed on its mirror image. (b) An alanine molecule, however, cannot. It is an example of a chiral molecule. One way of checking a molecule for chirality is to note whether all the groups attached to one carbon atom are different from each other (as they are in alanine but not in glycine).

▼ **EXAMPLE 24.7 Judging whether a compound is optically active**

Decide whether the amino acids (a) $CH_3CH_2CH(NH_2)COOH$ and (b) $(CH_3)_2C(NH_2)COOH$ are optically active.

STRATEGY To decide whether a compound is optically active, we must inspect it to see if it is chiral. A quick way of checking whether an organic compound is chiral, and hence optically active, is to note whether it contains a carbon atom to which are attached four *different* groups; if it does, then it cannot be the same as its mirror image.

SOLUTION The two (isomeric) amino acids are

$$
\text{(a) } CH_3CH_2-\overset{\overset{\displaystyle NH_2}{|}}{\underset{\underset{\displaystyle H}{|}}{C}}-COOH
\qquad
\text{(b) } CH_3-\overset{\overset{\displaystyle NH_2}{|}}{\underset{\underset{\displaystyle CH_3}{|}}{C}}-COOH
$$

We see that (a) contains a carbon atom to which four different groups are attached (the central C atom) and hence is chiral. However, (b) does not have such a carbon atom, is identical to its mirror image, is not chiral, and therefore is not optically active.

EXERCISE What is the simplest chlorofluorocarbon that is optically active?

[*Answer*: CH_3CHFCl]

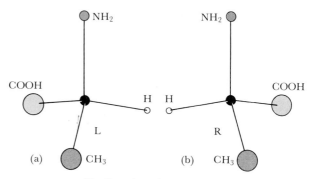

22 Enantiomeric amino acids

Ordinary laboratory syntheses of amino acids lead to mixtures of enantiomers (the two mirror-image optical isomers) in equal proportions. The isomers are labeled L, denoting a left-hand form, and D, denoting a right-hand form (**22**): the terms "left-hand" and "right-

TABLE 24.6 The naturally occurring amino acids

Formula	Name	Abbreviation	Formula	Name	Abbreviation
$\overset{\displaystyle NH_2}{\underset{\displaystyle H}{H-C-COOH}}$	Glycine	Gly	$\underset{\displaystyle CH_2CH_2-CO-NH_2}{\overset{\displaystyle NH_2}{H-C-COOH}}$	Glutamine	Gln
$\underset{\displaystyle CH_3}{\overset{\displaystyle NH_2}{H-C-COOH}}$	Alanine	Ala	$\underset{\displaystyle CH_2OH}{\overset{\displaystyle NH_2}{H-C-COOH}}$	Serine	Ser
$\underset{\displaystyle CH(CH_3)_2}{\overset{\displaystyle NH_2}{H-C-COOH}}$	Valine*	Val	$\underset{\displaystyle CH(OH)CH_3}{\overset{\displaystyle NH_2}{H-C-COOH}}$	Threonine	Thr
$\underset{\displaystyle CH_2CH(CH_3)_2}{\overset{\displaystyle NH_2}{H-C-COOH}}$	Leucine	Leu	$\underset{\displaystyle CH_2CH_2-NH_2}{\overset{\displaystyle NH_2}{H-C-COOH}}$	Lysine*	Lys
$\underset{\displaystyle CH(CH_3)CH_2CH_3}{\overset{\displaystyle NH_2}{H-C-COOH}}$	Isoleucine	Ile	$\underset{\displaystyle CH_2CH_2CH_2-NH-\underset{\displaystyle \|\!\!\;NH}{C}-NH_2}{\overset{\displaystyle NH_2}{H-C-COOH}}$	Arginine	Arg
$\underset{\displaystyle CH_2COOH}{\overset{\displaystyle NH_2}{H-C-COOH}}$	Aspartic acid	Asp	$\underset{\displaystyle CH_2}{\overset{\displaystyle NH_2}{H-C-COOH}}$ (imidazole ring)	Histidine*	His

* Essential amino acid.

hand" have a specific technical meaning in this context but are effectively defined by the two structures shown here. Hence, laboratory syntheses result in a *racemic mixture*, a mixture of equal numbers of L and D enantiomers. Since L and D enantiomers rotate the plane of polarization of light in opposite directions, a racemic sample is not optically active. However, biochemical reactions lead to only one enantiomer. It is a remarkable feature of nature (one that has not yet been fully explained) that all naturally occurring amino acids are L enantiomers.

Polypeptides and proteins. The central importance of amino acids is that they are the building blocks of proteins. In a sense, proteins are a very elaborate form of nylon. However, whereas nylon is a monotonous repetition of the same two monomers, proteins are copolymers of up to 20 different naturally occurring L-amino acids (Table 24.6). Our bodies can synthesize 15 of these acids in sufficient amounts for our needs. It is essential to ingest the other 5, and so they are known as the *essential amino acids*.

Formula	Name	Abbreviation
$H-\underset{CH_2CH_2COOH}{\overset{NH_2}{C}}-COOH$	Glutamic acid	Glu
$H-\underset{CH_2-SH}{\overset{NH_2}{C}}-COOH$	Cysteine	Cys
$H-\underset{CH_2CH_2-S-CH_2}{\overset{NH_2}{C}}-COOH$	Methionine	Met
$H-\underset{CH_2-CO-NH_2}{\overset{NH_2}{C}}-COOH$	Asparagine	Asn
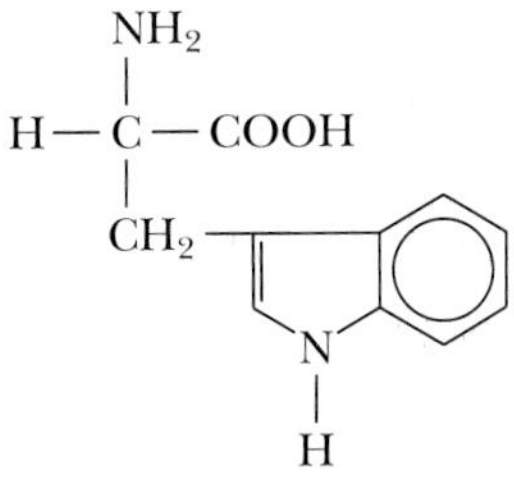	Tryptophan*	Trp
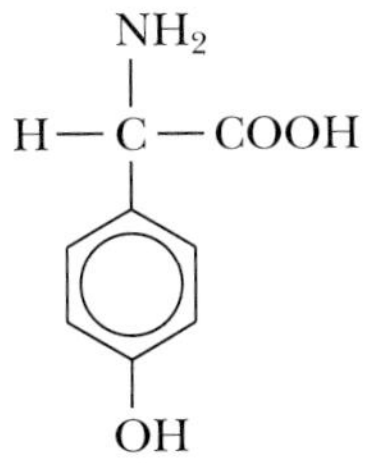	Tyrosine	Tyr
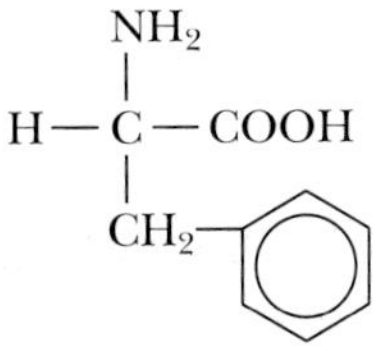	Phenylalanine*	Phe
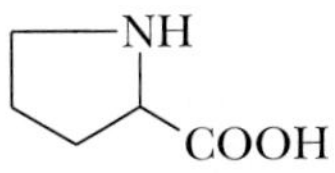	Proline	Pro

Leu-Ser-Pro-Ala-Asp-Lys-Thr-Asn-Val-Lys- . . . – – – – –

. . . -Val-Lys-Gly-Trp-Ala-Ala- . . . –

. . . -Ser-Thr-Val-Leu-Thr-Ser-Lys-Ser-Lys-Tyr-Arg

A molecule formed from two or more amino acids is called a *peptide*. An example is the combination of glycine and alanine, Gly-Ala:

$$NH_2CH_2—C \overset{O}{\underset{NH—CH(CH_3)COOH}{\Big\langle}}$$

The —CO—NH— link printed in red is called the *peptide bond*, and the amino acid in a peptide is called a *residue*. A typical protein is a *polypeptide chain* with more than about a hundred residues joined through peptide bonds and arranged in a strict order. A peptide chain with only a few amino acid residues is called an *oligopeptide*. The artificial sweetening agent aspartame is an oligopeptide with two residues; specifically, it is a *dipeptide*.

The sequence of amino acids in a peptide chain is the protein's *primary structure*. Aspartame consists of phenylalanine and aspartic acid, so its primary structure is Phe-Asp. The opposite arrangement, Asp-Phe, is a different primary structure and a different dipeptide. A fragment of the primary structure of human hemoglobin is given in Table 24.7. The determination of the primary structures of proteins is a very demanding analytical task, but many of these structures are now known.

Chemists have now also begun the even more demanding task of synthesizing proteins. Some syntheses have been completed, but so far the products have turned out to be biologically inactive. This is largely because a protein's ability to carry out its function (which may be to act as an enzyme) also depends on the shape adopted by the chain of residues, and synthetic proteins have not yet been produced with the same shapes as their biological counterparts.

The *α helix* (Fig. 24.13) is a specific conformation of a polypeptide chain held in place by hydrogen bonds between the residues and is one example of a *secondary structure*. An alternative secondary structure is the *β-pleated sheet* form of the protein we know as silk. In this structure, the protein molecules lie side by side to form nearly flat sheets. Many protein molecules consist of alternating regions of α helix and β-pleated sheet (Fig. 24.14).

Protein molecules also adopt a specific *tertiary structure*, the shape into which the α helix and β-pleated sheet sections are twisted as a result of interactions between residues lying in different parts of the primary structure. The globular form of the polypeptide in Fig. 24.14 is an example. One important type of link responsible for tertiary structure is the *disulfide link*, —S—S—, between residues containing sulfur (cys-

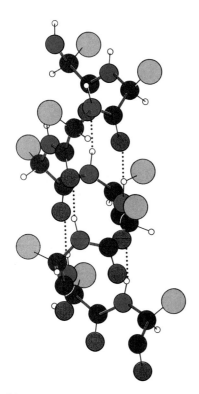

FIGURE 24.13 The α helix, one of the secondary structures adopted by polypeptide chains.

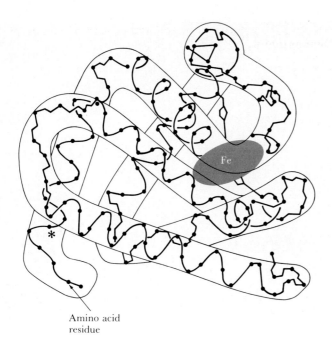

Amino acid
residue

FIGURE 24.14 One of the four polypeptide chains that make up the human hemoglobin molecule. The chain consists of alternating regions of α helix and β-pleated sheet.

teine and methionine). The overall shape of a protein is essential to its function. In Section 12.7, in discussing enzymes as catalysts, we saw how the action and malfunctioning of enzymes could be explained with a "lock-and-key" model. Here we see that the amino acid sequence, along with the secondary and tertiary structures to which this primary structure leads, account for the shape of the lock.

Proteins may also have a *quaternary structure*, when neighboring polypeptide units stack together in a specific arrangement. The hemoglobin molecule, for example, has a quaternary structure consisting of four polypeptide units like the one shown in Fig. 24.14. Any modification of the primary structure—the replacement of one amino acid residue by another—may lead to the malfunction we call congenital disease. Even one wrong amino acid in the chain can cause malfunction. In normal hemoglobin, the amino acid marked with an asterisk in Fig. 24.14 is glutamic acid; if it is replaced by valine, then the disease "sickle-cell anemia" may result. The affected person develops malformed, sickle-shaped red blood corpuscles (Fig. 24.15) that are poor transporters of oxygen.

A loss of structure in a protein is called *denaturation*. This may be a loss of quaternary, tertiary, or secondary structure, or even degradation of the primary structure by cleavage of the peptide bonds. Even mild heating can cause irreversible denaturation, as happens when we cook an egg and the albumin denatures into a white mass. The permanent waving of hair is a result of partial denaturation. Mild reducing agents are applied to the hair to sever disulfide links between the protein strands that make up hair. The links are then reformed by applying a mild oxidizing agent while the hair is stretched and twisted into the desired arrangement.

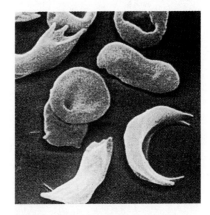

FIGURE 24.15 The sickle-shaped red blood cells that form when the glutamic acid marked with an asterisk in Fig. 24.14 is replaced by valine.

FIGURE 24.16 A computer graphics image of a short section of a DNA molecule, which consists of two entwined helices. In this illustration, the double helix is also coiled around itself, in a shape called a superhelix.

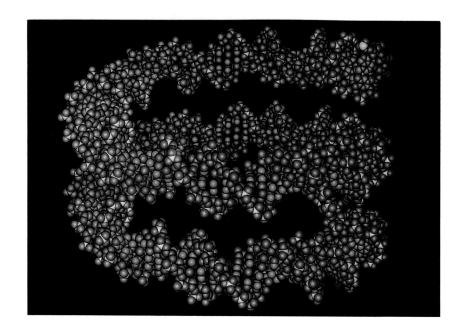

24.9 DNA AND RNA

Among the most complicated of all organic molecules are the deoxyribonucleic acids, DNA (Fig. 24.16). The nucleus of every living cell has at least one DNA molecule to control the production of proteins and carry genetic information from one generation of cells to the next. Human DNA molecules are immense: if one could be extracted without damage from a cell nucleus and straightened out from its highly coiled natural shape, it would be about 2 m long (Fig. 24.17). The ribonucleic acid (RNA) molecule is closely related to the DNA molecule but is smaller. One of its functions is to carry the information stored by DNA out of a nucleus and into a region of the cell where it can be used as a template for protein synthesis.

Nucleosides and nucleotides. The best way to understand the structure of DNA is to see how it gets its name. It is a polymer, and the repeating unit is a modification of a sugar molecule, ribose (**23**), in its furanose form. The modification is the absence of the oxygen atom at carbon atom 2; that is, the repeating unit—the monomer—is deoxyribose (**24**).

Attached by a covalent bond to carbon atom 1 of the ribose ring is an aminelike molecule (and therefore a base), which may be adenine (A; **25**), guanine (G; **26**), cytosine (C; **27**), or thymine (T; **28**). In RNA, uracil (U; **29**) occurs in place of thymine. The base bonds to carbon atom 1 of the deoxyribose through the nitrogen of the NH group (printed in red), and the compound so formed is called a *nucleoside*. All nucleosides have about the same structure, which we can summarize as the shape shown in (**30**).

The final detail in the structure of the DNA monomers is that a phosphate group ($-O-PO_3^{2-}$) is covalently bonded to carbon atom 5 to give a compound called a *nucleotide* (**31**). Since there are four possible nucleosides (one for each base), there are four possible nucleotides.

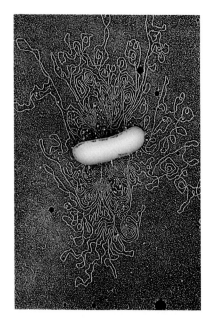

FIGURE 24.17 The DNA molecule is very large, even in bacteria. In this micrograph, a DNA molecule is seen to spill out of the damaged cell wall of a bacterium.

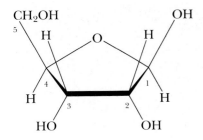

23 Ribose

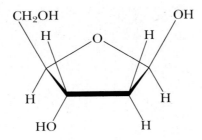

24 Deoxyribose

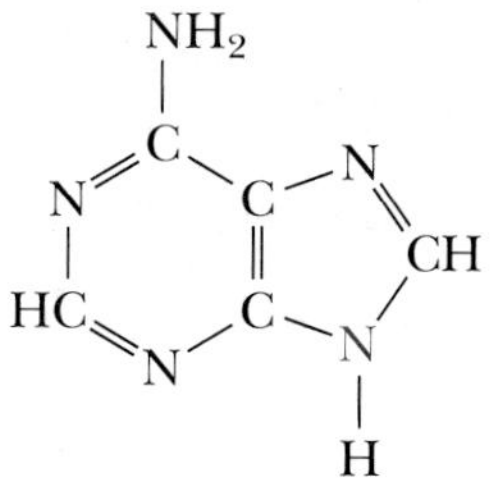

25 Adenine

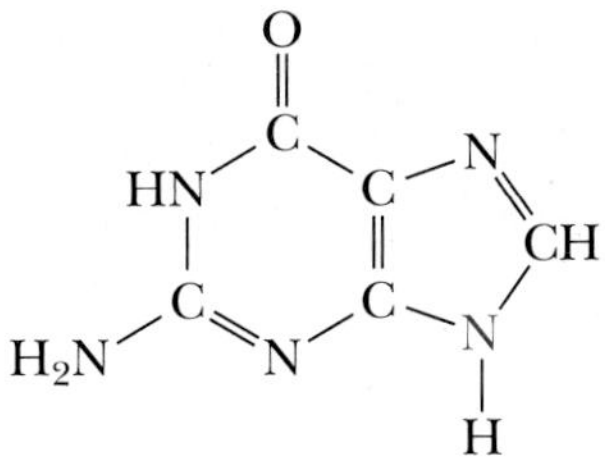

26 Guanine

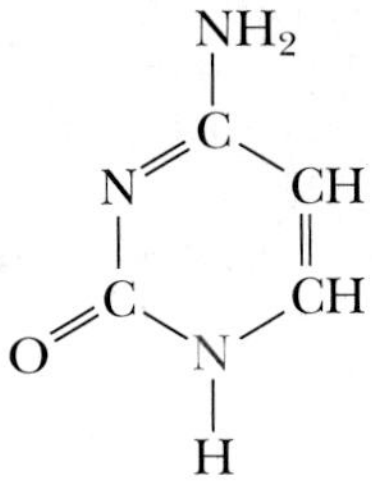

27 Cytosine

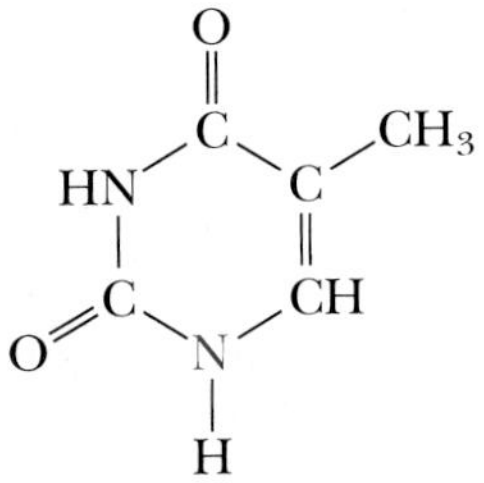

28 Thymine

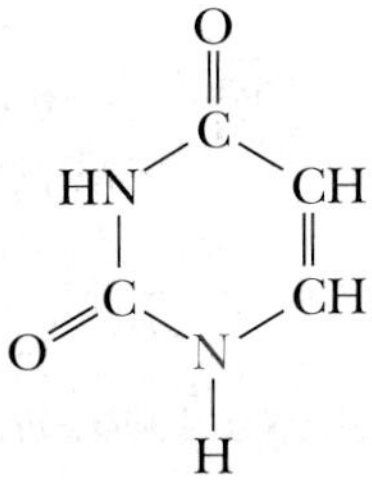

29 Uracil

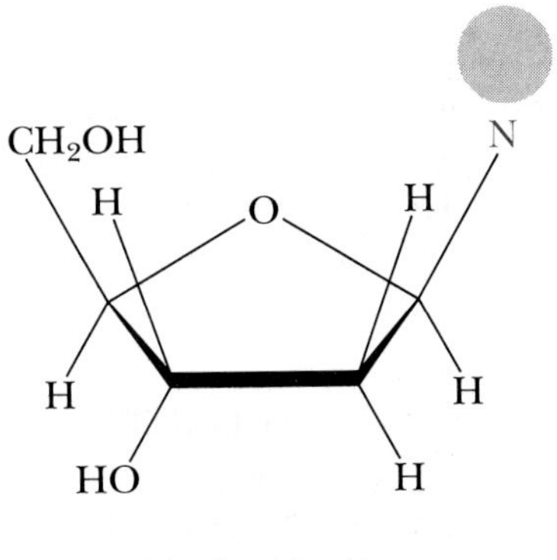

30 A nucleoside

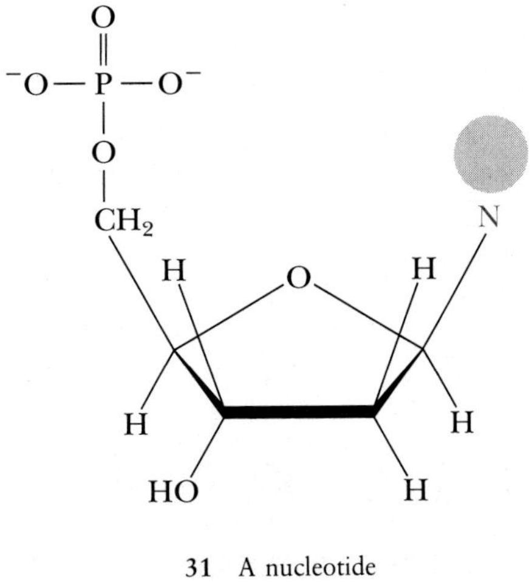

31 A nucleotide

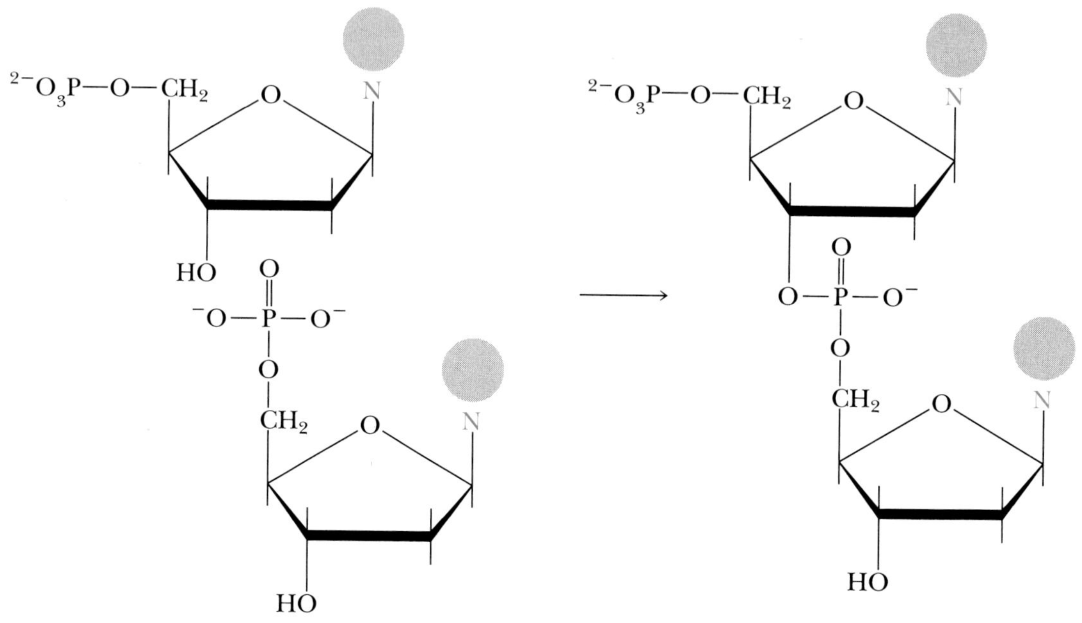

FIGURE 24.18 The condensation of nucleotides that leads to the formation of a nucleic acid—a polynucleotide.

Nucleic acids. DNA and RNA are *polynucleotides*, polymers built from nucleotide units. The polymers are formed in a reaction involving the phosphate group of one nucleotide and the remaining hydroxyl group of another. As this condensation continues, it results in a structure like that shown in Fig. 24.18, known as a *nucleic acid*. The DNA molecule itself is a double helix in which two long nucleic acid strands are wound around each other. The ability of DNA to replicate lies in its double helix, for there is a precise correspondence between the bases in the two strands; that is, the strands are held together by bonds between specific *base pairs*. Adenine in one strand always pairs with thymine in the other, and guanine always pairs with cytosine, so the base pairs are always AT and GC (Fig. 24.19).

The genetic code. Because of the base pairing, the base sequence in one strand of a helix reflects and is determined by the sequence in the other:

$$. . . \text{GATGTGCTCAG} . . .$$
$$. . . \text{CTACACGAGTC} . . .$$

Therefore, when the strands separate, each acts as a template for the growth of another that is identical to its original partner:

$$. . . \text{GATGTGCTCAG} . . .$$
$$. . . \text{CTACACGAGTC} . . .$$

. . . GATGTGCTCAG . . .   . . . CTACACGAGTC . . .

. . . GATGTGCTCAG . . .   . . . CTACACGAGTC . . .
. . . CTACACGAGTC . . .   . . . GATGTGCTCAG . . .

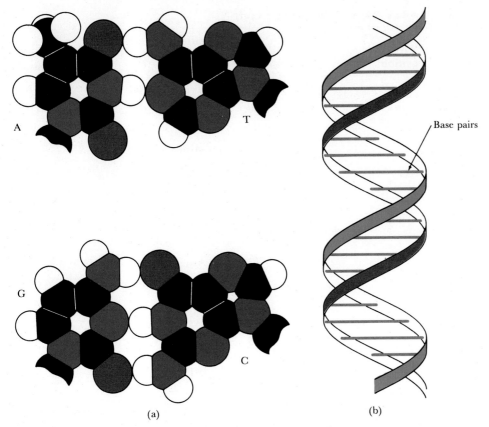

FIGURE 24.19 (a) The bases in the DNA double helix fit together by virtue of the hydrogen bonds that they can form. Once formed, the AT and GC pairs are almost identical in size and shape; as a result, the turns of the helix (b) are regular and consistent.

As well as replication—the production of copies of itself for reproduction and cell division—DNA governs the production of proteins. For this purpose, segments of the genetic message are carried out of the cell nucleus as RNA molecules, with U in place of T.

The precise sequence of bases in the nucleic acid constitutes the genetic information for the cell. The order of the bases controls the order in which amino acids condense together to form proteins. Since there are twenty amino acids but only four bases, each amino acid must be coded by a group of bases. These groups, which are called _codons_, are known to consist of three bases each; CAC, for instance, is the codon for histidine, and GCA is the codon for alanine. The complete list of codons in RNA (which is where the code is normally read and transcribed) is given in Table 24.8. The sequence CACGCA, for example, would result in the sequence His-Ala in the protein that was built to its design.

The principles of chemistry apply to RNA and DNA and to the proteins they code for, just as those principles apply to the very simple

TABLE 24.8   The genetic code

First position	Second position				Third position
	U	**C**	**A**	**G**	
U	Phe	Ser	Tyr	Cys	U
	Phe	Ser	Tyr	Cys	C
	Leu	Ser	STOP	STOP	A
	Leu	Ser	STOP	Trp	G
C	Leu	Pro	His	Arg	U
	Leu	Pro	His	Arg	C
	Leu	Pro	Glu	Arg	A
	Leu	Pro	Glu	Arg	G
A	Ile	Thr	Asn	Ser	U
	Ile	Thr	Asn	Ser	C
	Ile	Thr	Lys	Arg	A
	Met	Thr	Lys	Arg	G
G	Val	Ala	Asp	Gly	U
	Val	Ala	Asp	Gly	C
	Val	Ala	Glu	Gly	A
	Val	Ala	Glu	Gly	G

Note: AUG and GUG are also part of the START signal.

compounds we met earlier in the text. It is here, though, that chemistry becomes biology. We are at the point where the elements have completed their journey from the stars, have come alive (Fig. 24.20), and have begun to control their own destiny.

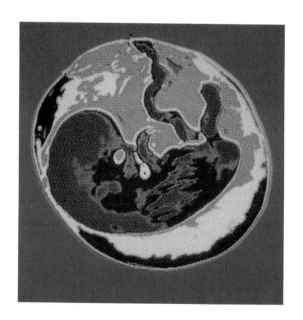

FIGURE 24.20 A human fetus in the first trimester (8–11 weeks). The ultrasound image has been computer enhanced for color.

SUMMARY

24.1 Aliphatic compounds containing the **hydroxyl group** are called **alcohols** (suffix -ol) and may be **primary** (with structure RCH_2—OH), **secondary** (R_2CH—OH), or **tertiary** (R_3C—OH). Alcohols are amphiprotic. They are very weak Brønsted acids, and when they act as bases they accept a proton to form an **oxonium ion.** They may be halogenated, dehydrated, and oxidized to aldehydes, ketones, and carboxylic acids. **Ethers** (R—O—R) are less reactive than alcohols, but can undergo acid-catalyzed substitution. Cyclic **crown ethers** are important complexing agents.

24.2 Compounds in which the hydroxyl group is attached directly to an aromatic ring are called **phenols;** they are weak acids. The benzene ring is much more reactive in phenol than in benzene, particularly at the *ortho* and *para* positions. A particularly important reaction of phenol is with formaldehyde, which leads to the formation of **phenolic resins.**

24.3 Compounds containing the **carbonyl group** include **aldehydes** (R—CO—H, more often denoted R—CHO; suffix -al) and **ketones** (R—CO—R; suffix -one). These are obtained by the oxidation of primary and secondary alcohols, respectively. Aldehydes are readily oxidized to carboxylic acids. Aldehydes reduce Fehling's solution and Tollens' reagent; ketones do not. The reactions of aldehydes (and some ketones) include **cyanohydrin** formation with cyanide ions. Aldehydes form **hemiacetals** and **acetals** by reaction with alcohols.

24.4 Carbohydrates may be aldehydes or ketones; they are often found in their hemiacetal **pyranose** (six-membered ring) or **furanose** (five-membered ring) forms. **Reducing sugars** are carbohydrates that reduce Fehling's solution and Tollens' reagent. **Monosaccharides** are the building blocks of **oligosaccharides,** including the **disaccharide** sucrose, and **polysaccharides,** such as **starch** and **cellulose.** Starch consists of two components—amylose and amylopectin.

24.5 Compounds containing the **carboxyl group**

(—COOH) are **carboxylic acids.** These are weak acids produced by the oxidation of hydrocarbons, primary alcohols, and aldehydes.

24.6 Carboxylic acids undergo a **condensation reaction** with alcohols to form **esters.** Dicarboxylic acids and diols undergo **condensation polymerization** to form **polyesters.**

24.7 Compounds containing the **amino group** (—NH_2 and its derivatives) not linked to a carbonyl group are called **amines** and may be **primary** (RNH_2), **secondary** (R_2NH), or **tertiary** (R_3N). Amines are Lewis and Brønsted bases. Alkylamines are stronger Brønsted bases than ammonia, arylamines weaker. Amines condense with carboxylic acids to form **amides;** some take part in condensation polymerization to form **polyamides.**

24.8 Amino acids are compounds containing both the amino group and the carboxyl group. Many have chiral molecules and hence are optically active, but only L-amino acids occur naturally. Proteins are **polypeptides,** copolymers of up to 20 naturally occurring amino acids. The **primary structure** of a polypeptide chain is its sequence of amino acids. Its **secondary structure** may be either an *α* **helix** or a *β***-pleated sheet.** The secondary structure is contorted into a **tertiary structure** by interactions between peptide units, some through **disulfide links** (—S—S—). Protein units may link together to give an overall **quaternary structure.**

24.9 The primary structure of proteins is controlled by the **nucleic acid** molecule **DNA,** a **polynucleotide.** A **nucleotide** is the combination of a phosphate group and a **nucleoside;** the latter is a combination of a base (A, G, C, or T; but U in place of T in RNA) and a **deoxyribose** molecule. The double helix of DNA has a structure that arises from specific **base pairing** between the two polynucleotide strands. **Codons,** which specify an amino acid and form the **genetic code,** consist of groups of three bases.

EXERCISES

Functional Groups

24.1 Explain what is meant by a functional group. Give three examples each of groups that contain (a) oxygen; (b) nitrogen; (c) neither oxygen nor nitrogen.

24.2 Write the formulas for the functional groups (a) amino; (b) bromo; (c) carboxyl; (d) cyano.

In Exercises 24.3 and 24.4, name each functional group.

24.3 (a) —OH; (b) $>$CO; (c) —NH_2.
24.4 (a) —COOH; (b) —NC; (c) —NO_2.

In Exercises 24.5 and 24.6, write the general formula for each family of compounds, using R— to denote an aliphatic group and Ar— to denote an aromatic group (if that needs to be specified). Give an example of each family.

24.5 (a) Alcohol; (b) carboxylic acid; (c) phenol; (d) primary amine.

24.6 (a) Aldehyde; (b) ketone; (c) ether; (d) tertiary amine.

In Exercises 24.7 and 24.8, classify each compound.

24.7 (a) CH_3OH; (b) $CH_3OC_6H_5$; (c) ArCHO; (d) $CH_3CH(NH_2)COOH$.

24.8 (a) CH_3COCH_3; (b) ArOH; (c) $CH_2(OH)CH_2(OH)$; (d) HCO_2^-.

In Exercises 24.9 and 24.10, list the functional groups that each compound contains.

24.9 (a) Vanillin **(10)**, the compound mainly responsible for the flavor of vanilla; (b) carvone **(11)**, which is mainly responsible for the flavor of spearmint.

24.10 (a) Zingiberone, a pungent, hot component of ginger:

(b) Monosodium glutamate, a flavor enhancer:

In Exercises 24.11 and 24.12, summarize the reactions typical of compounds that possess each functional group.

24.11 (a) A double bond; (b) a hydroxyl group.
24.12 (a) A carbonyl group; (b) an amino group.

The Hydroxyl Group

In Exercises 24.13 and 24.14, write the formula for each compound and classify it as a primary, secondary, or tertiary alcohol or as a phenol.

24.13 (a) Ethanol; (b) *p*-hydroxybenzoic acid; (c) 2-propanol; (d) 2-butanol.

24.14 (a) Methanol; (b) 1-butanol; (c) 2-methyl-2-propanol; (d) *p*-hydroxybenzaldehyde.

In Exercises 24.15 and 24.16, classify and name each alcohol.

24.15 (a) $CH_3(CH_2)_4OH$
(b) $CH_3CH(OH)(CH_2)_3CH_3$
(c) $CH_2(OH)CH_2CH_2OH$
(d) $(CH_3)_2C(OH)CH_2CH_2CH_3$

24.16 (a) $CH_3(CH_2)_6CH_2OH$
(b) $CH_2(OH)CH(CH_3)CH_2OH$
(c) $CH_2(OH)CH(OH)CH(OH)CH(CH_2CH_3)$-$CH_2OH$
(d) $C_6H_5CH_2OH$

In Exercises 24.17 and 24.18, suggest how you might prepare each alcohol, starting from (and naming) a bromoalkane.

24.17 (a) Ethanol; (b) 2-propanol; (c) 2-butanol.
24.18 (a) 1-propanol; (b) cyclohexanol; (c) 3-methyl-3-hexanol.

In Exercises 24.19 to 24.26, name the products of each reaction.

24.19 (a) Ethanol is heated with concentrated sulfuric acid.
(b) 2-Butanol is heated with concentrated sulfuric acid.
24.20 (a) 1-Propanol is heated with concentrated sulfuric acid.
(b) Isopropyl alcohol is heated with concentrated sulfuric acid.
24.21 (a) Methanol is heated with hydrochloric acid.
(b) 2-Propanol is heated with hydrobromic acid.
24.22 (a) 2-Butanol is heated with hydrobromic acid.
(b) *tert*-Butyl alcohol is heated with hydrobromic acid.
24.23 (a) Ethanol is heated with oxalic acid.
(b) Ethylene glycol is heated with stearic acid.
24.24 (a) 1-Butanol is heated with propionic acid.
(b) Glycerol is heated with benzoic acid.
24.25 Ethanol is heated with acidified sodium dichromate in an organic solvent.
24.26 2-Propanol is heated with acidified sodium dichromate in an organic solvent.

Aldehydes and Ketones

In Exercises 24.27 and 24.28, identify each compound as an aldehyde or a ketone, and give its systematic name.

24.27 (a) HCHO; (b) $C_6H_5COCH_3$; (c) $(CH_3CH_2)_2CO$; (d) $(CH_3)_2CHCHO$.
24.28 (a) CH_3CHO; (b) $CH_3CH_2CH_2COCH_2CH_3$; (c) $(CH_3CH_2)_2CHCH_2COCH_3$; (d) *p*-$CH_3C_6H_4CHO$.

In Exercises 24.29 and 24.30, write the formula for each compound.

24.29 (a) Ethanal; (b) propanone; (c) octanal; (d) 2-phenyl-3-hexanone.
24.30 (a) Methanal; (b) 2-heptanone; (c) 2-ethyl-2-methylpentanal; (d) 3,5-dihydroxy-4-octanone.

In Exercises 24.31 and 24.32, suggest an alcohol that would be a suitable starting material for the preparation of each compound, and indicate how the reaction would be carried out.

24.31 (a) Ethanal; (b) butanone; (c) 2-octanone; (d) 5-methyloctanal.

24.32 (a) Methanal; (b) propanone; (c) acetaldehyde; (d) 5-methyl-6-decanone.

24.33 By analogy with the behavior of glucose and fructose, suggest what product is formed when methanol and (a) acetaldehyde or (b) acetone are mixed.

24.34 Lithium aluminum hydride ($LiAlH_4$) is a reducing agent for the carbonyl group, but not for the $C{=}C$ double bond. Name the products formed (and give their structural formulas) when it is used to reduce (a) 3-hexanone; (b) 5-hexenal; (c) 5,7-octadienal; (d) ethyl-cyclohexane-2,6-dione.

24.35 Suggest a method for preparing acetaldehyde from limestone, water, and coke.

24.36 Compare and contrast the reactions of aldehydes and ketones, using as examples their oxidation and their reactions with HCN, Fehling's solution, and Tollens' reagent.

The Carboxyl Group

In Exercises 24.37 and 24.38, give the systematic name of each carboxylic acid.

24.37 (a) CH_3COOH; (b) $CH_3CH_2CH_2COOH$; (c) $(CH_3)_2CHCH_2COOH$; (d) $CH_2(OH)CH_2COOH$.
24.38 (a) $HCOOH$; (b) $CH_3(CH_2)_8COOH$; (c) $CH_2(NH_2)COOH$; (d) $(CH_3CH_2)_2CH(OH)COOH$.

In Exercises 24.39 and 24.40, write the structural formula for each carboxylic acid.

24.39 (a) Butanoic acid; (b) benzoic acid; (c) trichloro-ethanoic acid; (d) ethanedioic acid.
24.40 (a) Oxalic acid; (b) 2,3-diphenylbutanoic acid; (c) trifluoroacetic acid; (d) 2-hydroxypropionic acid.

In Exercises 24.41 and 24.42, suggest a method of preparing each acid from an alcohol or phenol.

24.41 (a) Acetic acid; (b) oxalic acid; (c) 2-methyl-butanoic acid.
24.42 (a) Formic acid; (b) benzoic acid; (c) 2-methylpropionic acid.

24.43 Acetic acid and its three chloro-substituted derivatives have the following pK_a values: CH_3COOH, 4.76; $CH_2ClCOOH$, 2.86; $CHCl_2COOH$, 1.29; CCl_3COOH, 0.65. Suggest an explanation for this trend.

24.44 You are given samples of an aldehyde, a ketone, and a carboxylic acid. State clearly, with equations, how you could distinguish the compounds.

24.45 Suggest an experimental method of showing that in ester formation, the oxygen atom in the eliminated water molecule comes from the carboxylate group. Does the process of amide formation shed any light on the details of the esterification reaction?

24.46 Explain the process of condensation polymerization. How might the polymer obtained from benzene-1,2-dicarboxylic acid differ from Dacron?

24.47 Suggest one reason why a condensation polymer (like any polymer) might not have a definite molecular weight. How would that affect the osmotic measurement of its apparent molecular weight?

Amines and Amides

24.48 Explain what is meant by the terms primary, secondary, and tertiary as applied to amines. In what way does the use of these terms for amines differ from their use for alcohols?

In Exercises 24.49 and 24.50, give the systematic name of each amine and classify it as primary, secondary, or tertiary.

24.49 (a) CH_3NH_2; (b) p-$CH_3C_6H_4NH_2$; (c) $NH_2CH_2CH_2NH_2$; (d) NH_2CH_2COOH.
24.50 (a) $(CH_3)_3N$; (b) $(CH_3CH_2)_4N^+$; (c) o-$CH_3C_6H_4NH_2$; (d) $(C_6H_5)_2NH$.

24.51 Write the structures of, classify, and give the names of the eight isomeric amines of formula $C_4H_{11}N$.

24.52 Give the structural formula for (a) 2-aminoethanol; (b) diphenylamine; (c) 2,4-dimethylaniline; (d) iso-propylammonium benzoate.

24.53 Suggest a method for the production of butylamine from (a) 1-bromobutane; (b) 1-butene; (c) 1-butanol; (d) ethanol.

24.54 Deduce the structure of putrescine ($C_4H_{12}N_2$) from the fact that it can be formed from 1,2-dibromo-ethane by reaction with KCN, which gives a compound of formula $C_4H_4N_2$, followed by reduction with sodium and ethanol.

24.55 Suggest the identity of the products of the reaction between aniline and (a) bromine; (b) hydrochloric acid.

24.56 The pK_a values of the conjugate acids of ammonia and methylamine are 9.25 and 10.64, respectively. What does this mean in terms of the relative strengths of the parent compounds as bases. Propose an explanation for the difference.

24.57 The substituents —CH_3, —NH_2, and —OCH_3 release electrons into the benzene ring, but —NO_2, —COOH, and halogen atoms withdraw them. Which set of substituents will increase the strength of aniline as a base?

24.58 Write equations showing how a polyamide can be prepared. Suggest how the following substance, called "caprolactam," might be used to prepare a version of nylon:

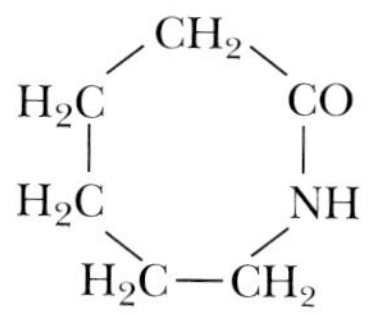

24.59 What is meant by an amide? Predict the products of the reaction between (a) acetic acid and ethylamine; (b) propionic acid and dimethylamine; (c) butane-1,4-diamine and oxalic acid; (d) 4-aminobutanoic acid and aminoethanoic acid.

Proteins and Nucleic Acids

24.60 Alanine is a chiral amino acid, but glycine is not. Explain this assertion. Is aspartic acid chiral?

***24.61** Explain what is meant by the primary, secondary, tertiary, and quaternary structures of proteins. Write the molecular formula of a fragment of a polypeptide chain with the composition (a) —Gly—Gly—Gly—Gly—; (b) —Gly—Leu—Ser—Ala—.

***24.62** The protein beef insulin consists of two chains in which the groups —(Leu)$_2$—Ile—(Cys)$_4$—Arg— and —Pro—(Phe)$_3$—(Cys)$_2$—Arg— occur. What could be the nature of the interaction between the chains. The chains can be separated by oxidation. Suggest a reason.

***24.63** The molecular weight of a protein extracted from salmon sperm is 10,000 amu. Analysis of 100 g of the protein gave the following masses of amino acids: Ile, 1.28 g; Ala, 0.89 g; Val, 3.68 g; Gly, 3.01 g; Ser, 7.29 g; Pro, 6.90 g; Arg, 86.40 g. What is the molecular formula of the protein?

***24.64** What amino acid residue sequence would be constructed from the RNA sequence UAUCUAUCUAU-CUAUCUAUCU?

***24.65** Suppose you wanted to produce a genetically engineered organism that could synthesize aspartame. What RNA sequences would you seek?

1

MATHEMATICAL INFORMATION

1A Scientific notation
1B Logarithms

1C Quadratic equations
1D Graphs

APPENDIX 1A: SCIENTIFIC NOTATION

In *scientific notation*, a number is written as $A \times 10^a$, where A is a decimal number with one digit in front of the decimal point and a is a whole number. For example, 333 is written 3.33×10^2 in decimal notation, since $10^2 = 10 \times 10 = 100$:

$$333 = 3.33 \times 10^2$$

We use

$$10^1 = 10$$

$$10^2 = 10 \times 10 = 100$$

$$10^3 = 10 \times 10 \times 10 = 1000$$

$$10^4 = 10 \times 10 \times 10 \times 10 = 10{,}000$$

and so on. Note that the number of zeros following 1 is equal to the power of 10. Thus, 10^6 is 1 followed by 6 zeros:

$$10^6 = 10 \times 10 \times 10 \times 10 \times 10 \times 10 = 1{,}000{,}000$$

The power of 10 shows us how many places the decimal point should be moved to the *right* to convert a number from scientific notation back to ordinary decimal notation. Thus to convert 3.33×10^2, we would move the decimal point 2 places to the right, to give 333. (or simply 333), and for 3.33×10^4 we move it 4 places to the right, giving 33,300. (or simply 33,300).

Numbers between 0 and 1 are expressed in the same way, but with a negative power of 10, and have the form $A \times 10^{-a}$, using $10^{-1} = \frac{1}{10} = 0.1$. Thus, 0.0333 in decimal notation is written 3.33×10^{-2}, since

$$10^{-2} = \frac{1}{10} \times \frac{1}{10} = \frac{1}{100}$$

and

$$0.0333 = 3.33 \times \frac{1}{100} = 3.33 \times 10^{-2}$$

We use

$$10^{-2} = 10^{-1} \times 10^{-1} = 0.01$$

$$10^{-3} = 10^{-1} \times 10^{-1} \times 10^{-1} = 0.001$$

$$10^{-4} = 10^{-1} \times 10^{-1} \times 10^{-1} \times 10^{-1} = 0.0001$$

When a negative power of 10 is written out as a decimal number, the number of zeros following the decimal point is one less than the number (disregarding the sign) to which 10 is raised. Thus, 10^{-6} is written as a decimal point followed by $6 - 1 = 5$ zeros and then a 1:

$$10^{-6} = 10^{-1} \times 10^{-1} \times 10^{-1} \times 10^{-1} \times 10^{-1} \times 10^{-1} = 0.000001$$

Note that the negative power tells us how many places to move the decimal point to the *left* when converting scientific notation to ordinary decimal notation. Thus to convert 3.33×10^{-2}, we would move the decimal point 2 places to the left, to give 0.0333, and for 3.33×10^{-4} we move it 4 places to the left, giving 0.000333.

To multiply numbers in scientific notation, the decimal parts of the numbers are multiplied, and the powers of 10 are added:

$$(A \times 10^a) \times (B \times 10^b) = (A \times B) \times 10^{a+b}$$

An example is

$$(1.23 \times 10^2) \times (4.56 \times 10^3) = 1.23 \times 4.56 \times 10^{2+3}$$
$$= 5.61 \times 10^5$$

This rule holds even if the powers of 10 are negative:

$$(1.23 \times 10^{-2}) \times (4.56 \times 10^3) = 1.23 \times 4.56 \times 10^{-2+3}$$
$$= 5.61 \times 10$$
$$(1.23 \times 10^{-2}) \times (4.56 \times 10^{-3}) = 1.23 \times 4.56 \times 10^{-2-3}$$
$$= 5.61 \times 10^{-5}$$

Note that the results of calculations are adjusted so that there is one digit in front of the decimal point:

$$(4.56 \times 10^3) \times (7.65 \times 10^6) = 34.88 \times 10^9 = 3.488 \times 10^{10}$$

When dividing two numbers in scientific notation, we divide the decimal parts of the numbers and subtract the powers of 10:

$$\frac{A \times 10^a}{B \times 10^b} = \frac{A}{B} \times 10^{a-b}$$

Two examples are

$$\frac{7.65 \times 10^8}{4.56 \times 10^3} = \frac{7.65}{4.56} \times 10^{8-3} = 1.68 \times 10^5$$

$$\frac{4.31 \times 10^{-5}}{9.87 \times 10^{-8}} = \frac{4.31}{9.87} \times 10^{-5-(-8)}$$
$$= 0.437 \times 10^3 = 4.37 \times 10^2$$

Before adding and subtracting numbers in scientific notation, we rewrite the numbers as decimal numbers multiplied by the *highest* power of 10 that appears:

$$1.00 \times 10^3 + 2.00 \times 10^2 = 1.00 \times 10^3 + 0.200 \times 10^3$$
$$= 1.20 \times 10^3$$
$$1.00 \times 10^{-3} - 2.00 \times 10^{-2} = 0.100 \times 10^{-2} - 2.00 \times 10^{-2}$$
$$= -1.90 \times 10^{-2}$$

Note that $10^{-2} = 0.01$ is a higher power of 10 than $10^{-3} = 0.001$.

When raising a number in scientific notation to a particular power, we raise the decimal part of the number to the new power and *multiply* the power of 10 by the new power:

$$(A \times 10^a)^b = A^b \times 10^{a \times b}$$

For example, 2.88×10^4 raised to the third power is

$$(2.88 \times 10^4)^3 = 2.88^3 \times (10^4)^3$$
$$= 2.88^3 \times 10^{3 \times 4}$$
$$= 23.9 \times 10^{12} = 2.39 \times 10^{13}$$

The rule follows from the fact that

$$(10^4)^3 = 10^4 \times 10^4 \times 10^4 = 10^{4+4+4} = 10^{3 \times 4}$$

APPENDIX 1B: LOGARITHMS

The *common logarithm* of a number x is denoted $\log x$ and is the power to which 10 must be raised to equal x. Thus, the logarithm of 100 is 2 (written $\log 100 = 2$) because $10^2 = 100$. The logarithm of 151 is 2.18 because $10^{2.18} = 151$. The *common antilogarithm* of a number x is the number that has x as its common logarithm: in practice, the common antilogarithm of x is simply another name for 10^x, so the common antilogarithm of 2 is $10^2 = 100$, and that of 2.18 is $10^{2.18} = 151$.

The logarithm of a number greater than 1 is positive, and that of a number smaller than 1 but greater than zero is negative:

$$\text{If } x > 1, \text{ then } \log x > 0$$

$$\text{If } x = 1, \text{ then } \log x = 0$$

$$\text{If } 0 < x < 1, \text{ then } \log x < 0$$

Logarithms are not defined for zero and negative numbers.

On an electronic calculator, the common logarithm of a number x is calculated by entering x and pressing the "$\log x$" key. Likewise, the common antilogarithm of x is found by entering x and pressing the "10^x" key.

The *natural logarithm* of a number x is denoted $\ln x$ and is the power to which the number $e = 2.718 \ldots$ must be raised to equal x. This number e may seem a peculiar choice, but it occurs naturally in a number of mathematical expressions, and its use simplifies many formulas. On an electronic calculator, the natural logarithm of x is calculated by entering x and then pressing the "$\ln x$" key. Thus, $\ln 10 = 2.303$, signifying that $e^{2.303}$ is equal to 10.

Common and natural logarithms are related by the expression

$$\ln x = \ln 10 \times \log x$$

$$= 2.303 \times \log x$$

The natural antilogarithm of x is normally called the *exponential* of e; it is equal to e raised to the power x. On a calculator, it is obtained by entering x and pressing the "e^x" key. Thus, the natural antilogarithm of 2.303 is $e^{2.303} = 10$.

The following relations between logarithms are useful (they are written here only for common logarithms, but they apply to natural logarithms as well):

$$\log x + \log y = \log (x \times y)$$

$$\log x - \log y = \log \frac{x}{y}$$

$$x \log y = \log y^x$$

$$\log \frac{1}{x} = -\log x$$

Logarithms are useful for solving expressions of the form

$$a^x = b$$

for the unknown x. To do so, we take the logarithms of both sides, obtaining

$$\log a^x = \log b$$

and then use the third relation above ($x \log y = \log y^x$) to find that

$$x \log a = \log b$$

Therefore,

$$x = \frac{\log b}{\log a}$$

Logarithms are also useful for evaluating awkward powers of numbers. Suppose, for example, that we want to find the value of $x = 6^{2.3}$. We begin by taking the logarithms of both sides, and then we use the relation $x \log y = \log y^x$ again:

$$\log x = \log 6^{2.3} = 2.3 \log 6 = 2.3 \times 0.778 = 1.79$$

Finally, to find x itself, we take the antilogarithm of 1.79 to find

$$x = 10^{1.79} = 61.6$$

APPENDIX 1C: QUADRATIC EQUATIONS

A *quadratic equation* is an equation of the form

$$ax^2 + bx + c = 0$$

Some quadratic equations can easily be factored into the form

$$(x - x_1)(x - x_2) = 0$$

which immediately shows that the *roots* of the equation, the values of x that satisfy the equation, are the values $x = x_1$ and $x = x_2$. For example, the equation

$$x^2 + 5x + 6 = 0$$

can be written

$$(x + 3)(x + 2) = 0$$

and hence is satisfied by $x = -3$ and $x = -2$.

When the factorization is not obvious, the roots can be found from the expressions

$$x_1 = \frac{-b + \sqrt{b^2 - 4ac}}{2a} \qquad x_2 = \frac{-b - \sqrt{b^2 - 4ac}}{2a}$$

When a quadratic equation arises from a real process, we accept only the root that leads to a physically plausible result. For example, if x is a concentration, then it must be a positive number, and a negative root can be discarded. However, if x is a *change* in concentration, then it may be either positive or negative. In such a case, we would have to determine which root led to an acceptable (positive) final concentration.

APPENDIX 1D: GRAPHS

Experimental data can often be analyzed by plotting a graph. In many cases, the best procedure is to find a way of plotting the data so that a straight line results. This is more useful than plotting a curve, because it is quite easy to tell whether or not the data do in fact fall on a straight line, whereas small deviations from a curve are harder to detect. Moreover, it is also easy to calculate the slope of a straight line; to *extrapolate*, or extend, a straight line beyond the range of the data; and to *interpolate* between the data—that is, find a value between two measured values.

The formula of a straight line graph of y (the vertical axis) plotted against x (the horizontal axis) is

$$y = a + bx$$

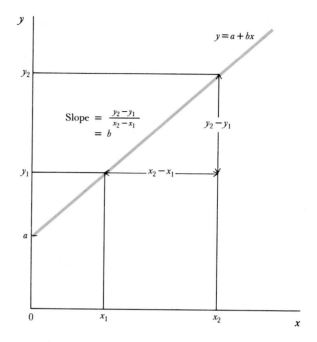

FIGURE A.1 The straight line $y = a + bx$, its intercept at $y = a$, and its slope, b.

Here, a is the *intercept* of the graph (Fig. A.1), the value of y where the graph cuts through the vertical axis at $x = 0$. The *slope* of the graph, its gradient, is b. The slope can be calculated by choosing two points x_1 and x_2 and their corresponding values on the y axis y_1 and y_2 and substituting the values into the formula

$$b = \frac{y_2 - y_1}{x_2 - x_1}$$

Because a is the intercept and b is the slope, the equation of the straight line is equivalent to

$$y = intercept + slope \times x$$

Many of the equations in the text can be rearranged so as to give a straight line when plotted. These include

	y	$= intercept$	$+ slope \times x$
Temperature-scale conversions:	$^\circ C$	$= -273.15 +$	$1 \times K$
	$^\circ F$	$= 32 +$	$\frac{9}{5} \times {}^\circ C$
Ideal gas law	P	$=$	$nRT \times \dfrac{1}{V}$
First-order rate law	$\ln [A] =$	$\ln [A]_0 -$	$k \times t$
Second-order rate law	$\dfrac{1}{[A]} =$	$\dfrac{1}{[A]_0} +$	$k \times t$
Arrhenius's law	$\ln k =$	$A -$	$\dfrac{E_a}{R} \times \dfrac{1}{T}$

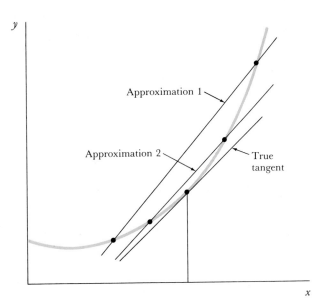

FIGURE A.2 Successive approximations to the true tangent are obtained as the two points defining the straight line come closer together and finally coincide.

We can speak of *the* slope of a straight line because the slope is the same at all points. On a curve, however, the slope changes from point to point. The *tangent* to a curve at a specified point is the straight line that has the same slope as the curve at that point. The tangent can be found with a series of approximations as shown in Fig. A.2. Approximation 1 is found by choosing a point of the curve on each side of the point of interest (corresponding to equal distances along the x axis) and joining them with a straight line. A better approximation (approximation 2) is obtained by moving the chosen points an equal distance closer to the point of interest, and drawing a new line. The "exact" tangent is obtained when the two points are virtually coincident with the point of interest. Its slope is then equal to the slope of the curve at the point. We use this technique in the text to measure the rate of a chemical reaction at a specified time.

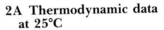

2

EXPERIMENTAL DATA

2A Thermodynamic data at 25°C
1 Inorganic substances
2 Organic compounds

2B Standard reduction potentials at 25°C
1 Potentials in electrochemical order
2 Potentials in alphabetical order

2C Ground-state electron configurations

2D The elements

2E The top fifty chemicals produced in the United States

APPENDIX 2A: THERMODYNAMIC DATA AT 25°C

1 INORGANIC SUBSTANCES

Substance	Atomic or molecular weight, amu	Enthalpy of formation ΔH_f°, kJ/mol	Free energy of formation ΔG_f°, kJ/mol	Entropy* S°, J/(K · mol)
Aluminum				
$Al(s)$	26.98	0	0	28.33
$Al^{3+}(aq)$	26.98	−524.7	−481.2	−321.7
$Al_2O_3(s)$	101.95	−1675.7	−1582.3	50.92
$Al(OH)_3(s)$	78.00	−1276		
$AlCl_3(s)$	133.24	−704.2	−628.8	110.67
Antimony				
$SbH_3(g)$	153.24	145.11	147.75	232.78
$SbCl_3(g)$	228.11	−313.8	−301.2	337.80
$SbCl_5(g)$	299.02	−394.34	−334.29	401.94
Barium				
$Ba(s)$	137.34	0	0	62.8
$Ba^{2+}(aq)$	137.34	−537.64	−560.77	9.6
$BaO(s)$	153.34	−553.5	−525.1	70.43
$BaCO_3(s)$	197.35	−1216.3	−1137.6	112.1
$BaCO_3(aq)$	197.35	−1214.78	−1088.59	−47.3
Boron				
$B(s)$	10.81	0	0	5.86
$B_2O_3(s)$	69.62	−1272.8	−1193.7	53.97
$BF_3(g)$	67.81	−1137.0	−1120.3	254.12
Bromine				
$Br_2(l)$	159.82	0	0	152.23
$Br_2(g)$	159.82	30.91	3.11	245.46
$Br^-(aq)$	79.91	−121.55	−103.96	82.4
$HBr(g)$	90.92	−36.40	−53.45	198.70
Calcium				
$Ca(s)$	40.08	0	0	41.42
$Ca(g)$	40.08	178.2	144.3	154.88
$Ca^{2+}(aq)$	40.08	−542.83	−553.58	−53.1
$CaO(s)$	56.08	−635.09	−604.03	39.75
$Ca(OH)_2(s)$	74.10	−986.09	−898.49	83.39
$Ca(OH)_2(aq)$	74.10	−1002.82	−868.07	−74.5
$CaCO_3(s)$ (calcite)	100.09	−1206.9	−1128.8	92.9

*The entropies of individual ions in solution are determined by setting the entropy of H^+ in water equal to zero and then defining the entropies of all other ions relative to this value; hence a negative entropy is one that is lower than the entropy of protons in water.

Substance	Atomic or molecular weight, amu	Enthalpy of formation ΔH_f°, kJ/mol	Free energy of formation ΔG_f°, kJ/mol	Entropy* S°, J/(K · mol)
$CaCO_3(s)$ (aragonite)	100.09	−1207.1	−1127.8	88.7
$CaCO_3(aq)$	100.09	−1219.97	−1081.39	−110.0
$CaF_2(s)$	78.08	−1219.6	−1167.3	68.87
$CaF_2(aq)$	78.08	−1208.09	−1111.15	−80.8
$CaCl_2(s)$	110.99	−795.8	−748.1	104.6
$CaBr_2(s)$	199.90	−682.8	−663.6	130
$CaC_2(s)$	64.10	−59.8	−64.9	69.96
$CaSO_4(s)$	136.14	−1434.11	−1321.79	106.7
$CaSO_4(aq)$	136.14	−1452.10	−1298.10	−33.1
Carbon†				
$C(s)$ (graphite)	12.011	0	0	5.740
$C(s)$ (diamond)	12.011	1.895	2.900	2.377
$CO(g)$	28.01	−110.53	−137.17	197.67
$CO_2(g)$	44.01	−393.51	−394.36	213.74
$CCl_4(l)$	153.82	−135.44	−65.21	216.40
$CS_2(l)$	76.14	89.70	65.27	151.34
$HCN(g)$	27.03	135.1	124.7	201.78
$HCN(l)$	27.03	108.87	124.97	112.84
Cerium				
$Ce(s)$	140.12	0	0	72.0
$Ce^{3+}(aq)$	140.12	−696.2	−672.0	−205
$Ce^{4+}(aq)$	140.12	−537.2	−503.8	−301
Chlorine				
$Cl_2(g)$	70.91	0	0	223.07
$Cl^-(g)$	35.45	−167.16	−131.23	56.5
$HCl(g)$	36.46	−92.31	−95.30	186.91
$HCl(aq)$	36.46	−167.16	−131.23	56.5
Copper				
$Cu(s)$	63.54	0	0	33.15
$Cu^+(aq)$	63.54	71.67	49.98	40.6
$Cu^{2+}(aq)$	63.54	64.77	65.49	−99.6
$Cu_2O(s)$	143.08	−168.6	−146.0	93.14

†For organic compounds, see the next table.

(continued)

Substance	Atomic or molecular weight, amu	Enthalpy of formation ΔH_f°, kJ/mol	Free energy of formation ΔG_f°, kJ/mol	Entropy* S°, J/(K · mol)
Copper (continued)				
$CuO(s)$	79.54	−157.3	−129.7	42.63
$CuSO_4(s)$	159.60	−771.36	−661.8	109
$CuSO_4 \cdot 5H_2O(s)$	249.68	−2279.7	−1879.7	300.4
Deuterium				
$D_2(g)$	4.028	0	0	144.96
$D_2O(g)$	20.028	−249.20	−234.54	198.34
$D_2O(l)$	20.028	−294.60	−243.44	75.94
Fluorine				
$F_2(g)$	38.00	0	0	202.78
$F^-(aq)$	19.00	−332.63	−278.79	−13.8
$HF(g)$	20.01	−271.1	−273.2	173.78
$HF(aq)$	20.01	−332.63	−278.79	−13.8
Hydrogen				
$H_2(g)$	2.016	0	0	130.68
$H^+(aq)$	19.03	0	0	0
$H_2O(l)$	18.02	−285.83	−237.13	69.91
$H_2O(g)$	18.02	−241.82	−228.57	188.83
$H_2O_2(l)$	34.02	−187.78	−120.35	109.6
Iodine				
$I_2(s)$	253.81	0	0	116.14
$I_2(g)$	253.81	62.44	19.33	260.69
$I^-(aq)$	126.90	−55.19	−51.57	111.3
$HI(g)$	127.91	26.48	1.70	206.59
Iron				
$Fe(s)$	55.85	0	0	27.28
$Fe^{2+}(aq)$	55.85	−89.1	−78.90	−137.7
$Fe^{3+}(aq)$	55.85	−48.5	−4.7	−315.9
$Fe_3O_4(s)$ (magnetite)	231.54	−1118.4	−1015.4	146.4
$Fe_2O_3(s)$ (hematite)	159.69	−824.2	−742.2	87.40
$FeS(s, \alpha)$	87.91	−100.0	−100.4	60.29
$FeS(aq)$	87.91		6.9	
$FeS_2(s)$	119.98	−178.2	−166.9	52.93
Lead				
$Pb(s)$	207.19	0	0	64.81
$PbBr_2(s)$	367.01	−278.7	−261.92	161.5
$PbBr_2(aq)$	367.01	−244.8	−232.34	175.3

Substance	Atomic or molecular weight, amu	Enthalpy of formation ΔH_f°, kJ/mol	Free energy of formation ΔG_f°, kJ/mol	Entropy* S°, J/(K · mol)
Magnesium				
Mg(*s*)	24.31	0	0	32.68
Mg^{2+}(*aq*)	24.31	−466.85	−454.8	−138.1
MgO(*s*)	40.31	−601.70	−569.43	26.94
MgCO$_3$(*s*)	84.32	−1095.8	−1012.1	65.7
Mercury				
Hg(*l*)	200.59	0	0	76.02
Hg(*g*)	200.59	61.32	31.82	174.96
HgO(*s*)	216.59	−90.83	−58.54	70.29
Hg$_2$Cl$_2$(*s*)	472.09	−265.22	−210.75	192.5
Nitrogen				
N$_2$(*g*)	28.01	0	0	191.61
NO(*g*)	30.01	90.25	86.55	210.76
N$_2$O(*g*)	44.01	82.05	104.20	219.85
NO$_2$(*g*)	46.01	33.18	51.31	240.06
N$_2$O$_4$(*g*)	92.01	9.16	97.89	304.29
HNO$_3$(*l*)	63.01	−174.10	−80.71	155.60
HNO$_3$(*aq*)	63.01	−207.36	−111.25	146.4
NO$_3^-$(*aq*)	62.01	−205.0	−108.74	146.4
NH$_3$(*g*)	17.03	−46.11	−16.45	192.45
NH$_3$(*aq*)	17.03	−80.29	−26.50	111.3
NH$_4^+$(*aq*)	18.04	−132.51	−79.31	113.4
NH$_2$OH(*s*)	33.03	−114.2		
HN$_3$(*g*)	43.03	294.1	328.1	238.97
N$_2$H$_4$(*l*)	32.05	50.63	149.34	121.21
NH$_4$NO$_3$(*s*)	80.04	−365.56	−183.87	151.08
NH$_4$Cl(*s*)	53.49	−314.43	−202.87	94.6
Oxygen				
O$_2$(*g*)	32.00	0	0	205.14
O$_3$(*g*)	48.00	142.7	163.2	238.93
OH$^-$(*aq*)	17.01	−229.99	−157.24	−10.75
Phosphorus				
P(*s*) (white)	30.97	0	0	41.09
P$_4$(*g*)	123.90	58.91	24.44	279.98
PH$_3$(*g*)	34.00	5.4	13.4	210.23
PCl$_3$(*l*)	137.33	−319.7	−272.3	217.18
PCl$_3$(*g*)	137.33	−287.0	−267.8	311.78

(continued)

Substance	Atomic or molecular weight, amu	Enthalpy of formation ΔH_f°, kJ/mol	Free energy of formation ΔG_f°, kJ/mol	Entropy* S°, J/(K · mol)
Phosphorus (*continued*)				
$PCl_5(g)$	208.24	−374.9	−305.0	364.6
$PCl_5(s)$	208.24	−443.5		
$H_3PO_3(aq)$	82.00	−964.8		
$H_3PO_4(l)$	98.00	−1266.9		
$H_3PO_4(aq)$	98.00	−1277.4	−1018.7	
Potassium				
$K(s)$	39.10	0	0	64.18
$K^+(aq)$	39.10	−252.38	−283.27	102.5
$KOH(s)$	56.11	−424.76	−379.08	78.9
$KOH(aq)$	56.11	−482.37	−440.50	−91.6
$KF(s)$	58.10	−567.27	−537.75	66.57
$KCl(s)$	74.56	−436.75	−409.14	82.59
$KBr(s)$	119.01	−393.80	−380.66	95.90
$KI(s)$	166.01	−327.90	−324.89	106.32
$KClO_3(s)$	122.55	−397.73	−296.25	143.1
$KClO_4(s)$	138.55	−432.75	−303.09	151.0
$K_2S(s)$	110.27	−380.7	−364.0	105
$K_2S(aq)$	110.27	−471.5	−480.7	190.4
Silicon				
$Si(s)$	28.09	0	0	18.83
$SiO_2(s, \alpha)$	60.09	−910.94	−856.64	41.84
Silver				
$Ag(s)$	107.87	0	0	42.55
$Ag^+(aq)$	107.87	105.58	77.11	72.68
$AgBr(s)$	187.78	−100.37	−96.90	107.1
$AgBr(aq)$	187.78	−15.98	−26.86	155.2
$AgCl(s)$	143.32	−127.07	−109.79	96.2
$AgCl(aq)$	143.32	−61.58	−54.12	129.3
$AgI(s)$	234.77	−61.84	−66.19	115.5
$AgI(aq)$	234.77	50.38	25.52	184.1
$Ag_2O(s)$	231.74	−31.05	−11.20	121.3
$AgNO_3(s)$	169.88	−124.39	−33.41	140.92
Sodium				
$Na(s)$	22.99	0	0	51.21
$Na^+(aq)$	22.99	−240.12	−261.91	59.0
$NaOH(s)$	40.00	−425.61	−379.49	64.46
$NaOH(aq)$	40.00	−470.11	−419.15	48.1
$NaCl(s)$	58.44	−411.15	−384.14	72.13

Substance	Atomic or molecular weight, amu	Enthalpy of formation ΔH_f°, kJ/mol	Free energy of formation ΔG_f°, kJ/mol	Entropy* S°, J/(K · mol)
NaBr(s)	102.90	−361.06	−348.98	86.82
NaI(s)	149.89	−287.78	−286.06	98.53
Sulfur				
S(s) (rhombic)	32.06	0	0	31.80
S(s) (monoclinic)	32.06	0.33	0.1	32.6
SO$_2$(g)	64.06	−296.83	−300.19	248.22
SO$_3$(g)	80.06	−395.72	−371.06	256.76
H$_2$SO$_4$(l)	98.08	−813.99	−690.00	156.90
H$_2$SO$_4$(aq)	98.08	−909.27	−744.53	20.1
H$_2$S(g)	34.08	−20.63	−33.56	205.79
H$_2$S(aq)	34.08	−39.7	−27.83	121
SF$_6$(g)	146.05	−1209	−1105.3	291.82
Tin				
Sn(s) (white)	118.69	0	0	51.55
Sn(s) (gray)	118.69	−2.09	0.13	44.14
SnO(s)	134.69	−285.8	−256.9	56.5
SnO$_2$(s)	150.69	−580.7	−519.6	52.3
Zinc				
Zn(s)	65.37	0	0	41.63
Zn^{2+}(aq)	65.37	−153.89	−147.06	−112.1
ZnO(s)	81.37	−348.28	−318.30	43.64

2 ORGANIC COMPOUNDS

Substance	Molecular weight, amu	Enthalpy of combustion ΔH_c°, kJ/mol	Enthalpy of formation ΔH_f°, kJ/mol	Free energy of formation ΔG_f°, kJ/mol	Entropy S°, J/(K · mol)
Hydrocarbons					
CH$_4$(g) (methane)	16.04	−890	−74.81	−50.72	186.26
C$_2$H$_2$(g) (acetylene)	26.04	−1300	226.73	209.20	200.94

(continued)

Substance	Molecular weight, amu	Enthalpy of combustion ΔH_c°, kJ/mol	Enthalpy of formation ΔH_f°, kJ/mol	Free energy of formation ΔG_f°, kJ/mol	Entropy S°, J/(K · mol)
Hydrocarbons (continued)					
$C_2H_4(g)$ (ethylene)	28.05	−1411	52.26	68.15	219.56
$C_2H_6(g)$ (ethane)	30.07	−1560	−84.68	−32.82	229.60
$C_3H_6(g)$ (propylene)	42.00	−2058	20.42	62.78	266.6
$C_3H_6(g)$ (cyclopropane)	42.08	−2091	53.30	104.45	237.4
$C_3H_8(g)$ (propane)	44.01	−2220	−103.85	−23.49	270.2
$C_4H_{10}(g)$ (butane)	58.13	−2878	−126.14	−17.03	310.1
$C_5H_{12}(g)$ (pentane)	72.15	−3537	−146.44	−8.20	349
$C_6H_6(l)$ (benzene)	78.12	−3268	49.0	124.3	173.3
$C_7H_8(l)$ (toluene)	92.15	−3910	12.0	113.8	221.0
$C_6H_{12}(l)$ (cyclohexane)	84.16	−3920	−156.4	26.7	204.4
$C_8H_{18}(l)$ (octane)	114.23	−5471	−249.9	6.4	358
Alcohols and phenols					
$CH_3OH(l)$ (methanol)	32.04	−726	−238.66	−166.27	126.8
$CH_3OH(g)$	32.04	−764	−200.66	−161.96	239.81
$C_2H_5OH(l)$ (ethanol)	46.07	−1368	−277.69	−174.78	160.7
$C_2H_5OH(g)$	46.07	−1409	−235.10	−168.49	282.70
$C_6H_5OH(s)$	94.11	−3054	−164.6	−50.42	144.0
Carboxylic acids					
$HCOOH(l)$ (formic)	46.03	−255	−424.72	−361.35	128.95
$CH_3COOH(l)$ (acetic)	60.05	−875	−484.5	−389.9	159.8
$CH_3COOH(aq)$	60.05		−485.76	−396.46	86.6
$(COOH)_2(s)$ (oxalic)	90.04	−254	−827.2	−697.9	120
$C_6H_5COOH(s)$ (benzoic)	122.13	−3227	−385.1	−245.3	167.6
Aldehydes and ketones					
$HCHO(g)$ (formaldehyde)	30.03	−571	−108.57	−102.53	218.77

Substance	Molecular weight, amu	Enthalpy of combustion ΔH_c°, kJ/mol	Enthalpy of formation ΔH_f°, kJ/mol	Free energy of formation ΔG_f°, kJ/mol	Entropy S°, J/(K · mol)
$CH_3CHO(l)$ (acetaldehyde)	44.05	−1166	−192.30	−128.12	160.2
$CH_3CHO(g)$	44.05	−1192	−166.19	−128.86	250.3
$CH_3COCH_3(l)$ (acetone)	58.08	−1790	−248.1	−155.4	200
Sugars					
$C_6H_{12}O_6(s)$ (glucose)	180.16	−2808	−1268	−910	212
$C_6H_{12}O_6(s)$ (fructose)	180.16	−2810	−1266		
$C_{12}H_{22}O_{11}(s)$ (sucrose)	342.30	−5645	−2222	−1545	360
Nitrogen compounds					
$CO(NH_2)_2(s)$ (urea)	60.06	−632	−333.51	−197.33	104.60
$C_6H_5NH_2(l)$ (aniline)	93.13	−3393	31.6	149.1	191.3
$NH_2CH_2COOH(s)$ (glycine)	75.07	−969	−532.9	−373.4	103.51
$CH_3NH_2(g)$ (methylamine)	31.06	−1085	−22.97	32.16	243.41

APPENDIX 2B: STANDARD REDUCTION POTENTIALS AT 25°C

1 POTENTIALS IN ELECTROCHEMICAL ORDER

Reduction half-reaction	E°, V
Strongly oxidizing	
$H_4XeO_6 + 2H^+ + 2e^- \longrightarrow XeO_3 + 3H_2O$	+3.0
$F_2 + 2e^- \longrightarrow 2F^-$	+2.87
$O_3 + 2H^+ + 2e^- \longrightarrow O_2 + H_2O$	+2.07
$S_2O_8^{2-} + 2e^- \longrightarrow 2SO_4^{2-}$	+2.05
$Ag^{2+} + e^- \longrightarrow Ag^+$	+1.98
$Co^{3+} + e^- \longrightarrow Co^{2+}$	+1.81
$H_2O_2 + 2H^+ + 2e^- \longrightarrow 2H_2O$	+1.78
$Au^+ + e^- \longrightarrow Au$	+1.69

(continued)

Reduction half-reaction	$E°$, V
$Pb^{4+} + 2e^- \longrightarrow Pb^{2+}$	$+1.67$
$2HClO + 2H^+ + 2e^- \longrightarrow Cl_2 + 2H_2O$	$+1.63$
$Ce^{4+} + e^- \longrightarrow Ce^{3+}$	$+1.61$
$2HBrO + 2H^+ + 2e^- \longrightarrow Br_2 + 2H_2O$	$+1.60$
$MnO_4^- + 8H^+ + 5e^- \longrightarrow Mn^{2+} + 4H_2O$	$+1.51$
$Mn^{3+} + e^- \longrightarrow Mn^{2+}$	$+1.51$
$Au^{3+} + 3e^- \longrightarrow Au$	$+1.40$
$Cl_2 + 2e^- \longrightarrow 2Cl^-$	$+1.36$
$Cr_2O_7^{2-} + 14H^+ + 6e^- \longrightarrow 2Cr^{3+} + 7H_2O$	$+1.33$
$O_3 + H_2O + 2e^- \longrightarrow O_2 + 2OH^-$	$+1.24$
$O_2 + 4H^+ + 4e^- \longrightarrow 2H_2O$	$+1.23$
$ClO_4^- + 2H^+ + 2e^- \longrightarrow ClO_3^- + H_2O$	$+1.23$
$Pt^{2+} + 2e^- \longrightarrow Pt$	$+1.20$
$Br_2 + 2e^- \longrightarrow 2Br^-$	$+1.09$
$Pu^{4+} + e^- \longrightarrow Pu^{3+}$	$+0.97$
$NO_3^- + 4H^+ + 3e^- \longrightarrow NO + 2H_2O$	$+0.96$
$2Hg^{2+} + 2e^- \longrightarrow Hg_2^{2+}$	$+0.92$
$ClO^- + H_2O + 2e^- \longrightarrow Cl^- + 2OH^-$	$+0.89$
$NO_3^- + 2H^+ + e^- \longrightarrow NO_2 + H_2O$	$+0.80$
$Ag^+ + e^- \longrightarrow Ag$	$+0.80$
$Hg_2^{2+} + 2e^- \longrightarrow 2Hg$	$+0.79$
$AgF + e^- \longrightarrow Ag + F^-$	$+0.78$
$Fe^{3+} + e^- \longrightarrow Fe^{2+}$	$+0.77$
$BrO^- + H_2O + 2e^- \longrightarrow Br^- + 2OH^-$	$+0.76$
$MnO_4^{2-} + 2H_2O + 2e^- \longrightarrow MnO_2 + 4OH^-$	$+0.60$
$MnO_4^- + e^- \longrightarrow MnO_4^{2-}$	$+0.56$
$I_2 + 2e^- \longrightarrow 2I^-$	$+0.54$
$Cu^+ + e^- \longrightarrow Cu$	$+0.52$
$I_3^- + 2e^- \longrightarrow 3I^-$	$+0.53$
$NiO(OH) + H_2O + e^- \longrightarrow Ni(OH)_2 + OH^-$	$+0.49$
$O_2 + 2H_2O + 4e^- \longrightarrow 4OH^-$	$+0.40$
$ClO_4^- + H_2O + 2e^- \longrightarrow ClO_3^- + 2OH^-$	$+0.36$
$Cu^{2+} + 2e^- \longrightarrow Cu$	$+0.34$
$Hg_2Cl_2 + 2e^- \longrightarrow 2Hg + 2Cl^-$	$+0.27$
$AgCl + e^- \longrightarrow Ag + Cl^-$	$+0.22$
$Bi^{3+} + 3e^- \longrightarrow Bi$	$+0.20$
$SO_4^{2-} + 4H^+ + 2e^- \longrightarrow H_2SO_3 + H_2O$	$+0.17$
$Cu^{2+} + e^- \longrightarrow Cu^+$	$+0.15$
$Sn^{4+} + 2e^- \longrightarrow Sn^{2+}$	$+0.15$

Reduction half-reaction	$E°$, V
$AgBr + e^- \longrightarrow Ag + Br^-$	$+0.07$
$NO_3^- + H_2O + 2e^- \longrightarrow NO_2^- + 2OH^-$	$+0.01$
$Ti^{4+} + e^- \longrightarrow Ti^{3+}$	0.00
$2H^+ + 2e^- \longrightarrow H_2$	0 (by definition)
$Fe^{3+} + 3e^- \longrightarrow Fe$	-0.04
$O_2 + H_2O + 2e^- \longrightarrow HO_2^- + OH^-$	-0.08
$Pb^{2+} + 2e^- \longrightarrow Pb$	-0.13
$In^+ + e^- \longrightarrow In$	-0.14
$Sn^{2+} + 2e^- \longrightarrow Sn$	-0.14
$AgI + e^- \longrightarrow Ag + I^-$	-0.15
$Ni^{2+} + 2e^- \longrightarrow Ni$	-0.23
$V^{3+} + e^- \longrightarrow V^{2+}$	-0.26
$Co^{2+} + 2e^- \longrightarrow Co$	-0.28
$In^{3+} + 3e^- \longrightarrow In$	-0.34
$PbSO_4 + 2e^- \longrightarrow Pb + SO_4^{2-}$	-0.36
$Ti^{3+} + e^- \longrightarrow Ti^{2+}$	-0.37
$In^{2+} + e^- \longrightarrow In^+$	-0.40
$Cr^{3+} + e^- \longrightarrow Cr^{2+}$	-0.41
$Fe^{2+} + 2e^- \longrightarrow Fe$	-0.44
$In^{3+} + 2e^- \longrightarrow In^+$	-0.44
$S + 2e^- \longrightarrow S^{2-}$	-0.48
$In^{3+} + e^- \longrightarrow In^{2+}$	-0.49
$O_2 + e^- \longrightarrow O_2^-$	-0.56
$U^{4+} + e^- \longrightarrow U^{3+}$	-0.61
$Cr^{3+} + 3e^- \longrightarrow Cr$	-0.74
$Zn^{2+} + 2e^- \longrightarrow Zn$	-0.76
$Cd(OH)_2 + 2e^- \longrightarrow Cd + 2OH^-$	-0.81
$2H_2O + 2e^- \longrightarrow H_2 + 2OH^-$	-0.83
$Cr^{2+} + 2e^- \longrightarrow Cr$	-0.91
$Mn^{2+} + 2e^- \longrightarrow Mn$	-1.18
$V^{2+} + 2e^- \longrightarrow V$	-1.19
$Ti^{2+} + 2e^- \longrightarrow Ti$	-1.63
$Al^{3+} + 3e^- \longrightarrow Al$	-1.66
$U^{3+} + 3e^- \longrightarrow U$	-1.79
$Be^{2+} + 2e^- \longrightarrow Be$	-1.85
$Mg^{2+} + 2e^- \longrightarrow Mg$	-2.36
$Ce^{3+} + 3e^- \longrightarrow Ce$	-2.48
$La^{3+} + 3e^- \longrightarrow La$	-2.52

(continued)

1 POTENTIALS IN ELECTROCHEMICAL ORDER *(continued)*

Reduction half-reaction	$E°$, V
$Na^+ + e^- \longrightarrow Na$	-2.71
$Ca^{2+} + 2e^- \longrightarrow Ca$	-2.87
$Sr^{2+} + 2e^- \longrightarrow Sr$	-2.89
$Ba^{2+} + 2e^- \longrightarrow Ba$	-2.91
$Ra^{2+} + 2e^- \longrightarrow Ra$	-2.92
$Cs^+ + e^- \longrightarrow Cs$	-2.92
$Rb^+ + e^- \longrightarrow Rb$	-2.93
$K^+ + e^- \longrightarrow K$	-2.93
$Li^+ + e^- \longrightarrow Li$	-3.05

Strongly
reducing

2 POTENTIALS IN ALPHABETICAL ORDER

Reduction half-reaction	$E°$, V
$Ag^+ + e^- \longrightarrow Ag$	$+0.80$
$Ag^{2+} + e^- \longrightarrow Ag^+$	$+1.98$
$AgBr + e^- \longrightarrow Ag + Br^-$	$+0.07$
$AgCl + e^- \longrightarrow Ag + Cl^-$	$+0.22$
$AgF + e^- \longrightarrow Ag + F^-$	$+0.78$
$AgI + e^- \longrightarrow Ag + I^-$	-0.15
$Al^{3+} + 3e^- \longrightarrow Al$	-1.66
$Au^+ + e^- \longrightarrow Au$	$+1.69$
$Au^{3+} + 3e^- \longrightarrow Au$	$+1.40$
$Ba^{2+} + 2e^- \longrightarrow Ba$	-2.91
$Be^{2+} + 2e^- \longrightarrow Be$	-1.85
$Bi^{3+} + 3e^- \longrightarrow Bi$	$+0.20$
$Br_2 + 2e^- \longrightarrow 2Br^-$	$+1.09$
$BrO^- + H_2O + 2e^- \longrightarrow Br^- + 2OH^-$	$+0.76$
$Ca^{2+} + 2e^- \longrightarrow Ca$	-2.87
$Cd(OH)_2 + 2e^- \longrightarrow Cd + 2OH^-$	-0.81
$Cd^{2+} + 2e^- \longrightarrow Cd$	-0.40
$Ce^{3+} + 3e^- \longrightarrow Ce$	-2.48
$Ce^{4+} + e^- \longrightarrow Ce^{3+}$	$+1.61$

Reduction half-reaction	$E°$, V
$Cl_2 + 2e^- \longrightarrow 2Cl^-$	$+1.36$
$ClO^- + H_2O + 2e^- \longrightarrow Cl^- + 2OH^-$	$+0.89$
$ClO_4^- + 2H^+ + 2e^- \longrightarrow ClO_3^- + H_2O$	$+1.23$
$ClO_4^- + H_2O + 2e^- \longrightarrow ClO_3^- + 2OH^-$	$+0.36$
$Co^{2+} + 2e^- \longrightarrow Co$	-0.28
$Co^{3+} + e^- \longrightarrow Co^{2+}$	$+1.81$
$Cr^{2+} + 2e^- \longrightarrow Cr$	-0.91
$Cr_2O_7^{2-} + 14H^+ + 6e^- \longrightarrow 2Cr^{3+} + 7H_2O$	$+1.33$
$Cr^{3+} + 3e^- \longrightarrow Cr$	-0.74
$Cr^{3+} + e^- \longrightarrow Cr^{2+}$	-0.41
$Cs^+ + e^- \longrightarrow Cs$	-2.92
$Cu^+ + e^- \longrightarrow Cu$	$+0.52$
$Cu^{2+} + 2e^- \longrightarrow Cu$	$+0.34$
$Cu^{2+} + e^- \longrightarrow Cu^+$	$+0.15$
$F_2 + 2e^- \longrightarrow 2F^-$	$+2.87$
$Fe^{2+} + 2e^- \longrightarrow Fe$	-0.44
$Fe^{3+} + 3e^- \longrightarrow Fe$	-0.04
$Fe^{3+} + e^- \longrightarrow Fe^{2+}$	$+0.77$
$2H^+ + 2e^- \longrightarrow H_2$	0 (by definition)
$2H_2O + 2e^- \longrightarrow H_2 + 2OH^-$	-0.83
$2HBrO + 2H^+ + 2e^- \longrightarrow Br_2 + 2H_2O$	$+1.60$
$2HClO + 2H^+ + 2e^- \longrightarrow Cl_2 + 2H_2O$	$+1.63$
$H_2O_2 + 2H^+ + 2e^- \longrightarrow 2H_2O$	$+1.78$
$H_4XeO_6 + 2H^+ + 2e^- \longrightarrow XeO_3 + 3H_2O$	$+3.0$
$Hg_2^{2+} + 2e^- \longrightarrow 2Hg$	$+0.79$
$Hg_2Cl_2 + 2e^- \longrightarrow 2Hg + 2Cl^-$	$+0.27$
$2Hg^{2+} + 2e^- \longrightarrow Hg_2^{2+}$	$+0.92$
$I_2 + 2e^- \longrightarrow 2I^-$	$+0.54$
$I_3^- + 2e^- \longrightarrow 3I^-$	$+0.53$
$In^+ + e^- \longrightarrow In$	-0.14
$In^{2+} + e^- \longrightarrow In^+$	-0.40
$In^{3+} + 2e^- \longrightarrow In^+$	-0.44
$In^{3+} + 3e^- \longrightarrow In$	-0.34
$In^{3+} + e^- \longrightarrow In^{2+}$	-0.49
$K^+ + e^- \longrightarrow K$	-2.93
$La^{3+} + 3e^- \longrightarrow La$	-2.52
$Li^+ + e^- \longrightarrow Li$	-3.05

(continued)

Reduction half-reaction	$E°$, V
$Mg^{2+} + 2e^- \longrightarrow Mg$	-2.36
$Mn^{2+} + 2e^- \longrightarrow Mn$	-1.18
$Mn^{3+} + e^- \longrightarrow Mn^{2+}$	$+1.51$
$MnO_4^- + 8H^+ + 5e^- \longrightarrow Mn^{2+} + 4H_2O$	$+1.51$
$MnO_4^- + e^- \longrightarrow MnO_4^{2-}$	$+0.56$
$MnO_4^{2-} + 2H_2O + 2e^- \longrightarrow MnO_2 + 4OH^-$	$+0.60$
$Na^+ + e^- \longrightarrow Na$	-2.71
$Ni^{2+} + 2e^- \longrightarrow Ni$	-0.23
$NiO(OH) + H_2O + e^- \longrightarrow Ni(OH)_2 + OH^-$	$+0.49$
$NO_3^- + 2H^+ + e^- \longrightarrow NO_2 + H_2O$	$+0.80$
$NO_3^- + 4H^+ + 3e^- \longrightarrow NO + 2H_2O$	$+0.96$
$NO_3^- + H_2O + 2e^- \longrightarrow NO_2^- + 2OH^-$	$+0.01$
$O_2 + 2H_2O + 4e^- \longrightarrow 4OH^-$	$+0.40$
$O_2 + 4H^+ + 4e^- \longrightarrow 2H_2O$	$+1.23$
$O_2 + e^- \longrightarrow O_2^-$	-0.56
$O_2 + H_2O + 2e^- \longrightarrow HO_2^- + OH^-$	-0.08
$O_3 + 2H^+ + 2e^- \longrightarrow O_2 + H_2O$	$+2.07$
$O_3 + H_2O + 2e^- \longrightarrow O_2 + 2OH^-$	$+1.24$
$Pb^{2+} + 2e^- \longrightarrow Pb$	-0.13
$Pb^{4+} + 2e^- \longrightarrow Pb^{2+}$	$+1.67$
$PbSO_4 + 2e^- \longrightarrow Pb + SO_4^{2-}$	-0.36
$Pt^{2+} + 2e^- \longrightarrow Pt$	$+1.20$
$Pu^{4+} + e^- \longrightarrow Pu^{3+}$	$+0.97$
$Ra^{2+} + 2e^- \longrightarrow Ra$	-2.92
$Rb^+ + e^- \longrightarrow Rb$	-2.93
$S + 2e^- \longrightarrow S^{2-}$	-0.48
$SO_4^{2-} + 4H^+ + 2e^- \longrightarrow H_2SO_3 + H_2O$	$+0.17$
$S_2O_8^{2-} + 2e^- \longrightarrow 2SO_4^{2-}$	$+2.05$
$Sn^{2+} + 2e^- \longrightarrow Sn$	-0.14
$Sn^{4+} + 2e^- \longrightarrow Sn^{2+}$	$+0.15$
$Sr^{2+} + 2e^- \longrightarrow Sr$	-2.89
$Ti^{2+} + 2e^- \longrightarrow Ti$	-1.63
$Ti^{3+} + e^- \longrightarrow Ti^{2+}$	-0.37
$Ti^{4+} + e^- \longrightarrow Ti^{3+}$	0.00
$U^{3+} + 3e^- \longrightarrow U$	-1.79
$U^{4+} + e^- \longrightarrow U^{3+}$	-0.61
$V^{2+} + 2e^- \longrightarrow V$	-1.19
$V^{3+} + e^- \longrightarrow V^{2+}$	-0.26
$Zn^{2+} + 2e^- \longrightarrow Zn$	-0.76

APPENDIX 2C: GROUND-STATE ELECTRON CONFIGURATIONS

Z	Symbol	Configuration	Z	Symbol	Configuration	Z	Symbol	Configuration
1	H	$1s^1$	36	Kr	$[Ar]3d^{10}4s^24p^6$	71	Lu	$[Xe]4f^{14}5d^16s^2$
2	He	$1s^2$	37	Rb	$[Kr]5s^1$	72	Hf	$[Xe]4f^{14}5d^26s^2$
3	Li	$[He]2s^1$	38	Sr	$[Kr]5s^2$	73	Ta	$[Xe]4f^{14}5d^36s^2$
4	Be	$[He]2s^2$	39	Y	$[Kr]4d^15s^2$	74	W	$[Xe]4f^{14}5d^46s^2$
5	B	$[He]2s^22p^1$	40	Zr	$[Kr]4d^25s^2$	75	Re	$[Xe]4f^{14}5d^56s^2$
6	C	$[He]2s^22p^2$	41	Nb	$[Kr]4d^45s^1$	76	Os	$[Xe]4f^{14}5d^66s^2$
7	N	$[He]2s^22p^3$	42	Mo	$[Kr]4d^55s^1$	77	Ir	$[Xe]4f^{14}5d^76s^2$
8	O	$[He]2s^22p^4$	43	Tc	$[Kr]4d^55s^2$	78	Pt	$[Xe]4f^{14}5d^96s^1$
9	F	$[He]2s^22p^5$	44	Ru	$[Kr]4d^75s^1$	79	Au	$[Xe]4f^{14}5d^{10}6s^1$
10	Ne	$[He]2s^22p^6$	45	Rh	$[Kr]4d^85s^1$	80	Hg	$[Xe]4f^{14}5d^{10}6s^2$
11	Na	$[Ne]3s^1$	46	Pd	$[Kr]4d^{10}$	81	Tl	$[Xe]4f^{14}5d^{10}6p^16s^2$
12	Mg	$[Ne]3s^2$	47	Ag	$[Kr]4d^{10}5s^1$	82	Pb	$[Xe]4f^{14}5d^{10}6p^26s^2$
13	Al	$[Ne]3s^23p^1$	48	Cd	$[Kr]4d^{10}5s^2$	83	Bi	$[Xe]4f^{14}5d^{10}6p^36s^2$
14	Si	$[Ne]3s^23p^2$	49	In	$[Kr]4d^{10}5s^25p^1$	84	Po	$[Xe]4f^{14}5d^{10}6p^46s^2$
15	P	$[Ne]3s^23p^3$	50	Sn	$[Kr]4d^{10}5s^25p^2$	85	At	$[Xe]4f^{14}5d^{10}6p^56s^2$
16	S	$[Ne]3s^23p^4$	51	Sb	$[Kr]4d^{10}5s^25p^3$	86	Rn	$[Xe]4f^{14}5d^{10}6p^66s^2$
17	Cl	$[Ne]3s^23p^5$	52	Te	$[Kr]4d^{10}5s^25p^4$	87	Fr	$[Rn]7s^1$
18	Ar	$[Ne]3s^23p^6$	53	I	$[Kr]4d^{10}5s^25p^5$	88	Ra	$[Rn]7s^2$
19	K	$[Ar]4s^1$	54	Xe	$[Kr]4d^{10}5s^25p^6$	89	Ac	$[Rn]6d^17s^2$
20	Ca	$[Ar]4s^2$	55	Cs	$[Xe]6s^1$	90	Th	$[Rn]6d^27s^2$
21	Sc	$[Ar]3d^14s^2$	56	Ba	$[Xe]6s^2$	91	Pa	$[Rn]5f^26d^17s^2$
22	Ti	$[Ar]3d^24s^2$	57	La	$[Xe]5d^16s^2$	92	U	$[Rn]5f^36d^17s^2$
23	V	$[Ar]3d^34s^2$	58	Ce	$[Xe]4f^15d^16s^2$	93	Np	$[Rn]5f^46d^17s^2$
24	Cr	$[Ar]3d^54s^1$	59	Pr	$[Xe]4f^36s^2$	94	Pu	$[Rn]5f^67s^2$
25	Mn	$[Ar]3d^54s^2$	60	Nd	$[Xe]4f^46s^2$	95	Am	$[Rn]5f^77s^2$
26	Fe	$[Ar]3d^64s^2$	61	Pm	$[Xe]4f^56s^2$	96	Cm	$[Rn]5f^76d^17s^2$
27	Co	$[Ar]3d^74s^2$	62	Sm	$[Xe]4f^66s^2$	97	Bk	$[Rn]5f^97s^2$
28	Ni	$[Ar]3d^84s^2$	63	Eu	$[Xe]4f^76s^2$	98	Cf	$[Rn]5f^{10}7s^2$
29	Cu	$[Ar]3d^{10}4s^1$	64	Gd	$[Xe]4f^75d^16s^2$	99	Es	$[Rn]5f^{11}7s^2$
30	Zn	$[Ar]3d^{10}4s^2$	65	Tb	$[Xe]4f^96s^2$	100	Fm	$[Rn]5f^{12}7s^2$
31	Ga	$[Ar]3d^{10}4s^24p^1$	66	Dy	$[Xe]4f^{10}6s^2$	101	Md	$[Rn]5f^{13}7s^2$
32	Ge	$[Ar]3d^{10}4s^24p^2$	67	Ho	$[Xe]4f^{11}6s^2$	102	No	$[Rn]5f^{14}7s^2$
33	As	$[Ar]3d^{10}4s^24p^3$	68	Er	$[Xe]4f^{12}6s^2$	103	Lr	$[Rn]5f^{14}6d^17s^2$
34	Se	$[Ar]3d^{10}4s^24p^4$	69	Tm	$[Xe]4f^{13}6s^2$	104	Rf	$[Rn]5f^{14}6d^27s^2$
35	Br	$[Ar]3d^{10}4s^24p^5$	70	Yb	$[Xe]4f^{14}6s^2$	105	Ha	$[Rn]5f^{14}6d^37s^2$

APPENDIX 2D: THE ELEMENTS

Element	Symbol	Atomic number	Atomic weight, amu	Normal state*	Density, g/cm³	Melting point, °C	Boiling point, °C
Actinium (Greek *aktis*, ray)	Ac	89	227.03	s, m	10.06	1230	3200
Aluminum [from alum, salts of the form $KAl(SO_4)_2 \cdot 12H_2O$]	Al	13	26.98	s, m	2.70	660	2350
Americium (The Americas)	Am	95	241.06	s, m	13.67	990	2600
Antimony (probably a corruption of an old Arabic word; Latin *stibium*)	Sb	51	121.75	s, m	6.69	631	1640
Argon (Greek *argos*, inactive)	Ar	18	39.95	g	1.66†	−189	−186
Arsenic (Greek *arsenikos*, male)	As	33	74.92	s, nm	5.78	613s‡	
Astatine (Greek *astatos*, unstable)	At	85	210.	s, nm		300	350
Barium (Greek *barys*, heavy)	Ba	56	137.34	s, m	3.59	710	1640
Berkelium (Berkeley, California)	Bk	97	249.08	s, m	14.79	986	
Beryllium (from the mineral beryl, $Be_3Al_2SiO_{18}$)	Be	4	9.01	s, m	1.85	1285	2470
Bismuth (German *weisse Masse*, white mass)	Bi	83	208.98	s, m	8.90	271	1650
Boron [Arabic *buraq*, borax, $Na_2B_4O_9H_4 \cdot 8H_2O$; *bor*(ax + carb)*on*]	B	5	10.81	s, nm	2.47	2030	3700
Bromine (Greek *bromos*, stench)	Br	35	79.91	l, nm	3.12	−7	59
Cadmium (Greek *Cadmus*, founder of Thebes)	Cd	48	112.40	s, m	8.65	321	770
Calcium (Latin *calx*, lime)	Ca	20	40.08	s, m	1.53	840	1490
Californium (California)	Cf	98	251.08	s, m			
Carbon (Latin *carbo*, coal or charcoal)	C	6	12.01	s, nm	2.27	3700s	
Cerium (the asteroid Ceres, discovered two days earlier)	Ce	58	140.12	s, m	6.71	800	3000
Cesium (Latin *caesius*, sky blue)	Cs	55	132.91	s, m	1.90	29	686

*The normal state is the state of the element at normal temperature and pressure (20°C and 1 atm): *s* denotes solid, *l* liquid, and *g* gas; m denotes metal, and nm nonmetal.
†The density quoted is for the liquid.
‡s means that the solid sublimes.

Element	Symbol	Atomic number	Atomic weight, amu	Normal state*	Density, g/cm³	Melting point, °C	Boiling point, °C
Chlorine (Greek *chloros*, yellowish green)	Cl	17	35.45	*g*	2.03†	−101	−34
Chromium (Greek *chroma*, color)	Cr	24	52.00	*s*, m	7.19	1860	2600
Cobalt (German *Kobald*, evil spirit; Greek *kobalos*, goblin)	Co	27	58.93	*s*, m	8.80	1494	2900
Copper (Latin *cuprum*, from Cyprus)	Cu	29	63.54	*s*, m	8.93	1085	2580
Curium (Pierre and Marie Curie)	Cm	96	247.07	*s*, m	13.30	1340	
Dysprosium (Greek *dysprositos*, hard to get at)	Dy	66	162.50	*s*, m	8.53	1410	2600
Einsteinium (Albert Einstein)	Es	99	254.09	*s*, m			
Erbium (Ytterby, a town in Sweden)	Er	68	167.26	*s*, m	9.04	1520	2600
Europium (Europe)	Eu	63	151.96	*s*, m	5.25	820	1450
Fermium (Enrico Fermi, an Italian physicist)	Fm	100	257.10	*s*, m			
Fluorine (Latin *fluere*, to flow)	F	9	19.00	*g*	1.14†	−220	−188
Francium (France)	Fr	87	223	*s*, m		30	650
Gadolinium (Johann Gadolin, a Finnish chemist)	Gd	64	157.25	*s*, m	7.87	1310	3000
Gallium (Latin *Gallia*, France; also a pun on the discoverer's forename, Le Coq)	Ga	31	69.72	*s*, m	5.91	30	2070
Germanium (Latin *Germania*, Germany)	Ge	32	72.59	*s*, m	5.32	959	2850
Gold (Anglo-Saxon *gold;* Latin *aurum*, gold)	Au	79	196.97	*s*, m	19.28	1064	2850
Hafnium (Latin *Hafnia*, Copenhagen)	Hf	72	178.49	*s*, m	13.28	2230	5300
Helium (Greek *helios*, the sun)	He	2	4.00	*g*	0.12†		−269
Holmium (Latin *Holmia*, Stockholm)	Ho	67	164.93	*s*, m	8.80	1470	2300

(*continued*)

Element	Symbol	Atomic number	Atomic weight, amu	Normal state*	Density, g/cm³	Melting point, °C	Boiling point, °C
Hydrogen (Greek *hydro* + *genes*, water-forming)	H	1	1.0080	*g*	0.089†	−259	−253
Indium (from the bright indigo line in its spectrum)	In	49	114.82	*s*, m	7.29	157	2050
Iodine (Greek *iodos*, violet)	I	53	126.90	*s*, nm	4.95	114	184
Iridium (Greek and Latin *iris*, rainbow)	Ir	77	192.2	*s*, m	22.55	2447	4550
Iron (Anglo-Saxon *iron;* Latin *ferrum*)	Fe	26	55.85	*s*, m	7.87	1540	2760
Krypton (Greek *kryptos*, hidden)	Kr	36	83.80	*g*	3.00†	−157	−153
Lanthanum (Greek *lanthanein*, to lie hidden)	La	57	138.91	*s*, m	6.17	920	3450
Lawrencium (Ernest Lawrence, an American physicist)	Lr	103	257	*s*, m			
Lead (Anglo-Saxon *lead;* Latin *plumbum*)	Pb	82	207.19	*s*, m	11.34	328	1760
Lithium (Greek *lithos*, stone)	Li	3	6.94	*s*, m	0.53	180	1360
Lutetium (*Lutetia*, ancient name of Paris)	Lu	71	174.97	*s*, m	9.84	1700	3400
Magnesium (Magnesia, a district in Thessaly, Greece)	Mg	12	24.31	*s*, m	1.74	650	1100
Manganese (Greek and Latin *magnes*, magnet)	Mn	25	54.94	*s*, m	7.47	1250	2120
Mendelevium (Dmitri Mendeleev)	Md	101	258.10	*s*, m			
Mercury (the planet Mercury; Latin *hydrargyrum*, liquid silver)	Hg	80	200.59	*l*, m	13.55	−39	357
Molybdenum (Greek *molybdos*, lead)	Mo	42	95.94	*s*, m	10.22	2620	4830
Neodymium (Green *neos* + *didymos*, new twin)	Nd	60	144.24	*s*, m	7.00	1024	3100
Neon (Greek *neos*, new)	Ne	10	20.18	*g*	1.44†	−249	−246
Neptunium (the planet Neptune)	Np	93	237.05	*s*, m	20.45	640	
Nickel (German *Nickel*, Satan)	Ni	28	58.71	*s*, m	8.91	1455	2150
Niobium (Niobe, daughter of Tantalus; see tantalum)	Nb	41	92.91	*s*, m	8.58	2425	5000
Nitrogen (Greek *nitron* + *genes*, soda-forming)	N	7	14.01	*g*	1.04†	−210	−196
Nobelium (Alfred Nobel, the founder of the Nobel Prizes)	No	102	255	*s*, m			

Element	Symbol	Atomic number	Atomic weight, amu	Normal state*	Density, g/cm³	Melting point, °C	Boiling point, °C
Osmium (Greek *osme*, a smell)	Os	76	190.2	*s*, m	22.58	3030	5000
Oxygen (Greek *oxys* + *genes*, acid-forming)	O	8	16.00	*g*	1.46†	−219	−183
Palladium (the asteroid Pallas, discovered at about the same time)	Pd	46	106.4	*s*, m	12.00	1554	3000
Phosphorus (Greek *phosphoros*, light-bearing)	P	15	30.97	*s*, nm	1.82	44	280
Platinum (Spanish *plata*, silver)	Pt	78	195.09	*s*, m	21.45	1772	3720
Plutonium (the planet Pluto)	Pu	94	239.05	*s*, m	19.81	640	3200
Polonium (Poland)	Po	84	210	*s*, m	9.40	254	960
Potassium (from potash; Latin *kalium* and Arabic *qali*, alkali)	K	19	39.10	*s*, m	0.86	63	777
Praseodymium (Greek *prasios* + *didymos*, green twin)	Pr	59	140.91	*s*, m	6.78	935	3000
Promethium (Prometheus, the Greek god)	Pm	61	146.92	*s*, m	7.22	1168	3300
Protactinium (Greek *protos* + *aktis*, first ray)	Pa	91	231.04	*s*, m	15.37	1200	4000
Radium (Latin *radius*, ray)	Ra	88	226.03	*s*, m	5.00	700	1500
Radon (from radium)	Rn	86	222	*g*	4.40†	−71	−62
Rhenium (Latin *Rhenus*, Rhine)	Re	75	186.2	*s*, m	21.02	3180	5600
Rhodium (Greek *rhodon*, rose; its aqueous solutions are often rose-colored)	Rh	45	102.91	*s*, m	12.42	1963	3700
Rubidium (Latin *rubidus*, deep red, "flushed")	Rb	37	85.47	*s*, m	1.53	39	705
Ruthenium (Latin *Ruthenia*, Russia)	Ru	44	101.07	*s*, m	12.36	2310	4100
Samarium (from samarskite, a mineral)	Sm	62	150.35	*s*, m	7.54	1060	1600
Scandium (Latin *Scandia*, Scandinavia)	Sc	21	44.96	*s*, m	2.99	1540	2800
Selenium (Greek *selene*, the moon)	Se	34	78.96	*s*, nm	4.81	220	685
Silicon (Latin *silex*, flint)	Si	14	28.09	*s*, nm	2.33	1410	2620
Silver (Anglo-Saxon *seolfor*; Latin *argentum*)	Ag	47	107.87	*s*, m	10.50	962	2160
Sodium (English soda; Latin *natrium*)	Na	11	22.99	*s*, m	0.97	98	900

(continued)

Element	Symbol	Atomic number	Atomic weight, amu	Normal state*	Density, g/cm³	Melting point, °C	Boiling point, °C
Strontium (Strontian, Scotland)	Sr	38	87.62	s, m	2.58	770	1380
Sulfur (Sanskrit *sulvere*)	S	16	32.06	s, nm	2.09	115	445
Tantalum (Tantalus, Greek mythological figure)	Ta	73	180.95	s, m	16.67	3000	5400
Technetium (Greek *technetos*, artificial)	Tc	43	98.91	s, m	11.50	2200	4600
Tellurium (Latin *tellus*, earth)	Te	52	127.60	s, nm	6.25	450	990
Terbium (Ytterby, a town in Sweden rich in minerals)	Tb	65	158.92	s, m	8.27	1360	2500
Thallium (Greek *thallos*, a green shoot)	Tl	81	204.37	s, m	11.87	304	1460
Thorium (Thor, Scandinavian god of war)	Th	90	232.04	s, m	11.73	1700	4500
Thulium (*Thule*, early name for Scandinavia)	Tm	69	168.93	s, m	9.33	1550	2000
Tin (Anglo-Saxon *tin*; Latin *stannum*)	Sn	50	118.69	s, m	7.29	232	2720
Titanium (Titans, Greek mythological figures, sons of the earth)	Ti	22	47.90	s, m	4.51	1670	3300
Tungsten (Swedish *tung + sten*, heavy stone)	W	74	183.85	s, m	19.25	3387	5420
Uranium (the planet Uranus)	U	92	238.03	s, m	19.05	1135	4000
Vanadium (Vanadis, Scandinavian mythological figure)	V	23	50.94	s, m	6.09	1920	3400
Xenon (Greek *xenos*, stranger)	Xe	54	131.30	g	3.56†	−112	−108
Ytterbium (Ytterby, a town in Sweden)	Yb	70	173.04	s, m	6.97	824	1500
Yttrium (Ytterby, a town in Sweden)	Y	39	88.91	s, m	4.48	1510	3300
Zinc (Anglo-Saxon *zinc*)	Zn	30	65.37	s, m	7.14	420	913
Zirconium (Arabic *zargun*, gold color)	Zr	40	91.22	s, m	6.51	1850	4400

APPENDIX 2E: THE TOP FIFTY CHEMICALS PRODUCED IN THE UNITED STATES

Rank	Name	Annual production, 10^9 kg	Comment on source*
1	Sulfuric acid	35.15	Contact process
2	Nitrogen	21.46	Fractional distillation
3	Ethylene	15.87	Thermal cracking
4	Ammonia	14.66	Haber process
5	Oxygen	14.64	Fractional distillation
6	Lime	13.77	Heating of limestone
7	Sodium hydroxide	10.45	Brine electrolysis
8	Chlorine	9.95	Electrolysis
9	Phosphoric acid	9.50	From phosphate rocks
10	Propylene	8.41	Thermal cracking
11	Sodium carbonate	8.07	Solvay process and mining
12	Urea	6.74	Ammonia + carbon dioxide
13	Nitric acid	6.44	Modern version of the Ostwald process
14	Dichloroethylene	6.26	Chlorination of ethylene
15	Ammonium nitrate	5.82	Ammonia + nitric acid
16	Benzene	5.29	Catalytic reforming
17	Carbon dioxide	4.46	Steam reforming of hydrocarbons
18	Ethylbenzene	4.27	Friedel-Crafts alkylation of benzene
19	Vinyl chloride	3.73	Dehydrochlorination of 1,2-dichloroethane
20	Terephthalic acid	3.68	Oxidation of *para*-xylene
21	Styrene	3.67	Dehydrogenation of ethylbenzene
22	Methanol	3.31	Hydrogenation of carbon monoxide (using synthesis gas)
23	Toluene	3.05	Catalytic reforming of naphtha†

(continued)

Production data are compiled annually by the American Chemical Society and published in *Chemical and Engineering News*. This table is based on 1987 data as listed in the April 1988 issue. Water, sodium chloride, and steel are traditionally not included but would outrank the rest if they were. Hydrogen is used heavily but almost always "on site," as soon as it is produced.

*In <u>cracking</u>, petroleum fractions of higher molecular weight than gasoline are converted to smaller molecules with a higher number of double bonds, as in

$$CH_3(CH_2)_9CH_3 \xrightarrow{\text{500°C, catalyst}} CH_3(CH_2)_6CH{=}CH_2 + CH_3CH_3$$

In <u>reforming</u>, the number of carbon atoms in the feedstock is left unchanged, but a greater number of double bonds and aromatic rings is formed, as in

†*Naphtha* is a petroleum fraction consisting of a mixture of C_4 through C_{10} aliphatic and cycloaliphatic hydrocarbons together with some aromatic compounds.

Rank	Name	Annual production, 10^9 kg	Comment on source*
24	Formaldehyde	2.76	Oxidation or dehydrogenation of methanol
25	Xylenes	2.59	Catalytic reforming of naphtha
26	Ethylene oxide	2.55	Addition of O_2 to ethylene
27	*para*-Xylene	2.34	Catalytic reforming of naphtha
28	Hydrochloric acid	2.26	By-product of hydrocarbon chlorination
29	Ethylene glycol	2.05	Hydration of ethylene oxide
30	Ammonium sulfate	1.98	Ammonia + sulfuric acid
31	Cumene (isopropyl benzene)	1.88	Friedel-Crafts alkylation
32	Methyl *tert*-butyl ether	1.53	Addition of methanol to 2-methylpropene
33‡	Phenol	1.47	Oxidation of cumene
33‡	Acetic acid	1.47	Monsanto process (reaction of CO with methanol)
35‡	Potash§	1.36	Mined (KCl); electrolysis (KOH); reaction of KCl with $NaNO_3$ (KNO_3)
35‡	Butadiene	1.36	Dehydrogenation of butane
37	Carbon black	1.23	Partial combustion of hydrocarbons
38	Propylene oxide	1.18	Oxidation of propylene
39	Acrylonitrile	1.16	Reaction of propylene with ammonia and oxygen
40	Vinyl acetate	1.14	Reaction of ethylene and acetic acid
41	Aluminum sulfate	1.11	Alumina + sulfuric acid
42	Cyclohexane	1.02	Hydrogenation of benzene
43	Acetone	0.94	Dehydrogenation of isopropyl alcohol
44‡	Titanium dioxide	0.86	Purification of rutile
44‡	Sodium silicate	0.86	Sodium carbonate and sand
46	Sodium sulfate	0.73	Mined, by-product of rayon manufacture
47	Adipic acid	0.71	Oxidation of cyclohexane
48	Calcium chloride	0.70	By-product of Solvay process
49	Isopropyl alcohol	0.59	Hydration of propylene
50	Caprolactam	0.53	From cyclohexane

‡Tied with one other chemical.
§Potash refers collectively to K_2CO_3, KOH, K_2SO_4, KCl, and KNO_3; its production is expressed in terms of the equivalent mass of K_2O.

GLOSSARY

absolute zero of temperature ($T = 0$) The lowest possible temperature ($-273.15°C$).

abundance (of an isotope) The percentage (in terms of numbers of atoms) of the isotope present in a sample of the element. The *natural abundance* of an isotope is its abundance in a naturally occurring sample.

accuracy Freedom from systematic (nonrandom) error; accurate measurements give a result close to the true value of the measured property.

acid See *Arrhenius acid, Brønsted acid,* and *Lewis acid.* Used alone, "acid" normally means Brønsted acid.

acidic oxide An oxide that reacts with water to give an acid; the oxides of nonmetallic elements generally form acidic oxides. *Examples:* CO_2, SO_3

acid ionization constant K_a The equilibrium constant for the ionization of a Brønsted acid; a measure of the strength of the acid. *Example:*

$$HF(aq) + H_2O(l) \rightleftharpoons H_3O^+(aq) + F^-(aq)$$

$$K_a = \frac{[H_3O^+][F^-]}{[HF]}$$

activated complex A combination of two reactant molecules that can either go on to form products or fall apart into the unchanged reactants.

activation energy E_a The minimum energy needed for reaction; the height of the activation barrier.

addition polymerization The polymerization, usually of alkenes, by an addition reaction.

addition reaction A reaction in which atoms or groups bond to the carbon atoms joined by a multiple bond. *Example*:

adhesive forces Forces that bind a substance to a surface.

adsorb To bind a substance to a surface; the surface *adsorbs* the substance.

alcohol An organic molecule containing an —OH group attached to a carbon atom that is not part of a carbonyl group or an aromatic ring. Alcohols are classified as primary, secondary, or tertiary according to the number of carbon atoms attached to the C—OH carbon atom. *Examples:* CH_3CH_2OH (primary), $(CH_3)_2CHOH$ (secondary), $(CH_3)_3COH$ (tertiary)

aldehyde An organic compound containing the —CHO group. *Examples:* CH_3CHO (acetaldehyde), C_6H_5CHO (benzaldehyde)

aliphatic compound An organic compound that does not include benzene rings in its structure.

alkali An aqueous solution of a base. *Examples:* aqueous NaOH, aqueous ammonia

alkali metals The elements of Group I of the periodic table (the lithium family).

alkaline earth metals The elements of Group II of the periodic table (the beryllium family).

alkanes A homologous series of hydrocarbons with formulas derived from CH_4 by inserting —CH_2— groups and thus having molecular formulas C_nH_{2n+2}. *Examples:* CH_4, CH_3CH_3, $CH_3(CH_2)_6CH_3$

alkenes A homologous series of unsaturated hydrocarbons with formulas derived from $CH_2{=}CH_2$ by inserting —CH_2— groups and thus having molecular formulas C_nH_{2n}. *Examples*: $CH_2{=}CH_2$, $CH_3CH{=}CH_2$, $CH_3CH{=}CHCH_2CH_3$

alkynes A homologous series of unsaturated hydrocarbons with formulas derived from $HC{\equiv}CH$ by inserting —CH_2— groups and thus having molecular formulas C_nH_{2n-2}. *Examples*: $CH{\equiv}CH$, $CH_3C{\equiv}CCH_3$

allotropes Alternative forms of an element that differ in the way the atoms are linked. *Examples*: O_2 and O_3; white and gray tin

alloy A solid homogeneous mixture of metals.

alpha (α) particles Helium nuclei ($^4_2He^{2+}$), the components of α rays.

amide A combination of an amine and a carboxylic acid and containing the group —CO—NR_2. *Example*: CH_3CONH_2 (acetamide)

amines Compounds derived from ammonia by replacing various numbers of H atoms with hydrocarbon groups. The number of hydrogen atoms replaced determines the classification of the amine as primary, secondary, or tertiary. *Examples*: $CH_3CH_2NH_2$ (primary), $(CH_3)_2NH$ (secondary), $(CH_3)_3N$ (tertiary); $(CH_3)_4N^+$ (quaternary ammonium ion)

amino acid A carboxylic acid that also contains an amino group. The *essential* amino acids are those that must be ingested as a part of the diet. *Example*: NH_2CH_2COOH (glycine, nonessential)

amorphous solid A solid in which the atoms, ions, or molecules lie together randomly.

amphiprotic substance A substance that can both donate and accept protons. *Examples*: H_2O, CH_3COOH

amphoteric substance A substance that reacts with both an acid and a base. *Examples*: Aluminum metal, Al_2O_3

analysis Determination of the composition of a substance.

analyte The solution of unknown concentration in a titration.

anhydride A compound that forms an oxoacid when it reacts with water. A *formal anhydride* is a compound that has the formula of an acid minus the elements of water but does not react with water to produce the acid. *Examples*: SO_3 (the anhydride of sulfuric acid), CO (the formal anhydride of formic acid, HCOOH)

anhydrous Lacking water. *Example*: $CuSO_4$ is anhydrous white copper(II) sulfate (in contrast to the hydrated blue solid $CuSO_4 \cdot 5H_2O$)

anion A negatively charged ion. *Examples*: F^-, SO_4^{2-}

anode The electrode at which oxidation occurs.

antibonding orbital (σ^* or π^*) A molecular orbital that, if occupied, raises the energy of a molecule and so destabilizes it.

aqueous solution A solution in which the solvent is water.

arene An aromatic hydrocarbon.

aromatic compound An organic compound that includes benzene rings in its structure. *Examples*: C_6H_6 (benzene), C_6H_5OH (phenol), $C_{10}H_8$ (naphthalene)

Arrhenius acid A compound that contains hydrogen and releases hydrogen ions in water. *Examples*: HCl, CH_3COOH (but not CH_4)

Arrhenius base A compound that produces hydroxide ions in water. *Examples*: NaOH, NH_3

Arrhenius behavior (for a reaction) The value of ln k (where k is the rate constant) is proportional to $1/T$.

Arrhenius parameters The preexponential factor A and the activation energy E_a, the parameters in the equation $k = Ae^{-E_a/RT}$ for the rate constant k.

atactic polymer A polymer in which substituents are attached to each side of the chain at random.

atom The smallest particle of an element that has the *chemical* properties characteristic of that element; a nucleus and the surrounding electrons.

atomic mass unit amu (sometimes u) Exactly $\frac{1}{12}$ the mass of one atom of carbon-12.

atomic nucleus The small, positively charged particle at the center of an atom that is responsible for most of its mass.

atomic number Z The number of protons in the nucleus of an atom of an element.

atomic orbital (s, p, d, f) A region of space in which there is a high probability of finding an electron in an atom (more specifically, a mathematical expression that gives the probability of finding an electron at each point in an atom).

atomic structure The arrangement of electrons around the central nucleus of an atom.

atomic weight The *average* mass of the atoms in a naturally occurring sample of an element.

***Aufbau* principle** See *building-up principle*.

autoionization A reaction in which a conjugate acid and a conjugate base are formed from the same substance; the equilibrium constant for the autoionization of water is the *water autoionization constant K_w*. *Example*: $H_2O(l) + H_2O(l) \rightleftharpoons H_3O^+(aq) + OH^-(aq)$

$$K_w = [H_3O^+][OH^-]$$

Avogadro's law The volume of a sample of gas at a given temperature and pressure is proportional to the number of moles of molecules in the sample.

Avogadro's number N_A The number of objects in one mole of objects ($N_A = 6.022 \times 10^{23}$).

axial bond A bond that is perpendicular to the molecular plane in a bipyramidal molecule.

back titration A titration procedure in which a measured excess of a reagent is added to an analyte, and the amount of that reagent remaining after reaction is measured.

Balmer series A series of spectral lines (some of which lie in the visible region) emitted by hydrogen atoms.

band gap The range of energies for which no orbitals exist in a solid and which therefore separates two bands of energy levels.

base See *Arrhenius base, Brønsted base*, and *Lewis base*. Used alone, "base" normally means Brønsted base.

base ionization constant K_b The equilibrium constant for the ionization of a Brønsted base; a measure of the strength of the base. *Example*:

$$NH_3(aq) + H_2O(l) \rightleftharpoons NH_4^+(aq) + OH^-(aq)$$
$$K_b = \frac{[NH_4^+][OH^-]}{[NH_3]}$$

base units The units of measurement in the International System (SI) in terms of which all other units are defined. *Examples*: kilogram for mass, meter for length, second for time, kelvin for temperature, ampere for electric current

basic oxide An oxide that is a Brønsted base. The oxides of metallic elements are generally basic. *Examples*: Na_2O, MgO

Bayer process The process for purifying alumina (Al_2O_3) for use in the Hall process.

beta (β) particles Electrons emitted by nuclei in radioactive decay.

bimolecular reaction An elementary reaction in which two molecules (or free atoms) come together to form a product. *Example*: $O(g) + O_3(g) \longrightarrow 2O_2(g)$

binary Consisting of two components, as in a binary mixture or binary compound. *Examples:* Acetone and water; HCl; Al_2O_3; C_6H_6

block (s, p, d, or f) A region of the periodic table containing elements for which, according to the building-up principle, the corresponding subshell is currently being occupied.

Bohr frequency condition The relation between the frequency of radiation emitted or absorbed by an atom or molecule and the resulting change in energy: $\Delta E E = h\nu$.

Bohr model (of the hydrogen atom) A model in which the electrons travel in a set of discrete orbits.

boiling point The temperature at which a liquid is in equilibrium with its vapor at the pressure of the surroundings; vaporization then occurs throughout the liquid, not only at the liquid's surface. The *normal boiling point T_b* is the boiling point under 1 atm pressure.

Boltzmann formula (for entropy) The formula $S = k \ln W$, where k is Boltzmann's constant and W is the number of atomic arrangements that correspond to the same energy.

bond A link between atoms (see *ionic bond* and *covalent bond*); a *single bond* consists of one shared electron pair, a *double bond* of two shared electron pairs, and a *triple bond* of three shared electron pairs.

bond angle In an X—Y—Z molecule or part of a molecule, the angle between the X—Y and Y—Z directions.

bond enthalpy $B(X—Y)$ The enthalpy change for the breaking of a bond. *Example*:

$$H_2(g) \longrightarrow 2H(g) \qquad B(H—H) = +436 \text{ kJ/mol}$$

bond length   The distance between the nuclei of two atoms joined by a bond.

bonding orbital   A molecular orbital which, if occupied, lowers the energy of a molecule and so stabilizes it.

bond order (BO)   The net number of bonds in a molecule, allowing for the cancellation of bonds by antibonds: $BO = \frac{1}{2} \times$ (number of electrons in bonding molecular orbitals − number of electrons in antibonding molecular orbitals)

bond parameters   The characteristics of a bond. *Examples*: bond length, bond enthalpy

Born-Haber cycle   A closed series of reactions used to express the enthalpy of formation of an ionic solid in terms of contributions that include the lattice enthalpy.

boundary surface   The surface showing the region of an orbital within which there is about 90% probability of finding an electron.

Boyle's law   At constant temperature, the volume of a fixed mass of gas is inversely proportional to the pressure.

branching (in a chain reaction)   A propagation step in a chain reaction in which more than one chain carrier is formed. *Example*: $\cdot O \cdot + H_2 \longrightarrow 2HO \cdot$

Brønsted acid   A proton donor (a source of hydrogen ions). *Examples*: HCl, CH_3COOH, HCO_3^-, NH_4^+

Brønsted base   A proton acceptor (a substance to which hydrogen ions can bond). *Examples*: OH^-, F^-, $CH_3CO_2^-$, HCO_3^-, NH_3

Brønsted equilibrium   The equilibrium $acid_1 + base_2 \rightleftharpoons base_1 + acid_2$. *Example*: $CH_3COOH(aq) + H_2O(l) \rightleftharpoons CH_3CO_2^-(aq) + H_3O^+(aq)$

buffer   A solution that resists changes in pH when small amounts of acid or base are added. *Example*: $CH_3COOH(aq)$ plus $CH_3CO_2^-(aq)$

buffer capacity   The amount of acid or base that may be added to a buffer before it loses its ability to resist changes in pH.

building-up principle   The procedure for arriving at the ground-state electron configurations of atoms and molecules.

bulk properties   Properties that depend on the collective behavior of large numbers of atoms. *Examples*: melting point, vapor pressure, internal energy

calibration   Interpretation of an observation by comparison with known information.

capillary action   The rising of a liquid up a narrow tube.

carbocation (carbonium ion)   A cation containing a carbon atom with only six valence electrons. *Example*: $CH_3CH_2CH_2^+$

carbohydrate   A compound of general formula $C_m(H_2O)_n$, although small deviations from this general formula are often encountered. *Examples*: $C_6H_{12}O_6$ (glucose), $C_{12}H_{22}O_{11}$ (sucrose)

carboxylic acid   An organic compound containing the carboxyl group (—COOH). *Examples*: CH_3COOH (acetic acid), C_6H_5COOH (benzoic acid)

catalyst   A substance that increases the rate of a reaction without being consumed in the reaction. A catalyst is *homogeneous* if it is present in the same phase as the reactants, and *heterogeneous* if it is in a different phase from the reactants. *Examples*: homogeneous—$Br_2(aq)$ in the decomposition of $H_2O_2(aq)$; heterogeneous—Pt in the Ostwald process

catenate   To form chains or rings of atoms, as in O_3 and S_8.

cathode   The electrode at which reduction occurs.

cation   A positively charged ion. *Examples*: Na^+, NH_4^+, Al^{3+}

cell diagram   A portrayal of the arrangement of electrodes in an electrochemical cell; a cell diagram is written with the cathode on the right and the anode on the left. *Example*: $Zn(s)|Zn^{2+}(aq) \,\|\, Cu^{2+}(aq)|Cu(s)$

cell potential E The potential difference between the electrodes of an electrochemical cell when it is producing no current; the cell potential is always positive.

chain reaction   A reaction in which an intermediate, the *chain carrier*, reacts to produce another intermediate. *Examples*:

$$Br \cdot + H_2 \longrightarrow HBr + H \cdot$$
$$H \cdot + Br_2 \longrightarrow HBr + Br \cdot$$

chalcogens   The elements of Group VI of the periodic table (the oxygen family).

change of state   The change from one physical state of a substance to another state of the same substance. *Example*: melting (solid $\longrightarrow$ liquid)

charge   A measure of the strength with which a particle can interact electrostatically with another particle.

Charles's law   The volume of a fixed mass of gas at constant pressure is proportional to its absolute temperature.

chelate   A complex containing at least one polydentate ligand forming a ring of atoms that includes the central metal atom. *Example*: $[Co(en)_3]^{3+}$, where en is ethylenediamine ($NH_2CH_2CH_2NH_2$)

chemical change   The conversion of one substance to another; a chemical reaction.

chemical kinetics   The study of rates of reactions and their dependence on the reaction conditions.

chemiluminescence   The emission of light by products that are in energetically excited states when they are formed by mixing reagents.

chemistry   The science of matter and the changes it can undergo.

chiral molecule   A molecule that is distinct from its own mirror image. *Examples*: $CH_3CH(NH_2)COOH$, $CHBrClF$

chromatography The separation of substances from each other by use of a surface that adsorbs them with different strengths.

cis-trans isomerization Conversion from one geometrical isomer into another. *Example*: *cis*-butene $\longrightarrow$ *trans*-butene

classical mechanics The laws of motion proposed by Isaac Newton, in which particles travel in definite paths in response to forces.

clathrate compound A solid compound in which a molecule of one component sits in a cage made up of molecules of a second component, typically water. *Example*: SO_2 in a water cage

Claus process A process for obtaining sulfur from the H_2S from oil wells by the oxidation of H_2S with SO_2.

closed shell (or subshell) A shell (or subshell) containing the maximum number of electrons allowed by the exclusion principle. *Example*: the neonlike core $1s^22s^22p^6$

close-packed structure A crystal structure in which atoms or ions occupy the smallest total volume with the least empty space. *Examples*: hexagonal close packing (ABAB . . .) and cubic close packing (ABCABC . . .) of identical spheres

cohesive forces Forces that bind the molecules of a substance together to form a bulk material and that are responsible for condensation.

coinage metals Copper, silver, and gold.

colligative property A property that depends only on the number of solute particles present in a solution and not on their chemical composition. *Examples*: elevation of boiling point, depression of freezing point, osmosis

collision theory The theory of elementary gas-phase bimolecular reactions in which it is assumed that molecules react only if they collide with at least enough kinetic energy for bonds to be broken.

colloidal dispersion A mixture in which tiny particles containing a few thousand atoms are dispersed through a liquid. *Example*: milk

combustion The reaction that occurs when a substance burns in oxygen. *Example*: $CH_4(g) + 2O_2(g) \longrightarrow CO_2(g) + 2H_2O(g)$

combustion analysis Determination of the composition of a sample by measuring the masses of the products of its combustion.

common-ion effect Reduction of the solubility of a salt by the presence of another salt with one ion in common. *Example*: the lower solubility of AgCl in NaCl(*aq*) than in pure water

common name An everyday name for a compound that may give little or no clue to the compound's composition. *Examples*: water, terephthalic acid

complementary color The color white light becomes when one of the colors present in it is removed.

complex The combination of a Lewis acid and a Lewis base; more specifically, a unit consisting of several ligands bonded to a single central metal ion. *Examples*: H_3N-BF_3, $[Fe(H_2O)_6]^{3+}$, $[PtCl_4]^-$

compound A substance made up of atoms of different elements bonded together in a definite ratio.

concentration See *molar concentration*.

condensation The formation of a liquid from a gas.

condensation polymer A polymer formed by a series of condensation reactions. *Examples*: polyesters, polyamides (nylon)

condensation reaction A reaction in which two molecules combine to form a larger one, and a small molecule is eliminated. *Example*: $CH_3COOH + C_2H_5OH \longrightarrow CH_3COOC_2H_5 + H_2O$

conduction band An incompletely filled band of orbitals in a metal.

configuration See *electron configuration*.

conformations Molecular shapes that can be interchanged by rotation about bonds without bond breakage and re-formation.

congeners Members of the same group in the periodic table. *Example*: calcium and beryllium

conjugate acid The Brønsted acid formed when a Brønsted base has accepted a proton. *Example*: NH_4^+ is the conjugate acid of NH_3.

conjugate base The Brønsted base formed when a Brønsted acid has donated a proton. *Example*: NH_3 is the conjugate base of NH_4^+.

contact process The production of sulfuric acid by the combustion of sulfur and the catalyzed oxidation of sulfur dioxide to sulfur trioxide.

conversion factor A factor that is used to convert measurements from one unit to another.

coordinate To use a lone pair to form a covalent bond. *Example*: $F_3B + :NH_3 \longrightarrow F_3B-NH_3$

coordination compound A neutral complex or a compound in which at least one of the ions is a complex. *Examples*: $Ni(CO)_4$, $K_2[Fe(CN)_6]$, chlorophyll

coordination number The number of nearest neighbors of an atom in a solid. In an ionic solid, the coordination number of an ion is the number of nearest neighbors of opposite charge.

coordination sphere The ligands directly attached to the central ion in a complex.

copolymer A polymer formed from a mixture of monomers.

core The inner closed shells of an atom.

couple The reduced and oxidized substances in a half-reaction; specified as oxidized form/reduced form. *Example*: Fe^{3+}/Fe^{2+} in $Fe^{3+} + e^- \longrightarrow Fe^{2+}$

covalent bond X—Y A pair of electrons shared between two atoms.

covalent radius The contribution of an atom to the length of a covalent bond.

cracking The process of converting petroleum fractions into smaller molecules with more double bonds. *Example*: $CH_3(CH_2)_6CH_3 \longrightarrow CH_3(CH_2)_3CH_3 + CH_3CH=CH_2$

critical mass The mass below which so many neutrons escape from a sample of nuclear fuel that a fission chain reaction is not sustained; a greater mass is *supercritical*, and a smaller mass is *subcritical*.

critical temperature The temperature above which a substance cannot exist as a liquid.

cryogenics The study of matter at very low temperatures.

cryoscopy The determination of the molecular weight of a substance by measuring the freezing-point depression it causes when dissolved in a solvent.

crystal field The electrostatic influence of the ligands (modeled as point negative charges) on the central ion of a complex.

crystalline solid (or crystal) A solid in which the atoms, ions, or molecules lie in an orderly array. *Examples*: NaCl, diamond, graphite

cycloalkane A saturated aliphatic hydrocarbon in which the carbon atoms form a ring. *Example*: C_6H_{12}(cyclohexane)

Dalton's law of partial pressures The total pressure of a mixture of gases is the sum of the pressures that each one would exert if it occupied the container alone.

de Broglie relation The proposition that every particle has wavelike properties and a wavelength λ that is related to its mass m and speed v by $\lambda = h/mv$.

decay constant k The rate constant for radioactive decay.

decomposition Breakdown into simpler substances; *thermal decomposition* is decomposition brought about by heat. *Example*:

$$CaCO_3(s) \xrightarrow{\Delta} CaO(s) + CO_2(g)$$

decomposition vapor pressure (dissociation pressure) The pressure of the gaseous decomposition product of a solid at equilibrium.

dehydrating agent A reagent that removes water, or the elements of water, from a compound. *Example*: H_2SO_4

delocalized orbitals Molecular orbitals that spread over an entire molecule.

delta Δ (in a chemical equation) A symbol signifying that a reaction occurs at temperatures above room temperature.

delta X ΔX The difference between the final and initial values of a property: $\Delta X = X_f - X_i$. *Examples*: ΔT, ΔE

denaturation The loss of structure of a protein.

density d Mass per unit volume: $d = m/V$.

derived unit A combination of base units. *Examples*: cubic centimeter (cm^3), joule ($kg \cdot m^2/s^2$)

diagonal relationship A similarity between diagonal neighbors in the periodic table, especially for elements on the left of the table. *Examples*: the similarity of lithium and magnesium, and of beryllium and aluminum

diamagnetic substance A substance that is pushed out of a magnetic field; composed of atoms or molecules with no unpaired electrons.

diatomic molecule A molecule that consists of two atoms. *Examples*: H_2, CO

diffraction The interference between waves that is caused by an object in their path; the basis of the technique of *x-ray diffraction*, used to determine the structures of crystals and molecules.

diffusion The spreading of one substance through another.

dilute To reduce the concentration of a solution by adding more solvent.

dimer The union of two identical molecules. *Example*: Al_2Cl_6, or $(AlCl_3)_2$

dipole See *electric dipole*.

dispersion force See *London force*.

disproportionation reaction A redox reaction in which a single substance is simultaneously oxidized and reduced. *Example*: $2Cu^+(aq) \longrightarrow Cu(s) + Cu^{2+}(aq)$

distribution (of molecular speeds) The percentage of gas molecules moving at each speed at any instant.

double bond Two shared electron pairs forming one σ bond and one π bond.

double replacement reaction (metathesis) A reaction of the form $MA + BX \rightarrow MX + BA$, in which two pairs of ions (M^+ and A^-; B^+ and X^-) change partners. *Example*: $AgNO_3(aq) + NaCl(aq) \longrightarrow AgCl(s) + NaNO_3(aq)$

Downs process The production of sodium and chlorine by the electrolysis of molten sodium chloride.

drying agent A substance that absorbs water and thus maintains a dry atmosphere. *Example*: P_2O_5

dynamic equilibrium A condition in which a forward process and its reverse are occurring at equal rates. *Examples*: vaporizing and condensing, chemical reactions at equilibrium

ebullioscopy The determination of the molecular weight of a substance by measuring the boiling-point elevation that it causes when dissolved in a solvent.

effective atomic number Z_{eff} An apparent atomic number experienced by an electron, that takes into account the repulsion from other electrons in the atom.

elastomer An elastic polymer. *Example*: rubber (polyisoprene)

electric dipole A positive charge next to an equal negative charge.

electric dipole moment μ The magnitude of the electric dipole (in debye).

electrochemical cell A cell for producing electricity from a chemical reaction. More specifically, a *galvanic cell* is an electrochemical cell used to produce electricity, and an *electrolytic cell* is an electrochemical cell in which an electric current is used to cause chemical change.

electrochemical series Redox couples arranged in order of oxidizing and reducing strengths, usually with strong oxidizing agents at the top of the list and strong reducing agents at the bottom.

electrochemistry The branch of chemistry that deals with the use of chemical reactions to produce electricity, the relative strengths of oxidizing and reducing agents, and the use of electricity to produce chemical change.

electrode One of two metallic contacts between an electrochemical cell and an external circuit; the *electrode compartment,* the part of the cell containing the electrode, is specified as reduced form|oxidized form. *Examples*: $Pt(s),H_2|H^+(aq)$; $Cu(s)|Cu^{2+}(aq)$

electrolysis A process in which a chemical change is produced by passing an electric current through a substance or its solution.

electrolyte A substance that dissolves to give an ionically conducting solution. A *weak electrolyte* is a solution in which only a small proportion of the solute molecules are ionized. *Examples*: $NaCl(aq)$ is an electrolyte solution; $CH_3COOH(aq)$ is a weak electrolyte solution; $C_6H_{12}O_6(aq)$ is a nonelectrolyte solution.

electromagnetic radiation A wave of oscillating electric and magnetic fields; includes light, x-rays, and gamma rays.

electron affinity The ability of a gas-phase atom to acquire an electron; more specifically, the negative of the electron-gain enthalpy.

electron configuration The list of occupied orbitals in an atom or molecule. *Examples*: N $1s^2 2s^2 2p^3$; N_2 $\sigma^2 \sigma^{*2} \pi^4 \sigma^2$

electron-deficient compound A compound with too few valence electrons for it to be assigned a Lewis structure. *Example:* B_2H_6

electronegativity χ The ability of an atom to attract electrons to itself when it is part of a compound.

electrophile A reactant that attacks an electron-rich region of a molecule. *Examples*: NO_2^+, Br_2

electroplating The deposition of a thin film of metal on an object by electrolysis.

electropositive Denotes an element with a low electronegativity (a metal, for instance).

element A substance that consists only of atoms with the same atomic number. *Examples*: hydrogen, gold, uranium

elementary reaction An individual reaction step in a mechanism. *Example*: $H\cdot + Cl_2 \longrightarrow HCl + Cl\cdot$

elimination reaction A reaction in which two atoms or groups of atoms on neighboring carbon atoms are removed from a molecule, leaving a multiple bond. *Example*:

empirical formula The simplest chemical formula that shows the relative numbers of atoms of each element in a compound. *Examples*: $NaCl$, P_2O_5, CH for benzene

enantiomers A pair of mirror-image optical isomers. *Examples*: L-alanine and D-alanine, $CH_3CH(NH_2)COOH$

endothermic reaction A reaction that takes in heat ($\Delta H > 0$). *Example*: $N_2O_4(g) \longrightarrow 2NO_2(g)$

end point The stage in a titration at which enough titrant has been added to take the indicator to a color halfway between its initial and final colors.

energy E The capacity to do work or supply heat. *Kinetic energy* is the energy of motion, and *potential energy* is the energy arising from position and from interactions between particles.

enthalpy change ΔH Heat output at constant pressure; *enthalpy* itself is $H = E + PV$, where E is the internal energy of a system.

enthalpy density (of a fuel) The enthalpy of combustion per liter (without the minus sign).

entropy S A measure of the disorder of a system. A change in entropy is equal to the heat added to a system divided by the temperature at which the transfer occurs.

enzyme A biological catalyst.

equation of state A mathematical relation linking the pressure, volume, and temperature of a substance and the amount present. *Example*: ideal gas law, $pV = nRT$

equatorial bond (in a bipyramidal molecule) A bond perpendicular to the molecular axis.

equilibrium constant K_c A characteristic of the equilibrium composition of a reaction mixture, with a form given by the law of mass action. *Example*:

$$N_2(g) + 3H_2(g) \rightleftharpoons 2NH_3(g) \qquad K_c = \frac{[NH_3]^2}{[N_2][H_2]^3}$$

equivalence point The stage in a titration when exactly the right volume of solution needed to complete the reaction has been added.

ester The product (other than water) of a reaction between a carboxylic acid and an alcohol. *Example*: $CH_3COOC_2H_5$ (ethyl acetate)

ether A compound of the form R—O—R. *Example*: $C_2H_5OC_2H_5$ (diethyl ether)

exclusion principle No more than two electrons may occupy any given orbital, and when two electrons do occupy one orbital, their spins must be paired.

exothermic reaction A reaction that gives out heat ($\Delta H < 0$). *Example*: $N_2(g) + 3H_2(g) \longrightarrow 2NH_3(g)$

expanded octet A valence shell containing more than eight electrons. *Examples*: the valence shells in PCl_5 and SF_6

exponential decay A variation with time of the form e^{-kt}.

extensive property A property that depends on the size (the "extent") of the sample. *Examples*: mass, internal energy, enthalpy, entropy

experiment A test carried out under carefully controlled conditions.

Faraday's law of electrolysis The number of moles of product formed by an electric current is chemically equivalent to the number of moles of electrons supplied by the current.

fats Esters of glycerol and carboxylic acids with long hydrocarbon chains; fats act as long-term energy storage for the body.

fatty acids Carboxylic acids with long hydrocarbon chains. *Example*: $CH_3(CH_2)_{16}COOH$ (stearic acid)

ferromagnetism The ability of some substances to be permanently magnetized. *Examples*: iron and magnetite (Fe_3O_4)

field An influence spreading over a region of space. *Examples*: electric field from a charge, magnetic field from a magnet

first law of thermodynamics The internal energy of an isolated system is constant.

fissile Having the ability to undergo fission induced by slow neutrons. *Example*: ^{235}U is fissile

fission The breakup of a nucleus into two smaller nuclei with similar masses. Fission may be *spontaneous* or *induced* (particularly by the impact of neutrons). *Examples*:

$$^{244}_{95}Am \longrightarrow {}^{134}_{53}I + {}^{107}_{42}Mo + 3n \text{ (spontaneous)}$$

$$^{235}_{92}U + n \longrightarrow {}^{142}_{56}Ba + {}^{92}_{36}Kr + 2n \text{ (induced)}$$

fixed nitrogen Nitrogen in combination with other elements. *Examples*: NH_3, $(NH_2)_2CO$

flammability The ability of a substance to burn in air.

force *F* An influence that can change the state of motion of an object. *Examples*: electrostatic force from an electric charge, mechanical force from an impact

formal charge FC = number of valence electrons in the free atom − number of lone-pair electrons − $\frac{1}{2} \times$ number of shared electrons.

formation constant (stability constant) K_f The equilibrium constant for complex formation. The overall formation constant is the product of stepwise formation constants.

formula unit A unit of a compound consisting of the same numbers of atoms as in the chemical formula. *Examples*: for NaCl, one Na^+ ion and one Cl^- ion; for CH_4, one CH_4 molecule

formula weight The mass of one formula unit.

fossil fuels The partially decomposed remains of vegetable and marine life; mainly coal, oil, and natural gas.

fractional distillation Separation of the components of a liquid mixture by repeated distillation, making use of their differing volatilities.

Frasch process A process for mining sulfur, using superheated water.

free energy *G* The energy of a system that is free to do work at constant temperature and pressure: $G = H - TS$, where *H* is the enthalpy and *S* the entropy of the system. The direction of spontaneous change (at constant pressure and temperature) is the direction of decreasing free energy.

freezing point The temperature at which a liquid freezes.

frequency (of radiation) ν The number of cycles (repeats of the waveform) per second.

Friedel-Crafts alkylation The attachment of alkyl groups to an aromatic ring, using aluminum chloride as catalyst. *Example*:

$$C_6H_6 + CH_3Cl \xrightarrow{AlCl_3} C_6H_5CH_3 + HCl$$

fuel cell An electrochemical cell in which the reactants are supplied from outside as the cell is used.

functional group A group of atoms that brings a characteristic set of chemical properties to an organic molecule. *Examples*: —OH, —Br, —COOH

furanose A five-member-ring form of a sugar molecule.

galvanize To coat a metal (usually steel) with an unbroken film of zinc.

gas A fluid substance that fills the container it occupies and can easily be compressed into a smaller volume.

glass electrode A thin-walled glass bulb containing an electrolyte; used for measuring pH.

Graham's law The rate of diffusion of a gas is inversely proportional to the square root of its molecular weight.

gravimetric analysis The analysis of composition by means of the measurement of mass.

greenhouse effect The blocking by some atmospheric gases (notably carbon dioxide) of the radiation of heat from the surface of the earth back into space, leading to a worldwide rise in temperature.

ground state (of an atom or molecule) The state of lowest energy.

groups (of the periodic table) The vertical columns of the periodic table. The number of a group (in the system we use) is equal to the number of electrons in the valence shell of its member atoms.

Haber process (Haber-Bosch process) The catalyzed synthesis of ammonia at high pressure and high temperature.

half-life $t_{1/2}$ In chemical kinetics, the time needed for the concentration of a substance to fall to half the initial value; in nuclear chemistry, the time needed for half the initial number of radioactive nuclei to disintegrate.

half-reactions Hypothetical reactions in which an oxidation and a reduction occur separately; the sum of their equations is the equation for the overall reaction. *Examples*: $Na(s) \longrightarrow Na^+(aq) + e^-$, $Cl_2(g) + 2e^- \longrightarrow 2Cl^-(aq)$

Hall process The production of aluminum by the electrolysis of aluminum oxide dissolved in molten cryolite.

halogens The elements of Group VII of the periodic table.

hard water Water that contains dissolved calcium and magnesium salts.

heat   The energy transferred as the result of a temperature difference between a system and its surroundings.

heat capacity   The constant of proportionality between the heat supplied to an object and the temperature rise it produces.

heating curve   A graph of the variation of the temperature of a sample as it is heated.

Henry's law   The solubility of a gas in a liquid is proportional to its partial pressure. That is, $S = k_H \times P$.

Hess's law   A reaction enthalpy is the sum of the enthalpies for any sequence of reactions (at the same temperature and pressure) into which the overall reaction may be divided.

heterogeneous equilibrium   An equilibrium in which at least one substance is in a different phase from the rest. *Example*: $AgCl(s) \rightleftharpoons Ag^+(aq) + Cl^-(aq)$

heterogeneous mixture   A mixture in which the individual components, though mixed together, lie in distinct regions, even on a microscopic scale. *Example*: a mixture of sand and sugar

high-spin complex   A d^n complex with the maximum number of unpaired electron spins.

homogeneous equilibrium   A chemical equilibrium in which all the substances are in the same phase. *Example*: $H_2(g) + I_2(g) \rightleftharpoons 2HI(g)$

homogeneous mixture   A mixture in which the individual components are uniformly mixed, even on an atomic scale. *Examples*: air, solutions

homologous series   A family of organic compounds with molecular formulas obtained by inserting a particular group (usually $-CH_2-$) repetitively into a parent structure. *Example*: CH_4, CH_3CH_3, $CH_3CH_2CH_3$, . . .

Hund's rule   If more than one orbital in a subshell is available, electrons occupy different orbitals of that subshell, with parallel spins.

hybrid orbital   A mixed orbital formed by blending atomic orbitals on the same atom. *Example*: sp^3 hybrid

hybridization   The formation of a hybrid orbital.

hydration   The attachment of water molecules to a central ion; or, the addition of water across a multiple bond in an organic compound (H to one carbon atom, OH to the other). *Example*: $CH_2{=}CH_2 + H_2O \longrightarrow CH_3CH_2OH$

hydrocarbons   Binary compounds of carbon and hydrogen. *Examples*: CH_4, C_6H_6

hydrogenation   The addition of hydrogen to multiple bonds. *Example*: $CH_3CH{=}CH_2 + H_2 \longrightarrow CH_3CH_2CH_3$

hydrogen bond   A link formed by a hydrogen atom lying between two strongly electronegative atoms (O, N, or F).

hydrogen economy   The widespread use of hydrogen as a fuel.

hydrohalogenation   The addition of a hydrogen halide across a carbon–carbon multiple bond. *Example*: $CH_3CH{=}CH_2 + HBr \longrightarrow CH_3CHBrCH_3$

hydrolysis   The reaction of water with a substance, resulting in the formation of a new element–oxygen bond. *Example*: $PCl_5(s) + 4H_2O(l) \longrightarrow H_3PO_4(aq) + 5HCl(aq)$

hydrometallurgy   The extraction of metals by dissolving them from their ores and then reducing the aqueous solutions of their ions. *Example*: $Cu^{2+}(aq) + Fe(s) \longrightarrow Cu(s) + Fe^{2+}(aq)$

hydrophilic   Water-attracting. *Example*: hydroxyl groups are hydrophilic

hydrophobic   Water-repelling. *Example*: hydrocarbon chains are hydrophobic

hypothesis   A suggestion put forward to account for a series of observations. *Example*: Dalton's atomic hypothesis

ideal gas   A gas that satisfies the ideal gas law, $PV = nRT$. All gases obey the law more and more closely as the pressure is reduced to very low values.

ideal solution   A solution that obeys Raoult's law at any concentration; all solutions behave ideally as the concentration approaches zero.

***i* factor**   A factor that takes into account the deviation from ideal behavior of ions in an electrolyte solution, particularly for the interpretation of colligative properties. *Example*: $i = 2$ (approximately) for $NaCl(aq)$

indicator   A substance that changes color when it goes from its acid to its base form (an *acid-base indicator*) or from its oxidized to its reduced form (a *redox indicator*).

indirect titration   A titration procedure in which the analyte is converted into another substance, which is then titrated.

inductive effect   The supply or withdrawal of electrons by the polarization of σ bonds (such as by an electronegative substituent on an aromatic ring).

inert pair   A pair of valence-shell s electrons that are tightly bound to the atom.

inhibition (of a chain reaction)   The removal of radicals other than by chain termination. *Example*: $H\cdot +$ other radicals $\longrightarrow$ unreactive substances

initial rate   The reaction rate at the start of a reaction, when no reverse reactions or other complications can affect the rate.

initiation (of a chain reaction)   The formation of radicals from a reactant. *Example*:

$$Br_2 \xrightarrow{\text{heat or light}} 2Br\cdot$$

insulator (electrical)   A substance that does not conduct electricity. *Examples*: nonmetallic elements, molecular solids

integrated rate law   An expression for the concentration of a reactant or product of a reaction in terms of time, obtained from the rate law for the reaction. *Example*:

$$\ln\frac{[A]_0}{[A]} = kt$$

intensive property   A property of a substance that is

independent of the size of the sample. *Examples*: density, molar volume, temperature

interference Interaction between waves, leading either to a greater amplitude (*constructive interference*) or to a smaller one (*destructive interference*).

interhalogens Binary compounds of the halogens among themselves. *Example*: IF_3

intermolecular Between molecules.

intermolecular forces Attractions and repulsions between molecules. *Examples*: ion–dipole, dipole–dipole, and London forces

internal energy E The total energy of all the particles in a system.

interstitial (atom) Occupying the gaps between other atoms.

intramolecular Within a molecule.

iodimetry Redox titration using iodine as the oxidizing agent.

ion An electrically charged atom or group of atoms. *Cations* are positively charged ions; *anions* are negatively charged ions. *Examples*: Na^+, NH_4^+, Al^{3+} (cations); Cl^-, SO_4^{2-} (anions)

ion–dipole interaction The interaction between an ion and the partial charges of an electric dipole.

ionic bond The interionic attraction arising from the opposite charges of a cation and an anion.

ionic compound A compound that consists of ions. *Examples*: $NaCl$, KNO_3

ionic radius The radius of an ion, measured from the distance between the centers of neighboring ions in a solid and setting the radius of the O^{2-} ion equal to 140 pm.

ionic solid A solid built from cations and anions. *Examples*: $NaCl(s)$, $KNO_3(s)$

ionization (of atoms and molecules) The conversion of atoms or molecules to ions by removing or adding electrons. *Examples*: $K(g) \longrightarrow K^+(g) + e^-(g)$, $Cl(g) + e^-(g) \longrightarrow Cl^-(g)$

ionization (of an acid) The donation of a proton from a neutral acid molecule to a base with the formation of the conjugate base (an anion in this instance) of the acid. *Example*: $CH_3COOH(aq) + H_2O(l) \longrightarrow CH_3CO_2^-(aq) + H_3O^+(aq)$

ionization energy I The minimum energy required to remove an electron from the ground state of a gaseous atom. The *second ionization energy* is the energy required to remove a second electron.

ion product Q The product of the ion concentrations in a solution, not necessarily a saturated solution. *Example*: $Hg_2Cl_2(s) \rightleftharpoons Hg_2^{2+}(aq) + 2Cl^-(aq)$ $Q = [Hg_2^{2+}][Cl^-]^2$

isoelectronic ions (or molecules) Ions (or molecules) with the same number of atoms and the same number of valence electrons. *Examples*: SO_2 and O_3; CN^- and CO; NO_3^- and CO_3^{2-}

isomers Compounds built from the same atoms in different arrangements. In *structural isomers*, the atoms have

different partners or lie in a different order; in *stereoisomers*, the atoms have the same partners but are in different arrangements in space. *Examples*: CH_3-O-CH_3 and CH_3CH_2-OH (structural); *cis* and *trans* 2-butene (stereo)

isomerization reaction A reaction in which a compound is converted into one of its isomers. *Example*: *cis*-butene $\longrightarrow$ *trans*-butene

isotactic polymer A polymer with all the substituents on the same side of the chain.

isotope Elements with the same atomic number but different atomic masses. *Example*: ^1H, ^2H, and ^3H

isotopic dating Determination of the age of an object by measuring the proportion of the radioactive isotopes it contains; particularly, *radiocarbon dating* using ^{14}C.

isotopic labeling The use of an isotope to follow the location of an element in a chemical reaction.

Joule-Thomson effect The cooling of a gas as it expands.

Kekulé structures Two Lewis structures of benzene, consisting of alternating single and double bonds.

ketones Organic compounds of the form $R-CO-R'$. *Example*: $CH_3-CO-CH_3$, acetone

kinetic theory (kinetic-molecular theory) A theory accounting for the properties of a gas and in which interactions between molecules are ignored.

labile complex A complex that survives for short periods only.

lanthanide contraction The reduction of metallic radii of the elements of the f block below the values that would be expected from a simple extrapolation of the trend up to Period 6 of the periodic table (and arising from the poor shielding ability of f electrons).

lattice enthalpy The standard enthalpy change that accompanies the conversion of a solid to a gas of ions.

law A summary of experience.

law of conservation of energy Energy can be neither created nor destroyed.

law of constant composition A compound has the same composition whatever its source.

law of mass action For a reaction of the form $aA + bB \rightleftharpoons cC + dD$, the concentrations at equilibrium satisfy

$$K_c = \frac{[C]^c[D]^d}{[A]^a[B]^b}$$

law of radioactive decay The rate of decay of a sample is proportional to the number of radioactive nuclides in the sample.

Le Chatelier's principle A dynamic equilibrium tends to oppose any change in the equilibrium conditions. *Example*: A reaction at equilibrium tends to proceed in the endothermic direction when the temperature is raised.

leveling The observation that strong acids all have the same strength in water, and all behave as though they were solutions of H_3O^+ ions.

Lewis acid An electron-pair acceptor. *Examples*: H^+, Fe^{3+}, BF_3

Lewis base An electron-pair donor. *Examples*: OH^-, H_2O, NH_3

Lewis structure A diagram showing how electron pairs are shared between atoms in a molecule. *Examples*:

$$H-\overset{..}{\underset{..}{Cl}}: , \quad \overset{..}{\underset{..}{O}}=C=\overset{..}{\underset{..}{O}}$$

ligands The groups attached to the central metal ion in a complex; *polydentate ligands* occupy more than one binding site.

ligand field splitting Δ The energy separation between the e and t orbitals in a complex.

limiting reagent The reactant that governs the maximum amount of product that can be formed in a particular experiment.

lipids Naturally occurring substances that dissolve in nonpolar solvents; more precisely, fatty acids and substances derived from fatty acids. *Examples*: fats, steroids, terpenes

liquid A fluid substance that takes the shape of the part of the container it occupies.

liquid crystal A substance that flows like a liquid but has molecules that lie in a moderately orderly array; a mesophase. Liquid crystals may be *nematic, smectic,* or *cholesteric,* depending on the arrangement of the molecules.

London forces Interactions between instantaneous electric dipoles on neighboring polar or nonpolar molecules.

low-spin complex A d^n complex with the minimum number of unpaired electron spins.

main-group elements The members of the s and p blocks of the periodic table. *Examples*: Li, Mg, C, Br

manometer An instrument used for measuring the pressure of a gas confined inside a container.

Markovnikov's rule In the addition of HX to an alkene, where X is an electronegative element or group, the H atom bonds to the C atom that already carries the greater number of H atoms. *Example*: $CH_3CH{=}CH_2 +$ $HBr \longrightarrow CH_3CHBrCH_3$

mass m The quantity of matter in a sample.

mass concentration The mass of solute per liter of solution.

mass number A The total number of nucleons (protons + neutrons) in the nucleus of an atom of an element. *Example*: $^{14}_{6}C$, with mass number 14, has 14 nucleons (6 protons and 8 neutrons)

mass percentage The mass of a substance present in a sample, expressed as a percentage of the total mass of the sample.

matter Everything that takes up space.

melting point The temperature at which a solid and liquid are in equilibrium.

meniscus The curved surface that a liquid forms in a narrow tube.

mesophase A state of matter showing some of the properties of both a liquid and a solid (liquid crystal).

metal A substance that conducts electricity, has a metallic luster, and is malleable and ductile; consists of cations held together by a sea of electrons.

metallic conductor An electronic conductor with a resistance that increases as the temperature is raised.

metallic radius (of a metallic element) Half the distance between the centers of neighboring atoms in a solid sample.

metathesis A reaction in which atoms, ions, or groups of atoms exchange partners, as in $MA + BX \longrightarrow$ $MX + BA$. *Example*: precipitation reactions

minerals Substances that are mined; more generally, solid inorganic substances.

mixture A blend of different substances that can be separated by physical methods.

moderator A substance that slows neutrons. *Examples*: graphite, heavy water

molality The number of moles of solute per kilogram of solvent.

molar Amount per mole. *Examples*: Molar mass is the mass per mole; molar volume is the volume per mole.

molar concentration [. . .] The number of moles of solute per liter of solution.

molarity Molar concentration.

mole The number of atoms in exactly 12 g of carbon-12.

molecular compound A compound that consists of molecules. *Examples*: water, sulfur hexafluoride, benzoic acid

molecular formula A combination of chemical symbols and subscripts showing the number of atoms of each element present in a molecule. *Examples*: H_2O, SF_6, C_6H_5COOH

molecularity The number of reactant molecules (or free atoms) taking part in an elementary reaction. *Examples*:

$$O_3 \longrightarrow O + O_2 \quad \text{(unimolecular)}$$

$$O + O_2 \longrightarrow O_3 \quad \text{(bimolecular)}$$

molecular orbital A region of space in which there is a high probability of finding an electron in a molecule (more specifically, a mathematical expression that gives the probability of finding an electron at each point in a molecule).

molecular potential-energy curve A graph showing the variation of the energy of a molecule as the bond length is changed.

molecular solid A solid consisting of a collection of individual molecules. *Examples*: glucose, aspirin, sulfur

molecular weight The average mass of one of the molecules of a molecular compound.

molecule A definite and distinct group of atoms bonded together; the smallest particle that has the characteristic chemical properties of a compound. *Examples*: H_2, NH_3, CH_3COOH

mole fraction x The number of moles (or atoms or ions) of a substance in a mixture, expressed as a fraction of the total number of moles in the mixture.

Mond process The purification of nickel by the formation and decomposition of nickel carbonyl.

monomer The small molecule from which a polymer is formed. *Examples*: $CH_2{=}CH_2$ for polyethylene, $NH_2(CH_2)_6NH_2$ for nylon

native Occurring in an uncombined state as the element itself.

Nernst equation An equation expressing cell potential in terms of the concentrations of the reagents in the cell reaction: $E = E° - (RT/nF) \ln Q$, where $E°$ is the standard potential and Q the reaction quotient.

net ionic equation The equation showing the net change, obtained by canceling the spectator ions in a chemical equation. *Example*: $Ag^+(aq) + Cl^-(aq) \longrightarrow AgCl(s)$

network solid A solid consisting of atoms linked together covalently throughout its extent. *Examples*: diamond, silica

neutralization The reaction of an acid with a base; specifically, $H_3O^+(aq) + OH^-(aq) \longrightarrow 2H_2O(l)$

noble gases The elements of Group VIII of the periodic table.

node A point or surface at which an electron occupying an orbital will not be found.

nonaqueous solution A solution in which the solvent is not water. *Example*: sulfur in carbon disulfide

nonelectrolyte A substance that dissolves to give a solution that does not conduct electricity. *Example*: sucrose

nonmetal A substance that does not conduct electricity very well and is neither malleable nor ductile. *Examples*: all gaseous elements, phosphorus, bromine

normal boiling point T_b The temperature at which the vapor pressure of a liquid is 1 atm.

NO$_x$ Nitrogen oxides, especially in the context of atmospheric chemistry and pollution.

nuclear binding energy The energy released when Z protons and $A - Z$ neutrons form a nucleus. The greater the binding energy, the lower the energy of the nucleus.

nuclear disintegration The partial breakup of a nucleus (including its fission). *Example*: $^{226}_{88}Ra \longrightarrow ^{222}_{86}Rn + ^4_2\alpha$

nuclear transmutation The conversion of one element into another. *Example*: $^{12}_6C + \alpha \longrightarrow ^{16}_8O + \gamma$

nucleon A proton or neutron; thus, either of the two principal components of a nucleus.

nucleophile A reactant that attacks an electron-poor region of a molecule. *Examples*: OH^-, Cl^-

nucleophilic substitution A substitution reaction that results from attack by a nucleophile. *Example*: $CH_3Br + OH^- \longrightarrow CH_3OH + Br^-$

nucleosynthesis The formation of elements.

nuclide A specific isotope of an element. *Examples*: ^2_1H, $^{16}_8O$

octet An s^2p^6 valence-electron configuration.

octet rule Atoms proceed as far as possible toward completing their octets by gaining or losing electrons or sharing electron pairs.

oligosaccharide A molecule consisting of a string of several bonded saccharide units (such as glucose molecules). *Examples*: cellulose, amylopectin

optical activity The ability of a substance to rotate the plane of polarization of polarized light passing through it.

order (of a reaction) The power to which the concentration of a substance is raised in a rate law. See also *overall order*. *Example*: rate $= k[SO_2][SO_3]^{-1/2}$, of first order in SO_2 and of order $-\frac{1}{2}$ in SO_3

ores The natural mineral sources of metals. *Example*: Fe_2O_3 (hematite)

organic compounds Compounds containing the element carbon and usually hydrogen.

osmometry The measurement of molecular weight from observations of osmotic pressure.

osmosis The passage of a solvent through a semipermeable membrane into a more concentrated solution.

osmotic pressure Π The pressure needed to stop osmosis.

Ostwald process The production of nitric acid by the catalytic oxidation of ammonia.

overall order (of a reaction) The sum of the powers to which individual concentrations are raised in the rate law for a reaction. *Example*: rate $= k[SO_2][SO_3]^{-1/2}$, of order $\frac{1}{2}$ overall

overlap The merging of orbitals belonging to different atoms in a molecule.

overpotential The additional potential difference that must be applied above the cell potential to cause appreciable electrolysis of a substance.

oxidation Electron loss. *Example*: $Mg(s) \longrightarrow Mg^{2+}(s) + 2e^-$

oxidation number The effective charge on an atom in a compound, calculated according to a prescribed set of rules. An increase in oxidation number corresponds to oxidation, and a decrease to reduction.

oxidizing agent A substance that causes oxidation (and is itself reduced in a redox reaction). *Examples*: O_2, O_3, MnO_4^-, Fe^{3+}

oxoacids Acids containing oxygen. *Examples*: H_2CO_3, HNO_3, HNO_2, $HOCl$

oxoanions Polyatomic anions that contain oxygen. *Examples*: HCO_3^-, CO_3^{2-}

oxonium ion An ion of the form ROH_2^+. The *hydronium ion*, H_3O^+, is a special case.

paired electrons   Two electrons with spins in opposite directions (denoted ↑↓).

parallel electrons   Two electrons with spins in the same direction (denoted ↑↑).

paramagnetic substance   A substance that is pulled into a magnetic field; composed of atoms or molecules with unpaired electrons. *Examples*: O_2, $[Fe(CN)_6]^{3-}$

partial pressure P_X The pressure a gas X in a mixture would exert if it alone occupied the container.

parts per million (ppm)   The number of solute particles among 1 million solution molecules. Parts per billion (ppb, 1 in 10^9) may also be used.

passivation   Protection of a metal from further reaction by a surface film. *Example*: the formation of an aluminum oxide coating on aluminum in air

Pauli exclusion principle   See *exclusion principle*.

percentage yield (of a reaction product)   The percentage of the theoretical yield that is actually achieved.

penetration   The possibility that an *s* electron may be found inside the inner shells of an atom and hence close to the nucleus.

peptide   A substance composed of two or more amino acids. A *dipeptide* consists of two amino acid residues bonded together by a peptide link, an *oligopeptide* consists of several residues, and a *polypeptide* consists of many residues. *Examples*: aspartame (dipeptide), proteins (polypeptides)

peptide link   The —CO—NH— group.

periods   The horizontal rows of the periodic table. The number of a period is equal to the principal quantum number of the valence shell of its members.

pH   The negative logarithm of the hydronium ion concentration in a solution: $pH = -\log[H_3O^+]$; $pH < 7$ indicates an acidic solution, $pH = 7$ a neutral solution, and $pH > 7$ a basic solution.

phase   A particular state of matter. A substance may exist in solid, liquid, and gas phases and, in certain cases, in more than one solid phase. *Examples*: White and gray tin are two solid phases of tin; calcite and aragonite are two solid phases of calcium carbonate

phase boundaries   The lines separating the areas in a phase diagram. The points on a phase boundary correspond to equilibria between the two adjoining phases.

phase diagram   A summary in graphical form of the conditions of temperature and pressure at which the solid, liquid, and gaseous phases of a substance exist.

pH curve   The graph of pH against volume of acid added during a titration.

phenol   A compound in which a hydroxyl group is attached directly to an aromatic ring (Ar—OH). *Example*: C_6H_5OH (phenol)

photochemical reaction   A reaction caused by light. *Example*:

$$H_2(g) + Cl_2(g) \xrightarrow{\text{light}} 2HCl(g)$$

photoelectric effect   The emission of electrons from a metal surface by the impact of incident ultraviolet radiation.

photon   A particle of electromagnetic radiation with energy $E = h\nu$, where ν is the frequency of the radiation.

physical state   The condition of being a solid, a liquid, or a gas.

pi (π) bond   Two electrons in a π orbital.

pi (π) orbital   A molecular orbital that looks like a p orbital when viewed along the internuclear axis.

pK_a and pK_b   The negative logarithms of the acid and base ionization constants: $pK = -\log K$. The larger the value of pK_a or pK_b, the weaker the acid or base, respectively.

plasma   An ionized gas; also, in biology, the colorless component of blood in which the corpuscles are dispersed.

polar bond   A covalent bond between atoms carrying partial electric charges. *Examples*: H—Cl, O—S

polarizable (of an atom or ion)   Easily polarized.

polarize   To distort the electronic charge cloud.

polarized light   Plane-polarized light is light in which the wave motion occurs in a single plane.

polar molecule   A molecule with a nonzero electric dipole moment. *Examples*: HCl, NH_3, C_6H_5Cl

polyamide   A polymer formed by the condensation polymerization of a dicarboxylic acid and a diamine. *Example*: nylon

polyatomic ion   A bonded group of atoms with an overall positive or negative charge. *Examples*: NH_4^+, NO_3^-, SiF_6^{2-}

polyatomic molecule   A molecule that consists of more than two atoms. *Examples*: O_3, $C_{12}H_{22}O_{11}$

polyester   A polymer formed by linking together dicarboxylic acids and diols by condensation polymerization.

polymer   A chain of covalently linked monomers. *Examples*: polyethylene, nylon

polypeptide chain   A polymer formed by the condensation of amino acids.

polyprotic acid or base   A Brønsted acid or base that can donate or accept more than one proton. A polyprotic acid is sometimes called "polybasic." *Examples*: H_3PO_4 (triprotic acid), N_2H_4 (diprotic base)

polysaccharide   A chain of many glucose units linked together. *Examples*: cellulose, amylose

potential difference   An electric potential difference between two electrodes is a measure of the work that can be done when electrons travel from one electrode of a cell to another: the greater the potential difference, the greater the work from a given transfer of electrons. Potential difference is measured in volts V.

precipitation   The process in which a product comes out of solution rapidly in the form of a finely divided powder, called a precipitate.

precipitation reaction   A reaction in which a solid product is formed when two electrolyte solutions are mixed.   *Example*: $KBr(aq) + AgNO_3(aq) \longrightarrow KNO_3(aq) + AgBr(s)$

precision Freedom from random error. Precise measurements have only small random error and are reproducible in repeated trials.

pressure P Force per unit area.

primary cell An electrochemical cell that produces electricity from chemicals sealed into the cell during manufacture.

primary structure The sequence of monomers in a biopolymer, such as the polypeptide chain of a protein.

propagation (of a chain reaction) Reaction of the chain carriers (for example, radicals) to produce more chain carriers. *Example*:

$$Br\cdot + H_2 \longrightarrow HBr + H\cdot$$

$$H\cdot + Br_2 \longrightarrow HBr + Br\cdot$$

properties Distinguishing features. *Examples*: vapor pressure, color, density, temperature

proteins The principal nitrogen-containing substances in living organisms; structurally, proteins are polypeptides. *Examples*: hemoglobin, enzymes

protonation Proton transfer to a Brønsted base. *Example*: $2H_3O^+(aq) + S^{2-}(aq) \longrightarrow H_2S(g) + 2H_2O(l)$

pseudo-first-order reaction A reaction with a rate law that is effectively first-order because one substance has a virtually constant concentration.

pseudo-noble gas configuration A noble-gas core surrounded by a completed d subshell. *Example*: Cu^+: $[Ar]3d^{10}$

purex process A process based on solvent extraction that is used to separate plutonium and uranium.

pyranose A six-membered-ring form of a sugar molecule.

pyrometallurgy The extraction of metals using high temperatures. *Example*:

$$Fe_2O_3(s) + 3CO(g) \xrightarrow{\Delta} 2Fe(l) + 3CO_2(g)$$

qualitative analysis Analysis of a sample to determine the identities of the elements it contains.

quantitative analysis Analysis of a sample to determine the amounts of the elements it contains.

quantization The restriction of a property to certain values. *Examples*: energy and the quantization of angular momentum

quantum A parcel of energy.

quantum mechanics The description of matter that takes into account the fact that the energy of an object may be changed only by discrete amounts.

quantum number A number that labels the state of an electron and specifies the value of a property. *Example*: principal quantum number n

quaternary structure The manner in which neighboring polypeptide units stack together to form a protein molecule.

R— (in organic chemistry) A hydrocarbon group; spe-

cifically, an alkyl group, when the distinction between aromatic and nonaromatic compounds is important. *Examples*: R—OH may stand for CH_3OH, C_2H_5OH, and so on

racemic mixture A mixture containing equal concentrations of two enantiomers.

radical A fragment of a molecule with at least one unpaired electron. *Examples*: $\cdot NO$, $\cdot O\cdot$, $\cdot CH_3$

radioactive series A stepwise nuclear decay path in which α and β particles are successively ejected and which terminates at a stable nuclide (often of lead).

radioactivity The spontaneous emission of radiation by nuclei.

random error An error that varies randomly from measurement to measurement, sometimes giving a high value and sometimes a low one.

Raoult's law The vapor pressure of a solution of a nonvolatile solute is proportional to the mole fraction of the solvent in the solution: $P = x_{solvent} \times P^*$, where P^* is the vapor pressure of the pure solvent.

rate constant k The constant of proportionality in a rate law.

rate-determining step The elementary reaction in a mechanism that is so much slower than the rest that it governs the rate of the overall reaction. *Example*: $O(g) + O_3(g) \longrightarrow 2O_2(g)$ in the decomposition of ozone

rate law An equation expressing the instantaneous reaction rate in terms of the concentrations of the substances taking part in the reaction. *Example*: rate = $k[NO_2]^2$

reaction enthalpy The change of enthalpy for the reaction exactly as the chemical equation is written. *Example*: $CH_4(g) + 2O_2(g) \longrightarrow CO_2(g) + 2H_2O(l)$ $\Delta H = -890$ kJ

reaction mechanism A series of elementary reactions that is proposed to account for the rate law of an overall reaction.

reaction profile The variation in potential energy that occurs as two reactants meet, form an activated complex, and separate as products.

reaction quotient Q_c The ratio of the product of the concentrations of the products to that of the reactants (defined like the equilibrium constant) at an arbitrary stage of a reaction. *Example*:

$$N_2(g) + 3H_2(g) \rightleftharpoons 2NH_3(g) \qquad Q_c = \frac{[NH_3]^2}{[N_2][H_2]^3}$$

reaction rate The rate of change of the concentrations of reactants and products at a specific instant.

reaction sequence A series of reactions in which products of one reaction take part as reactants in the next reaction. *Example*:

$$2C(s) + O_2(g) \longrightarrow 2CO(g)$$

$$2CO(g) + O_2(g) \longrightarrow 2CO_2(g)$$

recrystallization Purification by repeated dissolving and crystallization.

redox couple The reduced and oxidized substances taking part in a reduction or oxidation half-reaction. The couple is denoted Ox/Red, where Ox is the oxidized substance and Red the reduced. *Example*: Cu^{2+}/Cu for $Cu^{2+}(aq) + 2e^- \longrightarrow Cu(s)$

redox reaction A reaction in which oxidation and reduction occur. *Example*: $S(s) + 3F_2(g) \longrightarrow SF_6(g)$

reducing agent A substance that causes reduction (and is itself oxidized) in a redox reaction. *Examples*: H_2, H_2S, SO_3^{2-}

reduction Electron gain. *Example*: $Cl_2(g) + 2e^- \longrightarrow 2Cl^-(aq)$

reforming reaction A reaction in which the number of carbon atoms in the reactant hydrocarbon molecule is unchanged but more double bonds and aromatic rings are introduced. Also, in hydrogen manufacture, a reaction in which a hydrocarbon is converted to carbon monoxide and hydrogen over a nickel catalyst. *Examples*:

$$C_6H_{11}CH_3 \longrightarrow C_6H_5CH_3 + 3H_2$$

$$CH_4(g) + H_2O(g) \longrightarrow CO(g) + 3H_2(g)$$

resistance (electrical) A measure of the opposition to the flow of an electric current through a substance. The greater the resistance, the smaller the current for a given potential difference. Resistance is measured in ohms Ω.

resonance A blending of Lewis structures into a single, composite, hybrid structure. *Example*:

$$:\!\overset{\cdot\cdot}{O}\!-\!S\!=\!\overset{\cdot\cdot}{O}\!: \longleftrightarrow :\!\overset{\cdot\cdot}{O}\!=\!S\!-\!\overset{\cdot\cdot}{O}\!:$$

retardation (of a chain reaction) A reaction that diverts radicals from the formation of products. *Example*: $H\cdot + HBr \longrightarrow H_2 + Br\cdot$

reverse osmosis The passage of solvent out of a solution when a pressure greater than the osmotic pressure is applied on the solution side of a semipermeable membrane.

salt The product (other than water) of the reaction between an acid and a base. *Examples*: $NaCl$, K_2SO_4

salt bridge A tube containing a concentrated electrolyte, such as a salt (potassium chloride or potassium nitrate, for example) in a jelly and providing a conducting path between the two compartments of an electrochemical cell.

sample A representative part of a whole.

saturated hydrocarbon A hydrocarbon with no carbon-carbon multiple bonds. *Examples*: CH_4, CH_3CH_3

saturated solution A solution in which dissolved and undissolved solute are in dynamic equilibrium.

science The systematically collected and organized body of knowledge based on experiment, observation, and careful reasoning.

secondary cell An electrochemical cell that must be charged (or recharged) using another supply of electricity before it can be used.

secondary structure The manner in which a biopolymer is coiled.

second law of thermodynamics A spontaneous change is accompanied by an increase in the *total* entropy of the universe.

self-oxidation A reaction in which an anion oxidizes its accompanying cation. *Example*:

$$NH_4NO_3(s) \overset{\Delta}{\longrightarrow} N_2O(g) + 2H_2O(l)$$

semiconductor An electronic conductor with a resistance that decreases as the temperature is raised. In an *n*-type semiconductor, the current is carried by electrons in a largely empty band; in a *p*-type semiconductor, conduction is a result of electrons missing from otherwise filled bands.

semipermeable membrane A barrier that allows only certain types of molecules or ions to pass.

sequestration The formation of a complex between a cation and a bulky molecule or ion. *Example*: Ca^{2+} with $O_3POPO_2OPO_3^{5-}$

shell All the orbitals of a given principal quantum number. *Example*: A single $2s$ orbital and three $2p$ orbitals comprise the shell with $n = 2$.

shielding The repulsion of an electron in an atom by the other electrons present (particularly those in inner shells), which opposes the attraction exerted by the nucleus.

shift reaction A reaction between carbon monoxide and water that is used in the manufacture of hydrogen:

$$CO(g) + H_2O(g) \xrightarrow{400°C,\ Fe/Cu} CO_2(g) + H_2(g)$$

SI The International System of Units (*Système Internationale*), an extension and rationalization of the metric system.

sigma (σ) bond Two electrons in an orbital, having cylindrical symmetry with respect to the line joining two linked atoms.

significant figures (in a measurement) The number of digits in the measurement, up to and including the first uncertain digit in scientific notation. *Example*: 0.0206 mL, or 2.06×10^{-2} mL, a measurement with three significant figures

skeletal equation An unbalanced equation summarizing qualitative information about a reaction. *Example*:

$$H_2 + O_2 \longrightarrow H_2O \qquad \triangle$$

smelting Melting a metal ore with a reducing agent. *Example*:

$$CuS(l) + O_2(g) \overset{\Delta}{\longrightarrow} Cu(l) + SO_2(g)$$

solid A rigid substance with a shape that is independent of the shape of its container.

solubility S The molar concentration of a saturated solution of a substance.

solubility product K_{sp} The product of the ionic molar concentrations in a saturated solution; the dissolution equilibrium constant. *Example*: $Hg_2Cl_2(s) \rightleftharpoons Hg_2^{2+}(aq) + 2Cl^-(aq)$ $K_{sp} = [Hg_2^{2+}][Cl^-]^2$

solute   A dissolved substance.

solution   A homogeneous mixture, generally of a substance of smaller abundance (the solute) in one of greater abundance (the solvent).

solvated   Surrounded by and linked to solvent molecules. (Hydration is a special case in which the solvent is water.)

Solvay process   An industrial process for producing sodium carbonate in which ammonia is used at an intermediate stage.

solvent   The most abundant component of a solution.

specific   Per unit mass (typically per gram) *Example*: specific heat capacity (heat capacity per gram)

specific enthalpy (of a fuel)   The enthalpy of combustion of the fuel per gram (without the minus sign).

spectator ions   Ions that are present but remain unchanged during a reaction. *Example*: Na^+ and NO_3^- in $NaCl(aq) + AgNO_3(aq) \longrightarrow NaNO_3(aq) + AgCl(s)$

spectral line   Radiation of a single wavelength, emitted or absorbed by an atom or molecule.

spectrochemical series   Ligands ordered according to the strength of the ligand field splitting they produce.

spectroscopy   Analysis of the radiation emitted or absorbed by substances.

spin   The intrinsic angular momentum of an electron. The spin, which cannot be eliminated, may occur in only two senses, denoted ↑ and ↓.

spontaneous   Not needing to be driven. *Examples*: a gas expanding; a hot object cooling; methane burning

standard cell potential $E°$ The cell potential when the molar concentration of each ion taking part in the cell reaction is $1 M$ and all gases are at 1 atm pressure. The standard cell potential is the sum of two standard electrode potentials, $E° = E°(anode) + E°(cathode)$, and the difference of the two standard reduction potentials.

standard enthalpy of combustion $\Delta H_c°$ The change in enthalpy per mole of a substance when it burns completely in oxygen under standard conditions.

standard enthalpy of formation $\Delta H_f°$ The standard reaction enthalpy per mole of formula units for the synthesis of a compound from its elements in their most stable form at 1 atm pressure.

standard free energy of formation $\Delta EG_f°$ The standard reaction free energy for the formation of a compound from its elements.

standard hydrogen electrode (SHE)   The hydrogen electrode in its standard state (hydrogen ions at a concentration of $1 M$ and a hydrogen partial pressure of 1 atm), defined as having $E° = 0$.

standard molar entropy $S°$ (of a substance)   The entropy per mole of the pure substance at 1 atm pressure.

standard reaction enthalpy $\Delta H°$ The difference in enthalpy between the products of a reaction in their standard states and the reactants in their standard states; $\Delta H° = H°(reactants) - H°(products)$.

standard reaction entropy $\Delta ES°$ The difference in entropy between the products of a reaction in their standard states and the reactants in their standard states; $\Delta ES° = S°(reactants) - S°(products)$.

standard reaction free energy $\Delta EG°$ The difference in free energy between the products of a reaction in their standard states and the reactants in their standard states: $\Delta EG° = G°(products) - G°(reactants)$.

standard reduction potential $E°$ The standard electrode potential of a couple with the half-reaction taken as a reduction.

standard state   The pure form of a substance at 1 atm pressure.

standard temperature and pressure (STP)   0°C (273.15 K) and 1 atm.

state property   A property that is independent of how the sample was prepared. *Examples*: pressure, enthalpy, entropy, color

stepwise formation constants K_{f1}, K_{f2}, . . .   The formation constants for the successive addition of Lewis bases to a Lewis acid.

steric reasons   Reasons related to shape and size. *Example*: The nonexistence of NCl_5 is ascribed to the bulk of the chlorine atoms and the smallness of the nitrogen atom.

Stock number   A roman numeral equal to the number of electrons lost by the atom on formation of a compound and sometimes added in parenthesis to a symbol or name; more formally, the oxidation number of the element. *Example*: Cu(II) in compounds of copper(II) containing Cu^{2+}

stoichiometric coefficients   The numbers multiplying chemical formulas in a chemical equation. *Examples*: 1, 1, and 2 in $H_2 + Br_2 \longrightarrow 2HBr$

stoichiometric proportions   Reactants in the same molar proportions as their coefficients in the chemical equation. *Example*: equal numbers of moles of H_2 and Br_2 in $H_2 + Br_2 \longrightarrow 2HBr$

strong acid (or base)   An acid (or base) that is fully ionized in solution. *Examples*: HCl, $HClO_4$ (strong acids); NaOH, $Ca(OH)_2$ (strong bases)

subatomic particle   A particle that is smaller than an atom. *Examples*: electron, proton, neutron

sublimation   The direct conversion of a solid to a vapor without first forming a liquid.

subshell (of a given shell)   All the atomic orbitals with the same value of the quantum number l. *Example*: the five $3d$ orbitals of an atom

substances   Different kinds of matter.

substitution reaction   A reaction in which an atom (or group of atoms) is replaced by another atom or group. In complexes, a reaction in which one Lewis base expels another and takes its place. *Examples*:

$$C_6H_5OH + Br_2 \longrightarrow BrC_6H_4OH + HBr$$

$$[Fe(H_2O)_6]^{3+}(aq) + 6CN^-(aq) \longrightarrow$$
$$[Fe(CN)_6]^{3-}(aq) + 6H_2O(l)$$

substrate   The molecule on which an enzyme acts.

superconductor   An electronic conductor that conducts electricity with zero resistance.

superfluidity   The ability to flow without viscosity.

surface-active agent (surfactant)   A substance that accumulates at the surface of a solution (a component of detergents). *Example*: the stearate ion of soaps

surface tension γ The energy that is needed to create new surface area by moving molecules from the bulk of a liquid to the surface and that is responsible for the smooth surface of a liquid.

syndiotactic polymer   A polymer in which the substituents alternate on either side of the chain.

synthesis reaction   A reaction in which a substance is formed from simpler starting materials. *Example*: $N_2(g) + 3H_2(g) \longrightarrow 2NH_3(g)$

synthesis gas   A mixture of carbon monoxide and hydrogen.

systematic error   An error that persists in a series of measurements and does not average out.

systematic name   The name of a compound that reveals which elements are present (and, in its most complete form, how the atoms are arranged). *Example*: methylbenzene is the systematic name for toluene

temperature T How hot or cold a sample is; the property that determines the direction in which heat will flow between two objects in contact.

termination reaction (of a chain reaction)   A reaction in which radicals combine and no longer propagate the chain. *Example*: $2Br \cdot \longrightarrow Br_2$

tertiary structure   The shape into which the α helix and β-pleated sheet sections of a protein are twisted as a result of interactions between peptide groups lying in different parts of the primary structure.

theoretical yield   The maximum mass of product that can be obtained from a given mass of a specified reactant.

thermal motion   The random, chaotic motion of atoms.

thermal pollution   The damage caused to the environment by the waste heat of an industrial process.

thermite reaction   The reduction of a metal oxide with aluminum. *Example*: $2Al(s) + Fe_2O_3(s) \longrightarrow Al_2O_3(s) + 2Fe(l)$

thermodynamics   The study of the transformations of energy from one form to another

titrant   The solution added from a buret in a titration.

titration   The analysis of the composition of a solution by measuring the volume needed to react with another solution.

transition metals   The members of the d block of the periodic table. *Examples*: vanadium, iron, gold

triple bond   Three electron pairs shared (forming one σ and two π bonds).

triple point   A point at which three phase boundaries meet in a phase diagram; under the conditions represented by that point, all three phases coexist in dynamic equilibrium.

Trouton's rule $\Delta H^\circ_{vap}/T_b$ (in fact, the entropy of vaporization) is approximately equal to 85 J/(K · mol) for liquids in which hydrogen bonding is unimportant.

uncertainty principle   The more precisely we know the position of a particle, the less we can know about its speed (and vice versa); $\Delta x \times (m \times \Delta v) \geq h/4\pi$, where Δx is the uncertainty in position, Δv the uncertainty in speed, and m the mass of the particle.

unimolecular reaction   An elementary reaction in which a single reactant molecule changes into products. *Example*: $O_3(g) \longrightarrow O(g) + O_2(g)$

unit analysis   A technique for making conversions between units, using conversion factors and operating on units like arithmetical quantities.

unit cell   The smallest unit of a crystal that, when repeated, stacked together without gaps, can reproduce an entire crystal.

unsaturated hydrocarbon   A hydrocarbon with at least one carbon-carbon multiple bond. *Examples*: $CH_2{=}CH_2$, C_6H_6

valence   The number of bonds that an atom can form.

valence shell   The outermost shell of an atom. *Example*: the $n = 2$ shell of Period 2 atoms

valence-shell electron-pair repulsion theory (VSEPR theory)   A theory for predicting the shapes of molecules, using the fact that electron pairs repel each other.

van der Waals forces   Intermolecular forces.

van't Hoff equation   An equation for osmotic pressure Π in terms of molar concentration [solute]: $\Pi = i[solute]RT$; also (in an entirely different use) the equation that shows how the equilibrium constant varies with temperature.

vapor   The gaseous phase of a substance (specifically, of a substance that is normally a liquid or solid at the temperature in question).

vaporization   The formation of a gas or vapor.

vapor pressure   The pressure exerted by the vapor of a condensed phase (a liquid or a solid) when the vapor and the condensed phase are in dynamic equilibrium.

variable covalence   The ability to form different numbers of covalent bonds. *Example*: sulfur in SO_2 and SO_3

variable valence   The ability to form ions with different charge numbers. *Example*: indium as In^+ and In^{3+}

viscosity   The resistance of a fluid (a gas or a liquid) to flow: the higher the viscosity, the slower the flow.

volatility   The readiness with which a substance vaporizes.

volume V The amount of space a sample occupies.

volumetric analysis   chemical analysis by measurement of volume.

wavelength λ The peak-to-peak distance of a wave.
wave-particle duality   Combined wavelike and particlelike character.
weak acid (or base)   An acid (or base) for which K_a (or K_b) is small compared to 1 and which does not ionize to an appreciable extent in solution. *Examples*: HF, CH_3COOH (weak acids); NH_3, CH_3NH_2 (weak bases)
weight   The gravitational force on a sample.
work w The product of the distance that an object moves and the force opposing the motion.

x-ray diffraction   The analysis of crystal structures by studying the interference patterns they cause in a beam of x-rays.
x-rays   Electromagnetic radiation with wavelengths from about 10 pm up to about 1000 pm.

zero-order reaction   A reaction with a rate that is independent of the concentration of the reactant. *Example*: catalyzed decomposition of ammonia
zone refining   Purifying a solid by repeatedly melting a zone and moving it along the length of the sample.

ANSWERS TO ODD-NUMBERED NUMERICAL EXERCISES

CHAPTER 1

1.13 (a) 250 cm^3 (b) 0.250 L (c) 0.0283 kg
(d) 2.54 cm

1.15 100 mm^2, $1 \times 10^6 \text{ m}^3$, $1 \times 10^6 \text{ cm}^3$, $1 \times 10^{-5} \text{ m}^3$

1.17 (a) 0.155 in^2 (b) 0.836 m^2 (c) $4.047 \times 10^3 \text{ m}^2$
(d) 2.47 acre

1.19 (a) 1.0000×10^4 (b) 1.86000×10^5 (c) 1×10^{-9} (d) 5.5×10^{10} (e) 1×10^{-4}

1.21 (a) $1 \times 10^{-12} \text{ m}$ (b) $1 \times 10^{-7} \text{ cm}$ (c) $1 \times 10^4 \text{ mg}$ (d) $1.00 \times 10^{-3} \text{ kg}$

1.23 (a) $1 \times 10^{-18} \text{ m}^2$ (b) $1 \times 10^3 \text{ mm}^3$ (c) 1 dm^3
(d) 0.6 g/cm^2

1.25 (a) $37°\text{C} = 310 \text{ K}$ (b) $-40°\text{C} = 233 \text{ K}$
(c) $-459.4°\text{F}$ (d) 373 K

1.27 (a) $1945°\text{F}$, 1336 K (b) $833°\text{F}$, 718 K (c) 27 K
(d) $-270.5°\text{C}$

1.35 (a) 4.0 gal/s (b) 2 kg/ft^3 (c) $1.4 \times 10^2 \text{ lb/cm}^2$
(d) $0.3 \text{ in}^2/\text{s}$

1.37 (a) $5.4 \times 10^4 \text{ L/h}$ (b) $2 \times 10^2 \text{ lb/m}^3$, $2 \times 10^{-4} \text{ lb/cm}^3$ (c) $4.2 \times 10^2 \text{ kg/in}^2$ (d) $1.13 \times 10^7 \text{ g/m}^3$

1.39 (a) $9.2 \times 10^2 \text{ lb/in}^2$ (b) 705 lb/ft^3
(c) $1.4 \times 10^4 \text{ gal/h}$ (d) $6.7 \times 10^8 \text{ mi/h}$

1.41 $8.1 \times 10^3 \text{ mi}$

1.43 $5.6 \times 10^2 \text{ lb/ft}^3$

1.45 (a) $5.88 \times 10^{12} \text{ mi}$ (b, i) $3.00 \times 10^8 \text{ m/s}$
(b, ii) 0.300 m/ns (c) $1 \times 10^9 \text{ hr}$

1.47 1.8 m, 70 kg, 57 g

1.49 (a) 22 g (b) 0.77 oz

1.51 $6.2 \times 10^3 \text{ ton}$

1.53 (a) 3 (b) 3 (c) 3 (d) 3

1.55 6.60 mL, 30.0 mL

1.57 (a) 2.5 K (b) 2 K (c) $3.0 \times 10^8 \text{ m/s}$
(d) $1 \times 10^5 \text{ A} \cdot \text{s}$

1.59 3 g

1.61 103.6 cm

1.63 $1.76 \times 10^{-4} \text{ g}$

1.65 1799 ton

1.67 (a) $1.86 \times 10^{10} \text{ m}$ (b) $1.9 \times 10^{10} \text{ m}$

1.69 (a) 12.57 (b) 12.56637

1.71 13 mg

1.73 67.3% Cu and 32.7% Zn

1.75 62.9% Cu, 30.5% Zn, 6.6% Sn

1.83 1.59 g/cm^3

1.85 $9 \times 10^1 \text{ kg}$

1.87 $10 \text{ m}^2/\text{person}$

1.89 48 g Sn and 52 g Pb

1.91 (a) 28.6% alcohol and 71.4% water   (b) $2.9 \times 10^2 \text{ g}$ (c) $2.50 \text{ kg water/kg alcohol}$

1.93 (a) 9.49% salt, 90.5% water   (b) 94.9 g salt/kg solution   (c) $105 \text{ g salt/kg water}$

1.95 (a) 11.1 g salt (b) 1.52 g salt (c) 1.50 g salt

CHAPTER 2

2.15 (a) 0.456 g protons (b) 0.545 g neutrons
(c) $0.000248 \text{ g electrons}$ (d) 0.45 (or 45%)

2.17 (a) $3.1 \times 10^{18} \text{ atoms}$ (b) $5.3 \times 10^{22} \text{ atoms}$

2.19 (a) $2.99 \times 10^{-23} \text{ g}$ (b) $3.34 \times 10^{24} \text{ molecules}$

2.21 0.25 g Cu

2.23 (a) 56 g Fe (b) 27 g Fe

2.25 -0.03% of body mass

2.27 5, 0.055%

2.29 $5.01 \times 10^{-24} \text{ g}$, 2.0×10^{20} He-3 atoms

2.31 35.0 amu, $5.81 \times 10^{-23} \text{ g}$

2.33 (a) 1.00 mol H atoms (b) $2.0 \text{ mol electrons}$
(c) $4.98 \times 10^{-16} \text{ mol people}$ (d) $10^{-2} \text{ mol stars}$

2.35 (a) $6.02 \times 10^{23} \text{ O}_2 \text{ molecules}$ (b) 1.5×10^{23} $\text{SO}_4^{2-} \text{ ions}$ (c) $9.0 \times 10^{20} \text{ Al atoms}$ (d) 1.2×10^{24} glucose molecules

2.37 12.01 amu

2.39 6.94 amu

2.41 32.1, 1.88×10^{22} S atoms

2.43 (a) 12.0 g graphite (b) 18 g chlorine (c) 0.29 g platinum   (d) 321 g sulfur

2.45 $5.05 \times 10^{-21} \text{ dollars/atom}$

2.47 (a) 146.06 (b) 17.03 (c) 26.04 (d) 272.37

2.49 (a) 143.32 (b) 65.12 (c) 132.14 (d) 295.41

2.51 (a) $0.0128 \text{ mol C}_6\text{H}_6$ (b) 0.00292 mol $\text{C}_{12}\text{H}_{22}\text{O}_{11}$

2.53 (a) 0.0070 mol AgCl (b) 0.0070 mol Ag^+
(c) 0.0070 mol Cl^-

2.55 0.0068 mol Br^- ($4.1 \times 10^{21} \text{ ions}$)

2.57 (a) $18 \text{ g H}_2\text{O}$ (b) $1.1 \times 10^2 \text{ g Fe}$ (c) $4.3 \times 10^2 \text{ g UO}_3$ (d) $6.0 \text{ g C}_2\text{H}_4\text{O}_2$

2.59 (a) $55 \text{ mol H}_2\text{O}$ (b) $2.2 \text{ mol C}_2\text{H}_6\text{O}$ (c) $5.9 \times 10^{-3} \text{ mol AgNO}_3$ (d) $8.8 \times 10^{-3} \text{ mol C}_{12}\text{H}_{22}\text{O}_{11}$

2.77 10 g/cm^3; packed with much open space

CHAPTER 4

4.1 10 mol H_2

4.3 22 mol

4.5 (a) $1:1$ (b) $1:2$ (c) $3:2$ (d) $1:5$

4.7 8 mol CO_2

4.9 5 mol O_2

4.11 $0.10 \text{ mol N}_2\text{O}$

4.13 0.30 mol $Na_2S_2O_3$

4.15 4.40 g CO_2

4.17 189 g Al_2O_3

4.19 3.2×10^3 g (3.2 kg) H_2O, 1.3×10^4 g (13 kg) $Ca(OH)_2$

4.21 1.9 metric ton slag, 0.93 metric ton CaO, 1.7 metric ton $CaCO_3$

4.23 1.1 kg H_2O

4.25 4×10^3 g H_2O

4.27 35 metric tons O_2

4.29 0.25 g CH_4

4.31 9.4 g glucose

4.35 15 g H_3PO_3

4.37 96.8% yield

4.39 81.2%

4.41 (a) $C_7H_6O_2$ (b) C_6H_7N

4.43 (a) $C_{10}H_{20}$ (b) C_6H_6 (c) $C_6H_{12}O_6$
(d) $C_{10}H_{14}N_2$

4.45 CH

4.47 CHO_2

4.49 PCl_5

4.51 CH_2

4.53 $C_{10}H_8$

4.55 $C_{18}H_{22}O_2$

4.57 $C_{14}H_{18}N_2O_5$, empirical formula is the same

4.59 $C_4H_5N_2O$, $C_8H_{10}N_4O_2$

4.61 0.2105 M NaCl

4.63 0.850 g $AgNO_3$

4.65 0.00118 mol Cl^-

4.67 (a) 6.3 mL (b) 6.3 mL

4.69 57.5 mL

4.71 0.03123 M KCl

4.73 1.289 M HCl

4.75 0.7246 M H_2SO_4

4.77 9.43 M

4.79 0.216 M I_2, 5.48 g I_2/L

4.81 0.0204 M hydrazine sulfate

4.83 0.750 M HCN

4.85 4.24 M H_3PO_3

4.87 0.926 g vitamin C, about 10% less than claimed

CHAPTER 5

5.1 (a) 0.987 atm (b) 1 mmHg (c) 7.50×10^{-3} Torr (d) 133 Pa

5.3 (a) 1 atm (b) 100 kPa (c) 758 Torr (d) 96 kPa

5.5 (a) 0.097 atm, 74 Torr (b) 18 lb/in² 1.2 atm

5.7 0.048 atm

5.9 1.0×10^4 kg

5.11 (a) 763.8 Torr (b) 745.3 Torr

5.13 11.3 cm

5.15 (a) 0.500 L (b) 543 mL

5.17 (a) 0.500 atm (b) 977 Torr (c) 1.54 atm

5.19 (a) 98.7 mL (b) 1.01×10^4 mL

5.21 (a) 1.02 L (b) 1.43 L

5.23 (a) $-124.1°C$ (b) $14.3°C$

5.25 Sample 1: yes with absolute zero at $-276°C$
Sample 2: yes with absolute zero at $-272°C$

5.27 (a) 829.6 Torr (b) 0.0134 atm

5.29 (a) 7.56×10^3 Torr (b) 53 Torr

5.31 (a) 0.301 g (b) 4.73×10^{-3} g

5.33 875 Torr

5.35 68.3 mL

5.37 78.1 g/mol

5.39 (a) Van der Waals: 1.05 atm; ideal: 1.06 atm, about 1% less (b) Van der Waals: 1.00 atm; ideal: 1.00 atm, negligible

5.41 (a) 22.4 L (b) 44.8 L (c) 3.41 L (d) 2.5 L

5.43 (a) 0.700 L (b) 11.1 L (c) 2.55 L (d) 6.59 L

5.45 (a) 4.46×10^{-3} mol (b) 4.46×10^{-3} mol
(c) 45 mol (d) 4.5×10^{-5} mol

5.47 (a) N_2: 0.22 atm, H_2: 0.67 atm; total: 0.90 atm
(b) N_2: 2×10^1 Torr, H_2: 5×10^1 Torr, NH_3: 3×10^1 Torr; total: 1×10^2 Torr (c) Ar: 47.8 Torr, Ne: 94.5 Torr; total: 142.3 Torr (d) O_2: 0.804 Torr, CO: 0.614 Torr, CO_2: 0.405 Torr; total: 1.830 Torr

5.49 0.085 g H_2O

5.51 $P_{CO_2} = 4.6$ Torr, $P_{N_2} = 0.2$ Torr, $P_{Ar} = 0.1$ Torr

5.53 10.30 m

5.55 6.5 L H_2, 3.2 L O_2

5.57 1.1 L

5.59 2.33×10^4 L, 3.6×10^3 L

5.61 626 L

5.63 8.46×10^{-3} mol

5.65 6.17×10^{-3} mol

5.67 (a) 1840 m/s (b) 640 m/s (c) 392 m/s
(d) 215 m/s

5.69 $CO_2 < Ar < H_2$

5.71 (a) 3.51 times (b) 0.523 times (c) 0.949 times

5.73 (a) 154 s (b) 98.6 s (c) 205 s (d) 370 s

5.75 115 g/mol, $C_6H_{12}O_2$

5.77 220 g/mol, P_4O_6

5.79 324.3 g/mol, $C_{20}H_{24}N_2O_2$

CHAPTER 6

6.1 (a) 0.11 kJ (b) 6.0×10^2 kJ

6.3 (a) 14 J (b) 5.98×10^4 kJ

6.5 (a) 82 m (b) 0.7 m (c) 73 m

6.7 1.58 kJ

6.9 298 kJ, 89.6%

6.11 2.0 min

6.13 (a) 461 J/K (b) 43 J/K

6.15 43.5 kJ/mol

6.17 41.2 kJ/mol

6.19 226 kJ, 226 s

6.21 0.247 kJ, 0.247 s

6.23 2 g

6.25 1.90 kJ/mol

6.27 12.3 kJ

6.29 1.4 kJ, −57 kJ/mol

6.31 -1.20×10^3 kJ

6.33 -2.80×10^3 kJ/mol

6.35 (a) A (b) P (c) A (d) A

6.39 2.97×10^3 kJ

6.41 18 kg

6.43 Aniline, 36.43 kJ/g; phenol, 32.45 kJ/g

6.45 −802 kJ/mol

6.47 +607 kJ/mol

6.49 −1570 kJ/mol

6.51 −197.8 kJ

6.53 −1131 kJ

6.55 −444 kJ/mol

6.57 −676 kJ/mol

6.59 −137 kJ

6.61 −393.5 kJ/mol

6.63 (a) −124.7 kJ/mol (b) −1272 kJ/mol
(c) −533 kJ/mol

6.65 −44 kJ

6.67 −602 kJ/mol

6.69 −53 kJ mol

6.71 (a) +178.3 kJ/mol (for calcite) (b) +8.74 kJ/mol
(c) −480.30 kJ/mol (d) +246.5 kJ/mol

6.73 (a) −138.18 kJ (b) −36.03 kJ (c) −144.6 kJ
(d) +752.3 kJ

6.75 −11.3 kJ/mol

6.77 48.44 kJ/g, 3.3×10^4 kJ/L

6.79 (a) 37 kJ/L (b) 91 kJ/L (c) 3.7×10^4 kJ/L
(d) 1.8×10^4 kJ/L

6.81 24.75 kJ/g for Mg as compared to 31.05 kJ/g for
Al; Yes, Al is better

6.83 890 kJ, 684 kJ, 468 kJ

6.85 −34.02 kJ/mol, 2.47×10^4 kJ/L

6.87 688.1 kJ/mol

CHAPTER 7

7.1 (a) 3.1 m (b) 5.7×10^{-7} m (c) 150 pm

7.3 (a) 5.66×10^{14} Hz (b) 1.2 MHz (c) 2.01×10^{18} Hz

7.5 1.26 m

7.7 (a) 3.43×10^{-19} J, 2.06×10^5 J/mol
(b) 3.11×10^{-19} J, 1.87×10^5 J/mol
(c) 1.29×10^{-15} J, 7.78×10^8 J/mol

7.9 -4.08×10^{-19} J, -2.46×10^5 J/mol
-3.87×10^{-19} J, -2.33×10^5 J/mol
-3.66×10^{-19} J, -2.21×10^5 J/mol
-3.60×10^{-19} J, -2.17×10^5 J/mol
-3.44×10^{-19} J, -2.07×10^5 J/mol

7.11 2.9×10^{18} photons/s, 2.1×10^5 s

7.13 1.9×10^{18} photons/m² · s

7.15 600 nm, 4.63×10^{-19} J/electron

7.17 For $n_i = 3$, 656 nm; for $n_i = 4$, 486 nm; for $n_i = 5$, 434 nm; and so forth for $n_i = 6, 7, \cdots$

7.19 (a) For $n_i = 3$, 164 nm; for $n_i = 4$, 122 nm; for $n_i = 5$, 109 nm; and so forth for $n_i = 6, 7, \cdots$ (b) For $n_i = 3$, 18.2 nm; for $n_i = 4$, 13.5 nm; for $n_i = 5$, 12.1 nm; and so forth for $n_i = 6, 7 \cdots$

7.21 (a) 3.31×10^{-10} m (b) 0.0728 m (c) 3.96×10^{-5} m

7.23 The woman (the lighter)

7.25 (a) $\Delta v = 5.8 \times 10^5$ m/s (b) $\Delta x = 0.058$ m

7.27 (a) 5.3×10^{-21} m (b) 5.3×10^{-17} m/s

7.29 $\Delta x = 5.3 \times 10^{-45}$ m; no, you could not claim you were over 10 m away

7.31 (a) 1 (b) 4 (c) 9

7.33 (a) 1, 1s (b) 5, 3d (c) 5, 4d (d) 1, 3s

7.51 (a) $\psi^2 = 0.70$ at 10 pm; $\psi^2 = 0.10$ at 60 pm; it is seven times more probable to find the electron at 10 pm (b) Increase (c) Decrease, since the p orbital is shaped like a figure 8; the probability of finding an electron at the nucleus is zero for a p orbital

7.53 9.17×10^{-20} J (55.2 kJ/mol)

7.85 2.18×10^{-18} J/atom (1.31×10^6 kJ/mol)

7.87 (a) 6.94×10^{-19} J/atom (b) 287 nm

7.89 4.84×10^{-19} J/atom (2.91 kJ/mol)

7.93 -9.94×10^{-20} J (−59.9 kJ/mol)

7.113 (a) No (b) $n \geq 2$ (c) Maximum energy for $He^+ = 8.72 \times 10^{-18}$ J, which is less than the ionization energy of Li^{2+} from ground state, 19.6×10^{-18} J (d) $n \geq 3$ (e) $n = Z/\sqrt{3}$

7.119 (b) 1.10×10^5 cm^{-1} (c) 1.19×10^{-2} kJ/mol =
1 cm^{-1} (d) Li, 4.36×10^4 cm^{-1}; Na, $4.15 \times$
10^4 cm^{-1}; K, 3.51×10^4 cm^{-1}; Rb, 3.38×10^4 cm^{-1};
Cs, 3.16×10^4 cm^{-1}

CHAPTER 8

8.5 (a) $+2$ (b) -3

8.9 (a) $d < 959$ pm   (b) $d < 149$ pm

8.11 (a) $d < 511$ pm   (b) $d < 1500$ pm

8.15 AgI < AgBr < AgCl < AgF

8.17 1021 kJ/mol

8.19 3887 kJ/mol

8.21 NaCl$_2$(s) would be unlikely with ΔH_f°
approximately 2200 kJ/mol

8.45 (a) 926 kJ/mol   (b) 1486 kJ/mol   (c) 3150 kJ/mol
(d) 2317 kJ/mol

8.47 (a) -92 kJ/mol   (b) -151 kJ/mol

8.49 (a) -286 kJ/mol   (b) -257 kJ/mol   (c) -1 kJ/mol

8.51 (a) -997 kJ/mol   (b) -4170 kJ/mol
(c) -535 kJ/mol

8.53 (a) -2288 kJ/mol   (b) -1142 kJ/mol

8.75 (a) 1.4 D   (b) 0.5 D   (c) 1.0 D   (d) 1.0 D

CHAPTER 10

10.11 151 Torr

10.13 17 mg

10.15 1.8 kg

10.17 (a) 95°C   (b) 102°C   (c) 50°C

10.19 (a) 12 kJ/mol   (b) 30 kJ/mol   (c) 39 kJ/mol

10.21 (a) Abnormal, $\Delta H_{vap}/T_b = 125$ kJ/mol · K
(b) normal, $\Delta H_{vap}/T_b = 86$ kJ/mol · K

10.33 168 pm

10.35 (a) Ni radius = 124 pm   (b) K radius =
231 pm

10.37 (a) 405 pm   (b) 1.5×10^{22} unit cells/cm^3

10.39 (a) 1 atom/unit cell   (b) 6   (c) 280 pm

10.41 (a) 9.02 g/mL   (b) 1.77 g/mL

10.43 144 pm

10.45 132 pm

10.47 (a) 51 Na: 20 K   (b) 52 Zn: 100 Cu   (c) 7 Sn:
2 Pb: 111 Cu

10.49 (a) 49.29% Cu, 50.71% Zn   (b) 9.33% Zn,
90.67% Cu

10.61 10^{19} unit cells assuming an edge-length of
512 pm

10.63 (a) 4.4 g/mL   (b) 5.4 g/mL

10.67 178 kJ/mol of bonds

10.77 3.6×10^2 Torr

10.79 4.5 kg

10.81 21.1 g/mL

10.83 74.0%

10.85 Ne 170. pm, Ar 203 pm, Kr 225 pm, Xe 232 pm,
Rn 246 pm

10.87 (b) 690 nm

CHAPTER 11

11.1 (a) 0.34 M (b) 0.222 M

11.3 (a) 55.4 M (b) 53.2 M

11.5 (a) 1.5 g   (b) 2.8 g   (c) 45.0 g

11.7 (a) 0.68 m (b) 0.12 m (c) 0.029 m (d) 13 m

11.9 (a) 0.10 g   (b) 0.91 g   (c) 28 g   (d) 20 g

11.11 61 g

11.13 (a) 0.72 H$_2$O, 0.28 C$_2$H$_6$O   (b) 0.0018 glucose,
0.998 H$_2$O

11.15 (a) 0.0018 Na$^+$, 0.0018 Cl$^-$, 0.9964 H$_2$O
(b) 0.0036 Na$^+$, 0.0018 CO$_3^{2-}$, 0.9946 H$_2$O

11.17 (a) 0.90 m (b) 0.32 m

11.19 (a) 3.40 M (b) 5.41 M (c) 5×10^{-6} M
(d) 11.8 M

11.21 (a) 0.0013 M, 42 mg/L   (b) 2.7×10^{-4} M,
8.6 mg/L   (c) 0.023 M, 1.0×10^3 mg/L   (d) 0.0023 M,
10 mg/L

11.23 (a) 0.1 atm   (b) 0.5 atm

11.25 For each of the given solutions, it is assumed
that you have 110 g of solution with a specific heat of
4.0 J/gK: (a) -1.5 K   (b) $+1.3$ K   (c) $+56$ K
(d) -7.3 K

11.27 (a) Increase   (b) Decrease

11.29 (a) -440 kJ/mol   (b) -344 kJ/mol

11.31 (a) $+3$ kJ/mol   (b) $+16$ kJ/mol

11.33 (a) -8 kJ/mol   (b) -159 kJ/mol

11.35 (a) -306 kJ/mol   (b) -338 kJ/mol

11.37 (a) -387 kJ/mol   (b) -320 kJ/mol
(c) -1860 kJ/mol   (d) -4650 kJ/mol

11.39 (a) -512 kJ/mol   (b) -385 kJ/mol

11.41 (a) 179 pm   (b) 258 pm

11.43 (a) 684 Torr   (b) 622 Torr (*not 608* Torr)

11.45 (a) 55.1 Torr   (b) 20.9 Torr

11.47 (a) 343 Torr   (b) 337 Torr   (c) 349 Torr

11.49 1.1×10^2 g/mol

11.51 1.04%

11.53 (a) 0.051°C   (b) 0.010°C

11.55 0.091 K

11.57 (a) 0.34°C   (b) 0.29°C

11.59 168 g/mol

11.61 100.5°C

11.63 (a) 0.19°C   (b) 0.04°C

11.65 (a) 0.17 K   (b) 1.2 K   (c) 4 K

11.67 182 g/mol

11.69 About −2°C

11.71 (a) 2.4×10 kPa, 2.4 atm   (b) 49 kPa, 0.48 atm

11.73 (a) 2×10^{-2} kPa, 2×10^{-4} atm   (b) 1.6×10^3 kPa, 16 atm   (c) 4.9×10^3 kPa, 48 atm

11.75 (a) 12 m   (b) 0.50 m

11.77 2 mm

11.79 40 mm

11.81 46 mm

11.83 3.4×10^3 g/mol

CHAPTER 12

12.1 (a) 1.0 mmol/L · s   (b) 5.2×10^3 mmol/L · s   (c) 5.4 mol/L · s   (d) 0.28 mol/L · s

12.5

Time, s	Rate, *M*/s
0	7.3×10^{-8}
4000	6.4×10^{-8}
8000	5.6×10^{-8}
12000	4.9×10^{-8}
16000	4.3×10^{-8}

12.7

Time, s	Rate, *M*/s
0	1.3×10^{-5}
1000	7.6×10^{-6}
2000	1.0×10^{-6}
3000	5.9×10^{-7}
4000	3.4×10^{-7}
5000	2.2×10^{-7}

12.9 Rate = $k[\text{OH}^-][\text{CH}_3\text{Br}]$

12.11 Rate = $k[\text{ICl}][\text{H}_2]$

12.13 Rate = $k[\text{A}][\text{B}]^2[\text{C}]$

12.15 3.1×10^{-5} *M*/s

12.17 (a) Times 2   (b) No change

12.21 (a) Rate = $k[\text{A}][\text{B}]$ (b) Rate = $k[\text{A}][\text{C}]^{-1/2}$

12.23 (a) 400 s   (b) 800 s

12.25 (a) 700 s   (b) 3100 s

12.29 (a) 6.93×10^{-4}/s   (b) 1.39×10^{-2}/s   (c) 8.17×10^{-3}/s

12.31 (a) 4.00 *M* · s   (b) 7.5 *M* · s   (c) 0.0967 *M* · s

12.33 172 s

12.35 75 s

12.37 (a) 1.4×10^2 s   (b) 82 s

12.39 3.4×10^{-5}/s

12.41 4.4×10^{-4}/s

12.43 1.2×10^{-4}/*M* · s

12.45 1.3×10^3/*M* · s

12.47 Second order, 1.2×10^{-2}/*M* · s

12.49 Second order, 0.14/*M* · s

12.63 (a) 8.4×10^{-4}/s   (b) 2.6×10^9/*M* · s

12.65 2.4×10^2 kJ/mol

12.67 2.7×10^2 kJ/mol

12.69 92 kJ/mol, 23 mL/mol · s

12.71 7.6×10^{10}

12.73 300 kJ/mol

CHAPTER 13

13.3 (a) 0.024 atm^2 (b) 1.7×10^3 atm^{-4} (c) 6.4 atm^{-1}

13.5 48.8, 48.9, 48.9

13.7 5.84×10^{-3} *M*

13.9 0.0955 atm^2, 0.0939 atm^2, 0.0938 atm^2 (Average = 0.0944 atm^2)

13.11 6.6×10^{-3}

13.13 2.3×10^{-4} atm^3

13.15 (a) 0.024 *M*   (b) 2.5×10^{-3} *M*

13.17 7.1×10^{-5}/Torr^2

13.27 (a) 51 *M*   (b) 141 *M*   (c) 17.1 *M*   (d) 79.62 *M*

13.31 (a) 0.50, products   (b) 1, products

13.33 Yes

13.35 7.3×10^{-5} *M*

13.37 19 atm

13.39 $[\text{H}^+] = [\text{OH}^-] = 1 \times 10^{-7}$ *M*

13.41 (a) 1.9×10^{-3} g/L   (b) 2.6×10^{-7} g/L

13.45 (a) $[\text{Cl}_2] = 1 \times 10^{-3}$ *M*, $[\text{Cl}] = 1.1 \times 10^{-5}$ *M*   (b) $[\text{Cl}_2] = 9 \times 10^{-4}$ *M*, $[\text{Cl}] = 1.3 \times 10^{-4}$ *M*

13.47 $[\text{HBr}] = 0.0012$ *M*, $[\text{H}_2] = [\text{Br}_2] = 1.1 \times 10^{-8}$ *M*

13.49 $[\text{PCl}_5] = 8.2 \times 10^{-3}$; $[\text{PCl}_3] = [\text{Cl}_2] = 0.011$ *M*

13.51 2.8

13.57 (a) 0.95   (b) 0.91

13.65 4.2×10^{-5} *M*

13.67 44.2 kJ/mol, 305 K

CHAPTER 14

14.15 (a) 3.572   (b) 4.757   (c) 0.52   (d) 4.190

14.17 (a) 7.6×10^{-3} (b) 0.010   (c) 0.012   (d) 3.5×10^{-3}

14.19 (a) 4.747 (b) 4.96 (c) 5.77 (d) 7.971

14.21 (a) 4.3×10^{-10} (b) 7.2×10^{-11} (c) 1.4×10^{-10}

14.23 (a) 2.3×10^{-5}, 4.63 (b) 1.4×10^{-4}, 3.86 (c) 7.1×10^{-5}, 4.15

14.27 (a) 3.0% (b) 0.95%

14.33 (a) $0.1\ M$ (b) $1.0 \times 10^{-7}\ M$ (c) $3.2\ M$

14.35 (a) 4.70 (b) 0.00 (c) 13.30 (d) -0.70

14.37 (a) 2.0, 12.0 (b) 2.7, 11.3 (c) 12.0, 2.0

14.39 $2.0 \times 10^{-13}\ M$, 12.70

14.41 $3.0 \times 10^{-14}\ M$, 13.52

14.43 (a) 2.79 (b) 5.14 (c) 0.96

14.45 $1.6 \times 10^{-3}\ M$, 2.80, 11.20

14.47 2.0

14.49 $pK_b = 3.12$, $pK_a = 10.88$

14.51 0.80

14.53 (a) 3.94 (b) 4.68

14.55 (a) $[H^+] = [HC_2O_4^-] = 0.053\ M$; $[C_2O_4^{2-}] = 6.5 \times 10^{-5}\ M$ (b) $[H^+] = [H_2BO_3^-] = 2.7 \times 10^{-7}\ M$; all other ions negligible

14.57 0.125 kJ/mol

14.59 59 kJ/mol, 7.44

CHAPTER 15

15.3 (a) 9.02 (b) 5.13 (c) 10.6

15.5 (a) 9.18 (b) 4.61 (c) 7.00

15.7 (a) 8.40 (b) 9.00

15.9 2.4

15.11 (a) 9.55 (b) 8.95

15.13 (a) $pK_a = pH = 3.86$ (b) 3.55

15.15 (a) 4.75 (b) 8.95

15.17 2.21

15.19 11/1

15.21 acid/base = 2.5/1

15.23 18.3 mL, $0.0635\ M$ NaCl

15.25 423 mL, $0.142\ M$ NaCl

15.27 (a) 13.04 (b) 12.82 (c) 12.54 (d) 7.00 (e) 1.81 (f) 1.55

15.29 (a) 10.54 (b) 9.54 (c) 4.46 (d) 3.46

15.31 10.30 to 3.70

15.33 Initial pH = 2.87, pH at end point = 8.72

15.35 Initial pH = 11.22, end point pH = 5.24

15.37 (a) 11.15 (b) 9.68 (c) 9.17 (d) 5.22 (e) 1.81 (f) 1.55

15.39 (a) 8.01 (b) 6.99 (c) 3.09 (d) 2.11

15.41 (a) First end point at 1.54, second end point at 7.37 (b) First end point at 2.71, second end point at 8.51

15.49 (a) $pH \approx 3.8 - \log \dfrac{[\text{acid}]}{[\text{base}]}$

(b) $pH \approx 5.4 - \log \dfrac{[\text{acid}]}{[\text{base}]}$

(c) $pH \approx 7 - \log \dfrac{[\text{acid}]}{[\text{base}]}$

(d) $pH \approx 9.1 - \log \dfrac{[\text{acid}]}{[\text{base}]}$

15.51 (a) 7.0 to 2.3 (b) 4.75 to 4.71

15.53 (a) pK_a of lactic acid (b) 12.67 ± 1.00 (c) 7.21 ± 1.00

15.57 (a) 42 mL HCl, pH = 9.25 (b) 42 mL NaOH, pH = 11.25

15.59 pH = 7.78

15.63 (a) 7.7×10^{-13}, 12.11 (b) 1.7×10^{-14}, 13.77 (c) 5.3×10^{-3}, 2.28 (d) 2.5×10^{-13}, 12.60

15.65 (a) $1.2 \times 10^{-17}\ M$ (b) $9.2 \times 10^{-23}\ M$ (c) $9.3 \times 10^{-5}\ M$ (d) $1.6 \times 10^{-2}\ M$

15.67 (a) No, 5×10^{-13} does not exceed $K_{sp} = 4.0 \times 10^{-11}$ (b) $Ca_5(PO_4)_3OH$ solubility $= 3 \times 10^{-5}\ M$, $Ca_5(PO_4)_3F$ solubility $= 6 \times 10^{-8}\ M$

15.69 (a) $8 \times 10^{-10}\ M$ (b) $1.3 \times 10^{-16}\ M$ (c) $4.0 \times 10^{-4}\ M$

15.71 (a) $S = 10^{(42-3pH-pK_{sp})}$; (1) at pH = 7, $S = 10^{-12}\ M$; (2) at pH = 4.5, $S = 3 \times 10^{-5}\ M$ (b) $S = 10^{(28-2pH-pK_{sp})}$; (1) at pH = 7, $S = 2.0 \times 10^{-3}\ M$; (2) at pH = 6, $S = 0.20\ M$

15.73 (a) $1.7 \times 10^{-4}\ M$, $1.7 \times 10^{-20}\ M$ (b) $4.7 \times 10^{-17}\ M$, $4.7 \times 10^{-41}\ M$

15.75 (a) $2.1 \times 10^{-4}\ M$ (b) $3.6 \times 10^{-4}\ M$

15.77 (a) No, 1.4×10^{-6} does not exceed 1.4×10^{-4} (b) Yes, 6.8×10^{-14} exceeds 1.0×10^{-33} (c) Yes, 9.6×10^{-7} exceeds 1.6×10^{-10} (d) Yes, 3.7×10^{-5} just barely exceeds 2.4×10^{-5}

15.79 CuS

15.81 1st to last: Cu, Mn, Ca, Ba

15.83 Yes, since Hg^{2+} is down to about $10^{-25}\ M$ before PbI_2 begins to precipitate

15.85 $[Cu^{2+}] = 6.4 \times 10^{-14}\ M$

15.87 $[NH_3] = 0.09$, $S_{AgCl} = 4.6 \times 10^{-3}\ M$

CHAPTER 16

16.5 (a) 0 (b) 9.56×10^{-23} J/K

16.7 5.18 J/K

16.9 5.76 J/K · mol, indicates two orientations of N_2O_2 dimers in the solid

16.11 (a) -22.0 J/K   (b) $+45.3$ J/K

16.13 -326.68 J/K

16.15 (a) $+125.8$ J/K   (b) $+498.92$ J/K (assuming rhombic sulfur)   (c) -395.4 J/K   (d) $+366.9$ J/K

16.17 (a) 336 J/K   (b) 100 J/K   (c) 1×10^6 J/K

16.19 (a) 0.341 J/K · s   (b) 2.95×10^4 J/K · day

16.21 (a) 0.033 J/K   (b) 1 J/K   (c) 0.02 J/K

16.23

	Overall	Surroundings	Substance
(a)	0	-74 J/K	$+74$ J/K
(b)	0	-73 J/K	$+73$ J/K
(c)	0	-124 J/K	$+124$ J/K
(d)	0	-87 J/K	$+87$ J/K

16.25 (a) 22.0 J/K   (b) 28.9 J/K   (c) 14 J/K

16.27 1.7×10^2 K ($-107°$C)

16.29 80 K

16.31 30 kJ/mol

16.33 From 16.13: (a) -474.30 kJ   (b) $+124.5$ kJ   (c) -4.80 kJ   (d) $+34.61$ kJ   From 16.15: (a) -233.56 kJ   (b) $+97.82$ kJ   (c) -641.03 kJ   (d) $+642.03$ kJ

16.37 (a) -16.48 kJ/mol   (b) -237.12 kJ/mol   (c) -50.70 kJ/mol   (d) -120.32 kJ/mol

16.39 (a) 65.27 kJ/mol   (b) -300.19 kJ/mol   (c) -1128.9 kJ/mol   (d) -394.36 kJ/mol

16.41 (a) 3.40 kJ/mol   (b) -141.74 kJ/mol   (c) 130.41 kJ/mol (calcite)

16.43 (a) Stable   (b) Unstable   (c) Unstable   (d) Unstable

16.45 (a) More   (b) More   (c) More   (d) More

16.47 Yes

16.49 (a) Yes, for either CO or CO_2 as products   (b) No, for either CO or CO_2 as products

16.51 $+317.02$ kJ

16.53 (a) 0.25   (b) 7.2×10^{24} atm^{-1} (c) 1.4×10^{-23} atm

16.55 (a) 1.4×10^{83} (b) 6.75   (c) 8.4×10^{-7}

16.57 The reaction has a tendency to form reactants

16.59 (a) 1.7×10^{-10}, 1.3×10^{-5} M (b) 5.7×10^{-13}, 7.5×10^{-7} M

16.61 (a) -12.3 kJ, 6.2 kJ/mol   (b) $-60.$ kJ

16.63 $pK_{sp} = 16.08$

16.65 1.5 h

CHAPTER 17

17.21 (a) 0.34 V   (b) 0.77 V   (c) 0.00 V

17.23 (a) 0.53 V   (b) 0.27 V   (c) 0.07 V

17.25 (a) 0.18 V, anode: $Cu|Cu^{2+}$ (b) 0.83 V, anode: $I^-|I_3^-$ (c) 0.37 V, anode: $Ag|AgI|I^-$

17.27 (a) 0.52 V, anode: $Hg|Hg_2Cl_2|Cl^-$.   (b) 1.52 V, anode: $Sn^{2+}|Sn^{4+}$ (c) 0.83 V, anode: $Pt|O_2|OH^-$

17.49 (a) 1.10 V   (b) 1.23 V   (c) 1.21 V   (d) 0.58 V

17.51 -84.0 kJ/mol

17.55 (a) 0.018 V   (b) 0.059 V   (c) -0.40 V   (d) 0.51 V

17.57 $Q = 2.6 \times 10^6$

17.63 (a) 4.7×10^{-13} (b) 2.25×10^{51} (c) 1.2×10^5 (d) 14.02

17.65 Yes, $K = 2 \times 10^2$

17.71 (a) $\frac{1}{2}$ mol Cu   (b) $\frac{1}{3}$ mol Al   (c) $\frac{1}{2}$ mol H_2 (d) $\frac{1}{4}$ mol O_2

17.73 (a) 1×10^{-5} mol   (b) 0.003 mol   (c) 4 mol

17.75 0.67 mol

17.77 (a) 13 g Ne   (b) 15 g Zn   (c) 8.5 g F_2 (d) 18 g NaOH

17.79 (a) 0.418 L H_2 (b) 0.228 L O_2 (c) 0.409 L F_2

17.81 (a) 3.0 h   (b) 6.2 h   (c) 34 h   (d) 4.5 min

17.83 Ti^{2+}

17.85 0.9 g Zn

17.87 (a) 0.52 V   (b) -0.037 V

CHAPTER 18

18.11 (a) $+206.10$ kJ

18.13 1.5×10^{11} g H_2; 3.0×10^{12} g H_2 is needed, meaning additional H_2 needs to be imported

18.21 1.1 g/mL

18.33 7.15 g $Na_2CO_3 \cdot 10H_2O$

18.37 9 days

CHAPTER 19

19.17 35.5 kg Al, 68.4 kg Cr

19.25 Average B—H in BH_3, 372 kJ/mol; average B—H in B_2H_6, 400 kJ/mol; average B—H is 372 kJ/mol; average B—H—B = 228 kJ/mol; B—H—B are longer

19.27 -481 kJ/mol

19.75 2.7×10^2 g NaN_3

19.81 0.331 M NO_3^-

CHAPTER 20

20.35 $+553$ kJ

20.79 0.49 atm H_2, 0.49 atm I_2, 0.02 atm HI
20.85 0.117 M F^-
20.87 0.150 M ClO^-

CHAPTER 21

21.23 -0.74 V
21.37 $+0.52$ V
21.39 1.2×10^6

CHAPTER 22

22.7 (a) 15 pm, 0.01×10^{-12} J   (b) 0.06 pm, 3×10^{-12} J

22.9 (a) 6.2×10^{19} Hz, 4.8×10^{-12} m   (b) 2.4×10^{20} Hz, 1.2×10^{-12} m

22.11 (a) 3.2×10^{20} Hz, 9.3×10^{-13} m   (b) 2.6×10^{20} Hz, 1.1×10^{-13} m

22.33 (a) 1.0×10^{-4} Ci

22.35 (a) 3.7×10^{10} disintegrations   (b) 3.7×10^4 disintegrations

22.37 (a) 1.0×10^{-6} Ci   (b) 2.7×10^{-7} Ci

22.39 1.0×10^2 rad, 1.0×10^2 rem

22.41 100 days

22.43 (a) 0.0564/year   (b) 1.209×10^{-4}/year
(c) 5.3×10^{-10}/year   (d) 3.5×10^{-2}/s

22.45 (a) 1.60×10^3 years   (b) 10.52 years
(c) 3.78×10^9 years

22.47 0.0696 mCi

22.49 (a) 88.60%   (b) 95.54%   (c) 62.42%
(d) 91.75%

22.51 (a) 5.3×10^3 years   (b) 8.4×10^9 years
(c) 2 years

22.53 (a) 0.00100 Ci   (b) 1.8×10^{-6} Ci

22.55 1.15×10^4 years

22.57 1670 years

22.59 (a) 4.2×10^{-12} g, gain   (b) 66×10^{-12} g, loss
(c) 3.2×10^{-12} g, loss

22.61 (a) 9×10^{13} J, 6×10^{26} MeV   (b) 1.5×10^{-10} J,
9×10^2 MeV

22.63 (a) 1.14×10^{-12} J/nucleon, 7.12 MeV
(b) 1.23×10^{-12} J/nucleon, 7.69 MeV   (c) 1.41×10^{-12} J/nucleon, 8.81 MeV   (d) 1.22×10^{-12} J/nucleon, 7.62 MeV

22.65 (a) 7.8×10^{10} J/g D   (b) 5.9×10^{11} J/g ^{3}He
(c) 2.39×10^{11} J/g ^{7}Li   (d) 8.08×10^{11} J/g D

CHAPTER 23

23.25 C_2H_5, alkane
23.45 2.0×10^6 monomers/molecule

ILLUSTRATION CREDITS

All photos by Ken Karp except the following:

p. 2, Lick Observatory; p. 5, Edward J. Olsen; p. 7, The Aluminum Association; p. 8, C. F. Quate and Sang-il Park, W. W. Hansen Laboratories of Physics, Stanford University; p. 12, Champlin Petroleum Co.; p. 14, (Fig. 1.12) Paul Brierley; p. 21, S. Jonasson/Bruce Coleman; p. 33, Freer Gallery of Art, Smithsonian Institution; p. 40, Paul Brierley, RCA Ltd.; p. 46, Manchester Literary and Philosophical Society; p. 47, (Fig. 2.9) AT&T, (Fig. 2.10) Cavendish Laboratory, (Fig. 2.11) Donald Clegg; p. 48, Cavendish Laboratory; p. 57, (Fig. 2.22) Bettman Archives; p. 80, Woods Hole Oceanographic Institution; p. 106, F. S. Judd, Research Laboratories, Kodak Ltd.; p. 116, NASA; p. 119, The M. W. Kellogg Co.; p. 121, The Algoma Steel Corp. Ltd.; p. 150, NASA; p. 161, (Fig. 5.12) Jim Neill; p. 170, U. S. Department of the Interior, Bureau of Mines; p. 178, Martin Marietta Energy Systems, Oak Ridge, Tenn.; p. 186, Paul Brierley; p. 189, USDA Forest Service/Barry Nehr; p. 195, (Fig. 6.8b) NASA; p. 206, Gas Research Institute, Georgia Tech Laboratory; p. 217, Dr. Jeremy Burgess/Science Photo Library/Photo Researchers; p. 218, Atlas Power Co.; p. 235, Century Lubricant Specialists; p. 237, American Institute of Physics; p. 240, Science Museum, London; p. 241, Dublin Institute for Advanced Studies; p. 254, Edgar Fahs Smith Collection, ACS Center for History of Chemistry, University of Pennsylvania; p. 276, Paul Brierley; p. 291, University of California Archives, The Bancroft Library; p. 301, Allied-Signal Inc.; p. 303, (left) Photo Researchers, (right) National Astronomy Observatory; p. 318, Photo Researchers; p. 338, Donald Clegg; p. 350, Paul Brierley; p. 360, Robert Carlyle Day/Photo Researchers; p. 361, M. P. Allen, University of Bristol; p. 363, (Fig. 10.17) NASA; p. 371, Dr. Jeremy Burgess/Photo Science Library/Photo Researchers; p. 373, Argonne National Laboratory; p. 374, S. L. Craig/Bruce Coleman; p. 382, General Electric Company; p. 383, Chip Clark; p. 384, (Fig. 10.47) M. P. Allen, University of Bristol, (Fig. 10.48) J. R. Eyerman; p. 392, Paul Brierley; p. 397, adapted from P. A. Leighton, *The Photochemistry of Air Pollution*, Academic Press, New York, 1961; p. 402, (Fig. 11.6) Chilean Nitrate Corp., (Fig. 11.7) Runk/Schoenberger/Grant Heilman; p. 405, E. B. Smith, Physical Chemistry Laboratory, University of Oxford; p. 408, Airscam™ thermogram by Daedalus Enterprises Inc./National Geographic Magazine; p. 418, Wacker Siltronic Corp.; p. 423, (Fig. 11.32) from "Biological Membranes as Bilayer Couples," by M. Sheetz, R. Painter, and S. Singer, 1976, *Journal of Cell Biology*, 70:193, (Fig. 11.33) Paul Brierley; p. 430, Arthur T. Winfree, photographed by Fritz Goro; p. 453, Malcolm Lockwood, Geophysical Institute, University of Alaska—Fairbanks; p. 458, (Fig. 12.17) General Motors; p. 467,

Johnson Matthey; p. 468, Lubert Stryer; p. 491, NASA/Science Photo Library/Photo Researchers; p. 504, (Fig. 13.13, left) American Institute for Physics, Neils Bohr Library/Stein Collection, (Fig. 13.13, right) Bettman Archives, (Fig. 13.14) M. W. Kellogg Co.; p. 544, Photo Researchers; p. 560, Fisher Scientific; p. 561, Heather Angel/Biofotos; p. 586, Paul Brierley; p. 590, (Fig. 16.4) Dieter Flamm; p. 594, International Tin Research Institute; p. 601, Yale University Library; p. 616, Paul Brierley; p. 628, Bruce Coleman; p. 629, The National Museum of Science & Industry, London; p. 637, G. E. Thompson and G. C. Wood, Corrosion Protection Centre, UMIST, Manchester, England; p. 639, (Fig. 17.20) St. Joe Zinc Co./American Hot Dip Galvinizers Association; p. 650, R. H. Manley; p. 663, (Fig. 18.4) data from Martin Marietta Aerospace, New Orleans; p. 670, (Fig. 18.12) from P. P. Edwards and C. N. R. Rao, *The Metallic and Nonmetallic States of Matter*, Taylor & Francis, 1985; p. 672, R. H. Manley; p. 678, (Fig. 18.18) Houston Museum of Natural Science, (Fig. 18.19) AMAX Magnesium; p. 683, Bruce Coleman; p. 684, Field Museum of Natural History, Chicago; p. 686, Bruce Iverson; p. 690, Paul Brierley; p. 694, (Fig. 19.2) Paul Brierley, (Fig. 19.3) Waste Management, Inc.; p. 696, (a, c) Houston Museum of Science, (b) Lee Boltin, Inc.; p. 703, Field Museum of Natural History, Chicago; p. 706, Photo Researchers, Inc.; p. 708, The Rockwell Museum, Corning, N.Y.; p. 709, The Rockwell Museum, Corning, N.Y.; p. 710, (Fig. 19.16) Runk/Schoenburg/Grant Heilman, (Fig. 19.17) Lenox China; p. 711, Grant Heilman; p. 712, NASA; p. 715, Photo Researchers; p. 730, Paul Brierley; p. 734, Field Museum of Natural History, Chicago; p. 739, Lee Boltin; p. 744, (Fig. 20.13) Martin Marietta Energy Systems, Inc.; p. 750, Martin Thiokol; p. 753, Greater Pittsburgh Neon/Tom Anthony; p. 760, Photo Researchers; p. 767, (a) Field Museum of Natural History, Chicago, (b) Hagley Museum and Library, (c) Oremet Titanium; p. 768, Bruce Iverson; p. 770, Woods Hole Oceanographic Institute; p. 771, Paul Brierley; p. 772, U.S. Steel; p. 776, U.S. Department of the Interior, Bureau of Mines; p. 777, (Fig. 21.16) Lee Boltin Picture Library, (Fig. 21.17) Field Museum of Natural History, Chicago; p. 784, Concordia University, © Carol Moralejo 1988; p. 791, Concordia University, © Carol Moralejo 1988; p. 800, Paul Brierley; p. 803, Granger Collection; p. 812, Standard Linear Accelerator Center; p. 815, Granger Collection; p. 820, University of Arizona; p. 829, (Fig. 22.20) Princeton University, (Fig. 22.21) Harvard University; p. 830, Martin Marietta Energy Systems, Inc.; p. 831, Photo Researchers; p. 838, Paul Brierley; p. 850, American Petroleum Institute; p. 859, Photo Researchers; p. 870, Paul Brierley; p. 892, Hagley Museum and Library; p. 901, Photo Researchers; p. 902, Photo Researchers; p. 906, Stockmarket

INDEX

Page numbers followed by F denote figures, those followed by T denote tables.

A, mass number, 52
abrasive, carborundum, 711
absolute zero, 20
   entropy at, 591
abundance, isotopic, 54
accuracy, 31
acetal, 883
acetamide, 895
acetate ion
   as base, 519
   Lewis structure, 298
acetic acid, 887
   dimer, 361
   esterification
      equilibrium, 479
   Lewis structure, 296
   molecular model, 102, 529
   orbital structure, 335
   production, 512
acetone, iodination
   kinetics, 440
acetylene, 860
   orbital structure, 336
   preparation, 666
   production from
      calcium carbide, 124F
   VSEPR prediction, 327
acetylide ion, 711
acid, 73, 90, 512
   Brønsted, 92
   buffer, 564
   cation, 519
   conjugate, 517
   correlation with
      structure, 528
   hydrated cation, 548
   ionization constant,
      521, 524T
   ions as acids, 546
   leveled strength, 527
   Lewis, 514
   measurement of pK_a,
      551
   nomenclature, 74
   oxidizing agent, 99
   parent, 74
   production, 512
   reaction with gold, 778;
      mercury, 779; metals,
      636
   relative strength, 523
acid rain, 741
   nitrogen oxides, 720
   sulfur oxides, 740
acid-base titration, 139
acidic oxide, 93
acidities, in household
   products, 90
acidity constant, *see*
   ionization constant
acre, 36
acrylonitrile, 856
activated carbon, 702
activated complex theory,
   464
activation barrier, 464
activation energy, 459, 461T

catalyzed reaction, 466
   effect of temperature
      on equilibrium, 503
   interpretation, 464
   measurement, 460
   sensitivity of
      temperature
      dependence, 460
active site, 469
activity, radioactive, 815,
   819T
acyl chloride, 895
addition, 855
   polymerization, 856
additives, 457
   petroleum, 15
adenine structure, 903
adenosine
   diphosphate, 724
   triphosphate, molecular
      model, 724
adhesion, 363
ADP, adenosine
   diphosphate, 724
adsorb, 12
   in catalysis, 468
agate, 707F
air, composition, 152
alanine, 896, 898
albumin denaturation,
   901
alcohol, 873
   acidity, 529
   nomenclature, 873
   oxidation, 876
   primary, 873
   production, 873
   properties, 875
   reaction with
      aldehydes, 883
   reactions, 875
   secondary, 874
   tertiary, 874
aldehyde, 876, 879, 880T
   nomenclature, 881
   origin of name, 880
   preparation, 882
   reaction with alcohols,
      883
   reactions, 882
aliphatic, 862
alizarin, 563F
alkali, 90
   metals, 44, 265F, 666,
      667F
      compounds, 671T
      flame test, 232
      general chemical
         properties, 265
      physical properties,
         667
      reaction with
         nonmetals, 670;
         water, 44F
alkaline cell, 626
alkaline earth metals, 44,
   676, 677F
      compounds, 680, 681T
      flame test, 681F

introduction to
   chemical properties,
   265
   properties, 677T
   reaction with water,
      680
alkane, 840, 841
   branched, 844
   combustion, 850
   melting and boiling
      points, 848
   names, 845T
   nomenclature, 843
   properties, 841, 849
   unbranched, 844
alkene, 851
   melting points, 853
   nomenclature, 852
   reactions, 855
   role of double bond,
      853
alkyl
   group, 846, 845T
   halide, 846
alkylation, 865
alkyne, 860
   preparation, 861
   reactions, 861
allotrope, 381
   boron, 693
   phosphorus, 716
   sulfur, 734
   tin, 594F
alloy, 15, 377, 377T
   copper and beryllium,
      678
   formation by *d* metals,
      763
   nonferrous, 777
   sodium and potassium,
      668
alpha (α) decay,
   identifying daughter,
   806
alpha (α) disintegration,
   805
alpha (α) helix, 900
alpha (α) particles, 803
   discovery, 48
alpha (α) rays, 803
alum, 697
alumina, 695
   α and γ, 696
   gems, 696
   purification, 573
   solubility and pH, 571
aluminate ion, 572, 695,
   697
aluminosilicate, 710
   lapis lazuli, 739
aluminum
   amphoteric, 693
   anodized, 694
   aquaion, acidic, 515
   ion as acid, 548
   passivation, 637
   production, 694
   properties, 692
   similarity to Be, 266

smelting, 7
   solubility of oxide, 571
   structure, 375
   sulfate, 697
      in water purification,
         697
aluminum bromide,
   structure, 699
aluminum chloride
   catalyst, 850, 865
   Lewis structure, 302
   molecular model, 302
   structure, 699
aluminum halides
   production, 699
   reaction with water,
      699
aluminum hydride, 700
aluminum iodide,
   structure, 699
amethyst, 289, 707F
amide, 894
   preparation, 895
   ion, 64
amine, 872, 892
   nomenclature, 892
   preparation, 893
   primary, 892
   properties, 894
   reactions, 894
   secondary, 893
   tertiary, 893
amino acid, 896, 898T
   essential, 899
   sequence, 900
amino group, 893
ammonia
   anomalous boiling
      point, 360
   effect of pressure on
      equilibrium, 500
   equilibrium
      calculations, 497
      characteristics, 478
      constant, 487
      effect of
         temperature, 502
   Haber synthesis, 466
   Lewis base, 514, 515
   molecular model, 59
   oxidation,
      thermodynamic
      aspects, 607
   production, 153, 512,
      665, 716
   properties, 716
   reaction with boron
      trifluoride, 302; alkali
      metals, 668; hydrogen
      chloride, 177, 514
      in Solvay process, 674
   as solvent, 514
Ammonians, 717
ammonium hydroxide,
   716
ammonium ion, 717
   as acid, 519
   molecular model, 64
ammonium nitrate, self-

oxidation, 104
ammonium thiocyanate, endothermic reaction, 188
amorphous solid, 371
amphibole, 710
  structure, 709
amphiprotic, 519
amphoteric, 94
  aluminum, 266, 693
  beryllium, 266, 680
  character and oxidation number, 766
  oxides, solubility with pH, 571
  tin oxides, 267
  zinc, 778
amu, atomic mass unit, 54
amylopectin, structure, 886
amylose, structure, 886
analysis
  chemical, 10
  combustion, 127
  gravimetric, 127
  volumetric, 139
analyte, 139
anemia, 901
angular momentum, electrons in atoms, 244
anhydride, 695
  formal, 695
  formation, 705
anhydrous, 72
aniline, molecular model, 523, 893
anion, 62, 63
  basic, 519, 548
  basicity, 549T
  formed by elements, 64
anode, 620
  reaction, 621
  sacrificial, 639
  sludge, 734
anodic protection, 639
anodized aluminum, 694
antacid, 540F, 683
anthocyanidin, molecular model, 560
anthracene, 862
anthraquinone, molecular model, 738
antibonding orbital, 339, 340
antidote to poison, 782
antifreeze, 875
antilogarithm, 914
antimony, properties, 714
antioxidant, 457, 859
apatite, 402
Apollo mission, fuel, 717
apple odor, 890
apricot odor, 890
aqua regia, 580, 722, 778
aquaion, 289
  acidic, 515
aqueous solution, 15
aragonite, 353F, 684
arene, 862
  nomenclature, 863

argentite, 733
arginine, 898
argon, properties, 751
aroma, 890T
  cherries, 881
aromatic hydrocarbons, 862
  reactions, 865
arrangement, electron pairs, 323
Arrhenius
  acid, 90
  behavior, 458, 460
  parameter, 458, 461T
  plot, 459
arsenic
  as poison, 469
  properties, 714
asbestos, 710
asparagine, 899
aspartic acid, 898
asphalt, 12, 849
astatine, properties, 743
atactic polymer, 858
atmosphere
  composition, 152
  composition of polluted air, 397
  sulfur oxides, 740
  unit, 155
atom, 8, 8F
  configurations, box diagrams, 252
  orbitals, introduced, 241
  radii, periodicity, 256
  structure
    Bohr model, 238
    introduced, 232
    Thomson model, 48
atomic hydrogen spectrum, 236
atomic hypothesis, 46
atomic mass unit, 54
atomic nucleus, discovery, 49
atomic number, introduced, 49
atomic weight, 55, 920T
ATP, adenosine triphosphate, 724
Aufbau principle, 250
auroras, 453
autoionization, introduced, 519, 520
automobile battery, 627
average bond enthalpy, 304, 304T
  estimating reaction enthalpy, 305
Avogadro, Amedeo, 57F
Avogadro's law, 158
Avogadro's number, 57
axial position, 325
azide ion, 64
azimuthal quantum number, 242

back titration, 142
bacterial DNA, 902F

bactericide, 878
baking powder, 706
balance, 16F
balancing equations, 83
  net ionic equations, 88
  using half-reactions, 107
Balmer series, 236
banana odor, 890
band
  conduction, 379
  gap, 379
band of stability, 809
bar, 155
barium, 677F
  production, 679
barometer, 154
base, 512
  anion, 519
  Brønsted, 92
  conjugate, 517
  ionization constant, 522, 524T
  ions as bases, 546
  leveled strength, 527
  Lewis, 514
  origin of name, 95
  pair, 904
  production, 512
  relative strength, 523
  units, 16
basic copper carbonate, 777
basic oxide, 93
basic oxygen process, 771
battery, 620, 626
bauxite, 693
  purification, 573
Bayer process, 695
bcc, body-centered cubic, 375
becquerel, 815
Becquerel, Henri, 803
  discovery of radioactivity, 803
bends, 405
benzaldehyde, molecular model, 881
benzene, 862
  bonding, 862
  Lewis structure, 298
  orbital structure, 336
  thermodynamic stability, 606
  vapor pressure, 366
  viscosity, 362
benzenebergs, 383F
benzyl alcohol, 874
beryl, 676
beryllate ion, 680, 682
beryllium, 676, 677F, 678
  compounds, 680
  similarity to Al, 266
beryllium chloride
  Lewis structure, 310
  production, 682
  structure, 682
beryllium hydride, structure, 682
Bessemer converter, 771
beta (β) decay,

identifying daughter nuclide, 807
beta (β) disintegration, 807
beta (β) particles, 804
beta (β) pleated sheet, 900
beta (β) rays, 803
Bethe chain, 836
bidentate ligand, 782
bimolecular reaction, 450
binary compound, 72
binding energy, 825
biological cell walls, osmosis, 423
bismuth, properties, 714
blast furnace, 97F, 618, 771
bleach, 738, 744
  chlorine dioxide, 750
block, periodic table, 255
blood, 423
  buffer, 564, 566
  corpuscles, osmosis, 423
  plasma, pH, 546
boat conformation, 848
body-centered cubic, 375
Bohr
  frequency condition, 237
  model of atom, 238
  radius, 238
Bohr, Neils, 237F
boiling point, 6
  alcohols and ethers, 876
  alkanes, 848
  constant, 418T
  elements, 934T
  elevation, 417
  ethanol and water mixture, 30
  Group VI elements, 735
  hydrogen compounds, 360
  ionic compounds, 354T
  at low pressures, 367
  molecular compounds, 359
  normal, 366
  prediction from entropy, 598
  relation to vapor pressure, 366
Boltzmann
  constant, 590
  factor, 463
  formula for entropy, 590
Boltzmann, Ludwig, 590F
bomb calorimeter, 200
bond, 9
  angle, 322
  covalent, 291
  double, 293
  enthalpy, 304, 304T
    effect on acid strength, 528

bond *(continued)*
general rule for type of bond, 291
hydrogen, 360
ionic, 278
length, 306
multiple, 293
order, 342
parameters, 304
pi (π), 334
polar, 312
predicting type, 285
sigma (σ), 333
strength, 304, 850T
survey, 344
three-center, 344
triple, 293
bonding orbital, 339, 340
bone, 686
role of phosphate, 402
booster rockets, 750F
borane, 699
borazon, structure, 383
boric acid, 516, 695
Born equation, 413
Born-Haber cycle, 280
oxide formation, 737
borohydride ion, 699
boron
allotropes, 693
production, 693
properties, 692
structure, 382, 693
boron carbide, 697
structure, 382
boron halides, 698
Lewis acids, 699
boron nitride, 352F, 697
production, 383
structure, 383, 697
boron oxide, 695
boron trichloride,
production, 302, 698
boron trifluoride
formal charge, 308
hybridization, 331
Lewis acid, 514
Lewis structure, 301
polarity, 329
production, 301, 698
reaction with ammonia, 302
shape, 324
borosilicate glass, 708
Bosch, Carl, 504F
boundary surface, 244
d orbitals, 245
p orbitals, 245
box diagram, 249
Period-2 atoms, 252
Boyle, Robert, 157
Boyle's law, 157
Bragg, William and Lawrence, 372
Bragg equation, 372, 373
branched alkane, 844
nomenclature, 847
branching, 457
brass, 377, 777
breeder reactor, 824
brimstone, 733

brine electrolysis, 673
Brønsted acid, 92, 513
cation, 519
formation from Lewis acid, 516
in nonaqueous solvents, 513
Brønsted base, 513
anion, 519
Brønsted equilibria, 517
Brønsted-Lowry theory, 92, 513
bromate ions, reduction, 439
bromine
burning in hydrogen, 102F
catalyst for hydrogen peroxide decomposition, 467
production, 744
production from bromide, equilibrium, 483
properties, 743, 745
reaction with hydrogen, mechanism, 456; phosphorus, 99F water, 856
bromocresol green, 563F
bromomethane hydrolysis kinetics, 440
mechanism, 451
bromothymol blue, 563F
bronze, 377, 776, 777
manganese, 770
buffer, 562
blood, 564, 566
calculation of pH, 565
capacity, 565
explanation of action, 564
ocean, 564
building-up principle, 250
bulk property, 8
buret, 15F, 133, 133F
butane, 842
butene, 852
molecular model, 474
butyl rubber, 860

cadaverine, 892
cadmium
production, 778
properties, 775
calcite, 353F, 684
calcium
complex with EDTA, 782
production, 679
structural, 686
calcium carbide
production, 686
reaction with water, 124F
calcium carbonate, 684
calcite and aragonite, 353F
calculating

decomposition temperature, 603
decomposition equilibrium, 489
calcium chloride, energetics of formation, 282
calcium fluoride, 730F
solubility and pH, 573
calcium hydride, 662F
calcium hydroxide, 685
production, 512
calcium oxide, 685
structure, 380
calcium sulfate, 686
calibration, 201
calomel electrode, 622, 642
calorie, 190
dietary, 221
calorimeter, 200F
calorimetry, 199
capacity (buffer), 565
capillary action, 363
caprolactam, 910
carbide ion, 686, 861
Brønsted base, 666
carbides, 711
carbohydrates, 884
food, 217
carbon black, 702
carbon contrasted with silicon, 702, 713
diamond, 381
graphite, 381
phase diagram, 385
properties, 781
carbon dioxide, 706
atmospheric variation, 122
effervescence, 540
formal charge, 309
Lewis acid, 706
liquid, 371, 706
molecular model, 9, 60
phase diagram, 370
production, 153
reacting with potassium superoxide, 169F
supercritical, 368
carbon disulfide, 714
carbon-14 dating, 820
carbon monoxide
entropy at $T = 0$, 591
formal charge, 307
Lewis base, 516
molecular model, 60
preparation, 705
production, 153
properties, 705
reaction with iron, 773
reaction with nitrogen dioxide, mechanism, 451
reducing agent, 706
toxicity, 705
carbon suboxide, 705
carbon tetrachloride, density, temperature dependence, 369
carbon tetrafluoride, 359

carbonate
thermal decomposition, 82
on Venus, 491
carbonate ion, 706
action of acid, 706
calculation of concentration, 539
equilibrium in water, 536
thermal decomposition, 674, 706
molecular model, 64
carbonic acid, 706
anomalous acidity, 538
equilibrium in water, 536
as polyprotic acid, 536
carbonyl group, 879
carborundum, 711F
carboxyl group, 887
carboxylate group, 887
carboxylic acids, 887, 888T
acidity compared to alcohols, 529
nomenclature, 887
preparation, 888
strengths as acids, 889
as weak acids, 526
carnallite, 676
carotene, 352F
carrier gas, 13
carvone, 881
cassiterite, 704
catalysis, 465
catalyst, 466
Haber process, 467
heterogeneous, 467
homogeneous, 466
poison, 468
support, 13
vanadium pentoxide, 467
Ziegler-Natta, 858
catalytic converter, 468
catenate, 734, 739
cathode, 620
rays, 47
reaction, 621
cathodic protection, 639
cation, 62
acidic, 519, 548
acidity, 549T
formed by elements, 63T
caustic soda, 673
ccp, cubic close-packed, 375
cell
alkaline, 626
Daniell, 620
diagram, 623
diaphragm, 673
dry, 625
electrochemical, 620
fuel, 625, 628
galvanic, 946
lead-acid, 627
Leclanché, 625
mercury, 626

nickel-cadmium, 628, 775
potential, 623
  free energy, 629
  dependence on concentration, 639
  measurement, 624
  standard, 630
  thermodynamics, 628
primary, 625
reaction, deducing, 625
secondary, 625, 626
cell walls, osmosis, 423
cellulose, structure, 887
cellulose acetate membrane, 421
Celsius scale, 19
cement, 710
  high-alumina, 710
centimeter, 18
centrifugation, 11
  uranium enrichment, 831
ceramic, 710, 769
cesium, 667F
  chloride structure, 381
chain
  branching, 457
  carrier, 456
  reaction, 456
    nuclear, 822
    polymerization, 858
chair conformation, 848
chalcogens, *see* Group VI elements
chalcopyrite, 776
chalk, 684
champagne, 405
change
  free energy, 602
  spontaneous, 588
  state, 6
charge
  cloud, 244
  distribution, 307
  formal, 307
Charles, Jacques, 157
Charles's law, 158
chelate, 782
chemical analysis, 10
chemical bond, 9
  survey, 344
chemical change, 7
chemical equilibrium, 478
chemical kinetics, 432
chemical nomenclature, 67
chemical plating, 700
chemical symbols, 43
chemicals
  top 12 as acids and bases, 91
  top 50, 939T
chemiluminescence, 738F
chemistry, 5
cherry aroma, 881
chewing gum, 881
Chile saltpeter, 402F
China clay, 710
chiral, 784, 896
chlorates

preparation, 750
  thermal decomposition, 750
chlorinated water, 744
chlorine
  color, 152F
  hydrolysis, 736
  in drinking water, 143
  liquefaction, 179F
  production, 153, 649, 744
  properties, 743
  reaction with hydrogen, 665
  resonance, 312
chlorine dioxide, 750
chlorine trifluoride, formal charge, 309
chlorofluorocarbons, 712
chlorophyll
  chloroplasts, 217
  color, 791
  molecular model, 684
chloroplast, 217
cholesteric phase, 384
chromate ion, 70
  molecular model, 105
  solution, 105F
chromatogram, 13
chromatography, 11, 12
  gas-liquid, 13
chrome plating, 650, 769
chrome recording tape, 769
chrome yellow, 87F
chromite, 769
chromium
  plating, 650
  production, 769
  properties, 767
  steel, 772T
cinnabar, 733, 779
cinnamaldehyde, 881
cis, 854
cis-butene
  molecular model, 474
  structure, 854
cis-trans isomerization, 855
classical mechanics, 232
classification, of matter, 75F
classification of reactions by order, 439
clathrate, 740
Claus process, 104, 619, 733
Clausius-Clapeyron equation, 370
clay, 710
close-packed structure, 374
closed shell, 249
coal, 862
  dust, explosion, 170
  origin, 219
  structure, 863
cobalt, 774
  blue glass, 708
  properties, 767
cobalt(II) carbonate, 89F

cobalt chloride, 502F
code, genetic, 904, 906T
codon, 905
cohesion, 363
coinage metals, 45, 775
coke, 702
cold cream, 203
collagen, 770
colligative properties, 416
collision theory, 462
colloidal dispersion, 739
color
  complementary, 790
  complexes, 789
  dependence on ligands, 791
  relation to frequency, 234
  wheel, 790
combustion
  alkane, 850
  analysis, 127
  balancing, 84
  octane, 196
  pool of atoms, 213
common logarithm, 914
common molecular compounds, 73
common-ion effect, 570
competing oxidations, 645
competing reductions, 644
complete reaction, 126
complete shell, 249
complex, 289, 514
  chiral, 785
  color, 791
  concentration in solution, 579
  configuration, 793T
  d-metal complexes, 779, 780F
  electron configurations, 791
  formation, 780
  formation constant, 577
  formation, and solubility, 578
  gold, 580
  high-spin, 792, 794
  isomerism, 784
  low-spin, 792, 794
  magnetic properties, 795
  nomenclature, 782
  structure, 781
  theory of structure, 788
composition
  effect on free energy, 609
  mass percentage, 33
compound, 9, 59
  binary, 72
  common names, 73
  coordination, 781
  electron-deficient, 338
  inorganic, 67
  ionic, 59

molecular, 59
organic, 67, 840
compressibility, 154
concentration
  of complex, 579
  hydronium ions (pH), 530
  mass, 395
  measurement by redox titration, 142
  measurement by titration, 139
  molar, 133, 394
  pure solid, 489
  from rate law, 443
  second-order reaction, 447
  units, 395T
condensation, 6
condensation polymers, 891
condensation reaction, 890
conduction
  band, 379
  electronic, 377
  ionic, 377
configuration
  atomic, 933T
  d-metal complexes, 791, 793T
  deducing, 253
  diatomic molecules, 342
  electron, introduced, 249
  ions, 254
  Period-2 atoms, 252
  pseudonoble-gas, 285
  relation to group number, 256
conformation, boat and chair, 848
congener, 44
conjugate acids and bases, 517, 517T
conservation of energy, 191
constant
  Boltzmann, 590
  equilibrium, 479
  Faraday, 629
  Planck, 235
  rate, 437
  Rydberg, 237
constant composition law, 9, 127
constructive interference, 372
contact process, 741
controlled fission, 824
controlled fusion, 829
conversion
  factor, 24
  mass to molar concentration, 396
  mole fraction and molality, 399
  temperature scales, 19
coolant, 668

cooling as spontaneous process, 589
coordinate, 302
coordination
  compound, 781
  number, 374
  sphere, 781
copolymer, 860
copper
  disproportionation, 777
  electrolysis, 643
  electrolytic refining, 645F
  hydrometallurgy, 661
  production, 776
  properties, 775
  reaction with nitric acid, 100F, 637, 776; sulfuric acid, 742
  refining, 776
  structure, 375
  variable valence, 287
copper hydride, 666
copper(I)
  chloride, preparation, 288
  iodide, preparation, 288
  oxide, reaction with sulfuric acid, 287
copper(II)
  carbonate, 777
  ions
    complex formation with ammonia, 579F
    stepwise formation constants, 579T
  sulfate, 23F
    color, 791
    hydrous and anhydrous, 72F
    reaction with zinc, 619
    structure, 777
core, atomic, 250
cornflower color, 561
corrosion, 637
  mechanism, 638
  prevention, 638
  Titanic, 80
corrosive solution (EPA definition), 531
corundum, 696
coulomb, 624
  measuring number, 647
  potential, 279
couple, 620, 621
covalence, variable, 299
covalent bond, 291
covalent carbides, 711
covalent radii, 306, 306T
  Group VI elements, 735
  Group VII elements, 747
cracking, 939
crepe, 859
cresol red, 563F
cristobalite, 703F
  structure, 707

critical mass, 823
critical temperature, 367, 368, 368T
crown ether, 877
cryogenics, 752
cryolite, 694, 744
crystal
  examples, 381T
  face, 371
  field theory, 788
  fluorite, 389
  liquid, 384
  perovskite, 389
  structure
    rock-salt, 380
    rutile, 389
    superconductor, 389
crystalline solid, 371
crystallites, 768F
crystallize, 14
cube root, 569
cubic close-packed structure, 375
cupronickel, 377, 774, 776, 777
cured meat, 721
curie, 815
Curie, Marie, 815F
cyanate ion, 64
  formal charge, 309
cyanide ion, 713
  formal charge, 308
  Lewis base, 713
  process, 619
cyano group, 872
cyanohydrin, 882
cycles per second, 233
cyclic hemiacetal, 883
cycloalkanes, 847
cyclohexane, 847
cyclopentadienide ion, molecular model, 774
cyclopropane
  isomerization kinetics, 440
  molecular model, 443
cylohexane, structure, 848
cysteine, 899
cytosine, structure, 903

d-block elements, 256, 762
  alloy formation, 377
  complex formation, 779
  density, 764
  difficulty of extraction, 766
  general properties, 267
  oxidation numbers, 765
  physical properties, 767T
  properties, 763, 775
  trends in chemical properties, 765
  variable valence, 287
d orbital, 245
  boundary surface, 245
  hybrids, 331

occupation, 253
octet expansion, 292, 299
d subshell, 243
Dacron, 891, 892
dageurrotype, 46
Dalton, John, 46F
Dalton's law, 170
Daniell cell, 620
  free energy, 629
data, 31
dating, isotopic, 819
daughter nucleus, 804
Davisson-Germer experiment, 240
Davy, Humphry, 677
de Broglie relation, 239
debye, 311
Debye, Peter, 372
decay constant, 817
decomposition, 82
  equilibria, 492
  temperature, calculating, 603
  vapor pressure, 490
defoliant, 750
dehydration, producing acid anhydride, 705
dehydrogenation, 851
dehydrohalogenation reaction, 852
delocalized orbitals, 344
delta (Δ), in reaction equation, 82
denaturation, 901
density, 22
  calculating, 376
  in d block, 764
  elements, 934T
  enthalpy, 219
  gas, 160
  high-density solution, 745
  ice and water, 383
  variation with temperature, 369
deoxyribonucleic acid, 902
deoxyribose, 902
  structure, 903
depression of freezing point, 418
deprotonation, 666
derived unit, 21
destructive distillation, 862
destructive interference, 372
detergent, 403
  buffer in, 724
  phosphates, 723
  silicates, 708
detonator, 170
deuterium, 660
  as tracer, 478
  in nuclear fusion, 828
  properties, 53
development of image, 106
diagonal relationship, 266
  between Al and Be, 266

polarizing power, 311
diamagnetic, 795
diamines, 892
diamond, 702
  entropy, 593
  production, 702
  structure, 381, 703
  synthetic, 382F
diaphragm cell, 673
diatomic ion
  Lewis structures, 294
diatomic molecule, 60
  Lewis structures, 293
  orbital description, 341
diborane, 338
  electronic structure, 344
  preparation, 338, 700
  structure, 700
dichloroethylene isomers, polarity, 328
dichromate ion, 70, 769
  molecular model, 105
  solution, 105F
diffraction, 373
  electrons, 240
  pattern, 240, 373
diffusion
  Graham's law, 177
  uranium enrichment, 830
digestion, 886
dilution, 136
dimer, 699
dimethyl ether, molecular model, 366
dinitrogen oxide, preparation, 720
dinitrogen pentoxide, 433F
  decomposition, 433
  kinetics, 440
  preparation, 722
dinitrogen trioxide, 721F
  molecular model, 721
diol, 875
dipeptide, 900
dipole, 311
  instantaneous, 357
  moment, 311
  intermolecular forces, 354, 356
dipole-dipole
  force, 354
  interaction, distance dependence, 356
dipole-induced dipole force, 354
direct titration, 142
direction of reaction, 483
  and reaction quotient, 484
  spontaneous change, 588
disaccharide, 885
disease, and pH, 566
disorder
  entropy, 589
  solubility, 413
dispersion force, 354

of energy and matter, 357, 414
disproportionation, 103, 104
  copper, 777
  copper(I) compounds, 287
  in *d* block, 766
  gold, 778
  silver, 777
dissolve, 14
distillation, 11
  destructive, 862
distribution
  speed, 179
  speed, and reaction probability, 463
disulfide link, 900
disulfur dichloride, 743
DNA, deoxyribonucleic acid, 902
dolomite, 771
doping, 379
dose, radioactive, 816
dot diagram, 285
double bond, 293
  effect on shape, 853
  orbital structure, 334, 335
  properties, 336
  torsional rigidity, 337
double helix, 904
double replacement reaction, 88
double-action baking powder, 706
Downs process, 618, 619
  described, 649
drinking water, chlorine in, 143
dry cell, 625
dry ice, 179, 706
drying agent, 128, 723
duality, 239
ductile, 45
dynamic equilibrium, 365
  chemical, 454, 478
    Le Chatelier's principle, 407

e orbital, 788
earths, 676
ebonite, 859
EDTA, ethylenediaminetetraacetic acid, 781
eel, electric, 628
effective nuclear charge, 247
effervescence, 405
egg discoloration and smell, 739
Einstein, Albert, 825
Einstein's formula, 825
eka-silicon, 255
elastomer, 860
electric dipole, 311
  electronegativity dependence, 312
electric eel, 628

electric furnace, 618
electric organ, 628
electrical conduction, 377
  weak electrolyte, 527
electricity, measuring amount, 647
electrochemical cells, 619
electrochemical series, 98, 634
  predicting oxiding power, 635
  predicting spontaneous direction, 635
  writing cell diagram, 634
electrochemistry, 618
  thermodynamic aspects, 628
electrode, 620
  calomel, 622
  compartment, 622
  glass, 642
  hydrogen, 622
  potential, 630
    alkali metal trends, 668
  silver chloride, 622
  standard hydrogen, 630
electrolysis, 7, 643
  applications, 648
  brine, 673
  calculating
    mass produced, 648
    volume of gas produced, 646
  copper refining, 776
  extent, 646
  fluorine production, 744
  potential needed, 643
  production of magnesium, 678
  water, 644
electrolyte, 62
  weak, 93
electromagnetic radiation, 233
electromotive force, 623; *see also* cell potential
electron
  affinity, 262T
    periodicity, 262
  capture, 808
  configuration, 249
  configurations, 933T
  diffraction, 240
  discovery, 47
  dot diagram, 285
  fluctuation, 357
  gain, 262
  measurement of mass and charge, 48
  pair, explanation of role, 340
  properties, 48
  spin, 246
  transfer, introduced, 96
electron-deficient compound, 338, 700

orbital description, 344
electron-gain enthalpy, 262
electron-pair bond, 291
electron-pair repulsion, 322
electronegativity, 263, 264T
  dipole moment of bond, 312
  effect on type of bond, 291
  ionic character of bond, 312
  periodicity, 264
  strength of oxoacid, 529
electronic conduction, 377
electronic structure of Period-2 diatomic molecules, 341
electroplating, 650
elements, 8, 934T
  names, 42
  origin of names, 934T
  production, 618
  properties, 934
  symbols, 42
elementary reaction, 449
elevation, boiling point, 417
elimination, 852
Ellingham diagram, 756
emerald, 678F
EMF, electromotive force, 623; *see also* cell potential
empirical formula, 129, 130
emulsion, photographic, 105, 106
en, ethylenediamine, 781
enantiomer, 785
end point, 139
endothermic, 199
  reaction, origin of spontaneity, 600
  spontaneous process, 415
energy, 188
  activation, 459
  conservation of, 191
  dispersing, 414
  free, 601
  internal, 192
  kinetic, 189
  levels,
    hydrogen atom, 238
    orbitals of many-electron atoms, 248
  potential, 190
  quantization, introduced, 236
  quantum, 235
  total, 191
  transferred as heat, 194
  units, 189
energy-level diagram, molecular orbital, 340

enrichment, 829
enthalpy, 195
  bond, 304
  Born equation, 413
  change, definition, 197
  combustion, 209T, 925
  density, 219
  extensive property, 202
  food, 220
  for calculating entropy change, 595
  formation, 214, 215T, 920T
    combining, 215
  freezing, 204
  fusion, 204
  hydration, 410T
  ion hydration enthalpy, 411
  ionization, 258
  lattice, 281
  melting, 203
  neutralization, 513
  physical change, 202, 204T
  reaction, 205
  relation to internal energy, 198
  resources, 217
  solution, contributions, 409, 408T
  sublimation, 204
  transition, combining values, 205
  vaporization, 203
entropy, 588, 590
  calculated, 591
  carbon monoxide, 591
  change in
    surroundings, 595
    formula, 597
  $FClO_3$, 591
  hydrogen chloride, 591
  magnesium oxidation, 597
  measure of disorder, 589
  measurement, 591
  melting, 594
  nitrous oxide, 592
  physical change, 594
  prediction of boiling point, 598
  reaction, 594
  reaction, calculating, 595
  relation to equilibrium constant, 609
  solids, 590
  standard molar, 592, 593T
  and temperature, 593
  thermal motion, 592
  units, 592
  vaporization, 593
  water, 593
    freezing, 597
enzyme, 469, 901
EPA, Environmental Protection Agency, 532
Epsom salts, 683

equation of state, 159
equatorial position, 325
equilibrium
    autoionization, 520
    calculations, 491
        approximation, 498
        arbitrary initial
            concentrations, 495
        introduced, 491
        one unknown
            concentration, 492
        simplifications, 497
        starting from a
            mixture, 496
    stoichiometric
        proportions, 494
    chemical, 478
        equality of rates, 454
    constant, 479, 482T,
        487T
        ammonia, 487
        calculating from data
            481; electrode
            potentials, 639; free
            energy, 610
        heterogeneous
            reactions, 489
        hydrogen iodide, 487
        interpretation of size,
            483
        multistep reactions,
            485
        nitrogen dioxide
            dimerization, 487
        partial pressure, 486
        relating K_c and K_P,
            488
        relation to free
            energy, 609, 610;
            rate constant, 485
        sulfur dioxide
            oxidation, 487
        thermodynamic
            aspects, 610
    dynamic, 365, 478
    effect of temperature,
        502
    heterogeneous, 488
    homogeneous, 488
    in terms of entropy,
        597
    Le Chatelier's
        principle, 407
    proton transfer, 517
    response to added
        reagents, 499;
        conditions, 499;
        pressure, 500
    solubility, 566
    thermodynamic aspects,
        608
equivalence point, 139
    detection, 560
    strong-acid–strong-base
        titration, 555
    weak-acid–strong-base
        titration, 557
    weak-base–strong-acid
        titration, 559
erasers, 859
error

random, 30
    systematic, 31
essential, oil, 878
ester, 480, 889
    molecular model, 480
    occurrence, 889
    odors, 890T
esterification, 889, 890
    equilibrium data, 479
    mechanism, 891
    response to conditions,
        499
estradiol, molecular
    model, 60
etching, 708, 748F
ethane
    decomposition kinetics,
        440
    molecular model, 840
    structure, 841
ethanedioic acid, 888
ethanol
    bonding, 333
    molecular model, 59,
        102, 366, 529
    vapor pressure, 366
ethene, see ethylene
ether, 876
    crown, 877
ethoxide ion, 875
ethyl
    acetate, molecular
        model, 480
    group, 846
ethylene, 851
    bonding, 334
    hydrogenation,
        mechanism, 468
    shape, 326, 337
ethylene glycol
    antifreeze, 875
    molecular model, 875
    polyester formation,
        891
ethylenediamine, 781
    chelate formation, 782
    molecular model, 782
ethylenediaminetetraacetic
    acid, 781
    structure, 783
eugenol, 878
evaporation, 366
exclusion principle, 249
    role in bonding, 340
exothermic, 199
    reaction, origin of
        spontaneity, 600
expanded octet, 292, 299
expansion
    spontaneous, 589
    work, calculated, 196
experiment, 4
explosion
    chain branching, 457
    dust, 458
    nitrogen triiodide, 719
    nuclear, 823
    thermonuclear, 829
explosives, 722
exponential, 914
exponential decay, 443

extensive property, 22
extrapolate, 915

f orbital, 245
f subshell, 243
face, crystal, 371
face-centered cubic, 375
Fahrenheit scale, 19
fallout, 823
Faraday, constant, 629
Faraday, Michael, 629F
Faraday's law, 646
fat, 660
    introduced, 221
fatty acids, 889
fcc, face-centered cubic,
    375
Fehling's solution, 287
    with aldehydes and
        ketones, 882F
    with glucose, 885
feldspar, 710
fermentation, 873
ferricyanide ion, 783
ferroalloys, 769
ferrocene, 77
ferrochromium, 769
ferrocyanide ion, 780F,
    783
ferromagnetism, 772
ferromanganese, 770
ferrosilicon, 678
ferrovanadium, 769
fertilizer
    need for, 402
    phosphate, 720
    superphosphate, 723
fetus, 906F
fever, 782
fiber, 892F
filtration, 10, 11
fire retardants, 745
fireworks, 680, 750
first law of
    thermodynamics, 192
first-order reaction, 438
    effective, 448
    experimental test, 444
    half-life, 445
    time-dependence, 442
fissile, 822
fission, 822
    controlled, 824
    heat released, 826
fissionable, 822
fixed nitrogen, 715
flame front, 458
flame, sooty, 123
flame test
    alkali metals, 232F
    alkaline earth metals,
        681F
flares, 722
flavor, 889, 890
    synthetic, 891
flint, 707
floc, 697
fluorapatite, 402, 686,
    744
fluorescent, 405, 730

lighting, 753
fluoridation, 686
fluorides, special
    characteristics, 747
fluorinating agent, 754
fluorine
    production, 153, 649,
        744
    properties, 743
fluorite, 730F
    structure, 381, 389
fluorspar, 744
foaming agent, 720
food
    enthalpy, 220
    enthalpy aspects, 217
    flavor, 890
    thermochemical
        properties, 222T
fool's gold, 288, 771F
formal anhydride, 695
formal charge, 307
    plausible structure, 309
formaldehyde
    in wood smoke, 880
    phenolic resins, 878
    production, 661, 882
formalin, 880
formation
    constant, 577, 577T
        stepwise, 579
    enthalpy, 214, 215T
        combining, 215
    free energy, 605T
        calculation of, 605
    standard free energy,
        604
formic acid, 887
formula
    empirical, 129
    molecular, 131
    unit, 65, 66
    weight, 65, 66
fossil fuel, definition, 219
fraction, distillation, 12
fraction of successful
    collisions, 463
fractionating column, 12
francium, 667
Franklin, Rosalind, 372
Frasch process, 733
free energy, 601
    cell potential, 628
    composition
        dependence, 609
    electrical work, 629
    Ellingham diagram,
        756
    formation, 604, 605T,
        920T
        calculation of, 605
    reaction, 604
    relation to equilibrium
        constant, 610
    spontaneous changes,
        602
    why free, 601
freezing, 6
    enthalpy of, 204
    point, 6, 368
        constant, 419, 418T

depression, 418
effect of pressure, 369
Freon, 712
frequency, 233
  relation to color, 234
  units, 233
Friedel-Crafts alkylation, 865
frost, 371
froth flotation, 776
fructofuranose, 884
fructose, structure, 884
fuel
  booster rockets, 750
  cell, 194, 195F, 625, 628
  enthalpy aspects, 217
  fossil, 219
  heat output, 207
  nuclear, 829
  oil, 849
  thermochemical properties, 219T
functional groups, 872
  containing nitrogen, 892
furan, 884
furanose, 884
fusion
  enthalpy of, 204
  nuclear, 828

galena, 704, 733
gallium, properties, 692
galvanize, 638, 778
gamma (γ) ray, 803
  wavelength, 234
gangue, 776
gas, 6
  bottle, 172
  collection, 172
  constant, 159
  density, 160
  kinetic theory, 174
  liquefaction, 179
  molar volume, 159
  molecular model, 154
  molecular speed, 175
  preparation and properties, 153T
  real, 164
  solubility, 405
  volume occupied by given mass, 167
gas-liquid chromatography, 13
gasoline, 849
  lead-free, 468
Gay-Lussac, Joseph-Louis, 157
Geiger, Hans, 48
Geiger counter, 814
Geiger-Marsden experiment, 48
gem
  alumina, 696
  silica, 707
genetic code, 904, 906T
geometrical isomerism, 784

alkenes, 854
  nomenclature, 854
germanium
  production, 704
  properties, 701
getter, 679
Gibbs, Josiah, 601F
Gibbs free energy, 601
glass, 372, 707
  beer-bottle, 708
  borosilicate, 708
  cobalt-blue, 708
  colored, 708
  electrode, 642
  etching, 708
  production, 676
  reaction with hydrofluoric acid, 748
  soda-lime, 708
Glauber's salt, 673
glaze, 769
GLC, gas-liquid chromatography, 13
glucopyranose, 884
glucose
  concentration determination, 141
  determination, 143
  molecular model, 59
  reaction with Fehling's solution, 287F
  structure, 884
glutamic acid, 899
glutamine, 898
glycerol, molecular model, 875
glycine, 896, 898
goiter, 745
gold
  complex formation, 580
  disproportionation, 778
  extraction, 580
  foil, 45F
  properties, 775, 777
  reaction with acid, 778
  structure, 375
  variable valence, 288
  white, 777
Goodyear, Charles, 859
Goudsmit, Samuel, 246
graduated cylinder, 15F, 133, 133F
Graham, Thomas, 177
Graham's law, 177
  uranium enrichment, 830
grain, 106F
gram, 17
granite, 710
graphite, 702
  production, 382
  structure, 381, 702
graphs, 915
gravimetric analysis, 127
gray, 815
greenhouse effect, 220
ground state, 238
  configuration, 249
Group I elements, 667;
  see also alkali metals

general aspects of chemical properties, 265
Group II elements, 677;
  see also alkaline earth metals
  general aspects of chemical properties, 265
Group III elements
  oxides, 695
  properties, 692
Group IV elements, 702, 714
  halides, 712
  hydrides, 713
  oxides, 705
  properties, 701, 714
Group V elements
  hydrogen compounds, 716
  oxides, 719
  oxoacids, 719
Group VI elements, 732
  boiling points, 735
  compounds with hydrogen, 735
  electronegativities, 734
  properties, 732
Group VII elements
  oxides, 749
  oxoacids, 749
  properties, 743
  trends and differences, 746
Group VIII elements
  compounds, 753
  properties, 751
groups, 44
guanine, structure, 903
Guldberg, Cato, 479
gunpowder, 676
gutta-percha, 860
gypsum, 687, 710

H, see enthalpy
Haber, Fritz, 503, 504F
Haber process, 466, 665, 716
  catalyst, 467
  plant, 119F
hair, 901
  bleach, 738
half-life, 445
  nuclear, 819T
  radioactive, 818
  second-order reaction, 447
half-reaction, 105
  balancing reactions, 107
halide ions, 69
halides, 747
  hydration enthalpy, 410
Hall process, 618, 694
haloalkane, 846, 872
  production, 851
halogenation

alkene, 855
arenes, 865
halogens, 45; see also Group VII elements
ham, 721
hard water, 402, 684
  action of heat, 685
hcp, hexagonal close-packed, 375
heat, 194
  capacity, 200, 201T
    for measuring entropy, 591
    specific, 201
  constant pressure, 197
  detection of flow, 195
  enthalpy change, 197
  entropy change in surroundings, 596
  output
    from combustion enthalpies, 210
    fuel, 207
  produced by reaction, 188
  relation to temperature rise, 200
  released during fission, 826
heating curve, 205
heavy water, 54
Heisenberg, Werner, 240
helium
  phase diagram, 387, 752
  production, 153
  properties, 751, 752
helix
  alpha (α), 900
  double, 904
  polypeptide, 900
hematite, 771
hemiacetal, 883
  cyclic, 883
hemoglobin, 774
  amino acid sequence, 900
  structure, 901
Henderson-Hasselbalch equation, 550
Henry's constant, 405, 406T
Henry's law, 405
Hertz (unit), 233
Hess's law, 210
heterogeneous catalyst, 467
heterogeneous equilibria, 488
heterogeneous mixture, 14
hexagonally close-packed structure, 375
high-alumina cement, 710
high-spin complex, 792, 794
high-temperature superconductors, 378
histidine, 898
Hodgkin, Dorothy, 372

homogeneous catalyst, 466
homogeneous equilibria, 488
homogeneous mixture, 14
homologous series, 841
household products, acidities, 90
HRF products, highly radioactive fission products, 831
Hund's rule, 251
hunger, osmosis, 423
hybrid orbital, 330
hybridization, 330
hydrated cation, acidic properties, 548
hydrates, 71
  ion-dipole interaction, 355
hydration
  enthalpy, 410т
  ion-dipole interaction, 355
hydrazine, 717
hydride, 664
  distribution in periodic table, 664
  element-hydrogen bond strength, 304
  Group IV elements, 713
  ion
    reducing power, 665
    size, 665
  metallic, 666
  source of hydrogen, 220
hydrocarbon, 840
  aromatic, 862
  classification, 866т
  intermolecular forces, 359
  in petroleum, 849
  saturated, 840
  straight-chain, 844
  unsaturated, 840
  viscosity, 362
hydrofluoric acid, reaction with silica, 708, 748
hydrogen, 660
  abundance, 660
  atom
    Bohr model, 238
    orbitals, 243
    quantum numbers, 242
    structure, 232
    summary of structure, 246
  bond, 354, 360
    alcohols and ethers, 876
    effect on entropy, 599
    effects, 663
    glycerol, 875
    hydrogen halides, 747

phosphoric acid, 723
relevant elements, 361
role in action of solvent, 403
strength, 361; of nylon, 383
structure of ice, 383
burning in bromine, 102F
compounds
  boiling points, 360
  with Group VI elements, 735
ion
  oxidizing agent, 99
  small size, 663
isotopes, 660
liquid, 663
  as fuel, 220
molecular compounds, 665
molecule
  intermolecular forces, 358
  molecular orbitals, 339, 340
physical properties, 663
preparation, 662
production, 662
properties, 662
reaction with
  bromine, mechanism, 456
  chlorine, 665
  ethylene, mechanism, 468
  oxygen, mechanism, 457
reducing agent, 661
strength of element-hydrogen bonds, 305
sulfide
  preparation, 739
  production, 153
hydrogen bromide
  preparation, 747
  rate law, 442
hydrogen chloride
  dipole moment, 312
  entropy at $T = 0$, 591
  preparation, 747
  reaction with ammonia, 177, 514
hydrogen cyanide
  molecular model, 59
  production, 713
hydrogen economy, 220
hydrogen electrode, 622
hydrogen fluoride
  anomalous boiling point, 360
  preparation, 747
hydrogen halides, 748т
  preparation, 747
hydrogen iodide
  decomposition kinetics, 440, 447
  equilibrium constant, 487
  formation kinetics, 440

preparation, 747
reaction mechanism, 449
hydrogen peroxide
  catalyzed decomposition, 465; mechanism, 467
  chemiluminescence, 738
  decomposition, thermodynamic aspects, 607
  molecular model, 738
  preparation, 739
  production, 738
  properties, 738
hydrogenation, 661
  alkene, 855
hydrohalic acids, production, 512
hydrohalogenation, alkene, 856
hydrolysis, 736
  nonmetal halides, 718
hydrometallurgy, 618, 661, 776
hydronium ion
  molecular model, 92
  as strongest acid, 527
hydroperoxide ion, 739
hydrophilic, 404
hydrophobic, 404
hydroquinone, molecular model, 106
hydroxide ion
  Lewis base, 515
  as strongest base, 527
hydroxyapatite, 402, 686
hydroxyl group, 873
hypervalance, 299; see also expanded octet
hypo, 106, 577, 741
hypochlorous acid, preparation, 750
hypofluorous acid, preparation, 750
hypohalite ions, 750
hypohalous acids
  preparation, 750
  production, 512
hypothesis, 46
hypothetical reaction, 211

i factor, 418
ice, 68, 369
  structure, 383
icebergs, source of fresh water, 419
icosahedron, 693
ideal gas, 157, 159
  internal energy, 193
  law, 158
  volume occupied by given mass, 167
ideal solution, 417
ilmenite, 768
implosion, 823
inaccuracy, 31
incandescence, 123
incomplete octet, 292, 300

indicator, 139, 560
  choice, 561
  color change, 562т, 563F
indirect titration, 142
indium, properties, 692
induced nuclear fission, 822
inert gas, 751
inert pair, 261, 262
  growth of importance, 290
infrared radiation, 235
inhibition, 456
initial rate, 434
initiation, 456
inorganic compound, 67
instant cooling pack, 199
instantaneous
  dipole, 357
  reaction rate, 433
insulator, 377, 683
integers, significant figures, 29
integrated rate law, 443
intensive property, 22
intercept, 916
interference, 372
interhalogens, 745т
interionic forces, 354
intermediate reaction, 452
intermolecular forces, 164, 353, 354т
  effect on surface tension, 362
  effect on viscosity, 362
  role of molecular shape, 358
  variation with distance, 165
internal combustion engine, flame front, 458
internal energy, 192
  ideal gas, 193
  relation to enthalpy, 198
International Bureau of Weights and Measures, 17
International System, 16
interpolate, 915
interstellar molecules, 303
interstice, 666
interstitial carbides, 711
intramolecular hemiacetal formation, 883
iodimetry, 141т
iodine
  colors of solution, 745
  deficiency, 745
  production, 745
  properties, 743
  vapor color, 152F
iodized salt, 745
ion, 51
  acidic and basic properties, 547
  energetics of formation, 284

exchange, 735
forces, 354
hydrated, as Lewis
 acid, 515
hydration enthalpy,
 411, 412T
  pattern, 412
pair, 278
product, 574
ion-dipole force, 354
 distance dependence,
 355
ion-induced dipole force,
 354
ion-ion force, 354
ion configurations, 254
ionic bond, 278
 factors favoring, 284
ionic character,
 electronegativity, 312
ionic compound, 59
ionic conduction, 377
ionic radii, 257T
 periodicity, 256
ionic solid, 352, 379
 characteristics, 353
 energetics of
 formation, 280
 examples, 353
 melting and boiling
 points, 354
ionization, 61
 acid, 92
 atom, 247
 constant, 520, 521,
 524T
  acid, 521
  base, 522
  measurement, 551
  polyprotic acid, 537T
  relation to free
  energy, 611
 energy, 258, 260T
  core electrons, 259
  first, 258
  periodicity, 260
  second, 258
  spectroscopic
  measurement, 258
  trend in d block, 763
  variation across
  Period 2, 260
 enthalpy, 258
iron, 771
 color of solutions, 63,
 773
 corrosion, 638
 diet, 774
 pentacarbonyl, 773
 properties, 767
 structure, 375
 variable valence, 288
iron(II)
 compounds, color, 289
 ions, 773F
 sulfate, preparation,
 288
iron(III)
 compounds, color, 289
 ions, 773F
  acid character, 548

isooctane, molecular
 model, 846
isocyano group, 872
isoelectronic, 294
 molecules and shape,
 328
isoleucine, 898
isomer, 784
 geometrical, 784
 optical, 784
 steroisomer, 784
 structural, 784
isomerism, 784
 classification, 842
 d-metal complexes, 784
 geometrical in alkenes,
 853
 organic compounds,
 842
 structural, 842
 substituted benzenes,
 863
isomerization, 850
isonitrile, 872
isoprene, 859
isopropyl
 alcohol, 874
 group, 845
isotactic polymer, 858
isotope, 52, 802
 abundance, 54
 properties, 53T
isotopic dating, 819
isotopic tracing, 401
IUPAC, International
 Union of Pure and
 Applied Chemistry, 843

jade, 170
joule, 190
Joule-Thomson effect,
 180

karat, 777
Kekulé structure, 298,
 862
kelvin, 20
kernite, 693
kerosene, 849
ketone, 876, 879, 880T
 nomenclature, 881
 origin of name, 881
 preparation, 882
 reactions, 882
kilogram, 17
kinetic energy, 189
 needed for reaction,
 463
 temperature of gas,
 189
kinetic theory, 174
kinetics, reaction 432; see
 also chemical kinetics
Kroll process, 618
krypton
 difluoride, 753
 properties, 751

l, azimuthal quantum
 number, 242
 relation to angular
 momentum, 243
lability, 780
lactic acid, molecular
 model, 566
lanthanide contraction,
 764, 775
lapis lazuli, 739F
latex, 859F
 spheres, 363F
lattice enthalpy, 281,
 281T
law
 Avogadro's, 158
 Boyle's, 157
 Charles's, 158
 conservation of energy,
 191
 constant composition,
 9, 127
 Dalton's, 170
 Faraday's, 646
 Graham's, 177
 Henry's, 405
 Hess's, 210
 ideal gas, 158
 mass action, 480
 Ohm's, 377
 radioactive decay, 817
 Raoult's, 416
 rate, 437
lawrencium, 812
Le Chatelier's principle,
 407, 499
 chemical reactions, 499
 effect of pressure, 500
leaching, 832
lead, 267, 704
 entropy, 593
 in lead-acid battery,
 627
 oxidation, 290
 passivation, 704
 plumbing, 267
 poisoning antidote, 782
 production, 704
 properties, 701
lead azide, 170
lead halides, 712
lead nitrate, thermal
 decomposition, 721
lead oxide, 290F
lead sulfide, 89F
lead(II) chromate, 87F
 iodide, precipitation,
 14F
 oxide, 267, 711
lead(IV) oxide, 711
lead-acid cell, 627
lead-tin-bismuth alloy,
 377
leather, 770
Leclanché cell, 625
leguminous plant, 715
lemon juice, 532
leucine, 898
levelling effect, 527
levo, 898
Lewis, G. N., 291F

Lewis acids and bases,
 514
 boric acid, 695
 boron halides, 699
 carbon dioxide, 706
 cyanide ions, 713
 d-metal ions, 779
 failures of electron-pair
 model, 338
 structure for expanded
 octets, 300
 structures, 293
  rules for writing,
  294, 296
  and VSEPR theory,
  323
Libby, Willard, 821
ligand, 781T
 effect on color, 791
 field
  splitting, 788, 789
  theory, 788
 monodentate, 782
 polydentate, 782
 strong-field, 790
 weak-field, 790
light
 color, 234
 plane-polarized, 787
 properties, 233
 speed, 234
 visible, 234
 white, 234
light-year, 37
lightning, 715
like dissolves like, 403
lime
 slaked, 685
 water, 685
 for water softening,
 686
limestone, 684
limiting reagent, 123
 identification, 124
linear accelerator, 812F
linoleic acid, molecular
 model, 661
lipid, 890
liquefaction, 179
liquid, 6
 computer simulation,
 361F
liquid, 361
 crystal, 384
 hydrogen, fuel, 220
liter, 21
litharge, 290
lithium, 667F
 compounds, 672
 isotopic abundance,
 668
 slow reaction with
 water, 669
lithium aluminum
 hydride, 700
litmus, 560
lock-and-key model, 469,
 901
logarithm, 914
 properties, 914
London force, 354

London force (*continued*)
distance dependence, 357
lone pair
effect on shape, 324
introduced, 293
long period, 253
long-range interactions, 356
low-spin complex, 792, 794
lubricants, 849
Lucite, 856
lysine, 898
lysozyme, 468F

M, moles per liter, 395
magnesium, 677F, 678, 683
combustion, 97F, 590
compounds, 683
oxidation and entropy, 597
production, 678
magnesium hydroxide, 683
magnesium oxide, 683
Born-Haber cycle, 737
structure, 380
magnesium sulfate, 683
magnetic properties, 795
oxygen, 338
magnetic quantum number, 242, 243
magnetite, 771
main-group elements, 256
typical ions, 284
malleable, 45
malonic acid, 888
manganate ion, 70, 770
preparation, 771
manganese, 770
bronze, 770
nodules, 770F
production, 770
properties, 767
steel, 772T
manganese dioxide, 770
as catalyst, 466
Manhattan Project, 830
manic depression, 672
manometer, 155
many-electron atom, 247
orbital energy levels, 248
structure, 247
marble, 684
Mars, atmosphere, 371
mass, 15
atomic, 51
calculations, 120
concentration, 395
critical, 823
number, 52
percentage composition, 33, 395
produced in electrolysis, 648
mass action, 480

mass spectrometer, 51
mass spectrometry, molecular weight determination, 131
matches, 676, 722, 750
matter, 5
classification chart, 75
dispersing, 414
Maxwell distribution, 178
reactive collisions, 463
meat color, 721
mechanics, classical, 232
melting, 6
enthalpy of, 203
entropy, 594
point, 6
alkali metals, 667
alkane and alkene, 853
alkanes, 848
elements, 934T
for analysis, 7
ionic solids, 354T
molecular compounds, 359
Mendeleev, Dmitri Ivanovich, 254F
meniscus, 364
mercapto group, 872
mercury
cell, 626
production, 779
properties, 775
reaction with acids, 779
mercury(I) chloride, 779
mesophase, 384
metal, 45
characteristics, 265, 353
examples, 353
orbitals, 378
reaction with acids, 636; oxoacids, 637
structure, 352
metal-ammonia solution, 670F
metallic,
character, location in periodic table, 261
conductor, 377
hydride, 666
radii, 257T
measurement using x-rays, 373
periodicity, 256
trend in *d* block, 763
metallocene, 774
metals and nonmetals characteristics, 265T
location in periodic table, 46
metathesis, 88
meteorite, source of nickel-steel, 33
meter, 18
methane
flame color, 84F
molecular model, 9, 840
shape, 324
structure, 841
methanol, production, 661

methionine, 899
methyl methacrylate, 856
methyl orange, 563F
methyl radical, 303
methyl red, 563F
methylene group, 840
methylpropane, 842
metric system, 16
mica, 710F
micelle, 405
micropipet, 136F
milk of magnesia, 683
milligram, 17
Millikan, Robert, 48
mineral, 6
deposits, role of solubility, 402
minutes, 19
mixture, 9, 10
differences from compounds, 10
gas, 170
heterogeneous, 14
homogeneous, 14
racemic, 787, 899
separating, 10, 11T
m_l, magnetic quantum number, 242
mmHg, 154
moderator, 824
Moissan's method, 619
molality, 395, 398
molar concentration, 133, 394
molar mass, 57
molar volume, 166
gas, 159
molarity, 133; *see also* molar concentration
mole, 56
calculating number from mass, 58
calculations, 118
diagram, 120
elements, 57F
fraction, 173, 395, 396
ionic compound, 66F
molecular compounds, 60F
molecular compound, 59
melting and boiling points, 359T
molecular formula, 59, 131
molecular model, gas, 154
molecular orbital, 339
energy level diagram, 340
metals, 378
Period-2 diatomic molecules, 341
theory, 337
molecular potential energy curve, 341
molecular shape
effect on intermolecular forces, 358
hybridization, 330
VSEPR theory, 322

molecular solids, 352
characteristics, 353
molecular speed, 176
molecular structure, effect on vapor pressure, 366
molecular weight, 60, 920T
determination, 131
measurement by diffusion, 178; freezing point depression, 420; osmometry, 422
relation to speed, 176
molecularity, 450
molecule, 9, 59
diatomic, 60
polyatomic, 60
Mond process, 774
monodentate ligand, 782
monomer, 856
monosaccharide, 885
Monsanto process, 940
mortar, 686
Moseley, Henry, 49
m_s, spin-magnetic quantum number, 242
multiple bond, 293
bond length, 306
effect on shape, 326
strength, 304
multistep reactions, at equilibrium, 485
mustard gas, molecular model, 743

n-type semiconductor, 379
naphtha, 939
naphthalene, 862
natural
abundance, 54
gas, combustion, 206F
logarithm, 914
nematic phase, 384
neon
isotopes, 52
lights, 753
production, 153
properties, 75
neopentane, molecular model, 358
Nernst equation, 641
net ionic reaction, 87
network former, 831
network solids, 352
characteristics, 353
examples, 353
neutralization, 94
enthalpy, 513
in nonaqueous solvents, 514
proton transfer, 513
neutrino, 807
neutron, properties, 48
neutron-induced transmutation, 812
neutron-rich nuclei, 810
Newton, Isaac, 232

nickel
  properties, 767
  steel, 772T
nickel carbonyl, 774
  preparation, 705
nickel(II) sulfate, 352F
nickel-barium alloy, 377
nickel-cadmium cell, 628,
  775
nickel-steel, 33
nitrate ion, 722
  action of heat, 722
  Lewis structure, 297
nitric acid
  catalyst, 467
  oxidizing agent, 722
  production, 512, 721
  properties, 722
  reaction with copper,
    100, 776; mercury,
    779
nitric oxide
  formation equilibrium,
    494
  Lewis structure, 302
  preparation, 720
  production, 302, 720
  reaction in atmosphere,
    720
  reaction with oxygen,
    mechanism, 455
nitride, alkali metal, 670
nitrile, 872
nitrite, 721
  ion shape, 328
  preparation, 721
nitro compound, 872
nitrogen, 715
  contribution to aurora,
    453
  differences from
    congeners, 715
  fixing, 715
  Lewis structure, 293
  molecule configuration,
    342
  oxoacids, 719T
  production, 153
  properties, 714
nitrogen dioxide
  color, 152F
  decomposition kinetics,
    440
  dimerization equilibria,
    effect of pressure, 501
  equilibrium constant,
    487
  preparation, 721
  properties, 721
  reaction with carbon
    monoxide,
    mechanism, 451
nitrogen oxides, 719T
nitrogen triiodide
  explosion, 719
  molecular model, 719
nitroglycerin, explosion,
  170
nitrotoluene, molecular
  model, 864
nitrous acid, 721

nitrous oxide
  decomposition kinetics,
    440
  entropy, 592
  formal charge, 309
  preparation, 720
noble gas, 45, 751; see
  also Group VIII
  elements
    orbital explanation, 340
noble metals, origin of
  nobility, 764
nodal plane, 339
  p orbital, 244
nodule
  manganese, 770
  nitrogen fixing, 715F
nomenclature, 67
  acids, 74
  alcohols, 873
  aldehydes, 881
  alkanes, 843
  alkenes, 852
  amines, 892
  arenes, 863
  carboxylic acids, 887
  d-metal complexes, 782
  geometrical isomers,
    854
  ketones, 881
  organic, 843
  substituted benzenes,
    864
nonaqueous solution, 15
  Brønsted acids, 513
nonelectrolyte, 62
nonferrous alloy, 777T
nonmetal, characteristics,
  265
nonpolar solvent, 403
nonsparking metal, 678
nonstoichiometric
  compound, 10
normal boiling point, 366
northern lights, 453
NO$_x$, catalysis, 468
nuclear atom, 48, 49
nuclear binding energy,
  825
nuclear chemistry, 802
nuclear charge, effective,
  247
nuclear decay law, 817
nuclear disintegration,
  804
nuclear explosion, 823
nuclear fission, 822
nuclear fuel, 829
nuclear fusion, 828
  in stars, 836
nuclear power, 822
nuclear radiation, 802,
  805
nuclear reaction, 805
nuclear reactor, 824
nuclear stability, 809
nuclear structure, 802
nuclear transmutation,
  804
nuclear waste, 831
nucleic acid, 904

nucleon, 49, 802
nucleoside, 902
  structure, 903
nucleosynthesis, 812
nucleotide, 902
  structure, 903
nuclide, 802
nylon
  electrostatic charge,
    896
  preparation, 895
  salt, 895
  strength, 896

ocean
  composition, 394
  pH, buffer, 564
octahedral complex, 782
octane
  combustion, 196
  molecular model, 196,
    844
octet, 286
  expanded, 292, 299
  incomplete, 292
  rule, 292
odd-electron molecules,
  302
odor, 890T
ohm, 377
Ohm's law, 377
oil, 660
  essential, 878
oleum, 741
oligopeptide, 900
oligosaccharide, 885
onyx, 707F
optical activity, 787
  judging, 897
  organic compounds,
    896
optical isomer, 784
orange
  juice, 532
  odor, 890
orbital
  angular momentum,
    243
  antibonding, 340
  atomic, 241
  bonding, 340
  delocalized, 344
  e, 788
  hybrid, 330
  hydrogen atom, 243T
  molecular, 339
  order of occupation,
    251
  pi (π), 342
  sigma (σ), 339
  t, 788
order
  first, 438
  overall, 438
  second, 438
  zero, 438
ore, 7
organic
  chemistry, 840
  compound, 67

Orlon, 856
orthosilicates, 710
osmometry, 421
osmosis, 421
  blood corpuscles, 423
  reverse, 423
osmotic pressure, 421
Ostwald process, 721
  catalyst, 467
  thermodynamic aspects,
    607
-ous, 68
ovenware, 708
overall order, 438
overall reaction, 211
  mechanism, 451
overlap, 333
overpotential, 644
ox, oxalate ligand, 781
oxalic acid, 888
  molecular model, 99
oxidation, 97, 619
  alcohols, 876
  competing, in
    electrolysis, 645
  number, 100
    charge distribution,
      309
    of d metals, 765
    oxidizing ability, 104
    oxygen content, 102
    rule for calculating,
      101
  potential, 632
  production of
    elements, 619
oxide
  acidic, 93
    and oxidation
      number, 766
  alkali metal, 670
  basic, 93
  Group III elements,
    695
  Group IV elements,
    705
  Group V elements, 719
  halogen, 749
  ion
    basic character, 92
    Lewis base, 516
  thermodynamics of
    formation, 737
  xenon, 754
oxidizing ability
  oxidation number, 104
oxidizing agent, 98, 619
  hydrogen ion, 99
  identification by
    oxidation number,
    102
  strength, 634
  sulfur dioxide, 740
  water, 736
oxoacid, 74
  anhydride, 695
  Group V elements, 719
  halogen, strengths 749,
    749T
  reaction with metals,
    637

oxoacid *(continued)*
strengths, 529
sulfur, 740
oxoanions, 64, 70T
oxonium ion, 875
oxyacetylene welding, 861F
oxygen
atom, contribution to aurora, 453
characteristics, 734
differences from sulfur, 734
Lewis structure, 293
liquid, 338, 732
molecular configuration, 343
paramagnetism, 338, 343, 733
production, 153, 732
properties, 732
reaction with hydrogen, mechanism, 457; nitric oxide, mechanism, 455
ozone
decomposition equilibrium, 486
kinetics, 442
mechanism, 450
destruction, nitric oxide, 720
dipole moment, 313
hole, 712, 733F
polarity, 329
production, 733
reaction mechanism, 453
VSEPR prediction, 328
ozonide ion, 64

p-block elements, 255
properties, 266
role of inert pair, 261
variable valence, 289
p orbital, 244
boundary surface, 245
p subshell, 243
p–n junction, 379
p-type semiconductor, 379
paired electrons, 249
papermaker's alum, 697
paraffin wax, 849
paraffins, 842
parallel spins, 251
paramagnetic, 795
oxygen, 338F
parsec, 37
partial
charge, 311
intermolecular forces, 354
pressure, 171
equilibrium constant, 486
mole fraction, 173
particles and waves, 239
parts per million, 395

pascal, 155
passivation, 637, 694
lead, 704
path independence, state function, 193
patina, 777
Pauli, Wolfgang, 249
Pauli exclusion principle, 249
role in bonding, 340
Pauling, Linus, 264
pear odor, 890
pelargonidin, molecular model, 560
penetrating power, 814
penetration, 248
pentane, molecular model, 358
peptide, 899, 900
bond, 900
percentage composition, 33
percentage yield, 126
perchlorates, preparation, 751
period, 44
long, 253
periodic table, 43
location of acidic and basic oxides, 93; metals and nonmetals, 46
structure, 44; and the building-up principle, 255
periodicity, 50
electron affinity, 262
electrode potentials, 632
electronegativity, 264
polarizability, 311
survey of properties, 254
permanent waving, 901
permanganate
ion, 70
oxidizing agent, 766
oxidizing power, 771
titration, 140F, 141T
perovskite, 381
structure, 389
peroxide
alkali metal, 670
ion, 64
peroxyacetyl nitrate, molecular model, 739
perpetual motion, 191
Perutz, Max, 372
perxenates, 754
perylene, chemiluminescence, 738
petroleum, 12, 849T
pewter, 377, 777
pH, 530, 551F
blood, 564
calculations, 530
approximations, 534
buffer, 565
salt solution, 547
strong-acid–strong-base titration, 533, 553

strong-acid–weak-base salt solution, 547
sulfuric acid, 537
weak acid, 534
weak base, 535
weak-acid–strong-base titration, 556
weak-acid–strong-base solution, 549
weak-base–strong-acid titration, 559
common solutions, 531
curves, 552
strong-acid–strong-base titration, 553
weak-acid–strong-base titration, 556
weak-base–strong-acid titration, 559
disease, 566
electrical measurement, 642
meter, 532
mixed solution, 550
neutral solution, 532
qualitative analysis, 575
relation to pK_a, 551
salts, 546
shock, 566
solubility of amphoteric oxides, 571; salts, 573
stabilization, 564
strong acid or base, 533
phase, 353
boundary, 370
diagram, 369
carbon, 387
carbon dioxide, 370
helium, 387, 752
water, 369; with solute, 419
phenol, 877
molecular model, 878
preparation, 878
properties, 878
phenol-formaldehyde resins, 878
phenolic resins, 878
phenolphthalein, 561, 563F
phenoxide ion, 878
phenyl, 865
phenylalanine, 899
phonograph head, 769
phosphate, 723
buffer, 565
fertilizer, 720
ion, formal charge, 308
rock, 402F, 716, 723
solubility, 402
phosphine, 717
preparation, 666, 718
phosphoric acid, 719
molecular model, 723
production, 723
properties, 723
viscosity, 362
phosphorous acid
diprotic character, 529

molecular model, 529, 722
phosphorus
allotropes, 716
molecular model, 60
octet expansion, 299
production, 716
properties, 714
reaction with bromine, 99F
phosphorus halides, 718
phosphorus pentabromide, structure, 719
phosphorus pentachloride
Lewis structure, 299
preparation, 299, 718
shape, 324
structure, 299, 719
phosphorus pentafluoride, hybridization, 332
phosphorus trichloride
Lewis structure, 299
preparation, 299, 718
phosphorus(III) oxide
molecular model, 722
preparation, 722
phosphorus(V) oxide
drying agent, 128
molecular model, 723
preparation, 723
photochemical reaction, 106
photochromic sunglasses, 708
photocopying, 734
photoelectric effect, 235
photoelectron, 235
photography, 105
photon 235
energy and color, 234
photosynthesis, 684
energy storage, 189
food and fuel, 218
physical change
enthalpy of, 202
entropy, 594
physical state, 5
pi (π)
bond, introduced, 334
orbital, 342
piezoelectric, 768
pig iron, 771
pineapple flavor, 891
pipet, 15F, 133
pitchblende, 829F
Planck's constant, 235
plane-polarized light, 787
plasma, fusion research, 829
plaster of Paris, 687
plastic, 857
plating, 650
chemical, 700
pleated sheet, 900
Plexiglas, 856T
plutonium, critical mass, 823
pOH, 533

relation to pH, 533
poison
  arsenic, 469
  catalyst, 468
polar bond, 312
polar molecule, 328
polar solvent, 403
polarity, 328
  effect on acid strength,
  528
  experimental test, 328
polarizability
  intermolecular forces,
  359
  periodicity, 311
polarization, 310
polarizing power, 310
  periodicity, 311
pollution
  atmospheric, 397
  atmospheric sulfur, 740
  control, 643
  hydrogen peroxide,
  739
polonium, properties,
  732
polyacrylonitrile,
  molecular model, 857
polyamide, 894, 895
polyatomic ions, 64
polyatomic molecule, 60
  Lewis structures, 295
  molecular orbitals, 343
polycyclic, 862
polydentate ligand, 782
polyesters, 891
polyethylene, 856
  high-density, 858
  molecular model, 857
polyisoprene, 860
polymer, 856, 856T
  addition, 856
  atactic, 858
  condensation, 891
  isotactic, 858
  syndiotactic, 858
polymerization
  addition, 856
  radical, 858
  condensation, 891
polynucleotide, 904
polypeptide, 900
polyphosphate, 723
  in detergent, 404
  molecular model, 404
polyphosphoric acid, 723
polypropylene, 856
  molecular model, 857
polyprotic acid, 536
  ionization constants,
  537T
  strengths, 536
polysaccharide, 886
polysulfane, 739
polytetrafluoroethylene,
  inertness, 858
polyvinyl chloride, 856
  molecular model, 857
poppy color, 561
porcelain, 710
porphyrin, molecular

model, 684
positron, 805
  emission, 808
potash, 940
potassium
  compounds, 676
  production, 667
  structure, 375
potassium chlorate, 750
  catalyzed
  decomposition, 466
potassium chloride
  Born-Haber cycle, 280
  structure, 380
potassium iodide, catalyst,
  465
potassium nitrate, 676
potassium permanganate,
  667F
  production, 85F
potassium superoxide,
  671
  reacting with carbon
  dioxide, 169F
potential
  cell, 623
  coulombic, 279
  energy, 190
  curve, 341
  oxidation, 632
  reduction, 63
Powell, Herbert, 322
power units, 202
ppm, parts per million,
  395
precipitate, 14
precipitation
  reaction, 86
  qualitative analysis,
  574
  selective, 576
precision, 30, 31
preexponential factor,
  458, 461T
prefixes
  naming compounds, 71
  SI, 18
pressure, 154
  effect on equilibrium,
  500, 501
  effect on freezing
  point, 369
  height of liquid
  column, 154
  unit, 154
  units, 155
pressurized-water reactor,
  824
primary alcohol, 873
primary amine, 892
primary cell, 625
primary structure, 900
principal quantum
  number, 242
principle
  building-up, 250
  exclusion, 249
  Hund's, 251
  Le Chatelier's, 407, 499
  uncertainty, 240
probability, electron

position, 241
product, 82
proline, 899
prolyprotic base, 536
propagation, 456
propanedioic acid, 888
propene, 851
  molecular model, 443
property, 5
  bulk, 8
  extensive, 22
  intensive, 22
  periodicity, 254
  state, 192
propyne, 861
protein, 221, 899
  as polyprotic acid, 537
protium, 662
proton
  properties, 48
transfer, 92, 94, 513
  equilibrium, 517
proton-rich nuclei, 810
protonation, 666
pseudofirst-order
  reaction, 448
pseudonoble-gas
  configuration, 285
psi (ψ), wavefunction, 241
PTFE,
  polytetrafluoroethylene,
  856
purex process, 831
purgative, 683
purity test, 418
putrescine, 892
PVC, polyvinyl chloride,
  856
pX (as in pH, pK),
  $-\log X$, 521
pyran, 884
pyranose, 884
Pyrex, 708
pyrite, 288, 771F
pyrolusite, 770
pyrometallurgy, 776
pyroxene, 710
  structure, 709

quadratic equation, 496,
  915
qualitative analysis, 575,
  576F
  distinguishing mercury
  and silver, 779
  precipitation reactions,
  574
  role of pH, 575
  scheme, 576
quantitative reaction, 126
quantity calculus, 24; see
  also unit analysis
quantization, 237
  energy, 236
quantum
  light, 235
  number, 242
  azimuthal, l, 242
  magnetic, m_l, 242
  orbital, l, 242

principal, n, 242
  spin-magnetic, m_s,
  242
quartz, 703F
  structure, 372, 707
quartzite, 703F
quaternary
  ammonium ion, 893
  structure, 901
quicklime, 92, 685
quotient, reaction, 484

R
  gas constant, 159
  Rydberg constant, 237
racemic mixture, 787,
  899
rad, 815, 816
radiation, 805T
  effects of electric fields,
  803
  electromagnetic, 233
  health hazard, 817
  infrared, 235
  penetrating power, 814
  ultraviolet, 235
  units, 815
radical, 303
  chain reaction, 456
  polymerization, 858
radioactive, 802
  decay law, 817
  series, 810
radioactivity
  dose, 816
  survey, 813
radiocarbon dating, 820
radius
  atomic, 256
  hydrogen atom, 238
  ionic, 256
  metallic, 256
radon, 753
  properties, 751
random errors, 30
Raoult's law, 416
rate
  radioactive decay, 817
  relation to equilibrium
  constant, 485; half-
  life, 446
rate constant, 437
  at different
  temperatures, 462
rate law, 440T, 436, 437
  calculation of
  concentration, 443
  from mechanism, 455
  integrated, 443
  ozone decomposition,
  442
rate-determining step,
  452
  and reaction profile,
  465
rate reaction, 432
reactant, 82
reacting
  gases, 166

reacting (continued)
calculating volumes reacting, 168
volumes, solutions, 137
reaction
addition, 855
bimolecular, 450
condensation, 890
conditions favoring, 603
dehydrohalogenation, 852
elimination, 852
endothermic spontaneity, 600
exothermic, spontaneity, 600
hydrohalogenation, 856
intermediate, 452
isomerization, 850
nuclear, 805
precipitation, 574
spontaneous, 606
substitution, 851
complexes, 780
temperature required, 608
termolecular, 451
that "does not go," 483
that "goes," 483
unimolecular, 450
reaction direction, 483
reaction enthalpy, 205
combining, 211
developed, 206
estimation from bond enthalpy, 305
standard, 207, 209
reaction entropy, calculating, 594
reaction free energy, 604
reaction mechanism, 449
bromomethane hydrolysis, 451
deducing rate law, 455
hydrogen and bromine, 456
hydrogen iodide decomposition, 449
nature of proof, 455
nitric oxide and oxygen, 455
nitrogen dioxide and carbon monoxide, 451
ozone decomposition, 450, 453
reaction order, 436
reaction profile, 464
catalyzed reaction, 466
multistep reaction, 465
reaction quotient, 484
direction of reaction, 484
relation to free energy, 609
reaction rate, 432
collision theory, 462
dependence on concentration, 435
initial, 434
instantaneous, 433
relation to equilibrium

constant, 485
sensitivity of response to temperature, 460
temperature and equilibrium constant, 503
temperature dependence, 458
units, 433
reaction sequence, 210
reaction yield, 125
reactions at equilibrium, 478
reactor
coolant, 668
nuclear, 824
reagent, 82
real gas, 164
recrystallization, 11, 15
recycling aluminum, 694
red lead, 290
redox indicator, 140
redox reaction, 96, 98, 103T, 618
Ellingham diagram, 756
photography, 106
redox titration, 139, 635F
reducing agent, 98, 619
carbon monoxide, 706
hydrogen, 661
identification by oxidation number, 102
sulfur dioxide, 740
water, 736
reducing strength, 634
reducing sugar, 885
reduction, 97, 619
competing, in electrolysis, 644
elements prepared by, 618
potential, 632, 927T
alkali metal trends, 668
refining copper, 645
reforming, 850, 939
reaction, 662
relation between K_a and K_b, 523
relative biological effectiveness, 816
rem, roentgen equivalent man, 815T, 816
residual entropy, 591
residue, 900
resin, 878
resistance, 377
resonance, 297
effect on properties, 298; shape, 327
hybrid, 297
retardation, 456
retinal, molecular model, 337
reverse osmosis, 423
rhombic sulfur, 362, 734F
ribonucleic acid, 902
ribose, 902
structure, 903

RNA, ribonucleic acid, 902
rock-salt, 381T, 672
structure, 380
rocket fuel, hydrazine, 717
Roentgen, Wilhelm, 372, 817
roentgen equivalent man, 816
root-mean square speed, 175
roots of equation, 915
rubber, 859
butyl, 860
rubidium, 667F
ruby, 696F
rust, 638
Rutherford, Ernest, 48F
discovery of radioactivity, 803
rutile, 381T, 768
structure, 389
Rydberg constant, 237
Rydberg formula, 238

s-block elements, 255
introduction to chemical properties, 265
s orbital, 244
s subshell, 243
saccharide, 885
sacrificial anode, 639
sal ammoniac, 717
saline carbides, 711
saline hydrides, 664
salt, 94
mine, 672
predicting precipitation, 575
solution, calculation of pH, 547, 549, 551
salt bridge, 620
salt cake, 673
salt hydrates, 355
salted meat, 423
salts
pH of solutions, 546
solubility and pH, 573
sample, 5
sand, 707
sandwich compound, structure, 774
sapphire, 696F
saturated hydrocarbon, 840
saturated solution, 400, 401
scale, 402
scandium, properties, 767
scientific notation, 912
scintillation counter, 814
scum, 404
sea level, thermochemical, 214
sea of stability, 809
seawater
composition, 394T
origin of saltiness, 672

source of magnesium, 678
seaweed, 745
second, 19
second law, 588, 600
second-order reaction, 438
experimental test, 447
integrated rate law, 447
secondary alcohol, 874
secondary amines, 893
secondary cell, 625, 626
secondary structure, 900
selective precipitation, 576
selenium, 734
properties, 732
self-oxidation, 103, 104
self-reduction, 104
self-sustaining fission, 822
semiconductor, 374, 377, 379
semipermeable membrane, 421
sequester, 404
polyphosphate ions, 723
series
Balmer, 236
homologous, 841
radioactive, 810
spectrochemical, 790
serine, 898
SHE, standard hydrogen electrode, 631
shell
atom, 242
complete, 249
valence, 251
shielding, 248
short-range interactions, 356
shortening, 853
Shrödinger, Erwin, 241F
SI, international system of units, 16
prefixes, 18
sickle-cell anemia, 901
Sidgwick, Nevil, 322
sievert, 815
sigma (σ) bond, 333
sigma (σ) orbital, 339
significant figures, 28
counting, 29
rules in calculations, 31
silane, 713
silica
crystal forms, 703
gel, 708
gems, 707
reaction with hydrofluoric acid, 708, 748
structure, 372
silicates, structure, 707, 709
silicon
amorphous, 704
contrasted with carbon, 702

difference from carbon, 713
production, 703
properties, 701
single crystal growing, 704F
silicon carbide, 711
silicon halides, 713
silicone, 710
silver disproportionation, 777
photochromism, 708
plating, 650
properties, 775, 777
structure, 375
variable valence, 288
silver chloride
common-ion effect, 570
electrode, 622
solubility, 569
in ammonia, 577, 578
product calculation, 640
silver fluoride, solubility, 411
silver halide
covalent character, 311
photography, 106
solubility, 411
single bond, orbital structure, 335
skeletal equation, 84
slag, 685, 771
slaked lime, 92, 685
slope of graph, 916
smectic phase, 384
smelling salts, 717
smelting, 618, 776
smog, 720
smokescreen, 768
smokey flame, 123
smoking food, 880
soap, 403
production, 676
soda pop, 405
soda-ash, 674
soda-lime glass, 708
sodium, 667F
boiling point predicted, 598
production, 649
reaction with alcohol, 875
structure, 375
sodium bromide, 380
sodium carbonate
production, 674
water softening, 404
sodium chlorate, 750
sodium chloride
abundance in ocean, 672
crystals, 371F
production, 672
structure, 380
sodium chromate, 769
sodium fluoride, solubility, 411
sodium hydride, 665
sodium hydroxide,

production, 512, 673
sodium hypochlorite, production, 649
sodium peroxide, 671
sodium stearate, soap, 404
sodium sulfate, production, 673
sodium thiosulfate
iodine titrations, 143
preparation, 741
solder, 377
solidification, 368
solids, 6, 371
characteristics, 353
entropy, 590
solubility, 400, 401, 401T
amphoteric oxides, 571
calcium fluoride, 573
common-ion effect, 570
dependence on
solvent, 403
temperature, 407
effect of complex formation, 578
equilibria, 566
fluorides, 747
gas, 405
pressure dependence, 405
product, 567, 568T
calculation from electrode potentials, 640
measurement from solubility, 569
relation to K_{sp}, 569
rules, 89
salts, and pH, 573
solute, 14
solution, 14
aqueous, 15
enthalpy, 408T
contributions, 409
ideal, 417
nonaqueous, 15
pH of salt and acid, 551
solvation, 410
Solvay process, 674
solvent, 14
extraction, 831
soot, 702
sooty flame, 123
sound detection, 769
southern lights, 453
SO_x, 740
sp^2 hybrid, 330
sp^3 hybrid, 330
space shuttle
booster rockets, 750
fuel tanks, 663
spark plugs, 377
spearmint, 881
specific enthalpy density, 219
specific heat capacity, 201
spectator ions, 87
spectral line, 233
spectrochemical series, 790

spectrometer, 233F
spectroscopy, 232
spectrum, 236
atomic hydrogen, 236
speed
distribution, 178
gas molecules, 175
light, 234
molecular weight dependence, 176
of sound, 176
temperature dependence, 176
spent-fuel processing, 831
sphalerite, 778
spin, 246
electron, 246
quantum numbers, 242
spin-magnetic quantum number, 242, 246
spontaneous change, 414, 588
conditions, 603T
free-energy criterion, 602
and the second law, 600
spontaneous nuclear fission, 822
spontaneous reaction, 606
from electrochemical series, 635
stability, 606
nuclear, 809
sea of, 809
thermodynamic versus kinetic, 780
stainless steel, 377, 769
standard cell potential, 630
standard electrode potential, 630, 633T
standard enthalpy
combustion, 209
formation, 214
standard formation enthalpy, 215T
combining, 215
standard free energy of, 605T
calculation of, 605
formation, 604
reaction, 604
standard hydrogen electrode, 630
standard molar entropy, 592, 593T
standard reaction
enthalpy, 207
entropy, 594, 595
standard reduction potential, 632, 927T
standard state, 208
standard temperature and pressure, 166
Stanford linear accelerator, 812F
starch, 886
indicator, 141F

stars, energy source, 836
state, 5
change of, 6
equation of, 159
ground, 238
label, 83
property, 192
stearate ion, molecular model, 404
stearic acid, molecular model, 661
steel, 772T
stainless, 377
steelmaking, 771
stem, 68
stepwise formation constant, 579, 579T
stereoisomer, 784
Stern-Gerlach experiment, 246
Stock, Alfred, 338
Stock number, 68
relation to oxidation number, 101
stoichiometric coefficient, 83
interpreted, 118
stoichiometric proportions, 494
STP, standard temperature and pressure, 166
strength
oxoacid, and structure, 529
polyprotic acid, 536
relation to structure of acid, 528
strong acid, 525, 525T
calculation of pH, 533
strong base, 525, 525T
calculation of pH, 533
strong-field ligand, 790
strontium, 677F
production, 679
structural isomer, 784
structure
atomic, 232
many-electron atoms, 247
quaternary, 901
relation to strength of acid, 528
secondary, 900
tertiary, 900
styrene, 856
molecular model, 865
subatomic particle, 42
properties, 48T
subcritical mass, 823
sublimation, 6, 204
enthalpy of, 204
vapor pressure, 364
subshell
atom, 243
stability of half-full, 253
substance, 5
pure, 9
substituent, 844
substitution, 851

substitution (*continued*)
   predicting product, 877
   reaction, *d*-metal
      complex, 780
substrate, 469
sucrose
   hydrolysis kinetics, 440
   solubility, 400
   structure, 885
suffix, 68
sugar, 885
sulfate ion
   formal charge, 308
   Lewis structure, 297,
      300
   VSEPR prediction, 327
sulfates, 741
sulfide ions, in qualitative
   analysis, 576
sulfite ion, shape, 324
sulfites, 740
   oxidizing and reducing
      properties, 740
sulfo group, 872
sulfonic acid, 872
sulfur
   allotropes, 734
   atmospheric pollution,
      740
   combustion, 168F
   diatomic molecules,
      294
   differences from
      oxygen, 734
   dissolving in carbon
      disulfide, 403F
   effect of heating, 362
   octet expansion, 300
   oxoacids, 740T
   production, 733
   properties, 732
sulfur dichloride, 743
sulfur dioxide
   atmospheric, 396, 740
   equilibrium constant,
      487
   Lewis structure, 298
   oxidation, 439F
   oxidizing and reducing
      properties, 740
   production, 153, 740
   shape, 328
sulfur halides, 742
sulfur hexafluoride, 744
   bonding, 333
   hybridization, 332
   structure, 742
   molecular model, 60
sulfur oxides, 740T
sulfur tetrafluoride
   Lewis structure, 300
   structure, 742
   VSEPR prediction, 326
sulfur trioxide
   Lewis acid, 516
   preparation, 741
   production, 741
   structure, 741
sulfuric acid
   economic indicator, 91
   pH calculation, 537

as polyprotic acid, 536
production, 512, 741
properties, 741
reaction with copper,
   742
sulfurous acid, 740
sun, energy source, 836
sunglasses, 708
superconductor, 377
   high-temperature, 352
   structure, 389
supercritical, 368, 823
superfluidity, 752
superoxide ion, 64
superoxides, alkali metal,
   670
superphosphate, 723
support, catalyst, 13
surface tension, 361, 363
   dependence on
      intermolecular forces,
      362
surface tunneling
   microscopy, 47
surfactant, 404
surgical implants, 711
surroundings, 192
   entropy change, 595
swimming-pool water,
   735
sylvite, 676
syndiotactic polymer, 858
synthesis, 82
synthesis gas, 662
system, 192
systematic error, 31
systematic name, 67

t orbital, 788
talc, 710
tangent, 917
tanning, 770
Teflon, 744, 856, 858
tellurium, 734
   properties, 732
temperature, 15
   decomposition,
      calculating, 603
   effect on equilibrium,
      502; solubility, 407;
      reaction rate, 458
   kinetic energy of
      molecules, 189
   measurement, 19
   molecular speed, 176
   required for reaction,
      608
   scales
      Celsius, 19
      Fahrenheit, 19
      Kelvin, 19
terephthalic acid
   polyester formation,
      891
   production, 889
termination, 456
termites, 887
termolecular reaction,
   451
tert-butyl group, 845

tertiary alcohol, 874
tertiary amines, 893
tertiary structure, 900
tetrafluoroborate ion,
   preparation, 301
tetrafluoroethylene, 856
tetrathionate ion,
   molecular model, 143
thallium, properties, 692
theoretical yield, 125
thermal decomposition,
   carbonates, 82
thermal motion, 194
   contribution to
      entropy, 592
thermal pollution, 407
thermite reaction, 199,
   694
thermochemical
   properties
      food, 222T
      fuel, 219T
thermochemical sea level,
   214
thermochemistry, 188
thermodynamic altitude,
   604
thermodynamic cycle,
   reduction potential, 669
thermodynamic sea level,
   604
thermodynamic stability,
   606
thermodynamics, 188
   first law, 192
   second law, 588
thermometer, 15F
thermonuclear explosion,
   829
thioacetamide, 739
thiocyanate ion, 64
thiol, 872
thiosulfate ion
   complex formation,
      577
   molecular model, 143
Thomson, G. P., 240
Thomson, J. J., 47, 240
thorium-232 series, 810
three-center bonds, 344
threonine, 898
thymine, structure, 903
thymol, 878
thymol blue, 561, 563F
time, units, 19
time-dependence
   concentration, 447
   first-order reactions,
      442
tin, 267, 704
   allotropes, 594F
   plating, 267, 638, 704
   production, 704
   properties, 701
tin(II) oxide, 267, 290
   preparation, 711
tin(IV) oxide, 267
   preparation, 711
titanates, 768
Titanic, 76
titanium

production, 768
   properties, 767
titanium dioxide, 768
titanium tetrachloride,
   381, 768
titrant, 139
titration, 139
   apparatus, 560F
   back, 142
   choice of indicator, 561
   direct, 142
   indirect, 142
   strong-base–strong-
      acid, 553
   variation of pH, 552
   weak-acid–strong-base,
      556
   weak-base–strong-acid,
      558
TNT, trinitrotoluene, 864
Tokomak, 829
Tollens' reagent, 882
Tolu balsam, 863
toluene, molecular
   model, 863
tooth decay, 686
tooth enamel, 686
topaz, 696F
Torr, 154
torsional rigidity, 337
total energy, 191
toxicity
   carbon monoxide, 705
   cyanide ion, 713
tracer, ammonia
   equilibrium, 478
trans, 854
trans-butene
   molecular model, 474
   structure, 854
transferring solutions,
   135
transition metals, 45, 256,
   762
transmutation, 804, 812
tremolite, 710
trichlorophenol,
   molecular model, 878
tricine (buffer), 584
triiodide ion, 141, 745
trinitrotoluene, molecular
   model, 864
triol, 875
triolein, 890
triple bond, 293
   orbital structure, 335
triple point, 371
triple superphosphate,
   723
tris (buffer), 583
tristearin, 890
   molecular model, 221
tritium, 660
   in nuclear fusion, 828
   properties, 53
trona, 674
Trouton's rule, 367, 599
tryptophan, 899
tungsten steel, 772T
tungsten carbide, 712
turgidity, 423

tylenol, molecular model, 894
tyrosine, 899

u, atomic mass unit (alternative symbol), 54
Uhlenbeck, George, 246
ultraviolet radiation, introduced, 235
unbranched alkane, 844
uncertainty principle, stated, 240
unimolecular reaction, introduced, 450
unit, 16
  analysis, 24
  base, 16
  cancellation, 25
  concentration, 395
  debye, 311
  derived, 21
  energy, 189
  entropy, 592
  frequency, 233
  power, 202
  pressure, 154
  radiation, 815
  reaction rate, 433
  relation between, 17
  resistance, 377
unit cell, definition, 375
  counting ions, 380
universal indicator, 531
unsaturated hydrocarbon, 840
uracil, structure, 903
uraninite, 829F
uranium
  diffusion plant, 178F
  enrichment, 829
uranium hexafluoride, 744, 830
uranium tetrafluoride, 830
uranium-235 series, 810, 812
uranyl ion, 829
urea
  Lewis structure, 297
  molecular model, 221, 523

valence
  electron, 278
  origin of term, 278
  shell, 251
  variable, 286
valence-shell electron-pair repulsion, 322
valine, 898
van der Waals, Johannes, 165

van der Waals
  equation, 165
  forces, 353
    in hydrogen, 663
van't Hoff, Johannes, 418
van't Hoff
  equation for
    equilibrium constant, 503; osmotic pressure, 421
  i factor, 418
vanadium, 769
  colors of compounds, 43F
  properties, 767
  steel, 772T
vanadium pentoxide, 769
  catalyst, 467
vanadyl ion, 769
vanillin, 881
vapor, 6
vapor pressure, 172, 364, 365, 365T
  calculation, 611
  decomposition, 490
  lowering, 416
  molecular structure, 366
  relation to boiling point, 366
  temperature dependence, 370
vaporization, 6
  enthalpy of, 203
  entropy of, 593
  rate, 365
variable covalence, 299
variable valence, 286
  p-block, 289
vegetation, annual formation, 218
Venus, surface, 491
vinegar, 888
vinyl chloride, 856
viscosity, 361
  intermolecular forces, 362
visible light, 234
vision, 337
vitamin B$_{12}$, 774
volatile, 7
volcano reaction, 199
volt, 624
voltmeter, 624F
volume, 15
  moles of solute, 134
  produced by electrolysis, 646
volumetric analysis, 139
volumetric flask, 133F
von Laue, Max, 372
VSEPR, valence-shell electron-pair repulsion, 322

electron-pair
  arrangements, 323
  role of lone pairs, 324; multiple bonds, 326
vulcanize, 859

Waage, Peter, 479
washing soda, 404, 674
water, 735
  autoionization, 520
  boiling point, 735
  chlorinated, 744
  chlorine in, 143
  critical behavior, 368
  density, temperature dependence, 369
  dividing line between acids and vases, 528
  electrolysis, 644
  entropy, 593T
  formation, mechanism, 457
  freezing and entropy, 597
  freezing point and pressure, 369
  good solvent power, 415
  heavy, 54
  Lewis base, 515
  molecular model, 9
  oxidizing agent, 736
  phase diagram, 369
    with solute, 419
  purification, 697
  reaction with
    alkali metals, 668
    alkaline earth metals, 680
  reducing agent, 736
  special characteristics, 736
  swimming-pool, 735
  treatment with hydrazine, 717
  vapor pressure, 173, 366
water glass, 708
water-repelling organic groups, 711
water-splitting reaction, 662
watt (W), 202
wave properties of particles, 239
wave-particle duality, 239
wavelength, 234
  relation to color, 234
weak acid, 525
  calculation of pH, 533
weak base, 526
  calculation of pH, 533
weak electrolyte, 93

electrical conduction, 527
weak-field ligand, 790
weather chart, 155F
weed killer, 750
weight, distinction from mass, 15
welding, 861
whipped cream propellant, 720
whiskey, chromatogram, 13
white gold, 777
white light, 234
white paint, 768F
wine, 888
wood alcohol, 873
wood smoke, 880
work, 196
  done by reaction, 197
  electrical, 629

x-ray
  diffraction, 372
  diffractometer, 373
  wavelength, 234
xenon, 753
  properties, 751
xenon fluorides, 753
xenon oxides, 754
xenon tetrafluoride
  Lewis structure, 300
  VSEPR prediction, 326
xylene, 864
  oxidation, 889

yarn, 892
yeast, 706

Z, atomic number, 49
zeolite, 735
zero, as a significant figure, 28
zero order reaction, 438
Ziegler-Natta catalyst, 858
zinc
  amphoteric properties, 778
  plating, 638
  production, 778
  properties, 775
  reaction in copper(II) sulfate, 619
zinc acetate, common-ion effect, 570
zinc bromide, high-density solution, 745
zircon, 710
zone refining, 418F, 419

THE ELEMENTS

Element	Symbol	Atomic number	Atomic weight, amu	Element	Symbol	Atomic number	Atomic weight, amu
Actinium	Ac	89	227.03	Mercury	Hg	80	200.59
Aluminum	Al	13	26.98	Molybdenum	Mo	42	95.94
Americium	Am	95	241.06	Neodymium	Nd	60	144.24
Antimony	Sb	51	121.75	Neon	Ne	10	20.18
Argon	Ar	18	39.95	Neptunium	Np	93	237.05
Arsenic	As	33	74.92	Nickel	Ni	28	58.71
Astatine	At	85	210.	Niobium	Nb	41	92.91
Barium	Ba	56	137.34	Nitrogen	N	7	14.01
Berkelium	Bk	97	249.08	Nobelium	No	102	255.
Beryllium	Be	4	9.01	Osmium	Os	76	190.2
Bismuth	Bi	83	208.98	Oxygen	O	8	16.00
Boron	B	5	10.81	Palladium	Pd	46	106.4
Bromine	Br	35	79.91	Phosphorus	P	15	30.97
Cadmium	Cd	48	112.40	Platinum	Pt	78	195.09
Calcium	Ca	20	40.08	Plutonium	Pu	94	239.05
Californium	Cf	98	251.08	Polonium	Po	84	210.
Carbon	C	6	12.01	Potassium	K	19	39.10
Cerium	Ce	58	140.12	Praseodymium	Pr	59	140.91
Cesium	Cs	55	132.91	Promethium	Pm	61	146.92
Chlorine	Cl	17	35.45	Protactinium	Pa	91	231.04
Chromium	Cr	24	52.00	Radium	Ra	88	226.03
Cobalt	Co	27	58.93	Radon	Rn	86	222.
Copper	Cu	29	63.54	Rhenium	Re	75	186.2
Curium	Cm	96	247.07	Rhodium	Rh	45	102.91
Dysprosium	Dy	66	162.50	Rubidium	Rb	37	85.47
Einsteinium	Es	99	254.09	Ruthenium	Ru	44	101.07
Erbium	Er	68	167.26	Samarium	Sm	62	150.35
Europium	Eu	63	151.96	Scandium	Sc	21	44.96
Fermium	Fm	100	257.10	Selenium	Se	34	78.96
Fluorine	F	9	19.00	Silicon	Si	14	28.09
Francium	Fr	87	223.	Silver	Ag	47	107.87
Gadolinium	Gd	64	157.25	Sodium	Na	11	22.99
Gallium	Ga	31	69.72	Strontium	Sr	38	87.62
Germanium	Ge	32	72.59	Sulfur	S	16	32.06
Gold	Au	79	196.97	Tantalum	Ta	73	180.95
Hafnium	Hf	72	178.49	Technetium	Tc	43	98.91
Helium	He	2	4.00	Tellurium	Te	52	127.60
Holmium	Ho	67	164.93	Terbium	Tb	65	158.92
Hydrogen	H	1	1.0080	Thallium	Tl	81	204.37
Indium	In	49	114.82	Thorium	Th	90	232.04
Iodine	I	53	126.90	Thulium	Tm	69	168.93
Iridium	Ir	77	192.2	Tin	Sn	50	118.69
Iron	Fe	26	55.85	Titanium	Ti	22	47.90
Krypton	Kr	36	83.80	Tungsten	W	74	183.85
Lanthanum	La	57	138.91	Uranium	U	92	238.03
Lawrencium	Lr	103	257.	Vanadium	V	23	50.94
Lead	Pb	82	207.19	Xenon	Xe	54	131.30
Lithium	Li	3	6.94	Ytterbium	Yb	70	173.04
Lutetium	Lu	71	174.97	Yttrium	Y	39	88.91
Magnesium	Mg	12	24.31	Zinc	Zn	30	65.37
Manganese	Mn	25	54.94	Zirconium	Zr	40	91.22
Mendelevium	Md	101	258.10				

FUNDAMENTAL CONSTANTS

Name	Symbol	Value
Atomic mass unit	amu	1.6605×10^{-24} g
Avogadro's number	N_A	6.0221×10^{23}
Boltzmann's constant	k	1.3807×10^{-23} J/K
Elementary charge	e	1.6022×10^{-19} C
Faraday's constant	F	96.485 kC/mol
Gas constant	R	8.3145 J/(K · mol)
		0.08206 L · atm/(K · mol)
		62.37 L · Torr/(K · mol)
Mass of electron	m_e	9.1094×10^{-28} g
Mass of neutron	m_n	1.6749×10^{-24} g
Mass of proton	m_p	1.6726×10^{-24} g
Planck's constant	h	6.6261×10^{-34} J/Hz
Speed of light	c	2.9979×10^{8} m/s
Standard acceleration of free fall	g	9.8066 m/s^2

SI PREFIXES

p	n	μ	m	c	d	k	M	G
pico-	nano-	micro-	milli-	centi-	deci-	kilo-	mega-	giga-
10^{-12}	10^{-9}	10^{-6}	10^{-3}	10^{-2}	10^{-1}	10^{3}	10^{6}	10^{9}

RELATIONS BETWEEN UNITS

	Common unit	SI unit
Mass	1 ounce (oz)	28.35 grams (g)
	0.03527 oz	1 g
	1 pound (lb)	453.6 g
	2.205 lb	1 kilogram (kg)
	1 ton (2000 lb)	907.2 kg
Length	1 inch (in)	**2.54** centimeters (cm)
	0.3937 in	1 cm
	1 foot (ft)	**30.48** cm
	1 yard (yd)	0.914 meter (m)
	1.094 yd	1 m
	1 mile (mi)	1.6093 kilometers (km)
	0.6214 mi	1 km
Volume	1 liter (L)	**1000** cubic centimeters (cm^3)
	1 cubic foot (ft^3)	0.0283 cubic meter (m^3) (28.3 L)
	1 quart (qt)*	946 cm^3 (0.946 L)
	1 gallon (gal)*	3785 cm^3 (3.785 L)
Time	1 minute (min)	**60** seconds (s)
	1 hour (h)	**3,600** s
	1 day	**86,400** s

Temperature conversions

$°F = \mathbf{32} + \frac{9}{5} \times °C$ $°C = \frac{5}{9} \times (°F - \mathbf{32})$ $K = \mathbf{273.15} + °C$

Entries in bold type are exact.
*European quart and gallon are 1.201 times larger.